Analysis and Design of Autonomous Microwave Circuits

To my father Gerardo Suárez
and my mother Carmen Rodriguez

Analysis and Design of Autonomous Microwave Circuits

ALMUDENA SUÁREZ

IEEE

IEEE PRESS

WILEY

A JOHN WILEY & SONS, INC., PUBLICATION

Published by John Wiley & Sons, Inc., Hoboken, New Jersey.
Published simultaneously in Canada.

Library of Congress Cataloging-in-Publication Data:

Suárez, Almudena.
Analysis and design of autonomous microwave circuits / Almudena Suárez.
p. cm. – (Wiley series in microwave and optical engineering)
Includes bibliographical references and index.
ISBN 978-0-470-05074-3 (cloth)
1. Microwaves circuits–Mathematical models. 2. Oscillators, Microwave–Design and construction. 3. Oscillators, Microwaves–Automatic control. 4. System analysis. I. Title.
TK7876.S759 2008
621.381′32—dc22

2008007472

10 9 8 7 6 5 4 3 2 1

Contents

Preface

Autonomous circuits are capable of sustaining a steady-state oscillation at a frequency different from those delivered by input generators or their harmonic frequencies. The most obvious example is the free-running oscillator, generating a periodic solution from the energy delivered by direct-current (dc) sources only. Another example is the frequency divider, giving rise to a subharmonic frequency of the input periodic source. In injection-locked regimes, the oscillation frequency agrees with a multiple or submultiple of the input frequency, and this relationship is maintained within certain input frequency and input power intervals. Free-running oscillators and frequency dividers are used primarily in the frequency generation and frequency conversion stages of communication systems. Other applications of injection-locked oscillators take advantage of their high phase sensitivity with respect to their bias sources and component values to obtain phase shifters and phase-shift-keying modulators. In turn, the coupled-oscillator systems are composed of oscillator circuits connected through linear networks which operate in synchronous manner at a single fundamental frequency. They can be used for a variety of purposes. Multidevice oscillators with a global coupling network are applied for power combination at the fundamental frequency, or at a given harmonic component of this frequency. On the other hand, one- and two-dimensional oscillator systems with nearest-neighbor coupling can be used for beam steering in phased arrays. The beam steering capability comes from the fact that it is possible to synthesize a constant phase shift progression with a very simple tuning procedure by varying the tuning voltages of the peripheral elements only.

The autonomous circuits must contain amplitude-sensitive devices to enable the self-sustained oscillation: that is, an oscillation that does not grow unboundedly (which would be unphysical) or decays to zero. Thus, they must necessarily be nonlinear. The analysis of autonomous circuits is difficult due to this inherent nonlinearity and the usual coexistence of the oscillatory solution with a mathematical solution for which the circuit does not oscillate. As a simple example, consider the case of a free-running oscillator, which can always be solved for a dc solution even when the oscillatory solution is the only solution observed physically. The physical solutions are capable of recovering from the small perturbations, that are always present in real life, coming from noise or small fluctuations. They are robust versus small perturbations or *stable*. In fact, the stability analysis of a given mathematical solution is the verification of its physical existence. This analysis should be carried out in all circuits containing nonlinear devices and it is essential in autonomous circuits due to the typical coexistence of different steady-state solutions for the same values of the circuit elements.

Another undesired characteristic of autonomous circuits is the phase noise. In communication systems, the phase noise of the local oscillator corrupts the modulation signals and can give rise to demodulation errors. The phase noise of the free-running oscillator is due to the absence of a phase reference in this type of circuit: that is, a fixed phase value at a particular circuit location at the fundamental frequency or one of its harmonics. In forced periodic regimes phase reference is provided by the input periodic source. The free-running oscillator lacks this phase reference, and the solution is invariant versus phase shifts. Thus, the small perturbations coming from the circuit noise sources accumulate in the phase variable with a certain statistical variance. The resulting modulation of the oscillation carrier as well as other perturbation effects gives rise to an oscillator spectrum showing skirts about the fundamental and harmonic frequencies.

Another problem when dealing with autonomous circuits is the limited designer control over the autonomous solution and its characteristics. This is due to their inherently nonlinear behavior and to the strong dependence of the oscillation characteristics on the values of the circuit elements. The fundamental frequency of a free-running oscillator will vary under any change of these values. In the case of injection-locked oscillators, the phase shift and the operation bandwidth are also very sensitive to the component values. Usually, the oscillator circuits are designed in two steps. First a small-signal design is carried out to ensure the fulfilment of the oscillation startup conditions. Then the circuit is analyzed in its nonlinear steady-state regime to compare its actual performance with the original specifications. In general, the circuit is not designed in its nonlinear steady state, due to the difficulty in imposing the characteristics (frequency, power or bandwidth) of a fully grown oscillation.

The book has several objectives. One of them is to facilitate understanding of the free-running oscillation mechanism, startup from the noise level, and establishment of steady-state oscillation. The oscillation buildup is closely linked to stability concepts, which are presented in great depth. Other forms of oscillation are also treated in detail: for example, the superharmonically injection-locked oscillation, used for frequency division, the subharmonically injection-locked oscillation, used for frequency multiplication or phase noise reduction; and the parametric frequency division, due to the periodic variation of a nonlinear reactance. The causes of oscillator phase noise and its particular form of variation versus the offset from the carrier frequency are also studied. In each case the aim will be to unify or relate the various analysis approaches existing in the literature, from nonlinear dynamics, from simplified analytical formulations, or from accurate simulation techniques in the frequency domain. The various methodologies for stability analysis and for phase noise analysis are compared and related. Their degree of accuracy and their advantages or shortcomings, depending on the particular application, are discussed.

Nonlinear circuits can exhibit different types of steady-state regime, from dc to chaotic solutions. This variety of operational modes is most significant in the case of autonomous circuits. Generally, they behave only in the desired regime within certain intervals of their parameters, or magnitudes susceptible to be varied while maintaining the same circuit topology. Examples are the bias voltages, input generators, and linear-element values. For example, a frequency divider will operate as such only for certain intervals of the input power and frequency. Outside these intervals, the circuit self-oscillation will mix with the input source frequency or the oscillation will be extinguished. In both cases, the regime obtained will have no interest for the designer. The changes in the observed regime are due to bifurcations taking place in the circuit. A *bifurcation* is a qualitative change in the stability of a solution or in the number of solutions when a parameter is varied. Some bifurcations are natural in autonomous circuits, such as the oscillation extinction from a certain bias voltage or the loss of synchronization in an injection-locked oscillator. Other bifurcations, generally undesired, will depend on the particular design. Another objective of the book is to present a detailed and comprehensive classification of bifurcations to enable better understanding and more efficient design of such circuits as free and injection-locked oscillators or frequency dividers.

Realistic prediction of the behavior of nonlinear circuits requires the use of accurate simulation techniques. Analysis can be carried out in the time or frequency domain or using mixed time–frequency methods. The choice of one or another domain will generally depend on the type of circuit to be analyzed, the type of regime, and the information desired. For example, in most cases we are not interested in the transient response. However, this transient may be required in an investigation of the switching time of oscillator circuits, for instance. The frequency-domain techniques enable efficient simulation of circuits containing distributed elements, which are more easily described in this domain, usually by means of their frequency-dependent scattering parameters. In turn, time–frequency methods can be seen as an extension of the low pass equivalent of bandpass signals to solutions with multiple harmonic terms. They allow the analysis of microwave circuits containing modulations. These circuits cannot be simulated through standard time-domain integration, which is due to the requirement of a short integration step during a long simulation interval. They will also enable efficient determination of the envelope of the oscillation startup transient and the analysis of steady-state solutions with complex dynamics. Here, the main principles and properties of the various analysis methods are presented, as well as a detailed description of the algorithms and their most common options or improvements.

The simulation of autonomous circuits has added difficulties, especially when using frequency-domain methods, which is due to the existence of trivial solutions, with no oscillation. Frequency-domain methods such as a Fourier series representation of the circuit variables, provide only steady-state solutions, with no sensitivity to the stability or instability of these solutions. Error minimization techniques are used, which, by default, converge to the simplest steady state, for which the circuit exhibits no oscillation. Complementary techniques are required to avoid this undesired convergence. Another objective of the book is to provide techniques for simulation of the most usual types of autonomous regimes using in-house or commercial simulators. These techniques should be combined with a stability analysis of the solutions obtained. In the book, the main stability analysis methods in the frequency domain are presented and compared. The simpler implementation methods are applied to prevent instability during the design process, which will require practical and fast stability tests. The more involved methods are applied in a complementary manner for a final and rigorous stability analysis of the design developed, prior to manufacturing. When considering variations in the circuit parameters, accurate determination of the steady-state solutions, combined with a thorough stability analysis of these solutions, will provide great insight into circuit behavior. The book aims at extending knowledge from nonlinear dynamics, obtained from particular equations or simple topology circuits, to practical circuits with a lumped or distributed nature and containing one or several nonlinear devices.

As already stated, phase noise is an undesired characteristic of oscillator circuits, with negative implications on the performance of communication systems. Many different methods for phase noise analysis have been presented in the literature, using a time- or frequency-domain formulation of the oscillator equations. One objective of the book is the rigorous comparison of their capabilities and degree of accuracy, to facilitate designer decision as to the most convenient technique for the phase noise analysis of his or her particular circuit. The phase noise behavior of injection-locked oscillators is also presented in a comprehensive manner, giving insight into the effect of the input source noise and the circuit's own noise on the output phase noise spectrum.

The methods for stability and phase noise analysis constitute a compact set of tools for efficient and accurate prediction of autonomous circuit behavior. However, one more step must be taken, which is adapting these techniques to the optimized and accurate design of these circuits. This should make it possible to obtain maximum benefit of the circuit capabilities and saving *a posteriori* corrections. The designer has limited control over over the power and frequency of self-generated periodic oscillation, and the stable operation bands. In this book, an entire set of optimization techniques is presented for application to circuits with a given topology. The optimum topologies for oscillator or frequency-divider design with particular specifications have been investigated by other authors. Here, harmonic balance techniques are presented for the optimization of the circuit performance. Different design objectives will be considered: presetting the oscillation frequency and output power, increasing the efficiency, modifying the transient duration, or imposing operation bands. The techniques cover the three prinicipal operational modes of autonomous circuits—free-running, tuned, and synchronized—and can be applied externally by the user of commercial harmonic balance software standard library elements. Techniques for the reduction of oscillator phase noise are also be presented. These techniques are based on minimization of the coefficients that determine the variance of the phase deviation. The minimization is carried out by optimizing the values of the circuit elements of a given oscillator topology or by modifying this topology with an additional feedback loop.

Coupled oscillator systems can be used for power combination at the fundamental frequency, and beam steering in phased arrays. In the beam steering applications, the coupled-oscillator smaller system size, than that of a topology based on phase shifters, which requires individual control of the polarization and wiring for each phase shifter. An in-depth analytical study of the behavior of both types of coupled-oscillator systems is presented. The multidevice, multioscillator structure can give rise to various oscillation modes. Undesired modes, coexisting with the one desired by the designer, should be unstable. Techniques to obtain these modes systematically and to determine their stability properties are provided. The phase noise of the coupled-oscillator system is also studied. Also derived is a semianalytical formulation, which uses a perturbation model of the elementary oscillator in the coupled array, extracted with harmonic balance simulations. The semianalytical formulation combines this numerical model of the oscillators with an admittance description of the coupling networks. This provides a reduced-order nonlinear system describing the entire coupled-oscillator array. The greatest advantages of a numerical formulation are the low computational cost, even in the case of a large number of oscillator elements, and the higher accuracy than that of simple analytical oscillator models, often based on parallel/series resonators and cubic nonlinearities. The reduced cost will allow an optimized choice of the number of oscillator elements and an optimized synthesis of the coupling networks.

Oscillations are obtained not only in autonomous circuits. Nonlinear circuits that are not expected to oscillate, such as power amplifiers or frequency multipliers, often exhibit undesired instability phenomena, such as oscillations at incommensurable frequencies, frequency divisions, or jumps in amplitude and frequency. This type of behavior severely degrades circuit performance and in many cases prevents any practical application. Suppressing the undesired phenomena through trial-and-error procedures is inefficient and sometimes impossible, and the need to redesign a circuit increases the production cycles and the final manufacturing cost. As has been stated, techniques exist for accurate and complete stability analysis of the steady-state solutions of nonlinear circuits. However, the final goal is the efficient suppression of instability phenomena, with minimum degradation of the performance specified. We present systematic techniques to eliminate common types of undesired behavior, such as spurious oscillations, hysteresis, chaos, and sideband amplification. This requires a variety of considerations for a non autonomous circuit, such as a power amplifier or frequency multiplier, or for an autonomous circuit, such as an oscillator or a frequency divider. In the first case, characteristics such as output power and efficiency should be preserved. For free-running oscillators, the circuit autonomy constitutes an additional difficulty, since the oscillation frequency changes under any variation of circuit components. Thus, it will be affected by the introduction of the stabilization elements. Here techniques are presented for the stabilization of two main types of nonlinear circuits: power amplifiers and oscillator circuits. They are based, in each case, on in-depth analysis and understanding of the instabilization mechanism and the characteristics of the undesired solution that is to be suppressed. This will allow deriving an optimum stabilization strategy with minimum degradation of the performance specified.

The book is organized into twelve chapters. Each chapter starts with basic concepts and evolves from simple mathematical derivations to advanced theory. The chapters are closely related, but care has been taken to facilitate the independent study of a chapter, by a reader only interested in particular topics.

In Chapter 1 we analyze the dynamics of free-running oscillators. We show the mathematical conditions for oscillation startup and self-sustained steady-state oscillation and present the concept of stability, essential for an understanding of the oscillation mechanism. Emphasis is placed on the invariance of the oscillator solution versus time translations, which is the origin of the phase noise problem. The oscillator is analyzed in different manners: from a time-domain point of view, using simple analytical expressions, in the frequency domain, using impedance–admittance functions; and from the point of view of nonlinear dynamics. The results of these analyses and the stability conditions derived in each case are compared and related analytically.

Chapter 2 deals with oscillator phase noise. Time-domain methods for stochastic characterization of the phase noise spectrum are presented. They are based on determination of the variance of stochastic time deviation. The foundations of frequency-domain analysis, based on the carrier modulation approach, are also shown. Clear analytical relationships between the two methods are developed. The amplitude noise is analyzed and related to the common observations of resonances in the output noise spectrum.

In Chapter 3 we present a detailed classification of the primary bifurcations from th dc and periodic regimes. The meaning and implications of these bifurcations are discussed in detail with practical examples. The foundations of bifurcation detection in the frequency domain are also presented. They constitute the basis of the bifurcation analysis with harmonic balance presented in Chapter 6.

Chapter 4 deals with oscillators that have periodic forcing. Fundamentally synchronized oscillators, harmonic and subharmonic injection-locked oscillators, and parametric dividers are studied. Approximate analytical expressions are provided for steady-state solutions, their stability, and the limits of their operation bands in the desired mode. The phase noise spectrum of injection-locked oscillators is derived, analyzing the effect of the noise from an input synchronizing source and from the circuit noise sources on this spectrum.

In Chapter 5 we present the main analysis techniques for nonlinear circuits: time-domain integration, fast time-domain methods, harmonic balance, and envelope transient. Insight is given into the foundations of the various techniques, together with detailed descriptions of the algorithms used, and of their options and improvements. Numerous examples are provided. In-depth explanations, of the complementary techniques required for the analysis of autonomous circuits are included.

Chapter 6 covers the main harmonic balance techniques for the stability analysis of dc and periodic solutions. Emphasis is placed on the Nyquist criterion applied to the characteristic determinant of the harmonic balance system and pole–zero identification. Techniques for bifurcation detection from dc and periodic regimes are described in detail. These techniques can be implemented efficiently using in-house software. They can also be implemented externally by the user of commercial harmonic balance, using standard library elements.

Chapter 7 deals with the main harmonic balance techniques for phase noise analysis of oscillator circuits in the free-running and injection-locked regimes. The spectrum calculation from the variance of phase deviation is presented as well as the conversion matrix and the carrier modulation approaches. A detailed comparison of the techniques is presented, establishing their relationships, the degree of accuracy and ease of application. Expressions for amplitude noise calculation are derived and used for an analysis of noise spectrum resonances.

In Chapter 8 we present design techniques for oscillator circuits. An entire design procedure for free-running oscillators is presented, from an initial determination of the ideal feedback and termination element, using small-signal analysis to a final nonlinear- design stage, providing the circuit element values required for a specified oscillation frequency and output power. Techniques are also given for linearization of the frequency characteristic of voltage-controlled oscillators, for shortening the oscillation transient, and for phase noise reduction.

Chapter 9 covers stabilization techniques for phase noise reduction in oscillator circuits. Self-injection locking and low-frequency feedback are considered using delay lines and high-quality-factor resonators.

Chapter 10 is devoted to coupled oscillator systems. An in-depth analytical study of the operation of these systems is presented, considering aspects such as the coexistence of steady-state solutions, the stability of these solutions, and the phase noise. Practical techniques for the harmonic balance design of coupled-oscillator systems with global and nearest-neighbor coupling are also provided. For beam steering applications of coupled systems, the techniques will allow a simple synthesis of the constant phase shift progression. A semianalytical formulation for realistic prediction of the behavior of oscillator arrays with a large number of oscillator elements is presented. The technique is based on the extraction of a perturbation model of the oscillator elements by means of harmonic-balance simulations.

In Chapter 11 we present optimization procedures for analoge dividers. They allow presetting the operation band and avoiding variation of this band at different design stages. Techniques to broaden the division bandwidth are also provided. A simple semianalytical expression is used to evaluate the capability of a given free-running oscillator to operate as a harmonic injection divider by a different order N.

Chapter 12 deals with stabilization techniques for nonlinear circuits using harmonic balance simulations. Two principal types of circuits are considered: power amplifiers and free-running oscillators. Among the undesired phenomena suppressed are frequency division, incommensurate oscillations, chaos, hysteresis, and noise sideband amplification.

ACKNOWLEDGMENTS

The author would like to express her gratitude to the following: Dr. Sergio Sancho and Dr. Franco Ramírez, of the University of Cantabria, for their invaluable advice, support, and contribution in many of the analyses and results presented here and along many years of working together; Dr. Juan Mari Collantes and Dr. Aitziber Anakabe, of the University of the Basque Country, for insightful discussions; César Barquinero, Mabel Pontón, Elena Fernández, Jacobo Domínguez, Dr. Juan Pablo Pascual, Dr. Luisa de la Fuente, Dr. Amparo Herrera, of the University of Cantabria, and Dr. Victor Araña, of the University of Las Palmas de Gran Canaria, for their help in the revision of the manuscript; former members of the group, Dr. Samuel Ver Hoeye, Dr. Elena de Cos, and Dr. Ana Collado, for their help in the revision of the manuscript; Dr. Robert Melville, of the New Jersey Institute of Technology, and Dr. Christopher Silva, of the Aerospace Corporation, for interesting discussions; Prof. David Rutledge and Ms. Dale Yee, of Caltech, for the opportunity to visit Caltech and learn about power amplifier design; Dr. Sanggeun Jeon, Dr. Feiyu Wang, and Prof. David Rutledge, for their invaluable contributions to the techniques for power amplifier stabilization; Prof. Raymond Quere, of the University of Limoges, for his invaluable help and guidance at the beginning of the author's career; Prof. José Luis Garcia, for his support and help since the author joined the University of Cantabria; all the members of the Departamento Ingeniería de Communicaciones (DICOM); her family for their continuous support; and Angioline Loredo, of John Wiley & Sons, for her hard and careful work on the book.

Almudena Suárez

CHAPTER ONE

Oscillator Dynamics

1.1 INTRODUCTION

A well-designed free-running oscillator provides a periodic signal of constant amplitude and frequency f_o from the energy delivered by direct-current (dc) sources. This has an immediate application for the realization of local oscillators used in the frequency-conversion stages of communication systems [1]. In receivers, the modulated signal at radio-frequency (RF) f_{RF} is mixed with the output of a local oscillator at f_o, selecting the intermodulation product that corresponds to the frequency difference $f_{IF} = f_{RF} - f_o$, This allows down-conversion of the carrier frequency from f_{RF} to f_{IF}. An analogous procedure is followed in transmitters. The intermediate frequency f_{IF} is mixed with the output of the local oscillator, selecting the intermodulation product $f_{RF} = f_{IF} + f_o$. This allows up-conversion of the carrier frequency. The free-running oscillator is usually inserted into a phase-locked loop for this application [2].

A single oscillator having dc sources only is said to operate in free-running mode. However, other forms of behavior are possible. In injection-locked operation [3], the oscillation is synchronized with an independent periodic source, which means that the oscillation frequency, influenced by the input source, becomes equal to the input frequency $f_o = f_{in}$, with a constant phase shift between the oscillation and the input signal. The injection-locked mode is used for phase noise reduction, frequency division, or phase shifting. In coupled operation, several oscillators are interconnected by means of linear coupling networks [4] and oscillate in a synchronous manner. Coupled-oscillator systems can be used for power combination

Analysis and Design of Autonomous Microwave Circuits, By Almudena Suárez

or beam steering. In this chapter only the free-running mode of an oscillator circuit is considered. Familiarity with the behavior and properties of free-running oscillation is essential for an understanding of any other form of operation (e.g., injection-locked, coupled) treated in subsequent chapters.

Free-running oscillators have essential differences from other RF circuits, such as amplifiers, mixers, and frequency multipliers [5,6]. The operation frequency or frequencies (in the case of a mixer) of these circuits are determined by the input sources. In contrast, the fundamental frequency of an oscillator is self-generated or autonomous and depends on the values of the circuit elements. Thus, the circuit must be designed accurately to obtain the value desired for the oscillation frequency f_o. Due to the absence of time-varying sources, any free-running oscillator can be solved for a mathematical dc solution. The oscillation starts up from any small perturbation of this dc solution and must grow from noise level to a steady-state oscillatory solution with constant amplitude and period. As will be shown, the self-sustained oscillation is only possible in nonlinear, nonconservative systems. Stability concepts are also essential to the understanding of the oscillator behavior. The oscillation startup and the physical observation of the periodic solution are explained from the different stability properties of dc and the steady-state oscillation [7]. Because of the absence of an input periodic source establishing a time reference, arbitrary translations of the periodic waveform along the time axis give other solutions. There is an "irrelevance" with respect to time translations, or in the frequency domain, with respect to the phase origin. Thus, any phase-shifted solution constitutes a valid solution of the oscillator circuit. The absence of a restoring mechanism in the phase value gives rise to the phase noise problem in oscillator circuits [8,9].

In this chapter we deal with the main aspects of oscillator behavior. Oscillators are studied in the time domain and in the frequency domain, using impedance–admittance descriptions, which are very helpful for oscillator design, and the describing function approach, which allows nonlinear analysis at the fundamental frequency only. This one-harmonic approach will set the conceptual basis for harmonic balance analysis, covered in detail in Chapter 5. We relate various analysis techniques and unify concepts and properties, derived in the literature from very different viewpoints. Chapter 1 provides a general background for Chapter 2, which is devoted to phase noise analysis; Chapter 3, devoted to global stability analysis; and Chapter 4, devoted to an analysis of injection-locked oscillators and frequency dividers. The chapter is organized as follows. Section 1.2 provides intuitive explanations for oscillation startup and for the mechanism of self-sustained oscillation. In Section 1.3 we present the frequency-domain formulation based on the use of impedance or admittance functions, covering steady-state analysis and the stability of dc and periodic solutions. In Section 1.4 we extend the previous formulation to multiple harmonic components, for conceptual purposes, as this will be necessary for accurate stability analysis of oscillator circuit without limiting assumptions. In Section 1.5 we deal with oscillator circuits from the viewpoint of nonlinear dynamics, with the circuit described by a system of nonlinear differential equations. The main types of steady-state solutions and their properties are presented. In Section 1.6 we introduce formal mathematical procedures for the stability analysis of dc and periodic regimes and provide the necessary background for global stability analysis (i.e., versus variation in a circuit parameter), which is covered in Chapter 3. Finally, in Section 1.7 we emphasize the irrelevance of the oscillator solution versus time translations and show examples of phase shift response versus impulse perturbations. We establish the necessary background for Chapter 2, dealing with stochastic characterization of the spectrum of a noisy oscillator. Two different circuits are considered in this chapter: a parallel resonance oscillator with a two-terminal active element, and a FET-based oscillator at $f_o = 4.36$ GHz. The simplicity of the first circuit makes possible the derivation of meaningful analytical expressions. Comparison with a FET-based oscillator clarifies our understanding of deviations from ideal behavior in practical circuits.

1.2 OPERATIONAL PRINCIPLE OF FREE-RUNNING OSCILLATORS

An ideal circuit given by the parallel connection of an inductor L and a capacitor C, without resistance, will under any initial condition exhibit oscillation at the frequency $\omega_o = 1/\sqrt{LC}$, at which the average energies stored in the magnetic and electric fields are equal, so the sum of the inductor and capacitor susceptances is equal to zero [5]. The total energy in the circuit remains constant during the entire oscillation period, so it is a conservative system [10]. When the electrical energy stored in the capacitor is maximal, the magnetic energy stored in the inductor is zero, and vice versa. The energy displacement from one element to another gives rise to the oscillation observed in the node voltage and branch currents. By Kirchhoff's laws, the sum of the inductor plus capacitor current must be equal to zero, $i_C + i_L = 0$, which after some simple manipulations provides the linear differential equation

$$\frac{d^2v(t)}{dt^2} + \frac{1}{LC}v(t) = 0 \tag{1.1}$$

with $v(t)$ the node voltage. Equation (1.1) is a second-order differential equation with constant coefficients which can be transformed into two first-order equations by performing the variable change $x_1(t) = v(t), x_2(t) = dv(t)/dt$. Then, equation (1.1) becomes

$$\begin{bmatrix} \dot{x}_1(t) \\ \dot{x}_2(t) \end{bmatrix} = \begin{bmatrix} 0 & 1 \\ \frac{-1}{LC} & 0 \end{bmatrix} \begin{bmatrix} x_1(t) \\ x_2(t) \end{bmatrix} \tag{1.2}$$

System (1.2) belongs to the general class of linear differential equations with constant coefficients, which can be written in the general manner $\dot{\overline{x}}(t) = A\overline{x}(t)$, where $\overline{x}(t)$ is a vector of system unknowns and A is a constant matrix. For M variables in $\overline{x}(t)$, the general solution of $\dot{\overline{x}}(t) = A\overline{x}(t)$ has the form $\overline{x}(t) = c_1\overline{v}_1e^{\lambda_1 t} + c_2\overline{v}_2e^{\lambda_2 t} + \cdots + c_M\overline{v}_Me^{\lambda_M t}$, where the exponents λ_k are the eigenvalues of the matrix A, assumed different, and the vectors $\overline{v}_k$ are the eigenvectors of A. Because any physical variable $x(t)$ is real valued in the time domain, the constants c_k, v_k, and λ_k will be either real or complex conjugate. The constants c_k depend on the initial value $t_o, \overline{x}(t_o)$.

In the particular case of system (1.2), the eigenvalues of the 2×2 matrix A are $\lambda_{1,2} = \pm j\omega_o = \pm j1/\sqrt{LC}$ and the eigenvectors are given by $[1, j\omega_o]$ and $[1, -j\omega_o]$. Then the solution of (1.1) has, for $x_1(t) = v(t)$, the general form

$$v(t) = ce^{j\omega_o t} + c^* e^{-j\omega_o t} = 2(c_r \cos \omega_o t - c_i \sin \omega_o t) \tag{1.3}$$

with $c = c_r + jc_i$ being a complex constant, depending on the initial conditions $v(t_o)$ and $dv(t_o)/dt$. For a given initial value $v(t_o)$ and $dv(t_o)/dt$, this complex constant is calculated by means of the following system of boundary conditions:

$$\begin{aligned} v(t_o) &= 2(c_r \cos \omega_o t_o - c_i \sin \omega_o t_o) \\ \frac{dv(t_o)}{dt} &= -2(c_r \omega_o \sin \omega_o t_o + c_i \omega_o \cos \omega_o t_o) \end{aligned} \tag{1.4}$$

Thus, for each pair of possible initial conditions $v(t_o)$ and $dv(t_o)/dt$, an oscillatory solution with different amplitude would be obtained. This dependence of the oscillation amplitude on the initial conditions is unphysical and, of course, is never observed in the free-running oscillators measurements. An analogous situation would be found in an ideal pendulum with no friction in which the ball keeps oscillating at the amplitude of the initial elongation. In the case of the circuit described by (1.1), the unphysical situation is due to the absence of resistive elements in the ideal LC circuit. In practice it is not possible to have inductors or capacitors without resistive losses. Note that one of the solutions of (1.3) obtained from $v(t_o) = 0$ and $dv(t_o)/dt = 0$ is given by $v(t) = 0$ and $y(t) = 0$ $\forall t$. This solution, just one of the family $v(t) = ce^{j\omega t} + c^* e^{-j\omega t}$, provides no oscillation at all.

The eigenvalues λ_k of the matrix A in the general system $\dot{\overline{x}}(t) = A\overline{x}(t)$ can also be obtained from an application of the Laplace transform to this system, which provides $[sI_d - A]\overline{X}(s) = 0$, where I_d is the identity matrix and $\overline{X}(s)$ is the vector of the Laplace transforms of the different variables. (Note that the obtained system assumes a zero initial value $\overline{x}(0) = 0$, which otherwise should be taken into account in the transformation of the time-derivative to the Laplace domain.) The system $[sI_d - A]\overline{X}(s) = 0$ is a homogeneous linear system in $\overline{X}(s)$. Therefore, to obtain a solution $\overline{X}(s)$ different from zero, the matrix affecting $\overline{X}(s)$ must be singular. Thus, the condition $\det[sI_d - A] = 0$ must be fulfilled. The determinant introduced, known as the *characteristic determinant* of the linear system, provides a characteristic polynomial $P(s) = \det[sI_d - A] = 0$ of the same degree as the number of unknowns in $\overline{X}(s)$. In particular, application of the Laplace transform to (1.1) provides the characteristic polynomial $P(s) = s^2 + 1/LC = 0$. As can easily be seen, the roots of the characteristic polynomial agree with the eigenvalues $\lambda_{1,2} = \pm j\omega_o = \pm j1/\sqrt{LC}$ of matrix A of (1.2).

Now assume that a small-signal input $u(t)$ is introduced in the general linear system $\dot{\overline{x}}(t) = A\overline{x}(t)$. In the Laplace domain this will give rise to the equation $[sI_d - A]\overline{X}(s) = [G(s)]U(s)$, where $[G(s)]$ is a column matrix [11,12]. This matrix is necessary because we have not specified the nature of the input, so it may undergo time derivations when introduced in the system. Any possible output $Y(s)$ will be linearly related to the variable vector $\overline{X}(s)$, which in a general manner can be expressed as $Y(s) = [B][sI_d - A]^+$, where $[B][sI_d - A]^+$ is a row matrix. Thus, any possible single-input single-output transfer function will be written

$$H(s) = \frac{Y(s)}{U(s)} = \frac{[B][sI_d - A]^+[G(s)]}{P(s)} \tag{1.5}$$

with "+" being the transpose of the cofactor matrix. The roots of $P(s)$ will agree with the poles of the single-input single-output transfer function $H(s)$. Intuitively, the poles are associated with the zero-input solutions of the analyzed system, so they cannot depend on the particular input or output. However, pole–zero cancellations are possible due to the matrix product in the numerator, which will be different for different choices of the closed-loop transfer function. Pole–zero cancellations can be avoided through a suitable choice of $H(s)$. Provided that no pole–zero cancellations occur, it will be possible to calculate the roots λ_k of $P(s)$ indirectly from the pole analysis of a transfer function $H(s)$.

As an example, consider the connection of a small-signal current source $I_{\text{in}}(s)$ to the middle node of the LC resonator in Fig. 1.1. The input signal $U(s) = I_{\text{in}}(s)$ is the current introduced, and the output selected is the node voltage $Y(s) = V(s)$. Applying (1.5), the closed-loop transfer function is

$$Z(s) = \frac{V(s)}{I_{\text{in}}} = \frac{[1\ 0]\begin{bmatrix} s & -1/(LC) \\ 0 & s \end{bmatrix}\begin{bmatrix} 0 \\ s/C \end{bmatrix}}{P(s)} = \frac{Ls}{CLs^2 + 1} \tag{1.6}$$

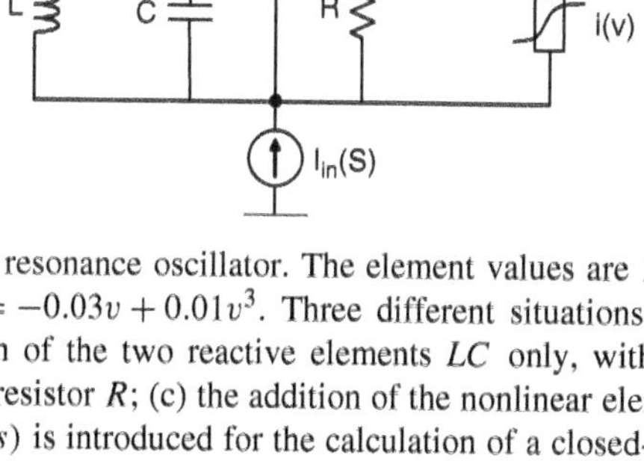

FIGURE 1.1 Parallel resonance oscillator. The element values are $L = 1$ nH, $C = 10$ pF, $R = 100\Omega$, and $i(v) = -0.03v + 0.01v^3$. Three different situations are considered in the text: (a) the connection of the two reactive elements LC only, without a resistor; (b) the inclusion of a positive resistor R; (c) the addition of the nonlinear element $i(v) = av + bv^3$. The current source $I_{\text{in}}(s)$ is introduced for the calculation of a closed-loop transfer function, defined as $Z(s) = V(s)/I_{\text{in}}(s)$.

with s the Laplace frequency. Clearly, the denominator agrees with the characteristic polynomial associated with (1.1), and the transfer function poles $p_{1,2} = \pm j\omega = \pm j\sqrt{1/LC}$ agree with the polynomial roots λ_1 and λ_2. The term *poles* is used often in the book to refer to the roots of the characteristic polynomial of a linear system, due to their equivalence. In the case of the second-order system (1.1), complex-conjugate poles are located on the imaginary axis. Therefore, the solution originating from given initial values $v(t_o)$ and $dv(t_o)/dt$ neither grows (which would correspond to poles on the right-hand side of the complex plane) nor vanishes (which would correspond to poles on the left-hand side of the plane), but remains with its initial amplitude. This is never observed in physical systems.

If a resistor is now introduced in the circuit of Fig. 1.1, the situation becomes totally different. The energy contained in the system no longer remains constant in time. The resistor dissipates energy as heat, at the rate $R\int_0^{t_{\max}} i_R^2(t)\,dt$, where $t_{\max}$ is the duration of the time interval considered and i_R is the current through the resistor. So the longer the time, the less energy is available for storage at the inductor and capacitor and the smaller is the oscillation amplitude. Thus, in an *RLC* circuit with $R > 0$, the oscillation amplitude decays to zero. When the resistor R is introduced in parallel, the circuit equations become

$$\frac{d^2v(t)}{dt^2} + \frac{1}{RC}\frac{dv(t)}{dt} + \frac{1}{LC}v(t) = 0 \tag{1.7}$$

Because of introduction of the resistor R, a new term has appeared in dv/dt, called the *damping term* [10]. The name *damping* indicates that the rate of extinction of the oscillation depends on the coefficient associated with dv/dt. The smaller the resistance value R, the higher its influence over the parallel resonator and the faster the oscillation extinction. As in the former case, equation (1.7) is a second-order linear system with constant coefficients. The associated characteristic polynomial $P(s)$ is obtained through application of the Laplace transform to (1.7). The two roots of this polynomial are given by

$$\lambda_{1,2} = -\frac{1}{2RC} \pm \sqrt{\frac{1}{(2RC)^2} - \frac{1}{LC}} \tag{1.8}$$

Provided that $1/LC > 1/(2RC)^2$, an exponentially decaying oscillation of the form $v(t) = e^{-(1/2CR)t}2(c_r\cos\omega t - c_i\sin\omega t)$ is obtained, with $\omega = \sqrt{1/LC - 1/(2RC)^2}$ and c_r and c_i constants that depend on the initial conditions. Note that the transient decay is ruled by the amplitude envelope $e^{(-1/2CR)t}$. The quality factor of the parallel resonance is given by $Q = RC\omega_o$, with $\omega_o = 1/\sqrt{LC}$. Therefore, the exponential transient can be described as $v(t) = e^{(-\omega_o/2Q)t}2(c_r\cos\omega t - c_i\sin\omega t)$, so the smaller the quality factor of the parallel circuit, the faster the oscillation extinction. Because the oscillation amplitude decays to zero for any initial value, the only steady-state solution of equation (1.7) is a dc regime with $v = 0$ and $dv/dt = 0$. This will be the only solution observed physically. The small noise perturbations will give rise to oscillatory transients, seen simply as noise about this dc regime.

As in the previous case, it is possible to define a closed-loop transfer function associated with the *RLC* circuit. This transfer function can be obtained by connecting a small-signal current source I_{in} in parallel and obtaining the ratio between the node voltage V and the current introduced, I_{in}. The poles of this transfer function, which agree with the roots of the characteristic polynomial $P(s)$, are located on the left-hand side of the complex plane. This indicates that whatever the initial condition, the linear system evolves to the steady state $v = 0$ and $dv/dt = 0$. The solution $v = 0$ and $dv/dt = 0$ also existed in the conservative system, but was just one of the infinite solutions in the family $v(t) = ce^{j\omega t} + c^*e^{-j\omega t}$. In contrast, the solution $v = 0$ and $dv/dt = 0$ is the only steady state of (1.7). Once this solution is reached, any instantaneous perturbation applied at a particular time value t_o only, and setting the initial values $v(t_o)$ and $dv(t_o)/dt$, will start a transient leading back to the dc solution $v = 0$ and $dv/dt = 0$. This dc solution is robust versus perturbations, or stable.

Clearly, to observe a steady-state oscillation, the effect of the resistor $R > 0$ must be compensated. Introduction of a negative-resistance element will provide an energy source to compensate for the energy loss in the resistor. This element can be a negative-resistance diode or a transistor under suitable configuration and bias conditions. The energy delivered will be taken from dc sources. Assuming a constant negative resistance R_N connected in parallel, the total resistance will be $R_T = 1/(G_N + G)$, with $G_N = 1/R_N$ and $G = 1/R$. The general circuit solution will be $v(t) = e^{[-(G+G_N)/2C]t} \cdot 2(c_r \cos \omega t - c_i \sin \omega t)$. For $G + G_N > 0$, which implies dominant positive resistance, the negative resistance introduced is not sufficient. The damping term will be positive and the oscillation amplitude will decay exponentially to zero from any initial condition. Thus, the dc solution will be the only one observable. For $G + G_N = 0$, a conservative *LC* circuit, with no effective resistance, is obtained again, which as discussed earlier, corresponds to a nonphysical situation. For $G + G_N < 0$, which implies dominant negative resistance, the damping term will be negative and the oscillation amplitude will increase exponentially ad infinitum. This is also nonphysical. The negative resistance cannot be insensitive to the growth of the node voltage. It has to depend on this voltage, or equivalently, it has to be nonlinear to enable saturation of the oscillation amplitude.

To illustrate the mechanism of self-sustained oscillation, a nonlinear element with the instantaneous characteristic $i(v)$ is introduced in the resonant circuit (see Figure 1.1). This provides the nonlinear differential equation

$$\frac{d^2v(t)}{dt^2} + \left[\frac{1}{RC} + \frac{1}{C}\frac{di}{dv}(v)\right]\frac{dv(t)}{dt} + \frac{1}{LC}v(t) = 0 \qquad (1.9)$$

To obtain sustained oscillation, the damping term affecting dv/dt must be nonlinear and thus sensitive to $v(t)$. A common example of nonlinearity in oscillator theory is $i(v) = av + bv^3$, with $a < 0$ and $b > 0$. This is an ideal element providing negative conductance at small signal $G_N = di(v = 0)/dv = a + 3bv^2|_{v=0} = a$ about $v = 0$.

In physical systems, bias sources delivering energy to the circuit will, of course, be required. Placing the derivative di/dv into (1.9) yields the following equation:

$$\frac{d^2v(t)}{dt^2} + \frac{1}{C}(a + G + 3bv^2)\frac{dv(t)}{dt} + \frac{1}{LC}v(t) = 0 \tag{1.10}$$

where $G = 1/R$. Thus, the nonlinear damping term is given by $\mu(v) = (a + G + 3bv^2)/C$. Equation (1.10) constitutes a good behavioral model of the oscillator circuit, with reduced analytical complexity. Clearly, equation (1.10) admits the steady-state solution $v(t) = 0, dv/dt = 0$, which corresponds to the constant or dc solution of the ideal circuit of Fig. 1.1. Note that any oscillator circuit can always be solved for a constant solution, even when it exhibits self-sustained oscillation. This can easily be verified by the reader and is due to the absence of time-varying generators. When the dc generators are first powered on, oscillation has not yet builtup and the circuit is at this dc solution, due to the existence of dc sources only. The reaction of the dc solution to small perturbations can be predicted by linearizing the nonlinear element $i(v) = av + bv^3$ about the dc solution $v = 0$. Thus, the nonlinear element is replaced by the constant conductance $G_N = di(v = 0)/dv = a$. This allows us to apply linear analysis techniques to the circuit constituted by the parallel connection of G, G_N, L, and C. The resulting poles [or roots of the characteristic determinant $P(s)$] are given by

$$p_{1,2} = -\frac{G_T}{2C} \pm \frac{\sqrt{G_T^2L^2 - 4LC}}{2LC} \tag{1.11}$$

with $G_T = G_N + G$. The two poles in (1.11) are associated with nonlinear circuit linearization about the dc solution, thus are often called poles of the dc solution. They determine the response of the dc solution of an oscillator circuit to a small instantaneous perturbation.

From an inspection of (1.11), to obtain an oscillatory transient with exponentially growing amplitude, the poles must be complex conjugate $p_{1,2} = \sigma \pm j\omega$, with $\sigma > 0$. The oscillatory transient requires a negative value of the term under the square root. Assuming that $4LC \gg G_T^2L^2$, the pole frequency will correspond approximately to the resonance frequency $\omega_o = 1/\sqrt{LC}$. For the oscillation amplitude to grow exponentially in time, the condition $\sigma = -G_T/2C > 0$ must be fulfilled, which in the circuit of Fig. 1.1 implies that $G_T = a + G < 0$. At small signal, we can consider the circuit of Fig. 1.1 (including the nonlinear element) as a feedback system with a direct-trajectory transfer function $Y_N = a$ and a feedback transfer function $Z(s) = (Cs + 1/(Ls) + G)^{-1}$. The combination of gain and feedback with a resonant network leads to a characteristic system with two complex-conjugate poles responsible for the oscillation startup. As in the case of a linear *RLC* circuit, the positive real part $\sigma = -G_T/2C$ can be expressed in terms of the quality factor $Q = C\omega_o/G$ of the linear part of the circuit as $\sigma = -\omega_o G_T/2GQ$. Thus, the duration of the startup transient depends on the quality factor and on the ratio between the total conductance G_T (with negative sign) and the load conductance G. The startup transient will be shorter for larger ratio G_T/G and smaller quality factor Q of the resonant circuit, which implies larger $\sigma > 0$. As the oscillation amplitude increases, the actual nonlinearity of the total conductance will give rise to continuous variation of σ, which must take a zero value at steady state.

The initial exponential growth of the oscillation amplitude is in agreement with the fact that for $v \cong 0$, the damping term $\mu(v) = (a + G + 3bv^2)/C$ is nearly constant and given by $\mu = (a + G)/C < 0$ and delivers energy continuously to the incipient oscillatory solution. Note, however, that this linearized analysis is valid as long as $|v(t)|$ is small enough for the linearization of $i(v)$ about the dc solution to be accurate. For not so small $|v(t)|$, the damping term $\mu(v)$ will no longer be constant and the oscillation amplitude will start to grow more slowly than the exponential prediction, until it reaches a constant value at the steady-state regime. None of this can be predicted with the linearization about the dc solution and the pole analysis. Evolution of the oscillatory solution to its steady-state regime has to be determined through numerical integration.

The results of the numerical integration of (1.10) are shown in Fig. 1.2, where the node voltage amplitude $|v(t)|$ and the associated evolution of the damping term $\mu(v) = [a + G + 3bv^2(t)]/C$ are represented. For small-signal $|v(t)|$, the damping term is nearly constant and negative, with the value $\mu(0) = (a + G)/C$. This negative damping term is responsible for the initial exponential growth of the oscillation amplitude as $e^{-[\mu(0)]/2t}$. As this amplitude increases, the nonlinearity of $\mu(v)$ starts to be noticeable. The nonlinear component of $\mu(v)$, given by $3bv^2(t)/C$, is always positive since $b > 0$, and constitutes a positive contribution to the damping term. For smaller amplitude $|v(t)|$, the damping term will be more negative than for larger amplitude $|v(t)|$, so more energy will be delivered to the oscillatory solution by the active element. Note that the damping term has oscillatory variation, as it is a function of the periodic $v(t)$. This can be seen in Fig. 1.2. The local

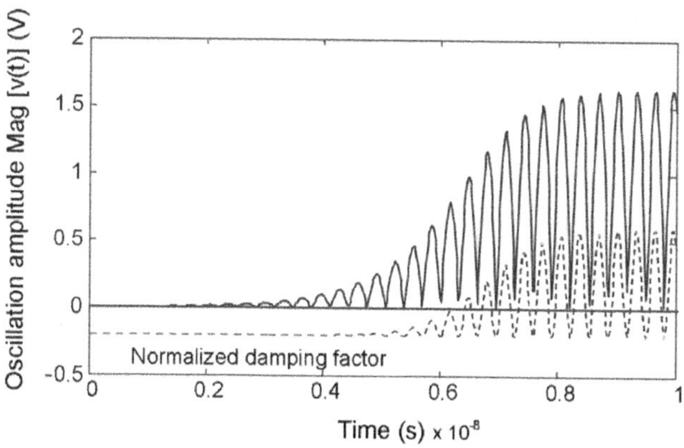

FIGURE 1.2 Analysis of the second-order oscillator of Fig. 1.1. Nonlinear equation (1.10) has been integrated for initial conditions different from $v = 0$ and $dv/dt = 0$. Both $|v(t)|$ and the normalized nonlinear term $(a + G + 3bv^2)/C$ have been represented.

maxima of $\mu(v)$ correspond to the local maxima of $|v(t)|$, and the local minima (most negative values) to the local minima of $|v(t)|$. The local maxima of $\mu(v)$ increase with $|v(t)|$ until steady state is reached. In steady state, both $v(t)$ and the damping term $\mu(v) = [a + G + 3bv^2(t)]/C$ exhibit periodic oscillation. As can be seen, the cubic nonlinearity provides a good model of the physical reduction of the device negative conductance when increasing the voltage amplitude across its terminals. This is why it is often chosen for a simple mathematical description of the oscillator behavior.

The circuit capability to self-sustain a steady-state oscillation is explained as follows. For small $|v(t)|$ during the oscillation period, the damping term $\mu(v)$ is negative (Fig. 1.2), so the energy delivered by the active element exceeds the resistor dissipation and makes $|v(t)|$ grow again. For large $|v(t)|$, a positive damping term is obtained and the dissipation exceeds the energy delivery, which makes $|v(t)|$ decrease. This mechanism allows sustaining the periodic oscillation with perfect balance between energy pumped in and energy dissipated over one cycle. Unlike the situation for a conservative system, with no energy dissipation at any time during the oscillation period, there is energy dissipation in the fraction of the oscillation period with $\mu(v) > 0$. Except in the case of coexistence of stable solutions (which has not yet been considered), the oscillation amplitude and frequency are independent of initial conditions. They are determined solely by the nonlinear characteristic of the damping term and the circuit topology and component values. Thus, for a system to exhibit sustained oscillation, it must be nonlinear and nonconservative.

Due to the nonexplicit time dependence of the differential equations describing an autonomous circuit, the time integration from different values at the same initial time t_o gives rise to time-shifted steady-state waveforms. This is illustrated in Fig. 1.3a, where different initial conditions have been considered in the time-domain integration of equation (1.10). The initial conditions are not known by the designers, as they come from noise or fluctuations at the experimental stage. Assuming that the voltage waveform $v(t)$ is a solution of (1.10), any time-delayed version of this voltage waveform $v(t - \tau)$ will be also a solution of (1.10). This is easily verified by defining the new time variable $t' = t - \tau$ and introducing $v(t')$ in (1.10). Note that the shape and period of the waveform are independent of these initial conditions. They satisfy the mathematical conditions for the self-sustained oscillation with zero net energy consumption. Nonautonomous circuits such as amplifiers and frequency multipliers are ruled by differential equations having coefficients with explicit time dependence. As an example, consider the parallel connection of an independent current generator $i_g(t)$ to a parallel *RLC* resonator. This circuit is governed by the linear equation $\ddot{v}(t) + G\dot{v}(t)/C + 1/(LC)v(t) - (di_g(t)/dt)/C = 0$, with the independent term $di_g(t)/dt$. This independent term establishes a time reference, so all the solutions obtained integrating the equations from different values v_o at the same initial time t_o converge to the same steady-state waveform. An example is shown in Fig. 1.3b, where the equations of a nonautonomous circuit have been integrated from totally different initial conditions $t = 0$, $v_o = -1V$, and $t = 0$, $v_o =$

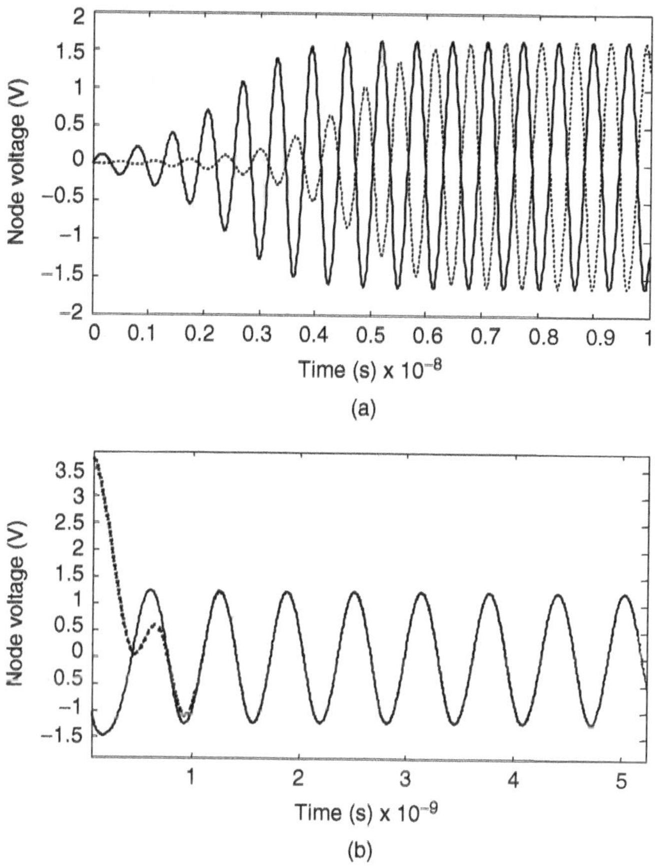

FIGURE 1.3 Time-domain integration of differential equations describing an autonomous and a forced circuit. (a) Integration of the nonlinear differential equation (1.10) describing the oscillator of Fig. 1.1, from different initial values at $t_o = 0$. This gives rise to time-shifted steady-state waveforms. (b) Integration of a forced circuit from different voltage values v_o at $t_o = 0$. The same waveform, without a time shift, is obtained for all the initial values.

4 V, obtaining the same steady-state waveform without a time shift. Compare with the situation shown in Fig. 1.3a.

Although the explanation above was based on a simple second-order nonlinear circuit, all the major conclusions are applicable to practical oscillators of much higher complexity. In a free-running oscillator, the oscillatory solution always coexists with a dc solution. When the dc generators are first powered on, the oscillation has not built yet up and the circuit is at the dc solution. In a well-designed oscillator, the dc solution is unstable and contains a pair of complex-conjugate poles on the right-hand side of the complex plane. This is due to the imbalance between the energy delivered by the active element and the energy dissipated by the resistors at the frequency of the poles. The unstable poles will give rise to oscillation startup under any small perturbation. For the circuit to be able to exhibit a self-sustained oscillation, the negative-resistance device must be nonlinear and thus sensitive to the oscillation amplitude. In the steady-stage regime, energy is alternately consumed and delivered during the oscillation period, so the system must be nonconservative (i.e., it must contain resistive elements).

1.3 IMPEDANCE–ADMITTANCE ANALYSIS OF AN OSCILLATOR

As noted in Section 1.2, an oscillator is ruled by a set of nonlinear differential equations that can only be accurately solved using numerical techniques. Time-domain analysis makes it possible to obtain the entire time evolution of the circuit variables, including transient and steady state. In frequency-domain analysis, each variable is represented by a Fourier series $v(t) = \sum_k V_k e^{jk\omega t}$, with constant complex coefficients V_k and constant ω, so only the steady-state regime can be determined. Note that due to the circuit nonlinearity, the saturation of the waveform amplitude gives rise inherently to some harmonic content. Due to the orthogonally of the Fourier basis, the circuit will be described by a set of equations, one at each harmonic frequency, relating the harmonic coefficients of the circuit variables. When limiting the analysis to one harmonic term (i.e., when assuming a sinusoidal oscillation), it will be possible to obtain meaningful analytical expressions for the oscillation frequency and amplitude. In what follows, an admittance–impedance analysis of the oscillator circuit is presented, assuming a sinusoidal waveform. This frequency-domain analysis offers a different viewpoint of the oscillator circuit and allows the derivation of useful design criteria. Note that the accuracy of the sinusoidal approach will be higher for a larger quality factor Q of the resonant circuit, due to the high attenuation of the harmonic frequencies.

As shown in Section 1.2, in a free-running oscillator, a negative-resistance element delivering energy to a resonator and a load or utilization resistance are necessary for oscillation buildup from the noise level. This negative resistance can be obtained from negative-resistance diodes, such as tunnel, Gunn, or Impatt [5], or by using transistors, which generally requires the introduction of suitable feedback between the two transistor ports [13,14]. Figure 1.4 shows a simple representation of an oscillator circuit. There are no periodic generators and the circuit is divided into a nonlinear block, providing the negative resistance, and a linear block, containing the output load. This block division is straightforward for a diode-based oscillator such as the one depicted in Fig. 1.1. For a transistor-based oscillator, the block division is more involved. In single-ended oscillators, the sketch shown in Fig. 1.5 is often used. Since there are no external RF sources, one of the transistor ports is ended by a given impedance (the termination), used only to obtain negative resistance at the other port. To avoid power loss, a reactive termination is often preferred. In addition to a proper choice of this termination and suitable biasing,

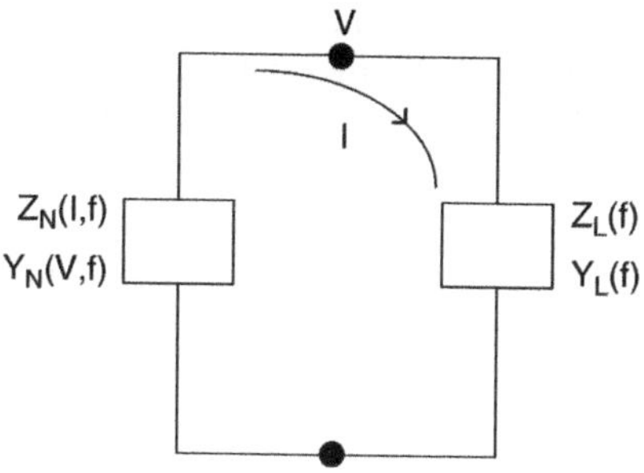

FIGURE 1.4 One-port representation of a free-running oscillator.

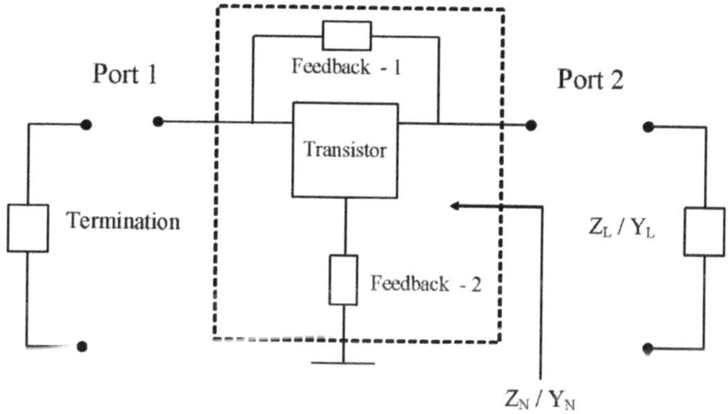

FIGURE 1.5 Schematic representation of a transistor-based oscillator. A one port description is used for the block, consisting of the transistor, its termination at port 1, and the feedback elements.

the transistor often requires an additional parallel or series feedback network to exhibit negative resistance about the oscillation frequency that is desired [15]. The transistor is loaded with an impedance Z_L containing the resistive load from which the oscillation output power is extracted.

A one-port definition of the subcircuit, consisting of the transistor, together with its termination at the other port and the series or parallel feedback network (the nonlinear block), is often assumed at the design stage. This allows modeling the transistor-based oscillator as in Fig. 1.4. Note that although this block contains nonlinear and linear elements, it is globally nonlinear. By taking into account the boundary condition imposed by the transistor termination, the admittance of the nonlinear block can be expressed as a function of the voltage V at the output port. This admittance will also depend on the frequency ω, due to the existence of reactive elements inside the nonlinear block. Thus, it is possible to define the function $Y_N(V, \omega)$. In turn, the load circuit exhibits the linear admittance $Y_L(\omega)$. This type of representation is not sufficient for an accurate analysis of transistor-based oscillators, which actually depend on the two state variables of the nonlinear model of the transistor (e.g., the gate-to-source voltage and the drain-to-source voltage of FET transistors). However, it will be very helpful for a general understanding of the oscillator behavior and for oscillator design.

Next, we analyze an oscillator in terms of general admittance–impedance functions from a single observation port, following Fig. 1.4.

1.3.1 Steady-State Analysis

When applying Kirchhoff's laws to the circuit of Fig. 1.4, either a series or a parallel connection may be considered between the linear and nonlinear blocks. For a series connection, an impedance analysis is carried out, in terms of the branch current, as this provides simpler equations. For a parallel connection, an admittance analysis is carried out in terms of the node voltage. Depending on the actual circuit topology, one or another analysis may be more convenient. Here, only the admittance analysis is considered. One based on an impedance description, in terms of the loop current, is totally analogous.

A steady-state oscillation with a sinusoidal node voltage $v(t) = V_o \cos(\omega_o t + \phi)$ will be assumed initially. In contrast to forced circuits, the fundamental frequency ω_o of the solution depends on the values of the circuit elements, bias sources, and other parameters, since it is not delivered to the circuit by an external source. Due to this fact, the oscillation frequency will be an unknown to be determined. Application of Kirchhoff's laws at the frequency ω_o provides the following complex equation, which relates the total branch current at ω_o to the node voltage at the same frequency:

$$Y_T(V, \omega_o)Ve^{j\phi} = [Y_N(V, \omega_o) + Y_L(\omega_o)]Ve^{j\phi} = 0 \tag{1.12}$$

where Y_L is the linear block admittance and Y_N the nonlinear block admittance, which, in general, will be frequency dependent, as it may contain reactive elements. Note that the nonlinear admittance function $Y_N(V, \omega_o)$ does not depend on the phase value of the periodic exciting signal $Ve^{j\phi}$. This is understood by comparison with the behavior of any circuit forced with a sinusoidal generator. A change $\Delta\phi$ in the phase of the periodic exciting source simply gives rise to the same phase increment in all the circuit variables. Thus, the solution phase shift with respect to this exciting source remains the same as before the application of $\Delta\phi$.

By inspecting (1.12), it is clear that at least two solutions coexist in the oscillator circuit. One is given by $V = 0$. This solution, with zero oscillation amplitude, is in fact the dc solution discussed in Section 1.2, for which any circuit with no time-varying external sources can be solved. The other solution is obtained from the nonlinear equation $Y_T(V_o, \omega_o) = 0$ and corresponds to a sinusoidal voltage $v(t) = \text{Re}\{V_o e^{j(\omega_o t+\phi)}\}$, as assumed when writing the admittance equation (1.12). Thus, the steady-state oscillation equation is written

$$Y_T(V_o, \omega_o) = [Y_N(V_o, \omega_o) + Y_L(\omega_o)] = 0 \tag{1.13}$$

The complex equation (1.13) can be split into two real equations in two real unknowns V_o and ω_o by considering the real and imaginary parts of Y_T: $\text{Re}[Y_T] = 0$ and $\text{Im}[Y_T] = 0$. It is actually the voltage dependence of $Y_N(V, \omega)$ (i.e., the circuit nonlinearity) which makes it possible to solve $Y_T = 0$ for the constant oscillation amplitude V_o. Note that any phase value ϕ provides a valid solution, as Y_T does not depend on ϕ. This is due to the absence of an independent periodic generator at the same frequency ω_o, establishing a phase reference. When this is the case, the coefficients of the differential equations ruling circuit behavior have no explicit time dependence, so any arbitrary time shift of the periodic waveform provides another solution. In the frequency domain, the different time shifts correspond to different phase origins, as $\Delta\phi = \omega_o(\tau' - \tau)$.

The complex equation $Y_T(V_o, \omega_o) = 0$ is in total agreement with the conclusions of Section 1.2. The first real equation, $\text{Re}[Y_T] = 0$, implies balance between average power delivered and consumed, as resulting from $\frac{1}{T}\int_0^T v(\lambda)i(\lambda)d\lambda = \frac{1}{2}\text{Re}[Y_T]V_o^2$, with T being the oscillation period $T = 2\pi/\omega_o$. The second equation, $\text{Im}[Y_T] = 0$, implies the existence of a resonance at the oscillation frequency.

The next objective is to obtain the nonlinear admittance function $Y_N(V, \omega)$, which constitutes the model of the active element in the approximate oscillator analysis. The model is based on use of the describing function. For a sinusoidal describing function [16], the input signal is represented by a sinusoid. Considering the nonlinearity $i(t) = i(v(t))$, the describing function will provide an admittance model $Y_N(V)$, depending on the voltage amplitude V. To obtain a sinusoidal describing function, the voltage $v(t) = V\cos(2\pi f_o t + \phi)$ is introduced into the nonlinearity $i(v)$, obtaining the ratio between the first harmonic of the resulting current and the voltage phasor $\frac{V}{2}e^{j\phi}$:

$$Y_N(V) = \frac{i(v)|_f}{Ve^{j\phi}} = \frac{2}{T}\frac{\int_0^T i(v(t))e^{-j\omega_o t}\,dt}{Ve^{j\phi}} \tag{1.14}$$

where $T = 2\pi/\omega_o$. Clearly, Y_N depends on the amplitude V of the voltage introduced but not on its phase ϕ, in agreement with previous discussions. To see this more clearly, a phase shift $\Delta\phi$ will be considered. This phase shift can be represented as $\Delta\phi = -\omega_o\tau$. Next, the variable change $t' = t - \tau$ is performed in (1.14), which provides

$$Y_N(V) = \frac{2}{T}\frac{\int_0^T i(v(t'))e^{-j\omega_o t'}e^{-j\omega_o\tau}\,dt'}{Ve^{j\phi}e^{-j\omega_o\tau}} \tag{1.15}$$

So the same nonlinear admittance $Y_N(V)$ is obtained. In polynomial nonlinearities, another way to obtain the same result would be to place $v(t) = V/2e^{j(\omega_o t+\phi)} + V/2e^{-j(\omega_o t+\phi)}$ into the nonlinear function $i(v)$, expand the function, and divide the resulting harmonic term at $j\omega_o$ by $V/2e^{j(\omega_o t+\phi)}$.

To illustrate the admittance analysis will be applied to the parallel resonance oscillator of Fig. 1.1. Using (1.14), the sinusoidal describing function associated with the constitutive relationship $i(v) = av + bv^3$ (with $a < 0$ and $b > 0$) is given by $Y_N(V) = a + 3/4bV^2$. From an inspection of this expression, the small-signal conductance is $Y_N(0) = a$, with a negative value. Because $b > 0$, the nonlinear conductance decreases with the voltage amplitude across the nonlinear element. Note that this physical behavior of the active element leads to an increase in the damping ratio $\mu(v)$ with the amplitude $|v(t)|$, discussed in Section 2.2, which allows us to reach a constant steady-state oscillation amplitude. Replacing the describing function obtained in (1.13) yields the following equations:

$$a + \frac{3bV^2}{4} + G_L = 0$$
$$C\omega_o - \frac{1}{L\omega_o} = 0 \tag{1.16}$$

Resolving equation (1.16), the oscillation amplitude is $V_o = \sqrt{(-a - G_L)/(3b/4)} = 1.64$ V and the oscillation frequency is $f_o = 1/(2\pi\sqrt{LC}) = 1.59$ GHz. From the expression of the oscillation amplitude, it is clear that the small-signal conductance $Y_N(V \cong 0) = a$ must have a larger absolute value than the positive linear conductance G_L to obtain a steady-state oscillation. Otherwise, the root of a negative value is obtained in $V_o = \sqrt{(-a - G_L)/(3/4b)}$. This agrees with the results of Section 1.2. The total circuit conductance G_T is negative in small-signal mode but equal to zero in the steady state, as $\mathrm{Re}[Y_T(V_o)] = 0$. To understand this, note that the negative conductance exhibited by the active element decreases with the voltage amplitude, as gathered from $Y_N(V) = a + 3/4bV^2$. The oscillation reaches steady state at the voltage amplitude for which $|Y_N(V_o)| = \mathrm{G}_L$.

It must be emphasized that the steady-state analysis (1.16) is very simplified, as it is limited to a single harmonic component. From (1.16), the oscillation frequency is given by the resonance frequency of the LC resonator. A time-domain simulation would show that depending on the quality factor, the oscillation frequency can differ noticeably from the resonance frequency. The one-harmonic limitation of (1.16) prevents prediction of this effect. To discuss the influence of the harmonic content, a voltage expression $v(t) = V_1 \cos \omega t + V_3 \cos(3\omega t + \phi)$ will be considered. In this expression it has been taken into account that in the circuit being analyzed with no dc sources, no dc or even harmonic components are generated by the cubic nonlinearity $i(v)$. To obtain the first- and third-harmonic admittance functions $Y_{N1}(V_1, V_3, \phi)$ and $Y_{N3}(V_1, V_3, \phi)$, the waveform $v(t)$ is introduced into the transfer characteristic $i(v)$. The admittance functions are calculated as

$$Y_{N1} = \frac{I_1(V_1, V_3, \phi)}{V_1}$$
$$Y_{N3} = \frac{I_3(V_1, V_3, \phi)}{V_3 e^{j\phi}} \tag{1.17}$$

Kirchhoff's laws are written at the first- and third-harmonic components, which provides a two-complex-equation system $Y_{T1} = 0$ and $Y_{T3} = 0$ in the four unknowns V_1, V_3, ϕ and, ω. Solving this system, the oscillation frequency does not exactly agree with the resonance frequency $1/\sqrt{LC}$. This is because unlike the case of the nonlinear admittance function $Y_N(V, \omega)$, in one-harmonic analysis, the imaginary part of Y_{N1} is different from zero. As an example, for $L = 4$ nH and $C = 2.5$ pF, $Y_{N1} = -0.01 + j0.002\Omega^{-1}$ and $Y_{N3} = -0.02 - j0.063\Omega^{-1}$, the oscillation frequency is $f_o = 1.52$ GHz instead of $1/(2\pi\sqrt{LC}) = 1.59$ GHz. The high discrepancy is due to the extremely low quality factor Q of the RLC resonator for these element values. This discrepancy is higher for a smaller quality factor Q, due to the lower filtering of the harmonic components $n\omega_o$ with $n > 1$.

1.3.2 Stability of Steady-State Oscillation

As already pointed out, for a given mathematical solution to be observable physically, it must be stable or robust versus small perturbations. Earlier we considered only the stability of the dc solution that coexists with steady-state oscillation. For the steady-state oscillation of (1.13), given by $v_o(t) = \text{Re}[V_o e^{j\omega_o t}]$, to be stable, the circuit must return to it exponentially under any small perturbation. To verify mathematically if this is the case, a small perturbation is applied at a given time instant t_o. This takes the circuit out of the steady state. However, because the perturbation is small at the beginning of the transient being generated, the circuit variables cannot differ much from their values in a steady-state regime. In the stability analysis proposed by Kurokawa [17], small variations are assumed in both the oscillation amplitude and frequency. The perturbation applied gives rise to time-varying amplitude, which can be expressed as $V_o + \Delta V(t)$. In turn, the frequency takes the time-varying value $\omega_o + \Delta\omega(t)$. Before continuing, the reader should be warned that the assumption of a small frequency variation $\Delta\omega(t)$ limits the validity of this analysis technique. This is because the small perturbation can actually have any frequency, not necessarily one fulfilling $\Delta\omega \ll \omega_o$. As an example, a common instability phenomenon is the onset of a subharmonic component at $\omega_o/2$, generated from a low-amplitude perturbation that clearly does not fulfill the assumption $\Delta\omega \ll \omega_o$. The stability analysis under the assumption $\Delta\omega \ll \omega_o$ is also called *quasistatic*. Despite of this limitation, the stability conditions obtained are extremely helpful use at the oscillator design stage.

Due to the use of instantaneous perturbation, the oscillator is no longer in steady state. The perturbed frequency is written as $j\omega_o + s$, where s is a complex frequency increment. Because the perturbation is small, the perturbed oscillation can be analyzed by performing a first-order Taylor series expansion of the total admittance function about the free-running solution (V_o, ω_o), fulfilling $Y_{To} = 0$. This provides the equation

$$Y_T[V_o + \Delta V(t)]e^{j\phi(t)} = \frac{\partial Y_{To}}{\partial V}\Delta V(t)[V_o + \Delta V(t)]e^{j\phi(t)} + \frac{\partial Y_{To}}{\partial j\omega}\frac{d[(V_o + \Delta V(t))e^{j\phi(t)}]}{dt} = 0 \tag{1.18}$$

with the increment s giving rise to a time derivation in the slow time scale of the perturbed voltage. After performing this derivation, equation (1.18) is written

$$\frac{\partial Y_{To}}{\partial V}\Delta V(t)V_o e^{j\phi(t)} + \frac{\partial Y_{To}}{\partial \omega}[\dot{\phi}(t)V_o e^{j\phi(t)} - j\Delta\dot{V}(t)e^{j\phi(t)}] = 0 \tag{1.19}$$

where higher-order terms have been neglected. Dividing by $V_o e^{j\phi(t)}$, equation (1.19) can be simplified to

$$\frac{\partial Y_{To}}{\partial V}\Delta V(t) + \frac{\partial Y_{To}}{\partial \omega}\left[-j\frac{\Delta \dot{V}(t)}{V_o} + \Delta\omega_o(t)\right] = 0 \tag{1.20}$$

where $\dot{\phi}(t)$ has been renamed $\dot{\phi}(t) = \Delta\omega_o(t)$. The complex nature of the frequency increment in (1.20) is due to the fact that the oscillator solution has been kicked out of the steady-state solution $V_o e^{j\omega_o t}$, so the amplitude must have an exponential variation associated with the imaginary term $-j[\Delta \dot{V}(t)/V_o]$. Splitting (1.20) into real and imaginary parts, the following linear system is obtained:

$$\begin{aligned} \frac{\partial Y_{To}^i}{\partial \omega}\frac{1}{V_o}\Delta \dot{V}(t) + \frac{\partial Y_{To}^r}{\partial \omega}\Delta\omega_o(t) &= -\frac{\partial Y_{To}^r}{\partial V}\Delta V(t) \\ -\frac{\partial Y_{To}^r}{\partial \omega}\frac{1}{V_o}\Delta \dot{V}(t) + \frac{\partial Y_{To}^i}{\partial \omega}\Delta\omega_o(t) &= -\frac{\partial Y_{To}^i}{\partial V}\Delta V(t) \end{aligned} \tag{1.21}$$

where the superscripts r and i indicate real and imaginary parts, respectively. Note that all the coefficients of (1.21) are constant and constitute the derivatives of the nonlinear function $Y_T(V, \omega)$, calculated at the free-running oscillation point, given by V_o and ω_o. By solving for $\Delta \dot{V}(t)$ in terms of $\Delta V(t)$, the following relationship is obtained:

$$\begin{aligned} \frac{d\Delta V(t)}{dt} &= \frac{-\left[(\partial Y_{To}^r/\partial V)(\partial Y_{To}^i/\partial \omega) - (\partial Y_{To}^i/\partial V)(\partial Y_{To}^r/\partial \omega)\right] V_o}{|\partial Y_{To}/\partial \omega|^2}\Delta V(t) \\ &= \sigma_o \Delta V(t) \end{aligned} \tag{1.22}$$

where the constant coefficient has been called σ_o. The amplitude increment $\Delta V(t)$ evolves according to $\Delta V(t) = \Delta V_o e^{\sigma_o t}$, where ΔV_o depends on the value of the initial instantaneous perturbation. The exponential reaction to small perturbations was also shown in Section 1.2 in the case of perturbed dc solutions. For the oscillation to be stable, the perturbation must vanish exponentially in time. Thus, the coefficient in (1.22) must fulfill $\sigma_o < 0$. Because the denominator of σ_o, given by $|\partial Y_{To}/\partial \omega|^2$, is necessarily positive, the stability condition is given by [17]

$$S = \frac{\partial Y_{To}^r}{\partial V}\frac{\partial Y_{To}^i}{\partial \omega} - \frac{\partial Y_{To}^i}{\partial V}\frac{\partial Y_{To}^r}{\partial \omega} > 0 \tag{1.23}$$

Expression (1.23) is very useful for oscillator design. Due to the physical reduction in negative resistance with signal amplitude, the factor $\partial Y_{To}^r/\partial V$ will generally have a positive sign. Then a sufficiently high value of $\partial Y_{To}^i/\partial \omega$ facilitates the oscillation stability. Actually, the second term, $(\partial Y_{To}^i/\partial V)(\partial Y_{To}^r/\partial \omega)$, is often small compared to the first term, which is explained as follows. The real part of Y_T usually has a small frequency dependence, because the dependence comes from the reactive elements. On the other hand, the imaginary part of Y_T usually has a small amplitude dependence, because the nonlinearities responsible for the free-running oscillation are usually voltage-controlled current sources.

The duration of the transient response to perturbation is considered next. Assuming that $\partial Y^i_{To}/\partial\omega \gg \partial Y^r_{To}/\partial\omega$, the denominator in (1.22) can be approached $|\partial Y_{To}/\partial\omega|^2 \cong (\partial Y^i_{To}/\partial\omega)^2$. A commonly used definition for the oscillator quality factor is $Q = (\omega_o/2G_L)(\partial Y^i_{To}/\partial\omega)$, with the derivative being evaluated at the oscillation frequency and G_L being the passive conductance. Thus, the coefficient σ_o is inversely proportional to the quality factor, meaning that the transient reaction to the perturbation will be slower for larger Q. A similar conclusion had been obtained in Section 1.2 for the oscillation startup transient. In the case of a stable steady-state oscillation, the system will return more slowly to this steady-state regime.

The nonlinear circuit of Fig. 1.1 fulfills the stability criterion. The real part of the total admittance is $\text{Re}[Y_T(V)] = a + 3/4bV^2 + G_L$, so the amplitude derivative in the first term is given by $\text{Re}[\partial Y_{To}/\partial V] = 3/2bV_o$. Note that V_o is the oscillation amplitude, defined as positive, so the term $3/2bV_o$ necessarily takes a positive value. On the other hand, the derivative of the imaginary part of the total admittance function, evaluated at the free-running oscillation, is given by $\text{Im}[\partial Y_{To}/\partial\omega] = 2C$. In turn, the derivatives in the second term of (1.23) are equal to zero. Thus, the condition $S > 0$ is satisfied and the oscillation is stable. Under any small perturbation, the amplitude increment of the perturbed oscillation evolves according to $\Delta V(t) = \Delta V_o e^{\sigma_o t}$, with $\sigma_o = -(3bV_o^2/4)(\omega_o/G_L Q)$ and $Q = C\omega_o/G_L$.

Note that condition (1.23) was derived under a quasistatic approximation, assuming a very small value of the perturbation frequency $\Delta\omega \ll \omega_o$ and using a single observation port. As already stated, this analysis is helpful for oscillation design, as it provides criteria for *likely* stable behavior from admittance functions accessible to the designer. However, the design procedure should be complemented by a rigorous verification of oscillator stability without the limiting assumption $\Delta\omega \ll \omega_o$ and taking into account the actual multidimensional nature of the circuit equations. Note that some unstable resonances may be hidden when inspecting the total impedance or admittance from a single observation port. At the end of the section, some hints about the basis for a more general stability analysis in the frequency domain are provided.

1.3.3 Oscillation Startup

As already known, stable oscillation, with steady-state amplitude V_o and frequency ω_o, must grow from the noise level. This growth is due to the instability of the dc solution, which under any small perturbation gives rise to an oscillatory transient. As shown in Section 1.2, the envelope of the initial transient follows an exponential law. From a certain oscillation amplitude, linearization is no longer valid and the device nonlinearity gives rise to saturation of the oscillation amplitude. When using admittance analysis, the instability of the dc solution is generally associated with fulfillment of the following conditions:

$$
\begin{aligned}
Y_T^r(V \cong 0, \omega_o') &< 0 \\
Y_T^i(V \cong 0, \omega_o') &= 0 \\
\frac{\partial(Y_T^i(V \cong 0, \omega_o'))}{\partial\omega} &> 0
\end{aligned}
\tag{1.24}
$$

where $V \cong 0$ refers to the admittance function evaluated in small-signal mode. Note that an analysis of conditions (1.24) constitutes a stability analysis of this dc solution. Actually, the small-signal admittance $Y_T(V \cong 0, \omega)$ depends on the dc solution about which the active element is linearized. That is, for different bias points, different $Y_T(V \cong 0, \omega)$ functions are obtained, fulfilling conditions (1.24) or not fulfilling them. The main point of conditions (1.24) is that they help in synthesizing a pair of complex-conjugate poles with positive σ at the desired oscillation frequency ω_o'. As shown in Section 1.2, this pair of complex-conjugate poles should give rise to an oscillatory transient of growing amplitude. To understand the relationship between (1.24) and the poles of the dc solution, consider the introduction of a small-signal current source $I_{in}(s)$ in parallel at the observation port. The ratio between the node voltage $V(s)$ and the current delivered, $I_{in}(s)$, provides the closed-loop transfer function $Z(s)$. Assuming that no pole–zero cancellations occur, the poles of $Z(s)$ will agree with the roots of the characteristic function $P(s)$ associated with circuit linearization about the dc solution. A pair of complex-conjugate poles $\sigma \pm j\omega_o$ provides a contribution of the form $Z_p(s) = A\omega_o^2/(s^2 + \sigma^2 - 2s\sigma + \omega_o^2)$, with A a constant value. We will assume that this is the dominant contribution of the pole-residue expansion of $Z_p(s)$ [16] from the observation port. Replacing s with $j\omega$, the impedance function becomes $Z_p(\omega) = A\omega_o^2/(\sigma^2 + \omega_o^2 - \omega^2 - 2\sigma j\omega)$. The property sign $(d\phi/dx) = sign(d\tan(\phi)/dx)$ is fulfilled for any angle ϕ and independent variable x. In the case of the impedance function $\tan(ang(Z_p(\omega)) = 2\sigma\omega/(\sigma^2 + \omega_o^2 - \omega^2)$ for positive σ, the phase associated with $Z_p(\omega)$ has positive slope at the resonance frequency ω_o. The function $Z(\omega)$ agrees with the inverse of the total admittance analyzed, $Y_T(\omega) = Y_T^r(\omega) + jY_T^i(\omega)$. In terms of $Y_T(\omega)$, it is possible to write $\tan(ang(Z_p(\omega))) = -Y_T^i(\omega)/Y_T^r(\omega)$. Assuming a small frequency variation of $Y_T^r(\omega)$, a resonance of the form $Y_T^r(\omega_o) < 0$, $Y_T^i(\omega_o) = 0$, $\partial(Y_T^i(\omega_o))/\partial\omega > 0$ will give rise to a positive slope of the phase associated with $Z(\omega)$, corresponding to a pair of unstable complex-conjugate poles. For a rigorous determination of the dc solution poles, pole–zero identification techniques [11] should be applied to the closed-loop transfer function $Z(\omega)$.

The result on the positive slope $\partial(Y_T^i(\omega_o))/\partial\omega > 0$ is in agreement with the preceding discussion on the stability conditions of steady-state oscillation. As already stated, the second term of (1.23) usually has little influence on the S value. The imaginary part, $Y_T^i(\omega)$, contributed primarily by the linear elements, is not typically very dependent on the oscillation amplitude. Thus, achieving the resonance condition at the oscillation frequency desired $Y_T^i(V \cong 0, \omega_o') = 0$, with positive slope $\partial(Y_T^i(V \cong 0, \omega_o')/\partial\omega > 0$, will facilitate stable oscillation at about ω_o'. Although small, there is usually a dependence of the susceptance Y_T^i on the signal amplitude. Therefore, the resonance frequency ω_o' under small-signal conditions will be similar to the oscillation frequency ω_o but generally not equal. In addition, the inherent nonlinearity of the oscillator circuit will generate a certain harmonic content that, as explained earlier, may give rise to a shift in the oscillation frequency.

The initial stage of oscillation startup will be ruled by the pair of unstable complex-conjugate poles $\sigma \pm j\omega$ of the dc solution, so the amplitude will grow according to $e^{\sigma t}$ from any small perturbation of this solution. The σ value is related linearly to $Y_T^r(\omega_o)$, fulfilling $Y_T^r(\omega_o) < 0$, and in general, σ will be more positive for larger absolute value $|Y_T^r(\omega_o)|$ [18]. This will imply a shorter initial transient. Actually, two different stages can be distinguished in the oscillation startup. In the initial stage, the oscillation amplitude is small and its variation can be predicted with circuit linearization about the dc solution. However, from a certain transient amplitude, the circuit will no longer be under small-signal conditions, and the real exponent σ will be different from the real part of the poles. In a simplified model it will exhibit an amplitude dependence $\sigma(V)$ coming from the amplitude dependence of the nonlinear conductance $G_N(V)$. The transient evolution depends on the function $\sigma(V)$. Usually, the positive exponent $\sigma(V)$ decreases monotonically to the value $\sigma = 0$, corresponding to steady state. However, in some cases, before reaching steady state, the positive σ increases, which is due to $G_N(V)$ becoming more negative versus V. After passing through a minimum, the conductance will increase (i.e., it will become less negative) until the steady-state condition $G_N(V) + G_L = 0$ is fulfilled. This type of behavior gives rise to an apparent delay in the startup transient as the amplitude growth becomes more noticeable for larger σ. It can be obtained in transistor-based oscillators that have power expansions of the nonlinear conductance of the form $G_N(V) = a_1 + a_2V^2 + a_3V^3 + \cdots$, with $a_1 < 0, a_2 < 0, a_3 > 0$ [18].

1.3.4 Formulation of Perturbed Oscillator Equations as an Eigenvalue Problem

For a better understanding of oscillator behavior, it will be convenient to formulate the perturbed oscillator equations as an eigenvalue problem. This will be done in terms of the amplitude and phase V and ϕ of the oscillator solution. The objective is to obtain a perturbed oscillator system of the form

$$\begin{bmatrix} \Delta\dot{V} \\ \Delta\dot{\phi} \end{bmatrix} = [M] \begin{bmatrix} \Delta V \\ \Delta\phi \end{bmatrix} \tag{1.25}$$

The matrix $[M]$ is derived directly from (1.21), taking into account that $\Delta\dot{\phi}(t) = \Delta\omega(t)$ and that $\partial Y_T/\partial\phi = 0$ due to the irrelevance with respect to the phase origin.

Thus, system (1.21) becomes

$$\begin{bmatrix} \Delta\dot{V}(t) \\ \Delta\dot{\phi}(t) \end{bmatrix} = \begin{bmatrix} \dfrac{1}{V_o}\dfrac{\partial Y^i_{To}}{\partial\omega} & \dfrac{\partial Y^r_{To}}{\partial\omega} \\ -\dfrac{1}{V_o}\dfrac{\partial Y^r_{To}}{\partial\omega} & \dfrac{\partial Y^i_{To}}{\partial\omega} \end{bmatrix}^{-1} \begin{bmatrix} -\dfrac{\partial Y^r_{To}}{\partial V} & \dfrac{\partial Y^r_{To}}{\partial\phi} \\ -\dfrac{\partial Y^i_{To}}{\partial V} & \dfrac{\partial Y^i_{To}}{\partial\phi} \end{bmatrix} \begin{bmatrix} \Delta V(t) \\ \Delta\phi(t) \end{bmatrix}$$
$$= \frac{\begin{bmatrix} -S & 0 \\ B & 0 \end{bmatrix}}{|\partial Y_{To}/\partial\omega|^2} \begin{bmatrix} \Delta V(t) \\ \Delta\phi(t) \end{bmatrix} \qquad (1.26)$$

where the coefficient B is deduced directly from the matrix product. One of the eigenvalues of the matrix on the right-hand side is $\lambda_1 = \sigma_o$. The second eigenvalue, $\lambda_2 = 0$, is due to the irrelevance of the oscillator solution versus any phase shift. From basic linear algebra [19,20], the general solution of linear differential equation system (1.26) with constant coefficients is

$$\begin{bmatrix} \Delta V(t) \\ \Delta\phi(t) \end{bmatrix} = c_1 e^{\sigma_o t} \begin{bmatrix} 1 \\ -\dfrac{B}{SV_o} \end{bmatrix} + c_2 \begin{bmatrix} 0 \\ 1 \end{bmatrix} \qquad (1.27)$$

Expression (1.27) evidences the irrelevance of the oscillator solution versus translations in the phase shift. Even if the oscillation is stable, which implies that $\sigma_o < 0$, the phase perturbation $\Delta\phi(t) = c_2$, with c_2 determined by the initial value, will remain in the steady-state solution. Equation (1.27) has a great conceptual interest. It enables the stability analysis of the steady-state oscillation, limited to two poles. Because one of the poles is necessarily zero, due to the autonomy of the free-running oscillator solution, the other pole must be real. In the case of an oscillator circuit with a single-resonant circuit, such as in Fig. 1.1, the system dimension (agreeing with the number of reactive elements) is $N = 2$. Thus, we only have two poles. The poles of the dc solution are complex-conjugate. The two poles of the steady-state oscillation are zero and real, respectively. The stability analysis derived by Kurokawa is limited to this real pole. In a "perfect" single-resonator oscillator, this should be sufficient. (A different problem is the limited accuracy of the analysis, considering only the fundamental frequency.) However, real-life oscillators, composed of several lumped reactive elements and distributed elements, will contain more poles. Therefore, the analysis from (1.27) will be unable to predict instabilities of the periodic solution coming from complex-conjugate poles or instabilities coming from two real poles on the right-hand side of the complex plane.

Clearly, for a free-running oscillator analyzed with one harmonic component and from a single observation port, the formulation above provides no advantage with respect to (1.23). However, when there is more than one state variable—two voltages, for instance, and/or several harmonic terms—phase variables will necessarily appear in the oscillator equations, as there is irrelevance with respect to the phase origin only. Then, use of a formulation of the type (1.26) will avoid a mixed system that includes the common frequency $\Delta\omega(t)$ in the set of circuit variables, together with the amplitudes and phases $\Delta V_n(t), n = 1$ to N, and $\Delta\phi_n(t), n = 2$ to N. An example of this type of formulation is the multiport stability analysis of a transistor-based oscillator, presented in the following. Other examples are shown throughout the book.

1.3.5 Generalization of Oscillation Conditions to Multiport Networks

As has been shown, in transistor-based oscillator design two of the transistor terminals are ended by particular immitance values, so it is possible to define the function $Y_N(V, \omega)$ depending only on the voltage amplitude at the reference plane. In turn, the load circuit exhibits the linear admittance $Y_L(\omega)$. Thus, fulfilment of the derived conditions (1.24) and (1.23) at the single-observation port considered will facilitate the stable oscillation. The same would be true for an impedance analysis in terms of the loop current. This is very helpful for circuit design, but does not fully guarantee stable operation of the oscillator circuit.

Alternatively, it is possible to use a generalization of the oscillation condition (1.13) in circuits containing multiport devices, which provides more accuracy and design flexibility. A brief explanation follows. The circuit is divided into two connected N-port networks, defined by their admittance matrixes $[Y_N(\overline{V}, \omega)]$ and $[Y_L(\omega)]$, with $\overline{V}$ the vector comprising voltage phasors at all N ports $\overline{V}^T$ with variables $[V_1, \ldots, V_N, \phi_1, \ldots, \phi_N]$. Note that the irrelevance with respect to the phase origin allows us to set any of the phase values to zero arbitrarily. Because there are no RF generators, the port voltages will be the same for the two connected multiport networks. The currents will have the same magnitude and opposite direction (or sign). Applying Kirchhoff's laws, it will be possible to write $([Y_N(\overline{V}, \omega)] + [Y_L(\omega)])\overline{V} = 0$. For $\overline{V}$ to differ from zero, the following oscillation condition must be fulfilled: $\det([Y_N(\overline{V}, \omega)] + [Y_L(\omega)]) = 0$. This condition generalizes (1.13) to multiport networks. Similar equations can be derived in terms of impedance or scattering matrixes.

The total admittance matrix of the circuit being considered can be defined as $[Y_T] = [Y_N(\overline{V}, \omega)] + [Y_L(\omega)]$. The circuit equations are written in matrix form as $\overline{H} = [Y_T]\overline{V} = 0$, the vector $\overline{V}$ comprising $[V_1, \ldots, V_N, \phi_1, \ldots, \phi_N]$. To balance the equation system, which must also be solved for ω, one of the phase variables is set arbitrarily to zero $\phi_k = 0$, which can be done due to the solution autonomy. For the quasistatic stability analysis of a given solution $\overline{V}_o^T = [V_1, \ldots, V_N, \phi_1, \ldots, \phi_N]$, the amplitudes and phases (except ϕ_k), as well as the frequency ω, must be perturbed about the steady-state values $\overline{V}_o^T$ and ω_o. Use of the frequency perturbation $\Delta\omega(t)$ leads to a mixed system, difficult to formulate in a compact manner. This can be avoided by considering the entire set of phase variables (including ϕ_k). Thus, the perturbations will be $\Delta V_1(t), \ldots, \Delta V_N(t), \Delta\phi_1(t), \ldots, \Delta\phi_N(t)$. The perturbed system will be derived by expanding the vector function $\overline{H}$ in a first-order Taylor series about the steady-state oscillation $(\overline{V}_o, \omega_o)$. Each H_k can be expressed as the product

$$H_k = [Y_T(\overline{V}, \omega)]_k^T \overline{V} \tag{1.28}$$

where $[Y_T]_k^T$ is a row matrix agreeing with the kth row of $[Y_T]$. The perturbed frequency will be given by $j\omega_o + s$. When performing the Taylor series expansion, it is taken into account that the multiplication by s acts like a time derivation of the perturbed variables, as shown in (1.18)–(1.20). Thus, the derivation of $[Y_T]_k^T$ with respect to frequency will give rise to terms of the form

$$\frac{\partial Y_{km}}{\partial \omega}\left(-j\frac{\Delta \dot{V}_m(t)}{V_m} + \Delta\dot{\phi}_m(t)\right) V_m e^{j\phi_m},$$

where k and m refer to the particular component of the kth row of the $[Y_T]$ matrix.

Using a development similar to the one in (1.18)–(1.20), each perturbed component H_k, with $k = 1, \ldots, N$, of the vector function $\overline{H}$ is given by

$$\frac{\partial H_k}{\partial V_1}\Delta V_1(t) + \cdots + \frac{\partial H_k}{\partial V_N}\Delta V_N(t) + \frac{\partial H_k}{\partial \phi_1}\Delta\phi_1(t)$$
$$+ \cdots + \frac{\partial H_k}{\partial \phi_N}\Delta\phi_N(t) + \frac{\partial Y_{k1}}{\partial \omega_o}\left(-j\frac{\Delta \dot{V}_1(t)}{V_1} + \Delta\dot{\phi}_1(t)\right) V_1 e^{j\phi_1}$$
$$+ \cdots \frac{\partial Y_{kN}}{\partial \omega_o}\left(-j\frac{\Delta \dot{V}_N(t)}{V_N} + \Delta\dot{\phi}_N(t)\right) V_N e^{j\phi_N} = 0 \quad (1.29)$$

with all the derivatives calculated at the steady-state oscillation. In expression (1.29) In matrix form it is possible to write

$$\left[\frac{\partial \overline{H}}{\partial \overline{V}_o}\right]\Delta\overline{V} + \left[\frac{\partial \overline{H}}{\partial \overline{\phi}_o}\right]\Delta\overline{\phi} + \left[\frac{\partial \overline{H}}{\partial \omega_o}\right]\left(-j\frac{\Delta\dot{\overline{V}}(t)}{V} + \Delta\dot{\overline{\phi}}(t)\right) = 0 \quad (1.30)$$

where $\Delta\overline{V}$ is the vector of amplitude increments, $\Delta\overline{\phi}$ the vector of phase increments, and $\Delta\overline{V}/V$ the vector of normalized amplitude increments. This notation indicates that each voltage increment is normalized by the corresponding steady-state value. On the other hand, $\left[\partial\overline{H}/\partial\omega_o\right]$ is a square matrix with k and m elements of the form $(\partial Y_{km}/\partial\omega)V_m e^{j\phi_m}$. Rearranging equation (1.30) it is possible to obtain a system of the form

$$\begin{bmatrix}\Delta\dot{\overline{V}} \\ \Delta\dot{\overline{\phi}}\end{bmatrix} = [JH]\begin{bmatrix}\Delta\overline{V} \\ \Delta\overline{\phi}\end{bmatrix} \quad (1.31)$$

Due to the irrelevance with respect to variations in the phase origin, the Jacobian matrix [JH] above must be singular. This is due to the fact that the solution remains the same if the phases of all the state variables are incremented by the same amount, $\Delta\alpha$. This means that for any H component,

$$\frac{\partial H_k^x}{\partial \phi_1}\Delta\alpha + \frac{\partial H_k^x}{\partial \phi_2}\Delta\alpha + \cdots + \frac{\partial H_k^x}{\partial \phi_N}\Delta\alpha = \left(\frac{\partial H_k^x}{\partial \phi_1} + \frac{\partial H_k^x}{\partial \phi_2} + \cdots + \frac{\partial H_k^x}{\partial \phi_N}\right)\Delta\alpha = 0 \tag{1.32}$$

where the superscript x refers to either a real or an imaginary part. From (1.32) it is clear that all the columns in (1.31) are related linearly, so the matrix $[JH]$ must be singular, with one eigenvalue, $\lambda_1 = 0$. This is in agreement with the fact that we are using one unnecessary phase variable that could have been set arbitrarily to any value. For the perturbation to vanish in time, all the rest of the eigenvalues of the matrix $[JH]$ must have a negative real part. As can be seen, this analysis generalizes the one-port analysis of (1.26) to multiple ports.

Analysis using (1.31) allows more insight into circuit behavior than does one-port analysis (1.26), as more observation ports are being considered. Actually, the analysis reflected in (1.26) is limited to one real eigenvalue, whereas (1.31) can provide a total of N eigenvalues, which can be real or complex conjugate. However, the analysis remains quasistatic, as a small frequency perturbation $\Delta\omega \ll \omega_o$ is still considered. Despite this, the formulation presented is helpful for understanding purposes and will be applied to some oscillator systems later in the book. It is particularly useful in the case of oscillator circuits composed of two or more suboscillator elements, as N-push oscillators [13] used for multiplication of the oscillation frequency, or in coupled-oscillator systems [4] used for beamsteering in phased arrays. However, for ordinary oscillator design, use of the total admittance function derived from a single sensitive port is more practical and intuitive.

1.3.6 Design of Transistor-Based Oscillators from a Single Observation Port

One-port analysis of transistor-based oscillators from a single observation port yields a simple oscillator design. It requires only the choice of a sensitive observation port and identification of a nonlinear active block and linear load network. As stated earlier, the steady-state oscillation condition $Y_T = 0$ is fulfilled at any possible observation node. However, the results of startup evaluation using $Y_T(V \cong 0, \omega)$ will depend on this observation node and may also be different for admittance analysis in terms of the node voltage, or impedance analysis in terms of the loop current. To show this more clearly, assume a parallel connection of the two blocks in Fig. 1.4. The total admittance is $Y_T = (G_N(\omega) + G_L(\omega)) + j(B_N(\omega) + B_L(\omega))$. Now, assume a series connection. The total impedance is $Z_T = (G_N(\omega) + jB_N(\omega))^{-1} + (G_L(\omega) + jB_L(\omega))^{-1}$. Developing this impedance function, it is easily seen that if the startup conditions (1.24) are fulfilled in terms of parallel admittance at ω_o', the equivalent conditions in terms of series impedance might be fulfilled at a different frequency, ω_o', or might never be fulfilled. Similar problems occur when changing the analysis port. A pure LC parallel (series) resonance will give a positive slope for admittance (impedance) analysis. Therefore, attention should be paid to the actual form of resonance of the circuit being analyzed. However, the nonlinear block of a practical circuit will generally contain several reactive and resistive elements, so the function Y_N cannot be modeled in a simple manner. In agreement with the discussion in Section 1.3.3, negative conductance $Y_T^r < 0$ at the resonance frequency ω_o' with negative slope of the susceptance $\partial(Y_T^i(V \cong 0, \omega_o')/\partial\omega < 0$ does not generally represent instability in the dc solution. It is advisable to perform an impedance analysis or to change the observation port until a positive slope is obtained. As a rule, the one-port conditions are helpful at the design stage. Then, a rigorous stability analysis of the steady-state solution obtained should be carried out. The use of numerical pole–zero identification [11] or other techniques, such as Nyquist criterion [16,21,22], will be necessary.

The topology of the FET-based oscillator of Fig. 1.6 matches the schematic representation of Fig. 1.5. The capacitor C_T, connected between the gate terminal and ground, constitutes a reactive "termination" at port 1. The capacitor C_{fb}, connected between the source terminal and ground, provides series feedback to the transistor in use. C_T and C_{fb} are both calculated to obtain negative resistance at the analysis port (port 2), defined between the drain terminal and ground, at the oscillation frequency desired, $f_o = 5.0$ GHz. The load circuit, with equivalent admittance Y_L (or impedance Z_L), is calculated to ensure the fulfilment of oscillation startup conditions at this specified oscillation frequency. The introduction of a parallel inductance provides a negative slope versus frequency of the small-signal susceptance $\partial Y_T^i(V \cong 0, \omega_o)/\partial\omega < 0$. A series inductance $L = 0.2$ nH, fulfilling $Z_N^i(\omega_o) + L\omega_o = 0$, is introduced instead, which also reduces the harmonic content due to lowpass filtering. An equivalent load resistance seen from the drain terminal is chosen: $R_L = 20\ \Omega$. This value provides an excess negative resistance $Z_N^r(\omega_o) + 20 = -45\Omega$, which should allow oscillation startup. Note that in the design discussed, the resonator is formed by the capacitive output of the nonlinear block containing the transistor and the series inductance introduced. It is possible to reduce the influence of the nonlinear block on the resonance frequency by adding a series capacitance C_s (not represented in the circuit schematic), such that the resonance frequency is determined primarily by the linear load circuit. Provided that C_s is small enough, the total capacitance (including the one at the output of the nonlinear block C_{out}) will be $C_T = C_s C_{\text{out}}/(C_s + C_{\text{out}}) \cong C_s$, much smaller than the original value C_{out}. Because $L_s = 1/(C_s\omega_o^2)$ to maintain the same resonance frequency, the quality factor Q of the series load resonator must increase significantly, giving a high value for the derivative $\partial Y_T^i(\omega_o)/\partial\omega$. For example, in the design discussed, the introduction of a series capacitance $C_s = 0.1$ pF requires an inductance of $L = 10.1$ nH to maintain the oscillation frequency at $f_o = 5$ GHz, which are quite extreme values. Thus, a high-Q resonator should be used. The derivative increases from $9.10 \times 10^{-9}\Omega^{-1}\cdot$ S in the original design to $2.9 \times 10^{-8}\Omega^{-1}$S, which is more than three times the original value. As will be shown, this increase in the frequency selectivity is very convenient for a low phase noise design, as well as the lower sensitivity (for $C_T \cong C_s$) to active device elements, subject to noise fluctuations.

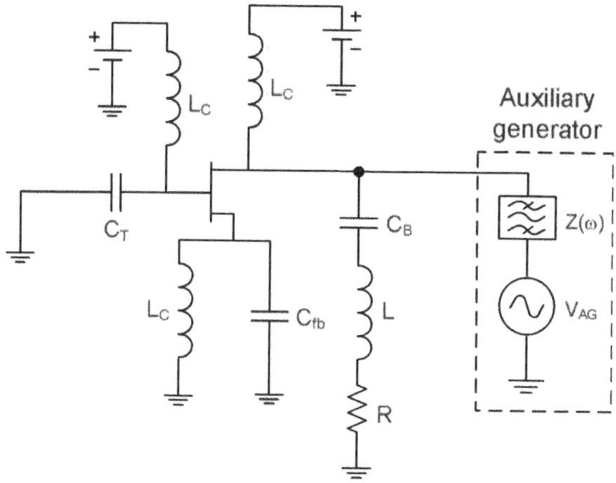

FIGURE 1.6 FET-based oscillator. The transistor is a NEC3210 biased at $V_{GS} = -0.25V$ and $V_{DS} = 2.25$ V. The capacitive termination C_T, together with the feedback capacitance C_{fb}, provide negative resistance at the drain port. The circuit topology matches the schematic representation of Fig. 1.5. The voltage auxiliary generator, connected in parallel at the transistor output node, is used for various analysis techniques presented in this chapter.

In the following, the load inductance $L = 0.2$ nH calculated originally will be considered instead of the high-Q load. This will allow a more general analysis of the oscillator dynamics, without the simplifications allowed by high-frequency selectivity. The circuit fulfills the oscillation startup conditions $Z_T^r(I \cong 0, \omega_o') < 0$, $Z_T^i(I \cong 0, \omega_o') = 0$, and $\partial(Z_T^i(I \cong 0, \omega_o')/\partial\omega > 0$ at the frequency $f_o = 5$ GHz. Evaluation of the admittance function $Y_T(V \cong 0, \omega) = 1/[Z_N^r(\omega) + jZ_N^i(\omega)] + 1/(R_L + jL\omega)$ shows a shift in the resonance frequency to the value $f_o = 5.2$ GHz (see Fig. 1.7). The total small-signal conductance has the negative value $Y_T^r(V \cong 0, \omega_o) < 0$, and there is a positive slope versus frequency of the susceptance $\partial Y_T^i(V \cong 0, \omega_o)/\partial\omega > 0$, with an excess of negative conductance in small-signal mode. Thus, the startup conditions are fulfilled. The pole analysis of this circuit, with a numerical technique, provides the unstable pair of complex-conjugate poles $2\pi(0.48 \pm j5.012) \times 10^9 s^{-1}$.

When using a commercial harmonic balance simulator, the nonlinear admittance function $Y_T(V, \omega)$ can be obtained with an auxiliary generator. The auxiliary generator is an artificial generator used for simulation purposes only. The voltage auxiliary generator at the frequency ω_{AG} is introduced in parallel at a circuit node (Fig. 1.6). Note that any voltage generator is a short circuit at any frequency different from the one that it delivers (ω_{AG}). To prevent the short circuiting of frequency components $\omega \neq \omega_{AG}$, the voltage generator is connected in series with an ideal bandpass filter, fulfilling $Z_f(\omega = \omega_{AG}) = 0$ and $Z_f(\omega \neq \omega_{AG}) = \infty$. The ratio between the current auxiliary generator current I_{AG} flowing into the circuit, and the voltage delivered, V_{AG}, provides the function $Y_{AG}(V_{AG}, \omega_{AG})$. This admittance function agrees with the total admittance $Y_T(V, \omega)$, depending on both the node

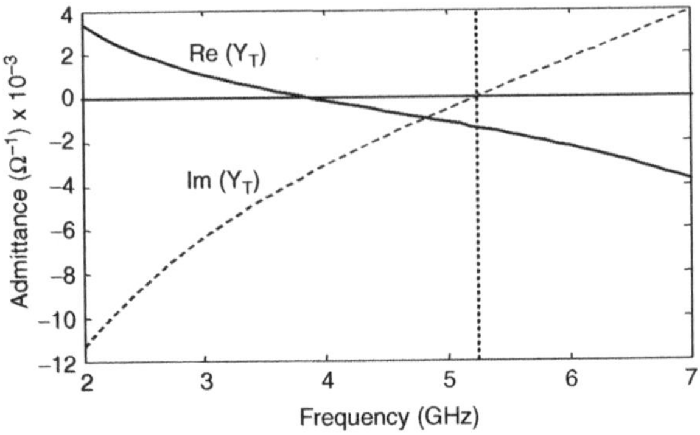

FIGURE 1.7 Small-signal admittance analysis of the FET-based oscillator of Fig. 1.6. There is excess negative conductance at the resonance frequency $f_o = 5.2$ GHz, so the startup of an oscillation at about this frequency can be expected for this particular design.

voltage amplitude $V = V_{AG}$ and the frequency $\omega = \omega_{AG}$ considered in all the preceding analyses. Thus, $Y_{AG} = Y_T$. An analogous procedure can be carried out to determine variations of the total impedance versus the branch current $Z_T(I, \omega)$. For this analysis, a current generator $I_{AG} = I$, at the frequency $\omega = \omega_{AG}$, is introduced in series at the circuit branch selected. To prevent open circuiting of frequency components $\omega \neq \omega_{AG}$, the current generator is connected in parallel with an ideal bandpass filter, fulfilling $Z_f(\omega = \omega_{AG}) = \infty$ and $Z_f(\omega \neq \omega_{AG}) = 0$. The input impedance function $Z_{AG} = Z_T(I, \omega)$ is given by the ratio between the voltage drop at the auxiliary generator V_{AG} and the current delivered, I_{AG}.

In the FET-based circuit considered here, the voltage auxiliary generator at the resonance frequency $f_o = 5.2$ GHz is connected in parallel at port 2 (see Figs. 1.5 and 1.6). By sweeping the auxiliary-generator amplitude V_{AG} from the small-signal conditions, it is possible to analyze the variation of the total admittance function Y_T versus the voltage amplitude at $f_{AG} = \text{f}_o$. The function $Y_{AG}(V_{AG}, f_o)$ is represented in Fig. 1.8. After a small-signal interval of nearly constant value, the conductance $\text{Re}[Y_{AG}]$ increases with the voltage amplitude, as expected in physical devices. On the other hand, the susceptance $\text{Im}[Y_{AG}]$ also varies with the voltage amplitude, due to the nonlinear behavior of the block containing the transistor. The susceptance $\text{Im}[Y_{AG}]$ increases with the amplitude. Thus, a smaller oscillation frequency than the one corresponding to the small-signal resonance of Fig. 1.7 should be expected. Optimization of the auxiliary generator voltage V_{AG} and frequency ω_{AG} to fulfill the goal $Y_{AG}(V_{AG}, \omega_{AG}) = 0$ allows us to obtain the steady-state oscillation amplitude $V_o = 4.4$ V and frequency $f_o = 4.4$ GHz. A multiharmonic analysis has actually been carried out for this calculation. More details are given in Chapter 5. The significant variation of the oscillation frequency is due to the low-frequency selectivity of the resonant consisting of the nonlinear block capacitance and the load inductance $L = 0.2$ nH. A high-quality factor load such as the one discussed at the beginning of the subsection would reduce the amplitude dependence of the imaginary part

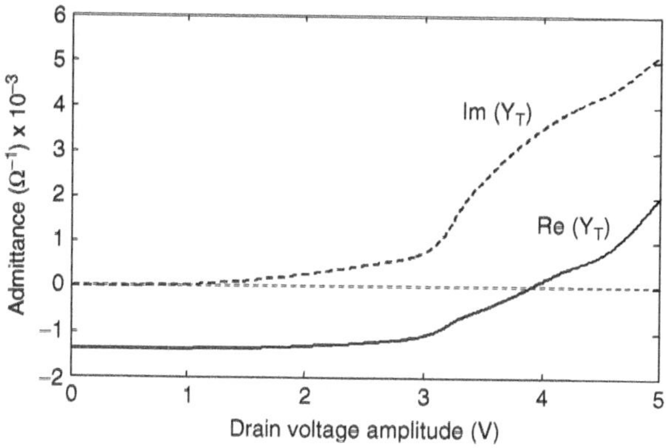

FIGURE 1.8 Variation of the admittance at port 2 versus the amplitude of a voltage auxiliary generator at the frequency $f_o = 5.2$ GHz introduced in parallel at the same port.

of the total admittance function $Y_T^l(V, \omega)$. Then the resonance frequency under small-signal conditions would be much closer to the actual oscillation frequency.

For validation, a time-domain simulation of the free-running oscillator has been carried out, with the results presented in Fig. 1.9. This shows that as predicted by the analysis of Fig. 1.7, the oscillation actually starts up. Steady state is reached after a transient. The envelope of the transient is initially exponential $e^{\sigma t}$ and then evolves gradually to the constant steady-state value. According to previous discussions, the exponent σ depends on the net negative conductance G_T and the quality factor of the resonant circuit. It is given by $\sigma = -\omega_o G_T / 2 G_L Q$. The steady-state oscillation obtained has the frequency $f_o = 4.4$ GHz and first-harmonic voltage amplitude $V = 4.4$ V at port 2. In agreement with the admittance function variation versus the voltage amplitude $Y_T(V, \omega_o)$, shown in Fig. 1.8, the steady-state oscillation frequency is smaller than the one predicted by small-signal analysis.

Note that the integration from a different initial condition provides a time-shifted steady-state solution with an identical waveform. In (1.12) it was shown that for a one-harmonic analysis of the oscillator, in terms of the node voltage $v(t) = \text{Re}[Ve^{j\omega_o t}]$, any phase shift $v(t) = \text{Re}[Ve^{j(\omega_o t + \phi)}]$ provides an equally valid solution. When considering several harmonic components, the solution will be invariant with respect to the phase of only one of these harmonic components. Otherwise, aside from the time shift, there would be a change in the waveform itself, which is not the case in periodic oscillation. In general frequency-domain analysis, considering two or more state variables, the solution will be invariant with respect to the phase of only one harmonic component of one of these state variables.

To illustrate we apply the stability condition (1.23), derived for a one-port and one-harmonic analysis, to our FET-based oscillator. The termination and feedback elements of the transistor were calculated to obtain negative resistance at the drain node, so this will be the reference node selected for stability analysis. Condition (1.23) is evaluated with the aid of the same auxiliary generator as that used to determine $Y_T(V_o, \omega_o)$ (Fig. 1.8). The derivatives about the free-running oscillation

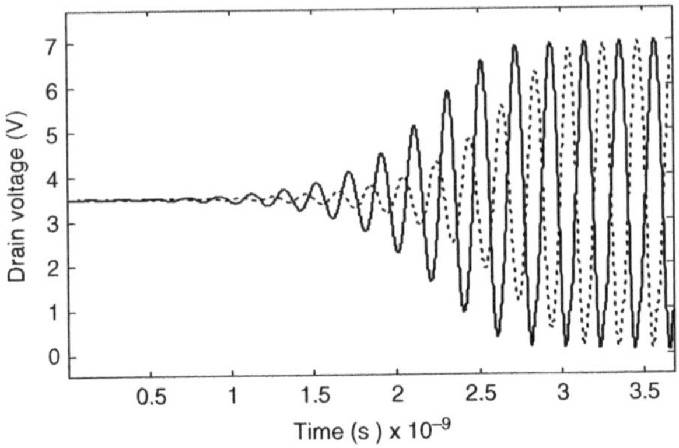

FIGURE 1.9 Time-domain analysis of the oscillator of Fig. 1.6. The envelope of the transient is initially exponential, evolving gradually to a constant steady-state value. The oscillation frequency is $f_o = 4.4$ GHz. Integration from different initial conditions gives rise to time-shifted steady-state waveforms which are equally valid oscillator solutions.

(V_o, ω_o) are obtained through finite differences. Initially, the generator amplitude is kept constant at the oscillation value $V_o = 4.4$ V, performing a frequency sweep about $f_o = 4.4$ GHz. The result is presented in Fig. 1.10. Note that the steady-state oscillation fulfills $\text{Re}[Y_T] = 0$ and $\text{Im}[Y_T] = 0$. Compared with the small-signal analysis of Fig. 1.7, the resonance frequency has decreased from 5.2 GHz to 4.4 GHz. On the other hand, the slope of $\text{Im}[Y_T]$ (the dashed line) remains positive at the resonance frequency, as in the small-signal analysis of Fig. 1.7. Next, the generator frequency is kept constant at f_o and the generator amplitude is swept about $V_o = 4.4$ V. When representing $Y_T(V, f_o)$ and $Y_T(V_o, f)$ in the plane defined by $\text{Re}[Y_T]$ and $\text{Im}[Y_T]$, Fig. 1.11 is obtained. The solid-line curve corresponds to the function $Y_T(V, f_o)$. The dashed-line curve corresponds to the function $Y_T(V_o, f)$.

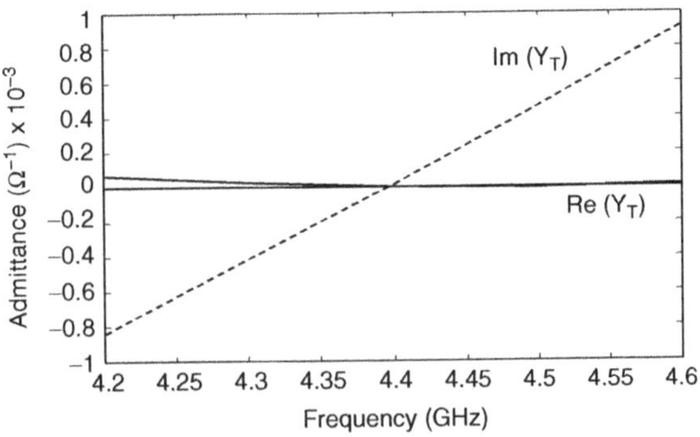

FIGURE 1.10 FET-based oscillator. Determination of the derivatives $\text{Re}[\partial Y_T/\partial f_o]$ and $\text{Im}[\partial Y_T/\partial f_o]$ about the free-running oscillation point V_o, f_o using a voltage auxiliary generator.

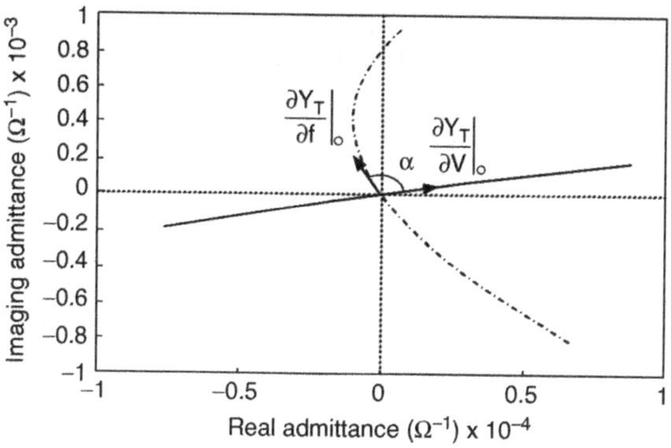

FIGURE 1.11 Representation of the curves $Y_T(V, f_o)$ and $Y_T(V_o, f)$ on the plane defined by $\mathrm{Re}[Y_T]$ and $\mathrm{Im}[Y_T]$. The origin, with $\mathrm{Re}[Y_T] = 0$ and $\mathrm{Im}[Y_T] = 0$, corresponds to the free-running oscillation $f_o = 4.4$ GHz and $V_o = 4.4$ V. The derivatives $\partial Y_{To}/\partial V$ and $\partial Y_{To}/\partial f$, agree, respectively, with the tangents to the origin of the curves $Y_T(V, f_o)$ and $Y_T(V_o, f)$.

The origin, with $\mathrm{Re}[Y_T] = 0$ and $\mathrm{Im}[Y_T] = 0$, corresponds to the free-running oscillation $f_o = 4.4$ GHz and $V_o = 4.4$ V. The derivatives $\partial Y_{To}/\partial V$ and $\partial Y_{To}/\partial f$ agree, respectively, with the tangents to the origin of the curves $Y_T(V, f_o)$ and $Y_T(V_o, f)$.

The derivative with respect to the voltage amplitude takes the value $\partial Y_T/\partial V_o = 5.529 \times 10^{-4} + j0.0012\Omega^{-1}/\mathrm{V}$. In turn, the derivative with respect to the frequency is $\partial Y_T/\partial \omega_o = -3.177 \times 10^{-14} + j6.929 \times 10^{-13}$. Thus, the term in (1.23) has the value $S = 4.2268 \times 10^{-16}$. Its positive sign indicates a stable solution. Note that the product in (1.23) will be positive for an angle $\alpha_{v\omega}$, defined as $\alpha_{v\omega} = \mathrm{ang}(\partial Y_{To}/\partial \omega) - \mathrm{ang}(\partial Y_{To}/\partial V)$, between 0 and π. Thus, the sign of σ_o can be determined graphically by tracing $\partial Y_{To}/\partial V$ and $\partial Y_{To}/\partial f$ in a polar plot (Fig. 1.11).

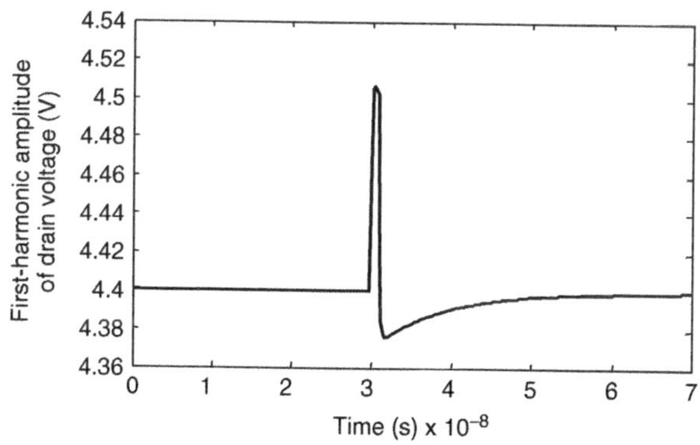

FIGURE 1.12 Oscillatory solution: reaction of the voltage amplitude at the drain node to an instantaneous perturbation applied at $t_p = 30$ ns. The oscillatory solution is stable, so the amplitude recovers its initial value after an exponential transient.

Figure 1.12 shows the effect of a perturbation on voltage amplitude at observation port 2. Instantaneous perturbation is applied at the time $t_p = 30$ ns. In agreement with the formulation (1.22), the amplitude perturbation follows an exponential transient $\Delta V(t) = \Delta V_o e^{\sigma_o t}$. Because the oscillatory solution is stable ($S > 0$), the exponent σ_o has a negative sign: $\sigma_o = -3.87 \times 10^9 S^{-1}$. Therefore, the transient leads back to the original value of the oscillation amplitude, $V_o = 4.4$ V.

1.4 FREQUENCY-DOMAIN FORMULATION OF AN OSCILLATOR CIRCUIT

The oscillator admittance–impedance analysis presented so far assumes a sinusoidal oscillation $v(t) = V_o \cos \omega_o t$. However, the inherent nonlinearity of the oscillator circuit will generate some harmonic content. As already stated, the relevance of the harmonic components will be higher for a smaller quality factor of the load circuit. The objective here is to derive the circuit equations when considering harmonic components up to a certain order N. This will show how the previous analysis at the fundamental frequency generalizes to N harmonic terms. The objective of introducing this formulation here is to provide the necessary background for the phase noise analysis in Chapter 2 and the analysis of frequency dividers in Chapter 4.

1.4.1 Steady-State Formulation

For the frequency-domain analysis of a given nonlinear circuit, the circuit variables are represented in a Fourier series. For simplicity, a single state variable $v(t)$ and a single nonlinearity of current type $i(v)$ are considered. The voltage variable is expressed as $v(t) = \sum_{k=-N}^{N} V_k e^{jk\omega_o t}$, with V_k complex coefficients. Note that because $v(t)$ is a real variable, the Fourier series contains both negative and positive harmonic frequencies $k\omega_o$, fulfilling $V_{-k} = V_k^*$. Due to the orthogonality of the Fourier frequency basis, a circuit of the form of Fig. 1.4 can be formulated by applying Kirchhoff's laws independently at the various harmonic frequencies $k\omega_o$. This provides a system of the form

$$\begin{aligned}
H_{-N} &= V_{-N} + Z_L(-N\omega_o)I_{-N}(V_{-N}, \ldots, V_o, \ldots, V_N) = 0 \\
&\vdots \\
H_o &= V_o + Z_L(0)I_o(V_{-N}, \ldots, V_o, \ldots, V_N) + E_{dc} = 0 \\
&\vdots \\
H_k &= V_k + Z_L(k\omega_o)I_k(V_{-N}, \ldots, V_o, \ldots, V_N) = 0 \\
&\vdots \\
H_N &= V_N + Z_L(N\omega_o)I_N(V_{-N}, \ldots, V_o, \ldots, V_N) = 0
\end{aligned} \tag{1.33}$$

where H_k are complex error functions. Note that the bias sources should be included in the dc term. As an example, a series voltage source E_{dc} has been considered in (1.33). The total number of equations is $2N + 1$, as each harmonic function H_k has real and imaginary parts, except the one corresponding to dc, given by H_o, which is real valued. The equation system (1.33) constitutes the harmonic balance formulation of the oscillator circuit, containing a single nonlinearity of current type only. As written in (1.33), it is valid only for current-type nonlinearities. It cannot be applied in the case of capacitive nonlinearities. As shown in Chapters 3 and 5, the capacitive nonlinearities are described in terms of the harmonic components of the corresponding nonlinear charge $q(v)$. Once the harmonic components $Q_{-N}, \ldots, Q_k, \ldots, Q_N$ are determined, the harmonics of the current through the capacitance are easily obtained from $-jN\omega_o Q_{-N}, \ldots, jk\omega_o Q_k, \ldots, jN\omega_o Q_N$.

As shown in (1.33), Kirchhoff's laws are fulfilled independently at each harmonic component. The key point is that each harmonic component of the nonlinear current depends on all the harmonic components of the node voltage, as they are linked through the constitutive relationship $i(t) = i(v(t))$. Analytically, the various harmonic terms of $i(t)$ would be obtained by calculating $i(t) = i\left(\sum_{k=-N}^{N} V_k e^{jk\omega_o t}\right)$. Note that use of the Fourier expansion from $k = -N$ to $k = N$ allows as to introduce $v(t)$ directly in the constitutive relationship $i(t) = i(v(t))$. This is why the harmonic balance system is generally expressed by considering positive and negative frequencies, even if we know that the harmonic terms at $k\omega_o$ and $-k\omega_o$ fulfill the Hermitian symmetry relationship $V_{-k} = V_k^*$. To understand the dependence $I_k(V_{-N}, \ldots, V_o, \ldots, V_N)$ of each harmonic component of $i(t)$ on all the harmonic components of $v(t)$, consider the particular case of a polynomial characteristic $i(t) = i(v(t))$. The expansion $i(t) = i\left(\sum_{k=-N}^{N} V_k e^{jk\omega_o t}\right)$ clearly gives rise to a mixed dependence of each I_k on different harmonic coefficients $V_{-N}, \ldots, V_o, \ldots, V_N$. In practice, the components $I_k(V_{-N}, \ldots, V_o, \ldots, V_N)$ are obtained numerically using inverse and forward Fourier transfers. Under any variation of $(V_{-N}, \ldots, V_o, \ldots, V_N)$, the waveform $v(t)$ is calculated with an inverse Fourier transform, then the waveform $i(t)$ is obtained from the relationship $i(v(t))$, and finally, the harmonic components I_k are calculated with a forward Fourier transform. Details of this calculation are given in Chapter 5.

System (1.33) is a nonlinear algebraic system which is usually resolved by employing the well-known Newton–Raphson algorithm. Note that in an oscillator circuit, the frequency ω_o is an unknown to be determined, so the system, in the form (1.33), is unbalanced, as it contains $2N + 1$ equations in $2N + 2$ unknowns, given by the real and imaginary parts of all the harmonic components of $v(t)$, plus the oscillation frequency ω_o. To solve this problem, either the real or imaginary part of one of the harmonic components of $v(t)$ will be set arbitrarily to zero, which is allowed by the autonomy of the steady-state oscillation.

As an example of the formulation (1.33), consider a case in which both the dc and first-harmonic component are taken into account in the oscillator solution, so the unknown voltage is expressed as $v(t) = V_o + V_1 e^{j\omega_o t} + V_{-1} e^{-j\omega_o t}$, with

$V_1 = V_{-1}^*$. The steady-state system is given by

$$\begin{aligned} H_o &\equiv V_o + R_L(0)I_o(V_o, V_1, V_{-1}) = 0 \\ H_1 &\equiv V_1 + Z_L(\omega_o)I_1(V_o, V_1, V_{-1}) = 0 \\ H_{-1} &\equiv V_{-1} + Z_L(-\omega_o)I_{-1}(V_o, V_1, V_{-1}) = 0 \end{aligned} \tag{1.34}$$

where, for simplicity, no bias sources are considered. This one-harmonic example is considered again later in the section.

The general system (1.33) can be written in matrix form as

$$\overline{H}_s = \overline{V}_s + [Z_L(k\omega_o)]\overline{I}_s(\overline{V}_s) = 0 \tag{1.35}$$

where the vector $\overline{V}_s$ is made up of the steady-state terms $\overline{V}_s = [V_o \quad V_1 \quad V_{-1} \quad \cdots \quad V_N \quad V_{-N}]$, the vector $\overline{I}_s$ is given by $\overline{I}_s = [I_o \quad I_1 \quad I_{-1} \quad \cdots \quad I_N \quad I_{-N}]$, and the linear matrix $[Z_L(k\omega_o)]$ is the diagonal matrix:

$$[Z_L(k\omega_o)] = \begin{bmatrix} R_L(0) & 0 & 0 & & 0 \\ 0 & \ddots & & & \\ & & Z_L(k\omega_o) & & \\ & & & \ddots & 0 \\ 0 & 0 & & 0 & Z_L(-N\omega_o) \end{bmatrix} \tag{1.36}$$

where k indicates the varying integer order of the harmonic coefficient, that is, $0, \ldots, k, -k, \ldots, N, -N$.

For conceptual purposes it is interesting to obtain the Jacobian matrix associated with system (1.33), which has the form

$$[JH] = [I_d] + [Z_L(k\omega_o)]\left[\frac{\partial \overline{I}}{\partial \overline{V}}\right]_s \tag{1.37}$$

with $[I_d]$ being the identity matrix and $\left[\partial \overline{I}/\partial \overline{V}\right]_o$ the Jacobian matrix of the nonlinear function, consisting of the derivatives of the various harmonic components of the current with respect to the harmonic components of the independent voltage. As an example, in the case of the system (1.34), comprising the dc and first-harmonic component, the Jacobian matrix $\left[\partial \overline{I}/\partial \overline{V}\right]_s$ is given by

$$\left[\frac{\partial \overline{I}}{\partial \overline{V}}\right]_s = \begin{bmatrix} \dfrac{\partial I_o}{\partial V_o} & \dfrac{\partial I_o}{\partial V_1} & \dfrac{\partial I_o}{\partial V_{-1}} \\ \dfrac{\partial I_1}{\partial V_o} & \dfrac{\partial I_1}{\partial V_1} & \dfrac{\partial I_1}{\partial V_{-1}} \\ \dfrac{\partial I_{-1}}{\partial V_o} & \dfrac{\partial I_{-1}}{\partial V_1} & \dfrac{\partial I_{-1}}{\partial V_{-1}} \end{bmatrix}_s \tag{1.38}$$

Determination of this matrix is much simpler than it seems. It is sufficient to take into account the fact that the derivative of the kth harmonic of the nonlinear current with respect to the mth harmonic of the voltage can be obtained as

$$\begin{aligned}\frac{\partial I_k}{\partial V_m} &= \frac{1}{T}\frac{\partial\left(\int_0^T i(t)e^{-jk\omega_o t}dt\right)}{\partial V_m} = \frac{1}{T}\int_0^T \frac{\partial i(t)}{\partial v(t)}\frac{\partial v(t)}{\partial V_m}e^{-jk\omega_o t}\,dt \\ &= \frac{1}{T}\int_0^T g(t)e^{jm\omega_o t}e^{-jk\omega_o t}\,dt \\ &= G_{k-m}\end{aligned} \tag{1.39}$$

with $g(t)$ being the time-domain derivative $g(t) = \partial i(t)/\partial v(t)$ and G_{k-m} being the $(k-m)$th harmonic component of $g(t)$. Taking the property above into account, the Jacobian matrix $\left[\partial \overline{I}/\partial \overline{V}\right]_s$ can be rewritten

$$\left[\frac{\partial I}{\partial V}\right]_s = \begin{bmatrix} G_0 & G_{-1} & G_1 \\ G_1 & G_0 & G_2 \\ G_{-1} & G_{-2} & G_0 \end{bmatrix}_s \tag{1.40}$$

The matrix (1.40), with equal diagonal elements G_0, is the conversion matrix associated with $g(t)$. Therefore, the matrix $[\partial \overline{I}/\partial \overline{V}]_s$ can be obtained from the Fourier series expansion of $g(t)$. Note that it is necessary to double the number of harmonic components considered in the Fourier series expansion of $g(t)$, which now goes from $-2\omega_o$ to $2\omega_o$. These results are easily extended to any number N of harmonic terms.

It has been shown in previous sections that an arbitrary variation of the phase origin of the oscillator solution provides another solution. Because of this, the Jacobian matrix associated with system (1.35), used in the Newton–Raphson algorithm, is singular at steady-state oscillation. To understand this singularity of $[JH]$, consider a time shift τ of the steady-state waveform. This will give rise to the phase shift $-k\omega_o\tau = k\alpha$ of the various harmonic terms, where $\alpha = -\omega_o\tau$ has been introduced. The phase-shifted solution must also be a solution of (1.35). Therefore, it is possible to write

$$\frac{\partial \overline{H}_s}{\partial \alpha} = \left[\frac{\partial \overline{H}_s}{\partial \overline{V}_s}\right]\frac{\partial \overline{V}_s}{\partial \alpha} = \overline{0} \tag{1.41}$$

Because the second factor of equation (1.41) is different from zero, the Jacobian matrix must be singular. In the numerical resolution of (1.35) with the Newton–Raphson algorithm, the singularity problem of the Jacobian matrix can be circumvented by arbitrarily setting to zero the imaginary part of one of the harmonic components (e.g., $V_1^i = 0$). As stated earlier, this also leads to a well-balanced system, with $2N+1$ equations in $2N+1$, given by the real and imaginary parts of all the harmonic components of $v(t)$ except $V_1^i = 0$ and the oscillation frequency ω_o.

1.4.2 Stability Analysis

The stability analysis presented in Section 1.3.2 assumed a small frequency perturbation $\Delta\omega \ll \omega_o$. However, the perturbation frequency is not necessarily small, as in the case of instabilities leading to a division by 2 of the oscillation frequency. For a more general stability analysis, the limitation $\Delta\omega \ll \omega_o$ must be eliminated. In the following, a small-amplitude perturbation of complex frequency $s = \sigma + j\omega$ is considered, with $\omega \in (0, \omega_o)$. Stability analysis is used once the steady-state oscillation has been determined by applying the Newton–Raphson algorithm to the system (1.33). The small perturbation at the initial time t_o will give rise to small amplitude increments in the voltage and current vectors, given by ΔV and ΔI, respectively. Thus, it will be possible to consider a first-order Taylor series expansion of the nonlinearity $\overline{I}(\overline{V})$ about the steady-state solution $\overline{V}_s$, so $\overline{I}(\overline{V})$ is replaced by $[\partial\overline{I}/\partial\overline{V}]_s\Delta\overline{V}$. The perturbed oscillator equations are written

$$[JH(jk\omega_o + s)]\Delta\overline{V}(s) = \left\{[I_d] + \left[Z_L(jk\omega_o + s)\right]\left[\frac{\partial\overline{I}}{\partial\overline{V}}\right]_s\right\}\Delta\overline{V}(s) = 0 \quad (1.42)$$

Note that the system (1.42) contains two different frequency variables, the steady-state frequency ω_o and the complex frequency s, generated as a result of perturbation at t_o. The formulation is similar to that used in the conversion matrix approach [23], although the small-signal frequency is complex in this case. The perturbation gives rise to a transient variation of the harmonic components that is taken into account by means of dependence on the complex frequency s. Note that unlike the analysis of Section 1.3, the imaginary part of the sideband frequency s is not limited to small values. It can take any value in the interval $(0, \omega_o)$. Note that particularizing s to the case of small frequency variations, the impedance matrix $[Z_L(jk\omega_o + s)]$ can be expanded in a Taylor series, so it is possible to write $[Z_L(jk\omega_o + s)] \cong Z_L(jk\omega_o) + (\partial Z_L/\partial(jk\omega_o))s$. It is easily seen that placing this in (1.42) and limiting analysis to the fundamental frequency ω_o, an equation equivalent to (1.26) is obtained. Therefore, the analysis technique (1.18 and 1.26) is a particularization of the more general stability analysis (1.42) to the case of a small perturbation frequency $\Delta\omega$.

System (1.42) is a homogeneous linear system, so for the perturbed solution $\Delta\overline{V}$ to differ from zero, the associated characteristic determinant must be zero $\det\{[I_d] + [Z_L(jk\omega_o + s)][\partial\overline{I}/\partial\overline{V}]_s\} = 0$, with $[I_d]$ the identity matrix. Note that the increment $\Delta\overline{V}(s)$ necessarily differs from zero since an instantaneous perturbation was actually applied at t_o. Because s is the complex frequency of the perturbation, evolution of this perturbation will depend on the roots s of the characteristic determinant $\det[JH(k\omega_o + s)]$. For $s = 0$, the characteristic determinant agrees with the determinant of the Jacobian matrix $\det[JH] = 0$ of the harmonic balance system [see (1.37)]. In (1.41) it was shown that this Jacobian matrix is singular, $\det[JH] = 0$. Therefore, one of the roots of the characteristic determinant will be $s = 0$. This zero root is due to the system autonomy. For this perturbation to vanish exponentially in time, all the rest of the roots must have a negative real part.

To transform the analysis of the characteristic system (1.42) into a pole analysis, a small-signal current source $I_{\text{in}}(s)$ is introduced in parallel with the nonlinear element. The current source at the frequency s will generate the sidebands $k\omega_o + s$. The original system (1.42), with the input $I_{\text{in}}(s)$, will be ruled by

$$\{[I_d] + [Z_L(jk\omega_o + s)] \left[\frac{\partial \overline{I}}{\partial \overline{V}} \right]_s \} \Delta \overline{V}(s) = \left[Z_L(jk\omega_o + s) \right] \begin{bmatrix} I_{\text{in}}(s) \\ \vdots \\ 0 \end{bmatrix} \quad (1.43)$$

Any output $Y(s)$ selected will be linearly related to the increment $\Delta\overline{V}(s)$, in the form $Y(s) = B\Delta\overline{V}(s)$, with B a row matrix. Unless pole–zero cancellations occur, all possible transfer functions will have the same denominator, due to the division by the characteristic determinant $\det\{[I_d] + [Z_L(jk\omega_o + s)][\partial\overline{I}/\partial\overline{V}]_s\}$. The system poles will agree with the roots of this determinant. In particular, it is possible to define the transfer function $Z_{\text{in}}(s) = \Delta V_1(s)/I_{\text{in}}(s)$, where $\Delta V_1(s)$ is the lowest sideband ($k = 0$) of node voltage perturbation. Clearly, this frequency-domain analysis is totally equivalent to the one presented for a dc solution in Section 1.2. However, unlike in dc analysis, two frequencies are involved in the linearization (1.43), one coming from the perturbation s and the other from ω_o, associated with the steady-state regime.

In the circuit shown in Fig. 1.1, with the cubic nonlinearity $i(v) = av + bv^3$, the terms in the Jacobian matrix (1.40) are given by $G_o = a + 3/2bV_o^2$, $G_1 = 0.0$, and $G_2 = 3/4bV_o^2$. The matrix $[Z_L(jk\omega_o + s)]$ is obtained directly from the inverse of the linear admittance $Z_L(\omega) - [G \mid j(C\omega - 1/(L\omega))]^{-1}$, with ω a generic frequency. The characteristic determinant is second order in s, so it has two different roots. Due to fulfilment of the oscillation condition (1.35), one root is $s = 0$, associated with the solution autonomy. The second real root is $-0.32 \times 10^9 s^{-1}$, so the oscillation is stable.

1.5 OSCILLATOR DYNAMICS

In this section the oscillator circuit is studied as a dynamic system [10,24] which will provide a geometric viewpoint of oscillator behavior and valuable background for an understanding of stability and phase noise. This study will be a general one, with no limiting assumptions in terms of state variables, harmonic content, or frequency of perturbations.

1.5.1 Equations and Steady-State Solutions

The nonlinear differential equations ruling circuit behavior are generally expressed in terms of a vector of state variables $\overline{x}$. This vector consists of the minimum number of variables such that its knowledge at time t_o together with that of the system input for $t > t_o$ determine the circuit response for $t > t_o$. Different choices are possible. As an example, the second-order nonlinear equation (1.9) can be split into two first-order equations by using the two state variables $x_1(t) = v(t)$ and $x_2(t) = dv/dt$. In lumped circuits, a common choice for the state variables in $\overline{x}$ is the set consisting of all inductor currents $i_{L1}, i_{L2}, \ldots$ and all capacitor voltages $v_{C1}, v_{C2}, \ldots$. The system order agrees with the number of reactive elements in the circuit. For circuits containing ideal transmission lines, the system order is ideally infinite, as the transmission lines are described with exponential terms of the form $\exp(A + sB)$, with s the Laplace frequency and A, B constant 2×2 matrixes (see Section 5.2, Chapter 5). A Taylor series expansion of this exponential would give rise to time derivatives of increasingly high order. If the time delay associated with each transmission line is not too high, it is possible to transform the differential equation system into a system of differential difference equations, due to the presence of the delayed variables $x_{T_1}(t + \tau_1) \cdots x_{T_M}(t + \tau_M)$, with M the number of transmission lines and $\tau_1, \ldots, \tau_M$ their corresponding time delays [25]. Other ways to tackle the simulation of distributed elements are presented in Chapter 5.

Because the main purpose of this section is to provide a general explanation of the oscillator dynamics, only the case of lumped-element circuits is considered. The vector containing the circuit state variables will be $\overline{x} \in R^N$. The time-domain equations will be written using Kirchhoff's laws together with the constitutive relationships of the nonlinear elements. This will provide a system of differential algebraic equations. In some cases, these time-domain equations can be expressed in *state form* [24]:

$$\begin{aligned} \dot{\overline{x}} &= \overline{f}(\overline{x}) \\ \overline{x}(t_o) &= \overline{x}_o \end{aligned} \tag{1.44}$$

Here $\overline{f}$ is a vector of nonlinear smooth functions (i.e., having continuous derivatives with respect to $\overline{x}$ up to infinite order). It must be noted that in free-running oscillators the function $\overline{f}$ does not depend explicitly on time. This is because it does not contain any time-varying external generators. As an example, in the parallel resonance oscillator of Fig. 1.1 the state variables are the voltage across the capacitor $v_c(t)$ and the current through the inductance $i_L(t)$. Thus, the state variable vector is defined as $\overline{x} = (v_c, i_L)^T$. Applying Kirchhoff's laws, it is possible to write

$$\frac{dv_c}{dt} = -\frac{i_L}{C} - \frac{v_c}{RC} - i_{nl}(v_c) = -\frac{i_L}{C} - \frac{v_c}{RC} - \frac{(av_c + bv_c^3)}{C} \qquad \text{(a)}$$
$$\frac{di_L}{dt} = \frac{v_c}{L} \qquad \text{(b)} \tag{1.45}$$

Clearly, equation system (1.45) is formally similar to the general equation (1.44), with a two-dimensional nonlinear function $\overline{f}$ that does not depend explicitly on time. A dc solution can generally be found for any circuit described by (1.44) by setting $\dot{\overline{x}} = 0$ and solving $\overline{f}(\overline{x}_{DC})$. This is the dc solution that always coexists with the oscillatory solution, as shown in previous sections. However, in a well-designed oscillator, this dc solution must be unstable. Thus, integration of the circuit differential equations from any initial condition $\overline{x}_0 \neq \overline{x}_{dc}$ must provide a transient leading to a periodic steady-state oscillation $\overline{x}_s(t)$.

Another property of autonomous systems already discussed is that any arbitrary time translation of the steady-state solution $\overline{x}_s(t)$ provides another valid solution $\overline{x}_s(t - \tau)$. Actually, the initial conditions $\overline{x}_o$ considered for the integration of (1.44) may be associated with any time value t_o, because the function $\overline{f}$ does not depend explicitly on time. When integrating the circuit equations (1.44) from different initial values $\overline{x}_o$, the same steady-state waveform, with different time shifts, is obtained (see Fig. 1.9).

The situation is different for circuits that have a time-varying independent source such as a sinusoidal generator. When employing Kirchhoff's laws, this source will give rise to an explicit time dependence of the nonlinear differential equation system, which in state form will be written $\dot{\overline{x}} = \overline{f}(\overline{x}, t)$. Note that in the common case of a periodic source with period T, the nonlinear function $\overline{f}$ will also be periodic, with the same period T.

For compactness of the formulation, the same formal equation (1.44) is often used for both autonomous and nonautonomous periodic systems. Actually, a nonautonomous system can be expressed as an autonomous system if the time t is included in the state variable vector, which becomes $\overline{x}'$. The new variable t is unbounded, as time tends to infinity. A different variable related to time can be chosen instead. For a periodic nonlinear function $\overline{f}$, the angle variable $\theta = (2\pi/T)t$ is used, with T being the independent source period [24]. The variation of this new variable can be limited to the range [0,2π). Defining the new state vector as $\overline{x}' = (\overline{x}, \theta)$, the equations of the nonautonomous periodic system are expressed as

$$\begin{aligned} \dot{\overline{x}} &= \overline{f}(\overline{x}, \theta) \\ \dot{\theta} &= \frac{2\pi}{T} \end{aligned} \tag{1.46}$$

As gathered from (1.46), the dimension of the nonautonomous periodic system increases in one respect to that of an autonomous system containing the same number of reactive elements. Note that this is merely a change in the formal expression of the system, since the dependence on the time reference is, of course, maintained under this change.

By *forced circuits* we mean circuits with an independent time-varying source that do not oscillate or circuits that exhibit an oscillation synchronized to the independent periodic source. When integrating the nonlinear differential equations that describe these circuits from the same initial time t_o and different initial values of the state vector $t_o, \overline{x}_o$ or $t_o, \overline{x}'_o$, the same steady-state waveform at the same time values (not shifted in time) will be obtained. Thus, the time-varying generator of the forced circuit prevents solution invariance versus time translations. This invariance is a property of autonomous circuits only. Here by the general term *autonomous circuit* we mean free-running oscillators or circuits containing oscillations that are not synchronized to the independent sources.

When analyzing a system such as (1.46), the designer usually performs a representation versus time of the solutions obtained. An alternative way to observe these solutions is by using the phase space [26]. In the phase space, each axis corresponds to a different state variable x_i. Then, an instantaneous representation of the time values of these variables $x_i(t)$ is carried out, as is done, for example, when tracing the load cycle of a transistor-based circuit. Plotting the numerical values of all the variables at a given time t provides a description of the state of the system at that time. When time evolves, the solution follows a "trajectory" or set of sequential points versus the implicit time variable. The evolution of the system is indicated by a path, or trajectory, in the phase space. The phase space enables a geometric and therefore comprehensible representation of complex behavior. In the case of a nonautonomous circuit, a time-related variable must be included, such as θ or the generator value $e_{\text{in}}(t)$. In practice, the phase space representation is limited to three state variables, so a projection of the phase space is actually obtained, which is usually enough to identify the most relevant properties.

As an example, Fig. 1.13 shows a phase space representation of the solutions of a FET-based oscillator. The variables chosen are the drain voltage v_D and the current through the load inductance i_L. The unstable dc solution, given by the constant voltage $v_D = 3.5$ V and current $i_L = 0$ (after the dc block), provides a point in this representation. The periodic steady-state solution gives rise to a closed trajectory termed a *cycle*, because the circuit variables repeat their values after one period. Actually, in a phase-space representation, the steady-state solutions give rise to bounded sets called *limit sets*. Dc solutions give rise to points called *equilibrium points*, and periodic solutions give rise to *cycles*. Other types of steady-state solutions give rise to other geometric figures.

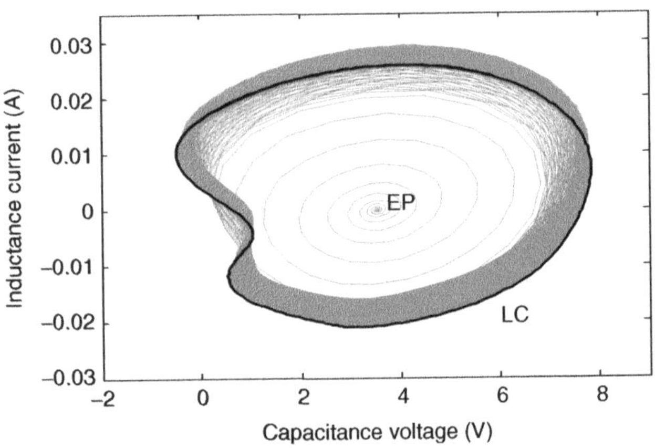

FIGURE 1.13 Phase space representation of the solutions of the FET-based oscillator of Fig. 1.6. The dc solution gives rise to the equilibrium point, indicated as EP. The steady-state oscillation gives rise to the limit cycle, LC. The spiral-like trajectory corresponds to the startup transient.

In a phase space representation [10], transients are open trajectories leading from one limit set to another. In the case of Fig. 1.13, the spiral trajectory from the equilibrium point (EP) to the cycle (LC) corresponds to the startup transient, leading from an unstable dc solution to steady-state oscillation. In a noiseless system, after reaching the cycle, the solution keeps turning in the cycle for a time tending to infinity. In practice, the stable cycle is continuously recovering from the small perturbations that are always present in real life. A single instantaneous perturbation kicks the system out of the cycle, but because the cycle is stable, an exponential transient leads the solution back to it. Due to the continuous noise influence, the solution trajectory will actually surround the cycle. The same is true for any other type of steady-state regime observed in real life.

As already indicated, when represented in phase space, steady-state solutions give rise to bounded sets. The type of limit set and its dimension depend on the particular type of steady-state solution. In general, the steady-state solutions of nonlinear systems can be classified into four principal types: dc solutions, periodic solutions, quasiperiodic solutions, and chaotic solutions. The main characteristics of each type of solution are summarized briefly next.

1.5.1.1 Constant Solution Constant solutions are only possible in circuits with no time-varying input generators. As already seen, they are obtained by imposing $\dot{\overline{x}} = 0$. Because there is no time variation of the state variables, the representation of a constant solution in phase space gives rise to a point called the *equilibrium point* (Fig. 1.13). The geometric dimension of the point is zero.

1.5.1.2 Periodic Solution The periodic solution, well known to designers, fulfills $\overline{x}(t + nT) = \overline{x}(t)$, with n an integer and T the solution period. The circuit variables can be expanded in a Fourier series with one fundamental frequency $\omega_0 = 2\pi/T$. For a free-running oscillator, the period T will depend on the values of the circuit elements and bias generators. In a forced circuit the period T is determined by the input generator. The periodic solution of a free-running oscillator gives rise to an isolated closed trajectory in the phase space (see the cycle in Fig. 1.13), known as the limit cycle. For a sinusoidal oscillator with no harmonic content, this cycle will be a circumference. Whatever the dimension of the system in R^N, the cycle will have one dimension because it is a line. The trajectories surrounding the limit cycle are open, corresponding to transients. This is why the limit cycle is an isolated closed trajectory in the phase space. In stable steady-state oscillation, all these neighboring trajectories lead to the cycle. The stable cycle must be attracting for all its surrounding neighborhood, which must have the same dimension as the entire phase space R^N. Note that the cycles of an ideal LC oscillator (with no resistance) are not isolated because any initial condition provides a different cycle (see Section 1.2). In a conservative oscillator, for arbitrarily close initial values, arbitrarily close cycles would be obtained. Thus, they are not limit cycles.

It must be remembered that the phase space of an autonomous system does not contain time or a time-related variable. Due to the invariance versus time translations of autonomous systems, all possible steady-state oscillations $\overline{x}_0(t - \tau)$

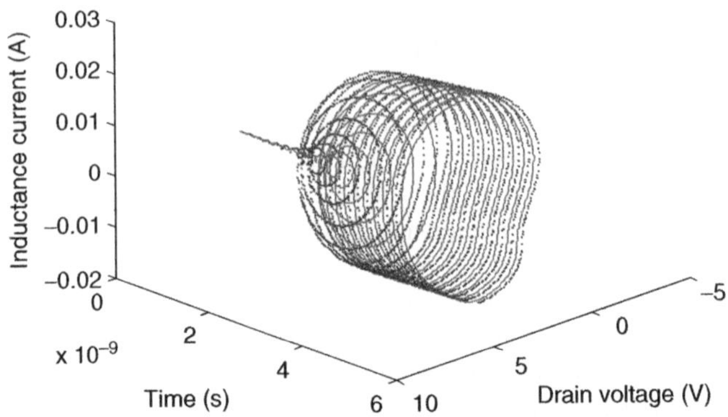

FIGURE 1.14 Solutions of the FET-based oscillator obtained when integrating the circuit equations for two different initial values. Although the solutions are time shifted, they provide the same limit cycle, which is obtained from the projection of these solutions over the plane defined by the drain voltage and inductance current.

lie on the same limit cycle. This is shown clearly in Fig. 1.14, where the solutions obtained when integrating the equations of the FET-based oscillator of Fig. 1.6 are shown for two different initial conditions. It is, in fact, the same analysis as that performed in Fig. 1.9. The difference is that two different state variables, the drain voltage v_D and the inductance current i_L, have been considered here for a three-dimensional representation versus time. The two time-shifted steady-state solutions give rise to the same limit cycle, obtained by projecting the figure over the plane defined by v_D and i_L.

The situation is different for the periodic solution of a forced system. For a given phase value of a forcing periodic source, the periodic solution is unique. The cycle is actually due to this source (not generated by the circuit). As shown in (1.46), time can be considered as a state variable of the nonautonomous system. Because time is unbounded, either the periodic source value $g_{\text{in}}(t)$ or the angle $\theta = \omega t$ should be assigned to one of the axis of the phase space representation. Therefore, a different cycle, with an identical shape, is obtained for each phase value of the input source.

1.5.1.3 Quasiperiodic Solution In an "almost-periodic" solution, no period can be defined [24]. However, for each $\varepsilon > 0$ there exists a time-interval length $l(\varepsilon)$ such that each real interval of length $l(\varepsilon)$ contains at least one number τ (the *translation number*) fulfilling $|\overline{x}(t+\tau) - \overline{x}(t)| < \varepsilon$. The quasiperiodic solutions can be expanded in a Fourier series with a finite number M of nonrationally related (*incommensurable*) fundamentals $\omega_{f1}, \omega_{f2}, \ldots, \omega_{fM}$ and are thus expressible as a sum of periodic waveforms [26]. Two frequencies, ω_{f1} and ω_{f2}, are incommensurable if $\omega_{f1}/\omega_{f2} \neq m/n$, with m and n integers. A key aspect of quasiperiodic solutions is that the number M of required fundamental frequencies is uniquely defined, but not the set of these fundamental frequencies. Actually, $\omega_{f1}, \omega_{f2} + \omega_{f1}$ span the same set of frequencies as ω_{f1} and ω_{f2}. For a simple explanation of why the solution cannot be periodic, note that it is not possible to obtain a time value T fulfilling $A\cos\omega_{f1}t + B\cos\omega_{f2}t = A\cos(\omega_{f1}t + \omega_{f1}T) + B\cos(\omega_{f2}t + \omega_{f2}T)$. Satisfying this would require that $\omega_{f1}T = n \cdot 2\pi$, $\omega_{f2}T = m \cdot 2\pi$, with m and n integers. Then the ratio ω_{f1}/ω_{f2} would be rational, which is against our initial assumption of incommensurable fundamentals.

A quasiperiodic solution with two fundamental frequencies is easily obtained when connecting a periodic generator at the frequency ω_{in} to an existing oscillator at the frequency ω_o. Although other regimes are possible (see Chapter 3), for a wide range of input generator frequency and power, mixer-like behavior, with two incommensurable fundamentals, ω_{in} and ω_o, will be observed, with the ω_o value influenced by the input generator. Note that it would be equally possible to consider the fundamental frequency basis ω_{in}, $|\omega_{in} - \omega_o|$. The circuit is said to operate in an autonomous quasiperiodic regime. An interesting aspect of this type of regime is that despite the existence of an independent periodic source connected to the circuit, different initial values $\bar{x}_o$ at the initial time t_o give rise to time-shifted solutions with the same pattern. This is due to the fact that the oscillation is not synchronized to the input source and can have any phase shift with respect to this source.

Figure 1.15 shows the quasiperiodic solution obtained when introducing a generator at $f_{in} = 6.33$ GHz in the oscillator of Fig. 1.6, with original free-running oscillation frequency $f_o = 4.4$ GHz. It is clear that when representing a quasiperiodic solution in the phase space, no closed cycle can be obtained because the solution is not periodic. In the particular case of two incommensurable fundamental frequencies, the steady-state trajectory lies on the surface of a 2-torus. This is in close relationship with the fact that two fundamental frequencies give two independent rotations in phase space. As an example, Fig. 1.16 shows the three dimensional representation of the quasiperiodic solution of the FET-based circuit. As time tends to infinity, the torus surface gets covered entirely by the solution trajectory. Because

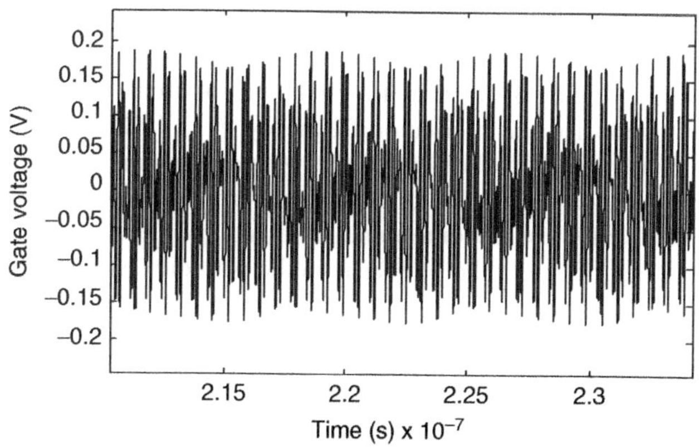

FIGURE 1.15 Time representation of the quasiperiodic solution of a FET-based oscillator at the original free-running frequency $f_o = 4.4$ GHz when a generator is introduced at $f_{in} = 6.33$ GHz.

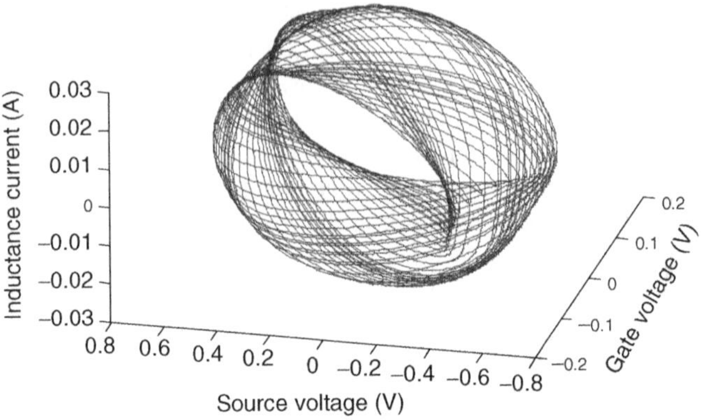

FIGURE 1.16 Phase space representation of the quasiperiodic solution of a FET-based oscillator. The steady-state solution lies on the surface of a 2-torus. This surface is covered entirely by the solution trajectory, as time tends to infinity.

the solution lies on the torus surface, it has two dimensions in the phase space. Note that a three-fundamental quasiperiodic solution would have three dimensions.

1.5.1.4 Chaotic Solution Chaotic solutions are neither periodic nor quasiperiodic [27]. Thus, they exhibit a continuous spectrum, at least for some frequency intervals. When performing frequency-domain measurements, chaotic solutions are often mistaken for noise or interference. However, the power measured is usually too high to be due to noise only. Chaotic solutions are quite common in practice. Actually, the minimum mathematical requirement for an autonomous circuit to exhibit this type of solution is that it contain at least three reactive elements plus one nonlinear element [28,29]. As an example, the commonly used Colpitts oscillator can exhibit chaotic solutions for some transistor bias voltages and linear element values.

Chaotic solutions are characterized by a sensitive dependence on initial conditions, meaning that solutions with arbitrarily close initial values diverge exponentially in time. Figure 1.17 presents simulations of a chaotic Colpitts oscillator, designed originally for the oscillation frequency $f_o = 1$ GHz. This figure shows the time evolution of the collector voltage when integrating the circuit equations from two close initial values. Initially the waveforms seem to overlap, but as time evolves they diverge and become quite different from each other. Compare the situation with that of a periodic or quasiperiodic periodic solution, giving the same steady-state waveform whatever the initial value is $\bar{x}_o$ at t_o. As we know, this waveform will be time-shifted for different $\bar{x}_o$ values in the case of autonomous behavior. Comparing Fig. 1.17, corresponding to a chaotic solution, with Fig. 1.15, corresponding to a quasiperiodic solution, it can be noted that the chaotic solution is nonperiodic and highly irregular. The quasiperiodic waveform usually looks like a periodic signal modulated with another periodic signal of incommensurate frequency. In the example in Fig. 1.18, the amplitude is clearly modulated and the frequency is modulated too, as the zero crossings are not uniformly spaced [26].

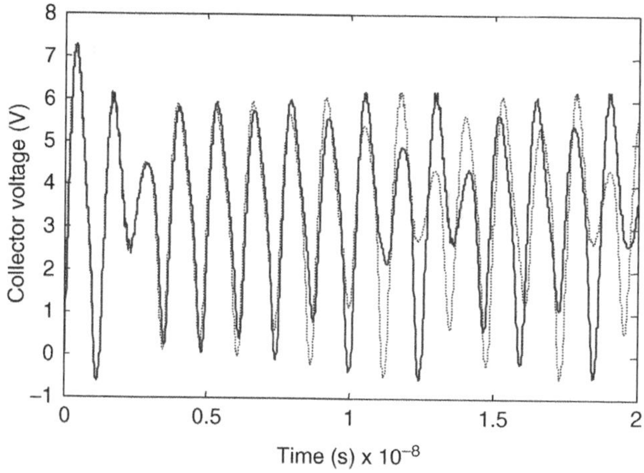

FIGURE 1.17 Time evolution of the collector voltage in a chaotic Colpitts oscillator when integrating the circuit equations from two close initial values. Initially, the waveforms overlap, then diverge, and after a little time become quite different from each other.

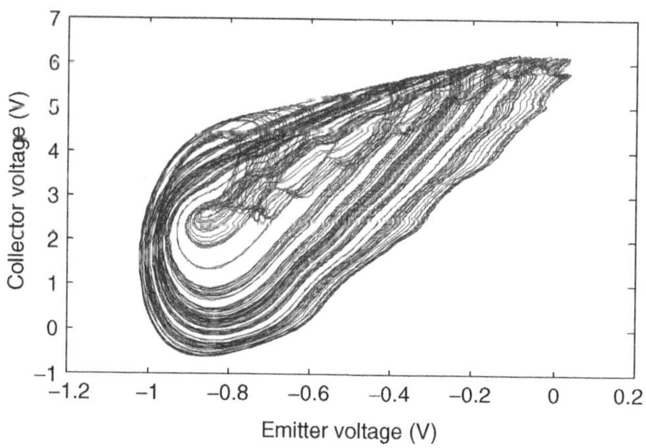

FIGURE 1.18 Phase space representation of a chaotic solution corresponding to a Colpitts oscillator. The bounded set obtained is not covered entirely by the trajectory; it has a fractal dimension. In close relation with this fractal dimension, the bounded set exhibits a self-similar structure.

When represented in phase space, the steady-state chaotic solution gives rise to a bounded figure, which unlike a limit cycle or a torus, is not entirely covered by the trajectory. As an example, Fig. 1.18 shows the phase space representation of a chaotic solution of a Colpitts oscillator. Some sections of the figure are not filled by the trajectory even when the simulation time tends to infinite. Owing to this fact, the dimension of the figure is fractal. The meaning of fractal dimension is explained in the following. A figure will have an entire dimension Dim if we can break it into an integer number N^{Dim} of self-similar figures. As an example, a line can be broken into N self-similar pieces. A square can be broken into N^2 self-similar pieces, and a cube can be broken into N^3 pieces. In each case the total number of pieces is N^{Dim} and the magnification factor of each piece, to recover the original figure, is N. As the reader can verify, the dimension of the figure can be obtained by setting Dim = log(number of pieces) − log(magnification of each piece). The chaotic bounded sets are characterized by a self-similar structure, meaning that they look the same for any scale of magnification. However, because some pieces are missing, as in Fig. 1.18, the definition of dimension introduced provides a fractional number.

1.5.2 Stability Analysis

As shown in previous sections, not all the steady-state solutions of a given circuit will be observable physically. To be observable, a given solution must be robust versus the small perturbations that are always present in real life (e.g., those coming from noise or any fluctuation of the bias sources). *Stable* means robust versus small perturbations. If a small perturbation is applied to a stable solution, the system will return to it exponentially in time. In contrast, if a small perturbation is applied to an unstable solution, the system will evolve to a different steady-state solution after an initially exponential transient. The solution obtained after the transient will be a stable solution and thus will be physically observable.

Note that for stability analysis, no assumption is made as to the value of the instantaneous perturbation applied. The only condition is that it has to be small. This is because two or more stable steady-state solutions may coexist and a large perturbation may lead the system to the a different stable solution. Thus, the stability definition is local in nature: it refers only to the system behavior near the steady-state solution [24]. The stability or instability of a given steady-state solution depends on the system and the particular solution, but not on the value of the applied small perturbation. This necessary restriction to small perturbations is advantageous, as it allows linearization of the circuit equations about the particular steady-state solution. Because an arbitrary perturbation will have components in any direction of an N-dimensional phase space, the stable steady-state solution must be attracting for all the neighboring trajectories. This is why stable solutions are also called *attractors*. An example of an attractor is the limit cycle of Fig. 1.13, which attracts, as can be seen, all its neighboring trajectories in the phase space.

We can express the solution of (1.44) in terms of its initial value as $\overline{x}(t, \overline{x}_o)$. The basin of attraction for a given steady-state solution $\overline{x}_s$ is the set of initial conditions $\overline{x}_o$ such that the system evolves to this solution as time tends to infinity: $\lim_{t\to\infty} \overline{x}(t, \overline{x}_o) = \overline{x}_s$ [24]. For an N-dimensional system with only one stable solution, the basin of attraction for this solution will be the entire space R^N. This single stable solution is said to be *globally asymptotically stable*. For a system with two or more coexisting stable solutions, each solution will have a different basin of attraction. The basins of attraction are disjoint because the solution of a system $\dot{\overline{x}} = \overline{f}(\overline{x})$, with $\overline{f}$ smooth, is unique, so the trajectories cannot intersect. If they did, using the intersection point $t_o, \overline{x}_o$ as the initial value for the system integration, the system might tend to either of the coexisting steady-state solutions, which is, of course, impossible. A key fact is that the dimension of each of the disjoint basins of attraction of the coexisting stable steady-state solutions agrees with the dimension N of the entire space. This is because each solution is stable. The union of the basins of attraction will be equal to R^N.

The circuit of Fig. 1.19a constitutes an example of bistable behavior. The nonlinearity is $i(v) = av + bv^3$. The circuit equations are obtained by adding the branch currents, which can be solved for three different dc solutions:

$$G_1 v + G_2 v + av + bv^3 = 0 \tag{1.47}$$

The three solutions are given by $V_{dc1} = 0$, which, as will be shown later, is unstable, and $V_{dc2} = 1$ V and $V_{dc3} = -1$ V, which are stable. Each of the stable solutions V_{dc2} and V_{dc3} has its own basin of attraction. Taking v and the current through

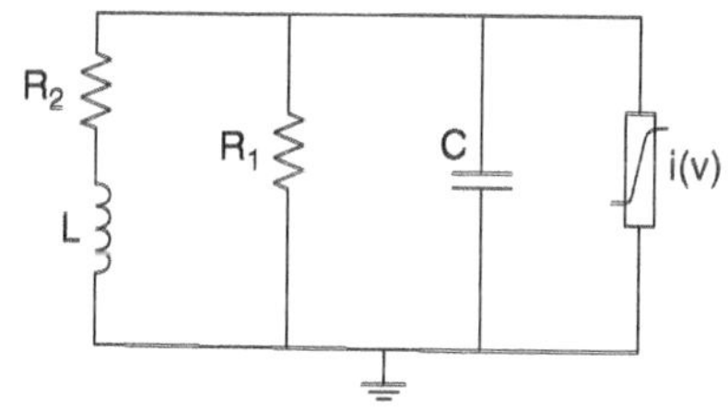

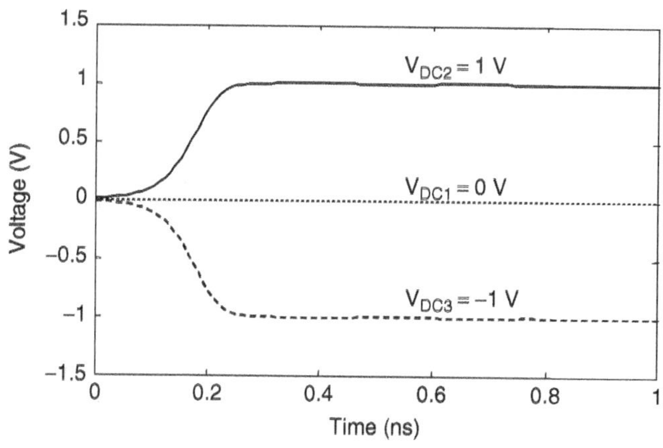

FIGURE 1.19 Cubic nonlinearity circuit with three dc solutions: $V_{dc1} = 0$, $V_{dc2} = 1$ V, and $V_{dc3} = -1$ V. The circuit element values are $L = 1$ nH, $C = 0.5$ pF, $R_2 = 100$ Ω, and $G_1 = 0.01$ Ω^{-1}. (a) Circuit schematic. (b) Solutions obtained for the two initial values: $i_{Lo} = 0$ and $v_o = 0.01$ V (solid line) and $i_{Lo} = 0$ and $v_o = -0.01$ V (dashed line).

the inductance i_L as state variables, the initial points $i_{Lo} = 0$ and $v_o > 0$ belong to the basin of attraction of $V_{dc2} = 1$ V, whereas the initial points $i_{Lo} = 0$ and $v_o < 0$ belong to the basin of attraction of $V_{dc3} = -1$ V. As an example, Fig. 1.19b shows the solution obtained when integrating the system equations from $i_{Lo} = 0$ and $v_o = 0.01$ V, which evolves to $V_{dc2} = 1$ V, and the solution obtained when integrating the system equations from $i_{Lo} = 0$ and $v_o = -0.01$ V, which evolves to $V_{dc2} = -1$ V.

For the stability analysis of a given steady-state solution $\overline{x}_s(t)$, either constant or time varying, a small perturbation is applied at a given time instant t_o, and from this value the system is allowed to evolve according to its own dynamics. Thus, beginning at this time value, the system analyzed is a perturbed system in which the stimulus that was applied is no longer present. Due to the effect of the instantaneous perturbation, the solution becomes $\overline{x}_s(t) + \Delta\overline{x}(t)$. Because the perturbation is small, it will be possible to expand the nonlinear equation system (1.47) in a Taylor series around $\overline{x}_s(t)$. The expansion is carried out only up to first order (higher order is rarely necessarily), which provides the following linear time-varying system:

$$\begin{aligned} \dot{\overline{x}}_s(t) + \Delta\dot{\overline{x}}(t) &= f(\overline{x}_s(t)) + Jf(\overline{x}_s(t))\Delta\overline{x}(t) && \text{(a)} \\ \Delta\dot{\overline{x}}(t) &= Jf(\overline{x}_s(t))\Delta\overline{x}(t) && \text{(b)} \end{aligned} \qquad (1.48)$$

where $Jf(\overline{x}_s(t))$ is the Jacobian matrix of the nonlinear function f in (1.48), evaluated at the steady-state solution $\overline{x}_s(t)$. Because $\overline{x}_s(t)$ fulfills (1.44), equation ((1.48)a) can be simplified to ((1.48)b). For the steady-state solution $\overline{x}_o(t)$ to be stable, the perturbation $\Delta\overline{x}(t)$ must vanish exponentially in time. This will depend on the properties of the Jacobian matrix $Jf(\overline{x}_s(t))$ evaluated at this particular solution. Because the Jacobian matrix is evaluated at the steady-state solution, it will have the same periodicity or nonperiodicity of this solution. Therefore, the difficulties of the stability analysis are totally dependent on the solution type. The simplest case will be that of a dc solution, providing a constant Jacobian $Jf(\overline{x}_{dc})$. For a periodic solution of period T, the Jacobian matrix will also be periodic, with the same period T.

1.5.2.1 *Stability Analysis of a dc Solution* In a dc solution $\overline{x}_{dc}$, the Jacobian matrix is constant. Then equation (1.44) becomes a time-invariant linear system given by $\Delta\dot{\overline{x}} = Jf(\overline{x}_{dc})\Delta\overline{x}(t)$, since $Jf(\overline{x}_{dc})$ is a constant matrix. The general solution of this linear system is [27]

$$\Delta\overline{x}(t) = \sum_{n=1}^{N} c_k e^{\lambda_k t}\overline{u}_k = c_{c1}e^{(\sigma_{c1}+j\omega_{c1})t}\overline{u}_{c1} + c_{c1}^* e^{(\sigma_{c1}-j\omega_{c1})t}\overline{u}_{c1}^* + c_{r1}e^{\gamma_{r1}t}\overline{u}_{r1} + \cdots \qquad (1.49)$$

where the exponents λ_k, $k = 1$ to N, which may be real or complex conjugate, are the eigenvalues of the Jacobian matrix $Jf(\overline{x}_{dc})$, the vectors $\overline{u}_k$ are the eigenvectors of this matrix, and c_k are constants that depend on the initial conditions, thus on the instantaneous perturbation applied. Note that for the expression (1.49) to be valid, all the eigenvalues $\lambda_k, k = 1$ to N of the Jacobian matrix $Jf(\overline{x}_{dc})$ are assumed different, which is the general case. For a double eigenvalue λ_j, the coefficient of the associated exponential term $e^{\lambda_j t}$ will depend linearly on time, as $(c_{oj} + c_{1j}t)e^{\lambda_j t}$. For three repeated real eigenvalues λ_j, the exponential term $e^{\lambda_j t}$ will have the quadratic dependence $(c_{oj} + c_{1j}t + c_{2j}t^2)e^{\lambda_j t}$. This is easily generalized to any number of repeated eigenvalues, whose presence will require calculation of generalized eigenvectors. Note, however, that the presence of repeated eigenvalues is rare in practical circuits unless they contain perfect symmetries. Thus, this possibility will be discarded in most derivations. Exceptions are the Rucker and N-push oscillators with symmetric topology that we study in Chapter 10.

The application of the Laplace transform to $\Delta\dot{\overline{x}} = Jf(\overline{x}_{dc})\Delta\overline{x}(t)$ provides the following system in the Laplace variable $s : (sI_N - Jf(\overline{x}_{dc}))\Delta\overline{X}(s) = 0$, with I_N the identity matrix in the space R^N. Clearly, the eigenvalues $\lambda_k, k = 1$ to N, of $Jf(\overline{x}_{dc})$ agree with the roots of the characteristic determinant $\det(sI_N - Jf(\overline{x}_{dc})) = 0$. It is possible to define a closed-loop transfer function associated with this linearized system as was done in Section 1.2. For that, an arbitrary input $U(s)$ is introduced into the system and an arbitrary output $Y(s)$ is selected:

$$\begin{aligned} (sI_N - Jf(\overline{x}_{dc}))\Delta\overline{X}(s) &= G(s)U(s) \\ Y(s) &= F\Delta\overline{X}(s) \end{aligned} \tag{1.50}$$

where $G(s)$ is a $N \times 1$ matrix and $F(s)$ is a $1 \times N$ matrix. The closed-loop transfer function will be

$$H(s) = \frac{F[sI_N - Jf(\overline{x}_{dc})]^+ G(s)}{\det\,[sI_N - Jf(\overline{x}_{dc})]} \tag{1.51}$$

Unless pole–zero cancellations occur, giving rise to changes in the numerator and denominator of (1.51), all possible closed-loop transfer functions that one may define in the linearized system will have the same denominator. This denominator agrees with the characteristic determinant, and its roots, which correspond to the system poles, agree with the exponents λ_k of (1.49). When dealing with a dc solution here, the exponents λ_k will be called, indistinctly, eigenvalues or poles.

It will be assumed that there are no repeated (multiple) eigenvalues and that no eigenvalue is zero. The time evolution of the perturbation $\Delta\overline{x}(t)$ will be determined by the eigenvalues λ_k, because c_k and $\overline{u}_k$ are constant. For stability, all the eigenvalues must have a negative real part. This means that the perturbation will vanish exponentially in time and the system will return exponentially to the original steady-state solution $\overline{x}_{dc}$. The linearized solution of a practical circuit will generally contain many eigenvalues (or poles), as many as the dimensions of the state variable vector $\overline{x}$, which in lumped circuits agrees with the total number of inductors plus capacitors. Typically, an unstable solution contains only a few unstable poles. Common situations are one real pole $\gamma > 0$ or a pair of complex-conjugate poles $\sigma \pm j\omega$, with $\sigma > 0$, with all the rest of the poles on the left-hand side of the complex plane.

From an inspection of (1.49), not all the eigenvalues λ_k will have the same weight on the transient response to the perturbation. This transient will be dominated by the eigenvalues with maximum real part σ_c or γ_r. In an unstable state, this real part will be positive. In a stable solution, the transient will be dominated by poles of smaller absolute value, abs(σ_c) or abs(γ_r). The associated frequencies will be observed during the transient response. In the common case of a single dominant pair of complex-conjugate poles $\sigma_c \pm j\omega_c$, an oscillation at the pole frequency ω_c, with amplitude decaying to zero will be observed during the transient response. Obviously, the response will be longer for a smaller absolute value of σ_c. As already shown, this can be due to a high quality factor for the resonance at ω_c. However, it can also be due to circuit operation close to instability, with the pair of complex-conjugate poles $\sigma_c \pm j\omega_c$ very near the imaginary axis. Actually, the observation (in simulation) of slower transients versus the variation in a circuit parameter such as bias voltage usually indicates that the circuit is approaching instability at the frequency of the transient.

According to (1.49), for any eigenvalue on the right-hand side of the complex plane, the perturbation $\Delta\overline{x}(t)$ tends to infinity over time. This unbounded growth of the perturbation is, of course, totally unrealistic. Expression (1.49) is a solution of the linearized system $\Delta\dot{\overline{x}} = Jf(\overline{x}_{\text{dc}})\Delta\overline{x}(t)$, which assumes a small perturbation $\Delta\overline{x}$. For any eigenvalue on the right-hand side of the complex plane, this assumption soon becomes invalid. After a very short time, the linearization is no longer applicable. The solution does not tend to infinity but to a different steady-state solution that cannot be predicted with the linearization.

Each eigenvalue of the Jacobian matrix $Jf(\overline{X}_{\text{dc}})$ is associated with a particular eigenvector, and the set of N vectors spans the space R^N. For illustration it will be assumed that m eigenvalues of a solution $\overline{x}_{\text{dc}}$ have a negative real part and $q = N - m$ eigenvalues have a positive real part. This type of solution, having stable and unstable eigenvalues or poles, is called a *saddle*. The solution is unstable because not all the eigenvalues have a negative real part. The m vectors associated with eigenvalues that have a negative real part $\overline{u}_1, \ldots, \overline{u}_m$ span, close to $\overline{x}_{\text{dc}}$, the stable eigenspace of this solution. The q vectors associated with eigenvalues that have a positive real part $\overline{u}_{m+1}, \ldots, \overline{u}_N$ span, close to $\overline{x}_{\text{dc}}$, the unstable eigenspace of this solution. Thus, unstable solutions may have stable eigenspaces, which is the most common situation in practical circuits, due to the usually high dimension of the system of differential equations determined by the number of reactive elements. For an unstable real pole, the unstable eigenspace will have one dimension and be defined by a single eigenvector, providing a straight line in the space R^N. In the case of two complex-conjugate poles, the unstable eigenspace will have two dimensions, defined by the two associated eigenvectors, corresponding to a plane in the space R^N.

Because the eigenvectors correspond to linearization of the original nonlinear system about the dc solution $\overline{x}_{\text{dc}}$, they are meaningful only in the neighborhood of this solution. At a larger distance from $\overline{x}_{\text{dc}}$, the stable and unstable eigenspaces become the stable and unstable manifolds associated with this solution. A manifold is a connected set in R^N instead of the disjoint junction of two or more nonempty subspaces [27]. All the points in the manifold have a continuous time derivative. The manifold can be closed, like a limit cycle, or it can be open, starting or ending in a steady-state solution or limit set. The stable manifold of $\overline{x}_{dc}$ is the set of initial values $\overline{x}_o$ such that $\lim_{t\to\infty} \overline{x}(t, \overline{x}_o) = \overline{x}_{dc}$. Close to $\overline{x}_{dc}$ the stable manifold is tangent to the stable eigenspace spanned by $\overline{u}_1, \ldots, \overline{u}_m$ [24]. The unstable manifold of $\overline{x}_{dc}$ is the set of initial values $\overline{x}_o$ such that $\lim_{t\to-\infty} \overline{x}(t, \overline{x}_o) = \overline{x}_{dc}$. Note the negative sign in the time limit, indicating that the system actually gets away from $\overline{x}_{dc}$ as time increases. Close to $\overline{x}_{dc}$ the unstable manifold is tangent to the unstable eigenspace spanned by $\overline{u}_{m+1}, \ldots, \overline{u}_N$. As already stated, because an arbitrary perturbation will have components in the N dimensions of the phase space, any dc solution with an unstable eigenspace or manifold will be unstable and physically unobservable. For the dc solution to be stable, the dimension m of its stable eigenspace must agree with the total system dimension N, so $m \equiv N$. Thus, the dc solution must be an attractor, that is, attracting for all the directions of the phase space.

Figure 1.20 shows an illustration of the eigenspaces of a dc solution in an R^3 system located in the plane in the center of the spiral [27]. Because it is an R^3 system, there are three eigenvalues associated with the dc solution. In this particular example, two eigenvalues are complex conjugate $\lambda_{1,2} = \sigma \pm j\omega$, with $\sigma < 0$. The associated eigenvectors define the stable eigenspace E_S associated with this dc solution. The third eigenvalue is real and positive, $\lambda_3 = \gamma > 0$. Its associated eigenvector defines a straight line that constitutes the unstable eigenspace of the dc solution. Note that due to the negative σ, the trajectories assume positions close to the straight line. The spirals shrink very quickly, so the system mostly evolves along a line corresponding to the unstable eigenspace E_U when getting away from the dc solution. Thus, the eigenvalue $\lambda_3 > 0$ totally dominates the transient behavior. However, as the distance to the dc point increases, the eigenspace evolves into a nonlinear manifold. It is no longer a straight line but generally a curve tending to a different steady-state solution that cannot be predicted using linear analysis.

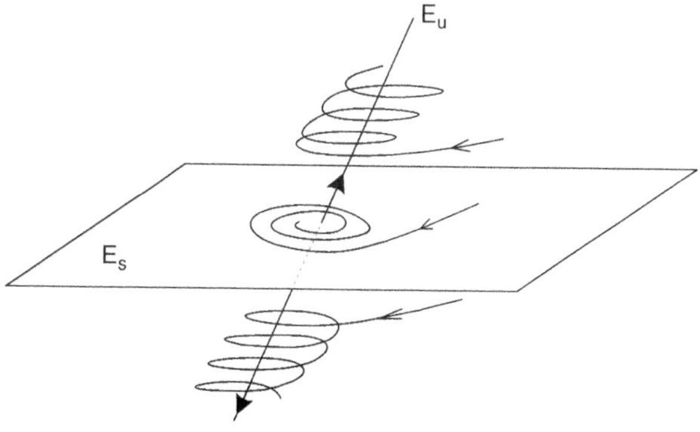

FIGURE 1.20 Eigenspaces of a dc solution in an R^3 system located in the plane in the center of a spiral. Because it is an R^3 system, there are three eigenvalues associated with the dc solution. The corresponding eigenvectors define stable and unstable eigenspaces.

Note that even though the new steady-state solution to which the solution evolves cannot be determined using the linearized analysis of (1.49), in most cases it will be possible to predict its constant or oscillatory nature and, in the latter case, the fundamental frequency of the oscillation. Exceptions can be encountered, however, because the linearization (1.49) has local validity only. As an example, for a dominant pair of complex-conjugate poles $\sigma \pm j\omega$, with $\sigma > 0$, the transient predicted by (1.49) will be oscillatory at the frequency ω, with exponentially growing amplitude. The system is expected to evolve to steady-state oscillation at about the frequency ω. In the case of a real pole γ on the right-hand side of the complex plane, no oscillation will be generally observed. The system is expected to evolve under a monotonic transient to a different dc solution (see Fig. 1.19b). Note that relaxation oscillations [24] are also possible in the presence of positive real poles.

To illustrate, the stability analysis described previously will be applied to the dc solution $v_{c,\text{dc}} = 0$, $i_{L,\text{dc}} = 0$ of the second-order nonlinear system (1.45). The linearized system is given by

$$\begin{bmatrix} \Delta\dot{v}_c(t) \\ \\ \Delta\dot{i}_L(t) \end{bmatrix} = \begin{bmatrix} \dfrac{-G_T + 3bv_{c,\text{dc}}^2}{C} & \dfrac{-1}{C} \\ 1/L & 0 \end{bmatrix} \begin{bmatrix} \Delta v_c(t) \\ \Delta i_L(t) \end{bmatrix} = \begin{bmatrix} \dfrac{-G_T}{C} & \dfrac{-1}{C} \\ 1/L & 0 \end{bmatrix} \begin{bmatrix} \Delta v_c(t) \\ \Delta i_L(t) \end{bmatrix} \tag{1.52}$$

where $G_T = 1/R + a$. As expected, the eigenvalues of the Jacobian matrix agree totally with the complex-conjugate poles $\sigma \pm j\omega$ calculated in (1.11), and are given by

$$\lambda_{1,2} = -\frac{G_T}{2C} \pm \frac{1}{2}\sqrt{\frac{G_T^2}{C^2} - \frac{4}{LC}} = 10^9 \pm j9.5 \times 10^9 \tag{1.53}$$

Because there are eigenvalues with a positive real part, the dc solution is unstable. The two complex-conjugate eigenvalues have two associated complex-conjugate eigenvectors spanning a two-dimensional space. The unstable manifold agrees, in this simple case, with the entire space R^2. If G_L is increased continuously, $G_T = a + G_L$ decreases, and at $G_T = 0$ the pair of poles crosses the imaginary axis to the left-hand side of the complex plane. The dc solution $v_{c\_\text{dc}} = 0, i_{L\_\text{dc}} = 0$ is stable for $G_T > 0$, unstable for $G_T < 0$, and undergoes a qualitative stability change at the critical value $G_L = |a|$. Note that the decaying transient, with oscillatory behavior, will be very slow for $G_L = |a| + \varepsilon$ with positive ε, that is, when approaching the critical value $G_L = |a|$. A qualitative change in the solution stability when a parameter is modified continuously is known as *bifurcation*. As has been shown, the values of the solution poles will generally change when varying a circuit parameter, such as G_L or L. Because of this, one real pole or a pair of complex-conjugate poles may cross the imaginary axis, giving rise to a qualitative change in the solution stability. Two complex poles may also turn into two real poles, or vice versa. This change in the nature of the poles may happen in either the left- or right-hand side of the complex plane, and does not give rise to a bifucation. However, the total pole number remains unchanged and is equal to the system order N. Note that for the large inductance value $L > 4C/G_T^2 = 0.1nH$, the linearized system will have two real eigenvalues of the same or different sign γ_1, γ_2, whose associated eigenvectors also span the entire space R^2.

As a second example, the stability of the three coexisting dc solutions of the circuit of Fig. 1.19b will be analyzed. These three dc solutions, calculated with (1.47), are $V_{dc1} = 0$, $V_{dc2} = 1$ V, and $V_{dc3} = -1$ V. For each stability analysis, the nonlinear element is replaced by its linearization about the corresponding dc solution $\partial i(V_{dci})/\partial v$, obtaining the linearized differential equation system

$$\begin{bmatrix} \Delta\dot{v}_c(t) \\ \Delta\dot{i}_L(t) \end{bmatrix} = \begin{bmatrix} \dfrac{-G'_T(V_{dci})}{C} & \dfrac{-1}{C} \\ \dfrac{1}{L} & \dfrac{-R_2}{L} \end{bmatrix} \begin{bmatrix} \Delta v_c(t) \\ \Delta i_L(t) \end{bmatrix} \tag{1.54}$$

with $G'_T(V_{dci}) = G_1 + a + 3bV_{dci}^2$ and $R_2 = 1/G_2$. $V_{dc2} = 1$ V and $V_{dc3} = -1$ V have the same two poles, which are complex conjugate: $p_{1,2} = -6 \times 10^{10} \pm j2 \times 10^{10}$. They fulfill $\text{Re}[p_{1,2}] < 0$, so V_{dc2} and V_{dc3} are stable. The two poles of $V_{dc2} = 0$ V are real, with the values $p_1 = -8.385 \times 10^{10}$ and $p_2 = 2.385 \times 10^{10}$. Thus, the solution $V_{dc2} = 0$ V is unstable and unobservable. The eigenvector associated with p_1, which is $\overline{u}_1 = (16.148, 1)$, spans the stable eigenspace of $V_{dc2} = 0$. The eigenvector associated with p_2, which is $\overline{u}_1 = (123.852, 1)$, spans the unstable eigenspace of $V_{dc2} = 0$. In terms of the state variables, this unstable eigenspace can be expressed as $\Delta v = 123.852\Delta i_L$. It separates the disjoint basins of attraction of the two stable solutions V_{dc2} and V_{dc3}.

1.5.2.2 Stability Analysis of a Periodic Solution Periodic solutions can be obtained in both autonomous and nonautonomous systems. For compactness, the two types of systems will be described as $\dot{\overline{x}} = f(\overline{x})$. However, in a nonautonomous system, the vector $\overline{x}$ will include $\theta = (2\pi/T)t$ as one of the state variables. In the following, the same dimension N of the vector $\overline{x}$ will be considered in the two cases, so the autonomous system should contain N reactive elements, whereas the nonautonomous system should contain $N-1$ reactive elements.

For the stability analysis of a periodic solution $\overline{x}_{sp}(t)$, with period T, a small perturbation will be applied at a particular time instant t_o, giving rise to the increment $\Delta\overline{x}(t)$ in the circuit state variables. Due to the small value of the increment $\Delta\overline{x}(t)$, the nonlinear system (1.55) can be linearized about $\overline{x}_{sp}(t)$. This leads to the time-varying linear system

$$\Delta\dot{\overline{x}}(t) = Jf(\overline{x}_{sp}(t))\Delta\overline{x}(t) \tag{1.55}$$

where the Jacobian matrix $Jf(\overline{x}_{sp}(t))$ is periodic with the same period T as the steady-state solution $\overline{x}_{sp}(t)$. The stability of the periodic solution will be determined by the time evolution of $\Delta\overline{x}(t)$. The general form of this perturbation is [24]

$$\begin{aligned} \Delta\overline{x}(t) &= \sum_{i=k}^{N} c_k e^{\lambda_k t}\overline{u}_k(t) \\ &= c_{c1}e^{(\sigma_{c1}+j\omega_{c1})t}\overline{u}_{c1}(t) + c_{c1}^* e^{(\sigma_{c1}-j\omega_{c1})t}\overline{u}_{c1}^*(t) + c_{r1}e^{\gamma_{r1}t}\overline{u}_{r1}(t) + \cdots \end{aligned} \tag{1.56}$$

where the complex vectors $\overline{u}_k(t)$, $k = 1$ to N, are periodic with the same period T as the periodic solution, and the complex exponents λ_k are constant. The complex constants c_k, $k = 1$ to N, depend on the initial conditions (i.e., on the applied instantaneous perturbation). Note the similarity with the general expression of the perturbation of a dc regime (1.49). The only difference is the periodicity of the vectors $\overline{u}_k(t)$. Because these vectors are periodic, the extinction (or not) of the perturbation will depend only on the real part of the exponents λ_k in (1.56). This transient will be dominated by the terms associated with exponents with maximum real part σ_c or γ_r. Calculation of the exponents λ_k requires a Floquet analysis of the time-periodic linear system (1.57). Note that the λ_k are constant and cannot be the eigenvalues of the periodic matrix $Jf(\overline{x}_{sp}(t))$. Their calculation is carried out in several steps, outlined below.

Because system (1.55) has N dimensions, it will also have N linearly independent solutions $\Delta\overline{x}_1(t), \Delta\overline{x}_2(t), \ldots, \Delta\overline{x}_N(t)$. This means that the solution obtained for any initial value $\Delta\overline{x}_o$ can be expressed as a linear combination of N independent solutions. However, different sets of N independent solutions can be chosen. A useful set is the one obtained by integrating the linear system (1.55) successively from initial condition vectors $\Delta\overline{x}_{ok}$ given by columns of the N-order identity matrix I_N. Each independent solution $\Delta\overline{x}_{ck}(t)$, with $k = 1$ to N, is determined by integrating (1.55) from $\Delta\overline{x}_k = [0, \ldots, 1, \ldots, 0]^T$, where the "1" is located at the kth position. A matrix $[W_c(t)]$ can then be defined by the N independent solutions $\Delta\overline{x}_{ck}(t)$ obtained in this manner. It is called a *canonical fundamental solution matrix*, $[W_c(t)] = [\Delta\overline{x}_{c1}(t), \Delta\overline{x}_{c2}(t), \ldots, \Delta\overline{x}_{cN}(t)]$. Note that $[W_c(t)]$ is not periodic since, as gathered from (1.56), it will contain products of periodic terms and exponential terms.

Any particular initial value vector $\Delta\overline{x}_o$ of an N-dimensional system can be written $\Delta\overline{x}_o = [I_N]\Delta\overline{x}_o$. Knowing the canonical matrix $[W_c(t)]$ and the initial value $\Delta\overline{x}_o$ at $t_o = 0$, the solution $\Delta\overline{x}(t)$ can be calculated through a simple matrix–vector product:

$$\Delta\overline{x}(t) = [W_c(t)]\Delta\overline{x}_o \tag{1.57}$$

To illustrate some other properties of $[W_c(t)]$, the auxiliary matrix $[V(t)] = [W_c(t + T)]$ will be defined, which is the same matrix $[W_c(t)]$ as that evaluated at the time incremented in one period of the steady-state solution: $t + T$. We emphasize that due to the exponential factors, $[W_c(t)]$ is not periodic. However, the Jacobian matrix $Jf(\overline{x}_{sp}(t))$ is indeed periodic. Because of this, $[V(t)]$ is also a fundamental solution matrix of (1.55), as can easily be verified:

$$[\dot{V}(t)] = [\dot{W}_c(t + T)] = [Jf(\overline{x}_{sp}(t + T))][W_c(t + T)] = [Jf(\overline{x}_{sp}(t))][V(t)] \tag{1.58}$$

Thus, all the components of $[V(t)]$ fulfill (1.55). Because $[V(t)]$ is a fundamental (but not canonical) solution matrix, its columns will be expressible as in (1.57). Therefore, the solution matrix $[V(t)] = [W_c(t + T)]$ can be expressed in terms of the canonical fundamental solution matrix as $[V(t)] = [W_c(t)][V(0)]$, where $[V(0)]$ is the initial condition matrix. Replacing $[V(t)]$ with its original expression $[V(t)] = [W_c(t + T)]$, the following relationship is obtained: $[W_c(t + T)] =$

$[W_c(t)][W_c(T)]$. Note that unlike $[W_c(t)]$, the matrix $[W_c(T)]$ evaluated at $t = T$ is constant and will have constant eigenvalues and eigenvectors. The eigenvalues, assumed different, will be $m_k, k = 1$ to N, and the associated eigenvectors will be $\overline{w}_k$.

The eigenvectors $\overline{w}_k$ of $[W_c(T)]$ are linearly independent, so when taking these vectors as initial values for linear system integration, a set of N independent solutions is obtained. For each $\overline{w}_k$, the following relationship is fulfilled:

$$\Delta\overline{x}_{fk}(t+T) = [W_c(t+T)]\overline{w}_k = [W_c(t)][W_c(T)]\overline{w}_k = [W_c(t)]m_k\overline{w}_k$$
$$= m_k[W_c(t)]\overline{w}_k = m_k\Delta\overline{x}_{fk}(t) \tag{1.59}$$

where m_k is the eigenvalue of $[W_c(T)]$ associated with the eigenvector $\overline{w}_k$. The N solutions $\Delta\overline{x}_{fk}(t)$ form a set of independent solutions in terms of which any general solution $\Delta\overline{x}(t)$ of (1.55) can be expressed. It is easily shown that each solution $\Delta\overline{x}_{fk}(t)$ fulfills $\Delta\overline{x}_{fk}(t+nT) = [W_c(T)]^n\Delta\overline{x}_{fk}(t) = m_k^n\Delta\overline{x}_{fk}(t)$. Due to this property, the solutions $\Delta\overline{x}_{fk}(t)$ are called *multiplicative*. The N eigenvalues m_k of $[W_c(T)]$ are known as the *Floquet multipliers* of the linearized system (1.55).

It is easily derived that a multiplicative solution fulfilling $\Delta\overline{x}_{fk}(t+nT) = m_k^n\Delta\overline{x}_{fk}(t)$ can be written as $\Delta\overline{x}_{fk}(t) = e^{\lambda_k t}\overline{u}_k(t)$, with $\overline{u}_k(t)$ a periodic vector. Taking into account the form of these independent solutions, the general solution $\Delta\overline{x}(t)$ is written

$$\Delta\overline{x}(t) = \sum_{k=1}^{N} c_k e^{\lambda_k t}\overline{u}_k(t) \tag{1.60}$$

which demonstrates (1.56). The N Floquet multipliers are related to the N exponents in (1.60) through the expressions [24]

$$m_k = e^{\lambda_k T} \qquad k = 1 \text{ to } N \tag{1.61}$$

The exponents λ_k are known as *Floquet's exponents*. At each time instant t, the periodic vectors $\overline{u}_k(t)$ provide N independent directions in which the perturbation is decomposed in a manner similar to the N eigenvectors associated with the linearization about a dc solution [see equation (1.49)]. Each vector $\overline{u}_k(t)$ is obtained by integrating the linearized system (1.55) from the eigenvector $\overline{w}_k$ of the constant matrix $[W_c(T)]$ and dividing by $e^{\lambda_k t}$. Note that the exponents λ_k are calculated directly from the eigenvalues $m_k, k = 1$ to N, of $[W_c(T)]$.

The Floquet multipliers m_k can be real or complex. The relation (1.61) between Floquet multipliers and Floquet exponents is not univocal. Actually, there is an infinite set of exponents $\lambda_k + jm(2\pi/T)$, with m an integer and T the solution period, associated with each multiplier m_k, as can easily be verified by introducing the exponent $\lambda_k + jm(2\pi/T)$ into (1.61).

Writing the time variable as $t = t' + nT$, with n a positive integer, it is possible to introduce the multipliers into the general expression (1.60)

$$\Delta\overline{x}(t' + nT) = \sum_{k=1}^{N} c_k m_k^n e^{\lambda_k t'} \overline{u}_k(t') \tag{1.62}$$

Remember that the objective is to determine the limit value of the perturbation when time tends to infinity. Whether the increment $\Delta\overline{x}(t)$ will decay to zero or grow unboundedly will depend solely on the limit value of m_k^n with n tending to infinity, as the vectors $\overline{u}_k(t)$ are periodic with the same period T as in the steady-state solution. Clearly, if any of the multipliers has a modulus larger than 1, the perturbation will tend to infinity and the solution will be unstable. For the periodic solution to be stable, all the multipliers must have modulus smaller than 1, except the one corresponding to variations tangent to the periodic cycle, with value $m = 1$. It is easily shown that in a nonautonomous circuit, this multiplier is associated with the extra variable θ. The case of a free-running oscillation is considered below. However, an exception is the periodic free-running oscillation, considered below.

As already shown, any arbitrary time shift τ of the periodic solution of an autonomous system $\overline{x}_{sp}(t)$ gives rise to a new solution $\overline{x}_{sp}(t - \tau)$. All the time-shifted solutions lie in the same limit cycle (see Fig. 1.14). Thus, we can assume that the periodic solution of an autonomous system is invariant under displacements along this cycle. The cycle has dimension 1, and at each time value, the tangent to the cycle can be considered as one of the N dimensions into which any small perturbation is decomposed. Perturbations tangent to the limit cycle will not vanish, as the solution is invariant under displacements along this cycle. Due to this invariance, one of the multipliers of the periodic solution will be $m_1 = 1$, which means that the perturbation neither grows nor decays. The associated vector $\overline{u}_1(t)$ is tangent to the cycle at each time value, and thus is equal to the time derivative of the periodic solution $\overline{u}_1(t) = \dot{\overline{x}}_{sp}(t)$. Therefore, $x_1(t) = e^{\lambda_1 t}\overline{u}_1(t) = \dot{\overline{x}}_{sp}(t)$, where the value $\lambda_1 = 0$ has been taken into account. Thus, $\overline{u}_1(t) = \dot{\overline{x}}_{sp}(t)$ must be an independent solution of the linearized system (1.55). This is easily demonstrated by deriving both sides of (1.55) with respect to time, which provides the equality $\ddot{\overline{x}}_{sp} = Jf(\overline{x}_{sp})\dot{\overline{x}}_{sp}$, so the vector $\dot{\overline{x}}_{sp}$, tangent to the limit cycle, fulfills (1.55).

The Floquet multiplier calculation has been used in the stability analysis of the steady-state oscillation at $f_o = 1.59$ GHz of the parallel resonance circuit shown in Fig. 1.1. Because it is a two-dimensional system, two different multipliers are obtained: $m_1 = 1$ and $m_2 = 0.2828$. As already explained, the first is associated with perturbations along the direction of the cycle. The second multiplier is real and has magnitude smaller than 1, which means that the steady-state oscillation is stable. The vector $\overline{u}_1(t)$ agrees with the time derivative of the periodic solution $\overline{u}_1(t) = \dot{\overline{x}}_{sp}(t)$. The vector $\overline{u}_2(t)$ can be calculated as $\overline{u}_2(t) = e^{-\lambda_2 t}\Delta\overline{x}_{f2}(t)$, where $\Delta\overline{x}_{f2}(t)$ is the fundamental solution obtained by integrating the linearized system from the initial condition $\overline{w}_2$, with $\overline{w}_2$ being the eigenvector of $[W_c(T)]$ associated with $m_2 = 0.2828$.

Because the steady solution $\overline{x}_{sp}(t)$ is periodic, it can be expressed in a Fourier series at the oscillation frequency $f_o : \overline{x}_{sp}(t) = \sum_{m=-M}^{M} \overline{X}_m e^{jm\omega_o t}$, where M is the number of harmonic terms considered. Note that the vectors $\overline{u}_k(t)$ in the general expression (1.56) for $\Delta\overline{x}(t)$ are also periodic, with the same fundamental frequency ω_o. Thus, considering two different time scales in $\Delta\overline{x}(t)$, one for the periodic $\overline{u}_k(t)$ and the other for the exponents $e^{\lambda_k t}$, it will be possible to decompose this perturbation in an M-order Fourier series, with time-varying harmonic terms $\Delta\overline{x}(t) = \sum_{m=-M}^{M} \Delta\overline{X}_m(t)e^{jm\omega_o t}$. Note that because N independent variables comprise $\Delta\overline{x}(t)$, each harmonic component $\Delta\overline{X}_m(t)$ will be an N-dimensional vector. Before continuing, note that the harmonic components of a time-domain product $c(t) = a(t)b(t)$ can be obtained as $\overline{C} = Toep(a)\overline{B}$, where $\overline{C}$ and $\overline{B}$ are vectors containing the $2M+1$ harmonic components of $c(t)$ and $b(t)$, respectively, and Toep(a) is a matrix composed of the Fourier coefficients of $a(t)$. The rows of this matrix are permutations of the harmonic components of $a(t)$, such that the product of row m by the harmonic vector $\overline{B}$ provides the mth harmonic component of $c(t)$. Note that the calculation is affected by the truncation error in the Fourier series. One example of this type of matrix was shown in (1.40). The same principle will be applied to the time-domain product $Jf(\overline{x}_{sp}(t))\Delta\overline{x}(t)$ in the system (1.55). On the other hand, the harmonic components of $\Delta\dot{\overline{x}}(t)$ can be related to the harmonic components of $\Delta\overline{x}(t)$. Note that $\Delta\dot{\overline{X}}(t)$ is given by $\Delta\dot{\overline{X}}(t) = \sum_{m=-M}^{M}(\Delta\dot{\overline{X}}_m(t) + jm\omega_o\Delta\overline{X}_m(t))e^{jm\omega_o t}$. Then it is possible to write in matrix form

$$\Delta\dot{\overline{X}}(t) + [jm\omega_o]\Delta\overline{X}(t) = Toep[Jf(\overline{x}_{sp})]\Delta\overline{X}(t) \tag{1.63}$$

with the components of the matrix $Toep[Jf(x_{sp})]$ being constant values, since $Jf(\overline{x}_{sp}(t))$ is periodic at ω_o. Note that the dimension of the system (1.63) is $(2M+1)N$ for N independent variables. Applying the Laplace transform to equation (1.63), the following system in the Laplace frequency s is obtained:

$$\left\{[s + jm\omega_o] - Toep\left[Jf(\overline{x}_{sp})\right]\right\}\Delta\overline{X}(s) = 0 \tag{1.64}$$

Note that (1.64) is the characteristic system associated with the system linearization about the periodic solution $\overline{x}_{sp}(t)$. Recent works [30, 31] have rigorously demonstrated that for $M = \infty$ the Floquet exponents λ_k agree with the poles associated with the harmonic linear system (1.64). Because of this, there will be a set of poles $\lambda_k + jm(2\pi/T)$, with $|m| \leq M$ and T the solution period associated with each multiplier m_k. In a free-running oscillator, one of these poles is $s = 0$, so the matrix $[jm\omega_o]-$ matrix $Toep[Jf(\overline{x}_{sp})]$ must be singular. The periodicity $\lambda_k + jm(2\pi/T)$ of the poles associated with a nonlinear system linearization about a periodic regime can be understood intuitively. Consider the particular case of an instability of the periodic regime at ω_o due to a pair of complex-conjugate poles $\sigma \pm j\omega$, with $\sigma > 0$. The instability will lead to the generation of an incommensurable frequency ω. This will give rise to sidebands of the form $m\omega_o \pm \omega$ in the oscillator spectrum, so the circuit reacts as if it had "sources" of instability at all the sidebands, originated by the periodic poles.

As an example, the analysis presented will be applied to the FET-based oscillator of Fig. 1.6. The system dimension is $N = 13$, due to a relatively large number of inductors and capacitors. The steady-state oscillation frequency is $f_o = 4.39$ GHz.

The two pairs of dominant poles, extracted through numerical calculation, are $p_{1,2} = 0 \pm j4.39$ GHz and $p_{3,4} = -0.071 \pm j0.404$ GHz. Note that the frequency of $p_{1,2}$ agrees with the oscillation frequency $f_o = 4.39$ GHz, so these poles correspond to the Floquet multiplier $m = 1$ and are due to the autonomy of the oscillator solution. On the other hand, the pair of poles $p_{3,4} = -0.071 \pm j0.404$ GHz correspond to the complex-conjugate multipliers $m_{1,2} = 0.9881 \pm j0.0503$, of absolute value $|m_{1,2}| = 0.9894$. Thus, the steady-state oscillation is stable.

There are three main types of instability of a periodic solution associated with the three different possible situations: one unstable real multiplier $m_u > 1$, one unstable real multiplier $m_u < -1$, and a pair of complex-conjugate multipliers m_u and m_u^*, with $|m_u| > 1$. The type of instability will generally determine the type of solution to which the system will evolve after a transient.

Instability Due to a Positive Real Multiplier $m_k > 1$ The case of a periodic oscillation with multipliers $m_1 = 1$, $|m_{j \neq k}| < 1$, $j = 1$ to N, and $m_k > 1$ will be considered. This oscillation will be unstable due to $m_k > 1$. The real multiplier $m_k > 1$ is associated with a real eigenvalue γ_k. Thus, the perturbed steady-state oscillation will have a transient dominated by $e^{\gamma_k t}\overline{u}_k(t)$, as gathered from (1.56). Because the real exponent does not introduce any new frequency components, this type of instability will generally lead to a different periodic solution. This type of instability is typically obtained in multivalued solution curves. An example is shown in Fig. 1.21, corresponding to a MOSFET-based oscillator at 0.4 GHz [32]. Variation of the oscillation amplitude has been represented versus the gate bias voltage. The curve is bi-valued in the interval represented, which is due to the existence of a turning point at $V_{GG} = -1.2$ V at which the solution curve folds over itself. The entire curve is composed of solution points of a free-running oscillator regime, so at all the solution points there is a multiplier $m_1 = 1$. The

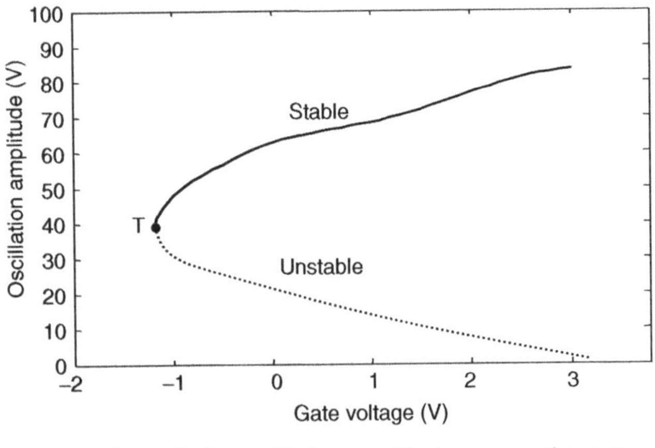

FIGURE 1.21 Variation of the oscillation amplitude versus the gate voltage in a MOSFET-based oscillator. The solution curve is bi-valued, due to the existence of a turning point.

upper section of the curve (the solid line) is stable, with all its Floquet multipliers, except $m_1 = 1$, having magnitude smaller than 1. However, a real multiplier $m_2 < 1$ increases its value when reducing V_{GG} and takes the critical value $m_2 = 1$ at the infinite slope point T. The lower section of the curve (the dashed line) is unstable with a real multiplier $m_2 > 1$. At the turning point T, $m_2 = 1$, which implies a real pole $\gamma_2 = 0$, due to the relationship between the multipliers and the roots of the characteristic determinant associated with (1.64). As demonstrated in Chapter 3, a pole at zero implies a singularity of the system at steady state, thus the infinite value of the curve slope at this point. This example shows again that the stability properties of a given steady-state solution vary when a parameter is modified. At the turning point, a qualitative stability change or bifurcation takes place in the system.

In Fig. 1.22a, the oscillator solutions at the particular bias voltage $V_{GG} = -1$ V have been represented in the phase space. The point designated EP corresponds to the coexisting dc solution. The stability of the dc solution has been analyzed independently and it has been found that this solution is stable. The limit cycle LC1 (the dashed line) has one multiplier, $m_1 = 1$, due to the solution autonomy, plus a second real multiplier, $m_2 = 1.0281$, so this limit cycle is unstable. The fundamental frequency of this solution is $f_o = 0.418$ GHz. The limit cycle LC2 has a multiplier $m_1 = 1$, plus a second, dominant multiplier $m_2 = 0.9863$, so this limit cycle is stable. Its fundamental frequency is $f_o = 0.414$ GHz.

The stable dc solution and the stable limit cycle coexist for the same values of the circuit elements (Fig. 1.22b). The unstable limit cycle is located between these two stable solutions. The unstable manifold of the unstable limit cycles separates their disjoint basins of attraction. To illustrate this idea, Fig. 1.22b shows the system behavior in a plane transversal to the unstable limit cycle LC1. The cycle intersection with this transversal plane gives rise to the point depicted. The stable manifold has dimension $N - 2$, with N the total system dimension. Note that one of the N dimensions corresponds to the cycle and is lost in the intersection with the transversal plane. The stable manifold of dimension $N - 2$ is simply sketched with two arrows pointing toward the cycle intersection point. The unstable manifold has one dimension, and depending on the initial conditions, leads to the stable limit cycle LC1 or the stable equilibrium point EP. According to Figs. 1.21 and 1.22, when reducing the gate bias voltage, the stable and unstable limit cycles LC1 and LC2 approach each other, overlap, and vanish at the turning point. For V_{GG} smaller than a value corresponding to the turning point, the dc solution is the only stable solution.

Instability Due to a Negative Real Multiplier $m_k < -1$ A periodic oscillation with associated Floquet multipliers $m_1 = 1$, $|m_{j\neq k}| < 1$, $j = 1$ to N, and a real multiplier $m_k < -1$ will be considered. This steady-state solution is unstable. Under small perturbations, the transient, ruled by (1.56), will be dominated by the real term $c_k e^{(\sigma + j(\omega_o/2))t} \overline{u}_k(t) + c_k^* e^{(\sigma - j(\omega_o/2))t} \overline{u}_k^*(t)$. To understand this, the relationship between Floquet multipliers and exponents $m_k = e^{\lambda_k T}$ must be taken into account. A real multiplier $m_k < -1$ can be expressed as $m_k = e^{(\sigma + j(1/2)(2\pi/T) + n(2\pi/T))T} =$

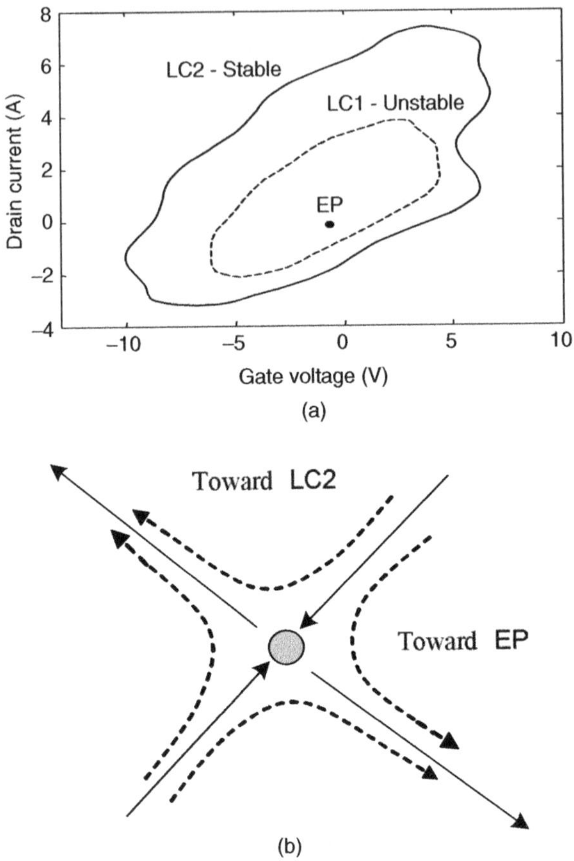

FIGURE 1.22 Phase space representation of coexisting solutions of a MOSFET-based oscillator for $V_{GG} = -1$ V. Both the equilibrium point EP and the outer limit cycle LC2 are stable. The inner limit cycle LC1 is unstable. Its unstable manifold behaves as a separator of the basins of attraction of EP and LC2.

$e^{(\sigma+j(\omega_o/2)+n\omega_o)T} = -e^{\sigma T}$, with n an integer. Because the vectors $\overline{u}_k(t)$ are periodic at the same frequency ω_o of the steady-state oscillation, the initial transient $c_k e^{(\sigma+j(\omega_o/2))t}\overline{u}_k(t) + c_k^* e^{(\sigma-j(\omega_o/2))t}\overline{u}_k^*(t)$ will correspond to an exponentially growing oscillation at the subharmonic frequency $\omega_o/2$. This transient will generally lead to a steady-state regime at the divided frequency $\omega_o/2$.

The frequency division by 2 has been observed in a Colpitts oscillator, discussed below. A stable periodic oscillation at $f_o = 1$ GHz is obtained for the original set of element values, with $L = 10$ nH. This solution has one multiplier, $m_1 = 1$, whereas the remaining multipliers have magnitude smaller than 1. Then the inductance L is swept, recalculating the steady-state solution for each L value. Note that due to the circuit autonomy, the oscillation frequency f_o will vary with L. After obtaining each steady-state solution, the corresponding Floquet multipliers are determined

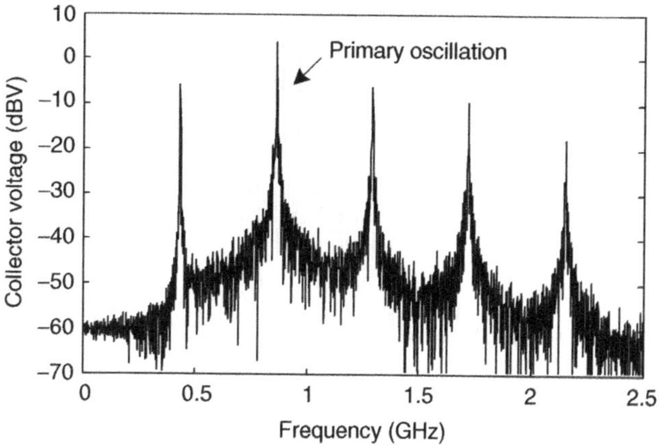

FIGURE 1.23 Subharmonic solution in a Colpitts oscillator. The oscillation at $f_o = 0.808$ GHz exhibites a real multiplier $m = -1.7340$, responsible for generation of the subharmonic frequency $f_o/2 = 0.404$ GHz.

numerically. It is found that when increasing the inductance value, one multiplier, m_2, crosses the circle through the point -1 at the value $L_o = 12.11$ nH. The periodic oscillation at f_o is unstable for $L > L_o$, so it is unobservable. For each $L > L_o$, the system evolves to a stable subharmonic solution at $f_o/2$. Figure 1.23 shows that the stable subharmonic solution emerged from the unstable periodic solution at $L = 16$ nH. This unstable solution has the Floquet multipliers $m_1 = 1$ and $m_2 = -1.734$. The voltage spectrum at the collector node is represented in Fig. 1.23. Both the primary oscillation at 0.870 GHz and the subharmonic components can be distinguished. This subharmonic solution is autonomous and periodic, so it will have a multiplier $m_1' = 1$. Because it is stable, the remaining multipliers will have magnitude smaller than 1. Note that the unstable nondivided solution coexists with the stable divided solution. It is a mathematical solution that cannot be observed physically.

Instability Due to a Pair of Complex-Conjugate Multipliers $m_k, m_{k+1} = m_k^*, |m_k| > 1$ A periodic solution with a pair of complex-conjugate multipliers $m_k, m_{k+1} = m_k^*$, $|m_k| > 1$ will be unstable and under small perturbations will generally lead the system to a quasiperiodic solution with two fundamental frequencies ω_o and $\omega_o' = \alpha\omega_o$, with $\alpha \in R$. To understand this, the relationship between Floquet multipliers and exponents $m_k = e^{\lambda_k T}$ must be taken into account. A Floquet multiplier $|m_k| > 1$ can be expressed as $m_k = e^{\sigma + j(\alpha(2\pi/T) + n(2\pi/T))T}$ $= e^{(\sigma + j\alpha\omega_o)T} = e^{(\sigma + j\omega_a)T}$. Because the multipliers m_k and m_{k+1} are the dominant ones, the transient after a small perturbation will initially evolve according to $c_k e^{(\sigma + j\omega_a)t}\overline{u}_k(t) + c_k^* e^{(\sigma - j\omega_a)t}\overline{u}_k^*(t)$. The vector $\overline{u}_k(t)$ is periodic at ω_o, so this transient will contain the two incommensurate frequencies ω_a and ω_o, which will generally lead to a quasiperiodic solution with a mixerlike spectrum at these two fundamental frequencies.

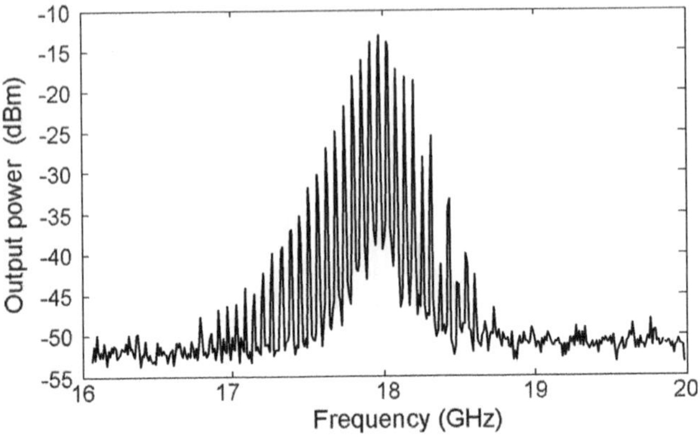

FIGURE 1.24 Output power spectrum of an oscillator at $f_o = 18$ GHz, with a second undesired oscillation at $f'_o = 8.989$ MHz.

As an example, Fig. 1.24 shows the output power spectrum of an oscillator at $f_o = 18$ GHz, with a second undesired oscillation at $f'_o = 8.989$ MHz. The unstable periodic solution had a pair of complex-conjugate multipliers $m_{1,2} = 1.0011 \pm j0.0077$. The nonharmonically related oscillation at $f'_o = \alpha f_o$ emerges from this solution. Mixing the two frequencies gives rise to the spectrum shown in Fig. 1.24. Note that the unstable periodic solution at f_o coexists with a quasiperiodic solution. It is a mathematical solution that cannot be observed physically.

1.6 PHASE NOISE

The phase noise problem in free-running oscillators is linked directly to the invariance of the steady-state periodic solution versus time translations. As shown in Section 1.2, in the frequency domain this gives rise to an irrelevance versus the phase origin. When a small impulse perturbation is applied to a stable periodic solution, the system will return to this solution (due to its stability) with a time shift τ (positive or negative) with respect to the original waveform. This gives rise to a shift $\Delta\phi = -\omega_o\tau$ in the phase origin. Note that the new phase value resulting from the perturbation corresponds to an equally valid oscillator solution. Because the linearized system $\Delta\dot{\overline{x}}(t) = Jf(\overline{x}_{sp}(t))\Delta\overline{x}(t)$ is a time-variant system, the time shift τ of the steady-state solution recovered depends on the particular time t_p of the solution period $(0, T]$ at which the perturbation is applied [8]. This is illustrated in the simulations of Fig. 1.25, which were carried out in the FET-based oscillator of Fig. 1.6. Figure 1.25a shows the steady-state waveform corresponding to the voltage across the gate capacitance $v_G(t)$. A short current pulse will be introduced at the gate node at different time values t_p, analyzing the effect on the drain voltage waveform. Figure 1.25b shows the original steady-state waveform (the solid line) and the time-shifted steady-state waveforms resulting from the use of an instantaneous perturbation of equal magnitude applied at different

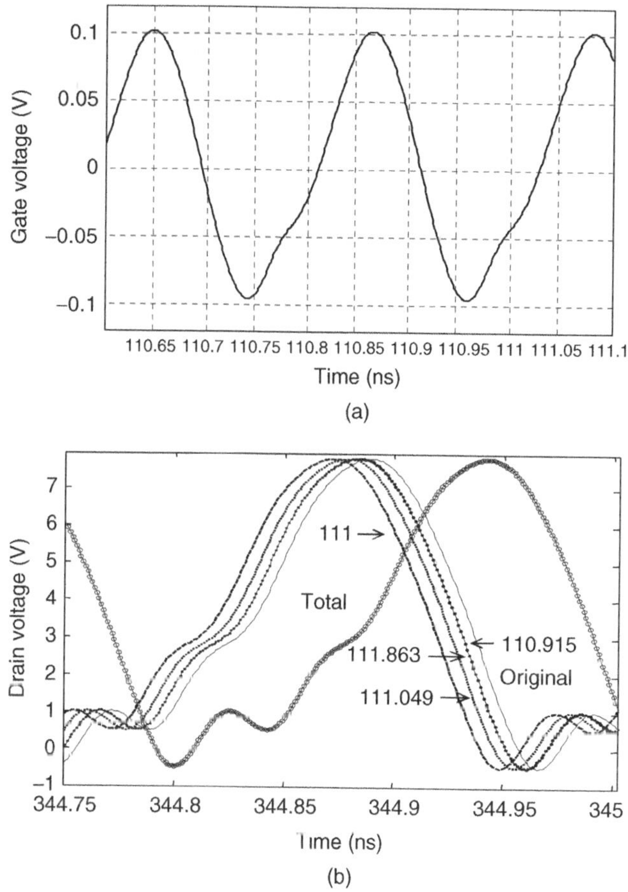

FIGURE 1.25 Time shift of the steady-state solution of the FET-based oscillator of Fig. 1.6 as a result of the introduction of short current pulses at different times. The current perturbations are introduced at the gate node. (a) Gate voltage waveform. (b) Drain voltage waveform. The waveform indicated by "total" is the result of three different perturbations applied at $t_p = 110.915$, 111, and 111.049 ns.

points in time. The curve corresponding to a perturbation applied at $t_p = 110.863$ ns is nearly overlapped with the curve corresponding to a perturbation applied at $t_p = 110.915$ ns, represented by diamonds. The dashed curve corresponds to a perturbation applied at $t_p = 111$ ns. The dotted curve corresponds to a perturbation applied at $t_p = 111.049$ ns. In agreement with Hajimiri and Lee [8], larger time shifts are obtained when a perturbation is applied at points of the waveform with a larger magnitude of the time derivative, due to rapid evolution of the system at these points. When applying several perturbations, the time shift accumulates. This is evidenced by the bold dotted curve, which is obtained as the result of three different perturbations applied at $t_p = 110.915$, 111, and 111.049 ns.

Unlike the test perturbations considered in the analysis shown in Fig. 1.25, the circuit noise sources are not deterministic. Thus, for phase noise analysis it will be necessary to obtain the stochastic characterization of the phase deviation in the presence of noise perturbations. The fundamental background for the understanding and analysis of oscillator phase noise is provided in Chapter 2.

REFERENCES

[1] A. B. Carlson, *Communication Systems*, McGraw-Hill, New York, 1986.

[2] U. L. Rohde, Nonlinear effects in oscillators and synthesizers, *IEEE MTT-S International Microwave Symposium*, Phoenix, AZ, pp. 689–692, 2001.

[3] K. Kurokawa, Injection locking of microwave solid state oscillators, *Proc. IEEE*, vol. 61, pp. 1386–1410, Oct. 1973.

[4] R. A. York, Nonlinear analysis of phase relationships in quasi-optical oscillator arrays, *IEEE Trans. Microwave Theory Tech.*, vol. 41, pp. 1799–1809, Oct. 1993.

[5] R. E. Collin, *Foundations for Microwave Engineering*, 2nd ed., Wiley, New York, 2001.

[6] P. F. Combes, J. Graffeuil, and J. F. Sautereau, *Microwave Components, Devices and Active Circuits*, Wiley, Chichester, UK, 1987.

[7] M. Odyniec, Oscillator stability analysis, *Microwave J.*, vol. 42, p. 6, 1999.

[8] A. Hajimiri and T. H. Lee, A general theory of phase noise in electrical oscillators, *IEEE J. Solid State Circuits*, vol. 33, Feb. 1998.

[9] F. X. Kaertner, Analysis of white and $f^{-\alpha}$ noise in oscillators, *Int. J. Circuit Theory Appl.*, vol. 18, pp. 485–519, 1990.

[10] J. M. T. Thompson and H. B. Stewart, *Nonlinear Dynamics and Chaos*, 2nd ed., Wiley, Hoboken, NJ, 2002.

[11] J. Jugo, J. Portilla, A. Anakabe, A. Suárez, and J. M. Collantes, Closed-loop stability analysis of microwave amplifiers, *IEE Electron. Lett.*, vol. 37, pp. 226–228, Feb. 2001.

[12] A. Anakabe, Detección y eliminación de inéstabilidades paramétricas en amplificadores de potencia para comunicaciones, Ph.D. Thesis, Universidad del Pais Vasco, 2003.

[13] U. L. Rohde, A. K. Poddar, and G. Bock, *The Design of Modern Microwave Oscillators for Wireless Applications*, Wiley, Hoboken, NJ, 2005.

[14] D. J. Vendelin, A. M. Pavio, and U. L. Rohde, *Microwave Circuit Design*, Wiley, New York, 1990.

[15] M. Odyniec (Ed.), *RF and Microwave Oscillator Design*, Artech House, Norwood, MA, 2002.

[16] K. Ogata, *Modern Control Engineering*, Prentice-Hall, Englewood Cliffs, NJ, 1980.

[17] K. Kurokawa, Some basic characteristics of broadband negative resistance oscillators, *Bell Syst. Tech. J.*, vol. 48, pp. 1937–1955, July–Aug. 1969.

[18] P. Gamand and V. Pauker, Starting phenomenon in negative resistance FET oscillators, *Electron. Lett.*, vol. 24, pp. 911–913, 1988.

[19] G. B. Arfken and H. J. Weber, *Mathematical Methods for Physicists*, Academic Press, San Diego, CA, 2001.

[20] J. H. Wilkinson, *The Algebraic Eigenvalue Problem*, Oxford University Press, New York, 1965.

[21] V. Rizzoli and A. Lipparini, General stability analysis of periodic steady-state regimes in nonlinear microwave circuits, *IEEE Trans. Microwave Theory Tech.*, vol. 33, pp. 30–37, Jan. 1985.

[22] S. Mons, J. C. Nallatamby, R. Quéré, P. Savary, and J. Obregón, A unified approach for the linear and nonlinear stability analysis of microwave circuits using commercially available tools, *IEEE Trans. Microwave Theory Tech.*, vol. 47, pp. 2403–2409, Dec. 1999.

[23] S. A. Maas, *Nonlinear Microwave Circuits*, Artech House, Norword, MA, 1988.

[24] J. Guckenheimer and P. Holmes, *Nonlinear Oscillations, Dynamic Systems, and Bifurcations of Vector Fields*, Springer-Verlag, New York, 1983.

[25] M. I. Sohby and A. K. Jastrzebsky, Direct integration methods of nonlinear microwave circuits, *European Microwave Conference*, Paris, pp. 1110–1118, 1985.

[26] T. S. Parker and L. O. Chua, *Practical Algorithms for Chaotic Systems*, Springer-Verlag, Berlin, 1989.

[27] S. Wiggins, *Introduction to Applied Nonlinear Dynamical Systems and Chaos*, Springer-Verlag, New York, 1990.

[28] L. Chua, Editorial in special issue, *IEEE Trans. Circuits Syst.*, vol. 30, pp. 617–619, 1983.

[29] C. P. Silva, Shil'nikov's theorem: a tutorial, *IEEE Trans. Circuits Syst. I Fundam. Theor. Appl.*, vol. 40, pp. 675–682, 1993.

[30] J. M. Collantes, I. Lizarraga, A. Anakabe, and J. Jugo, Stability verification of microwave circuits through Floquet multiplier analysis, *IEEE Asia-Pacific Proceedings on Circuits and Systems*, pp. 997–1000, 2004.

[31] F. Bonani and M. Gilli, Analysis of stability and bifurcations of limit cycles in Chua's circuit through the harmonic-balance approach, *IEEE Trans. Circuits and Syst. I*, vol. 46, no. 8, pp. 881–890, 1999.

[32] S. Jeon, A. Suarez, and D. B. Rutledge, Nonlinear design technique for high-power switching-mode oscillators, *IEEE Trans. Microwave Theory Tech.*, vol. 54, pp. 3630–3639, 2006.

CHAPTER TWO

Phase Noise

2.1 INTRODUCTION

In the frequency domain, a nonmodulated ideal oscillator is expected to provide pure spectral lines or impulses at the fundamental oscillation frequency and its harmonic terms $k\omega_o, k = 1$ to NH. In practice, the oscillator spectrum shows skirts about these central frequencies, associated with undesired modulations coming from the noise sources in the semiconductor devices and resistances contained in the circuit. In free-running oscillators, the phase noise dominates the amplitude noise, which is usually not too high, due to the limiter behavior of the oscillator circuit associated with its inherent nonlinearity [1]. The oscillator noise spectrum is frequency dependent. The highest values of noise spectral density are obtained near the carrier frequency, where this spectral density decreases more quickly versus the offset frequency Ω. Provided that all the noise spectrum can be attributed entirely to phase noise, the phase noise can be quantified by considering a unit bandwidth at a frequency offset Ω, calculating the noise power in this unit bandwidth, and dividing the result by the carrier power [2]. If only phase noise is present, the original oscillator power will spread around the steady-state frequencies $k\omega_o, k = 1$ to NH, with a total power obtained from integration of the power spectral density equal to that of a noiseless oscillator with impulses at $k\omega_o, k = 1$ to NH.

Oscillator circuits are often used as local oscillators in the frequency-conversion stages of communication systems, to up- or down-convert the carrier frequency of a modulated signal. At the mixing state, the phase noise of the local oscillator corrupts the modulation signal, which can give rise to demodulation errors. Other

Analysis and Design of Autonomous Microwave Circuits, By Almudena Suárez

undesirable situations have been pointed out by Razavi [2]. Assume a receiver of a weak signal at the frequency ω_1 and a large interferer in an adjacent channel, ω_2. If the local oscillator used for down-conversion has a high phase noise, the down-converted signal from ω_2 will have an increased bandwidth due to this phase noise. The down-converted signal desired from ω_1 may be corrupted by the overlapping tail of the down-converted interferer coming from ω_2. As another example, the weak signal of a noiseless receiver at ω_1 can also be corrupted with the phase noise tail of a high power transmitter at close frequency ω_2. Note that the power spectral density of the phase noise skirt may be relevant in a broader bandwidth than the difference between the two carrier frequencies.

To reduce phase noise, an oscillator circuit with a voltage control signal (voltage-controlled oscillator) is often introduced into a *phase-locked loop* [3], a feedback system with a voltage-controlled oscillator that is adjusted constantly to match the phase and frequency of a reference signal. The phase comparison with a low-noise reference signal provides an error current that after passing through a filter modifies the oscillator control voltage so as to reduce the phase error. The phase noise spectral density at a small frequency offset from the carrier can be reduced substantially with this technique. However, low spectral density at a higher frequency offset requires a low phase noise of the oscillator itself (i.e., of the voltage-controlled oscillator inserted into the loop). This phase noise depends on the active devices used, their bias conditions, and the particular design.

As introduced in Chapter 1, the phase noise in free-running oscillators is an undesired effect of the fact that any arbitrary time shift of the steady-state periodic solution provides another solution. When applying a small impulse perturbation to a stable periodic solution, the system returns, after a transient, to the steady state with a time shift with respect to the original unperturbed periodic solution. In the phase space, the stable oscillator returns to the limit cycle at a different cycle point. The instantaneous perturbation gives rise to a permanent shift of the solution in the cycle, whereas the increments $\Delta\overline{x}$ in the rest of the space directions vanish exponentially in time [4]. Thus, under continuous noise perturbations, the trajectory remains in the neighborhood of the limit cycle in the phase space, but the shifts along the cycle accumulate, due to the absence of a restoring mechanism in the direction tangent to the limit cycle.

Assuming a small impulse, the phase shift undergone by the steady-state solution depends greatly on the precise point of the periodic waveform at which this impulse is applied [5]. In the presence of noise sources, the oscillator solution is being perturbed continuously and the phase noise spectrum can be derived from the variance of the stochastic phase deviation. The variance depends on the spectral density and correlation of the noise sources and on the phase sensitivity functions, which are deterministic and periodic. The phase sensitivity function with respect to a particular noise source provides the phase shift resulting from application of a small-amplitude impulse at different times in the solution period. The impulse must be of the same type (current or voltage) as the noise source and must be applied from the same circuit location. As stated earlier, the phase noise spectrum can be calculated from the variance of the phase deviation, depending on the phase sensitivity functions.

Alternatively, the phase noise in oscillator circuits can be related to the fact that the frequency is a state variable of the oscillator circuit [6]. In a nonautonomous circuit (e.g., an amplifier), the noise sources give rise to perturbations in the amplitudes and phases of the harmonic components of all the circuit variables, voltages, and currents, but not to frequency perturbations, as the fundamental frequency of the solution is determined by the input periodic source. In the case of an oscillator, in addition to perturbations in the amplitudes and phases of the voltages and currents, the noise sources will give rise to perturbations in the fundamental frequency of the solution. Thus, the noise sources will give rise to a frequency modulation of the oscillator carrier. Because the phase is the integral of the frequency variable, these perturbations will be responsible for the undesired phase noise characteristic.

The aim of this chapter is to provide a conceptual background for an understanding of phase noise and its analysis techniques. In manner similar to the oscillator analysis of Chapter 1, the oscillator phase noise is studied in the time and frequency domains. Initially, the stochastic phase noise characterization of the oscillator spectrum, based on phase sensitivity functions, is presented, with basic and intuitive explanations. The most arduous mathematical details are omitted and readers are referred to fundamental references [4,7,8]. Then the frequency-domain analysis of phase noise is presented. Initially, the phase noise spectrum is derived from the oscillator frequency modulation and analyzed using an impedance–admittance description of the oscillator circuit. The results are related to those obtained using time-domain calculation. Next, the phase sensitivity functions used in the time-domain derivation are determined approximately using frequency-domain analysis limited to the fundamental frequency. The two types of frequency-domain analysis will establish a conceptual basis for the harmonic balance simulation of phase noise covered in Chapter 7. The amplitude noise in oscillator circuits is also covered, indicating the situations in which this type of noise constitutes a relevant contribution to the oscillator power spectrum.

The chapter is organized as follows. In Section 2.2 some generalities about random variables and random processes are presented, as a reminder. In Section 2.3 the types of noise sources in electronic circuits are defined. In Section 2.4 we present the time-domain derivation of the oscillator phase noise spectrum using phase sensitivity functions. Section 2.5 covers the frequency-domain analysis of the oscillator phase noise from modulation of the oscillator carrier and based on a calculation of the phase sensitivity functions. Amplitude noise is also discussed in Section 2.5.

2.2 RANDOM VARIABLES AND RANDOM PROCESSES

2.2.1 Random Variables and Probability

A real random variable X will take real values $x \in R$ according to a given probability distribution, depending on the value x. Thus, the probability that the variable X takes a value in the interval $[x - dx, x]$ is given by $p_X(x)dx$, where $p_X(x)$ is the probability density function (PDF). In turn, the distribution function $F_X(\alpha)$

provides the absolute probability P_X that a random variable X takes an equal or smaller value than a certain real number α; that is, $F_X(\alpha) = P_X[-\infty < x \leq \alpha]$. The probability density function agrees with the derivative of the distribution function $p_X(x) = dF_X(x)/dx$. The probability density function $p_X(x)$ is used for calculation of the mean value or expectation of any continuous function of x, given by $g(x)$:

$$E[g(x)] = \int_{-\infty}^{\infty} g(x)p_X(x)\,dx \tag{2.1}$$

For calculation of the mean value of X, the function $g(x) = x$ is used in (2.1). The mean value here will be called m_x. For calculation of the mean-square value of X, the function used is $g(x) = x^2$. The variance of a variable x is defined as the mean-square value of the variable deviation with respect to its mean $m_x : \sigma_x^2 = E[(x - m_x)^2] = E(x^2) - m_x^2$, where σ_x is called the *standard deviation* [9]. The nth-order moment of a random variable X is the expectation of the nth power of the variable $E[(x)^n]$. In turn, the nth-order central moment is the expectation $E[(x - m_x)^n]$. Thus, the variance σ_x^2 is the second-order central moment of the random variable X.

Generalizations of all the expressions above exist for functions depending on multiple variables. As an example, the joint probability density function $p_{XY}(x, y)$ provides the probability that the variables X and Y take values in the differential intervals $[x - dx, x], [y - dy, y]$. In the case of independent variables, the value taken by the variable x does not depend on the value taken by the variable y. Then the joint probability density fulfills $p_{XY}(x, y) = p_X(x)p_Y(y)$. This can be extended to any arbitrary number of variables. If the two variables are not independent, it will be possible to define a conditional probability density. This provides the probability that $X = x$ given that $Y = y$, calculated as

$$p_X(x|y) = \frac{p_{XY}(x, y)}{p_Y(y)} \tag{2.2}$$

where the vertical line indicates the condition $Y = y$. The probability $p_Y(y|x)$ would be calculated in a similar manner.

The characteristic function of a random variable X is given by

$$\phi_X(s) = E[e^{jsx}] = \int_{-\infty}^{\infty} e^{jsx} p_X(x)\,dx \tag{2.3}$$

which is a particular case of (2.1) with the function $g(x) = e^{jsx}$. Note that neither the expectation operation nor the integration affects the variable s. Thus, the expectation in (2.3) provides a function $\phi_X(s)$ of the variable s. From an inspection of (2.3), it can be gathered that the characteristic function and the probability density function constitute a Fourier transform pair [10]. Taking into account that $\partial^n e^{jsx}/\partial s^n = (j)^n x^n e^{jsx}$, the expectations values $E[(x)^n]$ can easily be obtained from the characteristic function of the random variable by setting

$$E[(x)^n] = (j)^{-n} \left. \frac{\partial^n \phi_x(s)}{\partial s^n} \right|_{s=0} \tag{2.4}$$

Later in the chapter we deal with partial differential equations in the probability density function. Use of the transformation (2.3) will allow a simpler resolution. This is because use of the dummy variable s transforms the derivation $\partial/\partial x$ into multiplication by js, in a manner similar to what happens when applying the Laplace transform to a system of linear differential equations.

The physical systems are modeled with different probability distributions, such as the binomial distribution, the Poisson distribution, or the Gaussian distribution [9]. The binomial distribution applies to integer random variables. The probability that a certain event A with probability p happens i times over n evaluations is the following: $P_A(i) = \binom{n}{i} p^i(1-p)^{n-i}$. The Poisson distribution arises as the limit of the binomial distribution for very large n and very small probability p. If the product np remains finite, the probability distribution can be approached as $P_A(i) = e^{-np}(np)^i/i!$. As an example, an event A with the probability of occurrence $P_A = \mu\Delta T \ll 1$ in the time interval ΔT will be considered [10]. If the occurrences are independent statistically, the probability that A occurs i times in the time interval T is $P_A(i) = e^{-\mu T}(\mu T)^i/i!$, with $np = \mu T$. A limit form of this probability distribution models the shot noise in electronic circuits. Details are given in section 2.3.2.

According to the central limit theorem, if X is the summation of many random components, and if each component represents a small contribution to this summation, the summation approaches a Gaussian probability distribution regardless of the probability distribution of the individual components. This is why the Gaussian probability distribution has great physical interest. The probability density function of a Gaussian random variable is given by

$$p_X(x) = \frac{1}{\sqrt{2\pi\sigma_x^2}} \exp\left[\frac{-(x-m_x)^2}{2\sigma_x^2}\right] \tag{2.5}$$

where m_x is the mean value of X and σ_x is its standard deviation. The probability distribution (2.5) is symmetrical about m_x. The larger values of this probability density are concentrated between $m_x - \sigma_x$ and $m_x + \sigma_x$. In fact, it is easily demonstrated that $P_X[|x - m_x| \leq \sigma_x] \cong 0.68$. As gathered from (2.5), the probability density function of a Gaussian variable X is totally determined by its mean value m_x and its standard deviation σ_x. Its statistical moments of order $n > 2$ are equal to zero. By replacing (2.5) in the integral expression (2.3), it is easily shown that the characteristic function associated with the Gaussian probability distribution is given by

$$\phi_X(s) = E[e^{jsx}] = \int_{-\infty}^{\infty} e^{jsx} p_X(x)\,dx = e^{jsm_x - s^2\sigma_x^2/2} \tag{2.6}$$

The multivariate probability density of N Gaussian random variables is an extension of (2.5). For a vector $\overline{x}$ of N Gaussian random variables, the covariance matrix is defined as

$$[\sigma_{\overline{x},2}] = E[(\overline{x} - \overline{x}_m)(\overline{x} - \overline{x}_m)^T] \tag{2.7}$$

where the vector $\overline{x}_m$ contains the mean values of the N random variables and $[\sigma_{\overline{x},2}]$ is the $N \times N$ covariance matrix whose elements are the second-order correlation functions of the variables in $\overline{x}$. It is a symmetric matrix, and if the variables are independent in pairs, it is a diagonal matrix [9]. From (2.7), the multivariate probability density function is written

$$p_{\overline{x}}(\overline{x}) = \frac{1}{((2\pi)^N \det[\sigma_{\overline{x},2}])^{1/2}} \exp\left(\frac{-1}{2}(\overline{x} - \overline{x}_m)^T [\sigma_{\overline{x},2}]^{-1}(\overline{x} - \overline{x}_m)\right) \tag{2.8}$$

which is a clear extension of expression (2.5).

2.2.2 Random Processes

As already stated, a real random variable X will take real values $x \in R$ according to a given probability distribution. A random or stochastic process is a function of a deterministic argument or index. This argument usually corresponds to the time variable, and the process is often known as a *time series*. Thus, a stochastic process is a collection of random variables $x(t)$. As an example, the noise sources $\overline{n}(t)$ in an electronic circuit are random processes. Two or more identical circuits with the same noise sources will have different perturbed values of their state variables $\overline{x}(t)$ along the same time interval. This is because for each different realization of the noise sources, according to their statistical characteristics, a different time variation of the circuit variables $\overline{x}(t)$ is obtained. If many identical circuits are evaluated for the same time interval, each $\overline{x}(t, s_i)$, with s_i referring to the particular circuit, is a sample. The set of different time functions is an *ensemble*. It can be said that a system of noiseless differential equations gives rise to a deterministic processes, whereas a system containing noise sources provides a stochastic process $\overline{x}(t)$.

The stochastic processes evolve probabilistically in time, so their probability density function is a function of time. Then the PDF of random process $x(t)$ will be written $p_X(x, t)$. If the variable X is measured at different time instants $t_1, t_2, t_3, \ldots$ the probability that this variable has followed the path $x_1, t_1; x_2, t_2; \ldots; x_n, t_n$ is determined by the corresponding joint probability density function

$$p_{\text{joint}}(x_n, t_n; x_{n-1}, t_{n-1}; \ldots; x_1, t_1) \tag{2.9}$$

Thus, the probability of having the state x_n at time t_n is determined by an entire set of joint probability density functions of the form (2.9), considering all possible values of the previous events $x_{n-1}, t_{n-1}; \ldots, x_1, t_1$. There are different types of processes, depending on the form of the joint probability density. If each instantaneous event t_n, x_n is independent of all previous or future events, the joint probability has the form $p_{\text{joint}} = \prod_i p(x_i, t_i)$. If the events are not independent, conditional probabilities such as the one defined in (2.2) must be taken into account. As an example, the case of three discrete time instants is considered in the following. The probability of the third measurement taking the value $x(t_3) = x_3$ under the condition $x(t_1) = x_1$ is given by

$$\begin{aligned} p(x_3, t_3|x_1, t_1) &= \int p(x_3, t_3; x_2, t_2|x_1, t_1)dx_2 \\ &= \int p(x_3, t_3|x_2, t_2; x_1, t_1)p(x_2, t_2|x_1, t_1)dx_2 \end{aligned} \tag{2.10}$$

Note that we are only interested in the probability measuring x_3 at t_3 under the condition that x_1 was measured at t_1, so x_2 is allowed to take any value. This is why the integral is carried out over all the possible x_2 values obtained under the condition $x(t_1) = x_1$.

If the probability of each event depends on the previous event only, we have a Markov process. Fortunately, most physical systems can be modeled approximately with this kind of process, characterized by a short-time memory. In a Markov process, the probability of having the state x_n at time t_n depends only on the previous state, x_{n-1}, t_{n-1}. This greatly simplifies expression of the joint probability, which can now be written in terms of conditional probabilities involving only two adjacent time instants:

$$\begin{aligned} &p_{\text{joint}}(x_n, t_n; x_{n-1}, t_{n-1}; \ldots; x_1, t_1) \\ &\quad = p(x_n, t_n|x_{n-1}, t_{n-1})p(x_{n-1}, t_{n-1}|x_{n-2}, t_{n-2}) \\ &\quad\quad \cdots p(x_2, t_2|x_1, t_1)p(x_1, t_1) \end{aligned} \tag{2.11}$$

For a Markov process, the probability of the third measurement taking the value $x(t_3) = x_3$ under the condition $x(t_1) = x_1$ is given by

$$p(x_3, t_3|x_1, t_1) = \int dx_2 \; p(x_2, t_2|x_1, t_1)p(x_3, t_3|x_2, t_2) \tag{2.12}$$

Compared to (2.10), expression (2.12) is simpler, as the conditional probability depending on two previous states $p(x_3, t_3|x_2, t_2; x_1, t_1)$ has been replaced by $p(x_3, t_3|x_2, t_2)$, depending only on the previous state.

The equality (2.12) is the well-known *Chapman–Kolmogorov equation* [9]. There is also a differential version of this equation. This differential version is essential in an analysis of stochastic processes. It will be the key element in deriving the partial differential equation that governs the time-varying PDF, $p_X(x, t)$, of a random process X ruled by a stochastic differential equation. It provides the time derivative of the probability of X having the value x at time t under the condition $X = x'$ at the previous time instant. Calculation of the time derivative of the PDF requires introduction of the transition rate w. This is the probability per unit time of transition from one state to another. The expression, written for continuous time, is the following [11]:

$$\frac{\partial p(x,t)}{\partial t} = \int w(x|x')p(x'|t)\,dx' - \int w(x'|x)p(x|t)\,dx' \tag{2.13}$$

So the time derivative of the probability of X having the value x at time t is the difference between the total probability of transition from any x' to the particular x minus the total probability of escape from the particular x to any x'. For a full demonstration the reader should check Gardiner's book [9].

Next, the following assumptions will be made: There are only small-amplitude jumps $|x - x'|$, and the functions w and p are slowly varying and sufficiently smooth (derivable) versus both arguments. To take advantage of these assumptions, the transition rate $w(x'|x)$ will be expressed in a different manner. It is possible to define $x' = x - r$ and make the change of notation $w(x|x') = w(x - r; r)$. Then equation (2.13) can be written

$$\frac{\partial p(x,t)}{\partial t} = \int w(x-r;r)p(x-r,t)\,dr - p(x,t)\int w(x;-r)\,dr \tag{2.14}$$

Note that when using r as the transition measure, the two integrals in (2.13) are performed in terms of r. This allows taking $p(x,t)$ out of the second integral. Assuming relatively small r, it will be possible to perform a Taylor series expansion of the first integral on the right-hand side about $r = 0$. This provides

$$\begin{aligned}
&\int w(x-r;r)p(x-r,t)\,dr \\
&\quad = \int_{-\infty}^{\infty} w(x;r)\,dr\;p(x,t) - \frac{\partial \int_{-\infty}^{\infty} r w(x;r)\,dr p(x,t)}{\partial x} \\
&\qquad + \frac{1}{2}\frac{\partial^2 \int_{-\infty}^{\infty} r^2 w(x;r)\,dr p(x,t)}{\partial x^2} + \text{higher-order terms}
\end{aligned} \tag{2.15}$$

If the transition rates w are slowly varying functions, it is possible to truncate the Taylor series expansion to the second order. Expression (2.15) must be placed into (2.14). The resulting equation is given by

$$\begin{aligned}
\frac{\partial}{\partial t}p(x,t) = {} & \underbrace{p(x,t)\int_{-\infty}^{\infty} w(x;r)\,dr - p(x,t)\int_{-\infty}^{\infty} w(x;-r)\,dr}_{\text{master}} \\
& - \frac{\partial}{\partial x}[a_1(x)p(x,t)] + \frac{1}{2}\frac{\partial^2}{\partial x^2}[a_2(x)p(x,t)]
\end{aligned} \tag{2.16}$$

with

$$\begin{aligned}
a_1(x) &= \int_{-\infty}^{\infty} r w(x;r)\,dr \\
a_2(x) &= \int_{-\infty}^{\infty} r^2 w(x;r)\,dr
\end{aligned} \tag{2.17}$$

There are three different terms on the right-hand side of equation (2.16). The first term, indicated as "master," governs jump phenomena and gives rise to discontinuous sample paths. In the master equation, which rules some stochastic processes (e.g., Poisson's process), the two additional terms on the right-hand side of (2.16) are equal to zero.

If the term denoted "master" in (2.16) is equal to zero, no discontinuous jumps occur versus the time variable [9,11]. For continuous paths, the relationship (2.16) simplifies to the *Fokker–Planck equation:*

$$\frac{\partial}{\partial t}p(x,t) = -\frac{\partial}{\partial x}[a_1(x)p(x,t)] + \frac{1}{2}\frac{\partial^2}{\partial x^2}[a_2(x)p(x,t)] \tag{2.18}$$

As can be seen, equation (2.18) is a partial differential equation in x and time t. In this equation, the term a_1 is called the *drift coefficient* and the term a_2 is called the *diffusion coefficient*. It can roughly be said that the drift term determines the variation of the mean value of the random process. The diffusion term determines the time evolution of its variance.

An example of the Markov process, ruled by the Fokker–Planck equation, is the Wiener process. In this process, the drift coefficient is equal to zero $a_1 = 0$ and the diffusion coefficient is equal to 1, $a_2 = 1$. The PDF obeys the partial differential equation

$$\frac{\partial}{\partial t}p(w,t|w_o,t_o) = \frac{1}{2}\frac{\partial^2}{\partial w^2}p(w,t|w_o,t_o) \tag{2.19}$$

with the initial condition $p(w,t_o|w_o,t_o) = \delta(w - w_o)$. The equation is solved easily using the characteristic function to transform the derivation $\partial/\partial x$ into multiplication by the dummy variable s. A detailed derivation has been provided by Gardiner [9]. The probability density function of the Wiener process obtained is given by

$$p(w,t|w_o,t_o) = \frac{1}{\sqrt{2\pi(t-t_o)}}e^{-(w-w_o)^2/2(t-t_o)} \tag{2.20}$$

Compared with (2.5), it is a Gaussian stochastic process with mean value w_o and variance $E[(w(t) - w_o)^2] = t - t_o$. Thus, the bell-shaped Gaussian distribution keeps centered about the initial value w_o but spreads in time. This means that the sample paths have great variation. The sample paths of the Wiener process $w(t)$ are continuous, in agreement with the fact that it is ruled by a Fokker–Planck equation. However, the sample paths are nondifferentiable, as the probability $P[(w(t+h) - w(t))/h] > k$ is different from zero in the limit $h \to 0$. Thus, the process is very irregular. The increments $w(t+h) - w(t)$, with $h > 0$, are Gaussian with zero mean and variance h, in agreement with (2.20). The increment $w(t+h) - w(t)$ is independent of $w(s)$ for $s \in [0,t)$. This means that the change of value in the interval $[t, t+h]$ is independent of what happened up to time t. Thus, it is possible to write $E[w(t)w(t')] = \min(t,t')$. The Wiener process has a notable implication in practical systems. It can be shown that the white noise $\varepsilon(t)$ is the generalized mean-square derivative of the Wiener process $w(t)$. Details are given later in the section.

2.2.3 Correlation Functions and Power Spectral Density

Provided that we know the time variation of the probability density function $p_X(x, t)$, the time-dependent mean value of $x(t)$ will be given by

$$E[x(t)] = \int_{-\infty}^{\infty} x(t) p_X(x, t)\, dx \tag{2.21}$$

Note that the time t is kept constant in the integral, so different mean values may be obtained for different t values. Thus, the mean value will generally depend on time. The autocorrelation function is the mean value of $x(t_1)x(t_2)$, with t_1 and t_2 being two different time instants. It is calculated as

$$R_x(t_1, t_2) = \int_{-\infty}^{\infty} \int_{-\infty}^{\infty} x(t_1)x(t_2) p_{X_1,X_2}(x_1, x_2) dx_1 dx_2 \tag{2.22}$$

where $x_1 = x(t_1)$ and $x_2 = x(t_2)$ and p_{X_1,X_2} is the joint probability density function, evaluated at the fixed time instants t_1 and t_2. The autocorrelation function gives a measure of the relatedness or dependence between the values of the variable x at the two different time instants t_1 and t_2, or equivalently, between the two variables $x(t_1)$ and $x(t_2)$ [10]. In the case of uncorrelated variables, the autocorrelation function (2.22) simplifies to $R_x(t_1, t_2) = E[x(t_1)]E[x(t_2)]$. Its value at t_1, t_2 will be zero if any of the mean values is zero. Note that two variables $x(t_1)$ and $x(t_2)$ may be uncorrelated but not statistically independent. Remember that for two variables $x(t_1)$ and $x(t_2)$ to be statistically independent, their joint probability density function must fulfill $p_{x_1x_2}(x(t_1), x(t_2)) = p_{x_1}(x(t_1))p_{x_2}(x(t_2))$, which is a more restrictive condition than $R[x(t_1), x(t_2)] = E[x(t_1)]E[x(t_2)]$. Thus, two uncorrelated variables may not be statistically independent. However, two independent variables are necessarily uncorrelated.

In a similar manner, the cross-correlation between two different variables $x(t)$ and $y(t)$ is the mean value of the product of the two different variables evaluated at two different time instants t_1 and t_2. It is given by

$$R_{xy}(t_1, t_2) = E[x(t_1)y(t_2)] \tag{2.23}$$

The processes are uncorrelated if the relationship $R_{xy}(t_1, t_2) = E[x(t_1)]E[y(t_2)]$ is fulfilled $\forall t_1$ and t_2. If any of the two processes has zero mean value, the cross-correlation is equal to zero.

The characteristics of a stationary process are invariant over all times, so any translation of the time origin along the ensemble does not affect the values of the ensemble averages [9]. The conditions for a wide-sense stationary process are less restrictive. A process is called *wide-sense stationary* if its mean value is time independent $E[x(t)] = m_x$ and its autocorrelation depends on the time difference only, $R_x(\tau) = E[x(t - \tau/2)x(t + \tau/2)]$. Note that the ensemble averages in the stationary process do not depend on time but do not necessarily agree with the time averages. In an ergodic process, the ensemble average at any time value t agrees with the time average. The following equalities are fulfilled:

$$\begin{aligned} \langle x(t)\rangle &= E[x(t)] = m_x \\ \langle x^2(t)\rangle &= E[x^2(t)] \\ \langle x(t-\tau/2)x(t+\tau/2)\rangle &= E[x(t-\tau/2)x(t+\tau/2)] \end{aligned} \tag{2.24}$$

Due to the ergodicity property, a single sample will be representative of the entire process. It is often considered that a wide-sense stationary function is also ergodic, if we can reasonably expect that a typical sample function exhibits the same statistical variations of the process [9].

The Gaussian process has great relevance in communication systems, as this model applies to many electrical phenomena [10]. A random process $x(t)$ is Gaussian if its associated time-dependent probability density function $p_X(x, t)$ is a Gaussian PDF for any time value t and, similarly, $p_{X_1X_2}(x(t_1), x(t_2), t_1, t_2)$ is a bivariate Gaussian PDF, which can also be extended to any other number n of considered time instants. As already stated, the probability density function of a Gaussian variable X is fully determined by its mean value m_x and its standard deviation σ_x [see (2.5)]. In turn, the Gaussian process is fully determined by its mean value $E[x(t)]$ and the correlation function $R_X(t_1, t_2)$. It is also shown that in case the property $R_X(t_1, t_2) = E[x(t_1)]E[x(t_2)]$ is fulfilled, the two variables $x(t_1)$ and $x(t_2)$ are uncorrelated and also statistically independent [12]. If the Gaussian process $x(t)$ is wide-sense stationary, it is also strictly stationary and ergodic. Any linear operation on a Gaussian variable $x(t)$ provides another Gaussian variable [10]. Many of the random processes that are dealt with in this book will be considered stationary, fulfilling (2.24), which will greatly simplify the calculations.

The Wiener–Kinchine theorem [10] allows calculation of the power spectral density $S(\Omega)$ of a stationary random variable from the Fourier transform of its autocorrelation function $R(\tau)$. Thus, for a given random process $x(t)$, the correlation function $R_x(\tau)$ and spectral density $S_x(\Omega)$ are related through

$$R_x(\tau) = E[x(t-\tau/2)x(t+\tau/2)] \qquad S_x(\Omega) = F[R_x(\tau)] = \int_{-\infty}^{\infty} R_x(\tau)e^{-j\Omega\tau}\,d\tau \tag{2.25}$$

where F is the Fourier transform. From inspection of (2.25), the mean-square value of the noise source $E[x^2(t)]$ agrees with its autocorrelation function, evaluated at $\tau = 0$; that is, $E[x^2(t)] = R_x(0)$. Application of the inverse Fourier transform to the power spectral density provides

$$R_x(\tau) = \int_{-\infty}^{\infty} S_x(f)e^{j2\pi f}\,df \tag{2.26}$$

where $\Omega = 2\pi f$. To obtain the mean-square value $E[x^2(t)] = R(0)$, it is possible to make $\tau = 0$ in expression (2.26):

$$E[x^2(t)] = R_x(\tau) = \int_{-\infty}^{\infty} S_x(f)\,df \tag{2.27}$$

Then the mean-square value of a stationary random variable agrees with the integral of its power spectral density. The equivalent bandwidth of the random process $x(t)$ is the bandwidth $[-W_x, W_x]$ which provides the same total available power $P_N = \int_{-\infty}^{\infty} S_x(f)df$ with a constant power density of the same value as the peak level $S_{\max}$ of the original distribution; that is, $P_N = 2S_{\max} W_x$.

2.2.4 Stochastic Differential Equations

The Langevin equation [9] constitutes a fundamental type of stochastic differential equations, ruling many physical random processes. It is given by

$$\frac{dx}{dt} = a(x, t) + b(x, t)\varepsilon(t) \tag{2.28}$$

where $a(x, t)$ and $b(x, t)$ are arbitrary functions and $\varepsilon(t)$ is the Gaussian white noise. The white noise $\varepsilon(t)$ is a stationary stochastic process with zero mean. It is rapidly varying and very irregular, so that $\varepsilon(t)$ and $\varepsilon(t')$ are statistically independent. Ideally, the autocorrelation function of white noise is $\langle\varepsilon(t)\varepsilon(t')\rangle = \delta(t - t')$ and its variance is infinite. Because the power spectral density is the Fourier transform of the autocorrelation function (2.25), the white noise will have a flat spectrum: thus, the adjective "white." It can be shown that the white noise $\varepsilon(t)$ is the generalized mean-square derivative of the Wiener process $w(t)$. The Wiener process considered is a continuous process defined for $t \geq 0$ with $w(0) = 0$. It has a Gaussian distribution with zero mean and variance $\sigma = t$. The time derivative of this autocorrelation function is given by

$$\frac{\partial^2}{\partial t \partial t'} \min(t, t') = \delta(t - t')$$

which agrees with the autocorrelation of the white noise.

It is possible to multiply both sides of (2.28) by dt. The resulting equation will depend on the differential element $dw(t) = \varepsilon(t)dt$ of the Wiener processes. This form of expression is more convenient, as the Wiener process is not instantaneously differentiable. The resulting equation is

$$dx = a(x, t)\,dt + b(x, t)\,dw \tag{2.29}$$

The integrals of stochastic functions admit different definitions, which unlike the case of deterministic functions, do not converge to the same result. This is because variations in the solution paths for different discretizations are too great. Two commonly used definitions are the following:

$$\int b(t)\,dw(t) = \begin{cases} \text{msl} \sum b(t_i)[w(t_{i+1}) - w(t_i)] & \text{Ito integral} \\ \text{msl} \sum \frac{b(t_{i+1}) + b(t_i)}{2}[w(t_{i+1}) - w(t_i)] & \text{Stratonovich integral} \end{cases} \tag{2.30}$$

where "msl" indicates minimum-square limit, as the number of considered points tends to infinity. The difference between the two definitions is the point of the $[w_i, w_{i+1}]$ interval where function b is calculated. In the Ito integral, b is calculated at the beginning of the interval, whereas in the Stratonovich integral b is calculated in the middle of this interval. The Ito integral definition allows taking advantage of the noncorrelation between the increments $w(t_{i+1}) - w(t_i)$, $w(t_i) - w(t_{i-1})$ of the Wiener process. The function $b(x, t)$ in (2.29) is nonanticipating if it is independent of the behavior of the Wiener process for $s > t$. If this is the case, Ito's integral allows us to write $\int b(t')\,dt' = \int b(t')[dw(t')]^2$ [9]. Note that the equivalence $dt = dw^2$ is a direct consequence of the variance of the Wiener process $\sigma = t$.

The probability density function $p(x, t|x_o, t_o)$ of the stochastic process x is derived from the Fokker–Planck equation associated with the stochastic differential equation (2.29). This equation is obtained in several steps. Initially, an arbitrary function of x, given by $f(x)$, is considered. Then, Ito's formula provides the following expression derived from the Taylor series expansion of $df(x)$:

$$\begin{aligned} df(x(t)) &= \frac{df}{dx}dx + \frac{1}{2}\frac{d^2 f}{dx^2}dx^2 \\ &= \left[a(x,t)\frac{df}{dx} + \frac{1}{2}b^2(x,t)\frac{d^2 f}{dx^2}\right]dt + b(x,t)\frac{df}{dx}dw \end{aligned} \tag{2.31}$$

where expression (2.29) for dx has been introduced and the relationship $[dw(t)]^2 = dt$ has been taken into account. Note that the expansion on the right-hand side has been limited to first order in the time increment dt. Dividing both terms of (2.31) by dt and obtaining the mean value with the conditional probability density $p(x, t|x_o, t_o)$, it is possible to derive

$$\begin{aligned} &\frac{d}{dt}\int f(x)p(x,t|x_o,t_o)\,dx \\ &= \int \left[a(x,t)\frac{\partial f}{\partial x} + \frac{1}{2}b^2(x,t)\frac{\partial^2 f}{\partial x^2}\right]p(x,t|x_o,t_o)\,dx \end{aligned} \tag{2.32}$$

where it has been taken into account that the mean of dw is equal to zero. Because the function $f(x)$ is arbitrary, it is possible to equate

$$\frac{\partial p(x,t|x_o,t_o)}{\partial t} = -\frac{\partial[a(x,t)p(x,t|x_o,t_o)]}{\partial x} + \frac{1}{2}\frac{\partial^2[b(x,t)^2 p(x,t|x_o,t_o)]}{\partial x^2} \qquad (2.33)$$

which constitutes the Fokker–Planck equation in p_X associated with differential equation (2.30) when using Ito's integral. The result (2.33) has a key relevance in the analysis of stochastic processes, as knowing a stochastic differential equation of the form (2.29) in the variable x allows us to obtain the partial differential equation that rules its probability density function $p_X(x,t)$. Note that this PDF will be necessary to determine essential magnitudes in circuit analysis, such as the autocorrelation (2.22) and the power spectral density.

Equation (2.33) is derived using Ito's definition of the stochastic integral, which is *nonanticipating*. A similar equation can be obtained using the Stratonovich integral. For that, we take into account that the discrete evaluations of the function $b(x,t)$ used in the summation of the Stratonovich integral (2.30) can be expanded as

$$b\left[\frac{x(t_{i-1})+x(t_i)}{2}, t_{i-1}\right] = b\left[x(t_{i-1}) + \frac{1}{2}dx(t_{i-1}), t_{i-1}\right] \qquad (2.34)$$

The Taylor series expansion of the above function, in combination with (2.31) [9], turns b into a nonanticipating function to which Ito's calculus can be applied. Thus, in some special cases it will be possible to transform one type of stochastic integral into another. Taking all the properties and definitions above into account, the partial differential equation in the probability density p_X associated with differential equation (2.28) is given by

$$\frac{\partial p_X(x,t)}{\partial t} = -\frac{\partial}{\partial x}\left[a(x,t)p_X(x,t) + \lambda\frac{\partial b(x,t)}{\partial x}b(x,t)p_x(x,t)\right] + \frac{1}{2}\frac{\partial^2}{\partial x^2}[b^2(x,t)p_X(x,t)] \qquad (2.35)$$

where the parameter λ takes the value $\lambda = 0$ for an Ito integral and $\lambda = \frac{1}{2}$ for a Stratonovich integral.

Because in general we deal with multiple state variables, it will be convenient to consider the extension of (2.35) to a vector $\overline{x} \in R^N$. This is given by

$$\frac{\partial \overline{p}_X(\overline{x},t)}{\partial t} = -\frac{\partial}{\partial \overline{x}}\left[[a(\overline{x},t)]\overline{p}_X(\overline{x},t) + \lambda\frac{\partial[b(\overline{x},t)]^{\mathrm{T}}}{\partial \overline{x}}[b(\overline{x},t)]\overline{p}_X(\overline{x},t)\right] + \frac{1}{2}\frac{\partial^2}{\partial \overline{x}^2}\left[[b(\overline{x},t)]^{\mathrm{T}}[b(\overline{x},t)]\overline{p}_X(\overline{x},t)\right] \qquad (2.36)$$

Note that the different components of equation (2.36) are matrixes and vectors.

The stochastic differential equation (2.28) is perturbed with Gaussian white noise associated with the Wiener process. It would also be possible to have perturbations associated with other processes. A stochastic process ruling some types of noise in electronic circuits is the *Orstein–Uhlenbeck process*. Its associated stochastic differential equation is

$$\frac{dy(t)}{dt} = -\gamma y(t) + \sqrt{D}\,\varepsilon(t) \tag{2.37}$$

with γ and D being constant and $\varepsilon(t)$ being Gaussian white noise. The square root is introduced for later notation convenience. Compared with the Langevin equation (2.29), it is possible to identify $a = -\gamma$ and $b = 1$. Also taking the general expression (2.35) for the partial differential equation in the PDF into account, it is possible to obtain

$$\frac{\partial p_Y(y,t)}{\partial t} = -\gamma \frac{\partial (y p_Y(y,t))}{\partial y} + \frac{1}{2} D \frac{\partial^2 p_Y(y,t)}{\partial y^2} \tag{2.38}$$

Thus, the Orstein–Uhlenbeck process is governed by a Fokker–Planck equation with nonzero drift term. Equation (2.38) can be solved with the aid of the characteristic function. A detailed derivation has been given by Gardiner [9]. Once the time-dependent PDF is known, the stationary time-correlation function can be calculated. It is given by

$$R(t, t-\tau) = \frac{D}{2\gamma} e^{-\gamma\tau} \tag{2.39}$$

As the time difference τ between samples increases, the correlation function decreases exponentially. The maximum correlation time is given approximately by $\tau_c = 1/\gamma$. Note that in the case of Gaussian white noise this correlation time is zero. The Fourier transform of a function of the form $A\exp(-B|t|)$ is $2AB/(B^2 + \Omega^2)$. Application of the Fourier transform to the stationary correlation function (2.39) provides the spectral density of the Orstein–Uhlenbeck process:)

$$S(\Omega) = \frac{D/2\gamma}{1 + (\Omega/\gamma)^2} \tag{2.40}$$

This type of spectrum is known as a *Lorentzian spectrum*. It is mathematically identical to the spectrum resulting from the introduction of white noise into a first-order lowpass Butterworth filter with cutoff frequency $\Omega_{3\text{dB}} = \gamma$. The spectrum is nearly flat for low-frequency Ω and drops -20 dB/dec above $\Omega_{3\text{dB}}$.

2.3 NOISE SOURCES IN ELECTRONIC CIRCUITS

The noise in electronic circuits is caused by fluctuations in the electric current generated by the movement of a discrete number of electrons. There are different types of noise sources according to the physical mechanism that causes the current fluctuations. For the analyses carried out in the book, the noise sources are considered stationary, fulfilling (2.24). In a first global classification, the noise sources are divided into white and colored sources [13–21]. The white noise sources have a flat spectral density, whereas the colored noise sources have a frequency-dependent density. A brief definition is presented in the following.

1. *White noise sources.* A white noise source will have the autocorrelation function $R_\varepsilon(\tau) = E[\varepsilon(t)\varepsilon(t-\tau)] = \Gamma_\varepsilon\delta(\tau)$ and the constant spectral density will be $S_\varepsilon(f) = \Gamma_\varepsilon$ for a single-sideband spectrum, or $S_\varepsilon(f) = \Gamma_\varepsilon/2$ for a double-sided spectrum (considering both negative and positive frequencies). The value Γ_ε depends on the type of noise perturbations and the type (current or voltage) of equivalent noise source considered. Note that this constant value of the spectral density is only ideal. Even the white noise sources must have a limited bandwidth. Otherwise, their mean-square value would tend to infinity, as derived from (2.27), which is physically impossible. The noise power of a white noise source in the bandwidth Δf is given by $\Delta P_\varepsilon = \Gamma_\varepsilon \Delta f$. Because the power agrees with the mean-square value of the normalized variable, it is possible to write $\langle\varepsilon^2(t)\rangle = \sigma^2 = \Gamma_\varepsilon \Delta f$, where $\langle\varepsilon(t)\rangle = 0$ has been taken into account. Then the probability density function associated with the Gaussian white noise in the bandwidth $\Delta\Omega = 2\pi\Delta f$ is given by

$$p_\varepsilon(\varepsilon) = \frac{1}{\sqrt{\Gamma_\varepsilon \Delta\Omega}} \exp\left(\frac{-\pi\varepsilon^2}{\Gamma_\varepsilon \Delta\Omega}\right) \tag{2.41}$$

The results above can be generalized to the case of several white–Gaussian noise sources coexisting in the same circuit. Any pair of samples of these sources is uncorrelated unless they are evaluated at the same time instant. Assuming zero average value for each of these sources,

$$\begin{aligned} E[\varepsilon_i(t)] &= 0 \\ E[\varepsilon_i(t)\varepsilon_j(t')] &= \Gamma_{ij}\delta(t-t') \end{aligned} \tag{2.42}$$

with Γ_{ij} being the correlation constants. For M different white noise sources coexisting in the circuit, an $M \times M$ correlation matrix $[\Gamma]$ can be defined. The joint probability density function is given by

$$p(\bar{\varepsilon}) = \frac{1}{\sqrt{(\Delta\Omega)^M \det[\Gamma]}} \exp\left(-\pi \frac{\bar{\varepsilon}^+[\Gamma]^{-1}\bar{\varepsilon}}{\Delta\Omega}\right) \tag{2.43}$$

2. *Colored noise sources.* If the white noise passes through a filter with transfer function $H(\Omega)$, colored noise is obtained. The colored noise sources have a frequency-dependent spectral density. Some physical mechanisms inherently give rise to colored noise. As some examples, the generation-recombination noise and the burst noise exhibit a Lorenzian spectrum like the one in (2.40) and are ruled by an Orstein–Uhlenbeck process with a frequency-dependent power spectral density. The flicker noise has a more difficult form of variation, proportional to $1/\Omega^{\alpha}$, with $\alpha \cong 1$. Its time variation can be modeled with an infinite sum of Ornstein–Uhlenbeck processes [4].

In a general manner, noise sources that have periodic statistical properties, depending on the periodic steady-state solution, are known as *cyclostationary*. A white cyclostationary source can be decomposed as $n(t) = n_o(t)\alpha(\omega_o t)$, where $n_o(t)$ is a white stationary process and $\alpha(\omega_o t)$ is a deterministic amplitude modulation, depending on the periodic steady-state solution [5]. The statistical properties of most noise sources discussed below depend on the periodic current through a device, thus are cyclostationary. This will give rise to mixing the stationary noise $n_o(t)$ with the large-signal current. A substantial amount of research work is being done on the modeling of cyclostationary noise sources, requiring the noise sideband spectra at all the harmonics as well as the interfrequency cross-correlation terms [13]. Due to these modeling difficulties, it is common practice to replace the periodic currents with their average or dc values, at the expense of a degradation of analysis accuracy. For simplicity, this is the type of modeling approach that will be considered here, although the analysis techniques presented in Sections 2.4 and 2.5 can be extended to cyclostationary models.

2.3.1 Thermal Noise

Let a conductor above the temperature $T = 0$ K be considered. From kinetic theory, the average energy of a particle at the absolute temperature T is kT, with k the Boltzmann constant. This energy facilitates the interaction of free electrons with other particles, which gives rise to random fluctuations of the electron movement. *Thermal noise*, also known as *Johnson* or *Nyquist noise*, is due to the perturbations that affect the trajectories of the charge carriers and give rise to a random current with zero average value. These charge carriers will be electrons and holes in semiconductor materials. The thermal noise exists even in the absence of an electric field applied to the material. Its spectral density has a Gaussian probability distribution, as expected from a phenomenon involving a large number of random events. The thermal noise does not depend on the periodic steady-state solution, so it is stationary.

Any resistive element R at a temperature T different from zero behaves as a source of noise power. The available power or power delivered to a resistance of the same value R in the bandwidth Δf is $P_N = kT\Delta f$. Thus, a noisy resistance can be represented with an equivalent model consisting of a noiseless resistance of the same value R in parallel with a noise current source with the same available power $P_N = \langle i_n^2(t)\rangle/4G$, with $G = 1/R$. Then the mean-square value of the current source is $\langle i_n^2(t)\rangle = 4GkT\Delta f$. The flat single-sideband spectral density associated with this current source is

$$S^T(\Omega) = 4GkT \qquad \text{A}^2/\text{Hz} \tag{2.44}$$

As can be expected, this constant value of the spectral density is actually an approximation, valid for frequencies below $0.1kT/h \cong 10^{12}$ Hz, with h being the Planck constant. It is also possible to consider an equivalent model given by the noiseless resistance R in series with a noise voltage source with mean-square value $\langle e_n^2(t)\rangle = 4RkT\Delta f\ V$.

2.3.2 Shot Noise

Shot noise is due to the discrete nature of an electric current, which cannot be considered as a uniform flow but as the superposition of a high number of elementary impulses. Shot noise is observed in currents generated by an electric field, unlike the case of thermal noise, which gives rise to current fluctuations without any applied voltage and with average current zero. Shot noise in semiconductor devices results from the passage of charged carriers across the potential barrier generated by semiconductor junctions. The shot noise current can be expressed $i_s(t) = q\sum_{k=-\infty}^{\infty}\delta(t - t_k)$, where q is the electron charge and the independent time instants t_k at which the pulses occur follow a Poisson law. The Poisson process has discontinuous sample paths, governed by the master equation, which is obtained by doing $a_1 = a_2 = 0$ in (2.16).

If the average value of events per second is $\overline{N}_\text{I}$, the normalized power spectral density (in A^2/Hz) associated with the shot noise will be given by

$$I_s^2(f) = \overline{N}_\text{I} E[|p_\text{I}(f)|^2] \tag{2.45}$$

where $p_\text{I}(f)$ is the Fourier transform of the current pulse shape. The mean $E[]$ is needed due to the variation in this pulse shape. The average value $\overline{N}_\text{I}$ can be obtained from the average current, I [14]. This average current is the number of events per second, $\overline{N}_\text{I}$, multiplied by the electron charge q; that is, $I = \overline{N}_\text{I} q$. Thus, it will be possible to write $\overline{N}_\text{I} = I/q$. The next objective will be to find the pulse spectrum $p_\text{I}(f)$. Assuming, for example, the case of a p-n junction, each electron traversing the depletion region causes a current pulse of height q/T_d, with T_d the drift time. If the mean velocity along the depletion region is v and the width of the depletion region is d, the drift time will be $T_d = d/v$. The current pulse can be assumed to have a square shape with value q/T_d for $-T_d/2 \le t \le T_d/2$ and zero otherwise. The corresponding Fourier transform is the sampling function or sinc function, with its main lobe of maximum value q cutting the frequency axis at $-1/T_d$ and $1/T_d$. For frequencies much smaller than the inverse of the carrier drift time $1/T_d$, the spectrum $p_\text{I}(f)$ is flat, with value q. Approaching $p_\text{I}(f) \cong q$ and substituting both $\overline{N}_I = I/q$ and $p_I(f) \cong q$ in (2.45), the double-sideband spectral density is given by

$$I_s^2(f) = qI \qquad \text{for} - 1/T_d \ll f \ll 1/T_d \tag{2.46}$$

For fluctuations in a frequency band Δf, the shot-noise normalized power will be

$$\langle i_n^2(t) \rangle = qI\Delta f \tag{2.47}$$

For a single-sideband spectrum, the power should be multiplied by 2. The current I is the steady-state current of the device. As an example, the shot noise across a p-n junction may be considered. The current through the diode is given by $I_d(v) = I_s(e^{(q/kT)v} - 1)$. The shot noise includes the contribution of the forward and reverse currents, which are statistically independent. Thus, the mean-square value of the noise current in the bandwidth Δf is given by

$$\langle i_n^2(t) \rangle = qI_s e^{(q/kT)v}\Delta f + qI_s\Delta f \tag{2.48}$$

For a single-sideband spectrum, the two terms should be multiplied by 2.

2.3.3 Generation–Recombination Noise

Generation–recombination noise is associated with the spontaneous fluctuation in the generation, recombination, and trapping of carriers in semiconductor devices, which gives rise to a fluctuation in the free carrier density. During the transition from the valence band to the conduction band, the carriers may stay at trap levels for a random time without contributing to the conduction. The spectral density of this noise is proportional to the square of the current traversing the semiconductor material. It is modeled with a current source of the spectral density:

$$S^{g-r}(f) = \frac{I_o^2}{N_o^2}\langle \Delta N^2 \rangle \frac{\tau}{1 + (2\pi f)^2\tau^2} \tag{2.49}$$

where I_o is the dc current through the semiconductor material, $\langle \Delta N^2 \rangle$ the constant mean-square value of the fluctuation in the total number of carriers, and τ the system time constant, given by the derivative of the difference between generation and recombination rates, $g(N) - r(N)$, with respect to the total charge number evaluated at the equilibrium. Note that the spectrum (2.49) has the form of (2.40), so the generation–recombination noise has nonzero correlation time.

2.3.4 Flicker Noise

Flicker noise is found in all physical systems. It has a power spectral density of the form $1/f^{\alpha}$, with $\alpha \cong 1$, and the name is associated with the fact that if a lamp had this distribution in its light intensity, we would perceive it as flickering [15]. In semiconductor devices it is believed to be caused by trap levels, due to contamination and crystal defects. At these trap levels, the charge carriers are captured and released in a random manner, and the associated time constants give rise to a noise signal with energy concentrated at low frequencies. The flicker noise has a spectral density that increases as frequency decreases, exceeding the thermal and shot noise in semiconductor devices. The power spectral density of the flicker noise is known to be proportional to the electrical current passing through a device and inversely proportional to the frequency $S^F(f) \propto 1/f^{\alpha}$, with α a constant close to 1, which depends on the particular device.

The $1/f$ characteristic of flicker noise has been measured up to 10^{-6} Hz. However, this characteristic implies infinite noise power at $f = 0$, which is not physical. It has been argued [16] that the flicker noise is actually a nonstationary process, and the nonphysical response at $f = 0$ arises when trying to model it as a stationary process. In an article by Kaertner [4], the flicker noise is modeled with an infinite sum of autocorrelation spectra of statistically independent Ornstein–Uhlenbeck processes. Each of these independent processes is ruled by the stochastic differential equation

$$\dot{y}_i(t) = -\gamma_i y_i(t) + \xi_i(t) \tag{2.50}$$

with $\xi_i(t)$ being white noise sources with the correlation function

$$\langle \xi_i(t)\xi_j(t') \rangle = \delta_{ij}\Gamma_i\delta(t - t') \tag{2.51}$$

δ_{ij} being the Kronecker delta. The damping constants γ_i are equipartitioned on a logarithmic scale and vary from $-\infty$ to ∞. Flicker noise is the result of the summation $y(t) = \sum_{i=-\infty}^{\infty} y_i(t)$. The values of the damping constants γ_i and the intensities Γ_i can be found from the approximation

$$\frac{1}{|2\pi f|^{\alpha}} = \lim_{\Delta\sigma \to 0} \sum_{i=-\infty}^{\infty} \frac{\Gamma_i}{\gamma_i^2 + (2\pi f)^2} \tag{2.52}$$

with $\gamma_i = e^{i\Delta\sigma}$. The Orstein–Uhlenbeck processes with $\gamma_i \to 0$ will have infinite correlation time and will give rise to the singularity of the noise spectrum for $f \to 0$. Due to the correlation time tending to infinity, it is not possible to neglect the effect of the finite time of the measurement. In Kaertner's article [4], expressions have been derived for the noise spectrum of the flicker noise, considering a finite measurement time interval. This finite time prevents the spectrum from tending to infinity when $f \to 0$.

A different way to solve this problem has been proposed by Demir [8] for $\alpha = 1$. The characteristic $1/|f|$ can be expressed in the integral form

$$\frac{1}{|f|} = 4\int_0^{\infty} \frac{1}{\gamma^2 + (2\pi f)^2}\, d\gamma \tag{2.53}$$

If instead of taking $\gamma = 0$ as the lower integration limit, a small cutoff value $f_{\min}$ (in rad/s) is used, a finite noise spectral density is obtained at $f = 0$. The resulting approximate model for the flicker noise is

$$S^F(f) = 4\int_{f_{\min}}^{\infty} \frac{1}{\gamma^2 + (2\pi f)^2}\, d\gamma = \frac{1}{|f|} - 4\frac{\arctan(f_{\min}/2\pi f)}{2\pi f} \tag{2.54}$$

Then a finite value of the noise spectral density is obtained at $f = 0$, given by $4/f_{\min}$.

A common model for the flicker noise at sufficiently high-frequency offset from the carrier is

$$S^F(f) = k\frac{I^a}{f} \tag{2.55}$$

where k is a constant depending on the particular device, I the current through this device, and a is a constant in the range 0.5 to 2.

For a small frequency f there will be both white and flicker noise in semiconductor devices. In some cases it will be possible to model contributions of both types of noise with a single low-frequency spectrum, expressed as

$$S(f) = N_o\frac{f_w + f}{f} \tag{2.56}$$

with N_o being the white noise spectral density. Note that the representation (2.56) will be valid only above the cutoff frequency $f_{\min}$. At the corner frequency f_w, there are equal contributions from the flicker and white noise. Below f_w, flicker noise dominates. Above f_w, white noise is the dominant contribution.

2.3.5 Burst Noise

Burst noise occurs in semiconductor devices and can be considered as a special form of generation–recombination noise, characterized by steplike transitions between two or more potential levels, occurring at time instants with a non-Gaussian distribution. There is no definitive explanation for the origin of burst noise, although it has been related to crystal imperfections and to the presence of heavy metal ion contamination. Burst noise can be modeled mathematically as a colored stochastic process with the Lorentzian spectrum

$$S^B(f) = k\frac{I^a}{1 + (f/f_c)^2} \tag{2.57}$$

where the constant k depends on the particular device, a is a constant in the range 0.5 to 2, I is the current through this device, and f_c is the 3-dB bandwidth. Note that the burst noise belongs to the class of Ornstein–Uhlenbeck processes.

2.4 DERIVATION OF THE OSCILLATOR NOISE SPECTRUM USING TIME-DOMAIN ANALYSIS

This section is based fully on the mathematical derivations of seminal work by F. Kaertner [4] and A. Demir, A. Mehrota, and J. Roychowdhury [8], which the author believes to be essential for an in-depth understanding of the oscillator phase noise.

For a derivation of the oscillator noise spectrum, a state-form representation of the nonlinear differential equations ruling the circuit behavior is assumed, for simplicity. However, the same analysis principles can be extended to the general circuit description with a system of differential algebraic equations [8]. The noiseless oscillator equations are

$$\dot{\overline{x}} = \overline{f}(\overline{x}) \tag{2.58}$$

where both $\dot{\overline{x}}$ and f are vectors in R^N. Next, noise sources will be introduced in (2.58). As we already know, the stochastic processes associated with white and colored (frequency-dependent) noise sources are different. This is why they are treated at two different stages in time-domain analysis.

2.4.1 Oscillator with White Noise Sources

2.4.1.1 Stochastic Differential Equations Let a circuit with L white noise sources $\varepsilon_i, i = 1$ to L, be considered, comprising the vector $\overline{\varepsilon}(t)$. The L noise sources fulfill $E[\varepsilon_i(t)\varepsilon_j(t+\tau)] = \Gamma_{ij}\delta(\tau)$. The vector of white noise sources $\overline{\varepsilon}(t)$ is introduced into the nonlinear differential equation system ruling circuit behavior

$$\dot{\overline{x}} = \overline{f}(\overline{x}, \overline{\varepsilon}(t)) \tag{2.59}$$

The noise sources will be of small amplitude, so the function $\overline{f}$ can be expanded in a Taylor series about $\overline{\varepsilon} = 0$, which provides the following system, linear with respect to the noise sources:

$$\dot{\overline{x}} = \overline{f}(\overline{x}) + g(x)\overline{\varepsilon} \tag{2.60}$$

where g is a matrix consisting of derivatives of the nonlinear function $\overline{f}$ with respect to the noise sources $\overline{\varepsilon}(t)$. As already stated, any small perturbation gives rise to a time deviation of the oscillator solution. Thus, the perturbed oscillatory solution can be expressed as

$$\overline{x}(t) = \overline{x}_{\mathrm{sp}}(t + \theta(t)) + \Delta\overline{x}(t + \theta(t)) \tag{2.61}$$

where $\bar{x}_{sp}$ represents the steady-state periodic waveform, θ the stochastic time deviation associated to perturbations tangent to the limit cycle, and $||\Delta\bar{x}(t)|| \ll ||x_{sp}(t)||$ the perturbation transversal to the cycle, here called *amplitude perturbation*. The vector $\Delta\bar{x}(t)$ contains small perturbations of all the circuit variables. On the other hand, the stochastic time deviation θ affects all the circuit variables in an identical manner and gives rise to phase modulation. Note that $\Delta\bar{x}(t)$ is small but θ might not be small, due to the absence of a restoring mechanism in the direction of the limit cycle. Thus, the perturbed oscillator system will be linearized with respect to $\Delta\bar{x}(t)$ but not with respect to θ. An auxiliary variable $y = t + \theta$ will be introduced, so the perturbed solution is compactly written $\bar{x}(t) = \bar{x}_{sp}(y) + \Delta\bar{x}(y)$. Using this expression and deriving $\bar{x}(t)$ with respect to the time t by parts yields

$$\begin{aligned}\dot{\bar{x}}(t) = [\bar{x}'_{sp}(y) + \Delta\bar{x}'(y)]\frac{dy}{dt} &= [\bar{x}'_{sp}(y) + \Delta\bar{x}'(y)](1 + \dot{\theta}) \\ &\cong \bar{x}'_{sp}(y) + \Delta\bar{x}'(y) + \bar{x}'_{sp}(y)\dot{\theta}\end{aligned} \tag{2.62}$$

where higher-order increments have been neglected. Introducing expression (2.61) and time derivative (2.62) into (2.60), the following equation is obtained:

$$\bar{x}'_{sp}(y) + \bar{x}'_{sp}(y)\dot{\theta} + \Delta\bar{x}'(y) = \bar{f}(\bar{x}_{sp}(\mathrm{y})) + \mathrm{Jf}(\bar{x}'_{sp}(y))\Delta\bar{x}(y) + \bar{g}(\bar{x}_{sp}(y)) \tag{2.63}$$

where $\bar{g}(x_{sp}(y), t) = [\partial\bar{f}(\bar{x}_{sp}(y))/\partial\bar{\varepsilon}]\bar{\varepsilon}(t)$. Equation (2.63) is linear in $\Delta\bar{x}$ but nonlinear in the stochastic time deviation θ.

The vector equation (2.63) is unbalanced, as it contains N equations in $N{+}1$ unknowns, given by the N components of $\Delta\bar{x}(t)$ plus the time deviation θ. An additional condition has to be imposed for its practical resolution. As shown by Sancho et al. [17], the different conditions will give rise to slightly different distributions of phase noise, coming from θ, and amplitude noise, coming from $\Delta\bar{x}(t)$, with the same total output noise power. Kaertner [7], uses the condition $\Delta\bar{x}^{\mathrm{T}}(t)\bar{u}_1(t) = 0$, that is, the zero value of the scalar product of the perturbation vector $\Delta\bar{x}(t)$ and the vector $\bar{u}_1(t)$, associated with the Floquet multiplier $m_1 = 1$ (Section 1.5.2.2). Remember that this multiplier is responsible for irrelevance of the oscillator solution versus translations along the limit cycle. As demonstrated in Chapter 1, the associated vector $\bar{u}_1(t)$ agrees with the time derivative of the oscillator periodic solution $\bar{u}_1(t) = \dot{\bar{x}}_{sp}(t)$. Thus, the condition $\Delta\bar{x}^{\mathrm{T}}(t)\bar{u}_1(t) = 0$ restricts the perturbation vector $\Delta\bar{x}(t)$ to the orthogonal complement space to the tangent space at the limit cycle. However, the additional condition $\Delta\bar{x}^{\mathrm{T}}(t)\bar{u}_1(t) = 0$ is not optimum for solving the mixed-variable system (2.63). Kaertner [4] proposed a different, more useful condition. The resulting perturbation $\Delta\bar{x}(t)$ is not orthogonal to the cycle but allows a very convenient uncoupling of the two dependences on θ and $\Delta\bar{x}(t)$ of system (2.63). This system is decomposed into two different subsystems, one depending only on the stochastic time deviation θ, which consists of a single scalar equation, and the other, containing $N{-}1$ equations, which depends only on $\Delta\bar{x}$. This decomposition facilitates phase noise analysis, as it provides a single scalar equation in the time deviation θ. Note that the two different additional conditions lead to different definitions of phase noise and thus to slightly different phase noise spectra.

The tools required for the decomposition of system (2.63) are related to those used for the stability analysis of periodic solutions and based on Floquet's analysis [4]. In Chapter 1 it was shown that the N independent solutions of the time-periodic linear system $\Delta\dot{\overline{x}}(t) = Jf(\overline{x}_{sp}(t))\Delta\overline{x}(t)$ are given by $\Delta\overline{x}_k(t) = e^{\lambda_k t}\overline{u}_k(t)$, with λ_k, $k = 1$ to N, being the Floquet exponents, fulfilling $m_k = e^{\lambda_k T}$, and m_k being the Floquet multipliers, with T as the solution period. Remember that the Floquet multipliers m_k are the eigenvalues of the monodromial matrix $[W(T)]$. The monodromial matrix is the canonical fundamental matrix of independent solutions of the linearized system, evaluated at constant time, equal to one period, $t = T$. In turn, the canonical fundamental matrix is obtained by integrating $\Delta\dot{\overline{x}}(t) = Jf(\overline{x}_{sp}(t))\Delta\overline{x}(t)$ from initial values agreeing with columns of the identity matrix. Each vector $\overline{u}_k(t)$ is obtained by integrating the linearized system $\Delta\dot{\overline{x}}(t) = Jf(\overline{x}_{sp}(t))\Delta\overline{x}(t)$ from an initial value given by the constant eigenvector $\overline{w}_k$ of the monodromial matrix $W(T)$. As already known, in the particular case of a free-running oscillator, one of the multipliers is $m_1 = 1$ and the associated periodic vector is $\overline{u}_1(t) = \dot{\overline{x}}_{sp}(t)$ (see Section 1.5.2.2).

Following a similar procedure, it is easily verified that the N independent solutions of the adjoint system $\Delta\dot{\overline{x}}^T(t) = -\Delta\overline{x}^T(t)Jf(\overline{x}_{sp}(t))$ are given by $\Delta\overline{x}_k^a(t) = e^{-\lambda_k t}\overline{v}_k(t)$, where the $\overline{v}_k(t)$ are periodic vectors, obtained in a manner similar to the $\overline{u}_k(t)$. These vectors fulfill the relations $\overline{v}_i^T(t)\overline{u}_j(t) = \delta_{ij}$ (Kronecker delta). Remembering that $\overline{u}_1(t) = \dot{\overline{x}}_{sp}(t)$, it will be possible to write

$$1 = \overline{v}_1^T(t)\dot{\overline{x}}_{sp}(t) \tag{2.64}$$

Because $\overline{v}_1(t)$ is associated with the multiplier $m_1 = 1$ (or $\lambda_1 = 0$), the corresponding solution of the adjoint system is given simply by $\Delta\overline{x}_1^a(t) = \overline{v}_1(t)$. Thus, the following relationship is fulfilled:

$$\dot{\overline{v}}_1^T(t) = -\overline{v}_1^T(t)Jf(\overline{x}_{sp}(t)) \tag{2.65}$$

Multiplying (2.63) by $\overline{v}_1^T(t)$, it will be possible to obtain a scalar equation depending only on θ. This multiplication provides

$$\begin{aligned}&\overline{v}_1^T(y)\overline{x}'_{sp}(y)\dot{\theta} + \overline{v}_1^T(y)\Delta\overline{x}'(y)\\&= \overline{v}_1^T(y)Jf(\overline{x}_{sp}(y))\Delta\overline{x}(y) + \overline{v}_1^T(y)\overline{g}(\overline{x}_{sp}(y))\end{aligned} \tag{2.66}$$

Next, relationships (2.64) and (2.65) are taken into account, so equation (2.66) is written

$$\dot{\theta} + \overline{v}_1^T(y)\Delta\overline{x}'(y) = -\overline{v}_1^{T\prime}(y)\Delta\overline{x}(y) + \overline{v}_1^T(y)\overline{g}(\overline{x}_{sp}(y)) \tag{2.67}$$

It is possible to move $-\overline{v}_1^{T\prime}(y)\Delta\overline{x}(y)$ to the left-hand side and write $d(\overline{v}_1^T\Delta\overline{x})/dy = \overline{v}_1^{T\prime}\Delta\overline{x} + \overline{v}_1^T\Delta\overline{x}'$. Thus, equation (2.67) becomes

$$\dot{\theta} + \frac{d(\overline{v}_1^T(y)\Delta\overline{x}(y))}{dy} = \overline{v}_1^T(y)\overline{g}(\overline{x}_{sp}(y)) \tag{2.68}$$

So far, no additional condition has been introduced in the perturbed oscillator system. The transformation from (2.63) to (2.68) has been carried out simply by using the properties (2.64) and (2.65). Thus, system (2.63) still has one degree of freedom, as the number of equations is N and the number of unknowns is $N+1$ (θ and $\Delta\overline{x}$). The additional condition on the vector $\Delta\overline{x}$ is introduced at this stage and is given by $\overline{v}_1^T(y)\Delta\overline{x}(y) = 0$. This condition eliminates the second term on the left-hand side of (2.67). Then the nonlinear equation in the time deviation θ is simplified to

$$\dot{\theta}(t) = \overline{v}_1^T(t+\theta)\overline{g}(\overline{x}_{sp}(t+\theta)) = \overline{v}_1^T(t+\theta)\frac{\partial\overline{f}}{\partial\overline{\varepsilon}}(\overline{x}_{sp}(t+\theta))\overline{\varepsilon}(t) \tag{2.69}$$

which can be written in a compact manner as

$$\dot{\theta}(t) = [b(t+\theta)]\overline{\varepsilon}(t) \tag{2.70}$$

The periodic row matrix b relates the white noise sources $\overline{\varepsilon}(t)$ directly to the time derivative of θ.

In summary, using Floquet decomposition, it has been possible to decouple the original equation (2.63), depending on both $\Delta\overline{x}(t)$ and $\theta(t)$, and obtain the scalar equation (2.70), depending only on $\theta(t)$. From inspection of (2.70), the dependence of the function $b(t)$ on θ, giving rise to nonlinearity, will be more relevant for larger θ values. These large values will be obtained at smaller noise frequencies. To understand this, for a moment neglect the nonlinear nature of (2.70). In the Fourier domain, $\theta(\Omega)$ would be determined by dividing the Fourier transform of the right-hand side of the equation by $j\Omega$, with Ω being the noise frequency. Thus, larger θ values would be obtained for smaller Ω values. Although this analysis is strictly invalid, it helps us to understand why the larger values of the time deviation $\theta(t)$ are obtained for the smaller noise frequencies. The nonlinearity of (2.70) will be more relevant at small offset frequency from the carrier.

Because the perturbation was expressed originally as $\Delta\overline{x}(t) = \sum_{k=1}^{N} c_k e^{\lambda_k t}\overline{u}_k(t)$ and the relation $\overline{v}_i^T(t)\overline{u}_j(t) = \delta_{ij}$ is fulfilled, the imposed condition $\overline{v}_1^T(y)\Delta\overline{x}(y) = 0$ allows a redefinition of $\Delta\overline{x}(t) = \sum_{k=1}^{N} c_k e^{\lambda_k t}\overline{u}_k(t)$ as $\Delta\overline{x}(t) = \sum_{k=2}^{N} c_k e^{\lambda_k t}\overline{u}_k(t)$, with the summation index starting at $k = 2$. As shown by Kaertner [4], complementing the scalar nonlinear equation (2.70), there is a linear equation system of $N-1$ dimensions in the amplitude perturbation $\Delta\overline{x}$. This system is obtained by multiplying the two sides of (2.63) by the projector matrix $P(y) = 1 - \overline{u}_1(y)\overline{v}_1^T(y)$. The resulting system allows calculation of the amplitude noise, affecting the waveform itself. In general, the contribution of the amplitude noise to the total noise power spectrum of the oscillator circuit will be much smaller than that of the phase noise. This type of noise is treated in Section 2.5.4, devoted to the frequency-domain analysis of the oscillator spectrum.

2.4.1.2 *Phase Noise Sensitivity* Hajimiri and Lee [5] demonstrated that the phase shift of the oscillator solution resulting from an impulse perturbation, applied at the time instant τ, takes the form of a step function. This function has been expressed as $h(t,\tau) = \Gamma(t)u(t-\tau)$, with $\Gamma(t)$ periodic with the same period T as the steady-state oscillation and $u(t)$ the unit step function. The function $h(t,\tau)$ is the time-variant impulse sensitivity function, which provides the phase response of the oscillator circuit with respect to a small impulse applied at the time instant τ. When considering an arbitrary noise input $\varepsilon(t)$, the phase shift $\phi(t) = \omega_o\theta(t)$ is determined by applying superposition in the noise time τ. The calculation is

$$\phi(t) = \int_{-\infty}^{\infty} h(t,\tau)\varepsilon(\tau)\,d\tau = \int_{-\infty}^{\infty} \Gamma(t)u(t-\tau)\varepsilon(\tau)\,d\tau$$
$$= \int_{-\infty}^{t} \Gamma(t)\varepsilon(\tau)\,d\tau \qquad (2.71)$$

Note that this calculation of the impulse response neglects the nonlinearity of (2.70), as the time deviation θ is not taken into account. To relate the function $\Gamma(t)$ with the coefficient $b(t)$ in (2.70), we derive (2.71) with respect to time, noting that $\phi(t) = \omega_o\theta(t)$. Thus, it will be possible to write

$$\omega_o\dot{\theta}(t) = \Gamma(t) \qquad (2.72)$$

Comparing (2.72) with (2.70), it is clear that each component of the row matrix $[b(t)]$ in (2.70) provides the phase sensitivity to a different white noise source existing in the circuit. Let a row matrix $[\Gamma(t)]$ be defined containing the phase-noise sensitivity functions $\Gamma_i(t)$, with $i = 1$ to L, to the L different white noise sources existing in the circuit. Then it will be possible to write

$$[\Gamma(t)] = \omega_o[b(t)] = \omega_o\overline{v}_1^T(t)\frac{\partial\overline{f}}{\partial\overline{\varepsilon}}(\overline{x}_{sp}(t)) \qquad (2.73)$$

Thus, the phase sensitivity to a given noise source $\varepsilon_i(t)$ will be given by

$$\Gamma_i(t) = \omega_o\left[v_1(t)\frac{\partial f_1(t)}{\partial\varepsilon_i} + v_2(t)\frac{\partial f_2(t)}{\partial\varepsilon_i} + \cdots + v_N(t)\frac{\partial f_N(t)}{\partial\varepsilon_i}\right] \qquad (2.74)$$

As can be seen, the phase noise sensitivity depends on the vector $\overline{v}_1^T(t)$ and increases with the magnitude of the derivatives of the vector function $\overline{f}$ with respect to the particular noise source $\varepsilon_i(t)$.

As an example, calculation of the phase-noise sensitivity has been applied to the parallel resonance oscillator of Fig. 1.1. A white noise current source $i_w(t)$ with spectral density $S_W(f) = 4kTG\,\text{A}^2/\text{Hz}$, with $G = 1/R$, connected in parallel, has been considered. The resulting stochastic differential equation system is

$$\begin{aligned}\frac{dv_c(t+\theta)}{dt} &= -\frac{i_L(t+\theta)}{C} - \frac{i_N(v(t+\theta))}{C} - \frac{i_w(t)}{C}\\ \frac{di_L(t+\theta)}{dt} &= \frac{v_C(t+\theta)}{L}\end{aligned} \tag{2.75}$$

which can be written in a compact manner as $\dot{\overline{x}} = \overline{f}(\overline{x}, \overline{\varepsilon}(t))$, with $\overline{f}$ being the vector of nonlinear functions on the right-hand side. The phase noise sensitivity function is given by $b_1(t) = \overline{v}_1^{\mathrm{T}}(t)[\partial\overline{f}/\partial i_W]$. As we already know, the vector $\overline{v}_1^{\mathrm{T}}(t)$ is a solution of the adjoint system $\Delta\dot{\overline{x}}^{\mathrm{T}}(t) = -\Delta\overline{x}^{\mathrm{T}}(t)Jf(\overline{x}_{\mathrm{sp}}(t))$ associated with the Floquet multiplier $m_1 = 1$ or the exponent $\lambda_1 = 0$. Clearly, the calculation of $v_1^{\mathrm{T}}(t)$ requires determination of the Jacobian matrix $Jf(\overline{x}_{\mathrm{sp}}(t))$ associated with system (2.75). This matrix is given by

$$Jf(\overline{x}_{\mathrm{sp}}(t)) = \begin{bmatrix} -\dfrac{1}{C}\dfrac{\partial i_N(v_s(t))}{\partial v} & -\dfrac{1}{C} \\ \dfrac{1}{L} & 0 \end{bmatrix} \tag{2.76}$$

The vector $\overline{v}_1^{\mathrm{T}}(t)$ is obtained by integrating the adjoint system from the left-hand-side eigenvector of the monodromial matrix, associated with $m_1 = 1$. Next, the Jacobian of the $\overline{f}$ vector with respect to the noise source $i_w(t)$ has to be calculated. From an inspection of (2.75), this is given simply by $\partial\overline{f}^{(t)}/\partial i_W = [-1/C, 0]^{\mathrm{T}}$.

The phase sensitivity to the white noise current source $i_w(t)$, given by $b_1(t) = \overline{v}_1^{\mathrm{T}}(t)[\partial\overline{f}(t)/\partial i_W]$, is represented in Fig. 2.1. The node voltage waveform has also

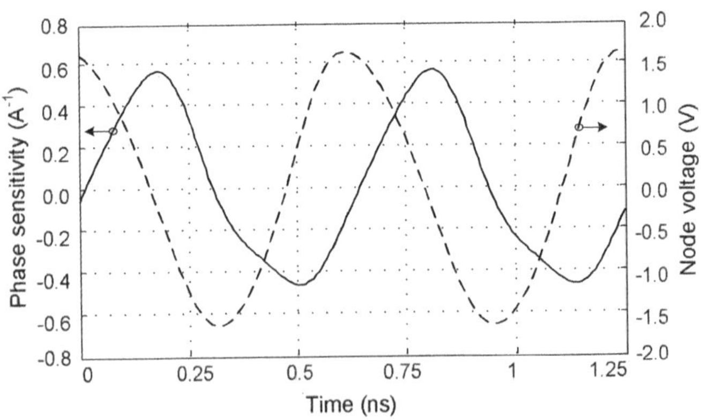

FIGURE 2.1 Phase noise sensitivity to a current source introduced in the parallel resonance oscillator in parallel with a cubic nonlinearity. The node voltage waveform is represented by the dashed line and the phase noise sensitivity is represented by the solid line.

been traced, for comparison. At each time value, this function provides the approximate magnitude of the phase step response to a current impulse of small amplitude applied at this particular instant of time. As demonstrated by Hajimiri and Lee [5], the phase noise sensitivity function depends on the conditions of the periodic oscillator solution when the current impulse is introduced. As gathered from Fig. 2.1, this sensitivity is larger for the larger absolute value of the time derivative of the voltage waveform and minimum for the zero-time derivative, obtained at the minima and maxima of the waveform. This can be understood intuitively from the fact that at points of the waveform with a large time derivative, the system evolves quickly along the unperturbed limit cycle. When applying a perturbation at these fast points, a larger phase shift is obtained, after the transient decay, with respect to the unperturbed solution.

2.4.1.3 Derivation of the Oscillator Spectrum Due to Phase Noise Although we have already derived the phase sensitivity functions, calculation of the time deviation $\theta(t)$ in a deterministic manner using these functions would be meaningless, as different random variations of the noise sources would give rise to different time functions $\theta(t)$. Instead, the objective will be to obtain its second-order magnitudes: for example, the autocorrelation $E[\theta(t)\theta(t+\tau)]$, with E the probabilistic expectation. The calculation of this expectation requires the time-varying probability density function p_θ associated with the variable θ, which is defined as $p_\theta(\eta, t) = \partial P[\theta(t) \leq \eta]/\partial\eta$, with $t \geq 0$ and P the probability measure. As shown in Section 2.2.4, some particular types of differential equations in a given variable x have an associated partial differential equation in the probability density of the variable $p_X(x,t)$. In fact, equation (2.70), given by $\dot{\theta} = b(t+\theta)\bar{\varepsilon}(t)$, is a particular case of the matrix form of the Fokker–Planck equation (2.36), with $a = 0$. Remember that p_θ is needed to determine the autocorrelation $E[\theta(t)\theta(t+\tau)]$. Applying (2.36), the differential equation in the probability density p_θ will be

$$\frac{\partial p_\theta(\theta,t)}{\partial t} = -\frac{\partial}{\partial\theta}\left[\lambda\frac{\partial b^T(t+\theta)}{\partial\theta}b(t+\theta)p_\theta(\theta,t)\right] + \frac{1}{2}\frac{\partial^2}{\partial\theta^2}[b^T(t+\theta)b(t+\theta)p_\theta(\theta,t)] \tag{2.77}$$

It is easily shown that when the probability density p_θ fulfills equation (2.77), the expectation of any smooth function $z(\theta)$ of the variable θ fulfills [8]

$$\frac{\partial E(z(\theta))}{\partial t} = -E\left[\lambda\frac{dz(\theta)}{d\theta}\frac{\partial b^T}{\partial\theta}b\right] + \frac{1}{2}E\left[\frac{d^2z(\theta)}{d\theta^2}b^Tb\right] \tag{2.78}$$

In particular, it is possible to choose z as $z = e^{j\omega\theta(t)}$. By definition, the mean value $E[e^{j\omega\theta(t)}]$ agrees with the characteristic function associated with θ, which is obtained as $F_\theta(\omega,t) = E[e^{j\omega\theta(t)}] = \int_{-\infty}^{\infty} e^{j\omega\eta}p_\theta(\eta,t)d\eta$. As shown in Section 2.2, the characteristic function $F_\theta(\omega,t)$ is similar to a Fourier transform from the θ domain to the ω domain. Note that time t remains as a variable after the transformation. The n-order derivative with respect to θ becomes a simple product by $(j\omega)^n$ in the ω domain. This is why it is simpler to solve (2.78) for $E(z) = E[e^{j\omega\theta(t)}] = F_\theta(\omega, t)$ than (2.77) for p_θ.

As shown in Section 2.2.1, a random variable x is Gaussian when its probability density function p_X is completely characterized by its mean value $\mu_x = \overline{x}$ and variance $\sigma_x^2 = E(x^2) - \mu_x^2$. When dealing with a stochastic process, the interest will be in the asymptotic behavior of the random variable after a long time. The characteristic function of a random variable that becomes Gaussian asymptotically in time fulfills [see (2.6)]

$$\lim_{t\to\infty} E[e^{j\omega\theta(t)}] = e^{j\omega\mu(t)-\omega^2\sigma^2(t)/2} \tag{2.79}$$

with μ being the average value and σ^2 the variance of the particular random variable.

In the particular case of the scalar nonlinear differential equation $\dot{\theta} = b(t + \theta)\overline{\varepsilon}(t)$ in the variable θ, with the matrix $[b]$ defined as $[b] = \overline{v}_1^{\mathrm{T}}\partial\overline{f}/\partial\overline{\varepsilon}$ and the white noise sources being stationary, the resolution of (2.78) for $z = e^{j\omega\theta(t)}$ provides a characteristic function $E[e^{j\omega\theta(t)}]$ that for a sufficiently large time fulfills (2.79), as demonstrated by Demir [8]. The demonstration is based on introduction of the expression $e^{j\omega\mu(t)-\omega^2\sigma^2(t)/2}$, as a test, in the equation of the characteristic function $F_\theta(\omega, t)$, associated with (2.78). By equating coefficients of the same order in the variable ω, it is shown that the expression indicated constitutes a solution of the characteristic function equation, provided that the time is large enough for terms of the form $e^{-1/2\omega_o^2(i-k)^2\sigma^2(t)}$ to be equal to 1 if $i = k$, and equal to zero otherwise. Note that we assume that the variance of the stochastic time deviation increases in time due to the absence of restoring mechanism in the phase variable. The "large time" depends on the oscillation frequency ω_o, so it will be smaller for larger ω_o. Thus, the stochastic time deviation θ becomes a Gaussian process asymptotically in time. Solving the equation in the characteristic function [8] we find that the mean takes the constant value $\mu_\theta(t) = m$, which should be determined numerically in each case, and the variance increases linearly in time as $\sigma_\theta^2 = ct$. Thus, the time deviation θ is a nonstationary process. The constant c is given by

$$\begin{aligned} c &= \frac{1}{T}\int_0^{\mathrm{T}} [b(t)]^{\mathrm{T}}\overline{\varepsilon}(t)\overline{\varepsilon}^{\mathrm{T}}(t)[b(t)]\,dt \\ &= \frac{1}{T}\int_0^{\mathrm{T}} \overline{v}_1^{\mathrm{T}}(t)\left[\frac{\partial f}{\partial\overline{\varepsilon}}\right]\overline{\varepsilon}(t)\overline{\varepsilon}^{\mathrm{T}}(t)\left[\frac{\partial f}{\partial\overline{\varepsilon}}\right]^{\mathrm{T}}\overline{v}_1(t)\,dt \\ &= \frac{1}{T}\int_0^{\mathrm{T}} \overline{v}_1^{\mathrm{T}}(t)\left[\frac{\partial f}{\partial\overline{\varepsilon}}\right][\Gamma_c]\left[\frac{\partial f}{\partial\overline{\varepsilon}}\right]^{\mathrm{T}}\overline{v}_1(t)\,dt \end{aligned} \tag{2.80}$$

with $[\Gamma_c]$ the correlation matrix of the L white noise sources, already defined as $E[\varepsilon_i(t)\varepsilon_j(t+\tau)] = \Gamma_{ij}\delta(\tau)$. Note that expression (2.80) provides a single scalar from the time averaging of a second-order function involving periodic phase sensitivity functions with respect to all the white noise sources constituting the matrix $\omega_o[b(t)]$ and the correlation matrix of these noise sources $[\Gamma_c]$.

The fact that σ_θ^2 increases linearly with time as $\sigma_\theta^2 = ct$ indicates that the bell of the probability density function flattens with time, with an associated increase in the interval of phase values fulfilling $P[|\theta - m| \leq \sigma_\theta^2] = 0.68$. This is in agreement with the invariance of the oscillator solution with respect to time translations or, equivalently, with the absence of a restoring mechanism in the direction of the limit cycle. In Demir et al. [8] it is shown that the variables $\theta(t), \theta(t+\tau)$ are jointly Gaussian, so it is derived easily that the autocorrelation of the stochastic time deviation θ is given by

$$E[\theta(t)\theta(t+\tau)] = m^2 + c\min(t, t+\tau) \tag{2.81}$$

Note that τ may be positive or negative. The need to take the minimum of t and $t + \tau$ in the second term of (2.81) comes from the fact that when using an Ito integral of the Langevin equation, the stochastic process $\theta(t)$ at time t is independent of the behavior of the Wiener process for $s > t$. The same result (2.81) is obtained with the Stratonovich integral [8].

The oscillator spectrum due to phase noise only is calculated as the Fourier transform of the autocorrelation function:

$$R(t, \tau) = E[x(t + \theta(t))x^*(t + \tau + \theta(t+\tau))] \tag{2.82}$$

with x any circuit state variable. Because the waveform itself is not affected by the noise perturbations, we can represent the time-shifted x variables using the same harmonic coefficients as those in its Fourier series expansion in the steady state $x(t) = \sum_{i=-\infty}^{\infty} X_i e^{ji\omega_o t}$. This provides

$$R(t, \tau) = \sum_{i=-\infty}^{\infty}\sum_{k=-\infty}^{\infty} X_i X_k^* e^{j(i-k)\omega_o t} e^{-jk\omega_o\tau} E[e^{j\omega_o(i\theta(t) - k\theta(t+\tau))}] \tag{2.83}$$

Note that the only stochastic terms are the exponentials of θ on the right-hand side, so the expectation of (2.82) has been shifted right in equation (2.83). Taking (2.79) and (2.81) into account and calculating the limit for a time tending to infinity,

$$\lim_{t\to\infty} E[e^{j\omega_o(i\theta(t) - k\theta(t+\tau))}] = e^{-1/2\omega_o^2[(i-k)^2 ct + k^2 c\tau - 2ikc\min(0,\tau)]} \tag{2.84}$$

Note that the first term of the exponential, involving $(i-k)^2 ct$, will tend to zero for $i \neq k$, due to the negative sign of the resulting real exponent. From (2.84), the autocorrelation of $x(t)$ is given by

$$\lim_{t\to\infty} R(t, \tau) = \sum_{i=-\infty}^{\infty} |X_i|^2 e^{-ji\omega_o\tau} e^{-1/2\omega_o^2 i^2 c|\tau|} \tag{2.85}$$

which only depends on the time difference τ, so $x(t)$ behaves, in the limit, as a stationary process. Finally, the oscillator spectrum is calculated from the transfer function of the autocorrelation function $S_o(\Omega) = F[R(\tau)]$. For that we take into account that the Fourier transform of $e^{a|\tau|}$ is given by $F(e^{a|\tau|}) = \frac{-2a}{a^2+\Omega^2}$. On the other hand, the exponentials $e^{-ji\omega_o\tau}$ give rise to a frequency shift, so the resulting single-sideband spectral density, considering all the harmonic terms, is given by

$$S_o(f_o, f) = 2\sum_{i=-\infty}^{\infty} |X_i|^2 \frac{i^2 f_o^2 c}{\pi^2 f_o^4 i^4 c^2 + (if_o + f)^2} \tag{2.86}$$

Note that $f = \Omega/(2\pi)$ is the offset frequency with respect to each harmonic frequency $i\omega_o = i2\pi f_o$. The spectrum above constitutes a Lorentzian line about each harmonic frequency $i\omega_o$, with the 3-dB cutoff frequency determined by the constant c. Note that at relatively large offset frequencies (far from the carrier) the spectrum drops as -20 dB/dec. The Lorentzian shape prevents the noise spectral density from tending to infinity as the offset frequency Ω tends to zero, which would be an unphysical situation. To see this, note that the frequency integral of the spectral density (2.86) provides

$$\int_{-\infty}^{\infty} S_o(f)\,df = \sum_{i=-\infty}^{\infty} |X_i|^2 \tag{2.87}$$

Thus, when only phase noise is considered, the oscillator spectrum spreads, but its total power remains equal to that of a noiseless oscillator. In contrast, the total power of an oscillator with amplitude noise is different from that of the noiseless oscillator.

As already stated, the single-sideband oscillator spectrum is defined for positive frequencies only and at these positive frequencies takes twice the value given by (2.86). This is because the circuit variables are real in the time domain and $X_{-1}^i = X_1^{i*}$. The oscillator phase noise at the frequency offset f is defined as the ratio between one-half the single-sideband power and the carrier power. At the fundamental frequency of the oscillator solution, this definition of the phase noise spectrum provides

$$S_{p1}(f) = \frac{S_{s1}(f)}{2|X_1|^2} = \frac{cf_o^2}{\pi^2 f_o^4 c^2 + f^2} \tag{2.88}$$

Note that the spectrum above is expressed in terms of the offset frequency from the fundamental frequency f_o. It is independent of $|X_1|^2$. Similar expressions hold for the phase noise about other harmonic components. According to (2.88), the oscillator phase noise due to white noise has a well-defined characteristic versus the offset frequency, initially flat and then dropping as -20 dB/dec. No other form of variation seems possible. However, the phase noise spectrum often tends to become flat versus the offset frequency in measurements or when using some types of frequency-domain simulation. The strict variation of (2.88) is due to the particular definition of phase noise resulting when the original perturbed system (2.63) is decoupled through multiplication by $\overline{v}_1^T(t)$ and use of the additional condition $\overline{v}_1^T(t)\Delta\overline{x}(t) = 0$. When using other constraints to solve system (2.63), the variation in $\theta(t)$ depends on the amplitude perturbation $\Delta\overline{x}(t)$, as demonstrated in [17]. The perturbed oscillator equation (2.62), has been solved by Kaertner [7] with the constraint $\Delta\overline{x}(t)\overline{u}_1(t) = 0$, which as already discussed, limits the perturbation to the orthogonal complement space to the tangent space at the limit cycle. In this resolution the perturbed N-equation system is also divided into two subsystems: a scalar equation and an $(N-1)$-equation subsystem. Unlike the preceding resolution, the resulting scalar equation contains both the stochastic time deviation $\theta(t)$ and the amplitude perturbation $\Delta\overline{x}(t)$ and can be expressed as

$$\dot{\theta}(t) = [\alpha(y)]\Delta\overline{x}(y) + [\beta(y)]\overline{\varepsilon}(t) \tag{2.89}$$

with $[\alpha(y)]$ and $[\beta(y)]$ being row matrixes depending on the Jacobian matrixes with respect to state variables and noise sources, respectively. When obtaining the phase noise spectrum from (2.89), the amplitude perturbation $\Delta\overline{x}(y)$ will have a direct influence over the stochastic time deviation $\theta(t)$. This amplitude perturbation is small compared with the phase perturbation, except at large offset from the carrier or when having a low damping resonance at a particular offset frequency Ω_r. In the absence of resonances, the amplitude perturbation $\Delta\overline{x}(t)$ will be significant only at a large frequency offset from the carrier, where its value will become comparable or even larger than that of the decreasing phase noise. The effect of the amplitude noise is discussed in some detail in Section 2.5.4. It is possible to suggest at this point that the common flattening of the phase noise spectrum for large Ω with some simulation techniques is due to the influence of the amplitude perturbation on the phase noise in the presence of white noise sources.

As an example, expression (2.88) has been used to calculate the phase noise spectrum of the circuit of Fig. 1.1, with oscillation frequency $f_o = 1.59$ GHz. The parallel white noise source has spectral density $S_W(\Omega) = 4\,kTG\text{A}^2/\text{Hz}$, with $G = 1/R$. The parameter c resulting from (2.80) is $c = 3.964 \times 10^{-23}\ \text{A}^{-2}/\ \text{Hz}$. This calculation is performed using the periodic phase sensitivity function of Fig. 2.1. As shown in Fig. 2.2, the phase noise spectrum resulting from the white noise source is a Lorentzian line with a corner frequency $f_{3\text{dB}} = c\omega_o^2/2/(2\pi) = 2 \times 10^{-3}$ Hz. The resulting corner frequency $f_{3\text{dB}}$ is below the smallest offset frequency considered in Fig. 2.2. Note, however, that the c value can be more significant in other circuits. From the corner frequency, the phase noise decreases as -20 dB/dec.

2.4.2 White and Colored Noise Sources

In this section, the oscillator spectrum in the presence of white and colored noise sources is determined. A single colored noise source $\gamma(t)$ with the spectral density $S_\gamma(f)$ is considered initially. The noise source $\gamma(t)$ is assumed to be a stationary Gaussian stochastic process with zero mean and autocorrelation $R_\gamma(\tau)$. Note that the spectral density of the noise source $S_\gamma(f)$ is the Fourier transform of its autocorrelation function $S_\gamma(f) = F(R_\gamma(\tau))$. Remember that in the case of a white noise source, this autocorrelation is simply $R_\gamma(\tau) = \varepsilon\delta(\tau)$, so the associated spectrum is flat. In contrast, the spectrum of the colored noise sources is not

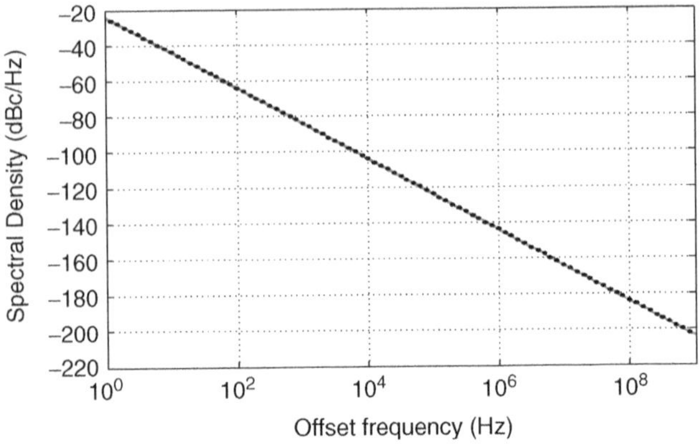

FIGURE 2.2 Phase noise spectral density resulting from the parallel connection of a white noise current source $i_w(t)$ to the parallel resonance oscillator of Fig. 1.1. It is a Lorentzian line with a very small cutoff frequency, given by $\Omega_{3dB} = 2 \times 10^{-3}\ s^{-1}$.

flat and their correlation time is different from zero. An example was shown in (2.39)–(2.40).

In the presence of the noise source $\gamma(t)$, the differential equation system ruling the circuit behavior becomes $\dot{\overline{x}} = \overline{f}(\overline{x}, \gamma(t))$. In a manner similar to (2.60), the nonlinear vector function $\overline{f}$ can be expanded in a Taylor series with respect to the noise source $\gamma(t)$, obtaining

$$\dot{\overline{x}} = \overline{f}(\overline{x}) + \frac{\partial \overline{f}(\overline{x})}{\partial \gamma}\gamma(t) \tag{2.90}$$

Following a procedure identical to (2.63)–(2.69), it is possible to obtain the stochastic differential equation in the time deviation θ:

$$\dot{\theta} = \overline{v}_1^{\mathrm{T}}(t+\theta)\frac{\partial \overline{f}}{\partial \gamma}(\overline{x}_{sp}(t+\theta))\gamma(t) = b_\gamma(t+\theta)\gamma(t) \tag{2.91}$$

Note that because only one colored noise source is being considered, the coefficient $b_\gamma(t)$ determining the phase sensitivity to this noise source is a scalar. Next, the equation in the probability density function $p_\theta(\theta, t)$ associated with the differential equation in θ (2.91) will be derived. This equation is a generalization of the Fokker–Planck equation in $p_\theta(\theta, t)$ to non-Markovian processes [8]. In the presence of a colored noise source, the stochastic process $\theta(t)$ will, in general, not be Markovian. This is due to the fact that the correlation time of the colored noise source can be very long, as in the case of flicker noise, so the short-memory assumption of Markovian processes becomes invalid.

The steps to determine the variance of stochastic time θ ruled by equation (2.91) are the following. Initially, the generalized Fokker–Planck equation in $p_\theta(\theta, t)$

associated with (2.91) is obtained. Next, the corresponding equation in the characteristic function $F_\theta(\omega, t) = E[e^{j\omega\theta(t)}]$ is derived. This is done by considering the function $z = e^{j\omega\theta(t)}$ and obtaining the equation that rules the time variation of its expectation $F_\theta(\omega, t) = E[e^{j\omega\theta(t)}]$, in a manner similar to (2.78). The expression $F_\theta(\omega, t) = e^{j\omega\mu(t) - \omega^2\sigma^2(t)/2}$ is introduced as a test in the equation in the characteristic function $F_\theta(\omega, t)$, equating equal powers of ω. It is seen that this expression constitutes a solution of the characteristic function equation, provided that the time is large enough for all terms of the form $e^{-1/2\omega_o^2(i-k)^2\sigma^2(t)}$ to be equal to 1 for $i = k$ and equal to zero otherwise. Thus, the characteristic function associated with $\theta(t)$ becomes, after a sufficiently long time t,

$$F_\theta(\omega, t) = \exp\left[j\omega\mu_\theta - \frac{\omega^2\sigma_\theta^2(t)}{2}\right] \tag{2.92}$$

with the mean μ_θ being a constant value. As already known, the characteristic function obtained agrees with the characteristic function of a Gaussian random variable. Thus, in the presence of a colored noise source, the stochastic time deviation θ becomes a Gaussian process asymptotically in time.

The scalar coefficient $b_\gamma(t)$ determining the phase sensitivity to the noise source is periodic and can be expanded in a Fourier series: $b_\gamma(t) = \sum_{i=-\infty}^{\infty} B_{\gamma,i} e^{ji\omega_o t}$. From an analysis of the equation in the characteristic function [derived from the generalized Fokker–Planck equation in $p_\theta(\theta, t)$] it is found in [8] that the variance σ_θ fulfills

$$\frac{d\sigma_\theta^2(t)}{dt} = \sum_{i=-\infty}^{\infty} 2|B_{\gamma,i}|^2 \int_0^t R_N(t-\tau)e^{ji\omega_o(t-\tau)}\,d\tau \tag{2.93}$$

For a baseband noise source, the autocorrelation function $R_N(t)$ has a slow time variation, so in the integral above, all terms different from the one corresponding to $i = 0$ will be divided approximately by $i\omega_o$. Therefore, the contribution of the integrals of the complex exponential functions for $i \geq 1$ is approximately zero. The only component of the phase sensitivity function $b_\gamma(t)$ that will contribute to the variance is the dc component $B_{\gamma,0}$, here renamed $b_{\gamma,\mathrm{dc}}$, for more clarity. This is different from the case of white noise sources. The reason is that unlike what happens with colored noise sources, the white noise sources are not at baseband, meaning that they contribute to all the harmonic components of the oscillator solution. This is why, as gathered from (2.80), all the harmonic components of the phase sensitivity matrix contribute to the variance of the stochastic time deviation $\sigma_\theta^2 = ct$ [see (2.80)].

Simplifying equation (2.93) according to the previous considerations, it is possible to obtain the following expression, depending on $b_{\gamma,\mathrm{dc}}$ and the autocorrelation $R_\gamma(\tau)$ of the colored noise source or, alternatively, on its spectrum $S_\gamma(f) = F(R_\gamma(\tau))$:

$$\sigma_\theta^2(t) = 2|b_{\gamma,\mathrm{dc}}|^2 \int_0^t (t-\tau)R_\gamma(\tau)d\tau = 2|b_{\gamma,\mathrm{dc}}|^2 \int_{-\infty}^{\infty} S_\gamma(f)\frac{1-e^{j2\pi ft}}{4\pi^2 f^2}df \tag{2.94}$$

The final value theorem allows establishing $\lim_{t\to\infty}\sigma_\theta^2(t)/t = b_{\gamma,\mathrm{dc}}^2 S_\gamma(0)$. This means that the variance $\sigma_\theta^2(t)$ resulting from any colored noise source with spectral density different from zero at $f = 0$ will tend to infinity as $t \to \infty$. The white noise sources, with $\sigma_\theta^2(t) = ct$, also fulfill this property.

So far, only the variance of the stochastic time deviation θ has been determined. However, the objective is to obtain the oscillator spectrum due to phase noise in the presence of the colored noise source $\gamma(t)$. As in the case of white noise sources, this requires the calculation of $\lim_{t\to\infty} R(t,\tau) = E[x(t+\theta(t))x^*(t+\tau+\theta(t+\tau))]$. The only stochastic terms remaining in the limit fulfill

$$\lim_{t\to\infty} E[e^{j\omega_o i(\theta(t)-\theta(t+\tau))}] = e^{-1/2\omega_o^2 i^2 \sigma_\theta^2(|\tau|)} \tag{2.95}$$

with i the harmonic order. In a manner similar to (2.84)–(2.85), the oscillator spectrum is determined from the Fourier transform of the autocorrelation function $R(t,\tau)$ when t tends to infinity, $S_o(f) = F[\lim_{t\to\infty} R(t,\tau)]$. As gathered from (2.95), this spectrum depends on the variance $\sigma_\theta(\tau)$ of the time deviation, which, in turn, depends on the spectral density $S_\gamma(f)$ or autocorrelation R_γ of the colored noise source (2.94). For some spectral densities $S_\gamma(f)$, no analytical forms of the Fourier transform $S_o(f) = F[\lim_{t\to\infty} R(t,\tau)]$ are available, so either limiting forms or a numerical calculation must be used. The following are some details.

A particular type of colored noise of great relevance to the performance of electronic oscillators is the flicker noise, with a spectral density exhibiting a frequency dependence of the form $1/f$. As already indicated, to avoid the singularity at $f = 0$, the use of a cutoff frequency $f_{\min}$ was proposed by Demir [8]. This cutoff frequency, which will typically be very small, can be related to the finite autocorrelation time T in the measurements [4]. The expression for the $1/f$ noise is $S^F(f) = 1/|f| - 4\arctan(f_{\min}/2\pi f)/2\pi f$. This expression should be introduced into (2.94) to obtain the variance of the time deviation due to one flicker noise source. Next, the noise spectrum about $i\omega_o$ is determined from the Fourier transform of the exponential (2.95). Clearly, the determination of this Fourier transform is not an easy task.

In the general case of L white noise sources and J colored noise sources, we must calculate the total variance $\sigma_\theta^2(t)$ due to all these sources. The total variance of the random variable θ, depending on several uncorrelated Gaussian processes, is calculated as the addition of the variances resulting from the individual processes [8]. We determine the parameter c, associated to the L white noise sources, and the parameters $b_{\gamma j,\mathrm{dc}}$, $j = 1\ldots J$, each associated to one of the colored noise sources. Then, the total variance is given by $\sigma_\theta^2(t) = \sigma^2(t) + \sum_{j=1}^{J}\sigma_{\gamma,j}^2(t) = ct + \sum_{j=1}^{J}\sigma_{\gamma,j}^2(t)$. The variance of each colored noise source γ_j is related to its spectral density $S_N^{\gamma,j}(f)$ source through expression (2.94). According to all the previous derivations, the phase noise spectrum about the oscillator carrier is calculated as

$$S(f) = F[R(\tau)] = F\left\{\exp\left[-\tfrac{1}{2}\omega_o^2\sigma_\theta^2(|\tau|)\right]\right\} \tag{2.96}$$

where F is the Fourier transform. Note that the phase noise spectrum does not depend on the considered circuit variable [see (2.88), for instance].

Due to the complexity of the expression (2.94) for the variance due to the colored noise sources, the determination of the Fourier transform in (2.96), providing the phase noise about the oscillator carrier, is not straightforward. The most accurate way to obtain the phase noise spectrum is to perform a numerical calculation of this. In fact, the major advantage of calculation of the phase noise spectrum from the variance of the phase deviation $\sigma_\theta^2(|\tau|)$ is the high accuracy that it provides at a small frequency offset from the carrier. Note that at these low offset frequencies, selection of the cutoff frequency f_{min} of the flicker noise sources will have a great influence on the phase noise spectrum.

Instead of the numerical calculation described, limiting forms of the Fourier transform [19] are used in Demir's article [8], which in the case of a single colored noise source $\gamma(t)$ provide the following spectrum:

$$S_{p1}(f) = \frac{S_{s1}(f)}{2|X_1|^2} = \begin{cases} \dfrac{|b_{\gamma,\text{dc}}|^2 S_\gamma(0)}{\pi^2 f_o^2 (|b_{\gamma,\text{dc}}|^2 S_\gamma(0))^2 + f^2} & f \approx 0 \quad \text{(a)} \\ |b_{\gamma,\text{dc}}|^2 S_\gamma(f) \dfrac{f_o^2}{f^2} & f \gg 0 \quad \text{(b)} \end{cases} \tag{2.97}$$

where the phase noise is defined as the ratio between one-half the single-sideband power and the carrier power. Note that for a low offset frequency, the phase noise spectrum is a Lorentzian line with the corner frequency determined by the dc component of the phase sensitivity function of the colored noise source $b_{\gamma_i,\text{dc}}$ and the constant value $S_\gamma(0)$, used to approximate the spectral density of the colored noise source at $f \to 0$. The colored noise source considered had a spectrum $S_\gamma(f) \propto 1/f$, so, for a larger frequency offset, the spectral density of the noisy oscillator will drop as -30 dB/dec.

In the case of J different Gaussian $1/f$ noise sources $\gamma_1(t), \gamma_2(t), \ldots, \gamma_J(t)$, uncorrelated with each other and with the L white noise sources, the phase noise spectrum at each harmonic component i is generalized as

$$S_{pi}(f) = \frac{S_{si}(f)}{2|X_i|^2} = \begin{cases} \dfrac{i^2 f_o^2 \left(c + \sum_{k=1}^{M} |b_{\gamma j,\text{dc}}|^2 S_{\gamma j}(0)\right)}{\pi^2 i^4 f_o^4 \left(c + \sum_{j=1}^{J} |b_{\gamma j,\text{dc}}|^2 S_{\gamma j}(0)\right)^2 + f^2} & f \approx 0 \quad \text{(a)} \\ \left(c + \sum\limits_{j=1}^{J} |b_{\gamma j,\text{dc}}|^2 S_{\gamma j}(f)\right) \dfrac{i^2 f_o^2}{f^2} & f \gg 0 \quad \text{(b)} \end{cases} \tag{2.98}$$

with $c = (1/T)\int_0^T \overline{b}^T(t)[\Gamma]\overline{b}(t)dt$, $b_{\gamma j,\text{dc}}$ the dc component of the phase sensitivity to the colored noise source $\gamma_j(t)$, with $j = 1$ to J, and $S_{\gamma j}(f)$ the spectral density of the colored noise source $\gamma_j(t)$. As in the case of a single colored noise source, for a small frequency offset, the spectrum can be approached by a Lorentzian line. Its bandwidth is determined by both the constant c, depending on the correlation of the white noise sources and the phase sensitivity to these sources, and on the summation, for all the colored noise sources, of the products $|b_{\gamma j,\text{dc}}|^2 S_{\gamma j}(0)$. These products involve the dc component of the phase sensitivity to the particular colored noise source $b_{\gamma j,\mathrm{dc}}$ and the value $S_{\gamma j}(0)$ by which its spectral density is approached in the limit $f \to 0$. In the case of the flicker noise sources, for a larger offset frequency, the spectrum will drop -30 dB/dec up to the corner frequency f_c, at which $\sum_{j=1}^{J} |b_{\gamma j,\mathrm{dc}}|^2 S_{\gamma j}(f_c) = c$. From this corner frequency it will drop -20 dB/dec.

Note that according to (2.98b), the oscillator phase noise due to white noise plus a flicker noise source has a very well defined characteristic, initially dropping -30 dB/dec and then dropping -20 dB/dec. No other form of variation seems possible. As in the case of white noise sources already discussed, this is due to the particular definition of phase noise, resulting when the original perturbed system (2.63) is decoupled through multiplication by $\overline{v}_1^{\mathrm{T}}(t)$ and use of the additional condition $v_1^{\mathrm{T}}(t)\Delta\overline{x}(t) = 0$. As shown by Sancho et al. [17], for other definitions of phase noise, the stochastic time deviation $\theta(t)$ and the amplitude deviation $\Delta\overline{x}(t)$ will be coupled, and other forms of variation will be possible.

The phase noise analysis above has been applied to a parallel resonant oscillator in Fig. 1.1. If a current flicker noise source is connected in parallel, the corresponding phase noise sensitivity $b_\gamma(t)$ has zero dc value: $b_{\gamma,\mathrm{dc}} = 0$. This means that there is no up-conversion to the carrier frequency of the flicker noise introduced. This is in agreement with the presence of the parallel inductance, exhibiting very low parallel impedance at baseband. Instead of the parallel flicker noise current source, a flicker noise voltage source $e_\gamma(t)$ is introduced in series with the nonlinear element. The spectral density of this voltage source is $S_F = 10^{-9}/f$ V^2/Hz. The resulting phase sensitivity function is represented in Fig. 2.3, where it can be compared with the waveform of the steady-state current through the nonlinear element. The maximum absolute values of the sensitivity function are obtained near the points of larger magnitude of the time derivative. The dc value of this phase sensitivity function is $b_{\gamma,\mathrm{dc}} = 1.6459 \times 10^{-4}$ A^{-1}/Hz.

The phase noise spectrum obtained when considering both the white noise current source $i_w(t)$ in parallel and the flicker noise voltage source $e_\gamma(t)$ in series

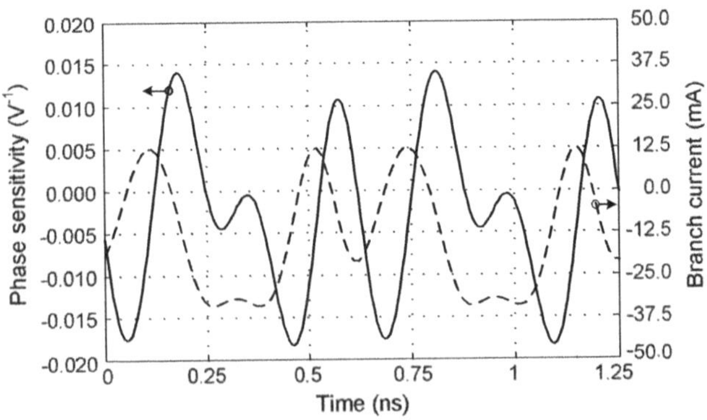

FIGURE 2.3 Phase sensitivity to a voltage noise source connected in series with a nonlinear element in a parallel resonance oscillator. The current waveform is shown by the dashed line and the sensitivity function in A^{-1}/Hz is shown by the solid line.

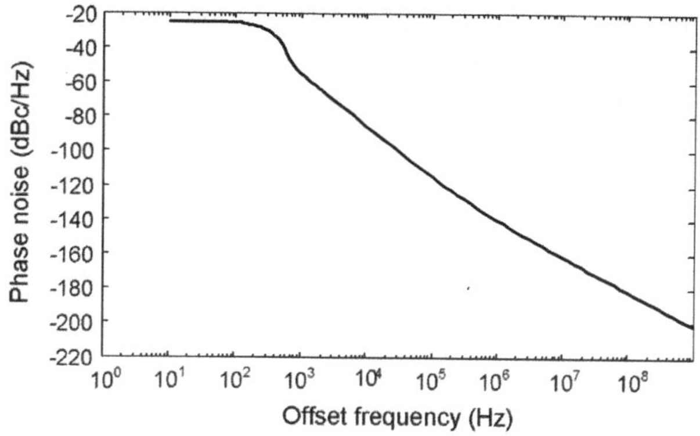

FIGURE 2.4 Numerical calculation of the phase noise spectral density of a cubic nonlinearity oscillator. A white noise current source in parallel with a nonlinear element and a flicker voltage source in series with this element have been considered.

with the nonlinear element is represented in Fig. 2.4. The spectrum has been calculated numerically assuming a cutoff frequency $f_{\min} = 10^{-5}$ Hz in the flicker noise source. In agreement with (2.98), the spectrum is initially flat, up to about 200 Hz, and then drops -30 dB/dec, up to the corner frequency at which $|b_{\gamma,\mathrm{dc}}|^2 S_\gamma(f_c) = c$. The value of this corner frequency is $f_c = 683.40$ kHz. From this corner frequency, the phase noise spectral density drops -20 dB/dec.

2.5 FREQUENCY-DOMAIN ANALYSIS OF A NOISY OSCILLATOR

In this section, two techniques for the frequency-domain analysis of the oscillator phase noise are presented. The first is based on a determination of the oscillator frequency modulation due to noise sources. This is done initially by means of an impedance–admittance formulation that is very helpful in circuit design. Then a harmonic balance formulation is used, limited here to dc and the first-harmonic term. Next, the frequency-domain equivalent of a time-domain analysis based on the phase sensitivity functions is presented.

2.5.1 Frequency-Domain Representation of Noise Sources

The noise sources in electronic circuits are either white, with a flat spectral density, as in the case of thermal and shot noise, or have a low-frequency spectrum, as in the case of flicker and burst noise. For the frequency-domain analysis of an oscillator at the steady-state frequency ω_o, the white noise is represented as flat narrowband spectra about the various harmonic terms $k\omega_o$. The limited bandwidth of these flat spectra is given by the maximum offset frequency from the carrier to be considered in the oscillator noise spectrum. Using a generalization of the lowpass equivalent of a bandpass signal, the white noise about each harmonic frequency k is represented as $N_k(t)e^{jk\omega_o t}$, where the $N_k(t)$ are slowly varying complex envelopes having the spectral density of flat white noise. In turn, low-frequency colored noise is represented as a baseband signal.

Next, the mean-square value of the real and imaginary parts of the complex envelopes $N_k(t)$ of a white noise source will be related to that of the mean-square value of the noise source in the time domain $n(t)$. The case of a single white noise current source is considered initially, with the analysis limited to the fundamental frequency ω_o. The current noise source is modeled as the summation $i_n(t) = \sum \varepsilon\delta(t - t_o)$, where both ε and t_o are independent random variables. The mean-square value of this noise current source will be [20]

$$\langle i_n(t)^2 \rangle = \frac{1}{2T} \int_{-T}^{T} \left[\sum \varepsilon\delta(t - t_o) \right]^2 dt \tag{2.99}$$

where T is a long time interval used for averaging. Assuming slow time variations about the oscillation frequency ω_o, the noise source can be written $i_n(t) = \text{Re}[I_n(t)e^{j\omega_o t}]$, where $I_n(t)$ is a slowly varying complex number constituting the envelope of the noise source about the carrier frequency ω_o. The real and imaginary parts of $I_n(t)$ will be calculated, respectively, as

$$\begin{aligned} I_n^r(t) &= \frac{2}{T_o} \int_{t-T_o}^{t} i_n(t) \cos \omega_o t \, dt \\ I_n^i(t) &= \frac{2}{T_o} \int_{t-T_o}^{t} i_n(t) \sin \omega_o t \, dt \end{aligned} \tag{2.100}$$

where T_o is the solution period. At each particular pulse $\varepsilon\delta(t - t_o)$, the integrals above become

$$\begin{aligned} \frac{2}{T_o} \int_{t-T_o}^{t} \varepsilon\delta(t - t_o) \cos(\omega_o t) \, dt &= \frac{2}{T_o} \varepsilon \cos(\omega_o t_o)[u(t - (t_o + T_o)) - u(t - t_o)] \\ &= \frac{2}{T_o} \varepsilon \cos(\omega_o t_o) p(t) \\ \frac{2}{T_o} \int_{t-T_o}^{t} \varepsilon\delta(t - t_o)(\sin \omega_o t) \, dt &= \frac{2}{T_o} \varepsilon \sin(\omega_o t_o)[u(t - (t_o + T_o)) - u(t - t_o)] \\ &= \frac{2}{T_o} \varepsilon \sin(\omega_o t_o) p(t) \end{aligned} \tag{2.101}$$

with $u(t)$ being the step function and $p(t)$ being a square pulse of unit amplitude, duration T_o, and initial time t_o. The mean-square value of $I_n^r(t)$ and $I_n^i(t)$ will be calculated as

$$\begin{aligned} \langle I_n^r(t)^2 \rangle &= \frac{1}{2T} \int_{-T}^{T} \left[\sum \frac{2}{T_o} \varepsilon \cos(\omega_o t_o) p(t) \right]^2 dt = 2\langle i_n(t)^2 \rangle \\ \langle I_n^i(t)^2 \rangle &= \frac{1}{2T} \int_{-T}^{T} \left[\sum \frac{2}{T_o} \varepsilon \sin(\omega_o t_o) p(t) \right]^2 dt = 2\langle i_n(t)^2 \rangle \end{aligned} \tag{2.102}$$

where T is a long time interval used for averaging. In the expressions above we have taken into account that the duration of the pulse $p(t)$ is so short compared to the long time interval considered in time averaging that the function $\varepsilon \sin(\omega_o t_o) p(t)/T_o$ is nearly a delta function. Because of the coefficient 2 in the calculation of the current components in $\cos \omega_o t$ and $\sin \omega_o t$ [see (2.100)], the mean-square value of either $I_n^r(t)$ or $I_n^i(t)$ is twice that of the time-domain current $i(t)$. On the other hand, because of the orthogonality between the sine and cosine functions, the noise envelopes $I_n^r(t)$ and $I_n^i(t)$ will be uncorrelated. This result, which can be generalized to multiple harmonic terms, will be essential for frequency-domain calculation of the noisy oscillator spectrum.

2.5.2 Carrier Modulation Analysis

The phase perturbations $\Delta\phi(t) = \omega_o \theta(t)$ give rise to a modulation of the oscillator carrier frequency [22], according to $\Delta\omega(t) = \omega_o \dot{\theta}(t)$. The phase noise analysis presented in this section is based on a determination of this carrier modulation. Initially, the analysis is performed at the fundamental frequency only (not including the dc term), so the noise perturbations will be limited to white noise about this frequency. Next, the more general case of flicker and white noise is considered.

As shown in Chapter 1, the oscillator circuit can be analyzed using the total admittance function $Y_T(V, \omega)$ calculated at a sensitive observation node. In the steady-state regime, corresponding to the node voltage $v(t) = \text{Re}[V_o e^{j\omega_o t}]$, the condition $Y_T(V_o, \omega_o) = 0$ is fulfilled. After the introduction of a white noise current source $i_n(t)$, perturbations will be obtained in the oscillator frequency and amplitude. This perturbed oscillator equation is formally identical to the equation used in Chapter 1 for the stability analysis of the steady-state oscillator solution. However, in contrast to this stability analysis, which assumes a small perturbation applied at the single time instant t_o, the random noise perturbations are present at all times. The noise source is considered a bandpass signal about the oscillation frequency ω_o. It is represented in terms of its complex envelope $I_n(t)$ as $i_n(t) = \text{Re}[I_n(t) e^{j\omega_o t}]$. Then the perturbed oscillator equation is given by

$$\frac{\partial Y_{To}}{\partial V} \Delta V(t+\theta) + \frac{\partial Y_{To}}{\partial \omega} \left(-j \frac{\Delta \dot{V}(t+\theta)}{V_o} + \omega_o \dot{\theta}(t) \right) = \frac{I_n(t)}{V_o} \tag{2.103}$$

where the stochastic time shift $\theta(t)$ of the node voltage has been taken into account for illustrative purposes. However, in the following derivation, based on [6], this dependence on θ will be neglected, which will reduce the accuracy at a small frequency offset from the carrier. A more complete analysis taking θ into account is presented later. Eliminating θ and splitting the complex equation (2.103) into real and imaginary parts, the following system is obtained:

$$\begin{aligned} \frac{\partial Y_{To}^i}{\partial \omega} \frac{1}{V_o} \Delta \dot{V}(t) + \frac{\partial Y_{To}^r}{\partial V} \Delta V(t) + \frac{\partial Y_{To}^r}{\partial \omega} \Delta\omega(t) &= \frac{I_n^r(t)}{V_o} = G_n(t) \\ -\frac{\partial Y_{To}^r}{\partial \omega} \frac{1}{V_o} \Delta \dot{V}(t) + \frac{\partial Y_{To}^i}{\partial V} \Delta V(t) + \frac{\partial Y_{To}^i}{\partial \omega} \Delta\omega(t) &= \frac{I_n^i(t)}{V_o} = B_n(t) \end{aligned} \tag{2.104}$$

where the noise conductance $G_n(t)$ and susceptance $B_n(t)$ have been introduced for compactness of the formulation. All the variables in (2.104) are real. Expressing these variables in the frequency domain, they will have Hermitian symmetry, so $\Delta V(\Omega) = \Delta V(-\Omega)^*$. Considering the positive-frequency sideband only, the following system is obtained:

$$\begin{aligned}
&\left(\frac{\partial Y_{To}^i}{\partial \omega}\frac{1}{V_o} j\Omega + \frac{\partial Y_{To}^r}{\partial V}\right)\Delta V(\Omega) + \frac{\partial Y_{To}^r}{\partial \omega} j\Omega\, \Delta\phi(\Omega) = G_n(\Omega) \\
&\left(-\frac{\partial Y_{To}^r}{\partial \omega}\frac{1}{V_o} j\Omega + \frac{\partial Y_{To}^i}{\partial V}\right)\Delta V(\Omega) + \frac{\partial Y_{To}^i}{\partial \omega} j\Omega\, \Delta\phi(\Omega) = B_n(\Omega)
\end{aligned} \tag{2.105}$$

where it has been taken into account that the derivation with respect to the slow time scale of the noise source is equivalent to multiplication by $j\Omega$, with $\Omega = 2\pi f$.

To obtain the phase perturbation, equation (2.105) must be solved for $\Delta\phi(\Omega)$, which can be done easily by using Kramer's rule. The phase noise spectrum is calculated by multiplying $\Delta\phi(\Omega)$ by $\Delta\phi^*(\Omega)$. It must be taken into account that $\langle G_n(t)\rangle^2 = 2\langle i_n(t)\rangle^2/V_o^2$, $\langle B_n(t)\rangle^2 = 2\langle i_n(t)\rangle^2/V_o^2$ and that $G_n(t)$ and $B_n(t)$ are uncorrelated, as demonstrated in Section 2.5.1. The mean-square value $\langle i_n(t)\rangle^2$ is related to the spectral density of the noise source through $\langle i_n(t)\rangle^2 = |I_n|_{\text{sd}}^2 \Delta f$, where the subscript sd indicates "spectral density." Then the phase noise spectral density is given by

$$|\Delta\phi(\Omega)|_{\text{sd}}^2 = \frac{\left|\frac{\partial Y_{To}}{\partial V}\right|^2 |V_o|^2 2|I_n|_{\text{sd}}^2 + \Omega^2 \left|\frac{\partial Y_{To}}{\partial \omega}\right|^2 2|I_n|_{\text{sd}}^2}{\Omega^4 \left|\frac{\partial Y_{To}}{\partial \omega}\right|^4 |V_o|^2 + \Omega^2 |V_o|^4 \left(\frac{\partial Y_{To}^r}{\partial V}\frac{\partial Y_{To}^i}{\partial \omega} - \frac{\partial Y_{To}^i}{\partial V}\frac{\partial Y_{To}^r}{\partial \omega}\right)^2} \tag{2.106}$$

For notational simplicity, the subscript sd will be dropped in the remainder of the book. However, the reader must bear in mind that we are dealing with spectral densities in units of Hertz^{-1} in the case of both noise sources and noise spectrum.

It can easily be seen that the time derivative $\Delta\dot{V}(t)$ gives rise to higher-order terms in the offset frequency Ω in both the numerator and denominator. Actually, when neglecting this derivative setting $\Delta\dot{V} = 0$ in (2.104), the following simpler expression is obtained:

$$S_\phi(\Omega) = |\Delta\phi(\Omega)|^2 = \frac{\left|\frac{\partial Y_{To}}{\partial V}\right|^2 2|I_n|^2}{\Omega^2 |V_o|^2 \left(\frac{\partial Y_{To}^r}{\partial V}\frac{\partial Y_{To}^i}{\partial \omega} - \frac{\partial Y_{To}^i}{\partial V}\frac{\partial Y_{To}^r}{\partial \omega}\right)^2} \tag{2.107}$$

The expression (2.106) approaches (2.107) at low-frequency offset. This is because the influence of the higher powers of Ω will be relevant only at relatively high

Ω values. This means that the time variation of the amplitude perturbation $\Delta V(t)$ will have an effect on the phase noise only at relatively high frequency offset from the carrier, where the phase noise takes lower values. On the other hand, at this relatively low frequency offset, the phase noise spectrum obtained with a white noise source will exhibit a $1/\Omega^2$ dependence, maintained up to the carrier frequency. This is different from the Lorentzian line resulting from the time-domain analysis of Section 2.4 in the presence of white noise sources. The difference comes from the fact that unlike system (2.104), the perturbed system considered in Section 2.4 is nonlinear in θ and the phase noise is extracted from the power spectrum of the perturbed variables x, which explains the saturation effect.

Coming back to the two frequency-domain expressions (2.106) and (2.107), and comparing the predicted phase noise spectrum, the results begin to differ as the frequency offset increases. Expression (2.107) maintains the $1/\Omega^2$ dependence, whereas expression (2.106) exhibits a different form of variation. As Ω increases, the numerator term $\Omega^2\,|\partial Y_{To}/\partial\omega|^2\,2|I_n|^2$ also increases and becomes equal to the constant value $|\partial Y_{To}/\partial V|^2\,|V_o|^2 2|I_n|^2$ at the corner frequency Ω_{c1}, given by

$$\Omega_{c1} = \frac{|\partial Y_{To}/\partial V||V_o|}{|\partial Y_{To}/\partial\omega|} \tag{2.108}$$

The corner frequency Ω_{c1} is lower for larger $|\partial Y_{To}/\partial\omega|$ compared to $|\partial Y_{To}/\partial V|$. For $\Omega > \Omega_{c1}$, the term $\Omega^2\,|\partial Y_{To}/\partial\omega|^2\,2|I_n|^2$ dominates the numerator. There is an intermediate range, between the frequencies Ω_{c1} and Ω_{c2}, for which the phase noise spectrum is flat. The frequency Ω_{c2} is the corner frequency at which the two contributions in the denominator of (2.106) have equal magnitude. This second corner frequency is given by

$$\Omega_{c2} = \frac{|V_o|\left(\frac{\partial Y^r_{To}}{\partial V}\frac{\partial Y^i_{To}}{\partial\omega} - \frac{\partial Y^i_{To}}{\partial V}\frac{\partial Y^r_{To}}{\partial\omega}\right)}{|\partial Y_{To}/\partial\omega|^2} \tag{2.109}$$

According to (2.106), from Ω_{c2} the phase noise spectrum will decrease again as -20 dB/dec. The flattening and 20 dB/dec drop (from Ω_{c2}) of the phase noise spectrum are due to the influence of the amplitude perturbations on the phase noise. Note that unlike the time-domain analysis of Section 2.4.1, the amplitude and phase perturbations have not been decoupled in equation (2.104). This is why this flattening is not obtained in the phase noise analysis of Section 2.4.1.

Note that, in general, the frequency dependence of the real part of the admittance will be low, so the term $|\partial Y_{To}/\partial\omega|$ will, in many cases, agree approximately with $\partial Y^i_{To}/\partial\omega$. Then the corner frequency Ω_{c1} between $1/\Omega^2$ and the flat spectrum sections will be smaller for a higher quality factor, $Q_L = \omega_o(\partial Y^i_{To}/\partial\omega)/2G$, in agreement with Leeson's model [21].

Expression (2.107) usually provides a good estimation of the phase noise spectrum up to a relatively high offset frequency from the carrier. The denominator can be written as

$$\frac{\partial Y_{To}^r}{\partial V}\frac{\partial Y_{To}^i}{\partial \omega} - \frac{\partial Y_{To}^i}{\partial V}\frac{\partial Y_{To}^r}{\partial \omega} = \left|\frac{\partial Y_{To}}{\partial V}\right|\left|\frac{\partial Y_{To}}{\partial \omega}\right| \sin \alpha_{v\omega} \tag{2.110}$$

where $\alpha_{v\omega} = \text{ang}\,(\partial Y_{To}/\partial \omega) - \text{ang}\,(\partial Y_{To}/\partial V)$. When introducing expression (2.110) into (2.107) it is clear that the phase noise will be minimized for $\alpha = \pi/2$ and will take lower value for higher $|\partial Y_{To}/\partial \omega|$. In general, the angle condition is true only for white noise perturbations. Note that the frequency dependence of the real part of the admittance will usually be small, so lower phase noise will be obtained for a higher quality factor. The phase noise will also be smaller for a larger oscillation amplitude, as gathered from (2.107).

The total power spectral density of the oscillator output signal is calculated by applying the Fourier transform to the autocorrelation of the perturbed voltage variable $S_{\overline{VV}}(\Omega) = F[E(\tilde{V}(t)\tilde{V}(t)^*(t-\tau))]$, with $\tilde{V}(t)$ the complete voltage envelope at the oscillation carrier ω_o, including the common phase perturbation $e^{j\theta(t)}$. This power spectral density can be decomposed into three contributions [17]: $S_{\overline{VV}}(\Omega) = S_\phi(\Omega) + S_{\Delta V}(\Omega) + 2S_{\phi\Delta V}(\Omega)$, where $S_\phi(\Omega)$ is the power spectral density due to phase noise, $S_{\Delta V}(\Omega)$ the spectral density due to the amplitude noise, and $S_{\phi\Delta V}(\Omega)$ the spectral density due to the correlation between phase and amplitude noise. This can be calculated from $\Delta\phi(\Omega)\Delta V^*(\Omega)$. To analyze the amplitude noise, the system (2.105) should be solved for the amplitude perturbation $\Delta V(\Omega)$. The amplitude–noise spectrum is obtained by multiplying this increment by its adjoint $\Delta V^*(\Omega)$:

$$S_{\Delta V}(\Omega) = |\Delta V(\Omega)|^2 = \frac{\Omega^2\,|\partial Y_{To}/\partial \omega|^2\,2|I_n|^2}{\Omega^4\left|\frac{\partial Y_{To}}{\partial \omega}\right|^4 + \Omega^2|V_o|^2\left(\frac{\partial Y_{To}^r}{\partial V}\frac{\partial Y_{To}^i}{\partial \omega} - \frac{\partial Y_{To}^i}{\partial V}\frac{\partial Y_{To}^r}{\partial \omega}\right)^2} \tag{2.111}$$

Note that because of the frequency dependence as Ω^2 in the numerator, the amplitude noise will be flat at low-frequency offset from the carrier. It will decay -20 dB/dec from the offset frequency at which the two terms in the denominator become equal, which agrees with the corner frequency Ω_{c2} defined in (2.109). It must be emphasized that this is true only for white noise perturbations. It is suggested that in the case of a flicker noise source, the amplitude noise decays -10 dB/dec closer to the carrier.

The admittance-based calculation above of the phase and amplitude noise spectrum has been used for a parallel resonance oscillator with a white noise current source connected in parallel. The two spectra are shown in Fig. 2.5. As can be seen, the amplitude noise is very low for this particular circuit, so it has little influence on the phase noise spectrum. This is why there is very good correspondence with the phase noise spectrum resulting from the time-domain analysis of Section 2.4 and represented in Fig. 2.2. In agreement with (2.111), the amplitude spectrum is initially flat, and from a large offset frequency, slightly smaller than the oscillation

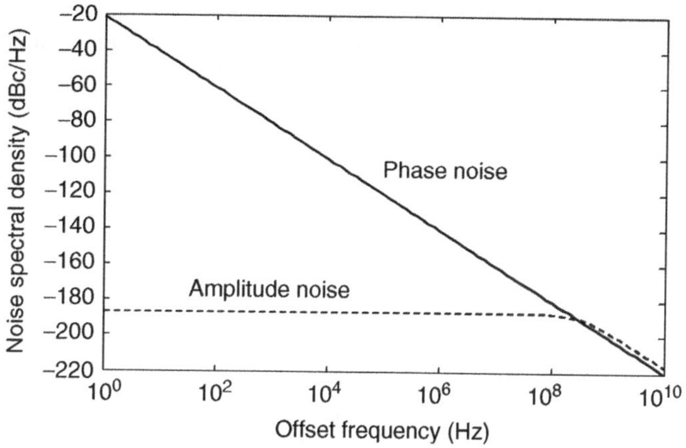

FIGURE 2.5 Frequency-domain calculation based on an admittance analysis of the phase and amplitude spectrum of a parallel resonance oscillator with a white noise current source.

frequency f_o, decays as -20 dB/dec. Note that the offset frequencies considered in the representation of Fig. 2.5 have been extended beyond reasonable values to illustrate the two noise characteristics.

To summarize, there are, in fact, two differences in calculation of the phase noise spectrum from (2.88) and (2.106). On the one hand, the time shift θ has been neglected in (2.104) in both ΔV and $\Delta \dot{V}$, whereas it is taken into account in all the variables of (2.69). This gives rise to different behavior close to the carrier. The nonlinearity with respect to θ in (2.69) and the spectrum calculation from the exponential in (2.95) give rise to the Lorentzian shape of the phase noise in expression (2.88), with a flat response for offset frequency $\Omega < \Omega_{3dB}$ and a $1/\Omega^2$ characteristic at higher-frequency offset. Thus, the phase noise obtained using (2.88) does not tend to infinity as the offset frequency tends to zero, $\Omega \to 0$. In (2.104), the phase perturbation has been decoupled from the amplitude perturbation through multiplication by the vector $\overline{v}_1^T(t)$ of the two sides of equation (2.63), using also $\overline{v}_1^T(t)\Delta\overline{x}(t) = 0$. Then the phase noise spectrum (2.88), with no influence from the amplitude perturbation, maintains the $1/\Omega^2$ characteristic of all Ω values. In contrast, the phase and amplitude perturbations are not decoupled in (2.104). Considering the time derivative $\Delta \dot{V}$ of the amplitude perturbation gives rise to the term in Ω^2 in the numerator of (2.106) and the term in Ω^4 in the denominator. Thus, the derivative $\Delta \dot{V}$ will have a greater effect at a larger frequency offset from the carrier. When the derivative $\Delta \dot{V}$ is neglected, the spectrum in (2.107) is obtained, which shows a dependence $1/\Omega^2$ at all values of the offset frequency from the carrier. As shown in the next section, for a sinusoidal steady-state oscillation the phase noise spectrum predicted by (2.88) and (2.107) will agree except at low offset frequencies.

The admittance analysis above considers white noise about the oscillator carrier frequency only. The inclusion of flicker noise or any other type of low-frequency noise requires an extension of the analysis technique. As shown in Section 2.5.1, the noise perturbations are represented in the frequency domain as bandpass signals about the corresponding harmonic terms $N_k(t)e^{jk\omega_o t}$, with $N_k(t)$ being the complex envelopes. Due to its low-frequency characteristic, the flicker noise, $S_F(f) = k/f^\gamma$, with $\gamma \cong 1$, constitutes a baseband perturbation. The aim here is to perform an oscillator analysis at the fundamental frequency only, so to take the flicker noise into account, both the dc and first-harmonic component of the node voltage, V_{dc} and V_o, must be considered. This will lead to a system of three equations in three unknowns, $\Delta V_{dc}(t)$, $\Delta V(t)$, and $\Delta\omega(t)$, where $\Delta V_{dc}(t)$ and $\Delta V(t)$ are, respectively, the time-varying increments of the dc component and the first-harmonic voltage amplitude.

Instead of using an admittance analysis, the oscillator circuit of Fig. 2.6 will be considered, containing a current nonlinearity and an impedance block with the total admittance Z. The circuit equations are obtained by applying Kirchhoff's laws to the network in Fig. 2.6, taking the nonlinearity $i(v)$ into account. These equations can be expressed in terms of the positive and negative spectra at dc, ω_o and $-\omega_o$, as was done in Chapter 1, or in terms of the real and imaginary parts of the various harmonic components. Because the voltage V is assumed real, the second type of representation will be more compact in this case. The steady-state equations are given by

$$\begin{aligned} H_{dc}^s &\equiv V_{dc}^s + R_L(0)I_{dc}^s(V_{dc}^s, V_1^s) = 0 \\ H_1^s &\equiv V_1^s + Z_L(\omega_o)I_1^s(V_{dc}^s, V_1^s) = 0 \end{aligned} \tag{2.112}$$

where, for simplicity, no dc bias sources have been considered. The unknowns of system (2.112), formulated in terms of the error functions H_{dc}^s (real) and H_1^s (complex), are V_{dc}^s, V_1^s, and the oscillation frequency ω_o. Note that the node voltage is assumed real, $V_1^s e^{j0}$, as in the case of admittance analysis. Next, a white noise current source $I_N(t)$ and a series flicker noise voltage source $V_N(t)$ are considered, as shown in Fig. 2.6. The small amplitude of the noise sources and the small value

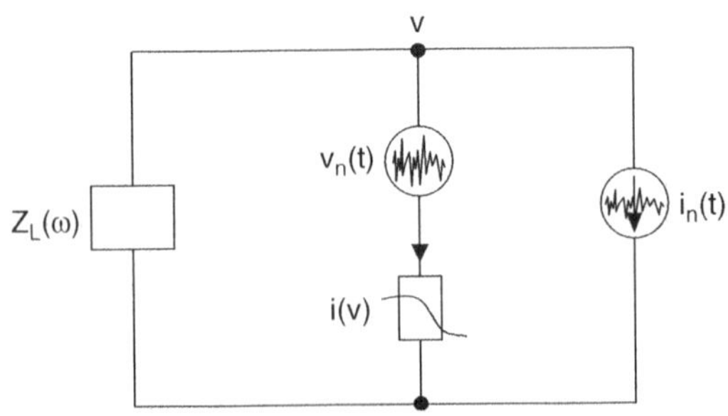

FIGURE 2.6 General representation of an oscillator circuit with a nonlinear current source $i(v)$ in parallel with the linear admittance $Z_L(\omega)$. Two different noise sources are considered: a parallel white noise current source and a flicker noise voltage source in series with the nonlinear element.

of the noise frequencies with respect to ω_o allow us to expand the error functions of (2.112) in a first-order Taylor series about V_{dc}^s, V_1^s, and ω_o. When doing so, the perturbed oscillator equations become

$$\begin{aligned}
&\frac{\partial H_{dc}}{\partial V_{dc}}\Delta V_{dc}(t)+\frac{\partial H_{dc}}{\partial V}\Delta V(t)+\left.\frac{\partial H_{dc}}{\partial s}\right|_{s=0}\frac{\Delta \dot{V}_{dc}(t)}{V_{dc}}=-V_N(t)\\
&\frac{\partial H_1^r}{\partial V_{dc}}\Delta V_{dc}(t)+\frac{\partial H_1^r}{\partial V}\Delta V(t)+\frac{\partial H_1^i}{\partial \omega}\frac{\Delta \dot{V}(t)}{V}\frac{\partial H_1^r}{\partial \omega}\Delta\omega(t)\\
&\qquad=-Z_L^r(\omega_o)I_N^r(t)+Z_L^i(\omega_o)I_N^i(t)\\
&\frac{\partial H_1^i}{\partial V_{dc}}\Delta V_{dc}(t)+\frac{\partial H_1^i}{\partial V}\Delta V(t)-\frac{\partial H_1^r}{\partial \omega}\frac{\Delta \dot{V}(t)}{V}+\frac{\partial H_1^i}{\partial \omega}\Delta\omega(t)\\
&\qquad=-Z_L^r(\omega_o)I_N^i(t)-Z_L^i(\omega_o)I_N^r(t)
\end{aligned} \tag{2.113}$$

Note that the baseband noise source $V_N(t)$ has been associated with the dc term. The current terms $I_N^r(t)$ and $I_N^i(t)$ are obtained from the lowpass equivalent of the white noise about the fundamental frequency ω_o. On the other hand, the perturbation of Z_L has been neglected in the terms affecting the noise source $i_N(t)$, as this would give rise to higher-order increments. The system (2.113) can be expressed in the matrix form

$$[JH_m]\begin{bmatrix}\Delta V_{dc}(t)\\ \Delta V(t)\\ \Delta\omega(t)\end{bmatrix}+\left[\frac{\partial H}{\partial \omega}\right]_o\begin{bmatrix}-j\dfrac{\Delta \dot{V}_{dc}(t)}{V_{dc}}\\ -j\dfrac{\Delta \dot{V}(t)}{V}\\ -j\dfrac{\Delta \dot{V}(t)}{V}\end{bmatrix}=-[Z_L]\overline{I}_N(t)-\overline{V}_N(t) \tag{2.114}$$

where $[\partial H/\partial\omega]_o$ is a diagonal matrix derived easily from the inspection of (2.113). The matrix $[JH_m]$ is a mixed-mode Jacobian, as it contains derivatives with respect to both the harmonic components of the node voltage and the oscillation frequency ω_o. This Jacobian matrix is not singular, since the system singularity was removed when using the additional condition $\phi = 0$. Remember that the node voltage has been assumed real, $V_1^s e^{j0}$. Neglecting the influence of the time derivative of the amplitude perturbation, system (2.114) can be rewritten in a more explicit manner:

$$\left[JH' \quad \left.\frac{\partial H}{\partial\omega}\right|_o\right]\begin{bmatrix}\Delta V_{dc}(t)\\ \Delta V(t)\\ \Delta\omega(t)\end{bmatrix}=-[Z_L]\overline{I}_N(t)-\overline{V}_N(t) \tag{2.115}$$

where the submatrix JH' of order 3×2 contains the derivatives of the three error functions H_{dc}, H_1^r, and H_1^i with respect to V_{dc} and the oscillation amplitude V. System (2.115) constitutes the formulation of the carrier modulation approach

[22], applied to the circuit in Fig. 2.6, considering only the fundamental frequency. The carrier modulation approach is, together with the conversion matrix approach [22–24], one of the most commonly used methods for the phase noise analysis of microwave oscillators. Usually, the methods are combined in the phase noise analysis of the same oscillator circuit. Details on the conversion matrix approach are given in Chapter 7.

The phase noise is calculated by solving the linear system (2.115) for the carrier modulation $\Delta\omega(t)$ and applying the same steps as in the preceding analysis, in terms of the admittance function. In this manner it is possible to determine the phase noise spectrum due to both white and flicker noise. The phase noise prediction will, of course, be limited to one harmonic only. It is easily demonstrated that as in the case of white noise only, the phase noise spectral density decreases with the oscillator quality factor.

2.5.3 Frequency-Domain Calculation of Variance of the Phase Deviation

In the time-domain analysis of Section 2.4 it was shown that the phase noise spectrum of the oscillator circuit can be determined accurately from the variance of the common phase deviation $\sigma_\theta^2(t)$. This variance is calculated from the phase sensitivity functions to the existing noise sources. As a reminder, the phase sensitivity with respect to multiple white noise sources is globally represented with the row matrix $b(t)$, whose elements provide the phase sensitivity to each of the white noise sources contained in the circuit. In the case of the colored noise sources, a different scalar function $b_{\gamma_i}(t)$ with $j = 1$ to J is used to represent phase sensitivity with respect to each colored noise source. In the following it is shown that these functions can also be determined from a frequency-domain analysis of the oscillator circuit. Once these functions are determined, the oscillator phase noise spectrum is calculated from the resulting variance of the phase deviation $\sigma_\theta^2(t)$.

For the frequency-domain derivation of the sensitivity functions, the same simplified circuit of Fig. 2.6, with a white noise source $I_N(t)$ and a flicker noise source $V_N(t)$, will be considered. As in the derivation of (2.112)–(2.113), the analysis will be limited to the dc and first-harmonic terms. However, an additional condition, different from $\Delta\phi = 0$, will be used to resolve the unbalanced perturbed oscillator equations. Because the first harmonic of the node voltage is considered complex in this case, it will be more convenient to formulate the system using both positive and negative spectra. The node voltage will be expressed as $v_s(t) = V_{dc}^s + V_1^s e^{j\omega_o t} + V_{-1}^s e^{-j\omega_o t}$, where V_1^s and V_{-1}^s are complex values fulfilling $V_1^s = V_{-1}^{s*}$. The application of Kirchhoff's laws to Fig. 2.6 provides the following steady-state system:

$$
\begin{aligned}
H_{dc}^s &\equiv V_{dc}^s + R_L(0) I_{dc}^s(V_{dc}^s, V_1^s, V_{-1}^s) = 0 \\
H_1^s &\equiv V_1^s + Z_L(\omega_o) I_1^s(V_{dc}^s, V_1^s, V_{-1}^s) = 0 \\
H_{-1}^s &\equiv V_{-1}^s + Z_L(-\omega_o) I_{-1}^s(V_{dc}^s, V_1^s, V_{-1}^s) = 0
\end{aligned}
\qquad (2.116)
$$

which can be rewritten in matrix form as

$$\overline{H} = \overline{V}_s + [Z_L(k\omega_o)]\overline{I}_s(\overline{V}_s) = 0 \tag{2.117}$$

with $\overline{V}_s = [V_{dc} \quad V_1 \quad V_{-1}]^T$ and $\overline{I}_s = [I_{dc} \quad I_1 \quad I_{-1}]^T$ and the linear matrix $[Z_L(k\omega_o)]$, with $k = 0, 1, -1$, defined as in expression (1.33).

In the presence of the noise perturbations, the voltage and current variables become

$$\begin{aligned} v(t+\theta) &= V_{dc}^s + V_1^s e^{j\omega_o(t+\theta)} + V_{-1}^s e^{-j\omega_o(t+\theta)} + \Delta V_{dc}(t+\theta) \\ &\quad + \Delta V_1(t+\theta)e^{j\omega_o(t+\theta)} + \Delta V_{-1}(t+\theta)e^{-j\omega_o(t+\theta)} \\ i(t+\theta) &= I_{dc}^s + I_1^s e^{j\omega_o(t+\theta)} + I_{-1}^s e^{-j\omega_o(t+\theta)} + \Delta I_{dc}(t+\theta) \\ &\quad + \Delta I_1(t+\theta)e^{j\omega_o(t+\theta)} + \Delta I_{-1}(t+\theta)e^{-j\omega_o(t+\theta)} \end{aligned} \tag{2.118}$$

The stochastic time deviation $\theta(t)$ with respect to the noise sources, neglected in (2.113), is now taken into account. Remember that unlike the case of (2.113), the condition $\Delta\phi = 0$ has not been imposed on (2.118), so the harmonic components of the perturbed voltage $\Delta v(t)$ contain both real and imaginary parts. The harmonic components of the perturbed voltage and current in (2.118) can be written, in vector form, as

$$\begin{aligned} \begin{bmatrix} V_{dc}(t) \\ V_1(t) \\ V_{-1}(t) \end{bmatrix} &= \begin{bmatrix} V_{dc}^s \\ V_1^s e^{j\omega_o\theta} \\ V_{-1}^s e^{-j\omega_o\theta} \end{bmatrix} + \begin{bmatrix} \Delta V_{dc}(t+\theta) \\ \Delta V_1(t+\theta)e^{j\omega_o\theta} \\ \Delta V_{-1}(t+\theta)e^{-j\omega_o\theta} \end{bmatrix} \\ \begin{bmatrix} I_{dc}(t) \\ I_1(t) \\ I_{-1}(t) \end{bmatrix} &= \begin{bmatrix} I_{dc}^s \\ I_1^s e^{j\omega_o\theta} \\ I_{-1}^s e^{-j\omega_o\theta} \end{bmatrix} + \begin{bmatrix} \Delta I_{dc}(t+\theta) \\ \Delta I_1(t+\theta)e^{j\omega_o\theta} \\ \Delta I_{-1}(t+\theta)e^{-j\omega_o\theta} \end{bmatrix} \end{aligned} \tag{2.119}$$

In the following, the stochastic time deviation $\theta(t)$ is assumed to be a baseband function. This assumption is not made in the analysis of Section 2.4, but it hardly limits the analysis accuracy. This is because the oscillator noise spectrum is analyzed up to certain frequency offset only, typically 100 MHz, which corresponds to slow variations of $\theta(t)$ with respect to the oscillator carrier. The perturbations of the harmonic components of the state variable $v(t)$ will have two contributions: one coming from $\Delta V_k(t+\theta)$, with $k = \text{dc}, 1, -1$, which in general is called *amplitude noise*, and the other coming from the common phase noise $e^{jk\omega_o(t+\theta)}$, associated to $\theta(t)$. Remember that the stochastic time deviation $\theta(t)$ is responsible for the modulation of the carrier frequency $\Delta\omega_o(t) = \omega_o\dot{\theta}(t)$. As in (2.113), the low value of the perturbation frequency Ω allows performing a Taylor series expansion of the linear matrix $[Z_L]$ about the steady-state frequencies $k\omega_o$. Then the perturbed oscillator equations, in matrix form, become

$$\begin{aligned}
&[e^{jk\omega_o\theta(t)}]\overline{V}_s + [e^{jk\omega_o\theta(t)}]\Delta\overline{V}(t+\theta) \\
&+ \left\{ \underbrace{\begin{bmatrix} Z_L(0) & 0 & 0 \\ 0 & Z_L(\omega_o) & 0 \\ 0 & 0 & Z_L(-\omega_o) \end{bmatrix}}_{[Z_L(k\omega_o)]} + \underbrace{\begin{bmatrix} \left.\dfrac{\partial Z_L(s)}{\partial s}\right|_{s=0} & 0 & 0 \\ 0 & \left.\dfrac{\partial Z_L(\omega_o+s)}{\partial s}\right|_{s=0} & 0 \\ 0 & 0 & \left.\dfrac{\partial Z_L(-\omega_o+s)}{\partial s}\right|_{s=0} \end{bmatrix}}_{[\partial Z_L/\partial(jk\omega_o)]} s \right\} \\
&\times \underbrace{\overline{I}([e^{jk\omega_o\theta(t)}](\overline{V}_s + \Delta\overline{V}(t+\theta)))}_{\overline{I}(t+\theta)} \\
&= -\begin{bmatrix} Z_L(0) & 0 & 0 \\ 0 & Z_L(\omega_o) & 0 \\ 0 & 0 & Z_L(-\omega_o) \end{bmatrix} \overline{I}_N(t) - \overline{V}_N(t) \qquad (2.120)
\end{aligned}$$

with s being a small frequency increment, acting as a derivation operator, to be applied to the time-varying quantities. Note that equation (2.120) still contains the steady-state terms, which will be suppressed at a later stage. The matrix $[e^{jk\omega_o\theta}]$ is organized as

$$[e^{jk\omega_o\theta}] = \begin{bmatrix} 1 & 0 & 0 \\ 0 & e^{j\omega_o\theta} & 0 \\ 0 & 0 & e^{-j\omega_o\theta} \end{bmatrix} \qquad (2.121)$$

Multiplication by s of the slowly varying phasors $\overline{I}(V(t))$ will be equivalent to a time derivation. To obtain the time derivatives, it is possible to apply the chain rule:

$$\begin{aligned}
\frac{d(\overline{I}(t+\theta))}{dt} &= \left[\frac{\partial \overline{I}}{\partial \overline{V}}\right]_s \frac{d([e^{jk\omega_o\theta}](\overline{V}_s + \Delta\overline{V}(t+\theta)))}{dt} \\
&= \left[\frac{\partial \overline{I}}{\partial \overline{V}}\right]_s [e^{jk\omega_o\theta}]([jk\omega_o]\dot{\theta}(t)\overline{V}_s + \Delta\dot{\overline{V}}(t+\theta)) \qquad (2.122)
\end{aligned}$$

The Jacobian matrix $[\partial\overline{I}/\partial\overline{V}]_s$ is calculated from the conversion matrix associated with $g(t) = \partial i(t)/\partial v$, as shown in Chapter 1. Suppressing the steady-state terms and neglecting increments of higher order yields

$$\underbrace{\left\{[I_d]+[Z_L(k\omega_o)]\left[\frac{\partial \overline{I}}{\partial \overline{V}}\right]_s\right\}}_{[JH]_s}\Delta\overline{V}(t+\theta)$$
$$+\underbrace{\left[\frac{\partial Z_L}{\partial(jk\omega_o)}\right]\left[\frac{\partial \overline{I}}{\partial \overline{V}}\right]_s}_{\partial[JH]_s/\partial(jk\omega_o)}([jk]\overline{V}_s\omega_o\dot{\theta}(t)+\Delta\dot{\overline{V}}(t+\theta))$$
$$=-\{[Z_L(k\omega_o)][e^{-jk\omega_o\theta}]\overline{I}_N(t)-[e^{-jk\omega_o\theta}]\overline{V}_N(t)\} \tag{2.123}$$

where some compact terms have been identified, taking into account the defined error vector $\overline{H}=[H_{dc}H_1H_{-1}]^T$ [see (2.116)]. For simplicity, a change in the time variable $t\to t+\theta$ has been considered in the exponential terms of the Fourier basis. On the other hand, due to the double time dependence of $t+\theta(t)$, it is possible to simply write $\Delta\overline{V}(t)$, $\Delta\dot{\overline{V}}(t)$. Thus, (2.123) can be written in compact form as

$$[JH]_s\Delta\overline{V}(t)+\left[\frac{\partial JH}{\partial(jk\omega_o)}\right][jk]\overline{V}_s\omega_o\dot{\theta}(t)+\left[\frac{\partial JH}{\partial(jk\omega_o)}\right]\Delta\dot{\overline{V}}(t)$$
$$=-[Z_L(k\omega_o)][e^{-jk\omega_o\theta}]\overline{I}_N(t)-[e^{-jk\omega_o\theta}]\overline{V}_N(t) \tag{2.124}$$

As in the case of time-domain analysis (Section 2.4), the system (2.124) is unbalanced. It contains three equations in four unknowns, given by $\Delta V_{dc}(t)$, $\mathrm{Re}[\Delta V_1(t)]$, $\mathrm{Im}[\Delta V_1(t)]$, and $\theta(t)$. Remember that unlike the calculation of (2.114), the phase of $\Delta V_1(t)$ has not been set to zero. Due to the irrelevance with respect to the phase origin, the matrix $[JH]_s=\left[\partial\overline{H}/\partial\overline{V}\right]_s$ is singular, as was shown in Chapter 1. Then the two sides of (2.124) can be multiplied by a row vector $\overline{V}_{\text{ker}}^T$, belonging to the kernel of $[JH]_s$, such that $\overline{V}_{\text{ker}}^T[JH]=\overline{0}$. All the vectors $\alpha\overline{V}_{\text{ker}}^T$, with α a scalar constant, equally fulfill $\alpha\overline{V}_{\text{ker}}^T[JH]=\overline{0}$. Then it is possible to choose a particular vector $\overline{V}_X^T$ such that it provides $\overline{V}_X^T[\partial JH/\partial(jk\omega_o)]\overline{U}_1=1$, with $\overline{U}_1=[jk]\omega_o\overline{V}_s$. Note that $\overline{U}_1$ contains the harmonic components of $\dot{v}_s(t)$ tangent to the limit cycle. As already known, the oscillator solution is invariant versus any shift in the phase origin. Thus, it is possible to write $\frac{\partial\overline{H}}{\partial\phi}=[JH]\cdot\frac{\partial\overline{V}_s}{\partial\phi}=0$, so the vector $\overline{U}_1=\omega_o\frac{\partial\overline{V}_s}{\partial\phi}=[jk]\omega_o\overline{V}_s$, in the direction of invariance, also fulfills the interesting property $[JH]_s\overline{U}_1=0$.

So far, no additional constraint has been imposed, so there is still one degree of freedom in (2.124). In particular, it is possible to choose $\Delta\overline{V}$ so that it fulfills $\overline{V}_X^T[\partial JH/\partial(jk\omega_o)]\Delta\overline{V}(t)=0$ [25]. This will be the additional condition used for the resolution of (2.124) instead of the condition $\Delta\phi(t)=0$ used in (2.114). The product $\overline{V}_X^T[\partial JH/\partial(jk\omega_o)]$ provides a constant row matrix, renamed here $\overline{A}^T=\overline{V}_X^T[\partial JH/\partial(jk\omega_o)]$, so the equality $A^T\Delta\dot{\overline{V}}(t)=d(A^T\Delta\overline{V}(t))/dt=0$

is satisfied. Thus, the increment vector $\Delta\overline{V}(t)$ must be orthogonal to $\overline{A}$ at all times t. Taking all this into account, the multiplication of both sides of (2.124) by $\overline{V}_x^{\mathrm{T}}$ provides

$$\Delta\omega(t) = \omega_o\dot{\theta}(t) = -\underbrace{\overline{V}_X^{\mathrm{T}}[Z_L(k\omega_o)][e^{-jk\omega_o\theta}]}_{\text{row matrix}=[B_W]}\overline{I}_N(t) - \underbrace{\overline{V}_X^{\mathrm{T}}[e^{-jk\omega_o\theta}]}_{\text{row matrix}=[B_Y]}\overline{V}_N(t) \tag{2.125}$$

Equation (2.125) should be compared with (2.69) and (2.91). In the case of a white noise current source, the time derivative $\dot{\theta}(t)$ is given by

$$\dot{\theta}(t) = \overline{v}_1^{\mathrm{T}}(t+\theta)\frac{\partial\overline{f}}{\partial i_N}[\overline{x}_{sp}(t+\theta)]i_N(t) = b(t+\theta)i_N(t) \tag{2.126}$$

To extract the sensitivity function $b(t)$ from system (2.125), we should take into account that any product $c(t) = d(t)a(t)$ of two time functions can be written in terms of the harmonic components of $d(t)$ and the Toeplitz matrix associated with $a(t)$. Limiting the analysis to the first harmonic term, the matrix–vector product is approached:

$$\begin{bmatrix} C_{dc} \\ C_1 \\ C_{-1} \end{bmatrix} = \begin{bmatrix} A_{dc} & A_{-1} & A_1 \\ A_1 & A_{dc} & A_2 \\ A_{-1} & A_{-2} & A_{dc} \end{bmatrix} \begin{bmatrix} D_{dc} \\ D_1 \\ D_{-1} \end{bmatrix} \tag{2.127}$$

The expression (2.127) is derived easily just by multiplying the harmonic expansions of $a(t)$ and $d(t)$ and assembling the harmonic components of the same order. The harmonic expression above can be applied to the time-domain product

$$\overline{v}_1^{\mathrm{T}}(t+\theta)\frac{\partial\overline{f}}{\partial i_N}[\overline{x}_{sp}(t+\theta)]$$

However, our frequency-domain analysis assumes slow time variations of the stochastic time deviation $\theta(t)$, limited by the maximum value of the noise frequency $\Omega = 2\pi f$. Thus, we have only the baseband component of $\dot{\theta}(t)$. Comparing (2.125) with (2.127), the harmonic components of the phase sensitivity to the white noise source $i_N(t)$ are given by

$$[b_{dc} \quad b_{-1} \quad b_1] = -\frac{\overline{V}_X^{\mathrm{T}}}{\omega_o}[Z_L(k\omega_o)] = \frac{[B_W]}{\omega_o} \tag{2.128}$$

Then the phase sensitivity function $b(t)$, limited to the first-harmonic term, is calculated as

$$b(t) = b_{dc} + b_1 e^{j\omega_o t} + b_{-1}e^{-j\omega_o t} \tag{2.129}$$

Compared to (2.125), there is some accuracy degradation due to the Taylor series expansion of $[Z_L(k\omega_o)]$ in (2.120). As will be shown in Chapter 7, the accuracy can be increased notably using a harmonic balance formulation of nodal type.

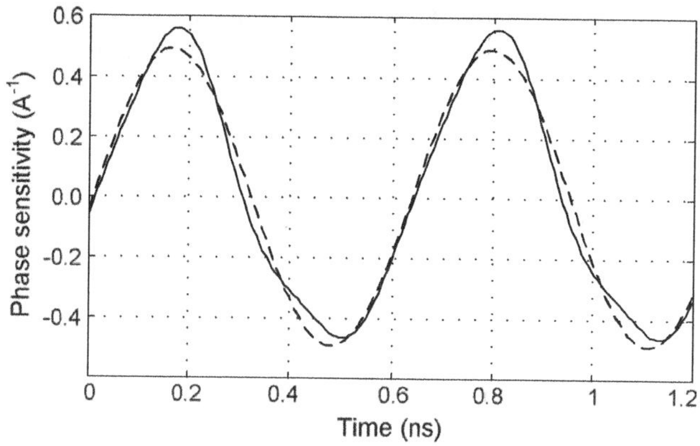

FIGURE 2.7 Phase sensitivity to the white noise current source of a parallel resonance oscillator. Comparison of the results obtained with the time-domain analysis of Section 2.4 (the solid line) and with the one-harmonic frequency-domain analysis of this section (the dashed line).

As an example, the analysis above has been applied to the parallel resonance circuit with cubic nonlinearity with respect to the parallel current noise source. The resulting harmonic terms are the following: $b_{dc} = 0$ and $\omega_o b_1 = -0.0424 + j0.4915$. Figure 2.7 presents the comparison between the phase sensitivity function $\omega_o b(t)$ obtained with (2.129) and with the time-domain calculation of Section 2.4. There is a slight disagreement attributed to the fact that only one harmonic component has been considered in the frequency-domain calculation.

As shown in (2.88), the oscillator phase noise spectrum due to white noise sources is a Lorentzian line with a 3-dB corner frequency, determined by the parameter $c = (1/T)\int_0^T \overline{b}^T(t)[\Gamma]\overline{b}(t)dt$. By taking into account the periodicity of $b(t)$ and the orthogonality of the Fourier transform, this parameter can be calculated directly in the frequency domain, doing

$$c = \frac{1}{\omega_o^2}[B_W][\Gamma_k][B_W]^+ = \frac{1}{\omega_o^2}\overline{V}_X^T[Z_L(k\omega_o)][\Gamma_k][Z_L(k\omega_o)]^+\overline{V}_X^* \tag{2.130}$$

where $[\Gamma_k]$ is the correlation matrix between the different harmonic terms 0, 1 and -1 of the white noise current source. As shown in Section 2.5.1, these harmonic components are uncorrelated, so for a single white noise source, this matrix is diagonal.

A similar analysis is performed to obtain the phase noise sensitivity to the flicker noise voltage source connected in series with the nonlinear element. The time-domain equation that relates the time derivative of the stochastic time deviation to the colored noise source $\gamma(t)$ is recalled here:

$$\dot{\theta} = \overline{v}_1^T(t+\theta)\frac{\partial \overline{f}}{\partial \gamma}[\overline{x}_{sp}(t+\theta)]\gamma(t) = b_\gamma(t+\theta)\gamma(t) \tag{2.131}$$

From an inspection of (2.125), the harmonic components of the phase sensitivity $b_\gamma(t)$ to the flicker noise source $v_N(t)$ in Fig. 2.6 must agree, in this particular case with the components of $[B_\gamma] = \overline{V}_{1x}^T/\omega_o$; that is,

$$[b_{\gamma,\mathrm{dc}} \quad b_{\gamma,-1} \quad b_{\gamma,1}] = \frac{-\overline{V}_X^T}{\omega_o} \tag{2.132}$$

When applying the calculation above to the parallel resonance oscillator, the resulting harmonic components of $b_\gamma(t)$ are $b_{\gamma,\mathrm{dc}} = -0.0029$ and $b_{\gamma,1} = -0.0011 + j0.0048$. As in the case of the white noise source, this provides just a one-harmonic approximation of the phase noise sensitivity function.

The phase noise spectrum in the presence of the colored noise source $\gamma(t)$ depends on the factor $b_{\gamma,\mathrm{dc}}$, which corresponds to the dc component of the periodic function $b_\gamma(t)$. For a direct calculation of $b_{\gamma,\mathrm{dc}}$, the following expression can be used:

$$b_{\gamma,\mathrm{dc}} = \frac{-1}{\omega_o}\overline{V}_X^T\overline{\Lambda}_\gamma \tag{2.133}$$

with $\overline{\Lambda}_\gamma$ being the vector $\overline{\Lambda}_\gamma = [1 \quad 0 \quad 0]^T$. Once the parameters $b_{\gamma,\mathrm{dc}}$ and c have been obtained, the phase noise spectrum can be determined numerically from the variance:

$$\sigma_\theta^2(t) = 2|b_{\gamma,\mathrm{dc}}|^2 \int_{-\infty}^{\infty} S_\gamma(f)\frac{1 - e^{j2\pi ft}}{4\pi^2 f^2}df + ct \tag{2.134}$$

which should be introduced in (2.96). The calculation above will require an accurate estimation of the cutoff frequency $f_{\min}$ of the flicker noise (2.54). Note that it is also possible to apply the approximate expressions (2.130).

2.5.4 Comparison of Two Techniques for Frequency-Domain Analysis of Phase Noise

Two different methods for the frequency-domain analysis of phase noise have been presented. The method of Section 2.5.2 provides the phase noise spectrum from the analysis of the carrier modulation $\Delta\omega(t)$. The carrier modulation approach in (2.115), which neglects $\Delta\dot{V}(t)$, provides at a sufficiently large distance from the carrier exactly the same phase noise variation as the time-domain calculation in (2.98). Near the carrier the accuracy degrades, due to the suppression in (2.114) of the stochastic time increment θ. The method of Section 2.5.3 enables a calculation of the variance of the phase deviation $\sigma_\theta^2(t)$ from a frequency-domain analysis of the circuit. As shown in Section 2.4, this variance depends on the periodic phase sensitivity functions relating $\Delta\omega(t) = \omega_o\dot{\theta}(t)$ to the noise sources. The time-domain analysis of Section 2.4 and the carrier modulation approach (2.114) exclusively provide the phase noise associated with the "common" phase perturbations $\omega_o\theta(t)$. This noise is common to all the oscillator variables and is simply multiplied by the harmonic order of the various harmonic terms $k\omega_o\theta(t)$.

In the following it will be shown that the variance $\sigma_\theta^2(t)$ can also be determined from the carrier modulation approach. Once this variance is known, it will be possible to apply the expressions derived in Section 2.4 for an accurate determination of the oscillator spectrum due to phase noise. Suppressing the time derivative $\Delta\dot{\overline{V}}(t) = 0$ in (2.124), the perturbed oscillator equation can be written, in a general manner, as

$$[JH]_s \Delta\overline{V}(t) + \left.\frac{\partial \overline{H}}{\partial \omega}\right|_o \Delta\omega(t) = [e^{-jk\omega_o\theta}]\overline{G}_N(t) \tag{2.135}$$

where the vector $\overline{G}_N(t) = -[Z_L]\overline{I}_N(t) - \overline{V}_N(t)$ accounts for the contribution of the noise sources. The following equality has also been taken into account:

$$\left.\frac{\partial \overline{H}}{\partial \omega}\right|_o = \frac{\partial Z_L}{\partial(jk\omega_o)}[jk]\overline{I}_s = \underbrace{\left[\frac{\partial Z_L}{\partial(jk\omega_o)}\right]\left[\frac{\partial \overline{I}}{\partial \overline{V}}\right]_s}_{\partial[JH]_s/\partial(jk\omega_o)}[jk]\overline{V}_s \tag{2.136}$$

where use has been made of the chain rule to relate the harmonic components $[jk]\omega_o\overline{I}_s$ of the time derivative of the steady-state current $i_s(t)$ to the harmonic components $[jk]\omega_o V_s$ of the time derivative of the steady-state voltage $v_s(t)$. The matrix $[JH]_s$ is singular at this stage, as no additional condition has been imposed. In agreement with the results of Section 2.4.1, suppressing the phase shift $[e^{-jk\omega_o\theta}]$ in (2.135) will not affect the periodic sensitivity functions, relating the carrier modulation $\Delta\omega(t)$ to the input noise sources. Due to the singularity of the Jacobian matrix $[JH]_s$, the increments $\Delta\overline{V}(t)$ are linearly related. Taking this into account, any possible additional condition can be expressed as $\overline{P}^T \Delta\overline{V}(t) = 0$, with $\overline{P}$ an arbitrary constant vector. Setting $\Delta\dot{\overline{V}}(t) = 0$ in (2.114) and combining the resulting equation with $\overline{P}^T \Delta\overline{V}(t) = 0$, the following matrix system is obtained:

$$\begin{pmatrix} [JH]_s \left.\frac{\partial \overline{H}}{\partial \omega}\right|_o & \left.\frac{\partial \overline{H}}{\partial \omega}\right|_o \\ \overline{P}^T & 0 \end{pmatrix} \begin{pmatrix} \Delta\overline{V}(t) \\ \Delta\omega(t) \end{pmatrix} = \begin{pmatrix} \overline{G}_N(t) \\ 0 \end{pmatrix} \tag{2.137}$$

To solve for the frequency perturbation $\Delta\omega(t)$, use is made of any vector $\overline{V}_{\text{ker}}^T$ belonging to the kernel of the singular matrix $[JH]_s$. The carrier modulation $\Delta\omega(t)$ is obtained by multiplying both sides of (2.137) by the row matrix $[\overline{V}_{\text{ker}}^T/(\overline{V}_{\text{ker}}^T(\partial\overline{H}/\partial\omega|_o)) \quad 0]$.

Clearly, the result is independent of any possible choice of the vector $\overline{P}$ when imposing the condition $\overline{P}^T \Delta V(t) = 0$. In particular, the condition $\phi = 0$ leads to the carrier modulation approach of (2.115). As a result, equations (2.115) and (2.125) will provide the same sensitivity functions, relating $\Delta\omega(t)$ to the noise sources. Because of this equivalence, we can apply the same identification techniques of (2.128)–(2.130) and (2.132)–(2.133) to a circuit analyzed using the carrier modulation approach.

As an example, considering only a white noise source about the carrier, it will be possible to obtain the phase sensitivity function from an admittance analysis of the circuit. Solving $\Delta\omega(t)$ from (2.104), with $\Delta\dot{V}(t) = 0$, yields

$$\Delta\omega(t) = \frac{\dfrac{\partial Y_{To}^r}{\partial V} I_n^i(t) - \dfrac{\partial Y_{To}^i}{\partial V} I_n^r(t)}{V_o S}$$

$$= \frac{-j\dfrac{1}{2}\dfrac{\partial Y_{To}^r}{\partial V} - \dfrac{1}{2}\dfrac{\partial Y_{To}^i}{\partial V}}{V_o S} I_n^1(t) + \frac{j\dfrac{1}{2}\dfrac{\partial Y_{To}^r}{\partial V} - \dfrac{1}{2}\dfrac{\partial Y_{To}^i}{\partial V}}{V_o S} I_n^{-1}(t) \quad (2.138)$$

with

$$S = \frac{\partial Y_{To}^r}{\partial V}\frac{\partial Y_{To}^i}{\partial \omega} - \frac{\partial Y_{To}^i}{\partial V}\frac{\partial Y_{To}^r}{\partial \omega} \quad (2.139)$$

where r and i stand for the real and imaginary parts and the general relationships $I_n^r(t) = (I_n^1 + I_n^{-1})/2$ and $I_n^i = -j(I_n^1 - I_n^{-1})/2$ have been taken into account. Therefore, through comparison with (2.129), the sinusoidal phase sensitivity function is given by

$$b(t) = \text{Re}\left[\frac{-j\frac{\partial Y_{To}^r}{\partial V} - \left(\frac{\partial Y_{To}^i}{\partial V}\right)}{V_o S} e^{j\omega_o t}\right] \quad (2.140)$$

There is a $T/4$ phase shift of $b(t)$ with respect to the sinusoidal waveform $v(t)$ of period T. Thus, the phase sensitivity is minimum at the maxima and minima of $v(t)$ and maximum at points with the highest $\dot{v}(t)$. To obtain the oscillator spectrum, the coefficient c, providing the cutoff frequency of the Lorentizan line, is calculated from $c = (1/T)\int_0^T \bar{b}^T(t)[\Gamma]\bar{b}(t)dt$. Note that this function will only allow determination of the phase noise spectrum due to white noise about the carrier, with one-harmonic accuracy. It is not possible to predict the effect of the flicker noise, as no dc component has been considered in the circuit equations.

2.5.5 Amplitude Noise

The objective of this section is to analyze the amplitude noise associated with the amplitude perturbation $\Delta\overline{V}(t)$. This amplitude noise in the absence of flicker noise had been studied in Section 2.5.2. The starting points were the equations (2.104), obtained from the additional condition $\phi = 0$. In system (2.104), the amplitude and phase perturbations are coupled and the resulting amplitude spectrum is given in (2.111). Here the amplitude noise is analyzed in an "isolated manner." It will be obtained by decoupling the phase and amplitude perturbations in the perturbed oscillator equation (2.124). The required additional condition for this decoupled analysis is $\overline{V}_X^T[\partial JH/\partial(jk\omega_o)]\Delta\dot{\overline{V}}(t) = 0$. For any other additional condition the amplitude noise $\Delta\overline{V}(t)$ will affect the phase noise $\Delta\phi(t)$. An example is the calculation of (2.113). We have seen that the variance of the phase deviation $\sigma_\theta^2(t)$ can be determined using the carrier modulation approach. Thus, the major interest of the more complex formulation of the perturbed oscillator circuit presented in Section 2.5.3 is this decoupled analysis of phase and amplitude noise.

To obtain a system in the amplitude perturbation $\Delta\overline{V}(t)$, both sides of equation (2.123) will be multiplied by the projector matrix $[P] = [I_d] - [\partial JH/\partial(jk\omega_o)]\overline{U}_1\overline{V}_X^T$, with $\overline{U}_1 = [jk]\omega_o\overline{V}_s$. Taking into account the normalization condition $\overline{V}_X^T[\partial JH/\partial(jk\omega_o)]\overline{U}_1 = 1$, the following differential equation in the uncoupled vector $\Delta\overline{V}(t)$ is obtained:

$$[JH]_s\Delta\overline{V}(t) + \left[\frac{\partial JH}{\partial(jk\omega_o)}\right]\Delta\dot{\overline{V}}(t) = [P][Z_L(k\omega_o)][e^{-jk\omega_o\theta}]\overline{I}_N(t) + [P][e^{-jk\omega_o\theta}]\overline{V}_N(t) \tag{2.141}$$

where the additional constraint used for the system decoupling has also been taken into account. System (2.141) will be solved neglecting the influence of the stochastic time shift θ. As already stated, this is acceptable at a relatively high-frequency offset from the carrier, where the amplitude noise is relevant. Applying the Fourier transform, the following system in $\Delta\overline{V}(\Omega)$ is obtained:

$$\left\{[JH]_s + \left[\frac{\partial JH}{\partial(jk\omega_o)}\right]j\Omega\right\}\Delta\overline{V}(\Omega) = [P][Z_L(k\omega_o)]\overline{I}_N(\Omega) + [P]\overline{V}_N(\Omega) \tag{2.142}$$

For a more compact expression, it is possible to use the definition

$$[JT(\Omega)] = \left\{[JH]_s + \left[\frac{\partial JH}{\partial(jk\omega_o)}\right]j\Omega\right\} \tag{2.143}$$

Replacing $j\Omega$ with the Laplace frequency s, it is easily seen that the matrix $[JT(s)]$ is, in fact, a first-order Taylor series expansion of the characteristic matrix $[JH(jk\omega_o + s)]$ derived in Chapter 1. One of the roots of the associated characteristic determinant is $s = 0$, due to the singularity of the Jacobian matrix $[JH]_s$ at the steady-state oscillation. Solving for the amplitude increment $\Delta\overline{V}(\Omega)$ requires multiplying both terms of (2.142) by $[JT(\Omega)]^{-1}$, which provides

$$\Delta\overline{V}(\Omega) = [JT(\Omega)]^{-1}[P][Z_L(k\omega_o)]\overline{I}_N(\Omega) + [JT(\Omega)]^{-1}[P]\overline{V}_N(\Omega) \tag{2.144}$$

Because of the singularity of $[JT(\Omega)]$ at $\Omega = 0$, equation (2.144) becomes ill conditioned near a small offset frequency from the carrier. However, as shown by Sancho et al. [17] the product $[JT(\Omega)]^{-1}[P]$ can be calculated in a way that inherently eliminates the pole at $\Omega = 0$. It is possible to express the matrix product $[JT(\Omega)]^{-1}[P]$ in terms of the eigenvalues of a constant matrix, defined as

$$[M] = \left[\frac{\partial JH}{\partial(jk\omega_o)}\right]^{-1}[JH]_s \tag{2.145}$$

In the simplified problem analyzed here, the dimension of the matrix $[M]$ is 3×3, so this matrix has three eigenvalues only. Note that $[JT(\Omega)]$ agrees with $[JH]_s$ at $\Omega = 0$, so $[M]$ must have a zero eigenvalue, denoted here as $\lambda_1 = 0$. After some algebraic manipulation, the inverse $[JT(\Omega)]^{-1}$ is given by

$$[JT(\Omega)]^{-1}[P] = \sum_{i=1}^{3} \frac{[M_i]}{j\Omega - \lambda_i} \tag{2.146}$$

Each matrix $[M_i]$ is calculated from the left and right kernels of the matrixes below, each obtained when replacing $j\Omega$ with the eigenvalue λ_j in (2.143). That is,

$$M_i = \overline{U}_i \overline{V}_i^T \tag{2.147}$$

with

$$\begin{aligned} &\left\{[JH]_s + \left[\frac{\partial JH}{\partial(jk\omega_o)}\right]\lambda_i\right\}\overline{U}_i = \overline{0} \\ &\overline{V}_i^T \left\{[JH]_s + \left[\frac{\partial JH}{\partial(jk\omega_o)}\right]\lambda_i\right\} = \overline{0} \\ &\overline{V}_i^T [\partial JH/\partial(jk\omega_o)]\overline{U}_j = \delta_{ij} \quad \text{Normalization condition} \end{aligned} \tag{2.148}$$

The eigenvector $\overline{U}_1$ associated with $\lambda_1 = 0$ agrees with $\overline{U}_1 = [jk]\omega_o \overline{V}_s$, due to the property $[JH]_s \overline{U}_1 = 0$ discussed, coming from the invariance of the oscillator solution versus time translations. On the other hand, the left eigenvector $\overline{V}_1^T$ associated with $\lambda_1 = 0$ agrees with the defined kernel $\overline{V}_X^T$ of $[JH]_s$. The imposed normalization condition $\overline{V}_X^T[\partial JH/\partial(jk\omega_o)]\overline{U}_1 = 1$ must also be taken into account [17]. Replacing the expression for the projector $[P]$ into (2.146) yields

$$\begin{aligned} [JT(\Omega)]^{-1}[P] &= \sum_{i=1}^{3} \frac{[M_i][P]}{j\Omega - \lambda_i} \\ &= \frac{\overline{U}_1 \overline{V}_X^T\{[I_d] - [\partial JH/\partial(jk\omega_o)]\overline{U}_1 \overline{V}_X^T\}}{j\Omega} + \sum_{i=2}^{3} \frac{[M_i][P]}{j\Omega - \lambda_i} \\ &= \sum_{i=2}^{3} \frac{[M_i][P]}{j\Omega - \lambda_i} \end{aligned} \tag{2.149}$$

Thus, the ill conditioning of $[JT(\Omega)]^{-1}$ near $\Omega = 0$ is avoided by obtaining $[JT(\Omega)]^{-1}[P]$ from the left and right kernels of the matrixes $[M_i]$ instead of performing the matrix inversion. As has been shown, the product $[JT(\Omega)]^{-1}[P]$ eliminates the pole $\Omega = 0$. However, the other two system poles are still present in (2.144). Note that the compact frequency-domain formulation used here, with only one state variable, severely limits the pole observability. The number of detectable poles increases with the number of independent voltages and currents considered in the frequency-domain formulation. As shown by Sancho et al. [17], the nodal harmonic balance, using all the node voltages and inductance currents as independent variables is best suited to analyze the frequency dependence of $[JT(\Omega)]$.

To determine the amplitude noise spectrum, $\Delta V(\Omega)$ in (2.144) should be multiplied by its adjoint $\Delta\overline{V}^{+}(\Omega)$. In the general case of multiple state variables, the dominant contribution to the amplitude spectrum will come from the poles with smaller absolute value of their real part. For a dominant real pole γ_i, and assuming a relatively large offset frequency such that the white noise is the dominant contribution, the amplitude spectrum will be flat up to the frequency of this dominant pole $\Omega_i = \gamma_i$. From this frequency, the amplitude spectrum will drop as of -20 dB/dec versus the offset frequency Ω. This is the case for the amplitude spectrum in Fig. 2.5. In the case of dominant complex-conjugate poles $\lambda_{i,i+1} = -\xi_i\omega_i \pm j\omega_i\sqrt{1-\xi_i^2} = \sigma_i \pm j\omega_i$, these poles can be combined to give rise to a single denominator term, $s^2 + 2\xi_i\omega_i s + \omega_i^2$, the damping factor fulfilling $\sigma_i = -\xi_i\omega_i$. Isolating the contribution of this pair of poles to the amplitude spectrum at the oscillator output node

$$|\Delta V_{\text{out}}(\Omega)|^2_{i,i+1} = A - 20\log\sqrt{\left(1-\frac{\Omega^2}{\omega_i^2}\right)^2 + \left(2\xi_i\frac{\Omega}{\omega_i}\right)^2} \tag{2.150}$$

with A as a constant coefficient. For small offset frequency, the contribution will be flat. For a large offset frequency, it will decay as -40 dB/dec. Provided that $0 < \xi_i < 1/\sqrt{2}$, there will be a resonance peak at $\Omega^2/\omega_i^2 = 1 - 2\xi_i^2$, near the pole frequency, with maximum amplitude $A - 20\log(2\xi_i\sqrt{1-\xi_i^2})$. Clearly, the damping factor ξ_i will be smaller for dominant complex-conjugate poles with smaller $|\sigma|$. This will also give rise to a higher resonance peak. Thus, the amplitude spectrum can exhibit resonance peaks or "noise bumps," due to the presence of stable complex-conjugate poles at a relatively small distance from the imaginary axis.

The noise bumps in the output spectrum can also be explained from the point of view of the oscillator dynamics. As shown in Chapter 1, the instantaneous perturbation of a stable periodic solution is extinguished according to $c_i\overline{u}_i(t)e^{(\sigma_i+j\omega_i)t} + c_i^*\overline{u}_i^*(t)e^{(\sigma_i-j\omega_i)t}$, where only the terms corresponding to the dominant poles $\sigma_i \pm j\omega_i$ have been retained, c_i is a constant-coefficient number, and $\overline{u}_i(t)$ is a complex periodic vector. Obviously, the smaller the absolute value of the negative σ_i, the slower the restoring transient that will lead the oscillator back to the steady state. The envelope of the transient waveform will have the frequency of the dominant poles ω_i. In real life the oscillator is being perturbed continuously by the noise sources. Thus, the oscillator is never able to fully return to the steady-state solution. For small $|\sigma_i|$ and due to the continuous noise perturbation, it will be possible to notice in the spectrum the modulation frequency ω_i. The smaller $|\sigma_i|$, the more noticeable will be the modulation frequency.

The noise bumps discussed are often observed in measurement. Figure 2.8 shows an example of this phenomenon in a FET-based oscillator at 5 GHz. The resonances

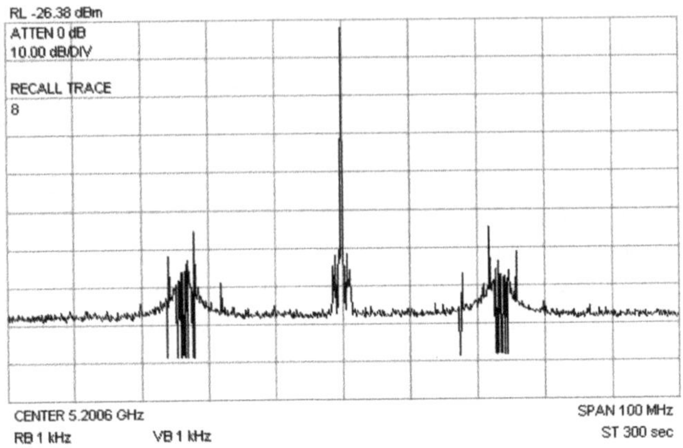

FIGURE 2.8 Resonances in the spectrum of a FET-based oscillator at 5.2 GHz. The resonances at about 250 MHz from the carrier frequency are due to a pair of complex-conjugate poles with negative σ of small absolute value.

are observed at about 250 MHz from the carrier frequency. If when varying a parameter η (e.g., a bias voltage), the complex-conjugate poles approach the imaginary axis, the height of the bumps increases due to the reduction of the absolute value of σ_i. If the complex-conjugate poles cross the imaginary axis, the noise bumps turn into distinct spectral lines, due to the onset of oscillation at ω_i. This is why the noise bumps are also called *noisy precursors* [26,27].

REFERENCES

[1] A. Demir, A. Mehrotra, and J. Roychowdhury, Phase noise in oscillators: a unifying theory and numerical methods for characterization, *IEEE Trans. Circuits Syst. Fundam. Theory Appl.*, vol. 47, May 2000.

[2] B. Razavi, Study of phase noise in CMOS oscillators, *IEEE J. Solid State Circuits*, vol. 31, pp. 331–343, 1996.

[3] U. L. Rohde, *Microwave and Wireless Synthesizers: Theory and Design*, Wiley-Interscience, New York, 1997.

[4] F. X. Kaertner, Analysis of white and $f^{-\text{alpha}}$ noise in oscillators, *Int. J. Circuit Theory Appl.*, vol. 18, pp. 485–519, 1990.

[5] A. Hajimiri and T. H. Lee, A general theory of phase noise in electrical oscillators, *IEEE J. Solid State Circuits*, vol. 33, Feb. 1998.

[6] K. Kurokawa, Some basic characteristics of broadband negative resistance oscillators, *Bell Syst. Tech. J.*, vol. 48, pp. 1937–1955, July–Aug. 1969.

[7] F. X. Kaertner, Determination of the correlation spectrum of oscillators with low noise, *IEEE Trans. Microwave Theory Tech.*, vol. 37, pp. 90–101, Jan. 1989.

[8] A. Demir, Phase noise in oscillators: DAEs and colored noise sources, *IEEE/ACM International Conference on Computer-Aided Design*, San Jose, CA, pp. 170–177, 1998.

[9] C. W. Gardiner, *Handbook of Stochastical Methods*, Springer-Verlag, New York, 1997.

[10] A. B. Carlson, *Communication Systems*, McGraw-Hill, New York, 1986.

[11] W. Paul and J. Baschnagel, *Stochastic Processes*, Springer-Verlag, New York, 1999.

[12] A. Papoulis, *Probability, Random Variables, and Stochastic Processes*, McGraw-Hill, New York, 1991.

[13] M. Rudolph, F. Lenk, O. Llopis, and W. Heinrich, On the simulation of low-frequency noise upconversion in InGaP/GaAs HBTs, *IEEE Trans. Microwave Theory Tech.*, vol. 54, pp. 2954–2961, 2006.

[14] T. Djurhuus, V. Krozer, J. Vidkjaer, and T. K. Johansen, AM to PM noise conversion in a cross-coupled quadrature harmonic oscillator, *Int. J. RF Microwave Computer-Aided Eng.*, vol. 16, pp. 34–41, 2006.

[15] H. Schmid, Aaargh! I just loooove flicker noise [Open Column], *IEEE Circuits Syst. Mag.*, vol. 7, pp. 32–35, 2007.

[16] M. S. Keshner, $1/f$ noise, *Proc. IEEE*, vol. 70, pp. 212–218, 1982.

[17] S. Sancho, A. Suárez, and F. Ramirez, Phase and amplitude noise analysis in microwave oscillators using nodal harmonic balance, *IEEE Trans. Microwave Theory Tech.*, vol. 55, pp. 1568–1583, 2007.

[18] S. Sancho, F. Ramirez, and A. Suárez, Analysis and reduction of the oscillator phase noise from the variance of the phase deviations, determined with harmonic balance, *IEEE MTT-S International Microwave Symposium Digest*, Atlanta GA, 2008.

[19] J. A. Mullen, Limiting forms of the FM spectra, *Proc. I.R.E.*, vol. 45, pp. 874–877, June 1957.

[20] K. Kurokawa, Injection locking of microwave solid state oscillators, *Proc. IEEE*, vol. 61, pp. 1386–1410, Oct. 1973.

[21] M. Odyniec (Ed.), *RF and Microwave Oscillator Design*, Artech House, 2002.

[22] V. Rizzoli, F. Mastri, and D. Masotti, General noise analysis of nonlinear microwave circuits by the piecewise harmonic-balance technique, *IEEE Trans. Microwave Theory Tech.*, vol. 42, pp. 807–819, May 1994.

[23] P. Bolcato, J. C. Nallatamby, R. Larcheveque, M. Prigent, and J. Obregón, A unified approach of PM noise calculation in large RF multitone autonomous circuits, *IEEE MTT-S International Microwave Symposium*, Boston, MA, pp. 417–420, 2000.

[24] M. Prigent and J. Obregón, Phase noise reduction in FET oscillators by low-frequency loading and feedback circuit optimization, *IEEE Trans. Microwave Theory Tech.*, vol. 35, pp. 349–352, Mar. 1987.

[25] A. Suárez, S. Sancho, S. Ver Hoeye, and J. Portilla, Analytical comparison between time- and frequency-domain techniques for phase-noise analysis, *IEEE Trans. Microwave Theory Tech.*, vol. 50, pp. 2353–2361, 2002.

[26] K. Taihyun and E. H. Abed, Closed-loop monitoring systems for detecting incipient instability, *Proc. 37th IEEE Conference on Decision and Control*, pp. 3033–3039, 1998.

[27] S. Jeon, A. Suárez, and D. B. Rutledge, Analysis and elimination of hysteresis and noisy precursors in power amplifiers, *IEEE Trans. Microwave Theory Tech.*, vol. 54, pp. 1096–1106, 2006.

CHAPTER THREE

Bifurcation Analysis

3.1 INTRODUCTION

The operation bands of autonomous circuits, or circuits exhibiting a self-sustained oscillation, are delimited by critical parameter values at which the circuit undergoes a qualitative change of behavior. A typical example is a voltage-controlled oscillator, in which the oscillation is extinguished from a given value of the varactor bias voltage. This is closely connected with the fact that as shown in Chapter 1, any free-running oscillator must fulfill particular mathematical conditions to be able to sustain the oscillation. On the other hand, when connecting a periodic source to an existing oscillator circuit, different operation modes are possible depending on the input power and input frequency. Injection-locked behavior is characterized by the existence of a rational relationship $\omega_a/\omega_{\text{in}} = m/k$ between the oscillation frequency ω_a and the input generator frequency ω_{in}, which can be maintained only for certain ranges of the input generator power and frequency [1]. Outside these ranges, the circuit will either behave as a self-oscillating mixer or the oscillation will be extinguished by the large power delivered by the input generator. The qualitative changes in the circuit solution observed are due to *bifurcations* or qualitative variations in the stability of a given circuit solution or in the number of solutions when a parameter is modified continuously [2,3].

The operation bands of autonomous circuits or circuits exhibiting oscillations are inherently delimited by bifurcation phenomena in which the oscillation is generated or extinguished. The situation is different in the case of nonoscillatory circuits such as amplifiers or mixers. As an example, a stable amplifier will have a continuous

Analysis and Design of Autonomous Microwave Circuits, By Almudena Suárez

gain curve, existing for all the frequency values. The operation band will only be defined from a quantitative criterion such as the 3-dB gain reduction. Note, however, that amplifier circuits may become unstable at some particular values of their parameters, which would also give rise to qualitative changes in the solution observed [4].

In this chapter we present a detailed classification of the most common types of bifurcations in practical circuits. In Section 3.2, two types of representation of the circuit solutions—the phase space, already introduced in Chapter 1, and the Poincaré map—are described briefly, as they will be helpful for an understanding of the bifurcation phenomena. The local bifurcations involve qualitative variations in the stability of a single solution. The local bifurcations from dc and periodic solutions are presented in Sections 3.3.1 and 3.3.2, respectively. Section 3.3.3 deals with global bifurcations or bifurcations involving qualitative changes in more than one steady-state solution. The mechanism and implications of each type of bifurcation are explained in detail, providing practical examples. The effects of the bifurcation are analyzed using time- and frequency-domain techniques. This chapter is connected closely to the more practical Chapter 4, in which the behavior of various types of autonomous circuits is studied in detail, with in-depth discussions of the stability aspects.

3.2 REPRESENTATION OF SOLUTIONS

3.2.1 Phase Space

Let a nonlinear system in state form $\dot{\overline{x}} = \overline{f}(\overline{x})$ in R^N be considered, with $\overline{f}$ being a continuous function. Note that the state form of the differential equations is not possible for all nonlinear circuits. A general representation would be constituted by a system of differential algebraic equations. However, all the conclusions of this chapter remain valid for that general case, so, for simplicity, only the state form will be considered. As shown in Chapter 1, the steady-state solution of the nonlinear system is generally independent of the initial value, but the transient is not. The system integration from different initial values t_o and $\overline{x}_o$ gives rise to different transient solutions, so to take this initial value into account, the solution is often expressed as $\overline{x}(t + t_o) = \overline{\phi}_t(\overline{x}_o)$, with the subscript t indicating the difference between the initial time and the present time $t + t_o$. In a phase space representation of the system solutions, we use orthogonal coordinate directions corresponding to each of the variables. Plotting the numerical values of all the variables at a given time provides a description of the state of the system at that time, and its dynamics, or evolution, is indicated by tracing a path, or trajectory, in that same space [2]. When using the phase space representation, the function $\overline{\phi}_t(\overline{x}_o)$ defines a trajectory based at $\overline{x}_o$. For some examples, see Fig. 1.14 and Fig. 1.19 in Chapter 1. For a continuous set of initial conditions $\overline{x}_o \in U$ at the same time t_o, the function $\overline{\phi}_t : U$ provides another continuous set V of image points obtained after the time t. The effect of $\overline{\phi}_t$ can be seen as a flow from the set U to the set V. This is why $\overline{\phi}_t$ is also called the *system flow*.

In the phase space, the steady-state solutions give rise to bounded sets or limit sets, as they are obtained doing $\lim_{t\to\infty} \overline{\phi}_t(\overline{x}_o)$. The four main types of limit set are the equilibrium point, corresponding to a dc solution; the cycle or limit cycle, corresponding to periodic solution; the M torus, corresponding to a quasiperiodic solution with M nonrationally related fundamental frequencies; or the fractal dimension bounded set, corresponding to a chaotic solution. Transients constitute open trajectories leading from a given $t_o, \overline{x}_o$ to a limit set. The stable limit sets are attracting for all their neighboring trajectories and are called *attractors*. Saddle-type solutions are attractive only for some of their neighboring trajectories. The dimension of their stable manifold is smaller than the dimension of the entire space R^N. Because the perturbation will generally have components in all the different directions, the saddle-type solutions are unstable and unobservable.

3.2.2 Poincaré Map

Let a periodic solution of a nonlinear system $\dot{\overline{x}} = \overline{f}(\overline{x})$ in R^N be assumed, giving rise to a cycle in the phase space. Then a transversal surface $\Sigma \in R^N$, of limited size, is considered. The surface dimension is $N - 1$ [2] and its size must be small enough for its intersection with the cycle to provide one point instead of two points. Then the cycle (with dimension 1) gives rise to a fixed point $\overline{x}_{ps}$ (with dimension zero) in this surface. If a small instantaneous perturbation is applied to the limit cycle, the transient trajectory generated will give rise to an ensemble of points in Σ. The intersections of the solution with the transversal surface will belong to a space of dimension $N - 1$ and here are denoted $\overline{x}_p^n$. The Poincaré map P is the ordered sequence of these intersections and can be expressed

$$\overline{x}_p^{n+1} = P\left(\overline{x}_p^n\right) = \overline{\phi}_\tau(\overline{x}_p^n) \tag{3.1}$$

where $\tau(\overline{x}_p^n)$ is the time taken for the trajectory $\overline{\phi}_t(\overline{x}_p^n)$ first to return to Σ, also called *time of flight*. The time of flight depends on $\overline{x}_p^n$ but approaches the period T of the cycle as $\overline{x}_p^n$ approaches $\overline{x}_{ps}$. This is due to the continuity of $\overline{\phi}_t$ with respect to the initial condition $\overline{x}_o$.

As gathered from (3.1), the Poincaré map relies on knowledge of the flow or solution of the nonlinear system, so it cannot be obtained unless general solutions of this system are available. Exceptions exist in specific cases and require the use of approximations. The Poincaré map has two main properties: It transforms the continuous flow $\overline{x}(t + t_o) = \overline{\phi}_t(\overline{x}_o)$ into an ensemble set of discrete points, and it reduces the dimension of the steady-state solutions or limit sets. As another example, a 2-torus (of dimension 2) will give rise to a cycle composed of discrete points. Due to the inherent reduction in the system dimension, the Poincaré map is a useful tool for the graphical analysis of the qualitative variations of the system steady-state solutions versus a parameter. An example is given later in this section.

When dealing with N-dimensional systems, the phase space representation is limited, for obvious reasons, to a maximum dimension $n = 3$. Then the Poincaré map can be obtained by choosing a particular value for one of the state variables represented, x_i, with $i = 1, 2$, or 3, and determining the intersection of the solution with the local surface Σ in R^N, defined by $x_i = x_{io}$ about the original steady-state solution. Note that the value x_{io} chosen must be contained within the range of variation of the particular variable x_i. In case of a nonautonomous circuit with a periodic input generator of period T, this intersection can be obtained in a very simple manner by sampling the steady-state solution at integer multiples nT of the input generator period, starting at the initial time t_o. As shown in Chapter 1, the phase $\theta = (2\pi/T)t$ can be considered as one of the state variables of a system containing a periodic generator. Thus, when performing the sampling at nT, we are actually obtaining the intersection of the solution with the surface $\theta_o = (2\pi/T)t_o$.

To illustrate, the quasiperiodic solution obtained when introducing a periodic generator with input frequency $f_{in} = 6.33$ GHz and input power $P_{in} = -15$ dBm into the FET-based oscillator circuit of Fig. 1.6 is considered. The steady-state solution is a quasiperiodic solution at the two fundamentals $f_{in} = 6.33$ GHz and $f_o = 4.7$ GHz, the latter being the oscillation frequency. The value of this oscillation frequency is slightly different from the frequency obtained in free-running conditions $f_o = 4.4$ GHz, due to the influence of the external generator at $f_{in} > f_o$. In the phase space, this quasiperiodic solution provides the 2-torus of Fig. 1.16. Due to the periodic input source, the Poincaré map can be obtained by sampling the steady-state solution at integer multiplies of $T_{in} = 1/f_{in} = 0.16$ ns, starting from a given time value t_o within the interval of steady-state behavior. The resulting map is represented in Fig. 3.1. In the case of a quasiperiodic solution considered, a cycle composed of discrete points is obtained. This cycle will eventually be filled as time tends to infinity. Note that the discrete points in the cycle are not consecutive. To see this, a Fourier expansion of the circuit variables can be considered,

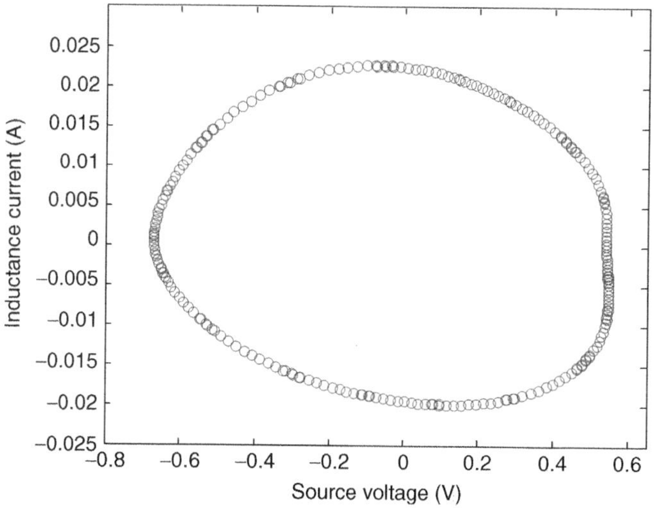

FIGURE 3.1 Poincaré map associated with the 2-torus of Fig. 1.16. This 2-torus corresponds to the quasiperiodic solution obtained when introducing a periodic generator at $f_{in} = 6.33$ GHz in the oscillator circuit of Fig. 1.6.

writing $x_i(t) = \sum_{m,k} X_{m,k} e^{j(m\omega_{\text{in}} + k\omega_a)t}$, with $1 \le i \le N$. By sampling these variables at integer multiples of $T_{\text{in}} = 2\pi/\omega_{\text{in}}$, the following set of discrete points will be obtained: $x_i(nT_{\text{in}}) = \sum_{m,k} X_{m,k} e^{jkn(\omega_a/\omega_{\text{in}})2\pi}$, with n an integer number. Thus, the different harmonic components evolve in the angle steps $k2\pi\omega_a/\omega_{\text{in}}$. The ratio $r = \omega_a/\omega_{\text{in}}$ is called the *rotation number*, as $r2\pi$ is the phase difference between two consecutive points of the discrete point cycle. Compare point n with the next point, $x_i((n+1)T_{\text{in}}) = \sum_{m,k} X_{m,k} e^{j2\pi(kn(\omega_a/\omega_{\text{in}})+kr)}$. In the quasiperiodic solution considered, the step r is an irrational number, and this is why the cycle is eventually filled. For $r = p/q$, with p and q integers and $p < q$, the steady-state solution will be periodic with the period $T = qT_{\text{in}}$, and the solution of the Poincaré map will consist of only q distinct points [3].

As a second example, the case of a frequency divider by 2 will be considered. The circuit contains a periodic generator at the frequency $f_{\text{in}} = 5$ GHz. The frequency division is obtained from a certain level of input power only. The Poincaré map can be obtained by sampling the steady-state solution at integer multiples of $T_{\text{in}} = 1/f_{\text{in}}$. As shown in Fig. 3.2, prior to this frequency division, sampling of the steady-state solution at nT_{in} provides one single point. When the circuit operates as a frequency divider, the sampling provides two points. This is because the steady-state solution is sampled at integer multiples of T_{in}, whereas the divided-solution period is $2T_{\text{in}}$, so two distinct points must be obtained. The frequency-divided regime starts at the point indicated as "bifurcation." At this point, the solution of the Poincaré map changes from one single point (no frequency division) to two points (frequency division by 2).

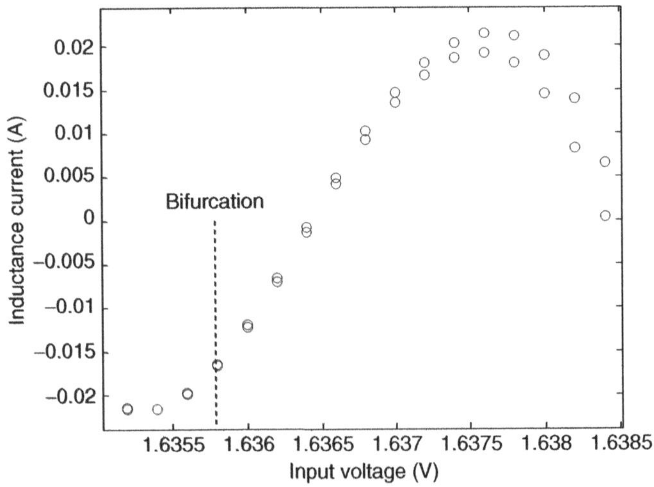

FIGURE 3.2 Evolution of the solutions of a Poincaré map applied to a frequency divider by 2 versus the input generator voltage. The map is obtained by sampling the steady-state solution at integer multiples of the input generator period T_{in}. Prior to the frequency division, a single point is obtained. After frequency division by 2, two points are obtained.

Intuitively, the steady-state discrete solutions of the Poincaré map $\overline{x}_p(n+1) = P(\overline{x}_p(n))$ will have the same stability properties as the corresponding continuous solutions of the original continuous-time system $\dot{\overline{x}} = f(\overline{x})$. As an example, a periodic steady-state solution $\overline{x}_s(t)$, which provides a cycle in the phase space, is considered. If the solution $\overline{x}_s(t)$ is stable, the cycle will attract all the neighboring trajectories in the phase space. The corresponding steady-state solution of the associated Poincaré map $\overline{x}_p(n+1) = P(\overline{x}_p(n))$ will be the single point $\overline{x}_{ps}$. When applying a small perturbation to the cycle in the phase space, a sequence of discrete points will be obtained in the Poincaré map P. Because the steady-state oscillation is stable, the sequence of discrete points will approach $\overline{x}_{ps}$ regardless of the value of the small perturbation applied.

As an example, Fig. 3.3 presents the Poincaré map resulting from the use of two different perturbations with a stable periodic solution $\overline{x}_s(t)$. The steady-state periodic solution corresponds to the single point of the map $\overline{x}_{ps}$. The response to two different perturbations is shown, one shown by dots and the other by circles. As can be seen, each perturbation provides a different sequence of discrete points, ending at the stable steady-state point $\overline{x}_{ps}$. The initial value considered in each case is slightly separated from the sequence. Then the points in each sequence approach $\overline{x}_{ps}$ in a flipping manner; that is, two consecutive points are at opposite sides of $\overline{x}_{ps}$. The reason for the flipping is the existence of a damped subharmonic component with doubled period $2T_{\text{in}}$ in the solution transient. More explanations are given later in this chapter.

Let the periodic solution of an autonomous system be considered. In the neighborhood of the limit cycle, the Poincaré map will have no component in the direction of this cycle, $\overline{u}_1(t) = \frac{d\overline{x}_{ps}(t)}{dt}$, as it is obtained from the solution

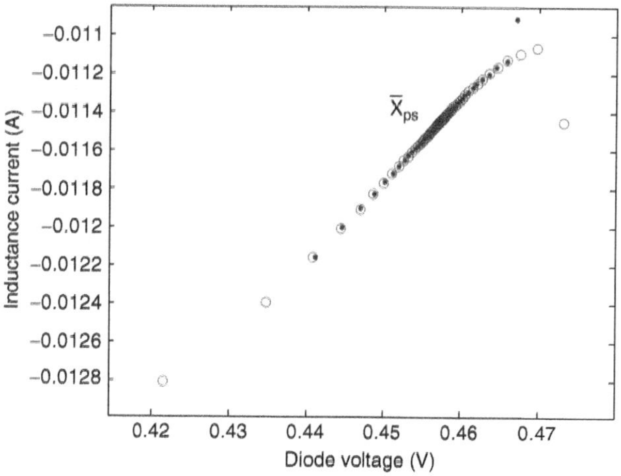

FIGURE 3.3 Perturbation of a stable periodic solution in a Poincaré map. The periodic steady-state solution provides the single point $\overline{x}_{ps}$. Two perturbations are applied, giving rise to two different sequences of discrete points, which flip from one side to another of $\overline{x}_{ps}$, ending at this point.

intersections with the transversal surface Σ. Considering a perturbation $\Delta\overline{x}_o$ at the initial time $t_o = 0$, the sequence of discrete points in the Poincaré map will evolve according to (see 1.56)

$$\Delta\overline{x}_p(n+1) = \sum_{k=2}^{N} c_k e^{\lambda_k \tau_{n+1}} \overline{u}_k(\tau_{n+1}) \tag{3.2}$$

where c_k are constants depending on the initial conditions, λ_k are the Floquet exponents of the periodic solution, $\overline{u}_k$ are periodic vectors, and τ_n is the time of the nth intersection with the surface. The component in $\overline{u}_1(t)$ corresponding to the direction of the cycle has been eliminated. Defining a special transversal surface [5], it will be possible to write

$$\Delta\overline{x}_p(n+1) = [JP(\overline{x}_{ps})]^n \Delta\overline{x}_p(1) = \sum_{k=2}^{N} c_k m_k^n \overline{u}_k \tag{3.3}$$

The matrix $[JP(\overline{x}_{ps})]$ is the Jacobian matrix of the Poincaré map. Remember that the time of flight $\tau_{n+1} - \tau_n$ is very close to the cycle period T for small $\Delta\overline{x}_p(n+1)$. Thus, it is easily derived that the Floquet multipliers $m_2, \ldots, m_N$ agree with the $N-1$ eigenvalues of $[JP(\overline{x}_{ps})]$ [2]. In a nonautonomous circuit, with a periodic input source of period T, the map is obtained generally from the intersections with the surface $\theta_o = \frac{2\pi}{T} t_o$. The perturbed Poincaré map fulfills (3.3), where the eliminated component $n = 1$ is the one associated to θ. A real Floquet multiplier m_k provides the amount of contraction ($m_k < 1$) or expansion ($m_k > 1$) near $\overline{x}_{ps}$ in the direction of $\overline{u}_k$. In turn, the contribution associated to each pair of complex-conjugate multipliers can be expressed as $c_k(m_k)^n \overline{u}_k + c_k^*(m_k^*)^n \overline{u}_k^* = 2\text{Re}[c_k(m_k)^n \overline{u}_k]$, which defines a spiral as the integer n increases. The magnitude $|m_k|$ provides the amount of contraction or expansion of this spiral [5].

3.3 BIFURCATIONS

A parameter is a relatively constant element or magnitude that determines the specific elements of an equation system but not its general nature. In circuit analysis it can be defined roughly as a magnitude susceptible to being varied while maintaining the same circuit topology. Examples of parameters are the linear component values, the values of the bias sources, or the amplitude or frequency of an input source. The continuous variation of a parameter η generates a set of steady-state solutions $\overline{x}(\eta)$ known as a *solution path*. This parameter variation ordinarily gives rise to a quantitative change in the circuit solution, such as the variation of its output power. However, in some cases a qualitative change may also be obtained at a particular parameter value η_b. This would be due to a *bifurcation*, defined as a qualitative change in the stability of a solution or in the number of solutions when a parameter is varied continuously. Figure 3.2 showed an example of bifurcation giving rise to a transition from a periodic regime at the input generator frequency f_{in} to a frequency-divided regime at $f_{in}/2$. Bifurcations can be classified as local or global. Local bifurcations are those involving variations in the stability properties of a single solution. They can be detected from the pole analysis of this single solution. Global bifurcations, roughly speaking, are qualitative variations in the phase space, involving intersections between the stable and unstable manifolds of one or more solutions [2].

3.3.1 Local Bifurcations

As already stated, local bifurcations are associated with qualitative change in the stability of a single steady-state solution $\overline{x}_s(t)$. This stability is determined by applying a small instantaneous perturbation to this solution and analyzing the evolution of the perturbed circuit variables. Due to the small value of the applied perturbation required, the circuit equations can be linearized about the particular steady-state solution $\overline{x}_s(t)$. When considering the continuous variation of a parameter η, the linearization must be applied about each steady-state solution obtained versus η:

$$\dot{\overline{x}} = \overline{f}(\overline{x}, \eta) \tag{3.4}$$

Then the circuit linearization about each steady-state solution $\overline{x}_s(t, \eta)$ is given by

$$\Delta\dot{\overline{x}}(t) = Jf(\overline{x}_s(t), \eta)\Delta\overline{x}(t) \tag{3.5}$$

Note that as η varies continuously, the poles associated with the system linearization about $\overline{x}_s(t)$ will also vary continuously. As shown in Chapter 1, by *solution poles* we refer to the poles of any or all possible closed-loop transfer functions that can be defined in a linearized system when introducing any small-signal input. As shown in Chapter 1, these poles, which are the same for all possible transfer functions, agree with the roots of the characteristic determinant associated with the particular linearized system in the frequency domain. Thus, they provide the stability of the solution $\overline{x}_s(t)$ about which the original nonlinear system is linearized.

For a particular parameter value η_b, a real pole γ or a pair of complex-conjugate poles $\sigma \pm j\omega$ may cross the imaginary axis, giving rise to a local bifurcation of the steady-state solution $\overline{x}_s(t, \eta)$ versus the parameter η. In a general manner, the crossing poles here are called *critical poles*. If the solution was originally stable, it will become unstable after the bifurcation point η_b. The bifurcations can be classified as direct or inverse. A direct bifurcation is obtained when the critical pole or poles cross the imaginary axis to the right-hand side of the complex plane. An inverse bifurcation is obtained when the critical pole or poles cross the imaginary axis to the left-hand side of the complex plane. As already known, transient behavior is dominated by the pole(s) with largest real-part value σ (or γ), which affects the envelope amplitude as $e^{\sigma t}$. As another general property, when approaching a bifurcation from a stable regime, the circuit transient response becomes progressively slower, due to the small magnitude of the negative σ (or γ).

Note that the original solution continues to exist after the bifurcation, as we can actually analyze its unstable poles. However, due to its instability, it will not be observable physically. Thus, the system will evolve to a different, stable steady-state solution after the bifurcation. This gives rise to the qualitative variation in the solution observed, associated with bifurcations. At a local bifurcation, one or more system solutions are created or extinguished. The local bifurcations are characterized by their continuity. All changes occur in an N-ball of radius $R = \varepsilon$ in the phase space. This means that the original and generated solutions overlap at the bifurcation point and diverge gradually from this point when varying the parameter further.

In nonlinear dynamics, the *center manifold theorem* provides a systematic way to reduce the dimension of the state spaces that have to be considered when analyzing a particular type of bifurcation [2]. As shown in Chapter 1, the stable manifold of a given steady-state solution consists of the set of space points for which this solution is attracting; that is, the set of points such that when used as initial conditions the system evolves exponentially in time to the particular steady state. The unstable manifold consists of the set of points for which the solution is repelling (the system evolves to it for $t \to -\infty$). For bifurcation points, the steady-state solution will also contain a center manifold. This manifold is associated with the eigenvalues located on the imaginary axis in a dc solution, or with the Floquet exponents located on the unit circle in a periodic solution. Note that the dynamics associated with the rest of poles is relatively simple and corresponds to expansions ($\sigma > 0$) or contractions ($\sigma < 0$). An example of the usefulness of the central manifold theorem is given later in the chapter.

In the following, a classification of the main types of local bifurcations is presented. Bifurcations from two different types of steady-state solutions are considered: from a dc solution, $\overline{x}_s(t) \equiv \overline{x}_{dc}$ and from a periodic solution of period T.

3.3.1.1 Bifurcations from a dc Solution

Different types of bifurcation may occur from a dc regime $\overline{x}_{dc}$ when a circuit parameter η is varied. For convenience, the general expression of the perturbation of a dc solution (1.42) is recalled here:

$$\Delta\overline{x}(t) = \sum_{k=1}^{N} c_k e^{\lambda_k t}\overline{u}_k = c_{c1}e^{(\sigma_{c1}+j\omega_{c1})t}\overline{u}_{c1} + c_{c1}^{*}e^{(\sigma_{c1}-j\omega_{c1})t}\overline{u}_{c1}^{*} + c_{r1}e^{\gamma_{r1}t}\overline{u}_{r1} + \cdots \tag{3.6}$$

where the exponents λ_k, $k = 1$ to N, which may be real or complex conjugate, are eigenvalues of the Jacobian matrix $Jf(\overline{x}_{dc})$, the vectors $\overline{u}_k$ are eigenvectors of this matrix, and the c_k are constants that depend on the initial conditions, thus on the instantaneous perturbation used. A local bifurcation will be obtained if at a certain parameter value η_b either a real pole γ or a pair of complex-conjugate poles $\sigma \pm j\omega$ crosses the imaginary axis of the complex plane. The two different situations are also described.

Bifurcations Associated with an Eigenvalue Passing Through Zero A real eigenvalue γ_k crosses the imaginary axis through the origin at the bifurcation parameter value η_b so the following conditions are fulfilled:

$$\begin{aligned} \gamma_k(\eta_b) &= 0 \\ \left.\frac{d\gamma_k}{d\eta}\right|_{\eta_b} &\neq 0 \end{aligned} \tag{3.7}$$

The second condition implies that the pole actually crosses the imaginary axis; that is, it is not tangent to this axis at η_b. Assuming that the dc solution was stable originally, it will become unstable with one real eigenvalue $\gamma_k > 0$ after the bifurcation. Note that the bifurcation condition (3.7) also applies to dc solutions that were originally unstable. In general, this condition states that if a dc solution originally has M eigenvalues (or poles) on the right-hand side of the complex plane, it will have $M \pm 1$ eigenvalues on this side of the plane after the bifurcation.

The existence of a zero eigenvalue $\gamma_k = 0$ gives rise to the singularity of the system Jacobian matrix at the bifurcation point, since $\det[\gamma U - Jf(\overline{X}_{\text{dc}})] = \det[Jf(\overline{X}_{\text{dc}})] = 0$, with U the identity matrix. Thus, the possible points fulfilling the bifurcation condition (3.7) can be determined from

$$\begin{aligned} \overline{f}(\overline{X}_{\text{dc}}, \eta_b) &= 0 \\ \det[Jf(\overline{X}_{\text{dc}}, \eta_b)] &= 0 \end{aligned} \tag{3.8}$$

Note that the bifurcation point must fulfill system (3.4). The same conditions (3.7) and (3.8) may actually correspond to three different types of bifurcations, giving rise to different qualitative changes in the circuit solution: the transcritical bifurcation, the pitchfork bifurcation, and the turning point. They can be distinguished using the *bifurcation coefficients* [2], obtained from a Taylor series expansion of the function $\overline{f}$ about the steady-state solution $\overline{X}_{\text{dc}}$, with order higher than 1. The bifurcations obtained most often in practical circuits are the pitchfork and turning-point bifurcations. These are the only types of bifurcations that are considered here. Note that condition (3.8) will be used to detect both types of bifurcations. As already noted, the distinction between them will require the use of the bifurcation coefficients or analysis of the solution paths about the bifurcation point.

PITCHFORK BIFURCATION In a pitchfork bifurcation, a dc solution $\overline{x}_{\text{dc}}$ gives rise to two new dc solution branches, $\overline{x}_{\text{dc}1}$ and $\overline{x}_{\text{dc}2}$ at the particular parameter value η_b, which arise from the system solution at the bifurcation point $\overline{x}_{\text{dc}}(\eta_b) = \overline{x}_{\text{dc}1} = \overline{x}_{\text{dc}2}$ (for an example, see Fig. 3.4b). The original path $\overline{x}_{\text{dc}}$ continues to exist after the bifurcation. If $\overline{x}_{\text{dc}}$ was stable, it will, of course, become unstable after the bifurcation, as it will have a real eigenvalue $\gamma_k > 0$. Note that three different dc solution branches merge at the bifurcation point, so the solution paths take the shape of a pitchfork about the bifurcation point: thus the name of this bifurcation. The occurrence of the pitchfork bifurcation requires the existence of odd symmetry in the system equations. This means invariance of the equations under a transformation of the type $(x_1, \ldots, x_i, \ldots, x_N) \rightarrow (x_1, \ldots, -x_i, \ldots, x_N)$.

The circuit of Fig. 1.19, defined by the perturbed linear system (1.45), is an example of this type of situation. This circuit is ruled by

$$\begin{aligned} \frac{dv_c}{dt} &= -\frac{G_1}{C}v_c - \frac{i_L}{C} - \frac{av_c + bv_c^3}{C} && \text{(a)} \\ \frac{di_L}{dt} &= \frac{v_c}{L} - \frac{R_2}{L}i_L && \text{(b)} \end{aligned} \tag{3.9}$$

Note that the above system is invariant under the transformations $v_c \rightarrow -v_c, i_L \rightarrow -i_L$. System (3.9) can be solved for the dc solutions

$$\begin{aligned} (a + bV_{\text{dc}}^2 + 1/R_2 + G_1)V_{\text{dc}} &= 0 \\ V_{\text{dc1}} &= 0 \\ V_{\text{dc2,3}} &= \pm\sqrt{\frac{-a - 1/R_2 - G_1}{b}} \end{aligned} \tag{3.10}$$

The solution $V_{\text{dc1}} = 0$ will exist for all possible values of the circuit elements. In contrast, the existence of the two dc solutions $V_{\text{dc2,3}}$ requires fulfillment of $-a - 1/R_2 - G_1 \geq 0$.

Next the poles associated with the system (3.9) linearized about the solution $V_{\text{dc1}} = 0$ are analyzed versus the parameter G_1. These poles are calculated as shown in Chapter 1. Because it is a second-order system, there are two poles, given by

$$p_{1,2} = \frac{-[(G_1 + a)L + R_2C] \pm \sqrt{[(G_1 + a)L + R_2C]^2 - 4LC[(G_1 + a)R_2 + 1]}}{2LC} \tag{3.11}$$

The pole evolution as G_1 increases is shown in Fig. 3.4a. The arrows indicate the sense of variation of these poles when G_1 is increased. For relatively small G_1, there are two real poles $\gamma_1 < 0, \gamma_2 > 0$, so the solution $V_{\text{dc1}} = 0$ is unstable. Increasing G_1, the poles approach each other, and at $G_1 = G_{1b} = 0.02\ \Omega^{-1}$, pole γ_2 crosses the imaginary axis to the left-hand side of the complex plane, so the solution $V_{\text{dc1}} = 0$ becomes stable. An inverse pitchfork bifurcation is obtained at the conductance value G_{1b}. As G_1 is increased further, the radicand in (3.11) becomes negative from $G_{1o} = 0.035\ \Omega^{-1}$, so the pair of negative real poles γ_1 and γ_2 turns into a pair of complex-conjugate poles $\sigma \pm j\omega$, with negative real part. Note that the number of poles must remain constant under any parameter variation, as this number of poles agrees with the system order N. As shown in Fig. 3.4a, the two real poles merge at G_{1o} and split into two complex-conjugate poles in a continuous fashion. Because this change in the nature of the poles takes place on the left-hand side of the complex plane, it does not have an influence on the steady-state solution observed.

Taking (3.10) into account, for $G_1 > G_{1b}$, the only circuit solution is $V_{\text{dc1}} = 0$ (see Fig. 3.4b). At $G_1 = G_{1b}$, the solution $V_{\text{dc1}} = 0$ becomes unstable and two

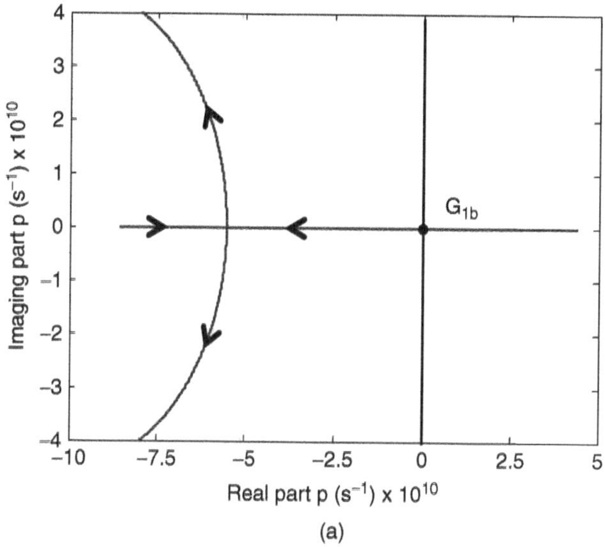

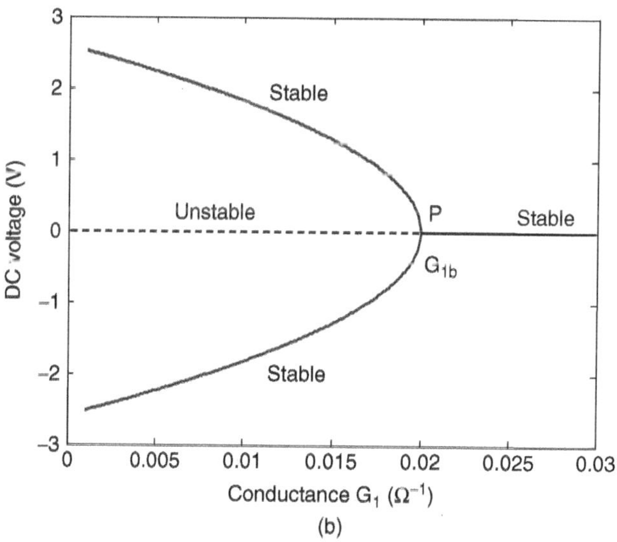

FIGURE 3.4 Pitchfork bifurcation in the circuit of Fig. 1.19: (a) evolution of the system poles versus the conductance G_1; (b) bifurcation diagram showing variation of the steady-state solutions versus G_1.

other dc solutions, $V_{dc2,3}$, are generated through a pitchfork bifurcation. The three solutions V_{dc1} and $V_{dc2,3}$ are overlapped at the bifurcation point G_{1b}, taking the same value $V_{dc} = 0$. This is in agreement with the continuity of the local bifurcations we have discussed. It is left to the reader to verify that the solutions generated are stable, which can be done through a pole analysis similar to the one carried out in (3.11). The reader can also verify that the Jacobian matrix associated with the system (3.9) is singular at $G_1 \equiv G_{1b}$ for $V_{dc} = 0$.

Due to the perfect odd-symmetry requirement in the circuit equations, the ideal pitchfork bifurcation is relatively rare. In a nearly symmetric system, an imperfect pitchfork bifurcation is obtained instead. To see an example, a dc current generator with a small value can be connected in parallel to the circuit of Fig. 1.19. This prevents the odd symmetry of equation (3.9a). For the generator value $I_{dc} = 1$ mA, the solution diagram of Fig. 3.4b turns into the one in Fig. 3.5. It is the typical diagram of an imperfect pitchfork bifurcation. The branching point P no longer exists. Due to addition of the nonsymmetric term, the solution diagram has split into two isolated curves, as can easily be verified by solving the cubic equation that provides the circuit dc solutions. Evolution from a diagram like Fig. 3.4b to the one shown in Fig. 3.5 is smooth versus the value of the dc generator that breaks the equation symmetry. One of two isolated curves exists for all the conductance values, and all its points are stable. The second curve exists below a certain conductance value $G_{1T} = 0.016\ \Omega^{-1}$ only. This maximum conductance value corresponds to an infinite slope point or turning point T, which separates the stable (lower) and unstable (upper) sections of the second curve. The turning point is a different type of bifurcation, which is discussed in detail next.

TURNING POINT As already stated, bifurcations from a dc regime associated with passing through zero of a real eigenvalue $\gamma_k = 0$ give rise to a singularity of the system Jacobian matrix $\det[Jf(\overline{x}_{dc}, \eta_b)] = 0$. Besides the pitchfork bifurcation, the turning point is another example of this situation. At turning points (e.g., T in Fig. 3.5) the solution curve $\overline{x}_{dc}(\eta)$ folds over itself, exhibiting an infinite slope

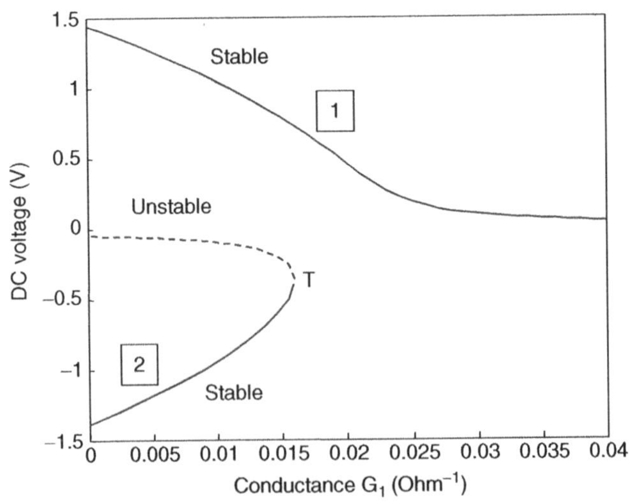

FIGURE 3.5 Imperfect pitchfork bifurcation obtained when connecting a dc current generator of small value to the circuit of Fig. 1.19. The resulting solution diagram should be compared with the one corresponding to the ideal pitchfork bifurcation in Fig. 3.4b.

$d\bar{x}_{dc}/d\eta_b = \infty$, which is due to the system singularity. Because one real eigenvalue γ_k crosses the imaginary axis through zero, if the solution curve had M unstable eigenvalues before the turning point, it will have $M \pm 1$ unstable eigenvalues after this point. Note that only two solutions, differing in one unstable eigenvalue, merge at the turning point (with infinite slope), unlike the case of pitchfork bifurcations, in which three solutions merge (see Fig. 3.4b). If the solution is originally stable, it will become unstable, with one unstable real eigenvalue, after the turning point. In this case, the folding of the curve is usually associated with a jump to a different stable solution. In Fig. 3.5, for $G_1 < G_{1T}$, operations in curve 1 or in the stable lower section of curve 2 are both possible. Provided that the circuit is operating initially in curve 2, it will remain there as G_1 increases until reaching point T. At this point, a jump to curve 1 will necessarily occur.

Turning points can also give rise to hysteresis when a circuit parameter is varied. An example of this phenomenon is obtained by keeping the conductance G_1 constant in the circuit of Fig. 1.19 at the initial value $G_1 = 0.01\ \Omega^{-1}$ and varying the dc current of the parallel current generator introduced. This provides the solution curve of Fig. 3.6. Two different turning points, T_1 and T_2, are obtained. The lower curve section (1) is stable up to point T_1. At this point, one of the two real eigenvalues crosses the imaginary axis to the right-hand side of the complex plane. This real eigenvalue remains on this side of the plane in the curve section (2), located between T_1 and T_2. At point T_2, the real pole again crosses the imaginary axis to the left-hand side of the complex plane, so the upper curve section (3) is stable. The two turning points give rise to hysteresis versus the dc current. To see this, let us assume that the dc current is increased from -0.1 A, for instance. The

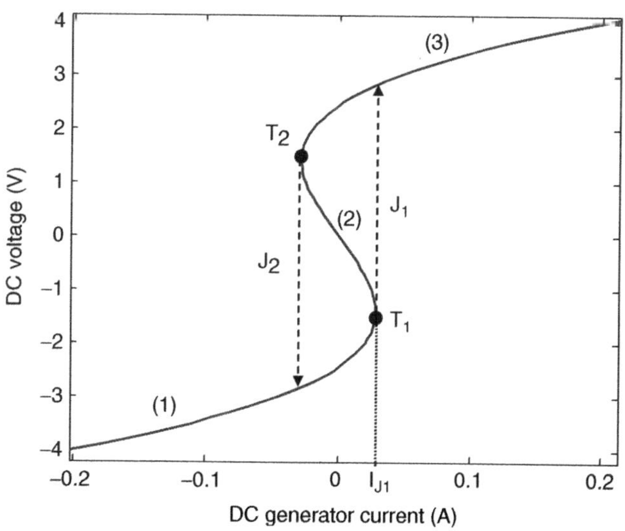

FIGURE 3.6 Hysteresis phenomenon in the circuit of Fig. 1.19 versus a dc current. The nonlinearity has been changed to $i(v) = -0.05v + 0.005v^3$. The phenomenon is due to turning points T_1 and T_2.

solution remains in section (1) until point T_1 is reached, which gives rise to the jump J_1 to section (3), occurring for the dc current value I_{J1}. Once in section (3), if we reduce the dc current from $I > I_{J1}$, the solution remains in section (3) when we pass through the value I_{J1}. This is because there is nothing anomalous happening at this current value in section (3). The system remains in section (3) until the turning point T_2 is reached, where the curve folds over itself. At this point, a second jump to section (1) J_2, occurs. Thus, a hysteresis cycle is observed.

In practical design, the occurrence of turning points in a dc solution curve requires the coexistence of some parameter ranges of two or more mathematical dc solutions. They are usually found in circuits with multiple transistors and no dc blocking [6].

Hopf Bifurcation A pair of complex-conjugate eigenvalues $\lambda_{k,k+1} = \sigma \pm j\omega$ of a dc solution $\overline{x}_{dc}$ cross the imaginary axis at the bifurcation parameter value η_b. Thus, the following conditions are fulfilled:

$$\begin{aligned} &\lambda_{k,k+1}(\eta_b) = \pm j\omega \qquad &\text{(a)} \\ &\left.\frac{d\sigma}{d\eta}\right|_{\eta_b} \neq 0 \qquad &\text{(b)} \end{aligned} \tag{3.12}$$

The second condition indicates that the pair of complex-conjugate poles actually cross the imaginary axis when varying the parameter η at the value of the particular η_b. Assuming that the dc solution $\overline{x}_{dc}$ was originally stable, it will become unstable after Hopf bifurcation. At the bifurcation point, the complex-conjugate eigenvalues will generate a limit cycle of frequency ω, agreeing with the imaginary part of the critical poles. Due to the continuity of the local bifurcation, this limit cycle will have zero amplitude at the bifurcation point and will be overlapped with the dc solution. Thus, it is a degenerate limit cycle. This is in agreement with the value $\sigma = 0$ of the real part of the critical poles of the dc solution at the bifurcation. Of course, the amplitude of the limit cycle will increase when further varying the parameter (in the same sense) from η_b, in agreement with the positive σ value of the poles of the dc solution after the bifurcation. Note that, intentionally, nothing is said about the stability or instability of the limit cycle generated. This aspect is treated later in this subsection.

An example of Hopf bifurcation from a dc regime is obtained in the circuit of Fig. 1.1. This circuit has only one dc solution, given by $V_{dc} = 0$. In terms of the circuit elements, the two poles associated with the system linearization about this solution are given by [see (1.46)]

$$\lambda_{1,2} = -\frac{G_T}{2C} \pm \frac{1}{2}\sqrt{\frac{G_T^2}{C^2} - \frac{4}{LC}} \tag{3.13}$$

with $G_T = G_L + a$. The poles are complex conjugate $\sigma \pm j\omega$ for $4/LC > G_T^2/C^2$. Assuming this situation, for $G_L > -a$ the poles will have $\sigma < 0$, and thus the dc solution will be stable. When reducing G_L continuously, the complex-conjugate poles approach the imaginary axis and cross this axis at $G_b = -a$. A periodic oscillation is generated at this conductance value. The evolution of the generated limit cycle versus the resistance $R_L = 1/G_L$ is shown in Fig. 3.7a. It has been obtained through numerical integration of the differential equation system (1.38). The limit cycle arises at the bifurcation point $R_b = 1/G_b = 33.33\ \Omega$, due to the instability of the equilibrium point (dc solution). It surrounds this unstable dc point for $R > R_b$. The fact that the limit cycle is generated in an N-ball of radius tending to zero at the bifurcation point can be noted clearly in Fig. 3.7a. As already stated, this is due to the continuity of the local bifurcation.

Taking advantage of the fact that we are dealing with the same circuit that was analyzed exhaustively in Chapter 1, we can apply the results of the admittance function analysis at the fundamental frequency performed in Section 1.3. This analysis relied on use of the describing function to model the nonlinear element. As gathered from system (1.14), the amplitude of the steady-state oscillation is given by $V_o = \sqrt{-(G_L + a)/(3b/4)}$. Replacing the conductance G_L with the bifurcation value $G_b = -a$, we obtain that the amplitude of the periodic oscillation takes zero value $V_o = 0$ V at this bifurcation point. This oscillation amplitude increases as G_L decreases from G_b, as shown in Fig. 3.7b, where the oscillation amplitude V_o has been represented versus the resistance $R_L = 1/G_L$.

EVOLUTION OF SOLUTION POLES At the bifurcation point, the dc solution gives rise to a degenerate limit cycle (oscillation) of zero amplitude. Because the two solutions are actually the same at the bifurcation point, the stability properties of the dc solution are transferred to the periodic solution. However, the stability of a dc solution is determined by the eigenvalues associated with this dc solution, whereas the stability of the periodic solution is determined by the Floquet multipliers associated with this periodic solution (see Section 1.5.2.2). To preserve the system dimension, the total number of Floquet multipliers of the periodic solution generated should agree with the total number of eigenvalues of the original dc solution. According to expression (1.62), the Floquet multipliers of the periodic solution generated can be expressed as $m = e^{\lambda T}$, with T the solution period $T = 2\pi/\omega$. The dc solution has the pair of critical eigenvalues $0 \pm j\omega$ at the bifurcation point, with ω being the frequency of the periodic solution generated. In the limit of zero oscillation amplitude, this pair of complex-conjugate critical eigenvalues is transformed into two real Floquet multipliers of value $+1$. Equivalently, they give rise to two infinite set of poles $\lambda_{1,n} = 0 \pm jn\omega$ and $\lambda_{2,n} = 0 \pm jn\omega$, with n an integer. This is due to the nonunivocal relationship between the Floquet multipliers and the Floquet exponents (which agree with the solution poles). We can also say that the two complex-conjugate eigenvalues of the dc solutions transform into two real poles $\lambda_{1,0} = \gamma = 0$ and $\lambda_{2,0} = \gamma\prime = 0$ of the periodic solution. Note that these values are, in fact, limit values, obtained in the limit of zero oscillation amplitude. As the amplitude of the oscillation generated increases, one of the poles, γ, remains on the imaginary axis, whereas the other pole, $\gamma\prime$, moves continuously away from this axis, either to the left-hand side of the complex plane (supercritical bifurcation)

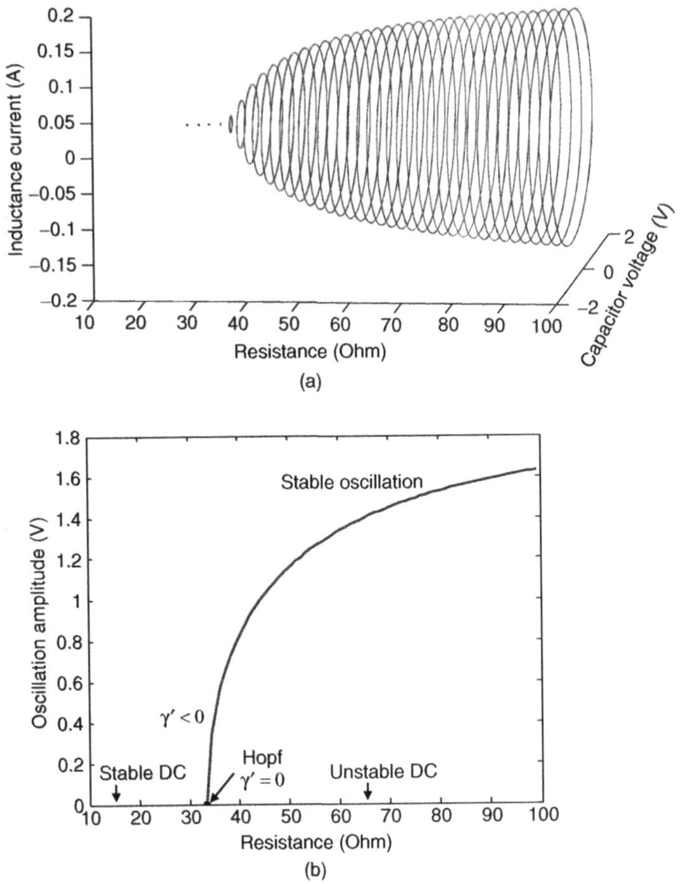

FIGURE 3.7 Hopf bifurcation in the circuit of Fig. 1.1. This bifurcation takes place at the resistance value $R_b = 33.33\ \Omega$ and gives rise to the onset of a limit cycle. (a) Evolution of the limit cycle generated versus the resistance R_L obtained through the numerical integration of the differential equation system. (b) Evolution of the limit cycle amplitude obtained through a first-harmonic admittance analysis, based on the describing function.

or to the right-hand side of this plane (subcritical bifurcation). These two different possibilities will give rise to very different qualitative behavior, discussed in more detail later in this section. Note that the presence of γ' in the neighborhood of the axis gives rise to a very high slope of the generated oscillatory solution versus the parameter (see Fig. 3.7), due to the nearly singular situation of the system. This is usually observed when tracing the oscillation amplitude or output power versus the parameter—a varactor bias voltage, for example.

BIFURCATION DETECTION The next aspect to be considered here is how to detect the Hopf bifurcation from the dc regime in practical circuit analysis. In the frequency domain this can be done in a very simple manner, by taking into account that the amplitude of the periodic oscillation tends to zero at the bifurcation point. Therefore, this point should fulfill the steady-state oscillation conditions for zero oscillation amplitude. When using an admittance (or impedance) analysis, assuming one harmonic component, the parameter value η_b giving rise to the Hopf bifurcation can be determined directly from the condition

$$\begin{aligned} Y_r(V_o = 0, \omega_o, \eta_b) &= 0 \\ Y_i(V_o = 0, \omega_o, \eta_b) &= 0 \end{aligned} \tag{3.14}$$

System (3.14) is a well-balanced system of two real equations in the two unknowns η_b and ω_o, which allows direct calculation of the bifurcation point η_b. The poles are purely imaginary $p = \pm j\omega_o$ at the bifurcation point η_b, and their frequency agrees with the oscillation frequency $\omega = \omega_o$ for V_o tending to zero. After the bifurcation, and due to the system nonlinearity, the frequency of the poles will generally be (slightly) different from the oscillation frequency. As an example, condition (3.14) will be applied to detect the Hopf bifurcation versus G_L in the parallel resonance oscillator of Fig. 1.1. Condition (3.14) becomes

$$\begin{aligned} Y_r(V_o = 0, \omega_o, G_L) &= \left(a + \tfrac{3}{4}bV_o^2 + G_L\right)_{Vo=0} = 0 \\ Y_i(V_o = 0, \omega_o, G_L) &= C\omega_o - \frac{1}{L\omega_o} = 0 \end{aligned} \tag{3.15}$$

which provides the same result, $G_L = -a$ and $\omega_o = 1/\sqrt{LC}$, as the former pole analysis.

In Fig. 3.7, as soon as the dc solution becomes unstable, the system evolves to a stable limit cycle, located in its immediate neighborhood. The situation is different in the case of Fig. 3.8. This corresponds to simulation of the MOSFET-based oscillator at 0.4 GHz [7], considered in Section 1.5.2. The amplitude of the first harmonic of the drain voltage has been represented versus the gate voltage V_{GG}, which constitutes the bifurcation parameter. The dc solution, for which no oscillation occurs, lies on the horizontal axis in this representation. Increasing V_{GG} from a very low value, this dc solution becomes unstable at the bifurcation point $V_{GGb} = 3.2$ V, where a pair of complex-conjugate eigenvalues of the dc solution crosses the imaginary axis to the right-hand side of the complex plane. An oscillation of zero amplitude is generated at the bifurcation point, in agreement with the discussion at the beginning of this section. However, the periodic solution path, represented in Fig. 3.8, goes backward; in other words, it coexists with the stable dc solution prior to the bifurcation (for $V_{GG} < V_{GGb}$). It is clear that for $V_{GG} > V_{GGb} = 3.2$ V, and just after the bifurcation, there are no stable oscillations in the neighborhood of the unstable dc solution. Note that for the steady-state oscillation to be located in the neighborhood of the dc solution, it must have very small amplitude. As seen in Fig. 3.8, for $V_{GG} > V_{GGb} = 3.2$ V, the only stable solution is the periodic solution in the solid-line section of the periodic path. Thus, when increasing V_{GG} the circuit oscillation arises in an abrupt, discontinuous manner,

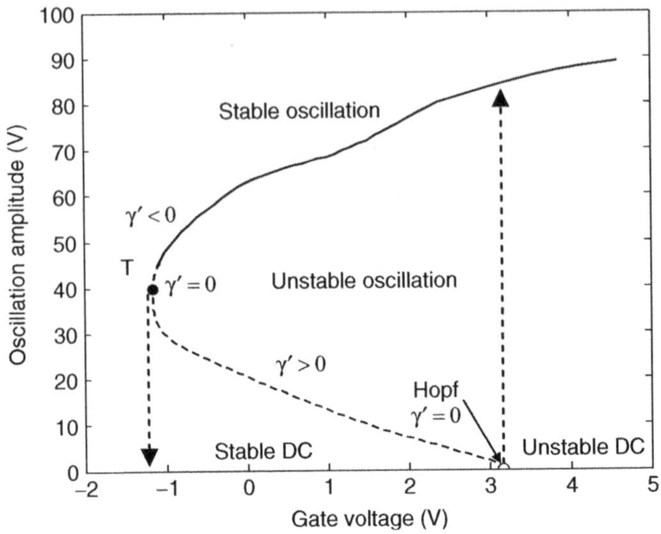

FIGURE 3.8 Subcritical Hopf bifurcation in a MOSFET-based oscillator.

unlike the smooth evolution of Fig. 3.7. In Fig. 3.8, the system goes to a limit cycle of large amplitude just after the bifurcation. Despite this, the periodic solution path does start from zero amplitude at the bifurcation point (although this amplitude increases in the opposite sense to the parameter), in agreement with the continuity of local bifurcations.

SUPERCRITICAL AND SUBCRITICAL BIFURCATIONS From the preceding discussion, the Hopf bifurcations can be divided into two types, according to the way in which the system evolves after the dc solution becomes unstable at the bifurcation point η_b. These two types are called supercritical and subcritical. As already stated, the degenerate periodic oscillation with zero amplitude that arises at the bifurcation point has two real poles of zero (limit) value, $\gamma = 0$ and $\gamma\prime = 0$ (belonging to the respective sets of poles $\lambda_{1,n} = 0 \pm jn\omega$ and $\lambda_{2,n} = 0 \pm jn\omega$, with n an integer). The real pole $\gamma = 0$ is due to the solution autonomy and will remain on the imaginary axis when the oscillation amplitude increases continuously from its original zero value at the bifurcation point. The second pole $\gamma\prime$ will either move to the left- or right-hand side of the complex plane, which will correspond to either a supercritical or subcritical bifurcation, respectively. The two types of bifurcation can also be distinguished geometrically by observing the variation of the oscillation amplitude versus the parameter η. In the following it is assumed that the bifurcation occurs when increasing the parameter η from dc regime, towards the bifurcation point η_b; that is, the dc regime is stable for $\eta < \eta_b$ and unstable for $\eta > \eta_b$. Then the two possible situations are:

1. *Supercritical Hopf bifurcation.* Just after the bifurcation, the pole $\gamma\prime$ shifts continuously to the left-hand side of the complex plane (see Fig. 3.7b). The generated oscillation is stable and its steady-state amplitude grows continuously from zero for $\eta > \eta_b$, with positive slope $dV/d\eta > 0$. Note that this slope tends to infinity at the bifurcation point, in agreement with the pole value $\gamma\prime = 0$. The limit cycle is generated after the critical parameter value, thus the term *supercritical*. The periodic solution path generated does not coexist with the stable dc regime (at least for small amplitude value).

2. *Subcritical Hopf bifurcation.* Just after the bifurcation, the pole $\gamma\prime$ shifts continuously to the right-hand side of the complex plane (see Fig. 3.8). The generated oscillation is unstable and its steady-state amplitude grows continuously from zero for $\eta < \eta_b$, with negative slope $dV/d\eta < 0$. The limit cycle exists before the critical parameter value, thus the term *subcritical*. The generated periodic solution path coexists with the stable dc regime. The subcritical bifurcation is often associated to a turning point of the periodic path, at which the pole $\gamma\prime$ passes through zero (see Fig. 3.8), so the periodic solution becomes stable.

The conditions on the derivative for the distinction between supercritical and subcritical bifurcations are the opposite ones if the dc regime is unstable for $\eta < \eta_b$ and stable for $\eta > \eta_b$. Note that the definition of supercritical or subcritical bifurcation is inherently local. We do not know how the periodic path generated will evolve away from the bifurcation.

The distinction between supercritical and subcritical is applicable to all the different types of *branching bifurcations* (i.e., bifurcations giving rise to the generation of new solution branches, like the pitchfork bifurcation). In the phase space, the supercritical bifurcation is characterized by the generation of a stable bifurcated solution in an N-ball of radius tending to zero $R = \varepsilon$, from an originally stable path, which loses its stability at the bifurcation point. The subcritical bifurcation is characterized by the generation of an unstable bifurcated solution in an N-ball of radius $R = \varepsilon$ from an originally stable path. The unstable solution generated coexists with the stable solution prior to the bifurcation. Subcritical bifurcations give rise to an abrupt change in the system state, as there is no stable solution in their neighborhood. This is why they are also called *hard-type bifurcations*. In turn, the supercritical bifurcations are also called *soft-type bifurcations*.

In nonlinear dynamics, the supercritical and subcritical bifurcations are distinguished by rewriting the system equations in *normal form* [2]. Use is made of the center manifold theorem, which, as already stated, provides a systematic way to reduce the dimension of the state spaces that have to be considered when analyzing a particular type of bifurcation. The Hopf bifurcation point contains a pair of complex-conjugate poles in the imaginary axis, so its stability cannot be determined with a first-order Taylor expansion of the nonlinear system, as in ordinary cases. The stability properties of the solution at the bifurcation point correspond to those of the generated limit cycle. For simplicity it will be assumed that the Hopf bifurcation takes place at $\overline{x} = \overline{X}_{\text{dc}} = 0$ for $\eta = 0$. The pair of critical eigenvalues will be $\lambda_{1,2} = \sigma(\eta) \pm j\omega(\eta)$, with $\sigma(0) = 0$. Note that for a system dimension N, the total number of eigenvalues will be $N = 2 + n_s + n_u$, with n_s the number of eigenvalues such that $\mathrm{Re}[\lambda_j] < 0$, $j = 3 \cdots 3 + n_s$, and n_u the number of eigenvalues such that $\mathrm{Re}[\lambda_j] > 0$, $j = 4 + n_s \cdots N$. Provided that two nondegeneracy conditions are fulfilled: $d\sigma(0)/d\eta \neq 0$ [in (3.12b)] plus an additional one (to be given later), the original nonlinear system $\dot{\overline{x}} = \overline{f}(\overline{x}, \eta)$ will be equivalent about the Hopf bifurcation to a much simpler system. This normal-form system is given by

$$\begin{aligned}
\dot{y}_1 &= \beta y_1 - y_2 + \alpha y_1(y_1^2 + y_2^2)\\
\dot{y}_2 &= y_1 + \beta y_2 + \alpha y_2(y_1^2 + y_2^2)\\
\dot{\overline{y}}_s &= -\overline{y}_s\\
\dot{\overline{y}}_u &= +\overline{y}_u
\end{aligned} \tag{3.16}$$

where a change of variables has been carried out from $\overline{x}$ to $\overline{y}$. The system (3.16) has the same dimension as the original one, $\dot{\overline{x}} = \overline{f}(\overline{x}, \eta)$. The vectors $\overline{y}_s$ and $\overline{y}_u$ have the dimensions n_s and n_u of the stable and unstable manifolds at the bifurcation point, respectively. Note that in situations of practical interest $n_u = 0$. The subsystem of dimension 2, in the variables y_1 and y_2, corresponds to the center manifold. The coefficients β and α are determined from a rather difficult function [2] obtained from the Taylor series expansion of the vector function $\overline{f}(\overline{x}, \eta = 0)$ about $\overline{x} = 0$ up to third order. The coefficients β and α are calculated by replacing the right and left eigenvectors of the Jacobian matrix $Jf(\overline{X}_{dc})$ into that function. These right and left eigenvectors are calculated from $[Jf(\overline{X}_{dc}, \eta = 0)]\overline{u} = j\omega\overline{u}$ and $[Jf(\overline{X}_{dc}, \eta = 0)]^T\overline{v} = -j\omega\overline{v}$, respectively.

The coefficient α can take the two possible values $\alpha = \pm 1$. It is given by $\alpha = \mathrm{sign}(L_1(\eta = 0))$, with L_1 being the *Lyapunov coefficient*. The second necessary condition for the existence of the normal form is $L_1(\eta = 0) \neq 0$. β and $L_1(\eta = 0)$ are both obtained from the Taylor series expansion of $\overline{f}(\overline{x}, \eta = 0)$, evaluated at $\overline{u}$ and $\overline{v}$. For $\alpha = 1$, the Hopf bifurcation is supercritical. For $\alpha = -1$, the Hopf bifurcation is subcritical. Extensions of the normal form exist for all the bifurcations of branching type, occurring from either the dc or the periodic regime. They enable a distinction between supercritical and subcritical type of bifurcations.

When using frequency-domain analysis, it will be possible to distinguish supercritical and subcritical bifurcations with a two-stage technique. First, the incipient periodic solution, near the bifurcation and with very small amplitude, is calculated. Then the poles associated with the circuit linearization about this solution are determined using a numerical pole–zero identification technique. In agreement with the preceding discussions, the incipient solution generated at a subcritical bifurcation will contain a real pole on the right-hand side of the complex plane, with all the rest of its poles on the left-hand side of this plane. The incipient solution generated at a supercritical bifurcation will contain one real pole on the left-hand side of the complex plane, with all the rest of its poles also located on the left-hand side of this plane. By means of this technique, there is no need to draw the solution curve versus the parameter to observe the slope of the subharmonic-component amplitude or to obtain the normal-form system, which would be virtually impossible in relatively large microwave circuits. On the other hand, the incipient solution is very near the actual bifurcation point, so this two-stage analysis provides the bifurcation point with sufficient accuracy, as well as information on the type of bifurcation: subcritical or supercritical. It must be noted that the real pole of the solution generated has zero value at the bifurcation point and varies continuously from this point. As an example, the foregoing technique has been applied to the practical microwave oscillator of Fig. 3.8. The amplitude considered for the incipient periodic solution is $V = 1$ V. Use of the numerical technique for pole calculation provides the following results, in Gigahertz:

$$\left.\begin{array}{l} -0.0000018 + j0.4197515 \\ -0.0000018 - j0.4197515 \end{array}\right\} \quad \text{oscillation autonomy}$$

$$\left.\begin{array}{l} 0.0019430 + j0.4196809 \\ 0.0019430 - j0.4196809 \end{array}\right\} \quad \text{unstable poles} \rightarrow \text{subcritical bifurcation}$$

The two pairs of poles have the same frequency, agreeing with the steady-state oscillation frequency. The small imaginary part in the first pair of poles is a numerical error. These poles are, in fact, located on the imaginary axis. Due to the periodicity of the poles of a periodic solution, the second pair of poles is equivalent to a positive real pole with the value $\gamma = 0.0019430 \times 10^9$.

As a second example, Table 3.1 presents the pole analysis of the solutions of the parallel resonance oscillator of Fig. 1.1. It is the same circuit as that considered in (3.13)–(3.15). This circuit has two reactive elements, so the dimension of the differential equation system describing its behavior is $N = 2$. Thus, the dc solution will contain two associated eigenvalues. In turn, the periodic solution generated at the Hopf bifurcation will contain two Floquet multipliers. The eigenvalues of the dc solution are complex conjugate. In turn, the two Floquet multipliers of the periodic solution are real and different. As we already know, the poles and Floquet multipliers are related through the nonunivocal relationship $m = e^{\lambda T}$. Because of that, the imaginary part $j\omega$ of the two pairs of poles is the same and agrees with the oscillation frequency 1.59 GHz. Table 3.1 shows the poles obtained for a different amplitude V of the incipient periodic solution, which confirm the supercritical nature of the Hopf bifurcation.

TABLE 3.1

Solution	Poles (GHz)
dc	$4.775 \times 10^{-7} \pm j1.5915494$
Periodic	
$V = 0.01$ V	$-0.0000012 \pm j1.5915499$
	$-0.0000132 \pm j1.5915501$
$V = 0.05$ V	$1.429 \times 10^{-7} \pm j1.59155$
	$-0.0002983 \pm j1.59155$

3.3.1.2 Bifurcations from a Periodic Solution In this section, evolution of a periodic solution $\overline{x}_s(t)$ of the nonlinear system (3.4) versus the continuous variation of a parameter η is analyzed. Because the periodic solution $\overline{x}_s(t)$ is given by an ordered set of time values, its representation versus the parameter η considered is not as straightforward as in dc regimes. It is convenient to represent each periodic solution with a single value, which can be done in different ways. When using a frequency-domain analysis, possible choices for the magnitude represented are the output power or the amplitude of a particular state variable at the fundamental frequency (as was done in Figs. 3.7b and Fig. 3.8). When using a time-domain analysis, the Poincaré map is very helpful, as the solution of the map associated with a periodic regime is a fixed point or M distinct fixed points, in the case of a frequency division by M. Thus, when limiting the representation to a one-state variable, each periodic regime can be represented by a limited number of discrete values. An example is shown in Fig. 3.2. Of course, the values will depend on the transversal surface $x_i = x_{io}$ selected, but this is not a problem because we are interested only in detecting the qualitative variations of the solution versus the parameter. Note that when using a Poincaré map to represent variations of the circuit solution versus a parameter, the transient must be totally extinguished at each step of this parameter. The variation in the number of discrete points obtained at a given parameter value η_b or a discontinuous jump in the fixed-point path will indicate a qualitative change in the solution or bifurcation (Fig. 3.2).

Because we are dealing with a periodic steady-state solution, its stability properties will be determined by the Floquet multipliers associated with the system (3.5), linearized about the periodic steady-state solution $\overline{x}_s(t)$, with fundamental frequency $\omega_o = 2\pi/T$. For each value of the analysis parameter η, the solution $\overline{x}_s(t)$ will have a given set of Floquet exponents, $\lambda_k, k = 1$ to N, which will evolve continuously versus the parameter. For convenience, the general expression of the perturbation of a periodic regime (1.50) in terms of these Floquet exponents $\lambda_k, k = 1$ to N, is recalled here:

$$\begin{aligned}\Delta\overline{x}(t) &= \sum_{i=k}^{N} c_k e^{\lambda_k t}\overline{u}_k(t)\\ &= c_{c1}e^{(\sigma_{c1}+j\omega_{c1})t}\overline{u}_{c1}(t) + c_{c1}^{*}e^{(\sigma_{c1}-j\omega_{c1})t}\overline{u}_{c1}^{*}(t) + c_{r1}e^{\gamma_{r1}t}\overline{u}_{r1}(t) + \cdots \end{aligned} \tag{3.17}$$

where the complex vectors $\overline{u}_k(t)$ are periodic with the same period T as the steady-state solution, and the complex constants c_k depend on the initial instantaneous perturbation. As shown in Chapter 1, the Floquet exponents λ_k agree with the system poles and are related to the Floquet multipliers through

$$m_k = e^{\lambda_k T} \qquad k = 1 \textit{ to } N \tag{3.18}$$

with T being the solution period. As shown in Chapter 1, the Floquet multipliers may be a real or complex conjugate. Note that there is a nonunivocal relationship between the poles and the Floquet multipliers. Associated with each multiplier is

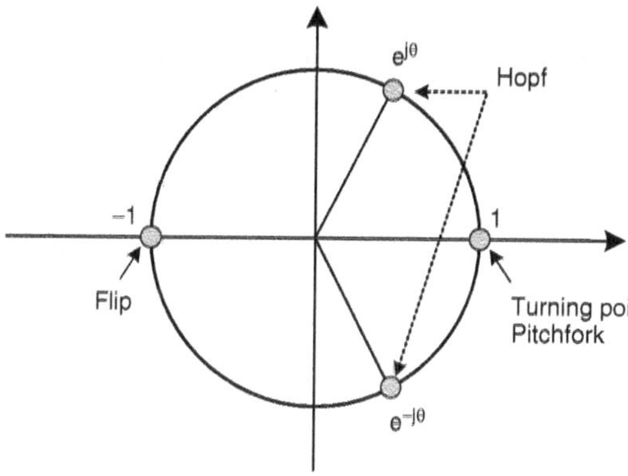

FIGURE 3.9 Bifurcations from a periodic regime. The three main types of local bifurcation are associated with the three ways that a real multiplier or pair of complex-conjugate multipliers can cross the unit circle.

an infinite set of poles of the form $\lambda_k \pm jn\omega_o$, with n a positive integer. Thus, there will be a different set of infinite poles associated with each single multiplier. Therefore, it will be more practical to define and classify the bifurcations in terms of the Floquet multipliers [8].

As is clear from (3.17) and (3.18), the N Floquet multipliers associated with a stable periodic solution $\overline{x}_s(t)$ will have a modulus smaller than 1 (i.e., they will be located inside a unit circle), except in the case of an autonomous solution, which will have one multiplier $m = 1$, with the rest inside the circle. The three main types of local bifurcation are associated with the three ways that a real multiplier or a pair of complex-conjugate multipliers can cross this circle (see Fig. 3.9). These three possibilities are as follows: (1) A real multiplier can cross the circle through the point (1,0), which will give rise to a D-type bifurcation; this crossing can have different effects and the general term *D-type* includes turning points and pitchfork bifurcations. (2) A real multiplier can cross the unit circle through the point $(-1, 0)$, which will give rise to a flip bifurcation. (3) A pair of complex-conjugate multipliers can cross the unit circle through $e^{\pm j\theta}$, which will give rise to a secondary Hopf bifurcation. The three types of local bifurcation from a periodic regime at $\omega_o = 2\pi/T$ are analyzed in detail next.

D-Type Bifurcation: Pitchfork and Turning-Point Bifurcations A real multiplier $m_k \in R$ crosses the unit circle through the point (1,0) at the parameter value η_b. The following conditions are fulfilled:

$$m_k(\eta_b) = 1$$
$$\left.\frac{dm_k}{d\eta}\right|_{\eta_b} \neq 0 \tag{3.19}$$

Taking the nonunivocal relationship (3.18) between the poles and the Floquet multipliers into account, it is possible to write $1 = e^{j(0+n\omega_o)T}$, with n an integer. Thus, when a multiplier $m_k = 1$ crosses the unit circle through the point (1,0), an infinite set of poles of the form $\gamma \pm jn\omega_o$, with the same real part $\sigma = \gamma$, cross the imaginary axis. They are all associated with the same real Floquet multiplier. Assuming that the periodic solution was originally stable, and taking (3.17) into account, after the bifurcation the perturbation will grow as $\Delta\overline{x}_k(t) = c_k\overline{u}_k(t)e^{\gamma t}$, with $\overline{u}_k(t)$ being the periodic vector (at the same frequency of the steady-state solution ω_o) associated with the unstable multiplier m_k, and $\gamma > 0$. Thus, there is no generation of new fundamental or subharmonic frequencies at the bifurcation point. Instead, a qualitatively different periodic solution arises at the bifurcation point.

Two different classes of D-type bifurcation can be distinguished, the pitchfork bifurcation and the turning point, which are analogous to those obtained from a dc regime. They are discussed next.

PITCHFORK BIFURCATION Assuming an initially stable periodic solution $\overline{x}_s(t)$, the periodic path $\overline{x}_s(t)$ versus η becomes unstable at the bifurcation point η_b and gives rise at this point to two new stable periodic branches. Three different periodic solutions merge at the bifurcation point, in a way similar to Fig. 3.4b. Thus, three periodic cycles will be overlapped at the bifurcation point. As in the case of a pitchfork bifurcation from a dc solution, the occurrence of this bifurcation from a periodic regime requires the fulfillment of certain symmetry conditions, which are rare in practical circuits. In nearly symmetric systems, imperfect pitchfork bifurcations like the one in Fig. 3.5 are obtained instead.

TURNING-POINT BIFURCATION The turning points of a solution path (traced in terms of the output power or the first-harmonic amplitude of a given state variable, for instance) are points of infinite slope versus the parameter η. The curve folds over itself, which is usually associated with a jump to a different stable solution. An example of this type of bifurcation can be seen in the periodic path of Fig. 3.8, showing the variation in the first-harmonic amplitude of the MOSFET-based oscillator [4] versus the gate voltage V_{GG}. As this voltage decreases from the upper branch (the solid line), a real multiplier escapes from the unit circle through the point (1,0) at the solution point corresponding to $V_{GG} = -1.2$ V, indicated by a "T." Thus, the upper section of the curve represented is stable, whereas the lower section is unstable. When varying the parameter toward the turning point, the two coexisting limit cycles (stable and unstable) approach each other, and finally, overlap at this point, in agreement with the continuity of the local bifurcations. This can be seen in Fig. 3.10, showing the coexisting stable (the solid line) and unstable (the dashed line) limit cycles for $V_{GG} = -1$ V in the MOSFET-based oscillator.

From the nonunivocal relationship (3.18) between the poles and the Floquet multipliers of the periodic solution, a multiplier crossing the unit circle through the point (1,0) implies an infinite set of poles crossing the imaginary axis of the complex plane at $\pm jn\omega_o$, n being a positive integer and ω_o the fundamental frequency of the periodic solution. Therefore, the turning-point bifurcation of a periodic solution can be detected by either the crossing of a real pole γ or the crossing of the pair

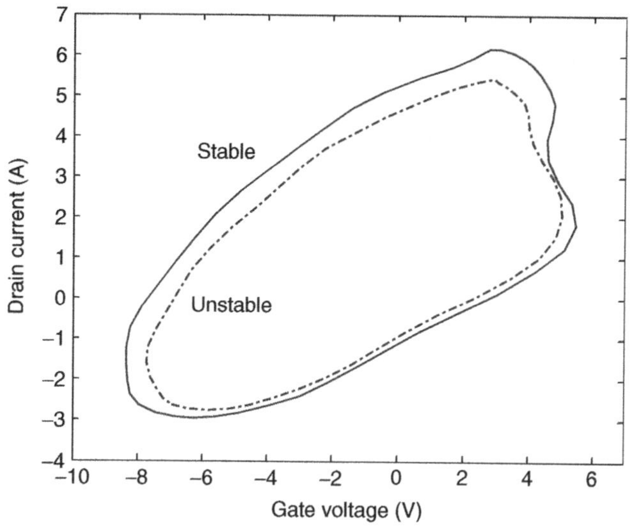

FIGURE 3.10 MOSFET-based oscillator. Stable (the solid line) and unstable (the dashed line) limit cycles near the turning point T of the bifurcation diagram of Fig. 3.8.

of complex-conjugate poles $\sigma \pm j\omega_o$ through the imaginary axis of the complex plane.

As already known, the steady-state solution of a free-running oscillator always contains a pair of complex-conjugate poles $\pm j\omega_o$, with ω_o the oscillation frequency. Thus, a possible turning point occurring in the periodic solution curve of a free-running oscillator (obtained versus a tuning voltage, for instance) would give rise to two overlapped pairs of poles $\pm j\omega_o$ at the oscillation frequency on the imaginary axis at the bifurcation point. An example can be seen in the pole locus of Fig. 3.11, corresponding to the MOSFET-based oscillator analyzed in Fig. 3.8. The diagram shows the variation of the poles closest to the imaginary axis (dominant poles) along the solution curve of Fig. 3.8 passing through the turning point. As can be seen, a pair of imaginary poles $\pm j\omega_o$ due to the solution autonomy are always located on the imaginary axis and slide slightly along the axis as the gate bias voltage is varied. Decreasing this voltage from the upper branch (the solid line), a second pair of poles, located initially on the left-hand side of the complex plane cross the imaginary axis at the gate bias value corresponding to the turning point T in the solution path of Fig. 3.8. The frequency of this second pair of poles agrees with that of the permanent pair $\pm j\omega_o$.

As we have already seen, subcritical Hopf bifurcations from a dc regime give rise to an unstable oscillation that coexists with the stable dc regime before the bifurcation is actually encountered. After the bifurcation takes place, there is no stable solution in the neighborhood of the unstable dc regime. The system cannot stay at the unstable dc solution, so it must evolve to some (distant) stable solution. The periodic path generated will usually exhibit a turning point, as the folding of the solution curve will enable the existence of a periodic solution (or other type of

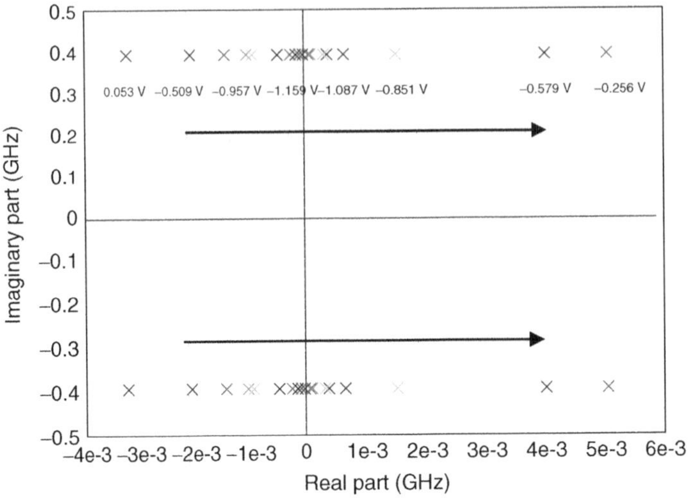

FIGURE 3.11 Evolution of the dominant solution poles along the periodic path of Fig. 3.8, corresponding to the MOSFET-based oscillator. Due to the autonomy of the solution, a pair of poles $\pm j\omega_o$ are located permanently on the imaginary axis. At the turning point, a second pair of complex-conjugate poles cross the imaginary axis at the same frequency.

time-varying solution) in the parameter interval for which the dc regime is unstable. See, for example, the upper section of the periodic solution curve in Fig. 3.8.

To get some physical insight into turning-point bifurcations, the behavior of a MOSFET-based oscillator [7] versus the gate bias voltage V_{GG} (Fig. 3.8) will be explained. For small V_{GG}, the transistor is cut off, so no oscillation can take place (Fig. 3.12). As V_{GG} increases, oscillation becomes possible at the bifurcation value V_{GGo}, corresponding to the conduction threshold. Due to the particular device operational conditions at this gate bias value, the energy absorbed from the dc sources and delivered to the oscillation is high enough to give rise to a large-amplitude limit cycle instead of a small one, which causes the subcritical Hopf bifurcation. When reducing the bias voltage, and due to the large amplitude of the oscillation, the transistor is on for a significant fraction of the period, even below V_{GGo}, due to the voltage peaks, so the oscillation persists for $V_{GG} < V_{GGo}$. The "on" fraction of the period decreases when reducing V_{GG}, so a value is reached, corresponding to the turning point T, from which it is impossible to maintain the oscillation. This is a general explanation of commonly observed hysteresis phenomena in oscillator circuits.

The bifurcation analysis of a periodic solution path requires the numerical determination of the parameter values η_b at which the critical Floquet multipliers or exponents (solution poles) are obtained. For that, the bifurcation conditions derived must be combined with the analysis techniques described in Chapter 1. However, as already shown, the bifurcations can also be detected through their effect on the circuit steady-state solutions. A pole at zero at the bifurcation point η_b will

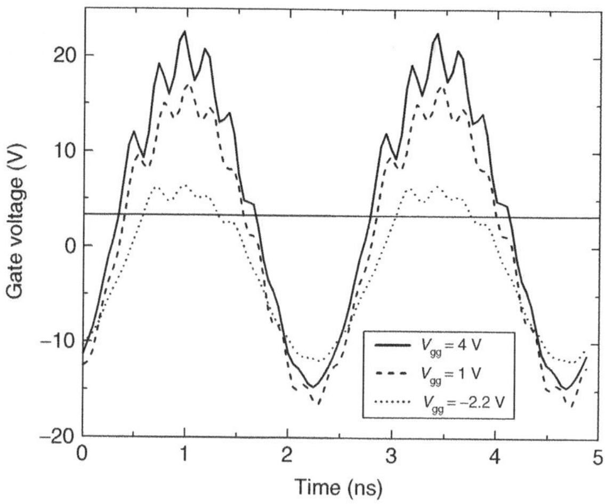

FIGURE 3.12 Gate voltage waveforms at various gate bias voltages. The threshold voltage is represented by a solid line. The drain bias considered is 25 V. (Reprinted with permission from IEEE.)

give rise to a singularity of the Jacobian matrix associated with the system linearization. Thus, the solution curve versus the parameter η will have infinite slope at η_b. Taking this into account, an approximate technique, based on admittance (or impedance) descriptions, will be presented in the following for turning-point detection.

When obtaining the solution path of a free-running oscillator versus a parameter η, due to the equation continuity, two consecutive points n and $n+1$ of this path will have relatively close values of the oscillation frequency ω_o and amplitude V_o. Then, the total admittance function Y_T at the point $n+1$, corresponding to η^{n+1}, can be estimated through the linearization of this function about the previous solution point n obtained for η^n. The admittance function is differentiated with respect to the oscillation amplitude V_o, frequency ω_o, and parameter η, which provides the following linearized system:

$$\overline{Y}_T(V_o^{n+1}, \omega_o^{n+1}, \eta^{n+1}) = \underbrace{\overline{Y}_T(V_o^n, \omega_o^n, \eta^n)}_{=0} + \Delta\overline{Y}_T$$

$$= \begin{bmatrix} \dfrac{\partial Y_T^{r,n}}{\partial V_o} & \dfrac{\partial Y_T^{r,n}}{\partial \omega_o} \\ \dfrac{\partial Y_T^{i,n}}{\partial V_o} & \dfrac{\partial Y_T^{i,n}}{\partial \omega_o} \end{bmatrix} \left(\begin{bmatrix} V_o^{n+1} \\ \omega_o^{n+1} \end{bmatrix} - \begin{bmatrix} V_o^n \\ \omega_o^n \end{bmatrix} \right)$$

$$+ \begin{bmatrix} \dfrac{\partial Y_{To}^r}{\partial \eta^n} \\ \dfrac{\partial Y_{To}^i}{\partial \eta^n} \end{bmatrix} (\eta_{n+1} - \eta_n) = 0 \qquad (3.20)$$

where $\overline{Y}_T$ is a column matrix consisting of the real and imaginary parts of the admittance function. Note that it has been taken into account that $Y_T(V_o^n, \omega_o^n) = 0$, as V_o^n, ω_o^n is the oscillatory solution already calculated for η^n. Thus, the point $n+1$ of the curve can be estimated from point n using a linear approach:

$$\begin{bmatrix} V_o^{n+1} \\ \omega_o^{n+1} \end{bmatrix} = \begin{bmatrix} V_o^n \\ \omega_o^n \end{bmatrix} - \begin{bmatrix} \dfrac{\partial Y_T^{r,n}}{\partial V_o} & \dfrac{\partial Y_T^{r,n}}{\partial \omega_o} \\ \dfrac{\partial Y_T^{i,n}}{\partial V_o} & \dfrac{\partial Y_T^{i,n}}{\partial \omega_o} \end{bmatrix}^{-1} \begin{bmatrix} \dfrac{\partial Y_{To}^r}{\partial \eta^n} \\ \dfrac{\partial Y_{To}^i}{\partial \eta^n} \end{bmatrix} (\eta_{n+1} - \eta_n) \tag{3.21}$$

Note that this linear analysis provides just an estimation of the next point of the solution curve. For an accurate determination of this point, a nonlinear analysis must be carried out using the estimated values as an initial guess. For a sufficiently small increment $\Delta\eta$, the slope of the solution curve versus the parameter will be given by the ratio $\Delta V_o/\Delta\eta$ or $\Delta\omega_o/\Delta\eta$, which is obtained from (3.21):

$$\begin{bmatrix} \dfrac{\Delta V_o^{n+1,n}}{\Delta\eta} \\ \dfrac{\Delta\omega_o^{n+1,n}}{\Delta\eta} \end{bmatrix} = - \begin{bmatrix} \dfrac{\partial Y_T^{r,n}}{\partial V_o} & \dfrac{\partial Y_T^{r,n}}{\partial \omega_o} \\ \dfrac{\partial Y_T^{i,n}}{\partial V_o} & \dfrac{\partial Y_T^{i,n}}{\partial \omega_o} \end{bmatrix}^{-1} \begin{bmatrix} \dfrac{\partial Y_{To}^r}{\partial \eta^n} \\ \dfrac{\partial Y_{To}^i}{\partial \eta^n} \end{bmatrix} \tag{3.22}$$

From an inspection of (3.22), the slope of the solution curve will tend to infinity at points where the Jacobian matrix of the admittance function becomes singular. Thus, the turning-point bifurcations can be detected from the following conditions:

$$\begin{aligned} Y_T(V_b, \omega_b, \eta_b) &= 0 \\ \det[JY(V_b, \omega_b, \eta_b)] &= \frac{\partial Y_T^r}{\partial V_o}\frac{\partial Y_T^i}{\partial \omega_o} - \frac{\partial Y_T^r}{\partial \omega_o}\frac{\partial Y_T^i}{\partial V_o} = 0 \end{aligned} \tag{3.23}$$

where it is taken into account that the turning point is also a steady-state solution of the nonlinear equation $Y_T = 0$. Note that the system above is a well-balanced system of three real equations in three unknowns V_b, ω_b, and η_b which allows direct determination of the bifurcation point. This singularity of the nonlinear system $Y_T = 0$ at the turning points is in total agreement with the existence of a real pole at zero at these bifurcation points.

For an analytical example of a turning point in the solution curve of an oscillator, the nonlinear function in the circuit of Fig. 1.1 will be modified with the inclusion of an odd-term polynomial of fifth order, leading to the describing function $Y_N(V) = a + bV^2 + cV^4$, with $a = -0.03$ A/V^{-1}, $b = 0.01$ A/V^{-3}, and $c = -0.001$ A/V^{-5}. Considering the conductance G_L as the analysis parameter, condition (3.23) can be used to detect possible turning points in the solution path. Use of this condition

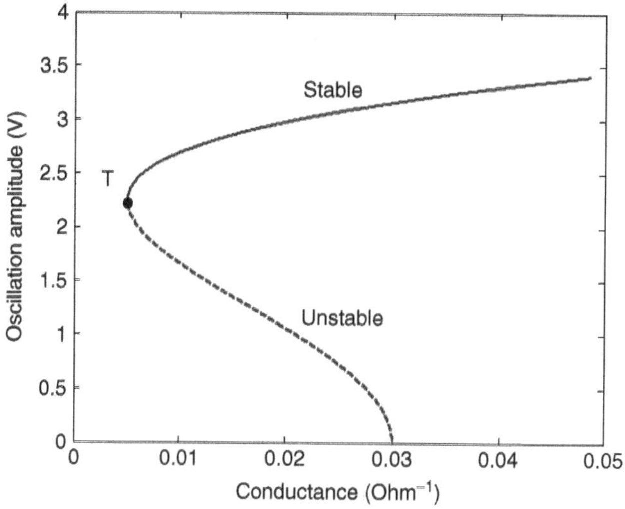

FIGURE 3.13 Turning-point bifurcation in the circuit of Fig. 1.1 versus the conductance G_L. The original nonlinear element has been replaced by a describing function of the form $Y_N(V) = a + bV^2 + cV^4$, with $a = -0.03$ A/V^{-1}, $b = 0.01$ A/V^{-3}, and $c = -0.001$ A/V^{-5}.

provides the system

$$\begin{aligned} Y_T(V_o, \omega_o, G_L) &= Y_N(V) + G_L + j\left(C\omega_o - \frac{1}{L\omega_o}\right) = 0 \\ \det[JY_o] &= \frac{\partial Y_T^{r,n}}{\partial V_o}\frac{\partial Y_T^{i,n}}{\partial \omega_o} = (bV + 2cV^2)2C = 0 \end{aligned} \tag{3.24}$$

Solving (3.24), a turning-point bifurcation is obtained at $G_b = 5.01 \times 10^{-3}\ \Omega^{-1}$ and $V_{ob} = 2.23$ V. This is confirmed by the simulation of Fig. 3.13, showing variation in the oscillation amplitude V_o versus the conductance G_L.

Note that the determinant in (3.23) agrees with the stability coefficient S defined in (1.20). In the one-harmonic, one-port approach presented in Chapter 2, the coefficient S must be positive for steady-state oscillation to be stable. If the coefficient S were negative, the steady-state oscillation would be unstable [9]. Thus, at points where the solution undergoes a qualitative change of stability, the coefficient S should take a zero value, $S = 0$, which is in agreement with (1.20). To understand this, note that if $S > 0$ is fulfilled before the turning point, then $S < 0$ must necessarily be fulfilled after the turning point. As shown in Section 3.3.1.1, at the Hopf bifurcation, two complex-conjugate poles of the dc solution become two real poles γ, γ' of the periodic solution. The real pole γ stays at zero due to the solution autonomy. The analysis of S aims at predicting the variation of the second pole γ'. However, this one-harmonic, one-port approach (presented in Section 1.3) is inherently limited. As shown in (1.22), it can be viewed as a one-pole description of the solution stability, strictly valid for dimension-2 systems only, as discussed in Chapter 1. The analysis of the coefficient S is unable to predict the transformation of an unstable solution with a pole on the right-hand side into a solution with two poles on the right-hand side. However, at the parameter value at which this second pole crosses the axis to the right-hand side of the complex plane, condition (3.23) would still be fulfilled. This is because it is actually evaluated from the system linearization about the steady-state solution V_o, ω_o, and the Jacobian matrix (3.22) is singular at this solution.

In the previous analyses, only turning points in solution curves of free-running oscillators versus a parameter η have been considered. In the expressions derived, the oscillation frequency ω_o is an unknown of the system, which varies with the parameter η. Generalization of the turning-point condition to nonautonomous systems is straightforward, and examples are presented in the next chapter.

Flip Bifurcation A real multiplier $m_k \in R$ crosses the unit circle through the point $(-1, 0)$ at the parameter value η_b. The following conditions are fulfilled:

$$\begin{aligned} &m_k(\eta_b) = -1 \\ &\left.\frac{dm_k}{d\eta}\right|_{\eta_b} \neq 0 \end{aligned} \tag{3.25}$$

Due to the relationship (3.18) between the poles of the periodic solution and the Floquet multipliers of this solution, it will be possible to write $m_k = -1 = e^{\pm j\pi} = e^{\pm j((\omega_o/2)+n\omega_o)T}$, with n an integer. Thus, the crossing of the unit circle by a Floquet multiplier through the point $(-1, 0)$ is equivalent to the crossing of an infinite set of complex-conjugate poles $\sigma \pm j(\omega_o/2 + n\omega_o)$, with n an integer, through the imaginary axis of the complex plane. Next, the general expression of the state-variable perturbation (3.17) is considered. Assuming that the periodic solution was originally stable, this perturbation will initially grow as $\Delta\overline{x}_k = c_k\overline{u}_k(t)e^{(\sigma+j(\omega_o/2))t} + c_k^*\overline{u}_k^*(t)e^{(\sigma-j(\omega_o/2))t}$, with $\overline{u}_k(t)$ a periodic vector at ω_o, after the bifurcation. Thus, the subharmonic frequency $\omega_o/2$ is generated at the bifurcation point η_b. It must be noted that the periodic solution at ω_o continues to exist after the flip bifurcation, although it is unstable and thus unobservable. From the point of view of the nonlinear system dimension, it must be kept in mind that the set of complex-conjugate poles $\sigma \pm j(\omega_o/2 + n\omega_o)$ corresponds to one single real multiplier, so they are associated with one system dimension only, defined by its associated periodic vector [see (3.17)].

RESONANCE AT THE DIVIDED-BY-2 FREQUENCY Figure 3.14 shows a simple circuit exhibiting a flip bifurcation. It is composed of a resistor, an inductance, and a varactor diode. Flip bifurcation occurs when increasing the input generator power, which leads to a nonlinear operation of the capacitances contained in the varactor diode. For analysis convenience a simpler circuit will be considered initially. It will be assumed that the resonant network is isolated from the driving source through appropriate filtering. Thus, we have an *RLC* resonator,

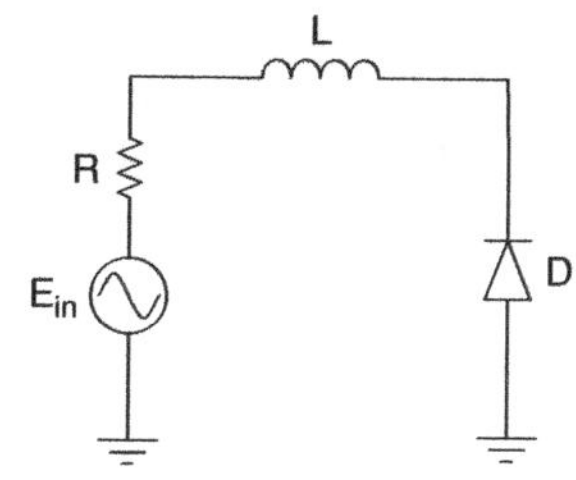

FIGURE 3.14 Varactor-based circuit exhibiting a flip bifurcation versus the input generator voltage.

with the capacitance (parameter) varying periodically at the frequency of the source. The nonlinear capacitance will be modeled using the well-known junction capacitance expression. Under the pumping of the input signal $v_{in}(t) = E_{in} \cos \omega t$ and assuming that the amplitude E_{in} is not too large, it will be possible to carry out a Taylor series expansion of this capacitance about the quiescent voltage V_o:

$$c(t) = \frac{c_{jo}}{\{1 - [V_o + v_{in}(t)]/\phi_o\}^{\gamma}} \cong \frac{C_o}{1 + m \cos \omega_{in} t} \tag{3.26}$$

with $C_o = c(V_o)$ and $m = v_{in}(t)\gamma/(-\phi_o + v_o)$. Therefore, $c(t)$ has a periodic variation. Next, expression (3.26) for $c(t)$ will be introduced in the differential equation ruling circuit behavior:

$$Ri + L\frac{di}{dt} + \frac{1}{C(t)}\int i(t)\,dt = \frac{d^2q}{dt^2} + \frac{R}{L}\frac{dq}{dt} + \omega_o^2(1 + m \cos \omega_{in} t)q = 0 \tag{3.27}$$

where q is the charge in the capacitor and $\omega_o = 1/\sqrt{LC_o}$. Assuming initial conditions different from zero and $m = 0$, the positive damping term R/L gives rise to the extinction of any oscillation at the natural frequency ω_o. Once this is known, to simplify the analysis of (3.27), the damping term R/L will be removed from (3.27), which provides the ideal equation:

$$\frac{d^2q}{dt^2} + \omega_o^2(1 + m \cos \omega_{in} t)q = 0 \tag{3.28}$$

Equation (3.28) is a linearized version of the well-known Mathieu equation [2]. For $m = 0$ the circuit would behave as an ideal conservative oscillator at the frequency ω_o. This is in agreement with the two roots of the associated characteristic system $s = \pm j\omega_o$.

We know that for $m = 0$ the oscillation will actually vanish in time, due to the existence of the positive damping term R/L in the physical equation (3.27). For $m \neq 0$ the system becomes a linear system with periodic coefficients, which can be solved with the aid of Floquet theory.

Renaming $x = q$, $y = dq/dt$, two independent solutions of the linear equation will be (x_1, y_1) and (x_2, y_2). The determinant associated with the fundamental solution matrix is $D = x_1 y_2 - x_2 y_1$. The time derivative of this determinant is calculated as $\dot{D} = \dot{x}_1 y_2 + x_1 \dot{y}_2 - \dot{x}_2 y_1 - x_2 \dot{y}_1$ and it is equal to zero in this particular case, $\dot{D} = 0$, as can easily be derived from equation (3.28). As already known, the canonical fundamental solution matrix $W_c(t)$ is obtained by integrating the linear system from two different initial vectors, given by $(x_{1o}, y_{1o})^T = (1, 0)^T$ and $(x_{2o}, y_{2o})^T = (0, 1)^T$. In turn, the monodromial matrix $W_c(T)$ agrees with the canonical fundamental matrix $W_c(t)$, evaluated at $t = T$, with T the period of the coefficients in the original linear system. In (3.28), this period is $T = 2\pi/\omega_{\text{in}}$. Clearly, the determinant associated with the initial value of $W_c(t)$ is $D(0) = D_o = 1$, because as already indicated, the used initial values are given by the two columns of the 2×2 identity matrix. The determinant does not depend on time in this particular case, so the determinant of the monodromial matrix $W_c(T)$ must also be $D = D_o = 1$. The determinant of a given matrix is equal to the product of its eigenvalues; thus, the product of the two Floquet multipliers (eigenvalues of the monodromial matrix) associated with equation (3.28) is $\mu_1\mu_2 = 1$. Depending on the values of m and ω_o, which are the two parameters of (3.28), the two multipliers can be complex-conjugate $\mu_{1,2} = e^{\pm j\alpha\omega_{\text{in}}}$, with $\alpha \in R$, they can be both equal to $+1$, both equal to -1, or real and reciprocal, with the same sign. The two multipliers $\mu_{1,2} = e^{\pm j\alpha\omega_{\text{in}}T}$ indicate neutral stability of an oscillation at $\alpha\omega_{\text{in}}$, coexisting with the input generator frequency ω_{in}. We will have this situation for $\omega_o \neq k/2\omega_{\text{in}}$, with k integer. Recalling that the positive damping term *R/L* has been suppressed for this simplified analysis, this oscillation will actually be extinguished. However, for $\omega_o = (2k+1)\omega_{\text{in}}/2$ and k different from zero, we can have $\mu_1 < -1 < \mu_2$. The multiplier $\mu_1 < -1$ will give rise to the onset of a frequency division. Note that in order to reach the steady-state subharmonic oscillation, a nonlinear model of the capacitance is necessary. In the nonlinear version of Mathieu's equation [2], this phenomenon occurs for a relatively large set of m or ω_{in} values, forming a resonance region in the plane defined by ω_{in} and m. In fact, there are multiple resonance regions about the input frequencies $\omega_{\text{in}} = 2\omega_o/(2k+1)\omega_{\text{in}} \geq 1$, although the one occurring for $k = 1$ is the most relevant. This explains the common occurrence of frequency division by 2 in parametric circuits [10,11] or circuits with a nonlinear reactance pumped by a periodic input source. Remember that ω_o is the resonance frequency of the varactor capacitance at its bias point c_o and the series inductor. If the drive frequencies deviates slightly from 2 ω_o, the average capacitance exhibited by the varactor shifts a little to keep the same frequency ratio 1/2. This locking phenomenon is enabled by the nonlinearity of the varactor capacitance.

As an example, the variation of the poles of the circuit of Fig. 3.14 versus the input amplitude has been analyzed with an accurate numerical technique based on pole–zero identification. Figure 3.15 shows the variation in the circuit-dominant poles (the poles closest to the imaginary axis) versus the input voltage amplitude E_{in} at the constant input frequency $f_{\text{in}} = 5.8$ GHz. The imaginary part of the pair of poles agrees approximately with the input frequency divided by 2. The pair of poles crosses the imaginary axis at $E_{\text{in}} = 1.64$ V, giving rise to a flip bifurcation, and as will be shown, to the generation of a subharmonic solution.

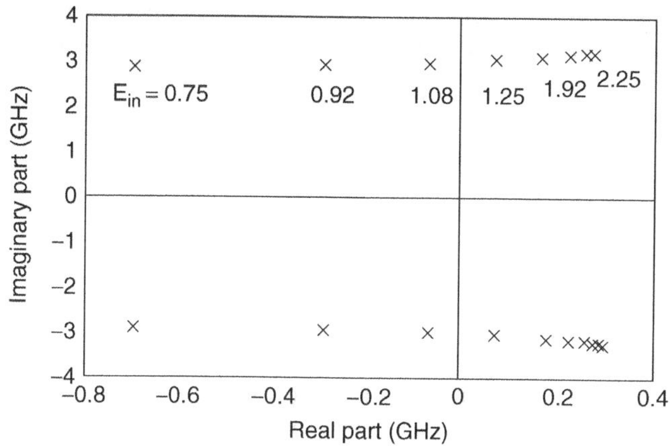

FIGURE 3.15 Evolution of the critical pair of complex-conjugate poles of the *RL* diode circuit of Fig. 3.14 versus the input generator amplitude E_{in} for constant input frequency $f_{in} = 5.8$ GHz.

Figure 3.16 presents the variation in steady-state solutions of the same circuit versus the input voltage E_{in} for constant input frequency $f_{in} = 5.8$ GHz. For low E_{in}, the only mathematical solution is a periodic solution at the generator frequency ω_{in}, which is represented by tracing the first-harmonic amplitude. This nondivided solution is stable up to the flip bifurcation, occurring for $E_{in} = 1.64$ V, where the subharmonic solution is generated. This solution is represented by tracing the

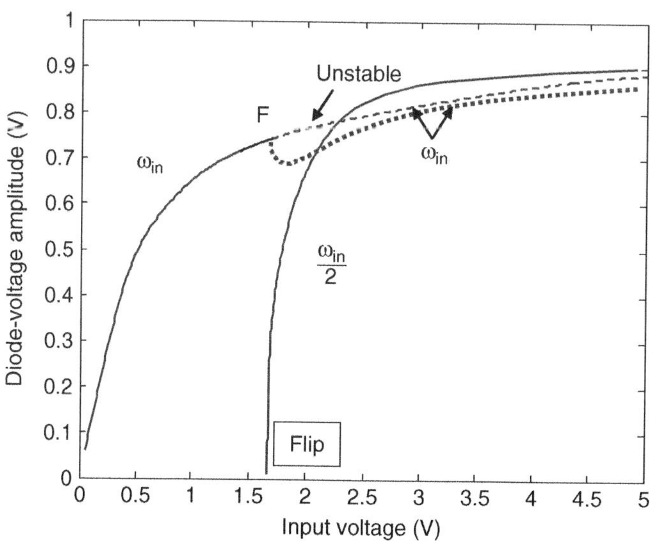

FIGURE 3.16 Bifurcation diagram of the circuit in Fig. 3.13 versus the input generator voltage E_{in}. A flip bifurcation occurs at the input voltage $E_{inb} = 1.64$ V.

amplitude of the diode voltage at the two frequency components ω_{in} and $\omega_{in}/2$. Note that the former periodic solution at ω_{in} continues to exist after the bifurcation, although it is unstable. After the flip bifurcation, the only observable solution is the frequency-divided solution, represented by the dotted line (the ω_{in} component) and the solid line at $\omega_{in}/2$. Due to the local nature of the flip bifurcation, the subharmonic amplitude tends to zero at the bifurcation point. After the flip bifurcation, there is quick growth of the subharmonic amplitude in order to balance the excess negative resistance exhibited by the device at $\omega_{in}/2$ with the positive resistance exhibited by the linear circuit. This is achieved through growth of the subharmonic amplitude. In this region, the phase values of the various harmonic components undergo a significant variation versus the input generator amplitude.

SUPERCRITICAL AND SUBCRITICAL FLIP BIFURCATIONS In a manner similar to the Hopf and pitchfork bifurcations from the dc regime, the flip bifurcations can be classified as supercritical or subcritical. Due to the continuity of the flip bifurcation, the bifurcating solution at ω_o gives rise to a degenerate frequency-divided solution at $\omega_o/2$ of subharmonic amplitude, tending to zero at the bifurcation point. This degenerate solution has a pair of complex-conjugate poles at $\pm j\omega_o/2$ on the imaginary axis. Because the fundamental frequency of this solution is $\omega_o/2$, the pair of poles $\pm j\omega_o/2$ will also be associated with a single real pole at zero $\gamma = 0$ due to the periodicity of the poles, in the form $\gamma \pm jk\omega_o/2$. With a supercritical bifurcation, these poles move to the left-hand side of the plane as the amplitude of the subharmonic increases continuously from zero. In the case of a subcritical bifurcation, they will move to the right-hand side of the plane. Note that unlike the case of Hopf bifurcations, there is no real pole staying at the origin for all the parameter values.

Unlike free-running oscillations away from the bifurcation point, the steady-state solution at the divided frequency $\omega_o/2$ does not contain a pair of imaginary poles located permanently on the imaginary axis. If this were the case, due to the nonunivocal relationship between the poles and the Floquet multipliers, the solution would also have a real pole at zero. Thus, the system would be singular, with invariance versus phase shifts, which is not true due to the rational relationship between the input frequency and the subharmonic oscillation frequency.

Observing the solution diagram versus the parameter, the supercritical and subcritical bifurcations can be distinguished from the same geometrical considerations as those discussed for Hopf bifurcations. Clearly, the pole $\gamma = 0$ at the bifurcation will give rise to a very high slope of the subharmonic amplitude versus the parameter immediately after this bifurcation. Assume that the parameter is varied in the sense for which the periodic solution at ω_o (from which the subharmonic emerges) evolves from *stable* to *unstable*. Then, in the case of a supercritical bifurcation (like in Fig. 3.16), the amplitude of the subharmonic component will exhibit positive slope versus the parameter immediately after the bifurcation and will never coexist with the stable solution at ω_o. For a subcritical bifurcation, it will exhibit negative slope and will coexist with the unstable periodic solution at ω_o prior to the bifurcation.

Without tracing the subharmonic solution curve, these two types of flip bifurcation can be distinguished, obtaining the normal form of the original nonlinear system about the bifurcation. Because the solution giving rise to the bifurcation is periodic instead of constant, the normal-form system will be a discrete system. The center manifold associated with the multiplier responsible for the bifurcation will have dimension 1, as the instability is associated with a single real multiplier escaping from the unit cycle through the point -1. In the frequency domain, the two types of bifurcation can be distinguished by obtaining the incipient divided-by-2 solution, with very small amplitude, and applying numerical techniques to obtain the poles associated with this solution. For a subcritical bifurcation, the incipient subharmonic solution at $\omega_{in}/2$ will contain an unstable real pole or, due to the periodicity of the poles, a pair of unstable complex-conjugate poles at the same frequency $\omega_{in}/2$.

FLIP BIFURCATIONS IN THE PHASE SPACE Figure 3.17 shows the qualitative variation of a cycle in the phase space when the system undergoes flip bifurcation. Figure 3.17a shows the cycle prior to the bifurcation ($E_{in} = 1.63$ V), and Fig. 3.17b shows this steady-state cycle just after the bifurcation ($E_{in} = 1.65$ V). The cycle doubles, as it takes the system twice the original period to return to the original values of the circuit variables. Initially, the doubled cycle overlaps with the original cycle, which is due to the continuity of the bifurcation. When drawing the amplitude of the subharmonic component $\omega_o/2$ of the steady-state solution versus the parameter η, it is seen that this subharmonic component arises from zero amplitude at the bifurcation point η_b.

FLIP BIFURCATIONS IN THE POINCARÉ MAP The Poincaré map gives additional insight into flip bifurcation. Remember that the map is obtained through intersection of the solution with a transversal surface Σ of small size. The map of Fig. 3.2 actually corresponds to the circuit of Fig. 3.14 considered in the simulations of Figs. 3.15 to 3.17. The map is obtained by sampling the steady-state solutions at integer multiples of the input generator period nT_{in}. As shown in Fig. 3.2 for a low input voltage, the map provides a single point. When flip bifurcation occurs, two points are obtained at the intersection with this surface. This can be seen as the result of cycle doubling, observed in Fig. 3.17b. It can also be related to the fact that due to the new periodicity $2T$ of the solution, it takes the system twice the original period to return to the same point on the map. Note that as shown in the bifurcation diagram of Fig. 3.16, the nondivided solution at the input generator frequency ω_{in} continues to exist after flip bifurcation. Due to the unstable poles $\sigma \pm j\omega_o/2$ with $\sigma > 0$, this solution cannot be obtained through standard time-domain integration, as the system gets away from it in simulation and converges to the frequency-divided steady-state solution. This is why it has not been represented in Fig. 3.2. However, the nondivided solution is still a valid mathematical solution which for each parameter value would give rise to a single point located between the pairs of points of the divided-by-2 regime (Fig. 3.2). The name *flip* comes from the fact that the steady-state period-doubled solution seems

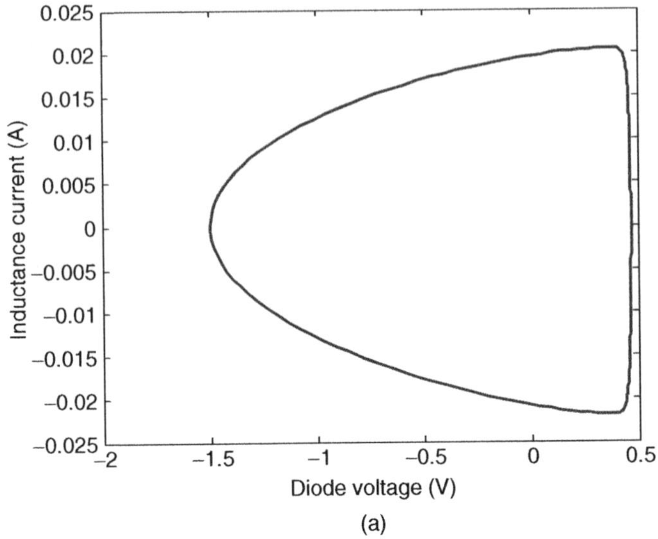

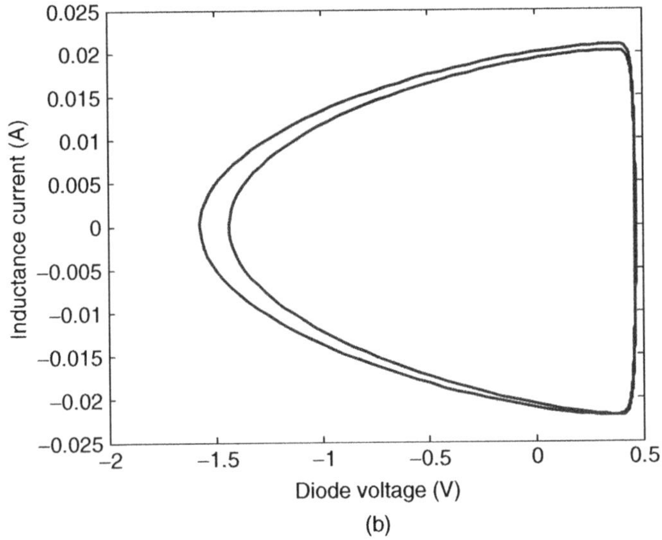

FIGURE 3.17 Qualitative variation of the cycle due to the flip bifurcation in Fig. 3.16: (a) periodic cycle corresponding to $E_{in} = 1.63$ V; (b) doubled periodic cycle corresponding to $E_{in} = 1.65$ V.

to bounce (flip) at each parameter value from one side to another of the unstable single point. Actually, when supercritical flip bifurcation occurs, two possibly stable period-doubled solutions are generated: one located at $t_o + nT_{in}$ at one of the two points and the other located at the other point. The two solutions have a time shift T_{in} when represented versus the time variable, but otherwise, they are totally identical, with the same stability properties. The convergence to one or the other will depend on the initial conditions. Note that the two solutions are overlapped in the bifurcation diagram of Fig. 3.2, and should give rise to the same period-doubled cycle in the phase space.

FREQUENCY-DOMAIN ANALYSIS OF FLIP BIFURCATIONS The analytical frequency-domain analysis of the parametric divider in Fig. 3.14 provides an alternative way to understand the natural frequency division by 2 or flip bifurcation. For simplicity, the diode of Fig. 3.14 is replaced by the instantaneous capacitance $c(t) = c_o + c_1 v(t)$, with $v(t)$ being the voltage across the diode. The usual procedure for the divider design is to obtain a series resonance R-L-c_o at the desired subharmonic frequency $\omega_{in}/2$. The input generator voltage is generally written as $e_{in}(t) = E_{in} \cos \omega_{in} t$, so the phase origin for the circuit analysis is established by this input voltage, with associated phase $\phi_{in} = 0$. In turn, the voltage across the nonlinear capacitance is written as $v(t) = 2V_1 \cos[(\omega_{in}/2)t + \phi_1] + 2V_2 \cos(\omega_{in} t + \phi_2)$, where the existence of a subharmonic solution is assumed. Because the two frequencies ω_{in} and $\omega_{in}/2$ are related harmonically, the phase of the subharmonic oscillation depends on the input generator phase. If the phase of the input generator is varied as $e_{in}(t) = E_{in} \cos(\omega_{in} t + \phi)$, the new solution will have the phase value $v(t) = 2V_1 \cos[(\omega_{in}/2)t + \phi_1 + \phi/2] + 2V_2 \cos(\omega_{in} t + \phi_2 + \phi)$, which corresponds to a time translation of the waveform τ, with $\phi = -\omega_{in}\tau$. To understand the coexistence of the two frequency-divided solutions, note that the phase shift $\phi/2 = 0$ and $\phi/2 = \pi$ at the subharmonic component gives rise to the same input generator phase in a 2π modulus. The two steady-state solutions will be the same except for a time shift $\tau = T_{in}$. This is in agreement with the observation of the two different fixed points on the Poincaré map. Convergence toward one or the other will depend on the initial conditions.

The current through the nonlinear capacitance will be calculated using $i(t) = c(v)dv/dt$. To obtain the harmonic expansion of this current, the waveform expression $v(t) = 2V_1 \cos[(\omega_{in}/2)t + \phi_1] + 2V_2 \cos(\omega_{in} t + \phi_2)$ is replaced into $c(v)$. For calculation simplicity it is convenient to express $v(t)$ as $v(t) = V_1 e^{j((\omega_{in}/2)t+\phi_1)} + V_1 e^{-j((\omega_{in}/2)t+\phi_1)} + V_2 e^{j(\omega_{in} t+\phi_2)} + V_2 e^{-j(\omega_{in} t+\phi_2)}$ and then replace this expression into $i(t) = c(v)dv/dt$, assembling the harmonic terms of the same order, $\omega_{in}/2$ and ω_{in}. In this manner it is possible to determine the harmonic terms at $\omega_{in}/2$ and ω_{in} of the current flowing through the nonlinear capacitance. They are given by

$$\tilde{I}_1(\tilde{V}_1, \tilde{V}_2) = c_o V_1 \frac{\omega_{in}}{2} e^{j(\phi_1+\pi/2)} + c_1 V_1 V_2 \frac{\omega_{in}}{2} e^{j(\phi_2-\phi_1+\pi/2)} = j\frac{\omega_{in}}{2} \tilde{Q}_1(\tilde{V}_1, \tilde{V}_2)$$
$$\tilde{I}_2(\tilde{V}_1, \tilde{V}_2) = c_o V_2 \omega_{in} e^{j(\phi_1+\pi/2)} + c_1 V_1^2 \frac{\omega_{in}}{2} e^{j(2\phi_1+\pi/2)} = j\omega_{in} \tilde{Q}_2(\tilde{V}_1, \tilde{V}_2) \tag{3.29}$$

where $\tilde{V}_1$ and $\tilde{V}_2$ are the first- and second-harmonic voltage phasors and $\tilde{Q}_1$ and $\tilde{Q}_2$ are the first- and second-harmonic phasors associated with the nonlinear charge. As seen in (3.29), the periodically pumped capacitance gives rise to a phase shift values different from $\pi/2$. Applying Kirchhoff's laws, the resulting frequency-domain equations are the following:

$$\left(R + jL\frac{\omega_{in}}{2}\right) j\frac{\omega_{in}}{2}\tilde{Q}_1(\tilde{V}_1, \tilde{V}_2) + \tilde{V}_1 = 0 \qquad \text{(a)}$$
$$(R + jL\omega_{in}) j\omega_{in}\tilde{Q}_2(\tilde{V}_1, \tilde{V}_2) + \tilde{V}_2 - \frac{E_{in}}{2} = 0 \qquad \text{(b)} \tag{3.30}$$

Note that (3.30) is a well-balanced system of two complex equations in two complex unknowns. In fact, it constitutes a harmonic balance formulation of the circuit in Fig. 3.13, limited to two harmonic terms. Note that as indicated in Chapter 1, the harmonic balance equations of circuits containing nonlinear capacitances are generally written in terms of the harmonic components of the corresponding nonlinear charges. Each harmonic component $\tilde{Q}_k$ is then multiplied by $jk\omega_o$ to obtain the harmonic component of the current at the same frequency $jk\omega_o$. System (3.30) is an example of this procedure.

It is interesting to observe that system (3.30) contains a homogeneous equation at the subharmonic frequency $\omega_{in}/2$. By *homogeneous*, here we mean that the system admits a zero solution. Thus, system (3.30) can be solved for a nondivided solution, even when frequency division actually takes place. This property can be generalized to all the systems exhibiting frequency division and justifies why the nondivided solution coexists with the divided solution in the bifurcation diagram of Fig. 3.16. The situation can be compared to that of free-running oscillators, which can always be solved for a dc solution.

After some manipulation, system (3.30a) can be rewritten

$$Y_{1/2}(\tilde{V}_1, \tilde{V}_2) = \frac{j(\omega_{in}/2)\tilde{Q}_1(\tilde{V}_1, \tilde{V}_2)}{\tilde{V}_1} + \frac{1}{R + jL(\omega_{in}/2)} = 0 \tag{3.31}$$

Equation (3.31) indicates that in order to get a frequency division, the circuit total admittance must be zero at the subharmonic frequency $\omega_{in}/2$. This is similar to the oscillation condition in free-running circuits, examined in Chapter 2. However, in the case of equation (3.30), the oscillation frequency is determined by the input generator, as it takes the subharmonic value $\omega_{in}/2$. Unlike the case of free-running oscillators, the phase origin cannot be fixed arbitrarily, which is due to the harmonic relationship between the input generator and oscillation frequencies that leads to a common period. The generator provides the phase reference of the circuit, and the two harmonic terms $\tilde{V}_1$ and $\tilde{V}_2$ must be solved in terms of both amplitude and phase.

For condition $Y_{1/2} = 0$ to be fulfilled, the capacitance must exhibit negative conductance. This can be possible due to its voltage dependence and its capability to provide phase shift between $\tilde{I}_1$ and $\tilde{V}_1$ at $\omega_{in}/2$ associated to negative conductance. In terms of the capacitance coefficients c_o and c_1, the input admittance exhibited by the capacitor, corresponding to the first term of (3.31), is given by

$$Y_{cap} = \frac{c_o V_1(\omega_{in}/2)e^{j(\phi_1+\pi/2)} + c_1 V_1 V_2 \frac{\omega_{in}}{2} e^{j(\phi_2-\phi_1+\pi/2)}}{V_1 e^{j\phi_1}} \tag{3.32}$$

where the two voltage phasors have been written in terms of amplitude and phase. Provided that the phase difference between the numerator and denominator of expression (3.32) falls between 90 and 270°, the nonlinear capacitance will exhibit negative resistance. From a certain amplitude of the input source voltage E_{in}, phase and amplitude solutions V_1, V_2, ϕ_1, and ϕ_2 will exist such that the real part of Y_{cap} has negative value and the imaginary part agrees exactly with the opposite of the susceptance exhibited by the inductive load. The second term in (3.32) adds real and imaginary contributions to the small-signal susceptance $c_o\omega_{\text{in}}/2$. Due to the phase shift dependence of (3.31), the condition for the existence of the subharmonic component will be fulfilled in a frequency band for a different phase value at each frequency. Thus, it is not only fulfilled at the frequency of the series resonance R-L-c_o. The parametric oscillation may also take place at a nonrational frequency $\omega = \alpha\omega_{\text{in}}$, $\alpha \in R$. Undesired parametric oscillations (either subharmonic or not) are often obtained in nonlinear circuits, such as power amplifiers and frequency multipliers, due to the nonlinear behavior of the device capacitances. The parametric oscillations are never observed for a low amplitude of the input signal. This is because at a small signal level, the capacitance behaves as a standard constant signal [see expression (3.32)], providing an ordinary 90° phase shift. To obtain negative resistance, a relatively high degree of pumping from the input generator is necessary. In summary, the nonlinear charges (as well as the nonlinear fluxes) have a phase-shifting capability that, from certain periodic-pumping amplitude, can give rise to negative resistance at a frequency different from that of the pumping source.

As an example, the element values $c_o = 1.1$ pF, $c_1 = 2.08$ pF/V, $R = 49\ \Omega$, and $L = 9.1$ nH will be considered. Solving system (3.30) for $E_{\text{in}} = 4.33$ V and $f_{\text{in}} = 4.62$ GHz, a subharmonic solution is obtained, with the following components of the voltage across the capacitance: $V_1 = 1.459e^{j86^\circ}$ and $V_2 = 0.638e^{-j22.9^\circ}$. The harmonic components of the current through the nonlinear capacitance are $I_1 = 10.32e^{-j163.7^\circ}$ mA and $I_2 = 13.56e^{-j75.6^\circ}$ mA. The admittance exhibited by the capacitance at the subharmonic frequency is $Y_{\text{cap}} = -0.0025 + j0.0066\ \Omega^{-1}$. The negative conductance, plus the resonance of the capacitive imaginary part with the circuit inductor, enables the subharmonic oscillation.

DETECTION OF FLIP BIFURCATIONS As already indicated, the flip bifurcations from the periodic regime at ω_{in} are associated with the crossing of a pair of complex-conjugate poles $\sigma \pm j\omega_{\text{in}}/2$ through the imaginary axis of the complex plane. The parameter values providing this type of bifurcation can be determined through root analysis of the characteristic determinant associated with the linearized system. A different way to detect flip bifurcation can be derived from the fact that the amplitude of the subharmonic $\omega_{\text{in}}/2$ tends to zero at the bifurcation point (Fig. 3.16). Thus, the flip bifurcation occurring at parameter value η_b can be detected by adding the following condition to the set of circuit equations in the frequency domain:

$$Y_{1/2}(\overline{V}, V_{1/2} = 0, \phi, \eta_b) = 0 \tag{3.33}$$

where $Y_{1/2}$ is the input admittance at a given observation port. The vector $\overline{V}$ consists of all the harmonic terms, $k\omega_o$, with k an integer, that are considered. Of the circuit state variables, $V_{1/2}$ is the subharmonic voltage amplitude at the observation port and ϕ is the corresponding subharmonic phase. In Chapter 6, details are provided about implementation of condition (3.33) on a frequency-domain simulator.

As an example, introduction of the condition $V_1 = 0$ in system (3.30) leads to the system

$$c_o\frac{\omega_{in}}{2}e^{j(\pi/2)} + c_1 V_2\frac{\omega_{in}}{2}e^{j(\phi_2-2\phi_1+\pi/2)} + \frac{1}{R + jL(\omega_{in}/2)} = 0$$
$$(R + jL\omega_{in})c_o V_2\omega_{in}e^{j(\phi_1+\pi/2)} + \tilde{V}_2 - \frac{E_{in}}{2} = 0 \quad (3.34)$$

which must be solved in terms of the four real unknowns V_2, ϕ_1, ϕ_2, and E_{in}. Resolution of this system allows direct calculation of the input generator voltage E_{in} at the flip bifurcation point, which is given by $E_{inb} = 4.83$ V. Considering (3.29), (3.30), and (3.32), we can gather that the phase values ϕ_1 and ϕ_2 will undergo significant variation after the flip bifurcation, due to the quick growth of the subharmonic amplitude V_1 versus E_{in} (see Fig. 3.16). From a certain E_{in} value, the squared amplitude V_1^2 tends to evolve E_{in} proportionally and the phase sensitivity is reduced [see (3.30b)].

The flip bifurcation can be obtained in forced circuits (containing a periodic generator) like the one in Fig. 3.14 or in free-running oscillators. In the latter case, the flip bifurcation gives rise to the division by 2 of the self-generated oscillation frequency. Thus, from the flip bifurcation, a periodic solution arises containing the self-generated oscillation frequency ω_o and the subharmonic $\omega_o/2$. An example of this type of regime was shown in the Colpitts oscillator spectrum of Fig. 1.23. For a free-running oscillator undergoing frequency division by 2, the oscillation frequency at the bifurcation point is an unknown to be determined. On the other hand, due to irrelevance versus phase shifts, it will be possible to assign a zero phase value arbitrarily to one of the harmonic components of one of the state variables. For detection of the flip bifurcation point at which the frequency division originates, the condition (3.33) should be replaced with

$$Y_{1/2}(\overline{V}', \omega_o, V_{1/2} = 0, \phi, \eta_b) = 0 \quad (3.35)$$

where ϕ stands for the phase shift between subharmonic and primary oscillations and the state-variable vector $\overline{V}$ has been replaced by the vector $\overline{V}'$, containing one less element. In exchange, the frequency of the primary oscillation is an unknown of the problem, because it is generated autonomously and varies with the parameter.

Secondary Hopf or Neimark Bifurcation A pair of complex-conjugate multipliers m_k and m_{k+1} fulfilling $m_k = m_{k+1}^* \in C$ cross the unit circle through the points $e^{\pm j\theta}$, with $\theta \neq 2n\pi, \theta \neq (2n+1)\pi$, and n an integer, at the parameter value η_b.

The following conditions are fulfilled:

$$
\begin{aligned}
& m_{k,k+1}(\eta_b) = e^{\pm j\theta} \\
& \left.\frac{dm_{k,k+1}}{d\eta}\right|_{\eta_b} \neq 0
\end{aligned}
\tag{3.36}
$$

In exponential form, the critical pair of complex-conjugate multipliers can be written $m_{k,k+1} = e^{\pm j\theta} = e^{\pm j\alpha\omega_o T \pm jn\omega_o T} = e^{\pm j\alpha\omega_o T}$, with $\alpha \in R$ and n a positive integer. Due to the relationship (3.18) between the poles and the Floquet multipliers, the condition above is equivalent to an infinite set of complex-conjugate poles of the form $\sigma \pm j(\alpha\omega_o + n\omega_o)$, with n an integer, crossing though the imaginary axis of the complex plane at the parameter value η_b. Assuming that the periodic solution was originally stable, and taking (3.17) into account, after the bifurcation the perturbed variables will initially grow as $\Delta\overline{x}_k = c_k\overline{u}_k(t)e^{(\sigma+j\alpha\omega_o)t} + c_k^*\overline{u}_k^*(t)e^{(\sigma-j\alpha\omega_o)t}$, with $\overline{u}_k(t)$ a periodic vector at ω_o. Thus, a second fundamental frequency $\omega_a = \alpha\omega_o$, related nonharmonically to ω_o, is generated at the bifurcation point η_b and gives rise to a quasiperiodic solution with the two fundamental frequencies ω_o and ω_a. In the phase space, a cycle becomes a 2-torus [8] at the Hopf bifurcation. In the Poincaré map, a point becomes a cycle of discrete points like the one presented in Fig. 3.1. At the bifurcation, the cycle has zero area about the fixed point corresponding to the periodic solution. Taking into account that the relationship between the Floquet multipliers and the Floquet exponents is not univocal, the critical poles $\sigma \pm j\omega_a$ can also be expressed in terms of the baseband difference frequency $\Delta\omega = \omega_o - \omega_a$ in the following manner: $\sigma \pm j(\Delta\omega + n\omega_o)$. From the point of view of the nonlinear system dimension, it must be kept in mind that the set of complex-conjugate poles $\sigma \pm j(\Delta\omega + n\omega_o)$ corresponds to a pair of complex-conjugate multipliers, so they are associated with two system dimensions, defined by their associated periodic vectors [see (3.17)].

Secondary Hopf bifurcation is very common in practical circuits. It is often found in power amplifiers or frequency multipliers when increasing the input power. The circuit behaves initially in a periodic regime at ω_{in}. Then, from a certain input power, an oscillation is generated at the frequency ω_a, which is often due to the nonlinear capacitances contained in these devices that exhibit negative resistance from certain amplitudes of the periodic pumping signal. The mechanism is very similar to the one explained in (3.29)–(3.32) concerning frequency division by 2. The only difference is that in the case of a Hopf bifurcation, the oscillation condition is fulfilled at an incommensurate frequency $\omega_a = \alpha\omega_o$ instead of the divided-by-2 frequency $\omega_o/2$, due to the particular circuit topology and element values. The Hopf bifurcation is also typical in injection-locked oscillators, as it is one of the mechanisms through which the oscillation loses its synchronized state [1].

Although the behavior of injection-locked oscillators is treated in detail in the next section, we deal with one of these circuits in the following example. At this point, only the geometric aspects of a secondary Hopf bifurcation are considered. An example concerning an injection-locked oscillator has been chosen to enable an immediate comparison with the local–global bifurcation (presented in the next subsection), which also leads from a periodic to a quasiperiodic regime, but through a different type of transformation with very different properties. Let a free-running oscillator at a frequency ω_o be considered. When a periodic input source of power P_{in} and frequency ω_{in}, relatively close to ω_o, is introduced, the oscillation frequency becomes equal to ω_{in} in a certain input frequency interval. When varying ω_{in}, the oscillation frequency ω_a varies, too, according to $\omega_a = \omega_{in}$. The equality $\omega_{in} = \omega_a$ is maintained within a certain synchronization band which is broader for higher power P_{in} delivered by the input source, due to the stronger influence of this source on self-oscillation. In synchronized state, there will be a phase relationship between the oscillation and the input source. As a simple example, consider the one-harmonic description of the parallel oscillator in which the free-running oscillation ω_o fulfills the resonance condition $jB(\omega) = j(C\omega - 1/L\omega) = 0$. If the oscillation is synchronized to an input current generator connected in parallel at the frequency ω_{in}, the susceptance will differ from zero at ω_{in}, $jB(\omega_{in}) \neq 0$, which will give rise to a certain phase relationship between the node voltage and the input generator. Note that this is just a very rough explanation. If the total susceptance jB depends on both ω and the voltage amplitude, we will have a phase shift different from zero even at $\omega_{in} = \omega_o$.

Depending on the input power level, the stable synchronization band is delimited by two different types of bifurcations. For the lower input power range, it is delimited by bifurcations of local–global type which are studied in the next section. For the higher input power range, it is delimited by secondary Hopf bifurcations. To illustrate the properties and implications of the secondary Hopf bifurcation, a periodic current generator $i_g(t)$ will be introduced in parallel in the cubic nonlinearity oscillator of Fig. 1.1. In the absence of this generator, this circuit exhibits self-oscillation at $f_o = 1.59$ GHz. For relatively high input generator current I_g, assumed constant, and input frequency ω_{in} within a certain interval $(\omega_{in}^{H1}, \omega_{in}^{H2})$ about the free-running oscillation value ω_o, the only stable solution will be a periodic solution at the same frequency of the input generator ω_{in}. To see this, the corresponding solutions of the Poincaré map will be represented versus the input frequency ω_{in}. The constant input current considered is $I_g = 25$ mA. Because the system is nonautonomous, the Poincaré map can be obtained by sampling the steady-state solutions at integer multiples of the input signal period $T_{in} = 2\pi/\omega_{in}$. This map is shown in Fig. 3.18. Within the interval of stable periodic operation at ω_{in}, a single point is obtained when sampling at nT_{in}. This interval is delimited by two Hopf bifurcations. At each of these bifurcations, an oscillation at the frequency ω_a is generated which gives rise to a self-oscillating mixer regime. In the phase space, the continuous cycle of period T_{in} turns into a 2-torus. The torus is overlapped with the cycle at the bifurcation point. In correspondence with this, the fixed point of the Poincaré map, corresponding to the periodic solution, becomes a cycle consisting of discrete points at the Hopf bifurcation. Because of the continuity of the local bifurcations, the discrete-point cycle arises at the bifurcation point with zero amplitude about the original fixed point of the map, so it can be seen as a degenerate cycle. After

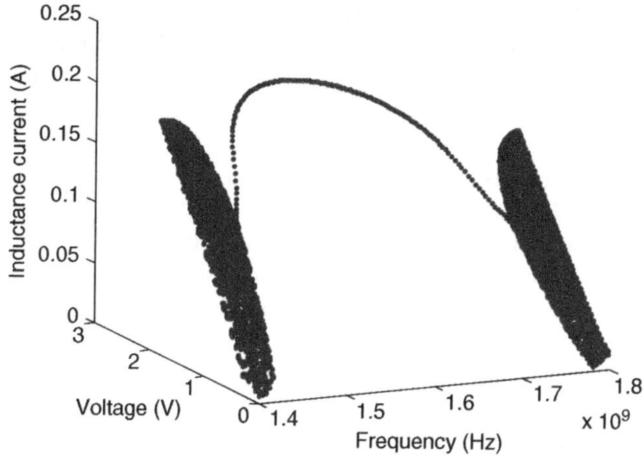

FIGURE 3.18 Analysis of the circuit of Fig. 1.1 when introducing a current generator in parallel. The generator amplitude considered is $I_g = 25$ mA. The analysis parameter is the input frequency ω_{in}. The Poincaré map represented has been obtained by sampling the steady-state solution at integer multiples of the period T_{in} for each parameter value.

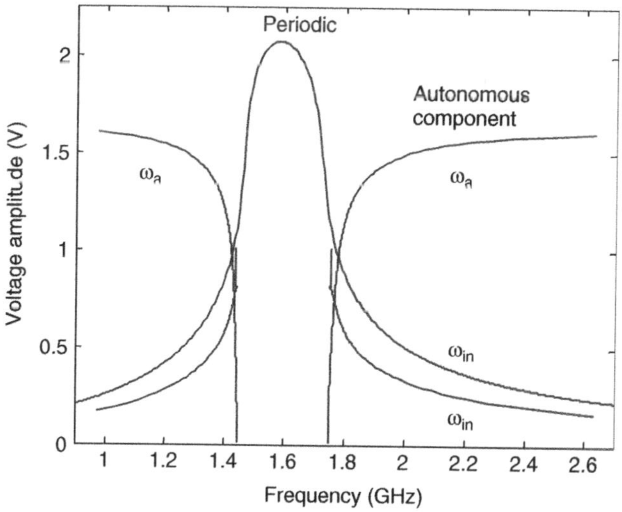

FIGURE 3.19 Frequency-domain analysis of the solutions of the circuit of Fig. 1.1 when introducing a current generator of $I_g = 25$ mA. The bifurcation diagram versus ω_{in} is the frequency-domain equivalent of the Poincaré map of Fig. 3.18.

the Hopf bifurcation, the solution at ω_{in} continuous to exist but is unstable, as it has two complex-conjugate poles on the right-hand side of the complex plane. In the Poincaré map this unstable solution corresponds to a point located inside the discrete cycle.

The circuit will also be analyzed in the frequency domain, considering, similarly, a constant input current amplitude $I_g = 25$ mA, taking the input frequency ω_{in} as a parameter. The results are shown in Fig. 3.19. The periodic solution is represented by means of the voltage amplitude at its fundamental frequency ω_{in}. This periodic solution exists for all the values of the parameter ω_{in} considered. However, it is only stable between the two Hopf bifurcations H_1 and H_2, that is, within the frequency interval 1.42 to 1.76 GHz. Outside this interval, the periodic solution has a pair of complex-conjugate poles on the right-hand side of the complex plane, so it is not physically observable. Instead, the circuit behaves as a self-oscillating mixer at the two fundamental frequencies ω_{in} and ω_a. This solution consists of intermodulation products of the form $k\omega_{\text{in}} + m\omega_a$, with k and m integers, in all the circuit variables. In Fig. 3.19 the self-oscillating mixer solution has been represented by drawing the voltage amplitude at the input frequency ω_{in} and oscillation frequency ω_a. As can be seen, the harmonic component at ω_a tends to zero at the two Hopf bifurcations. This is in agreement with the degenerate cycle of zero amplitude obtained at the Poincaré map. Because the frequency ω_a is generated autonomously, its value will change with the parameter ω_{in} along the solution curve.

Details on how to simulate the self-oscillating mixer solution in the frequency domain are given in Chapter 5. However, some brief hints will be given here. For an accurate frequency-domain analysis, a sufficiently high number of intermodulation terms $k\omega_{\text{in}} + m\omega_a$ must be taken into account. A criterion for the choice of the intermodulation products is provided by the *diamond truncation* [12,13], in which the intermodulation products selected must fulfill $|k| + |m| \leq N_L$, with N_L the nonlinearity order. Using the definition $\Delta\omega = \omega_{\text{in}} - \omega_a$, the two fundamental frequencies considered can be expressed as ω_{in} and $\omega_a = \omega_{\text{in}} - \Delta\omega$. If ω_{in} and ω_a have rather close values, the spectral lines $2\omega_{\text{in}} - \omega_a = \omega_{\text{in}} + \Delta\omega$ and $2\omega_a - \omega_{\text{in}} = \omega_{\text{in}} - 2\Delta\omega$ will be located in the immediate neighborhood of ω_{in}, and ω_a and will be relevant in the circuit operation, so the minimum N_L value should be $N_L = 3$.

In the circuit of Fig. 1.1, the independent variable is the node voltage. The nonlinear current depends on this voltage as $i(v)$. The frequency-domain equations for the self-oscillating mixer regime are obtained by defining the vectors $\overline{V}$ and $\overline{I}$, which contain the various harmonic terms of the corresponding variables. Then Kirchhoff's laws are applied at each harmonic frequency, in a manner similar to what was done in (3.30) when considering the defined vectors $\overline{V}$ and $\overline{I}(\overline{V})$. This leads to an equation system in matrix form. This system can be divided into two subsystems, one involving the harmonic terms of $k\omega_{\text{in}}$ only and the other involving the remainder of frequency terms, fulfilling the condition $|k| + |m| \leq N_L$. This separation provides the following system:

$$\begin{aligned}
\overline{V}^1(k\omega_{\text{in}}) + [Z(k\omega_{\text{in}})]\overline{I}^1(\overline{V}) &= [Z(k\omega_{\text{in}})]\overline{I}_g \\
\overline{V}^2(k\omega_{\text{in}} + m\omega_a) + [Z(k\omega_{\text{in}} + m\omega_a)]\overline{I}^2(\overline{V}) &= \overline{0} \qquad m \neq 0
\end{aligned} \tag{3.37}$$

where the vector $\overline{V}^1$ consists of the harmonic terms $V(k\omega_{\text{in}})$ and $\overline{V}^2$ consists of the remainder of the intermodulation products. Clearly, system (3.37) contains a homogeneous subsystem in $k\omega_{\text{in}} + m\omega_a, m \neq 0$, admitting a zero solution in $\overline{V}^2$. This explains why the solution with ω_{in} as the only fundamental is always a possible circuit solution, even when the self-oscillating mixer solution is the only stable solution. Because the frequency ω_a is generated autonomously, it constitutes an unknown of system (3.37). This frequency is related nonrationally to the input frequency ω_{in}, so the phase of the fundamental-frequency component at ω_a will have no influence on the components at $q\omega_{\text{in}}$, with q an integer, as the intermodulation products $k\omega_{\text{in}} + m\omega_a$, with $m \neq 0$, can never provide frequencies of the form $q\omega_{\text{in}}$. Thus, in the phase of a one-harmonic component, of the independent voltage v can be set arbitrarily to zero. If a different phase reference is chosen, the phase values of the intermodulation products $k\omega_{\text{in}} + m\omega_a$, with $m \neq 0$, will change so as to maintain the same relationships. If a phase shift $\Delta\phi_a$ is applied to the autonomous fundamental ω_a, the phase values of the intermodulation products become $\phi(k, m) + m\Delta\phi_a$. In a manner similar to free-running oscillations and frequency divisions, to be able to sustain the oscillation at ω_a, the device used must exhibit negative resistance and resonance at this frequency. Thus, the self-oscillating mixer solution of system (3.37) must fulfill the following oscillation condition, as is easily gathered from inspection of the system:

$$Y_a(\overline{V}, \omega_a) = 0 \tag{3.38}$$

In general circuit, Y_a is the admittance evaluated at the oscillation frequency ω_a at any observation port, and $\overline{V}$ is a vector that consists of the intermodulation products of the various state variables.

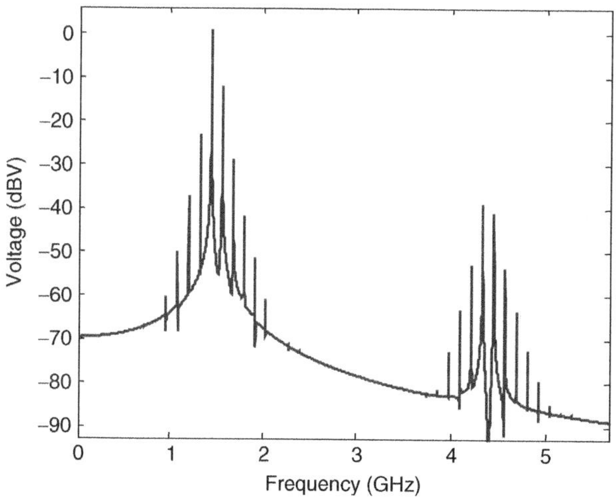

FIGURE 3.20 Spectrum of the quasiperiodic solution obtained for $I_g = 25$ mA and $f_{\text{in}} = 1.442$ GHz just after Hopf bifurcation, obtained using time-domain analysis. The autonomous frequency is $f_a = 1.561$ GHz.

As already seen, the input frequency interval for a stable periodic operation interval falls between 1.42 and 1.76 GHz. This interval is delimited by the two Hopf bifurcations. Immediately after the Hopf bifurcation, all the intermodulation products $k\omega_{in} + m\omega_a$, with $m \neq 0$, are of very small value, in correspondence with the low amplitude of the harmonic component at ω_a in Fig. 3.19. This can be seen in the spectrum of Fig. 3.20, obtained from a time-domain analysis of the circuit at $f_{in} = 1.442$ GHz. The autonomous frequency is $f_a = 1.561$ GHz. Note that the mixerlike spectrum arises with nonzero separation between the spectral lines, agreeing with the difference between the input and oscillation frequencies at which system (3.37) is fulfilled: $|\Delta\omega| = |\omega_{in} - \omega_a|$, equal to 119 MHz. This frequency difference is called *beat frequency*. At the Hopf bifurcation, it corresponds to the frequency $\Delta\omega$ of the pair of complex-conjugate poles $\sigma \pm j\Delta\omega$ crossing the imaginary axis. The fact that $|\Delta\omega| = |\omega_{in} - \omega_a|$ is different from zero at the secondary Hopf bifurcation explains why this type of bifurcation is also called *asynchronous*.

Direct calculation of Hopf bifurcations from a periodic regime at ω_o can be carried out by determining the parameter values at which a pair of complex-conjugate poles $\sigma \pm j\omega_a$, with $\omega_a/\omega_o \neq m/n$, crosses the imaginary axis. A different way to detect the Hopf bifurcation can be derived from the fact that the amplitude of the oscillation generated tends to zero as the bifurcation point is approached (Fig. 3.19). Taking this into account, the Hopf bifurcation, occurring at parameter value η_b, can be detected by adding the following oscillation condition to the set of circuit equations in the frequency domain:

$$Y_a(\overline{V}, V_a = 0, \omega_a, \eta_b) = 0 \tag{3.39}$$

where Y_a is the input admittance at a given observation port and V_a is the oscillation amplitude at the same node. Of course, equation (3.39) has to be combined with the general harmonic balance system (3.37) used to determine the steady-state solution. Because of the zero value of the oscillation amplitude V_a, condition (3.39) can be imposed, linearizing the circuit about the large-signal regime at $k\omega_o$, with k an integer. More details on the implementation of this Hopf bifurcation condition in frequency-domain simulations are provided in Chapter 4.

Although the preceding example corresponds to a Hopf bifurcation from a periodic regime occurring in a driven circuit at the frequency ω_{in}, this bifurcation can also take place from the periodic solution of a free-running oscillator at the frequency ω_o. In this case the Hopf bifurcation gives rise to a quasiperiodic solution at the frequencies ω_o and ω_a like the one presented in Fig. 1.24. This type of mixerlike regime often arises from the bias circuit instability in a free-running oscillator, so the oscillation frequency ω_o mixes with a rather low frequency ω_a generated from the resonance of the bias circuit elements. Note that both the frequency of the primary oscillation ω_o and that of the oscillation ω_a generated are autonomous and will constitute unknowns to be determined in the circuit analysis. In exchange, two phase values in one of the circuit state variables can be set arbitrarily to zero. Thus, the corresponding harmonic balance system will have the same number of equations and unknowns. The parameter value at the Hopf bifurcation can be detected using a condition analogous to (3.39).

3.3.2 Transformations Between Solution Poles

When analyzing a steady-state solution versus a given circuit parameter, the steady-state solution changes and so do its associated poles. Most of the stability analyses carried out in the book are based on calculation of these poles, either analytically or numerically [14,15]. Thus, some general comments on the types of transformations that we can expect in the structure of poles of a periodic solution will be of interest to understanding nonlinear circuit behavior.

1. In lumped circuits, the number of Floquet multipliers agrees with the system dimension given by the number of reactive elements contained in the circuit. Thus, the total number of multipliers is finite and constant, versus variations in any parameter. The number of poles associated with each multiplier is infinite, due to the nonunivocal relationship between the poles and the Floquet multipliers, $m = e^{[\sigma \pm j(\omega + n\omega_o)]T}$, with n an integer and $T = 2\pi/\omega_o$ the solution period. All the poles associated with the same multiplier have the same real part σ.

2. Due to the periodicity of the poles, pole analysis may be limited to the interval $(0, \omega_o]$. In case of a real multiplier of negative sign, we will find one pair of poles of the form $\sigma \pm j(\omega_o/2)$. For a pair of complex-conjugate multipliers, we will find the poles $\sigma \pm j\alpha\omega_o$ and $\sigma \pm j\omega_o(1 - \alpha)$, with $\alpha \in R$.

3. Consider a pair of complex-conjugate poles $\sigma \pm j\alpha\omega_o$ (associated with two complex-conjugate multipliers) which evolve versus a parameter so that α tends to $\alpha = 0$ (or to $\alpha = 1$). After merging at a particular parameter value, they will split into two different real poles γ and $\gamma\prime$, each associated with a different real and positive multiplier. Note that these poles can be expressed equally as $\gamma \pm jn\omega_o$ or $\gamma\prime \pm jn\omega_o$, due to the nonunivocal relationship between poles and multipliers. This is because the total number of multipliers must remain constant under the parameter variations. This pole transformation does not constitute a bifurcation, but if the transformed poles are the dominant poles of the periodic solution, it may have an influence on some circuit characteristics, such as the noise spectrum or the transient behavior (see Section 2.5.5). Examples are shown in Chapter 4 in an in-depth study of injection-locked oscillators and harmonic injection dividers.

4. Consider a pair of complex-conjugate poles $\sigma \pm j\alpha\omega_o$ (associated with two complex-conjugate multipliers) which evolve versus a parameter so that α tends to $\alpha = 1/2$. After merging at a particular parameter value, the poles will split into two different sets of poles, $\sigma \pm j\,((\omega_o/2) + n\omega_o)$ and $\sigma\prime \pm j\,((\omega_o/2) + n\omega_o)$, each associated with a different real and negative multiplier. Again, this pole transformation does not constitute a bifurcation but may have an influence on relevant circuit characteristics.

3.3.3 Global Bifurcations

As we already know, a saddle solution contains both stable and unstable poles and thus is attracting for a subset of the phase space R^N. This subset is the stable manifold of the saddle, which will also have an unstable manifold. Because an arbitrary perturbation will have components in all the directions of R^N, the saddle solution will be unstable. Despite this, the ability of the saddle solutions to attract some trajectories of the phase space can give rise to global bifurcations involving more than one solution. Global bifurcations are due to changes in the topological configuration of the stable and unstable manifolds of a solution of saddle type [16]. Unlike local bifurcations, they cannot be detected through pole analysis of a single steady-state solution. There are two main types of global bifurcation: saddle connection and saddle–node local–global bifurcation.

3.3.3.1 Saddle Connection Consider a dc solution constituting an equilibrium point in the phase space in R^N. The dc solution is assumed to be of saddle type, meaning, it has stable and unstable eigenvalues. In most cases, it will contain a stable manifold of N-1 (or N-2) dimension and an unstable manifold of dimension 1 (or 2). Considering variations in a given parameter η, a saddle connection will occur if the stable and unstable manifolds of the saddle-type solution intersect at η_o, giving rise to what is known as a *homoclinic orbit* (see Fig. 3.21). A transversal intersection requires the existence of vectors tangent to at least one of the manifolds that can span R^N at any intersection point [1]. The intersection of the stable and unstable manifolds of the saddle point is necessarily tangential. This is due to the fact that at any point $\overline{x}$ of the orbit, the vector $\dot{\overline{x}}$, determining the time evolution of the system, is tangent to both manifolds. Because of the tangential intersection, the homoclinic orbit is structurally unstable, which means that it will be destroyed under any slight perturbation with components in the full space R^N. However, under some circumstances, the breaking of this orbit can give rise to a stable limit cycle or even chaotic behavior (for $N \geq 3$). The mathematical conditions for the generation of a limit cycle and for the transition to chaotic behavior have been studied in low-dimension systems [1,2]. These conditions depend on the eigenvalues of the saddle point and some geometric characteristics of the homoclinic orbit.

Let us assume that a stable limit cycle has been generated from a homoclinic orbit. This cycle will persist when further varying the parameter η in the same sense. This oscillation will have amplitude different from zero and infinite period at the bifurcation point η_o. The oscillation amplitude is determined by the size of the manifold intersection of the saddle point. Just after the bifurcation, the periodic solution, though moving along the cycle, tends to spend a long time near the saddle point. This time tends to infinite at η_o, which justifies the infinite period of the generated cycle. Note that by definition the intersecting stable and unstable manifolds tend to the saddle for $t \to \infty$ and $t \to -\infty$, respectively, which also justifies the infinite period. The period will decrease quickly (and continuously) when moving away from the bifurcation point.

Saddle-connection bifurcations can also be obtained on a Poincaré map, and are associated to a saddle-type fixed point of this map. Depending on some mathematical conditions to be fulfilled by this saddle point, the bifurcation can give rise to the discontinuous generation of a quasi-periodic solution or to a transition to chaotic behavior. In the first case, a cycle composed of discrete points, corresponding to a 2-torus in the phase space (quasiperiodic solution), arises from the

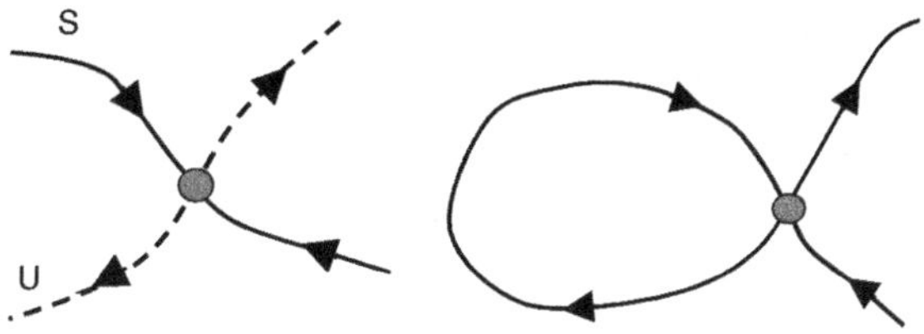

FIGURE 3.21 Saddle connection. The stable and unstable manifolds of an equilibrium point intersect, giving rise to a homoclinic orbit. Under some mathematical conditions, the homoclinic orbit transforms into a limit cycle for a further parameter variation.

intersection with the stable and unstable manifolds of the fixed point. This type of global bifurcation is found for some small parameter intervals in injection-locked oscillators and harmonic injection dividers, as discussed in Chapter 4.

3.3.3.2 Saddle–Node Bifurcations of Local–Global Type As discussed earlier, turning points are points of infinite slope of a given solution curve (dc or periodic) versus the analysis parameter. If the solution path is originally stable, it will become unstable after the turning point, due to the crossing of a real pole through the origin of the complex plane. The curve folds over itself at the turning point, as a zero pole implies an infinite slope of the solution curve versus the parameter. In all the cases considered so far, this gave rise to a jump to a different stable solution, due to the impossibility of remaining on a path section that, due to the folding, no longer exists (see Fig. 3.22a). However, a totally different phenomenon can also occur at the turning point, corresponding to a global bifurcation instead of a local one [16]. This phenomenon is described next.

Let the case of a turning point separating a stable and an unstable section of a dc solution path be considered (Fig. 3.22a). At a relatively short distance from the turning point, the unstable solution will have only one unstable pole, with all its other poles located on the left hand side of the complex plane. As already indicated, solutions having poles at both sides of the complex plane are called *saddles*. As the parameter is varied toward the turning point obtained at η_b, the two dc solutions approach each other and merge at this turning point. In some cases, before this turning point is reached, the unstable manifold of the saddle forms a closed connection, passing through the stable dc solution, also called a *node* (see Fig. 3.22b). The connection is not a steady-state solution, as the system does not turn around it in a unique sense, in contrast to what happens in a periodic cycle. Only the node is observed physically. However, this situation changes when the turning point is reached. The stable and unstable points merge and the loop gives rise to a stable cycle (Fig. 3.22c). This kind of bifurcation is also known as *saddle–node homoclinic bifurcation*.

The limit cycle has an infinite period, or zero frequency, at the bifurcation point. If a parameter is varied further in the same sense, only the limit cycle persists. Just after the bifurcation, the solution moving along the cycle tends to spend a long time

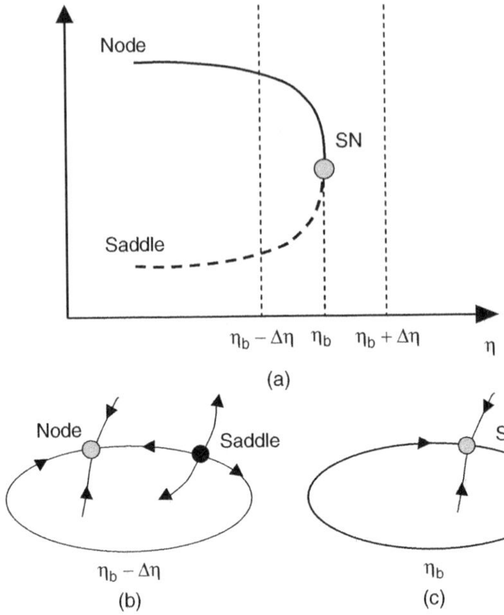

FIGURE 3.22 Limit cycle on a saddle–node. (a) Solution curve traced versus the parameter η. It exhibits a turning point SN. The upper section is composed of stable solutions or "nodes". The lower section is composed of unstable solutions or "saddles". (b) Near SN, the stable and unstable manifolds of a saddle point intersect, forming a loop. (c) As the parameter varies, the saddle–node approach [see (a)] and when they merge at the turning-point bifurcation, they give rise to a stable limit cycle.

near the place where the saddle–node point used to be. This justifies the infinite period of the cycle at the bifurcation point. This period decreases continuously when varying the parameter away from the bifurcation point. One essential property of this bifurcation is that the cycle is generated with amplitude different from zero at the bifurcation point. This amplitude is determined by the size of the manifold intersection of the saddle point. Both characteristics are opposite to those of limit cycles generated at Hopf bifurcations from dc solutions, which have zero amplitude and finite period at the bifurcation point. The discontinuous generation of the limit cycle is in correspondence with the global nature of this type of bifurcation. The bifurcation described is called a limit cycle on a saddle–node, although it is also known as local–global bifurcation, due to its occurrence in combination with a turning point of the dc solution path.

Local–global bifurcation can also occur on a Poincaré map. When this is the case, the stable and unstable points are fixed points of the map which correspond, in fact, to stable and unstable periodic solutions or cycles in the phase space. Prior to the turning point, the stable and unstable manifolds of the saddle solution intersect, forming a closed loop that contains a stable fixed point. At the turning point, the connection gives rise to a cycle comprised of discrete points which corresponds

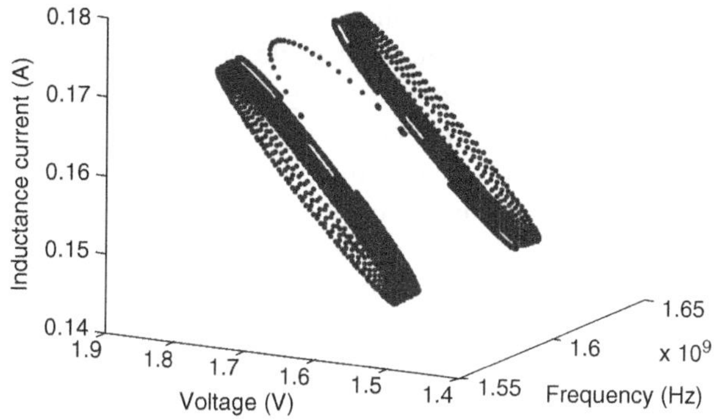

FIGURE 3.23 Poincaré map of the circuit of Fig. 1.1 when introducing a current generator in parallel. The generator amplitude considered is $I_g = 5$ mA. The analysis parameter is the input frequency ω_{in}. The stable range of periodic operation at the input generator frequency is delimited by two local–global bifurcations.

to a 2-torus in the phase space or a quasiperiodic solution. The two fundamental frequencies of this quasiperiodic solution will initially have the same value $\omega_{in} = \omega_a$, so $\Delta\omega = 0$, in agreement with the real pole at zero at the bifurcation point. This frequency difference increases quickly when moving the parameter slightly away from the bifurcation. The discrete-point cycle in the Poincaré map is generated with nonzero amplitude, also in correspondence with the discontinuous nature of the global bifurcations. Both characteristics are opposite those of quasiperiodic solutions generated at secondary Hopf bifurcations from a periodic regime.

The local–global bifurcation on the Poincaré map is found in all the injection-locked oscillators for a relatively small amplitude of the input generator. At this bifurcation, the oscillation synchronizes or desynchronizes to the input source. This is why it is also called *mode-locking bifurcation*. As an example, Fig. 3.23 presents an analysis of the circuit of Fig. 1.1 when introducing a periodic current generator in parallel, with amplitude $I_g = 5$ mA. We remind the reader that injection-locked oscillators are covered in detail in Chapter 4. The purpose of the example is just to illustrate the mathematical and geometrical aspects of local–global bifurcation. As in the Poincaré map of Fig. 3.18, this analysis has been carried out versus the input frequency ω_{in}. The behavior obtained is qualitatively very different from that shown in Fig. 3.18. In both cases the interval with stable periodic behavior is delimited by the frequency values at which the fixed point of the map turns into a discrete-point cycle. However, the way this cycle is generated is different in the two diagrams. In Fig. 3.18 the cycle is generated from zero amplitude with a nonzero difference between the fundamental frequencies of the corresponding quasiperiodic solution provided by the frequency of the crossing poles, which can be written $\sigma \pm j(\Delta\omega + k\omega_o)$. The single-point path continues to exist after bifurcation and is located inside the cycle.

In the map in Fig. 3.23, the discrete-point cycle is generated with relatively large amplitude at each end of the interval of stable periodic behavior. The single fixed point from which this cycle is generated is contained in the cycle at the bifurcation point, in agreement with the sketch shown in Fig. 3.22. Then the fixed point vanishes, leaving the discrete-point cycle only. Thus, the periodic solution from which the quasiperiodic regime was generated does not coexist with this regime (see Fig. 3.22a). The difference frequency between the two fundamentals of the quasiperiodic solution is zero at the bifurcation point, but increases quickly (and continuously) when varying the parameter away from this point.

For comparison, Fig. 3.24 presents the frequency-domain analysis of the parallel resonance oscillator with the current source $I_g = 5$ mA, equivalent to the time-domain analysis of Fig. 3.23. The periodic synchronized solution is represented by drawing the voltage magnitude at the fundamental frequency ω_{in} versus the input frequency. As can be seen, a closed-solution curve is obtained. The synchronized operation band is delimited at each end by a turning point. No synchronized solution exists beyond these turning points. Compared with Fig. 3.23, it is easily seen that the turning points agree with the points that in the Poincaré map give rise to the quasiperiodic solution (the discrete point cycle). Thus, the turning points in Fig. 3.24a are local–global bifurcations. The upper side of the closed curve consists of stable solutions or nodes. It corresponds to the single-point section of the map of Fig. 3.23. The lower side of the closed curve in Fig. 3.24a is composed of unstable solutions or saddles. These unstable solutions could not be obtained in the map of Fig. 3.23, due to their instability. Note that the map was generated though standard numerical integration of the nonlinear system ruling the circuit behavior. When varying ω_{in} in a synchronized regime toward any of the two turning points, the situation of the Poincaré map is depicted in Fig. 3.22a. The stable and unstable manifolds of the saddle point intersect, forming a closed connection. At the turning points, this connection becomes a cycle.

In Fig. 3.24b, simulations of the quasiperiodic solution outside the synchronization interval are also presented. As in the case of Fig. 3.19, the quasiperiodic solution has been represented tracing the node voltage amplitude at the input frequency ω_{in} and at the oscillation frequency ω_a. Remember that for the frequency-domain analysis of this mixerlike regime, the intermodulation products $k\omega_{\text{in}} + m\omega_a$ must be taken into account. The maximum order N_L of these products, defined as $|k| + |m| \leq N_L$, determines the degree of accuracy. As can be seen, the component at the autonomous frequency is extinguished in a discontinuous manner at the two ω_{in} values corresponding to the turning points of the closed synchronization curve. The discontinuity is in good agreement with the global nature of these bifurcations. This transformation from quasiperiodic to periodic should be compared with the one obtained in the case of an inverse Hopf bifurcation (see Fig. 3.19), showing a continuous extinction to zero of the oscillation amplitude.

The free-running frequency of the circuit in Fig. 1.1 is $f_o = 1.59$ GHz. After introduction of the current generator with amplitude $I_g = 5$ mA, the variation in the autonomous frequency with the input frequency is the one depicted in Fig. 3.25. Within the synchronization band, the relationship $\omega_a = \omega_{\text{in}}$ gives rise to a straight line of unity slope versus ω_{in}. When the synchronization is lost at each of the turning points, the oscillation frequency shows continuous behavior versus ω_{in}. The frequency difference $|\Delta\omega| = |\omega_{\text{in}} - \omega_a|$ (beat frequency) is zero at these turning

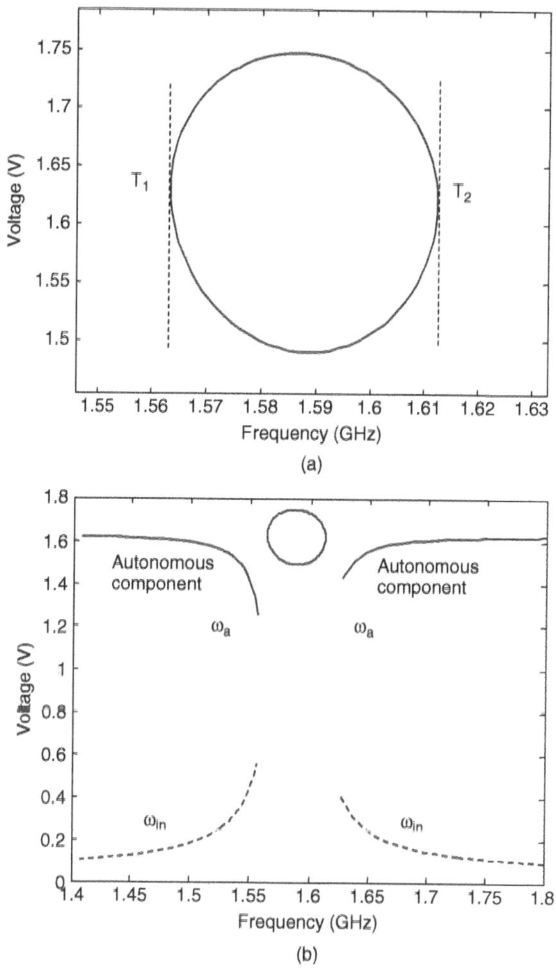

FIGURE 3.24 Frequency-domain simulation of the circuit of Fig. 1.1, with a current generator in parallel of amplitude $I_g = 5$ mA. (a) Synchronized periodic solution. The closed curve is delimited by two turning points. The upper side is stable, whereas the lower one is unstable. (b) The simulation of the quasiperiodic solution outside the synchronization interval has been added. This solution is represented by tracing the voltage amplitude at the input frequency ω_{in} and at the autonomous frequency ω_a. The path discontinuity at the two synchronization points is due to the global nature of the bifurcation.

points, in agreement with the properties described for local–global bifurcation. The oscillation frequency exhibits bigger variation near the turning points, which implies that the frequency difference $\Delta\omega$ increases quickly when moving the parameter away from these points. On the left-hand side of the turning point T_1, the autonomous frequency is smaller than the free-running frequency 1.59 GHz. On the right-hand side of the turning point T_2, the autonomous frequency is higher than the one obtained under free-running conditions. This is due to the influence

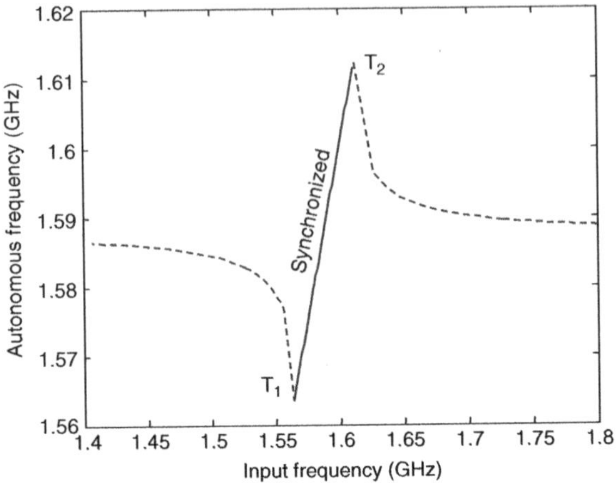

FIGURE 3.25 Evolution of autonomous frequency ω_a versus input generator frequency ω_{in}. The free-running frequency is $f_o = 1.59$ GHz. The straight line of unit slope between turning points T_1 and T_2 indicates the frequency variation within the synchronization interval. The dashed line indicates the frequency variation in the quasiperiodic (nonsynchronized) regime.

of the input generator. As can be seen, approaching the synchronization edges (turning points) from the quasiperiodic regime, there is a clear parameter interval (Fig. 3.25) in which the oscillation frequency varies very quickly to reduce the difference $|\Delta\omega|$. This interval is known as the *injection-pulling region*.

When represented in the time domain, the quasiperiodic solution generally looks like an amplitude-modulated waveform at the difference frequency $\Delta\omega = |\omega_{in} - \omega_a|$ (see Chapter 1). Because this difference frequency is so small near the turning points, a long simulation interval will be necessary to observe this modulation. An example is shown in Fig. 3.26, corresponding to the input frequency $f_{in} = 1.563$ GHz, near the turning point T_1. If only a short simulation interval is considered, the solution may look periodic. In a long interval, sudden bursts are observed, in agreement with the actual quasiperiodic nature of this solution. This type of behavior is also known as *quasiperiodic intermittence* or *quasilocking behavior*. The frequency of the bursts decreases gradually when approaching the turning point and tends to zero at this turning point. In quasilock conditions, the solution spectrum is extremely dense, in correspondence with the small difference between the two fundamental frequencies (Fig. 3.27). Because the waveform spends a long time in near periodic behavior at the frequency of the input source, the spectrum exhibits high power at the input-source frequency. During the bursts (Fig. 3.26), the instantaneous frequency tends to take value smaller than f_{in} (for $f_o < f_{in}$) or higher than f_{in} (for $f_o > f_{in}$), due to the influence of the self-oscillation. This justifies the spectrum triangular shape, with higher power to the opposite side of the input-drive frequency.

Note that in the measurements it is relatively easy to distinguish between oscillator synchronization due to an inverse Hopf bifurcation and that due to

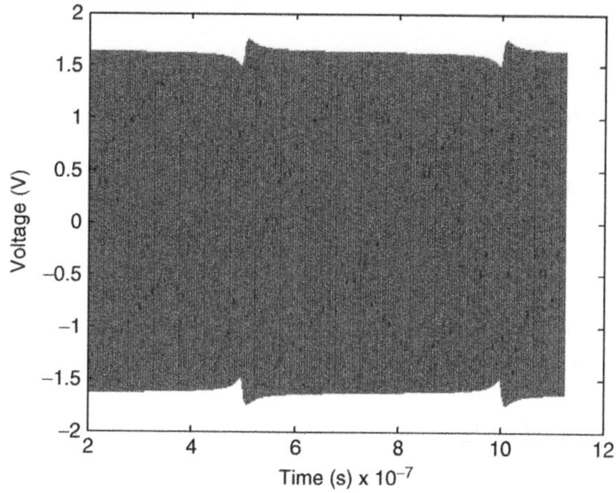

FIGURE 3.26 Voltage waveform for an input frequency $f_{in} = 1.563$ GHz near the turning point T_1. The waveform looks nearly periodic for long simulation intervals. Then a sudden burst occurs, in correspondence with the actual quasiperiodicity of the solution.

local–global bifurcation. In the former case, the spectral lines maintain a given distance $\Delta\omega$ as the power of the intermodulation products decreases continuously (see Fig. 3.20)—eventually to vanish at the bifurcation point, from which only the main spectral lines at $k\omega_{in}$ are left in the spectrum. In the case of local–global bifurcation, the spectral lines approach each other and the spectrum becomes very dense, like the one in Fig. 3.27. At the bifurcation point, this suddenly turns into a periodic spectrum, in a discontinuous manner.

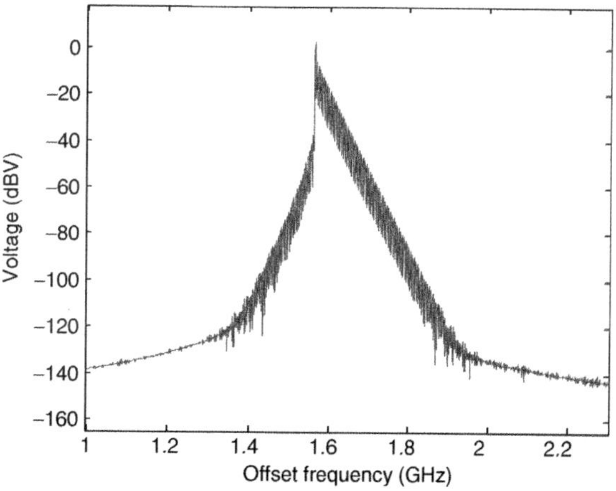

FIGURE 3.27 Voltage spectrum corresponding to the quasiperiodic waveform of Fig. 3.26. The spectrum is very dense, in correspondence with the small value of the difference between the two fundamental frequencies $|\Delta\omega| = |\omega_{in} - \omega_a|$.

REFERENCES

[1] A. Suárez, J. Morales, and R. Quéré, Synchronization analysis of autonomous microwave circuits using new global stability analysis tools, *IEEE Trans. Microwave Theory Tech.*, vol. 46, pp. 494–504, May 1998.

[2] J. Guckenheimer and P. Holmes, *Nonlinear Oscillations, Dynamic Systems, and Bifurcations of Vector Fields*, Springer-Verlag, New York, 1983.

[3] S. Wiggins, *Introduction to Applied Nonlinear Dynamical Systems and Chaos*, Springer-Verlag, New York, 1990.

[4] S. Jeon, A. Suárez, and D. B. Rutledge, Global stability analysis and stabilization of a class-E/F amplifier with a distributed active transformer, *IEEE Trans. Microwave Theory Tech.*, vol. 53, pp. 3712–3722, 2005.

[5] T.S. Parker and L.O. Chua, *Practical Algorithms for Chaotic Systems*, Springer-Verlag, Berlin, 1989.

[6] L. Trajkovic, R. C. Melville, and S. Fang, Finding dc operating points of transistor circuits using homotopy methods, *IEEE International Symposium on Circuits and Systems II*, Singapore, pp. 758–761, 1991.

[7] S. Jeon, A. Suárez, and D. B. Rutledge, Nonlinear design technique for high-power switching-mode oscillators, *IEEE Trans. Microwave Theory Tech.*, vol. 54, pp. 3630–3639, 2006.

[8] G. Iooss and D. D. Joseph, *Elementary Stability and Bifurcation Theory*, 2nd ed., Springer-Verlag, New York, 1990.

[9] K. Kurokawa, Some basic characteristics of broadband negative resistance oscillators, *Bell Syst. Tech. J.*, vol. 48, pp. 1937–1955, July–Aug. 1969.

[10] A. D'Ambrosio and A. Tattanelli, Parametric frequency dividers: operation and applications, *3rd European Microwave Conference*, pp. 1–5, 1973.

[11] G. Sarafian and B. Z. Kaplan, Dynamics of parametric frequency divider and some of its practical implications, *IEEE Convention of Electrical and Electronics Engineers*, Jerusalem, Israel, pp. 523–526, 1996.

[12] M. Gayral, E. Ngoya, R. Quéré, J. Rousset, and J. Obregón, Spectral balance: a general method for analysis of nonlinear microwave circuits driven by non-harmonically related generators, *IEEE MTT-S International Microwave Symposium*, Las Vegas, NV, pp. 119–121, 1987.

[13] K. S. Kundert, J. K. White, and A. Sangiovanni-Vincentelli, *Steady-State Methods for Simulating Analog and Microwave Circuits*, Kluwer Academic, Norwell, MA, 1990.

[14] J. Jugo, J. Portilla, A. Anakabe, A. Suárez, and J. M. Collantes, Closed-loop stability analysis of microwave amplifiers, *IEE Electron. Lett.*, vol. 37, pp. 226–228, Feb. 2001.

[15] A. Anakabe, J. M. Collantes, J. Portilla, et al., Analysis and elimination of parametric oscillations in monolithic power amplifiers, *IEEE MTT-S International Microwave Symposium*, Seattle, WA, pp. 2181–2184, 2002.

[16] J. M. T. Thompson and H. B. Stewart, *Nonlinear Dynamics and Chaos*, 2nd ed., Wiley, Chichester, UK, 2002.

[17] T. Endo and L. O. Chua, Chaos from phase-locked loops, *IEEE International Symposium on Circuits and Systems*, Espoo, pp. 1983–1986, 1988.

CHAPTER FOUR

Injected Oscillators and Frequency Dividers

4.1 INTRODUCTION

In Chapter 1 the operational principle and main characteristics of free-running oscillators were presented. The free-running oscillator, containing dc sources only, provides a self-sustained periodic oscillation from the energy delivered by these dc sources. However, a circuit can also oscillate in the presence of an input periodic source at the frequency ω_{in}. The oscillation frequency, with value ω_o in free-running conditions, will be influenced by the input source and take a different value ω_a. In the injection-locked mode, the two frequencies are rationally related $\omega_a/\omega_{in} = m/k$, with m and k positive integers, so the solution is periodic [1]. The oscillation signal is synchronized to the input source, so for any change in the input signal frequency, the oscillation frequency changes according to the relationship $\omega_a/\omega_{in} = m/k$. Due to the existence of frequency-selective elements in the circuit responsible for the original resonance at ω_o, the variation in the oscillation frequency will lead to a variation in the solution phase shift with respect to the input source. The synchronization is a complex nonlinear phenomenon, possible only for certain ranges of the input generator power and frequency, delimited, as shown in Chapter 3, by bifurcation phenomena. Outside these ranges, the oscillation will simply mix with the signal delivered by the input generator, showing *self-oscillating mixer behavior* [2]. Note that the extinction of the oscillation by the

Analysis and Design of Autonomous Microwave Circuits, By Almudena Suárez

input generator is also possible, especially when the generator has a high degree of power at quite a different frequency from that of the free-running oscillator.

Injection-locked operation has several applications. Synchronization at the fundamental frequency enables high-gain amplification [3] and can also be used for the implementation of phase shifters [4]. Synchronization of a given harmonic N of the oscillation signal to the input source, such that $\omega_a/\omega_{\text{in}} = 1/N$, is used for the implementation of frequency dividers [5,6]. On the other hand, the synchronization of the oscillation frequency with the Mth harmonic component of a periodic input source, such that $\omega_{\text{in}} = \omega_a/M$ allows high-gain frequency multiplication from very low input level at ω_{in}, due to the power contributed by the self-oscillation. Provided that a low phase noise source is used, it also enables phase noise reduction of the higher-frequency oscillator [7]. Finally, in the self-oscillating mixer mode, the oscillation frequency ω_a coexists and mixes with the frequency of the input source ω_{in}. With a suitable design, this type of operation can be used to implement a compact and low-power-consumption frequency converter, since the same nonlinear device enables local oscillation and performs frequency mixing [8].

In the operational modes discussed, the circuit exhibits free-running oscillation in the absence of input power from a periodic generator at ω_{in}. In other types of behavior, the circuit does not exhibit free-running oscillation. Instead, the oscillation arises from a certain power of this generator only. The frequency of the oscillation ω_a generated may be a subharmonic of the input frequency ω_{in}, giving rise to a frequency-divided regime, or may be related nonharmonically to this frequency, leading to a frequency mixer operation. An example is the parametric frequency division studied in Chapter 3, obtained when increasing the amplitude of the periodic pumping voltage of a nonlinear capacitance. Another type of divider with no self-oscillation in the absence of input signal is the regenerative divider [9], in which instability at ω_{in}/N is generated when increasing the input power through mixing and feedback effects. Oscillations obtained from a certain input power at ω_{in} can be a problem in forced nonlinear circuits such as frequency multipliers and power amplifiers [10], which are not expected to oscillate. These circuits are often stable at low signal levels but start to oscillate when the power is increased. The oscillation frequency ω_a can be a subharmonic of the input frequency $\omega_a = \omega_{\text{in}}/2$ or may not be related harmonically to this frequency. These undesired oscillations are often due to the nonlinear capacitances contained in active devices exhibiting negative resistance from a certain input power.

In this chapter we present the basic operational principles of the most relevant types of autonomous circuits, those with an input periodic source. The circuits considered are fundamentally synchronized oscillators, analog dividers of harmonic injection, regenerative and parametric types, subsynchronized oscillators, and self-oscillating mixers. The operational bands of all the circuits mentioned, in terms of input power and frequency, are delimited by bifurcations or qualitative stability changes. This chapter can be seen as a complement of Chapter 3, in which various types of local and global bifurcations were presented. Here, many practical examples of bifurcations are discussed in even more detail, due to their relevance in the operation of injected oscillators and frequency dividers among other valuable circuits. The phase noise spectrum of injection-locked oscillators is also derived analytically, to give insight into the magnitudes and parameters that determine its particular shape and corner frequencies. In summary, the chapter focuses on the operational principles, stability properties, and phase noise behavior of all the autonomous circuits mentioned, which, whenever possible, will be treated in an analytical manner. The numerical techniques for the simulation of these circuits are the object of Chapter 5.

4.2 INJECTION-LOCKED OSCILLATORS

In this section, use is made of the describing function introduced in Chapter 1 for a comprehensive study of the oscillator behavior in injection-locked conditions, and a determination of its stable operation ranges, in terms of the input generator frequency and power. Note that this approximate analysis is limited to the fundamental frequency of the circuit solution.

4.2.1 Analysis Based on Linearization About a Free-Running Solution

The injection-locked oscillator will be analyzed using admittance-function models. The analysis based on impedance functions would be analogous to this one. Let the circuit of Fig. 4.1 be considered. It is an equivalent circuit of an injection-locked oscillator, from a sensitive observation node. The admittance $Y_L(\omega)$ represents the linear block and $Y_N(V, \omega)$ is the current-to-voltage describing function associated with the nonlinear block, calculated as explained in Chapter 1. The variables V and ω of the general function Y_N represent the voltage amplitude at the analysis node and the excitation frequency, respectively. The current generator $i_{in}(t) = \text{Re}[I_{in}e^{j\omega_{in}t}]$ is the Norton equivalent of the input source, and the corresponding Norton impedance is included in $Y_L(\omega)$. It will be assumed that for $I_{in} = 0$, the circuit exhibits a free-running oscillation at the voltage amplitude V_o and frequency ω_o, expressed as $v_o(t) = \text{Re}(V_o e^{j\omega_o t})$. Remember that there is an irrelevance in the free-running oscillator solution versus the phase origin, so its phase will be set

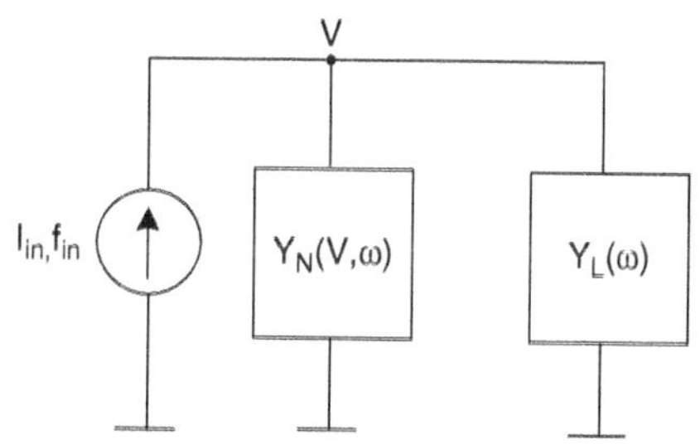

FIGURE 4.1 Admittance model of an injection-locked oscillator at a given observation port. The admittance associated with the Norton equivalent of an input generator has been included in the linear block, described by the admittance $Y_L(\omega)$.

arbitrarily to zero, $\phi_o = 0$. However, in the presence of the time-varying source $i_{in}(t) = \mathrm{Re}[I_{in} e^{j\omega_{in} t}]$, there will be a phase shift with respect to this source, so the node voltage will be represented by $v(t) = \mathrm{Re}(V(t)e^{j(\phi(t)+\omega_{in})t})$. For generality, no synchronized behavior is assumed, so $V(t)$ and $\phi(t)$ are time-dependent. We expect the circuit to operate near the synchronization band, where the oscillation frequency agrees with that of the synchronizing source, so we take ω_{in}, instead of ω_o, as the carrier frequency of $v(t)$. For small input amplitude I_{in} and frequency ω_{in} relatively close to the free-running frequency ω_o, the oscillator solution can be expressed as $v(t) = \mathrm{Re}[V_o + \Delta V(t)]e^{j(\phi(t)+\omega_{in} t)}$, where $\Delta V(t)$ and $\phi(t)$ are assumed to be slowly varying time functions. The circuit equations are written

$$Y_T(V(t), j\omega_{in} + s)V(t)e^{j(\phi(t)+\omega_{in} t)} = I_{in} e^{j\omega_{in} t} \tag{4.1}$$

where $Y_T = Y_L + Y_N$ and s is the complex frequency increment, acting (in an abuse of notation) as time derivator. Next, a Taylor series expansion of first order of the admittance function Y_T is carried out about the free-running solution with amplitude V_o and frequency ω_o, which fulfils $Y_T(V_o, \omega_o) = 0$. It is taken into account that the multiplication by s is equivalent to a time derivation in the slow time scale of the perturbed node voltage:

$$\begin{gathered}
\left.\frac{\partial Y_T}{\partial V}\right|_o (V(t) - V_o)V(t)e^{j\phi(t)} + \left.\frac{\partial Y_T}{\partial j\omega}\right|_o \frac{\mathrm{d}(V(t)e^{j\phi(t)})}{\mathrm{d}t} + \left.\frac{\partial Y_T}{\partial j\omega}\right|_o j(\omega_{in} - \omega_o) \\
V(t)e^{j\phi(t)} = I_{in} \\
\left[\left.\frac{\partial Y_T}{\partial V}\right|_o (V(t) - V_o) + \left.\frac{\partial Y_T}{\partial \omega}\right|_o \left(\omega_{in} + \dot{\phi}(t) - j\frac{\Delta\dot{V}(t)}{V_o} - \omega_o\right)\right] = \frac{I_{in}}{V_o} e^{-j\phi(t)}
\end{gathered} \tag{4.2}$$

where the approximation $I_{in}/(V_o + \Delta V) \cong I_{in}/V_o$ has also been considered. The complex equation (4.2) can be split into real and imaginary parts:

$$\begin{aligned}
\frac{\partial Y_{To}^r}{\partial V}\Delta V(t) + \frac{\partial Y_{To}^r}{\partial \omega}[\dot{\phi}(t) + \omega_{in} - \omega_o] &= \frac{I_{in}}{V_o}\cos\phi(t) \\
\frac{\partial Y_{To}^i}{\partial V}\Delta V(t) + \frac{\partial Y_{To}^i}{\partial \omega}[\dot{\phi}(t) + \omega_{in} - \omega_o] &= -\frac{I_{in}}{V_o}\sin\phi(t)
\end{aligned} \tag{4.3}$$

Note that the time derivative of the amplitude increment $\Delta\dot{V}(t)$ has been neglected in (4.3), as, due to the amplitude-limiting property of nonlinear elements, the magnitude of $\Delta V(t)$ is usually much smaller than that of $\phi(t)$. The coefficients in (4.3) are constant, as they are given by the derivatives of the admittance function, evaluated at the free-running solution. Thus, it will be possible to solve for $\dot{\phi}(t)$ through Kramer's rule. For notation convenience, the following vectors are defined:

$$\begin{aligned}\overline{Y}_{ToV} &\equiv (Y^r_{ToV}, Y^i_{ToV}) = \left(\frac{\partial Y^r_{To}}{\partial V}, \frac{\partial Y^i_{To}}{\partial V}\right)\\ \overline{Y}_{To\omega} &\equiv (Y^r_{To\omega}, Y^i_{To\omega}) = \left(\frac{\partial Y^r_{To}}{\partial \omega}, \frac{\partial Y^i_{To}}{\partial \omega}\right)\\ \tilde{e}^{j\phi} &\equiv (\cos\phi, \sin\phi)\end{aligned} \tag{4.4}$$

where the subscripts indicate the derivative with respect to the corresponding variable evaluated at the free-running oscillation V_o and ω_o, and the superscripts r and i indicate real and imaginary parts. Then it is possible to write

$$\begin{aligned}\frac{d\phi}{dt} &= \omega_o - \omega_{in} + \frac{I_{in}\overline{Y}_{ToV} \times \tilde{e}^{-j\phi}}{V_o(\overline{Y}_{ToV} \times \overline{Y}_{To\omega})}\\ &= \omega_o - \omega_{in} - \underbrace{\frac{I_{in}}{V_o\,|\partial Y_{To}/\partial\omega|\sin\alpha_{v\omega}}}_{F_o}\sin(\phi(t) + \alpha_v)\end{aligned} \tag{4.5}$$

where the multiplication sign implies the operation $a \times b = a^r b^i - a^i b^r = |a||b|\sin(\angle b - \angle a)$. Note that the defined product $a \times b$ provides a scalar number with either positive or negative sign. The angle α_v is the phase associated with $\overline{Y}_{ToV}$, and $\alpha_{v\omega}$ is the phase difference between $\overline{Y}_{To\omega}$ and $\overline{Y}_{ToV}$; that is, $\alpha_{v\omega} = \text{ang}(\overline{Y}_{To\omega}) - \text{ang}(\overline{Y}_{ToV})$. The maximum value of the phase derivative is determined by the injection frequency ω_{in} and the constant coefficient F_o, indicated in (4.5), and will typically be small, which justifies the approximations (4.2).

The frequency difference $\omega_o - \omega_{in}$ is known as the *frequency detuning*. If ω_o is higher than ω_{in}, the oscillation evolves more quickly than the input source, so the phase shift $\phi(t)$ will tend to increase due to the first term, $\omega_o - \omega_{in} > 0$, of the time derivative $\dot{\phi}(t)$. To achieve synchronization, the sinusoidal term $-F_o\sin(\phi(t) + \alpha_v)$ must oppose this phase growth. The opposite situation is obtained for $\omega_o - \omega_{in} < 0$. The circuit will achieve synchronization if, after a transient, the condition $\dot{\phi}(t) = 0$ is reached. To achieve this, the term $-F_o\sin(\phi(t) + \alpha_v)$ must have the sign opposite to $\omega_o - \omega_{in}$ and be large enough to cancel the frequency difference. Introducing the condition $\dot{\phi}(t) = 0$ into (4.5) gives us

$$\omega_{in} - \omega_o = -\frac{I_{in}\sin(\phi_s + \alpha_v)}{V_o\,|\partial Y_{To}/\partial\omega|\sin\alpha_{v\omega}} \tag{4.6}$$

Equation (4.6) is fulfilled only within the so-called frequency synchronization band $\omega_{in1}, \omega_{in2}$, to be analyzed later in this section. It indicates that the increment of the periodic oscillation frequency $\omega_{in} - \omega_o$ gives rise to a constant phase shift ϕ_s between the node voltage and the input source. This is due to the variation in total admittance Y_T in the presence of this periodic source, and the change in the periodic-solution frequency, from the original resonance frequency ω_o to ω_{in}. From inspection of (4.6) it is clear that synchronization will only be possible for ω_{in}

values provided that $|\sin(\phi_s + \alpha_v)| < 1$. Therefore, ω_{in} cannot be too different from ω_o.

For ω_{in} values leading to the impossible condition $|\sin(\phi_s + \alpha_v)| > 1$, there is no synchronized solution and the circuit operates in a quasiperiodic regime. The time-varying phase shift $\phi(t)$ can be calculated from (4.5) in different manners [11,12]. It can be expressed as $\phi(t) = \Delta\omega_b t + \sum_k \phi_k e^{jk\Delta\omega_b t}$, where $\Delta\omega_b$ is the so-called beat frequency, given by $\Delta\omega_b = \sqrt{(\omega_o - \omega_{in})^2 - F_o^2}$. More harmonic terms in the representation of $\phi(t)$ are typically necessary for smaller detuning $(\omega_o - \omega_{in})$ compared to F_o, due to the higher relevancy of the sinusoidal term in (4.5). The frequency $\Delta\omega_b$ corresponds to the spacing between the spectral lines of the quasiperiodic regime. We often consider that the quasiperiodic spectrum is spanned by the input frequency ω_{in} and an autonomous frequency ω_a. This autonomous frequency agrees with $\omega_{in} + \Delta\omega_b$ for $\omega_o - \omega_{in} > 0$ and with $\omega_{in} - \Delta\omega_b$ for $\omega_o - \omega_{in} < 0$. As an example, see the evolution of the frequency ω_a versus ω_{in} in Fig. 3.25. There is a region near the synchronization boundaries where the frequency ω_a of the steady-state quasiperiodic regime is highly influenced by ω_{in}. It is called the *injection-pulling region*. At a larger difference between the input source frequency ω_{in} and ω_o, the autonomous frequency ω_a is much less sensitive to ω_{in} variations, since $\Delta\omega_b$ tends to $(\omega_o - \omega_{in})$. For a detailed analysis of these variations, see the references [11,12].

We will now concentrate on the analysis of the synchronized solution. As gathered from (4.6), two phase values ϕ_s are possible for each ω_{in} within the synchronization band, determined by the condition (4.6). They correspond to the two different solutions of the arcsin(·) function. Thus, the synchronization bandwidth extends between the two ω_{in} values ω_{in1} and ω_{in2}, at which the sinus function is ± 1. An arcsin of 1 or -1 has only one solution, given by $\phi_s + \alpha_v = \pi/2, -\pi/2$, respectively, which means that the two synchronized solutions should merge in a single one at the frequency limits of the synchronization band. This in agreement with the closed shape of the synchronization curves discussed in Chapter 3 (see, e.g., Fig. 3.24). Note that a low I_{in} value gives rise to small amplitude and frequency increments ΔV and $\omega_{in} - \omega_o$. However, from (4.6), the phase shift ϕ_s will take all possible values between $-\pi$ and π along the closed solution curve. The phase values at band limits ω_{in1} and ω_{in2}, are given by $\phi_{s1} = -\alpha_v + \pi/2$ and $\phi_{s2} = -\alpha_v - \pi/2$, respectively. For small dependence of the imaginary part of $\overline{Y}_{ToV}$ on the oscillation amplitude V, the angle α_v will be close to zero, $\alpha_v \cong 0$, so the phase value at the frequency limits of the synchronization band will be $\phi_{s1} \cong \pi/2$, $\phi_{s2} \cong -\pi/2$.

To obtain the variation of the synchronization bandwidth versus the input generator amplitude I_{in}, the sinusoidal term $\sin(\phi_o + \alpha_v)$ is replaced by ± 1 in (4.6), solving for ω_{in}. This provides the following two straight lines I_{in}, ω_{in1} and I_{in}, ω_{in2}, merging at $I_{in} = 0$, ω_o:

$$\begin{aligned} \omega_{in1} &= \omega_o - \frac{I_{in}}{V_o\,|\partial Y_{To}/\partial\omega|\sin\alpha_{v\omega}} \\ \omega_{in2} &= \omega_o + \frac{I_{in}}{V_o\,|\partial Y_{To}/\partial\omega|\sin\alpha_{v\omega}} \end{aligned} \tag{4.7}$$

Equations (4.7) predict a totally symmetrical synchronization bandwidth for each value of the input amplitude I_{in}, which, in general, will only be true for very small I_{in}. Subtracting the two equations, the synchronization bandwidth $\Delta\omega_{max} = \omega_{in2} - \omega_{in1}$ varies with I_{in} according to

$$\Delta\omega_{max} = \frac{2I_{in}}{V_o\,|\partial Y_{To}/\partial\omega|\sin\alpha_{v\omega}} \tag{4.8}$$

Thus, in this linearized approach, the synchronization bandwidth is directly proportional to the input amplitude I_{in}. In the particular case of a purely resistive nonlinearity $Y_N \equiv G_N(V)$ and a linear admittance of the form $Y_L(\omega) = G_L + jB_L(\omega)$, equation (4.8) simplifies to

$$\Delta\omega_{max} = 2\frac{I_{in}}{V_o}\frac{1}{\partial B_T/\partial\omega|_o} = \frac{I_{in}}{V_o}\frac{\omega_o}{G_L Q} \tag{4.9}$$

where Q is the resonator quality factor, defined as $Q = \partial B_T/\partial\omega|_o\,\omega_o/2G_L$. From (4.9), broader synchronization bandwidth can be expected for a smaller quality factor of the resonator. Note that even in the general case (4.7), a low quality factor will enable a broader synchronization bandwidth [11], due to the usually much smaller value of $\partial G_{To}/\partial\omega$ than of $\partial B_{To}/\partial\omega$.

The next objective will be to obtain a variation of the periodic solutions $V_o + \Delta V$ and ϕ_s along the synchronization band delimited by (4.7). These solutions are determined by introducing the condition $d\phi/dt = 0$ into system (4.3) [13]:

$$\begin{aligned} Y^r_{ToV}(V - V_o) + Y^r_{To\omega}(\omega_{in} - \omega_o) &= \frac{I_{in}}{V_o}\cos\phi_s \\ Y^i_{ToV}(V - V_o) + Y^i_{To\omega}(\omega_{in} - \omega_o) &= -\frac{I_{in}}{V_o}\sin\phi_s \end{aligned} \tag{4.10}$$

where the subscripts V and ω indicate the variable with respect to which the derivative of the total admittance function Y_T is calculated. By squaring and adding the two equations, it is possible to obtain an approximate equation of the synchronized solution curve $V_s(\omega_{in})$, corresponding to each I_{in} value:

$$\begin{aligned} &|\overline{Y}_{ToV}|^2(V_s - V_o)^2 + |\overline{Y}_{To\omega}|^2(\omega_{in} - \omega_o)^2 \\ &\quad + 2\overline{Y}_{ToV}\cdot\overline{Y}_{To\omega}(V_s - V_o)(\omega_{in} - \omega_o) = \frac{I_{in}^2}{V_o^2} \end{aligned} \tag{4.11}$$

where the dot stands for the product $\overline{a}\cdot\overline{b} = a^r b^r + a^i b^i = |a||b|\cos\alpha_{ab}$, with $\alpha_{ab} = \text{ang}(b) - \text{ang}(a)$. Equation (4.11) defines a perfect ellipse in the plane V_s, ω_{in}, centered about the free-running solution V_o, ω_o. By expressing (4.11) in polar coordinates, we obtain the tilt angle of the ellipse, which is determined by the derivatives of the total admittance function, evaluated at the free-running oscillation. In the particular case of normal vectors $\overline{Y}_{ov}\cdot\overline{Y}_{o\omega} = 0$, with $\alpha_{v\omega} = \text{ang}(\overline{Y}_{To\omega}) - \text{ang}(\overline{Y}_{ToV}) = \pi/2$, the ellipse axes are parallel to axes ω_{in}, V_s of the coordinate system.

Calculating $\partial\phi/\partial\omega_{in}$ and $\partial V/\partial\omega_{in}$ from (4.10), it is easily seen that both derivatives tend to infinity at the two edges of the synchronization band, fulfilling $\cos(\phi_s + \alpha_v) = 0$ (see, e.g., Fig. 3.24). Thus, the edges of the synchronization band are given by turning points, in agreement with the conclusions reached in Chapter 3 and in previous discussions. At each turning point a real pole of the periodic solution crosses the imaginary axis, so when the upper section of the ellipse is stable, the lower section will be unstable. To see this in an intuitive manner, assume a small perturbation $\Delta\phi(t)$ of the phase ϕ_s corresponding to a particular steady-state solution within the synchronization bandwidth. For the perturbation $\Delta\phi(t)$ to vanish exponentially in time, the sinusoidal term $-\sin(\phi_s + \Delta\phi(t) + \alpha_v)/\sin\alpha_{v\omega}$ in (4.5) must have a sign opposite to $\Delta\phi(t)$. Due to the small value of $\Delta\phi(t)$, it will be possible to perform a Taylor series expansion of the sinusoidal term about ϕ_s, which provides the condition $\cos(\phi_s + \alpha_v)/\sin\alpha_{v\omega} > 0$. The limit stability condition is $\cos(\phi_s + \alpha_v) = 0$ and agrees with the turning-point condition, fulfilled at the edges of a closed synchronization curve. Therefore, in this linearized analysis, if the upper section of the synchronization ellipse is stable, the lower section will be unstable, and vice versa. In general, the upper section of the ellipse is the stable one. This can be explained roughly by the fact that it provides higher output power than the lower section, so it has less chance to exhibit "unused" negative resistance, associated with poles that have a positive real part.

As shown in Chapter 3, the turning points of the closed synchronization curves are usually mode-locking bifurcations (also known as *local– global bifurcations*), leading to the generation of a quasiperiodic solution (see Fig. 3.22). Particularizing the condition $\cos(\phi_s + \alpha_v)/\sin\alpha_{v\omega} > 0$ to the case of a purely resistive nonlinearity $Y_N \equiv G_N(V)$ and a linear admittance of the form $Y_L(\omega) = G_L + jB_L(\omega)$, the stable phase interval is $(-\pi/2, \pi/2)$. Note that this is just a particular case. In general, the stable phase interval depends on the angles α_v and $\alpha_{v\omega}$.

4.2.2 Nonlinear Analysis of Synchronized Solution Curves

The preceding analysis of an injection-locked oscillator linearized with respect to synchronizing source will be valid only for small input power values, as can be gathered from the fact that the nonlinear admittance function $Y_T(V, \omega)$ has been replaced by its first-order Taylor series expansion about the free-running solution V_o, ω_o. When the input power increases, the synchronized solution curve deviates from the perfect ellipse described by (4.11). Furthermore, linearized analysis is unable to provide open synchronization curves like the one represented in Fig. 3.19, obtained in a parallel resonance oscillator for the input current $I_{in} = 25$ mA. As shown in Chapter 3, secondary Hopf bifurcations, instead of turning points, delimit the stable synchronization band for higher values of the input generator amplitude. These bifurcations cannot be predicted using linearized analysis.

Using the describing function $Y_N(V, \omega)$ to model the nonlinear element [13], the equation ruling the circuit of Fig. 4.1 under synchronized conditions ($\omega_\partial = \omega_{in}$)

$$H_s = [Y_N(V, \omega_{in}) + Y_L(\omega_{in})]V - I_{in}e^{j\phi} = Y_s(V, \omega_{in})V - I_{in}e^{j\phi} = 0 \qquad (4.12)$$

where a compact error function H_s has been introduced and the total admittance function Y_s has been defined. For simplicity, the negative sign in the phase shift ϕ, affecting the input current I_{in}, has been suppressed. This will only give rise to a change of sign in the phase of the solutions obtained, with no effect regarding the synchronization curves or the stable sections of these curves. It must be emphasized that equation (4.12) assumes a periodic state and can only provide periodic solutions. However, as shown earlier, the injected oscillators have nonperiodic solutions outside the synchronization bandwidth, characterized by a time-varying phase shift $\phi(t)$. Note that for $I_{in} = 0$, equation (4.12) particularizes to the well-known free-running oscillator equation $Y_N + Y_L = 0$, satisfied by V_o, ω_o. This solution coexists with the trivial dc solution, with zero oscillation amplitude $V = 0$. For $I_{in} \neq 0$, the total admittance function will be different from zero and there may be one or several solutions, depending on the form of the nonlinear function $Y_N(V, \omega)$ and the input generator values I_{in} and ω_{in}.

As an example, the circuit in Fig. 1.1 will be considered. The corresponding nonlinear element has the instantaneous characteristic $i = av + bv^3$, with the associated describing function $Y_N(V) = a + 3/4bV^2$, with V the oscillation amplitude. This function and the linear network admittance $Y_L(\omega)$ will be substituted in the complex equation (4.12). Splitting this equation into real and imaginary parts, the following system of two real equations is obtained:

$$\begin{aligned} &\frac{3}{4}bV^3 + (G_L + a)V = I_{in}\cos\phi \\ &\left(C\omega_{in} - \frac{1}{L\omega_{in}}\right)V = I_{in}\sin\phi \end{aligned} \tag{4.13}$$

Provided that the input generator amplitude is held constant at I_{in} for each generator frequency ω_{in}, one or more solutions will be obtained in terms of the amplitude V and phase shift ϕ. Because of the cubic dependence on the amplitude V, one, two, or three different solutions may be found, depending on the generator values. To see this more clearly, the two real equations can be squared and added, which makes the phase disappear as a variable. This provides the following real equation:

$$(LC)^2V^2\omega_{in}^4 + \left[\left(\tfrac{3}{4}bV^3 + G_TV\right)^2L^2 - L^2I_{in}^2 - 2CLV^2\right]\omega_{in}^2 + V^2 = 0 \tag{4.14}$$

with $G_T = G_L + a$. As can be seen, it is a biquadratic equation in the frequency ω_{in}. The various coefficients will be renamed as follows:

$$\begin{aligned} A(V) &= (LC)^2V^2 \\ B(V, I_{in}) &= \left(\tfrac{3}{4}bV^3 + G_TV\right)^2L^2 - L^2I_{in}^2 - 2CLV^2 \\ D(V) &= V^2 \end{aligned} \tag{4.15}$$

Then the squared frequency ω_{in}^2 is given by

$$\omega_{in}^2 = \frac{-B(V, I_{in}) \pm \sqrt{B(V, I_{in})^2 - 4A(V)D(V)}}{2A(V)} \tag{4.16}$$

For each constant I_{in} value, the synchronized solution curve V versus ω_{in} can be obtained numerically in a very simple manner. The amplitude V is swept between zero and a few volts, calculating the coefficients in (4.15) and solving (4.16) at each V step. Only voltage values providing positive real values of ω_{in} are kept, storing the corresponding pairs ω_{in}, V. The results are shown in Fig. 4.2. Because the coefficients A and D are positive, when $B > 0$ there will be no solution, as this provides $\omega_{in}^2 < 0$. When $B < 0$ and $B^2 > 4AD$ there will be two ω_{in} solutions for each V value. When observing the ω_{in} variation versus V (the reader should turn the figure 90° clockwise), it is clear that two different regions can be identified in Fig. 4.2, depending on the I_{in} value. For relatively large I_{in}, the coefficient B is negative and fulfills $B^2 > 4AD$ in a single amplitude interval V_{min}, V_{max}. For a low I_{in} value, the coefficient B is negative and fulfills $B^2 > 4AD$ in two different V intervals: $(0, V_{max1})$ and (V_{min2}, V_{max2}). In the latter case, when tracing V versus ω_{in}, two different solution curves are obtained for the same I_{in} value (Fig. 4.2).

Figure 4.2 is very illustrative, as it shows the general evolution of periodic solutions of an injected oscillator versus the amplitude of the input generator. As stated earlier, the curves are represented in terms of the amplitude V at the fundamental frequency ω_{in}. The small circle indicates free-running oscillation corresponding to zero input amplitude $I_{in} = 0$. This solution coexists with a dc solution that in the representation of the figure would lie on the horizontal axis (zero oscillation amplitude). When the input generator power is injected, a synchronized solution

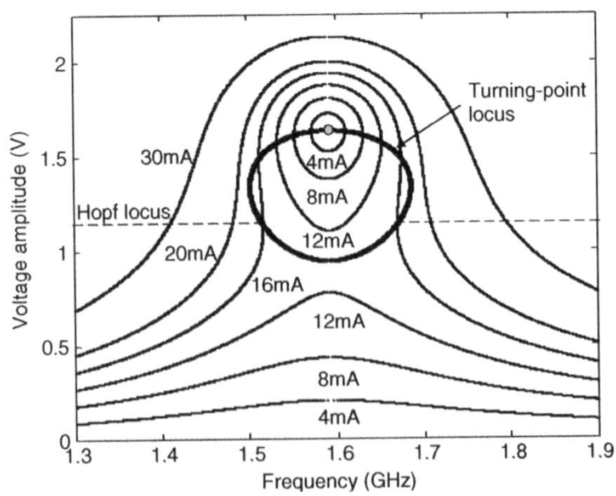

FIGURE 4.2 Periodic solutions of an injected cubic nonlinearity oscillator versus generator frequency for different values of the input generator current. The turning-point locus and Hopf locus are superimposed.

curve is obtained about the free-running oscillation point. For low input power, this synchronization curve is closed. See, for example, the curves corresponding to $I_{\text{in}} = 4$ mA, 8 mA, and 12 mA in the figure. Note that the closed solution curve coexists with a low-amplitude curve, which is present for all the frequency values. It is an unstable solution, equivalent to the dc solution of the free-running oscillator. In this low-amplitude solution, the circuit is not oscillating, which is why the solution has such a low amplitude. This curve provides a nonautonomous response of the circuit to the periodic input source. For low input current values, the limits of the synchronization band are given by the infinite slope points or turning points at each side of the closed curve. The closed solution curves are nearly perfect ellipses for low input power. As this power increases, the closed curves widen and become more irregular. The upper and lower curves merge at a certain input power, giving rise to a single solution curve. Then there is an intermediate range of input power for which the curve exhibits strong folding (see, e.g., the curve corresponding $I_{\text{in}} = 16$ mA). For the lower input power values, the turning points of the open curves are synchronization points (mode-locking bifurcations). For higher input power, they are simple jump points and the transition to quasiperiodic regime is due to direct Hopf bifurcations. As shown in the following section, the distinction between the two types of turning points requires a complementary stability analysis.

4.2.3 Stability Analysis

As indicated in Chapter 1, the stability of a given steady-state solution is determined from the poles associated with the circuit linearization about this particular solution. For an accurate analysis, the imaginary part ω of the poles $\sigma \pm j\omega$ must be allowed to take any value in the interval $[0, \omega_{\text{in}}/2)]$, with ω_{in} being the input frequency. However, the instabilities responsible for the desynchronization of the injected oscillator solution usually have a small pole frequency, $\omega = |\omega_{\text{in}} - \omega_a|$, with ω_a being the self-oscillation frequency. The following analytical derivation is restricted to small pole frequencies $\omega \ll \omega_o$. Although inherently limited, this analytical study is quite insightful and is compatible with a circuit description based on admittance functions.

Let a synchronized solution of the injected oscillator be considered, given by the amplitude V_s, frequency ω_{in}, and phase shift with respect to the input source $-\phi_s$. For a stability analysis of this solution, a small instantaneous perturbation is considered: This will give rise to a small increment in the node amplitude, $V_s + \Delta V(t)$, the node phase, $\phi(t) = -\phi_s + \Delta\phi(t)$, and the solution frequency $j\omega_{\text{in}} + s$, with s acting as a time derivator. Performing a first-order Taylor series expansion of Y_s and $e^{j\phi(t)}$ about the particular steady-state synchronized solution $V_s e^{-j\phi_s}$ at ω_{in}, it is possible to write

$$\left[\frac{\partial Y_s}{\partial V} V_s + Y_s\right] \Delta V(t) e^{-j\phi_s} + \frac{\partial Y_s}{\partial(j\omega_{\text{in}})} \frac{\mathrm{d}[V(t)e^{j\phi(t)}]}{\mathrm{dt}}$$

$$+ \underbrace{Y_s V_s}_{I_{\text{in}} e^{j\phi_s}} \Delta\phi(t) \frac{\partial e^{-j\phi_s}}{\partial \phi} = \left[\frac{\partial Y_s}{\partial V} V_s + Y_s \right] \Delta V(t)$$

$$+ \frac{\partial Y_s}{\partial(\omega_{\text{in}})} [-j \Delta \dot{V}(t) + V_s \Delta\dot{\phi}(t)] + j I_{\text{in}} e^{j\phi_s} \Delta\phi(t) = 0 \quad (4.17)$$

where all the higher-order terms have been neglected. Note that the phase derivative is calculated with respect to the node phase $\phi = -\phi_s$, instead of the input-source phase. It is possible to rewrite (4.17) in a more compact manner using a column vector composed of the real and imaginary parts of the error function H_s, defined in (4.12). This vector is given by $\overline{H}_s = (H^r, H^i)^T$. It is easily shown that in terms of this vector, equation (4.17) becomes

$$\overline{H}_V \Delta V(t) + \overline{H}_\omega \left(\Delta\dot{\phi}(t) - j \frac{\Delta \dot{V}(t)}{V_s} \right) - \overline{H}_\phi \Delta\phi(t) = \overline{0} \quad (4.18)$$

where the subindexes V, ϕ, and ω stand for derivatives of $\overline{H}_s$ with respect to the corresponding variables, evaluated at the particular synchronized steady-state solution given by V_s, ϕ_s, and ω_{in}. Splitting the complex equation (4.18) into real and imaginary parts, the following linear time invariant system is obtained:

$$\begin{pmatrix} \Delta \dot{V}(t) \\ \Delta\dot{\phi}(t) \end{pmatrix} = \left\{ \begin{pmatrix} H^i_\omega / V_s & H^r_\omega \\ -H^r_\omega / V_s & H^i_\omega \end{pmatrix}^{-1} \begin{pmatrix} -H^r_V & H^r_\phi \\ -H^i_V & H^i_\phi \end{pmatrix} \right\} \begin{pmatrix} \Delta V(t) \\ \Delta\phi(t) \end{pmatrix} \quad (4.19)$$

where the phase derivatives are calculated with respect to the input source phase. The solution poles will agree with the eigenvalues of the constant matrix within the braces $\{\cdot\}$. Note that the fact that the circuit is modeled using a two-dimensional system reduces the stability investigation to the two dominant poles. For stability, the two poles must be located on the left-hand side of the complex plane. Unlike what happens in the case of a free-running oscillator, neither of the two eigenvalues of (4.19) is intrinsically zero, as the periodic solutions of the injection-locked oscillator have no phase shift irrelevance. Remember that the phase shift $-\phi_s$ with respect to the input generator, together with the amplitude V_s, determines each solution within the synchronization bandwidth. When varying a circuit parameter, we can, of course, reach the conditions for a zero eigenvalue. Obviously, the constant matrix within the braces in (4.19) has a zero eigenvalue $\gamma = 0$ at points where the following matrix is singular:

$$[JH] = \begin{pmatrix} H^r_V & H^r_\phi \\ H^i_V & H^i_\phi \end{pmatrix} \quad (4.20)$$

This matrix, which contains derivatives of the real and imaginary parts of the error function H_s with respect to the amplitude V and phase ϕ, agrees with the Jacobian matrix associated with the nonlinear system (4.12). As demonstrated in Chapter 3, the infinite slope points of a solution curve versus a given parameter fulfill $\det[JH] = 0$, with JH the Jacobian matrix associated with the particular nonlinear system. Thus, the zero eigenvalue $\gamma = 0$ of the constant matrix in (4.19) will be responsible for the turning points of the synchronization curve.

The same stability formulation (4.19) is applicable when the synchronized circuit is analyzed using a linearization of the admittance function about the free-running solution, that is, when approaching $\overline{Y}(V, \omega_{in}) = \overline{Y}_{ToV}(V - V_o) + \overline{Y}_{To\omega}(\omega_{in} - \omega_o)$. In that case, the periodic synchronized solution fulfills (4.10) and the derivatives $\overline{H}_V$ and $\overline{H}_\omega$ in (4.19) can be approached by $\overline{H}_V = \overline{Y}_{Vo}V_o$ and $\overline{H}_\omega = \overline{Y}_{\omega o}V_o$. On the other hand, the phase derivative of $\overline{H}_s$ is calculated as $\overline{H}_\phi = (I_{in}\sin\phi, -I_{in}\cos\phi)$. It is left to the reader to compare the stability condition resulting from the eigenvalue analysis of (4.19) to the stability condition $\cos(\phi_s - \alpha_v)/\sin\alpha_{v\omega} > 0$ already obtained.

The stability analysis described has been applied along two representative solution curves of Fig. 4.2, obtained for constant I_{in} versus the input frequency ω_{in}. One of the two selected input current values is $I_{in} = 8$ mA, providing a closed curve and a low-amplitude curve. The second input current value is $I_{in} = 30$ mA, providing a single open solution curve. For each input current value I_{in}, stability analysis is applied versus ω_{in} in the following manner. In the first stage, the steady-state solution V_s, ϕ_s corresponding to each ω_{in} value is determined. Then the derivatives of the error function $\overline{H}_s$ are calculated at this solution V_s, ϕ_s, ω_{in}, using the expressions

$$\begin{aligned}\overline{H}_V &= \left(\frac{9}{4}bV^2 + (G_L + a), C\omega_{in} - \frac{1}{L\omega_{in}}\right)^T \\ \overline{H}_\omega &= V\left(0, C + \frac{1}{L\omega_{in}^2}\right)^T \qquad \overline{H}_\phi = I_{in}(\sin\phi, -\cos\phi)^T\end{aligned} \tag{4.21}$$

Next, the matrix within the braces in (4.19) is obtained from the real and imaginary parts of the vectors above. The two eigenvalues of this 2×2 matrix are calculated and stored. Then the next ω_{in} value is considered, following the same steps. Figure 4.3a shows the variation of the real part of the two dominant poles along the closed and low-amplitude curves obtained for $I_{in} = 8$ mA in Fig. 4.2. Note that for complex-conjugate poles $\sigma \pm j\omega$, a single value σ will be obtained, as the two poles have the same real part. As can be seen, the low-amplitude curve is always unstable, as its two poles have a positive real part for the entire input frequency interval. Even though the two poles belong to the low-amplitude curve without oscillation, their evolution versus the input frequency is related to the oscillation state: synchronized or not. Comparing Fig. 4.3a with Fig. 4.2, the two poles are complex-conjugate outside the synchronization band (one single curve), and they turn into two real poles near the edges of the synchronization band. Remember that the total number of poles agrees with the system dimension, which is 2 in this case. The dimension cannot change versus the parameter. A pair of unstable complex-conjugate poles indicates that oscillation at an incommensurable frequency $\omega_a/\omega_{in} \neq k/m$ is ready to start up. If the unstable poles are real, startup will take place at the frequency of the input source. On the other hand, the closed synchronization curve has two real poles, P_1 and P_2, each one describing, as can be expected, a closed path versus ω_{in},

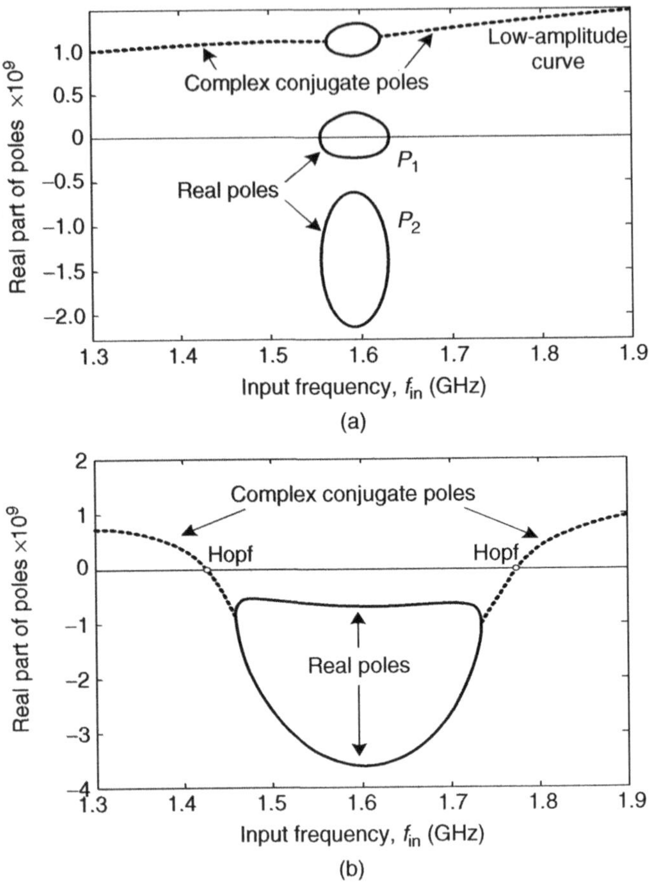

FIGURE 4.3 Stability analysis along the two solution curves of Fig. 4.2. The real part of the two poles calculated from (4.19) is represented versus the input frequency. Input amplitude (a) $I_{in} = 8$ mA and (b) $I_{in} = 30$ mA.

with the same two turning points as the synchronization curve. In the upper half of the synchronization curve, the two poles have a negative value. In the lower half, one of the poles is positive and the other is negative. As shown in the figure, one of the two real poles passes through zero at each of the two turning points.

Figure 4.3b shows variation in the real part of the two poles of the solution curve corresponding to $I_{in} = 30$ mA. The solution is stable in the interval (1.426 GHz, 1.776 GHz). Within the stable interval the two poles are real between 1.48 and 1.73 GHz. At these two frequency values, the two real poles merge into two complex-conjugate poles. Then at each edge of the stable synchronization interval, the real part of these complex-conjugate poles crosses through zero in a Hopf bifurcation.

The preceding analysis shows that for a low input amplitude, the stable synchronization ranges are delimited by the turning points at which a real pole crosses

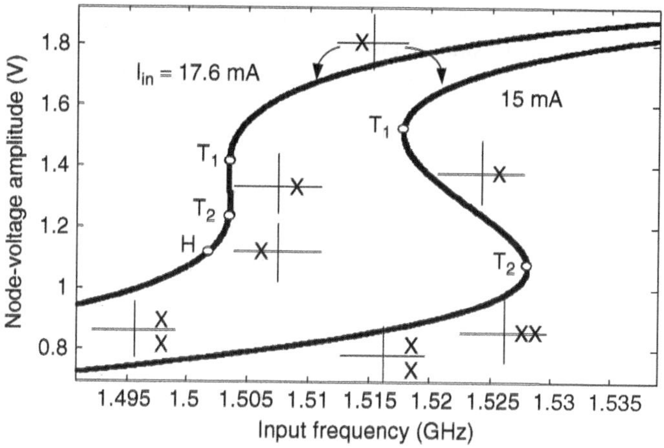

FIGURE 4.4 Pole variation along two periodic solution curves, in terms of voltage amplitude versus input frequency. Two different input current values are considered, $I_{\text{in}} = 15$ mA and $I_{\text{in}} = 17.6$ mA. In the first case, the turning point T_1 is a synchronization point. In the second case, the turning point T_2 is a jump point.

the zero value (Fig. 4.3a). For a relatively high input amplitude, the stable synchronization ranges are delimited by Hopf bifurcations, at which the real part of a pair of complex-conjugate poles crosses the zero value (Fig. 4.3b). The behavior is more complex in the intermediate range of input amplitude. For more insight into this behavior, pole analysis has also been used along two periodic solution curves, obtained for $I_{\text{in}} = 15$ mA and $I_{\text{in}} = 17.6$ mA, shown in Fig. 4.4. Both are open curves exhibiting two different turning points, T_1 and T_2. At these two points a real pole must necessarily cross the imaginary axis. The upper section of the two curves (starting from T_1 and increasing the input frequency) is stable for the two I_{in} values considered. When reducing the frequency from this upper section, one real pole γ_1 crosses the imaginary axis at T_1 in the two cases. For both solution curves, the section $T_1 - T_2$ is unstable, with one real pole on the right-hand side of the complex plane. At T_2, a real pole must also cross the imaginary axis, but the behavior is different for the two input current values.

When $I_{\text{in}} = 15$ mA, a second real pole, γ_2, different from γ_1, crosses the imaginary axis at T_2. Thus, after T_2, the solution keeps being unstable with two real poles γ_1 and γ_2 on the right-hand side of the complex plane. These two real poles merge and split into two complex-conjugate poles $\sigma \pm j\omega$ (Fig. 4.4) at about $f_{\text{in}} = 1.517$ GHz. Thus, the lower section of the periodic curve obtained for $I_{\text{in}} = 15$ mA is always unstable. For an input frequency below the value at which T_1 is obtained, that is, for $f_{\text{in}} < f_{\text{in}}(T_1)$, it has two unstable complex-conjugate poles. Thus, decreasing the input frequency from the stable periodic region (above T_1), the periodic solution turns quasiperiodic at point T_1. This point corresponds in this case to a local–global (mode-locking) bifurcation formally identical to those obtained at the turning points of the closed synchronization curves.

In the case of the input current $I_{in} = 17.6$ mA, the same real pole γ_1 that had crossed to the right-hand side of the complex plane at T_1 returns to the left-hand side at T_2 (see Fig. 4.4). The solution curve becomes stable at T_2. However, the stable interval is very short, as a pair of complex-conjugate poles $\sigma \pm j\omega$ crosses the imaginary axis at the input frequency $f_{in,H} = 1.502$ GHz, in a secondary Hopf bifurcation. Thus, for frequencies below $f_{in,H}$, the lower section of the curve will contain a pair of unstable complex-conjugate poles, and a quasiperiodic solution will be obtained. The two turning points T_1 and T_2 will give rise to a small hysteresis cycle, with jumps between the stable sections of the periodic solution curve.

4.2.4 Bifurcation Loci

As has been shown, the stable sections of the periodic solution curves of any injection-locked oscillators are delimited by two main types of bifurcations: turning points and secondary Hopf bifurcations, which are obtained at particular values of the amplitude I_{in} and frequency ω_{in} of the synchronizing source. Actually, the circuit will operate in a stable periodic regime only for certain input amplitude and frequency intervals. The set of ω_{in}, I_{in} values giving rise to stable operation is delimited by the bifurcation loci [14]. A *bifurcation locus* is a set of parameter values for which a given type of bifurcation takes place. In the following, equations are derived for the turning-point bifurcation locus and the secondary Hopf bifurcation locus using the describing function. To illustrate, calculations will be determined for a parallel resonance oscillator circuit, with the periodic solution curves represented in Fig. 4.2.

4.2.4.1 Turning-Point Locus The turning-point locus of an injection-locked oscillator in a periodic regime is the set of periodic solutions exhibiting infinite slope versus the input generator frequency or amplitude. To derive the locus, the equation system will be written in terms of the error function $H_s = Y_s V - I_{in} e^{j\phi}$ describing the circuit of Fig. 4.1. Let the solution curve V versus ω_{in} be considered. Assuming that the point n defined by ω_{in}^n, V^n, ϕ^n is known, the next point of the curve, ω_{in}^{n+1}, V^{n+1}, ϕ^{n+1}, corresponding to a frequency increment $\omega_{in}^{n+1} = \omega_{in}^n + \Delta\omega_{in}$, can be estimated by linearizing the function H_s about the previous point n. This provides the equation

$$[JH]_n \begin{bmatrix} \Delta V \\ \Delta\phi \end{bmatrix}_n + \begin{bmatrix} \dfrac{\partial H^r}{\partial \omega_{in}} \\ \dfrac{\partial H^i}{\partial \omega_{in}} \end{bmatrix}_n \Delta\omega_{in} = \underbrace{\begin{bmatrix} \dfrac{\partial H^r}{\partial V} & \dfrac{\partial H^r}{\partial \phi} \\ \dfrac{\partial H^i}{\partial V} & \dfrac{\partial H^i}{\partial \phi} \end{bmatrix}_n}_{[JH]_n} \begin{bmatrix} \Delta V \\ \Delta\phi \end{bmatrix}_n + \begin{bmatrix} \dfrac{\partial H^r}{\partial \omega_{in}} \\ \dfrac{\partial H^i}{\partial \omega_{in}} \end{bmatrix}_n \Delta\omega_{in} = 0 \qquad (4.22)$$

By solving for the vector $[\Delta V\ \Delta\phi]^T$, the infinite slope condition $dV/d\omega_{in} = \infty$ will be equivalent to the singularity of the Jacobian matrix $[JH]_n$. This is in agreement with the stability analysis of (4.19), as this condition implies the existence of a zero eigenvalue in the rightmost matrix of (4.19). Thus, the turning-point condition is $\det[JH] = 0$. This is equivalent to the condition derived in expression (3.21) for the case of free-running oscillators. The only difference is that in the case of an injection-locked oscillator, the phase variable replaces the free-running frequency.

For a more specific analysis, the Jacobian matrix $[JH]$ associated with the error function $H_s = Y_s V - I_{in} e^{j\phi}$ is

$$[JH] = \begin{bmatrix} Y^r(V, \omega_{in}) + \dfrac{\partial Y^r}{\partial V} V & I_{in} \sin\phi \\ Y^i(V, \omega_{in}) + \dfrac{\partial Y^i}{\partial V} V & -I_{in} \cos\phi \end{bmatrix} \tag{4.23}$$

Then the turning-point condition $\det[JH] = 0$ corresponds to

$$\left(Y^r + \frac{\partial Y^r}{\partial V} V\right) \cos\phi + \left(Y^i + \frac{\partial Y^i}{\partial V} V\right) \sin\phi = 0 \tag{4.24}$$

This equation must be combined with (4.12), as the turning points are also steady-state solutions of the synchronized system, so they fulfill $H_s = 0$. In terms of the variable V and the input frequency ω_{in}, the turning-point locus is given by

$$(Y^r)^2 + Y^r \frac{\partial Y^r}{\partial V} V + (Y^i)^2 + Y^i \frac{\partial Y^i}{\partial V} V = 0 \tag{4.25}$$

where the terms Y^r, Y^i, Y_v^r, and Y_v^i depend, in general, on both V and ω_{in}. As an example, equation (4.25) has been particularized to the parallel resonance oscillator, described by the steady-state system (4.13). This provides the following equation for the turning-point locus:

$$L^2C^2\omega_{in}^4 + \left(\frac{27}{16} b^2 V^4 L^2 + 3G_T b V^2 L^2 + G_T^2 L^2 - 2LC\right) \omega_{in}^2 + 1 = 0 \tag{4.26}$$

which is a biquadratic equation in ω_{in}, resolved in a manner similar to (4.14). The turning-point locus calculated with (4.26) has been superimposed on Fig. 4.2. The locus has an ellipsoidal shape and passes through all the points of infinite slope of the various solution curves. Sections of these curves inside the locus are unstable, as they are located between the two turning points T_1 and T_2. The unstable section $T_1 - T_2$ shrinks when increasing the input amplitude I_{in} and disappears at the value I_{in}^c, at which the solution curve is tangent to the turning-point locus. The tangency point is called the *cusp point* [1,16]. Note that at $I_{in} = I_{in}^c$, points T_1 and T_2 would overlap in a single point C. This is easy to figure out from the observation of Fig. 4.4. The real pole γ that for $I_{in} < I_{in}^c$ crossed the imaginary axis to the right-hand side of the complex plane at T_1 is tangent to the axis at the cusp

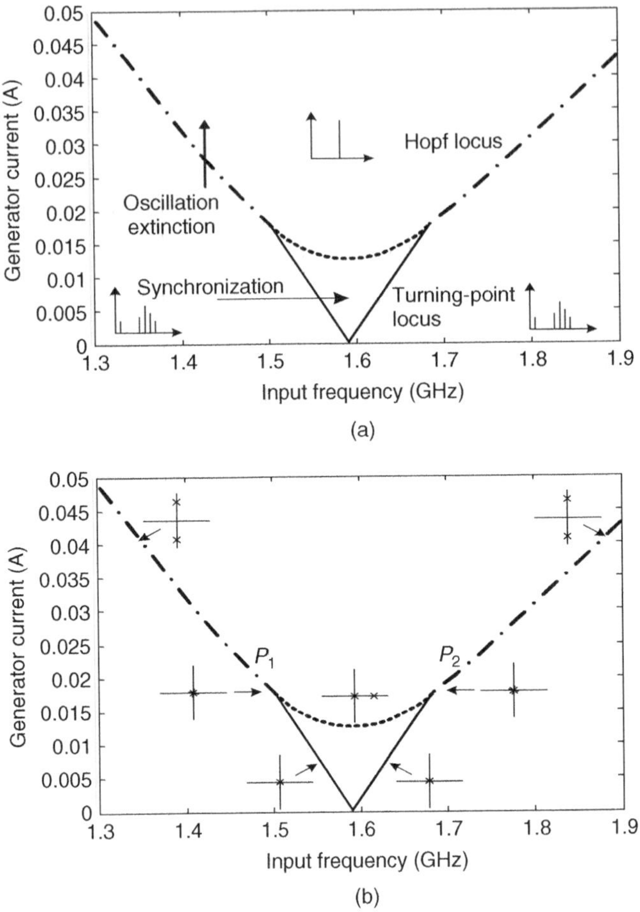

FIGURE 4.5 Bifurcation loci of an injected oscillator in the plane defined by the input generator frequency and current: (a) general view, with sketches of the solution spectrum at various operational regions delimited by the loci; (b) pole evolution along the solution points comprising the two loci.

point C. Therefore, this point does not fulfill the crossing condition $d\gamma/d\eta_b \neq 0$, so it is not actually a bifurcation point. In the analyzed circuit, two cusp points occur at $I_{\text{in}}^c = 17.78$ mA, and agree with the two infinite-slope points of the turning-point locus, at $f_{\text{inc1}} = 1.49$ GHz and $f_{\text{inc2}} = 1.69$ GHz (Fig. 4.2). For $I_{\text{in}} > I_{\text{in}}^c$, there are no turning points in the solution curves.

In Fig. 4.5a the turning-point locus has been drawn on a plane defined by ω_{in} and I_{in} where it corresponds to the trianglelike closed curve, composed of the solid line and dashed line sections. In this representation, the free-running solution, corresponding to $I_{\text{in}} = 0$, lies on the horizontal axis (zero oscillation amplitude) and is located on the lower vertex of the turning-point locus. Most of the points in the locus are actually local–global (mode-locking) bifurcations at which synchronization takes place. The Hopf locus has also been traced in Fig. 4.5, in dashed dotted line. Below the Hopf locus (discussed later) and outside the turning-point locus, the input frequency ω_{in} and the oscillation frequency ω_a coexist, giving rise to a quasiperiodic regime at the two fundamentals ω_{in} and ω_a. In terms of the input frequency ω_{in}, the synchronization band broadens with higher input power. The synchronization locus is called an *Arnold tongue*, due to its V shape. The top section of the turning-point locus (the curved zone between the two nearly straight lines) is the envelope of points at which a periodic solution with one unstable real pole transforms into a periodic solution with two unstable real poles. Thus, it has no physical effect. An example is the point T_2 of the curve corresponding to $I_{\text{in}} = 15$ mA in Fig. 4.4. For a better understanding, consider a straight line of constant amplitude $I_{\text{in}} = 15$ mA. This line crosses the turning-point locus at four different points. The outer points, corresponding to crossings with the solid line sections of the turning-point locus, are synchronization points. They delimit the stable synchronization range of the open solution curve (like T_1 in Fig. 4.4, for $I_{\text{in}} = 15$ mA). The two points at which the straight line crosses the dashed line section correspond to transitions between two unstable sections of the solution curve (like T_2 in Fig. 4.4, for $I_{\text{in}} = 15$ mA).

4.2.4.2 *Hopf Locus* As has been shown, secondary Hopf bifurcations are typically encountered in injection-locked oscillators for relatively high amplitude I_{in} values of the synchronizing source. At this type of bifurcation, an oscillation at an incommensurable frequency $\omega_a \neq (k/m)\omega_{\text{in}}$ is generated or extinguished. A general condition for the detection of this type of bifurcation in the frequency domain was given by expression (3.37). This condition, which should be combined with a full harmonic balance system, takes advantage of the fact that the oscillation is generated from zero amplitude value. After the Hopf bifurcation and in the immediate neighborhood of this bifurcation, the voltage waveform can be written

$$v(t) = V\cos(\omega_{\text{in}}t + \phi) + \Delta V\cos(\omega_a t + \theta) \tag{4.27}$$

where due to the nonrational relationship between ω_{in} and ω_a, θ is considered to be a uniformly distributed random variable. The expression (4.27) is introduced into the nonlinear function $i(v)$, which due to the small value of ΔV, can be expanded in a Taylor series about $v(t) = V\cos\omega_{\text{in}}t$. Next, the complex ratio between the output terms due to ΔV and the input term $\Delta V\cos(\omega_a t + \theta)$ is calculated and averaged with respect to phase θ. The result, independent of θ, provides an incremental describing function versus asynchronous inputs [14]:

$$Y_{\text{in,as}} = Y_N(V, \omega_a) + \frac{V}{2}\frac{dY}{dV} \tag{4.28}$$

The adjective *asynchronous* comes from the fact that the frequency ω_a is incommensurate with ω_{in}. In the particular case of the nonlinear element $i(v) = av + bv^3$, the incremental describing function $Y_{\text{in},as}$ is given by

$$Y_{\text{in},as} = a + \tfrac{3}{2}bV^2 \tag{4.29}$$

Applying Kirchhoff's laws, the oscillation condition at the frequency ω_a of the incipient quasiperiodic regime is given by

$$Y_{\text{in},as} + Y_L(\omega) = Y_{\text{in},as} + j\left(C\omega_a - \frac{1}{L\omega_a}\right) = G_T + \frac{3}{2}bV^2 = 0 \tag{4.30}$$

Note that the imaginary part of the total admittance function is equal to zero, as the circuit must resonate at the oscillation frequency generated. Condition (4.30) provides the secondary Hopf bifurcation locus. In this particular case, the incremental describing function does not depend on the input frequency, so the Hopf locus is determined by the constant oscillation amplitude:

$$V_h = \sqrt{\frac{-G_T}{\frac{3}{2}b}} = \frac{V_o}{\sqrt{2}} = 1.15\text{V} \tag{4.31}$$

The Hopf locus has also been superimposed on Fig. 4.2. When periodic curves cross this Hopf locus, they become unstable, due to the fact that a pair of complex-conjugate poles at the incommensurable frequency $\omega_a \neq (k/m)\omega_{\text{in}}$ cross through the imaginary axis. An autonomous oscillation is generated from zero amplitude, giving rise to a quasiperiodic regime (see Fig. 3.19). The quasiperiodic solution has two fundamental frequencies, the input frequency ω_{in} and the oscillation frequency ω_a. The oscillation generated is, in fact, the original circuit oscillation, reappearing in the circuit for I_{in} and ω_{in} values, for which the input generator has little influence over self-oscillation.

The turning-point and secondary Hopf bifurcation loci of injection-locked oscillators are very meaningful when traced in the plane defined by the input frequency and the input power or amplitude, as they provide a kind of "map" indicating the circuit operation mode for given generator values ω_{in} and I_{in}. To show this, sketches of the solution spectrum are included in Fig. 4.5a. The injected oscillator operates in a periodic regime at the input generator frequency ω_{in} inside the turning-point locus and above the Hopf locus. The circuit operates like a self-oscillating mixer at the two fundamental frequencies ω_{in} and ω_a outside the turning-point locus and below the Hopf locus. As already mentioned, the crossing of the upper section of the turning-point locus (shown dashed) has no physical implications. Note that solution curves and loci are symmetrical about the vertical axis passing through the free-running oscillation. This is true only in specific cases. Generally, the loci will not be symmetrical, due to the nonsymmetric response of the frequency-selective elements. An example is given later in the section.

For a better understanding of the meaning of loci, consider a particular value of the input amplitude I_{ino}, and trace a straight line $I_{\text{in}} = I_{\text{ino}}$ over the loci representation of Fig. 4.5a. Let us assume initially that for the selected value I_{ino}, the straight line $I_{in} = I_{ino}$ crosses the turning-point locus. The two frequency values at which the line $I_{in} = I_{ino}$ crosses this locus comprise the edges of the synchronization band $\omega_{in1}, \omega_{in2}$, which are determined by the two turning points of the corresponding closed solution curve. For zero input amplitude, the synchronization bandwidth degenerates in a single point, corresponding to the oscillation frequency ω_o. Thus, the lower vertex of the synchronization region corresponds to the free-running oscillation $I_{in} = 0$, $\omega_{in} = \omega_o$. As shown earlier, at the turning points that delimit the synchronization region, there is a transition from a quasiperiodic regime at ω_{in}, ω_a to a periodic regime at ω_{in}, or vice versa. When varying the parameter (input power or frequency) towards the turning point, the oscillation frequency ω_a of the quasiperiodic solution approaches continuously the input frequency ω_{in}. It becomes equal to this frequency at the synchronization point and the relationship $\omega_a = \omega_{in}$ is maintained within the entire synchronization band.

Repeating the procedure for a higher input amplitude I'_{ino} such that the turning-point locus is never traversed, the behavior is qualitatively different. Tracing the straight line $I_{in} = I'_{ino}$ over the loci diagram of Fig. 4.5a, the Hopf bifurcation is crossed twice. Increasing the frequency from a low value, the oscillation will be extinguished (instead of synchronized) at the first Hopf bifurcation. A transition from a quasiperiodic regime at ω_{in}, ω_a to a periodic regime at ω_{in} takes place at this point. If the input frequency is increased further, oscillation reemerges at the second Hopf bifurcation.

The turning-point and Hopf locus intersect at two points, P_1 and P_2, that are barely visible in Fig. 4.5b. Figure 4.6a shows an expanded view of Fig. 4.5b about the intersection point P_1. The intersection points between two different loci are very critical. It would be virtually impossible to obtain an intersection point varying one parameter only (either ω_{in} or I_{in}), with a constant value of the other. Points P_1 and P_2 are codimension 2 bifurcations [1], meaning that two different parameters must be varied simultaneously to obtain these points.

At points P_1 and P_2, the conditions for turning-point and Hopf bifurcation are fulfilled simultaneously. Hopf bifurcation implies two critical poles $\pm j\omega$, with zero real part $\sigma = 0$ and the turning point implies one zero real pole $\gamma_1 = 0$. These conditions are sketched in Fig. 4.5b. Because the poles must evolve in a continuous manner versus any parameter, the frequency of the critical poles $\pm j\omega$ decreases along the Hopf locus in the direction of each of points P_1 and P_2. Remember that $\omega = |\omega_{in} - \omega_a|$. At these intersection points, the pole frequency ω becomes zero, so P_1 and P_2 have two zero poles, $\gamma_1 = \gamma_2 = 0$. Next, the evolution of these two real poles along the turning-point locus will be discussed. From point P_1 (or P_2) one of the poles stays at zero $\gamma_1 = 0$, whereas the other shifts to either the left-hand side or the right-hand side of the complex plane and evolves along the turning-point locus in a continuous manner. From P_1 (or P_2) to the upper section of the turning-point locus, the pole γ_2 shifts to the right-hand side of the plane. Thus, all the points of the turning-point locus comprised between P_1 and P_2 (see Fig. 4.5b) have a real pole on the right-hand side of the complex plane $\gamma_2 > 0$ in addition to the real pole at zero $\gamma_1 = 0$. Due to the presence of an unstable pole, the crossing of the upper section of the turning-point locus gives rise to a transition

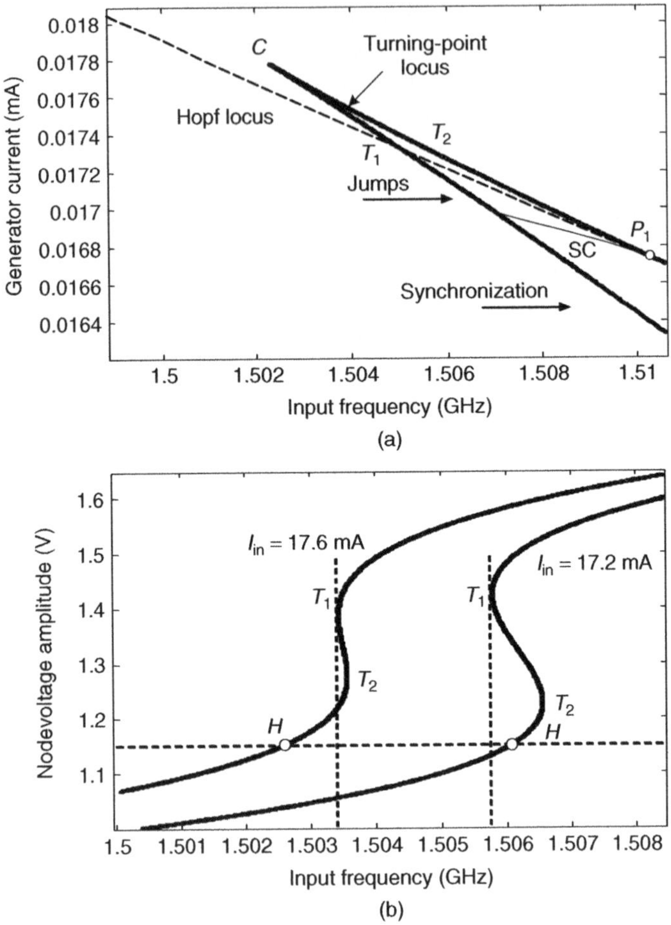

FIGURE 4.6 Behavior of an injection-locked oscillator near the intersection point of loci: (a) expanded view of the loci about intersection point P_1 between the turning-point and Hopf loci (a sketch of the saddle connection locus has been included); (b) solution curves with different behavior for $I_{in} = 17.6$ mA and $I_{in} = 17.2$ mA.

between two unstable solutions. Actually, the upper section of the locus contains turning points of identical class to point T_2 in the curve corresponding to $I_{in} = 15$ mA in Fig. 4.4. This is why this upper section has no physical meaning. In the rest of the turning-point locus (excluding the upper section), there is only one real pole at zero, with no poles on the right-hand side of the complex plane, so its crossing will give rise to transitions between stable and unstable sections of the solution curves.

For some more insight into circuit behavior, Fig. 4.6a shows an expanded view of the bifurcation loci about intersection point P_1. The turning-point locus has been divided into two sections. One section contains turning points T_1 obtained for lower values of input frequency, and the second section contains turning points T_2 obtained for higher values of this frequency. The two sections meet at cusp point C, at which the unstable section between T_1 and T_2 vanishes. The reasons for the name *cusp* become clear in the representation in the plane ω_{in}, I_{in}. The curve passing through the cusp point corresponds approximately to $I_{in} = 17.85$ mA. At this point, the real pole γ responsible for the instability of section $T_1 - T_2$ takes zero value but does not actually cross the imaginary axis. It fulfills $d\gamma/d\eta = 0$, with η being either the input generator amplitude or frequency, depending on the parameter varied, so the pole is tangent to the imaginary axis at the cusp point.

Besides the turning-point and Hopf loci, a sketch of a third bifurcation locus SC is represented in Fig. 4.5a. This locus, discussed later, consists of points at which saddle connection bifurcations occur in the Poincaré map. Above SC, sections T_1 and T_2 correspond to jump points. If the Hopf locus lies on the left-hand side of both the T_1 and T_2 sections, the jumps take place between two periodic solutions. As an example, note the curve corresponding to $I_{in} = 17.6$ mA in Fig. 4.4, redrawn for convenience in Fig. 4.6b. If the Hopf locus lies between the T_1 and T_2 sections, the jumps take place between periodic and quasiperiodic regimes. As an example, note the curve corresponding to $I_{in} = 17.2$ mA in Fig. 4.6b. When increasing the input frequency from low values, the quasiperiodic solution is extinguished at the Hopf bifurcation. Thus, a stable periodic regime is obtained from this point. If the frequency continues to be increased, a jump takes place at T_2 to the upper section of the periodic curve. When reducing the input frequency from this upper section, the system jumps to the coexisting quasiperiodic solution at point T_1. Note that T_1 is a jump point, not a synchronization point.

The circuit behavior when the SC locus is traversed is even more complex. The unstable periodic section $T_1 - T_2$ consists of saddle solutions that have an unstable pole, whereas the rest of their poles are on the left-hand side of the complex plane. At the SC locus, the saddle solution gives rise to a quasiperiodic solution QP2 through a global bifurcation termed *saddle connection* (see Chapter 3). Remember that at a saddle connection, a saddle-type fixed point of the Poincaré map gives rise to a discrete-point cycle corresponding to a quasiperiodic solution (Chapter 3). Considering a constant input current and increasing the frequency from a low value, the circuit exhibits a quasiperiodic solution QP1, which becomes synchronized at turning point T_1. However, at a higher input frequency, a new quasiperiodic solution QP2 is generated through a saddle connection when hitting the saddle connection locus. This quasiperiodic solution QP2 is stable and coexists with the stable periodic solution in the upper section of the periodic curve for a short frequency interval. The quasiperiodic solution QP2 is extinguished when crossing the Hopf locus in an inverse Hopf bifurcation. The saddle connection locus is difficult to obtain, as this requires detecting a collision between the discrete-point cycle and the saddle point of the Poincaré map. Fortunately, it is not very relevant in the behavior of injection-locked oscillators, as it occurs for a relatively small interval of input generator amplitude and frequency. However, transversal saddle connections can give rise to chaos near the codimension 2 bifurcations P_1 and P_2, which is often observed in practice. In general, the circuit behavior near the loci intersection is quite irregular. Though qualitatively similar to what has been presented in this section, it may vary substantially from circuit to circuit.

4.2.5 Phase Variation Along Periodic Curves

As shown in equation (4.12), when the oscillation is synchronized, there is a constant phase difference between this oscillation and the input source at ω_{in}. This phase shift, which varies with the generator frequency, can be determined by solving (4.12) for the phase ϕ_s:

$$\tan\phi_s = \frac{Y_N^i(V,\omega_{\text{in}}) + Y_L^i(\omega_{\text{in}})}{Y_N^r(V,\omega_{\text{in}}) + Y_L^r(\omega_{\text{in}})} \tag{4.32}$$

To make ϕ_s depend on the input frequency ω_{in} only, the relationship $V(\omega_{\text{in}})$, provided by (4.12), must also be taken into account. In the particular case of the parallel resonance oscillator, equation (4.32) becomes

$$\tan\phi_s = \frac{C\omega_{\text{in}} - 1/L\omega_{\text{in}}}{G_T + \frac{3}{4}bV^2(\omega_{\text{in}})} \tag{4.33}$$

with V related to ω_{in} through (4.14). Figure 4.7 shows the phase variation along the various types of periodic solution curves obtained in the simulations of Fig. 4.2. The case of small input power is considered first. We know that, in this case, for each I_{in} value we have two different solution curves: a closed curve and a small-amplitude curve. The phase shift along the closed-solution curves takes all possible values in the interval $(-180^\circ, 180^\circ)$. The corresponding phase curves are easily identified in Fig. 4.7, as they start and end at $\pm 180^\circ$ and exhibit two turning points $d\phi/d\omega_{\text{in}} = \infty$, which delimit the stable phase range, centered about 0°. The phase along the low-amplitude curve, coexisting with the closed synchronization curve, varies in a smaller range (see the dashed line curve in Fig. 4.7). The phase curve associated with the low-amplitude curve meets the one corresponding to the closed curve at the maximum phase values at $\pm 180^\circ$. Note that they are, in fact, two disjoint curves. In agreement with Fig. 4.2, at $I_{\text{in}} = 14$ mA, the two phase curves merge into a single curve. The total phase variation range (including stable and unstable sections) is now smaller than $(-180^\circ, 180^\circ)$. For $I_{\text{in}} > 14$ mA, the phase curves initially exhibit four turning points. At the cusp points, obtained in the two symmetrical sections at $I_{\text{in}} = 17.85$ mA, the two turning points on each side meet. From these I_{in} values, the phase curve does not exhibit any turning points. The stable phase shift range is delimited by the Hopf bifurcations presented in Fig. 4.2 and indicated in Fig. 4.7.

Note that the highest phase sensitivity to the input generator frequency ω_{in} is obtained (for all the input amplitude values) about the frequency of the free-running oscillation ω_o. This is due to the original circuit resonance at this frequency, $Y_{To}(V_o, \omega_o) = 0$, in the absence of input power. Far from the resonance, the phase

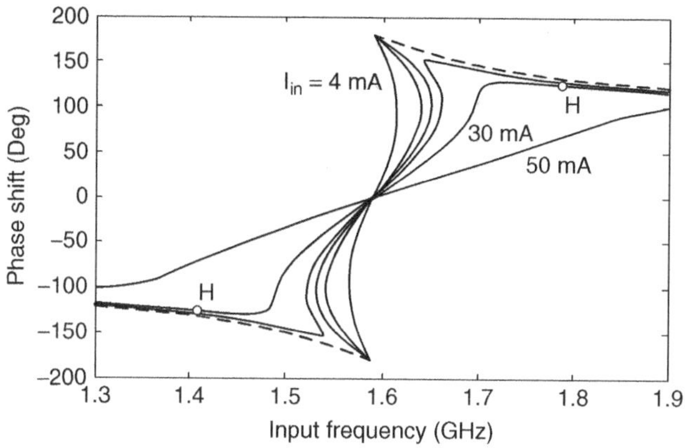

FIGURE 4.7 Variation of the solution phase ϕ versus the input frequency f_{in} for different values of the input generator amplitude.

sensitivity to the input frequency ω_{in} becomes substantially smaller. It is typically reduced from the Hopf bifurcations, at which the periodic solution becomes unstable.

As demonstrated with the linearized analysis of 4.2.1, for a low input amplitude I_{in}, the turning points of the synchronization curve, forming a perfect ellipse, correspond to the phase values $\phi_s - \alpha_v = -\pi/2, \pi/2$, where α_v is the angle associated with the phasor $\partial Y_{To}/\partial V$. On the other hand, according to (4.6), the phase at the same frequency ω_{in} as the original free-running oscillation $\omega_{in} = \omega_o$ is given by $\phi_s = \alpha_v$. For the small dependence of the imaginary part of Y_{To} on the oscillation amplitude, the associated phase will be $\alpha_v \simeq 0$. Then the phase value at the free-running frequency is $\phi_s(\omega_o) = 0$ and the phase shift at the turning points will be $\phi_{s1,s2} = -\pi/2, +\pi/2$, respectively. The validity of these results is restricted to small input amplitude, as confirmed by the simulations of Fig. 4.7.

4.2.6 Analysis of a FET-Based Oscillator

In this subsection, a synchronizing source is connected to an oscillator based on the FET transistor ATF26884. The oscillator schematic is shown in Fig. 4.8. The bias network consists of two dc sources, $V_{GG} = -1$ V and $V_{DD} = 5$ V, plus bias resistances. Series feedback is introduced at the source terminal. The gate subnetwork is given by an inductance and a resistance connected in series. The two elements have been implemented on a transmission line through a high-impedance line and a quarter-wave transformer. The transformer enables the connection of the synchronizing source, with a 50-Ω impedance, without extinguishing the oscillation. The used feedback capacitance, plus the gate subnetwork described, have been calculated to obtain negative resistance at the drain terminal at the desired oscillation frequency $f_o = 4.31$ GHz. The oscillator load at this drain terminal consists of an inductance and a resistance connected in series, also implemented through a high-impedance transmission line and a quarter-wave transformer.

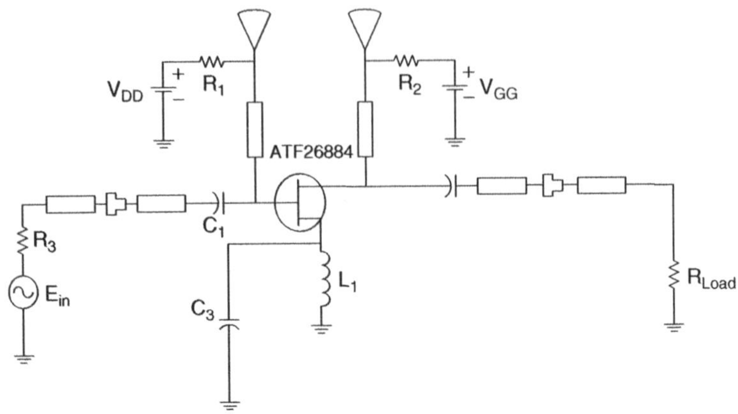

FIGURE 4.8 FET-based oscillator with series feedback at the source terminal. The circuit exhibits free-running oscillation at $f_o = 4.31$ GHz. An RF generator is introduced for injection locking.

The FET-based circuit of Fig. 4.8 exhibits free-running oscillation at $f_o = 4.31$ GHz. When introducing a periodic input source, the synchronized solution curves obtained for different values of the input amplitude E_{in} are as presented in Fig. 4.9. The curves are drawn in terms of the voltage amplitude at the drain terminal (V_{drain}). As can be seen, the behavior is qualitatively similar to that of the parallel resonance oscillator of Fig. 4.2. For a low input amplitude, the solution curves are nearly perfect ellipses, in agreement with the approximation (4.11). The values of the derivatives of the admittance function $Y(V, \omega)$, calculated at the free-running solution, are $Y_{ToV} = 0.002 + j0.011\ \Omega^{-1}$/V and $Y_{To\omega} = -7.16 \times 10^{-13} + j2.09 \times 10^{-12}\Omega^{-1}$/S respectively. As already shown, the ellipse axis in the coordinate system V_s, ω_{in} is defined by these derivatives. As in Fig. 4.2, for low input power, the ellipsoidal curve coexists with a low-amplitude curve. The two curves merge at the generator amplitude $E_{in} = 0.035$ V. After this merging, the single solution curve exhibits strong folding for a certain input amplitude interval. For a sufficiently high input amplitude, no turning points exist in the solution curve (see, e.g., the curve corresponding to $E_{in} = 0.37$ V).

As in the case of Fig. 4.2, the stable sections of periodic curves are delimited by the turning-point and Hopf loci. These two loci have been superimposed in Fig. 4.9. All sections of the periodic curves inside the turning-point locus and below the Hopf locus correspond to unstable behavior. The Hopf locus is nearly horizontal in the plane $f_{in} - V_{drain}$ except for a small section on the left-hand side. Note that it was perfectly horizontal in the parallel resonance oscillator of Fig. 4.2. In contrast with Fig. 4.9, neither the solution curves nor the loci are symmetrical with respect to the free-running oscillation, due to the frequency selectivity of the input/output filters and feedback network. The Hopf locus is crossed at lower power on the right-hand side of the diagram, which is due to the larger loss of the series feedback network versus the input frequency.

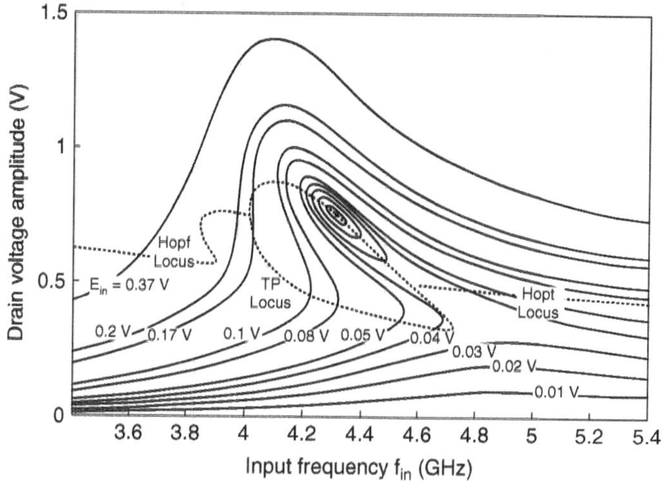

FIGURE 4.9 Periodic solutions of an injected FET-based oscillator for different values of the input generator voltage. The turning-point and Hopf loci are superimposed.

The turning-point and Hopf bifurcation loci have also been traced in the plane defined by the input frequency and input amplitude, with the results shown in Fig. 4.10. Sketches of the spectrum for different values of the input generator have been included. These loci should be compared with those represented in Fig. 4.5, corresponding to the parallel resonance oscillator. The turning-point locus is symmetrical about the free-running oscillation for very low input amplitude only. The Hopf bifurcation locus is also nonsymmetrical. In the higher-frequency range, the oscillation is extinguished from a much lower input generator amplitude,

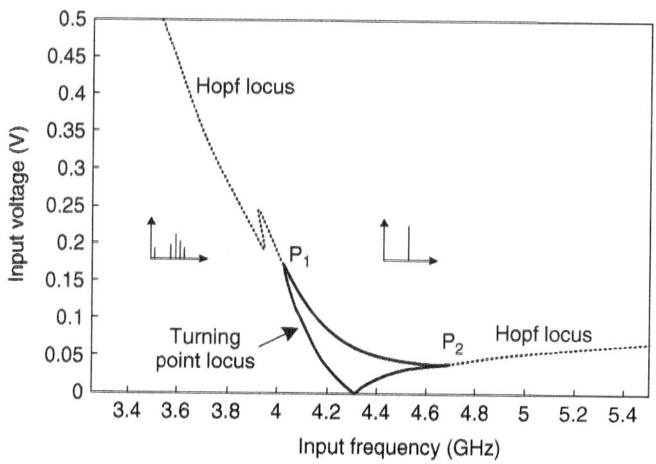

FIGURE 4.10 Bifurcation loci of an injected FET-based oscillator in the plane defined by the input generator frequency and voltage.

which, as already indicated, is due to the higher loss of the feedback network. The Hopf locus in the lower input frequency range exhibits two points of infinite slope. Folding of the Hopf locus will give rise to some irregularities in circuit behavior. As an example, consider the variations in the circuit solution when maintaining the constant input frequency $f_{in} = 3.8$ GHz and increasing the input power. For very low input power, the circuit behaves as a self-oscillating mixer. The oscillation is extinguished when crossing the Hopf locus for the first time but reemerges when crossing this locus a second time. Finally, when crossing the locus for the third time, the oscillation is definitively extinguished. This type of phenomenon is commonly observed in measurements.

The turning-point and Hopf loci intersect at points P_1 and P_2. As in the case of the loci in Fig. 4.5, the top line of the turning-point locus, located between P_1 and P_2, corresponds to the bifurcation of a periodic solution with one unstable real pole into a periodic solution with two unstable real poles (see Fig. 4.5b). Thus, it has no physical effect. Depending on the input amplitude, the stable operation range will be delimited at each side by either the turning-point locus sections P_1-O and $O-P_2$ or by the Hopf locus.

As an example, for the input voltage $E_{in} = 0.056$ V, there are two turning points, occurring at $f_{in,T1} = 4.2$ GHz and $f_{in,T2} = 4.5$ GHz, and a Hopf bifurcation, occurring at $f_{in,H} = 4.8$ GHz. Point T_2, which belongs to section $P_1 - P_2$ of the turning-point locus, will have no physical effect, so the operation bandwidth is delimited by T_1 and the Hopf bifurcation H. This is confirmed by simulations of Fig. 4.11, which include the quasiperiodic solutions outside the stable periodic operation band. The solutions are represented in terms of the drain voltage amplitude.

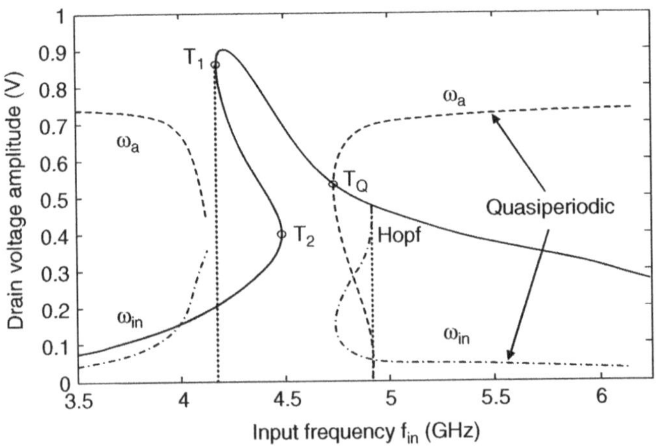

FIGURE 4.11 Solutions of a FET-based injection-locked oscillator for $E_{in} = 0.056$ V in terms of the drain voltage amplitude. The quasiperiodic solutions are represented by tracing the voltage amplitude at the input frequency ω_{in} and the oscillation frequency ω_a. The stable periodic range is delimited by the turning-point bifurcation T_1 (on the lower edge) and Hopf bifurcation (on the upper edge).

For quasiperiodic solutions, both the voltage amplitude at the input frequency ω_{in} and the oscillation frequency ω_a have been drawn. As can be seen, the lower edge of the stable synchronization band is determined by turning point T_1. The slight discontinuity in the generation of the quasiperiodic solution is associated with the global nature of turning points at which synchronization takes place. There is also limitted analysis accuracy because the spectrum becomes very dense near the synchronization point (see Chapter 3). Similar to Fig. 4.4, at the turning point T_2 of the periodic path, the periodic solution with two unstable real poles transforms into a periodic one with one unstable real pole, so this bifurcation has no physical effect. The upper edge of the synchronization band is determined by a subcritical Hopf bifurcation. The continuity of this bifurcation is in correspondence with the fact that a quasiperiodic solution is generated from zero oscillation amplitude. This amplitude grows in a continuous manner from the bifurcation point of subcritical type. Turning point T_Q in the quasiperiodic path generated will give rise to a hysteresis phenomenon in the transformation from a periodic to a quasiperiodic regime, and vice versa.

Figure 4.12 shows the variation in the solution phase at the drain terminal versus the input generator frequency for three input voltage values. These values, $E_{in1} = 0.01$ V, $E_{in2} = 0.02$ V, and $E_{in3} = 0.03$ V, correspond to the closed synchronization curves in the representation of Fig. 4.9. The phase corresponding to the frequency of the free-running oscillation, indicated as FO in the representation of Fig. 4.12, can be estimated from the linearized analysis of Section 2.3.1. Taking (4.6) into account, the relationship $\omega_{in} - \omega_o = \sin[\text{ang}(\overline{Y}_{ToV}, \phi_s)] = 0$ must be fulfilled. The phase value predicted is $\phi_{Fo} = 211^\circ$, quite close to the simulated value. According to the linearized analysis, for small values of the input amplitude E_{in}, the stable phase range is nearly independent of E_{in}. It is determined by the phase of the admittance derivative with respect to the amplitude $\overline{Y}_V$. This phase is given by $\text{ang}(\overline{Y}_{ToV}) = 211^\circ$. Thus, the stable phase shift range can be calculated approximately through $[-90^\circ + \text{ang}(\overline{Y}_{ToV}), 90^\circ + \text{ang}(\overline{Y}_{ToV})] = (121^\circ, 301^\circ)$. The stable interval simulated is $(127^\circ, 338^\circ)$, so the approximate calculation has a relative error of about 2%.

4.2.7 Phase Noise Analysis

For phase noise analysis of an injection-locked oscillator, the admittance model of Fig. 4.1 will be considered. The injection source $i_{in}(t)$ is initially assumed noiseless and for reasons given later, only white noise from the oscillator circuit is considered. This white noise contribution is modeled with a current generator in parallel $i_N(t)$ at the same observation node. The squared current spectral density of this noise source is $|I_N|^2$. The noiseless steady-state solution at the injection generator frequency ω_{in} fulfilling (4.12) has the node amplitude V_s and phase ϕ_s. For the admittance analysis, a complex envelope representation of the noise source $i_N(t)$ about the input frequency ω_{in} will be considered: $i_N(t) = \text{Re}[I_N(t)e^{j\omega_{in}t}]$. Due to the low amplitude of the noise current source, the amplitude, phase, and frequency of the solution will undergo only small variations with respect to the

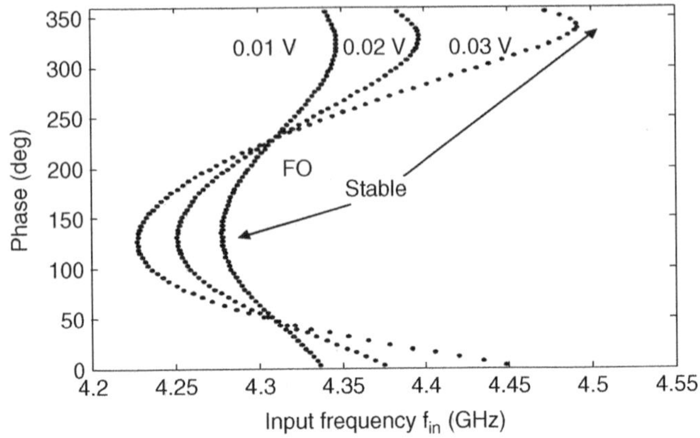

FIGURE 4.12 Phase variation ϕ of a FET-based oscillator versus the input frequency f_{in} for various small values of the input generator amplitude.

unperturbed synchronized oscillation, given by V_s, ϕ_s, and ω_{in}. In the presence of a noise source, the perturbed solution will have the amplitude $V(t) = V_s + \Delta V(t)$. In turn, the absolute phase of the node voltage will be expressed as $\phi(t) = -\phi_s + \Delta\phi(t)$ and the frequency will be incremented as $j\omega_{in} + s$. Performing a Taylor series expansion of the admittance function Y_s about the synchronized steady-state solution V_s, ϕ_s, ω_{in}, in a manner similar to what was done in (4.17), we obtain the following perturbed system in straightforward manner:

$$\left[\left.\frac{\partial Y_T}{\partial V}\right|_s \Delta V(t) + \left.\frac{\partial Y_T}{\partial \omega}\right|_s \left(\Delta\dot{\phi}(t) - j\frac{\Delta \dot{V}(t)}{V_s} \right) \right] = -\frac{I_{in}}{V_s} \left.\frac{\partial e^{j\phi_s}}{\partial \phi}\right|_s \Delta\phi(t) + \frac{I_N(t)}{V_s} \tag{4.34}$$

where the influence of the term $Y_T \Delta V(t)$ has been neglected. Note that the equation is essentially the same as in (4.17), except for the presence of the noise source. The derivative of the exponential is given by $\partial e^{j\phi}/\partial\phi\big|_s = -\sin\phi_s + j\cos\phi_s$. In an initial study, the time derivative $\Delta\dot{V}(t)$ will be neglected. As will be shown later, this variation will be significant only at the relatively large frequency offset from the carrier. Splitting (4.34) into real and imaginary parts yields

$$\begin{aligned} Y_{sV}^r \Delta V(t) + Y_{s\omega}^r \Delta\dot{\phi}(t) &= \frac{I_{in}}{V_s} \sin\phi_s \Delta\phi(t) + \frac{I_N^r(t)}{V_s} \\ Y_{sV}^i \Delta V(t) + Y_{s\omega}^i \Delta\dot{\phi}(t) &= -\frac{I_{in}}{V_s} \cos\phi_s \Delta\phi(t) + \frac{I_N^i(t)}{V_s} \end{aligned} \tag{4.35}$$

where the subscripts sV and $s\omega$ indicate derivatives of the total admittance function with respect to the amplitude and frequency, respectively, evaluated at the particular steady-state synchronized solution. System (4.35) is composed of real variables. Expressing these variables in the frequency domain and considering the positive-frequency sideband only, the following system is obtained [12]:

$$\begin{aligned} Y^r_{sV}\Delta V(\Omega) + Y^r_{s\omega} j\Omega\ \Delta\phi(\Omega) &= \frac{I_{\text{in}}}{V_s}\sin\phi_s \Delta\phi(\Omega) + \frac{I^r_N}{V_s} \\ Y^i_{sV}\Delta V(\Omega) + Y^i_{s\omega} j\Omega\ \Delta\phi(\Omega) &= -\frac{I_{\text{in}}}{V_s}\cos\phi_s \Delta\phi(\Omega) + \frac{I^i_N}{V_s} \end{aligned} \tag{4.36}$$

Grouping the terms in $\Delta\phi(\Omega)$ and solving for the phase perturbation $\Delta\phi(\Omega)$ yields

$$\Delta\phi(\Omega) = \frac{1}{V_s}\frac{Y^r_{sV}I^i_N - Y^i_{sV}I^r_N}{j\Omega(Y^r_{sV}Y^i_{s\omega} - Y^i_{sV}Y^r_{s\omega}) + (I_{\text{in}}/V_s)(Y^r_{sV}\cos\phi_s + Y^i_{sV}\sin\phi_s)} \tag{4.37}$$

Multiplying by the conjugate $\Delta\phi^*(\Omega)$ and taking into account that as shown in Chapter 2, I^r_N and I^i_N are uncorrelated and $|I^r_N|^2 = |I^i_N|^2 = 2|I_N|^2$, the phase noise spectrum is given by

$$|\Delta\phi(\Omega)|^2 = \frac{1}{V_s}\frac{2|Y_{sv}|^2|I_N|^2}{\Omega^2(Y^r_{sV}Y^i_{s\omega} - Y^i_{sV}Y^r_{s\omega})^2 + (I_{\text{in}}/V_s)^2\,(Y^r_{sV}\cos\phi_s + Y^i_{sV}\sin\phi_s)^2} \tag{4.38}$$

Note that expression (4.38) requires knowledge of the admittance function derivatives at the particular point ω_{in}, V_s, ϕ_s of the synchronized solution curve.

Now the case of a noisy injection source will be studied. Due to the much smaller value of the amplitude noise, only phase noise is considered, so the injection current source is expressed $i_{\text{in}}(t) = \text{Re}[I_{\text{in}}e^{j\psi(t)}]$, where $\psi(t)$ represents the phase perturbations introduced by this source. The phase perturbation of the input source does not alter directly the shift between the periodic-solution phase and the source phase. To understand this, the reader must remember that a shift α in the phase of the independent periodic source of a forced circuit gives rise to the same phase shift α in the circuit solution. Thus, the node voltage phase in the presence of the input phase noise $\psi(t)$ becomes $-\phi_s + \Delta\phi(t) + \psi(t)$, and the phase shift with respect to the input generator maintains the value $\phi(t) = -\phi_s + \Delta\phi(t)$. Performing a first-order Taylor series expansion similar to (4.17) and (4.34), the equations of the injection-locked oscillator in the presence of white noise from the oscillator circuit $I_N(t)$ and phase noise $\psi(t)$ from the injection source are written

$$\begin{aligned} Y^r_{sV}\Delta V(t) + Y^r_{s\omega}[\Delta\dot{\phi}(t) + \dot{\psi}(t)] &= \frac{I_{\text{in}}}{V_s}\sin\phi_s\ \Delta\phi(t) + \frac{I^r_N(t)}{V_s} \\ Y^i_{sV}\Delta V(t) + Y^i_{s\omega}[\Delta\dot{\phi}(t) + \dot{\psi}(t)] &= -\frac{I_{\text{in}}}{V_s}\cos\phi_s \Delta\phi(t) + \frac{I^i_N(t)}{V_s} \end{aligned} \tag{4.39}$$

Applying the Fourier transform in the slow time scale associated with the noise sources and solving for the phase perturbation $\Delta\phi(\Omega)$ yields

$$\begin{aligned}\Delta\phi(\Omega) &= \frac{-j\Omega\psi(\Omega)V_s(Y^r_{sV}Y^i_{s\omega} - Y^i_{sV}Y^r_{s\omega}) + Y^r_{sV}I^i_N - Y^i_{sV}I^r_N}{j\Omega V_s(Y^r_{sV}Y^i_{s\omega} - Y^i_{sV}Y^r_{s\omega}) + I_{\text{in}}(Y^r_{sV}\cos\phi_s + Y^i_{sV}\sin\phi_s)} \\ &= \frac{-j\Omega V_s\psi(\Omega)(\overline{Y}_{sV}\times\overline{Y}_{s\omega})}{j\Omega V_s(\overline{Y}_{sV}\times\overline{Y}_{s\omega}) + I_{\text{in}}(\overline{Y}_{sV}\cdot\tilde{e}^{j\phi})} \\ &\quad + \frac{\overline{Y}_{sV}\times\overline{I}_N}{j\Omega V_s(\overline{Y}_{sV}\times\overline{Y}_{s\omega}) + I_{\text{in}}(\overline{Y}_{sV}\cdot\tilde{e}^{j\phi})}\end{aligned} \tag{4.40}$$

where $\tilde{e}^{j\phi_s} = (\cos\phi_s, \sin\phi_s)$ and the symbols $\cdot$ and $\times$ indicate the products $\overline{a}\cdot\overline{b} = a^r b^r + a^i b^i$ and $\overline{a}\times\overline{b} = a^r b^i - a^i b^r$, respectively. Note that both products provide scalar numbers with either positive or negative sign.

The total phase perturbation of the node voltage is $\Delta\phi(t) + \psi(t)$, so the solution phase noise will have two contributions. The total phase perturbation in the frequency domain is given by [12]

$$\begin{aligned}\Delta\phi_T(\Omega) &= \Delta\phi(\Omega) + \psi(\Omega) = \psi(\Omega)\left[1 + \frac{-j\Omega V_s(\overline{Y}_{sV}\times\overline{Y}_{s\omega})}{j\Omega V_s(\overline{Y}_{sV}\times\overline{Y}_{s\omega}) + I_{\text{in}}(\overline{Y}_{sV}\cdot e^{j\phi_s})}\right] \\ &\quad + \frac{\overline{Y}_{sV}\times\overline{I}_N}{j\Omega V_s(\overline{Y}_{sV}\times\overline{Y}_{s\omega}) + I_{\text{in}}(\overline{Y}_{sV}\cdot e^{j\phi_s})}\end{aligned} \tag{4.41}$$

where the dot indicates the product $\overline{Y}_{sV}\cdot e^{j\phi_s} = Y^r_{sV}\cos\phi_s + Y^i_{sV}\sin\phi_s$. Expression (4.41) can be simplified as

$$\Delta\phi_T(\Omega) = \Delta\phi(\Omega) + \psi(\Omega) = \frac{I_{\text{in}}(\overline{Y}_{sV}\cdot e^{j\phi_s})\psi(\Omega) + \overline{Y}_{sV}\times\overline{I}_N}{j\Omega V_s(\overline{Y}_{sV}\times\overline{Y}_{s\omega}) + I_{\text{in}}(\overline{Y}_{sV}\cdot e^{j\phi_s})} \tag{4.42}$$

The phase noise spectral density is obtained by multiplying $\Delta\phi_T(\Omega)$ by its complex conjugate, $\Delta\phi^*_T(\Omega)$. It must be taken into account that the phase noise from the input source $\psi(\Omega)$ and the internal oscillator noise I_N are uncorrelated, as they have different origins. Then the phase noise spectrum of the injection-locked oscillator is given by

$$\begin{aligned}|\Delta\phi_T(\Omega)|^2 &= \frac{I^2_{\text{in}}(\overline{Y}_{sV}\cdot e^{j\phi_s})^2|\psi(\Omega)|^2 + 2|\overline{Y}_{sV}|^2|I_N|^2}{[I_{\text{in}}(\overline{Y}_{sV}\cdot e^{j\phi_s})]^2 + [V_s(\overline{Y}_{sV}\times\overline{Y}_{s\omega})]^2\Omega^2} \\ &= \frac{I^2_{\text{in}}\cos^2(\alpha_{sv} - \phi_s)|\psi(\Omega)|^2 + 2|I_N|^2}{I^2_{\text{in}}\cos^2(\alpha_{sv} - \phi_s) + V_s^2|Y_{s\omega}|^2\sin^2\alpha_{sv\omega}\Omega^2}\end{aligned} \tag{4.43}$$

where it has been taken into account that the products $\cdot$ and $\times$ can also be calculated as $\overline{a}\cdot\overline{b} = |a||b|\cos(\angle b) - \angle a)$ and $\overline{a}\times\overline{b} = |a||b|\sin(\angle b) - \angle a)$. As expected, for zero input current $I_{\text{in}} = 0$, expression (4.43) becomes the one corresponding to the phase noise spectral density of the free-running oscillator. The different angles are defined in a manner similar to the case of linearized analysis about the free-running oscillation in Section 4.2.1. However, the additional subscript s emphasizes the fact that they correspond to a linearization about a synchronized solution, so their values are different from those obtained from linearization of the free-running oscillation.

Expression (4.43) allows an intuitive understanding of the noise behavior of an injection-locked oscillator. As can be seen, expression (4.43) has two different inputs: one consisting of the source phase noise $|\psi(\Omega)|^2$ and the second consisting of the internal oscillator noise $|I_N|^2$. The numerator and denominator are both frequency dependent and are given by the summation of two different terms. This will give rise to two different corner frequencies when tracing the phase noise spectral density versus the offset frequency Ω. At low offset frequency Ω, the numerator term $I_{\text{in}}^2 \cos^2(\alpha_v - \phi_s)|\psi(\Omega)|^2$ will dominate over $2|I_N|^2$ because the input noise will generally grow as 30 dB/dec when approaching the carrier and thus will be much larger than the oscillator noise contribution. The difference in magnitude will be bigger for a higher input amplitude I_{in}. On the other hand, for low Ω, the denominator term $V_s^2|Y_{s\omega}|^2 \sin^2 \alpha_{v\omega}\Omega^2$ will be negligible compared with $I_{\text{in}}^2 \cos^2(\alpha_{sv} - \phi_s)$. Thus, the phase noise of the injection-locked oscillator can be approached: $|\Delta\phi_T(\Omega)|^2 \cong |\psi(\Omega)|^2$ regardless of the amplitude I_{in} of the injection source or the particular values of the steady-state synchronized solution, ω_{in}, V_s, and ϕ_s. Note that the offset frequency interval for which the equality $|\Delta\phi_T(\Omega)|^2 \cong |\psi(\Omega)|^2$ is fulfilled does depend on I_{in} and will be larger for higher I_{in}. This interval is limited due to the influence of the second input of (4.43), consisting of the internal oscillator noise $|I_N|^2$. This noise will dominate the numerator of (4.43) from a certain frequency value, determined by the condition

$$|\psi(\Omega_y)|^2 = \frac{2|Y_{sV}|^2|I_N|^2}{I_{\text{in}}^2(\overline{Y}_{sV} \cdot e^{j\phi_s})^2} = \frac{2|I_N|^2}{I_{\text{in}}^2 \cos^2(\alpha_{sv} - \phi_s)} \tag{4.44}$$

Note that the noise corner Ω_y will be larger for a higher input amplitude I_{in} (due to the decay of $|\psi(\Omega)|^2$ with the offset frequency) and, for constant I_{in}, will be maximum at the phase value fulfilling $\cos(\alpha_{sv} - \phi_s) = 1$. This should be at about the middle of the synchronization band. Remember that in linearized analysis, the middle of the synchronization band corresponds to input frequency equal to the free-running frequency $\omega_{\text{in}} = \omega_o$. The stable solution at this frequency has the phase value $\phi_s = \alpha_v$ (see Section 4.2.5), so $\cos(\alpha_v - \phi_s) = 1$ in (4.44). Note, however, that with increasing I_{in}, the angle α_{sv} deviates from α_v.

Condition (4.44) gives one of the two corner frequencies in the spectrum, here denoted Ω_y. The frequency Ω_y is usually well above the *flicker corner frequency*, the frequency value above which the spectral density of the circuit $1/f$ noise is below the density corresponding to the circuit white-noise contribution. This is why only white noise from the oscillator circuit $i_N(t)$ has been considered in the phase noise analysis of an injection-locked oscillator. For $\Omega > \Omega_y$ and up-to-the-second corner frequency, the numerator will be frequency independent or flat, and the constant phase noise spectral density will be given by

$$|\Delta\phi_T(\Omega)|^2 = \frac{2|\overline{Y}_{sV}|^2|I_N|^2}{[I_{\text{in}}(\overline{Y}_{sV}\cdot e^{j\phi_s})]^2} = \frac{2|I_N|^2}{I_{\text{in}}^2\cos^2(\alpha_{sv}-\phi_s)} \tag{4.45}$$

where it has been taken into account that $\overline{a}\cdot\overline{b} = |a||b|\cos(\angle b) - \angle a)$. In general terms, the flat spectral density will be smaller for larger I_{in}, but the value obtained depends also on the phase ϕ_s of the particular synchronized solution. It will increase when approaching the turning points of the closed synchronization curves, which delimit the stable operation band. These points are obtained from det[JH]= 0, with [JH] defined in (4.20) and $H_s = Y_s V - I_{\text{in}}e^{j\phi}$. If we write the error equation in terms of the total admittance, by doing $H_Y = Y_s - I_{\text{in}}e^{j\phi}/V = 0$, and we neglect the V variation in the denominator of the second term [as done in the admittance analysis of (4.34) to (4.35)], we clearly obtain that the turning points fulfill $\overline{Y}_{sV}\cdot e^{j\phi_s} = 0$, or equivalently, $\cos(\phi_s - \alpha_{sv}) = 0$. The phase spectral density $|\Delta\phi_T(\Omega)|^2$ tends to infinite at these turning points. This artificial result is due to the fact that the time derivative of the amplitude increment $\Delta\dot{V}(t)$ has been neglected in (4.43). It must also be noted that linearization used in (4.34) and (4.35) becomes invalid under this large-perturbation conditions.

Now inspecting the denominator of (4.43), when increasing the offset frequency Ω, the growing term $V_s^2|Y_{s\omega}|^2\sin^2\alpha_{sv\omega}\Omega^2$ will become equal to the constant term $I_{\text{in}}^2\cos^2(\alpha_{sv}-\phi_s)$ at a certain offset frequency, which will constitute the second corner frequency, $\Omega_{3\text{dB}}$. Actually, the injection-locked oscillator acts like a lowpass filter with respect to the injection source phase noise and the circuit noise, with a 3-dB cutoff frequency:

$$\Omega_{3\text{dB}} = \left|\frac{I_{\text{in}}\overline{Y}_{sV}\cdot e^{j\phi_s}}{V_s(\overline{Y}_{sV}\times\overline{Y}_{s\omega})}\right| = \frac{I_{\text{in}}|\cos(\alpha_{sv}-\phi_s)|}{V_s|\overline{Y}_{s\omega}||\sin\alpha_{sv\omega}|} \tag{4.46}$$

In general, the cutoff frequency $\Omega_{3\text{dB}}$ will be larger for higher input generator amplitude and smaller magnitude of the frequency derivative $|\overline{Y}_{s\omega}|$. The $\Omega_{3\text{dB}}$ value will also depend on the particular synchronized solution, given by $V_s, \phi_s, \omega_{\text{in}}$. As in the case of Ω_y, for given input amplitude I_{in}, the corner frequency $\Omega_{3\text{dB}}$ will be maximal at the phase shift ϕ_s fulfilling $\cos(\alpha_{sv}-\phi_s) = 1$. An interesting fact is that for a very small value of $|\overline{Y}_{s\omega}|$, the angle $\alpha_{sv}-\phi_s$ is always near 0° due to very small variations of ϕ_s along the solution curve. For a rough demonstration, consider the following linearization, used with small-signal amplitude:

$$\left[\left.\frac{\partial Y_T}{\partial V}\right|_o (V_s - V_o) + \left.\frac{\partial Y_T}{\partial\omega}\right|_o (\omega_{\text{in}} - \omega_o)\right] - \frac{I_{\text{in}}}{V_o}e^{j\phi_s} = 0 \tag{4.47}$$

where expression (4.10) has been taken into account to obtain the admittance function Y_s. Clearly, for small $\partial Y_T/\partial\omega|_o$, the following approximate relationship is fulfilled:

$$\frac{\partial Y_s}{\partial V} = \left.\frac{\partial Y_T}{\partial V}\right|_o \cong \frac{I_{\text{in}}}{V_o}\frac{e^{j\phi_s}}{(V_s - V_o)} \tag{4.48}$$

Therefore, the angle $\alpha_{sv} - \phi_s$ is about 0°. This is what we would get if we applied this analysis to an amplifier circuit instead of an injection-locked oscillator. As will be shown, this result is also interesting for understanding the phase noise behavior of different types of frequency dividers.

Finally, for offset frequency $\Omega > \Omega_{3\text{dB}}$, the oscillator phase noise spectral density $|\Delta\phi_T(\Omega)|^2$ will vary as

$$|\Delta\phi_T(\Omega)|^2 = \frac{2|\overline{Y}_{sV}|^2|I_N|^2}{[V_s(\overline{Y}_{sV} \times \overline{Y}_{s\omega})]^2\Omega^2} = \frac{2|I_N|^2}{V_s^2|\overline{Y}_{s\omega}|^2 \sin^2 \alpha_{sV\omega}\Omega^2} \tag{4.49}$$

So it will decrease as −20 dB/dec versus the offset frequency. Note that the phase noise spectral density in (4.49) is identical in form to that of a free-running oscillator under white noise perturbations. However, the values of the derivatives $\overline{Y}_{sV}$ and $\overline{Y}_{s\omega}$ and the steady-state amplitude V_s are not the same as under free-running conditions. The derivatives and the amplitude can approach those obtained under free-running conditions for low input power only. Then the phase noise spectrum obtained for $\Omega > \Omega_{3\text{dB}}$ agrees approximately with that corresponding to a free-running oscillator.

One may wonder if the values of the spectrum corner frequencies can be related to the synchronization bandwidth. This is more easily seen in the case of low input generator amplitude, due to the possibility of linearizing the equations of an injection-locked oscillator about a free-running solution. When this linearization is valid, that is, for low input amplitude, the derivatives Y_{sV} and $Y_{s\omega}$ can approach those obtained at free-running oscillation: Y_{ToV} and $Y_{To\omega}$. According to (4.7), the synchronization bandwidth is given by

$$\Delta\omega_{\text{max}} = \frac{2I_{\text{in}}}{V_o\,|\partial Y_{To}/\partial\omega|\,\sin\alpha_{v\omega}} = \frac{2I_{\text{in}}|Y_{T_oV}|}{V_o|Y_{ToV} \times Y_{To\omega}|} \tag{4.50}$$

Substituting $Y_{sV} \cong Y_{ToV}$, $Y_{s\omega} \cong Y_{To\omega}$, and $Y_{ToV} \times Y_{To\omega}$ from (4.50) into (4.43), it is possible to obtain the following expression, depending on the synchronization bandwidth:

$$|\Delta\phi_T(\Omega)|^2 = \frac{I_{\text{in}}^2 \cos^2(\alpha_v - \phi_s)|\psi(\Omega)|^2 + 2|I_N|^2}{I_{\text{in}}^2 \cos^2(\alpha_v - \phi_s) + 4I_{\text{in}}^2\,(\Omega/\Delta\omega_{\text{max}})^2} \tag{4.51}$$

From an inspection of (4.51), the expression for the corner frequency Ω_y agrees with expression (4.44), with the derivatives calculated at the free-running solution. On the other hand, the corner frequency $\Omega_{3\text{dB}}$ is given by

$$\Omega_{3\text{dB}} = \frac{\Delta\omega_{\text{max}}}{2}|\cos(\alpha_v - \phi_s)| \tag{4.52}$$

Thus, the corner frequency $\Omega_{3\text{dB}}$ is directly proportional to the synchronization bandwidth.

As a general conclusion, the phase noise reduction of the injection-locked oscillator with respect to its original free-running value comes from the phase relationship established between the oscillation and the input source. In the absence of perturbations, there is a constant phase shift $-\phi_s$ between the oscillation and the input generator. Now, consider input phase noise $\psi(t)$ only, without noise contributions from the oscillator circuit. From (4.43), the injection-locked oscillator tracks the low frequency variations of the input generator phase. However, from the frequency $\Omega_{3\text{dB}}$, the variations will be too fast and the circuit will be unable to maintain the locked behavior. Next, white noise perturbations from the oscillator circuit are considered $i_N(t)$. Provided that the offset frequency is smaller than $\Omega_{3\text{dB}}$, these perturbations will dominate the noise spectrum if they have larger spectral density than the contribution from the synchronizing source $I_{\text{in}}^2 \cos^2(\alpha_v - \phi_s)|\psi(\Omega)|^2$. This explains the flat region of the phase noise spectrum of injection-locked oscillators.

As an example, the analysis above has been applied to the parallel resonance oscillator of Fig. 1.1. In this case, the derivative vectors in (4.43), evaluated at the synchronized solution $V_s, \phi_s, \omega_{\text{in}}$ (instead of the free-running solution), take the values

$$\begin{aligned} \overline{Y}_{sV} &= \left[\frac{3bV_s}{2}, 0\right] \\ \overline{Y}_{s\omega} &= \left[0, C + \frac{1}{L\omega_{\text{in}}^2}\right] \end{aligned} \tag{4.53}$$

and expression (4.43) simplifies to

$$|\Delta\phi_T(\Omega)|^2 = \frac{I_{\text{in}}^2(\cos\phi_s)^2|\psi(\Omega)|^2 + 2|I_N|^2}{V_s^2(Y_{s\omega}^i)^2\Omega^2 + I_{\text{in}}^2(\cos\phi_s)^2} \tag{4.54}$$

The phase noise spectral density of the injection source considered is $|\psi(\Omega)|^2 = 10^5/(2\pi f)^3\text{Hz}^{-1}$. The spectral density of the oscillator white noise current source is $|I_N|^2 = 10^{-18}\text{A}^2/\text{Hz}$. For each input current amplitude, the value of the corner frequency Ω_y can be predicted using (4.44), which in this case simplifies to $|\psi(\Omega_y)|^2 = 2|I_N|^2/(I_{\text{in}}^2 \cos^2\phi_s)$. From (4.46), the second corner frequency $\Omega_{3\text{dB}}$ in this particular problem simplifies to $\Omega_{3\text{dB}} = I_{\text{in}} \cos\phi_s / V_s Y_{s\omega}^i$.

All the formulation presented previously is only approximate, since the second term of (4.34) has been divided by the unperturbed voltage amplitude V_s instead of the actual perturbed value $V_s + \Delta V_s$. Due to difficulties in solving the resulting nonlinear system, a different approach can be followed, based on use of the error function $H_s = Y_s V_s - I_{\text{in}} e^{j\phi_s}$, with current dimension. Next, a synchronizing source, with phase noise $\psi(t)$ only and a white noise current source $I_N(t)$, accounting for the noise contribution of the oscillator circuit, will be considered. The perturbed oscillator equations are

$$\begin{aligned} H_{sV}^r \Delta V(t) + H_{s\omega}^r[\Delta\dot{\phi}(t) + \dot{\psi}(t)] - H_{s\phi}^r \Delta\phi(t) &= I_N^r(t) \\ H_{sV}^i \Delta V(t) + H_{s\omega}^i[\Delta\dot{\phi}(t) + \dot{\psi}(t)] - H_{s\phi}^i \Delta\phi(t) &= I_N^i(t) \end{aligned} \tag{4.55}$$

By following steps identical to those in the preceding calculation based on the admittance function, the phase noise spectral density is given by

$$|\Delta\phi_T(\Omega)|^2 = \frac{|\overline{H}_{sV} \times \overline{H}_{s\phi}|^2 |\psi(\Omega)|^2 + 2|\overline{H}_{sV}|^2 |\overline{I}_N|^2}{|\overline{H}_{sV} \times \overline{H}_{s\phi}|^2 + |\overline{H}_{sV} \times \overline{H}_{s\omega}|^2 \Omega^2} \tag{4.56}$$

Now the phase noise corners, determined more accurately, are given by

$$|\psi(\Omega_y)|^2 = \frac{2|\overline{H}_{sV}|^2 |\overline{I}_N|^2}{|\overline{H}_{sV} \times \overline{H}_{s\phi}|^2} \qquad \Omega_{3\text{dB}} = \frac{|\overline{H}_{sV} \times \overline{H}_{s\phi}|}{|\overline{H}_{sV} \times \overline{H}_{s\omega}|} \tag{4.57}$$

The $|\cdot|$ notation indicates absolute value. Remember that the products $a \times b$ have been defined as scalar numbers. For a small offset frequency, fulfilling $\Omega < \Omega_{3\text{dB}}$, the phase noise will be maximum when $|\overline{H}_{sV} \times \overline{H}_{s\phi}| = 0$. This condition is fulfilled at the turning points of the solution curve [see (4.24)]. As seen already, it is equivalent to $\overline{Y}_{sV} \cdot e^{j\phi_s} = 0$ in the less accurate expression (4.43). Note that although less accurate, (4.43) is more useful for circuit design, as the total admittance function is more meaningful and easier to control than the error function H_s. This is why it has been derived here.

In the particular case of the parallel resonance oscillator, the derivatives in (4.56), in terms of the error function H_s instead of the total admittance Y_s, are given by

$$\begin{aligned} H_{sv} &= \frac{9}{4} bV^2 + j\left(C\omega_{\text{in}} - \frac{1}{L\omega_{\text{in}}}\right) \\ H_{s\omega} &= j\left(C + \frac{1}{L\omega_{\text{in}}^2}\right)V \\ H_{s\phi} &= I_{\text{in}} \sin\phi - jI_{\text{in}} \cos\phi \end{aligned} \tag{4.58}$$

The phase noise analysis has been carried out for two different values of the input current amplitude, $I_{\text{in1}} = 5$ mA and $I_{\text{in2}} = 50$ mA, at an input frequency agreeing with the free-running oscillation value $f_{\text{in}} = 1.59$ GHz. For $I_{\text{in1}} = 5$ mA and $f_{\text{in}} = 1.59$ GHz, the steady-state synchronized solution is given by the amplitude $V_s = 1.746$ V and phase $\phi_s = 0$. For $I_{\text{in1}} = 50$ mA and $f_{\text{in}} = 1.59$ GHz, the steady-state synchronized solution is given by the amplitude $V_s = 2.436$ V and phase $\phi_s = 0$. Due to the phase shift value $\phi_s = 0$, the expressions for the corners are given by $|\psi(\Omega_y)|^2 = 2|\overline{I}_N|^2/I_{\text{in}}^2$, $\Omega_{3\text{dB}} = I_{\text{in}}/2CV_s$. Thus, the Ω_y corner will be higher for larger I_{in} (which implies a smaller spectral density $|\psi(\Omega_y)|^2$. This can be verified through the comparison of the two phase noise spectra presented in Fig. 4.13. On the other hand, the second phase noise corner is $f_{3\text{dB}} = 45$ MHz for $I_{\text{in}} = 5$ mA and $f_{3\text{dB}} = 327$ MHz for $I_{\text{in}} = 50$ mA. Therefore, $f_{3\text{dB}}$ increases with I_{in}, due to the stronger influence of the input source over the self-oscillation for higher amplitude I_{in}.

Next, an analysis of variations in the phase noise spectral density along the synchronization curves versus the input generator frequency ω_{in} will be carried

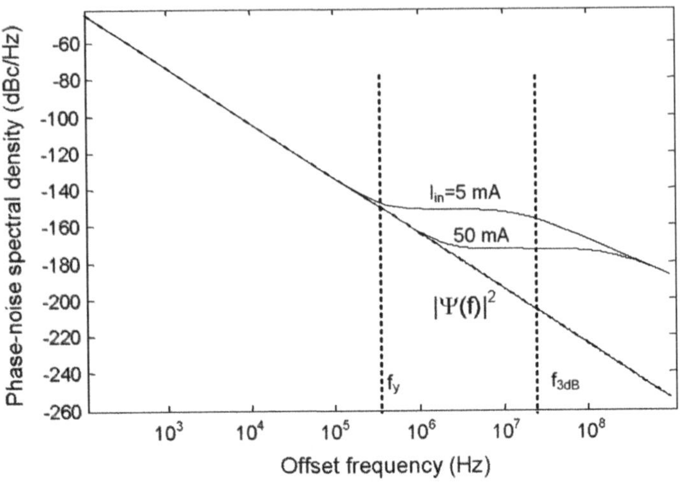

FIGURE 4.13 Comparison of the phase noise spectral density of an injection-locked parallel resonance circuit for two different values of the input current amplitude, $I_{in1} = 5$ mA and $I_{in2} = 50$ mA, at an input frequency agreeing with the free-running oscillation value $f_{in} = 1.59$ GHz. The phase noise spectral density of the injection source $|\psi(\Omega)|^2 = 10^5/(2\pi f)^3$ Hz^{-1} is also represented.

out. The analysis proceeds in two steps. In the first step, the steady-state solution, defined by V_s, ϕ_s and corresponding to the input frequency ω_{in}, is calculated from (4.12). In the second step, the derivatives of the error function H_s are evaluated at the particular solution, and the resulting values are introduced into the expression (4.56) for the phase noise spectral density. The procedure is repeated for the entire synchronization band.

According to (4.56), at low values of the offset frequency Ω, the phase noise spectral density will be constant and equal to the input phase noise $|\psi(\Omega)|^2$ for all the ω_{in} values. For a larger offset frequency, there will be phase noise variations along the synchronization band due to the different values of the two corner frequencies Ω_y and Ω_{3dB} at each point of this band. As an illustration, Fig. 4.14 shows the phase noise variations along the synchronization band of the parallel resonance oscillator for five input current amplitudes: $I_{in} = 4$ mA, $I_{in} = 12$ mA, $I_{in} = 16$ mA, $I_{in} = 20$ mA, and $I_{in} = 30$ mA, agreeing with the values considered in Fig. 4.2. The constant frequency offset is $f = 1$ MHz. At this offset frequency, the phase noise spectral density of the input source takes the constant value $|\psi(\Omega)|^2 = -159$ dBc/Hz.

As already discussed, for $\Omega > \Omega_y$, the phase noise spectral density will be maximum at the turning points of the solution curve, determined by the condition $|\overline{H}_{sV} \times \overline{H}_{s\phi}| = 0$. The value taken by the phase noise spectral density at these turning points depends on the second term in the denominator of $|\overline{H}_{sV} \times \overline{H}_{s\omega}|^2\Omega^2$. This term decreases with the offset frequency Ω and depends also on the particular steady-state solution ω_{in}, V_s, ϕ_s, at which the derivatives of H_s are calculated.

At input-power values for which the solution curves do not exhibit turning points, there will be maxima at the minima of the denominator in (4.56). These minima occur at input-frequency values near the Hopf bifurcations (compare Fig. 4.14 with Fig. 4.2). To see this, consider the characteristic system obtained when applying the Laplace transform to system (4.55) in the absence of noise perturbations $I_N(t)$, $\psi(t)$. The denominator in (4.56) agrees formally with the determinant of the characteristic matrix associated to this system when evaluated at $j\Omega$ instead of the Laplace frequency s. Clearly, the determinant should be zero for input-frequency values corresponding to the Hopf bifurcations and offset frequency agreeing with the frequency of the critical poles $\pm j\omega$; that is, $\Omega = \omega = |\omega_{in} - \omega_a|$. This explains the maxima of the phase-noise spectral density obtained for $I_{in} = 20$ mA and $I_{in} = 30$ mA in Fig. 4.14. The maxima are not infinite because the offset frequency does not agree with the pole frequency. For more accuracy in the determination of the spectrum resonances, the time derivative $\Delta\dot{V}(t)$ must be taken into account. The analysis considering this derivative is presented in the following.

As a final study, the influence of the time derivative of the amplitude perturbation $\Delta\dot{V}(t)$, neglected so far, will be analyzed. When this derivative is considered, the perturbation equation, in terms of the error function $\overline{H}_s$, is given by

$$\overline{H}_{sV}\Delta V(t) + \overline{H}_{s\omega}\left[\dot{\psi}(t) + \Delta\dot{\phi}(t) - j\frac{\Delta\dot{V}(t)}{V_s}\right] + \overline{H}_{s\phi}\Delta\phi(t) = \overline{I}_N(t) \qquad (4.59)$$

From (4.59), the oscillator phase noise spectral density is obtained in a straightforward, though cumbersome manner, following the same procedure as in

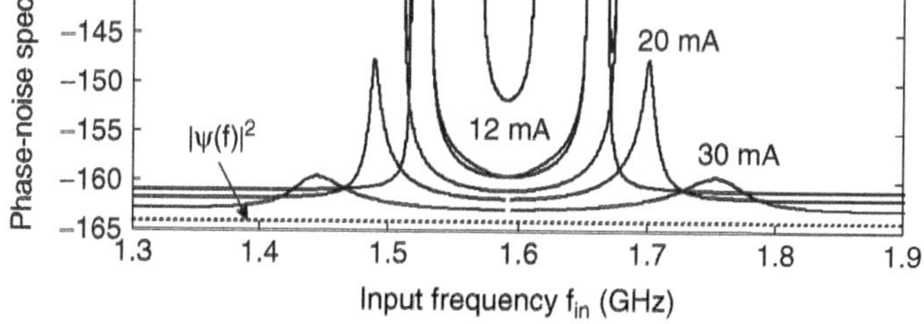

FIGURE 4.14 Evolution of the phase noise spectral density along the synchronization curves for five input current amplitudes: $I_{in} = 4$ mA, $I_{in} = 12$ mA, $I_{in} = 16$ mA, $I_{in} = 20$ mA, and $I_{in} = 30$ mA, agreeing with the values considered in Fig. 4.2. The constant frequency offset is $f = 1$ MHz.

(4.39)–(4.43). The final expression is

$$
|\Delta\phi_T(\Omega)|^2 = \frac{\left(2|\overline{H}_{sV}|^2 + \Omega^2 \frac{2|\overline{H}_{s\omega}|^2}{V_s^2}\right)|\overline{I}_N|^2 + \left(|\overline{H}_{sV} \times \overline{H}_{s\phi}|^2 + \Omega^2 \frac{|\overline{H}_{s\omega} \cdot \overline{H}_{s\phi}|^2}{V_s^2}\right)|\psi(\Omega)|^2}{|\overline{H}_{sV} \times \overline{H}_{s\phi}|^2 + \left[\left(\overline{H}_{sV} \times \overline{H}_{s\omega} - \frac{\overline{H}_{s\omega} \cdot \overline{H}_{s\phi}}{V_s}\right)^2 + 2\overline{H}_{sV} \times \overline{H}_{s\phi} \frac{|\overline{H}_{s\omega}|^2}{V_s}\right]\Omega^2 + \frac{|\overline{H}_{s\omega}|^4}{V_s^2}\Omega^4} \tag{4.60}
$$

Comparing (4.60) with (4.56), it is clear that the time derivative $\Delta\dot{V}(t)$ contributes the terms Ω^2 in the numerator and the terms Ω^4 in the denominator. It also modifies the coefficient affecting Ω^2 in the denominator. Furthermore, the near carrier phase noise does not totally agree with the input phase noise, but is influenced by the circuit noise through the term $2|\overline{H}_{sV}|^2|\overline{I}_N|^2$. In a manner similar to that of the noise analysis of the free-running oscillator (see Chapter 2), the terms introduced by $\Delta\dot{V}(t)$ will only be relevant at very high offset frequency from the carrier unless the circuit operates near a bifurcation. As an example, the phase noise of the parallel resonator for the input amplitude $I_{\text{in}} = 20$ mA and two values of input frequency, $f_{\text{in}} = 1.48$ GHz and $f_{\text{in}} = 1.475$ GHz, has been represented in Fig. 4.15. As can be verified from inspection of Fig. 4.2, the circuit is operating near a Hopf bifurcation, occurring for $f_{\text{in}} = 1.485$ GHz. The phase noise spectrum exhibits a resonance at an offset frequency agreeing with the frequency of the nearly-critical poles $\Omega = |\omega_{\text{in}} - \omega_a|$, about 80 MHz. The shift in the central frequency of the noise bump when f_{in} varies is due to the variation in pole frequency with the input frequency ω_{in}. Note that for each ω_{in}, the system operates at a different steady-state solution. The resonance is narrower and higher for a smaller distance to the bifurcation. Thus, the resonance is higher for the input frequency $f_{\text{in}} = 1.475$ GHz.

4.3 FREQUENCY DIVIDERS

There are three main types of analog frequency dividers: harmonic injection dividers, regenerative dividers, and parametric dividers. Considering a periodic input source of frequency ω_{in}, harmonic injection dividers are based on synchronization of the Nth harmonic component of the oscillation to the frequency of the input source $N\omega_a = \omega_{\text{in}}$. They exhibit a free-running oscillation in the absence of input power. In regenerative dividers a subharmonic oscillation is generated from a certain input power at ω_{in}, which requires suitable feedback and frequency mixing. These circuits do not oscillate in the absence of input power. Finally, the parametric frequency dividers give rise to frequency division from the negative conductance exhibited by a nonlinear capacitance pumped by a periodic source, which also resonates with an inductive element at the subharmonic frequency. A phase relationship is established with the input source, maintained in a certain frequenct band. No oscillation takes place in the absence of a pumping signal. In this section the three types of dividers are treated. We begin with a brief description of the general characteristics of any frequency-divided solution.

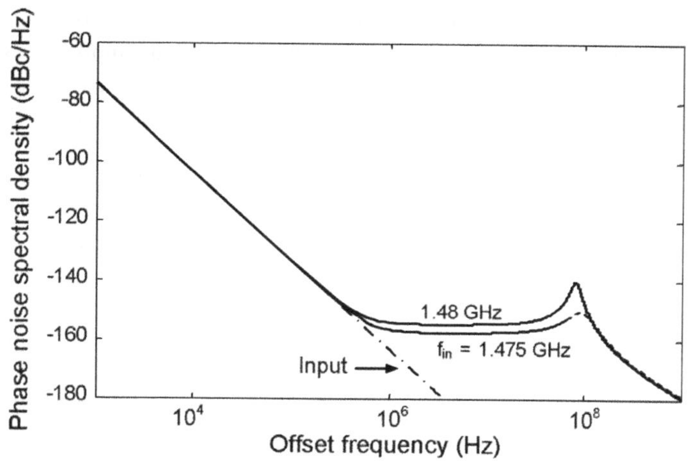

FIGURE 4.15 Phase noise spectral density of an injection-locked parallel resonance oscillator near a Hopf bifurcation. The input amplitude considered is $I_{in} = 20$ mA, with two values of input frequency: $f_{in} = 4.8$ GHz and $f_{in} = 4.75$ GHz (closer to the bifurcation).

4.3.1 General Characteristics of a Frequency-Divided Solution

Some general characteristics of the frequency-divided solution at ω_{in}/N in any type of divider circuit are discussed in the following.

4.3.1.1 Coexistence with a Nondivided Solution at ω_{in} Let a general frequency divider by N with input frequency ω_{in} and output frequency ω_{in}/N be considered. The divided solution always coexists with a nondivided mathematical solution at ω_{in}. This may be seen clearly when formulating the circuit equations in the frequency domain. For simplicity, a single nonlinear element depending on a single voltage variable is considered. The system is formulated in terms of the total branch current in a given analysis node. Considering NH harmonic terms of the divided frequency ω_{in}/N, this system is given by

$$\begin{aligned}
I_T^o(V_o, \tilde{V}_1, \ldots, \tilde{V}_N, \ldots, \tilde{V}_{NH}) &= I_o \\
\tilde{I}_T^1(V_o, \tilde{V}_1, \ldots, \tilde{V}_N, \ldots, \tilde{V}_{NH}) &= 0 \\
&\vdots \\
\tilde{I}_T^N(V_o, \tilde{V}_1, \ldots, \tilde{V}_N, \ldots, \tilde{V}_{NH}) &= I_{in}e^{j0} \\
&\vdots \\
\tilde{I}_T^{NH}(V_o, \tilde{V}_1, \ldots, \tilde{V}_N, \ldots, \tilde{V}_{NH}) &= 0
\end{aligned} \tag{4.61}$$

where I_o and I_{in} are the currents obtained from the Norton equivalents of the dc and injection sources, respectively, and $\tilde{V}_1, \ldots, \tilde{V}_{NH}$ are the phasors of the harmonic voltages at the observation node. Note that (4.61) is just a compact harmonic balance formulation description of the divider circuit, used simply for explanatory purposes. In general, the circuit will contain more than one state variable, so the system must be formulated in terms of the harmonic components of all these variables. The harmonic balance method is presented in detail in Chapter 5.

In (4.61), the input generator at ω_{in} plus the circuit nonlinearity naturally generate the harmonic frequencies $m\omega_{in}$, with m an integer. However, there are no generators at the subharmonic frequency ω_{in}/N, so the set of equations $I_T^k(\overline{V}) = 0$, with $\overline{V}$ the vector of voltage phasors, and $k \neq mN$, with m an integer, compose a nonlinear homogeneous subsystem $I_T^s(\overline{V}) = 0$, which admits a zero solution $\overline{V}^s = 0$. This solution corresponds to a periodic regime at the fundamental frequency ω_{in}, that is, with no frequency division. Thus, even if the circuit is actually performing a frequency division, the set of equations in the frequency domain that describe its behavior can be resolved for a nondivided solution.

4.3.1.2 *Phase Shift Variation* The frequency-divided solution of a given divider circuit will be expressed in the time domain in terms of its harmonic components as

$$v(t) = \mathrm{Re}\left[V_o + V_1 e^{j(\phi_1 + (\omega_{in}/N)t)} + V_2 e^{j(\phi_2 + 2(\omega_{in}/N)t)} + \cdots + V_N e^{j(\phi_N + \omega_{in} t)} + \cdots\right] \tag{4.62}$$

If a time shift τ is applied to this solution, giving the time value $t - \tau$, the phase of the frequency component at ω_{in} will vary as $\Delta\phi_N = -\omega_{in}\tau$. In turn, the phase shift of the subharmonic frequency will vary as $\Delta\phi_1 = -(\omega_{in}/N)\tau = (\Delta\phi_N/N)$. Thus, a phase shift of 2π radians at ω_{in} implies a phase shift of $2\pi/N$ radians at the divided frequency.

4.3.1.3 *Coexistence of N Stable Divided Solutions with Different Phase Shifts* Assuming that the input signal of a circuit behaving as a frequency divider by N is $v_{in}(t) = E_{in}\cos\omega_{in}t$, N stable divided solutions at ω_{in}/N will coexist, having the same waveform amplitude and N different phase shift values, given by

$$\phi_1 + k\frac{2\pi}{N} \qquad \text{with } k = 0 \text{ to } N - 1 \tag{4.63}$$

Note that they all give rise to the same phase value (in 2π modulus) at the input generator frequency ω_{in}, so there is an irrelevance with respect to phase shifts $k(2\pi/N)$, with k an integer. The existence of these N solutions is in agreement with the fact that the Poincaré map corresponding to a divided-by-N steady-state solution consists of N different fixed points $\overline{x}_{p1}, \ldots \overline{x}_{pN}$. Assume that the system is at one of these N points at time t_o. A time shift $kT_{in} = k(2\pi/\omega_{in})$, $k < N$, simply leads to a different point in the same set, $\overline{x}_{p1}, \ldots, \overline{x}_{pN}$. The only difference between the corresponding steady-state solutions is the phase shift $k(2\pi/N)$ or the time shift $\tau = kT_{in}$ in the time domain. Reaching one or another stable solution will depend only on the initial conditions. Therefore, for an analysis of the synchronization band of the frequency divider, the variation in phase shift ϕ between the divided frequency component and the input generator at ω_{in} can be limited to 0, $2\pi/N$. The waveform obtained for phase shift $2\pi/N + \Delta\phi$ is the same as the one obtained for $\Delta\phi$ with only a time shift $\tau = kT_{in}$.

4.3.1.4 *Bifurcations Leading to Frequency Division* In general, the behavior of frequency dividers is different for division by order $N = 2$ or by any other order $N \neq 2$. Note that a direct division by order $N \neq 2$ would imply the crossing of the unit circle by a pair of complex-conjugate multipliers at exactly $e^{\pm j(\omega_{in}/N)T}$, with T the period associated with the input frequency $\omega_{in} = 2\pi/T$. Due to the continuity of the unit circle, comprised of infinite points, the precise crossing of the pair of complex-conjugate multipliers through $e^{\pm(j\omega_{in}/N)T}$ is very unlikely. For instance, a direct division by $N = 3$ would require the crossing through the two precise points $e^{\pm j120^\circ}$. Thus, direct transition from a periodic regime at ω_{in} to a frequency-divided regime at ω_{in}/N with $N \neq 2$ will be rare. Instead, when varying the parameter, a quasiperiodic signal at $\omega_a \cong \omega_{in}/N$ generally arises in a secondary Hopf bifurcation, which, after further parameter variation, gets synchronized $\omega_a = \omega_{in}/N$ in a mode-locking bifurcation (at a turning point of the periodic curve at ω_{in}/N). The case of division by $N = 2$ is different. This type of division is associated with the crossing of a real multiplier through the point $(-1, 0)$ of the unit circle. The real multiplier cannot turn into a complex conjugate by itself, as this would require merging with another real multiplier. It simply slides along the axis, maintaining its real nature, so the crossing through $(-1, 0)$ will happen in a natural manner. Thus, this type of division is "very clean" versus the parameter, with no intermediate quasiperiodic signal.

4.3.2 Harmonic Injection Frequency Dividers

As shown in the preceding section, the oscillator synchronization at the fundamental frequency takes place for an input frequency interval about the free-running frequency. However, synchronization can also occur when the input frequency is close to a harmonic component, of N order, of the free-running frequency. Within the synchronization band, the relationship $\omega_{in} = N\omega_a$ will be fulfilled, so the oscillation frequency agrees with the subharmonic ω_{in}/N. The harmonic synchronization is used for the design of analog frequency dividers [5,17,18]. Due to the inherent nonlinearity of the self-oscillation, the presence of harmonic components $N\omega_a$ of significant amplitude capable of getting locked to the injection source will not be uncommon. However, as the harmonic order N increases, narrower synchronization bands will generally be obtained. The free-running oscillation of the harmonic injection dividers (at a frequency on the order of the desired output frequency $\omega_o \cong \omega_{in}/N$) enables frequency division from very low input power, which is a significant advantage with respect to other types of analog dividers.

First, a simple system model of harmonic injection dividers will be developed based on Fig. 4.16 [19,20]. Similar block diagrams will be considered for other types of oscillator circuits, as they enable an intuitive comparison of their behavior. It is assumed that the loop exhibits a free-running oscillation in the absence

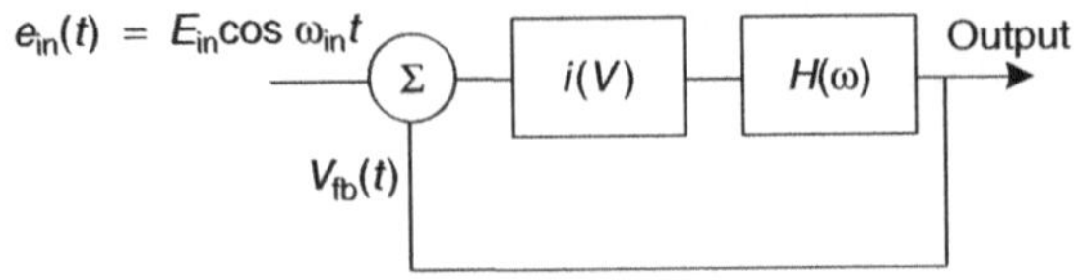

FIGURE 4.16 Harmonic injection divider by order N.

of input signal at $\omega_o \cong \omega_{in}/N$. The function $i(v)$ is the constitutive relationship of the nonlinear element and $H(\omega)$ is a resonant circuit with quality factor Q acting as a bandapass filter centered at ω_o. The nonlinear element $i(v)$ will be modeled with a power series: $i(v) = \sum_i \alpha_i v^i$. Assuming a sufficiently high quality factor Q of the bandpass filter $H(\omega)$, the output of the nonlinear element will be calculated from $i(v) = i(V_{fb}\cos\omega_o t)$. Assembling all the resulting terms at the first harmonic frequency, the frequency-domain equation ruling the free-running oscillation will be

$$F(V_{fb}^o, \omega_o) = V_{fb}^o - H(\omega_o)I_1(V_{fb}^o) = 0 \qquad (4.64)$$

where F is an error function and V_{fb}^o and ω_o are the free-running oscillation amplitude and frequency. Next, an analogous equation will be derived in the presence of the input signal $e_{in}(t) = E_{in}\cos\omega_{in}t$. Assuming that the synchronization has actually taken place, the input frequency will fulfill $\omega_{in} = N\omega_d$, with ω_d the subharmonic frequency, relatively close to ω_o. For a sufficiently high Q of $H(\omega)$, the output of the nonlinear element will be calculated from $i(v) = i(V_{fb}\cos\omega_d t + E_{in}\cos(N\omega_d t + \phi))$, where the phase origin $\phi_o = 0$ has been set at the first harmonic component of the feedback signal $v_{fb}(t)$. For a high Q value, it will be possible to limit the analysis to the terms of $i(v)$ at the divided frequency $\pm\omega_d$. These are generated from the intermodulation products $m\omega_d + k\omega_{in} = (-kN \pm 1)\omega_d + k\omega_{in} = \pm\omega_d$. The various contributions to the first harmonic of $i(v)$ can be shown explicitly by expressing $I_1 = I_{1,0} + \sum_{k\neq 0}|I_{-kN+1,k}|e^{jk\phi}$ [19], where only the components at the positive frequency ω_d are considered. The frequency division requires synchronous behavior; that is, the output of the nonlinear element, *i(v)*, at the divided frequency ω_d, must depend on the input generator phase ϕ. The equation at the divided frequency ω_d is given by

$$\begin{aligned} &V_{fb}^d - H(\omega_d)I_1 \\ &= V_{fb}^d - \frac{H_o}{1 + j2Q[(\omega_d - \omega_o)/\omega_o]}\left(I_{1,0} + \sum_{k\neq 0}|I_{-kN+1,k}|e^{jk\phi}\right) = 0 \end{aligned} \qquad (4.65)$$

The dominant terms in the summation will generally correspond to the lowest order $k = 1$; that is, $I_1 = I_{1,0} + |I_{-N+1,1}|e^{j\phi}$. Clearly, the synchronization bandwidth will increase with the mixing capability of the nonlinear element. Note that the division would be impossible if the subharmonic current were independent of the source phase ϕ or asynchronous.

For a small value of the input signal $e_{\text{in}}(t)$, it will be possible to expand the nonlinear function $i(v)$ in a first-order Taylor series about the free-running solution V_{fb}^{o}, ω_{o}, setting $\Delta i(t) = \partial i/\partial v|_{\text{o}}[\Delta v_{\text{fb}}(t) + e_{\text{in}}(t)]$. The terms contributed by $\partial i/\partial v|_{\text{o}}\Delta v_{\text{fb}}(t)$ are $\partial I_1/\partial V_1 \Delta V_{\text{fb}}/2 + \partial I_1/\partial V_{-1}\Delta V_{\text{fb}}^*/2 = (G_{\text{o}} + G_2)\Delta V_{\text{fb}}/2$, where the derivatives have also been replaced by G_{o} and G_2, which are the dc and second harmonic components of $g(t) = \partial i/\partial v|_{\text{o}}$, respectively. In turn, the input generator frequency is $N\omega_d$, so $e_{\text{in}}(t)$ contributes to the first harmonic of $i(t)$ through the Jacobian terms $\partial I_1/\partial V_N$ and $\partial I_1/\partial V_{-N}$. The corresponding dominant terms will be $(G_{1-N}e^{j\phi} + G_{N+1}e^{-j\phi})(E_{\text{in}}/2)$. Limiting the analysis to the lower-order term, the following linearized equation is obtained:

$$V_{\text{fb}}^{\text{o}} + \Delta V_{\text{fb}} - H(\omega_d)\left[I(V_{\text{fb}}^{\text{o}}) + G_{\text{o}}\Delta V_{\text{fb}} + G_{N-1}^* e^{j\phi}\frac{E_{\text{in}}}{2}\right] = 0 \tag{4.66}$$

Because equation (4.66) corresponds to the divided frequency $\omega_d = \omega_{\text{in}}/N$, the frequency increment is given by $\Delta\omega = \omega_d - \omega_{\text{o}}$. Performing a Taylor series expansion of $H(\omega_d)$ about $H(\omega_{\text{o}})$, equation (4.66) can be approximated as

$$[1 - H(\omega_{\text{o}})G_{\text{o}}]\Delta V_{\text{fb}} - \left.\frac{\partial H}{\partial \omega}\right|_{\text{o}} I(V_{\text{fb}}^{\text{o}})(\omega_d - \omega_{\text{o}}) = H(\omega_{\text{o}})G_{N-1}^* e^{j\phi}\frac{E_{\text{in}}}{2} \tag{4.67}$$

where the relationship (4.64) has been taken into account. Note that this linearized analysis is equivalent to the analysis performed in Section 3.2.1 for the fundamentally synchronized oscillator. The solution of (4.67) versus ω_{in} corresponds to a perfect ellipse in the plane defined by ω_{in} and ΔV_{fb}, as can easily be demonstrated by splitting the complex equation (4.67) into real and imaginary parts and making the phase ϕ disappear. Clearly, a broader bandwidth is obtained for the higher input amplitude E_{in} and larger magnitude of the derivative $\partial I_1/\partial V_N$, which increases with sensitivity to the input generator.

After the simple explanation above of the harmonic injection divider from a system point of view, an alternative explanation from a circuit point of view will be provided. The circuit exhibits a free-running oscillation at the frequency ω_{o}. The input source at about N times the free-running oscillation frequency $\omega_{\text{in}} \cong N\omega_{\text{o}}$ is expressed as $i_{\text{in}}(t) = \text{Re}[I_{\text{in}}e^{j\omega_{\text{in}}t}]$. When this generator is connected to the circuit, it is assumed that the circuit reaches a periodic steady-state solution at ω_{in}/N. In the frequency domain, the frequency divider is described by system (4.61). This complex system contains $2NH + 1$ unknowns, given by the real and imaginary parts of the NH harmonic components of the node voltage plus the dc term. To reduce the complexity of the following analyses, a single state variable $v(t)$ is assumed. As in (4.61), the harmonic balance system is formulated in a simplified manner in terms of the total branch current entering the analysis node. The dc bias current and injection current at ω_{in} are obtained through Norton equivalents. A two-tier resolution of the harmonic system is carried out. The $2NH - 1$ equations corresponding to the dc and harmonic terms $2 \cdots NH$ are used to express the complex state variables $V^{\text{dc}}, \tilde{V}^2, \ldots, \tilde{V}^{NH}$ in terms of the first-harmonic amplitude

$\tilde{V}^1 = V_s e^0$, taken as the phase origin. The substitution is performed in the following manner:

$$\left.\begin{array}{l}\left.\begin{array}{l}\tilde{I}_T^1(V_o, V_s e^{j0}, \tilde{V}_2, \ldots, \tilde{V}_N, \ldots, \tilde{V}_{NH}) = 0\\ I_T^0(V_o, V_s e^{j0}, \tilde{V}_2, \ldots, \tilde{V}_N, \ldots, \tilde{V}_{NH}) = I_o\\ \tilde{I}_T^2(V_o, V_s e^{j0}, \tilde{V}_2, \ldots, \tilde{V}_N, \ldots, \tilde{V}_{NH}) = 0\\ \vdots\\ \tilde{I}_T^N(V_o, V_s e^{j0}, \tilde{V}_2, \ldots, \tilde{V}_N, \ldots, \tilde{V}_{NH}) = I_{in} e^{j\phi_s}\\ \vdots\\ \tilde{I}_T^{NH}(V_o, V_s e^{j0}, \tilde{V}_2, \ldots, \tilde{V}_N, \ldots, \tilde{V}_{NH}) = 0\end{array}\right\} \Rightarrow \begin{array}{c}\text{inner tier}\\ \overline{V}'(V_s, I_{in}, \phi_s)\end{array}\end{array}\right\} \Rightarrow \begin{array}{c}\text{outer tier}\\ Y_T^1(\overline{V}'(V_s, I_{in}, \phi_s))V_s e^{j0} = 0\end{array} \tag{4.68}$$

where $\overline{V}'$ is the vector of state variables, obtained for fixed V_s and ϕ_s, and Y_T^1 is defined as the ratio $Y_T^1 = \tilde{I}_{T1}/V_s e^{j0}$. For notation simplicity, Y_T^1 is renamed Y_T here, so the outer-tier equation of the harmonic balance system (4.68) is given by

$$Y_T(V_s, \omega_{in}, I_{in} e^{j\phi_s})V_s e^{j0} = 0 \tag{4.69}$$

All the analysis techniques presented below are applied to the outer tier in equation (4.69), considering an absolute dependence on the first-harmonic voltage $V_s e^{j0}$, the phase shift ϕ_s, and the input generator frequency $\omega_{in} = N\omega_s$. This absolute dependence is allowed by the two-tier resolution of (4.69), which implies that under any variation of V_s, ω_{in}, $I_{in} e^{j\phi_s}$, the inner-tier system, consisting of the $2NH - 1$ harmonic equations in the leftmost bracket of (4.68), must be resolved. For given generator values I_{in}, ω_{in}, the harmonic balance system is solved in two different steps. In practice, this can be done using two nested Newton–Raphson algorithms. One is used to solve the outer-tier system in (4.69), which contains only two real equations in the two unknowns V_s, ϕ_s. The other is applied to obtain the derivatives of the function Y_T, required by the outer-tier Newton–Rahpson. These derivatives are calculated through finite differences, performing a Newton–Raphson iteration at each variable increment. This second Newton–Raphson algorithm is applied to the inner-tier system (4.68), with $2NH - 1$ equations.

Note that the subharmonic V_s at ω_s in (4.69) is related to the input generator $i_{in}(t) = \text{Re}[I_{in} e^{j(\omega_{in}t + \phi_s)}]$ at $\omega_{in} = N\omega_s$ through the inner tier. This point is essential for a good understanding of all the following derivations, written in terms of the outer tier only. Note that the same type of resolution is applicable to circuits containing more than one state variable, like those based on transistors, which are two-port devices. In these circuits it is also possible to obtain an outer tier of the same form as (4.69), with absolute dependence of the node voltage at a given observation port. Equation (4.69) and its corresponding inner tier allow nonlinear resolution of the frequency divider for any amplitude value I_{in} of the input generator. However, for low input generator amplitude I_{in}, it will be possible to expand (4.69) in a Taylor series about the free-running oscillation point (V_o, ω_o, $I_{in} = 0$).

This will allow an approximate analysis of the synchronized solutions for low input amplitude I_{in} [21]:

$$\frac{\partial Y_{To}}{\partial V}(V_s - V_o) + \frac{\partial Y_{To}}{\partial \omega}(\omega_{\text{in}} - \omega_o) = -\frac{\partial Y_{To}}{\partial I_{\text{in}}^r} I_{\text{in}} \cos \phi_s - \frac{\partial Y_{To}}{\partial I_{\text{in}}^i} I_{\text{in}} \sin \phi_s \tag{4.70}$$

Note that the inner tier is used for calculation of the various derivatives in equation (4.70). Due to the continuity of the entire complex system, the derivative of the complex admittance function Y_T with respect to $I_{\text{in}}e^{j\phi_s}$ must fulfill the Cauchy–Riemann relationships, so it is possible to write

$$\frac{\partial Y_{To}}{\partial I_{\text{in}}^r} = Y_{To,I_{\text{in}}}^r + jY_{To,I_{\text{in}}}^i \qquad \frac{\partial Y_{To}}{\partial I_{\text{in}}^i} = -Y_{To,I_{\text{in}}}^i + jY_{To,I_{\text{in}}}^r \tag{4.71}$$

where $Y_{To,I_{\text{in}}}^r$ and $Y_{To,I_{\text{in}}}^i$ are real values, with dimension V^{-1}. Taking the relationships above into account, the linearization of (4.71) about the steady-state solution is given by

$$\begin{aligned} Y_{ToV}^r(V_s - V_o) + Y_{To\omega}^r(\omega_{\text{in}} - \omega_o) + Y_{To,I_{\text{in}}}^r I_{\text{in}} \cos \phi_s - Y_{To,I_{\text{in}}}^i I_{\text{in}} \sin \phi_s = 0 \\ Y_{ToV}^i(V_s - V_o) + Y_{To\omega}^i(\omega_{\text{in}} - \omega_o) + Y_{To,I_{\text{in}}}^i I_{\text{in}} \cos \phi_s + Y_{To,I_{\text{in}}}^r I_{\text{in}} \sin \phi_s = 0 \end{aligned} \tag{4.72}$$

where the subscripts stand for the derivatives of the total admittance function Y_{To} of the free-running circuit, calculated with respect to the corresponding variables: amplitude V, frequency ω, and input generator $I_{\text{in}}e^{j\phi_s}$, operating at the harmonic component $N\omega_o$. The superscripts r and i stand for the real and imaginary parts of the derivatives of Y_{To}, respectively. Note that equations (4.72) involve relationships between the node voltage at the subharmonic frequency ω_s and the input generator at $\omega_{\text{in}} = N\omega_s$. Formulation (4.72) can also be applied to fundamentally injection-locked oscillators. It will be used when the synchronizing source is introduced at a distant circuit node or branch from the observation node considered and cannot be represented by means of its Norton equivalent seen from this node. An example is a transistor-based oscillator in which the synchronizing source is introduced at the gate terminal, whereas the observation node selected is the drain terminal.

The linearized equations above provide an ellipse in the plane defined by ω_s and V_s. This ellipse is obtained easily by separately squaring each of the two equations in (4.71), adding the two resulting expressions, and taking common factors. In a manner similar to what was done in the case of the fundamentally synchronized oscillators, the following vectors will be introduced:

$$\begin{aligned} \overline{Y}_{ToV} \equiv (Y_{ToV}^r Y_{ToV}^i) \qquad \overline{Y}_{To\omega} \equiv (Y_{To\omega}^r, Y_{To\omega}^i) \\ \overline{Y}_{To,I_{\text{in}}} \equiv (Y_{To,I_{\text{in}}}^r, Y_{To,I_{\text{in}}}^i) \end{aligned} \tag{4.73}$$

Then the ellipse equation takes the compact form

$$|\overline{Y}_{ToV}|^2(V_s - V_o)^2 + |\overline{Y}_{To\omega}|^2(\omega_{in} - \omega_o)^2 + 2\overline{Y}_{ToV} \cdot \overline{Y}_{To\omega}(V_s - V_o)(\omega_{in} - \omega_o)$$
$$= [(Y^r_{To,I_{in}})^2 + (Y^i_{To,I_{in}})^2]I^2_{in} \qquad (4.74)$$

where the dot stands for the product $\overline{a} \cdot \overline{b} = a^r b^r + a^i b^i$. Equation (4.74) constitutes the synchronized solution curve of the harmonic injection divider for low input generator amplitude. It is formally identical to equation (4.11), describing the synchronized solution curves of the injection-locked oscillator at the fundamental frequency. The derivatives of the total admittance function $\overline{Y}_{ToV}, \overline{Y}_{To\omega}$, evaluated at the free-running solution V_o, ω_o, are the same as those in (4.11). The difference between the two equations is in the independent term on the right-hand side. In (4.74), this term is $|Y_{To,I_{in}}|^2 I^2_{in}$. This means that the synchronization bandwidth is determined not only by the input generator amplitude I_{in} but also by the sensitivity of the total admittance function with respect to this generator. This is in total agreement with the results of the system-level analysis of (4.67).

It is possible to solve (4.72) in terms of $\Delta\omega_s$ by using Kramer's rule. This provides an expression with sinusoidal dependence on the phase shift ϕ_s. The division bandwidth $\Delta\omega_{1/N}$ by the order N is given by twice the maximum value of the increment $\Delta\omega_s$ versus the phase shift ϕ_s:

$$\Delta\omega_{1/N} \equiv 2|\Delta\omega_{in}|_{max} = 2I_{in}\frac{|\overline{Y}_{T_o,I_{in}}|}{|\partial Y_{To}/\partial\omega|\,|\sin(\alpha_{V\omega})|} \qquad (4.75)$$

where $\alpha_{V\omega}$ is the angle between the derivative vectors $\overline{Y}_{ToV}$ and $\overline{Y}_{To\omega}$. As can be seen, for constant I_{in} the division bandwidth increases with the magnitude of the derivative of the admittance function with respect to the input generator: in other words, with sensitivity to this generator. In addition, the magnitude of the frequency derivative $Y_{To\omega} = \partial Y_{To}/\partial\omega$ should be minimized. Because the imaginary part of the admittance function usually exhibits a higher frequency dependence on frequency variations than does the real part, a small value of $|\partial Y_{To}/\partial\omega|$ will be obtained for the low quality factor Q of the free-running oscillator circuit. The reduction in the quality factor will enable a general increase in the frequency-division bands at all orders N. However, the sensitivity to the input generator will change with the division order. This is because the magnitude of the derivative $|\overline{Y}_{To,I_{in}}|$ depends on the harmonic component $N\omega_o$ at which the input generator is introduced. In general, lower sensitivity is obtained at higher harmonic terms.

As an example, the behavior of the circuit of Fig. 1.1, operating as a harmonic injection divider by 3, has been studied. For this analysis the input current source about the free-running frequency ω_o considered previously is replaced with a current source about $N\omega_o$. For $N = 3$, the magnitude of $|\overline{Y}_{To,I_{in}}|$ in (4.75) is 0.046 V^{-1}. For $N = 5$, the magnitude of $|\overline{Y}_{To,I_{in}}|$ is 0.0017 V^{-1}. Because the derivatives Y_{ToV} and $Y_{To\omega}$ and the rest of the elements in expression (4.75) are the same for the two division orders, the division bandwidth will be much broader for $N = 3$

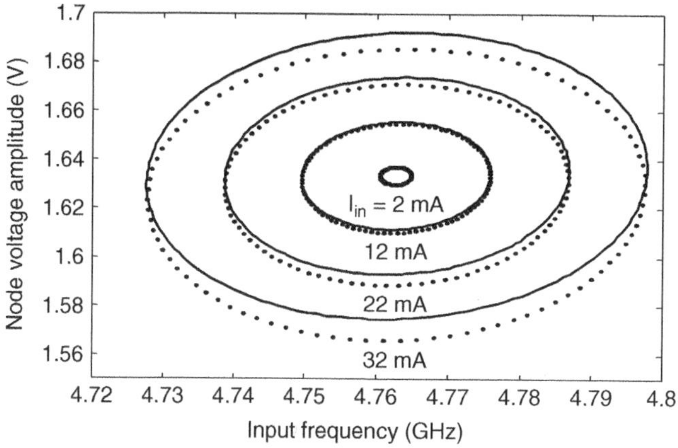

FIGURE 4.17 Synchronized solution curves of the circuit of Fig. 1.1 operating as a frequency divider by 3. Comparison of linearized analysis based on (4.74) with full nonlinear simulations.

than for $N = 5$. Note that the bandwidth prediction through (4.75) is valid only for low input generator amplitude, as it has been derived from the linearization of (4.69) with respect to this generator about the free-running regime. In the following analyses of this circuit, only the division order $N = 3$ is considered. Figure 4.17 compares solution curves divided by $N = 3$ obtained using linearized expression (4.70) (solid lines) with a full harmonic balance analysis of the form (4.68) (dotted lines). As can be seen, for low input current amplitude I_{in}, the curves are almost overlapped. For this small I_{in}, the circuit behaves in a linear manner with respect to the input generator, and the synchronization curves are perfect ellipses. As the input current grows, nonlinear effects become apparent, increasing the discrepancy between the linearization (4.74) and the nonlinear simulations with (4.69). The infinite slope points of the ellipse (one at each side) are local–global saddle–node bifurcations (see Chapter 3). Thus, only one section of the synchronization curve (either the upper or lower section) can be stable, similar to what was obtained in the analysis of an injection-locked oscillator at the fundamental frequency.

As already stated, frequency division by order $N \neq 2$ will generally take place from a quasiperiodic regime at local–global (mode-locking) bifurcations. As an illustration of the global behavior of dividers by $N \neq 2$, the bifurcation loci of the circuit of Fig. 1.1, operating as a harmonic injection divider by $N = 3$, have been determined. In Fig. 4.18 the corresponding secondary Hopf and turning-point loci have been represented in the plane defined by the input frequency ω_{in} and input current I_{in}. Sketches of the solution spectrum in the various operation regions, delimited by the loci, are included. The region of ω_{in}, I_{in} values for which the circuit behaves as a frequency divider by 3 is delimited by the turning-point locus, (consisting of) local–global bifurcations at which the synchronization $3\omega_s = \omega_{in}$ takes place. This locus is the envelope of the turning points of the solution curves in Fig. 4.17, obtained for different values of input generator amplitude I_{in}. For zero

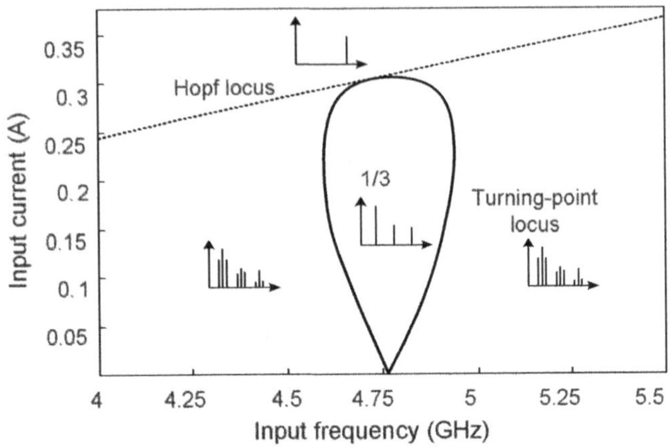

FIGURE 4.18 Bifurcation loci of the parallel resonance circuit of Fig. 1.1 operating as a harmonic injection divider by 3. The division region is delimited by the turning-point locus at which the synchronization $\omega_a/\omega_{in} = \frac{1}{3}$ takes place.

input amplitude, the division bandwidth degenerates to a single point, corresponding to the free-running oscillation. This point is given by $\omega_{in} = 3\omega_o$, $I_{in} = 0$.

The region delimited by the turning-point locus constitutes the Arnold tongue $\frac{1}{3}$, at which the synchronization $\omega_a/\omega_{in} = \frac{1}{3}$ takes place in the circuit. It is narrower than the Arnold tongue 1/1, represented in Fig. 4.5. In general, Arnold tongues become narrower for higher division order, due to the lesser influence of the input generator about a higher harmonic of the oscillation frequency $N\omega_o$.

As can be seen, the turning-point locus lies below the secondary Hopf locus, as the division through synchronization requires the existence of self-oscillation. The division region vanishes when the turning-point locus intersects the Hopf locus in a codimension 2 bifurcation, with two real poles of zero value. Outside the turning-point locus and below the Hopf locus, the circuit behaves as a self-oscillating mixer, with two fundamental frequencies: the input frequency ω_{in} and the oscillation frequency ω_a.

As already discussed, the behavior of the dividers $N = 2$ will generally be different from the behavior of dividers $N \neq 2$. This is due to the common occurrence of flip bifurcations at higher levels of input power, which lead directly from a periodic regime at ω_{in} to a regime at $\omega_{in}/2$, and vice versa. As we know, the three main types of local bifurcation from a periodic regime are the turning point, the flip bifurcation, and the Hopf bifurcation. Therefore, the flip bifurcation is a fundamental phenomenon in dynamical systems, leading to frequency division by $N = 2$. Figure 4.19a shows the typical bifurcation loci of frequency dividers by $N = 2$ in the plane defined by the two usual parameters P_{in} and ω_{in}.

For a harmonic injection divider by order $N = 2$, we will have three different loci: the secondary Hopf bifurcation locus, the turning-point locus, and the flip bifurcation locus. The turning-point locus is the envelope of infinite-slope points of closed synchronization curves obtained for different input power values. When

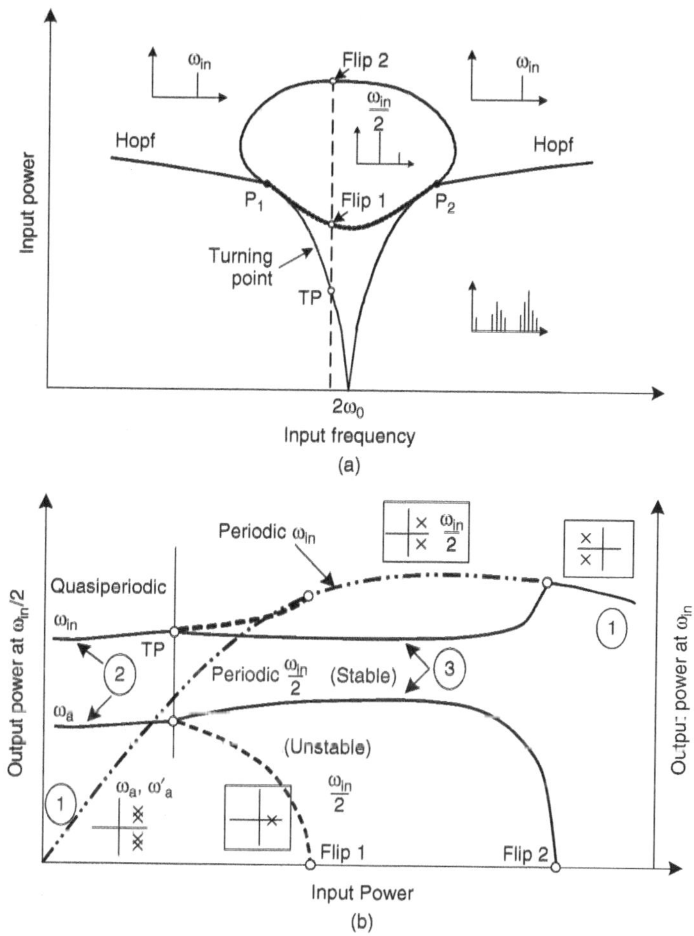

FIGURE 4.19 General behavior of a harmonic injection divider by $N = 2$: (a) bifurcation loci of a typical harmonic injection divider by order $N = 2$ (sketches of the solution spectrum at the different regions delimited by the loci are included); (b) stability versus the input power.

crossed from a quasiperiodic regime (below the Hopf locus), it gives rise to division through synchronization of the second harmonic component of the oscillation frequency $2\omega_a$ to the input source frequency ω_{in}. The synchronization region delimited by the turning-point locus in Fig. 4.19a constitutes an Arnold tongue of order $\frac{1}{2}$. For zero input amplitude, the division bandwidth degenerates to a single point, corresponding to the free-running oscillation. Thus, the lower vertex of the turning point locus is given by $\omega_{in} = 2\omega_o$ and $P_{in} = 0$W.

Above the Hopf locus and outside the synchronization region, the self-oscillation is extinguished and the nondivided solution at the input source frequency ω_{in} is stable. Crossing the flip bifurcation locus from this regime leads to a direct frequency division $\omega_{in} \rightarrow \omega_{in}/2$. This direct division does not involve the generation of any autonomous (nonsynchronized) frequency prior to the division itself. The subharmonic components $k\omega_{in}/2$, with k an integer, appear cleanly in the circuit output spectrum.

The turning-point locus consists of periodic solution points at frequency $\omega_{in}/2$ having one pole at zero. Due to the nonunivocal relationship between the poles and Floquet multipliers and the fact that the fundamental frequency is $\omega_{in}/2$, the solution will also have an infinite set of poles $\pm jk\omega_{in}/2$, with k an integer. Note that the turning-point locus, Hopf locus, and flip locus intersect at the two codimension 2 bifurcations P_1 and P_2. The frequency ω of the critical poles $\pm j\omega$ of the Hopf locus tends to $\omega_{in}/2$ when approaching these intersection points. It takes the value $\omega = \omega_{in}/2$ at P_1 and P_2. Thus, at these two points, there are two pairs of complex-conjugate poles with the same value $\pm j\omega_{in}/2$, one already existing in the turning-point locus and the second due to intersection with the Hopf locus. When moving away from P_1 or P_2 to the dotted section of the flip locus, one of the pairs of poles shifts to the right-hand side of the complex plane $\sigma \pm j\omega_{in}/2, \sigma > 0$, while the other remains at $\pm j\omega_{in}/2$. The dotted section of the flip locus indicates flip bifurcations giving rise to unstable divided solutions, so crossing this section when varying either P_{in} or ω_{in} will have no physical effect. In the remaining locus points, there is only a pair of poles at $\pm j\omega_{in}/2$ with all the other poles on the left-hand side of the complex plane.

Figure 4.19b helps make a distinction between the main types of bifurcations in a harmonic injection divider by $N = 2$. Note that there are two different output power axes: one corresponding to the subharmonic frequency $\omega_{in}/2$ and the other to the input frequency ω_{in} (which should have smaller values, so a different scale is considered in the right axis). The parameter considered is the input power P_{in}. When increasing this input power from near zero value at constant input frequency, the circuit traverses the regions indicated by the vertical line in Fig. 4.19a. For a good understanding of its behavior, both figures should actually be compared. For small P_{in}, the circuit behaves in a quasiperiodic regime at the two fundamental frequencies ω_{in} and ω_a. This is because the circuit exhibited a free-running oscillation prior to the connection of the periodic generator at ω_{in}, which for small P_{in}, persists in the circuit in a nonsynchronized manner. Because it is a harmonic injection divider by 2, we can expect the free-running oscillation frequency ω_o and the nonsynchronized frequency ω_a to be on the order of $\omega_{in}/2$. The autonomous quasiperiodic regime at ω_{in}, ω_a is indicated as (2) in the figure. The solution is represented by drawing the power values (in the output power spectrum) that correspond to the spectral lines at the two fundamental frequencies ω_{in} and ω_a.

The stable quasiperiodic solution at ω_{in}, ω_a coexists with a periodic solution at ω_{in}, indicated as (1). This periodic nondivided solution exists in any frequency divider, as shown at the beginning of the section. For the moment, we will concentrate on the poles of this periodic solution. For low P_{in}, the periodic solution at ω_{in} is unstable. This solution contains a pair of unstable complex-conjugate multipliers $m_{1,2} = e^{(\sigma \pm j\omega_a)T}$, with $\sigma > 0$ and T the input generator period $T = 2\pi/\omega_{in}$. To remind the Floquet multiplier theory, see Section 1.5.2. The pair of complex-conjugate $m_{1,2}$ multipliers is actually responsible for the circuit's self-oscillation at the frequency ω_a for low input power. Due to the nonunivocal relationship between the Floquet multipliers and the system poles [see equation 1.59], the periodic solution at ω_{in} will contain poles at $\sigma \pm j(\omega_a + k\omega_{in})$, with k an integer, all associated with the same pair of unstable multipliers $m_{1,2} = e^{(\sigma \pm j\omega_a)T}$ and thus having the same $\sigma > 0$. In particular, there will be two pairs of unstable poles located about the divided frequency $\omega_{in}/2$ and given by $\sigma \pm j\omega_a$ and $\sigma \pm j(\omega_{in} - \omega_a)$ (see the discussion of the pole structure in Section 3.3.2). Remember that ω_a is on the order of the divided-by-2 frequency. In Fig. 4.19b the difference frequency $\omega_{in} - \omega_a$ is denoted $\omega'_a = \omega_{in} - \omega_a$.

As the input power P_{in} increases, the frequencies ω_a and $\omega'_a = \omega_{in} - \omega_a$ tend to $\omega_{in}/2$. At a particular power value P_{ino} (not represented in Fig. 4.18b), the two pairs of $\omega_{in}/2$ poles $\sigma \pm j\omega_a$ and $\sigma \pm j\omega'_a$ merge into two pairs of poles at $\omega_{in}/2$ and split from this power value (see Section 3.3.2). Thus, for $P_{in} > P_{ino}$, they behave as two independent pairs of poles at the same frequency $\omega_{in}/2$, with a different real part. They can be expressed as $\sigma \pm j(\omega_{in}/2)$ and $\sigma' \pm j(\omega_{in}/2)$. From the point of view of Floquet multipliers, a pair of complex-conjugate multipliers transforms at $P_{in} = P_{ino}$ into two real multipliers m'_1, m'_2. Remember that the total number of multipliers agrees with the system dimension and cannot change under the variation of any system parameter.

For $P_{in} > P_{ino}$, one of the pairs of unstable poles at $\omega_{in}/2$, expressed as $\sigma \pm j\frac{\omega_{in}}{2}$, shifts leftward and crosses the imaginary axis at bifurcation flip 1, to the left-hand side of the complex plane. This bifurcation has no effect on divider behavior since the periodic solution is unstable before and after this bifurcation. This is because the second pair of poles $\sigma' \pm j\frac{\omega_{in}}{2}$ remains on the right-hand side of the complex plane after the bifurcation flip 1, which gives rise to a transition from a solution with two pairs of unstable complex-conjugate poles at $\omega_{in}/2$ to a solution with one pair of unstable complex-conjugate poles at $\omega_{in}/2$. The divided solution generated at $\omega_{in}/2$, indicated as (3), is initially unstable, due to the presence of the unstable poles $\sigma' \pm j\frac{\omega_{in}}{2}$. The fundamental frequency of the generated subharmonic solution is $\frac{\omega_{in}}{2}$, so this subharmonic solution will contain also a real pole γ' on the right-hand side of the complex plane. Note that γ' and $\sigma' \pm j\frac{\omega_{in}}{2}$ belong to the same set of poles, associated to the real Floquet multiplier $m = e^{\gamma' T} = e^{\gamma' 4\pi/\omega_{in}} < -1$.

The transformations of this divided-by-2 solution are now considered. When reaching the turning-point TP, the real pole γ' crosses the imaginary axis to the left-hand side of the plane, so from TP the divided-by-2 solution is stable. The turning point TP is a synchronization point. Actually, the oscillation frequency ω_a of the quasiperiodic solution (indicated with 2 in Fig. 4.18b) approaches continuously the divided frequency versus P_{in} and fulfills the synchronization relationship $\omega_{in} = 2\omega_a$ at the turning point TP. The divided-by-2 solution is maintained in a certain input power range. Then, it is extinguished to zero subharmonic amplitude at bifurcation flip 2, where the remaining pair of complex-conjugate poles at $\omega_{in}/2$ of the nondivided solution cross the imaginary axis to the left-hand side of the complex plane. From bifurcation flip 2, the periodic solution at ω_{in} is stable.

As a practical example, the circuit of Fig. 4.8 has been analyzed as a frequency divider by $N = 2$. The resulting bifurcation loci are shown in Fig. 4.20a together with sketches of the solution spectrum in the various sections delimited by the loci. The location and shape of these loci are variants of the ideal loci, represented in Fig. 4.19. As expected, for low input generator amplitude, the V-shaped turning-point locus delimits the frequency-division band through harmonic synchronization $\omega_{in} = 2\omega_a$. The flip bifurcation locus is located above this turning-point locus. The Hopf locus is observable on the left-hand side of the figure. Below this locus and outside the turning-point locus, the circuit behaves in a self-oscillating mixer regime at the fundamental frequencies ω_{in} and ω_a. On the right-hand side of the figure, the turning-point locus extends to quite high input amplitude and coexists with the flip bifurcation locus. The Hopf bifurcation locus is not reached for the values of input amplitude considered. This is due to the filtering action of the input network, which reduces the influence of the input signal at the higher frequency values, thus preventing oscillation extinction. Note that limitations in the precision of the device model can also degrade the accuracy in the prediction of the loci sections at high input generator amplitude. On the right-hand side of the turning-point locus, the circuit behaves in a self-oscillating mixer regime at the fundamental frequencies ω_{in} and ω_a. The division by 2 takes place at the turning points through harmonic synchronization $\omega_{in} = 2\omega_a$. Note that the flip bifurcation locus has no physical effect on the right-hand side of the figure, as it gives rise to unstable frequency-divided solutions that are not observable. To confirm this, Fig. 4.20b shows the evolution of the real part of the dominant poles of the nondivided solution at the constant input frequency $f_{in} = 9$ GHz versus the input amplitude E_{in}.

The periodic solution analyzed is analogous to solution 1 in Fig. 4.19b. For small E_{in} there are two pairs of unstable complex-conjugate poles at $\sigma \pm j\omega_a$ and $\sigma \pm j(\omega_{in} - \omega_a)$ at about the divided-by-2 frequency $f_{in}/2 = 4.5$ GHz. These poles have been calculated through a numerical technique [22], which explains the slight difference in the positive σ. As already stated, these two pairs of poles are associated with the same pair of complex-conjugate Floquet multipliers. The two pairs of poles merge at about $E_{in} = 0.37$ V and split into two pairs of poles at the divided-by-2 frequency $\sigma \pm j(\omega_{in}/2), \sigma' \pm j(\omega_{in}/2)$ associated with two different real Floquet multipliers fulfilling $m_1 < -1$ and $m_2 < -1$. At $E_{in} = 0.4$ V, one of the unstable pairs of poles crosses the imaginary axis to the left-hand side. Therefore, a flip bifurcation is obtained, in agreement with the loci of Fig. 4.20a. This bifurcation has no physical effect on circuit behavior. Actually, the circuit is already operating as a frequency divider when this flip bifurcation is obtained, since division by 2 took place through synchronization $2\omega_a = \omega_{in}$ at $E_{in} = 0.12$ V (see the loci of Fig. 4.20a). The situation is similar to the one in Fig. 4.19b.

Figure 4.21 shows the evolution of periodic solution curves at the divided frequency $\omega_{in}/2$, traced versus the input frequency ω_{in}, when the input generator amplitude increases. For a low input generator amplitude, the periodic curve is closed and coexists with a nondivided curve at f_{in}, with a much lower amplitude,

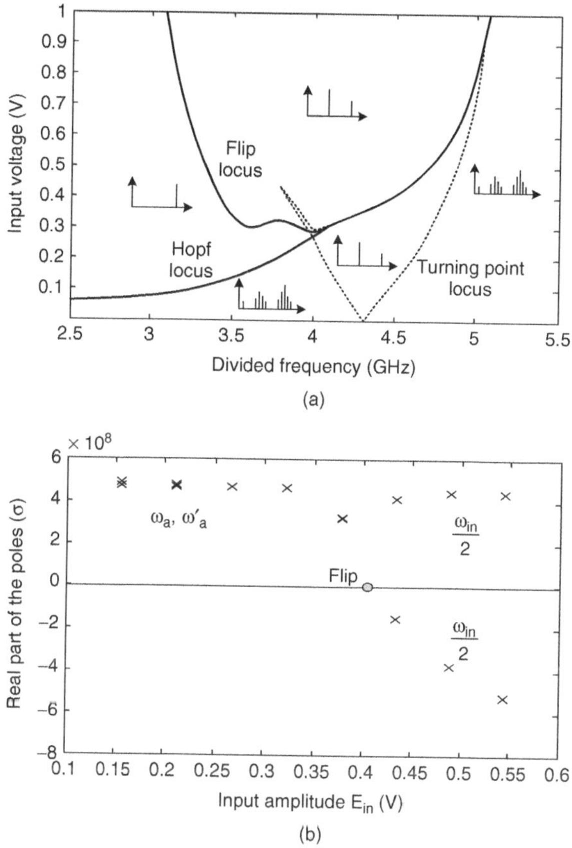

FIGURE 4.20 Bifurcation loci of the circuit of Fig. 4.8 operating as a frequency divider by $N = 2$: (a) representation of the loci in the plane defined by the input frequency ω_{in} and input voltage E_{in}; (b) evolution of the real part of the unstable poles of the periodic solution at $f_{in} = 9$ GHz versus the input amplitude E_{in}.

not represented in the figure. However, an example is shown in Fig. 4.22, corresponding to $E_{in} = 0.3$ V. Note the closed curve at $\omega_{in}/2$ and the second closed curve at f_{in}, corresponding to the fundamental and second harmonic components of the same solution. The nondivided solution is open and different from the divided solution. Figure 4.21 shows the evolution of the subharmonic amplitude $\omega_{in}/2$ versus the input power. The solution curves are oriented downward. As we know, for low input amplitude, the central axis of the ellipse in the coordinate system V_s, ω_{in} is determined by the derivatives of the total admittance function Y_{To} with respect to the amplitude and frequency Y_{ToV}, $Y_{To\omega}$, evaluated at the free-running oscillation. This is why this axis agrees with the axis that corresponds to the solution curves of Fig. 4.9, in which the same circuit was analyzed as a fundamentally synchronized oscillator. Note that the solution curves in Figs. 4.9 and 4.21 are traced in terms of the same node voltage.

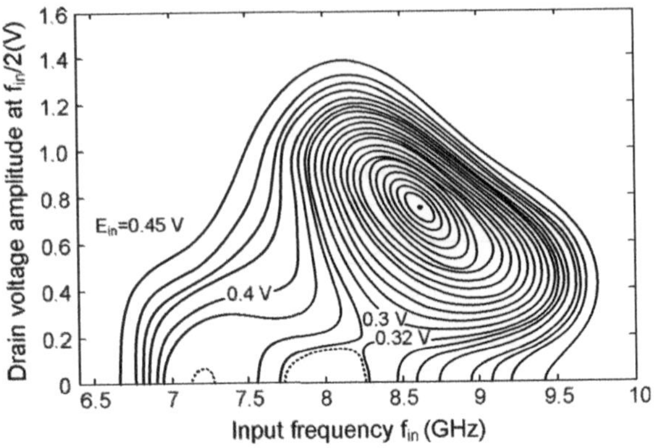

FIGURE 4.21 FET-based harmonic injection divider. Evolution of the solution curves at the divided frequency $\omega_{in}/2$ versus the input frequency ω_{in} for different values of input voltage amplitude E_{in}.

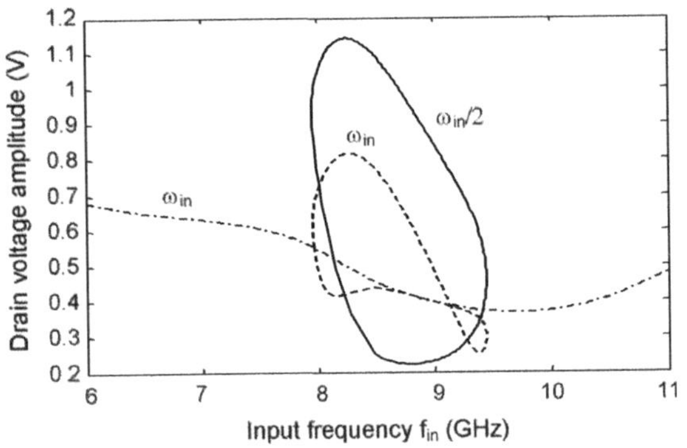

FIGURE 4.22 Solutions of the circuit of Fig. 4.8, with second harmonic injection for the input amplitude $E_{in} = 0.3$ V traced versus the input frequency. Within the frequency-division interval, the solution at $\omega_{in}/2$ coexists with a nondivided interval at ω_{in}. The divided solution is traced by representing the drain voltage amplitude at the divided frequency $\omega_{in}/2$ and at the second harmonic ω_{in}. The two closed curves correspond, in fact, to the same solution. The nondivided solution at ω_{in} provides an open curve.

The tuning-point locus of Fig. 4.20 is the envelope of all the turning points in the curves of Fig. 4.21. At approximately $E_{in} = 0.32$ V, the closed divided curve becomes open. The open solution curves obtained for $E_{in} > 0.32$ V are generated and extinguished at flip bifurcations. However, only flip bifurcations occurring above the Hopf bifurcation locus in Fig. 4.20 are physically meaningful. On the left-hand side, for E_{in} slightly higher than 0.32 V, they give rise to a transition between a stable periodic regime at ω_{in} and a stable divided-by-2 regime at $\omega_{in}/2$. Then the turning points of the divided curves lead to jumps between different sections of the divided solution curves. In turn, all the flip bifurcations on the right-hand side of Fig. 4.21 occur below the Hopf locus. They are unphysical, as division will take place through synchronization $2\omega_a = \omega_{in}$ at the turning point of each curve. As pointed out earlier, the circuit behavior is quite irregular in the neighborhood of the intersection points between different loci. As an example, for the divider analyzed, the flip locus exhibits local minima (Fig. 4.20), giving rise to low-amplitude divided solutions that are generated and extinguished in these zones, as confirmed through comparison of Fig. 4.20 and Fig. 4.21.

4.3.3 Regenerative Frequency Dividers

Unlike the case of harmonic injection dividers, a regenerative divider must not oscillate in the absence of an input generator signal. The oscillation should start from a certain level of this signal at the input frequency ω_{in}, and for that, a feedback loop is included in the system (see Fig. 4.23) [9]. The objective is to generate instability that the divided frequency ω_{in}/N (which is present in the circuit noise) through an increase in the feedback gain at the frequency $(N-1)\omega_{in}/N$. The nonlinear element mixes the feedback signal at $(N-1)\omega_{in}/N$ with the input signal at ω_{in}. The difference frequency ω_{in}/N is selected through a lowpass filter and amplified. Then it is introduced into the frequency multiplier by $N-1$ of the feedback branch. The instability is favored by the feedback at $(N-1)\omega_{in}/N$ and the mixing and amplification actions, which give rise to a gain increase versus the input amplitude E_{in} at the difference frequency ω_{in}/N. Note that the regenerative frequency division can also be achieved using feedback at ω_{in}/N plus a harmonic mixer, providing the component ω_{in}/N from the intermodulation product $\omega_{in} - (N-1)\omega_{in}/N$. This avoids the requirement for a frequency multiplier in the feedback branch.

A simplified analysis of a generic frequency divider based on the block diagram of Fig. 4.23 is presented next. Assuming an input signal $e_{in}(t) = E_{in}\cos(\omega_{in}t + \phi_{in})$ and a feedback signal $v_{fb}(t) = V_{fb}\cos[(N-1)(\omega_{in}/N)t + \phi]$, the mixer will provide the difference frequency $\omega_o = \omega_{in}/N$ and the summation frequency $2\omega_{in} - \omega_{in}/N$, which should be eliminated with the filter. For ideal filtering at ω_{in}/N, the system is ruled by the time-domain equation

$$\begin{aligned} v_{fb}(t) &= V_{fb}\cos\left[(N-1)\frac{\omega_{in}}{N}t + \phi\right] \\ &= A_T(\tilde{V}_{fb}, E_{in})\frac{E_{in}V_{fb}}{2}\cos\left[(N-1)\frac{\omega_{in}}{N}t + (N-1)(\phi_{in} - \phi + \gamma)\right] \end{aligned} \quad (4.76)$$

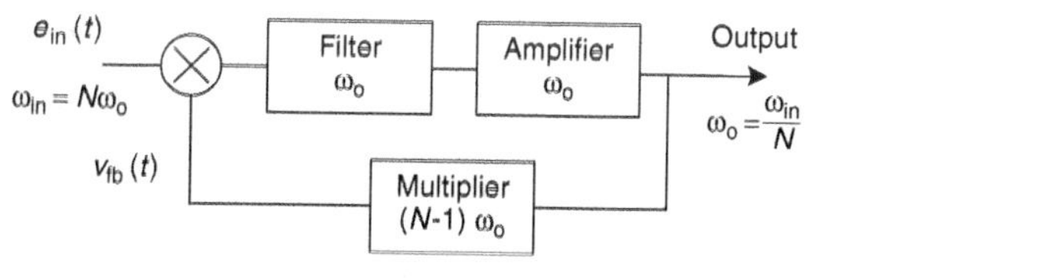

FIGURE 4.23 Operational principle of a regenerative frequency divider.

where $\tilde{V}_{fb}$ is the phasor associated with the feedback signal and γ is the phase shift contributed by the filter and amplifier. The nonlinear coefficient A_T affecting the feedback signal amplitude includes contributions from the mixer, filter, amplifier, and multiplier. For the subharmonic component ω_{in}/N to be self-sustained, the open-loop transfer function must fulfill

$$\begin{aligned} A_T(\tilde{V}_{fb}, E_{in})E_{in} &= 1 \\ \phi - (N-1)(\phi_{in} - \phi + \gamma) &= n2\pi \end{aligned} \tag{4.77}$$

where φ_T is the total open-loop phase shift, n is an integer, and for simplicity, A_T has been redefined to include the factor of $\frac{1}{2}$. The relationship (4.77) states clearly that the circuit cannot oscillate in the absence of input signal $E_{in} = 0$. For $E_{in} < E_{ino}$, the system will exhibit some linear gain at a subharmonic frequency, but the product $A_T(V_{fb} = 0, E_{in})E_{in} < 1$ will not be enough for oscillation startup. Note that the regime at ω_{in} is actually nonlinear and $A_T(V_{fb} = 0, E_{in})E_{in}$ is the gain obtained by linearizing the system in Fig. 4.23 about the steady-state regime at ω_{in}, due to the input generator $e_{in}(t)$ evaluated at the subharmonic frequency ω_{in}/N. From a certain input amplitude E_{ino}, the small-signal gain of the closed loop will be $A_T(V_{fb} = 0, E_{in})E_{in} > 1$, which provided that the phase condition $\varphi_T = 2n\pi(V_{fb} = 0, E_{in})$ is also fulfilled will give rise to oscillation startup. Note that the higher the coefficient $A_T(V_{fb} = 0, E_{in})$, the lower the amplitude threshold E_{ino} for the oscillation startup.

One disadvantage of frequency dividers based on the block diagram of Fig. 4.23 is the large number of building blocks required. As pointed out by Rauscher [23], a detailed analysis of the circuit reveals that many more functional blocks have to be added for proper operation, which would result in an expensive design. Another disadvantage is the need for relatively high input power to achieve frequency division. This is because the multiplier has to be driven hard by the mixer to deliver a sufficient output signal. In turn, the mixer can operate correctly only with relatively high amplitude at $(N-1)f_o$ provided by the multiplier. For a regenerative divider by 2, the schematic is simplified considerably, as the $N-1$ multiplier in the feedback branch can be replaced by a simple bandpass filter at f_o.

Instead of joining individual functional blocks, it is possible to perform a single circuit implementation of the regenerative divider. The schematic will consist of an input filter at $f_{in} = Nf_o$, a suitably biased transistor acting as both a mixer and an amplifier, an output filter at f_o, and a feedback block. The feedback block will be given by a second transistor at a convenient bias point. As an example, a frequency divider by $N = 4$ has been designed following the block diagram of Fig. 4.23. A MESFET transistor is used as an active device. It is biased near pinch-off for efficient frequency mixing, taking advantage of the quasiquadratic characteristic of the drain-to-source current i_{DS} versus the gate-to-source voltage v_{GS}. Input and output filters at the respective frequencies ω_{in} and ω_{in}/N are also introduced, together with a frequency multiplier by 3 in the feedback loop. Control of the phase shift introduced by the various linear elements is essential to fulfill the conditions $A_T(V_{fb} = 0, E_{in})E_{in} > 1$, $\varphi_T = 2n\pi$ from a certain input amplitude

E_{in}. Figure 4.24a shows the variation obtained for the magnitude and phase of the open-loop gain at $\omega_{in}/4$ versus the input power. As shown in the figure, the startup conditions of the subharmonic component $\omega_{in}/4$ are fulfilled at an input power of about 12 dBm. Figure 4.24b shows the evolution of the drain voltage at the subharmonic component versus the input generator amplitude E_{in}. The solution curve is quite regular, nearly dropping to zero when the input amplitude is reduced. However, the curve is unable to actually reach zero amplitude, unlike what happens in Hopf and flip bifurcations. Instead, a turning point T is obtained, at which the fourth harmonic component of an oscillation at $\omega_a \cong \omega_{in}/4$, generated for slightly lower input amplitude, synchronizes to the input signal. The evolution of the quasiperiodic solution near the turning point is similar to that sketched in Fig. 4.13.

As already stated, for $N = 2$ the frequency multiplier in the feedback loop of Fig. 4.23 can be replaced with a bandpass filter at $\omega_{in}/2$. The resulting configuration is similar to that of a harmonic injection divider, but the circuit must not oscillate

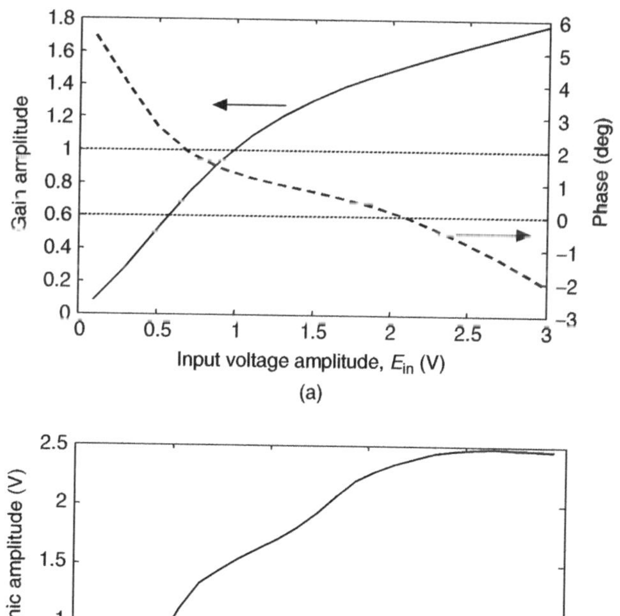

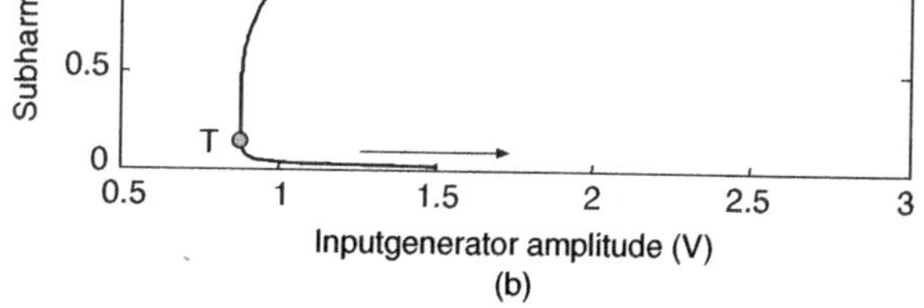

FIGURE 4.24 Regenerative frequency divider by $N = 4$: (a) variation of the magnitude and phase of the open-loop gain versus the input power; (b) evolution of the amplitude of the subharmonic component $\omega_{in}/4$ versus the input generator amplitude.

in the absence of an input signal. As an example, the circuit in Fig. 4.8, operating originally as a harmonic injection divider, can be transformed into a regenerative divider simply by reducing the gate bias voltage. By biasing the transistor below pinch-off, the originally existing free-running oscillation will be quenched. However, when increasing the input generator amplitude E_{in}, the gate voltage swing increases and makes the transistor conduct for a growing fraction of the input period. On the other hand, the quasiquadratic characteristic of the drain-to-source current i_{DS} versus the gate-to-source voltage v_{GS} will enable efficient mixing of the input signal at f_{in} and the feedback signal at $f_{in}/2$. The resulting increase in the open-loop gain at $f_{in}/2$ will lead to a frequency division by 2 from a certain input generator amplitude E_{ino}. The behavior is very sensitive to the gate bias voltage V_{GS}, so the input voltage E_{in} at which the flip bifurcation is obtained depends on this bias voltage. This is illustrated in Fig. 4.25, which shows the gate voltage waveform $v_{GS}(t)$ at the flip bifurcation for various V_{GS} values. For the transistor used, the pinch-off voltage is given by $V_{GS} = -1.8$ V. As can be seen, the input amplitude required is larger for lower gate bias voltage. Note that the waveform represented is periodic at f_{in}, as it corresponds to the instability threshold at which the subharmonic component $f_{in}/2$ is generated from zero amplitude. It is important to emphasize that the waveforms are calculated in the three cases at the flip bifurcation, obtained in each case for different values of V_{GG} and E_{in}.

In Fig. 4.26, the flip bifurcation locus has been traced in the plane defined by the gate bias and the input generator voltage. As can be seen, the locus has a negative slope, indicating that a lower input generator voltage is required for higher values of the gate bias. A second locus has been represented in the same plane. It is the Hopf bifurcation locus from the dc regime, which is traced in terms of the gate bias voltage and drain voltage amplitude at f_{in} when the oscillation is generated. The circuit oscillates in the absence of an input signal for gate bias voltage $V_{GG} > -1.2$ V. Thus, it can only behave as a regenerative divider for $V_{GG} < -1.2$ V. In the self-oscillation region, the circuit behaves as a harmonic injection divider, in a manner similar to the situation analyzed in Figs. 4.9 and 4.10.

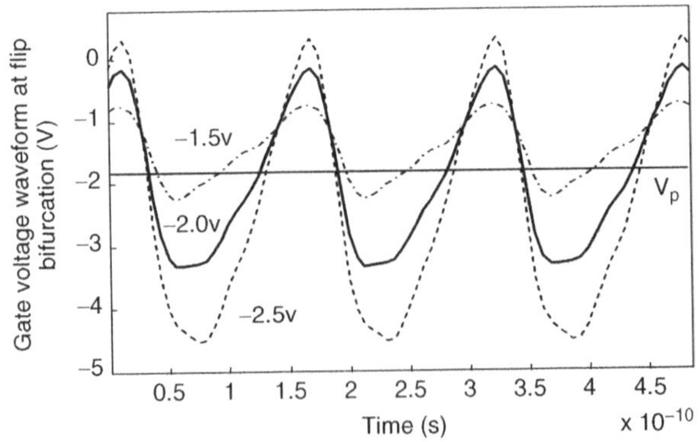

FIGURE 4.25 Gate voltage waveform at the flip bifurcation leading to a divided-by-2 regime for several values of gate bias voltage. The input amplitude required is larger for lower gate bias voltage.

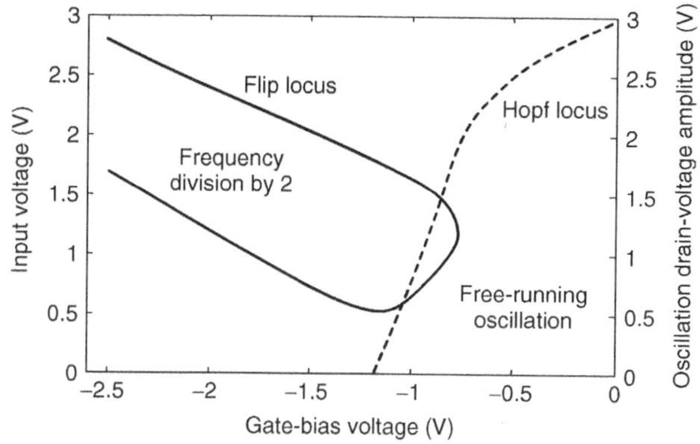

FIGURE 4.26 Flip bifurcation locus in a plane defined by the gate bias and input generator voltage. The locus has a negative slope, indicating that less input generator voltage is required for higher values of the gate bias. The Hopf bifurcation locus from the dc regime, drawn in terms of the gate bias voltage and second harmonic drain voltage amplitude, is also represented. The circuit oscillates in the absence of an input signal for gate bias voltage $V_{GG} > -1.2$ V.

Note that when considering a constant bias voltage and increasing the input generator amplitude from a low value, the flip bifurcation locus is crossed twice. At the first flip bifurcation, direct frequency division takes place from a periodic regime at f_{in}. The subharmonic component is generated from zero amplitude. At the second flip bifurcation, the subharmonic component at $f_{\text{in}}/2$ vanishes to zero.

In Fig. 4.27, the flip bifurcation locus of the circuit of Fig. 4.8 has been represented on the useful plane defined by the input frequency and the input generator amplitude. Because the circuit does not oscillate in the absence of input power, there is no oscillation outside the flip locus. Thus, unlike the case of harmonic

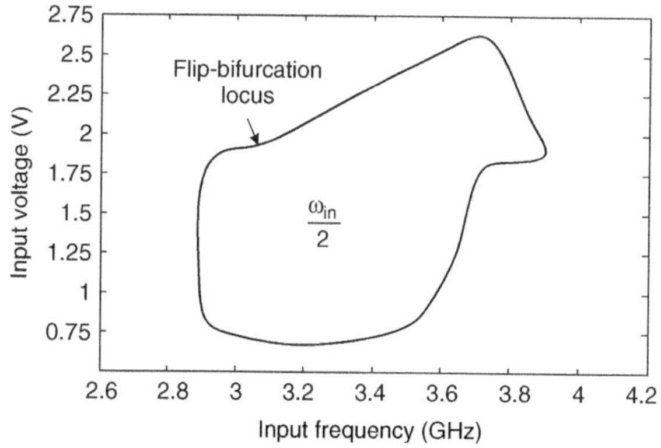

FIGURE 4.27 Flip bifurcation locus of the circuit of Fig. 4.8 in the plane defined by the input frequency and input generator amplitude. Because the circuit does not oscillate in the absence of input power, there is no Hopf locus or turning-point locus, corresponding to synchronization.

injection dividers, there is no Hopf locus for the input voltage and input frequency ranges considered. Turning points in the divided solution will generally occur for certain ranges of input generator amplitude and frequency, but they will correspond to jumps, giving rise to hysteresis. Significantly higher input generator amplitude is required for the frequency division than that shown in Fig. 4.10.

4.3.4 Parametric Frequency Dividers

In Section 3.3 it was shown how a circuit consisting of a varactor diode, biased at V_b, and an inductor can operate as frequency divider by 2 for input power above certain threshold. The inductor is calculated to fulfill, in the absence of this power, the resonance condition $2\pi f_o = 1/\sqrt{LC(V_b)}$. Then, for sufficiently high input power, the circuit will operate as a frequency divider by 2 in an input frequency f_{in} band about $2f_o$. For frequency divider design, the nonlinear capacitance of a varactor diode is usually employed. For the diode to deliver energy at a load at $f_{in}/2$, the diode loss must be less than this negative conductance [24]. Therefore, the diode quality factor at the subharmonic frequency must be relatively high. This quality factor is defined as the ratio between the intrinsic capacitance impedance at the selected bias voltage V_b and the series loss resistance R_s, due to the finite semiconductor conductivity, at the divided frequency $Q(V_b, f_o) = 1/[R_s C(V_b) 2\pi f_o]$. Note that the diode package introduces a parasitic capacitance C_p, which neglecting the loss R_s gives rise to the total capacitance $C_T = C_p + C(V_b)$.

The nonlinearity of the varactor diode capacitance is maximum about zero bias voltage. Thus, when biasing the diode about zero voltage, negative conductance is obtained from lower input power. However, the diode is likely to be driven into forward conduction, which usually results in high losses. Thus, a trade-off will be necessary when selecting this bias voltage. The practical divider design requires the addition of suitable resonant circuits or filters for frequency selection. Figure 4.28a shows a possible circuit schematic of a divider by 2. At the bias point selected for the diode, the total diode capacitance C_T resonates in series with inductor L_2 at a frequency between f_o and $2f_o$ the geometric ratio $\sqrt{2}f_o$ is convenient. The current flowing through the diode acts like a "pump," causing periodic variation in the capacitance. The input circuit $L_1 - C_1$ is selected to be parallel resonant at f_o and forms a series resonant circuit with L_2 and the diode at the input frequency f_{in}. Similarly, the output circuit $L_3 - C_3$ resonates in parallel at $f_{in} = 2f_o$ and forms a series resonant circuit with L_2 and the diode at f_o.

Parametric frequency division by an order different from $N = 2$ is also possible. To increase the efficiency of the division by $N > 2$, the load impedance at the undesired frequency components $k\omega_{in}/N$ with $k > 1$ should be made zero or infinite or purely reactive. Even though no output is desired at a frequency different from ω_{in}/N, these frequency components must exist inside the nonlinear reactance for efficient pumping, contributing to the negative resistance at ω_{in}/N through intermodulation. If the undesired frequencies are terminated in open or short circuits, it is ideally possible to achieve equality between delivered power at ω_{in} and consumed power at ω_{in}/N $-P(\omega_{in}) = P(\omega_{in}/N)$, as an ideal capacitance

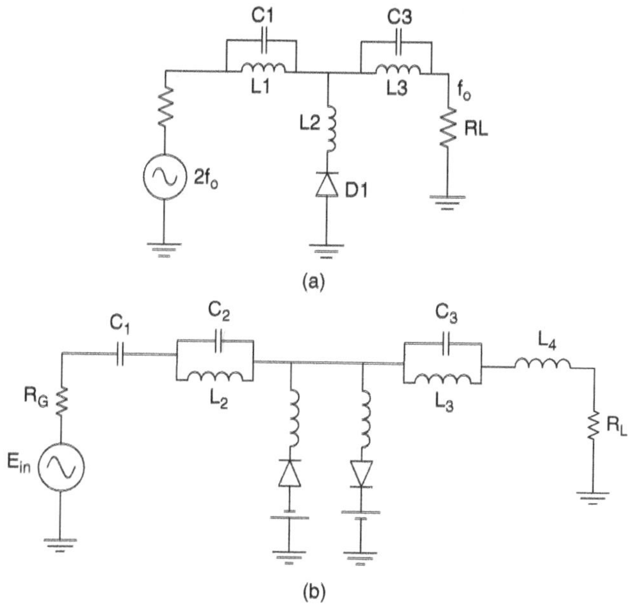

FIGURE 4.28 Circuit topology of a parametric frequency divider: (a) schematic of a parametric frequency divider by 2; (b) schematic of a frequency divider by 3, containing a resonant circuit at the idler frequency $2f_{\text{in}}/3$.

fulfills $\sum_k V_k I_k^* = 0$, with V_k being the harmonic terms $k\omega_{\text{in}}/N$ of the voltage across the nonlinear capacitance and I_k the harmonic terms of the current through this capacitance [25].

For division by $N = 3$, it is necessary to ensure the presence of the frequency component $2f_{\text{in}}/3$ across the diode, which will provide divided frequency through the difference term $f_{\text{id}} = f_{\text{in}} - 2f_{\text{in}}/3 = f_{\text{in}}/3$ (Fig. 4.28b). The frequency f_{id} is known as the *idler frequency*. The current generated by the diode at f_{id} will be unused in the sense that no power is extracted at this frequency. However, it is necessary to obtain the required pumping voltage at $2f_{\text{in}}/3$. The idler current is typically terminated in a short circuit so that no power is dissipated at the idler frequency. This has been implemented by Suárez and Melville [26] by connecting two symmetric legs, each containing an inductor and a diode, which are series resonant at twice the output frequency. Because of the orientation of the diodes, the frequency component $2f_{\text{in}}/3$ is evoked in antiphase and circulate insides the idler circuit only. It does not flow to the output or back to the source.

Figure 4.29 shows the simulation of a parametric divider by $N = 3$. The voltage amplitude between the diode terminals has been represented versus the input voltage at ω_{in}. An oscillation at $\omega_a \cong \omega_{\text{in}}/3$ is generated from $E_{\text{in}} = 1.4$ V in a direct Hopf bifurcation. The oscillation frequency is very close to the divided-by-3 frequency, but the solution is actually quasiperiodic, with the two fundamental frequencies ω_{in} and ω_a. This solution has been represented by means of the diode voltage amplitude at the spectral line corresponding to the oscillation frequency

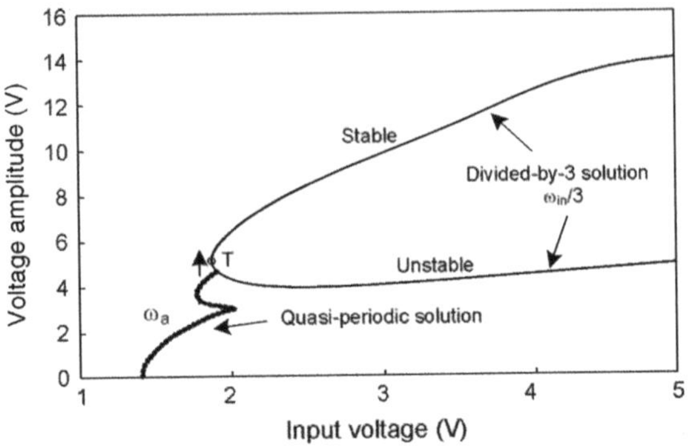

FIGURE 4.29 Variation in the solution of the frequency divider by 3 versus the input generator amplitude. The division occurs at a turning point of the periodic solution curve through synchronization of the oscillation for slightly lower amplitude. This oscillation gives rise to a quasiperiodic regime existing for a small input amplitude interval.

ω_a (the dashed line). The divided-by-3 solution is shown by the solid line. Division takes place through harmonic synchronization $3\omega_a = \omega_{in}$ at the turning point T of the divided-by-3 curve. This point is obtained for a slightly higher input amplitude than the Hopf bifurcation. Note that there is some inaccuracy about the turning point. The intermodulation spectrum in the nearly-synchronized (quasilocking) regime is very dense, and its frequency-domain analysis requires considering a large number of intermodulation products. See Fig. 3.26 as an example of the type of spectrum obtained in this operation mode. Here the same number of spectral lines has been considered along the entire curve. On the other hand, the synchronization is a local–global bifurcation giving rise to a discontinuous amplitude jump at the bifurcation point. The small jump takes place from the curve corresponding to the oscillation amplitude in the quasiperiodic regime to the turning point T of the curve in the divided-by-3 regime.

4.3.5 Phase Noise in Frequency Dividers

The objective of the phase noise analysis of frequency dividers presented in this section is to provide insight into the effect of the phase noise contributed by the input source at ω_{in} and the circuit noise sources on the output phase noise spectrum at ω_{in}/N. The amplitude noise introduced by the synchronizing source is neglected. Then the noise contributions will be the phase noise from this input source $\psi(t)$ and the different flicker and white noise sources contained in the divider circuit. To determine the phase noise spectrum of a divider by N, it is taken into account that according to Section 4.3.1.2, a phase shift $\psi(t)$ of the input generator at the input frequency ω_{in} gives rise to a phase shift $\psi(t)/N$ at the subharmonic frequency ω_{in}/N. The circuit will be described using the two-tier harmonic balance system (4.68). For a simple analytical derivation, the circuit noise contributions will be restricted to an equivalent white noise current source $I_N(t)$ at the divided frequency ω_{in}/N. Due to the small value of these perturbations, it will be possible to linearize the outer-tier equation $Y_s(V_s(t), \omega_s, I_{\text{in}}e^{j(\phi(t)+\psi(t))})V_s = I_N(t)$ about the particular frequency-divided solution $V_s, \omega_s, I_{\text{in}}e^{j\phi_s}$. Note that the frequency of the input generator is $\omega_{\text{in}} = N\omega_s$, so this generator is introduced at the Nth harmonic component in the system (4.68). Remember that the subharmonic voltage V_s at ω_s and the input generator $I_{\text{in}}e^{j\phi_s}$ at $N\omega_s$ are related through the inner tier of (4.68). Following steps similar to those in (4.39)–(4.56), it is easily shown that the output phase noise spectrum of a frequency divider by N is given by

$$|\Delta\phi_T(\Omega)|^2 = \frac{|\overline{Y}_{sV} \times \overline{Y}_{s\phi}|^2(|\psi(\Omega)|^2/N^2) + 2|\overline{Y}_{sV}|^2(|\overline{I}_N|^2/V_s^2)}{|\overline{Y}_{sV} \times \overline{Y}_{s\phi}|^2 + |\overline{Y}_{sV} \times \overline{Y}_{s\omega}|^2\Omega^2} \tag{4.78}$$

where the vectors $\overline{Y}_{sV}$, $\overline{Y}_{s\phi}$, and $\overline{Y}_{s\omega}$ are composed of the real and imaginary parts of the derivatives of the admittance function in (4.69) with respect to the variables V, ϕ, and ω, indicated by their corresponding subscripts. Note that the derivatives are calculated at the particular frequency-divided solution $V_s, \omega_s, I_{\text{in}}e^{j\phi_s}$. For the determination of these derivatives, the inner tier of the frequency-domain system must, of course, be taken into account.

As follows from (4.78), the structure of the phase noise spectrum of the frequency divider is similar to that of a fundamentally synchronized oscillator. Close to the carrier frequency, the phase noise spectrum approaches $|\psi(\Omega)|^2/N^2$. This output phase noise is maintained up to the corner frequency Ω_γ obtained when the two numerator terms become equal. From this corner frequency, the white noise contributions from the oscillator circuit become dominant. The second corner frequency $\Omega_{3\text{dB}}$ is obtained when the two denominator terms become equal. From $\Omega_{3\text{dB}}$, the divider is unable to track the fast noise perturbations of the oscillator circuit. Note that for $\Omega > \Omega_{3\text{dB}}$, the expression for the phase noise is similar to that corresponding to a free-running oscillator, with, of course, different values of the derivatives and oscillation amplitude.

As discussed earlier, the corner frequency $\Omega_{3\text{dB}}$ is inversely proportional to the magnitude of the derivative of the admittance function with respect to the frequency $|\overline{Y}_{s\omega}|$. This magnitude is usually smaller in parametric and regenerative dividers than in harmonic injection dividers. This is due to the higher frequency selectivity of harmonic injection dividers, based on an existing free-running oscillator with a pronounced frequency resonance. Therefore, the corner frequency $\Omega_{3\text{dB}}$ is usually higher in parametric and regenerative dividers, for which $|\overline{Y}_{s\omega}|$ generally takes smaller values. Remember that as shown in (4.48), the angle $\alpha_{sv} - \phi_s$ of $\overline{Y}_{sV} \times \overline{Y}_{s\phi}$ is about 90° for small $|\overline{Y}_{s\omega}|$.

Note that the phase noise spectrum in (4.78) refers to the common phase noise of the circuit variables, that is, the phase noise associated with time deviations (see Chapter 2). It does not take into account the phase and amplitude perturbation of the various harmonic components of the circuit variables. For a more detailed analysis of the divider phase noise, see an article by Rubiola et al. [27] describing the phase noise contributed by the various building blocks of a regenerative divider,

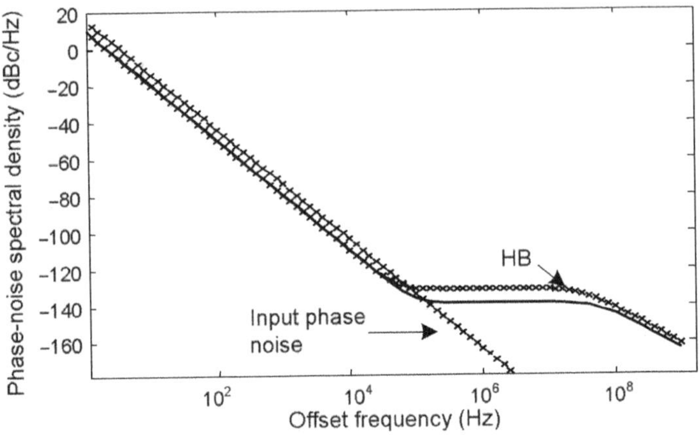

FIGURE 4.30 Phase noise spectral density of the circuit in Fig. 4.8 operating as a frequency divider by 2. The input voltage considered is $E_{in} = 0.1$ V. The results of (4.78), shown by the solid line, are compared with more accurate results obtained with a full harmonic balance simulation of the divider circuit (shown by crosses). The phase noise of the input source, with a slope of -30 dB/dec, is also represented.

or articles by Llopis et al. [28,29] analyzing the effect of the different nonlinearities and noise sources in a transistor-based divider.

As an example, Fig. 4.30 shows the application of expression (4.78) to calculation of the phase noise spectral density of the circuit in Fig. 4.8, operating as a frequency divider by 2. The considered input voltage is $v_{in} = 0.1$ V. The results of (4.78), shown by the solid line, are compared with more accurate results obtained with a full harmonic balance simulation of the divider circuit (shown by crosses). As can be seen, close to the carrier the output noise spectrum follows the input spectrum with a $20 \log N = 6$ dB reduction of the phase noise spectral density. Starting from the corner frequency f_y there is a flat region, and from the second corner frequency f_{3dB} there is a drop of -20 dB/dec of the spectral density, as in free-running conditions. The overall spectrum shows behavior similar to that of the fundamentally synchronized oscillator, with two different corners, which are well predicted by (4.78).

4.4 SUBHARMONICALLY AND ULTRASUBHARMONICALLY INJECTION-LOCKED OSCILLATORS

In a subharmonically injection-locked oscillator, the oscillation frequency gets locked to the mth harmonic of the input signal [7], which will require a sufficiently strong mth harmonic of this signal. Subsynchronization can be applied to improve the phase noise spectral density of a high-frequency oscillator. This is done by using the output of a lower-frequency oscillator, with low phase noise, as a synchronizing signal. The frequency of this oscillator will be on the order of ω_o/m, with ω_o

the free-running oscillation frequency, in the absence of input signal. Note that in synchronized operation the circuit output frequency is m times the frequency of the synchronizing signal $e_{in}(t) = E_{in} \cos \omega_{in} t$. The associated frequency multiplication gives rise to a multiplication of the phase perturbations $m\Delta\phi(t)$ and thus to an increase in the near-carrier phase noise spectral density in $20 \log m$ decibels with respect to the synchronizing signal; that is, $|\Delta\phi_{out}(\Omega)|^2 = |\Delta\phi_{in}(\Omega)|^2 + 20 \log m$. As for frequency dividers, this approximate relationship is true only at a relatively small frequency offset from the carrier. Thus, to obtain an actual phase noise reduction, the difference between the phase noise spectral density of the original oscillator and the synchronizing source must be larger than $20 \log m$ decibels. This is typically the case, as the higher the fundamental frequency of the oscillator, the lower the quality factor Q and the higher the phase noise spectral density. Thus, subsynchronization to a lower-frequency oscillator, with better spectral purity, enables a phase noise reduction in the higher-frequency oscillator.

As can be gathered, another possible application will be the multiplication by m of the frequency of the synchronizing source. Because the output (multiplied) frequency agrees with the oscillation frequency, the output power delivered will generally be higher than that obtained through standard frequency multiplication, using a transistor in nonlinear operation to generate the harmonic frequency $m\omega_{in}$ desired.

Next, an approximate analysis of the subsynchronized oscillator will be presented, from a system point of view. The analysis is similar to that used for harmonic injection dividers in Section 4.3.2 and based on the block diagram of Fig. 4.16. The system is assumed to exhibit a free-running oscillation at the frequency ω_o. Then an input signal is introduced at the frequency ω_{in}. Subsynchronization of the mth order is considered, so the circuit oscillates at $\omega_a = m\omega_{in}$. For this simplified analysis, the nonlinear element $i(v)$ will be represented in a power series as $i(v) = \sum_i \alpha_i v^i$. Assuming a high quality factor Q of the bandpass filter $H(\omega)$, centered at ω_o, the output signal of the nonlinear element can be obtained from $i(v) = i[V_{fb} \cos \omega_a t + E_{in} \cos(\omega_{in} t + \phi)]$, where the phase origin $\phi_o = 0$ is set at the harmonic component $m\omega_{in}$ of the feedback signal. For high Q it will be possible to limit the analysis to the terms of $i(v)$ at the subsynchronized oscillation frequency $\omega_a = m\omega_{in}$. These are generated from the intermodulation products $(-k \pm 1)\omega_a + km\omega_{in} = \pm\omega_a$. Considering only the positive frequencies, the different contributions to $\omega_a = m\omega_{in}$ can be expressed as $I_m = I_{1,0} + \sum_{k \neq 0} |I_{-k+1,km}| e^{jkm\phi}$. Then the system equation in the frequency domain is given by

$$V_{fb} - H(\omega_a) I_m = V_{fb} - \frac{H_o}{1 + j2Q[(\omega_a - \omega_o)/\omega_o]} \left(I_{1,0} + \sum_{k \neq 0} |I_{-k+1,km}| e^{jkm\phi} \right) = 0 \qquad (4.79)$$

Clearly, to get a phase relationship between the circuit oscillation and the input signal, the nonlinear element $i(v)$ must behave in a nonlinear manner with respect to $e_{in}(t)$, which will require a relatively large input power. The dominant terms in the summation will generally correspond to the lowest orders $k = 1; -1$; that is, $I_m = I_{1,0} + (I_{0,m} + I_{2,m})e^{jm\phi}$. From the trigonometric expansion of $i(v) = i[V_{fb} \cos \omega_a t + E_{in} \cos(\omega_{in} t + \phi)]$, Zhang et al. [7] have derived an expression of the form

$$I_m = A(E_{in}, V_{fb})V_{fb} + B(E_{in}, \phi, V_{fb})E_{in}^m \tag{4.80}$$

where the second term B can be seen as the response signal to E_{in}, sensitive to the input generator phase. Then the equation describing the closed-loop system is given by

$$V_{fb} - H(\omega_a)I_m = V_{fb} - \frac{H_o}{1 + j2Q[(\omega_a - \omega_o)/\omega_o]}[A(E_{in}, V_{fb})V_{fb} + B(E_{in}, \phi, V_{fb})E_{in}^m] = 0 \tag{4.81}$$

Splitting (4.81) into real and imaginary parts, Zhang et al. [7] demonstrated that the subharmonic injection-locking range increases with the second term of (4.80), representing the component of the system response that is sensitive to the input phase. The major application of the subsynchronized oscillators, which is the phase noise reduction of the original free-running oscillator, will be enabled by the phase relationship between the oscillation and the subharmonic injection source.

Note that synchronization can also occur at rational ratios $\omega_a/\omega_{in} = m/k$ between the oscillation frequency and the frequency of the synchronizing source. This is called *ultrasubharmonic synchronization*, in which the kth harmonic of the original oscillation gets locked to the mth harmonic of the input signal. Assuming that $m/k < 1$, it is easily shown that the solution will be periodic at the subharmonic frequency ω_{in}/k. However, the maximum output power will be obtained at the actual oscillation frequency, thus at the harmonic component $m\omega_{in}/k$. This harmonic component should be selected with the aid of a filter to obtain division by fractional order. For obvious reasons, the synchronization bandwidth is generally very narrow and will demand high levels of input power.

Figure 4.31 shows an Arnold tongue distribution in a general circuit. For each synchronization ratio $\omega_a/\omega_{in} = m/k$, a tongue, denoted $m : k$, is obtained in the plane defined by the input frequency and input power. Most tongues will have negligible width and will hardly be noticed in the measurements [29]. The main Arnold tongues are located about the harmonic components $k = 1, 2, 3, \ldots$ of the free-running oscillation frequency and correspond to the rational numbers $1 : k$, which implies fundamental harmonic synchronization (1 : 1) or frequency division ($1 : k, k \neq 1$). The subsynchronization tongues correspond to the rational numbers $m : 1$, and their width decreases quickly with the order m. Between two major tongues of the form $1 : k_o$ and $1 : (k_o + 1)$, the broadest Arnold tongue is the one corresponding to $2 : (2k_o + 1)$. In general, between two tongues with respective

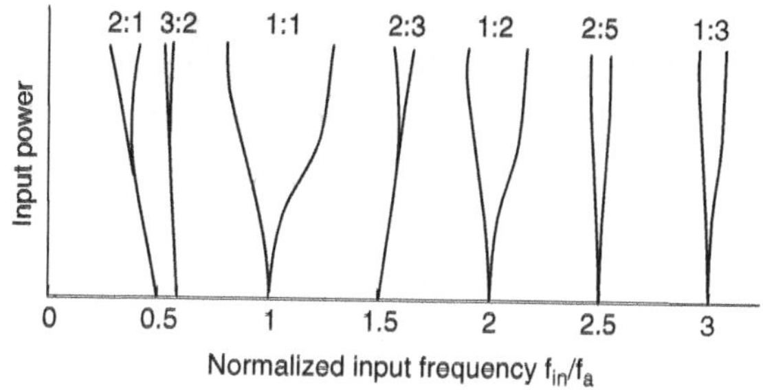

FIGURE 4.31 Arnold tongues in an injection-locked oscillator. They delimit the synchronization regions at different rational ratios $f_a/f_{in} = m/k$ between the oscillation and input frequencies.

ratios $m_1 : k_1$ and $m_2 : k_2$, the broadest tongue is the one corresponding to the ratio $(m_1 + m_2) : (k_1 + k_2)$.

For subsynchronization $\omega_a/\omega_{in} = m$ or ultrasubharmonic synchronization $\omega_a/\omega_{in} = m/k$, the circuit behavior is nonlinear with respect to the input generator, so the Arnold tongue bends with the input power. Therefore, the subsynchronization bandwidth is not centered about the free-running oscillation frequency (see the tongue 2 : 1 in Fig. 4.31). The bandwidth is negligible below a certain input power, and this is why these frequency divisions are rarely observed experimentally. For the analysis of these divisions, linearizations like the one discussed in Section 4.2.1 are not applicable.

To illustrate the behavior of subsynchronized oscillators, the parallel resonance oscillator of Fig. 1.1 with an input current source at about one-third of the oscillation frequency will be considered. To determine the input generator values providing a subsynchronized solution, the turning-point and Hopf loci are traced defined in the plane defined by the input frequency ω_{in} and input current I_{in} (Fig. 4.32). The

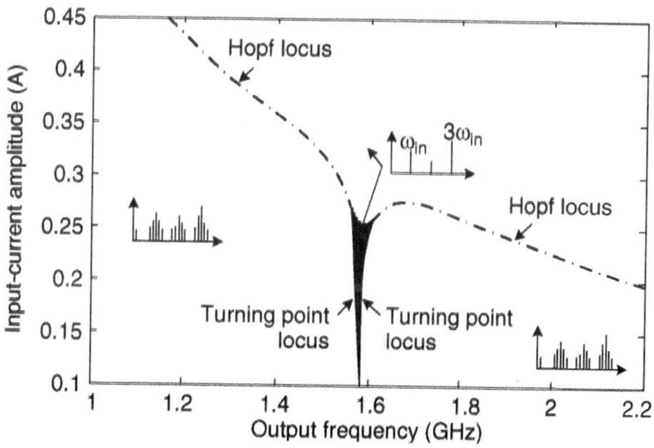

FIGURE 4.32 Bifurcation loci of a parallel resonance oscillator subsynchronized to an input source at about one-third of the oscillation frequency.

circuit exhibits a self-oscillation below the Hopf locus. The turning-point locus constitutes the Arnold tongue 3 : 1. Below the Hopf locus and outside the Arnold tongue, the circuit operates in the self-oscillating mixer regime. When entering the turning-point locus from this regime, the circuit oscillation synchronizes to the third harmonic of the input signal. Note that the input amplitude required for a noticeable synchronization bandwidth is quite high, in agreement with the previous discussion. Remember that the oscillation must synchronize to a harmonic component of the input signal, in this case to the third harmonic component. The Arnold tongue is very narrow. When increasing the input generator amplitude the oscillation is extinguished in an inverse Hopf bifurcation. However, despite the oscillation extinction, output power at $3\omega_{in}$ will still be obtained, due to the natural generation of this harmonic component of the input signal at ω_{in}. Furthermore, the input power is relatively high when the oscillation is extinguished, which justifies the significant harmonic amplitude at $3\omega_{in}$.

Next, evolution of the harmonic component $3\omega_{in}$ versus the input generator amplitude will be analyzed. Only the periodic solution with ω_{in} as fundamental has been considered in Fig. 4.33, where the voltage amplitude at $3\omega_{in}$ has been traced versus the input current. The input frequency considered is $f_{in} = 0.529$ GHz. The lower section of the curve is unstable. For low input amplitude it contains two complex-conjugate poles at about the free-running oscillation frequency ω_o, located on the right-hand side of the complex plane. As the input amplitude increases, the two complex-conjugate poles merge and split into two real poles on the right-hand side of the plane. One of the poles crosses the imaginary axis through zero at the turning point T_1, to the left-hand side of the complex plane. The section between T_1 and turning point T_2 is unstable, as the second real pole is still on the right-hand side of the complex plane. At T_2 this real pole crosses the imaginary axis to the left-hand side of the complex plane, so the upper section of the periodic curve, starting from T_2, is stable. The turning point T_2 is a synchronization point. For input power below T_2, the circuit operates in self-oscillating mixer regime, with two

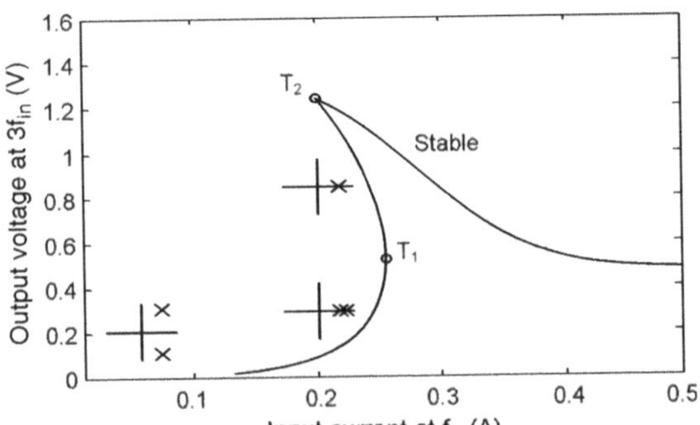

FIGURE 4.33 Evolution of the amplitude at $3\omega_{in}$ of the periodic solution of a subsynchronized parallel resonance oscillator. Only the upper section of the curve, starting from turning point T_2, corresponds to stable operation.

fundamental frequencies, the input frequency ω_{in} and the self-oscillation frequency ω_a. The shape of the curve in Fig. 4.33, providing evolution of the amplitude at $3\omega_{in}$ versus the input amplitude, is very meaningful. The stable section starts with an amplitude maximum coming from the synchronized oscillation. In plain words, the circuit oscillation gradually becomes less relevant versus the third harmonic of the input signal, so the amplitude at $3\omega_{in}$ decreases versus the input current. The reduction of this amplitude can also be attributed to the nonlinearity of the subsynchronized regime.

Figure 4.34 shows the family of synchronization curves versus the output frequency $3\omega_{in}$, obtained for increasing values of the input generator amplitude. In agreement with Fig. 4.32, the output amplitude decreases with the input signal. The Hopf bifurcation locus is also superimposed. Only the sections of the solution curves located above this locus correspond to stable behavior. For a small input current, the negligible synchronization band lies around the free-running solution, which is the point providing maximum voltage amplitude in Fig. 4.33. As the input power increases, the closed synchronization curve becomes noticeable. This curve coexists with an unstable curve at the same frequency $3\omega_{in}$, in which the circuit is not oscillating but simply responding to the input signal in a nonautonomous manner. This open solution curve is entirely unstable, as it lies below the Hopf locus. Its relatively high amplitude compared to an injection-locked oscillator at the fundamental frequency is due to the fact that the circuit must behave in a nonlinear regime with respect to the input source to achieve subharmonic synchronization, so high input power has been considered. Further increase in the input amplitude gives rise to wider synchronization curves with lower amplitude. At a certain amplitude value, the upper and lower curves merge. As already indicated, only the curve sections located above the Hopf locus correspond to stable behavior.

As another example, Fig. 4.35 demonstrates the phase noise reduction of an oscillator at 12 GHz by means of its ultrasubharmonic synchronization to a stable source at about 7.2 GHz. The ratio between the two frequencies is $f_a/f_{in} = 5/3$.

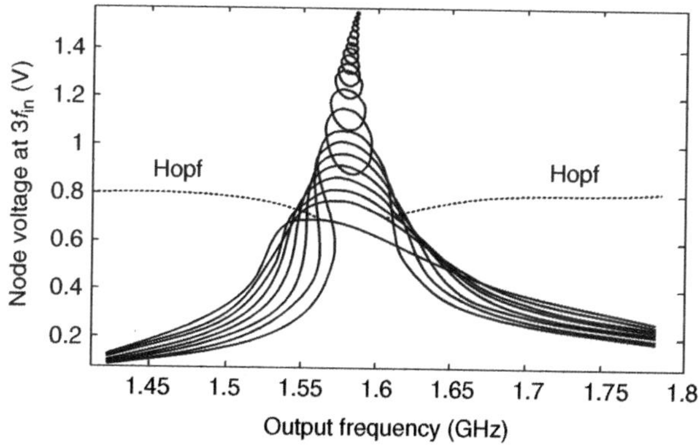

FIGURE 4.34 Evolution of subsynchronized solution curves versus the output frequency $3\omega_{in}$. Only the sections of the curves located above the Hopf locus correspond to stable behavior.

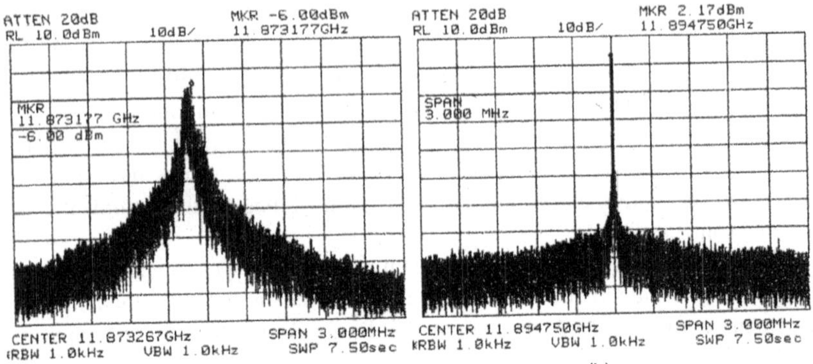

FIGURE 4.35 Ultrasubharmonic synchronization of a noisy oscillator at 12 GHz with a stable source at about $f_{in} = 7.2$ GHz. The synchronization ratio is $f_a/f_{in} = 5/3$. (a) Noisy spectrum prior to the synchronization. (b) Spectrum after synchronization with significant noise improvement.

Figure 4.35a shows the noisy spectrum of the free-running oscillator at $f_o = 11.87$ GHz. Figure 4.35b shows the oscillator spectrum after the ultrasubharmonic synchronization with significant noise reduction. Note that because of the bending of the Arnold tongue, the oscillator frequency after synchronization is not exactly the same as the one in free-running conditions.

4.5 SELF-OSCILLATING MIXERS

To obtain a self-oscillating mixer, an RF source is connected to an oscillator, avoiding oscillation synchronization to this source and oscillation extinction. The circuit operates in a quasiperiodic regime with two fundamental frequencies: one delivered by the input generator ω_{in} and one by the self-oscillation frequency ω_a. In the plane defined by ω_{in} and P_{in}, the circuit operates below the Hopf locus and outside the turning-point locus delimiting the synchronization region, which should be very narrow in this type of circuit. Advantage is taken of the mixing capabilities of the nonlinear device used to achieve frequency conversion. In down-conversion, the input frequency f_{in} mixes with the oscillation frequency f_a to provide the intermediate frequency $f_{IF} = |f_{in} - f_a|$. Thus, circuit self-oscillation plays the role of a local oscillator in standard mixers. The advantages of this type of circuit are the small size and low power consumption, since the same nonlinear device (a diode or transistor) behaves as a frequency mixer and sustains oscillation [8]. For good operation, the oscillation frequency must not be very sensitive to the input generator power and frequency. Otherwise, there will be undesired variations in the intermediate frequency $f_{IF} = f_{in} - f_a(P_{in}, f_{in})$, which is difficult for the designer to control. This can be solved by using a high-quality-factor resonator in the oscillator design [32] or by subsynchronizing the oscillation [33], which will totally prevent frequency shifts.

In a transistor-based self-oscillating mixer used as a down-converter, a lowpass filter is connected to the transistor output, which will generally correspond to the drain terminal. As an example, Fig. 4.36 shows the schematic of a self-oscillating mixer providing frequency down-conversion from $f_{in} = 5.5$ GHz to $f_{IF} = 0.5$ GHz. For the oscillator design, series feedback has been introduced at the source terminal, and the input network connected to the gate terminal is calculated so as to provide input matching at the RF frequency and to enable fulfillment of the oscillation condition at the required frequency $f_o = 5$ GHz. Both the oscillation and the signal delivered with the input generator will constitute transistor inputs. The nonlinear drain current will enable mixing of the two signals and provide an intermediate frequency $f_{IF} = f_{in} - f_a$ selected through the lowpass filter. The open-ended $\lambda/4$ transmission line at about the oscillation frequency enhances isolation of the higher-frequency components. To reduce oscillation frequency variations versus the input generator frequency or power, a dielectric resonator can be added to the input circuit.

Figure 4.37 shows the Hopf bifurcation locus delimiting the values of input power and frequency for self-oscillating mixer operation. The synchronization locus

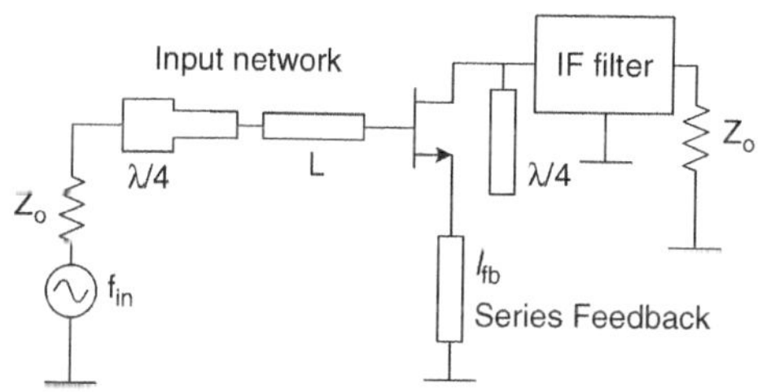

FIGURE 4.36 Self-oscillating mixer from frequency down-conversion $f_{in} = 5.5$ GHz to $f_{IF} = 0.5$ GHz. The higher oscillation amplitude is obtained at the gate port. The intermediate frequency is selected with a lowpass filter from the drain terminal.

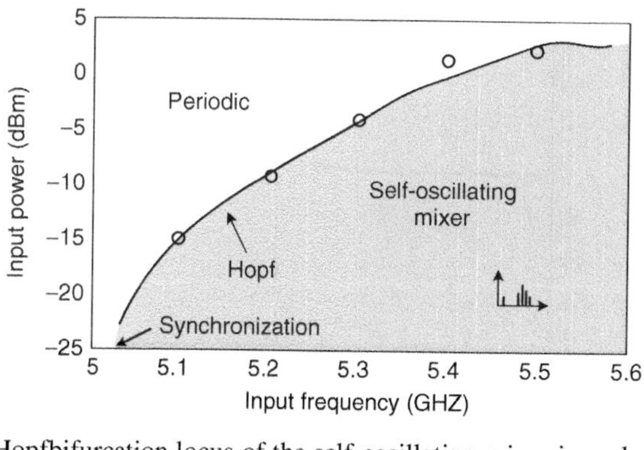

FIGURE 4.37 Hopfbifurcation locus of the self-oscillating mixer in a plane defined by the input frequency and power. The synchronization locus is obtained for very low input power, so it is not represented in the figure.

is obtained for very low input power values and is indicated in the figure simply by an arrow. Figure 4.37 shows the evolution of the conversion gain versus input power for a constant input frequency $f_{in} = 5.5$ GHz. As in a standard mixer, the conversion gain keeps constant for low input power because the oscillator circuit behaves linearly with respect to the RF source. From a certain input power, the nonlinear effects become apparent. The 1-dB gain compression point is obtained for $P_{in} = -3$ dBm. For slightly higher input power, the self-oscillation vanishes in an inverse Hopf bifurcation, in agreement with Fig. 4.38. Because no dielectric resonator has been used in the design, the oscillation frequency will vary with the input power and frequency affecting the intermediate frequency. These variations will, of course, be larger for higher input power. Figure 4.39 shows the oscillation

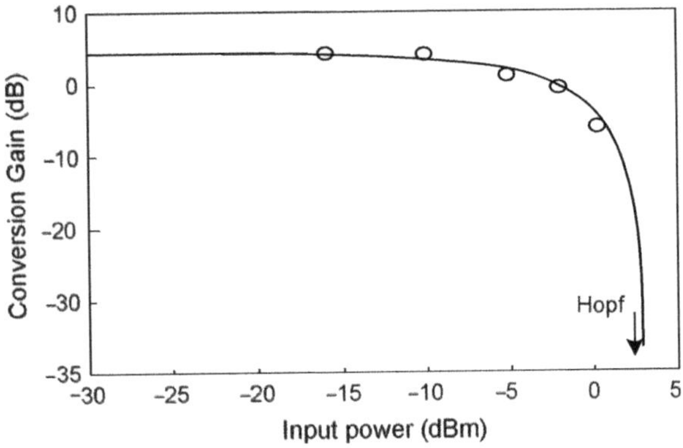

FIGURE 4.38 Variation in the conversion gain of a self-oscillating mixer versus the input power for constant input frequency $f_{in} = 5.5$ GHz. The 1-dB gain compression point is obtained for $P_{in} = -3$ dBm. The oscillation is extinguished for slightly higher power in an inverse Hopf bifurcation.

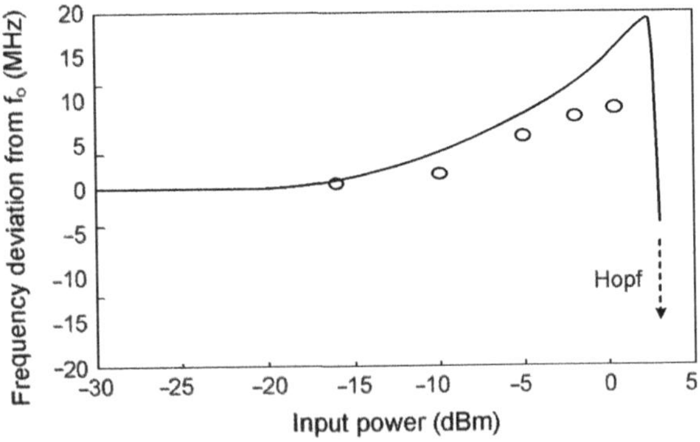

FIGURE 4.39 Deviations in the oscillation frequency with respect to the desired value $f_o = 5$ GHz versus the input power. As expected, the deviations increase with the input power.

frequency deviations with respect to the desired value $f_a = 5$ GHz when the input power increases.

REFERENCES

[1] J. Guckenheimer and P. Holmes, *Nonlinear Oscillations, Dynamic Systems, and Bifurcations of Vector Fields*, Springer-Verlag, New York, 1983.

[2] M. Tofighi and A. S. Daryoush, An IC based self-oscillating mixer for telecommunications, *IEEE Radio and Wireless Conference (RAWCON)*, Atlanta, GA, pp. 331–334, 2004.

[3] M. K. Kazimierczuk, V. G. Krizhanovski, J. V. Rassokhina, and D. V. Chernov, Injection-locked class-E oscillator, *IEEE Trans Circuits Syst. I Regul Pap.*, vol. 53, pp. 1214–1222, 2006.

[4] H. Grubinger, G. Von Buren, H. Barth, and R. Vahldieck, Continuous tunable phase shifter based on injection locked local oscillators at 30 GHz, *IEEE MTT-S International Microwave Symposium Digest*, pp. 1821–1824, 2006.

[5] R. Quéré, E. Ngoya, M. Camiade, A. Suárez, M. Hessane, and J. Obregón, Large signal design of broadband monolithic microwave frequency dividers and phase-locked oscillators, *IEEE Trans. Microwave Theory Tech.*, vol. 41, pp. 1928–1938, Nov. 1993.

[6] F. Giannini and G. Leuzzi, *Nonlinear Microwave Circuit Design*, Wiley, Hoboken, NJ, 2004.

[7] X. Zhang, X. Zhou, and A. S. Daryoush, A theoretical and experimental study of the noise behavior of subharmonically injection locked local oscillators, *IEEE Trans. Microwave Theory Tech.*, vol. 40, pp. 895–902, 1992.

[8] X. Zhou and A. S. Daryoush, Efficient self-oscillating mixer for communications, *IEEE Trans. Microwave Theory Tech.*, vol. 42, pp. 1858–1862, 1994.

[9] A. Safarian, S. Anand, and P. Heydari, On the dynamics of regenerative frequency dividers, *IEEE Trans. Circuits Syst. II Express Briefs*, vol. 53, pp. 1413–1417, 2006.

[10] S. Jeon, A. Suárez, and D. B. Rutledge, Global stability analysis and stabilization of a class-E/F amplifier with a distributed active transformer, *IEEE Trans. Microwave Theory Tech.*, vol. 53, pp. 3712–3722, 2005.

[11] R. Adler, A study of locking phenomena in oscillators, *Proc. IEEE*, vol. 61, pp. 1380–1385, Oct. 1973.

[12] B. Razavi, A study of injection locking and pulling in oscillators, *IEEE J Solid State Circuits*, vol. 39, pp. 1415–1424, 2004.

[13] K. Kurokawa, Injection locking of microwave solid state oscillators, *Proc. IEEE*, vol. 61, pp. 1386–1410, Oct. 1973.

[14] L. Gustafsson, G. H. Bertil Hansson, and K. I. Lundstrom, On the use of describing functions in the study of nonlinear active microwave circuits, *IEEE Trans. Microwave Theory Tech.*, vol. 20, pp. 402–409, 1972.

[15] K. Kurokawa, Some basic characteristics of broadband negative resistance oscillators, *Bell Syst. Tech. J.*, vol. 48, pp. 1937–1955, July–Aug. 1969.

[16] J. M. T. Thompson and H. B. Stewart, *Nonlinear Dynamics and Chaos*, 2nd ed., Wiley, Chichester, UK, 2002.

[17] P. Dorta and J. Perez, On the design of MESFET harmonic injection frequency dividers using the harmonic balance technique, *20th European Microwave Conference*, Budapest, Hungary, pp. 1730–1735, 1990.

[18] J. Perez, P. Dorta, A. Trueba, and F. Sierra, Application of harmonic injection dividers to frequency synthesizers in millimeter band, *Proc. of Meditteranean Electrotechnical Conference (MELECON '87)*, Rome, Italy, pp. 361–364, 1987.

[19] H. R. Rategh and T. H. Lee, Superharmonic injection locked oscillators as low power frequency dividers, *IEEE Symposium on VLSI Circuits*, Honolulu, HI, pp. 132–137, 1998.

[20] H. P. Moyer and A. S. Daryoush, Unified analytical model and experimental validations of injection-locking processes, *IEEE Trans. Microwave Theory Tech.*, vol. 48, pp. 493–499, 2000.

[21] F. Ramirez, E. de Cos, and A. Suárez, Nonlinear analysis tools for the optimized design of harmonic-injection dividers, *IEEE Trans. Microwave Theory Tech.*, vol. 51, June 2003.

[22] J. Jugo, J. Portilla, A. Anakabe, A. Suárez, and J. M. Collantes, Closed-loop stability analysis of microwave amplifiers, *IEE Electron. Lett.*, vol. 37, pp. 226–228, Feb. 2001.

[23] C. Rauscher, 16 GHz GaAs FET frequency divider, IEEE MTT-S International Microwave Symposium, Boston, MA, pp. 349–351, 1983.

[24] V. Manassewitsch, *Frequency Synthesizers: Theory and Design*, Wiley, New York, 1987.

[25] R. E. Collin, *Foundations for Microwave Engineering*, 2nd ed, Wiley, New York, 2001.

[26] A. Suárez and R. Melville, Simulation-assisted design and analysis of varactor-based frequency multipliers and dividers, *IEEE Trans. Microwave Theory Tech.*, vol. 54, pp. 1166–1179, 2006.

[27] E. Rubiola, M. Olivier, and J. Groslambert, Phase noise in the regenerative frequency dividers, *IEEE Trans. Instrum. Meas.*, vol. 41, pp. 353–360, 1992.

[28] O. Llopis, H. Amine, M. Gayral, J. Graffeuil, and J. F. Sautereau, Analytical model of noise in an analog frequency divider, *IEEE MTT-S International Microwave Symposium*, Atlanta, GA, pp. 1033–1036, 1993.

[29] O. Llopis, M. Regis, S. Desgrez, and J. Graffeuil, Phase noise performance of microwave analog frequency dividers application to the characterization of oscillators up to the MM-wave range, *IEEE MTT-S International Microwave Symposium*, Pasadena, CA, pp. 550–554, 1998.

[30] X. Zhang, X. Zhou and A. S. Daryoush, "A theoretical and experimental study of the noise behavior of subharmonically injection locked local oscillators," *IEEE Trans. Microwave Theory Tech.*, vol. 40, pp. 895–902, 1992.

[31] G. Iooss and D. D. Joseph, *Elementary Stability and Bifurcation Theory*, 2nd, ed., Springer-Verlag, New York, 1990.

[32] C. Tsironis, R. Stahlmann, and F. Ponse, Self-oscillating dual gate MESFET X-band mixer with 12dB conversion gain, *Proceedings of the European Microwave Conference*, 1979, pp. 321–325.

[33] X. Zhou, X. Zhang, and A. S. Daryoush, Phase controlled self-oscillating mixer, *IEEE MTT-S International Microwave Symposium*, San Diego, CA, pp. 749–752, 1994.

CHAPTER FIVE

Nonlinear Circuit Simulation

5.1 INTRODUCTION

To meet the high performance requirements of modern communication systems, accurate and efficient design tools are necessary. The design has added difficulties in the case of nonlinear circuits, in which the superposition principle does not hold, so their response depends on the input amplitude and there is a natural generation of harmonic frequencies [1,2]. The nonlinear circuits are also capable of exhibiting a self-sustained oscillation, which may be desired, as in the case of free-running oscillators or frequency dividers, or undesired, as in the case of power amplifiers and frequency multipliers. The oscillatory solution generally coexists with a mathematical solution for which the circuit does not oscillate, as has been shown in Chapters 3 and 4. Thus, the coexistence of steady-state solutions for the same values of the circuit elements is very common in nonlinear circuits. Only stable solutions are physically observable, so the stability analysis of the obtained steady-state solution is essential.

Different methods exist for the simulation of nonlinear circuits. The choice of one or another will depend on the type of circuit, lumped or distributed, with few or many active devices, and on its operational conditions: bias point and quality factor, for example, or the nature of the solution (e.g., periodic, quasiperiodic, modulated). In some simulation methods, the peculiarities of the autonomous solutions give rise to additional difficulties, so complementary techniques are necessary. The methods may be classified globally into analytical and numerical methods. Analytical methods such as the describing function [2,3] or Volterra series [4,5] are very well suited for circuit design since they provide insight into nonlinear behavior and enable an evaluation of its dependence on the circuit parameters. The describing function is widely used for oscillator design. Most of the analyses in previous chapters were based on this technique. It assumes a sinusoidal steady-state solution of the nonlinear circuit. Thus, its accuracy depends on the quality factor of this circuit. On the other hand, the Volterra series is a good approach for identifying the linearity limiting factor of a given transistor technology. It is also very well suited for the analysis of multitone signals and nonlinearities with memory. However, when the goal is to obtain an accurate solution, in terms of waveforms and spectral content, numerical iterative methods are generally preferred. These numerical simulation methods are the object of this chapter.

The numerical simulation methods can be classified into three main categories: time domain, frequency domain, and the more recent mixed time–frequency methods. In time-domain integration, the nonlinear circuit is described by a set of differential algebraic equations (DAEs) [6]. The nonlinear circuit is simulated by discretizing the time variable and applying a particular integration algorithm to the original DAE. This transforms the continuous system of DAEs into an algebraic system of nonlinear equations, depending on the discrete time samples of the circuit variables. The system is integrated from an initial condition at a constant or variable time step [7,8]. This method, used in programs such as SPICE [9], provides the entire evolution of the circuit solution from initial values to the steady state. Both the transient and steady states are simulated. However, the transient state, which usually has little interest for the designer, may be too long compared with the solution period. To cope with this problem, fast time-domain algorithms [7] such as the shooting [10] and finite-difference methods [11,12] perform time-domain analysis of the steady state only, avoiding the transient state. This is achieved through an additional constraint on the state of the solution. In the case of periodic regimes, the constraint imposes the equality of the circuit variables after one period. One advantage of the fast time-domain methods is their ability to simulate steady-state waveforms with sharp time transitions, which correspond to high harmonic content. Fast time-domain methods are more difficult to apply to quasiperiodic regimes [13].

Distributed elements such as transmission lines, stubs, coupled lines, or rings are often used in microwave circuit design. These elements, exhibiting loss and frequency dispersion, are difficult to model and analyze in the time domain. The most general approaches are based on a numerical calculation of the impulse response from the inverse Fourier transform of their transfer functions. It is also possible to use a Taylor series expansion of the transfer function in the Laplace domain, which is matched by a complex rational function in terms of pole–residue pairs [12,14]. The inverse Laplace transform of this type of function can be calculated analytically in a very simple manner and provides the impulse response associated with a particular distributed element. The distributed elements can be incorporated in differential equations by means of convolution products, requiring time-domain integration of the circuit equations from the initial time value. From an initial description in the Laplace domain, it is also possible to obtain a set of linear differential equations that describe the distributed element. These equations are combined through Kirchhoff's laws with the differential equation system accounting for the lumped section of the circuit.

In fact, the linear elements are more easily described in the frequency domain, since it is generally simpler to obtain their response using phasor analysis. However, the nonlinearities contained in transistors or diodes are naturally described in the time domain by means of their constitutive functions. These functions provide an instantaneous relationship between the particular nonlinear current, charge, or flux and its control voltages or currents. Examples of nonlinear elements are the voltage-controlled current $i(v)$ and junction capacitance $c_j(v)$ of a Schottky diode or the field-effect current $i_{ds}(v_{gs}, v_{ds})$ of a FET transistor, which is controlled by the gate-to-source voltage $v_{gs}(t)$ and drain-to-source voltage $v_{ds}(t)$. Taking these facts into account, the harmonic balance method [15,16] uses frequency-domain representation for the linear elements, lumped or distributed, maintaining the time-domain descriptions for the nonlinear devices. The circuit variables are represented by means of a Fourier series, with one or more fundamental frequencies. Because of this representation, only the steady state is simulated. On the other hand, use of a sinusoidal basis for the expression of circuit signals restricts the applicability of the method to circuits with relatively mild nonlinearities. Regimes with fast time transitions are better analyzed in the time domain using the shooting or finite-difference methods.

As shown in previous chapters, only stable steady-state solutions are observed physically. In the case of time-domain integration, provided that the integration step and algorithm are selected properly, the steady-state solution obtained will be stable. This is due to the fact that the integration process follows the actual time evolution of the circuit solution (transient) up to the steady state. If the initial value is close to a solution with unstable poles, the time-domain integration will initially follow an exponential transient, governed by the unstable dominant poles. Then the amplitude growth will progressively slow down until the system reaches a different, stable steady state. When using methods providing only steady-state solutions such as harmonic balance, it will be possible to obtain an unstable steady-state solution to which the circuit never evolves and is therefore never observed in practice. This situation is often faced in circuits such as power amplifiers. The designer simulates the periodic amplifier solution desired, which is actually unstable, and obtains a mixerlike spectrum in the measurement [17]. This stable solution is due to the mixing of the signal delivered by the input generator with self-oscillation. As can be gathered, verification of the solution stability will be essential for steady-state analysis methods, such as harmonic balance.

Due to the Fourier series expression required for the circuit variables, the harmonic balance technique cannot be applied to nonlinear circuits with modulated inputs. On the other hand, for a carrier frequency with much higher value than the modulation bandwidth, the time-domain simulation may be inefficient or even impossible. In most of these circuits, two different time scales may be distinguished: one associated with the modulation signal (slower time scale) and the other associated with the high-frequency carrier (faster time scale). Because the circuit is periodic with respect to the faster time scale, it will be possible to express the circuit variables in a harmonic series, with time-varying harmonic components, at the slower time scale. This is done in the envelope transient method [18–21]. The harmonic components become the unknowns of a system of nonlinear differential algebraic equations. The advantage over the time-domain integration is its lower computational cost, since the equation system is integrated at the time step of the slower time scale.

The objective of this chapter is to introduce and compare the various techniques for the numerical simulation of nonlinear circuits, showing the advantages and shortcomings of each of them when applied to different types of circuits and regimes. Special emphasis is placed on the simulation techniques for autonomous circuits, which are the focus in this book.

5.2 TIME-DOMAIN INTEGRATION

A key aspect for successful simulation of a nonlinear circuit is accurate modeling of the nonlinear devices used. Transistors and diodes are usually described through lumped electrical models, containing linear and nonlinear elements. The nonlinear elements are described through their constitutive relationships, which relate the instantaneous value of the nonlinear magnitude, current, charge, and flux with those of its control variables. An example is the well-known diode current model $i(t) = I_o(e^{\alpha v(t)} - 1)$, having $v(t)$ as control variable and I_o and α as parameters. For nonlinear capacitances, the model is usually written in terms of the associated nonlinear charge $q_j(v)$. For a junction capacitance, for example, the nonlinear charge is given by $q_j(v) = -2C_{jo}\phi\sqrt{1 - v/\phi}$, with ϕ the built-in potential and C_{jo} the capacitance for $v = 0$. For more complex nonlinearities, such as some of those contained in FET or bipolar transistors, great research effort has been necessary. Some of the most commonly used models can be found in books by Anholt [22] and Golio [23].

The most practical way to perform time-domain analysis of a nonlinear circuit is to formulate the circuit as a system of differential algebraic equations (DAEs). The case of a nonlinear circuit containing lumped elements only will be considered initially. In the nodal approach, the nonlinear system is derived by equating to zero the total current flowing into each node. The different node voltages constitute the system unknowns. This method has difficulty dealing with voltage sources, and the voltage drop at the various elements is not directly available. To cope with these problems, the modified nodal approach (MNA) is used instead. In this approach the equations are written in terms of both node voltages and inductance currents. The system of DAEs [7] describing a nonlinear circuit containing lumped elements is given by

$$\frac{d\overline{q}}{dt}(\overline{x}(t)) + \overline{f}(\overline{x}(t)) + \overline{g}(t) = 0 \tag{5.1}$$

where $\overline{q} \in R^P$ is a vector containing linear and nonlinear charges and fluxes of the circuit, $\overline{x} \in R^P$ is a vector of node voltages and inductance currents, $\overline{f} \in R^P$

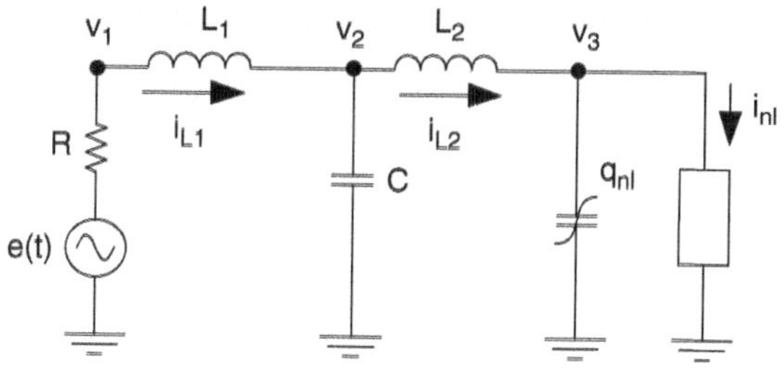

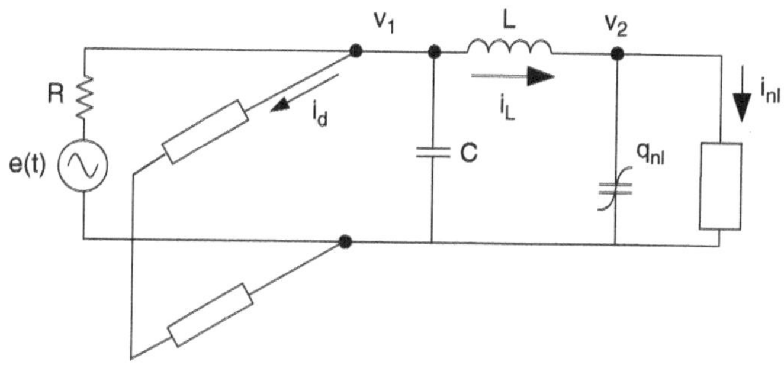

FIGURE 5.1 Circuits described as a system of nonlinear DAEs: (a) lumped circuit; (b) circuit containing a transmission line.

is a vector of sums of resistive currents (that enter each node) and loop voltages, and $\overline{g}(t) \in R^P$ is a vector that includes the input generators. Whenever the relationship $\overline{q}(\overline{v})$ is invertible, it will be convenient to use $\overline{q}$ as an unknown, since this ensures charge conservation [7]. As an example of how to describe a lumped-element circuit as a system of DAEs, the circuit of Fig. 5.1a is considered, The corresponding vector $\overline{x}$ consists of the two inductor currents i_{L1} and i_{L2} and the two node voltages v_2 and v_3; that is, $\overline{x} = [i_{L1} \quad i_{L2} \quad v_2 \quad v_3]$. The other vectors appearing in (5.1) are given by

$$\frac{d\overline{q}}{dt} = \begin{bmatrix} -L_1\dfrac{di_{L1}}{dt} \\ -L_2\dfrac{di_{L2}}{dt} \\ -C\dfrac{dv_2}{dt} \\ -\dfrac{dq_{nl}}{dt} \end{bmatrix} \qquad \overline{f}(\overline{x}) = \begin{bmatrix} -Ri_{L1} - v_2 \\ v_2 - v_3 \\ i_{L1} - i_{L2} \\ i_{L2} - i_{nl}(v_3) \end{bmatrix} \qquad \overline{g}(t) = \begin{bmatrix} e(t) \\ 0 \\ 0 \\ 0 \end{bmatrix} \tag{5.2}$$

Techniques to resolve the system of DAEs (5.1) are introduced later in the chapter.

5.2.1 Time-Domain Modeling of Distributed Elements

The aim of this subsection is just to provide the reader with some basic understanding of the time-domain simulation of distributed elements. For detailed explanations the reader should check [24–33]. The distributed elements are originally described through partial differential equations. One fundamental example is the telegrapher's equation describing a transmission line [24]:

$$\begin{aligned}\frac{\partial}{\partial x}v(x,t) &= -Ri(x,t) - L\frac{\partial}{\partial t}i(x,t)\\ \frac{\partial}{\partial x}i(x,t) &= -Gv(x,t) - C\frac{\partial}{\partial t}v(x,t)\end{aligned} \tag{5.3}$$

where x is the longitudinal coordinate and R, L, G, and C are the resistance, inductance, conductance, and capacitance per unit length, respectively. Our objective is to transform the partial differential equation system (5.3) into a system of ordinary equations. For the simplest case of a transmission line, several approaches have been proposed based on the discretization of (5.3) with respect to the longitudinal variable x. The line is divided into segments of length Δx, which must be a small fraction of the line length. In one possible approach, each segment has a lumped-network equivalent in terms of per-unit-length magnitudes, with element values $R\Delta x$, $L\Delta x$, $G\Delta x$, and $C\Delta x$. The problem is that an extremely high number of line segments might be necessary. In addition, this technique cannot model the frequency-dependent parameters directly. Other techniques, of more general application, have been proposed to incorporate distributed structures into time-domain equation systems. There are two primary approaches. The first is based on the generation of reduced-order models of the distributed elements in the Laplace domain, which are easily transformed to the time domain [25–27]. The second approach is based on computation of the impulse response of the distributed element, applying convolution at each time step to obtain the time-domain response of this element [28,29]. The two approaches, described briefly below, have advantages and limitations.

5.2.1.1 Reduced-Order Models Applying the Laplace transform, the partial differential equation system (5.3) is transformed into a system containing derivatives with respect to spatial coordinates only:

$$\begin{aligned}\frac{\partial}{\partial x}V(x,s) &= -RI(x,s) - LsI(x,s)\\ \frac{\partial}{\partial x}I(x,s) &= -GV(x,s) - CsV(x,s)\end{aligned} \tag{5.4}$$

Using the boundary conditions at $x = 0$, it is possible to integrate (5.4) with respect to x, which leads to the exponential relationship

$$\begin{bmatrix} V(d,s) \\ I(d,s) \end{bmatrix} = [e^{A+sB}] \begin{bmatrix} V(0,s) \\ I(0,s) \end{bmatrix} \tag{5.5}$$

where d is the line length and A and B are matrixes, given by

$$A = \begin{bmatrix} 0 & -R \\ -G & 0 \end{bmatrix} \qquad B = \begin{bmatrix} 0 & -L \\ -C & 0 \end{bmatrix} \tag{5.6}$$

We can easily transform the hybrid parameters in (5.5) into y-parameters, considering $x = 0$ as port 1 and $x = d$ as port 2. From knowledge of the terminal impedance at $x = 0$ it is possible to obtain the input admittance $Y(s)$ in the Laplace domain at $x = d$.

Generalizing the approach above, a distributed network will be described by the N-port admittance matrix

$$\begin{bmatrix} Y_{11}(s) & \cdots & Y_{1N}(s) \\ \vdots & \vdots & \vdots \\ Y_{N1}(s) & \cdots & Y_{NN}(s) \end{bmatrix} \begin{bmatrix} V_1(s) \\ \vdots \\ V_N(s) \end{bmatrix} = \begin{bmatrix} I_1(s) \\ \vdots \\ I_N(s) \end{bmatrix} \tag{5.7}$$

To obtain a time-domain description of the distributed element, the various components of the admittance matrix will be represented as complex rational functions in terms of pole–residue pairs. To obtain this description, each component Y_{ij} is expanded in a Taylor series about $s = 0$ [26]. This provides the *moments* of the Y_{ij} response. For simplicity, a one-port element is considered:

$$Y(s) \cong Y_r(s) = Y(0) + s\frac{dY(0)}{ds} + \frac{s^2}{2!}\frac{d^2Y(0)}{ds^2} + \cdots + \frac{s^M}{M!}\frac{d^MY(0)}{ds^M} \tag{5.8}$$

Note that the Taylor series expansion is limited to M order. Consistent with this, the subscript r in $Y_r(s)$ indicates that the M-order expansion is a reduced-order model of $Y(s)$. The coefficients $m_k = (1/k!)[d^kY(0)/ds^k]$ agree with the time-domain moments of the impulse response associated with $Y(s)$. Once the representation (5.8) has been obtained for the given function $Y(s)$, the next step will be to match this representation with a complex rational function [30]:

$$m_0 + m_1s + m_2s^2 + \cdots + m_Ms^M = \frac{a_0 + a_1s + a_2s^2 + \cdots + a_{M_1}s^{M_1}}{b_0 + b_1s + b_2s^2 + \cdots + b_{M_2}s^{M_2}} = \frac{P^{M1}(s)}{Q^{M2}(s)} \tag{5.9}$$

where $M_1 + M_2 = M$. The coefficients $a_1, \ldots, a_{M1}$ and $b_1, \ldots, b_{M2}$ are computed in terms of the known moments m_1 to m_M in two different steps. The b coefficients are obtained by cross-multiplying the left-hand side of (5.9) by the denominator of the rational function and equating the coefficients of powers of s from s^{M_1+1} to s^M.

In turn, the a coefficients are obtained by equating powers of s from s^0 to s^{M1}. This moment-matching technique is also known as a *Padé-based approximation* [25].

Once we have a representation of $Y(s)$ as a quotient of polynomials (5.9), it is possible to obtain the poles p_i of $Y(s)$ by applying a root-solving algorithm to $Q^{M2}(s)$. After determination of these poles, our objective is to find a pole–residue representation of the form

$$Y_r(s) = \alpha + \sum_{i=1}^{M_2} \frac{k_i}{s - p_i} \tag{5.10}$$

where α is the coupling factor and (p_i, k_i) are the pole–residue pairs. The residues k_i are calculated by equating the moment expansion $m_0 + m_1 s + m_2 s^2 + \cdots + m_M s^M$ to the Maclaurin series [30], which is a polynomial whose coefficients depend on the poles and residues of the rational function (5.9):

$$m_0 + m_1 s + m_2 s^2 + \cdots + m_M s^M = \alpha - \sum_{n=0}^{\infty} s^n \left(\sum_{i=1}^{M_1} \frac{k_i}{p_i^{n+1}} \right) \tag{5.11}$$

The residues k_i are computed by equating equal powers of s. The advantage of the pole–residue model is that it can be translated analytically to the time domain through calculation of the inverse Laplace transform of (5.10). This inverse transform provides the impulse response associated with the distributed element directly:

$$h(t) = \alpha\delta(t) + \sum_{i=1}^{M_2} k_i e^{p_i t} \tag{5.12}$$

Because of the exponential form of this impulse response, this method is called *asymptotic waveform evaluation* (AWE) [14,31,32]. Equation (5.12) describes the distributed element in terms of its impulse response. Integration of the impulse response into the system of nonlinear DAEs will require a convolution operation. However, a different analysis strategy is also possible, which consists of describing the distributed elements with a subset of linear differential equations. Integration of this system into the nonlinear system of DAEs will allow a *unified transient simulation* [30]. The differential equations describing the distributed element are obtained from the pole–residue model. A simple example is considered in the following. Assuming a single pole, we can write:

$$\left(\alpha + \frac{k}{s - p} \right) V(s) = I(s) \tag{5.13}$$

Performing the variable change

$$\frac{V(s)}{s-p} = Z(s) \tag{5.14}$$

and taking into account that $sZ(s) \leftrightarrow \dot{z}(t)$, it will be possible to write in the time domain,

$$\begin{aligned} \dot{z} &= v + pz \\ i &= \alpha v + kz \end{aligned} \tag{5.15}$$

It must be noted that the z is an implicit variable of system (5.15). However, this system is well balanced since the distributed element is linked to the rest of the circuit through common nodes, so either the current i or voltage v will be an input to (5.15).

The system obtained from the rational function (5.11) will be unstable if any poles of the pole–residue rational function used are located on the right-hand side of the complex plane. These unstable poles must be removed from the model. Besides the stability, the time-domain models of the distributed elements must fulfill two other essential characteristics: causality and passivity [27,33]. A *causal function* is nonanticipating; that is, it does not depend on future input. *Passive* means that the network does not generate more energy than it absorbs and cannot become unstable for any termination.

From (5.8) to (5.15) we have been dealing with a single-admittance function. In the case of a general network containing transmission lines, the moments would be obtained from the Taylor series expansion of an exponential matrix of the form $\exp(A + sB)$. Ill-conditioning may occur for relatively high orders of this expansion. As indicated by Achar and Nakhla [30], the number of accurate poles that can be extracted with this technique is generally less than 10. The situation is more complex in the general case of a multiport linear network such as (5.7). To cope with the accuracy problems, we can use moment-matching to multiple expansion points (complex frequency hopping). Once the dominant poles of Y_{ij} have been obtained, the residues are determined by taking for M data points, s_m, H_m, with $m = 1 \ldots M$ and solving a system of M linear equations in the form (5.10). A different class of algorithms, based on indirect moment-matching techniques, such as model reduction based on Krilov subspace techniques, can be more efficient [30].

Once we have obtained models of the form (5.10), in a general manner, a nonlinear circuit containing distributed elements can be modeled as

$$\begin{aligned} &\frac{d\overline{q}}{dt}(\overline{x}(t)) + \overline{f}(\overline{x}(t)) + [\Lambda]\overline{i}_d + \overline{g}(t) = 0 \\ &\left.\begin{aligned} \frac{d\overline{z}}{dt} - [p]\overline{z} - [d]\overline{v}_d &= 0 \\ \overline{i}_d - [k]\overline{z} - [\alpha]\overline{v}_d &= 0 \end{aligned}\right\} \text{distributed} \end{aligned} \tag{5.16}$$

where dimensions of the vector $\overline{z}$ depend on the number of ports and poles considered in the modeling of the distributed elements and the constant matrix $[d]$ depends on the particular system. The selector matrix $[\Lambda]$, with elements 1 or 0, links the two subsystems. As an example, the equations governing the circuit of Fig. 5.1b are presented in the following:

$$\begin{aligned} -\frac{dq_{nl}}{dt} - i_{nl}(t) + i_L(t) &= 0 \\ L\frac{di_L}{dt} + v_2(t) - v_1(t) &= 0 \\ C\frac{dv_1}{dt} + \frac{v_1(t) - e_g(t)}{R} + i_d(t) + i_L(t) &= 0 \\ \frac{d\overline{z}}{dt} - [p]\overline{z} - [d]v_1 &= 0 \\ i_d - [k]\overline{z} - [\alpha]v_1 &= 0 \end{aligned} \tag{5.17}$$

The short-circuit termination of the transmission line is taken into account in the derivation of the second subsystem. Thus, we can take the distributed element as a one-port network. The dimension of the vector $\overline{z}$ will be given by the number of considered poles. As shown in (5.17), the two subsystems are linked through the branch current i_d and can be solved in a simultaneous manner.

5.2.1.2 Impulse Response The distributed elements can be treated as black boxes, defined by the parameters S, Z, or Y, depending on the frequency ω. To facilitate the insertion of these elements into a nonlinear system of DAEs, they will be modeled by means of a matrix of transfer functions $H(\omega)$, with inputs belonging to the set of state variables $\overline{x}(t)$. This type of representation can easily be obtained from the original parameters S, Z, and Y. Then the impulse response is obtained from the inverse Fourier transform of $H(\omega)$ [29]. The frequency response $H(\omega)$ is computed in a frequency band $[0, \omega_m]$ such that the spectral content of $H(\omega)$ is negligible beyond $\omega_m = 2\pi f_m$. In practice, instead of abruptly cutting all frequency components above ω_m, which would be equivalent to multiplying the original transfer function $H(\omega)$ by a rectangle window, a smoothing window is generally used. Smoothing windows such as the Hanning window reduce ripple due to discontinuities in the frequency functions. It also eliminates noncausal and trailing effects.

The frequency ω_m is estimated from the input frequencies and expected rise times. For a rise time t_r, a conventional approximation of the equivalent bandwidth is $f_m = 2.2/t_r$ [26]. The maximum frequency ω_m determines the spacing of the time samples of the impulse response $h(t)$. In turn, the spacing between the frequency samples $\Delta\omega$ determined the time length of this impulse response. The impulse response $h(t)$ is usually defined in a time interval such that it has no energy in the second half of this interval. Note that aliasing will occur if the impulse response is indistinguishable from the time-domain samples provided by the inverse Fourier transform.

The convolution can be carried out using a lowpass description of the transfer function $H(\omega)$. However, it can also be performed in a discrete fashion, transforming the frequency response of the distributed element into a periodic function. This is done by forming a periodic extension of the function $H(\omega)$ over the entire frequency axis. Assuming that the function $H(\omega)$ has zero imaginary part at ω_m, the periodic extension gives rise to a smooth complex-valued function with period $2\omega_m$ [28]. Then the impulse response function becomes discrete and real valued and the convolution integrals can be replaced with summations in an exact manner. This analysis is faster than the one based on a lowpass description of the distributed element.

Defining a matrix $[h]$ with all the impulse responses associated with the various distributed elements, the modified nodal equations are written

$$\overline{f}(\overline{x}(t)) + \frac{d\overline{q}}{dt}(\overline{x}(t)) + \int_{-\infty}^{t} [h(t-\tau)]\overline{x}(\tau)d\tau + \overline{g}(t) = 0 \tag{5.18}$$

Modeling the distributed elements requires the calculation of convolution products when integrating system (5.18). Thus, the network response must be evaluated over the entire simulation interval (from the initial time value) at each time step. Some techniques [28] have been proposed to improve the interpolation quality of the model and thus enable a reduction of the required number of samples. Other techniques allow reducing the number of required numerical operations [33].

5.2.2 Integration Algorithms

To obtain the circuit solution, differential algebraic equation (5.17) or (5.18) must be integrated from an initial time point t_o. To do this, the continuous time variable t is discretized and replaced with the time points $[t_o, \ldots, t_n, \ldots, t_N]$. Initially, a constant time spacing $t_{n+1} - t_n = h$ will be considered. The original continuous system is transformed into a discrete system written in terms of the time samples at t_n. The derivative $d\overline{q}/dt$ can be approached in different manners in terms of $\ldots, \overline{q}_{n-1}, \overline{q}_n, \overline{q}_{n+1}$, and each approximation constitutes a different integration algorithm. The different approximations transform the original continuous system (5.1) into different discrete systems with different accuracy, efficiency, and stability properties [1]. The optimum algorithm depends on the particular problem.

The algorithms may be classified as either explicit or implicit. An algorithm is called *explicit* when each new point $\overline{q}_{n+1}$ is a function of previous solution points only: $\ldots, \overline{q}_{n-1}, \overline{q}_n$. An algorithm is called *implicit* when the point $\overline{q}_{n+1}$ is a function of itself; that is, it depends on $\ldots, \overline{q}_{n-1}, \overline{q}_n, \overline{q}_{n+1}$. In single-step algorithms, only one backward step of length h is used: $\overline{q}_n$. Multiple-step algorithms involve several past points $\ldots, \overline{q}_{n-1}, \overline{q}_n$ and tend to be more accurate. A third characteristic of the algorithm is the order. The *order* of a given algorithm is the number of required evaluations of the time derivative dq/dt.

Some of the most usual integration algorithms are described next.

5.2.2.1 *Forward Euler Algorithm* In the forward Euler algorithm, the derivative $d\overline{q}/dt$ is approached as

$$\frac{d\overline{q}(\overline{x}(t_n))}{dt} = \frac{\overline{q}(\overline{x}(t_{n+1})) - \overline{q}(\overline{x}(t_n))}{h} \tag{5.19}$$

where h is the time step, initially assumed constant and such that $t_n = t_o + nh$. As can be seen, the derivative at time t_n involves the variable evaluated at the next time point, t_{n+1}: thus the name *forward*. The sample $\overline{q}(\overline{x}(t_{n+1}))$ can be written explicitly in terms of the previous one, $\overline{q}(\overline{x}(t_n))$, as

$$\overline{q}(\overline{x}(t_{n+1})) = \overline{q}(\overline{x}(t_n)) + \frac{d\overline{q}(\overline{x}(t_n))}{dt} h \tag{5.20}$$

Applying the approach above to the system of DAEs (5.1), this continuous system turns into a discrete system:

$$\frac{\overline{q}(\overline{x}(t_{n+1})) - \overline{q}(\overline{x}(t_n))}{h} + \overline{f}(\overline{x}(t_n)) + \overline{g}(t_n) = 0 \tag{5.21}$$

Provided that the function $\overline{q}(\overline{x})$ is invertible, so that it is possible to write $\overline{x}(\overline{q})$, the system (5.21) becomes

$$\overline{q}_{n+1} = \overline{q}_n + h\overline{f}(\overline{q}_n) + \overline{g}(t_n)h \tag{5.22}$$

The forward Euler algorithm is an explicit approach. Provided that $\overline{q}_n$ is known, the value corresponding to the next time point t_{n+1} is obtained directly from (5.22). It is also clear that as $h \to 0$, the discrete system (5.21) tends to the continuous one (5.1).

5.2.2.2 *Runge–Kutta Family of Algorithms* From a single previous solution point $\overline{q}_n(t)$, the Runge–Kutta method generates a sequence of approximations in which $\overline{q}_{n+1}(t)$ is a linear combination of values of the function $d\overline{q}(t)/dt$ evaluated at time points in the interval $[t_n, t_{n+1}]$ and various arguments. It can be classified as a higher-order single-step method. It allows a larger step size and increased accuracy, but is expensive computationally. There are many different Runge–Kutta algorithms. The best known is the fourth-order Runge–Kutta algorithm, which uses an average of four different estimations of $d\overline{q}(t)/dt$ to obtain $\overline{q}(t)$. The Runge–Kutta algorithm is an explicit one because the point $\overline{q}_{n+1}$ is not used for derivative estimation. On the other hand, the estimations calculated within the interval $[t_n, t_{n+1}]$ are not reused, which gives rise to a relatively low efficiency.

5.2.2.3 Backward Euler Algorithm The backward Euler algorithm belongs to the general class of implicit algorithms. The derivative $d\overline{q}/dt$ is approached as

$$\frac{d\overline{q}(\overline{x}(t_n))}{dt} = \frac{\overline{q}(\overline{x}(t_n)) - q(\overline{x}(t_{n-1}))}{h} \tag{5.23}$$

Then the continuous equation (5.1), turns into the discrete equation

$$\frac{\overline{q}_{n+1} - \overline{q}_n}{h} + \overline{f}(\overline{q}_{n+1}) + \overline{g}(t_{n+1}) = 0 \tag{5.24}$$

From the inspection of (5.24), the backward Euler algorithm assumes that the solution is linear over one time step. The implicit equation (5.24) can only be solved through an error minimization algorithm such as the Newton–Raphson. Thus, implicit algorithms are generally more demanding from a computation point of view. However, compared to explicit algorithms, they offer better accuracy and improved stability properties [1]. This can be understood as a result of the dependence of the point $\overline{q}_{n+1}$ on itself, which gives rise to a feedback effect.

5.2.2.4 Trapezoidal Approximation The commonly used trapezoidal approximation estimates the derivative $d\overline{q}/dt$ using the average of its values at times t_n and t_{n+1}; that is,

$$\frac{d\overline{q}}{dt} = \frac{1}{2}\left(\frac{d\overline{q}_{n+1}}{dt} + \frac{d\overline{q}_n}{dt}\right) \tag{5.25}$$

The trapezoidal approximation is an implicit algorithm of order 2, as it uses two derivative evaluations. It is a single-step algorithm, as it uses one past step only. As gathered from (5.25), the trapezoidal rule estimates the area under $q(t)$ in the interval $t_n, t_n + \Delta t$ by a trapezium, the length of one side being Δt and the length of the other side being the average of the derivative $d\overline{q}/dt$, evaluated at $t_n, t_n + \Delta t$. From (5.25) it is possible to write

$$\frac{d\overline{q}_{n+1}}{dt} = \frac{2}{h}(\overline{q}_{n+1} - \overline{q}_n) - \frac{d\overline{q}_n}{dt} \tag{5.26}$$

Substituting (5.26) into (5.1), the following implicit equation is obtained:

$$2\frac{\overline{q}_{n+1} - \overline{q}_n}{h} - \frac{d\overline{q}_n}{dt} + \overline{f}(q_{n+1}) + \overline{g}(t_{n+1}) = 0 \tag{5.27}$$

The trapezoidal algorithm can give rise to artificial ringing in circuits with a small time constant compared with the time step. This can be related to the fact that (5.26) is a combination of the backward and forward Euler approaches and the latter has bad stability properties.

5.2.2.5 Gear Algorithms Before tackling the Gear algorithms, the multistep algorithms will be introduced. Multistep algorithms use several previous points, unlike all the algorithms based only on a single previous point $\overline{q}_{n-1}$. For the evaluation of $\overline{q}_{n+1}$, an m-step algorithm uses the inputs $\overline{q}_n, \overline{q}_{n-1}, \ldots, \overline{q}_{n-m+1}$. In general, an m-step algorithm can be written

$$\begin{aligned} q_{n+1} &= a_0 q_n + a_1 q_{n-1} + \cdots + a_{m-1} q_{n-m+1} \\ &\quad + h\left(b_{-1}\frac{dq_{n+1}}{dt} + b_0\frac{dq_n}{dt} + b_1\frac{dq_{n-1}}{dt} + \cdots + b_{m-1}\frac{dq_{n-m+1}}{dt}\right) \end{aligned} \tag{5.28}$$

Expression (5.28) has $2m+1$ coefficients. The a coefficients help predict $\overline{q}_{n+1}$ from the past values $\overline{q}_n \cdots \overline{q}_{n-m+1}$. The b coefficients add information from the time derivatives. Different types of multistep algorithms exist, depending on the values of the coefficients a_j and b_j. The approaches with $b_{-1} \neq 0$ will be algorithms of implicit type. All the integration algorithms with similar formal structure are grouped in families. A given algorithm has J order when the solution $\overline{q}(\overline{x}(t))$ and its J first time derivatives are continuous at the limits of the interval $[t_n, t_{n+1}]$. Thus, the integration is error-free for a system whose solution is a polynomial of the same order J or smaller [1].

The Jth-order Gear algorithm has $b_j = 0$ for $j = 0$ to J, and the reminder coefficients are chosen so that the algorithm is exact for polynomials of order J. The Jth-order Gear algorithm is a $(J-1)$-step algorithm of implicit type. Other criteria for the choice a_j and b_j provide other families of algorithms, such as the Adams–Bashforth and Adams–Moulton families. A detailed classification of the different integration algorithms is given by Parker and Chua [1].

The following expressions provide the q_{n+1} approximations in the first- to fourth-order Gear algorithms, commonly used in practice:

$$\begin{aligned} &\text{first order:} && q_{n+1} = q_n + h\frac{dq_{n+1}}{dt} \\ &\text{second order:} && q_{n+1} = \frac{1}{3}\left(4q_n - q_{n-1} + 2h\frac{dq_{n+1}}{dt}\right) \\ &\text{third order:} && q_{n+1} = \frac{1}{11}\left(18q_n - 9q_{n-1} + 2q_{n-2} + 6h\frac{dq_{n+1}}{dt}\right) \\ &\text{fourth order:} && q_{n+1} = \frac{1}{25}\left(48q_n - 36q_{n-1} + 16q_{n-2} - 3q_{n-3} + 12h\frac{dq_{n+1}}{dt}\right) \end{aligned} \tag{5.29}$$

It is clear that the Gear algorithm of first order is equivalent to the backward Euler approach.

In agreement with (5.29), the order of an algorithm increases with the number of time steps considered. The order of an M-step algorithm is generally either $J = M$ or $J = M + 1$. On the other hand, the order of single-step algorithms can be higher than 1, as in the case of the Runge–Kutta method. In a Jth-order single-step algorithm, J intermediate evaluations of the variable q and its derivative are carried out in time step h. The backward Euler algorithm adapts faster than the other algorithms to abrupt signal changes, but requires a shorter time step to maintain accuracy. It can also add artificial damping to the system solution. Algorithms of higher order use more information about the system and are generally more exact. They allow a longer time step without degrading accuracy and are convenient for smooth waveforms. On the other hand, they can give rise to instability on lightly damped circuits, that is, with dominant poles near the imaginary axis.

After reviewing the main integration algorithms, the case of circuits containing distributed elements, described through their impulse responses, will be considered. At each point t_n, the convolution products are calculated through a discrete sum:

$$\bar{i}(t_n) \cong \sum_{i=0}^{n-1} [h(t_n - t_i)]\bar{v}(t_i)\Delta t_i \tag{5.30}$$

where $\Delta t_i = t_{i+1} - t_i$. For simplicity, a rectangular rule has been considered in (5.30), although other, more efficient schemes of higher order are generally used in practice [33]. From (5.30) it is clear that evaluation of the circuit response over the total simulation interval $[t_o, t_N]$ requires $O(N^2)$ operations. An algorithm is proposed by Kapur et al. [33] to reduce the number of operations to $O(N\log(N))$.

Using, for instance, the trapezoidal approach, and taking the numerical convolution (5.30) into account, the following discretized version of the modified nodal equation is obtained:

$$\begin{aligned}\bar{e}(t_{n+1}) \equiv 2\frac{\bar{q}(\bar{x}(t_{n+1})) - \bar{q}(\bar{x}(t_n))}{t_{n+1} - t_n} - \frac{d\bar{q}(\bar{x}(t_n))}{dt} + \bar{f}(\bar{x}(t_{n+1}))\\ + \sum_{i=1}^{n} [h(t_{n+1} - t_i)]\bar{v}(t_i)\Delta t_i + \bar{g}(t_{n+1}) = 0\end{aligned} \tag{5.31}$$

where an error function $\bar{e}(t_{n+1})$ has been introduced. The process starts with a dc analysis of the circuit, providing the initial value $\bar{x}(0)$. This value can also be preset by the user in the form of node voltages or inductance currents. The integration algorithm is applied from $t_o = 0, \bar{x}(0)$. Initially, $\bar{q}(t_1)$ is obtained. Then the integration is applied recursively to determine $t_{n+1}, \bar{q}(t_{n+1})$ from a knowledge of $t_n, \bar{q}(t_n)$, and all the past points required for calculation of the convolution products. Note that in the case of multistep algorithms, the required m steps are not available in the first integration from the initial time value. To cope with this problem, the step number is gradually increased until the m previous points required are available.

The implicit system (5.31) (or, in general, the one resulting from the used integration algorithm) is usually resolved with the aid of the Newton–Raphson algorithm [34,35], which converts the nonlinear problem into a sequence of linear equations. At each time value, the unknown of the implicit algorithm is $q(t_{n+1})$. The Newton–Raphson algorithm requires computation of a Jacobian matrix of the error function with respect to the variable vector $\overline{q}(t_{n+1})$. The iteration $k+1$ of this algorithm is obtained as

$$\overline{q}_{k+1} = \overline{q}_k - \left[\frac{\partial \overline{e}}{\partial \overline{q}}\right]_k^{-1} \overline{e}_k \tag{5.32}$$

Compared to other techniques, the advantage of the Newton–Raphson algorithm comes from the large value of the step size $\Delta t_{n+1} = t_{n+1} - t_n$ that it allows. As time evolves, the Newton–Raphson algorithm uses the final value $\overline{q}(t_n)$ for the previous time point t_n as the initial guess for $\overline{q}(t_{n+1})$. At the beginning of the process, some tolerance limits must be imposed over the charge value or over the circuit voltages and currents. A maximum error value $e_{\max}$, below which the solution $\overline{q}(t_{n+1})$ is considered to be valid, must also be specified.

5.2.3 Convergence Considerations

Different causes may prevent convergence of the time-domain integration or degrade its accuracy. The main aspects are considered next.

1. *Noncausality.* As already stated, the response of a causal system at time t depends only on the inputs for time values less than or equal to t. Convergence problems often arise from the noncausality of ideal circuit components. Examples of noncausal components are constant complex impedances Z_o, or impedances with frequency-dependent real part and constant imaginary part $Z_r(\omega) + jZ_{io}$.

2. *Round-off and truncation errors.* The round-off error of a given algorithm depends on the number and type of arithmetical operations involved, so it will depend on the type of integration but not on the step size. The truncation error of a given integration algorithm is due to the particular discretization method and can be defined as the error obtained if the algorithm were implemented with infinite precision [34]. For a multistep algorithm of order J, it can be expressed as $\varepsilon_T = A_J h^{J+1}$, where A_J is a real number that does not depend on h, but depends on the order J, the number of utilized past points, the particular time-domain equation, and the particular point $n+1$ of the curve. In the trapezoidal rule ($J=2$), the error is proportional to h^3. In the backward Euler rule ($J=1$), the error is proportional to h^2. For sufficiently small h, the higher-order algorithms are more accurate than the lower-order algorithms. The situation can, however, be reversed if step h becomes too large. The global truncation error is the maximum accumulated truncation error. Circuits with large time constants (i.e., with long transients) are most sensitive to these errors.

3. *Selection of time step.* For maximum efficiency of the algorithm (i.e., the largest step size for a prefixed error tolerance), the step size must be adjusted during the integration process. The need for adjusting the step size comes from the fact that the coefficient A_J in the truncation error ε_T, which is independent of h, changes at each time step, as it depends on the particular point of the curve. Note, however, that, in multistep algorithms, the input past points must be spaced equally in time. Thus, when the step size is changed along the curve, the evenly spaced points must be calculated, which will require additional computational effort. The time step of the integration algorithm is generally determined from the current estimate of the truncation error, which provides a reasonable error bound $\varepsilon_T = A_J h^{J+1}$. It is also possible to use the *iteration-count technique*. This technique changes the time step according to the number of Newton–Raphson iterations that were required for convergence in the previous time point. If the number is larger than a given maximum value, the time step is divided by an integer factor. If it is smaller than a given minimum value, the time step is doubled. With this technique, the system may eventually diverge because it is not based on the actual rate of change of the circuit variables.

4. *Nonlinearity.* The time step must be smaller than the fastest rise time in the circuit solution. The Newton–Raphson algorithm may be unable to converge in the case of a very nonlinear behavior. Continuation algorithms, such as source stepping or $G_{\min}$ stepping, help resolve the initial value problem in very nonlinear situations. In the source stepping technique, the value of the input sources is reduced by multiplying them with a level η (source stepping). For very small η, the circuit is simpler to integrate due to the lower degree of nonlinearity. However, for this initial η value, the circuit is very different from the original one. The objective is to achieve the level $\eta = 1$ at which the circuit agrees with the original one. This is done by increasing η in discrete steps $\Delta\eta$ and using the final solution at η_n as the initial guess for the Newton–Raphson algorithm applied at η_{n+1}. In the $G_{\min}$ stepping technique, a small resistor is connected in parallel with the nonlinear devices terminals. Then the resistance value R is gradually increased in discrete steps, up to a very large value, for which the circuit containing the resistors is equivalent to the original circuit. The Newton–Raphson algorithm applied at each step R_{n+1} is initialized with the final solution obtained at the preceding step, R_n.

5. *Stability.* The stability of the integration algorithm is essential. Otherwise, the initial truncation errors will propagate through the integration and the artificial solution obtained may become unbounded. As already stated, the same continuous system gives rise to different discrete systems. For the same value of h, some integration algorithms may be unstable. The instability problems are mostly found in *stiff systems* or systems with very different time constants. To classify the various integration algorithms according to their stability properties, the very simple linear system $\dot{x} = \lambda x$ is considered. This autonomous continuous system has an equilibrium point at the origin, and for $\mathrm{Re}[\lambda] < 0$ the solution tends to $x = 0$. However, in the case of discretized systems, divergence may occur. Regions of stability are represented in the plane defined by $\mathrm{Re}[h\lambda]$ and $\mathrm{Im}[h\lambda]$. The regions of stability are determined in terms of λh, which means that in case the time

constant λ increases, a smaller time step h is necessary to maintain the same stability conditions. The implicit algorithms show the best stability properties, due to the inherent feedback. Among them, the backward Euler, trapezoidal, and Gear algorithms are stable for all the left-hand side of the plane defined, corresponding to $\text{Re}[h\lambda] < 0$, and are the most commonly used in practice. Although it is not possible to extrapolate the conclusions of this study to general differential equation systems, faster dynamics will obviously require smaller integration steps. Typically, the stability is not a critical issue for reasonably chosen step sizes. However, in *stiff systems*, with transient behavior ruled by very different time constants $|\text{Re}(\lambda_i)| \gg |\text{Re}(\lambda_j)|$, the stability considerations limit the step size more than do accuracy considerations [34].

As a first example, the parallel resonance oscillator of Fig. 1.1 has been solved with different integration methods. To reduce the quality factor, the circuit element values have been changed to $L = 1 \times 10^{-7}$ H, $C = 0.1$ pF (Fig. 5.2). The circuit operates like a relaxation oscillator. The behavior of this type of oscillator is characterized by sharp periodic transitions between two nearly constant states. In a first comparison of the integration methods, we have simulated the time interval 0–10 ns, with the same initial conditions and adjustable time step. Backward Euler requires 12904 points, with the highest CPU time. The trapezoidal rule requires 843 points; Gear 3, 1903 points; and Gear 4, 659 points. Thus, for higher-order polynomials, larger step can be used. Next, a fixed time step $t_{\text{step}} = 0.05$ ns has been considered in all cases (Fig. 5.2). The resulting waveforms are time shifted, which is attributed to the circuit autonomy and to the fact that each integration algorithm gives rise to a different system of discrete equations. The number of iterations required and the CPU time is similar in all cases. The Gear algorithms exhibit a numerical ringing after each sharp rise or fall which increases with the order considered.

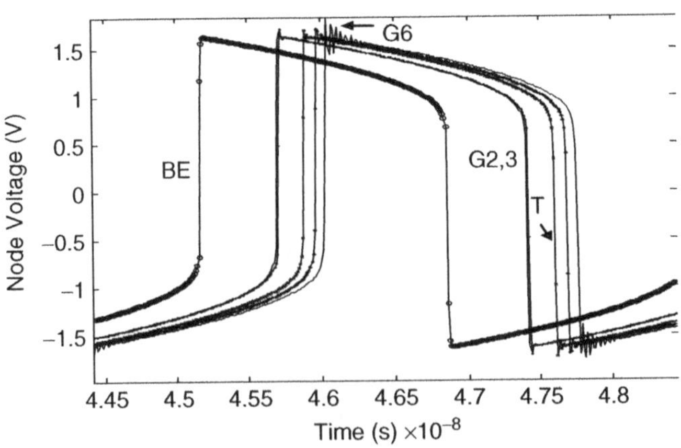

FIGURE 5.2 Time-domain integration of a parallel resonance oscillator for the reactive element values $L = 1 \times 10^{-7}$ H and $C = 0.1$ pF. BE stands for the backward Euler, T for the trapezoidal, and G for the Gear algorithm.

As a second example, a frequency divider by 2 similar to the one in Fig. 3.4 is considered. The input frequency is $f_{\text{in}} = 2.178$ GHz and the input amplitude $E_{\text{in}} = 4$ V. The lumped inductor is implemented with a microstrip line in CuClad $\varepsilon_r = 2.17$, which has width $w = 0.207$ mm and length $l = 6.8$ mm. The resistor value has been changed to $R = 33.3\ \Omega$. Different values are chosen for the maximum frequency f_{max} used for evaluation of the frequency response of the transmission line. Remember that the impulse response is calculated using the inverse Fourier transform. The impulse response should have no energy in the second half of the time interval determined by $1/\Delta f$. The point spacing is given by $1/f_{\text{max}}$. The circuit is analyzed with the trapezoidal integration rule in all cases. If the maximum frequency is below the actual system bandwidth, the results will be inaccurate. This is the case for the thin-dashed-line simulation in Fig. 5.3, which corresponds to a transmission-line characterization up to the fifth harmonic of the divided frequency, with 4096 sample points. It is also the case for the solid line, which corresponds to a line characterization up to the tenth harmonic component. For the bold dashed line, the transmission line is characterized up to the thirtieth harmonic term. Higher maximum frequency gives rise to no appreciable difference in the circuit solution.

The third example is based on the MESFET-based oscillator considered in Chapter 1. The nonlinear elements of the MESFET transistor used are the Schottky junction current $igs \equiv igs(vgs)$ and charge $qgs \equiv qgs(vgs)$, the drain-to-source current $ids \equiv ids(vgs, vds)$, which has been modeled through the Tajima equation [36], and the drain-to-gate current $idg \equiv idg(vgs, vds)$, which has been modeled through a diodelike equation [22,23]. The integration is performed with an adjustable time step. For the strict voltage tolerance 10^{-7} V the backward Euler

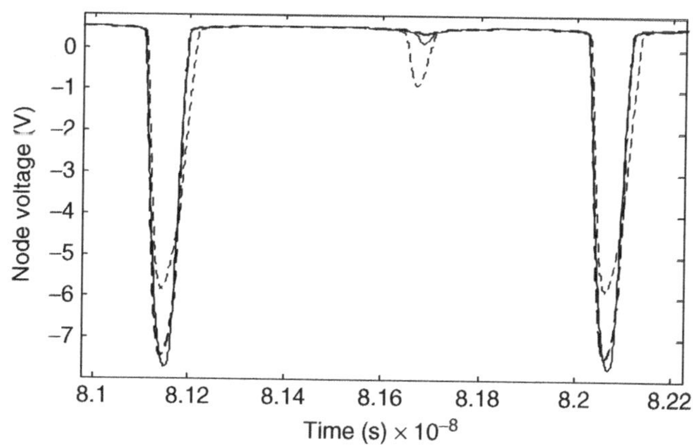

FIGURE 5.3 Simulation of a frequency divider by 2 with a microstrip line. The input frequency is $f_{\text{in}} = 2.178$ GHz and the input amplitude is $E_{\text{in}} = 4$ V. The thin dashed line corresponds to a transmission-line characterization up to the fifth harmonic of the divided frequency. The solid line corresponds to a line characterization up to the tenth harmonic component. The bold-dashed line corresponds to a characterization up to the thirtieth harmonic term.

method fails to converge from about 2.9 ns. The second-order Gear method fails to converge from about 3.2 ns. This can be seen in Fig. 5.4a. The trapezoidal rule and the third- and fourth-order Gear methods show good convergence properties. The trapezoidal rule requires 30.31 s of CPU time. The third- and fourth-order Gear methods require 23.95 and 16.93 s, respectively. The waveforms obtained using the five different methods overlap up to the loss of convergence. The duration of the transient is about 100 periods of the steady-state solution. Figure 5.4b shows the overlapped steady-state waveforms obtained using the trapezoidal rule and the fourth-order Gear method. For a voltage tolerance of 10^{-6} V, good convergence is achieved with the trapezoidal rule and all the Gear methods, whereas the backward Euler rule keeps failing to reach the steady state. For 10^{-6} V tolerance, the CPU

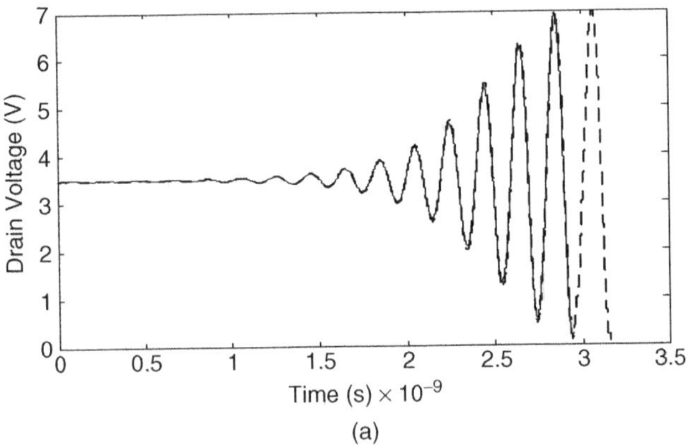

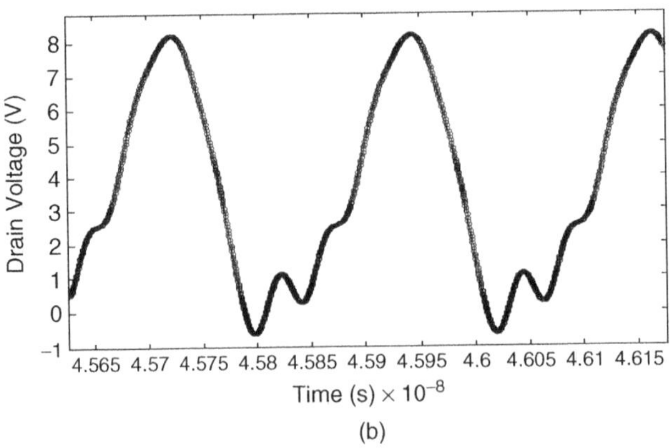

FIGURE 5.4 Integration of the MESFET-based oscillator considered in Chapter 2 with the strict tolerance 10^{-7} V and 10^{-15} C. (a) Divergence of the backward Euler algorithm (the solid line) and the second-order Gear algorithm (the dashed line). (b) Overlapped waveforms obtained for trapezoidal rule and fourth-order Gear algorithm.

time with the trapezoidal rule is 15.41 s, with second-order Gear it is 28.10 s, with third-order Gear it is 14.76 s, and with fourth-order Gear it is 12.33 s.

5.3 FAST TIME-DOMAIN TECHNIQUES

As has been shown, direct-integration methods provide the entire time evolution of the circuit solution from the initial value $t_o, \overline{x}_o$ to the steady state, including the transient. Generally, most simulation time is devoted to the transient. Examples of circuits with very long transients are high-quality-factor oscillators or circuits operating near a bifurcation. Circuit designers are usually interested in the steady-state solution only. Taking this into account, fast time-domain methods address the steady-state regime directly. The fast time-domain analysis is possibly the best option for the simulation of strongly nonlinear periodic regimes in lumped-element circuits. It is based on time-domain descriptions for both linear and nonlinear elements, so if the circuit contains the distributed elements, use of the harmonic balance method, presented in Section 5.4, could be more convenient. Two different fast time-domain techniques are outlined briefly next: the shooting methods and finite differences in the time domain.

5.3.1 Shooting Methods

The shooting methods [10,35,37,38] are applicable to circuits with periodic excitation. They are advantageous with respect to direct integration in circuits with slow transients. The shooting methods efficiently find, through an optimization technique, a vector of initial conditions $\overline{x}_o$ from which the circuit behaves in a periodic steady-state regime. The solution value at the end of the period must match the initial value $\overline{x}_o$. Assuming an initial time t_o, the circuit is evaluated for one period T and the following two-point constraint is imposed: $\overline{x}(t_o + T) - \overline{x}(t_o) = 0$, with T being the solution period. Because $\overline{x}_o \in R^P$, the two-point constraint provides a system of P equations in P unknowns. Thus, in the shooting methods, the differential equation integration is converted into a two-point boundary problem.

The shooting methods are iterative. They start with an estimation of the initial condition $\overline{x}_o$ desired. At each iteration, the final state is computed together with the sensitivity of the final state with respect to the initial state. Let the solution $\overline{x}(t)$ be expressed in terms of the initial value $\overline{x}_o = \overline{x}(t_o)$ as $\overline{x}(t) = \overline{\phi}(\overline{x}_o, t_o, t - t_o)$, where $\overline{\phi}$ is the state transition function [10,35,37,38]. Then the shooting equation system is given by $\overline{\phi}(\overline{x}_o, t_o, T) - \overline{x}_o = 0$, which contains P equations in P unknowns. A Newton–Raphson algorithm is used to solve the shooting equations $\overline{F}(\overline{x}_o) = \overline{\phi}(\overline{x}_o, t_o, T) - \overline{x}_o = 0$ in terms of the initial conditions $\overline{x}_o$. This requires determination of the Jacobian matrix $[JF] = \partial\overline{\phi}(\overline{x}_o, t_o, T)/\partial\overline{x}_o - I$, also called a *sensitivity matrix*. This matrix, together with the error in the periodicity, is used to compute a new initial condition.

The iterative algorithm is implemented as follows:

$$\overline{x}_o^{j+1} = \overline{x}_o^{j} - \left[\frac{\partial \overline{F}(\overline{x}_o, t_o, T)}{\partial \overline{x}_o}\right]_j^{-1} \overline{F}^j \tag{5.33}$$

where j indicates the iteration number. To obtain the transition matrix, a series of iterations are carried out at time points between t_o and $t_o + T$. Thus, recursively integrating equation (5.1) from $\overline{x}_o$, it will be possible to obtain the function $\overline{\phi}(\overline{x}_o, t_o, T)$ numerically. If the time interval $[0,T]$ is discretized in M values such that $t_o = 0$ and $t_{M-1} = T$, equation (5.1) will be fulfilled at each of these values, even when the shooting equations are not satisfied. Thus, the circuit is actually solved through a two-level Newton–Raphson algorithm. The outer level corresponds to the shooting equations. The inner level is applied to the integration of the system (5.1), which is required for determination of the transition matrix. Note that the computational expensiveness increases rapidly with the size of the circuit, so matrix-implicit iterative algorithms are generally used for the matrix inversion [10].

The shooting methods can deal with circuits that behave in a strongly nonlinear manner over a specific period. This is because the final state $\overline{x}(t_o + T)$is usually a nearly linear function with respect to the initial state $\overline{x}(t_o)$ even for very nonlinear periodic regimes. The nonlinearity can also be reduced by integrating over a number of periods, which implies delaying the starting time t_o. The convergence can be improved in this manner. As can easily be understood, it is better to start the process at a time when the solution waveform is varying slowly instead of using a point with a rapid time variation of this waveform.

The shooting methods require the circuit solution to be periodic. If there is more than one input source, their periods must be commensurable and the analysis period must be equal to the minimum common multiple of all the periods. In the case of a subharmonic regime, the analysis period will be that of the generated subharmonic. It must be noted that the boundary value constraint $\overline{x}(t_o + T) - \overline{x}(t_o) = 0$ cannot be applied to quasiperiodic regimes. The envelope-following method is a generalization of the shooting method, enabling simulation of these quasiperiodic regimes [13]. On the other hand, the initial conditions are difficult to establish for distributed devices, since these conditions must be specified throughout the devices, which is done through the use of special functions [10].

The shooting methods can be applied to the analysis of free-running oscillators. The same condition $\overline{x}(t_o + T) - \overline{x}(t_o) = 0$ is imposed. However, in an oscillator the period is an unknown of the problem, which depends on the values of the circuit elements and bias sources. Thus, the two-point constraint is a system of P equations and $P + 1$ unknowns: the P components of $\overline{x}(t_o)$ and the period T. An additional condition is necessary to balance the number of equations and unknowns. Taking into account the irrelevance of the circuit solution with respect to time translations, the increment applied to the initial values $\Delta\overline{x}(t_o)$ can be chosen to be orthogonal to the trajectory [1], which implies that

$$\frac{\partial\overline{F}(\overline{x}_o, 0, T)}{\partial T}\Delta\overline{x}_o = \frac{\partial\overline{\phi}(\overline{x}_o, 0, T)}{\partial T}\Delta\overline{x}_o = 0 \tag{5.34}$$

Adding this equation to the Newton–Raphson algorithm (5.33) provides the following system of $P + 1$ equations in $P + 1$ unknowns:

$$\begin{bmatrix}\overline{x}_o^{n+1}\\ T^{n+1}\end{bmatrix} = \begin{bmatrix}\overline{x}_o^{n}\\ T^{n}\end{bmatrix} - \begin{bmatrix}\left.\dfrac{\partial\overline{F}(\overline{x}_o, 0, T)}{\partial\overline{x}_o}\right|_n & \left.\dfrac{\partial\overline{\phi}(\overline{x}_o, 0, T)}{\partial T}\right|_n \\ \left.\dfrac{\partial\overline{\phi}(\overline{x}_o, 0, T)}{\partial T}\right|_n & 0\end{bmatrix}^{-1}\begin{bmatrix}\overline{F}^n\\ 0\end{bmatrix} \tag{5.35}$$

As an example, a parallel resonance oscillator with reactive element values $L = 1 \times 10^{-7}$ H, $C = 0.1$ pF will be considered. The shooting equation system has been solved considering two different initial points. In the first case, the initial point corresponds to one of the two nearly flat regions (those with a small time derivative) (Fig. 5.5a). The fundamental frequency obtained is 260.032 MHz, and the steady state is achieved in three iterations, with a total CPU time of 440 ms and 1153 steps. In the second case, the initial point corresponds to one of the two sections with a fast time variation (Fig. 5.5b). The fundamental frequency obtained is also 260.032 MHz. Steady state is achieved in five iterations, with a total CPU time of 540 ms and 1156 steps.

5.3.2 Finite Differences in the Time Domain

In a method based on finite differences in the time domain [38], the system (5.1) is combined with the periodicity condition $\overline{x}(t_o) - \overline{x}(t_{M-1}) = 0$ to obtain a global system whose unknowns are the M time samples $x_p(t_n)$ ($n = 0$ to $M - 1$) of each of the P state variables. For a circuit containing lumped elements only and assuming a backward Euler integration rule, the $P \times M$ equation system is written

$$\begin{aligned}
&\frac{1}{h}(\overline{q}(\overline{x}(t_1)) - \overline{q}(\overline{x}(0)) + \overline{f}(\overline{x}(t_1)) + \overline{g}(t_1) = 0\\
&\vdots\\
&\frac{1}{h}(\overline{q}(\overline{x}(t_{n+1})) - \overline{q}(\overline{x}(t_n)) + \overline{f}(\overline{x}(t_{n+1})) + \overline{g}(t_{n+1}) = 0\\
&\vdots\\
&\frac{1}{h}(\overline{q}(\overline{x}(T)) - \overline{q}(\overline{x}(t_{M-2})) + \overline{f}(\overline{x}(T)) + \overline{g}(T) = 0\\
&\overline{x}(0) - \overline{x}(T) = 0
\end{aligned} \tag{5.36}$$

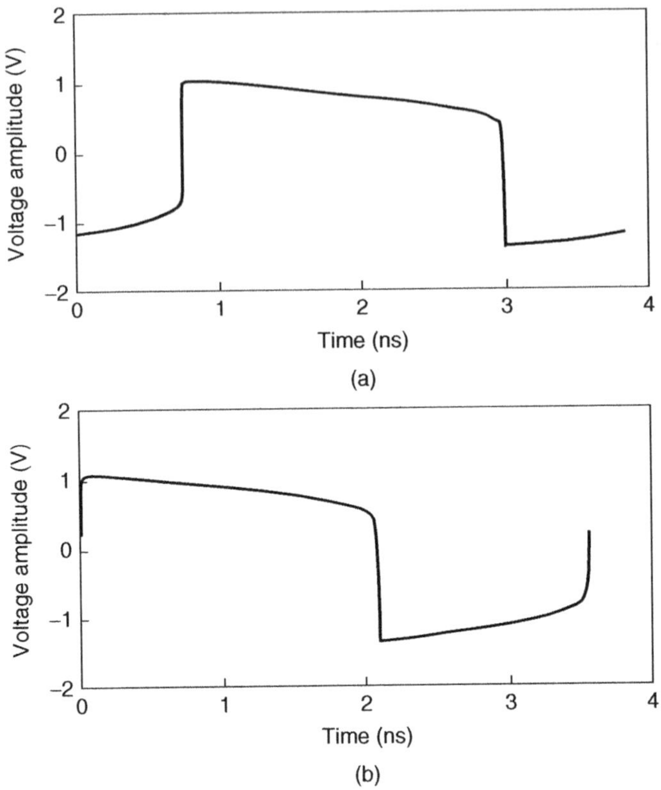

FIGURE 5.5 Simulation of a parallel resonance oscillator with reactive-element values $L = 1 \times 10^{-7}$ H and $C = 0.1$ pF, using the shooting method: (a) initial point located in a nearly flat region; (b) initial point located in a fast time variation region.

where equal spacing h between the time samples has been considered. In the subsystem consisting of the $P(M-1)$ first equations of (5.36), there are P additional unknowns, which correspond to the initial condition $\overline{x}(0)$. However, the periodicity condition $\overline{x}(0) - \overline{x}(t_{M-1}) = 0$ adds P more equations to the system. Thus, (5.36) is a well-balanced system of *PM* equations in *PM* unknowns, which is solved through the Newton–Raphson algorithm [38]. As in the case of the shooting methods, matrix-implicit iterative algorithms must be used for the analysis of large circuits.

In the case of a nonautonomous circuit, the time-varying input generator or generators $\overline{g}(t)$ set the conditions for the first point $t_o = 0$, as these generators take the specific value $\overline{g}(0)$. In the case of an autonomous circuit (free-running oscillator), there are no time-varying input generators. Thus, there is no external reference, providing the initial value of the circuit variables at $t_o = 0$. On the other hand, the solution period T is an unknown of the system, since the oscillation frequency is generated autonomously.

The autonomous circuit must be solved for T in addition to $\overline{x}(0), \ldots, \overline{x}(t_n), \ldots, \overline{x}(T)$. However, because no external reference exists, one of the components of $\overline{x}(0)$ may take any value in the expected variation range, $x^1(0) = x_0^1$, which is assigned arbitrarily [39]. Thus, in the case of an autonomous circuit, there are $MP - 1$ unknowns of the form $\overline{x}(t_n)$, with $n = 0$ to $M - 1$, plus the oscillation period T. So the system (5.36) is also a well-balanced system in the case of an autonomous circuit. It must also be taken into account that the steady-state free-running oscillation coexists with a dc solution. Thus, the initial values of the Newton–Raphson algorithm must be relatively close to the oscillating solution to avoid undesired convergence to the dc regime [39].

5.4 HARMONIC BALANCE

The harmonic balance method transforms the set of nonlinear differential algebraic equations that rule circuit behavior into a set of nonlinear algebraic equations in the frequency domain [15,16,40]. It uses frequency-domain descriptions for the linear elements, retaining the natural time-domain models of the nonlinear elements. The models of the different elements are given by instantaneous relationships with their control voltages. The circuit variables are represented in a Fourier series, so only periodic and quasiperiodic steady-state solutions can be simulated. Due to the difficulties in the representation of fast time variations in a sinusoidal basis, the application of harmonic balance is limited to relatively mild nonlinear regimes. In practice, the Fourier series must be truncated to a finite number N of harmonic components, so this type of representation might not be convenient for some periodic signals with low rise and fall times. As an example, the accurate harmonic balance simulation of the oscillator considered in Figs. 5.2 and 5.5 requires more than 30 harmonic terms. For a lower number of harmonic components, an artificial ringing is obtained which can be related to the Gibbs phenomenon [41]. Because of the use of Fourier series expansions for the circuit variables, harmonic balance analyzes only the steady-state solutions. Convergence may be obtained for either stable or unstable solutions, so a complementary stability analysis will be necessary.

5.4.1 Formulation of a Harmonic Balance System

For a formulation of the harmonic balance system, it will be assumed that the circuit variables can be expanded in a Fourier series, having a finite set of NF nonrationally related fundamentals F_1 to F_{NF}. In most practical circuits, the number of fundamentals is one or two. Examples of circuits with one fundamental frequency are amplifiers and oscillators. A circuit with two fundamental frequencies is a frequency mixer.

Physical circuits have intrinsic lowpass behavior, so it will be possible to truncate the Fourier series expansions of the circuit variables and keep only a certain number of harmonic terms. Let N be the total number of positive frequencies resulting from intermodulation products of the fundamental frequencies. The Fourier series expansions will have the general form

$$y(t) = \sum_{k=-N}^{N} Y_k e^{j\omega_k t} \qquad \omega_k \equiv \overline{\lambda}_k^t \overline{\Omega} \qquad \overline{\Omega}^t = (2\pi F_1, 2\pi F_2, \ldots, 2\pi F_{NF}) \quad (5.37)$$

where $y(t)$ stands for any of the circuit variables. The vector $\overline{\lambda}_k^t \in Z^{NF}$ contains the integer coefficients of the intermodulation product k. This intermodulation product is given by $\omega_k \equiv \overline{\lambda}_k^t \overline{\Omega} = \lambda_k^1 \Omega_1 + \lambda_k^2 \Omega_2 + \cdots + \lambda_k^{NF} \Omega_{NF}$, with λ_k^j integers. The superscript j in λ_k^j indicates the fundamental frequency ω_j that is affected by that particular integer. The subscript k places the resulting frequencies in increasing order $\omega_1 < \omega_2 < \cdots < \omega_N$. Note that since we are dealing with real variables (in time domain), the Fourier coefficients of (5.37) fulfill $Y_k = Y_{-k}^*$.

Different criteria can be used for truncation of the Fourier series representing a quasiperiodic signal. Two of the most common approaches are the box truncation and the diamond truncation [5,42]. In the box truncation, the intermodulation products $\omega_k \equiv \overline{\lambda}_k^t \overline{\Omega} = \lambda_k^1 \Omega_1 + \lambda_k^2 \Omega_2 + \cdots + \lambda_k^{NF} \Omega_{NF}$, with $-N \leq k \leq N$, fulfill $|\lambda_k^j| \leq nl_j$, with nl_j a constant positive integer. In the case of two fundamental frequencies, with $j = 1, 2$, representation of the selected pairs of coefficients λ_k^1 and λ_k^2 (one versus the other) provides a rectangle, which justifies the name *box truncation*. It is convenient in the case of quite different amplitudes at the various fundamental frequencies. In the diamond truncation, the criterion is $|\lambda_k^1| + |\lambda_k^2| + \cdots + |\lambda_k^{NF}| \leq nl$. For two fundamental frequencies, the representation of the selected pairs of coefficients λ_1 and λ_2 provides a diamond. It is easily shown that in this case, the total number of positive frequencies is $N = nl(nl + 1)$. Clearly, this form of truncation neglects intermodulation products components of higher order than the one corresponding to any fundamental, given by nl. These intermodulation products usually have much less power than the harmonics of the fundamentals $\lambda^j F_j$. Thus, for the same total number of analysis frequencies N, the diamond truncation is generally more efficient than the box truncation. Note that the a priori determination of the optimum truncation order is generally difficult. A saturation criterion can be used, increasing nl or (nl_j) until no appreciable changes are obtained in the circuit solution.

Other truncation schemes can also be used. For instance, it is possible to assign a different order nl_j, $1 \leq j \leq NF$, to each fundamental, while still imposing diamond truncation $|\lambda_k^1| + |\lambda_k^2| + \cdots + |\lambda_k^{NF}| \leq nl$ to the intermodulation products (involving two or more fundamentals). This technique is useful when there are significant differences in magnitude at the different fundamentals.

Two different harmonic balance formulations are possible and are presented in the following. The first formulation is obtained from direct introduction of the Fourier series expansions of the vectors $\overline{x}(t), \overline{q}(t), \overline{f}(t)$, or $\overline{g}(t)$ into equation (5.18). Taking into account the orthogonality of the Fourier basis, this leads to a nonlinear algebraic system in the Fourier frequency components of $\overline{x}$ (i.e., of the set of node voltages plus inductance currents). This formulation is known as *nodal harmonic balance* [43]. The size of the system is equal to the number of circuit nodes and inductance currents multiplied by the number $(2N+1)$ of spectral lines. Thus, a nonlinear system with a large number of unknowns may be obtained. The second formulation, known as *piecewise harmonic balance* [44], is based on a strict separation of the circuit elements into linear and nonlinear. This allows limiting the set of unknowns to the control variables of the nonlinear elements only (instead of having to determine all the node voltages). Compared with the nodal harmonic balance, the number of unknowns is reduced considerably at the expense of an increase in the complexity of the linear matrixes representing the linear embedding network, which will have higher order in the frequency ω.

5.4.2 Nodal Harmonic Balance

Let Fourier series expansions of the vectors $\overline{x}(t), \overline{q}(t), \overline{f}(t)$, and $\overline{g}(t)$, with the form (5.37), be considered. These expansions will be introduced into the modified nodal equation (5.18). Taking into account the orthogonality of the Fourier basis $e^{j\omega_k t}$, it will be possible to obtain a relationship between the harmonic components of $\overline{x}(t), \overline{q}(t), \overline{f}(t)$, or $\overline{g}(t)$. For a compact expression of this relationship, the set of Fourier coefficients will be written in the vector form

$$\overline{x}(t) \to \overline{X} = (\overline{X}_{-N}, \ldots, \overline{X}_k, \ldots, \overline{X}_N) \qquad \overline{X}_k = (X_k^1, \ldots, X_k^p, \ldots, X_k^P) \quad (5.38)$$

where p (from 1 to P) is the index of the state variable. Remember that the index k is used to rank the frequencies resulting from the different intermodulation products in increasing order. Similar expressions are used for the harmonic components of $\overline{q}(t), \overline{f}(t)$, and $\overline{g}(t)$. Then the relationship between the Fourier coefficients of the different sets of variables constituting the harmonic balance equation is the following:

$$\overline{E}(\overline{X}) \equiv \overline{F}(\overline{X}) + [j\omega]\overline{Q}(\overline{X}) + [H(j\omega)]\overline{X} + \overline{G} = 0 \quad (5.39)$$

where $[j\omega] = \text{diag}[(j\omega_{-N}) \cdots (j\omega_k) \cdots (j\omega_N)]$, with the $(j\omega_k)$ being diagonal matrixes of the form $j\omega_k[I_p]$, with $[I_p]$ being the identity matrix of P order. Comparing with (5.18), the convolution operation of the impulse responses associated with the distributed elements in (5.18) becomes a simple multiplication by the matrix $[H(j\omega)]$ in the frequency-domain formulation (5.39). This matrix contains the transfer functions of the various distributed elements. Finally, $\overline{E}(\overline{X})$ is an error function to be minimized in the solution process. Note that (5.39) is a well-balanced system of $(2N+1)P$ equations in $(2N+1)P$ unknowns.

In case the relationship $\overline{Q}(\overline{X})$ between the charge and state variables is invertible, it will be possible to express $\overline{F}(\overline{X}(\overline{Q})) \equiv \overline{F}(\overline{Q})$. Then the harmonic balance system can be reformulated using $\overline{Q}$ as an unknown. This technique ensures the charge conservation. Note that since the circuit variables are real, it is generally more convenient to limit the harmonic vectors $\overline{X}$, $\overline{F}$, $\overline{Q}$, and $\overline{G}$ to the positive-frequency spectrum only. Then the system is solved in terms of the real and imaginary parts of each Fourier coefficient of the independent variables. Considering the kth harmonic of the independent variable $x_p(t)$, the transformation from the complex components

X_k^p and X_{-k}^p of the double-sided spectrum to the real and imaginary parts $X_{k,r}^p$ and $X_{k,i}^p$ of the positive-frequency spectrum, and vice versa, are given by the matrix–vector products

$$\begin{bmatrix} X_k^p \\ X_{-k}^p \end{bmatrix} = \begin{bmatrix} 1/2 & j/2 \\ 1/2 & -j/2 \end{bmatrix} \begin{bmatrix} X_{k,r}^p \\ X_{k,i}^p \end{bmatrix} = [T_2] \begin{bmatrix} X_{k,r}^p \\ X_{k,i}^p \end{bmatrix}$$
$$\begin{bmatrix} X_{k,r}^p \\ X_{k,i}^p \end{bmatrix} = \begin{bmatrix} 1 & 1 \\ -j & j \end{bmatrix} \begin{bmatrix} X_k^p \\ X_{-k}^p \end{bmatrix} = [T_2]^{-1} \begin{bmatrix} X_k^p \\ X_{-k}^p \end{bmatrix} \tag{5.40}$$

where the subscript indicates the square-matrix dimension. The relationships above can be generalized to obtain the global transformation matrixes corresponding to the kth harmonic component:

$$\begin{bmatrix} \overline{X}_{k,r} \\ \overline{X}_{k,i} \end{bmatrix} = [T_{2xP}]^{-1} \begin{bmatrix} \overline{X}_k \\ \overline{X}_{-k} \end{bmatrix}$$
$$\begin{bmatrix} \overline{X}_k \\ \overline{X}_{-k} \end{bmatrix} = [T_{2xP}] \begin{bmatrix} \overline{X}_{k,r} \\ \overline{X}_{k,i} \end{bmatrix} \tag{5.41}$$

where, again, the subscript indicates the matrix dimension. These relationships can be generalized once more to obtain the global transformation matrixes associated with the different vectors in the harmonic balance system (5.39). Note that the total number of equations and unknowns remains $(2N+1)P$ after these transformations. For notational simplicity, the original complex system (5.39), with a double-sided spectrum, is considered in the reminder of the chapter.

As an example, the nodal harmonic balance formulation of the nonlinear circuit considered in Fig. 5.1a is presented. This circuit contains a nonlinear current, a nonlinear charge, and an input generator. It has four state variables, so $P = 4$. In turn, the subscript k goes from $-N$ to N, for N harmonic components. The number of harmonic balance unknowns is $4(2N+1)$. The following vectors are defined:

$$\overline{X}_k = \begin{bmatrix} I_{L1,k} \\ I_{L2,k} \\ V_{2,k} \\ V_{3,k} \end{bmatrix} \qquad \overline{Q}_k(\overline{X}) = \begin{bmatrix} -L_1 I_{L1,k} \\ -L_2 I_{L2,k} \\ -C V_{2,k} \\ -Q_{nl}((\overline{V}_3)) \end{bmatrix}$$
$$\overline{F}_k(\overline{X}) = \begin{bmatrix} -R I_{L1,k} - V_{2,k} \\ V_{2,k} - V_{3,k} \\ I_{L1,k} - I_{L2,k} \\ I_{L2,k} - I_{nl,k}(\overline{V}_3) \end{bmatrix} \qquad \overline{G}_k = \begin{bmatrix} E_{g,k} \\ 0 \\ 0 \\ 0 \end{bmatrix} \tag{5.42}$$

Once the vectors above have been determined, the equation system (5.39) is directly applicable for the circuit analysis.

Note that system (5.39) uses a description of the nonlinear elements in terms of their harmonic components, whereas these elements are originally described with time-domain relationships. For a given value of the state-variable vector $\overline{X}$, the corresponding $\overline{F}(\overline{X})$ and $\overline{Q}(\overline{X})$ are obtained indirectly through inverse (W^{-1}) and direct (W) Fourier transforms:

$$\overline{x} = W^{-1}(\overline{X}) \rightarrow \overline{f} = \overline{f}(\overline{x}), \overline{q} = \overline{q}(\overline{x}) \rightarrow \overline{F} = W(\overline{f}), \overline{Q} = W(\overline{q}) \qquad (5.43)$$

In practice, for calculation of the direct and inverse Fourier transforms, the time variable must be discretized. Discrete Fourier transforms (DFTs) must be used, implemented through matrix-vector products. Algorithms for the calculation of these transforms for periodic and quasiperiodic signals are presented in Section 5.5.1. The reader familiar with these algorithms can skip this section. The DFT algorithm requires a number M of discrete time samples of the vectors $\overline{x}(t), \overline{q}(t), \overline{f}(t)$, and $\overline{g}(t)$. The choice $M_{\min} = 2N + 1$, with N the number of positive frequencies, gives rise to square transformation matrixes (see Section 5.4.2). The square transformation of a representative variable $y(t)$ would take the form $\overline{Y} = W\overline{y}$, with $\overline{Y}$ being the vector containing the coefficients of the Fourier series representation of the signal $y(t)$, $\overline{y}$ being the vector containing the time samples chosen, and W being the constant transformation matrix. However, oversampling is commonly used to reduce aliasing that occurs when the continuous signal is indistinguishable from the samples obtained.

Note that according to the Nyquist sampling theorem, $2N + 1$ samples are needed to consider the highest harmonic component. Oversampling allows a more precise sampling of rapid transitions in the circuit waveforms. Then the number of samples M is chosen between $2M_{\min}$ and $10M_{\min}$ [7]. Due to the requirements of the fast Fourier transform, this number of samples is rounded up to the nearest power of 2. The excess of time samples with respect to the number $2N + 1$ of harmonic frequencies gives rise to a rectangular Fourier transformation matrix W to which pseudoinversion techniques are applied. In this case, the transformation between samples and Fourier coefficients will take the form $\overline{y} = (W^{*T}W)^{-1}W^{*T}\overline{Y}$.

The error function in (5.39) is usually minimized through the Newton–Raphson algorithm. This algorithm requires an initial value $\overline{X}^{0}$ which is usually generated with a dc analysis of the circuit. Then the system is resolved through an iterative process. The iteration $\overline{X}^{j+1}$ is obtained from a linearization of the nonlinear system about the point $\overline{X}^{j}$, resulting from the iteration j. This is equated as

$$[JE]_j(\overline{X}^{j+1} - \overline{X}^{j}) = -\overline{E}^{j}(\overline{X}^{j}) \Rightarrow \overline{X}^{j+1} = \overline{X}^{j} - [JE]_j^{-1}\overline{E}^{j}(\overline{X}^{j}) \qquad (5.44)$$

where $[JE]_j$ is the Jacobian matrix of the error function in (5.39) evaluated at the previous iteration j. The termination criterion depends on the desired relative (R) and absolute (A) tolerances. An efficient one is $|\overline{X}^{j+1} - \overline{X}^{j}| < R|\overline{X}^{j}| + A$. The Jacobian matrix $[JE]$ is given by

$$[JE] \equiv \left[\frac{\partial \overline{E}}{\partial \overline{X}}\right] = \left[\frac{\partial \overline{F}(\overline{X})}{\partial \overline{X}}\right] + [j\omega]\left[\frac{\partial \overline{Q}(\overline{X})}{\partial \overline{X}}\right] + [H(j\omega)] \qquad (5.45)$$

To obtain the components of the matrixes $\partial\overline{F}/\partial\overline{X}$ and $\partial\overline{Q}/\partial\overline{X}$, the Fourier transformation matrix W and the chain rule are used. As an example, let the element $f^{p1}(\overline{x})$, $1 \leq p_1 \leq P$, and the particular state variable x^{p2}, $1 \leq p_2 \leq P$, be considered. The box $\partial\overline{F}^{p1}/\partial\overline{X}^{p2}$, containing the derivatives of the harmonic components of f^{p1} with respect to each harmonic component of x^{p2}, can be computed as

$$\frac{\partial\overline{F}^{p1}}{\partial\overline{X}^{p2}} = [W]\left[\frac{\partial f^{p1}}{\partial x^{p2}}\right][W]^{-1} \tag{5.46}$$

which is easily demonstrated applying the chain rule to $\overline{F}^{p1} = [W]f^{p1}([W]^{-1}\overline{X})$. We can also take into account the following property:

$$\begin{aligned}\frac{\partial F_k^{p1}}{\partial X_m^{p2}} &= \frac{\partial\left(\frac{1}{T_o}\int_0^{T_o} f^{p1}(t)e^{-jk\omega_o t}dt\right)}{\partial X_m^{p2}} = \left(\frac{\partial f^{p1}}{\partial x^{p2}}\frac{\partial x^{p2}}{\partial X_m^{p2}}\right)_{\text{harm}.k} \\ &= \frac{1}{T_o}\int_0^{T_o}\frac{\partial f^{p1}}{\partial x^{p2}}e^{jm\omega_o t}e^{-jk\omega_o t}\,dt \\ &= \left.\frac{\partial f^{p1}}{\partial x^{p2}}\right|_{\text{harm}k-m}\end{aligned} \tag{5.47}$$

Thus, the derivative of the harmonic component k of the nonlinear function f^{p1} with respect to the harmonic component m of the variable x^{p2} is equal to the harmonic component $k-m$ of the derivative $\partial f^{p1}/\partial x^{p2}$. The box $\partial F^{p1}/\partial X^{p2}$ is full in the case of a nonlinear element $f^{p1}(x^{p2})$. In the case of a linear element f^{p1}, the box will be diagonal, since each harmonic component of order k of f^{p1} depends only on the harmonic components of the same order k of the state vector $\overline{x}$. Since the vectors $\overline{f}(\overline{x})$ and $\overline{q}(\overline{x})$ include both linear and nonlinear elements, calculations of the form (5.47) give rise to a Jacobian matrix (5.46) with many zero terms. A matrix of this type is called a *sparse matrix* [45,46].

Maintaining the organization (5.38) for the harmonic component of the circuit variables, the total Jacobian matrixes $\partial\overline{F}(\overline{X})/\partial\overline{X}$ and $\partial\overline{Q}(\overline{X})/\partial\overline{X}$ will be written

$$\left[\frac{\partial\overline{F}}{\partial\overline{X}}\right] = \begin{bmatrix}\frac{\partial\overline{F}_{-N}}{\partial\overline{X}_{-N}} & \cdots & \frac{\partial\overline{F}_{-N}}{\partial\overline{X}_N} \\ \vdots & \vdots & \vdots \\ \frac{\partial\overline{F}_N}{\partial\overline{X}_{-N}} & \cdots & \frac{\partial\overline{F}_N}{\partial\overline{X}_N}\end{bmatrix} \qquad \left[\frac{\partial\overline{Q}}{\partial\overline{X}}\right] = \begin{bmatrix}\frac{\partial\overline{Q}_{-N}}{\partial\overline{X}_{-N}} & \cdots & \frac{\partial\overline{Q}_{-N}}{\partial\overline{X}_N} \\ \vdots & \vdots & \vdots \\ \frac{\partial\overline{Q}_N}{\partial\overline{X}_{-N}} & \cdots & \frac{\partial\overline{Q}_N}{\partial\overline{X}_N}\end{bmatrix} \tag{5.48}$$

These matrixes should be introduced into (5.45) to obtain the Jacobian matrix of the error function $[JE]$. Note that the frequency-dependent matrix $[j\omega]$ is evaluated at the original steady-state frequencies ω_k.

As an example, the Jacobian matrix [JE] corresponding to the circuit of Fig. 5.1a, described by the harmonic balance equation (5.42), has been calculated here. The submatrix containing the derivatives of the kth harmonic component $\overline{E}_k$ of the error function with respect to the mth harmonic component of the state variables is given by

$$[JE]_{k,m} = \begin{bmatrix} -R\delta_{k,m} & 0 & -\delta_{k,m} & 0 \\ 0 & 0 & \delta_{k,m} & -\delta_{k,m} \\ \delta_{k,m} & -\delta_{k,m} & 0 & 0 \\ 0 & \delta_{k,m} & 0 & -\dfrac{\partial I_{nl,k}}{\partial V_{3,m}} \end{bmatrix} + [j\omega_k] \begin{bmatrix} -L_1\delta_{k,m} & 0 & 0 & 0 \\ 0 & -L_2\delta_{k,m} & 0 & 0 \\ 0 & 0 & -C\delta_{k,m} & 0 \\ 0 & 0 & 0 & -\dfrac{\partial Q_{nl,k}}{\partial V_{3,m}} \end{bmatrix} \quad (5.49)$$

where $\delta_{k,m}$ indicates the Kronecker delta. The $\delta_{k,m}$ terms come from the currents and voltages associated with the linear elements, which at each frequency ω_k depend only on the state variables at the same frequency. The high number of zero components in the Jacobian matrix can be noted.

As already shown, the Newton–Raphson algorithm provides the vector $\overline{X}^{j+1}$, corresponding to the iteration $j+1$, from the nonlinear system linearization about the point $\overline{X}^j$, resulting from the previous iteration j. This requires solving the linear system (5.44): $[JE]_j \Delta\overline{X}^{j+1} = -\overline{E}^j(\overline{X}^j)$, with $-\overline{E}^j(\overline{X}^j)$ being constant and the increment $\Delta\overline{X}^{j+1}$ being the unknown vector. It is a system of the general form $A\overline{x} = \overline{b}$, where A is a constant matrix and $\overline{b}$ is a constant vector. The linear system is generally solved through direct matrix factoring methods to invert the Jacobian matrix.

In the Gaussian elimination [47,48], the matrix $[A\overline{b}]$, obtained by adding the column $\overline{b}$ to the matrix $[A]$, is transformed into row-echelon form by means of row operations. For a general matrix $[M]$, with elements m_{ij}, the row-echelon form is such that the first nonzero element of the row m_i has a smaller index j than the first nonzero element of the next row, $i+1$. Then the system is solved through back-substitution, which is a classical way to resolve linear systems, through simple backward substitution of variables. An LU-factorization algorithm can also be employed, transforming the original system $A\overline{x} = \overline{b}$ into $LU\overline{x} = \overline{b}$. The two matrixes L and U are, respectively, lower and upper triangular matrixes. This allows breaking the original linear system into two successive systems: $L\overline{y} = \overline{b}$ and $U\overline{x} = \overline{y}$. The advantage is that the triangular systems obtained can be solved directly using forward and backward substitution. For an N-dimensional system, the total number of operations involved is $N^3/3$. Computation of the LU decomposition requires a Gaussian elimination process. The procedure becomes computationally expensive for a high order of the matrix A, that is, for a large number of state variables and/or frequency components, as the computation time increases with the cube of the matrix size.

As can be gathered, the matrix factorization required for the Jacobian inversion can be very demanding in terms of computing time and memory storage. This will be the case of large circuits, containing many active devices or quasiperiodic signals with a high number of intermodulation frequencies. However, the natural sparsity of the system (5.44) enables straightforward application of sparse matrix techniques for linear system solution. To give a brief explanation about these techniques [45,49,50], the notation $A\overline{x} = \overline{b}$ will be used for simplicity.

The system $A\overline{x} = \overline{b}$ can be solved iteratively starting from the initial value $\overline{x}_1 = \overline{b}$. The successive iterations would be obtained through $\overline{x}_{j+1} = \overline{x}_j + \overline{r}_j$, with $\overline{r}_j$ being the residue $\overline{r}_j = \overline{b} - A\overline{x}_j$. By combining the two expressions, it is possible to write $\overline{x}_{j+1} = (I_d - A)\overline{x}_j + \overline{b}$. By replacing the expressions of $\overline{x}_j, \overline{x}_{j-1}, \ldots$ recursively in terms of the residue, it is easily shown that $\overline{x}_{j+1}$ can be written as a combination of the vectors $\overline{b}, A\overline{b}, A^2\overline{b}, \ldots, A^j\overline{b}$. Each iteration j implies a new matrix–vector multiplication, with relatively low computational effort due to sparsity of the matrix A. The desired vector $\overline{x}$ is obtained when the residue $\overline{r}$ is below a certain imposed threshold. Thus, it is possible to calculate $\overline{x}$ without inverting the matrix A.

The vector set $\overline{b}, A\overline{b}, A^2\overline{b}, \ldots, A^j\overline{b}$ spans the Krylov subspace of jth-order K_j [46]. There are different approaches for the selection of $\overline{x}_{j+1}$. In the conjugate gradient approach the residual $\overline{r}_{j+1} = \overline{b} - A\overline{x}_{j+1}$ is made orthogonal to K_j. In the generalized minimal residual method (GMRES) the vector $\overline{x}_{j+1}$ is chosen to give the minimum norm of the residual $\overline{r}_{j+1}$. The GMRES is the most commonly used in the practical resolution of the harmonic balance equation [51,52].

One difficulty in the implementation of the iterative algorithm comes from the fact that the matrix V_j, comprised of the columns $\overline{b}, A\overline{b}, A^2\overline{b}, \ldots, A^j\overline{b}$, is ill conditioned [53]. The condition number of an arbitrary matrix M is defined as $N_M = \|M\|\|M^{-1}\|$, with $\|M\|$ being the infinite norm $\|M\| = \max_i \sum_j |m_{ij}|$. For an ill-conditioned matrix, the number N_M is much larger than 1. This is the case of the matrix V_j. As a result of this ill conditioning, the vectors $\overline{b}, A\overline{b}, A^2\overline{b}, \ldots, A^j\overline{b}$ tend very quickly to become almost linearly dependent, degrading the accuracy. To overcome this problem, the Arnoldi orthogonalization algorithm is applied to the Krylov basis $\overline{b}, A\overline{b}, A^2\overline{b}, \ldots, A^j\overline{b}$.

The Arnoldi algorithm provides an orthogonal basis consisting of the vectors $\overline{q}_1, \overline{q}_2, \ldots, \overline{q}_j$. These vectors span the same Krylov space K_j and the dimension j of this space increases in one at each iteration. The procedure starts by defining the normalized vector $\overline{q}_1 = \overline{b}/|\overline{b}|$. Then, each $\overline{q}_{j+1}$ is obtained by orthogonalizing the vector $A\overline{q}_j$ to the basis $\overline{q}_1, \overline{q}_2, \ldots, \overline{q}_j$, which is done by substracting the projections of $A\overline{q}_j$ over the basis $\overline{q}_1, \overline{q}_2, \ldots, \overline{q}_j$ and, thus, eliminating the components of $A\overline{q}_j$ in the directions of $\overline{q}_1, \overline{q}_2, \ldots, \overline{q}_j$. Remember that the subscript j indicates the iteration number in the iterative procedure for calculation of the unknown vector $\overline{x}$.

The orthogonalization of $A\overline{q}_j$ is carried out using the new elements $h_{i,j} = \overline{q}_i^{\mathrm{T}} A\overline{q}_j$ with $i \leq j$, in the following manner:

$$\overline{T}_j = A\overline{q}_j - (h_{1,j}\overline{q}_1 + h_{2,j}\overline{q}_2 + \cdots + h_{j,j}\overline{q}_j) \tag{5.50}$$

Because the $\overline{q}_j$ vectors have unit modulus, the vector $\overline{q}_{j+1}$ resulting from the orthogonalization procedure described will be obtained as

$$\overline{q}_{j+1} = \frac{\overline{T}_j}{\|\overline{T}_j\|} \tag{5.51}$$

For each $\overline{q}_j$ it is possible to consider an additional h element, given by $h_{j+1,j} = \|\overline{T}_j\|$. Introducing the relationship (5.51) into (5.50) with the change of variable $h_{j+1,j} = \|\overline{T}_j\|$, the following equality will be fulfilled:

$$A\overline{q}_j = h_{1,j}\overline{q}_1 + h_{2,j}\overline{q}_2 + \cdots + h_{j,j}\overline{q}_j + h_{j+1,j}\overline{q}_{j+1} \tag{5.52}$$

The matrix Q_j will be composed by the columns $\overline{q}_1, \overline{q}_2, \ldots, \overline{q}_j$. From the construction (5.52) we can also define a matrix $H_{j+1,j}$, of dimension $(j+1) \times j$, containing the h elements, and relating Q_{j+1} to Q_j. Using $H_{j+1,j}$, it is possible to express $AQ_j = Q_{j+1}H_{j+1,j}$. This matrix constitutes a representation of the orthogonal projection of A onto the Krylov subspace K_j in the basis formed by the Arnoldi vectors.

By the construction (5.52) the matrix $H_{j+1,j}$ is upper Hessenberg; that is, all its elements below the first subdiagonal are equal to zero [53]. Because of the orthogonality of Q, the relationship $QQ^{\mathrm{T}} = I$ is fulfilled. (The subindexes have been dropped for notation simplicity.) Then the matrix H can be written $H = Q^{\mathrm{T}}AQ$. Multiplying both terms of $A\overline{x} = \overline{b}$ by Q^{T} and also making use of $QQ^{\mathrm{T}} = I$, it is possible to express $Q^{\mathrm{T}}AQQ^{\mathrm{T}}\overline{x} = Q^{\mathrm{T}}\overline{b}$. The product $Q^{\mathrm{T}}\overline{b}$ is given simply by $Q^{\mathrm{T}}\overline{b} = [\|b\|0\cdots0]^{\mathrm{T}}$ due to the orthogonality of the base $\overline{q}_1, \overline{q}_2, \ldots, \overline{q}_j$ and the definition of the first vector $\overline{q}_1 = \overline{b}/|\overline{b}|$.

As shown at the beginning of this discussion, the residues in calculation of the unknown vector $\overline{x}$ are given by $\|r_j\| = \|\overline{b} - A\overline{x}_j\|$. Taking into account the equality $AQ_j = Q_{j+1}H_{j+1,j}$ and introducing the definition $Q^{\mathrm{T}}\overline{x} = \overline{y}$, it will be possible to write

$$\|r_j\| = \|\overline{b} - A\overline{x}_j\| = \left\| \overline{b} - \underbrace{AQ_j}_{Q_{j+1}H_{j+1,j}} \underbrace{Q_j^{\mathrm{T}}\overline{x}_j}_{\overline{y}_j} \right\| = \|\overline{b} - Q_{j+1}H_{j+1,j}\overline{y}_j\| \tag{5.53}$$

Because the norm $\|r_j\|$ does not change when multiplying by Q_{j+1}^{T}, the residue to minimize is $\|r_j\| = \|\overline{b} - A\overline{x}_j\| = \|[\|b\|\ 0\cdots0]^{\mathrm{T}} - H_{j+1,j}\overline{y}_j\|$. In conclusion, the system expansion on the Krylov subspace $\overline{q}_1, \overline{q}_2, \ldots, \overline{q}_j$, obtained through the Arnoldi algorithm, provides an orthogonal basis without ill-conditioning problems, and reduces the computational cost, due to the upper Hessenberg form of the matrix H.

In the practical resolution of $\|r_j\| = \|[\|b\| \ 0 \cdots 0]^{\mathrm{T}} - H_{j+1,j}\overline{y}_j\|$, a series of matrix transformations are initially carried out to reduce H to an upper triangular form. Then the optimal $\overline{y}$ is found through backward substitution. Some modifications of this technique have been proposed. For instance, a preconditioner can be applied before starting the entire process, multiplying both sides of the original system by a matrix P^{-1}, which provides $P^{-1}A\overline{x} = P^{-1}\overline{b}$. The two systems are equivalent. However, the preconditioner has the advantage of reducing the condition number of the system. The amount of iterations required increases with this condition number, so use of the preconditioner will improve the efficiency of the process [54,55].

5.4.3 Piecewise Harmonic Balance

In piecewise harmonic balance, the nonlinear elements of the circuit are identified initially [44]. These nonlinear elements are considered as dependent sources, and the set of all their control variables constitutes the set of unknowns to be determined. As an example, in the circuit in Fig. 5.1a, the only state variable is the voltage v_3, controlling the nonlinear current i_{nl}. Compared to the nodal harmonic balance, the number of state variables has been reduced from four to one. Clearly, the number of unknowns in piecewise harmonic balance will be much lower than the number in nodal harmonic balance. Remember that in nodal harmonic balance the unknowns consist of all the node voltages and inductance currents. In piecewise harmonic balance, the Q different variables that control the nonlinear elements form the set of unknowns x_q, with $1 \leq q \leq Q$. The P nonlinear elements form a second set of variables y_p, with $1 \leq p \leq P$. Finally, the S independent generators of the circuit form a third set g_s, with $1 \leq s \leq S$.

Once the Fourier basis is established, three vectors $\overline{X}, \overline{Y}$, and $\overline{G}$ containing the $2N + 1$ harmonic components of the variables x_q, y_p, and g_s, respectively, are defined. These vectors are organized similar to $\overline{x}(t)$ in (5.38). The piecewise harmonic balance system is easily obtained from the application of Kirchhoff's laws to linear networks connecting the three sets of elements $\overline{X}, \overline{Y}$, and $\overline{G}$ [40]:

$$\overline{E}(\overline{X}) = [A_x(\omega)]\overline{X} + [A_y(\omega)]\overline{Y}(\overline{X}) + [A_g(\omega)]\overline{G} = 0 \tag{5.54}$$

where $[A_x]$, $[A_y]$, and $[A_g]$ are frequency-dependent linear matrixes with a block diagonal structure composed of submatrixes at the different harmonic frequencies $-N \leq k \leq N$. The matrix $[A_x]$ contains the blocks $[A_x(\omega_k)]$ of order $Q \times Q$. The matrix $[A_y]$ contains the blocks $[A_y(\omega_k)]$ of order $Q \times P$. The matrix $[A_g]$ contains the blocks $[A_g(\omega_k)]$ of order $Q \times S$. The total number of equations is $(2N + 1)Q$ in $(2N + 1)Q$ unknowns. Equation (5.54) is, in fact, a nonlinear equation, since the functions $\overline{Y}$ and the state variables $\overline{X}$ are related nonlinearly through the constitutive relationships of the various nonlinear elements. The system (5.54) is solved numerically by minimizing the norm of the error function $\|\overline{E}\|$ through the Newton–Raphson algorithm.

As an example, the circuit of Fig. 5.1a will now be formulated through piecewise harmonic balance. The only state variable is the voltage v_3 across the nonlinear elements i_{nl} and q_{nl}. Thus, the state-variable and nonlinear element vectors are given by $\overline{x} = v_3$ and $\overline{y} = (q_{\rm nl}, i_{\rm nl})^{\rm T}$, respectively. At the frequency component ω_k, the different matrixes in (5.54) are given by

$$\begin{aligned} A_x(\omega_k) &= L_1 C\omega_k^2 - 1 - jC\omega_k R \\ A_y(\omega_k) &= [j\omega_k[R - L_2 CR\omega_k^2 + j((L_1 + L_2)\omega_k - L_1 L_2 C\omega_k^3)] \\ &\quad \times [R - L_2 CR\omega_k^2 + j((L_1 + L_2)\omega_k - L_1 L_2 C\omega_k^3)]] \\ A_g(\omega_k) &= -1 \end{aligned} \tag{5.55}$$

The vectors at ω_k are given by $\overline{X}_k = V_{3,k}, \overline{Y}_k = [Q_{{\rm nl},k}(\overline{X}), I_{{\rm nl},k}(\overline{X})]^{\rm T}$, and $\overline{G}_k = E_{g,k}$.

As can be seen, the degree in ω_k of the polynomials in $A_x(\omega_k)$ and $A_y(\omega_k)$ is two and four, whereas the degree of the linear matrixes in the nodal formulation (5.39)–(5.42) was only one. In piecewise harmonic balance a smaller number of unknowns are used at the expense of higher order in ω_k in the linear matrixes. One advantage of this type of formulation is that the complexity of the linear network or the accuracy in the description of its elements may be increased arbitrarily without increasing the number of unknowns of the nonlinear system. On the other hand, the Jacobian matrix of the piecewise formulation is generally dense (not sparse). Artificial sparsity may be created by setting to zero selected elements of the matrix according to a physical criterion [56,57]. However, a relatively high degree of sparsity is necessary for the sparse system solvers to operate efficiently. In a more recent work, an inexact Newton approach enables the use of GMRES for calculation of the inexact Newton update [52].

5.4.4 Continuation Techniques

As already stated, the initial value for the Newton–Raphson algorithm is generally provided by the circuit dc solution, obtained through a previous dc analysis of the circuit. The dc solution will constitute a valid starting point in the case of small-signal operation. However, for large-signal amplitude of the input generators, the actual solution will be quite different from the dc initial point, which will give rise to convergence problems in the Newton–Raphson algorithm. A simple way to cope with this problem is to apply the *source stepping technique*. In this technique a parameter η is introduced, expressing the generators as $\overline{G}_{\rm RF}(\eta) = \eta\overline{G}_{\rm RF}$. Then the parameter η is varied between a small initial value $\eta_{\rm o}$ and 1, in discrete steps $\eta_n = \eta_{\rm o} + n\Delta\eta$ [40,58]. When proceeding like this, the circuit operates in small-signal mode for the initial parameter value $\eta_{\rm o}$, so solving the harmonic balance system for this initial value will be straightforward. Then the next value, $\eta_1 = \eta_{\rm o} + \Delta\eta$, is considered. Due to the small value of the increment $\Delta\eta$, the final harmonic balance solution obtained for η_o will constitute a valid initial guess for η_1. The process is applied iteratively using the final harmonic balance solution at η_n as an initial guess for the harmonic balance calculation at η_{n+1}. The process repeats up to the final level $\eta = 1$, at which the circuit operates at the desired values of the input generators $\overline{G}_{RF}$. Note that a complete harmonic balance resolution, using the final solution for η_{n-1} as an initial guess, must be carried out at each η_n, with a successful Newton–Raphson convergence.

The technique described will fail if the solution path obtained as η increases exhibits a turning point versus the amplitude of any RF generator. As already known from numerous examples in Chapters 3 and 4, this is not an uncommon situation in nonlinear circuits. Even if the level step $\Delta\eta$ is reduced arbitrarily, convergence will be impossible because the curve folds over itself and evolves in an opposite sense to the parameter η. Because the turning point is associated with a qualitative change in the solution stability, techniques to cope efficiently with this problem will be provided in Chapter 6, devoted to stability analysis based on harmonic balance techniques.

The general idea behind the continuation technique also applies in making use of other parameters, with or without a physical meaning. Another example is the artificial introduction of resistances connected in parallel with nonlinear device ports, which are increased gradually in small logarithmic steps up to an infinite value. The circuit operates under nearly linear conditions for a low resistance value. Incrementing the resistance at small steps, the solution obtained at step n will constitute a valid initial guess for the next resistance value, at step $n+1$.

The continuation methods discussed will generally enable harmonic balance convergence of circuits in forced operation at the fundamental frequencies delivered by the input RF generators and its harmonic or intermodulation frequencies. However, in circuits with autonomous behavior, the steady-state solution may contain fundamental frequencies that are not delivered by any generator. This is the case for oscillation at the frequency ω_a, coexisting with the generator frequency ω_{in}, or a frequency-divided solution with the subharmonic frequency ω_{in}/N. Even if the designer is conscious about the existence of these autonomous frequencies and takes them into account when establishing the fundamental frequencies of the Fourier basis, the continuation techniques discussed will not be able to initialize frequency components of the form $n\omega_{in} + m\omega_a$ with $m \neq O$ (in the quasiperiodic regime) or $m\omega_{in}/N$, with $m \neq kN$ (in divided regime), with m an integer. As already known, coexisting with any oscillatory solution there is generally a mathematical solution for which the circuit exhibits no self-oscillation. This is because, as shown in previous chapters, the harmonic balance system contains a homogeneous nonlinear subsystem at the frequencies $n\omega_{in} + m\omega_a$ or $m\omega_{in}/N$, with m an integer, which admits a zero solution. On the other hand, the Newton–Raphson algorithm used for solution of the harmonic balance equations is very dependent on the initial value. Unless an accurate initial point is provided, this algorithm naturally converges to a solution having the input generator frequencies as the only fundamentals. This is because the generators naturally initialize these fundamental frequencies, and the nonlinearities naturally give rise to the harmonic components of these frequencies. Due to the absence of any generators at ω_a or ω_{in}/N, the iterative process will be unable to provide any values to the frequency components $n\omega_{in} + m\omega_a$ or $m\omega_{in}/N$. Complementary harmonic balance techniques for the analysis of autonomous circuits are presented in Section 5.5.

5.4.5 Algorithms for Calculation of Discrete Fourier Transforms

From the point of view of discrete Fourier transform (DFT) algorithms, two main types of signals can be distinguished: periodic and quasiperiodic. In the case of periodic signals, the techniques for DFT are well known. In particular, the fast Fourier transform (FFT) reduces the number of operations involved in the transformation. These algorithms cannot be applied directly to quasiperiodic signals, so complementary approaches will be needed [42,59,60]. A brief summary of the principal approaches follows.

5.4.5.1 DFT of Periodic Signals The case of a periodic signal $y(t)$ expressed in the form $y(t) = \sum_{k=-N}^{N} Y_k e^{jk2\pi f_o t}$ is considered first. The $2N + 1$ coefficients Y_k may be calculated from the linear equation system that is obtained when considering $2N + 1$ time points t_n [59]. The resulting square system is the following:

$$\begin{bmatrix} 1 & 1 & \cdots & 1 & 1 & \cdots & 1 \\ 1 & e^{j2\pi f_o t_1} & \cdots & e^{j2\pi N f_o t_1} & e^{-j2\pi N f_o t_1} & \cdots & e^{-j2\pi f_o t_1} \\ \vdots & \vdots & \cdots & \vdots & \vdots & \vdots & \vdots \\ 1 & e^{j2\pi f_o t_{2N}} & \cdots & e^{j2\pi N f_o t_{2N}} & e^{-j2\pi N f_o t_{2N}} & \cdots & e^{-j2\pi f_o t_{2N}} \end{bmatrix} \begin{bmatrix} Y_0 \\ Y_1 \\ \vdots \\ Y_N \\ Y_{-N} \\ \vdots \\ Y_{-1} \end{bmatrix}$$

$$= \begin{bmatrix} y(0) \\ y(t_1) \\ \vdots \\ y(t_{2N}) \end{bmatrix} \tag{5.56}$$

System (5.56) can be written in the compact form $[W]\overline{Y} = \overline{y}$. Provided that the time instants t_n are calculated such that $[W]$ is invertible, the Fourier transform of $y(t)$ will be given by $\overline{Y} = [W]^{-1}\overline{y}$. The time points t_n are

$$t_n = n\frac{1}{(2N+1)f_o} \equiv nT_s \tag{5.57}$$

The frequency $f_s = (2N + 1)f_o$ is clearly the sampling frequency of the time-domain signal. With the point selection (5.57), the matrix $[W]$ is invertible and well conditioned. For a matrix to be well conditioned, its condition number must be very close to unity. This number is calculated from the infinite norm of $[W]$ as $N_W = \|W\|\|W\|^{-1}$, with $\|W\| = \max_i \sum_j |W_{ij}|$.

From the time point choice (5.57), and taking into account the harmonic relationship $f_k = kf_o$, the harmonic components of $y(t)$ can be written

$$Y_k = \frac{1}{2N+1}\sum_{n=0}^{2N} y(n)e^{[-j2\pi/(2N+1)]nk} = \frac{1}{M}\sum_{n=0}^{2N} y(n)(\Phi_M)^{nk} \tag{5.58}$$

where $M = 2N + 1$ and $\Phi_M \equiv e^{-j2\pi/M}$. The FFT takes advantage of the periodicity of $(\Phi_M)^{nk}$, with respect to both n and k, reducing substantially the number of operations involved in calculation of the complete sequence Y_k. The $2N$-point DFT is subdivided into two N-point DFTs by splitting the input signal into odd- and even-numbered samples. The decimation process is continued until a series of DFTs with only two input samples is obtained. The FFT algorithm requires about $N \log_2 N$ operations instead of the N^2 operations needed for direct computation of the DFT. Efficient implementation of this technique requires a choice of the initial number of samples $2N$ as an integer power of 2.

5.4.5.2 DFT of Quasiperiodic Signals In the case of periodic signals, the sample points are spaced equally within the signal period. Then the DFT has great accuracy since the rows of the transformation matrix $[W]$ are orthogonal and the matrix is well conditioned. In the case of quasiperiodic signals, the equally spaced time points give rise to the ill conditioning of $[W]$. The truncation error is determined by the matrix condition number [59]. Thus, the DFT cannot be applied directly. Various methods for calculation of the Fourier transform of quasiperiodic signals have been presented in the literature. The most efficient are summarized below.

Almost-Periodic Fourier Transform In a paper by Kundert et al., the time points t_n are obtained randomly from a time interval equal to three times the period of the smallest nonzero frequency [59]. A number M' of time points, larger than M, is generated. However, instead of oversampling, which would give rise to an increase in the computational cost, only M time points are taken from the original set of M' points. A variation of the Gram–Schmidt orthogonalization procedure is used for this selection. The M time points selected are those giving rise to the nearest-to-1 condition number of the matrix $[W]$. However, the near orthogonality of $[W]$ can also be achieved using nonrandom selections of the sample points. Several strategies have been shown by Ngoya et al. [60].

Frequency Remapping In frequency remapping we take into account that the goal of Fourier transform in the context of the harmonic balance analysis is determination of the frequency components of nonlinear elements. Let the memory-less function $y \equiv y(\overline{x})$ be considered. It can easily be shown [58] that the Fourier coefficients of the nonlinear element Y_k, with $k = -N$ to N, depend only on the Fourier coefficients of the state variables, grouped in the vector $\overline{X}$, and on the integer vector $\overline{\lambda}_k$ that generates the particular intermodulation product f_k, but do not depend on the frequency basis $\overline{F}$. They can actually be written $Y_k \equiv Y_k(\overline{X}, \overline{\lambda}_k)$. Thus, it will not be necessary to use the actual waveforms of the state variables $\overline{x}(t)$ to obtain the Fourier coefficients of $y(\overline{x})$. It will be possible to use simpler artificial waveforms for the $\overline{Y}_k$ calculation. A convenient choice for the frequency basis $\overline{F}^d$ would be one providing periodic artificial waveforms to which the efficient FFT algorithm may be applicable. Thus, the actual fundamental frequencies $\overline{F}^t = (F_1 \cdots F_{NF})$ are remapped to the artificial frequencies $(\overline{F}^d)^t = (F_1^d \cdots F_{NF}^d)$. Once the harmonic values $\overline{Y}_k$ are determined, they will be assigned to the actual frequencies, given by $f_k = \overline{\lambda}_k^t \overline{F}$. For this calculation to be valid, the artificial basis $\overline{F}_d$ must generate $2N+1$ different frequencies $f_{kd} = \overline{\lambda}_k^t \overline{F}^d$. These artificial frequencies are the *remapped frequencies*.

The value of the artificial fundamentals depends on the truncation criterion that is used for the Fourier series. For $Q = 2$, the artificial fundamentals, in the two cases of box and diamond truncation, are given, respectively, by [42,58]

$$\left.\begin{aligned} F_1^d &= \Delta f \\ F_2^d &= (2nl+1)\Delta f \end{aligned}\right\} \text{box truncation} \qquad \left.\begin{aligned} F_1^d &= nl\Delta f \\ F_2^d &= (nl+1)\Delta f \end{aligned}\right\} \text{diamond truncation} \tag{5.59}$$

where $\Delta f$is an arbitrary frequency, different from zero, and nl is the nonlinearity order.

Multidimensional Fourier Transform Let a signal $x(t)$ and NF fundamental frequencies be considered. An artificial signal $\tilde{x}$ can be defined using a different time variable for each fundamental [i.e., $\tilde{x}(t_1, t_2, \ldots, t_{NF})$]. Then the following equality is fulfilled: $x(t) = \tilde{x}(t, t, \ldots, t)$. The artificial signal x is periodic in each time variable t_j. Considering a different truncation order nl_j $1 \le j \le NF$ for each fundamental frequency, the NF-dimensional DFT is written [42,61]

$$X(\lambda_1, \lambda_2, \ldots, \lambda_{NF}) = \frac{1}{N_{\text{tot}}} \sum_{n_1=0}^{2nl_1+1} \cdots \sum_{n_{NF}=0}^{2nl_{NF}+1} x_{n_1 \cdots n_{NF}} \times \exp^{-j2\pi\left(\frac{\lambda_1 n_1}{N_1} + \cdots + \frac{\lambda_{NF} n_{NF}}{N_{NF}}\right)} \tag{5.60}$$

where $N_j = 2nl_j + 1$, $1 \le j \le NF$, and $N_{\text{tot}} = \prod_{j=1}^{NF} N_j$. In expression (5.60) care must be taken to prevent aliasing when choosing the sampling orders. The multidimensional Fourier transform can be obtained through a sequential calculation of fast Fourier transforms in the various time variables. However, the computational cost of these operations can be greatly reduced through the use of special algorithms [62]. For a nonlinear relationship $y = f(x)$ it is easily shown [42,63] that samples of $y(t)$ can be obtained through $\tilde{y}_{n_1,\ldots,n_{NF}} = f(\tilde{x}_{n_1,\ldots,n_{NF}})$, which is easily generalized to a nonlinear dependence on multiple state variables. Then the Fourier coefficients of $y(t)$ can be determined through an expression formally identical to (5.60). The number of samples associated with each fundamental can be chosen individually. The total number of samples is equal to the product of the individual sample numbers.

5.5 HARMONIC BALANCE ANALYSIS OF AUTONOMOUS AND SYNCHRONIZED CIRCUITS

As has been shown, the harmonic balance technique requires the provision by the designer of the set of fundamentals $F_1, \ldots, F_{NF}$ to be used in the Fourier series expansion of the circuit variables. If the circuit exhibits self-oscillation, the corresponding frequency must be included in this basis. However, even when the entire frequency basis is provided correctly, there might still be problems for simulation of the oscillatory regime. The reason is that, coexisting with the oscillatory solution, there is generally a mathematical solution without oscillation. The most obvious example is the free-running oscillator, for which a dc solution always coexists with the oscillating solution. The Newton–Raphson algorithm used in resolution of the harmonic balance system is very dependent on the initial value. In an oscillatory regime, provision of an accurate starting point is not an easy task. The oscillation is self-generated, so it depends on the circuit element values and input sources. When this oscillation is not synchronized, a suitable guess of the oscillation amplitude and frequency, or amplitude and phase in the case of synchronization, will be necessary. Otherwise, we will obtain only the circuit nonautonomous response to the input sources. In case the oscillation frequency is different from that of the input source, frequency terms involving multiples of the self-generated fundamentals will tend to a zero value. The convergence process will provide the coexisting nonoscillatory steady state.

In the following, a distinction is made between regimes exhibiting oscillation at an incommensurable frequency (nonsynchronized) and regimes exhibiting oscillation at a frequency related rationally to that of the input source (synchronized). Here, the nonsynchronized regimes will also be called autonomous. In autonomous regimes, the oscillation frequency is an unknown to be added to the set of state variables of the harmonic balance system. Thus, the global set of unknowns will contain the ordinary voltage and current variables, plus the oscillation frequency. It is a mixed set and the associated harmonic balance technique is called *mixed harmonic balance* [15]. We present this mixed harmonic balance formulation in Section 5.5.2. This formulation requires a suitable initial value for the Newton–Raphson algorithm, to avoid convergence to a trivial, nonoscillatory solution.

The initial value problem can be circumvented through the use of complementary techniques. The aim is to initialize the oscillation in a systematic manner, with no need to provide a full initial value vector $\overline{X}_o$ as an initial guess. These techniques, presented in Section 5.6.1, are based on the introduction of an auxiliary generator into the circuit. This is an artificial generator used for simulation purposes only. The generator plays the role of the oscillation, so in harmonic balance, the oscillatory regime can be simulated as an ordinary forced one. The auxiliary generator technique can easily be implemented on either in-house or commercial harmonic balance software. The only requirements for commercial software would be the existence of ideal impedance elements and nonlinear optimization tools.

5.5.1 Mixed Harmonic Balance Formulation

For steady-state free-running oscillation or a self-oscillating mixer regime, the circuit oscillates at a self-generated frequency that depends on the circuit element values and the values of the dc generators (free-running oscillation) or dc and RF generators (self-oscillating mixer regime). Thus, the oscillation frequency F_a will be an unknown of the problem to be added to the set of harmonic balance state variables $\overline{X}$. In nodal harmonic balance, the set $\overline{X}$ consists of the $(2N + 1)$ harmonic components of the P state variables that constitute the node voltages and inductor currents. In piecewise harmonic balance, the set $\overline{X}$ consists of the $(2N + 1)$ harmonic components of the Q state variables that make up the control voltages of nonlinear devices. When adding the oscillation frequency to the set of state variables, a set of variables of different nature or mixed set is obtained. The number of equations remains equal to $P(2N + 1)$ (nodal) or $Q(2N + 1)$ (piecewise). However, inclusion of the oscillation frequency gives rise to one extra unknown, so the number of unknowns is, in each case, $\text{Dim}(\overline{X}) + 1$. Thus, the system is unbalanced. However, in an autonomous regime there is an invariance of the solution obtained with respect to time translations. Thus, it is possible to set arbitrarily to zero the phase of one of the harmonic components of one of the state variables. The first harmonic component of a given state variable is generally chosen as $\text{Im}[X_1^p] = 0$. Thus, there is one less unknown in the vector $\overline{X}$. A new vector $\overline{X}'$ is defined which is equal to $\overline{X}$ except that it does not contain $\text{Im}[X_1^p] = 0$. The new set of unknowns is given by $[\overline{X}', F_a]$, and the mixed harmonic balance equation is written $\overline{E}[\overline{X}', F_a] = 0$. As in standard harmonic balance, this equation is solved with the aid of the Newton–Raphson algorithm, which requires calculation of a mixed Jacobian matrix $[JE'] = \left[\partial\overline{E}/\partial\overline{X}', \partial\overline{E}/\partial F_a\right]$. Note that this Jacobian matrix is not singular, as its singularity has been removed when imposing the additional condition $\text{Im}[X_1^p] = 0$. Detailed explanations of this fact were given in Chapters 1 and 2.

For the reasons already discussed, unless a suitable initial point is provided for the Newton–Raphson algorithm, the mixed harmonic balance system will converge to the nonoscillatory mathematical solution that always coexists with the oscillating solution. In the literature [64–66], various techniques have been proposed for efficient initialization of the harmonic balance system. The oscillation frequency is often estimated from a small-signal analysis of the circuit, using the value at which the oscillation startup conditions are fulfilled. Some authors [65] propose adding the steady-state oscillation condition derived by Kurokawa to the set of mixed-mode harmonic balance equations. As already known, this condition would be $Y_T(\overline{X}', F_a) = 0$, with Y_T the total input current/voltage relationship at a given observation node. To avoid the trivial solution, other authors propose normalizing the error function $\overline{E}$ to the magnitude at the oscillation frequency of one of the state variables, setting $\overline{E}[\overline{X}', F_a]/\text{Mag}(X_1^p) = 0$ [66]. A different strategy is presented in the following.

5.5.2 Auxiliary Generator Technique

Auxiliary generators make it possible to force harmonic balance convergence toward special solutions that are not obtained in a default analysis. Examples are subharmonic or autonomous regimes or multivalued sections in the solution curves of power amplifiers or other forced circuits. Basically, the problem with oscillatory regimes derives from the fact that there is no generator at the oscillation frequency. In the auxiliary generator technique, an artificial generator is introduced at this frequency. The generator will play the role of the oscillation and will avoid the default convergence toward a nonoscillatory solution. In brief, the auxiliary generator technique takes advantage of the natural construction of the harmonic balance solution from the voltage or current of the existing generators.

Two different types of auxiliary generator are possible: voltage generators, connected in parallel at a circuit node, and current generators, connected in series at a circuit branch. The use of a voltage auxiliary generator is illustrated in Fig. 5.6. This figure shows the circuit page of the commercial harmonic balance software Advanced Design System (ADS). The simulated circuit is the oscillator in Fig. 1.6. In the schematic the auxiliary generator is separated from the circuit and connected in parallel at the drain node V_d. As stated earlier, the auxiliary generator operates at the autonomous or synchronized oscillation frequency, which is written $F_{AG} = F_a$, the subscript a standing, in general, for *autonomous*. It must also be taken into account that voltage generators and current generators are short circuits and open

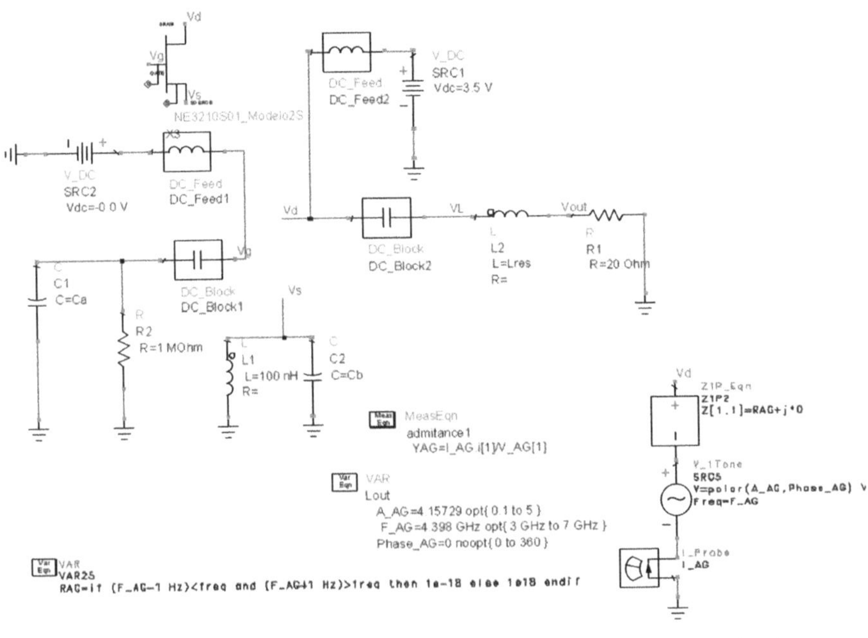

FIGURE 5.6 Oscillator circuit of Fig. 1.6 with an auxiliary generator of voltage type connected in parallel at the drain node.

circuits, respectively, at frequencies different from the ones that they deliver. Thus, an ideal filter is necessary in each case, to avoid a large perturbation of the solution due to the short circuiting or opening of the frequency components $F \neq F_{AG}$. For a voltage generator, the ideal filter is connected in series with this generator (see Fig. 5.6). The filter has a zero impedance value at the generator frequency $F_{AG} = F_a$ and infinite impedance at any other frequency (for which the generator will have no effect). In the case of a current auxiliary generator, the ideal filter is connected in parallel with this generator. In a manner analogous to the voltage auxiliary generator, the filter must exhibit an infinite impedance at the generator frequency $F_{AG} = F_a$ and zero impedance at any other frequency (for which the generator will have no effect).

As can be seen, the artificial voltage generator in Fig. 5.6 is connected in series with an ideal box, defined by its impedance $Z[1, 1] = R_{AG} + j0$. A conditional sentence is used to assign this impedance value. The resistance R_{AG} is equal to an arbitrary small value at the auxiliary generator frequency and near infinity at all the other frequencies. The conditional sentence is: *if freq* $=$ *FAG then RAG* $=$ *1E*$^{-18}$ *else RAG* $=$ *1E*^{18} *endif*.

As already stated, to be of any use in the determination of the oscillatory solution, the auxiliary generator must have no influence over this solution once the process is complete. To fulfill this condition, the voltage auxiliary generator must exhibit a zero value of its current-to-voltage relationship $Y_{AG} = 0$ at its operation frequency F_{AG}. In turn, the current auxiliary generator must exhibit a zero value of its voltage-to-current relationship $Z_{AG} = 0$ at F_{AG}. Although either a current or a voltage generator may be chosen, only voltage generators are considered in this book. As gathered from previous explanations, a current generator is the dual of a voltage generator. In most cases the oscillatory solution can be obtained with a voltage auxiliary generator. The current generator can be more effective for the analysis of series resonances, though it is rarely necessary.

The nonperturbation condition $Y_{AG} = 0$ introduces two additional real equations in the harmonic balance system: $\text{Re}[Y_{AG}] = 0$, $\text{Im}[Y_{AG}] = 0$, but there will also be two additional variables. These variables depend on the type of regime, autonomous or synchronized, to be analyzed. Free-running oscillators, injection-locked oscillators, and self-oscillating mixers are considered next.

5.5.2.1 Free-Running Oscillators

For the simulation of a free-running oscillator, an auxiliary generator is introduced at the oscillation frequency $F_{AG} = F_a$. Due to the fact that the frequency F_a is generated autonomously by the circuit, its value will depend on the values of the circuit elements and thus will be an unknown to be determined. The auxiliary generator will have amplitude A_{AG} and phase ϕ_{AG}. However, in an autonomous regime, the phase ϕ_{AG} can be of any value, due to the irrelevance of the steady-state periodic solution with respect to time translations. Any possible phase value ϕ_{AG} will give rise, after completing the harmonic balance simulation, to the same waveforms of the circuit variables. For simplicity, the value $\phi_{AG} = 0$ will be imposed. In the case of a voltage auxiliary generator introduced in parallel at a circuit node, this choice sets the solution phase reference at the node voltage at the oscillation frequency $F_{AG} = F_a$. Note that by means of this particular assignment, the phase shift irrelevance has been eliminated and the system is no longer singular. All the rest of harmonic components of the state variables must be fully determined as to amplitude and phase.

As already stated, the auxiliary generator must not have influence on the steady-state oscillatory solution, which is ensured by the condition $Y_{AG} = 0$, with Y_{AG} being the ratio between the auxiliary generator current and the voltage at the frequency F_{AG}. To fulfill this nonperturbation condition, the unknowns to be determined are the oscillation frequency, which is the frequency delivered by the auxiliary generator, and the amplitude A_{AG} of this generator. Defining a vector $\overline{Y}_{AG}$ composed of the real and imaginary parts of the current/voltage ratio, the generator nonperturbation condition is

$$\overline{Y}_{AG}(A_{AG}, F_{AG}) \equiv \begin{bmatrix} \text{Re}\left[\dfrac{I_{AG}}{A_{AG}}\right] \\ \text{Im}\left[\dfrac{I_{AG}}{A_{AG}}\right] \end{bmatrix} = \overline{0} \qquad V_{AG} \equiv A_{AG}e^{j0} \tag{5.61}$$

Note that the division by A_{AG} prevents the nonoscillatory solution $A_{AG} = 0$. Defining I_{AG} as the auxiliary generator current from ground to the connection node, the admittance function Y_{AG} agrees with the total input admittance of the circuit being analyzed observed from the node at which the auxiliary generator is connected. Thus, the condition $Y_{AG} = 0$ is equivalent to the steady-state oscillation condition derived in Chapter 1. Actually, the auxiliary generator is totally analogous to the voltage generator used in Section 1.3 to analyze the variations of the oscillator total admittance function $Y_T(V, \omega)$ versus the frequency ω and voltage amplitude V at the observation node selected. Here the amplitude and frequency that make the function $Y_T(V, \omega)$ equal to zero will be calculated directly through an error minimization algorithm or by using optimization tools in commercial harmonic balance software. The two types of resolution of the nonperturbation condition are described below.

Error Minimization in In-House Software The nonperturbation condition of the auxiliary generator $Y_{AG} = 0$ adds two more equations and two more unknowns to the harmonic balance system, which becomes

$$\begin{aligned} \overline{E}(\overline{X}, F_{AG}, A_{AG}) &= \overline{0} \qquad &\text{(a)} \\ \overline{Y}_{AG}(\overline{X}, F_{AG}, A_{AG}) &= \overline{0} \qquad &\text{(b)} \end{aligned} \tag{5.62}$$

The error function $\overline{E}$ is the error function of the standard harmonic balance system. The combined system (5.62) can be solved either in parallel or at two different levels.

In parallel resolution, a global error function is defined accounting for the two different subsystems: $\overline{E}' = [\overline{E}, \overline{Y}_{AG}]^{\text{T}}$. In the harmonic balance formulation, the auxiliary generator is introduced in the set of circuit generators $\overline{G}$. The auxiliary generator frequency F_{AG} is the fundamental frequency of the Fourier series. The new error function $\overline{E}'$, containing $\mathrm{Dim}(\overline{X})+2$ equations, is minimized through a global Newton–Raphson algorithm. The unknowns to be determined are the state variables $\overline{X}$, plus the generator amplitude A_{AG} and frequency F_{AG}. It is a mixed-variable system because the set of unknowns contains the auxiliary generator frequency in addition to the ordinary set of voltages and currents. Because of the arbitrary selection $\phi_{AG}=0$, the parallel connection of the auxiliary generator at a given circuit node will naturally force a zero value of the node voltage phase at F_{AG}. The parallel resolution of the two subsystems in (5.62) is very efficient in terms of computation time.

In the two-tier resolution of system (5.62), the nonperturbation equation $\overline{Y}_{AG}(A_{AG}, F_{AG})=0$, depending only on the auxiliary generator amplitude, and the frequency constitutes the outer tier. The pure harmonic balance system, solved as usual with the Newton–Raphson algorithm, constitutes the inner tier. In this inner-tier resolution, the auxiliary generator frequency F_{AG} and auxiliary generator amplitude A_{AG} are taken as constant values. For this inner-tier system, the auxiliary generator constitutes a simple forcing generator, like any of those composing the generator vector $\overline{G}$. In case the outer-tier equation $\overline{Y}_{AG}(A_{AG}, F_{AG})=0$ is also solved through Newton–Raphson, the corresponding Jacobian matrix is written

$$[JY_{AG}] - \begin{bmatrix} \dfrac{\partial Y^r_{AG}}{\partial A_{AG}} & \dfrac{\partial Y^r_{AG}}{\partial F_{AG}} \\ \dfrac{\partial Y^i_{AG}}{\partial A_{AG}} & \dfrac{\partial Y^i_{AG}}{\partial F_{AG}} \end{bmatrix} \tag{5.63}$$

The matrix (5.63) is calculated through finite differences. To determine the derivative of the complex admittance function Y_{AG} with respect to A_{AG}, a small increment is considered in the auxiliary generator amplitude $A_{AG}+\Delta A_{AG}$, whereas its frequency is maintained constant at F_{AG}. Next, a harmonic balance simulation is carried out for this new amplitude value, $A_{AG}+\Delta A_{AG}$. The derivative is obtained from the ratio $[Y_{AG}(A_{AG}+\Delta A_{AG}, F_{AG}) - Y_{AG}(A_{AG}, F_{AG})]/\Delta A_{AG}$. The derivative of the complex function Y_{AG} with respect to F_{AG} is calculated in a similar manner, considering a small frequency increment $F_{AG}+\Delta F_{AG}$ and maintaining the amplitude constant at A_{AG}. Then the outer-tier Newton–Raphson algorithm is formulated:

$$\begin{bmatrix} A_{AG} \\ F_{AG} \end{bmatrix}^{j+1} = \begin{bmatrix} A_{AG} \\ F_{AG} \end{bmatrix}^{j} - [JY_{AG}]_j^{-1} \begin{bmatrix} Y^r_{AG} \\ Y^i_{AG} \end{bmatrix}^{j} \tag{5.64}$$

where j indicates the iteration number.

Once convergence is achieved, through either a parallel or a two-tier resolution of system (5.62), the auxiliary generator will have no influence over the oscillatory solution. Its final value will agree with the connection-node voltage at the oscillation frequency F_{AG}, namely, $V(F_{AG}) = A_{AG}e^{j0}$.

Optimization Tools in Commercial Harmonic Balance Software The two real functions $|Y^r_{AG}|$ and $|Y^i_{AG}|$ can be minimized using the optimization tools of commercial harmonic balance software. In a manner similar to the two-tier resolution of system (5.62), this minimization is performed externally to the pure harmonic balance system $\overline{E}(\overline{X}) = 0$. The optimization variables will be A_{AG} and F_{AG} and the goals $|Y^r_{AG}| = 0$ and $|Y^i_{AG}| = 0$.

An example of this optimization procedure is shown in Fig. 5.7. The admittance function is defined as YAG = I_AG.i[1]/A_AG[1], with the symbol [k] indicating the frequency component selected, corresponding to the fundamental frequency. The optimization goals are Re(YAG) = 0, Im(YAG) = 0. In practice, the goals are written -1E^{-15} < real(YAG) < 1E^{-15} and -1E^{-15} < imag(YAG) < 1E^{-15}. As also shown in Fig. 5.7, the optimization variables are A_AG, taking values in the interval [0.1 V, 5 V] and F_AG, taking values in the interval [3 GHz, 7 GHz].

The gradient optimization is usually most efficient for the minimization of YAG, provided that convenient initial values for A_AG and F_AG are supplied to the simulator. In a general manner, the error of the optimization process is given by the

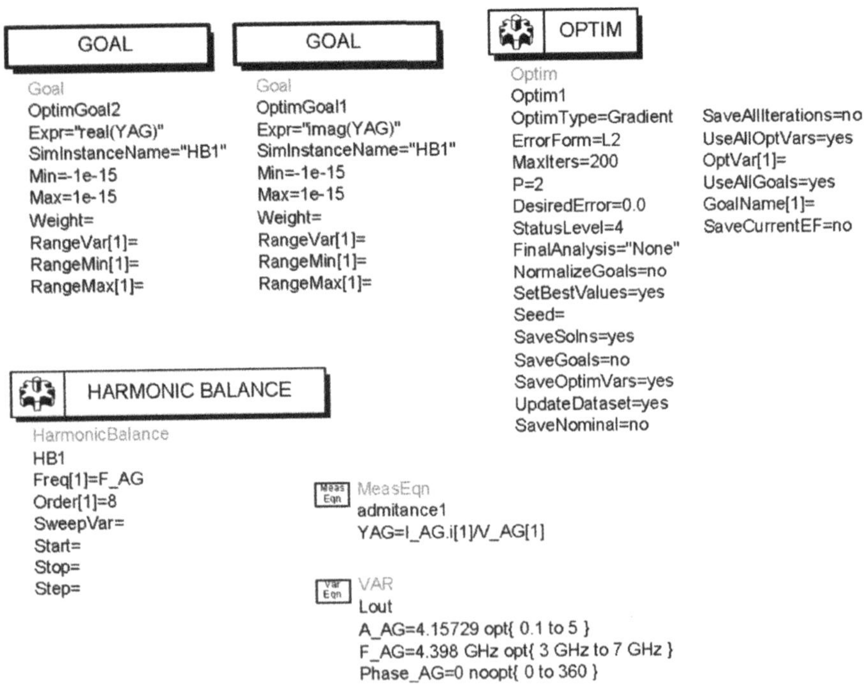

FIGURE 5.7 Detail of the circuit schematic page used for simulation of the circuit of Fig. 5.6 in the commercial harmonic balance software ADS. The outer-level equation $\overline{Y}_{AG} = 0$, corresponding to the *AG* nonperturbation condition, is solved through optimization with the goals real(YAG) = 0, imag(YAG) = 0. The optimization variables are the oscillation amplitude A_AG and frequency F_AG.

difference between the desired values or goals of the optimized functions and the values resulting after each iteration. In the gradient optimization, the real-valued error functions must be defined and differentiable in the neighborhood of each iteration point. The new values of the optimized variables are obtained by taking into account that the error function E decreases faster in the direction opposite the gradient of the error function with respect to these variables $-\nabla E$. The gradient is reevaluated after each iteration. In the particular case of oscillator analysis using an auxiliary generator, the real error functions agree with real(YAG) and imag(YAG), and the gradient is calculated with respect to A_AG and F_AG.

The gradient optimization can converge to a local minimum, not reaching the goals imposed on real(YAG) and imag(YAG). To avoid this problem, a previous random optimization can be carried out. Another possibility is to obtain the initial A_AG and F_AG values through a simple sweep technique. For example, a harmonic balance analysis of the oscillator of Figs. 5.6 and 5.7, using optimization to achieve $\overline{Y}_{AG} = 0$, is presented in the following.

For an analysis of the oscillator circuit in Figs. 5.6 and 5.7, the initial values of the auxiliary generator amplitude A_AG and frequency F_AG are estimated with a sweep technique. Three amplitude values of A_AG are selected: 0.01, 2, and 4 V, performing for each a sweep in the frequency F_AG from 3 to 7 GHz. The ratio between the current through the auxiliary generator (entering the circuit) and the voltage delivered is evaluated. This ratio, agreeing with YAG, is equal to the input admittance function observed from the node at which the auxiliary generator is connected. The results are shown in Fig. 5.8. The number of considered harmonic components is $N = 15$. As can be seen, for A_AG = 0.01 V (the dashed line), the input admittance function exhibits a negative real part and a resonance with positive slope at about 5.2 GHz. For A_AG = 2 V (the dotted line), there is little variation

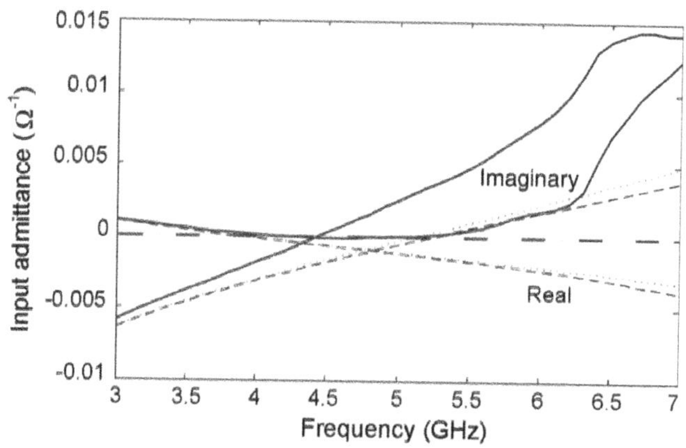

FIGURE 5.8 Estimation of a suitable initial condition for the minimization through gradient optimization of the admittance function YAG in terms A_AG and F_AG. The real and imaginary parts of the function YAG have been sketched versus the auxiliary generator frequency for three different amplitude values: A_AG = 0.01 V (dashed line), A_AG = 2 V (dotted line), and A_AG = 4 V (solid line).

of the negative conductance, with a small reduction of the resonance frequency. For A_AG = 4 V (the solid line) little negative conductance is observed, whereas the resonance frequency has decreased to 4.4 GHz. In view of these results, a good starting point for the iterative process could be the frequency F_AG = 4.5 GHz and the auxiliary generator amplitude A_AG = 3 V.

Figure 5.9 shows the evolution of the optimization process from the not so good initial value A_AG = 0.1 V and F_AG = 5 GHz. Nineteen iterations are necessary in order to reach the final amplitude and frequency values A_AG = 4.15 V and F_AG = 4.4 GHz, fulfilling the imposed goals |real(YAG)| < $1E^{-15}$, |imag(YAG)| < $1E^{-15}$. Figure 5.9a shows the variation of the amplitude and frequency versus the iteration number of the optimization process, and Fig. 5.9b shows the corresponding variation of the real and imaginary parts of YAG. When the admittance function is equal to zero, F_AG agrees with the free-running oscillation frequency F_a, and A_AG agrees with the amplitude of the first harmonic component of the voltage at the node at which the auxiliary generator is connected.

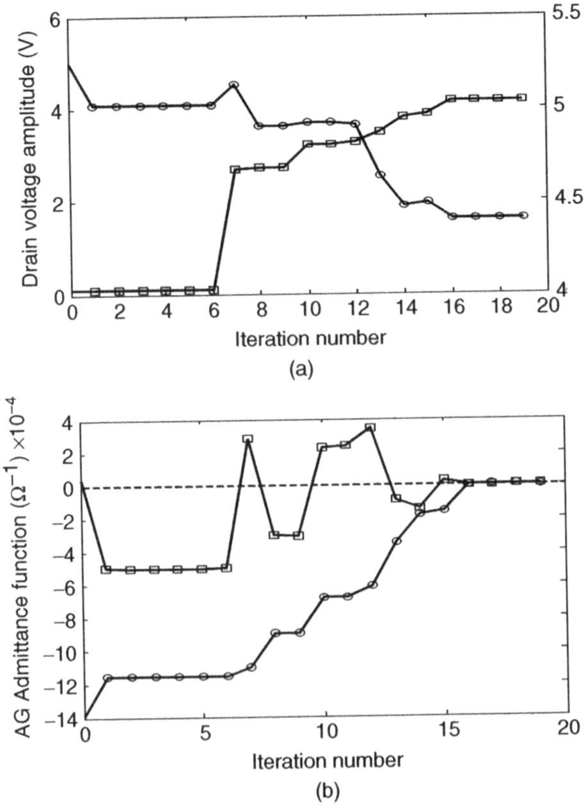

FIGURE 5.9 Use of the auxiliary generator technique. Evolution of the optimization process versus the iteration number. (a) Variation of the amplitude and frequency of an auxiliary generator. (b) Variation of the real and imaginary parts of the admittance function YAG.

5.5.2.2 Synchronized Regime In a synchronized regime, the self-oscillation frequency F_a is rationally related to the input generator frequency F_{RF}; that is, $F_a/F_{RF} = m/k$, with m and k integers. Because of this rational relationship, there will be constant phase shift between the oscillation and the input generator signal. Clearly, in this regime the oscillation frequency is not an unknown, as it is determined by that of the input generator as $F_a = mF_{RF}/k$.

In regard to the possible coexistence of the synchronized solution with a nonoscillatory solution, three cases are considered (see Chapter 4): fundamentally injection-locked oscillators, frequency dividers, and subsynchronized oscillators. For a fundamentally injection-locked oscillator, the coexistence occurs for low input power values, with the synchronized solutions located in a closed curve, and the nonoscillatory solutions located in a low-amplitude open curve. For larger input power, there is a unique solution curve. In the case of frequency dividers, the divided solutions at $F_a = F_{RF}/k$ always coexist with a nondivided solution at the input generator frequency F_{RF}. The case of subsynchronized oscillators is similar to that of fundamentally injection-locked oscillators. For lower input power values, the synchronized solutions are located in a closed curve and coexist with the nonoscillatory solutions, located in a low-amplitude open curve. For higher input power values, there is a unique solution curve.

In the coexistence of solutions, the harmonic balance method will provide by default to the nonoscillatory solution, which does not require a proper initial value and enables a simpler convergence due to its lower amplitude and lower degree of nonlinearity. A variant of the auxiliary generator technique presented in Section 5.5.2.1 can be used to avoid this undesired convergence. The cases of synchronization at $F_a = F_{RF}/k$, with $k \geq 1$ and $F_a = mF_{RF}$, with k and m integers, are considered separately.

Synchronization at $F_a = F_{RF}/k$ For the analysis of a synchronized regime at $F_a = F_{RF}/k$, with $k \geq 1$, an auxiliary generator is introduced at the frequency $F_{AG} = F_a$ at which the self-oscillation occurs. Since the oscillation frequency F_a is determined by F_{RF}, it will not be an unknown in the problem. In contrast, the phase ϕ_{AG} of the auxiliary generator is no longer irrelevant due to the presence of the input synchronizing source establishing the circuit phase reference. It will be an unknown to be resolved. This is due to the phase relationship between the oscillation and the input generator signal. The auxiliary generator must force convergence toward the oscillatory solution without affecting this solution. Thus, the nonperturbation condition $Y_{AG} = I_{AG}/V_{AG} = 0$ must be fulfilled.

For the harmonic balance simulation, the various circuit variables are expressed in a Fourier series with the frequency of the auxiliary generator $F_{AG} \equiv F_a = F_{RF}/k$ as fundamental. The harmonic balance system, including this generator, is the following:

$$\begin{aligned} \overline{E}(\overline{X}) &= 0 && \text{(a)} \\ Y_{AG}(\phi_{AG}, A_{AG}) &= 0 && \text{(b)} \end{aligned} \qquad (5.65)$$

When solving (5.65) through a two-level Newton–Raphson, the Jacobian matrix associated with the outer-level equation $\overline{Y}_{AG}(A_{AG}, \phi_{AG}) = 0$ is given by

$$[JY_{AG}] = \begin{bmatrix} \dfrac{\partial Y^r_{AG}}{\partial A_{AG}} & \dfrac{\partial Y^r_{AG}}{\partial \phi_{AG}} \\ \dfrac{\partial Y^i_{AG}}{\partial A_{AG}} & \dfrac{\partial Y^i_{AG}}{\partial \phi_{AG}} \end{bmatrix} \tag{5.66}$$

Matrix (5.66) is calculated through finite differences, performing a harmonic balance simulation for each increment of A_{AG} and ϕ_{AG}, as in the case of the matrix defined in (5.63). Then the outer-level Newton–Raphson algorithm is formulated:

$$\begin{bmatrix} A_{AG} \\ \phi_{AG} \end{bmatrix}^{j+1} = \begin{bmatrix} A_{AG} \\ \phi_{AG} \end{bmatrix}^{j} - [JY_{AG}]_j^{-1} \begin{bmatrix} Y^r_{AG} \\ Y^i_{AG} \end{bmatrix}^{j} \tag{5.67}$$

where j indicates the iteration number.

Due to the fact that the relevant variable in the synchronized oscillation is the phase shift between the oscillation and the input generator, it is equally possible to set the phase of the auxiliary generator to zero $\phi_{AG} = 0$ and solve system (5.67) in terms of the input generator phase ϕ_{RF} and the auxiliary generator amplitude A_{AG}.

When using the optimization tools of a commercial harmonic balance program, the optimization goals will be $|Y^r_{AG}| = 0$ and $|Y^i_{AG}| = 0$. In turn, the optimization variables can be (A_{AG} and ϕ_{AG}) or (A_{AG} and ϕ_{RF}). When analyzing a frequency divider by k, the optimization interval ϕ_{AG} considered can be limited to $2\pi/k$, due to the fact that as shown in Chapter 4, a phase shift of $2\pi/k$ in the divided-by-k solution simply gives rise to a time-shifted divided solution with exactly the same waveforms. Gradient optimization with a good starting point is usually convenient. The starting point is obtained through a couple of sweeps in the two variables A_{AG} and ϕ_{AG}. As an example, a fundamentally synchronized solution of the parallel resonance oscillator with the current generator values $I_g = 5$ mA and $F_{RF} = 1.59$ GHz will be analyzed here in the commercial harmonic balance software ADS. The admittance function is defined as YAG = I_AG.i[1]/A_AG[1]. Note that for both fundamentally synchronized oscillators and frequency dividers, the auxiliary generator operates at the fundamental frequency F_a; thus, its corresponding harmonic index is [1].

The optimization goals are written $-1\text{E}^{-15} < \text{real(YAG)} < 1\text{E}^{-15}$ and $-1\text{E}^{-15} < \text{imag(YAG)} < 1\text{E}^{-15}$. When the initial point is estimated, the fact that the phase variable is naturally bounded and limited to the interval 0 to 2π is taken into account. A sweep will be carried out in this phase for different values of the auxiliary generator amplitude A_{AG}. As already stated, it is the phase shift between the auxiliary generator and the synchronizing source that is actually relevant, so it is possible to set the auxiliary generator phase to zero and sweep (and later optimize) the input generator phase. This has been done in the case of the parallel resonance oscillator, with $I_g = 5$ mA at $F_{RF} = 1.59$ GHz. The three amplitude values considered are A_AG = 0.5, 0.75, and 1.5 V. The resulting function YAG

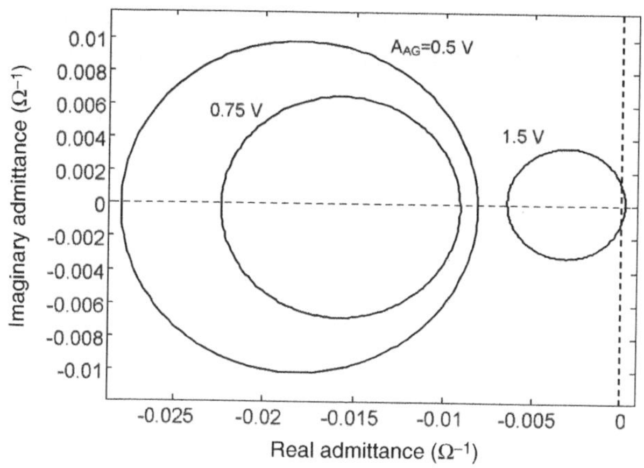

FIGURE 5.10 Estimation of a suitable initial condition for minimization through gradient optimization of the admittance function YAG in terms A_AG and the synchronizing generator phase ϕ. The function YAG has been represented in a polar plot, considering the three auxiliary generator amplitudes A_AG $= 0.5$, 0.75, and 1.5 V and sweeping the synchronizing generator phase between 0 and 2π. For each value of A_AG, a closed curve is obtained in this polar representation.

is shown in a polar diagram in Fig. 5.10. Because of the intrinsic periodicity of the admittance function with respect to the phase shift, a closed curve is obtained for each A_AG value. In view of the results of the phase sweep, the initial values selected for the gradient optimization are A_AG $= 1.5$ V and ϕ_AG $= 176^\circ$. They correspond to the point at which the closed curve for 1.5 V crosses the negative real semiaxis. Convergence with gradient optimization is achieved within four iterations, from an error of 4×10^{-8} to 5×10^{-19}. After the optimization, the resulting values of the auxiliary generator amplitude and input source phase were A_AG $= 1.49$ V and ϕ_AG $= 177.42^\circ$.

Synchronization at $F_a = mF_{\mathrm{RF}}$ For the analysis of a synchronized regime at $F_a = mF_{\mathrm{RF}}$, with $m > 1$, an auxiliary generator is introduced at the frequency $F_{AG} = F_a = mF_{\mathrm{RF}}$ at which the self-oscillation occurs. For harmonic balance simulation, the circuit variables are expressed in a Fourier series with the frequency of the input source F_{RF} as fundamental. Thus, the auxiliary generator will operate at the mth harmonic frequency $F_{AG} = mF_{\mathrm{RF}}$ of the Fourier series. Otherwise, the subsynchronized solution is determined from the same system (5.65), with the outer-level Newton–Raphson system (5.67). When using the optimization tools of a commercial harmonic balance program, the fundamental frequency of the Fourier series is set to F_{RF} and the admittance function is defined as YAG = I_AG.i[m]/A_AG[m]. Note that the auxiliary generator operates at the mth harmonic frequency, thus at the harmonic index $[m]$. This fact must also be taken into account in the definition of the ideal filter, connected in series with the voltage auxiliary generator, which must behave as a short circuit at mF_{AG} and as an open circuit at any other frequency.

As already known, the solution curves of synchronized oscillators versus the input generator frequency and other parameters are typically closed for low input power. The synchronized operation band is delimited by the turning points of this closed curve. One remarkable advantage of the auxiliary generator technique when dealing with synchronized circuits is the straightforward tracing of the closed solution curves. These curves would otherwise require the use of continuation methods, such as the one based on parameter switching presented in Chapter 6. Note that the simple sweeping of the input generator frequency will lead to convergence difficulties near the singular points and, obviously, cannot provide the entire solution curve because the frequency is swept in one sense only and the curve folds over itself at the turning points. To cope with these problems, instead of sweeping the input generator frequency, its phase ϕ_{RF} is swept between 0 and 2π in steps of $\Delta\phi$. At each step, an entire gradient optimization process, with the goal $Y_{AG} = 0$, is performed, using the optimization variables A_{AG} and F_{AG} instead of the original ones, A_{AG} and ϕ_{AG}. Proceeding like this, advantage is taken of the fact that neither the oscillation amplitude nor its frequency exhibit turning points versus the phase variable. This is illustrated in Fig. 5.11, corresponding to the analysis of the parallel resonance oscillator for the constant input generator current $I_g = 5$ mA versus the input frequency. Figure 5.11a shows the amplitude and frequency variation versus the input generator phase with periodic behavior and no turning points. It must be pointed out that the same result (except for a change of sign in the phase variable) should be obtained when sweeping the auxiliary generator phase instead of the phase of the input source. Figure 5.11b shows the closed synchronization curve versus the input frequency.

There are, however, some differences regarding the possible choice of the input generator phase, with $\phi_{AG} = 0$, or the auxiliary generator phase, $\phi_{RF} = 0$, as the swept variable. When dealing with a frequency divider by k, and sweeping ϕ_{AG}, the phase interval considered may be limited to 0 to $2\pi/k$, as the admittance at the divided frequency is periodic in phase with the period $2\pi/k$. If ϕ_{RF} is swept, the interval considered must be the ordinary one, 0 to 2π. In the case of a subsynchronized oscillator, if ϕ_{AG} is swept, the phase interval considered must be 0 to $2\pi m$. If ϕ_{RF} is swept, the ordinary interval 0 to 2π must be considered. In general, higher accuracy and better convergence properties are obtained when sweeping the input generator phase.

It must be taken into account that the described phase-sweeping technique cannot be applied to fully trace open solution curves. These open curves are found in injection-locked oscillators under relatively high input power (see, for example, Fig. 4.2 and Fig. 4.9 in Chapter 4). The phase-sweeping technique fails due to the lack of sensitivity of the solution versus the phase shift for input frequency too far from the free-running frequency (see Fig. 4.7 in Chapter 4). A parameter-switching continuation technique (described in Section 6.4, Chapter 6) should be used instead.

5.5.2.3 *Self-Oscillating Mixer Regime* In self-oscillating mixer operation, the circuit self-oscillation at the frequency F_a mixes with a periodic input signal at the frequency F_{RF} [67]. The circuit variables can be expanded in a Fourier

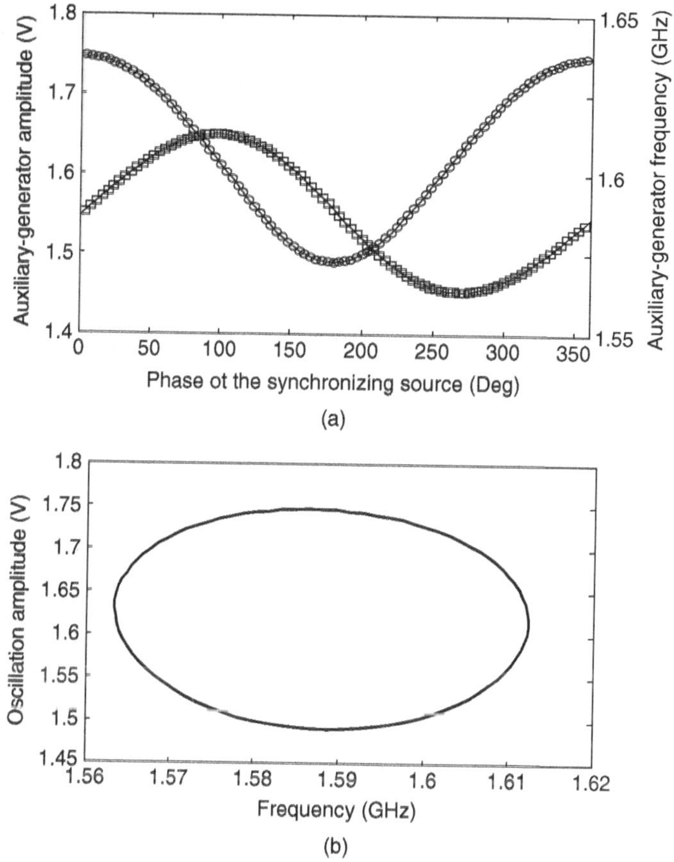

FIGURE 5.11 Synchronized solution curve of a parallel resonance oscillator with an input current generator $I_g = 5$ mA. (a) Variation of the auxiliary generator amplitude and frequency versus the synchronizing generator phase. (b) Closed synchronization curve resulting from the composition of the two curves represented in part (a).

series with two nonrationally related fundamentals: the input frequency F_{RF} and the frequency of the nonsynchronized oscillation F_a. The oscillation frequency, influenced by the input generator values, is an unknown to be determined. A major difficulty in the frequency-domain simulation of this type of regime is the fact that the self-oscillating-mixer solution at the two fundamentals F_{RF} and F_a always coexists with a nonoscillatory solution having the input frequency F_{RF} as the only fundamental. The harmonic-balance method converges by default to this periodic solution. To cope with this problem, the auxiliary generator technique can be used.

For the analysis of a self-oscillating mixer, an auxiliary generator operating at the oscillation frequency $F_{AG} = F_a$ is introduced into the circuit. Its amplitude will be A_{AG}. Its phase ϕ_{AG} can arbitrarily be set to zero because there is no phase relationship between the oscillation and the input RF generator. The auxiliary generator must fulfill the nonperturbation condition $Y_{AG} = I_{AG}/A_{AG} = 0$ at $F_{AG} = F_a$. This equation will be solved in terms of the auxiliary generator amplitude A_{AG} and frequency $F_{AG} = F_a$. The key difference with respect to the free-running oscillator analysis is that two independent fundamentals must be used in the Fourier series representation of the circuit variables. These two frequencies are given by F_{RF} and $F_{AG} = F_a$ or F_{RF} and $F_b = |F_{RF} - F_a|$. We should use $F_b = |F_{RF} - kF_a|$ in the case of input frequency about the harmonic frequency kF_a. Remember that as shown in Chapter 1, Section 1.5.1, the number of fundamental frequencies of a quasiperiodic solution is uniquely defined, but not the particular values of these fundamental frequencies. Here we will represent the circuit variables in the two-tone Fourier series as $x(t) = \sum_{k,m} X_{k,m} e^{j(k2\pi F_{RF} + m2\pi F_a)t}$, truncating the Fourier expansion to a certain number of harmonic terms (see Section 5.4.2).

As shown in Section 4.5, we can design a self-oscillating mixer to obtain a low-size, low-consumption frequency converter (from F_{RF} to $F_{IF} = |F_{RF} - F_a|$ or from F_{IF} to F_{RF}). This kind of design is usually based on a high Q oscillator. Other examples of circuits operating in a self-oscillating mixer regime are unstable power amplifiers and frequency multipliers, or injection-locked oscillators outside the synchronization bands. In the two cases, we will also have a mixerlike spectrum. Remember that two subsystems $Y_{AG} = 0$ and $\overline{H}(\overline{X}) = 0$ are resolved jointly when using the auxiliary technique. Thus, the A_{AG} and F_{AG} values that fulfill $Y_{AG} = 0$ will also lead to the intermodulation spectrum that satisfies $\overline{H}(\overline{X}) = 0$.

Note that for too close values of F_{RF} and F_a, differing, for instance, in just a few kilohertz, the multitone harmonic balance might not be accurate, since the frequency difference $|F_{RF} - F_a|$ is several orders of magnitude smaller than F_{RF} and F_a. In an injected oscillator, this situation is obtained for input frequencies outside the synchronization region but close to the edge of this region, delimited by the turning-point locus. As shown in Chapter 3, Section 3.3.3, due to the high density of the spectrum a large number of intermodulation terms must be considered in the Fourier series expansion of the circuit variables. This is because, as shown in Chapter 3, the frequency difference $|F_{RF} - F_a|$ tends to zero when approaching the turning-point locus. As shown in the next section, the envelope transient method is very well-suited for the simulation of self-oscillating mixer regimes with a very close value of the fundamental frequencies.

The auxiliary generator technique has been used for an analysis of the self-oscillating mixer in Fig. 4.30. Figure 5.12 shows the circuit description page used for simulation of the self-oscillation mixer regime in the commercial harmonic balance simulator ADS. In the absence of the RF input, the free-running oscillation frequency is $F_o = 5$ GHz. The circuit has been analyzed for constant input frequency $F_{RF} = 5.37$ GHz and input power $P_{RF} = -19$ dBm. Two-tone harmonic balance is used at the two fundamental frequencies F_{RF} and the auxiliary generator frequency F_AG. The frequency F_AG corresponds to the intermodulation product (1,0), whereas the input frequency corresponds to the intermodulation product (0,1). Thus, the nonperturbation condition is defined as the ratio YAG = I_AG.i[1,0]/A_AG[1,0]. The optimization goals are, as

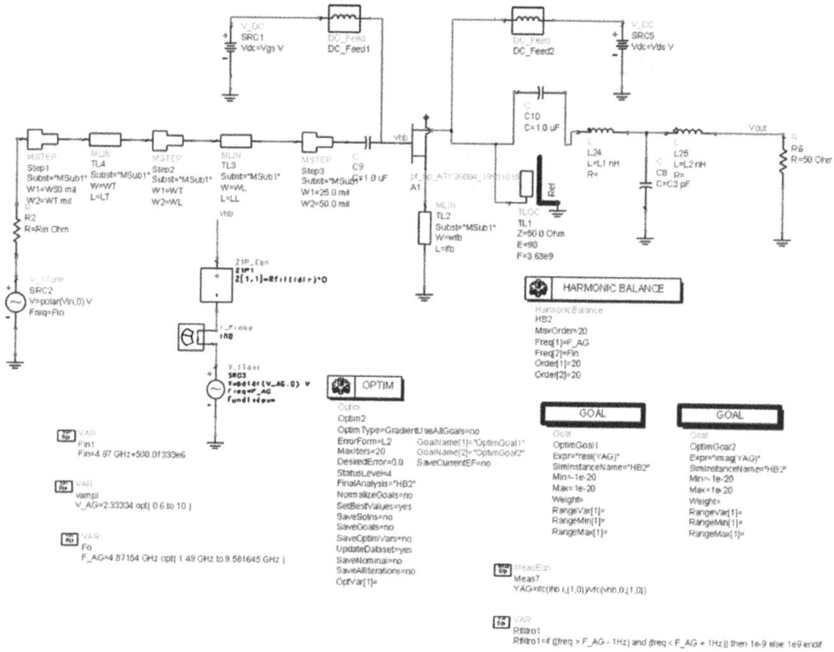

FIGURE 5.12 Circuit description page used for simulation of the self-oscillation mixer regime in the commercial harmonic-balance simulator ADS. Two-tone harmonic balance is used at the two fundamental frequencies F_{in} and the auxiliary generator frequency F_AG.

usual, real(YAG) = 0 and imag(YAG) = 0. The optimization is carried out in terms of A_AG and F_AG. For the circuit analysis a diamond truncation of the intermodulation products with nonlinearity order $nl = 20$ has been considered. The resulting spectrum is represented in Fig. 5.13. Due to the huge number of harmonic frequencies involved, $nl(nl + 1) = 420$, the Krylov subspace expansion has been used for resolution of the linear Newton–Raphson system. The auxiliary generator technique described was used to evaluate the conversion gain and oscillation frequency deviations versus the input power, presented in Figs. 4.32 and 4.33, respectively.

5.6 ENVELOPE TRANSIENT

In general, standard time-domain integration will not be applicable to an analysis of nonlinear circuits containing modulated signals. This would require integration of the system of nonlinear DAEs at a time step determined by the carrier frequency and its harmonics, and for a sufficiently long time interval to notice modulation effects. On the other hand, harmonic balance is unable to deal with modulated signals, due to the Fourier series expansion used for the circuit variables.

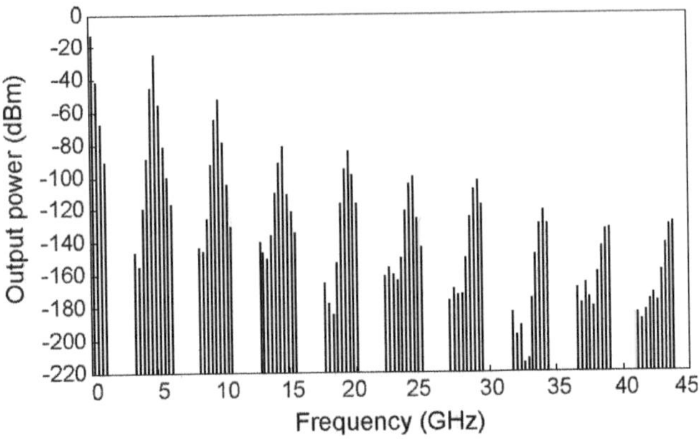

FIGURE 5.13 Harmonic balance analysis of the self-oscillating mixer of Fig. 4.36, for a constant input frequency $F_{in} = 5.37$ GHz. Output power spectrum for $P_{RF} = -19$ dBm. Due to the high number of spectral components, the Krylov subspace method has been used for this analysis.

In an analysis of communication systems, the problem of the two different time scales in the modulated signals is circumvented through the use of lowpass equivalents of bandpass signals and functions. The bandpass signals are expressed as $x_{bp}(t) = 2\text{Re}[x_{lp}(t)e^{j\omega_o t}]$, with $x_{lp}(t)$ being the complex lowpass equivalent and ω_o, the carrier frequency. In turn, each linear element is modeled by means of its lowpass impulse response $h_{lp}(t)$, related to the corresponding bandpass impulse response as $h_{bp}(t) = 2\text{Re}[h_{lp}(t)e^{j\omega_o t}]$. Then the envelope of the output signal $y_{bp}(t) = 2\text{Re}[y_{lp}(t)e^{j\omega_o t}]$of the particular linear block is obtained simply from the convolution $y_{lp}(t) = h_{lp}(t) * x_{lp}(t)$.

The recently introduced [18,21] envelope transient technique applies similar principles to nonlinear circuit analysis. The circuit variables are expressed in a Fourier series, with the carrier frequency ω_o as the fundamental and time-varying harmonic components $\overline{X}_k(t)$, with $-N \le k \le N$. These harmonic components vary at the slower time rate of the modulation signal. When these expressions are introduced in the system of nonlinear DAEs, in the state variables $\overline{x}(t)$, the orthogonality of the Fourier basis provides a differential equation system in the slowly varying harmonic components $\overline{X}_k(t)$ with $-N \le k \le N$. Due to this slow time variation rate, the system is integrated at a much larger time step than the one that would be required for the original system of full time-domain DAEs in $\overline{x}(t)$.

The envelope transient technique enables the efficient analysis of nonlinear circuits containing modulated signals such as amplifiers and mixers. One of the main applications of the technique is the accurate prediction of intermodulation distortion. The envelope transient can also be used for the simulation of autonomous circuits. This allows the analysis of modulated signals in voltage-controlled oscillators, injection-locked oscillators, or self-oscillating mixers. It is also a powerful tool to simulate dynamic behavior involving two different time scales, as in the near-synchronization regime, very difficult to analyze with standard time-domain integration or harmonic balance techniques.

We begin this section with a derivation of two common formulations of the envelope transient algorithm. Then autonomous circuits are analyzed. As in the case of harmonic balance, complementary techniques are necessary to avoid nonoscillatory solutions. The analysis techniques are particularized to free-running oscillators, injection-locked oscillators, and self-oscillating mixers. The main advantages and applications of the envelope transient simulation of autonomous circuits are highlighted.

5.6.1 Expression of Circuit Variables

For the envelope transient analysis of the circuit, two different time scales are considered. The faster time scale t_2 corresponds to the carrier, and the slower time scale t_1 corresponds to the modulation. The circuit is generally periodic in the "faster" time t_2. Then, any circuit variable $x(t)$ can be expanded in a harmonic series of the form [19,21]:

$$x(t_1, t_2) = \sum_{k=-N}^{N} X_k(t_1)e^{j\omega_k t_2} \tag{5.68}$$

where $X_k(t_1)$ are slowly varying envelopes. In an amplifier or an oscillator, the frequencies ω_k will be the harmonics of a single fundamental: $\omega_k = k\omega_o$. In a frequency mixer, the frequencies ω_k will be the intermodulation products of the RF/IF input frequency ω_{in} and the local oscillator frequency ω_a : $\omega_k = \overline{\lambda}_k^t(\omega_a, \omega_{in})^t$, with $\overline{\lambda}_k^t$ representing the different intermodulation coefficients. In both cases the ω_k frequencies will be ranged in increasing order: $\omega_{-N} < \cdots \omega_k \cdots < \omega_N$. According to (5.68), the circuit variables can be sampled using two different time rates, t_1 and t_2. Of course, since the two time scales are fictitious, the variables $x(t_1, t_2)$ will agree with $x(t)$ only when $t_1 = t_2$.

At each frequency ω_k associated to the fast time scale, we will have a vector $\overline{X}_k(t)$ of P elements, given by the time-varying harmonic components at ω_k of the P state variables $x^1(t), x^2(t), \ldots x^P(t)$. The harmonic components $\overline{X}_k(t)$ will vary at the "slower" modulation rate. The time- and frequency-domain expressions of the envelopes $\overline{X}_k(t)$ are related through

$$\overline{X}_k(t) = \frac{1}{2\pi}\int_{-B_k/2}^{B_k/2} \overline{X}_k(\Omega)e^{j\Omega t}\, d\Omega \tag{5.69}$$

where each vector $\overline{X}_k(\Omega)$ contains the spectra of the different state variables $x^1(t), x^2(t), \ldots, x^P(t)$ about the harmonic frequency ω_k. Note that the frequency Ω is, in fact, an offset frequency with respect to the corresponding harmonic frequency ω_k. For the Fourier series expansion to be unique, the bandwidth B_k associated with each $\overline{X}_k(t)$ must fulfill $B_k < (\omega_{k+1} - \omega_{k-1})/2$. On the other hand, the method will only be efficient in comparison with the full time-domain integration for relatively narrowband envelopes $\overline{X}_k(t)$.

5.6.2 Envelope Transient Formulation

When deriving an envelope transient equation system, two different cases may be considered according to the type of harmonic balance formulation, nodal or piecewise, in which the expansions (5.68) are introduced. The two cases are analyzed next.

5.6.2.1 *Nodal Harmonic Balance* The variables in the system of nonlinear DAEs (5.18) are expressed in a Fourier series with slowly varying harmonic terms, as shown in (5.68) [68]. Calculation of the time derivative of $\overline{q}(t_1, t_2) = \sum_{k=-N}^{N} Q_k(t_1)e^{j\omega_k t_2}$ will require two different derivative operators. Due to the solution periodicity with respect to t_2, for fixed t_1 the derivative with respect to t_2 will be obtained through multiplication of the various harmonic terms by $j\omega_k$. The full derivative of $\overline{q}(t)$ is given by $\dot{\overline{q}} = \sum_{k=-N}^{N} \dot{\overline{Q}}_k(t)e^{jk\omega_k t} + \sum_{k=-N}^{N} \overline{Q}_k(t)j\omega_k e^{j\omega_k t}$. Introducing this expression into the system of DAE (5.18), the time derivatives $\dot{\overline{Q}}_k(t)$ will lead to a nonlinear differential algebraic equation system in the harmonic components of the circuit variables. Taking into account the orthogonality of the various harmonic terms of the Fourier series expansions, the following relationship is obtained:

$$\overline{F}[\overline{X}(t)] + [j\omega]\overline{Q}[\overline{X}(t)] + \frac{d}{dt}\overline{Q}[\overline{X}(t)] + \int_{-\infty}^{t} \overline{H}(t-\tau)\overline{X}(\tau)d\tau + \overline{G}(t) = 0 \tag{5.70}$$

Note that the input–output relationship of the distributed elements is now written in terms of the convolution of the state-variable envelopes $\overline{X}(t)$ with the envelopes of the corresponding impulse responses $\overline{H}(t)$. As can be observed, (5.70) is a system of integrodifferential algebraic equations in the time-varying harmonic components $\overline{X}_k(t)$. To solve this system, the time variable must be discretized, as in the case of time-domain integration. This implies approaching the derivative $d\overline{Q}/dt$ in terms of the charge samples. As in the case of time-domain integration, implicit algorithms are the most efficient. When using backward Euler, the following discrete equation is obtained:

$$\overline{F}(\overline{X}(t_{n+1})) + [j\omega]\overline{Q}(\overline{X}(t_{n+1})) + \frac{\overline{Q}(\overline{X}(t_{n+1})) - \overline{Q}(\overline{X}(t_n))}{t_{n+1} - t_n} + \sum_{i=0}^{n} \overline{H}(t_{n+1} - t_i)\overline{X}(t_i)\Delta t_i + \overline{G}(t_{n+1}) = 0 \tag{5.71}$$

Note that at each integration step, the harmonic components $\overline{X}(t_{n+1})$, $\overline{X}(t_n)$, take a constant value. System (5.71) establishes an implicit nonlinear relationship between the unknown vector $\overline{X}(t_{n+1})$ and $\overline{Q}(\overline{X}(t_{n+1}))$, so an error minimization technique will be required in to integrate the system from the initial time value t_o. The Newton–Raphson algorithm is generally used for this purpose. The solution of a standard harmonic balance simulation with constant generator values $\overline{G}(t_o)$. This allows the systematic initialization of the integration procedure. The final value obtained after the Newton–Raphson convergence at the time step t_n is used as an initial guess for the next step t_{n+1}. This procedure is iteratively applied to obtain the envelope variation $\overline{X}(t)$ along the entire simulation interval $[0, T_s]$.

The major difficulty with the envelope transient, based on the nodal harmonic balance, comes from the need to compute the time-varying harmonic components of the impulse responses $\overline{H}(t)$, which will again need the use of Padé approximations, or numerical convolution as in the case of full time-domain analysis. However, computation of the convolution products is much less demanding than in standard time-domain integration, since the models of the distributed elements can be narrowband about the analysis frequencies ω_k.

5.6.2.2 *Piecewise Harmonic Balance* The time-varying harmonic components $\overline{X}_k(t)$, $\overline{Y}_k(t)$, and $\overline{G}_k(t)$ of the circuit variables can be introduced in the piecewise harmonic balance system (5.54). As shown in (5.69), these harmonic components have a continuous spectrum in the frequency Ω which must be relatively narrow about ω_k. Thus, the linear matrixes A_x, A_y, and A_g must be evaluated at the frequencies $\omega_k + \Omega$, where Ω represents the continuous frequency offset about the various harmonic frequencies ω_k. This provides the system

$$A_x(\omega_k + \Omega)\overline{X}_k(\Omega) + A_y(\omega_k + \Omega)\overline{Y}_k(\Omega) + A_g(\omega_k + \Omega)\overline{G}_k(\Omega) = \overline{0} \tag{5.72}$$

where $k = -N$, to N.

Under the assumption of slowly varying envelopes, the linear matrix may be expanded in a Taylor series about $\Omega = 0$. Assuming that a first-order development is sufficient, the following system is obtained [21]:

$$A_x(\omega_k)\overline{X}_k(t) + \frac{\partial A_x(\omega_k)}{\partial j\Omega}\dot{\overline{X}}_k(t) + A_y(\omega_k)\overline{Y}_k(t) + \frac{\partial A_y(\omega_k)}{\partial j\Omega}\dot{\overline{Y}}_k(t)$$
$$+A_g(\omega_k)\overline{G}_k(t) + \frac{\partial A_g(\omega_k)}{\partial j\Omega}\dot{\overline{G}}_k(t) = \overline{0} \qquad k = -N\text{to}N \tag{5.73}$$

where equivalences of the type $j\Omega\overline{X}_k(\Omega) = \dot{\overline{X}}_k(\Omega)$ have been taken into account. Due to the development of the linear matrixes in a first-order Taylor series about ω_k, equation (5.73) will only be valid for slowly varying variables $\overline{X}_k(t)$, $\overline{Y}_k(t)$, and $\overline{G}_k(t)$ (i.e., for "strictly" narrowband envelopes). This is a significant difference with the nodal formulation, which does not have this constraint. The problem can be circumvented by considering higher-order terms in the Taylor series expansions of the linear matrixes about the frequencies ω_k. One advantage of the piecewise formulation is that it requires neither calculation of the distributed element impulse responses nor computation of the convolution products.

For the solution of (5.73), a discrete equivalent of this system must be obtained, as in the case of standard time-domain integration. In the backward Euler approach, the time derivatives of the state variables and nonlinear sources at a given time value t_n are expressed as

$$\dot{\overline{X}}(t_n) \approx \frac{\overline{X}(t_n) - \overline{X}(t_{n-1})}{\Delta t}, \qquad \dot{\overline{Y}}(t_n) \approx \frac{\overline{Y}(t_n) - \overline{Y}(t_{n-1})}{\Delta t} \tag{5.74}$$

where $\Delta t = t_n - t_{n-1}$ is the time step selected. Introducing expressions (5.74) into system (5.73), an implicit equation in the unknowns $\overline{X}(t_n)$ is obtained:

$$\begin{aligned}\overline{H}_k[\overline{X}_k(t_n)] = {} & \left[A_x(\omega_k) + \frac{1}{\Delta t}\frac{\partial A_x(\omega_k)}{\partial j\Omega}\right]\overline{X}_k(t_n) + \left[A_y(\omega_k) + \frac{1}{\Delta t}\frac{\partial A_y(\omega_k)}{\partial j\Omega}\right] \\ & \times \overline{Y}_k[\overline{X}(t_n)] + A_g(\omega_k)\overline{G}_k(t_n) + \frac{\partial A_g(\omega_k)}{\partial j\Omega}\dot{\overline{G}}_k(t_n) \\ & - \frac{\partial A_x(\omega_k)}{\partial j\Omega}\frac{\overline{X}_k(t_{n-1})}{\Delta t} - \frac{\partial A_y(\omega_k)}{\partial j\Omega}\frac{\overline{Y}_k[\overline{X}(t_{n-1})]}{\Delta t} = 0\end{aligned} \tag{5.75}$$

where $k = -N$ to N.

The implicit system (5.75) is integrated with the Newton–Raphson algorithm. The system is integrated from the initial time t_o using the results of a preliminary harmonic balance analysis, with constant generator values $\overline{G}_o$ as initial guess. Depending on the particular circuit, the constant vector $\overline{G}_o$ may correspond to either the initial or average value of the modulated inputs. The solution of this standard harmonic balance simulation with constant generator values $\overline{G}_o$ is taken as the initial value at t_o. The final value obtained after Newton–Raphson convergence at time step t_n is used as an initial guess for the next step, t_{n+1}. This procedure is applied iteratively to obtain the envelope evolution $\overline{X}(t)$ along the entire simulation interval.

5.6.3 Extension of the Envelope Transient Method to the Simulation of Autonomous Circuits

The envelope transient method can be applied to autonomous circuits, but it generally requires complementary techniques to avoid convergence toward trivial, nonoscillatory solutions. Expressions (5.68) constitute a somehow artificial representation of the circuit variables, which often fails to follow the actual oscillator dynamics. Convergence is conditioned by the user choice of the Fourier frequency basis and the sampling rate of the time-varying harmonic components. For a systematic convergence toward the oscillatory solution, complementary techniques must be used. These techniques, described below, are particularized to free-running oscillators, injection-locked oscillators, and self-oscillating mixers. The most interesting applications of the envelope transient technique in these three main types of autonomous circuits will be presented.

5.6.3.1 Analysis of Free-Running Oscillations Ngoya et al. [69] have proposed a technique to avoid envelope transient convergence toward a trivial dc solution in free-running oscillators. To avoid the undesired convergence, an auxiliary generator or probe is introduced into the circuit. This generator is kept connected to the circuit during the entire integration interval $[0,T_s]$. The generator must fulfill the nonperturbation condition given by the zero value of the ratio between the generator current and the voltage delivered, $Y_{AG} = 0$. Due to the time-varying nature of the envelopes $\overline{X}_k(t)$, this nonperturbation condition must be fulfilled at each step t_n of the slow time variable. Thus, the amplitude and frequency of the auxiliary generator should also be time varying: $A_{AG}(t)$ and $\omega_{AG}(t)$. The equation $Y_{AG}(t) = 0$ is solved, together with the system (5.70) or (5.73), in terms of $\overline{X}_k(t)$, $A_{AG}(t)$, and $\omega_{AG}(t)$ at each step t_n of the time interval $[0,T_s]$ considered. This technique allows an efficient simulation of the oscillator transient response, with optimum adjustment of the integration time step.

A simpler analysis technique is also possible. In this technique, advantage is taken of the fact that the stable oscillatory solution behaves as an attractor of the neighboring transient trajectories. The oscillation occurs in the fast time scale, so it is possible to initialize the oscillation disregarding the influence of the modulations. The oscillatory solution at the initial time value $t = t_o$ is obtained with standard harmonic balance using an auxiliary generator. The resulting constant solution is stored as $\overline{X}_o$ and supplied to the envelope transient simulator as an initial condition. From this initial value, the system is allowed to evolve according to its own dynamics. Assuming that the oscillatory solution is stable, the system will naturally tend to it, with no need to keep the auxiliary generator connected to the circuit and solve the nonperturbation $Y_{AG}(t) = 0$ at each time step. Note that an envelope transient is available in some commercial harmonic balance simulators, but in most of these simulators, it lacks a complementary technique for the robust analysis of oscillatory solutions. The initialization technique indicated can be applied externally by the users of commercial harmonic balance. The analysis procedure is described below.

For the envelope transient analysis of free-running oscillations, a preliminary harmonic balance simulation is carried out, disregarding the possible modulation of the input sources. The auxiliary generator technique will be used for this simulation. Its amplitude A_{AGo} and frequency ω_{AGo} must be calculated to fulfill the nonperturbation condition, $Y_{AG}(A_{AGo}, \omega_{AGo}) = 0$. The resulting frequency ω_{AGo} will be used as the fundamental frequency of the Fourier series expansions, with time-varying envelopes. Thus, the circuit variables are written

$$\overline{x}(t) = \sum_{k=-N}^{N} \overline{X}_k(t) e^{jk\omega_{AGo}t} \tag{5.76}$$

The auxiliary generator is used to initialize the envelope transient system

$$\begin{aligned} &\overline{F}[\overline{X}(t)] + [j\omega]\overline{Q}[\overline{X}(t)] + \frac{d}{dt}\overline{Q}[\overline{X}(t)] + \int_{-\infty}^{t} \overline{H}(t-\tau)\overline{X}(\tau)d\tau + \overline{G}(t) = 0 \\ &V_a^m = A_{AGo}e^{j0} \qquad t = t_0 \end{aligned} \tag{5.77}$$

where the vector $\overline{G}(t)$ contains the dc sources and possible modulation inputs and V_a^m is the voltage at the autonomous fundamental ω_a at the node m where the auxiliary generator is connected. Note that this generator forces a constant value at the harmonic component V_a^m only. The auxiliary generator must be disconnected from the circuit for $t > t_o$ once the circuit variables have been initialized. Thus, for $t > t_o$ the circuit is allowed to evolve according to its own dynamics, without the auxiliary generator. When using commercial software in which the envelope transient is available, this disconnection may be carried out with the aid of a time-varying resistor, $R_{AG}(t)$, in series with the voltage auxiliary generator. The condition on this resistance will simply be

$$R_{AG} = \begin{cases} 0 & t = t_o \\ \infty & t > t_o \end{cases} \tag{5.78}$$

When modulation signals are introduced into the free-running oscillator, both harmonics values of the circuit voltages and currents and the oscillation frequency will exhibit time variations. In the analysis method proposed here, these variables will be expressed as

$$\begin{aligned} \overline{X}_k(t) &= \overline{X}_k^0 + \Delta\overline{X}_k(t) \qquad k = -N \text{ to } N \\ \omega_a(t) &\equiv \omega_{AGo} + \Delta\omega_a(t) \end{aligned} \tag{5.79}$$

where $\Delta\overline{X}_k(t)$ and $\Delta\omega_a(t)$ are the time variations of the harmonics components and oscillation frequency, respectively, due to the influence of the modulation. The circuit variables can be expressed as

$$\overline{x}(t) = \sum_{k=-N}^{N} [\overline{X}_k^0 + \Delta\overline{X}_k(t)] e^{jk\int_0^t \Delta\omega_a(s)ds} e^{jk\omega_{AGo}t} \equiv \sum_{k=-N}^{N} \overline{X}_k(t) e^{jk\omega_{AGo}t} \tag{5.80}$$

As gathered from (5.80), because the frequency basis is kept constant in the Fourier series expansions of the circuit variables, the frequency modulation is transformed into a phase modulation. This modulation is added to the inherent modulation of the various envelopes, since each $\overline{X}_k(t)$ will generally exhibit both amplitude and phase modulation.

To clarify (5.80), we will particularize this variable representation to the simplest case of a nonmodulated free-running oscillator. In steady state, the oscillation will have constant frequency ω_a and constant envelopes. In the case of an error in the estimation of the oscillation frequency $\omega_{AGo} \neq \omega_a$, the fundamental frequency ω_{AGo} of the Fourier series expansions $\sum_{k=-N}^{N} \overline{X}_k(t) e^{jk\omega_{AGo}t}$ will not agree with the actual oscillation frequency ω_a. Due to the imposed fundamental frequency ω_{AGo}, the envelopes $\overline{X}_k(t)$ must artificially oscillate at the difference frequency $|\omega_a - \omega_{AG}|$ to compensate the frequency error. This can be seen more clearly from the following relationship:

$$\overline{x}(t) = \sum_{k=-N}^{N} \overline{X}_k^{\text{o}} e^{jk\omega_a t} = \sum_{k=-N}^{N} \overline{X}_k(t) e^{jk\omega_{AG\text{o}} t} \equiv \sum_{k=-N}^{N} \overline{X}_k^{\text{o}} e^{jk(\omega_a - \omega_{AG\text{o}})t} e^{jk\omega_{AG\text{o}} t} \tag{5.81}$$

From the equality above, it will be possible to write

$$\overline{X}_k(t) = \overline{X}_k^{\text{o}} e^{jk(\omega_a - \omega_{AG\text{o}})t} \tag{5.82}$$

So the real and imaginary parts of $\overline{X}_k(t)$ will oscillate at $|\omega_a - \omega_{AG}|$. By inspecting (5.82) it is clear that the harmonics $\overline{X}_k^{\text{o}}$ can easily be extracted from $\overline{X}_k(t)$. Due to the unit magnitude of the complex exponential, they will both have the same magnitude $|X_k^p| = |X_k^{p\text{o}}|$, with p indicating the particular state variable. On the other hand, the phase of $X_k^{p\text{o}}$ can be obtained by subtracting the sawtooth function $\text{Mod}_{2\pi}[(\omega_a - \omega_{AG})t]$ from the phase of $X_k^p(t)$. Because the envelopes oscillate at $|\omega_a - \omega_{AG}|$, the time step used for the integration of the envelope transient system must be small enough to sample accurately the solution variations associated with this frequency. Thus, the frequency error will have the penalty of requiring a smaller time integration step, with increased computational effort.

The frequency spectrum associated with the envelopes $\overline{X}_k(t)$ is now considered. If the fundamental frequency of the Fourier series ω_{AG} agrees exactly with the oscillation frequency $\omega_{AG} = \omega_a$, the spectrum of $X_k^p(\Omega)$ will be centered about $\Omega = 0$. In case there is an error in the estimation of the oscillation frequency $\omega_{AG} \neq \omega_a$, the spectrum of $X_k^p(\Omega)$ will be shifted $k(\omega_a - \omega_{AG})$ in the Ω axis. As an example, Fig. 5.14 shows the output power spectrum of the free-running oscillator of Fig. 1.6. It is the output power spectrum about the fundamental frequency $P_{\text{out}}[1](\Omega)$.

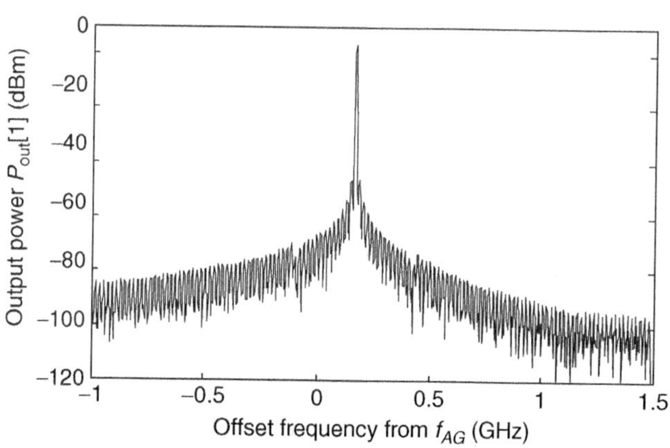

FIGURE 5.14 Output power spectrum about the fundamental frequency of the free-running oscillator of Fig. 1.6.

The spectral line is shifted $\Delta f = 0.2$ GHz to the right, which indicates that the actual oscillation frequency F_a is 0.2 GHz higher than the fundamental frequency $F_{AG} = 4.39$ GHz used.

The two different applications of the envelope transient analysis of free-running oscillators are described in the following: the analysis of the oscillator startup transients and the analysis of modulated oscillators.

Oscillator Transient The envelope transient formulation (5.77) is quite limited for the transient analyses of free-running oscillators. As already indicated, a constant frequency basis $\omega_k = k\omega_{AG}$ is considered for this formulation, with ω_{AG} being the frequency resulting from a preliminary harmonic balance simulation. The limitations for the startup transient analysis come from the fact that during this transient, the actual oscillation frequency ω_a may undergo significant variations. Thus, a very small time integration step Δt will be necessary to account for these time variations in the circuit envelopes $X_k(t)$. Ngoya et al.'s technique [69], using a probe with time-varying values of amplitude and frequency, avoids this problem, as the probe is connected to the circuit during the entire simulation interval $[0,T_s]$, and its amplitude and frequency are updated at each time step. On the other hand, most commercial harmonic balance simulators offer the envelope-transient analysis for forced (nonoscillatory) circuits only. The technique based on the time-varying probe cannot be applied by users of these simulators. In contrast, the initialization method of (5.77) requires only standard library elements and can be applied in a very simple manner.

As an example, the technique in (5.77) has been applied to simulation of the startup transient of the oscillator of Fig. 1.6. The initial harmonic balance analysis provides the oscillation frequency $F_{ao} = 4.4$ GHz, which corresponds to the steady-state oscillation frequency. Next, envelope transient analysis is carried out using this constant value as the fundamental frequency of the Fourier series. Because of the relatively large variation in the oscillation frequency during the startup transient, the largest time step allowed for the integration of the envelope transient equations is $\Delta t = 0.2$ ns. For a larger time step, the oscillation envelope decays to zero, so the simulator converges to the unstable dc solution. Figure 5.15 shows the time evolution of the magnitude of the first-harmonic component of the drain voltage Mag($V_{\text{drain}}[1]$) for the integration step $\Delta t = 0.2$ ns. Note the initially exponential growth and saturation of the oscillation amplitude.

Modulated Oscillator An envelope transient can be used for simulation of frequency-modulated oscillators. As an example, a voltage-controlled oscillator whose frequency varies under the action of a digital control signal $V_P(t)$ has been considered. The control signal $V_P(t)$ consists of a pulse train varying between the two values $V_{P1} = 2.75$ V and $V_{P2} = 6.2$ V, with a period of 50 ns and a duty cycle of 40%. The circuit schematic is shown in Fig. 5.16a. For the constant bias voltage $V_{P1} = 2.75$ V, the harmonic balance analysis provides the oscillation frequency 3.8 GHz. For $V_{P2} = 6.2$ V, the oscillation frequency obtained is 5 MHz above this value. The fundamental frequency considered is

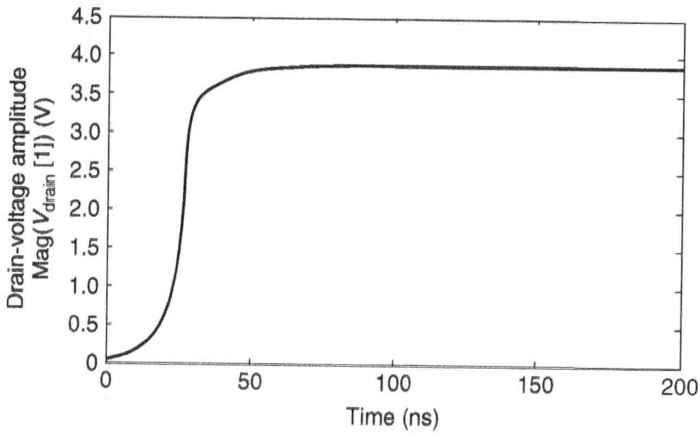

FIGURE 5.15 Envelope transient simulation of startup of the free-running oscillation of the circuit of Fig. 1.6.

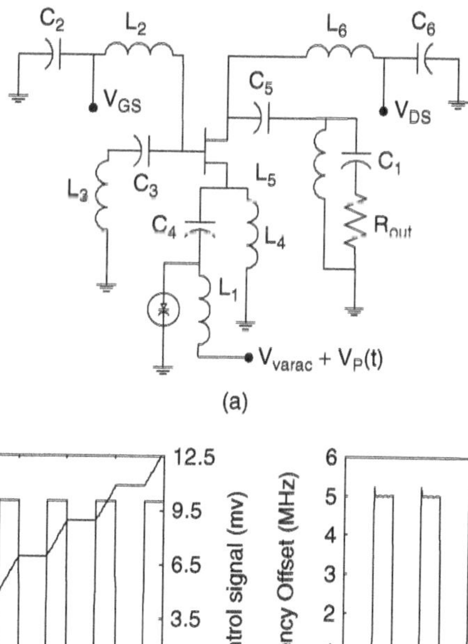

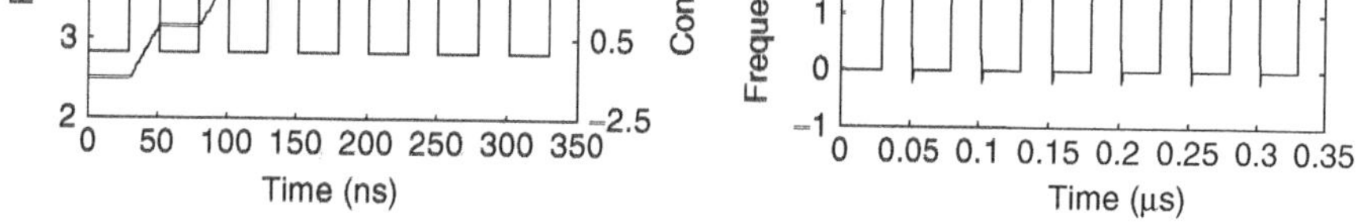

FIGURE 5.16 Voltage-controlled oscillator at 3.8 GHz modulated with a digital signal $V_P(t)$, consisting of a pulse train varying between the two values $V_{P1} = 2.75$ V and $V_{P2} = 6.2$ V, with a period of 50 ns and a duty cycle of 40 %: (a) circuit schematic; (b) time variation of the envelope frequency and phase.

$F_{AG} = 3.8$ GHz. The phase of the harmonic components follows the time integral of the control signal $V_P(t)$. Actually, a ramp is obtained in the phase of the various harmonic components. The instantaneous frequency offset $\Delta\omega_a(t)$ with respect to the fundamental frequency ω_{AG} is obtained through derivation of the phase modulation. The phase and the associated instantaneous frequency are represented in Fig 5.16b. When the control signal decreases again to $V_{P1} = 2.75$ V, the frequency value returns to ω_{ao}. However, the phase increments produced by the control signal are accumulated, due to the autonomy of the oscillator solution.

5.6.3.2 Analysis of Injected Oscillators For an injection-locked oscillator, the oscillation frequency ω_a will be determined by that of the external generator ω_{RF}. The two frequencies will fulfill a rational relationship of the form $\omega_a = (1/k)\omega_{\text{RF}}$ or $\omega_a = m\omega_{\text{RF}}$, with k and m integers, and there will be a constant phase relationship between the oscillation and the input signal. As already known, injection-locked solutions generally coexist with a solution in which the circuit self-oscillation is not excited, so it simply responds to the input periodic source in a nonautonomous manner. The envelope transient usually converges to this nonoscillatory solution, which is due to the limitations of the variable representation (5.68) to follow the actual oscillator dynamics. Thus, the envelope transient analysis of injected oscillators will require a complementary technique to initialize the circuit oscillation.

For the envelope transient analysis of injection-locked oscillators, a single fundamental frequency will be considered in the Fourier series expansion of the circuit variables $\sum_{k=-N}^{N} \overline{X}_k(t)e^{jk\omega_{\text{RF},f}t}$. For a fundamentally synchronized oscillator or a subsynchronized oscillator, the fundamental frequency $\omega_{\text{RF},f}$ will be the input generator frequency ω_{RF}; that is, $\omega_{\text{RF},f} = \omega_{\text{RF}}$. For a frequency divider by k, the fundamental frequency will be $\omega_{\text{RF},f} = \omega_{\text{RF}}/k$ for a frequency divider. Hereafter the additional subindex f will be dropped for notation simplicity. To initialize the oscillation, a standard harmonic balance simulation will be carried out, with constant values of the input generators $\overline{G}_o$. An auxiliary generator is used for this simulation. The auxiliary generator frequency ω_{AG} will be $\omega_{AG} = \omega_{\text{RF}}/k$ in a frequency divider or $\omega_{AG} = m\omega_{\text{RF}}$ in a subsynchronized oscillator. The nonperturbation condition $Y_{AG} = 0$ is solved in terms of the auxiliary generator amplitude A_{AG} and phase ϕ_{AG}; that is, $Y_{AG}(A_{AG}, \phi_{AG}) = 0$. The resulting solution $\overline{X}_o$ is taken as the initial value of the integration algorithm of the envelope transient system. An alternative way to initialize the integration algorithm will be the connection of the auxiliary generator to the circuit at the initial time t_o only. The generator values will be those resulting from the preliminary harmonic balance simulation. Then the envelope transient system is expressed as

$$\overline{F}[\overline{X}(t)] + [j\omega]\overline{Q}[\overline{X}(t)] + \frac{d}{dt}\overline{Q}[\overline{X}(t)] + \int_{-\infty}^{t} \overline{H}(t-\tau)\overline{X}(\tau)d\tau + \overline{G}(t) = \overline{0}$$
$$V_a^m = A_{AG}e^{j\phi_{AG}} \qquad t = t_o \tag{5.83}$$

with V_a^m being the voltage at the node m at the oscillation frequency. Note that for a relatively low power of the synchronizing source, the preliminary harmonic balance simulation using the auxiliary generator may be carried out in free-running conditions. Then the values $A_{AG} = A_{AGo}, \phi_{AG} = 0$ will be used in (5.83) at the frequency $\omega_{AG} = \omega_{RF}/k$ in a frequency divider, or $\omega_{AG} = m\omega_{RF}$ in a subsynchronized oscillator. For higher input power, the influence of the input generator will be more relevant, so the initial harmonic balance analysis must necessarily be performed under synchronized conditions. The nonperturbation equation $Y_{AG} = 0$ is solved in terms of A_{AG} and ϕ_{AG}.

The envelope transient allows the simulation of synchronized oscillators containing modulation signals, such as phase and frequency modulators based on injection locking. However, in the absence of modulations, the envelope transient simulation also enables an efficient and insightful analysis of near-synchronization states or a straightforward determination of the limits of the synchronization band when variations in a given circuit parameter are considered. The main applications are presented next.

Analysis of Synchronized Oscillator Dynamics As we already know, the synchronization phenomenon is inherently bandlimited. Thus, the synchronized solutions will exist only within certain ranges of the input generator frequency and power. As an example, see the closed synchronization curve obtained in the parallel resonance oscillator for the input current $I_g = 5$ mA represented in Fig. 5.11b. The circuit behaves in a periodic synchronized regime within the input frequency interval delimited by the two turning points of this closed curve. Assuming a representation of the solution as $\overline{x}(t) = \sum_{k=-N}^{N} \overline{X}_k(t)e^{jk\omega_{RF}t}$, the envelopes will tend to a constant steady-state value for generator frequencies in this interval. An example

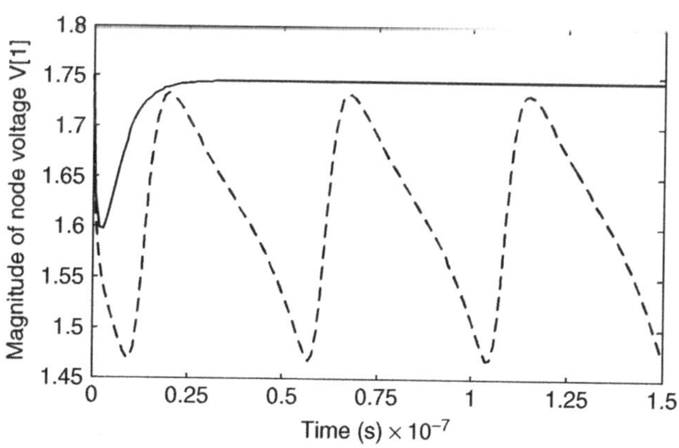

FIGURE 5.17 Envelope transient simulation of a parallel resonance oscillator for the input generator amplitude $I_g = 5$ mA. (a) Simulation for the input frequency $F_{RF} = 1.59$ GHz belonging to the synchronization band of Fig. 5.11b. The magnitude of the first-harmonic component of the node voltage tends to a constant value in steady state. (b) Simulation for the input frequency $F_{RF} = 1.62$ GHz outside the synchronization band. The magnitude of the first-harmonic component of the node voltage oscillates at the beat frequency $\omega_{IF} = |\omega_a - \omega_{RF}|$ in the steady state.

is shown in Fig. 5.17, where the initial value is purposely different from the value obtained with the harmonic balance simulation. As can be seen, the amplitude of the first-harmonic component of the node voltage Mag($V[1]$) tends, after a transient, to a constant value. This value agrees with the one resulting from the standard harmonic balance simulation of Fig. 5.11b.

As shown in Chapters 3 and 4, outside the frequency range delimited by the two turning points of the closed synchronization curves, the circuit behaves in a self-oscillating mixer regime. The envelope transient analysis of this type of regime is presented in the following. For simplicity, the case of a fundamentally synchronized oscillator has been considered, although the same derivation can easily be applied to frequency dividers and subsynchronized oscillators. When analyzed using standard harmonic balance, the steady-state solution will be expressed as

$$\overline{x}(t) = \sum_{n,m} \overline{X}_{n,m} e^{j[n\omega_{RF}+m\omega_a]t} \tag{5.84}$$

where the coefficients $\overline{X}_{n,m}$ are complex constant values. Making the expression (5.84) equal to $\sum_k \overline{X}_k(t)e^{jk\omega_{RF}t}$, it will be possible to obtain the time variation of the envelopes $\overline{X}_k(t)$ outside the synchronization band:

$$\begin{aligned}\overline{x}(t) &= \sum_{n,m} \overline{X}_{n,m} e^{j[n\omega_{RF}+m\omega_a]t} = \sum_{n,m} \overline{X}_{n,m} e^{jm[\omega_a-\omega_{RF}]t} e^{j(n+m)\omega_{RF}t} \\ &= \sum_{k,m} \overline{X}_{k-m,m} e^{jm[\omega_a-\omega_{RF}]t} e^{jk\omega_{RF}t} \\ &= \sum_k \left\{ \sum_m \overline{X}_{k-m,m} e^{jm[\omega_a-\omega_{RF}]t} \right\} e^{jk\omega_{RF}t} = \sum_k \overline{X}_k(t) e^{jk\omega_{RF}t}\end{aligned} \tag{5.85}$$

where $k = n + m$. Equation (5.85) indicates that outside the synchronization band, the time-varying harmonics oscillate periodically at the beat frequency $\omega_{IF} = |\omega_a - \omega_{RF}|$. Thus, when using an envelope transient, the quasiperiodic regime obtained outside the synchronization bands can be simulated with a single fundamental frequency in the Fourier series representation of the circuit variables. The efficiency of this simulation depends on the value of the difference frequency ω_{IF}. For high value, a small time step should be used, increasing the computational cost in comparison with a standard harmonic balance simulation at the two fundamentals ω_{RF} and ω_a. On the other hand, the simulation will fail if the integration step selected is not small enough to sample accurately the envelopes $\overline{X}_k^0(t)$, varying at the difference frequency $\omega_{IF} = |\omega_a - \omega_{RF}|$. If this is the case, the system will converge toward the unstable nonoscillatory solution at the input generator frequency ω_{RF} that coexists with the stable quasiperiodic solution.

As an example, Fig. 5.17 shows the envelope transient simulation (the dashed line) of the parallel resonance oscillator for $I_g = 5$ mA and an input frequency outside the synchronization interval of Fig. 5.11b. The input frequency selected is $F_{RF} = 1.62$ GHz. After a transient, the steady state is reached in which the magnitude of the harmonic component Mag($V[1]$) oscillates at the beat frequency $\omega_{IF} = |\omega_a - \omega_{RF}|$.

As gathered from the previous paragraphs, provided that no modulation is considered, the synchronized or nonsynchronized state of an injected oscillator can be distinguished from inspection of the envelope magnitude $|\overline{X}_k(t)|$. In a synchronized regime, the magnitude $|\overline{X}_k(t)|$ will take a constant value. Outside the synchronization band, the magnitude $|\overline{X}_k(t)|$ will exhibit a periodic variation at the beat frequency $\omega_{IF} = |\omega_a - \omega_{RF}|$, which increases with the parameter distance to the edges of the synchronization band. This significant difference between the nature of the envelopes provides a straightforward manner to determine the synchronization bands versus a given parameter η.

To determine the oscillator synchronization band versus a parameter η, a simple direct sweep is carried out in this parameter. The oscillation is initialized at the first point η_o of the parameter sweep only. For this initialization, the auxiliary generator is connected to the circuit at the initial time t_o and disconnected for $t > t_o$. Remember that the values of this generator are those resulting from a preliminary harmonic balance simulation. Starting from η_o, the parameter η is swept, performing an envelope transient simulation at each step $\Delta\eta$. The envelope transient equations are integrated for a sufficiently long interval t_o to t_{end} to ensure that the envelopes have reached the steady-state regime. Only results corresponding to the final fraction of the simulation interval t_{start} to t_{end} are stored at each step. This interval must correspond to circuit operation in the steady-state regime. The still-in-memory harmonic values $\overline{X}_k^n(t_{end})$ at the parameter value η_n are used as an initial guess for the next point, η_{n+1}, in a continuation technique. After the simulation is completed, the set of time values stored in the interval t_{start} to t_{end} (corresponding to the magnitude of a representative variable $|V_1^{out}(t, \eta)|$) is represented versus the parameter η. When the oscillation is synchronized, the magnitude takes a constant value, so a single point is obtained at the particular generator frequency. Outside the synchronization band, the solution is quasiperiodic at the frequencies ω_{RF} and ω_a. Therefore, $|V_1^{out}(t)|$ oscillates at the frequency difference ω_{IF}. The projection of $|V_1^{out}(t)|$ over the vertical axis provides a segment with a length determined by the oscillation swing.

As an example, the envelope transient analysis described has been used to obtain the synchronization band of the parallel resonance oscillator for the input generator amplitude $I_g = 5$ mA. Figure 5.18 shows the resulting variation of the magnitude of the first harmonic of the node voltage Mag($V[1]$) versus the input frequency. The results of this frequency sweep should be compared with the closed synchronization curve of Fig. 5.11b, obtained with harmonic balance. As can be observed, there is an excellent agreement of the single-point interval of Fig. 5.18 with the upper section of the closed synchronization curve, which corresponds to stable behavior. The solutions in the lower section of this curve are unstable, as they contain a real pole on the right-hand side of the complex plane. This real pole crosses the imaginary axis at each of the two turning points of the periodic solution curve.

Envelope transient analysis is highly valuable for simulation of nearly synchronized solutions. This analysis is difficult with either time-domain integration or

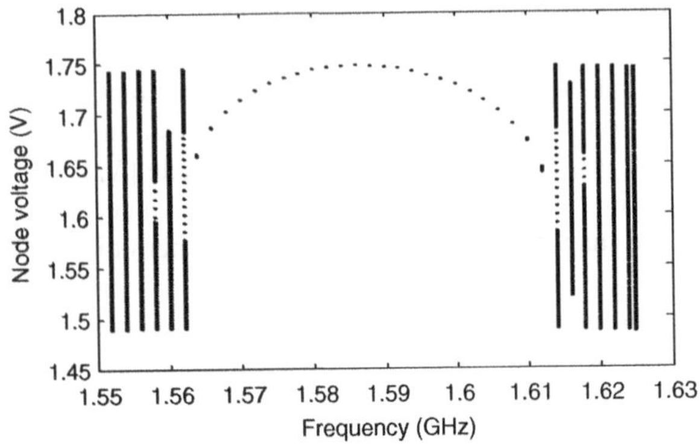

FIGURE 5.18 Determination of the synchronization band of a parallel resonance oscillator through a sweep of the input generator frequency, performing an envelope transient simulation at each sweep step. The variation of the magnitude of the first harmonic of the node voltage Mag($V[1]$) has been represented versus the input frequency.

harmonic balance. Actually, for small ω_{IF}, two different time scales can be clearly distinguished in the circuit solution: one corresponding to ω_{RF} and the other corresponding to the low beat frequency ω_{IF}. Considering, for instance, the harmonic balance simulation of Fig. 5.11b, the circuit will operate in this near-synchronization regime for frequencies outside the ellipsoidal curve, but quite close to any of its two turning points. Remember that the turning points of the synchronization curve are actually mode-locking bifurcations (also called local–global bifurcations). When reaching these points from a periodic synchronized regime, a transition to a quasiperiodic regime takes place. This is due to generation of an oscillation of infinite period at the turning point. Remember that at this kind of bifurcation, a discrete-point cycle, passing through the turning point arises in the Poincaré map (see Chapter 3, Section 3.3.3.2). The infinite period corresponds to the zero value of the difference frequency $\omega_{IF} = |\omega_a - \omega_{RF}|$. Therefore, near the turning points, the two fundamental frequencies of the quasiperiodic solution will be very close. When using time-domain integration, the simulation interval, with the sampling rate determined by the oscillation frequency, will have to be extremely long to take into account the small frequency difference $\omega_{IF} = |\omega_a - \omega_{RF}|$. The standard harmonic balance simulation will also be demanding and often inaccurate, due to the similar value of the two fundamental frequencies. Actually, the closer the value of the two fundamental frequencies ω_a and ω_{RF}, the higher the nonlinearity order nl required for an accurate simulation in Fourier series representation of the circuit variables. (see Section 5.4.2). In contrast, the envelope transient analysis of near-synchronization solutions is straightforward.

In the envelope transient analysis at the fundamental frequency ω_{RF}, the envelopes $\overline{X}_k(t)$ oscillate at the difference frequency $|\omega_a - \omega_{RF}|$. Near synchronization, this frequency is very small, so the envelope integration can be performed efficiently with a large time step, and a long simulation interval can also be considered without great computational effort. As an example, the technique has been used with a parallel resonance oscillator operating at the input generator amplitude $I_g = 5$ mA and frequency $F_{RF} = 1.563$ GHz. For these input generator values the circuit behaves in *quasilocked mode*, exhibiting quasiperiodic intermittency (Chapter 3). Figure 5.19a shows the time variation of the magnitude of the first harmonic of the node voltage Mag($V[1]$). Due to the proximity to the turning point, the envelope oscillates at a very small frequency. The resulting dense spectrum about the first-harmonic component, calculated with an envelope transient, is shown in Fig. 5.19b. The envelope variation of Fig. 5.19a should be compared with the waveform of Fig. 3.26, obtained from a time-domain simulation of high computational cost. An apparently periodic waveform is observed for long time intervals. Then the envelope variations associated with the actual quasiperiodic nature of the solution are noted, a phenomenon known as *synchronization intermittency*. The practical

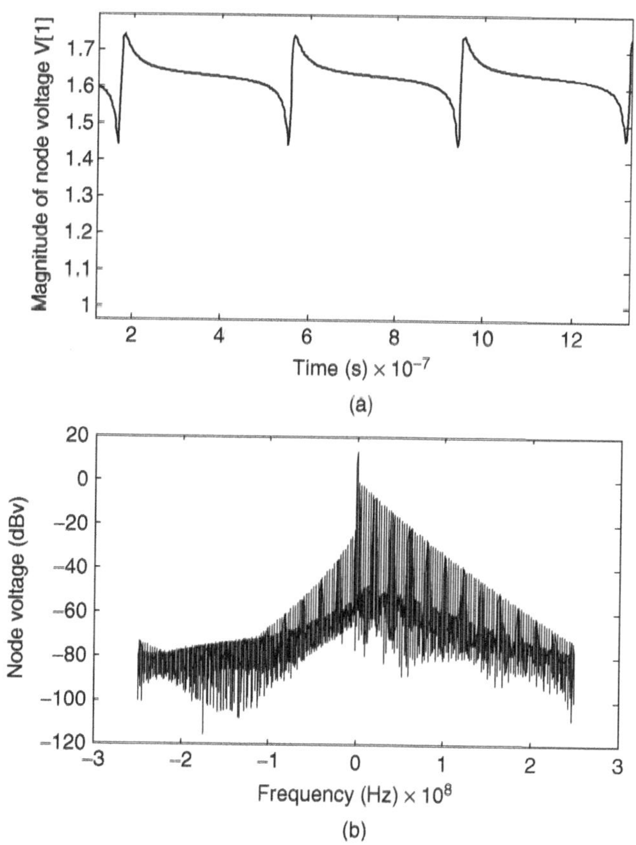

FIGURE 5.19 Envelope transient simulation of quasiperiodic solution of the parallel resonance oscillator $F_{RF} = 1.563$ GHz near the synchronization edge, determined by turning point T_1 of the synchronized solution curve of Fig. 5.11b: (a) low-frequency oscillation of the magnitude of the first harmonic component of the node voltage; (b) dense spectrum about the first-harmonic component.

limitation of the simulation time interval in time-domain integration may lead to the erroneous conclusion that the circuit is synchronized. In contrast, when using an envelope transient, the system can be integrated for a long time interval, which allows noticing the slow solution variations due to the low frequency ω_{IF}.

Injection-Locked Oscillators Containing Modulated Signals Some recent works [70,71] have shown the possibility to use the synchronization principle for the implementation of active antennas. The objective is to obtain low-cost PSK modulators and demodulators. The carrier frequency will agree with that of the synchronizing source ω_{RF}, and the phase modulation will be due to the time variation of a bias voltage. In the absence of modulation, the phase ϕ_k of the various harmonic components of any circuit variable will take a constant value, determined by the input frequency and bias sources. When a modulation signal $v_m(t)$ is introduced through one of the bias sources, the phase ϕ_k becomes time varying and can be expressed as

$$\phi_k(t) = \phi_k^0 + \Delta\phi(t) \tag{5.86}$$

where ϕ_k^0 is the harmonic phase, in the absence of modulations. The circuit variables can be written as:

$$\overline{x}(t) = \sum_{k=-N}^{N} [\overline{X}_k^0 + \Delta\overline{X}_k(t)]e^{jk\Delta\phi(t)}e^{jk\omega_{RF}t} \equiv \sum_{k=-N}^{N} \overline{X}(t)e^{jk\omega_{RF}t} \tag{5.87}$$

The modulation signal gives rise to both amplitude and phase variations. However, in a synchronized oscillator, the phase modulation will generally be more relevant than the amplitude modulation, although the stable phase-shift range obtained with a single oscillator may be insufficient for a practical phase modulator. To overcome this problem, two oscillator stages can be combined, adjusting the bias voltages to obtain the total phase variation required: for example, $-135^\circ, -45^\circ, 45^\circ, 135^\circ$ in a QPSK modulation. Using a similar principle, it is also possible to obtain a phase demodulator. The oscillator circuit will be synchronized to the modulated input signal. The low-frequency modulation signal will be extracted using a typical bias filter [70,71].

As an example, the envelope transient has been applied to analyze an injection-locked FET-based oscillator at 2.72 GHz, with a modulation voltage signal $v_m(t)$ introduced in the gate bias line. As can be expected, the oscillation synchronization to the input source will only be maintained for a certain interval of the bias voltage. Note that the oscillation frequency depends on the bias conditions, so for some values of the bias voltage, it may become too different from that of the input generator to maintain the synchronized state. In the absence of modulation, the synchronization interval has been determined with an envelope transient technique, described in the previous section a). A sweep has been carried out versus the bias voltage V_{GS} using the final results of each analysis as an initial guess for the next V_{GS} value. Figure 5.20 shows

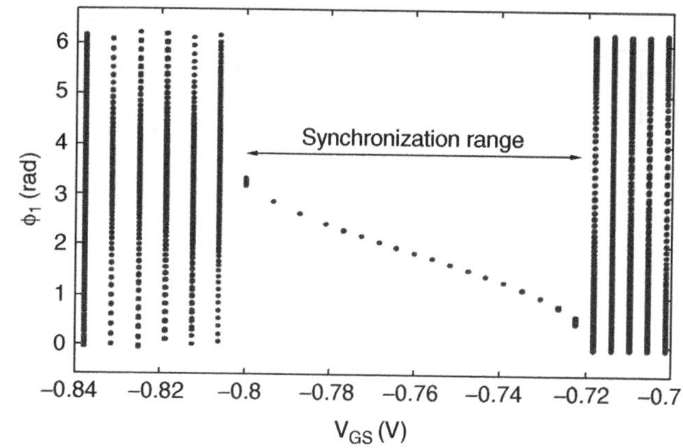

FIGURE 5.20 Determination of the synchronized operation interval of a FET-based oscillator at 2.72 GHz versus the bias voltage V_{GS}.

the variation phase $\phi_1(t)$ of the first harmonic of the output voltage $v_{out}(t)$ versus the bias voltage V_{GS}. The representation has been confined to the phase interval 0 to 2π, in radians. Within the synchronization interval, the circuit variables are periodic, and therefore the phase of the harmonic V_1^{out} is given by a constant value $\phi_1 = \phi_1^0$. When synchronization is lost, the circuit variables become quasiperiodic and the harmonic V_1^{out} oscillates periodically at the difference frequency $|\omega_a - \omega_{RF}|$. As can be observed, the constant phase shift can be varied between $\phi_1 = 0.5$ rad $= 28^\circ$ and $\phi_1 = 3$ rad $= 172^\circ$.

Next, a square-pulse periodic modulation $v_m(t)$ of amplitude $V_m = 40$ mV and frequency $f_m = 2.5$ MHz is added to the bias voltage $V_{GS} = -0.75$ V. The amplitude of this signal has been selected to maintain the circuit in the synchronized regime for all $v_{GS}(t)$ values. The results of the envelope transient analysis in terms of the phase $\phi_1(t)$ are shown in Fig. 5.21a. The input modulation signal is superimposed, for comparison. It is possible to notice the rise and decay times of the modulated phase $\phi_1(t)$. For V_{GS} values near the edges of the synchronization band in Fig. 5.20, the modulation signal might lead the circuit to a nonsynchronized state. This is the case of the simulation in Fig. 5.21b, showing transitions between synchronized and nonsynchronized behavior. At the minima of the modulation signal $V_{GS} + v_m(t)$, the waveform exhibits an oscillation at the difference frequency $|\omega_a(t) - \omega_{RF}|$. It must also be taken into account that the modulation signal influences the system dynamics and can give rise to a shift in the average V_{GS} values at which the bifurcations, delimiting the synchronization band, are obtained. Note that in the presence of modulations, the circuit is ruled by the time-varying system (5.70) instead of the static harmonic balance system. Thus, the edges of a synchronization band in the presence of a modulation band may be slightly different from those predicted with the static simulation of Fig. 5.20.

One difficulty in the design of phase modulators based on injection-locked oscillators is the limited range of stable phase shift that can be achieved with

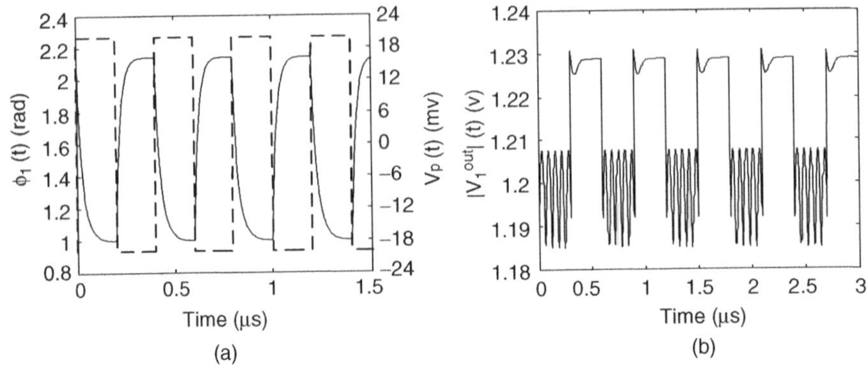

FIGURE 5.21 Envelope transient simulation of the phase modulation in a FET-based injection-locked oscillator at 2.72 GHz. The modulation signal $v_m(t)$ is a square pulse of amplitude $V_m = 40$ mV and frequency $f_m = 2.5$ MHz and the bias voltage $V_{GS} = -0.75$ V. The time variation of the phase of the first-harmonic component of the output voltage has been represented. (a) Operation within the synchronization band. The modulation signal is represented for comparison. (b) Operation near the band edges. For about half the period of the modulation signal, the circuit behaves in a nonsynchronized regime.

a single-stage circuit. In the case of PSK modulators, a stable phase-shift range of about 2π is required versus the bias voltage. The chain connection of two injection-locked oscillators has been proposed [70,71] in the literature. Only the first oscillator is connected to the synchronizing source. To achieve the 2π stable phase shift with constant input frequency, two bias voltages, one at each oscillator, must be varied. This is convenient for the parallel introduction of the four different bit pairs used in the QPSK modulation. Note that the minima and maxima of the introduced pulse chains must be adapted to the bias voltages required for the four phase values $-135^\circ, -45^\circ, 45^\circ$, and 135°.

The technique described above has been applied to obtain a QPSK modulator based on the use of two transistor-based oscillators. A pulsed signal is applied to the bias voltage of each transistor. The amplitude of the two voltage pulses is adjusted so as to obtain the combinations that provide the phase shift values required: $-135^\circ, -45^\circ, 45^\circ$, and 135°. The envelope transient simulation in Fig. 5.22a shows the transition between the four phase shift values. If the bit rate is too high, the modulator may not be able to reach the steady-state phase value. Figure 5.22b shows the constellation corresponding to 2 and 5 Mbps. For 5 Mbps, the system is failing, due to slow dynamics in comparison with the high rate of the modulation signal.

5.6.3.3 Analysis of Self-Oscillating Mixers For the envelope transient analysis of the self-oscillating mixer [72], the circuit variables will be represented in a two-fundamental Fourier series, with time-varying harmonic components. The two frequencies will be the carrier of the RF/IF input signal ω_{RF}/ω_{IF} and the oscillation frequency ω_a. The autonomous fundamental is initialized with the aid of an

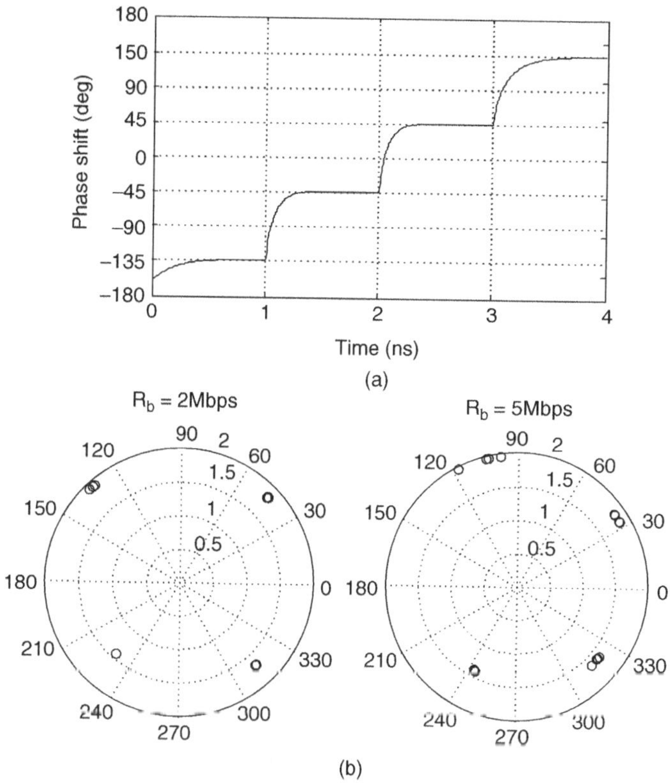

FIGURE 5.22 Transition between the four values of constant phase shift in the QPSK modulation: (a) time variation of the phase shift; (b) constellations for 2 and 5 Mbps.

auxiliary generator connected to the circuit at the initial time t_o only. The amplitude A_{AG} and frequency ω_{AG} of this auxiliary generator are obtained from a preliminary harmonic balance simulation (with constant harmonic terms). The auxiliary generator introduced must fulfill the nonperturbation condition $Y_{AG} = 0$, solved in terms of its amplitude A_{AG} and frequency ω_{AG}. The nonperturbing values resulting from this initial simulation are denoted here as A_{AGo} and ω_{AGo}. The circuit variables are expressed in a general manner as

$$\begin{aligned}\overline{v}(t) &= \sum_{k,m} \overline{V}_{k,m}(t)e^{jk\int_0^t \Delta\omega_a(s)ds}e^{j(k\omega_{AG}+m\omega_{\text{in}})t} \\ &= \sum_{k,m} \overline{V}_{k,m}(t)e^{jk\phi_a(t)}e^{jk\Delta\omega_a t}e^{j(k\omega_{AG}+m\omega_{\text{in}})t}\end{aligned} \tag{5.88}$$

where ω_{AG} is the frequency resulting from the preliminary harmonic balance simulation with the auxiliary generator. On the right-hand side, the frequency integral in (5.88) has been separated into a time-varying phase $k\phi_a(t)$ and a term $\Delta\omega_a t$

resulting from the possible static frequency shift. For the initialization of the oscillatory solution, the auxiliary generator, with the values A_{AGo} and ω_{AGo}, is connected to the circuit at the initial time t_o and is disconnected afterward. As already known, this can be done in a very simple manner with the aid of a time-varying resistor in series with auxiliary generator.

The envelope transient can be used for the analysis of intermodulation distortion in self-oscillating mixers. This analysis is commonly carried out by considering two closely spaced tones about the RF carrier: $\omega_{in} - \Omega/2$ and $\omega_{in} + \Omega/2$. As gathered from (5.88), the frequency of the self-oscillating mixer will be modulated by the input signal, due to its autonomy. Assuming a Fourier series expansion in $n\Omega$ for this modulated frequency, with harmonic coefficients W_n, the state variables can be expressed as

$$\overline{x}(t) = \sum_{k,m,n} \overline{X}_{k,m,n} e^{jn\Omega t} e^{jk\Delta\omega_a t} e^{jm \sum_{n\neq 0} \frac{W_n}{jn\Omega} e^{jn\Omega t}} e^{j(k\omega_{AG} t + m\omega_{in})t} \tag{5.89}$$

with k, m, and n integers. Due to the oscillation autonomy, the frequency modulation gives rise to an additional exponential term that will increase the harmonic content at $k\omega_{in} + m\omega_{ao} + n\Omega$ and might expand the modulation bandwidth. The intermodulation distortion will decrease with the oscillator quality factor Q_f. This is due to the smaller sensitivity of the oscillation frequency ω_a to any perturbation when Q_f increases. The modulation of the input signal will have less influence over the oscillator frequency for higher Q_f.

The envelope transient has been applied to the 5.5- to 0.5-GHz down-converter of Fig. 4.30. For simulation tests, two input tones, with 10-MHz frequency spacing and power −6 dBm, have been considered. Using a lowpass equivalent, the input signal is represented as $E_{in}(t) = \text{Re}[E_{lp}(t)e^{j\omega_{in}t}]$, with $E_{lp}(t) = E_{ino}(e^{j\Omega t/2} + e^{-j\Omega t/2})$. To initialize the solution, simulation with a nonmodulated input is carried out initially. The input generator amplitude considered delivers the same power as the modulated signal, so its amplitude is $E'_{ino} = \sqrt{2}E_{ino}$. The resulting auxiliary generator frequency is slightly different from the one obtained in the absence of input generator power.

The modulation spectrum around the IF frequency is obtained from the Fourier transform of the corresponding envelope $V^{out}_{1,-1}(t)$ (see Fig. 5.23a). The frequency shift is due to slight difference between the frequency ω_{AG} and the actual oscillation frequency ω_{ao}. Because of the low quality factor, a relatively broadband spectrum is obtained as a result of the modulation of the oscillation frequency $\omega_a \equiv \omega_a(t)$ at $\Delta f = 1$ MHz. The time variation of the magnitude of the first harmonic of the node voltage at the drain terminal $\text{Mag}(V_{drain}[1, -1])$ is shown in Fig. 5.23b, where 1-MHz modulation can be noted.

5.7 CONVERSION MATRIX APPROACH

The conversion-matrix approach provides the linearized response of a circuit in large-signal periodic regime at a frequency ω_o versus small-signal inputs at

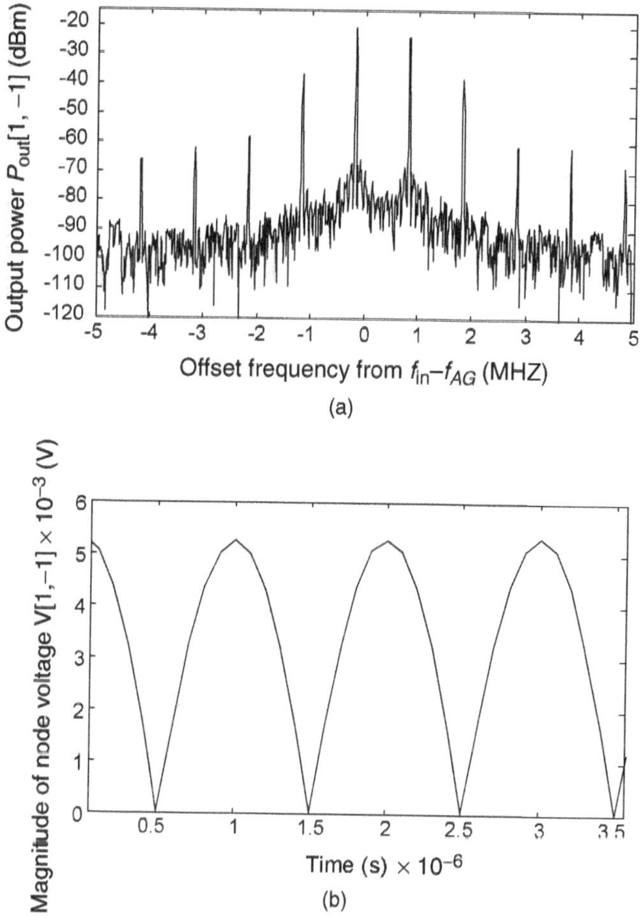

FIGURE 5.23 Self-oscillating mixer with autonomous oscillation: (a) output power spectrum of the envelope about the intermediate frequency (i.e., about the harmonic component $f_{in} - f_{AG}$) for two input tones with frequency spacing $\Delta f = 10$ MHz and total power $P_{in} = -19$ dBm; (b) variation of the magnitude of the first harmonic of the node voltage at the drain terminal.

one or more incommensurable frequencies, represented as $k\omega_o + \Omega$ [73,74]. The large-signal regime may be due to the power delivered by the input generator or to a self-oscillation. The conversion matrix approach is obtained by linearizing the harmonic balance formulation about the large-signal steady-state regime ω_o when the small-signal inputs at $k\omega_o + \Omega$ are considered. The conversion matrix approach will be invaluable for stability and phase noise analyses based on harmonic balance, covered in Chapters 6 and 7. These two types of analysis consider small signal perturbations of a large-signal periodic regime.

The periodic solution obtained with harmonic balance will be represented with the state-variable vector $\overline{X}_o$. This vector contains the harmonic components $k\omega_o$, with $k = 0, \pm 1, \ldots, \pm N$, of the different state variables of the harmonic balance equations. One or more small-signal sources at one or several of the frequencies $k\omega_o + \Omega$ will now be introduced into the circuit. Clearly, the nonlinear circuit will behave linearly with respect to these sources. Thus, to obtain the circuit response to these sources it will be possible to linearize the harmonic balance equation about the large-signal solution at $k\omega_o$. The nonlinear elements will be approached with their derivatives with respect to the control variables, evaluated at the periodic steady-state solution $\overline{X}_o$. Due to the presence of the small-signal sources at any $k\omega_o + \Omega$, the solution will contain the frequency components $\pm\Omega, \pm\omega_o \pm \Omega, \pm|k|\omega_o \pm \Omega$, and $\pm N\omega_o \pm \Omega$. Note that only coefficients ± 1 are considered in the frequency Ω as the circuit behaves in small-signal mode with respect to the sources at $k\omega_o + \Omega$.

Due to the Hermitian symmetry of real variables, the terms at $-|k|\omega_o \pm \Omega$ will be complex conjugates of the components at $|k|\omega_o \mp \Omega$. Thus, it is sufficient to consider the frequency components $\Omega, \omega_o + \Omega, -\omega_o + \Omega, \ldots, \pm N\omega_o + \Omega$, retaining only the positive sign for Ω. The sideband vector $\Delta\overline{X}_{-|k|}$ at $-|k|\omega_o + \Omega$ will be the complex conjugate of the sideband vector $\Delta\overline{X}_{|k|}$ at $|k|\omega_o - \Omega$, so it will be indicated as $\Delta\overline{X}_k^*$. As an example, the terms at $\omega_o + \Omega$ will agree with upper sidebands $\Delta\overline{X}_1 = \Delta\overline{X}_u$ about the fundamental frequency. In turn, the terms at $-\omega_o + \Omega$ will agree with the complex conjugate of the lower sideband about the fundamental frequency $\Delta\overline{X}_{-1} = \Delta\overline{X}_l^*$.

The conversion matrix approach will be initially particularized to the piecewise harmonic balance formulation. Note that the frequencies of the linearized system given by $k\omega_o + \Omega$ are different from those of the original periodic solution $k\omega_o$. Thus, the linear matrixes must be evaluated at the sideband frequencies $k\omega_o + \Omega$. This leads to the following linear system:

$$\left\{[A_x(k\omega_o + \Omega)] + [A_y(k\omega_o + \Omega)]\left[\frac{\partial \overline{Y}}{\partial \overline{X}}\right]_o\right\}\Delta\overline{X} = [A_g(k\omega_o + \Omega)]\Delta\overline{G} \quad (5.90)$$

with $k = -N$ to N in the linear matrixes. The generator vector $\Delta\overline{G}$ in (5.90) contains small-signal inputs at the frequencies $k\omega_o + \Omega$. The vector of sidebands $\Delta\overline{X}$ is solved through inversion of the matrix in the braces on the left-hand side. The derivative matrix $\left[\partial\overline{Y}/\partial\overline{X}\right]_o$ is the same as that calculated for implementation of the Newton–Raphson algorithm that provides the large-signal periodic solution at $\overline{X}_o$ at $k\omega_o$. Maintaining the organization (5.38) for the harmonic components of the circuit variables, the total Jacobian matrix of the nonlinear elements with respect to the state variables will be written

$$\left[\frac{\partial \overline{Y}}{\partial \overline{X}}\right] = \begin{bmatrix} \dfrac{\partial \overline{Y}_{-N}}{\partial \overline{X}_{-N}} & \cdots & \dfrac{\partial \overline{Y}_{-N}}{\partial \overline{X}_{N}} \\ \vdots & \vdots & \vdots \\ \dfrac{\partial \overline{Y}_{N}}{\partial \overline{X}_{-N}} & \cdots & \dfrac{\partial \overline{Y}_{N}}{\partial \overline{X}_{N}} \end{bmatrix} \quad (5.91)$$

The submatrix containing the derivatives of the harmonic components of order k of all the nonlinear elements (contained in the time-domain vector $\overline{y}$) with respect to the harmonic components of order m of all the state variables (contained in the time-domain vector $\overline{x}$) is given by

$$\frac{\partial \overline{Y}^k}{\partial \overline{X}^m} = \left[\frac{\partial \overline{y}}{\partial \overline{x}}\right]_{harmonic\ k-m} \tag{5.92}$$

The conversion matrix approach can also be applied to nodal harmonic balance. This provides the system

$$\left\{ \left.\frac{\partial \overline{F}}{\partial \overline{X}}\right|_{o} + [j(k\omega_o + \Omega)]\frac{\partial \overline{Q}}{\partial \overline{X}} + [H(k\omega_o + \Omega)] \right\} \Delta\overline{X} = \Delta\overline{G} \tag{5.93}$$

Note that the conversion matrix approach derived is a multiharmonic generalization of the linearized analysis approach discussed in Section 1.4 and based on the describing function.

One essential characteristic of the conversion matrix approach is that the linearization used regarding the periodic solution $\overline{X}_o$ applies to the control variables, generally voltages, of the nonlinear elements, but not to the small-signal frequency ω. This frequency Ω, incommensurable with ω_o, can take any value in the interval $0 < \Omega < \omega_o/2$. The restriction to this interval simply comes from the fact that the sidebands about major spectral lines $k\omega_o$ overlap at $\Omega = \omega_o/2$. This degeneracy leads to a singular linear matrix affecting $\Delta\overline{X}$ in (5.93), which cannot be inverted to obtain the state-variable increments $\Delta\overline{X}$.

The conversion matrix analysis has been applied to the self-oscillating mixer of Fig. 4.30 with constant input frequency $F_{RF} = 5.37$ GHz and RF power $P_{RF} =$

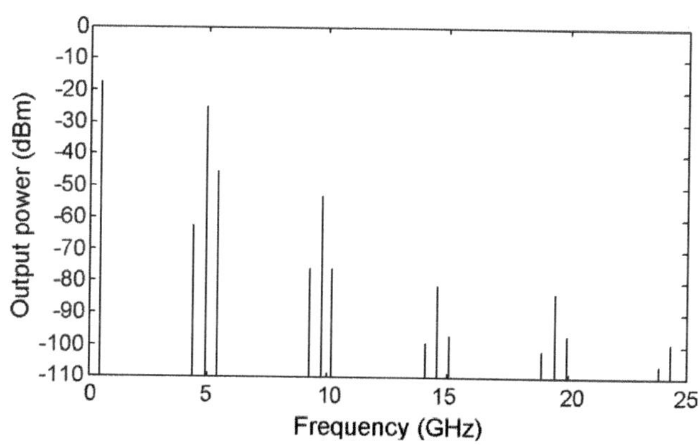

FIGURE 5.24 Conversion matrix analysis of the self-oscillating mixer of Fig. 5.13. Output power spectrum for constant input frequency $F_{in} = 5.37$ GHz and RF power $P_{RF} = -19$ dBm.

−19 dBm. The circuit is linearized about its periodic free-running oscillation, obtained with harmonic balance. This harmonic balance analysis provides the oscillation frequency $f_o = 4.87$ GHz and output power $P_{out} = -24$ dBm. For conversion matrix analysis, the periodic input source is introduced at $F_{RF} = 5.37$ GHz, which gives rise to the sideband frequency $F_{RF} - F_o = \Omega/2\pi = 0.5$ GHz. (The auxiliary generator can be kept connected to the circuit to sustain the oscillation during this analysis.) The spectrum obtained, presented in Fig. 5.24, should be compared with the one in Fig. 5.13, obtained using two-tone harmonic balance. Due to the relatively low value of the RF input power, there is good agreement in output power prediction at the intermediate frequency $f_{IF} = 0.5$ GHz. The conversion matrix analysis is inherently linear, so it is unable to predict variations in the conversion gain versus input power such as those shown in Fig. 4.32. Thus, it is unable to predict the 1-dB gain compression point or the oscillation extinction. It is equally unable to predict variations of the oscillation frequency due to the influence of the input power such as those shown in Fig. 4.33.

REFERENCES

[1] T. S. Parker and L. O. Chua, *Practical Algorithms for Chaotic Systems*, Springer-Verlag, Berlin, 1989.

[2] K. Ogata, *Modern Control Engineering*, Prentice-Hall, Englewood Cliffs, NJ, 1980.

[3] L. Gustafsson, B. Hansson, G. H. and K. I. Lundstrom, On the use of describing functions in the study of nonlinear active microwave circuits, *IEEE Trans. Microwave Theory Tech.*, vol. 20, pp. 402–409, 1972.

[4] S. A. Maas, *Nonlinear Microwave Circuits*, Artech House, Norword, MA, 1988.

[5] J. C. Pedro and N. B. Carvalho, *Intermodulation Distortion in Nonlinear Microwave Circuits*, Artech House, Norwood, MA, 2003.

[6] U. M. Ascher and L. R. Petzold, Computer methods for ordinary differential equations and differential-algebraic equations, in *Computer Methods for Ordinary Differential Equations and Differential-Algebraic Equations*, SIAM, Ed. 1998.

[7] K. Kundert, Introduction to RF simulation and its application, pp. 67–78, 1998.

[8] M. I. Sohby and A. K. Jastrzebsky, Direct integration methods of nonlinear microwave circuits, *European Microwave Conference*, pp. 1110–1118, 1985.

[9] L. W. Nagel, SPICE 2: a computer program to simulate semiconductor circuits, Ph.D. Thesis, University of Berkeley, 1975.

[10] K. S. Kundert and A. Sangiovanni-Vincentelli, Finding the steady-state response of analog and microwave circuits, *Proceedings of the IEEE 1988 Custom Integrated Circuits Conference*, pp. 6–1, 1988.

[11] J. Bonet, P. Pala, and J. M. Miro, Discrete-time approach to the steady state analysis of distributed nonlinear autonomous circuits, *IEEE International Symposium on Circuits and Systems*, pp. 460–463, 1998.

[12] I. Maio and F. G. Canavero, Differential-difference equations for the transient simulation of lossy MTLs, *IEEE International Symposium on Circuits and Systems*, pp. 1412–1415, 1995.

[13] K. Kundert, J. White, and A. Sangiovanni-Vincentelli, Envelope-following method for the efficient transient simulation of switching power and filter circuits, *IEEE International Conference on Computer-Aided Design*, pp. 446–449, 1988.

[14] L. T. Pillage and R. A. Rohrer, Asymptotic waveform evaluation for timing analysis, *IEEE Trans. Comput. Aided Des. Integrated Circuits Syst.*, vol. 9, pp. 352–366, 1990.

[15] V. Rizzoli and A. Neri, State of the art and present trends in nonlinear microwave CAD techniques, *IEEE Trans. Microwave Theory Tech.*, vol. 36, pp. 343–356, Feb. 1988.

[16] C. Camacho-Peñalosa, Numerical steady-state analysis of nonlinear microwave circuits with periodic excitation, *IEEE Trans. Microwave Theory Tech.*, vol. 31, pp. 724–730, Sept. 1983.

[17] S. Jeon, A. Suárez, and D. B. Rutledge, Global stability analysis and stabilization of a class-E/F amplifier with a distributed active transformer, *IEEE Trans. Microwave Theory Tech.*, vol. 53, pp. 3712–3722, 2005.

[18] H. G. Brachtendorf, G. Welsch, and R. Laur, Time-frequency algorithm for the simulation of the initial transient response of oscillators, *IEEE International Symposium on Circuits and Systems*, pp. 236–238, 1998.

[19] J. C. Pedro and N. B. Carvalho, Simulation of RF circuits driven by modulated signals without bandwidth constraints, *IEEE MTT-S International Microwave Symposium Digest*, pp. 2173–2176, 2002.

[20] J. Roychowdhury, Efficient methods for simulating highly nonlinear multi-rate circuits, *Proceedings of the 1997 34th Design Automation Conference*, pp. 269–274, 1997.

[21] E. Ngoya and R. Larcheveque, Envelope transient analysis: a new method for the transient and steady-state analysis of microwave communication circuits and systems, *IEEE Microwave Theory and Techniques Symposium*, pp. 1365–1368, 1996.

[22] R. Anholt, *Electrical and Thermal Characterization of MESFETs, HEMTs and BTs*, Artech House, Norwood, MA, 1994.

[23] J. M. Golio, *Microwave MESFETs and HEMTs*, Artech House, Norwood, MA, 1991.

[24] R. E. Collin, *Foundations for Microwave Engineering*, 2nd ed., Wiley, New York, 2001.

[25] P. K. Gunupudi, M. Nakhla, and R. Achar, Simulation of high-speed distributed interconnects using Krylov-space techniques, *IEEE Trans. Comput Aided Des. Integrated Circuits Syst.*, vol. 19, pp. 799–808, 2000.

[26] A. Dounavis, X. Li, M. Nakhla, and R. Achar, Passive closed-form transmission line model for general purpose circuit simulators, *IEEE Trans. Microwave Theory Tech.*, vol. 47, pp. 2450–2459, Dec. 1999.

[27] R. Mohan, M. J. Choi, S. E. Mick, et al., Causal reduced-order modeling of distributed structures in a transient circuit simulator, *IEEE Trans. Microwave Theory Tech.*, vol. 52, pp. 2207–2214, 2004.

[28] T. J. Brazil, Causal-convolution: a new method for the transient analysis of linear systems a microwave frequencies, *IEEE Trans. Microwave Theory Tech.*, vol. 43, p. 315, 1995.

[29] B. Yang and J. Phillips, Time-domain steady-state simulation of frequency-dependent components using multi-interval Chebyshev method, *39th Design Automation Conference*, pp. 504–509, 2002.

[30] R. Achar and M. Nakhla, Simulation of high speed interconnects, *Proceedings of the IEEE*, vol. 48, no. 5, pp. 693–728, 2001.

[31] P. Feldmann and R. W. Freund, Efficient linear circuit analysis by Padé approximation via the Lanczos process, *Proceedings of the 1994 European Design Automation Conference* pp. 170–175, 1994.

[32] L. T. Pillage, X. Huang, and R. A. Rohrer, AWEsim: asymptotic waveform evaluation for timing analysis, *26th ACM/IEEE Design Automation Conference*, pp. 634–637, 1989.

[33] S. Kapur, D. E. Long, and J. Roychowdhury, Efficient time-domain simulation of frequency-dependent elements, *Proceedings of the 1996 IEEE/ACM International Conference on Computer-Aided Design*, pp. 569–573, 1996.

[34] T. S. Parker and L. O. Chua, *Practical Algorithms for Chaotic Systems*, Springer-Verlag, New York, 1989.

[35] K. S. Kundert, Introduction to RF simulation and its application, *IEEE J. Solid State Circuits*, vol. 34, pp. 1298–1319, Sept. 1999.

[36] Y. Tajima, B. Wrona, and K. Mishima, "GaAs FET large-signal model and its application to circuit designs, *IEEE Trans. Electron Devices*, vol. 28, pp. 171–175, 1981.

[37] K. S. Kundert, Simulation methods for RF integrated circuits, *Proceedings of the IEEE International Conference on Computer-Aided Design*, pp. 752–765, Nov. 1997.

[38] R. Telichevesky, K. Kundert, I. Elfadel, and J. White, Fast simulation algorithms for RF circuits, *Proceedings of the 1996 IEEE Custom Integrated Circuits Conference*, pp. 437–444, 1996.

[39] J. Bonet-Dalmau and P. Pala-Schonwalder, Discrete-time approach to the steady-state and stability analysis of distributed nonlinear autonomous circuits, *IEEE Trans. Circuits Syst. I Fundam. Theor. Appl.*, vol. 47, pp. 231–236, 2000.

[40] R. Quéré, E. Ngoya, M. Camiade, A. Suárez, M. Hessane, and J. Obregón, Large signal design of broadband monolithic microwave frequency dividers and phase-locked oscillators, *IEEE Trans. Microwave Theory Tech.*, vol. 41, pp. 1928–1938, Nov. 1993.

[41] A. B. Carlson, *Communication Systems*, McGraw-Hill, New York, 1986.

[42] P. J. C. Rodrigues, Computer aided analysis of nonlinear microwave circuits, 1997.

[43] K. S. Kundert and A. Sangiovanni-Vincentelli, Simulation of nonlinear circuits in the frequency domain, *IEEE Trans. Comput. Aided Des. Integrated Circuits Syst.*, vol. 5, p. 1985, 1986.

[44] V. Rizzoli, A. Lipparini, A. Costanzo, et al., State-of-the-art harmonic-balance simulation of forced nonlinear microwave circuits by the piecewise technique, *IEEE Trans. Microwave Theory Tech.*, vol. 40, pp. 12–28, 1992.

[45] R. W. Freund, Krylov-subspace methods for reduced-order modeling in circuit simulation, *J. Comput. Appl. Math.*, vol. 123, pp. 395–421, 2000.

[46] W. M. Coughran, Jr., and R. W. Freund, Recent advances in Krylov-subspace solvers for linear systems and applications in device simulation, *Proceedings of the 1997 International Conference on Simulation of Semiconductor Process and Devices, SISPAD'97*, pp. 9–16, 1997.

[47] P. Misra and K. Naishadham, Order-recursive Gaussian elimination (ORGE) and efficient CAD of microwave circuits, *IEEE Trans. Microwave Theory Tech.*, vol. 44, pp. 2166–2173, 1996.

[48] K. Naishadham and P. Misra, Order recursive Gaussian elimination and efficient CAD of microwave circuits, *IEEE MTT-S International Microwave Symposium Digest*, pp. 1435–1438, 1995.

[49] A. Dounavis, E. Gad, R. Achar, and M. Nakhla, Passive model-reduction of distributed networks with frequency-dependent parameters, *IEEE MTT-S International Microwave Symposium Digest*, vol. 3, pp. 1789–1792, 2000.

[50] W. T. Beyene and J. E. Schutt-Aine, Krylov subspace-based model-order reduction techniques for circuit simulations, *Midwest Symposium on Circuits and Systems*, pp. 331–334, 1996.

[51] J. Wang, X. Zeng, W. Cai, C. Chiang, J. Tong, and D. Zhou, Frequency domain wavelet method with GMRES for large-scale linear circuit simulation, *2004 IEEE International Symposium on Circuits and Systems-Proceedings*, pp. 321–324, 2004.

[52] V. Rizzoli, F. Mastri, C. Cecchetti, and F. Sgallari, Fast and robust inexact Newton approach to the harmonic-balance analysis of nonlinear microwave circuits, *IEEE Microwave Guided Wave Lett.*, vol. 7, pp. 359–361, 1997.

[53] O. Axelsson, *Iterative Solution Methods*, Cambridge University Press, New York, 1994.

[54] Y. Cao and G. Wang, An efficient preconditioner for RFICs simulation using harmonic balance method, International Conference on Wireless Communications, Networking and Mobile Computing (WiCOM 2006), Wuhan, China, pp. 4149339, 2007.

[55] M. M. Gourary, S. G. Rusakov, S. L. Ulyanov, M. M. Zharov, K. K. Gullapalli, and B. J. Mulvaney, Adaptive preconditioners for the simulation of extremely nonlinear circuits using harmonic balance, *IEEE MTT-S International Microwave Symposium Digest*, vol. 2, pp. 779–782, 1999.

[56] V. Rizzoli, A. Lipparini, F. Mastri, A. Neri, F. Sgallari, and V. Frontini, Intermodulation analysis of microwave mixers by a sparse-matrix method coupled with the piecewise harmonic-balance technique, *20th European Microwave Conference, Budapest, Hungary*, pp. 189–194, 1990.

[57] V. Rizzoli, F. Mastri, F. Sgallari, and V. Frontini, Exploitation of sparse-matrix techniques in conjunction with the piecewise harmonic-balance method for nonlinear microwave circuit analysis, *IEEE MTT-S International Microwave Symposium Digest*, vol. 3, pp. 1295–1298, 1990.

[58] D. Hente and R. H. Jansen, Frequency domain continuation method for the analysis and stability investigation of nonlinear microwave circuits, *IEE Proc. H Microwaves Antennas Propag.*, vol. 133, pp. 351–362, 1986.

[59] K. S. Kundert, G. B. Sorkin, and A. Sangiovanni-Vincentelli, Applying harmonic balance to almost-periodic circuits, *IEEE Trans. Microwave Theory Tech.*, vol. 36, pp. 366–378, 1988.

[60] E. Ngoya, J. Rousset, M. Gayral, R. Quéré, and J. Obregón, Efficient algorithms for spectra calculations in nonlinear microwave circuits simulators, *IEEE Trans. Circuits Syst.*, vol. 37, pp. 1339–1355, 1990.

[61] P. L. Heron and M. B. Steer, Jacobian calculation using the multidimensional fast Fourier transform in the harmonic balance analysis of nonlinear circuits, *IEEE Trans. Microwave Theory Tech.*, vol. 38, pp. 429–431, 1990.

[62] V. Rizzoli, C. Cecchetti, A. Lipparini, and F. Mastri, General-purpose harmonic balance analysis of nonlinear microwave circuits under multitone excitation, *IEEE Trans. Microwave Theory Tech.*, vol. 36, pp. 1650–1660, 1988.

[63] B. Troyanovsky, Frequency-domain algorithms for simulating large-signal distortion in semiconductor devices, 1997.

[64] A. Suárez, J. Morales, and R. Quéré, Synchronization analysis of autonomous microwave circuits using new global stability analysis tools, *IEEE Trans. Microwave Theory Tech.*, vol. 46, pp. 494–504, May 1998.

[65] Y. Xuan and C. M. Snowden, New generalised approach to the design of microwave oscillators, pp. 661–664, 1987.

[66] D. Elad, A. Madjar, and A. Bar-Lev, New approach to the analysis and design of microwave feedback oscillators, pp. 369–374, 1989.

[67] X. Zhou and A. S. Daryoush, Efficient self-oscillating mixer for communications, *IEEE Trans. Microwave Theory Tech.*, vol. 42, pp. 1858–1862, 1994.

[68] H. G. Brachtendorf, G. Welsch, and R. Laur, Novel time-frequency method for the simulation of the steady state of circuits driven by multi-tone signals, pp. 1508–1511, 1997.

[69] E. Ngoya, J. Rousset, and D. Argollo, Rigorous RF and microwave oscillator phase noise calculation by envelope transient technique, *IEEE MTT-S International Microwave Symposium Digest*, pp. 91–94, 2000.

[70] L. Dussopt and J. Laheurte, BPSK and QPSK modulations of an oscillating antenna for transponding applications, *IEE Proc. Microwaves Antennas Propag.*, vol. 147, pp. 335–338, 2000.

[71] X. Liu, C. L. Law, Z. Shen, A. Sheel, C. Qian, and Z. Sun, New approach for QPSK modulation, *IEEE VTS 53rd Vehicular Technology Conference (VTS SPRING 2001)*, pp. 1225–1228, 2001.

[72] E. De Cos, A. Suárez, and S. Sancho, Envelope transient analysis of self-oscillating mixers, *IEEE Trans. Microwave Theory Tech.*, vol. 52, pp. 1090–1100, 2004.

[73] J. C. Nallatamby, M. Prigent, J. C. Sarkissian, R. Quéré, and J. Obregón, A new approach to nonlinear analysis of noise behaviour of synchronized oscillators and analog-frequency dividers, *IEEE Trans. Microwave Theory Tech.*, vol. 46, pp. 1168–1171, Aug. 1998.

[74] J. M. Paillot, J. C. Nallatamby, M. Hessane, R. Quéré, M. Prigent, and J. Rousset, A general program for steady state, stability, and FM noise analysis of microwave oscillators, *IEEE MTT-S Int. Microwave Symp. Dig.*, pp. 1287–1290, 1990.

CHAPTER SIX

Stability Analysis Using Harmonic Balance

6.1 INTRODUCTION

When using frequency-domain analysis techniques, the solution transient is not simulated, so there is no information about how the steady-state regime obtained reacts to perturbations. Thus, frequency-domain techniques such as harmonic balance or linear analysis based on scattering parameters provide no information about solution stability or, equivalently, about its physical existence. To verify the physical existence of the solutions obtained, a complementary stability analysis method must be used. In this chapter, the main stability analysis techniques implementable in in-house and commercial harmonic balance simulators are described.

Two different types of stability analyses are considered: local stability analysis, applied to a single steady-state solution, obtained for particular values of the circuit parameters, such as the input sources or circuit element values; and global stability analysis, used when considering a certain variation range of one or more of the circuit parameters [1,2]. For local stability analysis, existing techniques applicable to small- and large-signal regimes are reviewed briefly. For global stability analysis it is necessary to obtain the variation of the steady-state solution versus the parameter considered, which might require the use of continuation methods. This is due to the common multivalued response of nonlinear circuits that are autonomous in nature. Efficient continuation methods are presented together with techniques for detection of the most common types of bifurcations in electronic circuits occurring

Analysis and Design of Autonomous Microwave Circuits, By Almudena Suárez from dc and periodic regimes. The meaning and implications of these bifurcations were studied in detail in Chapter 3, so here only the techniques for their detection, using harmonic balance, are shown.

6.2 LOCAL STABILITY ANALYSIS

For the stability analysis of a small-signal regime, circuit equations are linearized about the dc solution, neglecting the influence of the small-signal generators. For the stability analysis of a large-signal periodic regime, the circuit equations are linearized about this large-signal periodic regime. The two cases are studied in the following.

6.2.1 Small-Signal Regime

Let a circuit containing one or more small-signal independent sources be considered such that the solution is linear with respect to these input sources. The stability properties of the small-signal regime are the same as those of the dc solution obtained by setting all the input sources to zero value. This is easily derived from the fact that the circuit is linear with respect to these sources. Superposition is fulfilled, so they cannot have any influence on the circuit linearization about the dc solution used for the stability analysis. Any possible oscillation comes from the energy delivered by the bias sources. Thus, stability analysis of small-signal solutions can be performed by suppressing the time-varying input sources. The circuit response versus the perturbation frequency ω (which may take any value) is analyzed by linearizing this circuit about the dc solution. An example is the stability analysis of small-signal amplifiers based on the Rollet factor and stability circles [3], usually based on a scattering-matrix description of the active two-port network. Note that this scattering matrix constitutes a linearization of the active device about the particular dc operation point. The main techniques for small-signal stability analysis are summarized below.

6.2.1.1 Rollet Stability Analysis The Rollet stability analysis, based on the k factor and the stability circles, is applicable to two-port networks that are intrinsically stable [4]. This means that when unloaded, or loaded with infinite impedances, the two-port network does not contain any poles on the right-hand side of the complex plane. This condition (known as *Rollet's condition*) is relatively easy to fulfill when the two-port network contains one transistor only. However, it may not be fulfilled if the two-port network contains more than one transistor or if it includes the transistor(s) plus additional feedback elements or bias paths. In these situations, the two-port network may contain unstable loops that cannot be detected from the analysis of its input or output impedance. As will be shown later in this section, we will have a feedback loop per active element contained in the analyzed circuit. This active element is generally constituted voltage-controlled current source. The feedback path is given by linear embedding network connecting the control voltage and controlled source. If we cannot guarantee that a two-port network is intrinsically stable, stability analysis based on the k factor and stability circles is not applicable.

To describe the Rollet stability analysis briefly, an intrinsically stable two-port network, described with its scattering matrix $[S(\omega)]$, will be assumed. As already discussed, all the periodic input sources are set to zero value. Then we take into account that the input (output) impedance of the two-port network depends on the scattering matrix and the load impedance connected to its output (input) [3]; that is, $Z_{in}([S], Z_L)$ and $Z_{out}([S], Z_S)$, with Z_{in} and Z_{out} being the input and output impedances, Z_L and Z_S being the load and source impedances, and $[S(\omega)]$ being the frequency-dependent scattering matrix. Variations in the analysis frequency are now considered in the entire frequency interval $(0,\omega_{max})$. The frequency ω_{max} is the maximum frequency up to which any of the devices included in $[S(\omega)]$ exhibits gain. Note that the frequency ω is not delivered by any existing source. It is, instead, a perturbation frequency, used to analyze the circuit response under small perturbations coming from noise or fluctuations. The two-port is said to be unconditionally stable if $\text{Re}[Z_{in}(\omega)] > 0$ for whatever passive Z_L and $\text{Re}[Z_{out}(\omega)] > 0$ for whatever passive Z_S in the entire frequency interval $(0,\omega_{max})$. This is fulfilled if the two following conditions are satisfied [4]:

$$\begin{aligned} k &= \frac{1 - |S_{11}(\omega)|^2 - |S_{22}(\omega)|^2 + |\Delta(\omega)|^2}{2|S_{12}S_{21}|} > 1 \\ |\Delta(\omega)| &= |S_{11}(\omega)S_{22}(\omega) - S_{12}(\omega)S_{21}(\omega)| < 1 \end{aligned} \tag{6.1}$$

for all frequencies in the interval $(0,\omega_{max})$. This means that when connecting any passive load to a two-port network, it will not exhibit negative resistance at either its input or output. Thus, it will not be able to oscillate. Note that the negative resistance of the device $\text{Re}[Z_{in}(\omega)] < 0$ ($\text{Re}[Z_{out}(\omega)] < 0$) is a necessary condition for oscillation startup, but it is not sufficient. As shown in Chapter 1, the fulfillment of oscillation startup conditions also depends on the values and frequency variation of the impedance Z_S in series with Z_{in} or the impedance Z_L in series with Z_{out}.

Fulfillment of the conditions for unconditional stability (6.1) depends on the particular values of the scattering matrix parameters and therefore on the particular active device and its bias point. If these conditions are not fulfilled, for a stable small-signal behavior, the impedances Z_L and Z_S should be restricted to certain values. The impedances should be chosen so as to prevent the two-port network from exhibiting negative resistance at any frequency. On the contrary, if the aim is to design a free-running oscillator, the impedance Z_S (or Z_L) should be chosen so as to obtain negative resistance at the output (input) of the two-port network at the desired oscillation frequency. Once the negative resistance is achieved, the circuit will have to be loaded suitably to fulfill the conditions for oscillation startup.

Use of the reflection coefficient is very convenient in establishing a boundary between the load impedances demonstrating $\text{Re}[Z_{in}] > 0$ and $\text{Re}[Z_{in}] < 0$ on the Smith chart [3]. Note that the reflection coefficient associated with an impedance fulfilling $\text{Re}[Z_{in}] < 0$ and referred to any passive reference impedance Z_c satisfies

$|\Gamma_{in}| > 1$. In turn, the reflection coefficient associated with an impedance that has $\text{Re}[Z_{in}] > 0$ satisfies $|\Gamma_{in}| < 1$. Thus, the boundary between the two cases is given by $|\Gamma_{in}| = 1$. Equivalently, the boundary between source impedances such that $\text{Re}[Z_{out}] > 0$ or $\text{Re}[Z_{out}] < 0$ is given by $|\Gamma_{out}| = 1$, with Γ_{out} being the reflection coefficient at the output of the two-port network. The input and output reflection coefficients of the two-port network, referred to the same impedance Z_c as that of the scattering matrix, usually $Z_c = 50\ \Omega$, are given by

$$\Gamma_{in} = S_{11} + \frac{S_{12}S_{21}\Gamma_L}{1 - S_{22}\Gamma_L} \qquad \text{(a)}$$
$$\Gamma_{out} = S_{22} + \frac{S_{12}S_{21}\Gamma_S}{1 - S_{11}\Gamma_S} \qquad \text{(b)} \qquad (6.2)$$

where the reflection load and source coefficients Γ_L and Γ_S are also referred to the characteristic impedance Z_c. At a constant frequency value, the scattering parameters in (6.2) will also be constant, so when Γ_{in} describes a circle $\Gamma_{in} = 1e^{j\varphi}$, with φ varying between 0 and 360° in (6.2a), Γ_L describes a circle, too. Thus, the $\Gamma_{in} = 1e^{j\varphi}$ limiting "stable" and "unstable" behavior is mapped to a circle in the plane Γ_L. It can be shown that this circle delimits the Γ_L values, providing $|\Gamma_{in}| > 1$ and $|\Gamma_{in}| < 1$ [3]. When traced on the Smith chart associated with Γ_L, this circle will delimit the stable and unstable load impedances Z_L. Expressions for the centers and radius of this circle are given in many microwave books [3]. From equation (6.2b) it is possible to obtain the circle in Γ_S associated with the circle $\Gamma_{out} = 1e^{j\varphi}$, with φ varying between 0 and 360°. When traced on the Smith chart associated with Γ_S, this circle will delimit the stable and unstable source impedances Z_S.

Whether the stable impedance region corresponds to the inside or outside of the circle obtained in the plane Γ_L depends on the particular case. This can easily be determined by taking into account that the input reflection coefficient obtained for $Z_L = 0$ agrees with the two-port scattering parameter S_{11}. Thus, for $|S_{11}| < 1$, the center point of the load Smith chart Γ_L belongs to the stable region. If this point is inside the stability circle, all the internal points will be stable, as the stability circle is the border between stable and unstable loads. Similarly, if the center point is outside the stability circle, all the external points will be stable. Following identical reasoning, the stability of the center of the Γ_S chart will depend on the modulus of S_{22}.

After discussing the boundary between stable and unstable impedances in the plane Γ_L (or Γ_S), it is possible to introduce the μ-factor [5] as an alternative criterion for the unconditional stability of the two-port network (provided that this network contains no poles on the right-hand side of the complex plane). The advantage of this criterion lies in the fact that it is based on a single condition. The stability factor μ provides the distance from the center of the unit Smith chart and the nearest point of the load stability circle in the plane corresponding to Γ_L. It is given by

$$\mu = \frac{1 - |S_{11}|^2}{|S_{22} - S_{11}^*\Delta| + |S_{12}S_{21}|} \qquad (6.3)$$

The single necessary and sufficient condition for the two-port network (with no unstable poles) to be unconditionally stable is $\mu > 1$. Note that in case of stable behavior, the μ factor provides an estimation of the stability margin. There is an alternative parameter μ' that provides the minimum distance between the center of the unit Smith chart and the stability circle in the source plane Γ_S.

Rollet's theory may be difficult to apply to multistage amplifiers since it is not easy to determine the limits between the different stages, which may also be ended in active loads [15]. However, the theory is very useful for the design of one-stage amplifiers and also for oscillator design. As an example, the Rollet criteria will be applied to obtain a free-running oscillator at 3.0 GHz using the MESFET transistor MGF135. The bias point considered is $V_{GS} = -0.5$ V and $V_{DS} = 3$ V. The stability factor obtained at 3.2 GHz for these bias conditions is $\mu = 0.182$. The transistor is conditionally stable, so for some passive source (load) impedances, negative resistance will be obtained at the transistor output (input). The source and load stability circles are represented in Fig. 6.1. Because the scattering parameters of the transistor fulfill $|S_{11}| < 1$ and $|S_{22}| < 1$, the unstable regions correspond to the inside of the stability circle in the plane Γ_L and the outside of the stability circle in the plane Γ_S.

In oscillator design, the size of the unstable regions in the Smith chart may be increased by changing the bias point or adding a feedback network to the transistor. As an example, a parallel resonant network at the desired oscillation frequency

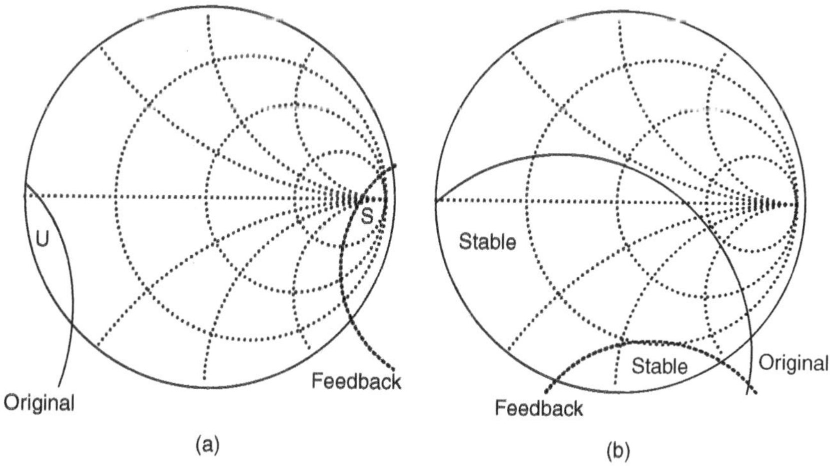

FIGURE 6.1 Stability circles corresponding to the transistor CFY30 at the bias point $V_{GS} = -0.5$ V and $V_{DS} = 3$ V at $f_o = 3.2$ GHz, before and after introducing feedback networks. (a) Stability circle traced on the Smith chart corresponding to the source impedance Z_S. Without additional feedback, the outside of the circle is the stable region. After the introduction of feedback, the inside of the circuit is the stable region. (b) Stability circle traced on the Smith chart corresponding to load impedance Z_L. Without additional feedback, the inside of the circle corresponds to the stable region. After the introduction of feedback, the inside of the circuit is the stable region.

$f_o = 3.0$ GHz has been connected to the source terminal of the transistor considered here. Then the scattering matrix of the two-port network is redefined to include the additional feedback resonant tank. By proceeding like this, the instability regions increase substantially, as shown in Fig. 6.1. For the oscillator design, a source impedance Z_s in the unstable region of the Smith chart is selected. The value chosen is $Z_S = 13.6 + j21\ \Omega$. This provides negative conductance at the output of a two-port network (including the feedback resonant tank). The impedance value is $Z_{out} = -80 - j31\ \Omega$. Then a load admittance Z_L is selected so as to fulfill the condition $\text{Re}[Z_T(f_o)] = \text{Re}[Z_N(f_o) + Z_L(f_o)] < 0$. The load impedance chosen is $Z_L = 26 + j31\ \Omega$. The impedances Z_L and Z_S selected have to be implemented with lumped or distributed circuit elements. After this implementation, the following conditions should be fulfilled to facilitate oscillation startup at $f_o = 3.0$ GHz:

$$\begin{aligned} \text{Re}[Z_T(f_o)] &= \text{Re}[Z_N(f_o) + Z_L(f_o)] < 0 \\ \text{Im}[Z_T(f_o)] &= \text{Im}[Z_N(f_o) + Z_L(f_o)] = 0 \\ \frac{\partial \text{Im}[Z_T(f_o)]}{\partial f} &> 0 \end{aligned} \tag{6.4}$$

As shown in Chapter 1, fulfillment of the conditions above generally implies the existence of a pair of complex-conjugate poles at the frequency f_o on the right-hand side of the complex plane. However, a more rigorous stability analysis based on pole–zero identification or the Nyquist criterion is advisable. These techniques are presented later in this section.

6.2.1.2 *Calculation of the Input Impedance and Admittance* The fulfillment of the oscillation startup conditions (6.4) at any frequency f_o generally means that the dc solution is unstable. For an impedance-based stability analysis, using (6.4), a circuit loop branch is broken to introduce a low-amplitude voltage source E_n at a frequency ω in series with this branch (see Fig. 6.2). The ratio between the voltage E_n and the current I through the voltage source agrees, by Kirchhoff's laws, with the total small-signal impedance of the circuit loop considered: $Z_{Tn}(\omega) = E_n(\omega)/I$. To have good sensitivity, the loop should include one of the active device ports. Next, an ac analysis is applied to obtain the circuit response to the small-signal input $E_n(\omega)$. For an admittance-based stability analysis at the circuit node n, a small-signal current source I_n at a frequency ω should be connected in parallel to this node. The ratio between the current introduced (entering the circuit) and the node voltage V provides the total input admittance seen from this node: $Y_{Tn}(\omega) = I_n/V$ (see Fig. 6.2). Convenient nodes for good analysis sensitivity are those corresponding to the transistor terminals. Note that the existing small-signal generators (such as the input generator of a linear amplifier) cannot influence the stability or instability of the dc solution because the circuit is linear with respect to these generators. Thus, these generators can be eliminated. In this kind of analysis, we are reducing the circuit model to just one port, described with a single impedance/admittance function. To compensate for the loss of information due to this model reduction, it is convenient to repeat the analysis at different

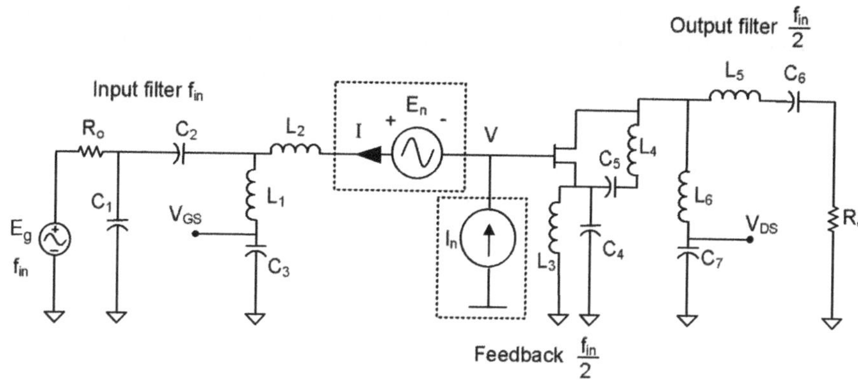

FIGURE 6.2 Oscillator circuit based on the MESFET transistor MGF135. The bias point is $V_{GS} = -0.5$ V and $V_{DS} = 3$ V. The auxiliary sources used for stability analysis are represented inside the dashed-line squares. They are not used simultaneously.

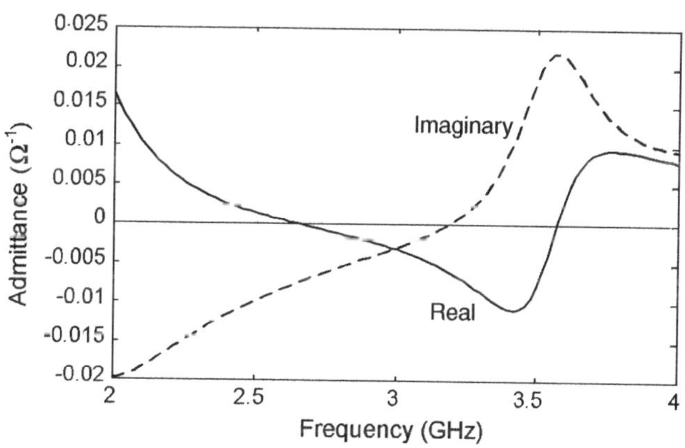

FIGURE 6.3 Admittance-based stability analysis of the dc solution of the circuit shown in Fig. 6.2, applied to a gate terminal, showing fulfillment of the oscillation startup condition at 3.22 GHz.

observation points. In transistor-based designs, we should consider all the terminals, for instance, gate, drain, and source in a FET-based circuit. The circuit will generally oscillate, provided the oscillation startup conditions (6.4) are fulfilled at any of the considered observation points. As an example, Fig. 6.3 shows the small-signal admittance analysis of the circuit in Fig. 6.2, performed at the gate terminal. As can be seen, the oscillation startup conditions are fulfilled at the frequency $f_o' = 3.2$ GHz. When using an impedance analysis, these conditions are fulfilled at $f_o = 3$ GHz.

The stability analysis based on impedance and admittance diagrams is easily implementable on any simulator by using a simple ac analysis to obtain the circuit response to the small-signal source I_n or E_n. However, the circuit may be unstable, and despite this, the analysis may provide an incorrect conclusion regarding stability. This will happen if the observation branch (node) is far from the "source" of negative resistance. It must also be noted that the observation branch (node) reduces the analysis of a usually multiresonant circuit to the analysis of one impedance (admittance) only. Analysis at a particular observation port may be unable to detect internal unstable resonances. The choice of observation port may be difficult in large multidevice circuits. However, the technique usually provides good results in small circuits.

6.2.1.3 Nyquist Stability Analysis Applied to the Characteristic Determinant of a Harmonic Balance System The stability of a given steady-state solution can be determined from the analysis of the perturbed harmonic balance system, linearized about this solution. The analysis presented here is based on a piecewise harmonic balance formulation more compact than the nodal formulation. However, the principles of stability analysis are applicable equally to both types of formulation. In the piecewise formulation (Section 5.4.3), the set of state variables considered is composed of all the control variables of the various nonlinear elements. The dc solution is given by the vector of dc components of the state variables $\overline{X}_{\text{dc}}$. For the stability analysis of this dc solution $\overline{X}_{\text{dc}}$, a small instantaneous perturbation will be introduced into the circuit, which will give rise to state-variable increments of low-amplitude and complex frequency s. In the practical description of the techniques, this complex frequency will be expressed explicitly as $s = \sigma + j\omega$. Because of the low amplitude of the perturbation, the circuit state variables will undergo a small increment, so the nonlinear elements $\overline{Y}(\overline{X})$ can be expanded in a first-order Taylor series about $\overline{X}_{\text{dc}}$. The linear matrixes A_x, A_y must be evaluated at the frequencies $\sigma \pm j\omega$ of the perturbed solution. The perturbed system can be split into two linear subsystems, one at the frequency $\sigma + j\omega$, formulated in terms of the state-variable increments at the same frequency $\Delta\overline{X}(\omega - j\sigma)$, and another subsystem at $\sigma - j\omega$, formulated in terms of $\Delta\overline{X}(-\omega - j\sigma)$. The relationship $\Delta\overline{X}(\omega - j\sigma) = \Delta\overline{X}^*(-\omega - j\sigma)$ is fulfilled, so the analysis can be limited to the following characteristic system:

$$\left\{[A_x(\omega - j\sigma)] + [A_y(\omega - j\sigma)]\left.\frac{\partial\overline{Y}}{\partial\overline{X}}\right|_{\text{dc}}\right\}\Delta\overline{X} = [JH(\omega - j\sigma)]\Delta\overline{X} = 0 \qquad (6.5)$$

where $\partial\overline{Y}/\partial\overline{X}\big|_{\text{dc}}$ is the dc-component matrix obtained from the derivation of nonlinear instantaneous functions with respect to the independent variables $[\partial\overline{y}/\partial\overline{x}]_{\text{dc}}$. For the variable increment $\Delta\overline{X}$ to be different from zero, the associated characteristic matrix [in braces in (6.5)] must be singular, fulfilling

$$\det[JH(\omega - j\sigma)] = 0 \qquad (6.6)$$

The poles associated with the dc solution $\overline{X}_{\text{dc}}$ are given by the roots of the characteristic determinant (6.6). For stability, all these poles must be located on the left-hand side of the complex plane. This means that the perturbation $\Delta\overline{x}(t)$ will vanish exponentially in time (due to the negative sign of σ), in agreement with the definition of stability. In contrast with an analysis based on the Rollet factor and the stability circles (which considers the input and output of a two-port network) or with the impedance–admittance analysis (from a particular circuit location), the analysis (6.6) globally takes into account all the circuit state variables.

Due to the usually high order of the linear matrixes A_x and A_y in the piecewise harmonic balance system, direct calculation of the complex roots $\sigma \pm j\omega$ is a nearly impossible task, so the Nyquist criterion, to be described later in this section, is applied instead. In the case of a nodal harmonic balance formulation of a circuit containing lumped elements only, the root calculation becomes a simple eigenvalue analysis. This is easily seen from the form of the Jacobian matrix in (5.45) in Chapter 5. The roots $\sigma \pm j\omega$ are given by the eigenvalues of the characteristic system, $s\left[\partial \overline{Q}/\partial \overline{X}\right]\Delta\overline{X}(s) = -\left[\partial \overline{F}/\partial \overline{X}\right]\Delta\overline{X}(s)$, with s the Laplace variable. Note that the matrix $\left[\partial \overline{Q}/\partial \overline{X}\right]$ is not always invertible, so generalized eigenvalues-search algorithms must be used [9]. This eigenvalue calculation will not be possible if the circuit constrains distributed elements, described by the transfer functions $[H_d(\omega - j\sigma)]$. Instead, the Nyquist stability criterion can be applied.

For a brief description of the Nyquist stability criterion, assume a complex function $F(s)$, which can be represented as a quotient of polynomials, such that $\lim_{s\to\infty} F(s) = \text{constant}$. Now, consider the plot resulting from evaluating the complex function $F(s)$ along a closed contour Γ of the complex plane, in a clockwise sense [10]. Note that the considered contour Γ cannot pass through any pole or zero of $F(s)$. The number N of clockwise encirclements of the plot $F(\Gamma)$ around the origin of the complex plane is equal to the difference between the number of zeros and poles of complex function F contained inside the contour Γ. For the complex function $\det[JH(\omega - j\sigma)] = 0$, the interest is in the number of zeros and poles on the right-hand side of the complex plane. This region of the plane is bounded by the entire imaginary axis $j\omega$ and a semicircular trajectory of infinite radius $s \to \infty$. The evaluation of $\det[JH]$ over the semicircular trajectory will provide a constant value, as for application of the Nyquist criterion we required $\lim_{s\to\infty} \det(s) = \text{constant}$. Thus, it will be sufficient to evaluate $\det[JH]$ along the imaginary axis $j\omega$, with ω going from $-\infty$ to ∞. Because the matrix terms corresponding to the negative frequency $-\omega$ are not considered in (6.5), the determinant will be a complex function. The Nyquist plot is obtained by sweeping ω and tracing $\text{Im}\{\det[JH(\omega)]\}$ versus $\text{Re}\{\det[JH(\omega)]\}$. The resulting number of clockwise encirclements of the origin is given by [6]

$$N = Z - P \tag{6.7}$$

with Z and P the number of zeros and poles of the analyzed function $\det[JH(\omega)]$, located on the right-hand side of the complex plane. From the inspection of (6.5), the poles of the perturbed harmonic balance system can only come from linear matrixes. These matrixes will not introduce any unstable poles in the determinant function $\det[JH(\omega - j\sigma)]$, as they come from the impedance or admittance of the passive linear elements [6], which are positive real. On the other hand, to have a bounded determinant for $\omega \to \infty$ we should redefine the linear matrixes according to the "feedback formulation" $\overline{X} + A_y \overline{Y}(\overline{X}) = A_g \overline{G}$, with A_x the unit matrix [9]. For example, in the case of the dc solution of the circuit in Fig. 1.1, the function evaluated is $\det(\omega) = 1 + Z_T(\omega)a$, with Z_T the passive-network impedance and "a" the device small-signal conductance.

From the discussion above, the number of clockwise encirclements Z around the origin of the complex function $\det[JH(\omega)]$, evaluated from $\omega = -\infty$ to $\omega = \infty$, will directly provide the number of unstable roots of $\det[JH(\omega - j\sigma)]$ corresponding to the poles associated to the dc solution. It is easily shown that the function $\det[JH(\omega)]$ is symmetrical for $\omega > 0$ and $\omega < 0$. This is because of the Hermitian symmetry of the perturbed-system equations in the frequency domain, so that $\det[JH(\omega)] = \det^*[JH(-\omega)]$. In practical applications it will be sufficient to perform a frequency sweep between $\omega = 0$ and $\omega = \omega_{\text{max}}$ in order to obtain the Nyquist plot. Emphasis should be placed on the fact that the Nyquist stability analysis described takes into account the actual multivariable nature of this circuit, so in principle allowing detection of any unstable loop. As already known, the dimension of the characteristic matrix is higher for a nodal harmonic formulation than for a piecewise one, which usually implies stronger numerical difficulties in the evaluation of the determinant $\det[JH(\omega)]$. Alternatively, we can obtain the generalized eigenvalues of the characteristic matrix $JH(\omega)$. Note that these eigenvalues are calculated in the ω domain instead of the s domain. For a Q dimension system, we will have Q eigenvalues. When sweeping ω, the variation of the eigenvalues will exhibit branching points. Thus, it is not possible to follow the Q eigenvalues independently. Instead, we can apply the Nyquist criterion to the so-called characteristic loci $L_j(\omega)$, $j = 1 \ldots Q$, in which the evaluated eigenvalues are exchanged at the branching points [9,19].

As an example, the Nyquist stability analysis described above has been applied to the dc solution of the circuit in Fig. 6.4a. The Nyquist criterion has been applied to the characteristic determinant resulting from a piecewise formulation of this circuit. As shown, the Nyquist plot encircles the origin and the dc solution is unstable. The crossing of the negative real semiaxis takes place at the frequency $f_c = 3.02$ GHz. When reducing the bias voltage to $V_{GS} = -1.5$ V, the origin is no longer enclosed and the dc solution is stable (Fig. 6.4b). The frequency ω_c at which the Nyquist plot crosses the negative real semiaxis provides an estimation of the oscillation frequency. This is in close relationship with the fact that at steady-state oscillation, the Jacobian matrix of the harmonic balance system is singular and fulfills $\text{Re}\{\det[JH(\overline{X}, \omega_o)]\} = 0$, $\text{Im}\{\det[JH(\overline{X}, \omega_o)]\} = 0$. This has been shown in Chapters 1 and 2 and is due to the irrelevancy of the oscillator solutions versus variations of the phase origin. Because the imaginary part mainly depends on the reactive elements, with most of them being linear, the frequency ω_c at which the condition $\text{Im}\{\det[JH(\omega_c)]\} = 0$ is fulfilled will be relatively close to the actual oscillation frequency ω_o.

6.2.1.4 *Normalized Determinant Function* The application of the Nyquist stability analysis to the characteristic determinant (6.6) requires an in-house harmonic balance formulation. A different technique allows the stability analysis of

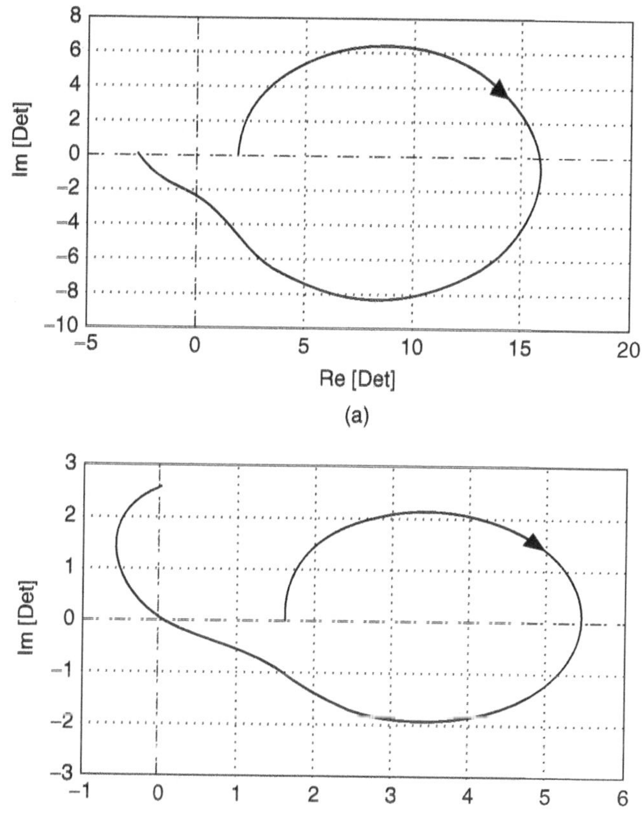

FIGURE 6.4 Stability analysis of the circuit in Fig. 6.2 based on use of the Nyquist criterion to the characteristic determinant of the harmonic balance system, linearized about the dc solution. (a) Analysis for $V_{GS} = -0.5$ V; the origin is enclosed and the dc solution is unstable. (b) Analysis for $V_{GS} = -1.5$ V; the origin is not enclosed and the dc solution is stable.

dc solutions using commercial software in which the Jacobian matrix $\partial\overline{Y}/\partial\overline{X}\big|_{dc}$ is not accessible to the designers. The technique provides a normalized version of the determinant det$[JH(\omega)]$ that does not alter the information contained in this determinant. Furthermore, the normalization used reduces the complexity of the Nyquist plot, often intricate in high-order systems. As outlined briefly in the following, the normalized determinant is obtained indirectly from open-loop transfer functions that can be calculated with commercial harmonic balance [11,12].

The normalized determinant function (NDF) is given by

$$\text{NDF}(\omega) = \frac{\det(\omega)}{\det_o(\omega)} \tag{6.8}$$

where $\det_o(\omega)$ is the determinant obtained when all the active devices are switched off. Any linear network parameters Y, Z, etc. can be used for the calculation of the determinants in (6.8) [11]. Obviously, the determinant $\det_o(\omega)$ will have the same denominator as $\det(\omega)$, so it cannot introduce any additional roots in the NDF(ω). On the other hand, division by $\det_o(\omega)$ cannot give rise to any unstable poles in NDF(ω), since all the active elements are switched off in $\det_o(\omega)$. Next, we obtain an expression for the normalized determinant NDF(ω) in terms of the open-loop transfer functions associated with the various active elements contained in the circuit.

Let a feedback system like the one depicted in Fig. 6.5 be assumed. The corresponding transfer function is given by $H(f) = A/(1 - AB)$. The product AB constitutes the open-loop transfer function of the system. The *return ratio* is defined as $\mathrm{RR} = -AB$ [11,12]. In terms of the return ratio RR, the denominator of the closed-loop transfer function $H(\omega)$ is given by $F = 1 + \mathrm{RR}$. Assume a circuit containing a controlled source, e.g., a voltage-controlled current source $i = g_m v$, plus a linear network. This circuit, shown in Fig. 6.6a, can be seen as a feedback system. The entire passive network constitutes the feedback loop of the active element $i = g_m v$. To obtain the return ratio, the closed loop will be broken, making the current depend on an external voltage V_{ext} and obtaining the voltage drop at the original location of the control voltage V' (Fig. 6.6b). The ratio V'/V_{ext} provides the open-loop transfer function $-\mathrm{RR} = V'/V_{ext}$. In [12] it is demonstrated that $F = 1 + \mathrm{RR}$ agrees with the normalized determinant function associated with the system of Fig. 6.6a. Then, for a single nonlinear element, it is possible to write $\mathrm{NDF}(\omega) = 1 + \mathrm{RR}$.

Obtaining the normalized determinant function NDF for multiple active elements is more involved. It requires the calculation of one open-loop transfer function per active element. Note that in the presence multiple feedback loops, we cannot determine the solution stability from a single open-loop transfer function. We cannot guarantee that the number of unstable poles of the resulting function $1 - AB$ is equal to zero, so the application of the Nyquist criterion would require a previous evaluation of these poles to obtain the number of roots of $\det(\omega)$ from $Z = N + P$. A different technique can be used instead. For M active elements, M open-loop transfer functions must be calculated successively in different operation conditions for the M elements. The calculation starts obtaining the return ratio associated with

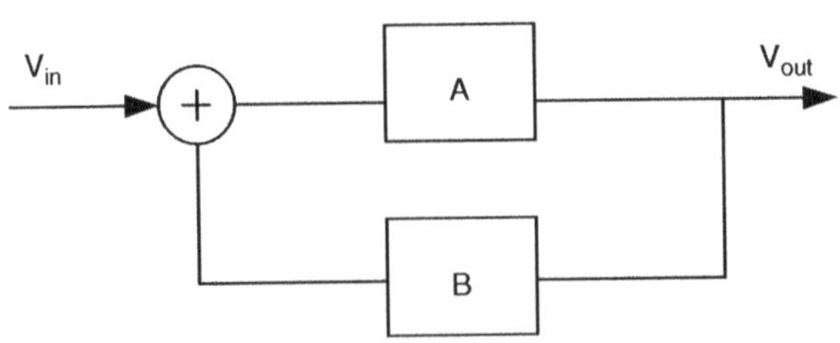

FIGURE 6.5 Feedback loop. The return ratio is equal to the open-loop transfer function with negative sign: $\mathrm{RR} = -AB$.

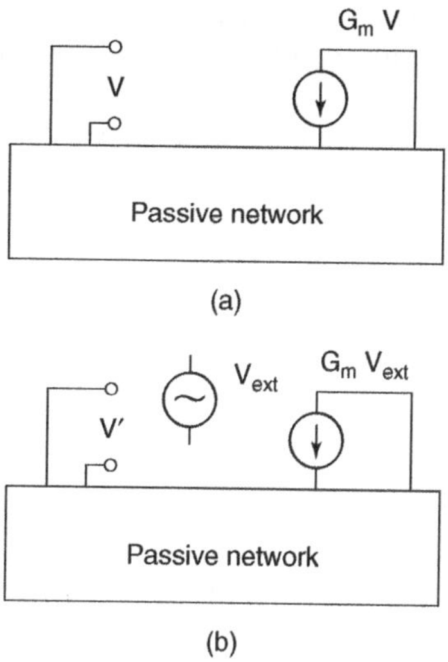

FIGURE 6.6 Circuit analysis in open-loop conditions for determination of the return ratio: (a) original circuit containing a single active device; (b) after opening the loop, the ratio V'/V_{ext} provides the open-loop transfer function $-\text{RR} = V'/V_{ext}$. The return ratio is given by RR.

the first active element RR_1. This is determined by breaking the loop associated with this controlled element, as in the single-element case, and keeping the rest of active elements 2 to M in nominal operation. Once RR_1 is known, it is possible to obtain the ratio $\det(\omega)/\det_1(\omega) = 1 + RR_1$, where $\det_1$ is the determinant obtained when the active element 1 is turned off and the remaining $M-1$ active elements are still on. Note that the determinant $\det_1(\omega)$ can still have roots on the right-hand side of the complex plane, since the remaining $M-1$ elements are in nominal operation. These roots will give rise to unstable poles of the normalized determinant $\det(\omega)/\det_1(\omega)$. Remember that the poles associated with the dc solution examined agree with the zeros of the determinant $\det(\omega)$.This determinant is the same as the one analyzed in Section 6.2.1.3. As shown in this subsection, the function $\det(\omega)$ cannot have any pole on the right-hand side of the complex plane ($P = 0$). However, when calculating $1 + RR_1$ in a circuit with multiple active elements, we are limiting the circuit analysis to a single active element, so the rest of the active elements can give rise to right-hand-side poles in $\det(\omega)/\det_1(\omega)$. The situation is similar to the admittance or impedance analysis from a single observation port, which may not be able to detect instabilities inside some circuit blocks.

With one more iteration, the determinant $\det_1(\omega)$ can be expressed as $\det_1(\omega) = \det_2(\omega)(1 + \text{RR}_2)$, where RR_2 is the return ratio associated with the second active element. The return ratio RR_2 is calculated by turning off the first active element and keeping the elements 3 to M in nominal operation. After this second iteration, the normalized determinant is $\det(\omega)/\det_2(\omega) = (1 + \text{RR}_1)(1 + \text{RR}_2)$. This determinant can still contain unstable poles due to nominal operation of elements

3 to M. To suppress all the possible unstable poles, the procedure is applied successively up to the final return ratio RR_M, which is the open-loop transfer function associated with the final active element M calculated with all the active elements 1 to $M-1$ turned off. Then the normalized determinant function (6.8) of the circuit containing M active elements is given by

$$\mathrm{NDF}(j\omega) = \frac{\det(j\omega)}{\det_o(j\omega)} = (1+\mathrm{RR}_1)(1+\mathrm{RR}_2)\cdots(1+\mathrm{RR}_M) \tag{6.9}$$

Note that the NDF function in (6.9) cannot contain any right-hand-side poles. The main advantage of the method is that it allows the use of Nyquist analysis of the circuit characteristic determinant without actually calculating this determinant. The NDF can be evaluated in commercial software by obtaining the return ratio functions RR_1 to RR_M successively, which requires access to all the different active elements. These elements must alternately in nominal operation, open-loop configuration, and turned off. One advantage of the gradual introduction of the factors $(1+\mathrm{RR}_m)$, $m = 1$ to M is that it can help detect the unstable loop existing in the circuit. The analysis of the factors $(1+\mathrm{RR}_m)$ can allow the identification of the active device involved in the unstable feedback loop.

6.2.1.5 Pole–Zero Identification Stability analysis of the dc solution of a given circuit can be carried out with pole–zero identification. This analysis relies on the fact that as shown in Chapter 1, any closed-loop transfer function that can be defined in a linear system can be expressed with the same denominator [13,14]. However, the numerator depends on the particular transfer functions, so pole–zero cancellations can occur. Pole–zero identification requires the definition of a particular closed-loop transfer function associated with a circuit linearization about the dc solution $\overline{X}_{dc}$ being examined. To obtain this transfer function in practical circuit analysis, a small-signal input current source I_n at the frequency ω will be introduced into the circuit, connected in parallel at a circuit node. Due to the small amplitude of the current generator, the nonlinear devices can be linearized about the dc solution, as in the case of the stability analysis (6.5). This provides the linear system

$$\left\{[A_x(\omega)] + [A_y(\omega)]\left.\frac{\partial \overline{Y}}{\partial \overline{X}}\right|_{dc}\right\}\Delta\overline{X} = [A_{gI}(\omega)]I_n \tag{6.10}$$

where A_{gI} is the matrix relating the circuit variables to the current generator introduced. Equations (6.10) and (6.5) can now be compared. Both equations constitute linearizations of the original harmonic balance system about the dc solution. In (6.5) the perturbation considered was applied at the initial time only, so it does not appear on the linearized equation. The system is in a transient state as a result of this perturbation, so the frequency is complex. In (6.10), a small-signal generator is connected permanently to the circuit, which behaves linearly with respect to this generator. The circuit is in steady state, so the frequency is real.

Let us continue with a definition of the closed-loop transfer function. Clearly, any output variable that the designer may choose will be a linear combination of

$\Delta\overline{X}$, so it will be expressed as $W = [C]\Delta\overline{X}$, with $[C]$ being a row matrix. From an inspection of (6.10), any possible transfer function $H = W/I_n$ will share the same denominator. This denominator is obtained when inverting the matrix between the symbols {} in (6.10), with the following determinant:

$$\det\left\{[A_x(\omega)] + [A_y(\omega)]\left.\frac{\partial\overline{Y}}{\partial\overline{X}}\right|_{\mathrm{dc}}\right\} \tag{6.11}$$

Note that the determinant above is the one used for the Nyquist stability analysis in previous subsections. The poles of the system are the roots of this determinant when expressed in terms of $\omega - j\sigma$ instead of ω. Operations in the numerator (associated to the product of $[C]$ by the transpose of the cofactor matrix) may give rise to pole–zero cancellations but cannot introduce new poles. For a definition of the transfer function, a very convenient output is the voltage V_n at the node to which the current generator is connected [13]. The proximity between the input and output provides good analysis sensitivity. Thus, the transfer function considered is

$$Z_{\mathrm{in}}(\omega) = \frac{V_n(\omega)}{I_n(\omega)} \tag{6.12}$$

The objective will be to determine the poles of the function $Z_{\mathrm{in}}(\omega - j\sigma)$, which agree with the roots of $\det[JH(\omega - j\sigma)]$ and thus provide information on the stability of the dc solution $\overline{X}_{\mathrm{dc}}$. Using pole–zero identification, the complex function $Z_{\mathrm{in}}(\omega)$ is fitted through a least minimum squares method, with a quotient of polynomials of the form [15]

$$Z_{\mathrm{in}}(\omega) = \frac{V_n}{I_n}(\omega) = A\frac{(j\omega - z_1)\cdots(j\omega - z_m)}{(j\omega - p_1)\cdots(j\omega - p_n)} \tag{6.13}$$

Note that the zeros and poles of $Z_{\mathrm{in}}(\omega - j\sigma)$ depend on the constants z_1 to z_m and p_1 to p_n, as the Z_{in} expression would be formally identical in terms of $\omega - j\sigma$. Thus, from the pole–zero identification of $Z_{\mathrm{in}}(\omega)$ it is possible to obtain the poles and zeros of the function $Z_{\mathrm{in}}(\omega - j\sigma)$. The identification should be performed between $\omega = 0$ and the frequency $\omega_{\max}$ up to which any of the active devices contained in the circuit exhibits gain. Note that accurate identification in a very wide frequency interval may require a high order n of the denominator polynomial in (6.13). Proceeding like this, the quality of the identification may degrade. As a matter of fact, the required order would be theoretically infinite for circuits containing distributed elements. However, because the transfer function $Z_{\mathrm{in}}(\omega)$ is linear, the total frequency interval 0 to $\omega_{\max}$ may be divided into subintervals, which will allow accurate identification with a small order n of the denominator polynomial in (6.13).

As already stated, all the closed-loop transfer functions that can be defined in a given linear system share the same denominator, corresponding to its characteristic determinant evaluated at the frequency ω. In contrast, the numerator depends on the particular definition of the transfer function. Thus, it will depend on the node selected for connection of the current source. Because of this, cancellations of unstable poles with zeros on the right-hand side of the complex plane may take place at particular locations. If unstable poles are canceled, an incorrect conclusion may be drawn about the stability of the solution [13,14]. This is why the pole–zero identification should be performed for different locations of the current source. The terminal nodes of the active devices are the most convenient for this analysis, due to their proximity to the potential sources of instability. In the case of series resonances, better sensitivity may be obtained using a small-signal voltage source (instead of a current source), introduced in series at a circuit branch. In a totally dual manner, the closed-loop transfer function would be defined as the ratio between the current through the voltage source and the introduced voltage.

Note that the possible existence of zeros on the right-hand side of the complex plane does not have any implications regarding system stability. Systems that do not have zeros in the right-half plane, called *minimum phase systems*, are characterized by the fact that their phase response $\angle Z_{in}(\omega)$ can be calculated from their amplitude response $\mathrm{Mag}(Z_{in}(\omega))$ and conversely. If a transfer function has poles and/or zeros in the right-half s-plane, this system shows *non-minimum-phase behavior*. The interval of phase variation of the phase response is then always larger than that for a system with minimum phase behavior with the same amplitude response.

The transfer function (6.13) agrees exactly with the inverse of the total input admittance at the node n, defined in Section 6.2.1.4 as $Y_{in}(\omega) = I_n(\omega)/V_n(\omega)$. However, there is one essential difference: The function $Z_{in}(\omega)$ corresponds to the ratio between the single circuit output $V_n(\omega)$ resulting from the single input $I_n(\omega)$. It is, in fact, a single-input single-output transfer function. In contrast, the input admittance $Y_{in}(\omega)$ is not a transfer function, and pole–zero identification cannot be applied to this function. Provided that $Y_{in}(\omega)$ is evaluated at a sensitive location, in some cases it is possible to relate the oscillation startup conditions $Y_{in}^r(\omega_o) < 0$, $Y_{in}^i(\omega_o) = 0$, and $\partial Y_{in}^i(\omega_o)/\partial\omega > 0$ (with r and i indicating the real and imaginary parts) to the existence of a pair of unstable complex-conjugate poles in the closed-loop transfer function $Z_{in}(\omega)$, as discussed in Chapter 1.

Use of the foregoing technique for stability analysis of the dc solution of the circuit in Fig. 6.2 provides the pole–zero locus of Fig. 6.7. The frequency interval considered in obtaining this particular locus was (2.5 GHz, 3.5 GHz). The pair of unstable complex-conjugate poles has the value $0.28 \pm j3.015$ GHz. As gathered from the example above, the pole–zero identification technique gives very complete information on the stability properties of the steady-state solution obtained. Compared with the Nyquist analysis, it provides the values of the poles and zeros of the transfer function $Z_{in}(\omega)$ obtained instead of simply the number of unstable poles.

6.2.2 Large-Signal Regime

The preceding stability analysis, based on the circuit linearization about the dc solution $\overline{X}_{dc}$, is applicable to circuits operating in a linear regime with respect

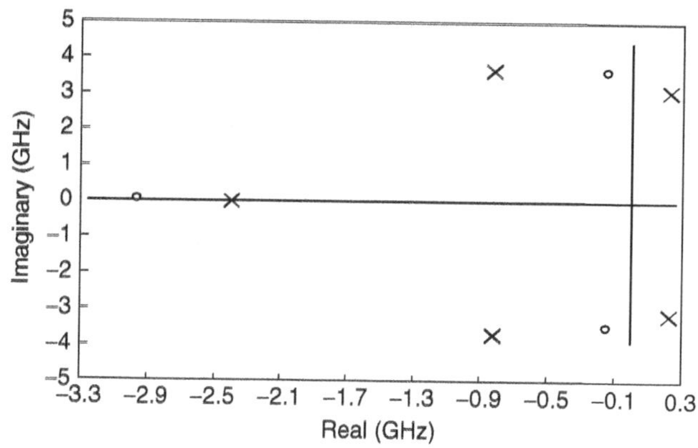

FIGURE 6.7 Pole–zero locus corresponding to the dc solution of the circuit in Fig. 6.2.

to the input sources. If the circuit is nonlinear with respect to these sources, its equations must be linearized about the particular time-varying steady-state regime obtained in the presence of the sources. As an example, a power amplifier may be stable in small-signal mode, fulfilling the Rollet stability criteria or other more complete criteria described in Section 6.2.1. However, as the input power increases, instability phenomena are often obtained from a certain value of this power, which cannot be predicated using small-signal stability analysis. This is because the dc solution is actually stable, and the instability originates from the large-signal regime resulting from the input drive [16]. As shown in previous chapters, a common cause of this instability is the negative resistance exhibited by the nonlinear capacitances contained in the active devices under strong periodic pumping at the input generator frequency.

In the following, stability analysis techniques of a large-signal periodic solution $\overline{X}_s$ are presented. This will require linearization of the harmonic balance equations about the steady-state solution $\overline{X}_s$, having the harmonic components $k\omega_o$ with $k = 0, \pm 1, \pm NH$. The techniques are extensions of those already presented for a dc regime to the case of a periodic steady state.

6.2.2.1 *Admittance and Impedance Diagrams* The oscillation startup from a dc regime can be related intuitively to the existence of negative resistance (energy unbalance) at a resonance frequency. As shown in Chapters 3 and 4, the negative resistance can also arise from a periodic large-signal steady-state regime at the harmonic frequencies $k\omega_o$, with k an integer. A spurious oscillation generated from this regime would give rise to the sidebands $k\omega_o + \omega_b$. Thus, in principle, it would be sufficient to analyze the stability of the periodic regime in the frequency interval $[0, \omega_o/2]$.

The admittance–impedance analysis, used in Section 6.2.1 for dc solutions, can be extended to periodic regimes. To calculate the total input admittance (impedance) at ω, a small-signal current (voltage) source in introduced at a given observation node (branch). Emphasis is placed on the fact that the circuit is linear with respect to this small-signal source at ω, used for the input admittance (impedance) calculation, but it is nonlinear with respect to the existing input sources or with respect to the self-generated oscillation in the case of a free-running oscillator. The small-signal source introduced must not alter the steady-state periodic regime at ω_o. Thus, the admittance–impedance analysis must be carried out with the input generator at its original value or in the presence of the nonperturbing auxiliary generator that sustains the steady-state oscillation in the case of an oscillator or a frequency divider. On the other hand, instability may occur at any frequency in the interval $[0, \omega_o/2]$, so the perturbation frequency ω must be swept, to obtain the frequency variation of the input admittance (impedance) [17,18]. Due to possible lack of sensitivity of this single node (branch) analysis, the frequency interval considered should be quite large, going from $\omega \cong 0$ to ω_o or even beyond this value.

For more clarity, an admittance analysis will be described in detail. A small-signal current source I_n at the frequency ω is introduced in parallel at a sensitive circuit node n. Because the circuit is linear with respect to I_n, its response to this source can be obtained by linearizing the harmonic balance equations about the steady-state large-signal regime at ω_o with respect to this source [17,18]. This can be done using the conversion matrix approach, which, as shown in Chapter 5, replaces the nonlinear elements with their conversion matrixes and evaluates the linear network at the sidebands $k\omega_o + \omega$. The input admittance at the frequency ω is given by the ratio between the current introduced and the node voltage V: $Y_{in}(\omega) = I_n/V$. The oscillation startup conditions at the frequency $\omega = \omega_a$ will be given by $\text{Re}[Y_{in}(\omega_a)] < 0$, $\text{Im}[Y_{in}(\omega_a)] = 0$, with positive slope $\partial(\text{Im}[Y_n])/\partial\omega_a$. This technique generalizes the impedance–admittance analysis to check the oscillation startup conditions from a dc regime, to a circuit in a large-signal periodic regime at ω_o.

The admittance–impedance analysis of periodic solutions described above suffers from the same limitations as the one used for the stability analysis of dc solutions. Even if the periodic solution is unstable, the oscillation startup conditions will not be observed at insensitive locations of the voltage (current) source introduced. As already pointed out, maximum sensitivity is obtained at the device terminals. However, even when only these points are considered, the selection of the test node (branch) may be arduous in the case of large multitransistor topologies. In the case of small circuits with few devices, the technique is very simple to apply and very intuitive. It can be used in commercial harmonic balance, relying on either the conversion matrix approach or two-tone analysis at the two fundamental ω_o and ω to obtain $Y_{in}(\omega)[Z_{in}(\omega)]$. If two-tone analysis is used, a very small amplitude value must be given to the generator at ω, to ensure linear behavior with respect to this generator. However, an excessively small amplitude may give rise to inaccuracy due to lack of sensitivity of the simulator to the generator at ω.

For illustration, the analysis technique described has been applied to the circuit of Fig. 6.2, biased at $V_{GS} = -0.5$ V and $V_{DS} = 3$ V and driven with input power $P_{in} = 12$ dBm. Two different input frequency values $f_{in} = 5$ GHz and $f_{in} = 6$ GHz have been considered. In the two cases, the harmonic balance system converges by default to a periodic solution at f_{in}, with some harmonic content, due to the

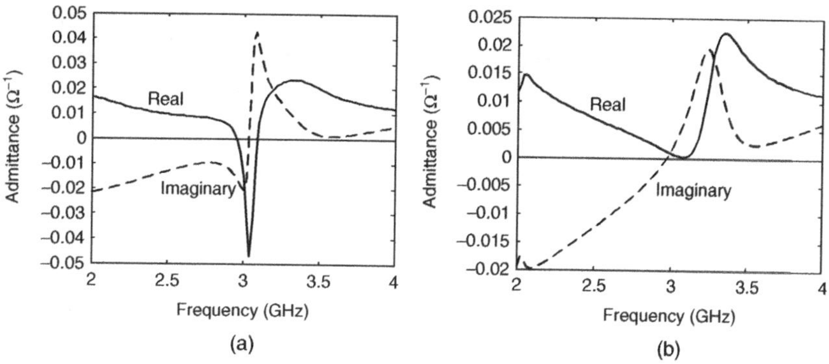

FIGURE 6.8 Admittance diagram for the circuit of Fig. 6.2 when introducing an input generator of power $P_{in} = 12$ dBm, considering two different input frequencies. (a) Input frequency $f_{in} = 6$ GHz; the oscillation startup conditions are fulfilled. (b) Input frequency $f_{in} = 5$ GHz; the oscillation startup conditions are not fulfilled.

relatively high power $P_{in} = 12$ dBm. To analyze the stability of this solution, a small-signal current generator $I_n(\omega)$ is connected to the gate terminal. Let the case of the input frequency $f_{in} = 6$ GHz be considered initially. The circuit is linearized about the periodic regime at $f_{in} = 6$ GHz with the conversion matrix approach, which enables the calculation of $Y_{in}(\omega) = I_n/V$, with V being the gate-node voltage at the frequency ω. The resulting frequency variation in input admittance is shown in Fig. 6.8a. The oscillation startup conditions are fulfilled at 3 GHz. Note that this is the same instability as already detected in analysis of the dc solution (Fig. 6.3). However, the diagram is different because the linearization is now carried out about the large-signal periodic regime at $f_{in} = 6$ GHz. In this particular case, the oscillation existed prior to the input power injection and is not extinguished for $f_{in} = 6$ GHz and $P_{in} = 12$ dBm. Next, a second test has been performed, changing the input frequency to $f_{in} = 5$ GHz, with the same input power $P_{in} = 12$ dBm. The resulting admittance diagram is shown in Fig. 6.8b. The oscillation startup conditions are not fulfilled. At this frequency, $f_{in} = 5$ GHz, the power delivered by the input generator extinguishes the original free-running oscillation.

Before continuing, it will be helpful to provide some additional information about the circuit behavior. As already stated, the circuit of Fig. 6.2 was designed to operate as a harmonic injection divider by 2. For the relatively high input power $P_{in} = 12$ dBm considered here, the frequency-division band is delimited by two flip bifurcations, F_1 and F_2 (Fig. 6.9). Increasing the input frequency, a divided regime at $\omega_{in}/2$ is generated at $f_{in} = 5.8$ GHz in a flip bifurcation. The nondivided regime at ω_{in} continues to exist after the bifurcation. It is the solution to which harmonic balance converges by default. Both the divided and nondivided solutions have been represented in Fig. 6.9a versus the input frequency f_{in}. To reduce the number of curves, the average value of the harmonic components of the independent variables has been traced in each case. This average value changes when frequency division takes place, due to the generation of subharmonic oscillation.

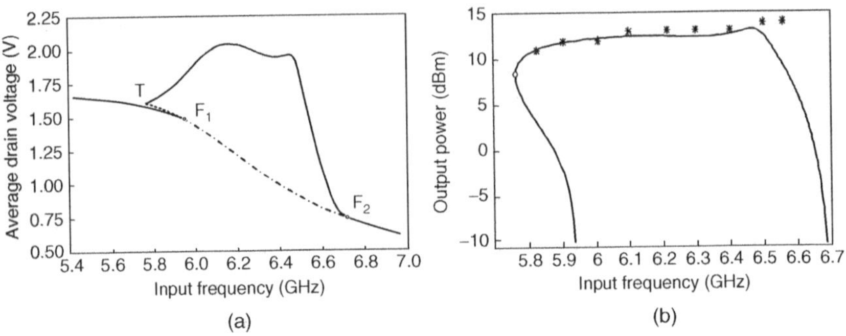

FIGURE 6.9 Solution curves of the circuit in Fig. 6.2 for constant input power $P_{in} = 12$ dBm versus the input frequency f_{in}: (a) periodic solutions at f_{in} and $f_{in}/2$ traced in terms of the average value of the harmonic components of the independent variables; (b) output power at the divided frequency $f_{in}/2$ versus the input frequency. This output power is zero outside the frequency-division band.

Figure 6.9b shows the output power at $f_{in}/2$. This power is different from zero only within the frequency-division band. The results of the admittance diagrams of Fig. 6.8 are in agreement with the solution curves represented in Fig. 6.9. For $f_{in} = 6$ GHz, the nondivided solution is unstable. The solution observed physically is the frequency-divided solution at $f_{in}/2 = 3$ GHz. For $f_{in} = 5$ GHz, the original self-oscillation is extinguished, whereas frequency division has not taken place yet. Thus, the nondivided solution at $f_{in} = 5$ GHz is stable.

6.2.2.2 Nyquist Stability Analysis Applied to the Characteristic Determinant of a Linearized Harmonic Balance Equation Let the piecewise harmonic balance system be considered, with a periodic solution $\overline{X}_s$ at the frequency ω_o:

$$\overline{H}(X_s) = [A_x(k\omega_o)]\overline{X}_s + [A_y(k\omega_o)]\overline{Y}(\overline{X}_s) = [A_g(k\omega_o)]\overline{G} \tag{6.14}$$

with $\overline{X}_s$ being the vector containing the $2N + 1$ harmonic components $k\omega_o$ of the Q control voltages and currents of the nonlinear elements, Y_s being the vector of harmonic components of the nonlinear elements, and G the vector of input sources. The matrixes $A_x(k\omega_o)$, $A_y(k\omega_o)$, and $A_g(k\omega_o)$ are obtained from the application of Kirchoffs laws to the embedding network at the different harmonic frequencies $k\omega_o$, with $k = -N$ to N, which is expressed with the general dependence on $(k\omega_o)$. For the stability analysis of the steady-state solution $\overline{X}_s$, a small instantaneous perturbation is considered. The complex frequency $\sigma + j\omega$ associated to this perturbation will give rise to a small increment of the circuit variables $\Delta\overline{X}$ at the intermodulation frequencies $k\omega_o \pm \omega - j\sigma$. The harmonic balance system can be linearized about the steady-state solution $\overline{X}_s$. This provides the following characteristic equation [6,7]:

$$\left\{[A_x(k\omega_o+\omega-j\sigma)]+[A_y(k\omega_o+\omega-j\sigma)]\left.\frac{\partial\overline{Y}}{\partial\overline{X}}\right|_s\right\}\Delta\overline{X}$$
$$=[JH(k\omega_o+\omega-j\sigma)]\Delta\overline{X}=0 \qquad (6.15)$$

where $\partial\overline{Y}/\partial\overline{X}\big|_s$ is the Jacobian of the nonlinear elements, with respect to the state variables, evaluated at the steady-state solution. A detailed derivation of this matrix was given in Section 5.6. Note that only the set of linear equations involving the frequency components $k\omega_o+\omega-j\sigma$ has been considered in (6.15). Because (6.15) is a linear homogeneous system, in order to have $\Delta\overline{X}\neq 0$, the associated characteristic matrix $[JH(k\omega_o+\omega-j\sigma)]$ must be singular.

The poles associated with the periodic solution $\overline{X}_s$ will agree with the roots of the characteristic determinant $\det[JH(k\omega_o+\omega-j\sigma)]=0$. For stability, all the solution poles must be located on the left-hand side of the complex plane. Because the steady-state solution $\overline{X}_s$ is periodic in time with the period $T=2\pi/\omega_o$, the solution poles will have the form $\sigma\pm j\omega+jk\omega_o$ or $\gamma+jk\omega_o$, with $-N\leq k\leq N$. Thus, the entire pole structure, with identical distances to the imaginary axis, will be repeated at ω_o intervals. Direct calculation of these roots is a nearly impossible task in the case of piecewise harmonic balance. The calculation is simpler in the case of nodal harmonic balance, due to the first order of the linear matrix $[j\omega]$ [see equation (5.45) in Chapter 5]. As in the case of the stability analysis of a dc solution, it becomes a simple eigenvalue calculation for circuits containing lumped elements only. However, this calculation is not possible in the case of circuit containing distributed elements. Then, the Nyquist stability criterion is usually applied to the characteristic determinant of the harmonic balance system, as in the case of the piecewise formulation.

Note that the function evaluated in order to obtain the Nyquist plot is the determinant $\det[JH(k\omega_o+\omega)]$. This is a complex function because the frequencies considered, $k\omega_o+\omega$, are not symmetric about the origin, as only the "plus" sign is considered in the perturbation frequency ω. The Nyquist plot is obtained by sweeping ω and tracing $\text{Im}\{\det[JH(k\omega_o+\omega)]\}=0$ versus $\text{Re}\{\det[JH(k\omega_o+\omega)]\}=0$. Taking into account the aforementioned periodicity of the poles, the ω sweep can be reduced to the interval $(0,\omega_o)$. The number N_T of clockwise encirclements of the Nyquist plot associated with the complex function $\det[JH(k\omega_o+\omega)]$ will correspond to $N_T=Z-P$, with Z being the number of right-hand-side zeros of the determinant $\det[JH(k\omega_o+\omega-j\sigma)]$ and P being the number of right-hand-side poles. However, the determinant $\det[JH(k\omega_o+\omega-j\sigma)]$ cannot have any unstable poles, due to the fact that the only denominators in (6.15) come from the linear matrixes A_x and A_y, which cannot contain any unstable poles, due to their inherent passivity. Thus, the number of clockwise encirclements N of the Nyquist poles will agree with the number of right-hand zeros of the determinant function $\det[JH(k\omega_o+\omega-j\sigma)]$. Remember that the zeros of this function constitute the poles associated with the circuit linearization about the periodic steady-state solution $\overline{X}_s$.

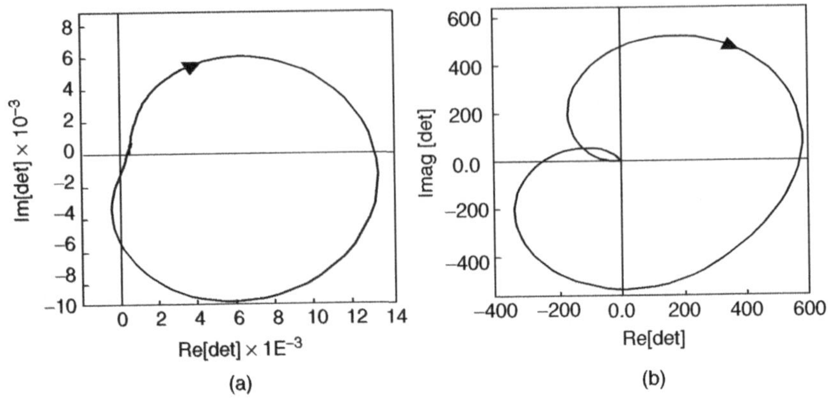

FIGURE 6.10 Stability analysis through the Nyquist criterion of the nondivided solution of the circuit in Fig. 6.2 for the input power $P_{in} = 12$ dBm. (a) Input frequency $f_{in} = 5$ GHz below the flip bifurcation F_1; the nondivided solution is stable. (b) Input frequency $f_{in} = 6$ GHz, belonging to the interval F_1, F_2; the nondivided solution is unstable. The physically observed solution is the frequency-divided solution.

To illustrate, the Nyquist stability criterion has been applied to the nondivided solution (with f_{in} as the fundamental) of the circuit in Fig. 6.2, at the two same f_{in} values considered in the admittance analysis of Fig. 6.8. For $f_{in} = 5$ GHz, below the flip bifurcation F_1 of Fig. 6.9, the corresponding Nyquist plot (shown in Fig. 6.10a) does not encircle the origin, so the nondivided solution is stable, in agreement with the fact that the divided path has not yet been generated (Fig. 6.9). For $f_{in} < F_1$ and $f_{in} > F_2$, all the points of the nondivided solution curve are similarly stable. However, in the frequency interval $[F_1, F_2]$ between the two flip bifurcations, the nondivided solution is unstable, as shown by the plot of Fig. 6.10b obtained for $f_{in} = 6$ GHz. This plot crosses the negative real semiaxis at about the divided frequency $f_{in}/2$.

The stability of the frequency-divided solution has also been analyzed. This solution, with $f_{in}/2$ as the fundamental, is qualitatively different from the nondivided solution at f_{in}. The frequency-divided steady state $\overline{X}_{s,1/2}$ is obtained with the aid of an auxiliary generator operating at the frequency $f_{AG} = f_{in}/2$, with amplitude A_{AG} and phase ϕ_{AG}. Once the steady state is determined accurately, the generator fulfills $Y_{AG}(A_{AG}, \phi_{AG}) = 0$ and thus has no influence over the steady-state solution. For the stability analysis, the harmonic balance system is linearized about the solution $\overline{X}_{s,1/2}$, determined with the aid of the auxiliary generator (see Section 5.5.2.2, Chapter 5). The Nyquist stability criterion is applied to the associated determinant $\det[JH(k\omega_{in}/2 + \omega)]$. The Nyquist plot obtained for $P_{in} = 12$ dBm and $f_{in} = 6$ GHz is shown in Fig. 6.11. The plot does not actually encircle the origin, as confirmed by the expanded view of Fig. 6.11b. The division path is stable between the turning point and F_2, as shown by a similar Nyquist analysis. The division path is unstable between F_1 and the turning point. The output power measurements

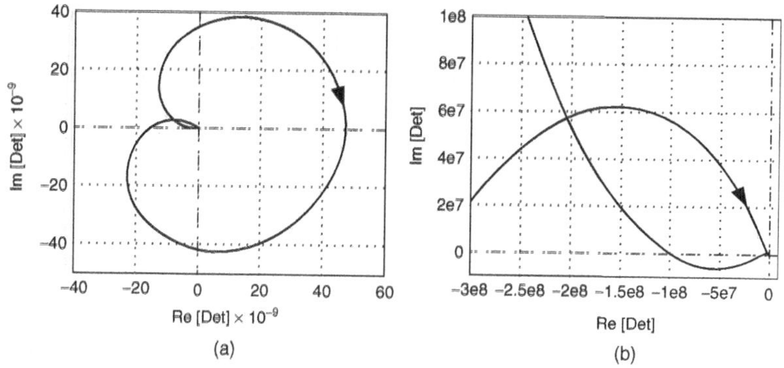

FIGURE 6.11 Stability analysis through the Nyquist criterion of the divided solution of the circuit in Fig. 6.2 for the input power $P_{in} = 12$ dBm and the input frequency $f_{in} = 6$ GHz: (a) Nyquist plot; (b) expanded view about the origin, showing stable behavior.

superimposed on Fig. 6.10b confirm the stability of the frequency-divided section between the turning point and F_2.

6.2.2.3 Normalized Determinant Function The normalized determinant function can be extended to the stability analysis of periodic solutions $\overline{X}_s$. This function is given by the ratio between the characteristic determinant associated with the particular periodic solution and the determinant $\det_o$ obtained when all the nonlinear dependences contained in the original circuit are eliminated; that is, $\text{NDF}(\omega) = \det(k\omega_o + \omega)/\det_o(k\omega_o + \omega)$. As in the case of dc regimes, instead of performing a direct calculation, the NDF(ω) function will be obtained from the recursive calculation of a series of open-loop transfer functions associated with circuit linearization about the periodic steady-state solution [19].

As already known, the different nonlinearities contained in the solid-state devices can be considered as nonlinear controlled sources. Calculation of the open-loop transfer functions requires breaking their dependence on their corresponding control variables on their corresponding control variables. However, applying variable substitution to the characteristic system [9] we can eliminate some internal variables. In practice, we consider one fundamental current and one control voltage per transistor element. Let a periodic steady-state solution with N harmonic components be assumed. A small perturbation at any frequency $\sigma + j(n\omega_o + \omega)$ with $-N < n < N$ will give rise to all the sideband frequencies $\sigma + j(m\omega_o + \omega)$ with $-N < m < N$. Thus, when breaking the circuit loop at a particular nonlinear element, $2N + 1$ different inputs and $2N + 1$ different outputs should be considered. Taking all the existing semiconductor devices into account, an open-loop transfer function must be calculated per controlled current and pair of sidebands. Each scalar transconductance is replaced [9] with a conversion matrix of dimension $(2N + 1) \times (2N + 1)$. Assuming, as an example, the nonlinear dependence $i(v)$, the external stimulus V_{ext} must be introduced recursively at each sideband $n\omega_o + \omega$, with n varying between $-N$ and N, considering, for each n value, all the possible outputs $m\omega_o + \omega$, with $-N < m < N$ [19]. These outputs are calculated, in each case, at the location of the original control variable v.

Note that the described analysis must be carried out with no alteration of the steady-state solution, which requires linear behavior with respect to the input stimulus V_{ext}. Thus, the open-loop transfer functions are determined with the conversion matrix approach. As n varies, the loop is open only to calculate the transfer functions from the stimulus at $n\omega_o + \omega$ to the various outputs at $m\omega_o + \omega$, with $-N < m < N$. It must remain closed at all the frequencies $k\omega_o$ of the steady-state regime that must not be affected by this loop opening. Furthermore, the input stimulus must be zero at all sidebands different from $n\omega_o + \omega$. It is necessary to define an ideal filter that ensures two conditions: the control voltage is unaltered at $k\omega_o$ and the input voltage signal V_{ext} of the conversion matrix has only one nonnull component at $n\omega_o + \omega$ [19]. All the different transfer functions obtained per nonlinear current existing in the circuit are placed in a submatrix of dimension $(2N + 1) \times (2N + 1)$. This constitutes the return ratio matrix associated with a particular fundamental current. Each element $R_{m,n}$ in the submatrix is obtained by setting the input stimulus at the sideband $n\omega_o + \omega$ and obtaining the output $m\omega_o + \omega$ at the location of the control variable. The position of the stimulus V_{ext} varies between $N\omega_o + \omega$ and $N\omega_o + \omega$ in the following manner:

$$\begin{bmatrix} \Delta V_{-N} \\ \vdots \\ \Delta V_0 \\ \vdots \\ \Delta V_N \end{bmatrix} = - \begin{bmatrix} RR_{-N,-N} & \cdots & RR_{-N,0} & \cdots & RR_{-N,N} \\ \vdots & \vdots & \vdots & \vdots & \vdots \\ RR_{0,-N} & \cdots & RR_{0,0} & \cdots & RR_{0,N} \\ \vdots & \vdots & \vdots & \vdots & \vdots \\ RR_{N,-N} & \cdots & RR_{N,0} & \cdots & RR_{N,N} \end{bmatrix} \begin{bmatrix} 0 \\ \vdots \\ V_{ext} \\ \vdots \\ 0 \end{bmatrix} \tag{6.16}$$

To obtain the different matrix elements, the stimulus V_{ext} is introduced successively at the sidebands $n\omega_o + \omega$, with $n = -N$ to N. The terms ΔV_m refer to the output (at the location of the control voltage) at the different sidebands $m\omega_o + \omega$, with $-N < m < N$. The nth position of V_{ext} in the right-hand-side vector makes possible determination of the nth column of the return-ratio submatrix. The submatrix in (6.16) is the equivalent, for periodic solutions, of the single-element return-ratio function associated with each active device, calculated for dc regimes. Note that the matrix (6.16) must be totally filled by changing from $n = -N$ to $n = N$ the sideband $n\omega_o + \omega$ at which the stimulus V_{ext} is introduced.

For the stability analysis of the periodic solution, the perturbation frequency ω at which each submatrix $[RR_i]$ is calculated must be swept between $\omega = 0$ and $\omega = \omega_o/2$. This is done by sweeping the excitation frequency ω in this same interval at each sideband $n\omega_o + \omega$. To obtain the normalized determinant function NDF(ω) from the M different nonlinear dependences existing in the circuit, the procedure is analogous to the one used for dc solutions. The return-ratio matrix $[RR_1]$ is calculated by opening the loop corresponding to the first nonlinear element, with the remaining elements 2 to M at their nominal operation. The matrix $[RR_2]$ is calculated by opening the loop corresponding to the second nonlinear element, with the first element turned off and the remaining elements, 3 to M, in nominal operation. Finally, the matrix $[RR_M]$ is calculated with the last element M in open-loop conditions and all the previous elements, 1 to $M - 1$, turned off.

Once the different return-ratio matrixes have been obtained, the normalized determinant function associated with the periodic regime analyzed is given by the following product [19]:

$$\mathrm{NDF}(\omega) = \det\left(\prod_{i=1}^{M}[\mathrm{RR}_i] + [U]\right) \tag{6.17}$$

where M is the fundamental currents and $[U]$ is the identity matrix.

The stability analysis may be performed by obtaining the Nyquist plot associated with the normalized determinant function (6.17). Because the determinant has been normalized, the complexity of the resulting Nyquist plot will be reduced significantly compared to the one resulting from direct evaluation of the original characteristic determinant obtained from (6.15). Alternatively, the eigenvalues of the matrixes $[\mathrm{RR}_i]$ can be calculated using algorithms such as the ones described in [20]. The Nyquist criterion can be applied individually to these eigenvalues. The matrixes $[\mathrm{RR}_i]$ have a relatively low order, given by the number of harmonic terms considered $(2N+1)\times(2N+1)$, so the calculation is not computationally very expensive. On the other hand, this individual evaluation of the various return-ratio matrixes allows us to identify the circuit elements at which the unstable behavior originates.

Use of the foregoing technique requires some form of automatic implementation of the input stimulus and transfer function calculation, to avoid user error. The nonlinearities contained in the devices used must also be fully accessible, as they have to be manipulated to break the various closed loops.

6.2.2.4 Pole–Zero Identification The pole–zero identification can also be generalized to the stability analysis of a periodic steady-state solution $\overline{X}_s$, at a frequency ω_o, obtained using harmonic balance [13,15,21]. In this case the closed-loop transfer function is determined by linearizing the harmonic balance system about the periodic solution $\overline{X}_s$, with respect to a small-signal input at the incommensurable frequency ω. As in the case of dc regimes, the small-signal input will consist of a small-signal current source I_n, at the frequency ω. This source will be connected in parallel at a sensitive circuit node n, like those corresponding to the transistor terminals. This part of the procedure is identical to the one used for the determination of admittance diagrams, described earlier in this section. The circuit operates in a linear regime with respect to $I_n(\omega)$, so at each frequency ω, its solution may be determined using the conversion matrix approach. The following linearized system is obtained:

$$\left\{[A_x(k\omega_o+\omega)] + [A_y(k\omega_o+\omega)]\left.\frac{\partial\overline{Y}}{\partial\overline{X}}\right|_s\right\}\Delta\overline{X} = [A_{gI}(k\omega_o+\omega)]I_n(\omega) \tag{6.18}$$

where the vector $\Delta\overline{X}$ contains sidebands at $k\omega_o+\omega$, with $-N\le k\le N$, of the circuit state variables and A_{gI} is the matrix relating $\Delta\overline{X}$ to the small-signal current source. Clearly, any circuit output W at any sideband frequency $n\omega_o + \omega$, with n a particular integer $-N \leq n \leq N$, will be a linear combination of $\Delta\overline{X}$, so it can be written $W(n\omega_o + \omega) = [C]\Delta\overline{X}$. Thus, all possible closed-loop transfer functions of the linearized system will share the same denominator, as resulting from the inversion of the matrix on the left-hand side of (6.18), with associated determinant:

$$\det\left\{[A_x(k\omega_o + \omega)] + [A_y(k\omega_o + \omega)]\left.\frac{\partial\overline{Y}}{\partial\overline{X}}\right|_s\right\} \tag{6.19}$$

The poles associated with the periodic solution $\overline{X}_s$ are the roots of the above determinant, evaluated at the complex perturbation frequency $\omega - j\sigma$ instead of the real frequency ω. Thus, any possible closed-loop transfer function will have implicit information on the solution stability. A convenient choice for the output of the transfer function is the lower-frequency sideband $V_n(\omega)$ of the voltage at node n at which the current generator $I_n(\omega)$ is connected [13,14] due to the proximity between input and output. Thus, the single input–single output transfer function considered is $Z_{in}(\omega) = V_n(\omega)/I_n(\omega)$. Note that it is possible to choose other sideband frequencies for the input current and/or output voltage, although the accuracy will generally degrade. As already stated, all possible closed–closed loop transfer function share the same denominator and thus will have the same poles. In contrast, the zeros of the closed-loop transfer function will depend on the particular definition of this function. Because of this, pole–zero cancellations may be obtained at some locations of the current source. If the canceled poles are unstable, an incorrect conclusion will be reached about the solution stability.

To extract the stability information from the transfer function $Z_{in}(\omega)$, the perturbation frequency ω is swept, applying pole–zero identification to $Z_{in}(\omega)$. This complex function is modeled with a quotient of polynomials, in a manner similar to (6.13). Note that the zeros and poles of $Z_{in}(\omega - j\sigma)$ depend on the constants z_1 to z_m and p_1 to p_n. Thus, it is possible using pole–zero identification to obtain explicit poles and zeros of the function $Z_{in}(\omega - j\sigma)$. These poles may be real or complex conjugate. As already known, the poles of a periodic solution at ω_o are also periodic, structured in sets of the form $\gamma_i + jk\omega_o$ or $\sigma_m \pm j\omega_m + jk\omega_o$, where k is an integer number. Because of the periodicity of the poles (see Chapter 1), the sweep interval in the frequency ω may be limited to $(0, \omega_o/2 + \varepsilon)$ or $(\omega_o/2 - \varepsilon, \omega_o + \varepsilon)$. Both subintervals must actually be considered, due to the possible lack of sensitivity. The extra frequency margin ε enables the inclusion of the edges of each subinterval in the polynomial fitting for the pole–zero identification. Practical circuits contain many poles which would require a high order n in the denominator of $Z_{in}(\omega)$ to match the complex function $Z_{in}(\omega)$ with the quotient of polynomials in (6.13). Better accuracy is obtained by dividing the frequency interval $(0, \omega_o)$ in subintervals of smaller length using a lower order n in the identification procedure. Note that this division in subintervals is valid because we are dealing with a linearized system and superposition applies. For stability, all the circuit poles must be located on the left-hand side of the complex plane, which implies an exponential extinction in time of any small perturbation. Note that the conversion

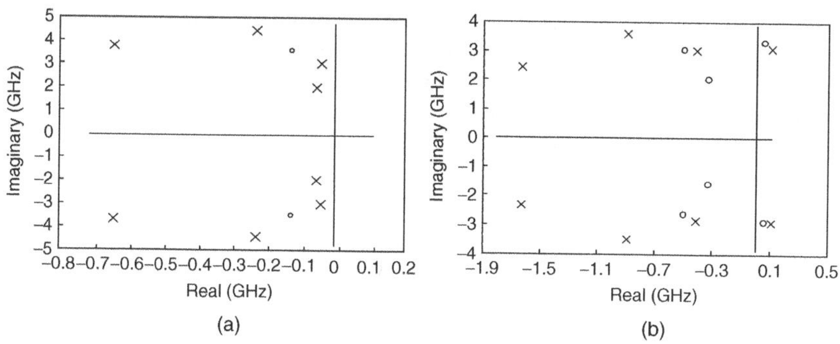

FIGURE 6.12 Stability analysis with pole–zero identification of the nondivided solution of Fig. 6.9 for two values of the input frequency of a steady-state periodic solution at f_{in}, obtained with harmonic balance: (a) input frequency $f_{in} = 5$ GHz; (b) input frequency $f_{in} = 6$ GHz.

matrix approach is ill-conditioned in the case of frequency division $\omega_o/2$, as there is an overlapping of the lowest analysis frequency ω and the lower carrier sideband $\omega_{in} - \omega$ for $\omega = \omega_{in}/2$, which leads to system degeneracy. Despite this, the pole–zero identification usually provides the existing pairs of poles at the divided frequency with enough accuracy, as it is based on the fitting of an impedance function in a frequency band instead of a single frequency point.

The pole–zero identification technique has been applied to the nondivided solution of the circuit of Fig. 6.2, obtained for the input power $P_{in} = 12$ dBm. The same two input frequency values taken for the analyses through admittance diagrams (Fig. 6.8) and the Nyquist plot (Fig. 6.10) have been considered: $f_{in} = 5$ GHz and $f_{in} = 6$ GHz (Fig. 6.12). For $f_{in} = 5$ GHz (Fig. 6.9a), all the solution poles are located on the left-hand side of the complex plane, so this solution is stable. However, there are two pairs of complex-conjugate poles relatively close to the imaginary axis, one of them at about $f_{in}/2$. For $f_{in} = 6$ GHz, there is a pair of complex-conjugate poles located on the right-hand side of the complex plane, so this solution is unstable. The frequency of this pair of poles, $f = 3$ GHz, agrees with the input frequency divided by 2. This is in correspondence with the existence for the considered circuit element values of a frequency-divided solution at $f_{in}/2$.

6.3 STABILITY ANALYSIS OF FREE-RUNNING OSCILLATORS

The characteristic system obtained from the small perturbation of a free-running oscillator about its periodic steady-state solution $\overline{X}_s$ at the frequency ω_o contains a pair of complex-conjugate poles on the imaginary axis at this same frequency $\pm j\omega_o$. This is in close relationship with the invariance of the oscillatory solution $\overline{X}_s$ versus any shift in the phase origin. As shown in Chapters 1 and 3, this gives rise to the singularity of the Jacobian matrix of the harmonic balance equation, evaluated at the steady-state free-running oscillation $\det[JH]_s = 0$. The singularity implies a zero pole $\gamma = 0$ of the associated characteristic determinant. Due to the multiplicity of the poles, there will also be an infinite set of poles $\pm jn\omega_o$, with n a positive integer located on the imaginary axis.

For the stability analysis, the harmonic balance system must be linearized about the free-running steady-state oscillation. It the oscillation was obtained with the aid of an auxiliary generator, this generator may be kept connected to the circuit for subsequent stability analysis. Actually, once convergence has been achieved, the auxiliary generator fulfills $Y_{AG} = 0$ and does not affect the oscillatory solution. This generator sustains the large-signal periodic regime about which the circuit will be linearized. The fact that the steady-state solution contains intrinsically the set of poles $\pm jn\omega_o$, with n a positive integer, gives rise to some peculiarities in the application of the different stability analysis methods.

When using the Nyquist plot for the stability analysis of a free-running regime it must be taken into account that the Jacobian matrix $[JH(k\omega_o + \omega)]$ is singular for $\omega = n\omega_o$, so the Nyquist plot must necessarily start from the origin at $\omega = 0$. Due to numerical inaccuracies, this will not be the case in a practical application. To cope with this problem, a normalization technique has been proposed [22]. The Nyquist criterion is applied to $\det[JH(k\omega_o + \omega)] - \det[JH(k\omega_o)]$ instead of the original function $\det[JH(k\omega_o + \omega)]$. The complex value $\det[JH(k\omega_o)] \neq 0$, due to numerical inaccuracy, is subtracted from the original determinant function, so the Nyquist plot, obtained by sweeping ω, necessarily starts from the origin at $\omega = 0$.

The pole–zero identification for the stability analysis of the free-running oscillation is performed in the frequency interval $[(0, \omega_o)]$, which for improved accuracy should be divided in subintervals. Due to the mentioned autonomy of the steady-state solution, the lower-frequency interval will contain the real pole $\gamma = 0$. The higher-frequency interval must contain the complex-conjugate poles $\pm j\omega_o$. For stability, all the rest of the poles must be located on the left-hand side of the complex plane. The poles $\gamma = 0$ and $\pm j\omega_o$ will suffer from some accuracy limitations. To understand this, remember that the closed-loop transfer function $Z_n(\omega) = V_n(\omega)/I_n(\omega)$ is calculated from the inversion of the Jacobian matrix $[JH(k\omega_o + \omega)]$ [see (6.18)]. In the case of a free-running oscillator, this Jacobian matrix approaches singularity for ω tending to zero. This will degrade the accuracy in the pole–zero identification about $jn\omega_o$. Anyway, knowing that these poles are intrinsic to the free-running oscillation, the user will generally recognize them without difficulty. Due to accuracy limitations, the poles associated with the oscillation autonomy may be located slightly to the left or to the right of the complex plane. The stability or instability of the obtained oscillator solution will depend on the location of the rest of the poles.

As an example, pole–zero identification has been applied to the free-running solution of the circuit in Fig. 6.2 (in the absence of input generator power) at the gate bias voltage $V_{GS} = -0.8$ V [16]. At this particular voltage, two periodic solutions coexist, one stable and the other unstable. Both of them have a pair of complex-conjugate poles at $\pm j\omega_o$ with different ω_o values, as they have different steady-state oscillation frequency. For the stable solution, all the rest of the poles are located on the left-hand side of the complex plane. In the case of the

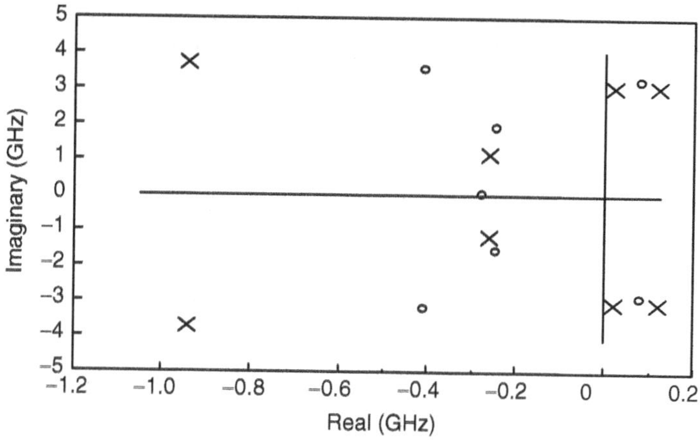

FIGURE 6.13 Pole locus of one of the two coexisting free-running solutions in the circuit of Fig. 6.2 for $V_{GS} = -0.8$ V. Besides the two poles on the imaginary axis at $\pm j\omega_o$ due to the solution autonomy, there is a pair of complex-conjugate poles at the same frequency, located on the right-hand side of the complex plane, which is equivalent to an unstable real pole.

unstable solution (Fig. 6.13), a pair of complex-conjugate poles $\sigma \pm j\omega_o$, at the same frequency ω_o of the poles on the imaginary axis, is located on the right-hand side of the complex plane. Because the frequency of the unstable poles is equal to the oscillation frequency, the unstable pair of poles is equivalent to a real pole $\gamma' > 0$. This suggests the existence of a turning point in the solution curve versus V_{GS}. Remember that at turning points a real pole passes through zero. This will be clarified in the next section.

6.4 SOLUTION CURVES VERSUS A CIRCUIT PARAMETER

Usually, the interest of the designer is not limited to obtaining a circuit solution just for a particular set of values of the input generator or generators, or the circuit elements. Instead, the designer generally wishes to know the circuit response versus one or more parameters (e.g., the input generator power or frequency, a bias voltage, or the value of a linear element). However, when varying a parameter, a stable regime may become unstable because the steady-state solution changes with this parameter and so do its associated poles. Many examples were shown in Chapter 4, devoted to injection-locked oscillators and frequency dividers. Thus, the capability to analyze the variation of the steady-state solution and its stability properties versus any circuit parameter will be essential.

Let a periodic solution $\overline{X}_s$ of a given circuit, obtained with harmonic balance, be considered. The straightforward way to determine the evolution of this solution versus a parameter η is to carry out a sweep in this parameter, performing a sequence of harmonic balance simulations and using the final solution obtained at a given step η_n as the initial guess for the next step, η_{n+1}. However, this direct sweep will lead to unsolvable convergence problems if the curve contains turning points. At the turning points (or infinite-slope points) the curve folds over itself, so before the turning point, there is a multivalued solution section versus the parameter (see, for instance, Fig. 4.4 in Chapter 4). As shown in Chapter 4, the turning points are intrinsic to the performance injection-locked oscillators and frequency dividers. They are also commonly obtained in voltage-controlled oscillators versus the bias voltage and in power amplifiers and frequency multipliers versus the input power.

The turning points of the solution curve can be circumvented using continuation methods. However, the continuation methods required are more involved than the simple parameter sweep (described so far) which uses the final harmonic balance solution at a given sweep step to initialize the circuit variables at the next step. Two of the most common continuation methods used to obtain multivalued curves are the prediction–correction technique with parameter switching, and arc-length continuation.

6.4.1 Parameter Switching Applied to Harmonic Balance Equations

The parameter switching technique, with prediction and correction stages, is summarized in the following [23–25]. The prediction–correction technique is described first. Let variations in a parameter η be considered. The technique is applied from an initial solution $(\eta_0, \overline{X}_0)$, which must be easy to determine with the harmonic balance method. Once a solution point $(\eta_n, \overline{X}_n)$ has been calculated, the next one $(\eta_{n+1}, \overline{X}_{n+1})$ is estimated through linearization of the harmonic balance system about the preceding point, $(\eta_n, \overline{X}_n)$. This constitutes the prediction stage of the technique. The estimated point will be called here $(\eta_{n+1,p}, \overline{X}_{n+1,p})$. It is calculated by differentiating the error function $\overline{H}$ of the harmonic balance system:

$$\begin{aligned} \Delta\overline{H} &= [JH]_n(\overline{X}_{n+1,p} - \overline{X}_n) + \left.\frac{\partial\overline{H}}{\partial\eta}\right|_n (\eta_{n+1} - \eta_n) = 0 \rightarrow \overline{X}_{n+1,p} \\ &= \overline{X}_n - [JH]_n^{-1}\left.\frac{\partial\overline{H}}{\partial\eta}\right|_n (\eta_{n+1} - \eta_n) \end{aligned} \tag{6.20}$$

where the Jacobian matrix $[JH]_n$ and the derivative of $\overline{H}$ with respect to the parameter are evaluated at the preceding solution point n. The predicted point $(\eta_{n+1,p}, \overline{X}_{n+1,p})$ is not the actual solution point since the actual system $\overline{H}(\overline{X}, \eta) = 0$, unlike equation (6.20), is nonlinear. The point obtained through (6.20) is a linear estimation (prediction) of the solution for η_{n+1}.

At the *correction stage* of the continuation technique, the Newton–Raphson algorithm is used for correction of the estimated point n, which is done for constant parameter value $\eta = \eta_{n+1,p}$. The input of this algorithm is the estimated solution $\overline{X}_{n+1,p}$, which is used as the starting point: $\overline{X}^{o}_{n+1} \equiv \overline{X}_{n+1,p}$. Unlike $\overline{X}_{n+1,p}$, the point resulting from the Newton–Raphson convergence $\overline{X}_{n+1} = \overline{X}^{c}_{n+1}$ is an actual solution point.

It is easily seen from (6.20) that the Jacobian matrix $[JH]$ becomes singular at any point of the solution curve with infinite slope or turning point. Because of this, the Newton–Raphson algorithm fails to converge in the neighborhood of the turning point. To cope with this problem, a parameter-switching algorithm [8,23] can be applied in combination with the prediction–correction technique. In this algorithm the continuation parameter is at some sections of the curve different from the actual circuit parameter η. In fact, at each prediction stage $(\eta_{n+1,p}, \overline{X}_{n+1,p})$, the increments predicted for the different variables (including the parameter η), $|\mathrm{Re}[\Delta X^1_{0n+1,p}]|$, $|\mathrm{Im}[\Delta X^1_{0n+1,p}]|, \ldots, |\mathrm{Re}[\Delta X^Q_{NHn+1,p}]|$, $|\mathrm{Im}[\Delta X^Q_{NHn+1,p}]|$, $|\Delta\eta_{n+1,p}|$ are compared, taking the variable with the biggest increment as the new analysis parameter.

If the new parameter is, for instance, $\mathrm{Re}[X^j_k]$, the increment h is assigned to this variable. Therefore, this variable is not considered as an unknown. Its value is incremented as $\mathrm{Re}[X^j_{kn+1}] = \mathrm{Re}[X^j_{kn}] + \mathrm{sign}(\mathrm{Re}[X^j_{kn+1,p})]h$. This assignment preserves the natural sense of variation of the variable $\mathrm{Re}[X^j_k]$ by affecting step h with the sign of the predicted increment: $\mathrm{sign}(\mathrm{Re}[X^j_{kn+1,p}])$. Thus, the parameter increment $\Delta\eta_{n+1}$, as well as the increments of all the circuit variables except $\mathrm{Re}[X^j_k]$, are the unknowns to be determined at the correction stage of the step $n+1$. Note that taking the variable with the biggest increment as a parameter prevents any unbounded growth of the circuit variables. Through this technique, points of infinite slope (turning points) become points of zero slope.

The described above technique has been used to obtain the multivalued curve, corresponding to the frequency-divided regime, represented in Fig. 6.10. The technique allows us to circumvent the turning point (TP) efficiently without convergence problems.

6.4.2 Parameter Switching Applied to an Auxiliary Generator Equation

The nonperturbation equation associated with the auxiliary generator $Y_{AG} = 0$ enables a simple derivation of a continuation technique to obtain multivalued solution curves versus a circuit parameter η. The major interest of the continuation techniques applied to $Y_{AG} = 0$ is the possibility of user implementation on commercial harmonic balance software using standard library elements. Unlike the case of continuation techniques applied to the harmonic balance equation, there is no need for user access to the harmonic balance equations or its Jacobian matrix.

As has already been shown, when using an auxiliary generator, the harmonic balance system can be solved in a two-tier procedure. The auxiliary generator will have amplitude A_{AG}, frequency ω_{AG}, and phase ϕ_{AG}. The nonperturbation equation $Y_{AG} = 0$ is the outer tier, and the pure harmonic balance system $\overline{H}(\overline{X}) = 0$ is the inner tier. The inner tier is solved for fixed values of the auxiliary generator: $\overline{X}(A_{AG}, \omega_{AG})$ with $\phi_{AG} = 0$ for an autonomous oscillation, or $\overline{X}(A_{AG}, \phi_{AG})$ in the case of a forced regime. An autonomous oscillation will be considered initially. For mathematical convenience, a vector consisting of the real and imaginary parts of Y_{AG} is defined: $\overline{Y}_{AG} \equiv [\mathrm{Re}(Y_{AG}), \mathrm{Im}(Y_{AG})]^{\mathrm{T}}$. At each steady-state solution versus a parameter η, the following equation is fulfilled:

$$\overline{Y}_{AG}(A_{AG}, \omega_{AG}, \eta) = 0 \tag{6.21}$$

with the pure harmonic balance equations as inner tier. The solution curve versus the parameter η is obtained with the following continuation technique. Assume that the auxiliary generator values that fulfill $\overline{Y}_{AG} = 0$ are known at an initial parameter value $\eta_0 : A^0_{AG}, \omega^0_{AG}$. Starting from this point, it is possible to apply an iterative procedure to predict the auxiliary generator values at the next point, $n + 1$, through linearization of the outer-tier equation $Y_{AG}(V_{AG}, \omega_{AG}) = 0$ about the preceding point, n:

$$\begin{bmatrix} A^{n+1}_{AG} \\ \omega^{n+1}_{AG} \end{bmatrix}_p = \begin{bmatrix} A^{n}_{AG} \\ \omega^{n}_{AG} \end{bmatrix} - [JY_{AG}]_n^{-1} \left.\frac{\partial \overline{Y}_{AG}}{\partial \eta}\right|_n (\eta_{n+1} - \eta_n) \tag{6.22}$$

where the subscript p indicates "predicted" and $[JY_{AG}]$ is the Jacobian matrix of the function $\overline{Y}_{AG}$ with respect to the two variables A_{AG} and ω_{AG}:

$$[JY_{AG}] = \begin{bmatrix} \dfrac{\partial \mathrm{Re}(Y_{AG})}{\partial A_{AG}} & \dfrac{\partial \mathrm{Re}(Y_{AG})}{\partial \omega_{AG}} \\ \dfrac{\partial \mathrm{Im}(Y_{AG})}{\partial A_{AG}} & \dfrac{\partial \mathrm{Im}(Y_{AG})}{\partial \omega_{AG}} \end{bmatrix} \tag{6.23}$$

Next, the absolute values of the increments $\Delta A_{AG} = A^{n+1}_{AG} - A^{n}_{AG}$ and $\Delta\omega_{AG} = \omega^{n+1}_{AG} - \omega^{n}_{AG}$ are compared with the increment applied to the parameter $|\Delta\eta| = |\eta_{n+1} - \eta_n| = h$. In contrast to the continuation technique applied to the harmonic balance equations, only increments of three variables are compared here: $\{|\Delta\eta| = h, |\Delta A_{AG}|, |\Delta\omega_{AG}|\}$. Note that ω_{AG} should be normalized for a proper comparison. If the variable with the largest increment agrees with the actual parameter η, the process continues to the correction stage. The Newton–Raphson algorithm is applied to solve $Y^{n+1}_{AG}(A^{n+1}_{AG}, \omega^{n+1}_{AG}) = 0$ at the constant parameter value $\eta_{n+1} = \eta_n + \Delta\eta$, with $|\Delta\eta| = h$. If the element with the largest increment happens, for instance, to be the amplitude A_{AG}, a new variable prediction is carried out, in terms of ω_{AG}, η, using A_{AG} as the parameter instead of η:

$$\begin{bmatrix} \omega^{n+1}_{AG} \\ \eta^{n+1} \end{bmatrix}_p = \begin{bmatrix} \omega^{n}_{AG} \\ \eta^{n} \end{bmatrix} - \begin{bmatrix} \dfrac{\partial \mathrm{Re}(Y_{AG})}{\partial \omega_{AG}} & \dfrac{\partial \mathrm{Re}(Y_{AG})}{\partial \eta} \\ \dfrac{\partial \mathrm{Im}(Y_{AG})}{\partial \omega_{AG}} & \dfrac{\partial \mathrm{Im}(Y_{AG})}{\partial \eta} \end{bmatrix}_{\mathrm{n}}^{-1} \left.\frac{\partial Y_{AG}}{\partial A_{AG}}\right|_n (A^{n+1}_{AG} - A^{n}_{AG}) \tag{6.24}$$

with $A^{n+1}_{AG} = A^{n}_{AG} + \mathrm{sign}(\Delta A_{AG})h$. After the prediction (6.24), the values estimated are corrected through a Newton–Raphson algorithm used to solve $Y^{n+1}_{AG}(\omega^{n+1}_{AG}, \eta^{n+1}) = 0$ for the constant amplitude value $A^{n+1}_{AG} = A^{n}_{AG} + \Delta A_{AG}$, with $|\Delta A_{AG}| = h$. A similar procedure would be followed if the element with the largest increment were the frequency ω_{AG}, exchanging the roles of ω_{AG} and A_{AG} in (6.24).

When using a commercial harmonic balance, the nonperturbation $Y_{AG} = 0$ equation is solved through optimization. The optimization goals are $\text{Re}[Y_{AG}] = 0$ and $\text{Im}[Y_{AG}] = 0$. In commercial harmonic balance, it is not generally possible to implement an automatic algorithm for parameter switching. This switching has to be performed manually, so it is better to minimize the number of parameter changes. One significant advantage of the auxiliary generator technique is that only three magnitudes must be considered—A_{AG}, ω_{AG}, and η—so it is possible to follow a very simple criterion for the choice of the analysis parameter. This criterion is illustrated in the following. It has been used to obtain the variation of the free-running oscillation of the circuit in Fig. 6.2 versus the gate bias voltage V_{GS}. The results are shown in Fig. 6.14. The circuit parameter is $\eta = V_{GS}$. Two different situations may be distinguished in the solution curve: operation far from any turning point, with a relatively small slope of the solution curve versus the parameter $\eta = V_{GS}$ (the nonproblematic region), and operation close to a turning point, with high slope versus η. In the nonproblematic region, the parameter η will be swept, optimizing A_{AG} and ω_{AG} to fulfill the optimization goals $\text{Re}[Y_{AG}] = 0$ and $\text{Im}[Y_{AG}] = 0$. This is the case for the flat section of the curve in Fig. 6.14, where the swept parameter is V_{GS}. In the regions of high slope versus η, the amplitude A_{AG} (or exceptionally, the frequency ω_{AG}) should be increased or decreased, optimizing η and ω_{AG} (or η and A_{AG} when ω_{AG} is swept). With this technique, the section of the oscillation curve in Fig. 6.14 containing the turning point is obtained in a very simple manner. We start from a solution in the nonproblematic upper section. Then we sweep down to zero the auxiliary generator amplitude A_{AG}. Due to the turning point T, the solution is multivalued in the V_{GS} interval 0.88 V, 0.72 V, which is in agreement with the two different steady-state solutions obtained for $V_{GS} = -0.8$ V in the analyses of Section 6.2.1. The unstable solution analyzed with pole–zero identification in Fig. 6.13 corresponds to the lower section. As already known, the turning point T_P in the solution curve will be responsible for a hysteresis phenomenon in the transformation from dc to oscillatory regime, and vice versa.

The continuation technique described, based on the auxiliary generator, is equally valid for forced periodic regimes, with an input generator at the frequency ω_{in}. For this use the auxiliary generator frequency ω_{AG} must be replaced, in all the expressions (6.21)–(6.24), with the auxiliary generator phase ϕ_{AG}. Note that turning points are not found only in the periodic solution curves of injection-locked oscillators or frequency dividers but can also be found in the solution curves of nonlinear circuits without oscillations, such as power amplifiers or frequency multipliers [26]. These turning points give rise to anomalous discontinuous jumps when tracing the power transfer curve (P_{out} versus P_{in}), for example. The jumps are often erroneously attributed to discontinuities in the device models.

To circumvent the turning points in a nonoscillatory curve, an auxiliary generator is connected to a sensitive circuit node n (e.g., active device terminals). Its frequency is made equal to the input frequency $\omega_{AG} = \omega_{\text{in}}$. Simulation with an auxiliary generator should start from a solution point in the neighborhood of the jump,

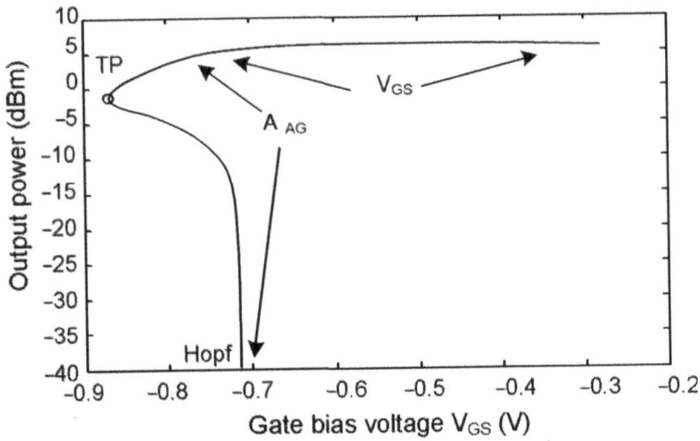

FIGURE 6.14 Variation in the output power of the circuit in Fig. 6.2 versus the gate bias voltage when analyzed as a free-running oscillator. The curve has been traced with the parameter-switching technique based on the use of an auxiliary generator. The parameter swept in each section of the curve is indicated in the figure.

obtained with default harmonic balance. Of course, this default harmonic balance simulation must have good convergence properties at the chosen initial point. After this initial analysis, the amplitude and phase of the auxiliary generator are made equal to the amplitude and phase of the node voltage V_n selected, at ω_{in}, resulting from the default harmonic balance convergence; that is, $A_{AG}e^{j\phi_{AG}} = A_n e^{j\phi_n} = V_n$. Then the amplitude of the auxiliary generator is increased or decreased to circumvent the turning point, solving the nonperturbation equation $Y_{AG} = 0$ in terms of ϕ_{AG} and the analysis parameter η, which may correspond to the input power, for instance. This allows the implementation of a switching parameter algorithm used in a commercial harmonic balance. Several examples of the application of this technique to power amplifiers will be presented in Chapter 12.

6.4.3 Arc-Length Continuation

Arc-length continuation [27] is based on the artificial introduction in the harmonic balance system of a second parameter α (in addition to the actual parameter η) to obtain the solution curve $\overline{X}(\eta)$. Both the circuit variables and the parameter itself are made to depend on the auxiliary parameter α: $\overline{X}(\alpha)$ and $\eta(\alpha)$. To obtain the solution curve $\overline{X}(\eta)$, instead of the actual parameter η, the auxiliary parameter α will be swept in steps $\Delta\alpha$, calculating $\overline{X}(\alpha)$ and $\eta(\alpha)$. This change of parameter will be advantageous if neither $\overline{X}(\alpha)$ nor $\eta(\alpha)$ exhibit turning points versus α. The turning points will only be obtained when composing $\overline{X}(\alpha)$ versus $\eta(\alpha)$. The dependence on α introduces one more variable in a harmonic balance system, so one more equation is required. The additional equation will be a normalization condition. The norm of the vector containing the derivatives with respect to α, at each point n of the solution curve, is made equal to unity $\left|d\overline{X}/d\alpha, d\eta/d\alpha\right| = 1$.

When this condition is imposed, the norm of the increment vector $[\Delta\overline{X} \quad \Delta\eta]^T$ agrees with the increment $\Delta\alpha$ or distance between two consecutive points, thus the name *arc-length continuation*. This is easily seen by writing $[\Delta\overline{X} \quad \Delta\eta]^T = \left[\partial\overline{X}/\partial\alpha \quad \partial\eta/\partial\alpha\right]^T \Delta\alpha$ and taking into account that the norm of the derivative vector is equal to 1.

6.5 GLOBAL STABILITY ANALYSIS

A solution may be stable for given values of circuit parameters. However, under the continuous variation of one of these parameters, η, the circuit steady-state solution changes and a real pole or a pair of complex-conjugate poles associated with the circuit linearization about this solution may cross the imaginary axis to the right-hand side of the complex plane. A local bifurcation occurs in the system at the particular parameter value η_b at which the crossing takes place. From this parameter value, the solution becomes unstable and thus unobservable. The characteristics of the new observed solution generally depend on the type (dc, periodic, quasiperiodic) of the original steady state and the nature (real or complex) and frequency value of the crossing pole(s). Thus, different types of transformations in the solution observed may be obtained, corresponding to the different types of local bifurcations. This was studied in Chapter 3.

As an example, Fig. 6.15 shows the variation in the stability properties of the free-running solution of the circuit in Fig. 6.2 when the gate bias voltage V_{GS} is modified. The pole evolution versus the gate bias voltage V_{GS} is obtained using a two-stage procedure at each V_{GS} value [14]:

1. *Determination of the steady-state solution with an auxiliary generator.* The auxiliary generator operates at the oscillation frequency $\omega_{AG} = \omega_o$. Either

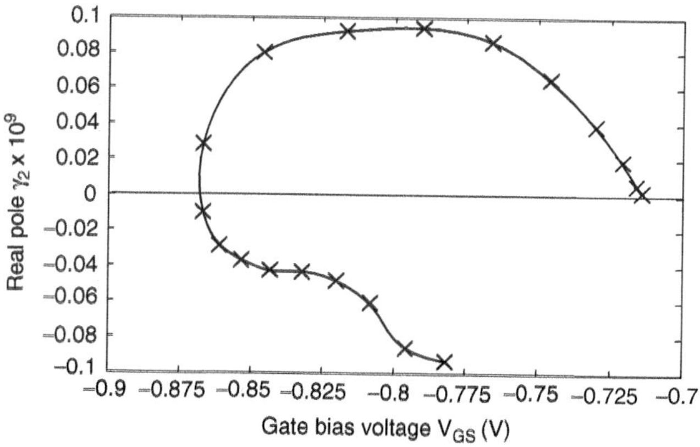

FIGURE 6.15 Variation in the dominant real pole γ_2 along the free-running oscillation curve of Fig. 6.14.

$\overline{Y}_{AG}(A_{AG}, \omega_{AG}) = 0$ for constant V_{GS}, or $\overline{Y}_{AG}(V_{GS}, \omega_{AG}) = 0$ for constant A_{AG}, is solved, depending on the curve section (see Fig. 6.14). The resolution for constant A_{AG} is carried out in the problematic section of the solution curve.

2. *Calculation of the closed-loop transfer function $Z_n(\omega)$, to be analyzed with pole–zero identification.* A small-signal current generator $I_n(\omega)$, at the frequency ω nonrationally related to ω_o, is introduced into the circuit. The auxiliary generator with the nonperturbing values A_{AGo} and ω_{AGo}, obtained at the preceding stage, is maintained connected to the circuit. This avoids losing the oscillatory steady-state solution when the current generator is introduced. The frequency ω is swept to calculate the closed-loop transfer function $Z_n(\omega)$. The results corresponding to this V_{GS} value are stored as $Z_n(V_{GS}, \omega)$. The next V_{GS} value is considered, returning to stage 1.

The application of pole–zero identification to the resulting function $Z_n(V_{GS}, \omega)$ shows that each point of the curve has two dominant real poles. One of them is $\gamma_1 = 0$. As already known, this real pole is due to the solution autonomy. Because of the periodicity of the poles of periodic solutions associated with this pole, there is also the set $\pm jk\omega_o$, with k a positive integer, and ω_o the oscillation frequency [21]. Note that the oscillation frequency varies with V_{GS}, so the poles $\pm jk\omega_o$ shift along the imaginary axis, whereas the real pole γ_1, belonging to the same set, remains at zero. The second dominant real pole is γ_2, with its associated set of complex-conjugate poles $\gamma_2 \pm jk\omega_o$. When varying V_{GS}, the pole γ_2 shifts along the real axis. The variation γ_2 near the turning point of Fig. 6.14 is represented in Fig. 6.15. Starting from the upper section of the curve in Fig. 6.14, the pole is originally negative and decreases in magnitude when approaching the turning point. At the turning point it crosses the imaginary axis through zero, so the solution curve in Fig. 6.12 becomes unstable at this point. From the turning point, this positive real pole increases up to a maximum value, then decreases and tends to zero at the $V_{GS} = -0.71$ V, where the oscillation is generated through a subcritical Hopf bifurcation. As shown in Chapter 3, Section 3.3.1.1, the two eigenvalues of the dc solution that becomes unstable at the Hopf bifurcation transform into two real poles, γ_1, γ_2, of the generated periodic solution, with zero oscillation amplitude at the bifurcation point. As the oscillation amplitude grows continuously from zero, the pole γ_1 stays at zero, whereas the pole γ_2 shifts to the left-hand side of the complex plane in a supercritical bifurcation and to the right-hand side in a subcritical bifurcation. In agreement with this, the Hopf bifurcation in Fig. 6.15 is clearly subcritical.

Next, two methodologies for bifurcation detection with harmonic balance are presented. The first is based on an analysis of the characteristic determinant of the harmonic balance system and is intended for in-house simulators, with access to harmonic balance equations and their Jacobian matrix [7,28]. The second methodology is based on the use of auxiliary generators and can be applied to bifurcation analysis using a commercial harmonic balance, with standard library elements.

6.5.1 Bifurcation Detection from the Characteristic Determinant of a Harmonic Balance System

As has been shown, the poles of a dc solution $\overline{X}_{dc}$ or a periodic solution $\overline{X}_s$ are given by the roots of the characteristic determinant of the harmonic balance equation, linearized about this solution $\det[JH(\omega - j\sigma)] = 0$ [7]. Analysis of this determinant will make it possible to detect the possible bifurcations from dc and periodic regimes.

6.5.1.1 *Bifurcations from a dc Regime* Let a dc solution $\overline{X}_{dc}$ of a given circuit be considered. If a circuit parameter η is varied continuously, this solution will (generally) change as well as its associated poles. A bifurcation will be obtained if at a particular parameter value η_b, a real pole or a pair of complex-conjugate poles crosses the imaginary axis. This is expressed mathematically as

$$\begin{aligned} \overline{H}(\overline{X}^b_{dc}, \eta_b) &= 0 \\ \det[JH(\omega_b, \eta_b)] &= 0 \\ \left.\frac{d\sigma}{d\eta}\right|_{\eta=\eta_b} &\neq 0 \end{aligned} \tag{6.25}$$

The condition on the derivative is essential, as it guarantees that the critical pole or poles actually cross the imaginary axis. Otherwise, the real pole could just be tangent to the imaginary axis at η_b without giving rise to a qualitative stability change versus the parameter. The main types of bifurcation from a dc regime depend on the ω_b value.

$\omega_b = 0$, Turning Point or Pitchfork Bifurcation Let the dc solution $\overline{X}_{dc}$ be analyzed versus a parameter η. A turning point or pitchfork bifurcation is obtained when conditions (6.25) are fulfilled for $\omega_b = 0$ at the parameter value η_b. This implies that a real pole crosses the imaginary axis of the complex plane at η_b. For $\omega_b = 0$, the determinant $\det[JH]$ in (6.25) takes a real value and adds a real equation $\det[JH] = 0$ to the harmonic balance system at the bifurcation point. Thus, the bifurcation point $\overline{X}^b_{dc}, \eta_b$ is obtained by solving a system of $Q + 1$ real equations in $Q + 1$ unknowns, Q being the number of state variables in $\overline{X}_{dc}$. Due to the fact that a real pole at zero implies a singularity of the system $\overline{H}(\overline{X}_{dc}, \eta) = 0$, the solution curve $\overline{X}_{dc}(\eta)$ will exhibit an infinite slope at η_b, which can be associated to a turning point or to a pitchfork bifurcation. As shown in Chapter 3, in the case of the pitchfork bifurcation from the dc regime, three different dc paths merge at the bifurcation point: the original dc path, plus two new dc paths, generated with infinite slope at the bifurcation point (for example, see Fig. 3.4b). The original dc path continues to exist after the bifurcation. If this path was stable prior to the bifurcation at η_b, it will become unstable after this bifurcation, due to the crossing of the real pole to the right-hand side of the complex plane. The occurrence of a pitchfork bifurcation requires odd symmetries in the circuit equations, which is

relatively uncommon. In the case of a turning point, two dc paths merge at η_b at a solution point with infinite slope versus the parameter η. This will give rise to a discontinuous jump to a different solution or (rarely) to generation of an oscillation through a global bifurcation (see Section 3.3.3.2 in Chapter 3).

$\omega_b \neq 0$, Hopf Bifurcation For $\omega_b \neq 0$ in (6.25), a Hopf bifurcation takes place at the parameter value $\eta = \eta_b$, which will give rise to the onset of an oscillation at the frequency ω_b. The original dc path versus the parameter η continues to exist after the bifurcation. Note that for $\omega_b \neq 0$, the determinant in (6.25) is complex, as only the positive frequency ω is considered, as shown in (6.5). Thus, the system (6.25) consists of $Q + 2$ real equations in $Q + 2$ unknowns, given by $\overline{X}^b_{\text{dc}}$, η_b, and ω_b. For a supercritical (subcritical) Hopf bifurcation, a stable (unstable) oscillation is generated (from zero amplitude) from a stable dc regime which becomes unstable after the bifurcation. As shown in Fig. 6.14, in the case of a subcritical bifurcation, the oscillation amplitude grows continuously from zero when varying the parameter in opposite sense. Techniques to distinguish between supercritical and subcritical bifurcations will be provided in Section 6.5.2. When further varying the parameter η from η_b, in a continuous manner, the oscillation frequency will no longer agree exactly with the frequency of the unstable pair of complex-conjugate poles.

As an example, a bifurcation analysis has been applied to the dc solution of the circuit in Fig. 6.2 (without periodic source). The parameter η considered is the bias voltage V_{GS}. Solving the Hopf bifurcation condition (6.25), in terms of $\overline{X}^b_{\text{dc}}$, V_{GSb}, and ω_b, the results are $f_b = 2.964$ GHz and $V_{GSb} = -0.782$ V, which provide the Hopf bifurcation of the solution curve in Fig. 6.14 directly.

6.5.1.2 Bifurcations from a Periodic Regime

A periodic solution $\overline{X}_s$ at the fundamental frequency ω_o is considered. When varying a parameter η continuously, the crossing through the imaginary axis of a real pole or a pair of complex-conjugate poles for a particular parameter value η_b will give rise to a bifurcation. This is expressed mathematically as

$$
\begin{aligned}
\overline{H}(\overline{X}^b_s, \eta_b) &= 0 \\
\det[JH(0 + j\omega_b, \eta_b)] &= 0 \\
\left.\frac{d\sigma}{d\eta}\right|_{\eta=\eta_b} &\neq 0
\end{aligned}
\tag{6.26}
$$

There are three main types of local bifurcations from a periodic regime at the fundamental frequency ω_o. The type of bifurcation depends on the frequency ω_b of the crossing poles.

$\omega_b = 0$, Turning Point or Pitchfork Bifurcation For $\omega_b = 0$ in (6.26), a real pole crosses the imaginary axis of the complex plane, which will correspond to either a turning point or a pitchfork bifurcation. With $\omega_b = 0$, the determinant in (6.26) takes a real value, because all the frequency components at $k\omega_o$ considered have

their complex-conjugate counterparts (at $-k\omega_o$) included in the calculation. The bifurcation point $\overline{X}_s^b, \eta_b$ is obtained by solving a system of $(2N+1)Q+1$ real equations in $(2N+1)Q+1$ unknowns, given by $\overline{X}_s^b, \eta_b$. Note that the bifurcation condition (6.26) with $\omega_b = 0$ implies the singularity of the Jacobian matrix of the harmonic balance system $\det[JH(0+j\omega_b, \eta_b)] = 0$ at the bifurcation point. Thus, the solution curve $\overline{X}_s(\eta)$ will exhibit infinite slope at η_b. The pitchfork bifurcations require the existence of odd-symmetry conditions in the circuit equations. As already known, the turning points can be associated with either jumps or hysteresis or synchronization phenomena.

As an example, a bifurcation analysis of the divided-by-2 solution of the circuit in Fig. 6.2 has been carried out. The divided-solution curve, analyzed with the aid of an auxiliary generator at the frequency $\omega_{AG} = \omega_{in}/2$, is represented in Fig. 6.9. It has been obtained for the constant input power $P_{in} = 12$ dBm, taking the input frequency f_{in} as parameter. Solving (6.26) with $\omega_b = 0$ provides the values of the input frequency and the circuit solution $\overline{X}_s^b$ directly at the turning point of the curve, obtained at $f_{TP} = 5.76$ GHz. This turning point is responsible for a hysteresis phenomenon in the transformation from nondivided to divided regime, and vice versa.

The technique to detect turning point or pitchfork bifurcations is slightly different in the case of free-running oscillators. In this type of circuit, the oscillation frequency is a state variable of the circuit, which changes under any variation of the parameter considered. Thus, the vector $\overline{X}_s$ in (6.26) must be replaced with the mixed vector $\overline{X}_s'$, in which the imaginary part of one of the harmonic variables has been suppressed and replaced with the oscillation frequency ω_o. The suppressed imaginary part is arbitrarily set to zero, which can be done due to the autonomy of the oscillator solution. The Jacobian matrix associated with the harmonic balance system resulting from this change of variables is no longer intrinsically singular (for all the parameter values). Then, turning points or pitchfork bifurcations can be detected from conditions (6.26) with $\omega_b' = 0$.

As an example, the technique for the turning-point detection has been applied to the circuit of Fig. 6.2 operating as a free-running oscillator. The free-running oscillator solution is analyzed versus the gate bias voltage V_{GS}, taken as parameter. The singularity condition of the Jacobian matrix [equations (6.26) with $\omega_b' = 0$] is solved in terms of $\overline{X}_s'^b$ and V_{GSb}. This provides directly the gate bias value $V_{GSb} = -0.867$ V at the steady-state oscillation frequency $f_b = 3.043$ GHz, corresponding to the point *TP* in Fig. 6.15. As already known, the turning point in the periodic solution curve of Fig. 6.15 is responsible for a hysteresis phenomenon in the transformation from the dc to the oscillatory regime, and vice versa.

$\omega_b = \omega_o/2$, Flip Bifurcation For $\omega_b = \omega_o/2$ in (6.26), a pair of complex-conjugate poles at $\omega_o/2$ cross the imaginary axis of the complex plane at the parameter value η_b. This gives rise to a division by 2 of the fundamental frequency ω_o of the periodic solution. The subharmonic components $(2k+1)\omega_o/2$, with k an integer, are generated from zero amplitude at the bifurcation point. For $\omega_b = \omega_o/2$, the determinant in (6.26) takes a real value, because the frequency components

$|k|\omega_o + \omega_o/2$ have their complex-conjugate counterparts $-|2k|\omega_o + \omega_o/2$ included in the calculation. Actually, to detect the flip bifurcation we can use a periodic frequency basis, with $\omega_o/2$ as fundamental, in (6.26), with the determinant $\det[JH]$ of the linearized system evaluated at the harmonic frequencies $k\omega_o/2$. The bifurcation point $\overline{X}_s^b, \eta_b$ is obtained by solving a system of $(2N+1)Q+1$ real equations in $(2N+1)Q+1$ unknowns, given by $\overline{X}_s^b, \eta_b$. Techniques to distinguish between supercritical and subcritical bifurcations are given in Section 6.5.2.

As an example, condition (6.25), with $\omega_b = \omega_{in}/2$, has been applied to detect the flip bifurcations of the circuit of Fig. 6.2. This time, the circuit is analyzed in a nondivided regime at the frequency ω_{in} of the input generator. As already known, the nondivided solution is the default solution provided by a harmonic balance simulator when no auxiliary generator is used or no accurate initial value (for the subharmonic solution) is provided. In this example, a constant input power $P_{in} = 12$ dBm has been considered, taking the input frequency f_{in} as parameter. The solution curve is represented in Fig. 6.9. Application of (6.25) with $\omega_b = \omega_{in}/2$ directly provides the two flip bifurcation points in Fig. 6.9 at the respective frequency values $f_{F1} = 5.94$ GHz and $f_{F2} = 6.68$ GHz.

$\omega_b = \alpha\omega_o$, with $\alpha \in R$, Secondary Hopf or Neimark Bifurcation For $\omega_b = \alpha\omega_o$, with $\alpha \in R$, in (6.26), a pair of complex-conjugate poles at a frequency ω_b nonharmonically related to that of the original periodic regime crosses the imaginary axis at the parameter value η_b. This gives rise to the onset of an oscillation at the frequency $\omega_a = \omega_b$ or $\omega_a = \omega_o \pm \omega_b$. The secondary Hopf bifurcation leads to a quasiperiodic regime at the two incommensurable fundamentals ω_o and ω_a. For $\omega_b = \alpha\omega_o$, the determinant in (6.26) takes a complex value, because the frequency components considered do not appear in complex-conjugate pairs. The bifurcation point $\overline{X}_s^b, \eta_b$ is obtained by solving a system of $(2N+1)Q+2$ real equations in $(2N+1)Q+2$ unknowns, given by $\overline{X}_s^b$, η_b, and ω_b. For a supercritical Hopf bifurcation, a stable quasiperiodic solution at ω_o, ω_b is generated from a stable periodic regime that becomes unstable after the bifurcation. The frequency ω_b agrees with the oscillation frequency at the bifurcation point. When varying the parameter η further from η_b in a continuous manner, the oscillation frequency will no longer agree exactly with the frequency of an unstable pair of complex-conjugate poles.

6.5.2 Bifurcation Detection Using Auxiliary Generators

For bifurcation detection with the aid of auxiliary generators, it is convenient to classify the bifurcations into two main geometrical types: bifurcations associated with a singularity of the original harmonic balance system, and bifurcations that give rise to the generation of new frequency components. The first type includes turning points and pitchfork bifurcations. The second type includes the flip bifurcation, which gives rise to the subharmonic $\omega_o/2$, and the Hopf bifurcation, which gives rise to an oscillation at the incommensurable frequency ω_b. As already known, the Hopf bifurcation may take place from a dc regime or from a periodic regime at ω_0. The amplitude of the autonomous (Hopf) or subharmonic components (flip) generated tends to zero at the bifurcation point, which is due to the continuity of the local bifurcations (see Figs. 6.9 and 6.14). Detection using an auxiliary generator of flip and Hopf bifurcations will rely on this property.

Below we describe the use of auxiliary generators for the analysis of Hopf, flip, and turning-point bifurcations.

6.5.2.1 Hopf Bifurcation For the detection of Hopf bifurcation it is taken into account that the amplitude of the newly generated oscillation tends to zero at the bifurcation point. Two different cases are distinguished here: Hopf bifurcations from a dc regime, leading to an oscillation at the frequency ω_b, and Hopf bifurcations from a periodic regime at ω_o leading to a quasiperiodic regime at the two fundamental frequencies ω_o and ω_b.

Hopf Bifurcation from a dc Regime As shown in Chapter 1, when a circuit exhibits a steady-state oscillation, the total current-to-voltage ratio is equal to zero at any circuit node $Y_T = 0$. At the Hopf bifurcations from the dc regime, this condition is fulfilled for an oscillation amplitude tending to zero. The Hopf bifurcation is detected by making the small-signal input admittance equal to zero. To obtain this input admittance, a small-signal current source $I_n(\omega)$, like the one used for the pole–zero identification, is connected in parallel at a sensitive circuit node. Then the input admittance is given simply by the ratio between the introduced current $I_n(\omega)$ and the node voltage $V_n(\omega)$ as $Y_T(\omega) = I_n(\omega)/V_n(\omega)$. To detect the Hopf bifurcation, the input admittance is set to zero, solving for the parameter value η_b and the oscillation frequency ω_b at the bifurcation point. Thus, the bifurcation condition is given by

$$Y_T(\omega_b, \eta_b) = 0 \qquad \text{with small-signal analysis} \tag{6.27}$$

The complex equation (6.27) is composed of two real equations in two unknowns, corresponding to the oscillation frequency and the parameter value η_b at the bifurcation point. Because the oscillation amplitude tends to zero at this bifurcation point, the input admittance is calculated with a small-signal analysis. The fulfillment of condition (6.27) implies the occurrence of a Hopf bifurcation at the parameter value η_b, leading to the onset of an oscillation at the frequency ω_b. Note that the input admittance can be determined equally using an auxiliary generator of low voltage amplitude connected in parallel at the sensitive circuit node. Because the oscillation amplitude tends to zero at the bifurcation point, this input admittance can be calculated with a small-signal analysis.

As shown in Chapter 3, the Hopf bifurcations can be classified into supercritical and subcritical. In the first case, a stable solution is generated at the bifurcation point. In the second case, an unstable periodic solution is generated at this point. The two types of bifurcation can be distinguished, obtaining the incipient oscillatory solution with the aid of an auxiliary generator and applying pole–zero identification to this solution. To obtain the incipient solution, an auxiliary generator of very small amplitude $A_{AG} = \varepsilon$ is introduced into the circuit. Due to the high slope of the solution curve versus the parameter in the neighborhood of the bifurcation point, the accuracy in the bifurcation detection is not very dependent on the value chosen for the small amplitude $A_{AG} = \varepsilon$. Then, a zero value of the ratio Y_{AG} between the auxiliary generator current and voltage is imposed, solving $Y_{AG} = 0$ in terms of the parameter and the oscillation frequency, in an manner analogous to condition (6.27). In the case of a subcritical bifurcation, the incipient solution will contain a real pole or a pair of complex-conjugate poles at about $\sigma \pm j\omega_b$ on the right-hand side of the complex plane. As an example, see the evolution of the real dominant pole in Fig. 6.15, corresponding to the subcritical bifurcation of the free-running oscillator solution in Fig. 6.14. The pole–zero identification can be performed at baseband, to detect the unstable real pole or about ω_b, in order to detect the unstable poles at $\sigma \pm j\omega_b$.

If two parameters η_1 and η_2 were considered, a curve composed of Hopf bifurcations, called a Hopf bifurcation locus, would be obtained. This curve would be defined by the same condition (6.27), replacing the single parameter η with the two parameters η_1 and η_2. The Hopf bifurcation locus consists of the pairs of parameter values η_{1b} and η_{2b} at which an oscillation is generated or extinguished. Examples of useful pairs of parameters are the two transistor bias voltages, one transistor bias voltage and a linear element value or one transistor bias voltage and the bias voltage of a varactor diode. The Hopf bifurcation locus will provide a contour enclosing the parameter values for which the circuit behaves as a free-running oscillator. An example of this locus was shown in Fig. 4.26.

Hopf Bifurcation from a Periodic Regime Assume the small perturbation at the frequency ω of a large-signal periodic regime $\overline{X}_s$ at $k\omega_o$, with k an integer. The frequency ω is incommensurate with ω_o. The small-signal input admittance $Y_T(\omega) = I_n(\omega)/V_n(\omega)$ is calculated through linearization of the harmonic-balance equations about the periodic solution $\overline{X}_s$ at ω_o. At the secondary Hopf bifurcation, the steady-state oscillation condition $Y_T(\omega_a) = 0$ is fulfilled with an oscillation amplitude tending to zero. This condition can be implemented easily on commercial harmonic balance software. For the detection of secondary Hopf bifurcations, a small-signal current source at the frequency ω is connected in parallel at a sensitive circuit node. The secondary Hopf bifurcation condition is thus given by

$$\overline{Y}_T(\omega_a, \eta_b) = 0 \qquad \text{with the conversion matrix approach} \tag{6.28}$$

where ω_a is the value of the incommensurable oscillation frequency generated at the parameter value η_b. Note that the secondary Hopf bifurcations can also be detected by introducing into the circuit an auxiliary generator of voltage type, with small-signal amplitude at the incommensurable frequency ω_{AG}, which should take the value $\omega_{AG} = \omega_a$ at the bifurcation point. When using either a small-signal current source or a small-signal auxiliary generator, it is possible to obtain the total admittance function Y_T with two-tone harmonic balance instead of the conversion matrix approach. The two fundamental frequencies considered are ω_o and ω (or

ω_{AG}). This, of course, requires very low amplitude in the auxiliary generator, as the circuit must operate in a linear regime with respect to this generator. The results should be the same as those obtained using the conversion matrix approach.

If the two parameters η_1 and η_2 are considered in (6.28), a curve is obtained consisting of the periodic solution points at which an incommensurable oscillation is generated or extinguished. This curve, called a secondary Hopf bifurcation locus, would be defined by the same condition (6.28), replacing the single parameter η with the two parameters η_1 and η_2. The Hopf bifurcation locus provides a contour of parameter values for which the circuit operates in the quasiperiodic regime, with two fundamental frequencies. As will be shown, this is very helpful for the efficient determination of the unstable behavior regions of power amplifiers and other nonlinear circuits. On the other hand, secondary Hopf bifurcations are found intrinsically in injection-locked oscillators for some values of the frequency and power of the synchronizing source (see Chapter 4).

The technique presented for determination of the Hopf bifurcation locus from a periodic regime has been applied to a frequency divider by 2 similar to the one shown in Fig. 6.2. The Hopf locus has been obtained with a commercial harmonic balance simulator. The circuit page used is shown in Fig. 6.16. The

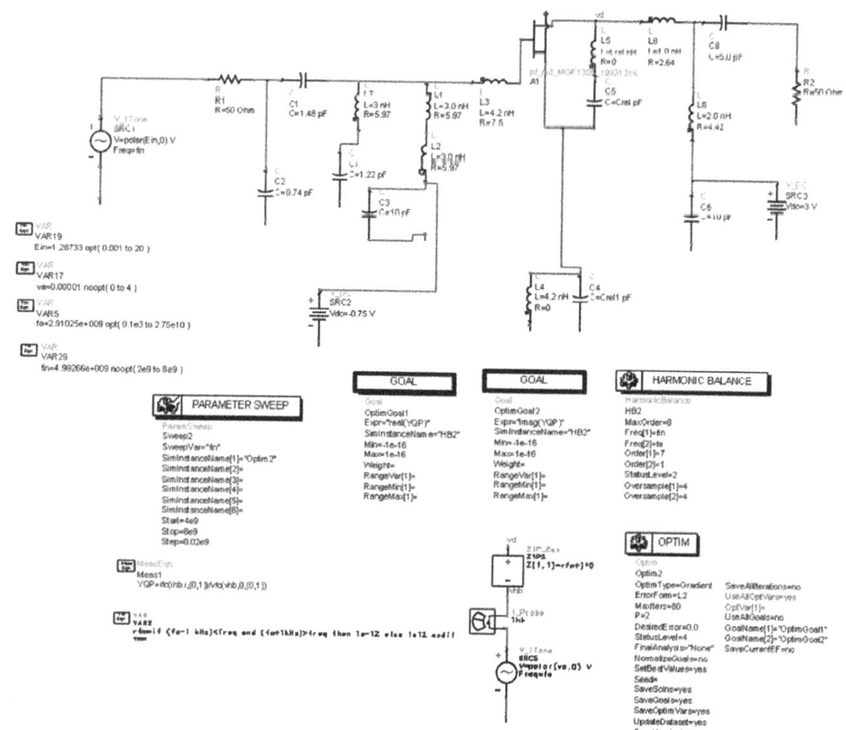

FIGURE 6.16 Circuit description for the determination of the Hopf bifurcation locus in a commercial harmonic balance simulator.

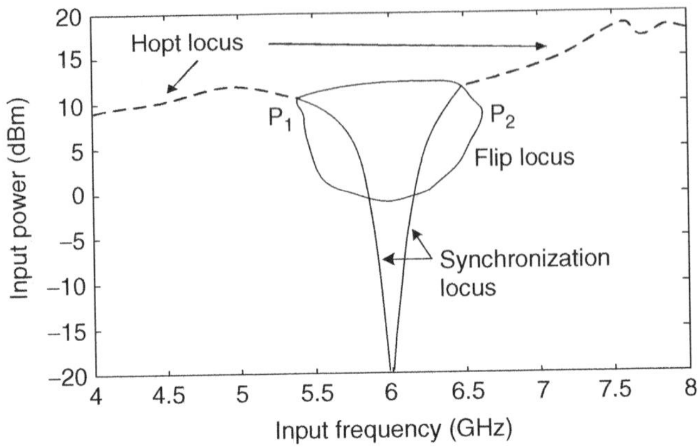

FIGURE 6.17 Bifurcation loci of the frequency divider of Fig. 6.2 in the plane defined by the input frequency and input power. The input frequency interval considered is about twice the oscillation frequency.

parameters considered are the input generator power P_{in} and frequency ω_{in}. The bifurcation condition $\overline{Y}_T(\omega_a, P_{in}, \omega_{in}) = 0$ is solved through optimization with a two-tone harmonic balance. As can be seen, the auxiliary generator is connected to the drain terminal. This generator operates at the oscillation frequency $\omega_{AG} = \omega_b$, with very small amplitude $V_{AG} = 10^{-3}$ V. The Hopf locus in the plane is defined by the input frequency ω_{in} and the input amplitude E_{in} obtained by sweeping the input frequency ω_{in} and calculating E_{in} and the oscillation frequency ω_{AG} at each step, by solving $\overline{Y}_{AG}(E_{in}, \omega_{AG}) = 0$ through optimization. The optimization goals are $\text{Re}[Y_{AG} = 0]$ and $\text{Im}[Y_{AG}] = 0$ and the optimization variables E_{in} and ω_{AG}. If the Hopf locus exhibits points of infinite or nearly infinite slope versus ω_{in}, the swept parameter must be changed to E_{in}, solving $\overline{Y}_{AG}(\omega_{in}, \omega_{AG}) = 0$. The resulting locus is shown in Fig. 6.17. Note that the oscillation at ω_a is generated or extinguished when crossing this locus (depending on the sense of variation of the parameters), and the frequency ω_a varies along the locus. This is shown in Fig. 6.18, where the frequency ω_a has been represented versus the input frequency ω_{in}. The section between points P_1 and P_2, fulfilling $\omega_b = \omega_{in}/2$, does not belong to the Hopf locus. This section corresponds, in fact, to the flip bifurcation locus, which provides the generator values for the generation (or extinction) of the subharmonic component $\omega_{in}/2$. Below the Hopf locus and outside the synchronization locus, also represented in Fig. 6.17, the circuit behaves as a self-oscillating mixer at the two fundamental frequencies ω_{in} and ω_a, with ω_a the self-oscillation frequency.

6.5.2.2 *Flip Bifurcations*

The flip bifurcations lead from a periodic solution at a frequency ω_o to a frequency-divided solution at $\omega_o/2$, or vice versa. The amplitude of the subharmonic component at $\omega_o/2$ tends to zero at the bifurcation point. Therefore, to detect the flip bifurcations, an auxiliary generator at the subharmonic

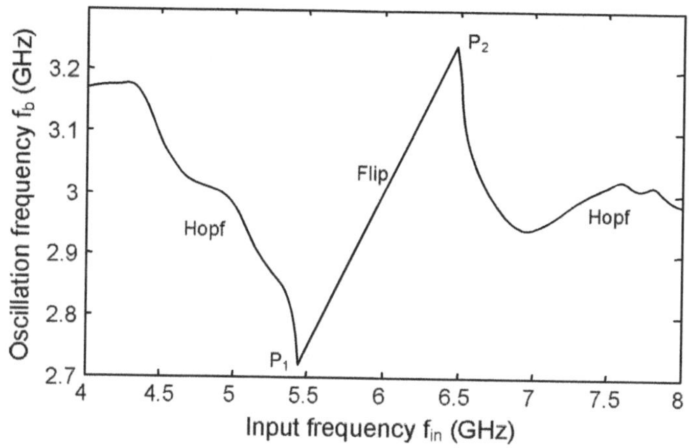

FIGURE 6.18 Variation in the oscillation frequency f_b along the Hopf bifurcation locus. The middle section, located between the two codimension 2 bifurcations P_1 and P_2 is part of the flip locus (see Fig. 6.17). Along the flip locus, the oscillation frequency is a subharmonic of the second order of the input frequency. The phase shift with respect to the input generator is constant for each input frequency value, but varies along the locus.

frequency $\omega_o/2$ is introduced into the circuit, setting its amplitude to a very small value:

$$Y_{AG}(\phi_{AGb}, \eta_b) = 0 \qquad \text{with} \quad \omega_{AG} = \omega_o/2 \text{ and } A_{AG} = \varepsilon \tag{6.29}$$

A suitable value for the amplitude of the auxiliary generator may be $A_{AG} = 10^{-3}$ V. The system (6.29) is solved using single-tone harmonic balance at the fundamental frequency $\omega_o/2$. The unknowns of the complex equation $Y_{AG} = 0$ are the phase of the auxiliary generator ϕ_{AG} and the parameter value η_b at the bifurcation point. This provides the parameter values at which the flip bifurcations are obtained. As already indicated, the simulation of Fig. 6.9 corresponds to the circuit of Fig. 6.2, operating with constant input power $P_{in} = 12$ dBm. The two flip bifurcations, F_1 and F_2, can be obtained directly by solving the complex equation (6.29) in terms of the input frequency ω_{in} and the auxiliary generator phase ϕ_{AG}. Supercritical and subcritical bifurcations can be distinguished by applying pole–zero identification to the incipient solution, with $A_{AG} = \varepsilon$. The subcritical (supercritical) bifurcation will contain a real pole or, equivalently, a pair of complex-conjugate poles at $\sigma \pm j\omega_{in}/2$ on the right-hand (left-hand) side of the complex plane. The identification can be performed at baseband in order to detect the unstable real pole or about $\omega_{in}/2$ in order to detect the unstable poles at $\sigma \pm j\omega_{in}/2$.

One may wonder why the detection of flip bifurcations, leading to a frequency division by 2, is not carried out with the conversion matrix approach. The reason is that the standard conversion matrix approach is inapplicable for coincident values of the sideband frequencies, as the matrix becomes singular in this situation, due to the system degeneracy. In the case of a frequency division by 2, the upper sideband of each harmonic frequency $k\omega_o + \omega = k\omega_o + \omega_o/2$ agrees exactly with the lower sideband $(k+1)\omega_o - \omega = (k+1)\omega_o - \omega_o/2$ of the next harmonic term. The frequencies are overlapped and the matrixes become singular. An alternative formulation has been proposed [29] to overcome this problem in frequency dividers. The technique is based on a judicious suppression of the overlapping sideband frequencies.

If two parameters η_1 and η_2 are considered in the complex equation (6.29), a curve comprised of flip bifurcations, called a *flip bifurcation locus*, is obtained. This locus is very interesting for the design of frequency dividers by 2. As shown in Chapter 4, the frequency-divided solution is not the only possible solution of frequency dividers. Depending on the input generator values, they can also operate in the nondivided regime, at the input frequency ω_{in}, or in the quasiperiodic regime, at ω_{in} and a self-generated oscillation frequency ω_a. The flip bifurcation locus delimits the frequency-division region in the plane defined by the input frequency and power, constituting the two analysis parameters. Using (6.29), the flip bifurcation locus can be determined with a commercial harmonic balance simulator. As an example, Fig. 6.19 presents the circuit description to obtain the flip bifurcation locus of the divider considered in Fig. 6.17. An auxiliary generator is

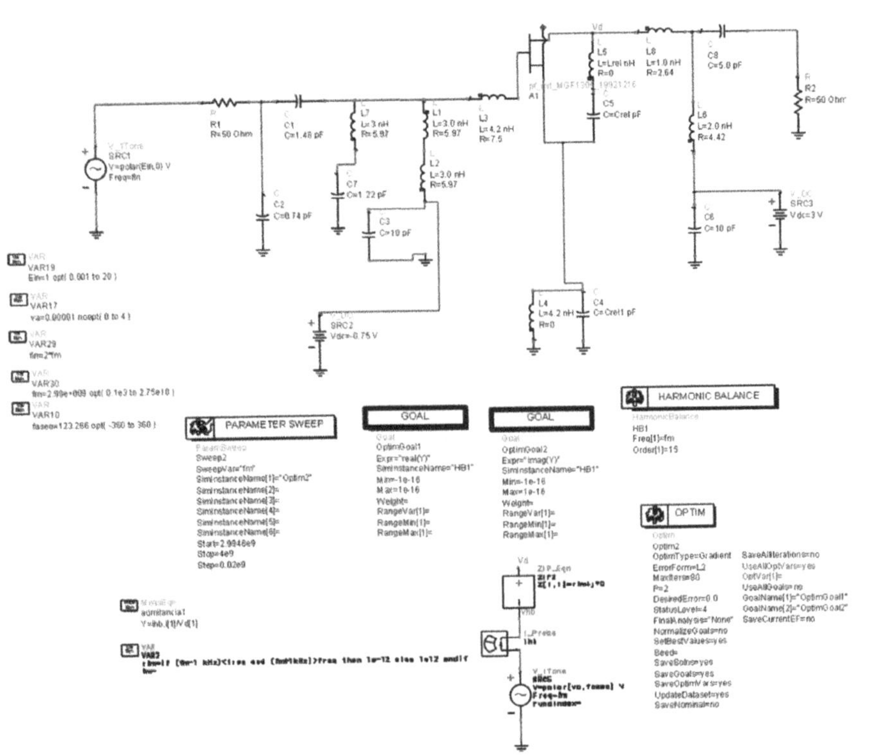

FIGURE 6.19 Circuit description for determination of the flip bifurcation locus in a commercial harmonic balance simulator.

introduced at a sensitive circuit node. In this case the node chosen is the transistor drain terminal. This generator operates at the divided frequency $\omega_{in}/2$ with very small amplitude $A_{AG} = 10^{-3}$ V and phase ϕ_{AG}. The frequency ω_{in} is swept, and at each step, the equation $Y_{AG}(P_{in}, \phi_{AG}) = 0$ is solved through optimization, with the goals $\text{Re}[Y_{AG}] = 0$ and $\text{Im}[Y_{AG}] = 0$ and the optimization variables P_{in} and ϕ_{AG}. The resulting flip bifurcation locus has been added to Fig. 6.17. The phase ϕ_{AG} varies along the locus. Note that the flip bifurcation locus exhibits turning points. Thus, to obtain the locus, parameter switching has to be used. In the zones of high slope versus the input frequency ω_{in}, the input generator amplitude E_{in} is swept, calculating ω_{in} and the auxiliary generator phase ϕ_{AG} at each step, so as to fulfill $Y_{AG}(\omega_{in}, \phi_{AG}) = 0$. Sometimes it can also be useful to sweep the phase ϕ_{AG} and solve $Y_{AG} = 0$ for P_{in} and ω_{in}, with $A_{AG} = \varepsilon$.

As shown in Fig. 6.17, the flip and Hopf bifurcation locus intersect at the two codimension bifurcation points P_1 and P_2. At these two points the oscillation frequency ω_a generated at the Hopf locus, which varies along this locus, becomes equal to $\omega_a = \omega_{in}/2$. This relationship is fulfilled along the entire flip bifurcation locus. Note that the flip bifurcation locus provides the pairs of ω_{in} and P_{in} values at which a frequency division is generated from a nondivided regime at ω_{in}. As shown in Chapter 4, Section 4.3.2, only the section of the flip bifurcation locus located above the Hopf locus corresponds to physically observable transformations. Below the Hopf locus, the unstable nondivided regime generates an unstable divided solution, so this transformation is not observed physically. From the inspection of Fig. 6.17, division through flip bifurcation has little relevance in the performance of this particular divider.

6.5.2.3 *Turning Points* Assume the case of a free-running oscillator with the periodic steady-state solution $\overline{X}_s$ at $k\omega_o$. The oscillation frequency ω_o will change under any variation of the parameter η. For the circuit analysis, an auxiliary generator at the oscillation frequency $\omega_{AG} = \omega_o$ will be introduced into the circuit. For the parameter increment $\Delta\eta$, the variation of the auxiliary generator values, A_{AG} and ω_{AG}, can be estimated using the linearization (6.22). Then the slope of these two variables versus this parameter is estimated from

$$\begin{bmatrix} \dfrac{A_{AG}^{n+1} - A_{AG}^{n}}{\Delta\eta} \\ \dfrac{\omega_{AG}^{n+1} - \omega_{AG}^{n}}{\Delta\eta} \end{bmatrix}_p = -[JY_{AG}]_n^{-1} \left.\frac{\partial Y_{AG}}{\partial\eta}\right|_n \tag{6.30}$$

In the limit $\Delta\eta \to 0$, the terms on the left-hand side agree with the derivatives of the oscillation amplitude and the oscillation frequency versus the parameter: $\partial A_{AG}/\partial\eta$ and $\partial\omega_{AG}/\partial\eta$. From an inspection of (6.30), the condition for infinite slope (or turning point) of A_{AG} or ω_{AG} versus the parameter η is given by

$$\begin{aligned} Y_{AG}(A_{AGb}, \omega_{AGb}) &= 0 \\ \det[JY_{AG}(A_{AGb}, \omega_{AGb}, \eta_b)] &= 0 \end{aligned} \tag{6.31}$$

with the Jacobian matrix $[JY_{AG}]$ being the one defined in (6.23). Note that the turning point of the curve is also a solution of $Y_{AG} = 0$. The determinant $\det[JY_{AG}] = 0$ is real, so system (6.31) is a well-balanced system of three real equations in three real unknowns. System (6.31) can be used to detect directly the turning points of the solution curve of a free-running oscillator, traced versus a bias voltage or versus a circuit element value. As an example, it can be used to detect directly the turning point of the curve of Fig. 6.14, providing variation the of the free-running oscillation amplitude of the circuit of Fig. 6.2 versus the gate bias voltage V_{GS}. In this case, the parameter η in (6.31) corresponds to the bias voltage.

Condition (6.31) can also be used to detect the possible turning points in autonomous quasiperiodic solutions, such as those corresponding to self-oscillating mixers. These quasiperiodic regimes have two fundamental frequencies: the input generator frequency ω_{in} and the self-oscillation frequency ω_a. As shown in Chapter 5, they are simulated with two-tone harmonic balance at ω_{in} and ω_a using an auxiliary generator at the oscillation frequency $\omega_{AG} = \omega_a$, with amplitude A_{AG}. The auxiliary generator must fulfill $Y_{AG}(A_{AG}, \omega_{AG}) = 0$. The Jacobian matrix associated with Y_{AG} is formally identical to the one defined in (6.23). At turning points of the quasiperiodic solution curve versus any parameter η, this Jacobian matrix becomes singular. Thus, the turning points of the quasiperiodic solution versus a parameter η can be detected using the same conditions (6.31). The only difference is that these conditions require using a two-tone harmonic balance simulation of the circuit at the two fundamental frequencies ω_{in} and $\omega_o = \omega_{AG}$. The number of analysis frequencies and thus harmonic balance variables increases substantially in the quasiperiodic regime. However, the Jacobian matrix used for the bifurcation detection with an auxiliary generator keeps the same order (2×2) that it had in the periodic regime.

As an example, Fig. 6.20 shows the variation in the conversion gain of a self-oscillating mixer versus the input power. The circuit being analyzed is similar to the one in Fig. 4.36. The input frequency $f_{\text{RF}} = 5.5$ GHz and output frequency $f_{\text{IF}} = f_{\text{RF}} - f_a = 0.5$ GHz. As can be seen in Fig. 6.20, the quasiperiodic solution exhibits a turning point T at the input power $P_{\text{in}} = 2$ dBm. This bifurcation is not the only one in Fig. 6.20. The self-oscillation at ω_a is extinguished at an inverse Hopf bifurcation of subcritical type at the input power $P_{\text{in}} = 0$ dBm. The two bifurcations give rise to a hysteresis phenomenon in the transformation from quasiperiodic regime to periodic regime, and vice versa.

A condition analogous to (6.23) can be used to detect the turning points of solution curves corresponding to circuits operating in a forced regime, such as injection-locked oscillators, frequency dividers, or power amplifiers. In this case, the auxiliary generator operates at a frequency related harmonically to that of the input generator. In the case, for example, of a power amplifier or a fundamentally synchronized oscillator, this frequency will be $\omega_{AG} = \omega_{\text{in}}$. In the case of a frequency divider by N, it will be $\omega_{AG} = \omega_{\text{in}}/N$. For turning-point detection, the variable ω_{AG} must be replaced with auxiliary generator phase ϕ_{AG}, in both equation (6.22) and the Jacobian matrix (6.23). Thus, the turning-point condition

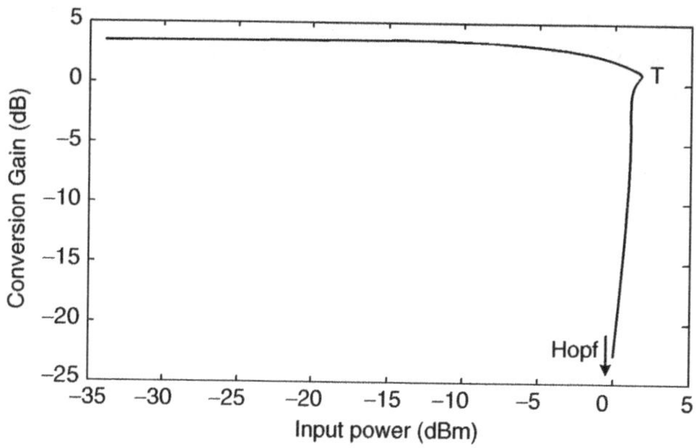

FIGURE 6.20 Variation of the conversion gain versus the input power in a self-oscillating mixer similar to the one shown in Fig. 4.36. The solution curve exhibits a turning point at $P_{in} = 2$ dBm. There is also a subcritical Hopf bifurcation occurring at about $P_{in} = 0$ dBm.

is given by

$$\begin{aligned} Y_{AG}(A_{AGb}, \phi_{AGb}) &= 0 \\ \det[JY_{AG}(A_{AGb}, \phi_{AGb}, \eta_b)] &= 0 \end{aligned} \tag{6.32}$$

with the Jacobian matrix

$$[JY_{AG}] = \begin{bmatrix} \dfrac{\partial \text{Re}[Y_{AG}]}{\partial A_{AG}} & \dfrac{\partial \text{Re}[Y_{AG}]}{\partial \phi_{AG}} \\ \dfrac{\partial \text{Im}[Y_{AG}]}{\partial A_{AG}} & \dfrac{\partial \text{Im}[Y_{AG}]}{\partial \phi_{AG}} \end{bmatrix} \tag{6.33}$$

As has been shown, the Jacobian matrixes used for the bifurcation detection with auxiliary generators have order 2 × 2. This reduced order enables an accurate determination of the parameter values at which the associated determinant takes a zero value. This calculation can be more accurate than the one based on an analysis of the high-order harmonic balance Jacobian matrix due to the common ill-conditioning problems of this matrix. However, this technique is computationally more expensive, since the numerical calculation of the derivatives in the Jacobian matrix of the admittance function requires several harmonic balance executions (one for each variable increment: amplitude and frequency or phase). A harmonic balance simulation is performed at each increment of the variables considered.

The turning-point locus in the plane defined by two parameters η_1 and η_2 is traced by sweeping one parameter η_1 and solving the following three-equation system at each η_1 step:

$$\begin{aligned} \overline{Y}_{AG}(A_{AG}, \xi_{AG}, \eta_2) &= 0 \\ \det[JY_{AG}(A_{AG}, \xi_{AG}, \eta_2)] &= 0 \end{aligned} \tag{6.34}$$

where ξ_{AG} can be either the auxiliary generator frequency or its phase, this depending on the type of regime to be analyzed, autonomous or forced. The resolution above provides a curve in the plane defined by η_1 and η_2. The locus is obtained by sweeping η_1 and solving (6.34) for A_{AG} and η_2. Parameter switching may be necessary in case of an infinite or near-infinite slope of the locus versus η_1. The turning-point locus can be obtained with commercial harmonic balance, but the procedure is more complex than the one applied to obtain the flip and Hopf bifurcation loci. This procedure is described in the following.

The case of a turning-point locus in a forced regime using ϕ_{AG} instead of ω_{AG} will be considered initially. When using commercial harmonic balance, the derivatives in $[JY_{AG}]$ are obtained by analyzing three copies of the circuit simultaneously. Each copy must have an auxiliary generator at the fundamental frequency of the periodic regime connected to the same circuit node. A sketch is shown in Fig. 6.21. Circuit 1 is simulated at the actual operation point, corresponding to A_{AG} and ϕ_{AG}. This circuit must fulfill the nonperturbation condition $Y_{AG}(A_{AG}, \phi_{AG}) = 0$. Circuit 2 is simulated at A_{AG} and $\phi_{AG} + \Delta\phi_{AG}$. Circuit 3 is simulated at $A_{AG} + \Delta A_{AG}$ and ϕ_{AG}. To solve the system (6.34), optimization is used, with the goals $Y_{AG} = 0$ (to be fulfilled by circuit 1) and $\det[Y_{AG}] = 0$. Circuits 2 and 3 enable the calculation through increments of the derivatives required for the determination of $\det[JY_{AG}]$. The derivative $\partial Y_{AG}/\partial\phi_{AG}$ is obtained from the ratio $\Delta Y_{AG}/\Delta\phi_{AG}$, with ΔY_{AG} being the admittance difference between circuits 2 and 1. In turn, the derivative $\partial Y_{AG}/\partial A_{AG}$ is calculated from the ratio $\Delta Y_{AG}/\Delta A_{AG}$, with ΔY_{AG} being the admittance difference between circuits 3 and 1. Practical values of $\Delta\phi_{AG}$ and ΔA_{AG} are 2° and 0.01 V. As already stated, the three circuits are solved in a single harmonic balance simulation. The technique can also be applied to obtain the turning-point locus of a free-running oscillator. However, one of the circuit copies must operate at the incremented frequency $\omega_{AG} + \Delta\omega_{AG}$, which requires using two incommensurable fundamental frequencies in the harmonic balance simulation at ω_{AG} and $\omega_{AG} + \Delta\omega_{AG}$. At the time to truncate the Fourier series for the two-tone simulation, all the intermodulation

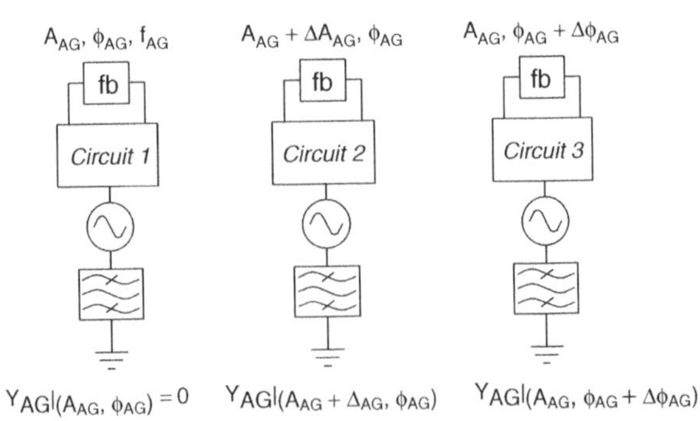

FIGURE 6.21 Three replicas of the circuit, used for determination of the derivatives container in the Jacobian matrix, associated with the admittance function Y_{AG}, in a single harmonic balance simulation.

products of the two frequencies, meaningless for this analysis, must be suppressed. This can be done by assigning the nonlinearity order $nl_1 = N$ to ω_{AG}, $nl_2 = N$ to $\omega_{AG} + \Delta\omega_{AG}$, and zero to the intermodulation products.

In injection-locked oscillators, the turning-point locus that provides variation in the synchronization bandwidth can be obtained in a much simpler manner. Advantage is taken of the closed form of the synchronized solution curves and the regular variation of the solution phase shift with respect to the input generator frequency. As an example, the synchronization locus in the plane defined by the input generator frequency ω_{in} and the input generator power P_{in} will be considered. Only one circuit has to be analyzed, with an auxiliary generator connected to a sensitive node. The free-running solution of the circuit, obtained for zero input power P_{in}, is determined initially. At this free-running solution, the auxiliary generator fulfills $Y_{AG}(A_{AG}, \omega_{AG}) = 0$. Next, to obtain the synchronization locus, the frequency ω_{AG} is written in terms of ω_{in}, according to their rational relationship $\omega_{AG} = m/k\omega_{in}$, and the auxiliary generator phase (or input generator phase) is taken as a new variable. Two nested sweeps are performed in the commercial harmonic balance simulator. In the external sweep, P_{in} is swept. In the internal sweep, ϕ_{AG} is swept between 0 and $2\pi/k$, with k the division order. It is also possible to set the phase of the auxiliary generator to zero $\phi_{AG} = 0$ and to sweep the phase of the input generator between 0 and 2π. At each P_{in}, ϕ_{AG} step, the auxiliary generator amplitude A_{AG} and the input frequency ω_{in} are calculated through optimization, so as to fulfill $Y_{AG}(A_{AG}, \omega_{in}) = 0$.

Assuming that all of the curves obtained are closed, the edges of the synchronization band will be given, at each P_{in} value, by the two turning points T_1 and T_2 of the corresponding solution curve. These are, in turn, the two outermost points of the curve. Thus, the synchronization locus can be obtained simply by projecting the collection of closed curves over the plane defined by the input frequency ω_{in} and input power P_{in}. Note that the projection will only be valid provided that the synchronization curves are closed and relatively regular, which happens for low and intermediate input power. Remember that the turning-point locus and Hopf bifurcation locus intersect at the two codimension 2 bifurcation points P_1 and P_2. In the region about these two points, the curves become quite irregular. Above this region the curves open up and the technique described provides confusing results. The technique is computationally expensive, as each entire solution curve is calculated rather then just the turning points of each curve. However, it is of very simple application and allows obtaining the synchronization locus of a given oscillator or frequency divider in a straightforward manner.

As an example, the technique described above has been applied to determine the synchronization locus of the circuit in Fig. 6.2, behaving as a harmonic injection divided by $N = 2$. The collection of solution curves resulting from the two nested sweeps is shown in Fig. 6.22. A three-dimensional representation has been carried out in the space defined by input frequency, the input voltage, and the drain voltage amplitude. The locus resulting from the projection of the curves in Fig. 6.22 over the plane defined by the input power and frequency is superimposed in Fig. 6.17.

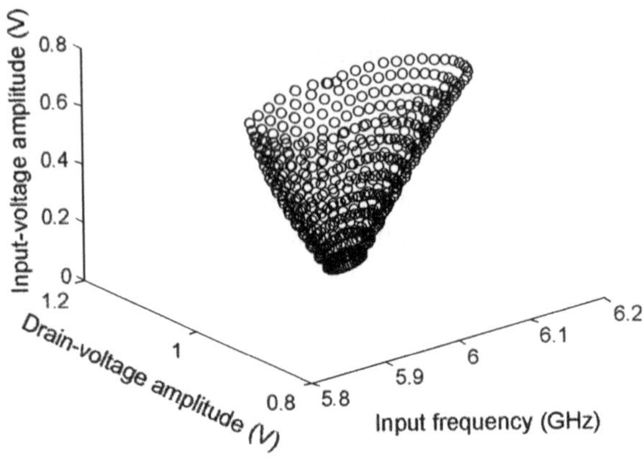

FIGURE 6.22 Three-dimensional representation of the results of the two nested sweeps used for the determination of the synchronization locus.

6.6 BIFURCATION SYNTHESIS AND CONTROL

The techniques presented so far enable local and global stability analysis of a given nonlinear circuit. They allow the detection of different types of bifurcations with in-house and commercial harmonic balance simulators. These bifurcations delimit the operation bands of circuits of autonomous nature such as voltage-controlled oscillators, injection-locked oscillators, or frequency dividers. Therefore, the possibility of presetting the parameter values at which these bifurcations take place will be of great interest for autonomous circuit design. Various techniques have been proposed in the literature, aiming at the bifurcation synthesis or control. The principles of these techniques are outlined briefly in the following.

6.6.1 Bifurcation Synthesis

The bifurcation synthesis modifies an initial circuit design so as to generate a particular type of bifurcation at the parameter value η_b. As an example, a method is presented [29] to synthesize the startup conditions of a subharmonic oscillation of order N. The method is based on a modified conversion matrix approach and leads to a variation of the circuit topology and its element values so as to fulfill the subharmonic startup conditions without modifying the steady-state solution at $k\omega_{\text{in}}$.

6.6.2 Bifurcation Control

In bifurcation control, the properties or location (in terms of the parameter) of a bifurcation existing originally in the system are modified. Example of control actions are the shifting of an existing bifurcation originally occurring at the parameter value η_b to η_b' or the transformation of a subcritical bifurcation into a supercritical bifurcation. In bifurcation control, the bifurcation properties of a given

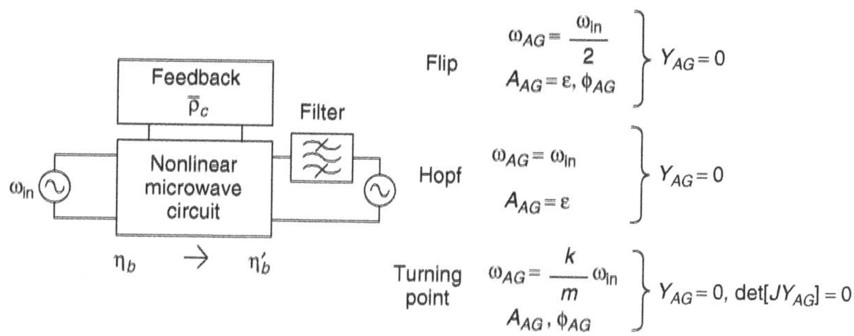

FIGURE 6.23 Auxiliary generator technique for bifurcation control. The conditions for the control of various types of bifurcations are shown on the right-hand side.

nonlinear system are modified through the synthesis of specific control functions [30–32], which introduce feedback in the system equations. The feedback functions contain additional parameters, which did not exist in the original circuit, so its response can be modified by changing the values of these feedback parameters. In the frequency domain this can be done using a describing function model for the nonlinear element. The bifurcation condition is written in terms of the characteristic system obtained from perturbation of the (nonbifurcated) steady-state solution in the presence of the feedback functions.

As will be shown in this section, with the aid of auxiliary generator(s), we can derive an alternative "black-box" technique, with no requirement for determination of the characteristic system associated with the circuit nonlinear equations. The introduction of a feedback network provides additional parameters to perform the control action and reduces the disturbance of the original circuit response. Note that, in general, the bifurcation can also be imposed by modifying element values of the original circuit, with no need of additional feedback elements. However, this can give rise to significant changes with respect to the original performance.

The principle of bifurcation control proposed here is a very simple one. The bifurcation is imposed at the desired parameter values by means of an auxiliary generator. The bifurcation condition used will depend on the type of bifurcation to be controlled. Assuming that a feedback network is introduced for bifurcation control, the element values of this network, contained in the vector $\overline{p}_c$, will be calculated to fulfill the particular bifurcation condition at the desired parameter value η'_b. The procedure is illustrated in Fig. 6.23. The auxiliary generator imposes the required oscillation conditions at the particular type of bifurcation: flip, Hopf, or turning point. The controller consists of a feedback network with a set of parameters $\overline{p}_c$. These parameters are calculated so as to preset the bifurcation at the desired parameter value η'_b.

Several application examples of the bifurcation control are presented in Chapter 11. Here an analytical demonstration of the technique is shown. The aim is to shift the resistance value at which a Hopf bifurcation occurs from the dc regime in the parallel resonance oscillator of Fig. 6.24a. The dc solution of this circuit is given by $v = 0$. Its stability is determined by considering a

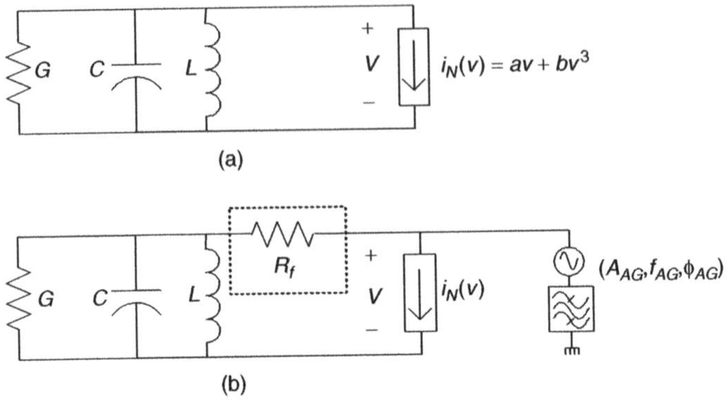

FIGURE 6.24 Parallel resonance oscillator with a cubic nonlinearity $i = av + bv^3$, $a < 0, b > 0$: (a) uncontrolled circuit; (b) addition of a resistance for bifurcation control.

low-amplitude perturbation of complex frequency p and linearizing the nonlinear element about the dc solution. This provides the following characteristic equation of the uncontrolled circuit (Fig. 6.24):

$$(a + G)Lp + (LCp^2 + 1) = 0 \tag{6.35}$$

The corresponding poles $p = \alpha \pm j\omega$ are given by

$$\alpha = \frac{-(a+G)}{2C}$$
$$\omega = \sqrt{\frac{1}{LC} - \frac{(a+G)^2}{4C^2}} \tag{6.36}$$

Considering the linear conductance G as a parameter, a Hopf bifurcation takes place for $G_b = -a$ and $\omega_b = 1/\sqrt{LC}$. A possible control objective may be the shifting of this bifurcation to the new conductance value G'_b. To do so, in this simple example, the resistance R_f (Fig. 6.5b) is introduced. Then the roots of the characteristic equation, or system poles, become

$$\alpha = \frac{-(a + G + aGR_f)}{2(C + aR_fC)}$$
$$\omega = \sqrt{\frac{1}{LC} - \frac{(a + G + aGR_f)^2}{4C^2(1 + aR_f)^2}} \tag{6.37}$$

which, as expected, agree with the poles of the original system for $R_f = 0$. From (6.37) it is possible to calculate R_f so as to obtain the Hopf bifurcation at the conductance value G'_b desired. This is given by

$$R_f = -\frac{1}{a} - \frac{1}{G'_b} \tag{6.38}$$

As can easily be verified, the frequency value at the bifurcation point remains at $\omega_b = 1/\sqrt{LC}$.

Next, the bifurcation control technique based on the auxiliary generator will be used to shift the Hopf bifurcation from G_b to G'_b. The auxiliary generator allows us to impose the parameter value for the incipient solution just after the bifurcation. Therefore, the oscillator steady-state equation should be written under conditions of very small oscillation amplitude $A_{AG} = \varepsilon$. Assuming one harmonic component, the nonlinear element can be modeled with its describing function. The steady-state equation of the circuit, after the introduction of R_f, is given by

$$\begin{aligned}
&\text{Re}[Y_{AG}] = \left(a + \tfrac{3}{4}bA_{AG}^2\right)(1 + R_f G'_b) + G'_b = 0 \\
&\text{Im}[Y_{AG}] = \left(C\omega - \frac{1}{L\omega}\right)\left[\left(a + \frac{3}{4}bV^2\right)R_f + 1\right] = 0 \\
&A_{AG} = \varepsilon
\end{aligned} \tag{6.39}$$

where G'_b is the conductance value at which the bifurcation is desired. From (6.39), the R_f value, to have a Hopf bifurcation at G'_b, should be

$$R_f = \frac{-a - \frac{3}{4}bA_{AG}^2 - G'_b}{\left(a + \frac{3}{4}bA_{AG}^2\right)G'_b} = \frac{-a - \frac{3}{4}b\varepsilon^2 - G'_b}{\left(a + \frac{3}{4}b\varepsilon^2\right)G'_b} \tag{6.40}$$

which agrees with (6.38) for $\lim \varepsilon \to 0$. In turn, the auxiliary generator frequency fulfills $\omega_{AG} = 1/\sqrt{LC}$. Thus, the two methods agree to an accuracy degree $A_{AG} = \varepsilon$.

As has been shown in the example, the control conditions, based on use of the auxiliary generator, are valid mathematical conditions. The aim of the control will be to modify the circuit bifurcation properties versus a given parameter with minimum variation of the controlled-circuit response compared with the original one. In the literature [30,32], equality is imposed at the bifurcation point between solutions of the uncontrolled and controlled systems. However, in autonomous circuits such as forced oscillators and frequency dividers, the designer is interested in the solution after the bifurcation, so the condition mentioned will not be imposed.

As already known, *branching bifurcations* are those giving rise to new solution paths growing from zero amplitude at newly generated frequency components. Examples are the pitchfork bifurcations (with no new frequency components), the flip bifurcations (with the generation of $k\omega_{\text{in}}/2$, with k odd) and the Hopf bifurcations, with the generation of $k\omega_{\text{in}} + m\omega_a$, with $m \neq 0$. The branching bifurcations may be supercritical or subcritical. As shown in Chapter 3, subcritical bifurcations are associated with hysteresis in the transformation from the original regime to the bifurcated regime. To avoid this possible hysteresis, it is possible to force the supercritical nature of the bifurcation with mathematical conditions [30]. Examples will be provided in Chapter 11. Finally, it must be taken into account that bifurcation control is inherently local. Away from the bifurcation, the variation of the circuit response will depend on the particular choice of feedback networks and control parameters.

REFERENCES

[1] J. Guckenheimer and P. Holmes, *Nonlinear Oscillations, Dynamical Systems and Bifurcations of Vector Fields*, Springer-Verlag, New York, 1990.

[2] J. M. T. Thompson and H. B. Stewart, *Nonlinear Dynamics and Chaos*, 2nd ed., Wiley, Chichester, UK, 2002.

[3] R. E. Collin, *Foundations for Microwave Engineering*, 2nd ed., Wiley, New York, 2001.

[4] J. M. Rollett, Stability and power-gain invariants of linear twoports, *IRE Trans. Circuit Theory*, vol. 9, pp. 29–32, Mar. 1962.

[5] M. L. Edwards and J. H. Sinsky, A new criterion for linear 2-port stability using a single geometrically derived parameter, *IEEE Trans. Microwave Theory Tech.*, vol. 40, pp. 2303–2311, Dec. 1992.

[6] V. Rizzoli and A. Lipparini, General stability analysis of periodic steady-state regimes in nonlinear microwave circuits, *IEEE Trans. Microwave Theory Tech.*, vol. 33, pp. 30–37, Jan. 1985.

[7] V. Rizzoli and A. Neri, State of the art and present trends in nonlinear microwave CAD techniques, *IEEE Trans. Microwave Theory Tech.*, vol. 36, pp. 343–356, Feb. 1988.

[8] R. Quéré, E. Ngoya, M. Camiade, A. Suárez, M. Hessane, and J. Obregón, Large signal design of broadband monolithic microwave frequency dividers and phase-locked oscillators, *IEEE Trans. Microwave Theory Tech.*, vol. 41, pp. 1928–1938, Nov. 1993.

[9] A. Suárez and R. Quéré, *Stability Analysis of Nonlinear Microwave Circuits*, Chapter 3, Artech House, Norwood, MA, Jan. 2003.

[10] K. Ogata, *Modern Control Engineering*, Prentice-Hall, Englewood Cliffs, NJ, 1980.

[11] W. Struble and A. Platzker, Rigorous yet simple method for determining stability of linear N-port networks, *Proceedings of the 15th GaAs IC Symposium*, San Jose, CA, pp. 251–254, 1993.

[12] A. Platzker and W. Struble, Rigorous determination of the stability of linear N-node circuits from network determinants and the appropriate role of the stability factor K of their reduced two-ports, *Integrated Nonlinear Microwave and Millimeterwave Circuits, Third International Workshop (INMMIC)*, pp. 93–107, 1994.

[13] A. Anakabe, J. M. Collantes, J. Portilla, et al., Analysis and elimination of parametric oscillations in monolithic power amplifiers, *IEEE MTT-S International Microwave Symposium Digest*, pp. 2181–2184, 2002.

[14] J. Jugo, J. Portilla, A. Anakabe, A. Suárez and J. M. Collantes, Closed-loop stability analysis of microwave amplifiers, *IEE Electron. Lett.*, vol. 37, pp. 226–228, Feb. 2001.

[15] A. Anakabe, Detección y eliminación de inestabilidades paramétricas en amplificadores de potencia para comunicaciones, Ph.D. Thesis, Universidad del Pais Vasco, 2003.

[16] S. Jeon, A. Suárez, and D. B. Rutledge, Global stability analysis and stabilization of a class-E/F amplifier with a distributed active transformer, *IEEE Trans. Microwave Theory Tech.*, vol. 53, pp. 3712–3722, 2005.

[17] A. Suárez, V. Iglesias, J. M. Collantes, J. Jugo, and J. L. García, Nonlinear stability analysis of microwave circuits using commercial software, *IEE Electron. Lett.*, vol. 34, June 1998.

[18] V. Iglesias, A. Suárez, and J. L. García, New technique for the determination through commercial software of the stable-operation parameter ranges in nonlinear microwave circuits, *IEEE Microwave Guided Wave Lett.*, vol. 8, Dec. 1998.

[19] S. Mons, J. C. Nallatamby, R. Quéré, P. Savary, and J. Obregón, A unified approach for the linear and nonlinear stability analysis of microwave circuits using commercially available tools, *IEEE Trans. Microwave Theory Tech.*, vol. 47, pp. 2403–2409, Dec. 1999.

[20] Sheng-Guo Wang, Robust Schur stability and eigenvectors of uncertain matrices, *Proceedings of 16th American CONTROL Conference*, pp. 3449–3454, 1997.

[21] J. M. Collantes, I. Lizarraga, A. Anakabe, and J. Jugo, Stability verification of microwave circuits through Floquet multiplier analysis, *IEEE Asia-Pacific Proceedings on Circuits and Systems*, pp. 997–1000, 2004.

[22] V. Rizzoli, A. Neri, and D. Masotti, Local stability analysis of microwave oscillators based on Nyquist's theorem, *IEEE Microwave and Guided Wave Letters*, vol. 7, no. 10, pp. 341–343, Oct. 1997.

[23] L. O. Chua and A. Ushida, Switching-parameter algorithm for finding multiple solutions of nonlinear resistive circuits, *Int. J. Circuit Theory Appl.*, vol. 4, pp. 215–239, 1976.

[24] D. Hente and R. H. Jansen, Frequency domain continuation method for the analysis and stability investigation of nonlinear microwave circuits, *IEE Proc. H: Microwaves Antennas Propag.*, vol. 133, pp. 351–362, 1986.

[25] E. Ngoya, A. Suárez, R. Sommet, and R. Quéré, Steady state analysis of free or forced oscillators by harmonic balance and stability investigation of periodic and quasi-periodic regimes, *Int. J. Microwave Millimeter-Wave Comput Aided Eng.*, vol. 5, pp. 210–223, 1995.

[26] S. Jeon, A. Suárez, and D. B. Rutledge, Analysis and elimination of hysteresis and noisy precursors in power amplifiers, *IEEE Trans. Microwave Theory Tech.*, vol. 54, pp. 1096–1106, 2006.

[27] J. Wu, G. W. H. Hui, and H. Ding, A kind of arc-length method for ordinary differential equations, *Commun. Nonlinear Sci. Numer. Simul.*, vol. 2, pp. 145–150, 1997.

[28] V. Rizzoli and A. Neri, Use of Hopf bifurcations loci for spurious-free nonlinear microwave circuit design, *Proceedings of the IEEE MTT-S International Microwave Symposium*, Orlando, FL, pp. 1089–1092, 1995.

[29] F. Di Paolo, G. Leuzzi, D. Schreurs, and A. Serino, Theoretical investigation and experimental verification of the nonanalytic form of the conversion equations in a frequency divider by two, *Int. J. RF Microwave Comput. Aided Eng.*, vol. 16, pp. 42–58, 2006.

[30] T. Kim and E. H. Abed, Stationary bifurcation control of systems with uncontrollable linearization, *Int. J. Control*, vol. 74, pp. 445–452, 2001.

[31] G. Chen, K. C. Yap, and J. Lu, Feedback control of Hopf bifurcations, *IEEE International Symposium on Circuits and Systems*, pp. 639–642, 1998.

[32] E. H. Abed and J. Fu, Bifurcation control and feedback stabilization, *Proceedings of the 1986 American Control Conference*, pp. 387–391, 1986.

CHAPTER SEVEN

Noise Analysis Using Harmonic Balance

7.1 INTRODUCTION

In this chapter, the main techniques for harmonic balance analysis of phase and amplitude noise in oscillator circuits are presented for both free-running and injection-locked operation. The chapter begins with a brief description of the most commonly used noise models in FET and bipolar transistors and varactor diodes. In the case of a free-running regime, four different techniques will be shown. The first two techniques, which resolve the perturbed harmonic balance system in two different manners, can be seen as the generalization of the frequency-domain analyses presented in Chapter 2 to multiple state variables and harmonic components. Then the classical carrier modulation [1] and conversion matrix approaches [1,2] are described in detail. In the case of injection-locked oscillators, two different techniques are presented: the conversion matrix approach and a two-tier harmonic balance analysis of the locked oscillator perturbed by the noise sources. More details on the foundations of various simulation methodologies covered in this chapter are given in the following.

In free-running operation, the oscillation frequency, acting as a state variable, becomes modulated by noise perturbations. This gives rise to a common phase noise, affecting all the harmonic components of the voltages and currents as

$k\omega_o\theta(t)$, with $\theta(t)$ being the stochastic time deviation, k the harmonic order, and ω_o the oscillation frequency. Besides this "common" phase noise, the free-running oscillator is affected by the amplitude noise associated with the amplitude and phase perturbations of the harmonic components of the state variables $\Delta\overline{X}_k(t)$. The common phase noise is the dominant contribution up to relatively high frequency offset from the carrier. This is due to the absence of a restoring mechanism in the phase variable, so the perturbations accumulate in phase, according to a certain time-varying probability distribution. The stochastic characterization of the oscillator spectrum presented by Kaertner [3] and Demir [4] shows that the spectrum due to phase noise can be calculated from knowledge of the variance of the phase deviation $\sigma_\theta^2(t)$. Here a technique is provided to determine this variance using parameters extracted from harmonic balance simulations. On the other hand, the effect of the amplitude noise can be isolated from that of the phase noise by splitting the perturbed oscillator harmonic balance equation into one scalar equation, depending only on $\theta(t)$ and a subsystem depending only on $\Delta\overline{X}_k(t)$. As shown in Chapter 2, when decoupling the amplitude and phase perturbation, the possible circuit resonances will affect only the amplitude noise.

After the decoupled analysis of phase and amplitude noise, a coupled resolution of the perturbed oscillator harmonic balance system is presented, which will provide a valid prediction of the phase and amplitude noise spectra, except at a very small frequency offset from the carrier. This analysis generalizes the perturbation analysis proposed by Kurokawa [5] to circuits with any number of state variables and multiple harmonic terms. The formulation obtained will be capable of predicting the possible spectrum resonances. Because the amplitude and phase perturbations are not decoupled, these resonances will affect both the phase and amplitude spectra, as shown in Chapter 2. Next, the classical techniques for the harmonic balance simulation of the noisy oscillator spectrum—carrier modulation and conversion matrix—are described in detail and compared with the previous methodologies based on an analysis of $\theta(t)$ and $\Delta\overline{X}_k(t)$.

In injection-locked operation, the input synchronizing source provides a phase reference to the periodic oscillation, so no frequency modulation $\omega_o\dot{\theta}(t)$ associated with the stochastic phase deviation will exist in the perturbed system. The injection-locked oscillator is linear with respect to the noise sources and can be analyzed in similar manner to a frequency mixer with the conversion matrix approach. Although very accurate, this method provides little insight into the conversion mechanism of the noise from the synchronizing source and the circuit noise sources to the output phase noise spectrum. For a better understanding, a phase noise analysis technique based on the two-tier formulation of the noisy oscillator will also be presented, having one single complex equation as the outer tier and the pure harmonic balance system as the inner tier. This formulation will enable an understanding of the general structure of the phase noise spectrum, having sections of different slope versus the offset frequency, and identification of the magnitudes and parameters that determine the frequency corners.

7.2 NOISE IN SEMICONDUCTOR DEVICES

Transistors and diodes contain different types of noise sources. Here the physical origin, location, and spectral density of these noise sources are presented briefly, considering three device types: field-effect transistor, bipolar transistor, and varactor diode. Schematics including the noise sources discussed can be found in books by Odyniec [6] and Rohde et al. [7].

7.2.1 Noise in Field-Effect Transistors

The main noise sources in field-effect transistors are attributed to thermal noise, shot noise, and flicker noise [8,9].

7.2.1.1 Thermal Noise in a Channel The thermal agitation gives rise to fluctuations of the instantaneous velocity of the carriers traversing the channel. This causes perturbations of both the gate current and the drain current, which are correlated, as they have the same origin. The fluctuation of the drain current is modeled with a current generator, placed in parallel between the drain and source intrinsic terminals. The mean-square value of this generator is the following:

$$\langle i_{Td}^2 \rangle = 4kTPg_m\,\Delta f \tag{7.1}$$

where P is a coefficient depending on the bias conditions and the transistor geometry, g_m is the transconductance $\partial I_{ds}/\partial V_{gs}$, k is the Boltzmann constant, and Δf is the noise bandwidth considered. In a periodic regime, the transconductance varies in time at the solution period. Thus, this thermal noise has a periodic modulation in oscillator circuits. However, it is often assumed that the periodic transconductance can be approached with its average value over one oscillation period g_{mo}.

The transistor gate and the channel are coupled capacitively through the gate-to-source capacitance C_{gs}. The noise due to this capacitive coupling can be modeled with an equivalent noise current i_{Tg} placed in parallel between the intrinsic gate and source terminals. Because the capacitance gives rise to frequency-dependent impedance, the mean-square value of this noise current will be frequency dependent [10,11]:

$$\langle i_{Tg}^2 \rangle = 4kT\Delta f\left[1+\left(\frac{f}{f_o}\right)^2\right]L_g(\alpha W_g+\beta I_{dso})\frac{g_m}{C_{gs}}L_g \tag{7.2}$$

where L_g is the gate length, W_g the gate width, I_{dso} the drain saturation current, and α and β are constants with respective values $\alpha = 2\times 10^{-3}$ and $\beta = 1.25\times 10^{-3}$ in the international system. On the other hand, the frequency f_o is given by

$$f_o = \frac{G_d}{2\pi C_{gd}} \tag{7.3}$$

with G_d being the drain conductance and C_{gd} the gate-to-drain capacitance. As in the previous case, g_m is periodic at the oscillation period, but in practice it is usually assumed constant and equal to its average value over one period g_{mo}.

The spectral densities of the two thermal noise sources i_{Tg} and i_{Td} depend on the bias point selected. Because the two sources have the same origin, corresponding to the thermal fluctuations, they are correlated and the correlation coefficient C takes the value

$$C = j\frac{\langle i_{Tg}^* i_{Td}\rangle}{\sqrt{i_{Tg}^2 i_{Td}^2}} \tag{7.4}$$

7.2.1.2 Thermal Noise Associated with Terminal Resistances The terminal resistances of a FET transistor at the gate, drain, and source, R_g, R_d, and R_s, generate thermal noise. When using equivalent current noise sources in parallel with these resistances, the mean-square values of these sources are given by

$$\langle i_{rg}^2\rangle = \frac{4kT\Delta f}{R_g} \qquad \langle i_{rd}^2\rangle = \frac{4kT\Delta f}{R_d} \qquad \langle i_{rs}^2\rangle = \frac{4kT\Delta f}{R_s} \tag{7.5}$$

with Δf being the noise bandwidth.

7.2.1.3 Excess Noise As indicated in Chapter 2, the excess noise is due to defaults in the semiconductor material, which give rise to electrical fluctuations with long time constants and thus with a low-frequency spectrum. The main sources of excess noise in the field-effect transistor are the flicker noise and the generation–recombination noise. The flicker noise has been associated with trap effects on the semiconductor surface, trap effects in the overall semiconductor volume, and fluctuations in carrier mobility. The generation–recombination noise originates in the depleted regions between the gate and the channel and between the channel and the substrate. The total noise spectrum depends on the device and its operating conditions. In the region dominated by flicker noise, the spectral density of the equivalent voltage noise source at the gate terminal is given by

$$S_e(f) = \frac{V_p^2 \alpha_c}{2W_g L_g N_d a}\frac{1}{f} \tag{7.6}$$

where V_p is the pinch-off voltage, α_c a characteristic parameter of the flicker noise in the field-effect transistor depending on the bias point, N_d the carrier density at 300 K, and a the width of the active section between the drain and the source. As can be gathered from an inspection of (7.6), the flicker noise depends on the technological process used for the manufacturing of the transistor. It decreases when increasing the gate width and the gate length, so it is smaller for power devices. For the circuit analysis, the excess noise is usually modeled with a voltage generator connected in series with the gate terminal.

7.2.1.4 Shot Noise The leakage current through the reverse-biased Schottky junction between gate and source gives rise to shot noise, modeled with a parallel current generator, with mean-square value

$$\langle i_{s-g}^2 \rangle = 2qI_g\ \Delta f \tag{7.7}$$

with q the electron charge. The shot noise depends on the gate current. Its spectral density is not very high in field-effect transistors, because these transistors contain no p-n junctions.

7.2.2 Noise in Bipolar Transistors

The bipolar transistors have three main types of noise sources: thermal noise, shot noise, and flicker noise [7,12].

7.2.2.1 Thermal Noise The terminal resistances of the bipolar transistor, at the base, collector, and emitter, R_b, R_c, and R_e, generate thermal noise that can be modeled by means of current noise sources in parallel with these resistances, with the respective mean-square values

$$\langle i_{rb}^2 \rangle = \frac{4kT\Delta f}{R_g} \qquad \langle i_{rc}^2 \rangle = \frac{4kT\Delta f}{R_d} \qquad \langle i_{re}^2 \rangle = \frac{4kT\Delta f}{R_s} \tag{7.8}$$

where Δf is the noise bandwidth.

7.2.2.2 Shot Noise The bipolar transistor is composed of two p-n junctions. For modeling shot noise, two different current noise sources are considered. The current noise source $i_{s,be}$ is connected between the intrinsic base and emitter terminals, whereas the noise source $i_{s,bc}$ is connected between the intrinsic base and collector terminals. Their respective mean-square values are given by

$$\langle i_{s,be}^2 \rangle = 2qi_{be}\Delta f \qquad \langle i_{s,bc}^2 \rangle = 2qi_{bc}\Delta f \tag{7.9}$$

7.2.2.3 Excess Noise The excess noise in bipolar transistors is due primarily to the flicker and burst noise generated by the dc base current. It can be modeled with a singe current noise source, connected in parallel at the base terminal, with the following spectral density:

$$S_{f,b}(\Omega) = K_f \frac{I_{be}^{A_f}}{f^{F_{fe}}} + K_b \frac{I_{be}^{A_b}}{1 + (f/f_b)^2} \tag{7.10}$$

where K_f is the flicker noise coefficient; A_f the flicker noise exponent, with value around 1; F_{fe} the flicker noise frequency exponent, with value around 1; A_b the burst noise exponent; and f_b the burst noise corner frequency. The flicker noise spectral density of silicon bipolars is lower than in FET transistors.

7.2.3 Noise in Varactor Diodes

The two main noise sources in the varactor diode are the thermal noise associated with the loss resistance and the shot and flicker noise due to current flowing through the diode.

7.2.3.1 ***Thermal Noise Associated with Loss Resistance*** The loss resistance R_s due to the limited conductibility of the neutral regions gives rise to thermal noise with the following mean-square value:

$$\langle i_{rs}^2 \rangle = \frac{4kT\Delta f}{R_s} \tag{7.11}$$

7.2.3.2 ***Shot and Flicker Noise*** The current through the diode gives rise to both shot and flicker noise, which can be modeled with a single current source, in parallel with the model of a diode junction.

$$S_{f,b}(f) = 2qI_d + K_f \frac{I_d^{A_f}}{f^{F_{fe}}} \tag{7.12}$$

where I_d is the current through the diode; K_f the flicker noise coefficient; A_f the flicker noise exponent, with value around 1; and F_{fe} the flicker noise frequency exponent, with value around 1.

7.3 DECOUPLED ANALYSIS OF PHASE AND AMPLITUDE PERTURBATIONS IN A HARMONIC BALANCE SYSTEM

In this section, the perturbed oscillator system in the presence of noise sources is derived using harmonic balance. This system has one degree of freedom, due to the addition of the frequency perturbation $\Delta\omega(t)$ to the set of perturbed state variables. Thus, for its practical resolution, it will require an additional constraint. By choosing this constraint suitably, it is possible to obtain two decoupled subsystems in the phase and amplitude perturbations, respectively. Then the phase noise spectrum, isolated from the influence of the amplitude noise, is determined from the variance of the phase deviation $\sigma_\theta^2(t)$ [13]. Next, the decoupled amplitude noise spectrum is derived by applying a projector matrix to the perturbed harmonic balance system, using an additional constraint [14].

7.3.1 Perturbed Oscillator Equations

For the harmonic balance description of the noiseless oscillator, either a nodal or a piecewise formulation may be chosen. However, for the noise analysis presented in Sections 7.3.3 and 7.4, the two formulations provide a different degree of accuracy, at relatively high frequency offset from the carrier. An explanation for this is given later in this section. For the moment, the two types of formulation will only be rewritten and compared:

$$\overline{H}_{nod}(\overline{X}) = \overline{F}(\overline{X}) + [jk\omega]\overline{Q}(\overline{X}) + [H_D(jk\omega)]\overline{X} + \overline{G}_{dc} = 0 \text{ nodal}$$
$$\overline{H}_P(\overline{X}) = [A_x(jk\omega)]\overline{X} + [A_y(jk\omega)]\overline{Y}(\overline{X}) + [A_g(jk\omega)]\overline{G}_{dc} = 0 \text{ piecewise} \tag{7.13}$$

where $\overline{G}_{dc}$ is the vector of bias sources. Each formulation uses a different set of state variables $\overline{X}$ which can be organized in subvectors $\overline{X} = (\overline{X}_{-N} \ldots, \overline{X}_k, \ldots \overline{X}_N)$, each composed of the kth harmonic term of the different state variables (see Chapter 5, Section 5.4). In *nodal formulation*, $\overline{X}$ contains the harmonic components of the node voltages and inductance currents, the vector $\overline{Q}$ contains the harmonic components of the linear and nonlinear charges and fluxes, the vector $\overline{F}$ contains the harmonic components of the sums of resistive currents (that enter each node) and loop voltages, and $[H_D]$ is a matrix containing the transfer functions of the distributed elements. The matrixes $[jk\omega_o]$ and $[H_D(jk\omega_o)]$ are block diagonal, composed of submatrixes evaluated at the various harmonic terms $k\omega_o$, with $k = -N$ to N. In *piecewise formulation*, the vector $\overline{X}$ contains the harmonic components of the control voltages or currents of the various nonlinearities $\overline{Y}(\overline{X})$ and the matrixes $[A_x]$, $[A_y]$, and $[A_g]$ are linear matrixes obtained from analysis of the linear embedding network. Similar to the nodal case, these matrixes are composed of blocks at the harmonic terms $k = -N$ to N.

In brief, the nodal harmonic balance formulation uses a higher number of state variables than the piecewise formulation. Fewer state variables in the piecewise formulation leads to a higher degree of linear matrixes in the frequency $k\omega_o$ (see Chapter 5). In the following derivations, and for the sake of compactness, the harmonic balance system will simply be written $\overline{H}(\overline{X}) = 0$ when corresponding to either a nodal or a piecewise formulation. The dimension of the vector $\overline{X}$ will be $Q(2N + 1)$, assuming Q state variables and $2N + 1$ harmonic terms. Note that generally Q will be much smaller for a piecewise formulation.

Next, the various noise sources contained in the circuit are considered, expressed in vector form as $\overline{g}_{noi}(t)$. This vector will consist of L white noise sources and J colored sources. Because the noise analysis will only be performed up to a certain frequency offset, the noise sources can be considered to be narrowband about the harmonic terms of the steady-state solution. Thus, the vector $\overline{g}_{noi}(t)$ can be expressed in a Fourier series with the oscillation frequency ω_o as fundamental and slowly varying harmonic terms:

$$\overline{g}_{noi}(t) = \sum_{k=-N}^{N} \overline{G}_{noi}(t)e^{jk\omega_o t} \tag{7.14}$$

The time-varying harmonic terms of the noise generators $\overline{G}_{noi}(t)$ give rise to a modulation of the circuit variables and oscillation frequency. Then the perturbed harmonic balance equation is formulated as in the envelope transient method. The circuit state variables undergo a time shift $\theta(t)$ and an increment $\Delta\overline{X}(t+\theta)$. Our frequency-domain analysis assumes slow time variations of the stochastic time deviation $\theta(t)$, limited by the maximum value of the noise frequency $\Omega = 2\pi f$. Then the perturbed circuit variables $\overline{x}'$ can be expressed as

$$\overline{x}'(t+\theta) = \sum_{k=-N}^{N} [\overline{X}_s^k + \Delta\overline{X}^k(t+\theta)]e^{jk\omega_o\theta}e^{jk\omega_o t} \tag{7.15}$$

where $\overline{X}_s$ is an unperturbed steady-state vector. The stochastic time shift will give rise to a common phase perturbation affecting all the harmonic terms of all the state variables. The increment $\Delta\overline{X}(t+\theta)$, called the *amplitude perturbation*, is actually a complex term and thus will affect the phase of the solution but much less than will θ. The term $\Delta\overline{X}$, the amplitude perturbation, is responsible for the slight difference between the phase-noise measurements at different circuit nodes.

For the frequency domain analysis, we consider that the noise sources $\overline{G}_{noi}(t)$ give rise to the perturbation frequencies $jk\omega_o + s$. Then a first-order Taylor series expansion of the linear matrixes in (7.13) is performed, with s acting as a time derivator. Taking exactly the same steps as in the derivation (2.116)–(2.120), the following relationship is obtained:

$$\begin{aligned}\frac{\partial[JH]_s}{\partial(jk\omega_o)}[jk]\overline{X}_s\Delta\omega_o(t) + [JH]_s\Delta\overline{X}(t+\theta) + \frac{\partial[JH]_s}{\partial(jk\omega_o)}\Delta\dot{\overline{X}}(t+\theta)\\ = [D_G(k\omega_o)][e^{-jk\omega_o\theta}]\overline{G}_{noi}(t)\end{aligned} \tag{7.16}$$

where $\Delta\omega_o(t) = \omega_o\dot{\theta}(t)$ and $[JH]_s$ is calculated as shown in the derivation of (5.45) to (5.48) in Chapter 5, and $[D_G]$ is a linear matrix relating the circuit variables to the $2N+1$ harmonic terms of the $L+J$ noise sources. The vector $\overline{G}_{noi}$ is organized as subvectors $\overline{G}_{noi} = (\overline{G}_{noi,-N}, \ldots, \overline{G}_{noi,N})$. The matrix $\partial[JH]_s/\partial(jk\omega_o)$ contains the derivatives of the Jacobian matrix $[JH]$ with respect to $jk\omega_o + s$, evaluated at $s = 0$, with $k = -N$ to N. The derivatives of each $\overline{H}_k$ depend only on $jk\omega_o$. The matrixes $[jk]$ and $[e^{-jk\omega_o\theta}]$ are composed of diagonal boxes of order $Q \times Q$ and $(L+J) \times (L+J)$, respectively, with the integer k varying from $-N$ to N. The rather complex first term on the left side can be simplified, since the following relationship is fulfilled:

$$\left.\frac{\partial\overline{H}}{\partial\omega_o}\right|_s = \frac{\partial[JH]_s}{\partial(jk\omega_o)}[jk]\overline{X}_s \tag{7.17}$$

Equation (7.17) was demonstrated in Chapter 2 [equation (2.136)]. The common phase noise associated with $\theta(t)$ will be determined from the frequency perturbation $\Delta\omega_o(t)$, and the amplitude noise will be determined from the state-variable perturbations $\Delta\overline{X}(t)$.

System (7.16) contains $Q(2N+1)$ real equations in $Q(2N+1)+1$ unknowns, given by $\Delta\overline{X}(t)$ and $\Delta\omega_o(t)$. However, the Jacobian matrix of the vector of harmonic balance error functions $\overline{H}$ with respect to the full set of harmonic components of the state variables $[JH]_s = \left[\partial\overline{H}/\partial\overline{X}\right]_s$ is singular when evaluated at noiseless steady-state oscillation; that is, $\det[JH]_s = 0$. As already known, this singularity is due to the irrelevance of the oscillatory solution with respect to the phase origin. As a result, there is one degree of freedom in the vector $\Delta\overline{X}(t)$, so an additional condition is required to resolve the perturbed oscillator system (see Chapter 2). In this section, the condition used will lead to two decoupled subsystems, one consisting of a single real equation depending on $\Delta\omega_o(t)$ only and the other composed of $Q(2N+1)$ equations, depending exclusively on $\Delta\overline{X}(t)$.

We recall at this point that the number of state variables Q in the perturbed oscillator system (7.16) is smaller for a piecewise formulation than for a nodal formulation. In exchange, the order of the linear matrixes is higher for the piecewise technique. Thus, the Taylor series expansion of the linear matrixes will differ for the two formulations. Note that in the case of lumped elements, these matrixes are simply first order $(jk\omega_o)$ for a nodal formulation in Chapter 5 [see (5.39)], so no Taylor series expansion is actually required when considering the perturbed frequencies $jk\omega_o + s$. Thus, even though the two formulations provide the same steady-state oscillatory solution (except for tolerances and round-off errors), their associated perturbed systems are expected to provide slightly different results. More details on the implications of these system differences are given in Section 7.5.2.

7.3.2 Phase Noise

As already stated, the Jacobian $[JH]_s$ is a singular matrix, due to the oscillator autonomy. Let the vector $\overline{V}_{1x}$ belong on the left-hand-side kernel of the Jacobian matrix $[JH]_s$ such that $\overline{V}_{1x}^T[JH]_s = \overline{0}$. This vector can be chosen so that it fulfills the normalization condition $\overline{V}_{1X}^T[\partial JH/\partial(jk\omega_o)]\overline{U}_1 = 1$, with $\overline{U}_1 = [jk]\omega_o\overline{X}_s$. Note that $\overline{U}_1$ contains the harmonic components of $\dot{\overline{x}}_s(t)$ in the direction tangent to the limit cycle. As already known, the oscillator solution is invariant versus translations along the limit cycle, so the vector $\overline{U}_1 = [jk]\omega_o\overline{X}_s$ also fulfills the interesting property $[JH]_s\overline{U}_1 = 0$ [3,14], as demonstrated in Chapter 2 [3,14].

Due to the singularity of the perturbed system, there is one degree of freedom in $\Delta\overline{X}(t)$. Then it is also possible to impose $\overline{V}_{1x}^T(\partial[JH]_s/\partial(jk\omega_o))\Delta\dot{\overline{X}}(t) = 0$. Because $\overline{V}_{1x}^T$ is constant, the condition imposed restricts the variation of $\Delta\overline{X}(t)$ to the hyperplane orthogonal to the vector $\overline{V}_{1x}^T(\partial[JH]_s/\partial(jk\omega_o)) = 0$. Taking these properties into account, the system (7.16) can be divided into an equation depending only on θ and a subsystem depending only on $\Delta\overline{X}(t)$, which will be considered later. The left multiplication of (7.16) by the vector $\overline{V}_{1x}^T$ provides

$$\omega_o\dot{\theta}(t) = \overline{V}_{1x}^T[D_G(k\omega_o)][e^{-jk\omega_o\theta}]\overline{G}_{noi}(t) \tag{7.18}$$

Clearly, equation (7.18) is nonlinear in the stochastic time deviation θ.

Before continuing, some general aspects of the calculation will be noted:

1. *Independency of* $\Delta\omega(t) = \omega_o\dot{\theta}(t)$ *on the used additional constraint in static conditions* $\Delta\dot{\overline{X}}(t) = 0$. If $\Delta\overline{X}(t)$ is made zero in (7.16), no matter what additional

constraint we use, the row matrix relating $\Delta\omega(t) = \omega_o\dot{\theta}(t)$ to the noise sources $\overline{G}_{noi}(t)$ will agree with $\overline{V}_{1x}^{T}[D_G(k\omega_o)][e^{-jk\omega_o\theta}]$. This property was demonstrated in Section 2.5.4. Thus, in case we are interested only in the phase noise, we can obtain the relationship (7.18) in a much simpler manner: for example, by setting the phase value of one of the first-harmonic components of one of the state variables to zero arbitrarily, as is done in the carrier modulation approach (Section 7.4). This must be seen only as a helpful shortcut to get the relationship (7.18). It cannot provide accurately the amplitude noise, which requires considering the term in $\Delta\dot{\overline{X}}(t)$ in (7.16). When this term is taken into account the relationship between $\Delta\omega(t)$ and $\overline{G}_{noi}(t)$ depends on the additional constraint used [14]. The advantage of the constraint $\overline{V}_{1x}^{T}(\partial[JH]_s/\partial(jk\omega_o))\Delta\dot{\overline{X}}(t) = 0$ comes from the fact that it provides two decoupled subsystems in $\Delta\omega(t)$ and $\Delta\dot{\overline{X}}(t)$ without neglecting the system dependence on $\Delta\dot{\overline{X}}(t)$. The subsystem in $\Delta\overline{X}(t)$ will enable accurate prediction of the noise spectrum due to amplitude perturbations only (Section 7.5.2).

2. Sidebands due to the phase and amplitude modulations. From an inspection of (7.18) it is clear that if instead of the noise sources, a small-signal deterministic periodic source at a frequency Ω is introduced into the circuit, the oscillation carrier will become modulated at the same frequency Ω. Thus, both the phase and amplitude perturbation of the spectrum will contribute to the same sideband frequencies, $k\omega_o \pm \Omega$. The conversion matrix approach (Section 7.6) provides the phase and amplitude noise spectra from an analysis of the sidebands of the output voltage (or current). Because the contributions from the common phase noise and harmonic perturbations $\Delta\overline{X}_k$ are considered jointly in the conversion matrix approach, the resulting phase and amplitude noise spectra will show a slight dependence on the analysis node considered.

7.3.2.1 Phase-Sensitivity Functions At this point, the frequency-domain equation (7.18) in the stochastic time deviation will be compared with the one obtained in the time domain. The original unperturbed time-domain equation is $\dot{\overline{x}} = \overline{f}(\overline{x})$. In the presence of a set of L white noise sources $\overline{\varepsilon}_w(t)$ and J colored sources $\gamma_j(t)$, $j = 1$ to J, the resulting equation in the stochastic time deviation θ is

$$\begin{aligned}\dot{\theta}(t) &= \overline{v}_1^{T}(t+\theta)\frac{\partial \overline{f}}{\partial \overline{\varepsilon}_w}(\overline{x}_{sp}(t+\theta))\overline{\varepsilon}_w(t) + \sum_j \overline{v}_1^{T}(t+\theta)\frac{\partial \overline{f}}{\partial \gamma_j}(\overline{x}_{sp}(t+\theta))\gamma_j(t)\\ &= [b(t+\theta)]\overline{\varepsilon}_w(t) + \sum_j b_{\gamma j}(t+\theta)\gamma_j(t)\end{aligned} \tag{7.19}$$

From equation (7.19) the phase sensitivity to white noise sources is given by the periodic row matrix $[b(t)]$. In turn, the phase sensitivity to each colored source $\gamma_j(t)$ is given by the periodic scalar $b_\gamma(t)$. Next, a Fourier series expansion of $\theta(t)$ in (7.19) will be considered: $\theta(t) = \sum_{k=-N}^{N}\theta_k(t)e^{jk\omega_o t}$. The formulation in (7.18) assumes slowly varying envelopes about the harmonic frequencies $k\omega_o$. Thus, the time deviation considered in (7.18) should actually agree with the low-frequency component $\theta_o(t)$, as was demonstrated rigorously in Chapter 2. For convenience, expression (7.18) can be rewritten as

$$\Delta\omega_o(t) = \overline{V}_{1x}^{\mathrm{T}}[D_G][e^{-jk\omega_o\theta}]\overline{G}_{noi}(t) = [C_M]^{\mathrm{T}}[e^{-jk\omega_o\theta}]\overline{G}_{noi}(t) \tag{7.20}$$

where for notation simplicity the row matrix $[C_M]^{\mathrm{T}}$ has been introduced. As discussed earlier (and demonstrated in Section 2.5.4), when doing $\Delta\dot{\overline{X}} = 0$ in (7.16), the same matrix $[C_M]^{\mathrm{T}}$ is obtained, regardless of the additional constraint selected. Next, the vector $\overline{G}_{noi}(t)$ will be split into a vector containing the white noise sources $\overline{G}_w(t)$ and several vectors of a single nonzero element $\overline{G}_i^{\gamma}(t)$, each associated with one of the J colored noise sources. This will provide

$$\Delta\omega_o(t) = [C_M]_W^{\mathrm{T}}[e^{-jk\omega_o\theta}]_w\overline{G}_w(t) + \sum_{j=1}^{J}[C_M]_{\gamma,j}^{\mathrm{T}}[e^{-jk\omega_o\theta}]_{\gamma}\overline{G}_j^{\gamma}(t) \tag{7.21}$$

where the row matrix $[C_M]_W^{\mathrm{T}}$ relates $\Delta\omega_o(t)$ to the white noise sources and $[C_M]_{\gamma,j}^{\mathrm{T}}$ relates $\Delta\omega_o(t)$ to the colored noise source j. Now, comparing terms affecting the L white noise sources in (7.19) and (7.20) yields

$$\omega_o\overline{B}_W^T = [C_M]_W^T \tag{7.22}$$

where $\overline{B}_W^T = [B_1^N \quad \cdots \quad B_L^N \quad \cdots \quad B_1^{-N} \quad \cdots \quad B_L^{-N}]$. The term B_l^k refers to the harmonic k affecting the white noise source l (with $l = 1$ to L). As explained in the derivation (2.122)–(2.125), each subset $[B_l^N \quad \cdots \quad B_l^{-N}]$, with $l = 1$ to L, must agree with the central row of the Fourier coefficient matrix associated with $b_l(t)$. This central row contains all the harmonic components of the $b_l(t)$. Taking this into account, it is possible to obtain the time domain expressions of all the periodic sensitivity functions $b_l(t)$, $l = 1$ to L, by introducing the harmonic terms identified from (7.22) and (7.21) into the corresponding Fourier expansions:

$$[b_1(t)\cdots \quad b_L(t)] = \left[\sum_{k=-N}^{N} B_1^k e^{jk\omega_o t}\cdots \quad \sum_{k=-N}^{N} B_L^k e^{jk\omega_o t}\right] \tag{7.23}$$

Thus, it is possible to determine the phase sensitivity function to each of the white noise sources from harmonic balance.

On the other hand, for each colored noise source, there is a relationship of the form

$$\omega_o\overline{B}_{\gamma_j}^T = [C_M]_{\gamma,j}^{\mathrm{T}} \tag{7.24}$$

and $\overline{B}_{\gamma_j}^{\mathrm{T}} = [B_{\gamma_j}^N \quad \cdots \quad B_{\gamma_j}^{-N}]$ and $B_{\gamma_j}^k$ refers to the harmonic k affecting the colored noise source j. Once these harmonic components are known, it is possible to write the phase sensitivity function to each colored noise source j as

$$b_j^{\gamma}(t) = \sum_{k=-N}^{N} B_{\gamma_j}^k e^{jk\omega_o t} \qquad j = 1 \text{ to } J \tag{7.25}$$

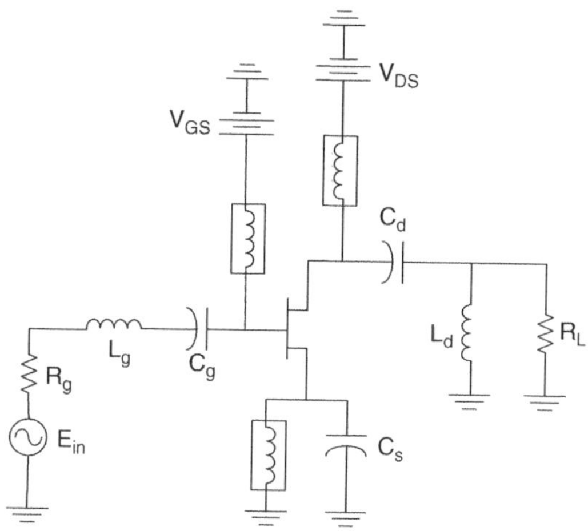

FIGURE 7.1 FET-based injection-locked oscillator using the transistor NE3210S01 at the bias point $V_{GS} = -0.5$ V, $V_{DS} = 2.5$ V.

The technique described above has been applied to the oscillator circuit in Fig. 7.1 under free-running conditions. It is a FET-based oscillator with the steady-state oscillation frequency $f_o = 4.9$ GHz. The gate terminal is loaded with an inductor and a 50-Ω resistance for possible injection. The noise sources considered account for the channel thermal noise, the thermal noise associated with the terminal resistances R_g, R_d, and R_s; the shot noise associated with the reverse-biased Schottky junction between gate and source; and the flicker noise, modeled as a voltage source in series with the intrinsic gate terminal. The transistor is biased with the gate-to-source voltage $V_{GS} = -0.5$ and the drain-to-source voltage $V_{DS} = 2.5$ V. The inductance L_g and the feedback capacitance C_s have been determined so as to obtain negative resistance at the drain terminal. The phase noise sensitivity function with respect to a current noise source introduced in parallel with intrinsic gate terminal is represented in Fig. 7.2. Note that this periodic function depends only on the steady-state solution and the type and location of the noise source. The voltage waveform at the node where the current source is introduced has also been represented. As can be seen, the sensitivity function takes larger absolute values in the time intervals with faster variation of the steady-state solution (a larger time derivative).

In Fig. 7.3, the phase noise sensitivity of the oscillator circuit has been evaluated with respect to a voltage noise source connected in series with the intrinsic gate terminal. The intrinsic current waveform is also superimposed, for comparison. As can be seen, the largest absolute values of the phase noise sensitivity function are obtained in the sections with a faster variation of the steady-state solution.

7.3.2.2 Phase Noise Spectrum As shown in Chapter 2, the phase noise spectrum can be obtained from the variance of the stochastic time deviation $\sigma_\theta^2(t)$. The calculation of this variance requires the determination of one coefficient c,

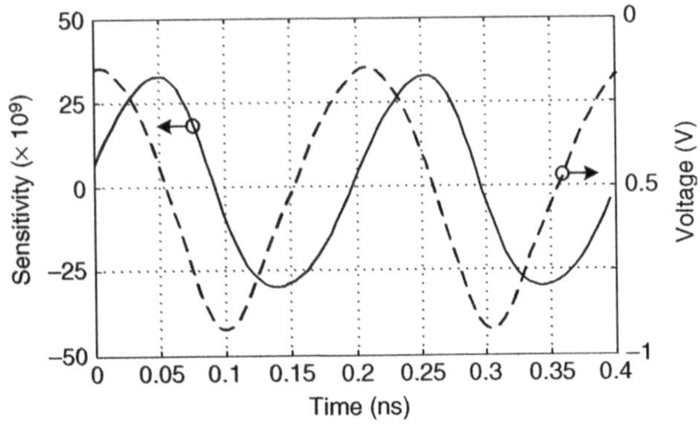

FIGURE 7.2 Time variation of the phase noise sensitivity of the oscillator in Fig. 7.1 with respect to a current noise source introduced in parallel with the gate terminal. The voltage waveform at this terminal is shown by the dashed line.

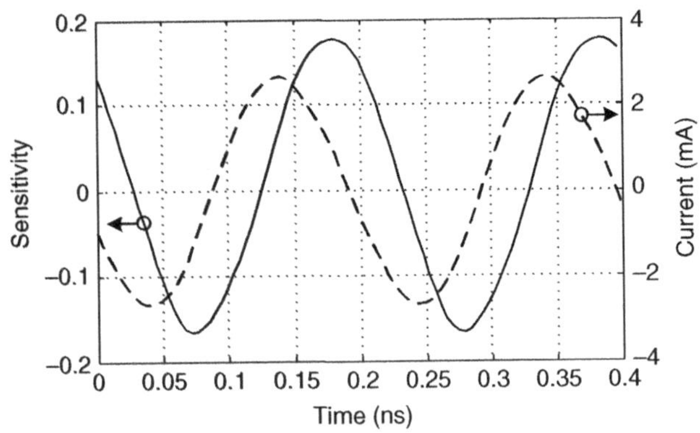

FIGURE 7.3 Time variation of the phase noise sensitivity of the oscillator in Fig. 7.1 with respect to a voltage noise source connected in series with the gate terminal. The current waveform through this same terminal is shown by the dashed line.

accounting globally for all the white noise sources, and J coefficients $b_{\gamma j,\mathrm{dc}}$, with $j = 1$ to J, one for each colored noise source. These coefficients are given by

$$
\begin{aligned}
c &= \frac{1}{T}\int_0^T \bar{b}^T(t)[\Gamma]\bar{b}(t)\,dt \\
b_{\gamma j,\mathrm{dc}} &= \frac{1}{T}\int_0^T b_{\gamma j}(t)\,dt \quad \text{with } j = 1 \text{ to } J
\end{aligned}
\tag{7.26}
$$

with $[\Gamma]$ the autocorrelation matrix of the white noise sources, calculated from $\langle \varepsilon_i(t)\varepsilon_j(t+\tau)\rangle = \Gamma_{ij}\delta(\tau)$. The coefficients c and $b_{\gamma j,\mathrm{dc}}$, with $j = 1$ to J, can be determined from (7.21). To obtain c, only the white noise sources are considered, using the vector $\overline{G}_W(t)$. Taking into account the orthogonality of the components of the Fourier series, the coefficient c is given by

$$c = \frac{1}{\omega_0^2}[C_M]_w^T[G_{n,w}][C_M]_w^*\overline{V}_1^* \tag{7.27}$$

where $[G_{n,w}]$ is the correlation matrix between the various harmonic terms of the white noise sources. On the other hand, each $b_{\gamma j,\mathrm{dc}}$ is calculated from the dc component $C_{\gamma,j}^0$, extracted directly from inspection of the row matrix $[C_M]_{\gamma,j}^T$ as

$$b_{\gamma j,\mathrm{dc}} = \frac{1}{\omega_0}C_{\gamma,j}^0 = \frac{1}{\omega_0}[\mathrm{C_M}(N+1)]_{\gamma,\mathrm{j}}^T \tag{7.28}$$

Once the parameters c and $b_{\gamma j,\mathrm{dc}}$, with $j = 1$ to J, have been determined, the phase noise spectrum is obtained from the total variance of the phase deviation [4], which has contributions from the L white noise sources and the set of J colored noise sources. The total variance is calculated as the addition of the variances due to the different uncorrelated processes [4]

$$\sigma_\theta^2(t) = \sigma_w^2(t) + \sum_{j=1}^{J}\sigma_{\gamma,j}^2(t) = ct + \sum_{j=1}^{J}\sigma_{\gamma,j}^2(t) \tag{7.29}$$

As shown in Chapter 2, the spectrum due to the common phase noise is independent of the observation node. It is calculated from the Fourier transform of the following autocorrelation function:

$$R(\tau) = E[e^{j\omega_0\theta(t)}e^{j\omega_0\theta(t-\tau)}] \tag{7.30}$$

where $e^{j\omega_0\theta(t)}$ is the characteristic function associated with the time deviation $\theta(t)$ and τ is the correlation time. Autocorrelation function (7.30) can be related analytically to the variance of the phase deviation (see Section 2.4.2), which leads to the expression

$$S(f) = F[R(\tau)] = F\left\{\exp\left[-\tfrac{1}{2}\omega_0^2\sigma_\theta^2(|\tau|)\right]\right\} \tag{7.31}$$

As shown in Chapter 2, the variance of the phase deviation due to a colored noise source γ_j is related to the spectral density $S_{\gamma j}(f)$ of this noise source through the expression

$$\sigma_{\gamma,j}^2(t) = 2|b_{\gamma j,\mathrm{dc}}|^2\int_{-\infty}^{\infty}S_{\gamma j}(f)\frac{1-\exp(j2\pi ft)}{4\pi^2 f^2}\,df \tag{7.32}$$

Thus, one expression of the form (7.32) should be considered for each of the colored noise sources contained in the circuit. These expressions must be combined with (7.29) and (7.31) to obtain the phase noise spectrum. The advantage of this calculation is the accuracy in the prediction of the spectral density at very small frequency offset from the carrier. In flicker noise sources, the spectral density $S_N(f)$ is approached by

$$S_N(f) = 4\alpha\int_{f_{\min}}^{\infty}\frac{1}{\lambda^2+(2\pi f)^2}\,d\lambda = \frac{\alpha}{|f|} - 4\frac{\alpha\arctan(f_{\min}/2\pi f)}{2\pi f} \tag{7.33}$$

Introduction of the cutoff frequency $f_{\min}$ avoids the singularity at $f = 0$. Provided that it is small, the particular value of this cutoff frequency has little relevance on the noise spectral density at a relatively high frequency offset from the carrier. However, at a small frequency offset, the phase noise spectral density will obviously depend on $f_{\min}$. Thus, the cutoff frequency in (7.33) will have to be selected carefully according to the measurement time of the oscillator circuit.

As can be gathered, analytical determination of the Fourier transform (7.30) in the case of white and colored noise sources is virtually impossible. Instead, a numerical calculation can be carried out, as has been done by Sancho et al. [15]. The accuracy of the phase noise spectrum depends on the estimation of $f_{\min}$ and the numerical technique used to determine the Fourier transform in (7.31). Successive application of (7.33), (7.32), (7.29), and (7.31) enables the most realistic determination of the phase noise spectrum in oscillator circuits, but, as already stated, requires a good estimation of $f_{\min}$. When we do not need such a detailed calculation, the approximate expressions provided by Demir [4] can be used. They have been obtained using approximate limiting forms of the Fourier transform when considering white and colored noise sources. They provide a phase noise spectrum about the different harmonic components if_o given by

$$S_{pi}(f) = \frac{S_{si}(f)}{2|X_i|^2} = \begin{cases} \dfrac{i^2 f_o^2\left(c+\sum_{j=1}^{J} |b_{\gamma j,\mathrm{dc}}|^2 S_{\gamma j}(0)\right)}{\pi^2 i^4 f_o^4\left(c+\sum_{j=1}^{J} |b_{\gamma j,\mathrm{dc}}|^2 S_{\gamma j}(0)\right)^2+f^2} & f \approx 0 \text{ (a)} \\ \left(c + \sum\limits_{j=1}^{J} |b_{\gamma j,\mathrm{dc}}|^2 S_{\gamma j}(f)\right) \dfrac{i^2 f_o^2}{f^2} & f \gg 0 \text{ (b)} \end{cases} \tag{7.34}$$

with $f = \Omega/(2\pi)$. The expression for low offset frequency depends on the dc component of the phase sensitivity function to the colored noise source $b_{\gamma j,\mathrm{dc}}$ and the constant values $S_{\gamma j}(0)$. These values correspond to the spectral density of the colored noise sources in the limit $f \to 0$. If the colored noise sources considered have a spectrum $S_\gamma(f) \propto 1/f$ for a larger frequency offset, the spectral density of the noisy oscillator would drop as -30 dB/dec. Then it will drop -20 dB/dec from the corner frequency defined by $c = \sum_{j=1}^{J} |b_{\gamma j,\mathrm{dc}}|^2 S_{\gamma j}(f_c)$.

The calculation above has been used to determine the phase noise spectrum due to white noise sources of the oscillator in Fig. 7.1. The white noise sources considered are located inside the transistor and in the resistances of the linear embedding network. The white noise contributions from the FET transistor are the channel thermal noise, the thermal noise generated by the terminal resistances R_g, R_d, and R_s, and the shot noise of the gate-to-source junction. The phase noise spectrum obtained is represented in Fig. 7.4. The cutoff frequency of the

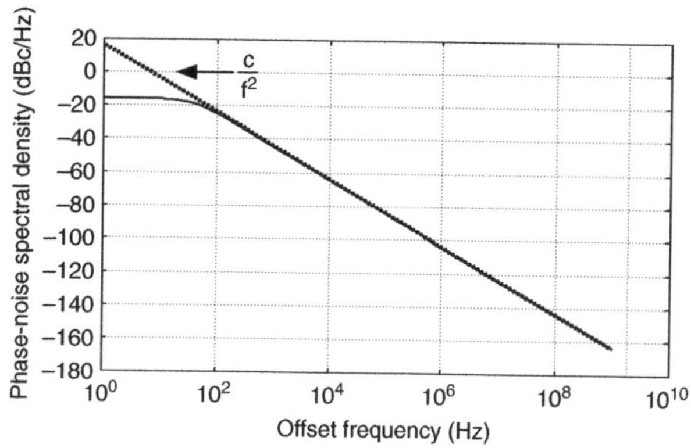

FIGURE 7.4 Phase noise spectrum of the oscillator in Fig. 7.1 when only white noise sources are considered. It is a Lorentzian line with cutoff frequency $c/2\pi$. The asymptote with constant slope −20 dB/dec is also shown.

Lorenzian line is $f_{3\text{dB}} = 100$ Hz, a rather low value. From about $f_{3\text{dB}}$, the phase noise spectral density decreases as 20 dB/dec. The asymptote with this constant slope is also shown in Fig. 7.4.

Next, the phase noise spectrum due to flicker and white noise of the oscillator in Fig. 7.1 has been determined using the approximations (7.34). The calculation has been carried out for the same set of white noise sources considered in the previous analysis, plus a flicker noise source of voltage type in series with the intrinsic gate terminal. This voltage source has the spectral density $S_F(f) = 1.5 \times 10^{-11}/f$. The spectrum resulting from the application of (7.34b) is shown in Fig. 7.5. The spectrum of the flicker noise source introduced is also shown.

7.3.3 Amplitude Noise

The determination of the amplitude noise requires deriving a subsystem in the amplitude perturbation $\Delta\overline{X}(t)$ from the global perturbed oscillator equation (7.16). This subsystem in $\Delta\overline{X}(t)$ is obtained by multiplying (7.16) by the projection matrix $[P] = [I_d] - [\partial JH(k\omega_o)/\partial(jk\omega_o)]\overline{U}_1\overline{V}_{1X}^T$, with $\overline{U}_1 = [jk]\omega_o\overline{X}_s$. Taking into account the normalization condition $\overline{V}_X^T[\partial JH(k\omega_o)/\partial(jk\omega_o)]\overline{U}_1 = 1$, this provides the following differential equation in the uncoupled vector $\Delta\overline{X}(t)$:

$$[JH]_s\Delta\overline{X}(t+\theta) + \frac{\partial[JH]_s}{\partial(jk\omega_o)}\Delta\dot{\overline{X}}(t+\theta) = [P][D_G(k\omega_o)][e^{-jk\omega_o\theta}]\overline{G}_{noi}(t) \quad (7.35)$$

As shown in Chapter 2, the stochastic time deviation θ is more relevant at small frequency offset from the carrier, where the phase noise totally dominates the amplitude noise. Thus, the time shift θ will be neglected in the calculation of $\Delta\overline{X}(t)$. Applying the Fourier transform to the resulting expression, it will be possible

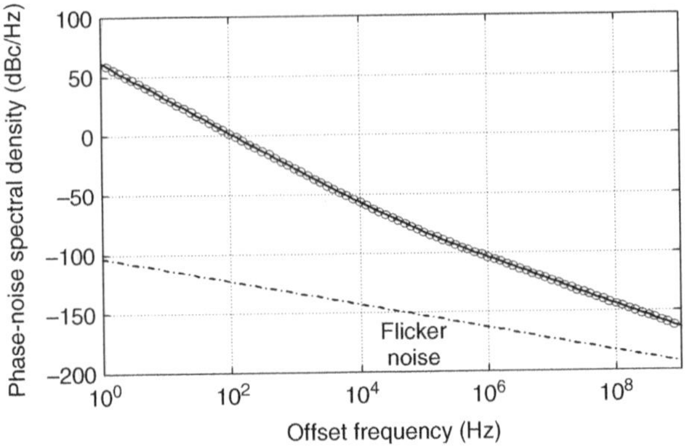

FIGURE 7.5 Phase noise spectrum of the oscillator in Fig. 7.1 when white noise sources plus a flicker noise source are considered. The flicker noise source is a voltage source connected in series with the intrinsic gate terminal. The spectrum of the flicker noise source considered is also shown.

to write

$$\left[[JH]_s + \frac{\partial [JH]_s}{\partial (jk\omega_o)} j\Omega \right] \Delta \overline{X}(\Omega) = [P][D_G(k\omega_o)]\overline{G}_{\text{noi}}(\Omega) \tag{7.36}$$

Calculation of the amplitude increment $\Delta \overline{X}(\Omega)$ requires inversion of the matrix

$$[JT(j\Omega)] = \left[[JH]_s + j\Omega \frac{\partial [JH]_s}{\partial (jk\omega_o)} \right] \tag{7.37}$$

Matrix (7.37) is the first-order Taylor series expansion of the characteristic matrix $[JT(jk\omega_o + s)]$, with s particularized to $s = j\Omega$. At this point it is interesting to compare the two formulations nodal and piecewise, in (7.16). The characteristic determinant $\det[JT(jk\omega_o + s)] = 0$ will have the same degree in both types of formulation, as the smaller number of state variables in the piecewise formulation leads to a higher degree of the linear matrixes. However, the perturbation analysis performed in (7.16) is based on a Taylor series expansion of the linear matrixes. The degree in $j\Omega$ of the expanded determinant in (7.37) will agree with $Q(2N + 1)$, with Q being the number of steady-state variables considered. Thus, the number of poles associated with the perturbed piecewise system will be smaller than the number of poles associated with the nodal system. As shown in Chapter 2, accuracy in the analysis of the amplitude noise is highly dependent on the number of poles of the associated linearized system, which determines the number of detectable resonances [14]. In the following, the amplitude spectrum will be derived in a general manner, using compact expressions valid for nodal and piecewise harmonic balance.

The matrix in (7.37) becomes singular at $\Omega = 0$, as can easily be gathered from the fact that it agrees with $[JH]_s$ at zero offset frequency. However, the right-hand multiplication of $[JT(j\Omega)]^{-1}$ by the projector matrix $[P]$ removes the singularity in an analytical manner, which is very convenient for reasons of numerical accuracy.

Clearly, the roots of the determinant $\det[JT(\lambda)]$ are given by the opposite of the $Q \times (2N+1)$ eigenvalues λ_{qk}, with $q = 1$ to Q and $k = -N$ to N, of the following matrix:

$$[M] = \left[\frac{\partial[JH]_s}{\partial(jk\omega_o)}\right]^{-1} [JH]_s \tag{7.38}$$

Due to the periodicity of poles of the characteristic matrix associated with a periodic steady-state solution, the eigenvalues of (7.38) will have the form $\lambda_{qk} = \lambda_{q0} + k\omega_o$, with λ_{q0} being the canonical poles with imaginary part (frequency) restricted to the interval $[(0,\omega_o)]$. Note that the number of canonical eigenvalues is determined by the number of state variables in the formulation used and is given by Q. In a piecewise harmonic balance formulation, this number is limited to the number of control voltages in the nonlinear elements. In a nodal formulation, it agrees with the number of node voltages and inductance currents.

The $(2N+1)Q$ eigenvalues of (7.38) can be determined in a simple manner. Because $\overline{V}_{1x}^T$ belongs to the left-side kernel of $[JH]_s$, it is associated to the canonical pole $\lambda_{10} = 0$. Using linear algebra relationships [14], the inverse matrix $[JT(j\Omega)]^{-1}$ can be expressed as

$$[JT(j\Omega)]^{-1} = \sum_{q=1}^{Q} \sum_{k=-N}^{N} \frac{M_{qk}}{j\Omega - \lambda_{qk}} \tag{7.39}$$

where the matrixes $[M_{qk}]$ are calculated from the left- and right-hand-side kernels of the constant matrixes obtained when replacing the eigenvalues λ_{qk} in $[JT]$, doing $[JT(\lambda_{qk})]$. The procedure is the following:

$$M_{qk} = \overline{U}_{qk}\overline{V}_{qk}^T \tag{7.40}$$

with

$$\begin{aligned} &[JT(\lambda_{qk})]\overline{U}_{qk} = \overline{0} \\ &\overline{V}_{qk}^T[JT(\lambda_{qk})] = \overline{0} \\ &\overline{V}_i^T \frac{\partial JH}{\partial(jk\omega_o)}\overline{U}_j = \delta_{ij} \quad \text{Normalization condition} \end{aligned} \tag{7.41}$$

The Kronecker delta will be zero if the two indexes qk are not identical. Clearly, the vector $\overline{V}_{1x}^T = \overline{V}_{10}^T$, fulfilling $\overline{V}_{1x}^T[JH]_s = 0$, is the left eigenvector associated with the eigenvalue $\lambda_{10} = 0$. On the other hand, the vector $\overline{U}_{10} = [jk]\omega_o\overline{X}_s$, containing the harmonic components of the time derivative $\dot{\overline{x}}_s$, is the right-hand-side eigenvector of the matrix $[JT(\lambda_{10})]$, with $\lambda_{10} = 0$. Therefore, the right-hand multiplication of $[JT(\Omega)]^{-1}$ by the projector matrix $[P]$ will eliminate the term of (7.39) associated with $\lambda_{10} = 0$. To see this more clearly, the term of summation (7.39) associated with $\lambda_{10} = 0$ is considered. This term is given by

$$[JT(j\Omega)]_{\lambda_{10}}^{-1} = \frac{M_{10}}{j\Omega} = \frac{\overline{U}_{10}\overline{V}_{1x}^T}{j\Omega} \tag{7.42}$$

The multiplication by the projector $[P]$ provides

$$[JT(j\Omega)]^{-1}_{\lambda_{10}}[P] = \frac{\overline{U}_{10}\overline{V}^T_{10}}{j\Omega}\left[[I_d] - \frac{\partial [JH]_s}{\partial jk\omega_o}\overline{U}_{10}\overline{V}^T_{1x}\right]$$
$$= \frac{1}{j\Omega}\left[\overline{U}_{10}\overline{V}^T_{1x} - \overline{U}_{10}\underbrace{\overline{V}^T_{1x}\frac{\partial [JH]_s}{\partial jk\omega_o}\overline{U}_{10}}_{1}\overline{V}^T_{1x}\right] = [0] \tag{7.43}$$

Thus, the singularity is removed from the system. The amplitude noise spectrum will depend on

$$[JT(\Omega)]^{-1}[P] = \sum_{q,k/1,0} \frac{M_{qk}}{j\Omega - \lambda_{qk}} \tag{7.44}$$

where $q, k/1, 0$ indicates that the summation is performed for all the combinations of q and k values except $q = 1$ and $k = 0$. Note that by performing the calculation (7.44), the matrix $[JT(\Omega)]$ is not actually inverted, which avoids any accuracy problems due to the ill conditioning at small offset frequency Ω from the carrier. Associated with the pole $\lambda_{10} = 0$ there are other imaginary poles $\lambda_{1k} = jk\omega_o$ which are not removed by the product $[JT(\Omega)]^{-1}[P]$. These poles will give rise to singularity at the offset frequencies $jk\omega_o$, with $k \neq 0$, which are never considered in the noise analysis. Next, the expression for $[JT(\Omega)]^{-1}[P]$ will be replaced into the general equation for the amplitude noise. This provides

$$[S_{\Delta\overline{X}}(\Omega)] \equiv \Delta\overline{X}(\Omega)\Delta\overline{X}^+(\Omega)$$
$$= \sum_{q,k/1,0} \frac{[M_{qk}]}{j\Omega - \lambda_{qk}}[D_G(k\omega_o)][S_{noi}(\Omega)][D_G(k\omega_o)]^+ \sum_{q,k/1,0} \frac{[M_{qk}]^+}{-j\Omega - \lambda^*_{qk}} \tag{7.45}$$

with $[S_{noi}(\Omega)]$ being the correlation matrix of the noise sources, given by

$$[S_{noi}(\Omega)] \equiv \overline{G}_{noi}(\Omega)\overline{G}^+_{noi}(\Omega) \tag{7.46}$$

The above square matrix is composed of submatrixes containing the correlation and intercorrelation terms between the different harmonic components of the noise sources. Assuming a stable oscillation, the poles with the greatest influence on the noise spectrum will be those with a smaller absolute value of the real part, also called *dominant poles*. If these poles are relatively far from the imaginary axis, the shape of the amplitude spectrum will be determined by the spectrum of the input noise sources. At low offset frequency Ω from the carrier, this spectrum will have a $1/\Omega$ characteristic. At larger offset frequency Ω, the spectrum will be flat. The case of dominant poles with a small absolute value of the negative real part $|\sigma|$ is treated in the following.

In case the solution at $k\omega_o$ contains a pair of (canonical) complex-conjugate poles $\sigma_d \pm j\omega_d$ at small distance from the imaginary axis, negative resistance will surely exist in the circuit at the pole frequency, susceptible to give rise to noise amplification. The circuit will behave in a similar manner to a negative-resistance amplifier at the frequencies $k\omega_o \pm \omega_d$. Note that the poles of a periodic solution are actually periodic, so the complete set of poles is given by $\sigma_d \pm j\omega_d \pm jk\omega_o$, with k an integer. Unlike negative-resistance amplifiers, exhibiting a single amplification band, an oscillator with a small stability margin will exhibit multiple amplification bands centered about the frequencies of its multiple dominant poles. The only circuit inputs in these frequency bands are the noise sources, which give rise to noise bumps in the oscillator spectrum. The central frequencies of the bumps will be $\omega_d, \omega_o - \omega_d, \omega_o + \omega_d, 2\omega_o \pm \omega_d$, and so on. When calculating the amplitude spectrum from (7.45), about each spectral line $k\omega_o$, with $k \neq 0$, it will exhibit maxima at the offset frequencies $\Omega = \omega_d$ and $\Omega = -\omega_d$. Considering the contribution of the dominant poles only and neglecting the cross-correlation between various terms of $\Delta\overline{X}(\Omega)$, the amplitude spectrum about the harmonic frequency $k\omega_o$ can be approached [3]:

$$[S_{\Delta\overline{X},k\omega_o}(\Omega)] \equiv \frac{C(\Omega)_k^{lsb}}{\sigma_d^2 + (\Omega + \omega_d)^2} + \frac{C(\Omega)_k^{usb}}{\sigma_d^2 + (\Omega - \omega_d)^2} \tag{7.47}$$

where the numerator terms $C(\Omega)_k^{lsb}$ and $C(\Omega)_k^{usb}$, corresponding to the lower and upper sidebands, respectively, are obtained directly from expression (7.45). Note that the simplified expression (7.47) will not be valid if there is more than one pair of poles close to the imaginary axis, as their joint contribution should be analyzed. In (7.47) a double sideband is considered about each harmonic frequency $k\omega_o$, in terms of the positive and negative offset frequency Ω. Assuming white noise inputs, the expression (7.47) predicts a pair of Lorentzian lines, centered about the resonance frequencies $k\omega_o \pm \omega_d$. Note that the amplitude spectrum will not generally be symmetric about $k\omega_0$, as the numerator coefficients $C(\Omega)_k^{lsb}$ and $C(\Omega)_k^{usb}$ corresponding to the upper and lower sidebands can be different. The spectrum is not necessarily symmetric about the central frequencies $k\omega_o \pm \omega_d$ either.

From an inspection of (7.47), the height of the two Lorentzian lines about $k\omega_o \pm \omega_d$ will increase for a lower absolute value of σ_d, which means a shorter distance from the critical poles to the imaginary axis. This distance will vary when modifying a circuit parameter η, such as a bias voltage. Higher power is also obtained, as the frequency approaches the critical values $k\omega_o \pm \omega_d$, that is, if the offset frequency Ω approaches $\pm\omega_d$ about each harmonic frequency. If when varying η, the poles $\sigma_d \pm j\omega_d \pm jk\omega_o$ happen to cross the axis to the right-hand side of the complex plane, a Hopf bifurcation will be obtained and the bumps will turn into distinct spectral lines. This is why the noise bumps have been called *noisy precursors*. It is also possible to have a pair of dominant poles at one-half the oscillation frequency, given by $\sigma_d \pm j\omega_o/2 + jk\omega_o$. In this case, the upper and lower sidebands of (7.47) will merge into a single sideband and the noisy precursors will be centered about the resonance frequencies $jk\omega_o + \omega_o/2$. If the poles cross the imaginary axis, a flip bifurcation will be obtained and the bumps will turn into distinct subharmonic spectral lines. Note that the oscillator analysis presented, which is linear in terms of the perturbation $\Delta\overline{X}(\Omega)$, becomes invalid in the immediate neighborhood of the bifurcation.

The influence of the dominant poles depends on the coefficients $C(\Omega)_k^{lsb}$ and $C(\Omega)_k^{usb}$, which can be related to the conversion gain from the noise sources to the oscillator output about the large-signal regime at $k\omega_o$. The relevance of the precursors in the output spectrum will also depend on the form of variation of the dominant poles versus the parameter η. If under any slight variation of this parameter, the poles evolve quickly, the precursors will hardly be observable, as, in practice, it will be unlikely to have the particular circuit element values at which the poles are close to the imaginary axis. It the poles keep close to the axis when the parameter is varied, the precursors will be observable. As a general conclusion, the operation under a small stability margin (small $|\sigma|$) should be avoided, not only for the risk of getting into experimental instability under slight parameter variation, but also for the possible observation of an anomalous noise spectrum under these conditions.

The technique described for analysis of the noise amplitude spectrum in oscillator circuits has been applied to a push-push oscillator with output frequency $f_{\text{out}} = 18$ GHz. More details on the operation of this oscillator configuration are given in Chapter 10. The white noise sources considered in each bipolar transistor are three voltage sources, accounting for thermal noise associated with the terminal resistances R_b, R_c, and R_e and two parallel current sources accounting for the shot noise associated with each of the two p-n junctions. The excess noise, with contributions from the flicker and burst noise generated by the dc base current, is modeled with a single current noise source with the spectral density in (7.10) connected to the base terminal. The steady-state oscillation has a pair of complex-conjugate poles a short distance from the imaginary axis, with the values $2\pi \times 0.01 \times 10^7 \pm j2\pi \times 2 \times 10^6$. The perturbed oscillator equation has been separated into two equations, one scalar equation depending only on $\Delta\theta$ and one vector equation depending only on $\Delta\overline{X}$. Thus, the complex-conjugate poles $\sigma_d \pm j\omega_d$ affect only the amplitude noise, with no influence over the phase noise. The phase and amplitude noise spectra can be seen in Fig. 7.6. The phase noise asymptote is given by $S_\phi(\Omega) = |b_{\gamma,\text{dc}}|^2 S_\gamma(\Omega)/\Omega^2$. For a different resolution of the perturbed oscillator system, the dominant poles will also affect the phase noise, as shown in the following sections.

7.4 COUPLED PHASE AND AMPLITUDE NOISE CALCULATION

As we already know, the perturbed oscillator system has one degree of freedom. It contains $Q(2N+1)$ equations in $Q(2N+1)+1$ unknowns, so it requires an additional constraint for its practical resolution in terms of $\Delta\omega_o$ and $\Delta\overline{X}$. The harmonic increments $\Delta\overline{X}$ are linearly related, due to the singularity of the Jacobian matrix $[JH]_s$. Thus, any arbitrary condition $\overline{P}^T\Delta\overline{X} = 0$, with $\overline{P}$ an arbitrary

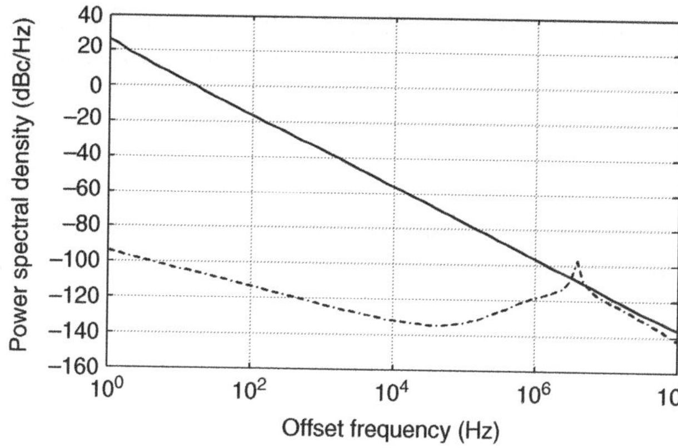

FIGURE 7.6 Phase and amplitude noise of the oscillator circuit in Fig. 7.5. As can be seen, the amplitude spectrum exhibits a local maximum at the frequency of the dominant poles, $f_d = 5$ MHz. The phase noise spectrum is unaltered by these complex-conjugate poles.

constant vector, can be used. One possibility is to set to zero the imaginary part of one of the harmonics components of $\Delta\overline{X}$: for example, $\text{Im}[\Delta X_1^1] = 0$. Note that this condition is analogous to the one considered in mixed-mode harmonic balance, in which the singularity of the free-running oscillator equations is removed by arbitrarily imposing $\text{Im}[X_1^1] = 0$.

By setting $\text{Im}[\Delta X_1^1] = 0$, the number of components of the state-variable vector $\Delta\overline{X}$ is reduced by one, so it will have the dimension $(2N+1)Q-1$. This reduced-order vector will be called $\Delta\overline{X}'$. Using the new constraint in the general system (7.16), the noisy oscillator is formulated as

$$\left.\frac{\partial\overline{H}}{\partial\omega}\right|_s \Delta\omega_o(t) + [JH]_s'\Delta\overline{X}'(t) + \frac{\partial[JH]_s'}{\partial(jk\omega_o)}\Delta\dot{\overline{X}}'(t) = [D_G(k\omega_o)][e^{-jk\omega_o\theta}]\overline{G}_{noi}(t) \tag{7.48}$$

where $[JH]_s'$ is the nonsquare Jacobian matrix of $\overline{H}$ with respect to $\overline{X}'$, with dimensions $(2N+1)Q$ and $(2N+1)Q-1$, and the vector $\Delta\overline{X}'(t)$ contains the perturbations in all the state variables except the suppressed variable.

System (7.48) is nonlinear in the stochastic time deviation θ, due to the relationship $\Delta\omega_o(t) = \omega_o\dot{\theta}(t)$ and the presence of exponential term $[e^{-jk\omega_o\theta}]$ on the right-hand side. As shown in Chapter 2, this nonlinearity will be relevant at a small frequency offset from the carrier. In the following derivation, the influence of the term $[e^{-jk\omega_o\theta}]$ will be neglected, so the resulting phase spectrum will be inaccurate at very small offset frequency from the carrier. Applying the Fourier transform to the slowly time-varying variables in system (7.48), the following perturbation equation is obtained in terms of the offset frequency Ω:

$$\begin{bmatrix} \left.\dfrac{\partial\overline{H}}{\partial\omega}\right|_s & [JH]_s' + \dfrac{\partial[JH]_s'}{\partial(k\omega_o)}j\Omega \end{bmatrix}\begin{bmatrix} \Delta\omega_o(\Omega) \\ \Delta\overline{X}'(\Omega) \end{bmatrix} = [D_G(k\omega_o)]\overline{G}_{noi}(\Omega) \tag{7.49}$$

where a global Jacobian matrix, affecting both $\Delta\overline{X}'(\Omega)$ and $\Delta\omega_o(\Omega)$, has been used. Note that the reduced-dimension Jacobian matrix $[JH]_s$ can be expressed

$$[JH]'_s = [JH]_s[Id_{V\times V-1}] \tag{7.50}$$

where $[Id_{V\times V-1}]$ is a nonsquare version of the identity matrix, containing $V = (2N+1)Q$ rows and $V-1$ columns. The missing column corresponds to the suppressed variable $\mathrm{Im}[X_1^1] = 0$. The submatrix affecting $\Delta\overline{X}'(\Omega)$ has the same form as the already defined matrix $[JT(j\Omega)]$. Actually, it is possible to express

$$[JT(\Omega)]' = \left[[JH]'_s + j\Omega\frac{\partial[JH]'_s}{\partial(k\omega_o)}\right] = [JT(\Omega)][Id_{V\times V-1}] \tag{7.51}$$

Using this definition, (7.49) becomes

$$\left[\left.\frac{\partial\overline{H}}{\partial\omega}\right|_s \quad [JT(\Omega)]'\right]\begin{bmatrix}\Delta\omega_o(\Omega)\\ \Delta\overline{X}'(\Omega)\end{bmatrix} = [D_G(k\omega_o)]\overline{G}_{noi}(\Omega) \tag{7.52}$$

Note that the matrix on the left-hand side of (7.52) is not singular, due to the fact that one of the state variables, $\mathrm{Im}[X_1^1]$, has been not been considered in calculation of the Jacobian matrix $[JH]'_s$. Solving for $\Delta\omega_o(\Omega)$ and multiplying by the adjoint $\Delta\omega_o^*(\Omega)$, the following compact expression is obtained for the phase noise spectrum:

$$S_{\Delta\phi}(\Omega) = \frac{[P_N(\Omega)][D_G(k\omega_o)][S_{noi}(\Omega)][D_G(k\omega_o)]^+[P_N(\Omega)]^+}{\Omega^2} \tag{7.53}$$

with $[P_N(\Omega)]$ being the following row matrix:

$$[P_N(\Omega)] = [1 \quad 0 \quad \cdots \quad 0]\left[\left.\frac{\partial\overline{H}}{\partial\omega}\right|_s \quad [JT(\Omega)]'\right]^{-1} \tag{7.54}$$

Note that the phase noise calculation above, unlike expression (7.34), includes the effect of the amplitude perturbations. It can be seen as a multiharmonic generalization of expression (2.112). For the considered phase noise definition, the $(2N+1)Q-1$ poles of $[JT(\Omega)]'$ will obviously affect the phase noise spectrum. Thus, for a small stability margin in steady-state oscillation, with stable poles near the imaginary axis, bumps can be observed in the phase noise spectrum, unlike what happens when using the phase noise definition (7.34).

As an example, the formulation described above has been used in the phase noise analysis of a push-push oscillator, with a noisy precursor at the offset frequency 5 MHz. The resulting spectral density is shown by the dashed line in Fig. 7.7. In this case the dominant complex-conjugate poles have an influence on the phase noise. This is gathered from expression (7.53), which depends on the matrix $[P_N(\Omega)]$. Actually, the solution poles are the roots of the common denominator of this matrix. For comparison, the phase noise characteristic obtained with (7.34)

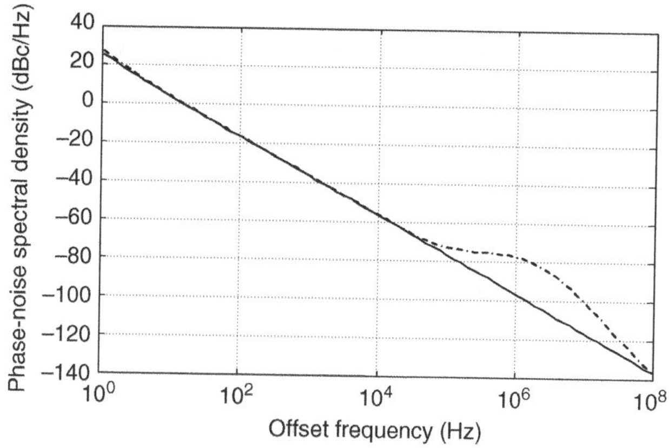

FIGURE 7.7 Phase noise spectrum of a bipolar-based push-push oscillator obtained using a phase sensitivity function (solid line) and the global analysis (dashed line) of Section 7.4. The noise bump due to the dominant complex-conjugate poles is noticeable only in the spectrum resulting from the global calculation.

for $\Omega \gg$ is shown by the solid line. The two different calculations for the phase noise spectrum provide approximately the same result except around the frequency of the dominant poles, ω_d. This is due to the significant influence of $\Delta\dot{\overline{X}}(t)$ in the perturbed oscillator equation for poles at small distance from the imaginary axis.

The amplitude noise is obtained by solving (7.49) for $\Delta\overline{X}(\Omega)$. This will require a definition of the compact matrix:

$$[A_N(\Omega)] = [0 \quad 1 \quad \cdots \quad 1]\left[\left.\frac{\partial\overline{H}}{\partial\omega}\right|_s \quad [JT(\Omega)]'\right]^{-1} \tag{7.55}$$

Then the amplitude noise spectral density is given by

$$[S_{\Delta\overline{X}'}(\Omega)] = [A_N(\Omega)][D_G(k\omega_o)][S_{noi}(\Omega)][D_G(k\omega_o)]^+[A_N(\Omega)]^+ \tag{7.56}$$

where $[S_{\Delta\overline{X}}(\Omega)]$ is the noise correlation matrix of the state variables. Remember that the global matrix $[\partial\overline{H}_s/\partial\omega \quad [JT(\Omega)]']$ is not singular at $\Omega = 0$, so no ill-conditioning problems will arise at small Ω. On the other hand, the poles of $[JT(\Omega)]'$ will affect the amplitude noise spectrum, and bumps can be observed for a small stability margin.

7.5 CARRIER MODULATION APPROACH

In the carrier modulation approach implemented on commercial harmonic balance simulators, the time variation of the amplitude perturbations is neglected in (7.49), setting $\Delta\dot{\overline{X}} = \overline{0}$. The time shift θ is not taken into account either. The technique can be applied for direct calculation of the phase and amplitude noise spectra or can be used to extract the parameters that provide variance in the stochastic phase deviation $\sigma_\theta^2(t)$. The latter possibility would allow a very accurate numerical determination of the phase noise spectrum through (7.29)–(7.33). The two options are discussed below.

7.5.1 Direct Calculation of Phase and Amplitude Noise Spectra

Doing $\Delta\dot{\overline{X}} = \overline{0}$ in (7.48), the following simplified expression is obtained for the perturbed oscillator equations in the frequency domain [1]:

$$\begin{bmatrix} \left.\dfrac{\partial \overline{H}}{\partial \omega}\right|_s & [JH]'_s \end{bmatrix} \begin{bmatrix} \Delta\omega_o(\Omega) \\ \Delta\overline{X}(\Omega) \end{bmatrix} = [D_G(k\omega_o)]\overline{G}_{noi}(\Omega) \tag{7.57}$$

where the matrix on the left-hand side agrees totally with the Jacobian matrix of the mixed harmonic balance formulation $[JH_{\text{mix}}]$. Note that the system (7.57) is well balanced, as the imaginary part of one of the harmonic components of one of the state variables in $\Delta\overline{X}(\Omega)$ has been set arbitrarily to zero, so this vector has the dimension $(2N+1)Q-1$. Then the frequency perturbation is calculated directly through matrix inversion. This provides

$$\Delta\omega_o(\Omega) = [1 \quad 0 \quad \cdots \quad 0][JH_{\text{mix}}]^{-1}[D_G(k\omega_o)]\overline{G}_{noi}(\Omega) \tag{7.58}$$

Note that the row matrix of dimension $Q(2N+1)$, relating $\Delta\omega_o(\Omega)$ to $\overline{G}_{noi}(\Omega)$, is constant, with no dependence on Ω.

As already stated, when neglecting the time derivative $\Delta\dot{\overline{X}}(t)$ in the perturbed oscillator equations (7.16), the carrier modulation obtained, $\Delta\omega_o(t)$, is independent of the choice of the additional constraint. Thus, the row matrix affecting $\overline{G}_{noi}(\Omega)$ in (7.58) must agree totally with $[C_M]^T$ in (7.20). Then it is possible to equate

$$\begin{aligned} \Delta\omega_o(\Omega) &= \overline{V}_{1x}^{\text{T}}[D_G(k\omega_o)]\overline{G}_{\text{noi}}(\Omega) \\ &= [1 \quad 0 \quad \cdots \quad 0][JH_{\text{mix}}]^{-1}[D_G(k\omega_o)]\overline{G}_{noi}(\Omega) \end{aligned} \tag{7.59}$$

and the phase noise calculated with the carrier modulation approach is given by

$$\begin{aligned} S_{\Delta\phi}(\Omega) = \frac{1}{\Omega^2}&[1 \quad 0 \quad \cdots \quad 0][JH_{\text{mix}}]^{-1}[D_G(k\omega_o)] \\ &[S_{noi}(\Omega)]\{[1 \quad 0 \quad \cdots \quad 0][JH_{\text{mix}}]^{-1}[D_G(k\omega_o)]\}^+ \end{aligned} \tag{7.60}$$

The phase noise spectrum obtained from the carrier modulation approach (7.60) agrees exactly with the spectrum resulting from the expression (7.34b), valid from relatively high frequency offset from the carrier. For white noise sources only, equation (7.60) predicts a constant phase noise reduction of −20 dB/dec versus Ω. In the case of a single flicker noise, there will be a constant phase noise reduction of −30 dB/dec at $\Omega \gg f_{\text{min}}$. In the case of both white and colored noise sources, there will be different sections in the spectrum, separated by corner frequencies, depending on the spectral density of these sources and their correlation spectrum (see (7.34)).

The amplitude noise spectrum is obtained directly by solving (7.57) for $\Delta\overline{X}(\Omega)$. Due to the quasistatic nature of this system, in which the time derivative $\Delta\dot{\overline{X}}(t)$ has been neglected, the shape of the spectrum will be determined by the correlation matrix of the noise sources. The poles associated with circuit linearization about the large-signal regime are not taken into account in the calculation, so it is not possible to predict any low-damping resonances.

7.5.2 Calculation of Variance of the Phase Deviation $\sigma_\theta^2(t)$

One advantage of the carrier modulation approach is that it allows a very simple extraction of the coefficients c and $b_{\gamma j,\mathrm{dc}}$, with $j = 1$ to J, providing the variance of the phase deviation $\sigma_\theta^2(t)$ from (7.32) and (7.29). This is because this approach is implemented on commercial software, so the accurate numerical procedure for determination of the phase noise spectrum, based on successive applications of (7.33), (7.32), (7.29), and (7.31), can be used with no need for dedicated in-house software. The way the coefficients c and $b_{\gamma j,\mathrm{dc}}$, with $j = 1$ to J, are extracted is explained below.

1. *Extraction of coefficient c.* To extract the coefficient c, all the colored noise sources contained in the circuit must be suppressed. This will provide a phase noise spectrum with the characteristic $S_{\Delta\phi}(f) = c/f^2$, with f being the offset frequency. Then the coefficient c can be extracted from one or two points of the phase noise spectrum, setting $c = S_{\Delta\phi}(f_o)f_o^2$.

2. *Extraction of coefficients $b_{\gamma j,\mathrm{dc}}$.* To extract each component $b_{\gamma j,\mathrm{dc}}$, all the white noise and colored sources are suppressed except the source, $\gamma_j(t)$. Then the coefficient $b_{\gamma j,\mathrm{dc}}$ can be extracted from one or two points of the phase noise characteristic $S_{\Delta\phi}(f) = |b_{\gamma j,\mathrm{dc}}|^2 S_{\gamma j}(f)/f^2$, doing $|b_{\gamma j,\mathrm{dc}}| = \sqrt{S_{\Delta\phi}(f_o)f_o^2/S_{\gamma j}(f_o)}$. The procedure is applied recursively for all the colored noise sources with $j = 1$ to J. Note that in the case of flicker noise, simple models of the form $S_{\gamma j}(f) = \alpha_j/f$ are used for the coefficient extraction.

Once the coefficients c, $b_{\gamma j,\mathrm{dc}}$, $j = 1 \ldots J$ have been extracted, the phase noise spectrum can be determined accurately with a numerical calculation of the Fourier transform in (7.31), as shown in [15]. Note that this spectrum will depend on the cutoff frequency $f_{\min}$ of the flicker noise sources, which has great relevance to the phase noise analysis from very small frequency offset from the carrier. Remember that the flicker noise sources are described through expressions of the form (7.33).

7.6 CONVERSION MATRIX APPROACH

In the techniques presented so far, the noise perturbations give rise to modulation of the oscillator carrier $\Delta\omega(t)$ and modulation of the harmonic components of the various state variables $\overline{X}_s + \Delta\overline{X}(t)$. In the conversion matrix approach, the joint effect of the two modulations is considered by expressing the state variables as

$$\overline{x}(t) = \sum_k (\overline{X}_{k,s} + \Delta\overline{X}_k(t))e^{jk\int_0^t \Delta\omega(s)\,ds} e^{jk\omega_o t} = \sum_k (\overline{X}_{k,s} + \Delta\overline{X}_{Tk}(t))e^{jk\omega_o t} \tag{7.61}$$

The phase and amplitude perturbations will be determined from the total sidebands. Actually, in the conversion matrix approach [2,16,17], the phase and amplitude noise spectra are calculated in two stages. First, the complex sidebands $\Delta\overline{X}_{Tk}$ are determined, and next, the phase and amplitude perturbations at a given node voltage (or branch current) are extracted from the corresponding voltage (or current) sidebands. The details of the formulation are presented next.

7.6.1 Calculation of complex sidebands $\Delta\overline{X}_T$

So far, the noise perturbations have been represented by the low-pass equivalents $\overline{G}_{noi,k}(t)$ about the main spectral frequencies $k\omega_o$. In the conversion matrix approach, the oscillator circuit is analyzed similar to a frequency mixer, with the small-signal inputs comprised by the noise sources. The noise inputs are described by means of a summation of pseudosinusoids with random amplitude and phase. In the perturbed-oscillator equation they will be represented as $\overline{G}_{noi}(k\omega_o \pm \Omega)$. Due to the Hermitian symmetry of real variables, the terms at $-|k|\omega_o \pm \Omega$ will be complex conjugates of the components at $|k|\omega_o \mp \Omega$. Thus, it is sufficient to consider the frequency components $\Omega, \omega_o + \Omega, -\omega_o + \Omega, \ldots, \pm N\omega_o + \Omega$, retaining only the positive sign for Ω [17,18].

The conversion matrix approach, based on a piecewise harmonic balance formulation, provides the following perturbed oscillator equations:

$$\left\{[A_x(k\omega_o + \Omega)] + [A_y(k\omega_o + \Omega)]\left[\frac{\partial\overline{Y}}{\partial\overline{X}}\right]_s\right\} \Delta\overline{X}_T(k\omega_o + \Omega) = [D_G(k\omega_o + \Omega)]\overline{G}_{noi}(k\omega_o + \Omega) \tag{7.62}$$

Note that the linear matrixes are evaluated accurately at the sidebands $k\omega_o + \Omega$. This is different from the approximate analysis based on the first-order Taylor series expansion of these matrixes about the steady-state frequencies $k\omega_o$ utilized in all the previous techniques. This constitutes an essential difference with respect to those techniques. On the other hand, the frequency modulation $\Delta\omega(t)$ is not taken as an independent variable of system (7.62). Its effect is included in the total perturbation vector $\Delta\overline{X}_T(k\omega_o + \Omega)$. Note that increments are considered in the real and imaginary parts of all the state variables, so system (7.62) is a well-balanced system of $(2N + 1)Q$ equations in $(2N + 1)Q$ unknowns [see (7.61)]. For the compactness of formulation, it is convenient to define the following Jacobian matrix:

$$[JH(k\omega_o + \Omega)] = [A_x(k\omega_o + \Omega)] + [A_y(k\omega_o + \Omega)]\left[\frac{\partial \overline{Y}}{\partial \overline{X}}\right]_s \tag{7.63}$$

The organization of the components at the various perturbation frequencies of each state variable X_q and each noise source G^g_{noi} may be the following:

$$\Delta\overline{X}^q_T(k\omega_o + \Omega) = \begin{bmatrix} \Delta X^q_T(-N\omega_o + \Omega) \\ \vdots \\ \Delta X^q_T(\Omega) \\ \vdots \\ \Delta X^q_T(N\omega_o + \Omega) \end{bmatrix}$$
$$\overline{G}^g_{noi}(k\omega_o + \Omega) = \begin{bmatrix} G^g_{noi}(-N\omega_o + \Omega) \\ \vdots \\ G^g_{noi}(\Omega) \\ \vdots \\ G^g_{noi}(N\omega_o + \Omega) \end{bmatrix} \tag{7.64}$$

On the other hand, the Jacobian matrix $[\partial\overline{Y}/\partial\overline{X}]_s$ contains the Toeplitz matrixes associated with the different time derivatives $j^p_q(t) = \partial y^p/\partial x^q$, x^q being the qth state variable and y^p being the pth nonlinearity. These matrixes have the form

$$[J^p_q] = \begin{bmatrix} \dfrac{\partial Y^p_{-N}}{\partial X^q_{-N}} & \dfrac{\partial Y^p_{-N}}{\partial X^q_{-N+1}} & \cdots & \dfrac{\partial Y^p_{-N}}{\partial X^q_{N-1}} & \dfrac{\partial Y^p_{-N}}{\partial X^q_{N}} \\ \dfrac{\partial Y^p_{-N+1}}{\partial X^q_{-N}} & \dfrac{\partial Y^p_{-N+1}}{\partial X^q_{-N+1}} & \cdots & \dfrac{\partial Y^p_{-N+1}}{\partial X^q_{N-1}} & \dfrac{\partial Y^p_{-N+1}}{\partial X^q_{N}} \\ & & \cdots & & \\ \vdots & \vdots & \vdots\ \vdots & \vdots & \vdots \\ \dfrac{\partial Y^p_{N-1}}{\partial X^q_{-N}} & \dfrac{\partial Y^p_{N-1}}{\partial X^q_{-N+1}} & \cdots & \dfrac{\partial Y^p_{N-1}}{\partial X^q_{N-1}} & \dfrac{\partial Y^p_{N-1}}{\partial X^q_{N}} \\ \dfrac{\partial Y^p_{N}}{\partial X^q_{-N}} & \dfrac{\partial Y^p_{N}}{\partial X^q_{-N+1}} & \cdots & \dfrac{\partial Y^p_{N}}{\partial X^q_{N-1}} & \dfrac{\partial Y^p_{N}}{\partial X^q_{N}} \end{bmatrix} \tag{7.65}$$
$$= \begin{bmatrix} J^p_{q,0} & J^p_{q,-1} & J^p_{q,-2} & \cdots & J^p_{q,-2N+1} & J^p_{q,-2N} \\ J^p_{q,1} & J^p_{q,0} & J^p_{q,-1} & & J^p_{q,-2N+2} & J^p_{q,-2N+1} \\ J^p_{q,2} & J^p_{q,1} & J^p_{q,0} & & J^p_{q,-2N+3} & J^p_{q,-2N+2} \\ & & & & & \\ J^p_{q,2N-1} & J^p_{q,2N-2} & J^p_{q,2N-3} & & J^p_{q,0} & J^p_{q,-1} \\ J^p_{q,2N} & J^p_{q,2N-1} & J^p_{q,2N-2} & & J^p_{q,1} & J^p_{q,0} \end{bmatrix}$$

where $J^p_{q,k}$, with $k = -2N$ to $2N$, are the different harmonic components of the derivative $j^p_q(t) = \partial y^p/\partial x^q$. As stated in Chapter 5, calculation of the Jacobian matrix inherently requires twice the number of harmonic components as those considered for steady-state analysis. That is, for N frequency terms in harmonic balance analysis, $2N$ terms are necessary in the derivatives $j^p_q(t) = \partial y^p/\partial x^q$ to obtain the matrix (7.65). Taking all the expressions into account, the increment of the state variables $\Delta\overline{X}_T(k\omega_o + \Omega)$ is given by

$$\Delta\overline{X}_T(k\omega_o + \Omega) = [JH(k\omega_o + \Omega)]^{-1}[D_G(k\omega_o + \Omega)]\overline{G}_{noi}(k\omega_o + \Omega) \qquad (7.66)$$

Note that as long as the matrix $[JH(k\omega_o + \Omega)]$ is evaluated for $\Omega \neq 0$, it will be nonsingular. However, at $\Omega = 0$ it agrees totally with the full Jacobian matrix of the harmonic balance system (with respect to the complete vector of state variables $\overline{X}$), evaluated at the steady-state oscillation at $k\omega_o$. As already known, this Jacobian matrix is singular for a steady-state free-running oscillation, since there is an irrelevance of the solution versus variations in the phase origin. Thus, as Ω decreases, the linear system (7.66) approaches singularity. At small Ω, the accuracy in the sideband calculation is determined by that of the matrix inversion. As demonstrated by Bolcato et al. [2], the accuracy can be increased by using oversampling and a high number of harmonic terms in the calculation of the steady-state oscillator solution $\overline{X}_s$, about which the circuit is linearized by means of (7.62).

For the actual determination of the oscillator noise spectrum, the incremental vector $\Delta\overline{X}(\Omega)$ in (7.66) must be multiplied by its adjoint $\Delta\overline{X}^+(\Omega)$. This will provide the correlation matrix of the perturbed state variables, which, in turn, depends on the correlation matrix of the noise sources:

$$\begin{aligned}[\Delta\overline{X}(\Omega)\quad \Delta\overline{X}(\Omega)^+] = {} & [JH(k\omega_o + \Omega)]^{-1}[D_G(k\omega_o + \Omega)] \\ & \times [S_{noi}(k\omega_o + \Omega, m\omega_o + \Omega)] \\ & \times \{[JH(k\omega_o + \Omega)]^{-1}[D_G(k\omega_o + \Omega)]\}^+ \end{aligned} \qquad (7.67)$$

where S_{noi} is a matrix containing the correlation spectra of the input noise sources.

7.6.2 Determination of Phase and Amplitude Noise Spectra

The phase and amplitude noise at a particular circuit node about a harmonic frequency $k\omega_o$ is obtained from the complex voltage sidebands corresponding to the node voltage selected, $\Delta V(k\omega_o - \Omega)$ and $\Delta V(k\omega_o + \Omega)$. The sidebands are calculated from the linear matrixes relating the selected voltage to the state variables $\Delta\overline{X}(\Omega)$. For a brief explanation of the calculation procedure, consider the following representation of the perturbed node voltage in a single-harmonic analysis ($N = 1$):

$$v(t) = Re\left[V_o\left(1 + \frac{\Delta v(t)}{V_o}\right)e^{j(\omega_o t + \phi_o + \Delta\phi(t))}\right] \qquad (7.68)$$

where V_o, ω_o, and ϕ_o are the steady-state amplitude, frequency, and phase at that particular node and $\Delta v(t)$ and $\Delta\phi(t)$ are the amplitude and phase perturbations. For a small-signal sinusoid at a frequency Ω, the perturbations $\Delta v(t)$ and $\Delta\phi(t)$ will also be sinusoids, expressed as

$$\begin{aligned} \Delta v(t) &= \text{Re}[\Delta V e^{j\Omega t}] \qquad \Delta V \in C \\ \Delta\phi(t) &= \text{Re}[\Delta\phi e^{j\Omega t}] \qquad \Delta\phi \in C \end{aligned} \tag{7.69}$$

Performing a first-order Taylor series expansion of (7.68) about the steady-state solution $Re[V_o e^{j(\omega_o t+\phi_o)}]$ in both the amplitude and phase variables, the following expression is obtained:

$$\begin{aligned} v(t) = \text{Re}\Bigg[& V_o e^{j(\omega_o t+\phi_o)} + V_o e^{j(\omega_o t+\phi_o)} \frac{1}{V_o}\left(\frac{\Delta V}{2} e^{j\Omega t} + \frac{\Delta V^*}{2} e^{-j\Omega t}\right) \\ & + V_o e^{j(\omega_o t+\phi_o)} j\left(\frac{\Delta\phi}{2} e^{j\Omega t} + \frac{\Delta\phi^*}{2} e^{-j\Omega t}\right)\Bigg] \end{aligned} \tag{7.70}$$

where the higher-order increments have been neglected. Grouping terms at the same frequency, it is possible to write

$$v(t) = \text{Re}[V_o e^{j\omega_o t + \phi_o} + V_L e^{j\omega_L t} + V_U e^{j\omega_U t}] \tag{7.71}$$

where ω_L and ω_U refer to the lower and upper frequency sidebands, respectively. The complex terms V_L and V_U are given by

$$\begin{aligned} V_U &= V_o e^{j\phi_o}\left(j\frac{\Delta\phi}{2} + \frac{\Delta V}{2V_o}\right) \\ V_L &= V_o e^{j\phi_o}\left(j\frac{\Delta\phi^*}{2} + \frac{\Delta V^*}{2V_o}\right) \end{aligned} \tag{7.72}$$

which is a system of two complex equations in two complex unknowns. Solving for ΔV and $\Delta\phi$, the following expressions are obtained:

$$\begin{aligned} \Delta V &= e^{j\phi_o} V_L^* + e^{-j\phi_o} V_U \\ \Delta\phi &= j\frac{e^{j\phi_o} V_L^* - e^{-j\phi_o} V_U}{V_o} \end{aligned} \tag{7.73}$$

The relationships above are used to extract the phase and amplitude perturbations of the node voltage from knowledge of the complex sidebands $\Delta\overline{X}_T(k\omega_o + \Omega)$. Note that calculation of the sidebands $\Delta V(k\omega_o - \Omega)$ and $\Delta V(k\omega_o + \Omega)$, corresponding to any node voltage, is immediate using the linear embedding matrixes that connect the voltage selected to the circuit state variables. Generalizing the expressions (7.73) to the various harmonic terms $k\omega_o$ and multiplying by the adjoint, the phase noise about the harmonic $k\omega_o$ is given by

$$S_{k\phi}(\Omega) = \frac{|\Delta V(k\omega_o - \Omega)^* e^{j\phi_{k,o}} - \Delta V(k\omega_o + \Omega)e^{-j\phi_{k,o}}|^2}{|V_{k,o}|^2} \tag{7.74}$$

where $|V_{k,o}|$, $\phi_{k,o}$ are the steady-state voltage amplitude and phase at the harmonic component $k\omega_o$ of the node selected. Note that the sideband $\Delta V(k\omega_o - \Omega)$ is obtained directly from the conjugation: $\Delta V(k\omega_o - \Omega) = \Delta V^*(-k\omega_o + \Omega)$. The amplitude noise at this same node is given by

$$S_{k\Delta V}(\Omega) = \frac{|\Delta V(k\omega_o - \Omega)^* e^{j\phi_{k,o}} + \Delta V(k\omega_o + \Omega)e^{-j\phi_{k,o}}|^2}{|V_{k,o}|^2} \tag{7.75}$$

As shown in (7.66), the perturbation vector $\Delta\overline{X}(k\omega_o + \Omega)$ is obtained from inversion of the matrix $[JH(k\omega_o + \Omega)]$, defined in (7.63). Replacing $j\Omega$ with s, this matrix agrees with the characteristic matrix associated with the system linearization about the steady-state oscillation; that is, $[JH(jk\omega_o + s)]$. The associated determinant $\det[JH(jk\omega_o + s)] = 0$ will have $Q \times (2N + 1)$ roots λ_{qk}, with $q = 1$ to Q and $k = 0$ to N, with the structure $\lambda_{qk} = \lambda_{q0} + kj\omega_o$. One of the roots is $\lambda_{10} = 0$, due to the solution autonomy. This zero pole has not been removed from expression (7.67), unlike the analysis methods described in Sections 7.3 and 7.4.

In the presence of white and flicker noise sources, the phase noise spectrum obtained from (7.74) will generally have three different sections versus the offset frequency, an initial section of −30 dB/dec near the carrier, a section of −20 dB/dec, and a flat section, corresponding to the higher offset frequencies. This is cumbersome to demonstrate, except for very simple oscillator models [18], due to the difficulty in dealing analytically with the sidebands. Provided that an accurate calculation is carried out, it is the pole at the origin ($\Omega = 0$) of $\det[JH(jk\omega_o + j\Omega)] = 0$ that actually gives rise to the typical phase noise characteristic described for the oscillator circuit. A small error in the prediction of this pole coming from a possible inaccuracy in the matrix inversion will lead to a miscalculation of the phase noise spectrum at small offset frequencies. This is similar to the problem of identifying accurately the zero pole in the stability analysis, through pole–zero identification. If instead of the zero pole a real pole of very small value $\gamma = \omega_r$ is obtained, the phase noise spectrum will become flat below ω_r, as expected from the frequency response associated with a real pole. Note that this flattening is due to a calculation error and has nothing to do with the flattening of the phase noise characteristic predicted by the accurate phase noise characterization in the time domain. The spectrum flattening at very small offset frequencies, predicted by (7.34), cannot be obtained using the conversion matrix approach, since this approach is based on a linearized analysis of the noisy oscillator in terms of the entire set of perturbation variables $\Delta\overline{X}(\Omega)$.

As already stated, the phase and amplitude noise calculation with the conversion matrix approach takes into account all the poles associated with the system

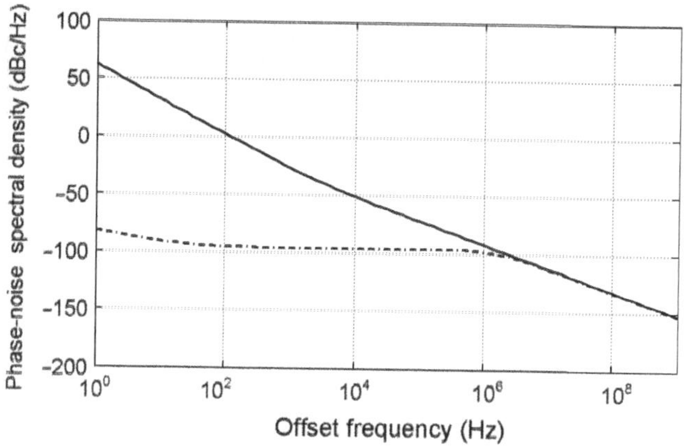

FIGURE 7.8 Phase noise analysis of the oscillator circuit in Fig. 7.1 with the conversion matrix approach (dashed line) and the carrier modulation approach (solid line).

linearization about the steady-state solution at $k\omega_o$. The possible spectrum resonances due to a small stability margin can be predicted with more accuracy than by using the methods described in Sections 7.3 and 7.4. This is because the linear matrixes are not expanded in Taylor series, but are calculated accurately at the sidebands $k\omega_o + \Omega$. However, the accurate implementation of the method is demanding from a computational point of view, due to the ill conditioning of the matrix $[JH(k\omega_o + \Omega)]$ at small Ω. Remember that the resonances and flattening of the phase noise spectrum (from certain offset frequency) are due to joint consideration of the common phase perturbation $\omega_0\theta(t)$ and the phase perturbation associated with the harmonic components $\Delta\overline{X}_k(t)$. Note that this does not mean lower accuracy; it is just the result of a different form of calculation in which the common phase noise $\theta(t)$ and the amplitude noise $\Delta\overline{X}(t)$ are not decoupled.

The conversion matrix approach has been used for the phase noise analysis of the oscillator in Fig. 7.1, containing white noise and flicker noise voltage sources in series with the intrinsic gate terminal. In Fig. 7.8 the results are compared with those obtained with the carrier modulation approach. As can be seen, the conversion matrix approach seems to fail for offset frequencies below approximately 1 MHz, due to numerical accuracies in the Jacobian matrix inversion. Note that no extra technique has been used to increase the accuracy of the calculation. On the other hand, the carrier modulation approach, which is accurate at a relatively short frequency distance from the carrier, is unable to detect the possible resonances at a longer frequency offset from the carrier.

7.7 NOISE IN SYNCHRONIZED OSCILLATORS

In an injection-locked oscillator, the synchronizing source establishes a phase origin for the noiseless steady-state solution, so there is no longer irrelevance of the periodic solution versus phase shifts. Therefore, there is no stochastic time deviation that originates from the noise sources. The phase noise will be due to the phase perturbation of the various harmonic terms coming from the phase and amplitude noise of the synchronizing source and from the noise sources contained in the circuit. In the frequency domain, the conversion matrix approach will be the most accurate technique for phase noise analysis. There is no ill-conditioning problem for low offset frequencies, since the Jacobian matrix $[JH]_s$ evaluated at the synchronized steady-state oscillation is not singular. Thus, the conversion matrix approach will be the most suitable simulation technique. Recently, a semianalytical formulation has also been proposed [19]. Although less accurate than the conversion matrix formulation, it enables a good understanding of the synchronized oscillator response to the various noise sources and provides design criteria for phase noise reduction. The semianalytical approach is based on the phase noise formulation derived in Chapter 4, relying on an admittance analysis of the oscillator circuit. The phase noise depends on derivatives of the admittance function with respect to the frequency, amplitude, and phase at the observation node. These derivatives can be extracted from a harmonic balance simulation of the circuit using an auxiliary generator, as shown in Chapter 5. The two analysis methods are presented in the following sections.

7.7.1 Conversion Matrix Approach

The harmonic balance formulation used for determination of the steady-state solution of the injection - locked oscillator is the following:

$$\overline{H}(\overline{X}) = [A_x(k\omega_o)]\overline{X} + [A_x(k\omega_o)]\overline{Y}(\overline{X}) + [A_g(k\omega_o)]\overline{G} = 0 \tag{7.76}$$

where $\overline{G}$ is the generator vector containing the bias sources and the input periodic generator G_s at its corresponding frequency $\omega_s = N\omega_o$ in the case of a frequency divider by N or a submultiple of the oscillation frequency in the case of a sub-synchronized oscillator $\omega_s = \omega_o/L$. Note that in the latter case, the fundamental frequency for the harmonic balance analysis will be ω_s instead of ω_o, so in (7.76), the harmonic terms $k\omega_o$ indicated should be replaced with $k\omega_s$.

Considering the vector of circuit noise sources $\overline{G}_{noi}(k\omega_o \pm \Omega)$, as well as the noise contributed by the input source at ω_s, perturbations will be generated about the harmonic terms. The sidebands $\Delta\overline{X}(k\omega_o \pm \Omega)$ are calculated using a expression similar to the one in the case of free-running oscillators:

$$\begin{aligned}\Delta\overline{X}(k\omega_o + \Omega) = [JH(k\omega_o + \Omega)]^{-1}\big\{&[D_G(k\omega_o + \Omega)]\overline{G}_{noi}(k\omega_o + \Omega)\\ &+ [A_g(k\omega_o + \Omega)]\overline{G}(k\omega_o + \Omega)\big\}\end{aligned} \tag{7.77}$$

where the Jacobian matrix is calculated at the synchronized steady-state solution. The vector $\overline{G}(k\omega_o + \Omega)$ represents the noise from the synchronizing source. In general, the only terms different from zero will be those corresponding to $\omega_s + \Omega, -\omega_s + \Omega$. The steady-state value G_s of this source is not included because the perturbed equation system involves the sidebands only. The noise spectrum is calculated multiplying (7.7) by ΔX^+. The required noise sidebands of the input source can be determined from its phase, amplitude, and phase-amplitude correlation spectrum using expression (7.72). Unlike the case of the free-running regime, the Jacobian matrix associated with the steady-state solution at $\Omega = 0$ is not singular, because there is no irrelevancy versus phase shifts. Thus, there are no accuracy problems at small values of the offset frequency Ω.

The phase and amplitude noise at a given observation node are determined from the same expressions as those used for the free-running oscillator. The corresponding spectra at the harmonic frequency $k\omega_o$ are calculated as

$$
\begin{aligned}
S_{k\phi}(\Omega) &= \frac{|\Delta V(k\omega_o - \Omega)^* e^{j\phi_{k,o}} - \Delta V(k\omega_o + \Omega)e^{-j\phi_{k,o}}|^2}{|V_{k,o}|^2} \\
S_{k\Delta V}(\Omega) &= \frac{|\Delta V(k\omega_o - \Omega)^* e^{j\phi_{k,o}} + \Delta V(k\omega_o + \Omega)e^{-j\phi_{k,o}}|^2}{|V_{k,o}|^2}
\end{aligned}
\tag{7.78}
$$

The characteristic determinant $\det[JH(k\omega_o + \lambda)]$ associated with the linearized system (7.77) will contain $Q \times (2N + 1)$ poles λ_{qk}, with $q = 1$ to n and $k = -N$ to N. The poles will have the structure $\lambda_{qk} = \lambda_{q0} + kj\omega_o$. Unlike the case of free-running oscillators, the synchronized oscillations have no intrinsic pole at zero. Of course, the solution can have a pole at zero if it corresponds to a bifurcation point, i.e., a point of qualitative stability change. Assuming stable behavior, all the poles $\lambda_{10}, \ldots, \lambda_{Q0}$ will be located on the left-hand side of the complex plane. The possible poles near the imaginary axis will give rise to local noise maxima, as in the case of free-running oscillators. A key fact in injection-locked oscillators is that their operation bands are always limited by bifurcations. Thus, solutions near the band edges will necessarily have a small stability margin and poles near the imaginary axis.

7.7.2 Semianalytical Formulation

The semianalytical formulation is applicable to injection-locked oscillators solved with a two-tier harmonic balance simulation. The outer tier is an admittance function conceptually similar to the one used in the analyses in Chapter 4, and the inner tier is the pure harmonic balance formulation. We will have this kind of resolution in the case of an injection-locked oscillator simulated with the aid of an auxiliary generator at the oscillation frequency $\omega_{AG} = \omega_o$. Note that for injection locking, the fundamental frequency ω_o agrees with the input-generator frequency $\omega_o = \omega_{in}$. When using an auxiliary generator of voltage type, the outer tier consists of the total admittance function $Y_T = Y_{AG}(A_{AG}, \phi_{AG}) = 0$, and, with A_{AG} and ϕ_{AG} being the auxiliary generator amplitude and phase, respectively. The inner tier is the pure harmonic balance system $\overline{H}(\overline{X}) = 0$. When convergence is achieved, the amplitude A_{AG} and phase ϕ_{AG} agree with those corresponding to the voltage component at ω_o at the node where the auxiliary generator at $\omega_{in} = \omega_o$ is connected (see Chapter 5). Considering the input generator value $G_s e^{j0}$ as an analysis parameter, the following condition is fulfilled:

$$Y_T(V_s, \omega_s, G_s e^{j\phi_s}) = 0 \tag{7.79}$$

In the above equation, it is considered that the auxiliary generator is connected to a node different from that of the synchronizing source. For convenience, the phase origin of the oscillator solution is taken at the observation node voltage, so the phase difference between the node and the generator is $-\phi_s$. Remember that only this phase difference is actually relevant for the analysis of synchronized circuits. The following derivation is identical to the one presented in Chapter 4, but the analysis in Chapter 4 was applied to injection-locked oscillators containing a nonlinear device modeled with its describing function. The key difference is that here we are dealing with the outer-tier equation of the harmonic balance system, so equation (7.79) implicitly contains the pure harmonic balance system $\overline{H}(\overline{X}) = 0$ as the inner tier. The two subsystems must be satisfied simultaneously by the synchronized solution.

As already stated, the edges of the synchronization band of an injection-locked oscillator are determined by bifurcation phenomena. Thus, close to the band edges, we will have near-critical poles at short distance from the imaginary axis. These poles will have a crucial influence on the oscillator phase noise spectrum. In the following, this phase noise spectrum will be related to the location of the dominant poles. The analysis, applied to the outer tier $Y_T = 0$ of the harmonic balance system, is limited to relatively small values of the pole frequency ω. However, the ω values of dominant poles $\sigma \pm j\omega$ of injection-locked oscillators are typically small. This is because the limits of the stable synchronization bands at which the dominant poles cross the imaginary axis are given by turning-point bifurcations ($\omega = 0$) or Hopf bifurcations ($\omega = |\omega_{in} - \omega_a|$, with ω_{in} the input frequency and ω_a the unlocked self-oscillation frequency). For more details on the stability properties of injection-locked oscillators, see Section 4.2.3. Note that the purpose of this simplified stability analysis is to relate the phase noise spectrum to the stability properties of the steady-state solution of the injection-locked oscillator.

7.7.2.1 Stability Analysis Initially, the dominant poles of the injection-locked oscillator will be determined. As shown in (4.17) in Chapter 4, the stability analysis based on the admittance function is carried out by considering the perturbed frequency $j\omega_o + s$. A first-order Taylor series expansion of $Y_T(V_s, j\omega_o + s, \phi_s)$ is performed about the synchronized solution, with s acting as a time derivator in $Y_T V_s(t) e^{j\Delta\phi(t)} = 0$. In this manner, it is possible to write

$$\overline{Y}_{TV}\Delta V(t) + \overline{Y}_{T\omega}\left(\Delta\dot{\phi}(t) - j\frac{\Delta\dot{V}(t)}{V_s + \Delta V(t)}\right) - \overline{Y}_{T\phi}\Delta\phi(t) = \overline{0} \tag{7.80}$$

where the term between the parentheses represents complex instantaneous frequency and, as usual, the different vectors are composed of the real and imaginary parts of the complex magnitudes. The subindexes V, ϕ, and ω stand for derivatives of Y_T with respect to the corresponding variables, evaluated at the particular steady-state solution, given by V_s, ϕ_s, and ω_s. Splitting the complex equation into real and imaginary parts, the following linear-time invariant (LTI) system is obtained:

$$\begin{pmatrix} \Delta\dot{V}(t) \\ \Delta\dot{\phi}(t) \end{pmatrix} = \left\{ \begin{pmatrix} Y^i_{T\omega}/V_s & Y^r_{T\omega} \\ -Y^r_{T\omega}/V_s & Y^i_{T\omega} \end{pmatrix}^{-1} \begin{pmatrix} -Y^r_{TV} & Y^r_{T\phi} \\ -Y^i_{TV} & Y^i_{T\phi} \end{pmatrix} \right\} \begin{pmatrix} \Delta V(t) \\ \Delta\phi(t) \end{pmatrix} = [M] \begin{pmatrix} \Delta V(t) \\ \Delta\phi(t) \end{pmatrix} \tag{7.81}$$

Note that the increment $\Delta V(t)$ in the denominator of (7.80) has been neglected, as it gives rise to second-order terms. System (7.81) is formally identical to the one derived in Chapter 4 for the analysis of injection-locked oscillators based on the describing function. As already stated, the only difference is that (7.81) refers to the outer-tier equation (7.79), which implicitly contains the harmonic balance system as the inner tier. The derivatives in (7.81) are easily calculated in commercial harmonic balance software with the aid of an auxiliary generator, as described in Chapter 6. The auxiliary generator must fulfill the steady-state condition $\overline{Y}_{AG} = 0$, agreeing with the total admittance Y_T, calculated through increments. Provided that the steady-state values are V_s, ϕ_s, and ω_s, the derivative $\overline{Y}_{TV}$ is calculated by considering the amplitude increment $V_s + \Delta V_s$, while the other two variables are kept constant at their steady-state values ϕ_s and ω_s. The derivatives $\overline{Y}_{T\phi}$ and $\overline{Y}_{T\omega}$ are calculated in an analogous manner.

The poles of the perturbed system will be given by the eigenvalues of the constant matrix $[M]$ in (7.81). Because the LTI system has order 2, only the two dominant poles of the synchronized solution can be detected. Note that the capability to detect instabilities is limited by the first-order Taylor series expansion in the perturbation frequency. However, as already stated, the most common instabilities in synchronized oscillators take place at a relatively small frequency difference with respect to the steady-state fundamental [20]. The two poles detected can be real or complex conjugate, this depending on the input power level. For relatively small input power, the two poles are real and the limit of the stable synchronization range is determined by a real pole crossing through zero at a turning point of the solution curve. From an inspection of (7.81), a real pole at zero implies fulfillment of the following condition:

$$\det \begin{bmatrix} Y^r_{TV} & Y^r_{T\phi} \\ Y^i_{TV} & Y^i_{T\phi} \end{bmatrix} = 0 \tag{7.82}$$

For higher input power, generally the two dominant poles of (7.81) will be complex conjugate, and the edges of the synchronization band will be given by Hopf-type bifurcations.

As an example, the stability analysis technique has been applied to the FET-based oscillator of Fig. 7.1. The synchronizing source is connected to the gate subnetwork. The synchronized solution curves for various input power values are shown in Fig. 7.9. The solution curves are closed for small input power and become open for a certain value of this power. To illustrate the effects of the two different types of bifurcation, two input power values have been considered:

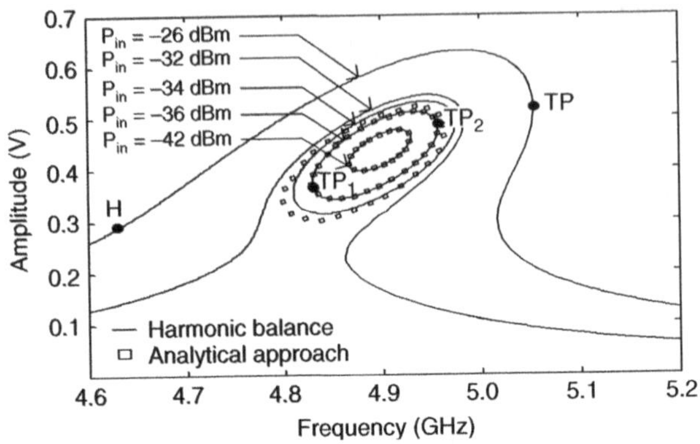

FIGURE 7.9 Synchronization curves versus the input frequency of the oscillator of Fig. 7.1 for various input power values.

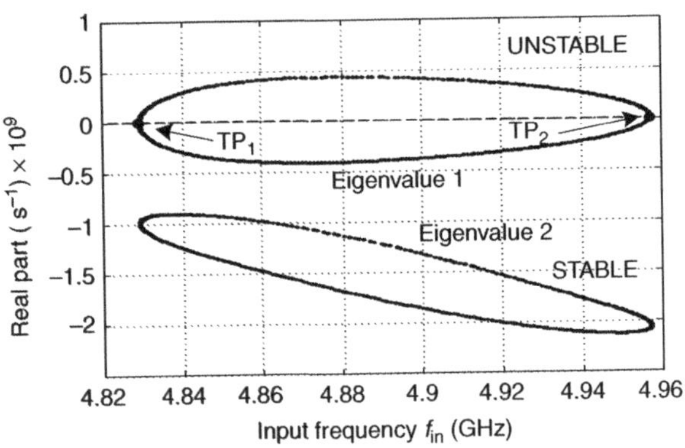

FIGURE 7.10 Variation of the two dominant poles of the closed synchronization curve of the circuit in Fig. 7.1, obtained for $P_{in} = -36$ dBm and represented in Fig. 7.9.

$P_{in1} = -36$ dBm and $P_{in2} = -26$ dBm. The corresponding synchronization curves can be seen in Fig. 7.9.

For $P_{in1} = -36$ dBm, the solution curve is closed and the operation band is delimited by two turning points, one at each edge. A stability analysis has been performed along this solution curve using the formulation (7.81). At each point of the closed curve, there are two real dominant poles. Their evolution is shown in Fig. 7.10. As can be seen, one of the poles γ_1 is always negative, whereas the other one γ_2 changes sign at each of the two turning points. The region of the synchronization curve with $\gamma_2 < 0$ is the upper section of the closed curve in Fig. 7.9, so this is the stable section.

For $P_{in2} = -26$ dBm, the synchronization curve (Fig. 7.9) is open and exhibits two turning points on its right-hand side. The variation in the real and imaginary parts of the two dominant poles has been represented in Fig. 7.11, with the input frequency as parameter. As can be seen, the poles are complex conjugate for

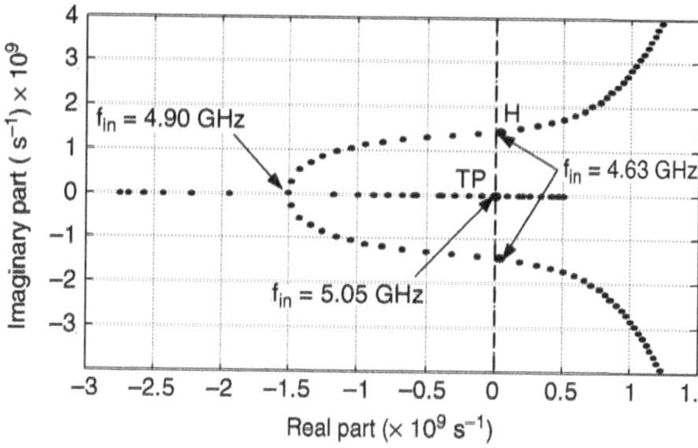

FIGURE 7.11 Pole locus with the input frequency as a parameter of the open solution curve of the circuit in Fig. 7.1, obtained for $P_{in} = -26$ dBm. At the Hopf bifurcation the frequency of the crossing poles is $f_{pH} = 230$ MHz.

low input frequency. They are located initially on the right-hand side of the complex plane, so the solution is unstable initially. As the input frequency increases, the pair of poles approach the imaginary axis and cross this axis at $f_{in} = 4.63$ GHz. It is an inverse Hopf bifurcation versus the input frequency, and the periodic solution becomes stable from this frequency value. At the Hopf bifurcation, the frequency of the complex-conjugate poles is $f_{pH} = 230$ MHz. As the input frequency continues to increase, the two stable complex-conjugate poles approach each other, merge, and split into two real poles, which happens at $f_{in} = 4.9$ GHz. From this frequency value, one of the real poles moves leftwards whereas the other pole moves rightward and crosses the imaginary axis through zero at $f_{in} = 5.05$ GHz. This frequency agrees with that of the turning point TP and the periodic solution curve becomes unstable at this frequency value.

7.7.2.2 Phase Noise Analysis Phase noise analysis based on the semianalytical formulation is also conceptually identical to that presented in Chapter 4. Two different noise inputs will be considered: phase noise from the synchronizing source $\psi(t)$ and an equivalent noise current generator at the observation node $i_n(t)$. This equivalent noise current model can be obtained from a phase-noise analysis in free-running conditions, with an accurate technique, based on harmonic balance. The spectral density of the noise source $|I_N|^2$ is fitted until the phase noise spectrum obtained with this single source agrees with the one obtained with the entire set of circuit noise sources [19]. The circuit noise sources only become relevant (versus the input phase noise $\psi(t)$) from certain offset frequency, as was shown in Chapter 4. Thus, the fitting can be carried out using the conversion matrix approach. The value of the spectral density of $i_n(t)$ is modified until the phase noise spectrum of the free-running oscillator matches the one obtained with multiple noise sources [19].

Let us begin by analyzing the influence of the phase perturbations from the synchronizing source $\psi(t)$ on the phase of the node voltage analyzed. A change in the phase of the synchronizing source $G_s e^{j(\omega_s+\phi_s)}$ gives rise to a shift of equal value in the phase of the first harmonic component of all the circuit variables, and thus in the phase value of the voltage at the observation node. If only the phase noise coming from the synchronizing source were considered, the phase of the voltage at the observation node N would be $\phi_n = \psi(t)$ instead of $\phi_n = 0$. Because noise contributions are also coming from the oscillator circuit, modeled using the equivalent source $i_n(t)$, the total phase value of the voltage at the observation node N will be $\Delta\phi_T(t) = \phi_n = \psi(t) + \Delta\phi(t)$. Note that the phase perturbation $\psi(t)$ does not affect the phase shift between the observation node N (reference node) and the input generator, which is given by $-\phi_s + \Delta\phi(t)$. In turn, the node amplitude, under the influence of $\psi(t)$ and $i_n(t)$, becomes $V_s + \Delta V(t)$, with V_s being the amplitude of the steady-state synchronized solution. As demonstrated in Chapter 4, the instantaneous complex frequency of the perturbed regime can be expressed as

$$\omega_s \mapsto \omega_s + (\dot{\psi} + \Delta\dot{\phi}) - j\frac{\Delta\dot{V}}{V_s + \Delta V} \tag{7.83}$$

Performing a first-order Taylor series expansion of the total node current about the particular steady-state synchronized solution, V_s, ϕ_s, and ω_s, it is possible to write

$$\overline{Y}_{TV}\Delta V(t) + \overline{Y}_{T\omega}\left[(\dot{\psi}(t) + \Delta\dot{\phi}(t)) - j\frac{\Delta\dot{V}(t)}{V_s + \Delta V(t)}\right] - \overline{Y}_{T\phi}\Delta\phi(t) = \frac{\overline{I}_N(t)}{V_s} \tag{7.84}$$

where the subscripts stand for derivatives of $\overline{Y}_T$ with respect to the corresponding variables V, ϕ, and ω, and $\overline{I}_N(t)$ is the envelope of the current noise source $i_n(t)$ about the fundamental frequency of the synchronized steady-state solution ω_s. Next, the complex equation (7.84) is split into real and imaginary parts and the Fourier transform is applied in the slowly varying time scale of the noise perturbations, with associated frequency Ω. Solving for $\Delta\phi_T(\Omega) = \Delta\phi(\Omega) + \psi(\Omega)$ and multiplying by the adjoint, the phase noise spectral density is given by

$$|\Delta\phi_T(\Omega)|^2 = \frac{\begin{matrix}(2|\overline{Y}_{TV}|^2 + \Omega^2(2|\overline{Y}_{T\omega}|^2/V_s^2))\,(|\overline{I}_N|^2/V_s^2) \\ + (|\overline{Y}_{TV} \times \overline{Y}_{T\phi}|^2 + \Omega^2(|\overline{Y}_{T\omega}\cdot\overline{Y}_{T\phi}|/V_s^2))\,|\psi(\Omega)|^2\end{matrix}}{\begin{matrix}\Big\{|\overline{Y}_{TV}\times\overline{Y}_{T\phi}|^2 + \Big[(\overline{Y}_{TV}\times\overline{Y}_{T\omega} - (\overline{Y}_{T\omega}\cdot\overline{Y}_{T\phi}/V_s))^2 \\ + 2\overline{Y}_{TV}\times\overline{Y}_{T\phi}(|\overline{Y}_{T\omega}|^2/V_s\Big]\Omega^2 + (|\overline{Y}_{T\omega}|^4/V_s^2)\Omega^4\Big\}\end{matrix}} \tag{7.85}$$

where the complex-number products of the form $a \times b$ are real and are defined as $a \times b = \text{Re}[a]\text{Im}[b] - \text{Re}[b]\text{Im}[a] = |a||b|\sin(\angle b - \angle a)$. To derive the expression (7.85), we have taken into account that the real and imaginary parts of the equivalent noise source $\overline{I}_N$ are uncorrelated (see Chapter 2, Section 2.5.1) and the noise from the synchronizing source is uncorrelated with the oscillator noise. Expression (7.85) is formally identical to the expression derived in Chapter 4 using the describing function. However, expression (7.85) is based on the perturbation analysis of a two-tier harmonic balance formulation of the oscillator circuit. Thus, it enables higher accuracy and can be applied to any injection-locked oscillator. The only requirement is the availability of an accurate equivalent noise model $\overline{I}_N(t)$ at the observation port.

The higher-order terms in the numerator and denominator of (7.85) will be relevant in only two situations: at relatively high frequency offset from the carrier or where the circuit is operating near the edges of the synchronization band (this case will be considered later). Thus, in most practical situations, the phase noise spectral density can be approached as

$$|\Delta\phi_T(\Omega)|^2 = \frac{2|\overline{Y}_{TV}|^2(|\overline{I}_N|^2/V_s^2) + |\overline{Y}_{TV} \times \overline{Y}_{T\phi}|^2|\psi(\Omega)|^2}{|\overline{Y}_{TV} \times \overline{Y}_{T\phi}|^2 + |\overline{Y}_{TV} \times \overline{Y}_{T\omega}|^2\Omega^2} \tag{7.86}$$

Two different frequency corners can be identified from an inspection of (7.86), in total agreement with the analysis presented in Chapter 4. The first is due to the decrease with Ω of the numerator term $|\overline{Y}_{TV} \times \overline{Y}_{T\phi}|^2|\psi(\Omega)|^2$, which will become equal to $2|\overline{Y}_{TV}|^2|\overline{I}_N|^2/V_s^2$ at the offset frequency Ω_y, defined by the condition

$$|\psi(\Omega_y)|^2 = \frac{2|\overline{Y}_{TV}|^2|\overline{I}_N|^2}{V_s^2|\overline{Y}_{TV} \times \overline{Y}_{T\phi}|^2} = \frac{2|\overline{I}_N|^2}{V_s^2|\overline{Y}_{T\phi}|^2 \sin^2\alpha_{v\phi}} = \frac{2|\overline{I}_N|^2}{|\overline{Y}_{TG}|^2 V_s^2 G_s^2 \sin^2\alpha_{v\phi}} \tag{7.87}$$

where $\alpha_{v\phi}$ is the angle $\alpha_{v\phi} = \text{ang}(\overline{Y}_{T\phi}) - \text{ang}(\overline{Y}_{TV})$. The chain rule has also been used to express $\overline{Y}_{T\phi}$ as $\overline{Y}_{T\phi} = \overline{Y}_{TG} jG_s e^{j\phi_s}$, where $\overline{Y}_{TG}$ is the derivative of the total admittance $\overline{Y}_T$ with respect to the input generator at the particular synchronized solution. The corner frequency Ω_y will increase with the generator amplitude G_s or the sensitivity of the total admittance to this generator, which is achieved by improving the input matching of the oscillator circuit.

The second slope change predicted by (7.86) occurs at the frequency $\Omega_{3\text{dB}}$, at which the two denominator terms become equal. This frequency is defined by the condition

$$\Omega_{3\text{dB}} = \frac{|\overline{Y}_{TV} \times \overline{Y}_{T\phi}|}{|\overline{Y}_{TV} \times \overline{Y}_{T\omega}|} = \frac{|\overline{Y}_{T\phi}||\sin\alpha_{v\phi}|}{|\overline{Y}_{T\omega}||\sin\alpha_{v\omega}|} \tag{7.88}$$

where $\alpha_{v\omega}$ is the angle $\alpha_{v\omega} = \text{ang}(\overline{Y}_{T\omega}) - \text{ang}(\overline{Y}_{TV})$. To maximize $\Omega_{3\text{dB}}$, the term $|\overline{Y}_{T\omega}||\sin\alpha_{v\omega}|$ must be minimized. Note that minimizing the magnitude $|\overline{Y}_{T\omega}|$ will imply generally reducing the oscillator quality factor. As in the case of Ω_y, the frequency $\Omega_{3\text{dB}}$ increases with the magnitude $|\overline{Y}_{T\phi}|$, so higher values will be obtained for larger generator amplitude and better input matching.

As gathered from (7.86), the output phase noise spectrum will follow the input spectrum for offset frequency $\Omega < \Omega_y$. It will be flat in the interval $\Omega_y < \Omega < \Omega_{3\text{dB}}$ and it will drop -20 dB/dec for $\Omega > \Omega_{3\text{dB}}$. An injection-locked oscillator is able to track input phase perturbations below the frequency $\Omega_{3\text{dB}}$. Perturbations above this frequency are too fast to be tracked by the oscillator circuit. Below the corner frequency Ω_y, the dominant perturbations are due to the phase of the synchronizing source $\psi(t)$. Above this frequency, the dominant perturbations are due to the circuit noise.

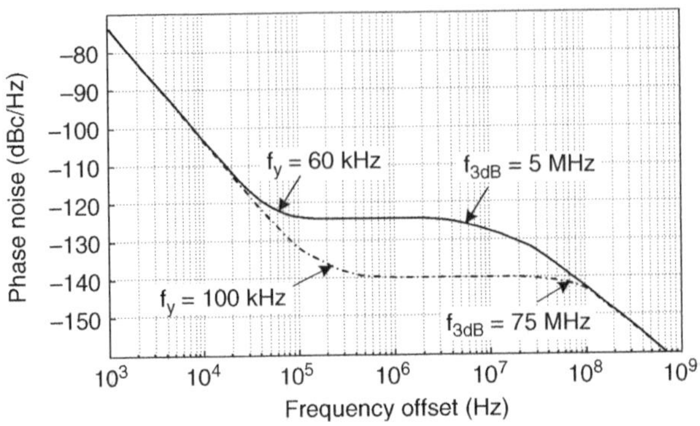

FIGURE 7.12 Phase noise spectra obtained with the conversion matrix approach for input power $P_{\rm in} = -36$ dBm and two values of the synchronization frequency, $f_s = 4.9$ GHz and $f_s = 4.958$ GHz.

As an example, phase noise calculation using both the conversion matrix approach and the semianalytical formulation (7.85)–(7.86) has been applied to the FET-based oscillator of Fig. 7.1. In an initial simulation, the phase noise of the injection-locked oscillator has been calculated for the input power $P_{\rm in} = -36$ dBm and two different values of input frequency, $f_s = 4.9$ GHz and $f_s = 4.958$ GHz. The phase noise spectrum of the synchronizing source is assumed to have a constant -30 dB/dec ($1/\Omega^3$) slope. The results are shown in Fig. 7.12, where the calculations have been carried out at two different nodes: the gate node and the drain node. Spectra determined at the two nodes, with the conversion matrix approach, agree totally with the semianalytical model in (7.86). The output phase noise follows that of the input source up to the first corner frequency, corresponding to Ω_y. It becomes flat from this corner frequency up to the second frequency at $\Omega_{3\rm dB}$. From the offset frequency $\Omega_{3\rm dB}$, the phase noise spectral density drops as -20 dB/dec ($1/\Omega^2$), in agreement with the prediction of (7.86).

Next, the phase noise variation along the synchronization band obtained for $P_{\rm in1} = -36$ dBm will be analyzed. The corresponding evolution of the two dominant poles is shown in Fig. 7.10. In Fig. 7.13, the phase noise spectral density has been represented versus the input frequency for different values of the offset frequency Ω. In agreement with the semianalytical formulation (7.86), for very small offset, the circuit phase noise replicates the noise introduced by the synchronized source, with no influence of the noise sources and the dynamics of the oscillator circuit, except at the band edges, due to the pole at zero. Note that at these band edges, the circuit behaves, at low-frequency offsets, in a nearly singular manner. This is because the denominator in (7.86) is nearly zero, since at turning points the condition $|\overline{Y}_{TV} \times \overline{Y}_{T\phi}| = 0$ (see Chapter 4, Section 4.2.7) is satisfied and Ω takes a very small value. As can be seen, the noise amplification at the two band edges decreases with the offset frequency. This is because the term $|\overline{Y}_{TV} \times \overline{Y}_{T\omega}|^2\Omega^2$ in the denominator of (7.86) gets larger and reduces the influence of the singularity at $\Omega = 0$, due to $|\overline{Y}_{TV} \times \overline{Y}_{T\phi}| = 0$. At larger offset frequency, the circuit will be operating in different regions of the phase noise spectrum (of similar form to the

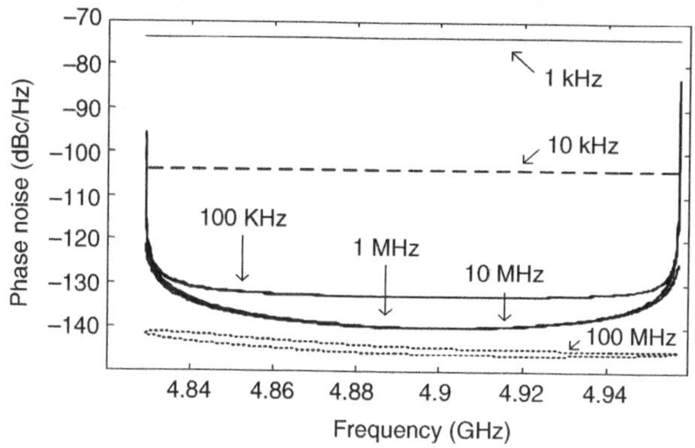

FIGURE 7.13 Variation in the phase noise spectral density of the oscillator circuit in Fig. 7.1 along the synchronization band obtained for $P_{in} = -36$ dBm. Different values of the constant offset frequency have been considered.

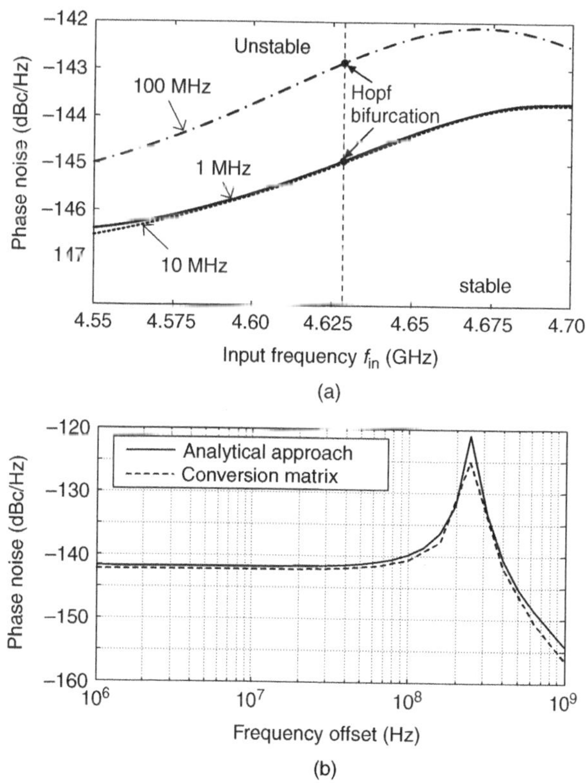

FIGURE 7.14 Phase noise analysis of the oscillator circuit of Fig. 7.1 for $P_{in} = -26$ dBm: (a) variation of the phase noise spectral density at different offset frequencies versus the input generator frequency; (b) phase noise spectrum for $f_{in} = 4.6$ GHz.

one in Fig. 7.12) for the different f_{in} values considered in the representation of Fig. 7.13. Therefore, there is a slight variation of the phase noise spectral density versus the input generator frequency.

Finally, the phase noise variation along the synchronization band obtained for $P_{in2} = -26$ dBm will be analyzed. The corresponding evolution of the two dominant poles is shown in Fig. 7.10. As can be seen, the upper edge of the stable synchronization band is delimited by a turning point, whereas the lower edge is delimited by a Hopf bifurcation. Fig. 7.14a shows the variation of the phase noise spectral density at different offset frequencies versus the input frequency f_{in} in the neighborhood of the Hopf bifurcation. As can be seen, the phase noise undergoes amplification about the f_{in} value corresponding to the Hopf bifurcation ($f_{inH} = 4.63$ GHz). The maximum amplification will occur for an offset frequency agreeing with the frequency of the unstable pair of poles. This is due the fact that the denominator of (7.85) agrees with the characteristic determinant, associated to the perturbed system (7.81), when evaluated at $j\Omega$ instead of the Laplace frequency s. At the Hopf bifurcation point, the roots of this demominator agree with the frequencies of the critical poles $s = \pm j\Omega = \pm j|\omega_{in} - \omega_a|$. An expanded view of the phase noise spectrum obtained for $f_{in} = 4.6$ GHz is shown in Fig. 7.14b. A sharp increase at the offset frequency 230 MHz, agreeing with the frequency of the unstable poles, can be observed. This is due to the proximity of the input frequency $f_{in} = 4.6$ GHz to that corresponding to the Hopf bifurcation, $f_{inH} = 4.63$ GHz.

REFERENCES

[1] V. Rizzoli, F. Mastri, and D. Masotti, General noise analysis of nonlinear microwave circuits by the piecewise harmonic-balance technique, *IEEE Trans. Microwave Theory Tech.*, vol. 42, pp. 807–819, May 1994.

[2] P. Bolcato, J. C. Nallatamby, R. Larcheveque, M. Prigent, and J. Obregón, A unified approach of PM noise calculation in large RF multitone autonomous circuits, *IEEE MTT-S International Microwave Symposium*, Boston, MA, pp. 417–420, 2000.

[3] F. X. Kaertner, Analysis of white and $f^{-\text{alpha}}$ noise in oscillators, *Int. J. Circuit Theory Appl.*, vol. 18, pp. 485–519, 1990.

[4] A. Demir, Phase noise in oscillators: DAEs and colored noise sources, *IEEE/ACM International Conference on Computer-Aided Design*, pp. 170–177, 1998.

[5] K. Kurokawa, Some basic characteristics of broadband negative resistance oscillators, *Bell Syst. Tech. J.*, vol. 48, pp. 1937–1955, July–Aug. 1969.

[6] M. Odyniec, *RF and Microwave Oscillator Design*, Artech House, Norwood, MA, 2002.

[7] U. L. Rohde, A. K. Poddar, and G. Bock, *The Design of Modern Microwave Oscillators for Wireless Applications*, Wiley, Hoboken, NJ, 2005.

[8] D. J. Vendelin, A. M. Pavio, and U. L. Rohde, *Microwave Circuit Design*, Wiley, New York, 1990.

[9] S. A. Maas, *Noise in Linear and Nonlinear Circuits*, Artech House, Boston, 2005.

[10] A. Cappy and W. Heinrich, High-frequency FET noise performance: a new approach, *IEEE Trans. Electron Devi.*, vol. 36, pp. 403–409, 1989.

[11] A. Cappy, F. Danneville, and G. Dambrine, Noise analysis in devices under nonlinear operation, *Proc. of the 27th European Solid-State Device Research Conference*, pp. 117–124, 1997.

[12] J. Nallatamby, M. Prigent, M. Camiade, A. Sion, C. Gourdon, and J. J. Obregón, An advanced low-frequency noise model of GaInP-GaAs HBT for accurate prediction of phase noise in oscillators, *IEEE Trans. Microwave Theory Tech.*, vol. 53, pp. 1601–1611, 2005.

[13] A. Suárez, S. Sancho, S. Ver Hoeye, and J. Portilla, Analytical comparison between time- and frequency-domain techniques for phase-noise analysis, *IEEE Trans. Microwave Theory Tech.*, vol. 50, pp. 2353–2361, 2002.

[14] S. Sancho, A. Suárez, and F. Ramirez, Phase and amplitude noise analysis in microwave oscillators using nodal harmonic balance, *IEEE Trans. Microwave Theory Tech.*, vol. 55, pp. 1568–1583, 2007.

[15] S. Sancho, F. Ramirez, and A. Suárez, Analysis and reduction of the oscillator phase noise from the variance of the phase deviations, determined with harmonic balance, *IEEE MTT-S International Microwave Symposium Digest*, Atlanta GA, 2008.

[16] M. Prigent and J. Obregón, Phase noise reduction in FET oscillators by low-frequency loading and feedback circuit optimization, *IEEE Trans. Microwave Theory Tech.*, vol. 35, pp. 349–352, Mar. 1987.

[17] J. M. Paillot, J. C. Nallatamby, M. Hessane, R. Quéré, M. Prigent and J. Rousset, A general program for steady state, stability, and FM noise analysis of microwave oscillators, *IEEE MTT-S International Microwave Symposium Digest*, vol. 3, pp. 1287–1290, 1990.

[18] J. M. Paillot, C.A.O des circuits analogiques non-linéaires: réalisation d'un simulateur pour l'analyse des spectres de bruit des oscillateurs, Ph.D. Dissertation, University of Limoges, France, 1991.

[19] F. Ramirez, M. Ponton, S. Sancho, and A. Suárez, Phase noise of injection-locked oscillators and frequency dividers, *IEEE Trans. Microwave Theory Tech.*, vol. 56, Feb. 2008.

[20] A. Suárez and R. Quéré, *Stability Analysis of Nonlinear Microwave Circuits*, Artech-House, Norwood, MA, 2003.

CHAPTER EIGHT

Harmonic Balance Techniques for Oscillator Design

8.1 INTRODUCTION

The accurate design of free-running oscillators is not a simple task. They behave in a nonlinear regime, as required to sustain constant amplitude oscillation, and the designer has little control over the oscillation frequency and power, which depend on the circuit element values and bias conditions. In general, the oscillator design is performed in two stages. First, the feedback, termination, and load elements of the active devices are calculated with linear simulations to fulfil the oscillation startup conditions. Because this design is linear, it is not possible to impose the desired output power value at the oscillation frequency. At the second stage, a nonlinear analysis is performed to determine the steady-state oscillation frequency and output power. A frequency shift with respect to the original design is generally found, due to nonlinear effects.

As shown in previous chapters, an oscillator analysis by means of an auxiliary generator operating at the oscillation frequency efficiently avoids the default harmonic balance convergence toward the dc solution that coexists with the oscillatory solution. The auxiliary generator plays the role of the oscillation, so it can also provide the designer with privileged control over this oscillation. It can be used for a direct nonlinear oscillator design, imposing the values of the steady-state amplitude and frequency. Note that this only guarantees the existence of the oscillatory

Analysis and Design of Autonomous Microwave Circuits, By Almudena Suárez

solution as a valid mathematical solution of the circuit. The stability considerations will be essential, as otherwise the oscillation will be physically unobservable, will fail to start up, or its spectrum will present undesired frequency components. In this chapter, techniques are provided for an efficient oscillator design. An auxiliary generator will be used to preset the amplitude and frequency of the steady-state oscillation. This technique requires a valid oscillator topology as a starting point, so the initial circuit design, to obtain the suitable feedback, termination, and load elements, is also presented at the beginning of the chapter. Only single-ended topologies are considered, although the techniques can be applied equally to differential oscillators or other configurations. Other critical aspects are the oscillation efficiency, the transient duration, and the phase noise. Techniques to enhance designer control over these oscillator characteristics, which are briefly discussed in the following, are also presented in this chapter.

The efficiency is the ratio between the useful RF power delivered to the load at the fundamental frequency and the dc power consumed [1,2]. Low-power consumption is of great interest in modern communication systems, as it increases the duration of the batteries, reduces the cost of thermal management, and reduces the risk of component damage due to excessive heating. Here three different techniques for increasing the oscillator efficiency without affecting the steady-state oscillation frequency are presented: class E design [3], class F design [4], and a multiharmonic load pull [5] of general application. Reduction in the transient duration will be convenient for power saving: the faster the system can be powered up, perform its function, and powered down, the less power will be needed. Fast starting times are also very important in pulsed systems such as ultra wideband communications (UWB) or radar systems, because narrower pulse widths can be used. The startup time of the oscillator circuits will be reduced here through the increase of the positive real part σ of the pair of complex-conjugate poles $\sigma \pm j\omega$ of the dc solution [6] responsible for the onset of the oscillation. Another essential oscillation characteristic is the phase noise, which can substantially degrade the communication signal through an increase in the signal-to-noise ratio in the case of analog signals, or to an increase in the bit error rate in digital communications, due to misinterpretation of the symbols transmitted. Here techniques are provided to minimize the phase noise spectral density of a given oscillator topology through variations of its circuit elements. The objective will be to reduce the variance of the phase deviation without affecting the oscillation frequency.

As one step forward in the designer control over the oscillation, techniques will be provided to preset the oscillator response over a frequency band instead of a single operation point. It will be possible to impose a linear frequency response of a voltage-controlled oscillator (VCO) versus the bias voltage applied. In phase-locked loops, the linear characteristic avoids variations in the loop bandwidth. In chirp signal generators, even when using a linearization loop, the oscillator itself must originally exhibit a good degree of linearity, as demonstrated by Kulpa [7]. In contrast to all the design objectives discussed previously, the linearization of the VCO characteristic implies dealing with an entire set of steady-state oscillator solutions obtained versus the bias voltage used.

The chapter is organized as follows. In Section 8.2 a simple technique for the initial small-signal design of the oscillator circuit is presented, starting from the Rollet factor and stability circles [8]. Then the nonlinear design technique, based on the use of auxiliary generator, is shown. In the next two sections we cover maximization of the oscillator efficiency and reduction of the startup transient. Next, the techniques for the optimized design of voltage-controlled oscillators are presented. Finally, a technique for the phase noise reduction of a given oscillator topology is described. The objective is not to present an exhaustive review of the possible circuit configurations. Instead, the principles and implementation of modern simulation techniques, aiming at an efficient oscillator design, are proposed. The techniques are illustrated by means of their application to simple, though practical microwave circuits.

8.2 OSCILLATOR SYNTHESIS

Oscillator circuits are designed most efficiently in the frequency domain, where impedance–admittance analysis can easily be applied for evaluation of the oscillation startup and steady-state conditions. The distributed elements of the linear networks are also described more easily in this domain. As shown in Chapter 1, the oscillator circuit is often modeled as the connection of a linear and a nonlinear network (Fig. 1.4), which is conceptually useful for the oscillator design [8,9]. The oscillator design requires an initial choice of the active device. This device may be a negative-resistance diode or a transistor [9]. The transistor is a two-port device, which to provide negative resistance at one port will usually require external feedback together with a proper termination at the other port (Fig. 1.5). The nonlinear network contains active device, feedback, and terminal elements, whereas the output resistive load is included in the linear network. With respect to this output load, the remainder of the circuit must behave as an energy source at the desired oscillation frequency, corresponding to a circuit resonance. This energy is taken from the dc sources. Thus, the nonlinear network must exhibit negative resistance at the oscillation frequency, which is determined by the circuit resonance.

Next, a systematic design procedure for oscillator circuits is presented. It is illustrated by means of its application to the design of an oscillator, based on the MESFET CFY30, at the frequency $f_o = 4.5$ GHz.

8.2.1 Oscillation Startup Conditions

The evaluation of the oscillation startup conditions is, in fact, an evaluation of the stability of its dc solution. As already known, this solution has to be unstable, with two poles $\sigma \pm j\omega_o$ on the right-hand side of the complex plane. An excess negative resistance is necessary to have $\sigma > 0$ at the oscillation frequency ω_o. As shown in Chapter 1, when using a transistor, this negative resistance is provided by a combination of gain and feedback. Any transistor is actually bilateral $S_{12} \neq 0$, so it inherently contains feedback. Analysis of the transistor μ factor at the desired oscillation frequency ω_o allows us to determine whether it is possible to obtain negative resistance looking into the transistor input or output port through suitable loading at the opposite port. For $\mu < 1$ at ω_o, the transistor will be conditionally stable at this frequency and negative resistance will be obtained for some passive terminations, without adding any external feedback elements to the transistor. As shown in Chapter 6, the μ factor [10] depends on the transistor scattering parameters and thus on its bias point. It provides the distance from the center of the Smith chart to the nearest point of the output (load) stability circle. It is given by

$$\mu = \frac{1 - |S_{11}|^2}{|S_{22} - S_{11}^* \Delta| + |S_{12} S_{21}|} \tag{8.1}$$

For $\mu > 1$, provided that the unloaded transistor is itself stable, no passive loads connected to either its input or output will give rise to instability. No negative resistance will ever be observed at the transistor input or output and the oscillator design will be impossible in these conditions. If the transistor is unconditionally stable ($\mu > 1$) at the oscillation frequency desired, a different bias point can be tried. However, it is usually more convenient to add a feedback network and reevaluate the μ factor. Because it is a different two-port network, with gain and enhanced feedback, the μ factor may now be smaller than 1 for some implementations of this feedback network. Reactive elements are generally chosen to avoid the introduction of unnecessary loss. Different feedback networks are possible. A simple topology is obtained through series feedback, with either a capacitance or an inductance, as shown in Fig. 8.1a. Topologies such as those used in Colpitts and Clapp–Gouriet oscillators [8,11] reduce the influence of the elements of the transistor model on the oscillation frequency.

A simple series feedback at the transistor source is considered here (Fig. 8.1a). When doing so, a new two-port can be defined, globally including the transistor and the feedback elements. The two ports of the active network (including the feedback) are denoted port 1 and port 2. The input impedance Z_{in} at port 1 depends on the scattering parameters S, the feedback elements, and the circuit load at the other end, Z_2. In turn, the output impedance Z_{out} at port 2 depends on S, the feedback elements, and Z_1 [8,12]. For $\mu < 1$, it will be possible to obtain negative resistance for some passive impedances Z_1 and Z_2. The oscillator synthesis can be carried out at either port 1 or port 2. The choice will depend on the transistor type and the particular oscillator configuration. Without loss of generality, port 2 will be selected here (see Fig. 8.1b).

The μ factor will now be analyzed in the presence of series feedback at the desired oscillation frequency ω_o. To consider all the positive and negative reactive values globally, an ideal feedback element, defined by its reflection coefficient $\Gamma_{fb} = 1e^{j\theta_{fb}}$, will be used. Then a new two-port network is defined, containing the transistor and the feedback element $\Gamma_{fb} = 1e^{j\theta_{fb}}$. For some values of θ_{fb}, a stability coefficient $\mu < 1$ will generally be obtained. Figure 8.2a shows the variation of μ versus θ_{fb} for the transistor CFY30 when considering different values of gate bias. The drain voltage is in all cases $V_{DS} = 2.5$ V.

For $\mu < 1$ it will be possible to have source impedances such that $\text{Re}[Z_{out}] < 0$. These are determined by tracing the source stability circle on the Smith chart

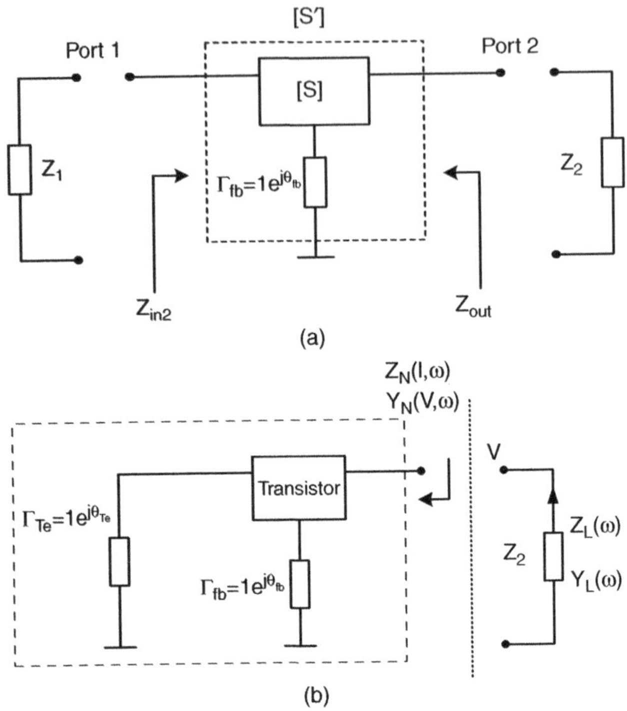

FIGURE 8.1 Transistor with series feedback to achieve negative resistance at the drain port: (a) two-port analysis; (b) one-port analysis, with a reactive terminal element at the gate port.

corresponding to $\Gamma_1 = (Z_1 - Z_c)/(Z_1 + Z_c)$, with Z_c the reference impedance of the scattering matrix. This circle constitutes the border between the source impedances Z_1 for which $\text{Re}[Z_{\text{out}}] < 0$ and those giving $\text{Re}[Z_{\text{out}}] > 0$. The condition $\text{Re}[Z_{\text{out}}] < 0$ is equivalent to an output reflection coefficient with modulus larger than 1: $|\Gamma_{\text{out}}| > 1$. For $\Gamma_1 = 0$, this output reflection coefficient agrees with the scattering parameter S_{22}. Thus, provided that the scattering matrix of the two-port network considered (referred to 50 Ω and including the feedback element $\Gamma_{\text{fb}} = 1e^{j\theta_{\text{fb}}}$) fulfills $|S_{22}| < 1$, the origin of the Smith chart will belong to the zone giving $\text{Re}[Z_{\text{out}}] > 0$. The opposite is true for $|S_{22}| > 1$. Figure 8.2b shows the location of the stability circles, at the oscillation frequency desired, $f_o = 4.5$ GHz, for some θ_{fb} values with $\mu < 1$.

In most cases, the impedance Z_1 is used to achieve the negative resistance only. It is often called the terminal impedance, Z_{Te}. To avoid additional loss, a reactive value X_{Te} is usually chosen (Fig. 8.1). However, in some cases it will be necessary to consider a real part in Z_{Te}: for example, if the oscillator is going to be injection locked at a later stage, since the forcing generator, connected at port 1, will necessarily introduce a resistive part $\text{Re}[Z_{Te}] > 0$. For a free-running oscillator, only reactive Z_{Te} values will generally be considered. These values will

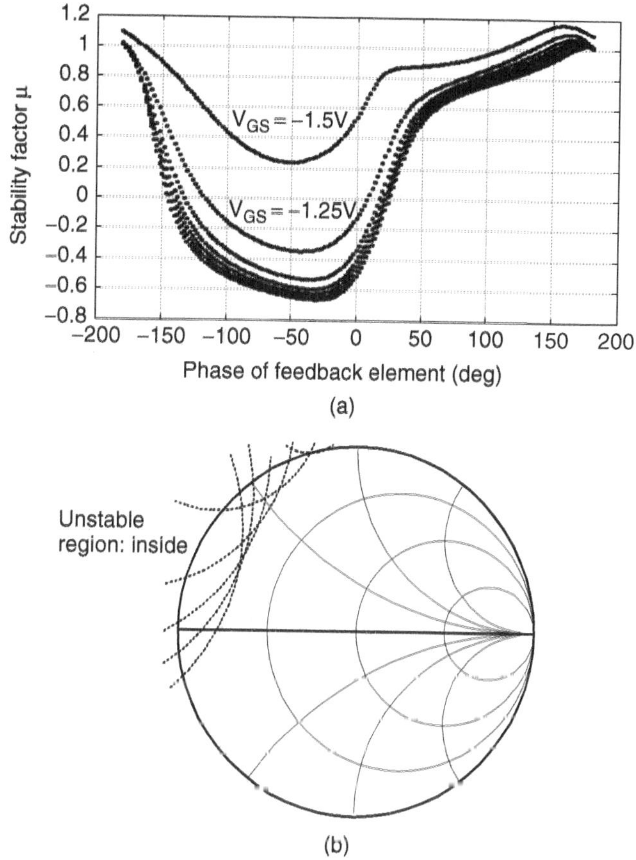

FIGURE 8.2 Analysis of potential instability of the series feedback topology in Fig. 8.1: (a) variation of the stability factor μ versus the phase θ_{fb} for different values of the gate bias voltage; (b) regions in the Z_1 plane, providing negative resistance at the drain port, obtained θ_{fb} values for which $\mu < 1$.

be located on the border of the Smith chart. Although a single terminal element (inductance or capacitance) will be chosen here, it is also possible to use a resonator with a high quality factor. As has been shown in Chapter 2, this will reduce the phase noise spectral density.

As stated earlier, to facilitate oscillation startup, excess negative resistance and resonance at the desired oscillation frequency ω_o must be achieved at port 2 in Fig. 8.1. The oscillation startup conditions can be evaluated in terms of the total impedance or the total admittance at port 2 (Fig. 8.1b). Actually, at this early stage, connection of the *transistor block* (including the transistor itself, the feedback elements, and the termination impedance) and the load Z_2 can be regarded as a series connection, which would require the sum $Z_{out} + Z_2$, or as a parallel connection, which would require the sum of the respective admittances $Y_{out}(V, \omega) = 1/Z_{out}$

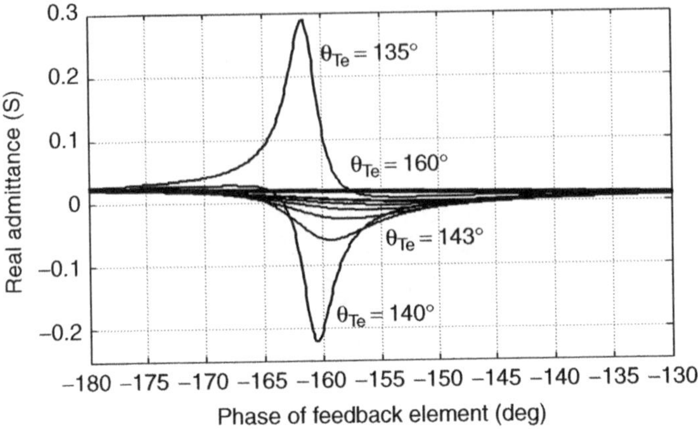

FIGURE 8.3 Conductance variation versus the phase of the reactive terminal and feedback elements, given by θ_T and θ_{fb}, respectively.

and $1/Z_2(\omega)$ [i.e., $Y_T(V, \omega) = 1/Z_{out} + 1/Z_2$], with V being the node voltage. The choice of admittance functions is generally more convenient due to the common use of voltage-dependent models of the nonlinear devices. Thus, only admittance functions are considered in the remainder of the chapter.

The analysis of Fig. 8.2b provides values of the terminal impedance X_{Te} (on the border of the Smith chart) for which negative resistance (or conductance) is obtained, at port 2, at different θ_{fb} values, with $\mu < 1$. However, it does not show the actual values of negative resistance or conductance. This requires an additional analysis that will allow the optimum selection of both the feedback and terminal element values. Again, to consider both inductive and capacitive values globally, this terminal impedance will be written $\Gamma_{Te} = 1e^{j\theta_{Te}}$ (Fig. 8.1b). Renaming $Y_N \equiv 1/Z_{out} = G_N + jB_N$, the variation of the function $G_N(\theta_{fb}, \theta_{Te})$ at the desired oscillation frequency ω_o will be analyzed by means of a double sweep in the two phases θ_{fb} and θ_{Te}.

Figure 8.3 shows the results of the application of the double-sweep technique to the MESFET oscillator, biased at $V_{GS} = -0.5$ V and $V_{DS} = 2.5$ V. Note that the analysis frequency agrees with the desired oscillation frequency $f_o = 4.5$ GHz. Phase values between 0 and 180° will correspond to inductances, whereas phase values between 180 and 360° will correspond to capacitances. This distinction is useful if a feedback or terminal element of a particular type is desired, as in the case of voltage-controlled oscillators (VCOs), requiring the inclusion of a tuned capacitance either as terminal or feedback element. In the case of Fig. 8.3, the maximum conductance is obtained for the phase values $\theta_{fb} = 200^\circ$ and $\theta_{Te} = 130^\circ$. Thus, a capacitive feedback element $C = 4$ pF and an inductive terminal element $L = 0.82$ nH will be used. Note that it would be possible to impose a different combination of linear elements (i.e., inductive feedback and capacitive termination) by choosing a different bias point or a different transistor.

After the choice of the optimum terminal and feedback elements, a load admittance $Y_L \equiv G_L + jB_L$ is connected to the transistor block. To facilitate the oscillation startup, this load impedance must fulfill (Chapter 1)

$$\begin{aligned} G_T(V \cong 0, \omega_o) &\equiv G_N(V \cong 0, \omega_o) + G_L(\omega_o) < 0 \\ B_T(V \cong 0, \omega_o) &\equiv B_N(V \cong 0, \omega_o) + B_L(\omega_o) = 0 \\ \frac{\partial B_T(V \cong 0, \omega_o)}{\partial \omega} &> 0 \end{aligned} \tag{8.2}$$

where ω_o is the desired oscillation frequency and $V \cong 0$ indicates small-signal voltage. To maximize the output power $P_o = \frac{1}{2} G_L V_o^2$, an empirical criterion is $G_L = -1/3G_N(V \cong 0, \omega_o)$. Conditions (8.2) are extremely helpful for circuit design. They generally allow the synthesis of a pair of complex conjugate poles of the dc solution with positive σ at the oscillation frequency ω_o. As discussed in Chapter 1, due to the relationship between the admittance and the linear system dominant poles, the condition on the positive sign of the derivative $\partial B_T/\partial \omega|_o$ is essential. However, to minimize the risk of nonoscillation, it is also convenient to apply more rigorous stability analysis techniques to the dc solution, based on pole–zero identification, for example.

In a MESFET-based oscillator, $B_N(4.5 \text{ GHz}) = 0.069$ S. Thus, a parallel inductance of value $L = 0.511$ nH should be introduced to obtain $B_T = 0$. The oscillator resonance is formed between the inductance and the output capacitance of the device used. Figure 8.4 shows the fulfillment of oscillation startup conditions at $f_o = 4.5$ GHz. The total conductance is shown by the solid line, the total susceptance by the dashed line. As can be seen, the frequency derivative of the susceptance $\partial B_T/\partial \omega|_o$ is rather small, so the design will have low frequency selectivity. The low selectivity is associated with high sensitivity to perturbations of the oscillation frequency ω_o, as shown in Chapter 2. Thus, higher phase noise should be expected for lower $\partial B_T/\partial \omega|_o$. On the other hand, the device capacitance, resonating with the parallel inductance L introduced, depends on the oscillation amplitude, the temperature, and the bias supply, which also contributes to increase the oscillator phase noise.

Proper choice of the load impedance is essential for a low-phase-noise design [13]. As indicated in Chapter 1, the dependence on the device capacitance can be reduced considerably by introducing a high-quality factor resonator. Assuming parallel resonance, the output resonance pair C_{load} and L_{load} will consist of a large capacitance and a small inductance. One advantage of this procedure is that the oscillation frequency is nearly independent of the device capacitance, since $C_{\text{load}} + C_d \cong C_{\text{load}}$. This reduces the influence of the device noise. A second advantage is the increase in quality factor due to the large $C_{\text{load}}/L_{\text{load}}$ ratio of the parallel resonator. Thus, the oscillator phase noise should decrease. Note that the changes have been performed without affecting the negative conductance exhibited by the nonlinear block, due to the parallel connection of C_{load} and L_{load}. Exactly the same procedure could have been used in the case of a series resonance, connecting a small capacitance in series to the transistor output, such that $C_{\text{load}}C_d/(C_{\text{load}} + C_d) \cong$

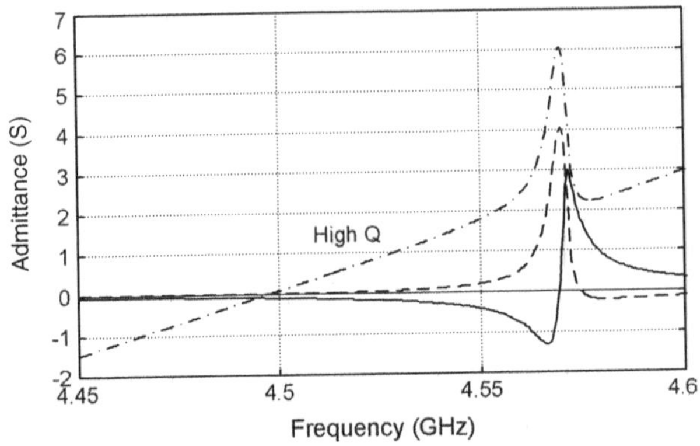

FIGURE 8.4 Fulfillment of oscillation startup conditions for $f_o = 4.5$ GHz. The total conductance is shown by the solid line, and the total susceptance is shown by the dashed line. Introduction of the LC parallel resonator $C = 0.34$ nH, $L = 5.14$ pF does not change the real part of the total admittance (solid line), but the frequency derivative of the imaginary part at f_o is remarkably higher (dotted-dashed line).

C_{load}. In this case, a large load inductance L_{load} would be required to maintain the same value of the oscillation frequency, which leads equally to a high-quality-factor resonator.

As an example, the original load inductance $L_{\text{load}} = 0.511$ nH of the MESFET-based oscillator has been replaced with the LC parallel resonator $C = 0.34$ pF, $L = 5.14$ nH, fulfilling $\omega_o = 1/\sqrt{LC}$, with ω_o being the desired oscillation frequency. Introduction of this reactive subcircuit does not change the real part of the total admittance (the solid line), but the frequency derivative of the imaginary part at ω_o is remarkably higher (the dotted-dashed line). Thus, higher-frequency selectivity and less sensitivity to fluctuations of the device capacitance are obtained.

From the discussion above, the connection of a high-quality-factor resonator to the output of the nonlinear block will reduce the dependence of the resonance frequency ω_o on the device capacitance and will lead to phase noise reduction. This can be done in three different manners [13]:

1. *Dielectric resonator.* Dielectric resonators are made of ceramic material, which is stable in temperature, with low loss and high dielectric constant [8,14]. The resonator is placed in the neighborhood of a transmission line and is coupled to this line. Its equivalent circuit is a parallel resonant circuit, in series with the line, whose element values R, L, and C depend on the distance of the resonator from the transmission line and on the length of this transmission line that is actually coupled to the resonator [14]. The dielectric resonators cannot be implemented in planar technology.

2. *Transmission-line resonators.* A distributed resonator, based on a short-circuited/open-circuited transmission line can be used instead of a dielectric resonator. As shown in the appendix, an entire series or parallel R-L-C

resonator can be implemented with a single properly ended transmission line [15]. As an example, a transmission line with physical length l ending in a short circuit will give rise to a parallel resonance at the frequencies fulfilling $\beta_n l = (\omega_{o,n}/v_g)l = (2n+1)(\pi/2)$. To implement a particular lumped parallel resonator L-C at ω_o, the physical length of the line must agree with $l = (2n+1)\lambda_o/4$, with λ_o being the wavelength at ω_o. The frequency derivative of the line susceptance at the resonance frequency is given by (see Section A.4)

$$\left.\frac{\partial B(l)}{\partial \omega_o}\right| = \frac{2n+1}{4 f_o Z_c} = 2C \tag{8.3}$$

where Z_c is the required characteristic impedance. For the resonator design, the lowest resonance $n = 0$ is generally chosen. As gathered from (8.3), we can use a long transmission line to increment the spectral purity of the oscillator signal. Generally this line will be part of a feedback loop added to an existing oscillator circuit (Chapter 9). The "external" use of the long transmission line avoids the problem associated to its multiple resonances at $f_o/(2n+1)$.

Alternatively, the parallel resonator of the oscillator circuit can be constituted by a lumped capacitance and an inductance, implemented on a transmission line, using the expression derived in the appendix. The total input susceptance will be $B_T(\omega) = C\omega - \{Z_c \tan[(\omega/v_g)l]\}^{-1}$. At resonance, the condition $B_T(\omega_o) = 0$ is fulfilled. The frequency selectivity is $\partial B_T(\omega_o)/\partial\omega = C + l/Z_c v_g \sin^2\left[\omega_o l/v_g\right]$. Thus, higher selectivity is obtained for a longer physical length of transmission line.

3. *Two-section transmission line.* The frequency selectivity of the distributed resonator can be increased using a two-section line [13]. The expression (A.1) in the appendix provides the input impedance of a transmission line of characteristic impedance Z_c, ended with the load impedance Z_L. Assume two-line sections of the same electrical length θ and characteristic impedances Z_1 and Z_2. The Z_2 section is ended in a short circuit. Using (A.1) we find that this section exhibits the input impedance $Z_{L1} = jZ_2 tg\theta$, which in turn constitutes the load impedance of the Z_1 section. Using (A.1) again, we obtain the total input impedance $Z_{in} = jZ_1(1+M)tg\theta/(1 - Mtg^2\theta)$, with M being the ratio between the two characteristic impedances $M = Z_2/Z_1$. Provided that the Z_{in} zeros are obtained at the electrical lengths $k\theta_z$, with k an integer, the poles are located at $\theta_p = k\theta_z - 2\tan^{-1}(1/M)$. By suitably choosing the ratio M we can make the poles approach the zeros, thus increasing the positive slope of the reactance at the resonance frequencies. Note that inverting (A.1), we obtain an identical result for the case of a *parallel* resonator, implemented with a two-section transmission line, ended in an open circuit.

8.2.2 Steady-State Design Using One-Harmonic Accuracy

Assuming an ideal waveform $v(t) = V_o \cos\omega_o t$, the steady-state oscillation condition showing the balance between the delivered and dissipated energy in one oscillation period is given by

$$\begin{aligned} G_T(V_o, \omega_o') &\equiv G_N(V_o, \omega_o') + G_L(\omega_o') = 0 \\ B_T(V_o, \omega_o') &\equiv B_N(V_o, \omega_o') + B_L(\omega_o') = 0 \end{aligned} \tag{8.4}$$

As shown in equations (8.4), the steady-state oscillation frequency will generally be different from the desired value ω_o at which the startup condition (8.2) is evaluated. This is due to the nonlinear dependence of both G_N and B_N on the amplitude V. Unlike the startup condition (8.2), the steady-state equation (8.4) requires a nonlinear model $Y_N(V, \omega)$ for admittance of the active device. This equation will enable approximate calculation (with one harmonic component) of the oscillation output power and frequency. After the initial design, based on (8.4), a rigorous nonlinear analysis should be carried out, taking into account a sufficient number of harmonic terms, which will have an influence on both the oscillation frequency and power.

The nonlinear function $Y_N(V, \omega)$ can be determined empirically, analytically, or numerically. The empirical determination is carried out from measurements of the scattering parameters versus the input amplitude V in a frequency band about the oscillation frequency expected. The analytical determination of Y_N relies on the *describing function*, discussed exhaustively in Chapters 1 and 4. When multiple nonlinearities are present, as is the case with the transistor devices, this analytical determination is demanding. It requires an individual describing function for each nonlinear element having either one or two input variables. Due to the generalized use of commercial harmonic balance simulations these days, only numerical determination of Y_N will be considered here, This will be done simply with a one-harmonic analysis of the nonlinear block, containing the transistor, plus the feedback and termination elements, by limiting the harmonic balance simulation to the fundamental frequency only. It is assumed that values of the terminal and feedback element, ensuring that $G_N(V \cong 0, \omega_o) < 0$, have already been determined using the technique in Section 8.2.1. For these particular values of the feedback elements and termination, the block can be made dependent on the design port voltage only. A simple harmonic balance technique can be used for an initial approximate design. This technique is described below.

Evaluation of the admittance function associated with the transistor block will be carried out with an auxiliary generator of voltage type. An auxiliary generator of amplitude $V_{AG} = V$ and frequency $\omega_{AG} = \omega_o$ (equal to the desired oscillation frequency) will be connected to the design port (port 2 of Fig. 8.1). A high-valued resistor should also be added in parallel to avoid any convergence problem of the harmonic balance algorithm. Then, through Kirchhoff's laws, the ratio $Y_{AG} = I_{AG}/V_{AG}$, with I_{AG} entering the circuit, agrees with the sum $Y_{AG} = Y_N(V_{AG}, \omega_{AG}) + 1/R \cong Y_N(V, \omega)$. Then a harmonic balance sweep in the auxiliary generator amplitude $V_{AG} = V$ from a small value to a few volts is carried out. Both the real and imaginary parts of Y_N are traced versus V. The resulting curve will allow estimating the required values of the load admittance to obtain an oscillation at V_o, ω_o. One selects the voltage amplitude V_o on the abscissas and reads the

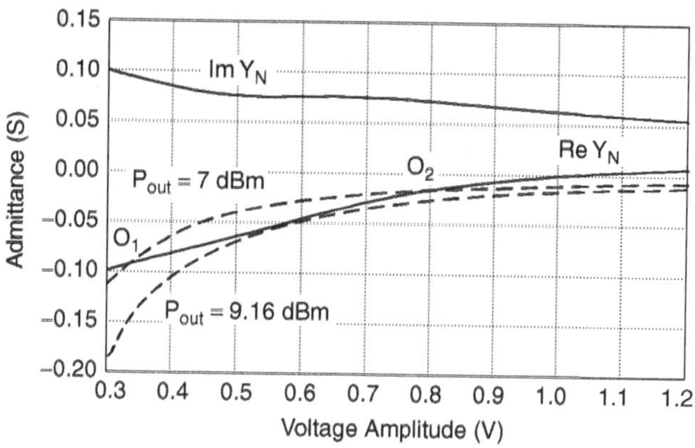

FIGURE 8.5 Approximate nonlinear design for a specified output power based on an amplitude sweep.

values of the block admittance $Y_N(V_o, \omega_o)$. To complete the design, a load admittance Y_L such that $G_L(\omega_o) = -G_N(V_o, \omega_o)$ and $B_L(\omega_o) = -B_N(V_o, \omega_o)$ should be chosen. Note that this corresponds to the steady-state oscillation condition, evaluated with one-harmonic accuracy, given in (8.4).

It is also possible to impose the output power desired. This power depends on both the voltage amplitude at the observation port and the real part of the load impedance: $P_o = G_L(V^2/2)$. Thus, there are infinite pairs of values V and G_L providing the same output power. This pair of values comprises the curve $G_L = 2P_o/V^2$. To design for the output power P_o, the opposite curve, $-G_L = -2P_o/V^2$, must be superimposed over those providing a variation of $G_N(V, \omega_o)$ and $B_N(V, \omega_o)$. For an approximate design of the oscillator circuit for the output power P_o, a suitable intersection point of the curves $G_N(V, \omega_o)$ and $-G_L(V)$ should be chosen. The values at ω_o of the load admittance selected will be $Y_L = G_{Lo} + jB_{Lo}$, with G_{Lo}, B_{Lo} being the opposite of the admittance values of the nonlinear block at the intersection point selected.

In Fig. 8.5, the technique is applied to a MESFET oscillator. The design for specified output power P_o is carried out by adding the curve $-G_L = -2P_o/V^2$ to the representation of Fig. 8.5. Two different curves, $-G_{L1}(V)$ and $-G_{L2}(V)$, have actually been considered, corresponding to the output power $P_o = 7$ dBm and $P_o = 9.16$ dBm, respectively. In the case of $-G_{L1}(V)$, two different intersection points are obtained with $-G_N(V_o, \omega_o)$. The first intersection point, with $V = 0.33$ V, requires the load resistance $R_L = 1/G_L = 10.87\ \Omega$. This low value is impractical for circuit implementation. The second intersection point, with $V = 0.78$ V, requires a load resistance $R_L = 62.5\ \Omega$, which is much better suited for implementation. Thus, the second intersection point is kept for the design.

The nonlinear susceptance varies from $B_N = 0.11$ S, in small-signal mode, to $B_N = 0.073$ S, at the oscillation point desired. Thus, the required susceptance value of the oscillator load will be $B_L = -0.073$ S, which can be obtained with a parallel inductance, of value $L = 0.4845$ nH. Of course, for this design to be valid, the oscillation startup conditions must still be fulfilled with these load-element values, and the steady-state oscillation must be stable.

8.2.3 Multiharmonic Steady-State Design

The accurate oscillator design using admittance equations requires an auxiliary generator of voltage type. This generator is connected in parallel at the observation node. For the oscillator of Fig. 8.1, it would be connected between port 2 and ground, as shown in Fig. 8.6. The auxiliary generator frequency is equal to the oscillation frequency $\omega_{AG} = \omega_o$. Its amplitude is A_{AG} and its phase is arbitrarily set to zero: $\phi_{AG} = 0$. The voltage auxiliary generator has an ideal bandpass filter in series, which exhibits zero impedance at $\omega = \omega_{AG}$ and an infinite impedance at $\omega \neq \omega_{AG}$. The objective of this filter is to prevent the voltage generator from short-circuiting frequency components different from $\omega = \omega_{AG}$. Two applications of the auxiliary generator technique to oscillator synthesis are presented next.

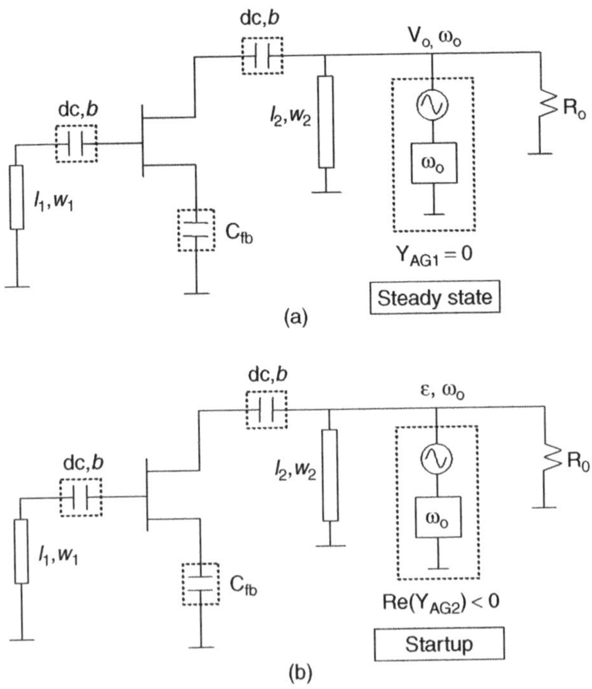

FIGURE 8.6 Oscillator circuit implementation on a microstrip line using the auxiliary generator technique. Accurate models have also been introduced for packaged elements. The circuit is duplicated during the optimization procedure. (a) Circuit analyzed at the steady-state oscillation, with the auxiliary generator fulfilling $Y_{AG1} = 0$. (b) Circuit analyzed in small-signal mode used to facilitate fulfillment of oscillation startup conditions. Three optimization goals are used: $\mathrm{Re}[Y_{AG1}] = 0$, $\mathrm{Im}[Y_{AG1}] = 0$, and $\mathrm{Re}[Y_{AG1}] < 0$.

8.2.3.1 Nonlinear Analysis After implementation of the load circuit resulting from the approximate technique of Section 8.2.2, the accuracy of the oscillator design is verified through a harmonic balance simulation. A sufficiently high number of harmonic components must be considered. The ratio Y_{AG} between the current through the generator entering the circuit and the voltage delivered agrees, by Kirchhoff's laws, with the total admittance function seen from the observation node at the frequency ω_{AG}: that is, $Y_{AG} = Y_N + Y_L$. If good accuracy has been obtained in the oscillator design, the ratio Y_{AG} will be close to zero, $Y_{AG}(V_o, \omega_o) \cong 0$. Note that the harmonic content will influence the admittance values exhibited by the nonlinear block at ω_o, so the condition $Y_{AG}(V_o, \omega_o) = 0$ is not likely to be fulfilled at a first attempt.

As presented in Chapter 5, to fulfill $Y_{AG} = 0$, the actual oscillator solution is obtained through the calculation/optimization of V_{AG}, ω_{AG}. This multiharmonic analysis will provide the oscillation frequency ω_o' and node voltage amplitude V_o'. The primes represent the fact that the actual oscillation amplitude and frequency are different from those imposed at the design stage, limited to the fundamental frequency. For a MESFET oscillator, when considering 8 harmonic terms, the oscillation frequency changes from $f_o = 4.5$ GHz to $f_o' = 4.513$ GHz, and the output power changes from $P_o = 7$ dBm to $P_o' = 7.8$ dBm.

8.2.3.2 Nonlinear Optimization The auxiliary generator can be used for a nonlinear optimization of the oscillator circuit, avoiding the shift from (V_o, ω_o) to (V_o', ω_o') when all the harmonic components are taken into account. To achieve this, the auxiliary generator amplitude and frequency are set to the desired values $A_{AG} = V_o$ and $\omega_{AG} = \omega_o$ respectively. Then two circuit elements, η_1 and η_2, are optimized or calculated so as to fulfill $Y_{AG}(\eta_1, \eta_2) = 0$. In a circuit output load that has a parallel topology, setting the amplitude $A_{AG} = V_o$ will allow imposing the output power value, since $P_o = V_o^2/2R_L$. In the case of a series resonance, the output power can be preset using a current auxiliary generator rather than a voltage generator (Chapter 5). However, even in the case of a series resonance, it is often possible to connect the voltage auxiliary generator in parallel with the output load resistance and still have good convergence properties. This will allow presetting the output power through $V_o = \sqrt{2P_oR_L}$.

The auxiliary generator technique is very useful for the sequential introduction of accurate models of the oscillator elements. It can be applied for the transmission-line implementation of lumped elements, avoiding any undesired shift of the oscillation frequency. In the appendix, expressions are derived for the transmission-line implementation of different types of lumped elements [15]. The lumped elements considered are series and parallel inductors and parallel capacitors, series and parallel resonators, and load resistances. The expressions allow calculation of the physical length l and/or characteristic impedance Z_c for given values of the lumped element or resonator circuit. Note that the expressions are approximate. In resonator circuits they are valid in a certain bandwidth about the resonance frequency. This comes from the fact that inherently, the lines and lumped elements have a different frequency response and the equivalences are obtained by equating the frequency derivative of the reactance/susceptance at the resonance frequency. On the other hand, the load resistance R_L can be implemented with a quarter-wavelength line $l = \lambda_o/4$, with characteristic impedance $Z_c = \sqrt{R_L 50}\ \Omega$. In the case of microstrip line, the line width w providing the necessary Z_c and the effective dielectric constant ε_{eff}, determining $\lambda_o = c/f\sqrt{\varepsilon_{\text{eff}}}$, can be obtained with a line calculation program.

The transmission-line implementation of an initial oscillator design is carried out element by element. This is illustrated in Fig. 8.6, corresponding to the FET-based oscillator at 4.5 GHz. Because the relationships in the appendix are only approximate, the geometrical dimensions of each new distributed element, length (l) and width (w), for instance, are optimized or calculated so as to fulfill $Y_{AG}(l, w) = 0$, with the auxiliary generator operating at the desired values $A_{AG} = V_o$ and $\omega_{AG} = \omega_o$. The implementation modifies the linear network, so the oscillation startup conditions might not be fulfilled after the optimization procedure. After implementation of the various elements, microstrip discontinuities (steps, tees, crosses, etc.) are also considered in the linear circuit. After the introduction of each discontinuity, two circuit elements must be reoptimized in order to fulfill $Y_{AG} = 0$ at $A_{AG} = V_o$ and $\omega_{AG} = \omega_o$.

To facilitate fulfillment of oscillation startup conditions during nonlinear optimization, a simple simulation technique can be used. The oscillator is duplicated in the circuit page. The two circuit copies are simulated with the same harmonic balance analysis at the same frequency, ω_o. An auxiliary generator is connected to each circuit at identical nodes. The first circuit (Fig. 8.6a) operates in large-signal mode at the steady-state values desired, $A_{AG} = V_o$ and $\omega_{AG} = \omega_o$. The second circuit (Fig. 8.6b) operates at very small amplitude $A_{AG} = \varepsilon$ and frequency $\omega_{AG} = \omega_o$. The second circuit is used for calculation of the total small-signal admittance $Y_{AG}(A_{AG} = \varepsilon, \omega_o)$ at the observation node. All the element values are the same in the two circuit copies, which are analyzed simultaneously. Note that even though the auxiliary generator of the second circuit has small-signal values, it is analyzed with harmonic balance, together with the first circuit, in a single harmonic balance simulation.

At each element implementation, the line length l and width w (or any other parameters) are optimized so as to fulfill three different goals: $\text{Re}[Y_{AG1}] = 0$, $\text{Im}[Y_{AG1}] = 0$, and $\text{Re}[Y_{AG2}] < 0$. The first two goals refer to the circuit that operates at the steady-state regime $A_{AG} = V_o$ and $\omega_{AG} = \omega_o$. The third condition refers to a circuit that operates at small-signal mode. It forces the circuit to exhibit negative resistance under small-signal conditions. Note that for oscillation startup, resonance with the positive frequency slope of the imaginary part is also necessary. However, the amplitude dependence of the imaginary part of the circuit admittance/impedance is usually much lower than that of the real part. Thus, the steady-state condition $\text{Im}[Y_{AG1}] = 0$ at ω_o will generally enable a nearby resonance, also at small signal. Note that the startup frequency is usually different from the steady-state frequency. This is why no condition is imposed on the imaginary part of the small-signal admittance.

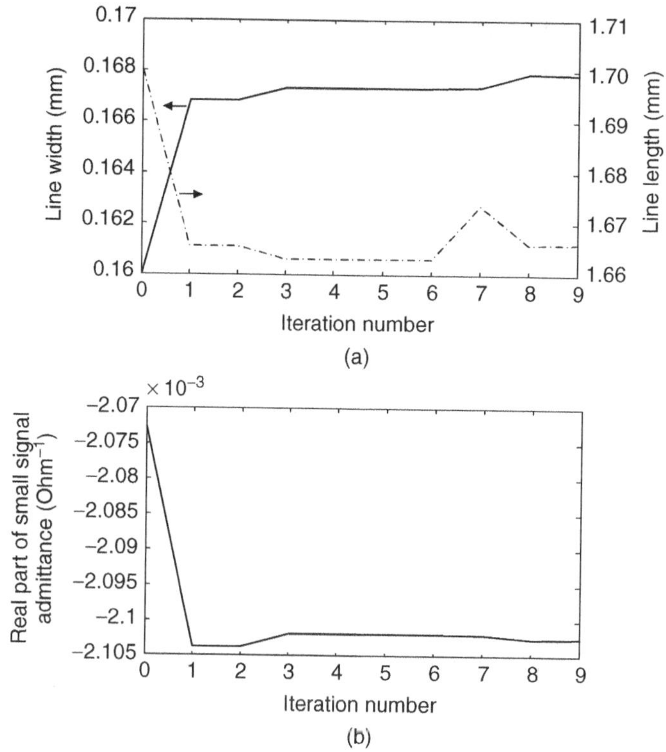

FIGURE 8.7 Implementation of the resonance inductance $L_{res} = 0.8$ nH: (a) variation in the transmission-line width and length during the optimization process with constant oscillation amplitude $V_o = 1$ V and frequency $f_o = 1$ GHz; (b) variation in the real part of the small-signal admittance during the same optimization procedure.

The technique described has been applied to microstrip implementation of the various elements of a FET-based oscillator. As already stated, this implementation is carried out element by element. Implementation of the parallel resonance inductor, $L_{res} = 0.8$ nH, connected to the drain output is shown here as an example. A line with the high characteristic impedance $Z_c = 110\ \Omega$ is used. For CuClad 2.17 at 4.5 GHz, this corresponds to a line width of $w = 0.16$ mm and an effective dielectric constant $\varepsilon_{eff} = 1.693$. Applying the expression (A.1), the required line length for $L_{res} = 0.8$ nH is $l = 1.7$ mm. Figure 8.7a shows the variation of the two variables l and w during the optimization procedure, with constant $f_{AG} = 4.5$ GHz and $V_{AG} = 1$ V. Figure 8.7b shows the variation of the real part of the small-signal admittance $\mathrm{Re}[Y_{AG2}]$. The negative value $\mathrm{Re}[Y_{AG2}] < 0$ is maintained during the entire optimization procedure. In some cases it may also be necessary to allow the variation of other circuit elements. Here, besides l and w, the length of the transformer line was taken as an optimization variable. The final line length and width values are $w = 0.17$ mm and $l = 16.6$ mm, quite close to the initial ones.

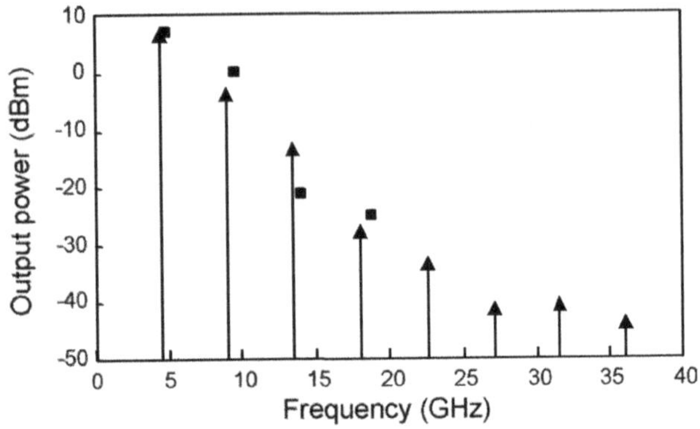

FIGURE 8.8 Output spectrum of the final oscillator circuit, with inclusion of experimental results.

The same procedure can also be used for the progressive introduction of accurate models for varactor diodes and packaged inductors or capacitors. For example, in the MESFET oscillator considered, it has been used for introduction of the dc block and feedback capacitor models, to avoid any shift of the oscillation frequency. After concluding the oscillator design, the stability of the steady-state oscillation and fulfillment of the startup conditions must be verified, through an accurate stability analysis technique such as the pole–zero identification [16] or the Nyquist criterion [17,18] (see Chapter 6). In the case of instability, techniques can still be applied to remove undesired spurious frequencies. These techniques are shown in Chapter 12. The final oscillator spectrum at $f_o = 4.5$ GHz is shown in Fig. 8.8, together with the experimental values.

8.3 DESIGN OF VOLTAGE-CONTROLLED OSCILLATORS

In a voltage-controlled oscillator (VCO), the oscillation frequency is tuned by modifying the bias voltage of a varactor diode. The reverse-biased diode is essentially a voltage-dependent junction capacitance $C_j(v)$, in series with a loss resistance R_s. The junction capacitance varies according to $C_j(v) = C_{jo}/(1 - v/\phi_o)^{\gamma}$, where ϕ_o is the built-in potential, C_{jo} is the capacitance value at zero bias voltage, and the parameter γ is $\gamma = 1/2$ for an abrupt junction and $\gamma = 1/3$ for a gradual one. The largest capacitance value is $C_{max} \cong C_{jo}$, and the region of maximum sensitivity corresponds to relatively low reverse-bias voltages, as the capacitance characteristic $C_j(v)$ saturates versus this bias voltage. The practical variation range is limited to C_{min} to C_{max}.

8.3.1 Technique for Increasing Oscillation Bandwidth

A simple technique can be applied to maximize the variation of the oscillator frequency [14] for a given capacitance range C_{min} to C_{max}. The technique is illustrated by means of its application to the MESFET oscillator circuit of Fig. 8.6. In the VCO design, the varactor diode can be connected to a transistor terminal different from the one at which the series feedback is introduced. However, the varactor can also be used as the feedback element. This will be done in the case of the oscillator of Fig. 8.6, which has a single capacitive element connected to the source terminal. The transistor, together with the feedback and terminal elements constituting the nonlinear block, must exhibit negative conductance in a frequency interval containing the oscillation band desired (Fig. 8.9). For the approximate determination of this band, we analyze the small-signal susceptance $B_N(V \cong 0, \omega)$ seen from the varactor terminals. These terminals correspond to the transistor source terminal and ground. The small-signal susceptance $B_N(V \cong 0, \omega)$ is calculated from these terminals, looking into the transistor. This susceptance $B_N(V \cong 0, \omega)$ must be traced versus ω, together with the two straight lines $-C_{min}\omega$ and $-C_{max}\omega$. Then

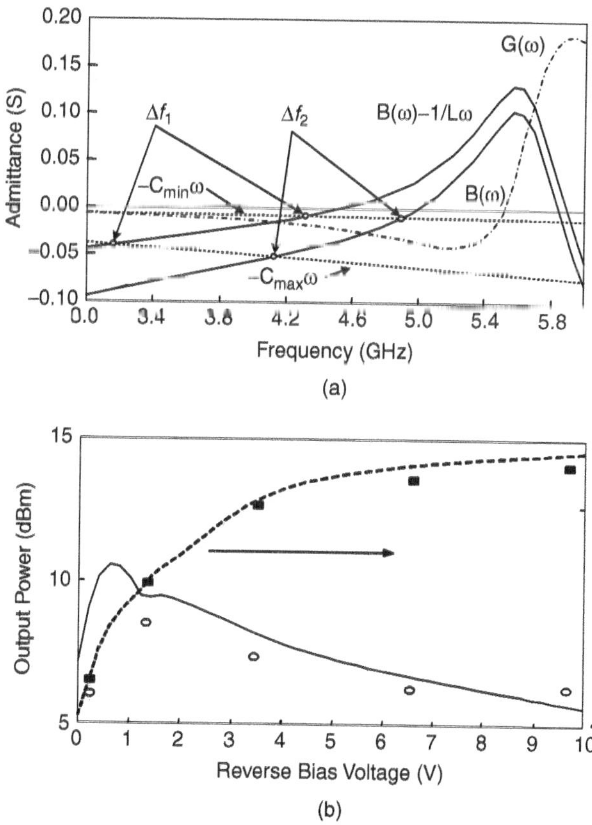

FIGURE 8.9 Design of a voltage-controlled oscillator: (a) method to increase the frequency variation band versus the varactor capacitance through the introduction of an inductive element in parallel; (b) frequency and output power variation versus the reverse bias voltage obtained through a harmonic balance analysis. Experimental points are included.

the oscillation frequency interval (indicated in the figure as Δf_2) is delimited by the two intersection points of these two straight lines with the curve $B(\omega)$. This is shown in Fig. 8.9, where the varactor diode has the capacitance variation range $C_{\min} = 0.3$ pF to $C_{\max} = 2$ pF.

As shown by Soares et al. [14], the introduction of an inductor in parallel with $B(\omega)$ allows flattening of the susceptance curve seen from the transistor source terminal (and looking into the transistor), which becomes $B'(\omega) = B(\omega) - 1/L\omega$. The inductance will increase the frequency separation between the two intersection points. Provided that the two points belong to the negative-resistance interval, a broader frequency tuning range will be obtained. In Fig. 8.9a, the initial frequency bandwidth is $\Delta f = 734$ MHz. After the introduction of parallel inductance, the oscillation interval becomes Δf_1 and the bandwidth increases to $\Delta f = 1.286$ GHz. There is a reduction of the operation frequency band which could be mitigated with the technique to be presented in the next section. Nonlinear analysis of Fig. 8.9b, showing the variation in oscillator frequency and power versus the varactor bias, confirms the validity of the technique. There is, however, a shift of the oscillation band predicted, due to nonlinear effects. The experimental values are in good agreement with the simulations.

8.3.2 Technique to Preset the Oscillation Band

As shown in Section 8.2, it is possible to impose the frequency ω_o and output power P_o of the steady-state oscillation using an auxiliary generator in harmonic balance. When dealing with voltage-controlled oscillators, an entire operation band $\omega_1 - \omega_2$ must be considered, with all the oscillator steady-state solutions located between these two frequencies. Here a technique is provided to preset the operation band of the voltage-controlled oscillator. This technique should be applied after the initial selection of a convenient topology for the oscillator circuit.

To modify the oscillation band, two points of the VCO characteristic are set to the tuning voltage and frequency values desired. The corresponding oscillation frequencies, ω_1 to ω_2, will agree with the limits of the operation band or will be contained within this band. For control over a broad band, the tuning voltages considered should be those providing the limit capacitance values $C_{\min}(V_{t2})$ and $C_{\max}(V_{t1})$. Although only two points are considered, the VCO characteristic is typically regular, so smooth behavior is expected between the two points imposed. The oscillation frequency must be ω_1 if the varactor diode is biased at V_{t1} and ω_2 if the varactor diode is biased at V_{t2}. Introducing an auxiliary generator into the circuit, the oscillation condition $Y_{AG} = 0$ must be fulfilled simultaneously at the two points (V_{t1}, ω_1) and (V_{t2}, ω_2) for the same circuit element values (except the tuning voltage V_t). The auxiliary generator will operate, at each turning-voltage value, at the desired oscillation frequency, so $\omega_{AG} = \omega_1$ for V_{t1} and $\omega_{AG} = \omega_2$ for V_{t2}. To preset the oscillation band, two circuit copies, operating at V_{t1} and V_{t2}, may be resolved in harmonic balance simultaneously to fulfill $Y_{AG1}(\overline{p}) = 0$ and $Y_{AG2}(\overline{p}) = 0$, with $\overline{p}$ being the vector containing the optimization parameters selected: some circuit-element values and the auxiliary-generator amplitude A_{AGi}, which, in general, will not be prefixed. If multipoint harmonic balance optimization is available, the following goal will also be possible:

$$H(\overline{p}) = \sum_{i=1}^{2} |Y_{AGi}(V_{ti}, \omega_{AGi}, \overline{p})|^2 = 0 \tag{8.5}$$

where V_{ti} and ω_{AGi} have the desired turning voltage and frequency values. To solve this equation, a gradient method is used. At each iteration k, the optimization circuit parameters are set to the value $\overline{p} = \overline{p}_k$ starting from an initial guess $\overline{p}_o$. Then the error function $H(\overline{p}_k)$ is constructed by setting the VCO tuning voltage V_t and the auxiliary generator frequency ω_{AG} sequentially to the imposed values (V_{ti}, ω_i) and calculating the admittance $Y_{AGi}(V_{ti}, \omega_{AGi}, \overline{p})$. Due to the quadratic nature of $H(\overline{p})$, the solution $\overline{p} = \overline{p}_s$ will be obtained for the zero gradient $\partial H(\overline{p})/\partial\overline{p} = 0$.

The CAD technique described is illustrated in Fig. 8.10 by means of its application to an existing MESFET-based VCO at 5 GHz. The oscillation frequency is changed by sweeping the tuning voltage V_t that is applied to the varactor diode D_1 connected to node 1. To obtain the steady-state oscillation, an auxiliary generator has been used. The auxiliary generator at the oscillation frequency $\omega_{AG} = \omega_o$ is connected to the gate node. The original frequency–voltage characteristic is obtained by sweeping V_t and calculating the auxiliary generator amplitude V_{AG} and frequency ω_{AG} to fulfill $Y_{AG} = 0$. This original characteristic is represented by the curve (1) of Fig. 8.11. As can be seen, the curve saturates for the tuning voltage value $V_s = 2.5$ V, so the VCO operating band is limited to the range 4.3 to 5.34 GHz.

To show the capabilities of the technique, the original VCO band has been modified in different ways using the optimization parameters $\overline{p} = [L_1, L_2, L_3, C_1]$. In the characteristics (2) and (3) of Fig. 8.11, the technique has been applied to impose the upper limit of the band while keeping the lower limit to the fixed

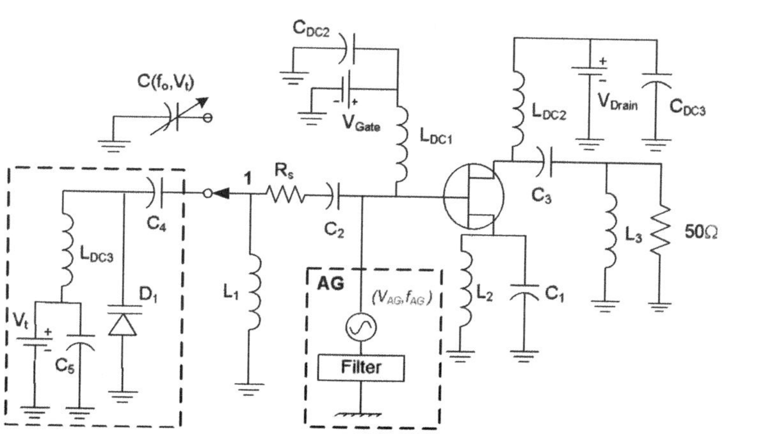

FIGURE 8.10 Schematic of the FET-based VCO. The auxiliary generator is introduced for different optimization purposes.

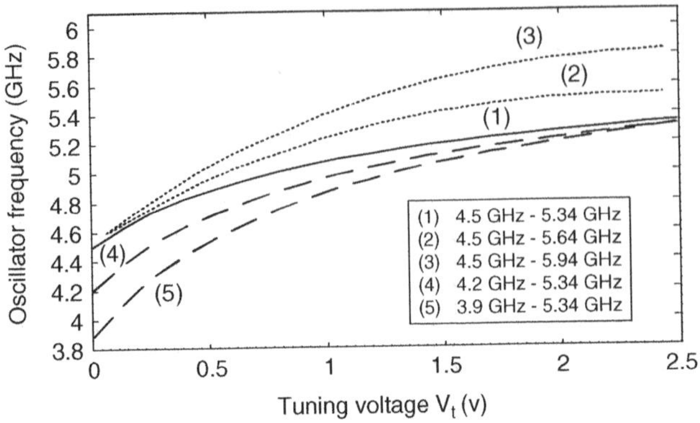

FIGURE 8.11 Technique to impose the limits of the VCO operation band. In the dotted-line curves, f_1 is kept constant and f_2 changes in Δf; in the dashed-line curves, f_2 is kept constant and f_1 changes in Δf.

value $f_1 = 4.5$ GHz. The use of a continuation technique is generally convenient, sweeping f_2 from the original value in curve (1) to the desired value and solving equation (8.5) at each step. In curve (2) the upper limit has been increased in $\Delta f =$ 300 MHz. In curve (3) this limit has been increased in $2\Delta f$. In the characteristics (4) and (5), the lower limit has been reduced in Δf and $2\Delta f$ while keeping the upper limit at the original value $f_2 = 5.34$ GHz. The regular behavior of the solution curve expected between the two imposed points f_1 and f_2 is confirmed in all cases, while the oscillation amplitude does not vary appreciably with respect to that of the original characteristic.

If a broader oscillation band is desired, with the capability to preset the limits of this band accurately, an inductor can be connected in parallel with the varactor diode, as shown in Section 8.3.1. Then the two-point optimization described would be used in an identical manner.

8.3.3 Technique to Linearize the VCO Characteristic

A linear frequency–voltage characteristic is desirable in most VCO applications. In phase-locked loops, the linear characteristic avoids incorrect estimations of the loop bandwidth and unexpected reductions of the phase-locked band. In chirp signal generators, the original oscillator circuit must exhibit a good degree of linearity, even when using a linearization loop, as demonstrated by Kulpa [7]. Thus, the availability of a simulation procedure for linearization of the frequency–voltage characteristic will be of interest for VCO design.

The linearization technique will be based on the use of an auxiliary generator in harmonic balance [19]. It represents an advance with respect to the preceding techniques, in which a single operation point (Section 8.2) or two operation points (Section 8.3.2) were preset. In the linearization technique, the auxiliary generator will impose a full frequency–voltage characteristic $f_o(V_t)$. As will be shown, the circuit is optimized in the entire oscillation band, in terms of an ideal element, with a frequency-dependent value. This ideal element is implemented at a later stage. The principle of the technique is based on the work of Marazzi and Rizzoli [20]. This work presents a methodology for the experimental determination of the linear impedance required at the varactor connection node to obtain a linear frequency–voltage characteristic in an interval f_1 to f_2. This linear impedance is synthesized by embedding the varactor in a linear network and fitting the elements of this network to match the impedance variation required.

In the simulation technique, the varactor diode and its bias circuit are provisionally replaced with an ideal capacitance C. See, for example, the replacement carried out in the oscillator of Fig. 8.10. The ideal capacitance C will later be implemented with the same varactor diode, embedded in a linear network. This ideal capacitance (used for simulation only) is independent of the bias voltage. The linear characteristic $f_o = k_v V_t + f_{oo}$ desired is imposed with an auxiliary generator, doing $f_{AG}(V_t) = k_v V_t + f_{oo}$. Then V_t is swept, which only affects the value of the auxiliary generator frequency. The oscillation condition $Y_{AG} = 0$ must be fulfilled at each imposed frequency $f_{AG}(V_t)$. The equation $Y_{AG} = 0$ is solved in terms of C and the auxiliary generator amplitude A_{AG} at each V_t value: $Y_{AG}(C, V_{AG}) = 0$. Solving this equation for each V_t value provides the capacitance function $C(f_o, V_t)$ required to obtain the linear characteristic $f_o = k_v V_t + f_{oo}$.

In general it will not be possible to implement the ideal characteristic $C(f_o, V_t)$ using a varactor diode only. Instead, the varactor, with all its parasitic elements, is embedded in a linear network, as suggested by Marazzi and Rizzoli [20]. The network elements are fitted so as to obtain the same variation of the reactance at the fundamental frequency versus the tuning voltage: that is, $X_{\text{net}}(f_o, V_t) = -1/[C(f_o, V_t)2\pi f_o]$. Note that the fitting is limited to the fundamental frequency, so further corrections will be necessary to take higher-order harmonics into account. Actually, the reactance exhibited by the subcircuit at higher harmonic frequencies will surely differ from the one presented by the ideal capacitance. The correction of the embedding network used is carried out with a multipoint harmonic balance optimization using the auxiliary generator. The goal will be

$$
\begin{aligned}
H(\overline{p}_L) &= \sum_{i=1}^{N} |Y_{AGi}(\overline{p}_L)|^2 = 0 \\
f_{AGi} &\equiv f_{oi} = k_v V_{ti} + f_{oo}
\end{aligned}
\tag{8.6}
$$

where (V_{ti}, f_{oi}) are points of the imposed linear characteristic $f_o(V_t)$, N is the number of points considered, and $\overline{p}_L$ contains the linearization network parameters to be optimized, as well as the auxiliary generator amplitude. The linear characteristic desired is imposed with the auxiliary generator by setting $f_{AG} = k_v V_t + f_{OO}$. At each point of this characteristic, the oscillation condition $Y_{AGi} = 0$ must be fulfilled so that a multipoint harmonic balance optimization is performed.

For illustration, the technique has been applied to the voltage-controlled oscillator of Fig. 8.10. The aim is to linearize the oscillation band between 4.5 and 5.34 GHz with the imposed sensitivity $k_v = 336$ MHz/V. Solving $Y_{AG} = 0$ for

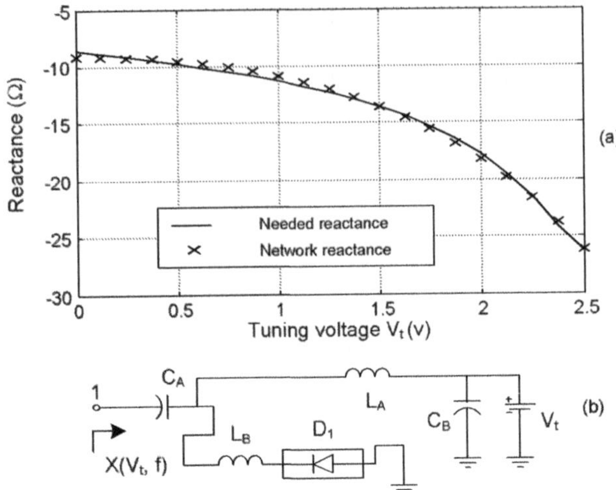

FIGURE 8.12 Implementation of the required first-harmonic reactance for a linear frequency–voltage characteristic: (a) comparison between the required reactance and the reactance provided by the implemented circuit, versus the tuning voltage V_t; (b) schematic of the linearizing network. The boxed diode represents the varactor together with its parasitics.

$f_{AG} = k_v V_t + f_{00}$ versus V_t provides the required ideal capacitance C at each oscillation frequency $f_{AG} = f_o$. Its associated first-harmonic reactance $1/(2\pi f_o C)$ is shown in Fig. 8.12a. The network used to synthesize this first-harmonic reactance is represented in Fig. 8.12b. The elements of the embedding network are fitted to match the ideal reactance, which provides the reactance X_{net} superimposed in Fig. 8.12a, with good agreement. In a second step, the reactance at higher-order harmonics is adjusted through harmonic balance multipoint optimization. For this particular application, the number of points considered is $N = 21$. The original and

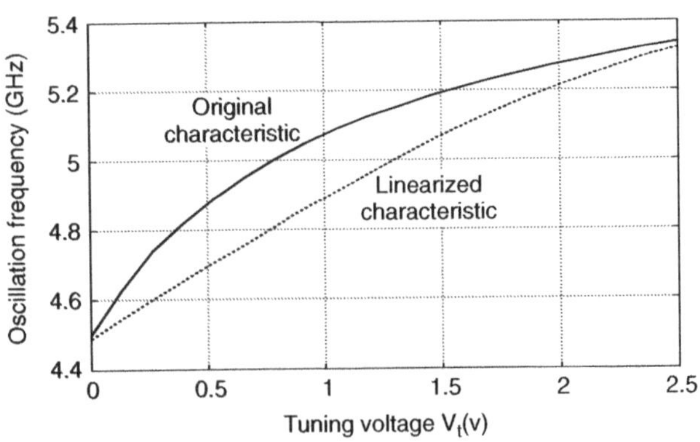

FIGURE 8.13 Comparison of the frequency–voltage characteristics of the original and linearized VCOs.

linearized characteristics are compared in Fig. 8.13. For the linearized characteristic, a full circuit implementation, including parasitics, has been considered.

8.4 MAXIMIZATION OF OSCILLATOR EFFICIENCY

In power oscillators, the high dc power consumption may lead to component damage due to excessive heating. It may also degrade the circuit response, reduce the design reliability, and increase the cost of the thermal management. The oscillator efficiency is defined as the ratio between the output power at the fundamental frequency ω_o and the dc consumption: $E_{ff} = P_o/P_{dc}$, so an oscillator with lower efficiency will require a higher power supply P_{dc} for the same output power P_o. The consumed power P_{dc} is mostly dissipated at the transistor output [2]. In field-effect transistors, the gate dc current is nearly zero and the power is dissipated at the drain–source port. In bipolar transistors most of the power is dissipated at the collector–emitter port. The instantaneous power is given by $P(t) = v(t)i(t)$, with $v(t)$ and $i(t)$ being the voltage and current waveforms, respectively. Ideally, the power consumed by the transistor will tend to zero if the output voltage and current waveforms do not overlap. In fact, the reduction of the overlapping of $v(t)$ and $i(t)$ has been a crucial design criterion for high-efficiency oscillators [21]. This is achieved more easily if the transistor behaves as a switch instead of as a voltage-controlled current source. For the switching behavior, the oscillation amplitude should be large enough to make the transistor operate in a saturated manner for about one-half the period. The two main types of switching oscillators are class E oscillators and class F oscillators. In a *class E oscillator*, a capacitor and an inductive reactance are connected to the transistor output, with their values determined mathematically so as to prevent overlapping of the voltage and current waveform [22]. In a *class F oscillator* [4], a square output voltage waveform with short rise and decay times reduces overlapping with the transistor output current. The square waveform is achieved through control of the harmonic components of the output voltage, which requires a careful implementation of the various harmonic loads. Class E operation allows for a simpler design; class F operation has the advantages of low peak voltage and low root-mean-square current. The oscillator efficiency depends heavily on load impedance at the various harmonic components. Thus, the availability of a general-purpose multiharmonic load–pull technique is of interest for high-efficiency design.

The in-depth study of high-efficiency design is beyond the scope of the book. Instead, the aim here is to show how harmonic balance techniques that allow oscillation control can be combined with different methodologies for the design of high-efficiency oscillators.

8.4.1 Class E Design

In a class E oscillator, the transistor ideally operates as a switch between the triode and cutoff regions, and conducts for approximately half of the oscillation period. This type of operation is achieved with a resonant output network, with precisely calculated element values [23]. A capacitor is introduced in parallel with the transistor output and the load impedance is constituted by an inductance, in series with the output resistance. The switching operation of the transistor requires a sufficiently large amplitude of the periodic signal between its input terminals. When the switch is open, with infinite impedance, the current flows through the capacitor. When the switch is closed, all the current flows through the transistor. The value of the parallel capacitor and that of the inductive load are calculated so as to fulfill the zero-voltage switching condition [24,25]. This condition is given by the zero value of the voltage across the capacitance and its time derivative when the transistor starts to conduct. This prevents energy stored in the capacitance from being discharged through the active device, which would give rise to an overlapping of voltage and current waveforms.

The following class E design was carried out in collaboration with Caltech. The co-authors were S. Jeon and D. Rutledge and was published in S. Jeon et al. [26]. A three-stage technique was used. In the first stage, a class E amplifier is designed at the target oscillation frequency, with the values of output power and efficiency desired. At the second stage, the transistor input and output networks are replaced by a single feedback network exhibiting the same impedance values at the transistor ports. According to the substitution theorem, the transistor in an oscillator operates in the same way as in an amplifier under the same set of terminal voltages and currents. Due to the difficulty in synthesizing the feedback network at the various harmonic terms, this synthesis is performed at the fundamental frequency only. Thus, at the third stage, the oscillator is analyzed and optimized with harmonic balance, considering the entire harmonic content. The optimization technique will enable an efficiency and/or output power increase, with the circuit in the actual oscillatory stage and taking all the harmonic components into account. The technique will be illustrated by means of its application to the design of a class E oscillator, based on the use of a MRF183 LDMOS from Freescale Semiconductor. The oscillation frequency desired is $f_o = 410$ MHz.

8.4.1.1 Amplifier Design As already stated, a class E amplifier at the desired oscillation frequency $f_{in} = f_o$ is designed initially. The amplifier schematic is shown in Fig. 8.14. Note that the role of the input source at f_{in} is just to provide sufficient input amplitude so as to enable the transistor operation as a switch, with the load cycle mostly located between the cutoff and triode regions. Because this is just a virtual amplifier, not to be implemented in practice, there is no 50-Ω resistance associated with the RF voltage generator. The key elements for class E operation are the parallel capacitance C_{out} and the detuning inductance L_{detune}. These two elements allow fulfillment of zero-voltage switching conditions and are calculated using the well-known expressions for class E design [27]:

$$\begin{aligned} C_{out} &= \frac{0.1836}{2\pi f_{in} R_L} \\ L_{detune} &= \frac{1.1525 R_L}{2\pi f_{in}} \end{aligned} \tag{8.7}$$

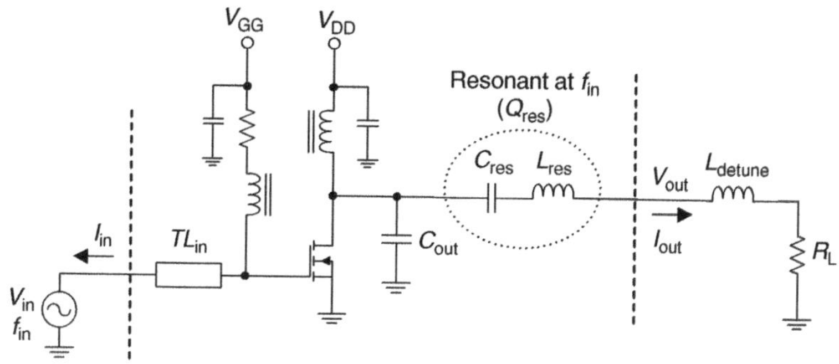

FIGURE 8.14 Schematic of a class E amplifier. The input drive frequency is set to the oscillation frequency. TL_{in} is a transmission line added at the gate to facilitate the layout of the feedback network that will be synthesized in the next sections. Dashed lines represent the reference planes for the synthesis. (Reprinted with permission of IEEE.)

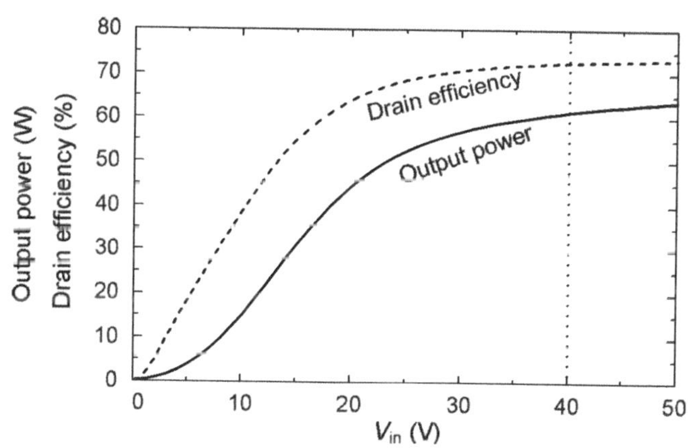

FIGURE 8.15 Simulated output power and drain efficiency as a function of the input drive level. L_{detune} and R_L are tuned to the maximum drain efficiency point. The dotted line at 40 V represents the determined input drive level for saturated operation. (Reprinted with permission of IEEE.)

Assuming 2 Ω for R_L, the equations give $C_{out} = 36$ pF and $L_{detune} = 0.9$ nH. Since this transistor already has an output capacitance near 36 pF, C_{out} is completely absorbed into the transistor. In between C_{out} and L_{detune}, a series LC tank is introduced, which resonates at the input drive frequency f_{in}, equal to the oscillation frequency f_o. This tank provides the required frequency selectivity of the oscillator. The input drive level (V_{in}) and the output circuit parameters (C_{out}, Q_{res}, L_{detune}, and R_L) are optimized so that the amplifier operates at high drain efficiency. Next, the influence of the input drive level on the power and efficiency is analyzed, which provides the curves shown in Fig. 8.15. The input-voltage level selected is $V_{in} = 40$ V, which provides an output power of 58 W and a drain efficiency of 73%.

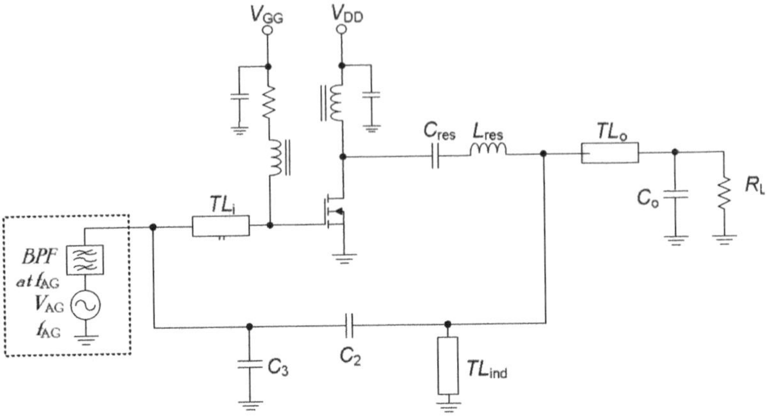

FIGURE 8.16 Complete schematic for a class E oscillator. The embedding network is implemented by capacitors (C_2, C_3, C_o), transmission lines (TL_{ind}, TL_o), and a 50-Ω load (R_L). An auxiliary generator, consisting of a voltage source and an ideal bandpass filter inside the dashed box, is used for nonlinear simulation. (Reprinted with permission of IEEE.)

8.4.1.2 Oscillator Design In the second step an embedding network for the transistor and some terminal elements is synthesized using the substitution theorem [28]. The network must provide the same source and load impedance as that seen by the transistor in a class E amplifier. Because the synthesis will be performed only at the fundamental frequency, it will be helpful to choose reference planes that minimize the influence of the harmonic terms. These reference planes are shown by a dashed line in Fig. 8.14. The input reference plane is given by the node at which the RF voltage source is connected. The output reference plane is placed after the LC resonant tank, which should reduce the harmonic influence significantly. After the synthesis, the terminal voltages and currents must be the same as those of the optimized amplifier: V_{in}, I_{in}, V_{out}, and I_{out}. The embedding network can be configured as a T- or a Π-network [29]. Here, a Π-network is chosen, with three reactive elements (jB_1, jB_2, jB_3) and a resistive element (G_1) representing the load resistance. The network Y matrix is written at the fundamental frequency in terms of these four elements. This matrix relates the network terminal currents $-I_{in}$ and $-I_{out}$ to the terminal voltages V_{in} and V_{out}. Imposing these voltage and currents at the network terminals, it is possible to solve the two complex equations provided by the Y-matrix parameters in terms of the four elements B_1, B_2, B_3, and G_1. The resulting network, introduced into the oscillator circuit, is shown in Fig. 8.16. Its element values are $R_1 = 3.3\ \Omega$, $L_1 = 2.6$ nH, $C_2 = 2.6$ pF, and $C_3 = 23.5$ pF.

The synthesis technique above is applied at the fundamental frequency only. Extension of the technique to several harmonic terms would be involved, as the feedback network should fulfill simultaneously the various Y-matrix relationships resulting from all the harmonic terms considered. As already stated, judicious choice of the reference planes for the feedback network connection reduces the harmonic influence. Despite this, full nonlinear analysis of the oscillator obtained will be necessary. In the next section, a nonlinear technique will also allow improvements in this original oscillator design.

8.4.1.3 Nonlinear Optimization For the nonlinear analysis and optimization of the obtained oscillator, an auxiliary generator has been connected to the transistor gate terminal. The auxiliary generator frequency is made equal to the desired oscillation frequency $f_{AG} = 410$ MHz and its amplitude is made equal to the amplitude value of the RF voltage source in the original amplifier $V_{AG} = V_{in}$. After imposing the amplitude and frequency values of the auxiliary generator, two circuit elements will be calculated in order to fulfill the non-perturbation condition of the oscillatory steady state. The circuit elements chosen are the two capacitors in the feedback network, C_2 and C_3. The final values are $C_2 = 2.6$ pF and $C_3 = 23.5$ pF. The nonlinear simulation with 11 harmonic components predicts 61 W of output power with 71% dc-to-rf conversion efficiency. These values are quite close to those obtained in the original amplified design, with an output power of 58 W and a drain efficiency of 73%.

To investigate the influence of the feedback element values on the oscillator performance, two nested sweeps are carried out in C_2 and C_3. For each pair of capacitance values (C_2, C_3), the auxiliary generator amplitude and the capacitance of the series resonant tank are calculated so as to fulfill nonperturbation condition $Y_{AG}(V_{AG}, C_{res}) = 0$ [26]. The output power and efficiency are evaluated at each step of the double sweep. It is important to note that the frequency of the auxiliary generator keeps constant during the entire parameter sweep at the oscillation value desired, $f_{AG} = 410$ MHz. To enhance the convergence of the simulation, other circuit elements can be included in a set of optimization variables to fulfill the nonperturbation condition $Y_{AG} = 0$ taken as the optimization goal.

With the results of this double sweep, contours of constant output power and efficiency have been plotted in the plane (C_2, C_3). The efficiency in Fig. 8.17 exhibits its maximum value near the point obtained from the quasinonlinear analysis, corresponding to $C_2 = 2.6$ pF and $C_3 = 23.5$ pF (marked by a star). It means that the effect of harmonic components on the efficiency is not very significant in this class E oscillator. This is partly due to the series LC tank with the high Q-factor, which prevents the harmonics generated at the drain from affecting the output load. It is also interesting to see that the contour plot in Fig. 8.17a has a narrow ridge of efficiency along the dashed line. By changing the parameters of (C_2, C_3) along the line, other characteristics of the oscillator performance could be improved, maintaining the high efficiency of more than 67%. As shown in Fig. 8.17b, the output power keeps increasing along the dashed line toward the square. Hence, the feedback elements can be modified from the original synthesized values to obtain the higher output power without degrading the efficiency significantly. With the parameters at the square point, the output power predicted is 85 W with 68% dc-to-RF conversion efficiency.

Remember that for a valid oscillator design, the steady-state solution obtained must be stable. As shown in previous chapters, this can be done using the pole–zero identification technique. The oscillation startup conditions must also be analyzed.

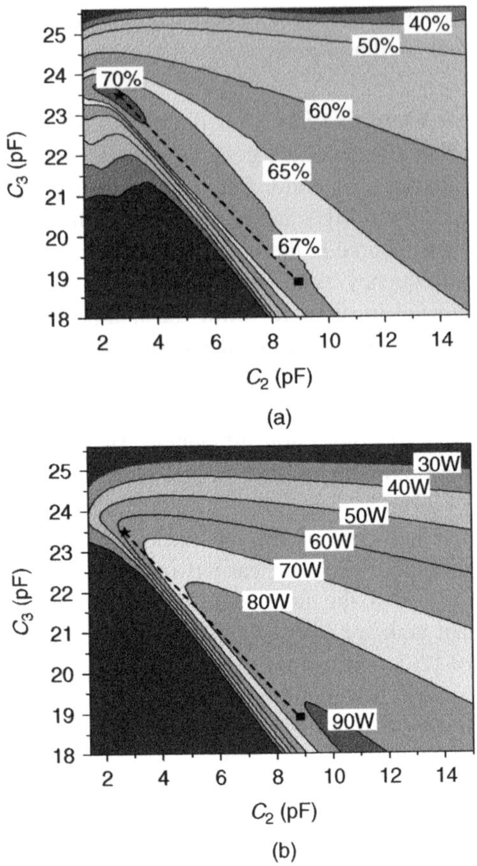

FIGURE 8.17 Contour plots of the simulated dc-to-rf conversion efficiency (a) and output power (b) in the plane of (C_2, C_3). For the entire solution, the oscillation frequency is fixed to 410 MHz. The points of a star and a square represent the original values of (C_2, C_3) obtained from the feedback network synthesis and the new values nonlinearly optimized in the output power and efficiency, respectively. (Reprinted with permission of IEEE.)

Typically, class E amplifiers and oscillators exhibit higher efficiency for gate bias below the threshold voltage. However, the oscillation does not start up from the dc regime for a gate bias below this threshold, which is due to the stability of the dc solution. When the transistor starts to conduct, the changes in the operation conditions associated with the oscillation start up are so strong that an oscillation with large steady-state amplitude must be generated in order to balance the energy delivery and consumption. Once this large-amplitude solution has been established, it is possible to reduce the bias voltage below the threshold. This is explained by the fact that the large-signal gate voltage keeps the transistor conducting for a fraction of the oscillation period even when the gate bias is reduced below the threshold voltage. Thus, a hysteresis phenomenon is observed [26].

8.4.2 Class F Design

In class F operation the harmonic loads are tuned so as to constitute an open circuit at several odd harmonics of low order and a short circuit at even harmonic terms. The voltage waveform is compressed, resembling a square waveform, since the harmonic components flatten the upper and lower sections of the waveform. The quasisquare shape has a very low rise and decay time. The output current is half sinusoidal and there is little overlapping between the two waveforms. Maximizing the oscillator efficiency implies maximizing the ratio $r = V_1/V_{dc}$ between the first-harmonic amplitude and dc component of the output voltage. It has been shown [30–32] that considering three harmonic components of the output voltage waveform, the harmonic components that maximize r are $V_1 = 2V_{dc}/\sqrt{3}e^{j180^\circ}$, $V_2 = 0$, and $V_3 = 2V_{dc}/3\sqrt{3}$, with V_{dc} the particular bias voltage. Note that the phase associated with the first-harmonic component has been set arbitrarily to 180°, as allowed by the solution autonomy. Here a method will be presented to impose the harmonic values indicated for the output voltage waveform, while maintaining the oscillator in its nonlinear steady state. The technique, applied successfully to bipolar-based oscillators, is described next.

The oscillation frequency specified is f_o. The bias voltage V_{CEo} is selected according to the allowed minimum and maximum of the collector–emitter voltage $v_{CE}(t)$, setting $V_{CEo} = (V_{CEMax} + V_{CEMin})/2$. To obtain negative resistance at the collector terminal, two reactive elements X_2 and X_1 are connected to the emitter terminal (enabling series feedback) and the gate terminal, respectively (see Fig. 8.18). Note that in a standard bipolar oscillator, the ranges of these elements providing negative resistance will decrease when reducing the bias voltage V_{BEo}. In most cases, both X_1 and X_2 will be varied during the nonlinear optimization process, described in the next paragraphs. To avoid hysteresis in the oscillation startup due to a stable dc solution, the technique described in Section 8.2.3 can be applied. Just as a brief reminder, two copies of the circuit, one in steady-state oscillatory regime and the other in small signal about the coexisting dc solution, are simultaneously simulated (see Fig. 8.6). An excess negative conductance in small signal is imposed as an additional optimization condition.

For the class F design the objective will be to impose the harmonic voltage values $V_1 = 2V_{dc}/\sqrt{3}e^{j180^\circ}$, $V_2 = 0$, and $V_3 = 2V_{dc}/3\sqrt{3}$ required for a quasisquare waveform at the collector node of the bipolar transistor. In fact, these voltage values should be imposed between the intrinsic terminals of the transistor used. Otherwise, there will be power dissipation at the transistor parasitic elements. Note that this will not be possible when using a black-box transistor model. The analysis is performed with an auxiliary generator at the oscillation frequency f_o (Fig. 8.18) connected to the collector node, so its amplitude will agree with that of the ideal quasisquare waveform at the fundamental frequency; that is, $V_{AG} = V_{C1} = 2V_{CEo}/\sqrt{3}e^{j180^\circ}$. The load required at this frequency component will be described with its reflection coefficient $\Gamma_1 \equiv |\Gamma_1|e^{j\varphi_l}$, with $|\Gamma_1| < 1$ to ensure passivity. An ideal bandpass filter at f_o precedes this ideal load to avoid influencing other harmonic components. Note that Γ_1 will be one of the unknowns to be determined to fulfill the nonperturbation

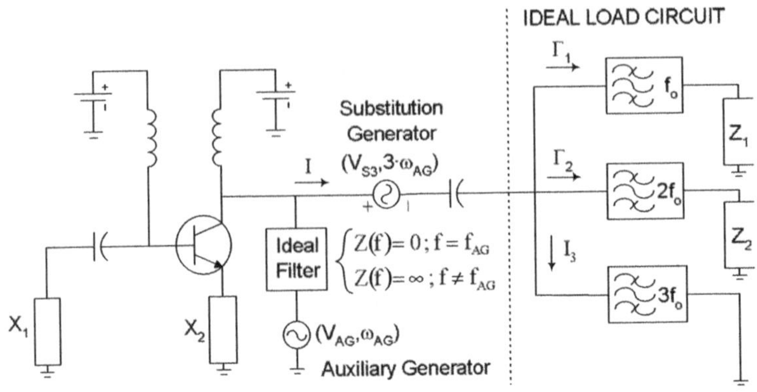

FIGURE 8.18 Circuit schematic for class F oscillator design using a nonperturbing generator plus a substitution generator to obtain an ideal quasisquare output waveform.

condition of the auxiliary generator $Y_{AG} = 0$. At $2f_o$, requiring zero voltage value $V_{C1} = 0$, an ideal short-circuit ($\Gamma_2 \equiv 1e^{j180^\circ}$) is introduced, preceded by an ideal bandpass filter at $2f_o$. At $3f_o$, an ordinary voltage generator with the required value $V_{C3} = 2V_{CEo}/3\sqrt{3}$ is connected between the collector terminal and ground, also using an ideal filter at $3f_o$ (see Fig. 3.18). Unlike the auxiliary generator, this artificial source perturbs the circuit conditions. The voltage source with the value $V_{C3} = 2V_{CEo}/3\sqrt{3}$ at $3f_o$ is actually used to determine the load impedance required at $3f_o$, from the ratio between its voltage and current. It is another example of the use of the substitution theorem. If the generator at $3f_o$ is replaced with a linear element providing the same voltage-to-current ratio, no variation in the circuit solution will be obtained. Thus, the substitution generator adds no new unknowns to the nonlinear system. This system should be solved in terms of Γ_1 and the two reactive elements X_1 and X_2 to fulfill the nonperturbation condition $Y_{AG} = 0$. Note that use of the auxiliary generator ensures the actual oscillation of the circuit, at the desired frequency $\omega_{AG} \equiv \omega_o$, with passive load $\Gamma_1, \Gamma_2, \Gamma_3$. On the other hand, for the resulting third harmonic load to be implementable, the complex relationship $Y_3 = I_3/V_{S3}$, with I_3 the third-harmonic component of the source current (Fig. 8.18), must have a positive real part.

The technique has been applied to the design of an oscillator at $f_o = 6$ GHz, based on the bipolar transistor BFP420 [33]. The reactive elements X_1 and X_2 at the base and emitter, providing negative resistance, are two inductors. Figure 8.19a shows the family of quasisquare waveforms obtained at the extrinsic collector terminal, obtained for four different values of the collector bias voltage between $V_{CEo} = 0.57$ V and $V_{CEo} = 1.39$ V. In all the cases, the substitution voltage generator has already been replaced with a passive load Γ_3. Only three harmonic components have been taken into account for this analysis, which justifies perfect agreement with the theoretical quasisquare waveforms. Note that the oscillation frequency is the same for each solution, which is due to the use of auxiliary generator, forcing, in all cases, the same oscillation frequency $f_o = 6$ GHz. For $E_{ff} = 61.5\%$,

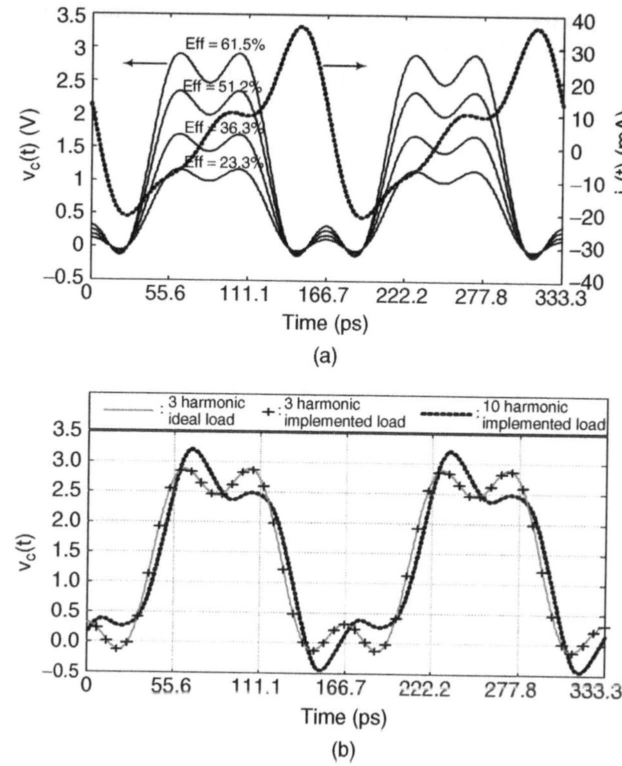

FIGURE 8.19 Class F oscillator design through the synthesis of a multiharmonic load providing a quasisquare waveform at the transistor output: (a) ideal quasisquare voltage waveforms at the extrinsic collector terminal $v_C(t)$ for different values of the bias voltage V_{CEo} and thus different efficiency values $E_{ff} = 61.5\%$; (b) comparison of the ideal quasisquare waveform for $E_{ff} = 61.5\%$. With the waveforms obtained after transmission-line implementation of the ideal load, considering three and 12 harmonic components.

the resulting first harmonic load is $\Gamma_1 = 0.8\angle 295^\circ$. The extrinsic collector current is also superimposed for this efficiency value. The first-harmonic output power is $P_{\text{out}} = 5.46$ dBm.

The next step will be the transmission-line implementation of the ideal multiharmonic load from impedance values at the three first-harmonic frequencies, resulting from the optimization procedure. This can be done in a straightforward manner using standard impedance synthesis techniques [12,15]. Figure 8.19b compares the voltage waveform, corresponding to $E_{ff} = 61.5\%$, obtained with the ideal load and with the load implemented on the transmission line. To evaluate the degree of accuracy, the circuit with the implemented load has been simulated considering three and 12 harmonic components. The two simulations with three harmonic components overlap; they correspond to the ideal and implemented loads, respectively. Simulation with 12 harmonic components deviates slightly from the ideal semisquare waveform. Improvement in the accuracy of the calculation, due to an increased number of harmonic components, gives rise to slightly different values of efficiency and output power, $E_{ff} = 62.7\%$ and $P_{\text{out}} = 4.89$ dBm.

Note that the design is not finished yet. As already known, verification of the stability of the steady-state oscillation is essential. Restriction of the values of the feedback and termination elements to intervals providing negative resistance in small-signal mode should facilitate the condition of the oscillation startup. Nevertheless, a rigorous stability analysis is necessary.

8.4.3 General Load–Pull System

The class F design described in the preceding section might not be possible for some active devices or bias conditions of these devices. This is because it requires quite a high ratio between the third and first harmonic amplitudes of self-sustained oscillation. A less restrictive technique can be applied instead, based on the load–pull system of Fig. 8.20a, implemented on harmonic balance software [30]. For this example, a design based on a FET transistor will be considered. An auxiliary generator at the desired oscillation frequency ω_o maintains a fixed value of this frequency during the entire optimization process. The load at each harmonic component $n = 1$ to N consists of an ideal bandpass filter centered at $n\omega_o$ and terminated in a reflection coefficient $\Gamma_n = \rho_n e^{j\theta_n}$. In the example, the number of harmonic terms will be limited to $N = 3$ (Fig. 8.19a). The use of reflection coefficients is more convenient than the use of impedances or admittances, since all possible passive terminations can be taken into account with just two bounded sweeps in the amplitude and phase intervals $\rho_n = 0$ to 1 and $\theta_n = 0$ to 360°.

The process starts with determination of the Γ_1 value, for maximum efficiency, when short circuiting the higher harmonic terms by making $\Gamma_n = 1e^{j\pi}, n = 2$ to N. Note that even when using these short-circuit terminations, the harmonic current still circulates inside the transistor device, as the i_{ds} current source is not short-circuited. The optimum Γ_1 is determined through a double sweep in ρ_1 and θ_1. The reactive feedback and terminal elements are calculated or optimized for each Γ_1 variation in order to fulfill $Y_{AG} = 0$. Of course, due to the autonomous nature of the circuit, no solution will exist for some Γ_1 values. Once the optimum Γ_1 has been determined, the influence of the higher-harmonic loads is analyzed. These loads will be reactive, so the amplitude of their associated refection coefficients is kept at 1, performing consecutive sweeps in the phases $\theta_2, \theta_3, \ldots$. An example of application is shown in Fig. 8.20b, where the arrows indicate the order of the sweeps. The multiharmonic load will have to be implemented with a single passive network, using standard implementation techniques. Both the fulfillment of the oscillation startup conditions and the stability of the steady-state oscillation must be verified. As in previous examples, to facilitate oscillation startup, it is possible to use a circuit copy with a low-amplitude auxiliary generator. The negative value of the total small-signal conductance, seen from the transistor output, will be imposed as an additional optimization goal (see Fig. 8.6).

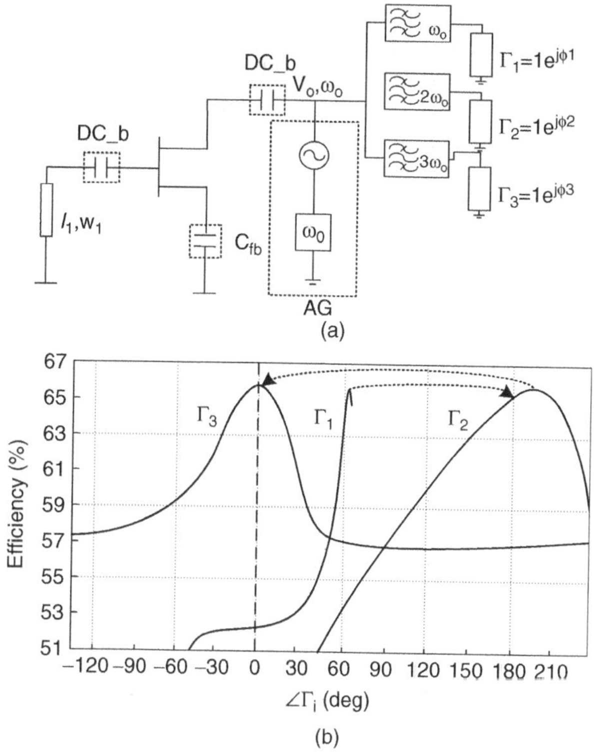

FIGURE 8.20 Load–pull system for maximization of oscillator efficiency: (a) circuit schematic with an ideal harmonic load; (b) maximization of the oscillator efficiency through optimum selection of the harmonic loads. The arrows indicate the order of the optimization procedure. Initially, the amplitude and phase of Γ_1 are varied in a double sweep, until a maximum is reached. Then the phase of Γ_2 is swept, and finally, the phase of Γ_3 is swept.

8.5 CONTROL OF OSCILLATOR TRANSIENTS

As stated in Section 8.1, several applications require oscillator circuits having fast startup times. As an example, the capability to turn the oscillator on when needed and power it down when idle is of great interest for power saving in mobile communication devices. The faster the system can be powered up, perform its function, and be powered down, the less power will be needed. Fast starting times are also very important in pulsed systems such as ultrawideband communications (UWB) or radar systems because narrower pulse widths can be used. Actually, the maximum usable modulation frequency is determined by the ability of the oscillation to build up in a given time interval.

Another use requiring control of the oscillator transient will be that of a modulated oscillator. This modulation can be achieved by introducing a baseband signal in the voltage input of a voltage-controlled oscillator. This scheme can be used to generate a linear frequency-modulated (LFM) chirp signal, which requires the introduction of a sawtooth or triangular waveform between the diode terminals. This waveform typically has a high amplitude and a fast falling edge, so the quality of the generated chirp signal will depend critically on the VCO dynamics. Even when the steady-state oscillatory solution is stable and all the poles are located on the left-hand side of the complex plane, a small stability margin at a frequency ω_a, different from ω_o, will give rise to a slowly decaying transient at this frequency. Mathematically, this means that although all the solution poles have $\sigma < 0$, there is a pair of complex-conjugate poles at ω_a with small $|\sigma_a|$. If the decay time $\tau = 1/|\sigma_a|$ is relatively low compared to the period of the applied sawtooth signal, the frequency ω_a will appear in the modulated output of the VCO, giving rise to an undesired response. In this section a technique is provided to suppress spurious components from modulated oscillator circuits. These spurious frequencies will be eliminated by increasing the stability margin of the oscillator circuit in its periodic steady-state regime.

8.5.1 Reduction of Oscillator Startup Time

For the startup of a free-running oscillation at frequency ω_o from the noise level, the circuit dc solution must be unstable, with a pair of complex-conjugate poles $\sigma \pm j\omega$ such that $\sigma > 0$ and $\omega \cong \omega_o$. This pair of poles dominates all the rest of poles and rules the exponential transient, which will evolve as $\Delta\overline{x}(t) \cong ce^{(\sigma+j\omega)t}\overline{v} + c^*e^{(\sigma-j\omega)t}\overline{v}^*$, where the c and $\overline{v}$ are a constant scalar and a constant vector, respectively, with the same dimension as that of the system analyzed. This pair of dominant poles will be called *critical poles*. Note that the exponential increase in the oscillation amplitude will only be true as long as the perturbation is sufficiently small for the linearization to be valid. From a certain amplitude of the growing oscillation, the circuit will behave in a nonlinear manner and the envelope of the transient solution will no longer correspond to an exponential $e^{\sigma t}$. Roughly, the exponent real part σ must decrease to the value $\sigma = 0$, which implies a nonlinear evolution of the system [34–36]. Therefore, two stages are distinguished in the oscillator startup. The initial stage can be modeled using an exponential of constant complex frequency, corresponding to the dominant poles of the dc solution. The second stage exhibits nonlinear behavior and the complex frequency varies with the oscillation amplitude. The objective here will be reduction of the initial transient, with no attempt to control the nonlinear transient stage.

From previous considerations, oscillators with critical poles having larger $\sigma > 0$ will exhibit shorter initial transients [37]. All the pole values and, in particular, the real part $\sigma > 0$ of the critical poles will depend on the circuit parameters. Thus, for a given technology and circuit topology it will be possible to tune the circuit element values so as to increase the positive real part $\sigma > 0$ of the critical poles. This will reduce the duration of the initial transient. However, when dealing with autonomous circuits, variation of the circuit element values will also give rise to an undesired variation of the oscillation frequency. Thus, the technique presented here aims at shifting the critical poles of the unstable dc solution without affecting the frequency and first harmonic output power of the steady-state oscillatory solution.

For practical implementation of the technique, the oscillator circuit is duplicated, to be analyzed at two different levels (dc and periodic regime), similar to what is shown in Fig. 8.6. However, unlike the technique described in Section 8.2.3, the two circuit copies are analyzed not simultaneously but sequentially. The first circuit copy is simulated at the steady-state oscillation, which is maintained through the connection of an auxiliary generator at the desired values of oscillation frequency $\omega_{AG} = \omega_o$ and first-harmonic amplitude $V_{AG} = V_o$. The second circuit copy contains the small-signal current source $I_n(\omega)$. A linear analysis is carried out to extract the complex transfer function $Z_n(\omega)$ sweeping the frequency ω. This second copy is used to determine the critical poles $\sigma \pm j\omega$ of the coexisting dc solution through the application of pole–zero identification to $Z_n(\omega)$.

The objective is to increase the value of $\sigma > 0$. Thus, a sensitive circuit element or bias voltage η_1 is varied by means of a sweep. At each η_1 step, two different simulations are carried out. The first is a harmonic balance analysis of the steady-state oscillation, with fixed values $\omega_{AG} = \omega_o$ and $V_{AG} = V_o$ of the auxiliary generator The aim is to avoid the undesired shift of the oscillation frequency or amplitude due to the η_1 variation. Because ω_{AG}, V_{AG} are kept fixed, two other elements, η_2 and η_3, must be optimized to fulfill $Y_{AG} = 0$. At each η_1 step, the achievement of the optimization goal provides the three-element set η_1, η_2, η_3, where η_2 and η_3 are the parameter values resulting from optimization at the constant value η_1. In a second simulation for the same η_1 value, the other circuit copy, with η_2, η_3 values resulting from the previous simulation, is analyzed at small signal. The frequency of the small-signal current source $I_n(\omega)$ contained in this circuit is swept to obtain the closed-loop transfer function $Z_n(\omega) = V(\omega)/I_n$. Note that this ω sweep affects the second circuit copy only. The values $Z_n(\eta_1, \eta_2, \eta_3, \omega)$ are stored for the subsequent application of pole–zero identification.

To summarize, the sweep procedure is the following:

For $k = 1$ to N

$$\eta_1 = \eta_1^k$$

Circuit 1 in the steady-state regime with an auxiliary generator :

Harmonic balance optimization $Y_{AG}(\eta_2, \eta_3) = 0$,

with $\omega_{AG} = \omega_o$, $V_{AG} = V_o$

Final values : η_2^k, η_3^k

$$\eta_1 = \eta_1^k, \eta_2 = \eta_2^k, \eta_3 = \eta_3^k$$

Circuit 2 with small-signal current source I_n :

Small-signal analysis; sweep in the I_n frequency ω

Calculation of the closed-loop transfer function $Z_n(\omega)$

The closed-loop transfer functions $Z_n^k(\omega)$, $k = 1$ to N, obtained for each of the N parameter sets $\eta_1^k, \eta_2^k, \eta_3^k$, are exported and pole–zero identification is applied to

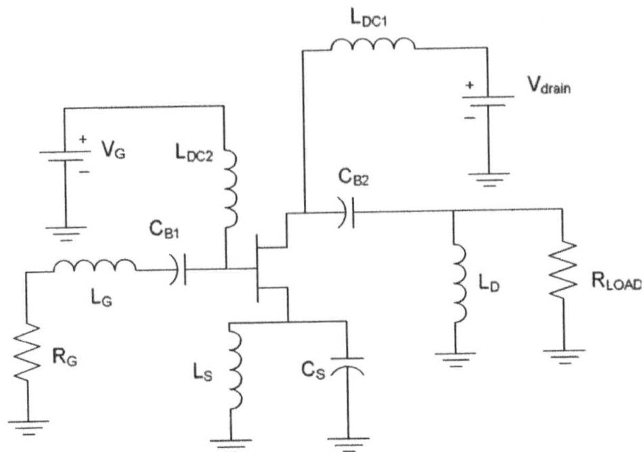

FIGURE 8.21 FET-based oscillator circuit at 2.4 GHz.

these functions. By observing the pole evolution versus η_1, η_2, η_3, it is possible to choose the set of parameter values providing the largest $\sigma > 0$ without altering the steady-state oscillation given by ω_o and V_o. Note that the technique above enables shortening of the initial startup transient of the oscillator circuit when this circuit still behaves in a linear manner. The final stage of the transient is nonlinear and no control over the behavior can be performed through the technique described. Thus, a final evaluation of the transient, including linear and nonlinear effects, will also be required. This can be done in an efficient manner through the envelope transient method [6] (Chapter 5).

For illustration, the technique described has been applied to the FET-based oscillator of Fig. 8.21, with gate bias voltage $V_{\text{gate}} = -0.25$ V. The objective has been to reduce the duration of the startup transient while maintaining the steady-state oscillation frequency and output power at the desired values $f_o = 2.4$ GHz and $P_{\text{out}} = 0$ dBm. In a first analysis, the drain bias V_{drain} is swept and the values of the two elements C_S and L_D are calculated at each point of the sweep in order to satisfy the nonperturbation condition $Y_{AG} = 0$ for $f_o = 2.4$ GHz and $P_{\text{out}} = 0$ dBm. As described earlier, for each set of values V_{drain}, $C_D(V_{\text{drain}})$, $L_D(V_{\text{drain}})$, the circuit is also linearized about its dc solution, applying pole–zero identification to the impedance function $Z_n(\omega)$, obtained using small-signal analysis. The resulting variation C_S and L_D versus V_{drain} is shown in Fig. 8.22a. Figure 8.22b shows the evolution of the critical pair of complex-conjugate poles versus the parameter set V_{drain}, $C_D(V_{\text{drain}})$, $L_D(V_{\text{drain}})$. For some combinations of these element values the critical poles are located on the left-hand side of the complex plane and they have not been represented. As can be seen, the positive real part σ increases versus V_{drain} up to a maximum value. This maximum is due to the limitation in the negative resistance that the device can provide versus variations in V_{drain}.

For a further reduction of the transient, a second parametric analysis has been carried out. The drain voltage is set to $V_{\text{drain}} = 4.5$ V, with the shortest startup time. The swept parameter is now the source capacitance C_S, whereas the inductors

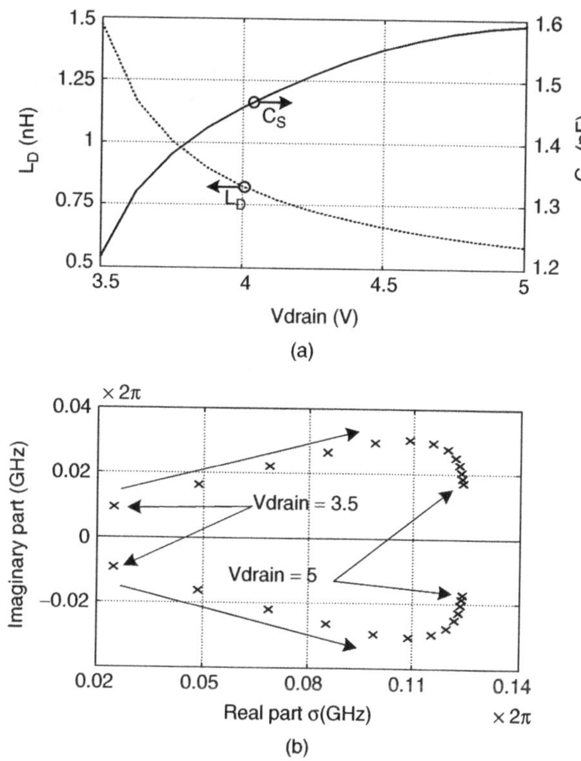

FIGURE 8.22 Parametric analysis versus V_{drain}: (a) solution curves showing the variation of C_s and L_d in order to keep the oscillation frequency at $f_o = 2.4$ GHz and the output power at $P_{\text{out}} = 0$ dBm; (b) evolution of the critic poles versus the resulting sets $V_{\text{drain}}-C_s-L_d$.

L_G and L_D are calculated in order to satisfy $Y_{AG} = 0$ for $f_o = 2.4$ GHz and $P_{\text{out}} = 0$ dBm. A diagram similar to Fig. 8.22b has been obtained when sweeping C_s. This diagram, which can be seen in [6], has been obtained by applying the same technique, described in the preceding paragraphs, with a capacitance sweep from $C_s = 2$ pF to $C_s = 0.86$ pF. The maximum σ value obtained versus the set C_S, $L_G(C_S)$, $L_D(C_S)$ is $\sigma = 2\pi \times 0.17 \times 10^9 \text{S}^{-1}$, corresponding to $C_S = 0.86$ pF. The slowest transients correspond to the largest value $C_S = 2$ pF.

To verify the validity and accuracy of the technique, transient simulation has been used for the analysis of two different oscillator designs, corresponding to $C_S = 2$ pF and $C_S = 0.86$ pF respectively. According to the predictions of the previous analyses, they should enable the longest and shortest transient duration, respectively. The results of the time-domain integration (Fig. 8.23a) confirm these predictions. There is noticeable peak value difference between the two waveforms, as only the first-harmonic power has been preset. Figure 8.23b shows the measured startup time. To verify that the output power and frequency remain at the desired values, the spectra of two different steady-state oscillations (the initial oscillation and the one obtained after the transient shortening) are compared in Fig. 8.23c.

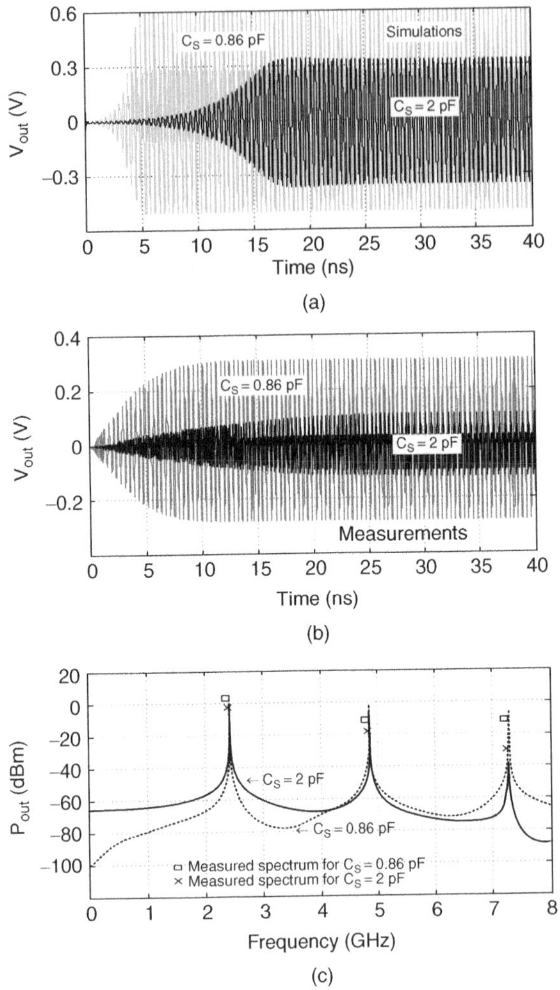

FIGURE 8.23 Validation in the transient-shortening technique (time-domain analysis): (a) simulated transient duration for $C_S = 0.86$ pF and $C_S = 2$ pF; (b) measured transient duration for the same capacitance values; (c) comparison of the simulated and measured output power spectrum. Measurements were obtained using the HP-83480 digital communications analyzer.

Both oscillations have the same fundamental frequency 2.4 GHz and output power 0 dBm.

The on–off capability of the two oscillator designs has also been analyzed through the introduction of a square pulse in a varactor diode. To avoid the high computational cost of the time-domain integration, the envelope transient method has been used. Envelope transient simulation of oscillator switching requires following the entire startup transient along which the oscillation frequency can

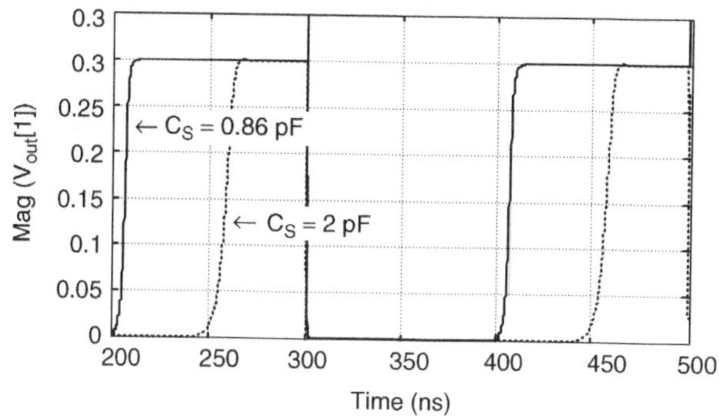

FIGURE 8.24 Envelope transient simulation of the on–off process. As can be noted, the switching time for $C_S = 0.86$ pF is much shorter than for $C_S = 2$ pF.

undergo significant variations. Therefore, the oscillation frequency has been used as one of the state variables of envelope transient integration as proposed by Ngoya and described in Section 5.6.3. This requires a time-varying auxiliary generator with amplitude $A_{AG}(t)$ and frequency $\omega_{AG}(t)$, where t is the slow-time scale associated with the harmonic components of the circuit variables. The values $A_{AG}(t)$ and $\omega_{AG}(t)$ have to be determined at each time step of the integration process, in order to fulfill the nonperturbation condition $Y_{AG}(t) = 0$. Figure 8.24 shows a comparison of the switching times of the two designs. By choosing the linear element values corresponding to the rightmost pair of complex-conjugate poles, the oscillator switching time has been reduced considerably.

8.5.2 Improvement in the Modulated Response of a Voltage-Controlled Oscillator

As stated Section 8.1, a VCO can be used to generate a linear frequency-modulated (LFM) chirp signal, which requires the introduction of a sawtooth or triangular waveform between the diode terminals [19]. As an example, the VCO linearized in Section 8.3.3 has been used to generate a chirp signal with an input sawtooth signal $V_t(t)$ of period $T_s = 50$ ns covering the range 0 to 2.5 V (Fig. 8.25). As shown by the time-domain simulation superimposed on the same figure, the oscillator response is corrupted by a spurious frequency modulation. The modulation appears during the transient response to the steep falling edge of the sawtooth waveform. During this transient, the nearly critical frequencies with small $|\sigma|$ become observable. The stability of the steady-state oscillation at $V_t = 0$ V has been analyzed with pole–zero identification. In addition to the pair of imaginary poles at the oscillation frequency $\pm j\omega_o$ due to the solution autonomy, there is a pair of nearly critical complex-conjugate poles at $f_a \approx 10$ MHz, with very small absolute value of the negative real part $|\sigma_a|$. This will be responsible for a slowly vanishing oscillation at f_a, which is attenuated with the long time constant $\tau = 1/|\sigma_a|$. As shown in Fig. 8.25, this slowly decaying oscillation significantly degrades the chirp signal generated. To eliminate the spurious modulation, the real part σ_a has been

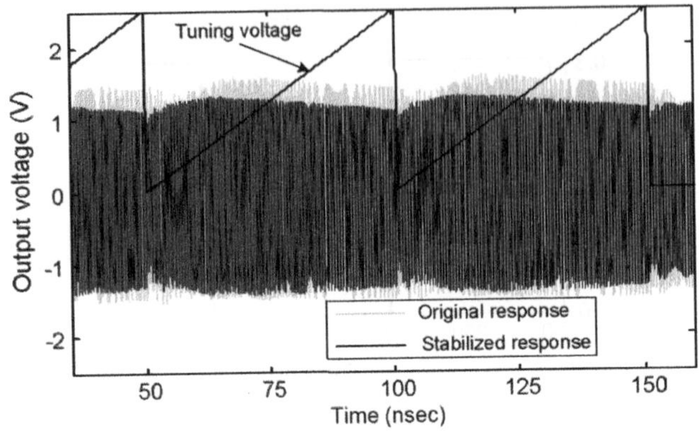

FIGURE 8.25 Chirp signal generated with the linearized VCO (time-domain simulation). The spurious modulation is due to a small stability margin at $f_a = 10$ MHz. The original and improved responses are compared.

made more negative by sweeping a circuit resistance R_s. To maintain the original values of the oscillation amplitude $V_o = 0.7$ V and frequency $f_o = 4.5$ GHz, an auxiliary generator at $A_{AG} = V_o$ and $f_{AG} = f_o$ is kept connected to the circuit during the resistance sweep. At each R_s value, two circuit elements C_1 and L_3 are optimized to fulfill the oscillation condition $Y_{AG}(C_1, L_3) = 0$. Next, the stability of the periodic solution obtained with element values R_s, C_1, and L_3 is analyzed with pole–zero identification. Note that it is not necessary to duplicate the circuit, as the stability analysis is applied to the same periodic regime as that analyzed in the first simulation. The auxiliary generator that sustains the oscillation is kept connected to the circuit and the function $Z_n(\omega)$ is calculated with the conversion matrix approach. The procedure of general application to modulated oscillators is summarized as follows:

For $k = 1$ to N

$R_s = R_s^k$

Simulation 1 : Harmonic balance analysis

Optimization $Y_{AG}(C_1, L_3) = 0$, with $\omega_{AG} = \omega_o$, $V_{AG} = V_o$

Final values : C_1^k, L_3^k

$R_s = R_s^k, C_1 = C_1^k, L_3 = L_3^k$

Simulation 2 : Conversion matrix approach with small-signal source I_n

Sweep in the I_n frequency ω

Calculation of the closed-loop transfer function $Z_n(\omega)$

The pole locus versus R_s–C_1–L_3 is shown in Fig. 8.26. For $R_s = 5\ \Omega$ (corresponding to the most negative σ_a), the time constant τ of the spurious component

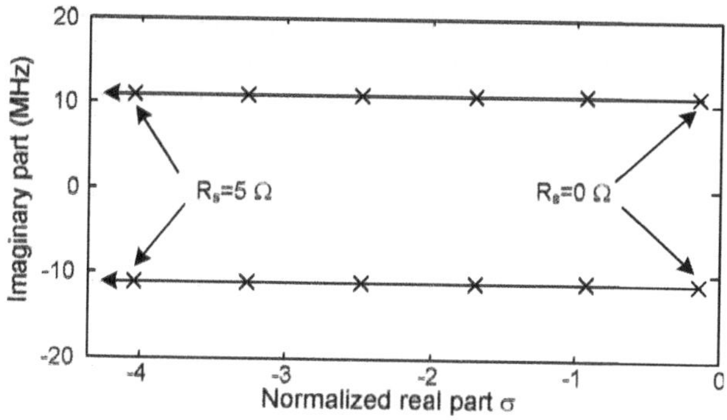

FIGURE 8.26 Stability analysis of the steady state oscillation for $V_t = 0$; pole locus when increasing R_s. The elements C_1 and L_3 are also varied at each R_s step, to maintain oscillation at the original frequency and amplitude values.

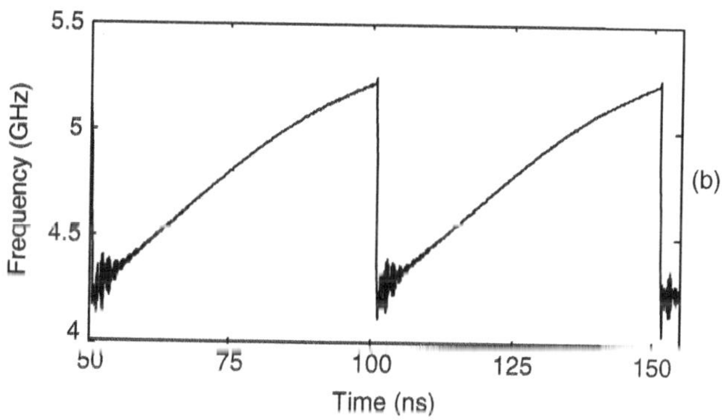

FIGURE 8.27 Instantaneous frequency variation of the chirp signal after elimination of the spurious frequency.

is small compared to the sawtooth period T_s. Therefore, the influence of the spurious frequency should be reduced significantly. This is shown in the superimposed waveform of Fig. 8.25. Figure 8.27 shows the instantaneous frequency variation of the chirp signal obtained with the corrected circuit. The small transient originated by the ideal infinite slope falling edge of the sawtooth signal is unavoidable. Actually, the falling edge of the physical sawtooth signals is high but different from infinite.

8.6 PHASE NOISE REDUCTION

Noise perturbations originated at the circuit resistive elements and semiconductor devices give rise to phase and amplitude noise in the free-running oscillator signal. As shown in Chapters 2 and 3, the phase noise will have the greatest influence on the output spectrum due to the solution irrelevance versus variations of the phase origin. Different criteria can be applied for low-phase-noise oscillator design. Once a convenient configuration has been selected, the circuit element values can be optimized for further reduction of the phase noise. In the following, the three major strategies that do derive a low-phase-noise design are summarized briefly [13,38]. Then emphasis is placed on simulation techniques for phase noise minimization of a given oscillator topology.

1. *Bias conditions.* As shown in [41], the limiting mechanism of the oscillator circuit influences the output power and the noise behavior. In FET devices, for small drain source voltage at the transition between saturation and linear region, the output admittance increases with decreasing source voltage, which causes a reduction of voltage gain. Other limiting mechanisms are the forward biasing of the gate-source junction, the channel cut-off, and the gate-drain breakdown. In [41] they show that to obtain low phase noise at low offset frequencies with dominant 1/f noise, the steady state should be limited by increasing output admittance at the transition between the saturation region and the linear region, as well as the channel cut-off. The limiting mechanism associated to the breakdown of the gate-drain diode should be avoided.

2. *Enhancement of the frequency selectivity of the resonant circuit.* Increasing the positive slope of the load reactance/susceptance versus frequency implies enhancing the frequency selectivity of oscillator circuit. This will reduce the oscillation-frequency sensitivity to the device output capacitance, which depends on the oscillation amplitude, the temperature, and the bias supply. Various ways to increase the frequency selectivity of the load circuit were discussed in Section 8.2.1.

3. *Low-frequency loading.* The up-conversion of low-frequency noise to the carrier frequency can be reduced through suitable low-frequency loading of the active device, in order to reduce the low-frequency noise voltages. As shown in [42], in FET, with an equivalent voltage noise source in series with the gate terminal, this can be done by open-circuiting the gate and short-circuiting the drain port at low frequencies. To achieve this practically, a series branch, composed of resistance R_{in} and a blocking capacitance C_b, is connected between the input terminal and ground. A second identical branch, with the resistance R_{out}, is connected between the output terminal and ground. Then the resistances should be optimized to minimize the noise perturbations of the oscillator circuit.

4. *Low-frequency feedback.* Negative feedback at low frequency, implemented through the parallel connection of a resistance in series with a blocking capacitor, allows low-frequency and thermal noise reduction at low frequencies. Feedback reduces the transistor gain, so this gain must have a sufficiently high value for the technique to be applicable. It is also possible to use active feedback at low frequencies [11]. The device low-frequency output signal is sampled, amplified, and subtracted from the input signal. For the subtraction, a 180° phase shift is achieved with an inverter transistor. Other forms of low-frequency feedback include the use of long transmission lines. The two main configurations are the one based on a self-injected oscillator and the one based on a stabilization loop, with a frequency discriminator. They are presented in detail in Chapter 9.

After this brief review, a technique will be proposed to reduce the oscillator phase noise without further modification of its original topology or inclusion of additional elements. Only values of the existing circuit elements will be varied. Let an oscillator circuit containing M white noise sources and J colored noise sources be assumed. As a brief reminder, the oscillator output power spectrum at the fundamental frequency due to phase noise at sufficiently high offset frequency Ω from the carrier is given by [39]

$$S(\Omega) = \frac{c + \sum_{j=1}^{M} |b_{\gamma j,\mathrm{dc}}|^2 S_{\gamma j}(\Omega)}{\Omega^2} \tag{8.8}$$

where the different constant parameters are calculated as

$$\begin{aligned} c &= \frac{1}{T}\int_0^T \bar{b}^{\mathrm{T}}(t)[\Gamma]\bar{b}(t)\,dt \\ b_{\gamma j,\mathrm{dc}} &= \frac{1}{T}\int_0^T b_{\gamma j}(t)\,dt \qquad \text{with} j = 1 \text{to} J \end{aligned} \tag{8.9}$$

with $\bar{b}^{\mathrm{T}}(t)$ being the $M \times 1$ row matrix containing the periodic noise sensitivity functions to the M white noise sources, $b_{\gamma j}(t)$ the periodic noise-sensitivity function to the colored noise source $\gamma_j(t)$ with $j = 1$ to J, and $[\Gamma]$ the autocorrelation matrix of the white noise sources calculated from $\langle e_i(t)e_j(t+\tau)\rangle = \Gamma_{ij}\delta(\tau)$. Note that all the constant parameters in (8.9) can be determined from the analysis of the perturbed harmonic balance system, as was shown in Chapter 7.

In most single-transistor oscillators, only one flicker noise source is considered, in addition to the set of white noise sources, including shot and thermal noise. The largest contribution at relatively low frequency offset from the carrier is due to this single flicker noise source. According to (8.8), the phase noise spectral density close to the carrier could be reduced by decreasing the magnitude of the coefficient $b_{\gamma,\mathrm{dc}}$ associated with the flicker noise source. At a larger frequency offset, the phase noise spectrum will be dominated by the white noise perturbations, and the phase noise is determined mainly by the value of the parameter c in (8.9).

The aim here will be to reduce $|b_{\gamma,\mathrm{dc}}|$ and c without affecting the oscillation frequency and first-harmonic amplitude. To maintain the oscillation frequency and amplitude at the desired values V_o and ω_o, an auxiliary generator at $V_{AG} = V_o$ and $\omega_{AG} = \omega_o$ is introduced into the circuit. At each variation of a given parameter η_1, two other parameters, η_2 and η_3, are optimized in order to fulfill $Y_{AG}(\eta_2, \eta_3) = 0$. To reduce the near carrier noise, the magnitude $|b_{\gamma,\mathrm{dc}}|$ is evaluated versus the three-parameter set η_1, η_2, η_3, choosing the point $\eta_{1o}, \eta_{2o}, \eta_{3o}$ that provides the minimum value of $|b_{\gamma,\mathrm{dc}}|$. Note that more analysis parameters η_k can also be considered. A similar technique can be applied to noise reduction at higher offset frequencies. In this case we will minimize the coefficient c by means of a parameter sweep.

The technique described has been applied to the FET-based oscillator of Fig. 6.1 at 4.9 GHz [40]. The analysis parameters are L_g, C_d, C_s, and L_d. An auxiliary generator with $f_{AG} = 4.9$ GHz and $V_{AG} = 0.45$ V is introduced into the circuit so as to maintain the original values of the oscillation frequency and first harmonic amplitude. A sweep in the inductance L_g is carried out, calculating C_d, C_s, and L_d at each step so as to fulfill the nonperturbation condition $Y_{AG}(C_d, C_s, L_d) = 0$. The amplitude V_{AG} is also allowed to vary slightly about $V_{AG} = 0.45$ V. Figure 8.28 shows the representation of $b_{\gamma,\text{dc}}$ versus one of the parameters in the set, L_g, although the other two, C_d and C_s, also vary along the curve traced. As can be seen, there is a zero minimum of $|b_{\gamma,\text{dc}}|$ obtained for $L_g = 3.2$ nH. Ideally, this should give rise to a zero contribution of the flicker noise source to the oscillator phase noise. Note that this is impossible in practice, due to the distributed nature of the noise sources. Figure 8.29 shows the phase noise variation versus the analysis

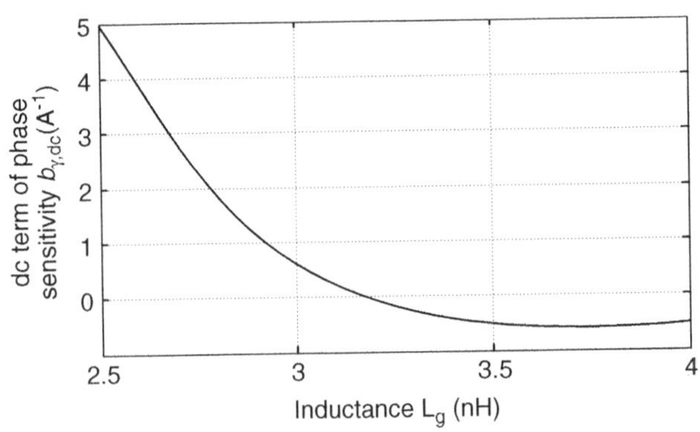

FIGURE 8.28 Variation in the dc component $b_{\gamma,\text{dc}}$ of the phase sensitivity function to the flicker noise of a FET-based oscillator of Fig. 6.1 at 4.9 GHz versus the analysis parameters L_g, L_d, C_d, and C_s.

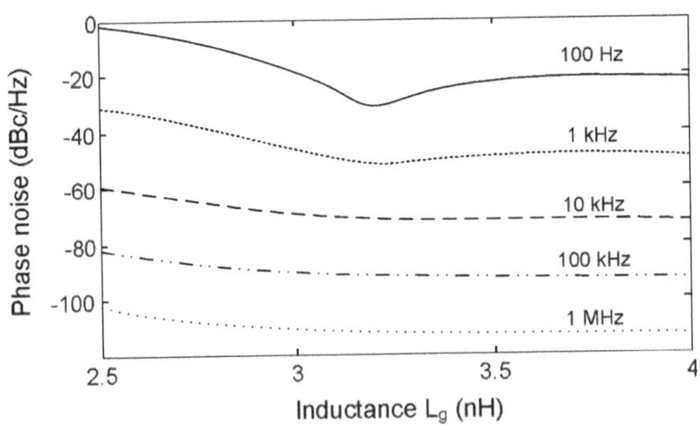

FIGURE 8.29 Variation in the phase noise spectral density of the oscillator in Fig. 6.1 for the same element values considered in the analysis of Fig. 8.28.

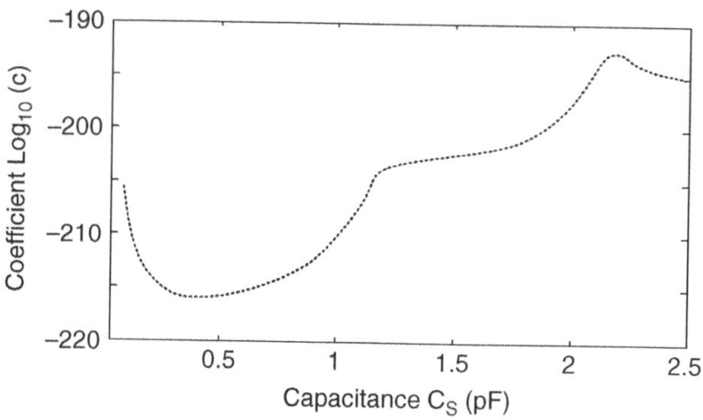

FIGURE 8.30 Variation in log(c) versus the source capacitance C_s.

parameters at a different frequency offset from the carrier. The phase noise spectral density of the original oscillator is also given, for comparison. Again, only the inductance L_g is considered in the representation. As can be seen, the phase noise minimum agrees with the point at which the magnitude $|b_{\gamma,\mathrm{dc}}|$ takes a zero value. As the offset frequency increases, the white noise sources have a greater influence on the phase noise spectrum, as can be gathered from (8.8). In agreement with this, the improvement in the phase noise spectral density decreases with the offset frequency.

A similar technique can be used for reduction of the phase noise spectral density due to white noise sources. A sweep in the capacitance C_s is carried out, calculating C_d and L_d at each step so as to fulfill the nonperturbation condition $Y_{AG}(C_d, C_s, L_d) = 0$. Figure 8.30 shows the variation in the parameter log(c) versus the capacitance C_s. The white noise spectral density is minimized for the capacitance value $C_s = 0.4$ pF. Figure 8.31 shows a comparison between the original phase noise spectrum (the dotted line) and those resulting from minimization of $|b_{\gamma,\mathrm{dc}}|$ (the solid line) and c (the dotted-dashed line). As can be seen, minimization of $|b_{\gamma,\mathrm{dc}}|$ eliminates the section of the spectrum with slope −30 dB/dec. The minimization of c associated with the white noise contribution gives rise to a noise reduction at the higher offset frequencies.

The advantage of the proposed noise minimization technique is that it does not require a double simulation, harmonic balance plus phase noise analysis, at each parameter step. Instead, the coefficients $b_{\gamma,\mathrm{dc}}$ and c can be determined directly from the Jacobian matrix associated with the steady-state solution and inherently evaluated during harmonic balance simulation. On the other hand, when using commercial software, the parameters c and $b_{\gamma,\mathrm{dc}}$ can be obtained from a phase noise simulation with the carrier-modulation approach. The circuit is first simulated with the flicker noise source off, and all the white noise sources and the parameter c is determined from only one or two points $[f, S(f)]$ of the phase noise spectrum, as explained in Chapter 7, Section 7.5. Then, the circuit is simulated with the flicker noise source on and all the white noise sources off, using only one or two points of the phase noise spectrum $[f, S(f)]$ to obtain the parameter $b_{\gamma,DC}$. Because only one or two points are required in the offset frequency sweep for the phase noise

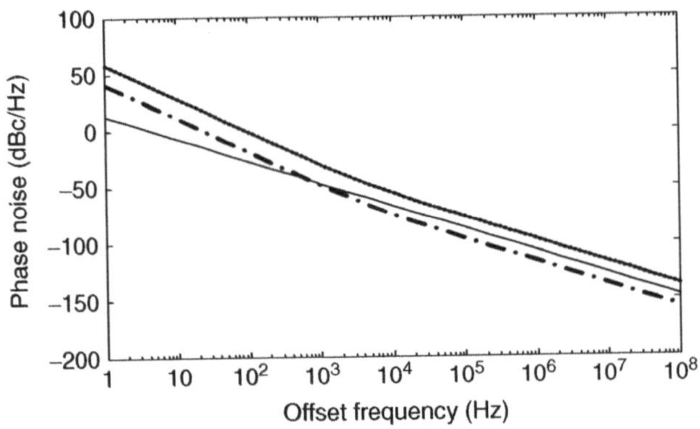

FIGURE 8.31 Comparison between the original phase noise spectrum (dotted line) and those resulting from minimization of $b_{\gamma,\mathrm{dc}}$ (solid line) and c (dotted-dashed line).

calculation, this alternative technique for the determination of the parameters c and $b_{\gamma,dc}$ has also a low computational cost.

APPENDIX

As a brief remainder, the transmission-line equivalences of different types of lumped elements are obtained from the general expression of the input impedance of a transmission-line section of characteristic impedance Z_c, ended with load impedance Z_L:

$$Z(l) = Z_c \frac{Z_L + jZ_c \tan \beta l}{Z_c + jZ_L \tan \beta l} \tag{A.1}$$

where l the transmission-line length and $\beta = 2\pi/\lambda_g$, with λ_g the transmission-line wavelength. Implementation of several types of lumped elements is considered next.

A.1 Inductive Impedances

For small βl such that $\beta l \cong \tan \beta l$ and $Z_L \ll \beta l Z_c$, expression (A.1) can be approached:

$$Z(l) = jZ_c\beta l = j\frac{Z_c l}{v_g}\omega \tag{A.2}$$

where the relationship $\beta = \omega/v_g$, with v_g the wave velocity, has been taken into account. Thus, the line section of length l is equivalent to an inductor of value $L = Z_c l/v_g = Z_c l\sqrt{\varepsilon_{\mathrm{eff}}}/c$, with c the speed of light and $\varepsilon_{\mathrm{eff}}$ the effective dielectric constant of the transmission line. To ensure the condition $Z_L \ll \beta l Z_c$, a line with high characteristic impedance is chosen, which implies a relatively narrow line width w.

A.2 Parallel Capacitive Impedances

For small βl and $Z_L \gg \beta l Z_c$, expression (A.1) can be approached:

$$Z(l) = -j\frac{Z_c}{\beta l} = -j\frac{Z_c v_g}{l\omega} \tag{A.3}$$

Thus, the line section is equivalent to a capacitor of value $C = l/Z_c v_g = l\sqrt{\varepsilon_{\text{eff}}}/Z_c c$. To ensure the condition $Z_L \gg \beta l Z_c$, a line with low characteristic impedance is chosen, which implies a relatively large line width w usually ended in an open circuit.

A.3 Series Resonant Circuit

According to expression (A.1), a transmission line with electrical length βl, ended with resistive load impedance $Z_L = R$, will give rise to a series resonance, with input impedance $Z = R$, at the frequencies $\omega_{o,n} = (2n+1)\omega_o$ fulfilling $\beta_n l = (\omega_{o,n}/v_g)l = (2n+1)\pi$, with n an integer. Thus, for a series resonance at ω_o, the physical length of the line must agree with $l = (2n+1)\lambda_o/2$, with λ_o being the wavelength at ω_o. In order to implement a particular lumped series resonator at ω_o with reactive elements $L, C, R = 0$, the frequency derivative of the transmission-line reactance at the resonance frequency will be made equal to that of the reactance of the lumped circuit [14]. By deriving expression (A.1) with respect to the frequency and particularizing to $n = 0$ and $l = \lambda_o/2$, the condition will be

$$\left.\frac{\partial X(l)}{\partial \omega_o}\right| = \frac{Z_c}{2f_o} = 2L \tag{A.4}$$

Thus, for an approximate implementation of the series resonant circuit R-L-C, the transmission line must fulfill

$$\begin{aligned} Z_L &= R \text{ with } Z_c > Z_L \\ l &= \frac{\pi v_g}{\omega_o} = \frac{c}{2f_o\sqrt{\varepsilon_{\text{eff}}}} \\ Z_c &= 4Lf_o \end{aligned} \tag{A.5}$$

Note that the frequency variation of the lumped reactance is inherently different from that of the $\lambda_o/2$ transmission line. Therefore, the equivalency is valid only in a relatively narrow band about the resonance frequency. The characteristic impedance Z_c of the transmission line is determined by the inductance value L of the resonant circuit and the resonance frequency f_o. The line width w required for this characteristic impedance must be obtained with transmission-line synthesis programs.

Following similar reasoning, it is also possible to implement the series resonant *RLC* circuit with a $\lambda_o/4$ transmission line. The conditions will be

$$\begin{aligned} Z_L &= \frac{Z_c^2}{R} \text{ with } Z_c < Z_L \\ l &= \frac{\pi v_g}{2\omega_o} = \frac{c}{4 f_o \sqrt{\varepsilon_{\text{eff}}}} \\ Z_c &= 8 L f_o \end{aligned} \tag{A.6}$$

As in the previous case, the line width w required for the Z_c value obtained must be determined with transmission-line synthesis programs.

A.4 Parallel Resonant Circuit

The inversion of expression (A.1) provides the input admittance of a transmission line ended with load admittance $Y_L = 1/Z_L$. It is easily demonstrated that transmission lines with electrical length βl ended in the conductance $Y_L = G$ give rise to parallel resonance at the frequencies $\omega_{o,n} = (2n+1)\omega_o$ fulfilling $\beta_n l = (\omega_{o,n}/v_g)l = (2n+1)(\pi/2)$. Thus, for a parallel resonance at ω_o, the physical length of the line must agree with $l = (2n+1)\lambda_o/4$, with λ_o being the wavelength at ω_o. To implement a particular lumped parallel resonator at ω_o, with elements $G = 0, L, C$, the frequency derivative of the transmission-line susceptance at the resonance frequency will be made equal to that of the susceptance of the lumped circuit [14]. By deriving the inverse of expression (A.1) with respect to the frequency, and particularizing to $n = 0$ and $l = \lambda_o/4$, the condition will be

$$\left.\frac{\partial B(l)}{\partial \omega_o}\right| = \frac{1}{4 f_o Z_c} = 2C \tag{A.7}$$

Thus, for an approximate implementation of the parallel *GLC* circuit, the transmission line must fulfill

$$\begin{aligned} Y_L &= \frac{Y_c^2}{G} \text{ with } Y_c < Y_L \\ l &= \frac{\pi v_g}{2\omega_o} = \frac{c}{4 f_o \sqrt{\varepsilon_{\text{eff}}}} \\ \frac{1}{Z_c} &= 8 C f_o \end{aligned} \tag{A.8}$$

As in the previous case, the equivalency is valid only in a relatively narrow band about the resonance frequency. Following similar reasoning, it is also possible to implement the parallel *GLC* circuit with a $\lambda_o/2$ transmission line ended in an open circuit $Y_L = 0$. The conditions will be

$$Y_L = G \text{ with } Y_c > Y_L$$
$$l = \frac{\pi v_g}{\omega_o} = \frac{c}{2f_o\sqrt{\varepsilon_{\text{eff}}}} \tag{A.9}$$
$$\frac{1}{Z_c} = 4Cf_o$$

REFERENCES

[1] A. Hajimiri, I. Aoki, S. Kee, and D. Rutledge, A fully-integrated 1.8-V. 2.8-W, 1.9-GHz, CMOS power amplifier, pp. 199–202, 2003.

[2] S. D. Kee, I. Aoki, A. Hajimiri, and D. Rutledge, The class-E/F family of ZVS switching amplifiers, *IEEE Trans. Microwave Theory Tech.*, vol. 51, pp. 1677–1690, 2003.

[3] H. Oh, T. Song, E. Yoon, and C. Kim, A power-efficient injection-locked class-E power amplifier for wireless sensor network, *IEEE Microwave Wireless Components Lett.*, vol. 16, pp. 173–175, 2006.

[4] M. Prigent, M. Camiade, G. Pataut, D. Reffet, J. M. Nebus, and J. Obregón, High efficiency free running class F oscillator, *IEEE Radio Frequency Integrated Circuits (RFIC) Symposium*, Philadelphia, PA, pp. 1317–1320, 1995.

[5] S. Lingling and C. Xiane, Load-pull simulation for large signal GaAsFET, *IEEE MTT-S International Microwave Symposium*, Orlando, FL, pp. 601–603, 1995.

[6] F. Ramirez, A. Suárez, and S. Sancho, Harmonic-balance technique for the shortening of the initial transient of microwave oscillators, *IEEE MTT-S International Microwave Symposium Digest*, p. 4, 2005.

[7] K. S. Kulpa, Novel method of decreasing influence of phase noise on FMCW radar, *CIE International Conference of Radar*, Beijing, China, pp. 319–323, 2001.

[8] D. J. Vendelin, A. M. Pavio, and U. L. Rohde, *Microwave Circuit Design*, Wiley, New York, 1990.

[9] K. Kurokawa, Some basic characteristics of broadband negative resistance oscillators, *Bell Syst. Tech. J.*, vol. 48, pp. 1937–1955, July–Aug. 1969.

[10] M. L. Edwards and J. H. Sinsky, A new criterion for linear 2-port stability using a single geometrically derived parameter, *IEEE Trans. Microwave Theory Tech.*, vol. 40, pp. 2303–2311, Dec. 1992.

[11] U. L. Rohde, A. K. Poddar, and G. Bock, *The Design of Modern Microwave Oscillators for Wireless Applications*, Wiley, Hoboken, NJ, 2005.

[12] R. E. Collin, *Foundations for Microwave Engineering*, 2nd ed., Wiley, New York, 2001.

[13] A. Grebennikov, Noise reduction in transistor oscillators: Part 1– Resonant circuits, *High Freq. Electron.*, vol. 4, no. 5, pp. 16–31, 2005.

[14] R. Soares, J. Graffeuille, and J. Obregón, *Applications des transistors a effet de champ en arseniure de gallium*, Eyrolles, Paris, 1984.

[15] P. F. Combes, J. Graffeuil, and J. F. Sautereau, *Microwave Components, Devices and Active Circuits*, Wiley, Chichester, UK, 1987.

[16] A. Anakabe, J. M. Collantes, J. Portilla, et al. Analysis and elimination of parametric oscillations in monolithic power amplifiers, *IEEE MTT-S International Microwave Symposium*, Seattle, WA, pp. 2181–2184, 2002.

[17] V. Rizzoli and A. Lipparini, General stability analysis of periodic steady-state regimes in nonlinear microwave circuits, *IEEE Trans. Microwave Theory Tech.*, vol. 33, pp. 30–37, Jan. 1985.

[18] R. Quéré, E. Ngoya, M. Camiade, A. Suárez, M. Hessane, and J. Obregón, Large signal design of broadband monolithic microwave frequency dividers and phase-locked oscillators, *IEEE Trans. Microwave Theory Tech.*, vol. 41, pp. 1928–1938, Nov. 1993.

[19] J. Dominguez, S. Sancho, and A. Suárez. VCO linearization using harmonic balance, *IEEE MTT-S International Microwave Symposium Digest*, pp. 864–867, 2006.

[20] E. Marazzi and V. Rizzoli, Design of linearizing networks for high-power varactor-tuned frequency modulators, *IEEE Trans. Microwave Theory Tech.*, vol. 28, pp. 767–773, 1980.

[21] E. W. Bryerton, W. A. Shiroma, and Z. B. Popovic, A 5-GHz high-efficiency class-E oscillator, *IEEE Microwave Guided Wave Lett.*, vol. 6, pp. 441–443, Dec. 1996.

[22] J. Ebert and M. Kazimierczuk, Class E high-efficiency tuned power oscillator, *IEEE J. Solid State Circuits*, vol. 16, pp. 62–66, 1981.

[23] J. F. Davis and D. B. Rutledge, Low-cost class-E power amplifier with sine-wave drive, *IEEE MTT-S International Microwave Symposium*, Baltimore, MD, pp. 1113–1116, 1998.

[24] N. O. Sokal, Class-E switching-mode high-efficiency tuned RF/microwave power amplifier: improved design equations, *IEEE MTT-S International Microwave Symposium Digest*, vol. 2, pp. 779–782, 2000.

[25] N. O. Sokal, Class E high-efficiency power amplifiers, from HF to microwave, *IEEE MTT-S International Microwave Symposium*, Baltimore, MD, pp. 1109–1112, 1998.

[26] S. Jeon, A. Suárez, and D. B. Rutledge, Nonlinear design technique for high-power switching-mode oscillators, *IEEE Trans. Microwave Theory Tech.*, vol. 54, pp. 3630–3639, 2006.

[27] F. H. Raab, Idealized operation of the class E tuned power amplifier, *IEEE Trans. Circuits Syst.*, vol. 24, pp. 725–735, 1977.

[28] M. Lee, S. Yi, S. Nam, Y. Kwon, and K. Yeom, High-efficiency harmonic loaded oscillator with low bias using nonlinear design approach, *IEEE Trans. Microwave Theory Tech.*, vol. 47, pp. 1670–1679, Sept. 1999.

[29] Y. Xuan and C. M. Snowden, A generalized approach to the design of microwave oscillators, *IEEE Trans. Microwave Theory Tech.*, vol. 35, pp. 1340–1347, 1987.

[30] D. Barataud, M. Champovecchio, and J. Nebus, Optimum design of very high-efficiency microwave power amplifiers based on time-domain harmonic load-pull measurements, *IEEE Trans. Microwave Theory Tech.*, vol. 49, pp. 1107–1112, June 2001.

[31] C. Duvanaud, P. Bouysse, S. Dietsche, J. M. Nebus, J. M. Paillot, and D. Roques, A design method for high efficient power amplifiers, *Int. J. Microwave Millimeter Wave Comput. Aided Eng.*, vol. 6, pp. 205–210, July 1996.

[32] C. Duvanaud, P. Bouysse, S. Dietsche, J. M. Nebus, J. M. Paillot, and D. Roques, Design method for highly efficient power amplifiers: application to class F amplifiers, *Int. J. Microwave Millimeter Wave Comput. Aided Eng.*, vol. 6, pp. 288–293, 1996.

[33] S. Ver Hoeye, F. Ramirez, and A. Suárez, Nonlinear optimization tools for the design of high-efficiency microwave oscillators, *IEEE Microwave Wireless Components Lett.*, vol. 14, pp. 189–191, 2004.

[34] P. Gammand and V. Pauker, Starting phenomenon in negative FET oscillators, *IEE Electron. Lett.*, vol. 24, pp. 911–913, July 1988.

[35] B. Voigt, D. Seefeldt, and E. H. Horneber, A new approach to determine the start-up time of sinusoidal oscillators, *Proceedings of the 37th Midwest Symposium on Circuits and Systems*, vol. 2, pp. 1119–1122, Aug. 1994.

[36] T. Tsuzuki, T. Adachi, and J. Zhang, Formulation of nonlinear negative resistance for calculation of start-up characteristics of crystal oscillators, *Proceedings of the IEEE International Frequency Control Symposium*, pp. 710–713, June 1996.

[37] J. Jugo, J. Portilla, A. Anakabe, A. Suárez, and J. M. Collantes, Closed-loop stability analysis of microwave amplifiers, *IEE Electron. Lett.*, vol. 37, pp. 226–228, Feb. 2001.

[38] A. Grebennikov, Noise reduction in transistor oscillators: 2. Low frequency loading and filtering, *High Freq. Electron.*, pp. 24–36, June 2005.

[39] A. Demir, Phase noise in oscillators: DAEs and colored noise sources, *IEEE/ACM International Conference On Computer-Aided Design*, San Jose, CA, pp. 170–177, 1998.

[40] S. Sancho, F. Ramirez, and A. Suárez, Analysis and reduction of the oscillator phase noise from the variance of the phase deviations, determined with harmonic balance, *IEEE MTT-S International Microwave Symposium Digest*, Atlanta, GA, 2008.

[41] V. Gunderich, J. Deuringer, W. Anzill, and P. Russer, Phase-noise reduction of microwave oscillators by optimization of the dynamic behaviour, *IEEE MTT-S Int. Microwave Symposium*, pp. 953–956, 1994.

[42] M. Prigent and J. Obregon, Phase noise reduction in FET oscillators by low-frequency loading and feedback circuitry optimization, *IEEE Trans. Microwave Theory Tech.*, vol. 35, no. 3, pp. 349–352, March 1987.

CHAPTER NINE

Stabilization Techniques for Phase Noise Reduction

9.1 INTRODUCTION

The phase noise of an existing oscillator circuit can be reduced by introducing this oscillator in a feedback loop. This loop may contain either a long delay line or a dielectric resonator coupled to a short line. In both cases, the aim is to obtain a feedback signal with high spectral purity, enabling the phase noise reduction of the closed-loop configuration. The long transmission line stabilizes the feedback signal due to its high frequency selectivity [see equation (8.3) in Chapter 8]. It also reduces the correlation of the feedback signal with the oscillator signal, as this correlation decreases with the time shift. Note that at the output $v(t-\tau)$ of the delay line is re-injected into the circuit and affects its different variables. The line can be a coaxial cable suitably connected to the circuit, or a microstrip line with multiple bends for a reasonable small size. In turn, the high-Q resonator increases the spectral purity of the feedback signal of the closed-loop configuration. A different technique is based on the use of active feedback at low frequencies. In active feedback, the device low-frequency output signal is sampled, amplified, and subtracted from the input signal [1,2]. For the subtraction, 180° phase shift is achieved with an inverter transistor. In this chapter the first two options are considered: feedback with a long delay line or a high-quality-factor resonator. The main objective will be to provide insight into phase noise reduction mechanisms

Analysis and Design of Autonomous Microwave Circuits, By Almudena Suárez

when using any of these two feedback elements. A second objective will be detailed analysis of the instability phenomena obtained for some values of the feedback elements.

Let an existing free-running oscillator be assumed. The feedback can be introduced at the oscillation frequency or at baseband. The two main configurations are represented in Fig. 9.1. Figure 9.1a shows a self-injection topology. A circulator is used to reinject the oscillator signal to the circuit after passing through the long delay line or a high-quality-factor Q resonator. Thus, the feedback is introduced at the oscillation frequency. In stable periodic operation the frequency of the feedback signal will agree with the oscillation frequency. Chang [3,4] explains the phase noise reduction as a result of the oscillator synchronization to the reinjected RF signal. As already known, the synchronization to a low-noise signal enables the reduction of the oscillator phase noise, due to the phase relationship established with the low-noise synchronizing signal (see Section 4.2.7 in Chapter 4). Figure 9.1b shows a stabilization loop [5], based on the use of a frequency discriminator. The output signal of a voltage-controlled oscillator (VCO) is coupled to a frequency discriminator using the baseband output signal of the discriminator as the control voltage of the oscillator circuit. The discriminator [5–7] converts the frequency fluctuations into baseband voltage fluctuations which provide an error signal that

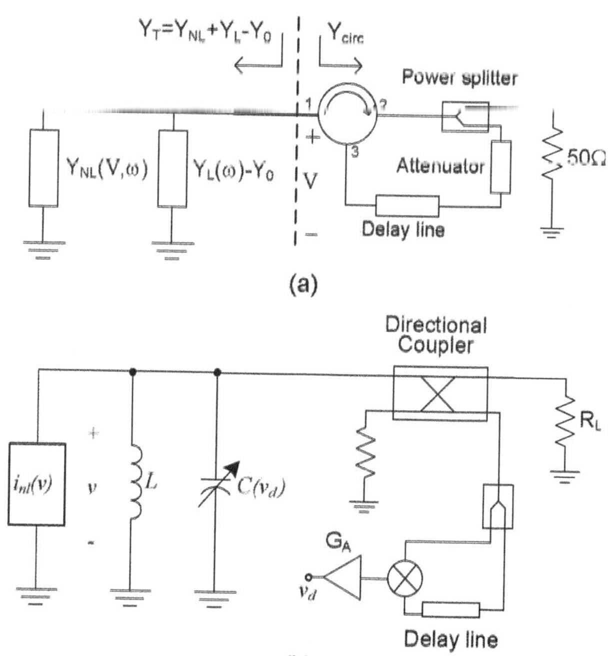

FIGURE 9.1 Feedback topologies for the phase noise reduction of an existing oscillator: (a) self-injection topology; (b) use of a stabilization loop, with a frequency discriminator.

readjusts the oscillator solution. In stable noiseless operation, the discriminator output will be a dc signal and the oscillator will be in the periodic regime. Although the overall circuit is actually a free-running oscillator, the phase noise reduction may be understood intuitively as the result of the oscillator phase locking to a stabilized signal. In the discriminator, this stabilized signal is achieved by passing the sampled signal through either a long delay line or a high-Q resonator.

As demonstrated experimentally, the two feedback configurations mentioned enable the phase noise reduction of an existing oscillator. However, use of these phase-reduction systems has two associated problems. One of them is the undesired variation in the oscillation frequency and amplitude. In fact, the overall system, including the particular feedback loop, is also a free-running oscillator. As we know, the oscillation frequency is a state variable of the free-running oscillator that depends on the values of the circuit elements. Introduction of the feedback element may thus give rise to variation in the original free-running frequency. The second problem is even more serious. The feedback loop introduces a change in the circuit topology and adds new circuit elements. Even if the original free-running oscillator is stable, after inclusion of the feedback, the new global circuit solution loop, may be unstable for some values of the loop elements.

In this chapter we present an analytical study of the self-injection topology and the topology based on introduction of a low-frequency feedback loop. In each section three aspects are considered: variation of the steady-state solution versus the loop elements, variation of the stability properties, and the mechanism for phase noise reduction. Practical transistor-based oscillators are considered and the results of an analytical study will be compared with the harmonic balance simulations of these circuits. Additionally, a semianalytical formulation will be presented. This formulation is similar to the one introduced in Chapter 7, Section 7.7, for the stability and phase noise analysis of injection-locked oscillators. This formulation is based on a perturbation model of the original oscillator extracted from harmonic balance simulations. It allows the application of the expressions derived from the analytical study to transistor-based oscillators. As an illustration, the two feedback techniques are applied to a MESFET-based VCO at 5 GHz.

9.2 SELF-INJECTION TOPOLOGY

9.2.1 Steady-State Solution

Analytical study of the self-injected oscillator is carried out in terms of admittance functions at the fundamental frequency. We take into account that the original oscillator (prior to the introduction of the feedback elements) fulfills the steady-state oscillation condition

$$Y_T(V_o, \omega_o) = 0 \tag{9.1}$$

Next, the self-injection topology of Fig. 9.1 is considered. The reflection structure, comprising the circulator, the power splitter, and the delay line, replaces the original load $Z_o = 50\ \Omega$. Thus, the output load admittance, given initially by $Y_o = 1/50\ \Omega^{-1}$, becomes $Y_{circ} = Y_o[(1+\Gamma)/(1-\Gamma)]$, where Γ is the reflection coefficient at the circulator port 1. Taking this into account, the steady-state equations of the self-injected topology will be

$$Y'_T(V,\omega) = Y_T(V,\omega) - Y_o + Y_o\frac{1+\Gamma(\omega)}{1-\Gamma(\omega)} = 0 \tag{9.2}$$

where the characteristic admittance Y_o is subtracted from the total admittance $Y_T(V,\omega)$ of the original oscillator, as it has been replaced with the input admittance at port 1 of the circulator. Note that a self-synchronized state is actually assumed in (9.2), as only one fundamental frequency is considered. On the other hand, the admittance function $Y_T(V,\omega)$ is now different from zero, as due to Kirchhoff's laws, it must take a value opposite to the input admittance exhibited by the reflection structure minus Y_o. Due to the changes in the load network introduced, one can expect variations in the oscillation frequency ω and amplitude V with respect to the original values V_o and ω_o.

Assuming an ideal circulator, the reflection coefficient Γ used in (9.2) is given by the ratio between the power injected into the circulator port 3 and the power injected into the circulator port 1. The case of a long delay line of electrical length θ, plus a total attenuation A_L decibels, is considered initially. Then the reflection coefficient may be expressed as $\Gamma = 10^{-A_L/20}e^{j\theta}$. For a physical line length l, the electrical length θ is given by $\theta = (2\pi/\lambda_g)l = (\omega/v_g)l$, where ω is the angular frequency and v_g is the wave velocity. Thus, it is possible to write $\theta = \omega\,\Delta T$, with ΔT the delay associated with the long transmission line. In the following analysis, the line dispersion associated with the frequency dependence $\lambda_g(\omega)$ is neglected, so ΔT will be assumed constant regardless of the ω value.

To avoid a big perturbation of the oscillator solution, high attenuation is usually introduced between ports 2 and 3 of the circulator. Thus, the modulus of the reflection coefficient will be close to $\Gamma = 0$. This will allow expanding $Y_{circ} = Y_o[(1+\Gamma)/(1-\Gamma)]$ in a first-order Taylor series about $\Gamma = 0$, which provides $Y_{circ} \cong Y_o(1-2\Gamma)$. Because of the relatively high attenuation, the power of the reinjected signal will be rather small, so only relatively small changes are expected in the oscillation amplitude V and frequency ω. It will also be possible to consider a first-order Taylor series expansion of the admittance function $Y_T(V,\omega)$ about the free-running solution V_o, ω_o. Then the steady-state equation of the self-injected oscillator becomes

$$Y_{TVo}(V - V_o) + Y_{T\omega o}(\omega - \omega_o) - Y_o + Y_o(1 - 2\rho e^{-j\omega\Delta T}) = 0 \tag{9.3}$$

where $\rho = 10^{-A_L/20}$ is the magnitude of the reflection coefficient and Y_{TVo} and $Y_{T\omega o}$ stand for the Y_T derivatives with respect to the voltage amplitude and frequency, respectively, evaluated at the steady-state solution V_o, ω_o. It has also been taken into account that the total admittance function of the original oscillator fulfills

$Y_T(V_o, \omega_o) = 0$. Splitting (9.3) into real and imaginary parts, the following system of two real equations in the two unknowns V and ω is obtained:

$$\begin{aligned} G_{vo}(V - V_o) + G_{\omega o}(\omega - \omega_o) - 2\rho Y_o \cos(\omega\,\Delta T) = 0 \\ B_{vo}(V - V_o) + B_{\omega o}(\omega - \omega_o) + 2\rho Y_o \sin(\omega\,\Delta T) = 0 \end{aligned} \tag{9.4}$$

where G and B stand for the real and imaginary parts of the admittance function Y_T, respectively. System (9.4) provides the variation of the self-injected oscillator solution in terms of its amplitude and frequency versus the time delay ΔT. The linearization about the free-running oscillation will be valid for high A_L, which is the usual case. For low values of A_L, the linearization is not valid, so the nonlinear admittance function $Y_T(V, \omega)$ modeling the response of the original free-running oscillator should be introduced in (9.4) instead of its first-order Taylor series expansion. This function would be obtained from nonlinear element models based on the function or by means of a suitable numerical approach. In the following discussion only high attenuation values are considered, so it will be assumed that $Y_T(V, \omega)$ can be linearized about the original steady-state solution V_o, ω_o. Despite this linearization, system (9.4) is actually nonlinear in ω, due to the cos and sin terms in ωT. These terms are associated with the self-injected signal and evidence the feedback action of the topology in Fig. 9.1a. System (9.4) is solved with a simple root-finding routine. When considering variations in the time delay $\Delta T$of the transmission line, the linear plus sinusoidal dependence of system (9.4) on the oscillation frequency ω will lead to multivalued curves for large ΔT. This is due to the faster variation of the sinusoidal terms versus ω for larger ΔT.

For illustration, the system (9.4) has been particularized to the case of a parallel resonance oscillator with a cubic nonlinearity. The circuit elements are $i(v) = av + bv^3$, with $a = -0.037A/V$ and $b = 0.021A/V^3, C = 0.715pF, L = 1.328\ nH$. Variation in the steady-state solution in terms of V and ω versus the time delay ΔT is shown in Fig. 9.2 for the attenuation value $A_L = 30$ dB. As can be seen, only small deviations from the free-running amplitude and frequency are obtained in the entire ΔT interval, which confirms the validity of the linearization in (9.3) and (9.4). The curves are single-valued for relatively small ΔT. However, from certain ΔT value, they start to exhibit points of infinite slope (turning points). When tracing the oscillation amplitude versus ΔT, the lower sections of these curves fold over themselves, giving rise to the loops that are shown in the inset of Fig. 9.2a. The turning points give rise to low-amplitude jumps between different sections of the multivalued curve. The effect of the tuning points is more significant in the oscillation frequency. The frequency deviation with respect to the original value ω_o (without the self-injection loop) is represented in Fig. 9.2b. The unstable sections located between the turning points are indicated with a U. Note that in order to take the line dispersive effects into account, the physical length l should be used as the parameter (instead of ΔT), as will be done for the transistor-based oscillator of Section 9.5.

In Fig. 9.2, the results obtained by solving the nonlinear admittance equation (9.4) are compared with those provided by harmonic balance, considering one and

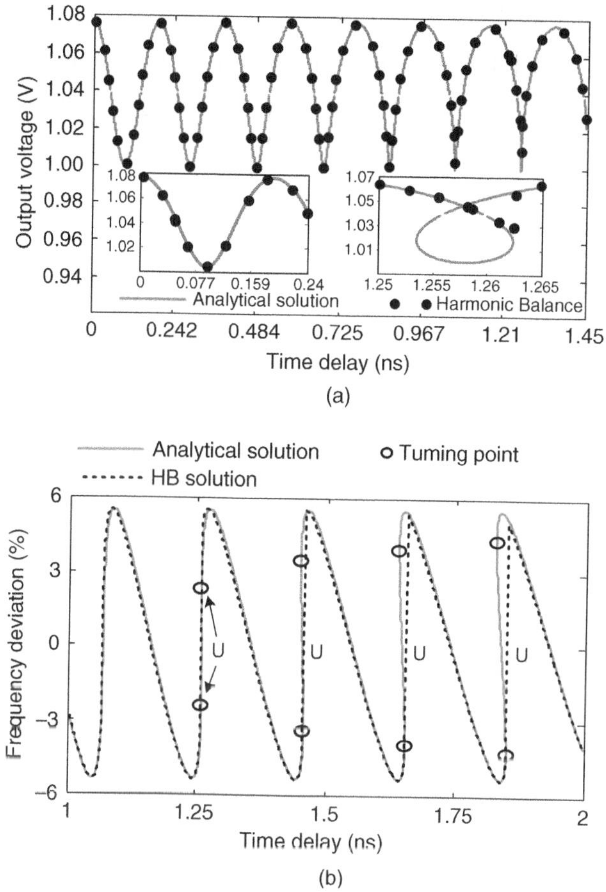

FIGURE 9.2 Parallel topology oscillator based on the cubic nonlinearity $i(v) = av + bv^3$, with $a = -0.037$ A/V and $b = 0.021$ A/V^3. The linear element values are $C = 0.715$ pF and $L = 1.328\ nH$. Variation of the oscillator solution is shown versus the time delay ΔT. (a) Oscillation amplitude: comparison between the results obtained with expression (9.2) and those using harmonic balance with one harmonic component. (b) Frequency deviation from the free-running value: comparison between the results obtained with expression (9.2) and those using harmonic balance with one and 15 harmonic components. The unstable sections located between the turning points are indicated with a U.

15 harmonic components. A default harmonic balance simulation unable to pass through the turning points has been used, so discontinuous jumps are obtained at the ΔT values corresponding to the turning points. Note that to be able to pass through the turning points, the harmonic balance simulation should have been combined with a parameter-switching continuation technique, such as those described in Chapter 6, Section 6.4. As can be expected, for 15 harmonic components (Fig. 9.2b), some quantitative discrepancies exist between the curves, but the qualitative behavior is still well predicted.

9.2.2 Stability Analysis

In Chapter 1, Section 1.3, an approximate stability analysis of the periodic solution of free-running oscillators was presented assuming a small frequency perturbation $\Delta\omega \ll$ limited to the detection of a real dominant pole. The same approximate analysis can be applied to the self-injected oscillator, since the overall circuit, including the feedback loop, behaves as a free-running oscillator. The only requirement is, of course, replacing the total admittance function of the free-running oscillator Y_T with the global admittance function Y'_T, including the feedback loop. Then the stability condition, under the assumptions indicated, is given by [8]

$$S = \frac{\partial Y'^r_{To}}{\partial V}\frac{\partial Y'^i_{To}}{\partial \omega} - \frac{\partial Y'^i_{To}}{\partial V}\frac{\partial Y'^r_{To}}{\partial \omega} > 0 \tag{9.5}$$

where $Y'_{To} = 0$ is the total admittance function of the oscillator circuit containing the feedback loop, particularized at a given steady-state regime. Note that the factor S agrees with the determinant of the Jacobian matrix associated with the global admittance function Y'_T.

When modifying ΔT, the self-injected oscillator solution must be recalculated at every ΔT step, using (9.4), so the determinant $\det[JY'_T(V, \omega)]$ will vary versus this parameter. At turning points of the oscillator solution curve versus a particular parameter, the Jacobian matrix associated with $Y'_T(V, \omega) = 0$ must become singular. Thus, the turning-point condition is $\det[JY'_T(V, \omega)] = 0$. This determinant is given by

$$\begin{aligned}\det[JY'_T(\omega, \Delta T)] &= G_{vo}[B_{\omega o} + 2\rho Y_o \Delta T \cos(\omega\,\Delta T)] \\ &\quad - B_{vo}[G_{\omega o} + 2\rho Y_o \Delta T sin(\omega\,\Delta T)] \\ &= \det[JY_T]_o + 2\rho Y_o \Delta T[G_{vo}\cos(\omega\,\Delta T) - B_{vo}\sin(\omega\,\Delta T)]\end{aligned} \tag{9.6}$$

where $\det[JY_T]_o$ is the constant determinant of the Jacobian matrix evaluated at the original free-running oscillator given by $\det[JY_T]_o = G_{vo}B_{\omega o} - G_{\omega o}B_{vo}$. As can be seen, the determinant exhibits a double dependence on ΔT, affecting both the angle and amplitude of the sinusoidal terms. Limiting the analysis to only one real pole, the oscillator is stable for $\det[JY'_T(V, \omega)] > 0$ and unstable for $\det[JY'_T(V, \omega)] < 0$. The solution will exhibit a turning point for $\det[JY'_T(V, \omega)] = 0$. Note that the original oscillator to which the self-injection loop is added must itself be stable. Thus, the condition $\det[JY_T]_o > 0$ should be fulfilled. Turning points can only be obtained when the amplitude of the sinusoidal term in (9.6), directly proportional to ΔT, becomes comparable to $\det[JY_T]_o$.

Figure 9.3 shows the variation of $\det[JY'_T(V, \omega)]$ versus ΔT for a cubic nonlinearity oscillator. Note that (9.6) must be resolved in combination with the steady-state system (9.4). The difference between the positive maxima and negative minima of this determinant increases with ΔT, as gathered from (9.6). As can be observed, the determinant zeros accurately predict the turning points. Unstable behavior is obtained for the ΔT values at which $\det[JY'_T(V, \omega)] < 0$. This has

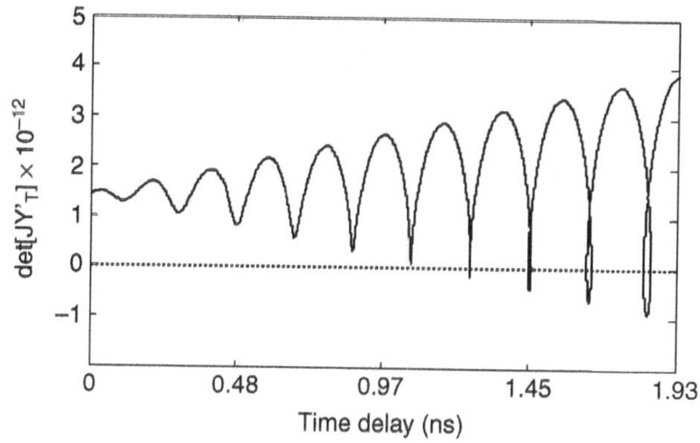

FIGURE 9.3 Variation of $\det[JY'_T]$ versus the time delay ΔT. The turning-point condition is $\det[JY'_T] = 0$.

been confirmed with an accurate stability analysis based on pole–zero identification. Points with unstable behavior predicted by Nyquist analysis [9] and pole–zero identification [10] are indicated in the solution curves of Fig. 9.2b. It is also possible to calculate the turning points of these curves (versus ΔT) directly through simultaneous resolution of (9.4) and (9.6). The results are shown by circles in Fig. 9.2b. Each unstable section is located between two consecutive turning points.

To summarize, the behavior of the self-injected oscillator is not periodic versus the time delay. For relatively large ΔT, solution of the self-injected oscillator becomes multivalued in small ΔT intervals, due to the fast variation of the sinusoidal terms versus ΔT, as gathered from inspection of (9.4). The instability is associated with turning points in the solution curve, which give rise to jumps to other curve sections. The solution remains periodic after the jump. However, a sudden shift is obtained in the oscillator amplitude and frequency. It must be noted that the stability analysis presented here is limited to one real dominant pole, assuming a small perturbation frequency. For particular circuit topologies, other instabilities may also be obtained. Thus, the stability of the oscillator after addition of the feedback loop should be investigated with more accurate and complete techniques, such as pole–zero identification or the Nyquist criterion.

9.2.3 Phase Noise Analysis

The objective of this section is the analytical study of mechanisms for phase noise reduction in a self-injected oscillator. For simplicity, the analysis is carried out at the fundamental frequency, so only white noise perturbations about the oscillator are considered. The constant spectral density of this white noise source is S_w. The analytical study will be based on the admittance function $Y'_T(V, \omega) = 0$. Variations in phase noise spectral density versus ΔT are analyzed through successive linearizations of the system equations about each steady-state oscillation obtained versus ΔT. The noise perturbations are modeled with a current noise source connected parallel to the oscillator output. Note that despite the presence of the self-injection loop, the circuit still behaves like a free-running oscillator, as no external time reference is present. Thus, Kurokawa's formulation for the noise analysis at the fundamental frequency, presented in Chapter 2, is applicable. As already known, under the influence of noise, the perturbed oscillator amplitude is given by $V + \Delta V(t)$ and its frequency becomes $\omega + d\phi(t)/dt - j[(d\frac{\Delta V}{V_o}(t))/dt]$. Then the noisy oscillator equations are written as (see Chapter 2, Section 2.5.2)

$$\left[Y'_{TV}\Delta V(t) + Y'_{T\omega}\left(\frac{d\phi(t)}{dt} - j\frac{d\Delta V(t)}{V_o\,dt}\right)\right]V = I_n(t) \tag{9.7}$$

where $I_n(t)$ is the complex envelope of the current noise source about the oscillator's fundamental frequency. Note that the admittance derivative with a prime refers to the derivative of the admittance function of the self-injected oscillator evaluated at its steady-state solution. However, due to the first-order Taylor series expansion, Y'_{TV} actually agrees with Y_{TV_o}, as can be verified from inspection of (9.4). In contrast, $Y'_{T\omega}$ is different from $Y_{T\omega_o}$, due to the influence of the self-injection structure containing the delay line. Neglecting the time variation of V on the right-hand side of the brackets and splitting the term into real and imaginary parts, the following system is obtained:

$$\begin{aligned} G_{vo}\Delta V(t) + G'_{\omega o}\frac{d\phi(t)}{dt} + \frac{B'_{\omega o}}{V_o}\frac{d\Delta V(t)}{dt} &= G_n(t) \\ B_{vo}\Delta V(t) + B'_{\omega o}\frac{d\phi(t)}{dt} - \frac{G'_{\omega o}}{V_o}\frac{d\Delta V(t)}{dt} &= B_n(t) \end{aligned} \tag{9.8}$$

where $G_n(t)$ and $B_n(t)$ are the equivalent conductance and susceptance of the white noise current obtained by dividing the real and imaginary parts of $I_n(t)$, respectively, by the oscillation amplitude. Applying the Fourier transform to (9.8), the time derivatives become products by $j\Omega$. Solving for $j\Omega\,\Delta\phi$, the following expression is obtained:

$$j\Omega\,\Delta\phi = \frac{(G_{VO}B_n - G_nB_{VO})V_o + j\Omega(B'_{\omega_o}B_n + G'_{\omega_o}G_n)}{(G_{VO}B'_{\omega_o} - B_{VO}G'_{\omega_o})V_o + j\Omega(B'^2_{\omega_o} + G'^2_{\omega_o})} \tag{9.9}$$

Multiplying by the adjoint and taking into account that as shown in Chapter 2, G_n and B_n are uncorrelated, we derive the expression

$$|\Delta\phi(\Omega)|^2 = \frac{1}{\Omega^2V_o^2}\frac{|Y_{V_o}|^2V_o^2 + \Omega^2|Y'_{\omega_o}|^2}{(G_{VO}B'_{\omega_o} - B_{VO}G'_{\omega_o})^2V_o^2 + \Omega^2|Y'_{\omega_o}|^2}2S_w \tag{9.10}$$

The expression is formally identical to the expression derived for standard free-running oscillators (without feedback). However, the presence of the feedback elements gives rise to a change in the frequency derivative of the admittance function Y'_{ω_o} with respect to the derivative corresponding to the original oscillator. From an inspection of (9.10), at high-frequency offset from the carrier the term

$\Omega^2|Y'_{\omega_o}|^2$ will dominate both the numerator and denominator, and the influence of the self-injection topology will be negligible. Thus, the effect of self-injection will be noticeable at relatively low frequency offsets. At these offsets, equation (9.9) can be simplified to

$$|\Delta\phi(\Omega)|^2 \cong \frac{|Y_{V_o}|^2 2S_w}{[\det[JY_T]_o + 2\rho Y_o \Delta T(G_{vo}\cos(\omega\,\Delta T) - B_{vo}\sin(\omega\,\Delta T))]^2 V_o^2 \Omega^2} \tag{9.11}$$

where $|Y_{V_o}|^2 = G_{V_o}^2 + B_{V_o}^2$. As already indicated, the expression (9.11) neglects the time variation of the amplitude perturbation ΔV. Through the inspection of (9.11), it is possible to obtain and understand the variation of the phase noise spectral density versus the time delay ΔT. The phase noise reduction comes from the large magnitude reached by the denominator of (9.11) for some ΔT values. Because the numerator does not depend on ΔT, the minima of (9.11) will be obtained at the maxima of the denominator. For constant offset frequency Ω_m, the phase noise minima versus ΔT will be given by

$$|\Delta\phi(\Omega)|^2 \cong \frac{|Y_{V_o}|^2 2S_w}{[\det[JY_T]_o + 2\rho Y_o \Delta T]^2 \Omega_m^2} \tag{9.12}$$

Clearly, the lowest minima are obtained for higher ΔT, thus the need to use long delay lines. The coefficient in the denominator of (9.12) agrees with the square of $\det[JY'_T(V,\omega)]$[see (9.6)]. Therefore, only denominator values with positive sign will correspond to stable solutions. Divisions by zero are obtained at the turning points. The phase noise will tend to infinity at those points. However, the more complete expression (9.10) prevents the phase noise from actually growing to infinity. In addition, the system linearization that is always used in phase noise analysis is no longer valid in the neighborhood of the turning points, since the amplitude perturbation ΔV grows rapidly as these points are approached.

For illustration, expression (9.12) has been used for the phase noise analysis of the parallel topology oscillator. In Fig. 9.4, the phase noise spectral density at fixed frequency offset $f_m = 1$ kHz, with $f_m = \Omega_m/2\pi$, has been traced versus the time delay ΔT. As discussed previously, the phase noise tends to infinity at the turning points of the solution curve. On the other hand, the phase noise minima lie in the stable sections, as can be verified through comparison with Figs. 9.2 and 9.3. For validation, harmonic balance simulations using the carrier modulation approach [8,11] have been superimposed. An improvement of about 10 dB is obtained for the time delay $\Delta T = 2.33$ ns. Clearly, the asymptotic reduction of the spectral density at these minima (provided that the condition $2\rho Y_o \Delta T \gg \det[JY_T]$ is fulfilled) will be $-20\log(\Delta T)$. Thus, a 20-dB phase noise reduction is obtained for a 10-fold longer delay line. Extremely long lines will, of course, be unimplementable. On the other hand, the spectral density tends to the value obtained with the original oscillator for time delay ΔT tending to zero.

As already indicated, the phase noise minima correspond to the maxima of $G_{vo}\cos(\omega\,\Delta T) - B_{vo}\sin(\omega\,\Delta T)$ versus ΔT. The resulting phase

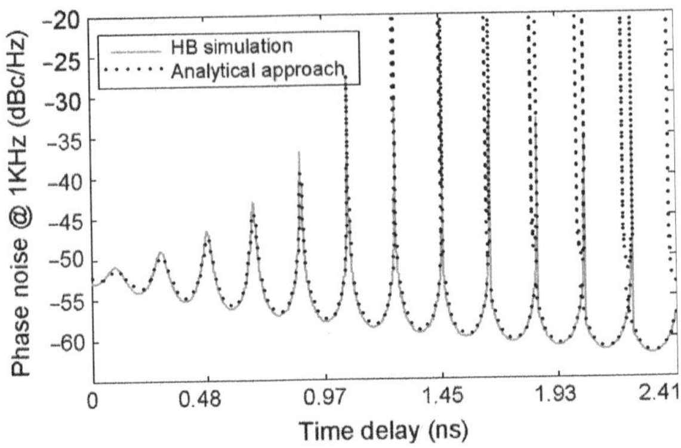

FIGURE 9.4 Phase noise variation versus time delay ΔT at a fixed frequency offset from the carrier, $f_m = 1$ kHz. The analytical results are compared with commercial harmonic balance simulations using the carrier modulation approach.

minimum condition is $\tan(\omega\,\Delta T) = -B_{VO}/G_{VO}$, with $\cos(\omega\,\Delta T) > 0$. For $2\rho Y_o \Delta T \gg \det[JY_T]$, the maximum phase noise reduction with respect to the free-running value tends to $\Delta S = 20\log(2\rho Y_o \Delta T |Y_{TVo}|/\det[JY_T]) = -A_L + 20\log(2Y_o\Delta T|Y_{TVo}|/\det[JY_T])$. As can be seen, a larger reduction is obtained for a smaller attenuation A_L. However, for small A_L the accuracy of the linearization (9.4) will degrade.

The ΔT values providing the phase noise minima have the additional advantage of corresponding to zero frequency deviation with respect to the original oscillation frequency ω_o, obtained before insertion of the oscillator into the feedback loop. This can be demonstrated easily by taking (9.4) into account. Actually, using this equation, it is possible to express the frequency deviation as

$$\omega - \omega_o = \frac{-G_{vo}2\rho Y_o \sin(\omega\,\Delta T) - B_{vo}2\rho Y_o \cos(\omega\,\Delta T)}{G_{vo}B_{\omega o} - G_{\omega o}B_{vo}} \tag{9.13}$$

The phase noise minima condition $\tan(\omega\,\Delta T) = -B_{VO}/G_{VO}$ provides a zero value in the numerator of (9.13). Therefore, at points with minimum phase noise, no frequency deviation is obtained with respect to the original circuit, which is very convenient for the oscillator design. The points with minimum phase noise should also lie on the stable sections of the solution curves. As already stated, they correspond to the positive maxima of $G_{vo}\cos(\omega\,\Delta T) - B_{vo}\sin(\omega\,\Delta T)$, so taking (9.6) into account, they should fulfill $\det[JY'] > 0$.

In the analytical formulation derived, only white noise perturbations about the oscillator carrier have been taken into account. To include the influence of the flicker noise, the oscillator equations should be written at both dc and the first harmonic term. This would provide a system of three real equations in three real unknowns, given by V_{dc}, V, and ω. As discussed in Chapter 2, this requires system formulation at a node where the dc components are not blocked. The extended system has the form

$$
\begin{aligned}
G_{T\mathrm{dc,\,dco}}(V_{\mathrm{dc}} - V_{\mathrm{dco}}) + G_{T\mathrm{dc},Vo}(V - V_o) + G_{T\mathrm{dc},\omega o}(\omega - \omega_o) &= G_{n,\mathrm{dc}}(t) \\
G_{T,\mathrm{dco}}(V_{\mathrm{dc}} - V_{\mathrm{dco}}) + G_{vo}(V - V_o) + G_{\omega o}(\omega - \omega_o) - 2\rho Y_o \cos(\omega\,\Delta T) &= G_n(t) \\
B_{T,\mathrm{dco}}(V_{\mathrm{dc}} - V_{\mathrm{dco}}) + B_{vo}(V - V_o) + B_{\omega o}(\omega - \omega_o) + 2\rho Y_o \sin(\omega\,\Delta T) &= B_n(t)
\end{aligned}
\tag{9.14}
$$

Note that the time delay does not appear explicitly in the dc equation. However, the three equations are coupled, so an overall phase noise reduction will take place. To see this, the carrier frequency perturbation due to the noise sources will be obtained simply by following the same steps in (9.8)–(9.11). The expression for the time derivative of the phase increment is $\Delta\dot{\phi}(t)$:

$$
\begin{aligned}
\Delta\dot{\phi}(t) &= \frac{\begin{vmatrix} G_{T\mathrm{dc,\,dco}} & G_{T\mathrm{dc,\,Vo}} & G_{n,\mathrm{dc}}(t) \\ G_{T,\mathrm{dco}} & G_{VO} & G_n(t) \\ B_{T,\mathrm{dco}} & B_{VO} & B_n(t) \end{vmatrix}}{\begin{vmatrix} G_{T\mathrm{dc,\,dco}} & G_{T\mathrm{dc},Vo} & G_{T\mathrm{dc},\omega o} \\ G_{T,\mathrm{dco}} & G_{VO} & G_{\omega o} + 2\rho Y_o \Delta T \sin(\omega\,\Delta T) \\ B_{T,\mathrm{dco}} & B_{VO} & B_{\omega o} + 2\rho Y_o \Delta T \cos(\omega\,\Delta T) \end{vmatrix}} \\
&= \frac{G_{n,\mathrm{dc}}(t)\alpha + G_n(t)\beta + B_n(t)\gamma}{[G_{\omega o} + 2\rho Y_o \Delta T \sin(\omega\,\Delta T)]\beta + [B_{\omega o} + 2\rho Y_o \Delta T \cos(\omega\,\Delta T)]\gamma}
\end{aligned}
\tag{9.15}
$$

where α, β, and γ are constants obtained directly from expansion of the two determinants. Clearly, the self injection topology reduces the low frequency and white noise equally. The amount of reduction will, of course, differ quantitatively from the predictions of (9.11). Note that it is also possible to rewrite (9.15) in terms of the determinant associated with the original free-running oscillator solution:

$$
\Delta\dot{\phi}(t) = \frac{G_{n,\mathrm{dc}}(t)\alpha + G_n(t)\beta + B_n(t)\gamma}{\det[JY_T]_o + 2Y_o\rho\Delta T[\beta\sin(\omega\,\Delta T) + \gamma\cos(\omega\,\Delta T)]}
\tag{9.16}
$$

which is an expression similar to (9.11).

9.3 USE OF HIGH-Q RESONATORS

The use of a dielectric resonator increases the spectral purity of the reinjected signal due to its high quality factor and thus to its high-frequency selectivity. A dielectric resonator is a rectangular or cylindrical dielectric block, with high permittivity and small loss. It behaves in similar manner to a cavity resonator, with boundaries at the interface between the air and the dielectric material. The resonance frequency depends on the dielectric constant, the block dimensions, and the operation conditions [12,13]. In the self-injected oscillator, the resonator is coupled to a transmission line in the circulator loop [3,4]. Note that unlike the case of the delay-line method, the transmission line is used only to couple the resonator to the feedback loop, so it will be of short length. However, the stability of the solution is very sensitive to the length of this coupling line [12]. Even if the original oscillator solution is stable, it will become unstable for some values of the coupling-line length. This instability effect, as well as the mechanism for phase noise reduction in a high-Q resonator oscillator, is analyzed next.

For a description of the resonator-based self-injected topology, the same expression (9.2) will be used. The dielectric resonator is coupled to a microstrip transmission line connected between ports 2 and 3 of the circulator. The amount of coupling is determined by the lateral distance between the resonator and the microstrip line. Use of an attenuator of A_L decibels will also be necessary to limit the feedback amount. The formulation (9.2) requires determination of the reflection coefficient Γ at port 1 of the circulator. In a first approximate calculation, this will be given by the scattering parameter S_{21} of the linear network connected between ports 2 and 3 of the circulator (see Fig. 9.5). The dielectric resonator coupled to a transmission line is modeled with a lumped parallel RLC resonator in series with the line. The element values of this equivalent resonator depend on the resonator characteristics (dimensions and permittivity) and the coupling strength. Details on the modeling procedure are given in [13,15]. The total transmission-line length gives rise to the phase shift $e^{-j\theta}$, with $\theta = \omega\,\Delta T$ [12], and the attenuator provides the attenuation factor $\gamma = 10^{-A_L/20}$. The reflection coefficient Γ at port 1 is given by the parameter S_{21} of the scattering matrix associated with the chain connection of the three elements:

$$S_{21} = \frac{2Z_o Y_{\rm res}}{1 + 2Z_o Y_{\rm res}} \gamma e^{-j\omega\,\Delta T} \tag{9.17}$$

where $Y_{\rm res}$ is the total admittance of the equivalent RLC resonator. This model is only approximate, since the phase shift provided by the transmission line is equal to its electrical length $\theta = \omega\,\Delta T$ only for matched terminations [13,14], which is not the case when the line is ended in the resonator admittance $Y_{\rm res}$. For the determination of the steady-state solutions versus ΔT and their stability and phase noise analysis, expression (9.17) should be replaced into (9.2), following identical steps to the case of the long delay line.

The reflection coefficient Γat port 1 of the circulator can be calculated in a more accurate way from multiplication of the chain matrices associated with a parallel RLC resonator, in series with the length of transmission line and the attenuator. The total chain matrix of the three elements between ports 2 and 3 of the circulator is

$$[C_T] = \begin{bmatrix} \dfrac{1+\gamma^2}{2\gamma} & \dfrac{Z_o(1-\gamma^2)}{2\gamma} \\ \dfrac{1-\gamma^2}{2Z_o\gamma} & \dfrac{1+\gamma^2}{2\gamma} \end{bmatrix} \begin{bmatrix} \cos(\omega\,\Delta T) & Z_o j\sin(\omega\,\Delta T) \\ Y_o j\sin(\omega\,\Delta T) & \cos(\omega\,\Delta T) \end{bmatrix} \begin{bmatrix} 1 & \dfrac{1}{Y_{\rm res}} \\ 0 & 1 \end{bmatrix} \tag{9.18}$$

where the matrix depending on $\omega\,\Delta T$ is the chain matrix of the transmission line, and the third matrix is the one associated with the resonator. The next step is the transformation of the total chain matrix into the total scattering matrix $[S_T]$ using

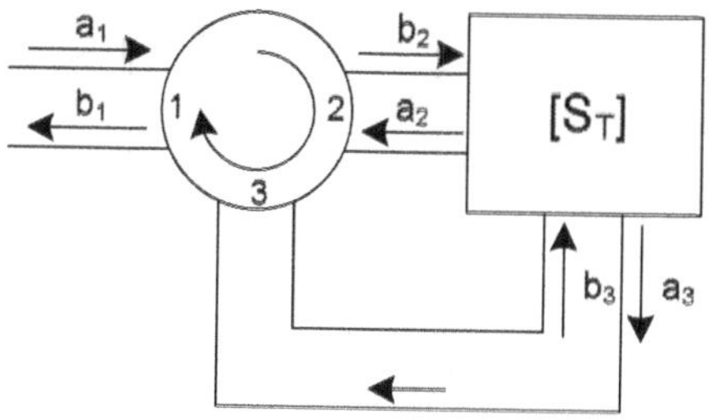

FIGURE 9.5 Sketch used for determination of the reflection coefficient at port 1 of the circulator when the total scattering matrix is determined from the chain matrixes of the attenuator, the delay line, and the resonator.

transformation expressions [15]. Following Fig. 9.5, the scattering matrix provides the following relationship between the incident and reflected power waves at ports 1 and 3 of the circulator:

$$\begin{aligned} b_3 &= S_{11}^{\mathrm{T}} a_1 + S_{12}^{\mathrm{T}} b_3 \\ b_1 &= S_{21}^{\mathrm{T}} a_1 + S_{22}^{\mathrm{T}} b_3 \end{aligned} \tag{9.19}$$

Thus, the reflection coefficient at port 1 is given by

$$\Gamma = \frac{b_1}{a_1} = S_{21}^T + \frac{S_{11}^{\mathrm{T}} S_{22}^T}{1 - S_{12}^T} \tag{9.20}$$

Formulation (9.20) is introduced into in-house software for numerical determination of the reflection coefficient.

As illustration, a dielectric resonator with $Q = 1000$ and resonance frequency $f_0 = 5.16$ GHz, coupled to a short transmission line, has been introduced in the self-injection loop of the parallel topology oscillator. Figure 9.6 shows the variation of oscillation frequency obtained versus the time delay ΔT when both the simplified and accurate models are used. Two different attenuation values have been considered: $A_L = 30$ dB (Fig. 9.6a) and $A_L = 15$ dB (Fig. 9.6b). There are discrepancies between the two models which increase as the attenuation is reduced.

As shown in Fig. 9.6, there is great sensitivity to the length of the short transmission line, which gives rise to significant variations of the solution and its stability properties in a short time-delay interval. This is associated to the high quality factor of the resonator, which, as gathered from (9.16), gives rise to a substantial variation in the reflection coefficient Γ at port 1 of the circulator versus any small change in transmission-line length. As shown in Fig. 9.6, there are some intervals with three coexisting solutions delimited by turning points. For maximum rigor, the stable regions of the curves in Fig. 9.6a have been determined through sequential application of the pole–zero identification technique [10]. Figure 9.7 shows the dominant complex-conjugate poles corresponding to the three coexisting solutions obtained for the same time-delay value, $\Delta T = 0.1$ ns. The solutions in sections A

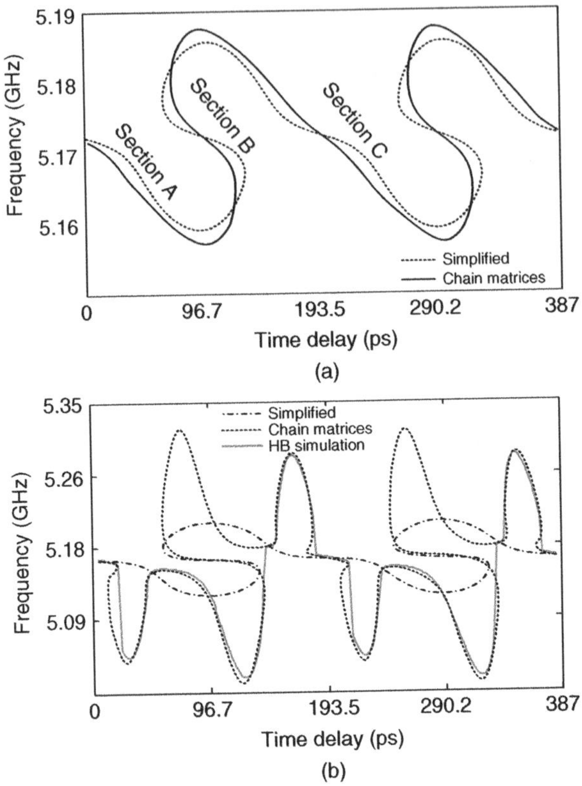

FIGURE 9.6 Self-injected oscillator with a dielectric resonator in the circulator loop. Variation of the oscillation frequency versus the time delay ΔT. Comparison of the results obtained using simplified and accurate coupled resonator models for two different values of the loop attenuation: (a) $A_L = 30$ dB; (b) $A_L = 15$ dB.

and C are stable, whereas the solution in section B is unstable. For each of the three solutions, there is also a pair of complex-conjugate poles on the imaginary axis at the oscillation frequency, inherent in the oscillator's autonomy.

As already mentioned, for smaller attenuation A_L, greater discrepancy is obtained between the results obtained with the two coupled-resonator models (see Fig. 9.6b). When rigorous calculation of Y_{circ} is carried out, a second multivalued region is obtained, not predicted by the simplified expression. Figure 9.6b shows a comparison of the results of the describing function analysis and those obtained with commercial harmonic balance. There is excellent agreement with the solution curve obtained with an accurate resonator model. As already known, with a default harmonic balance simulation, it is not possible to pass through the solution turning points, and a jump is obtained for a different curve section. Because the sections are very close, discontinuity of the response might be difficult to note, and a continuous curve may be wrongly assumed.

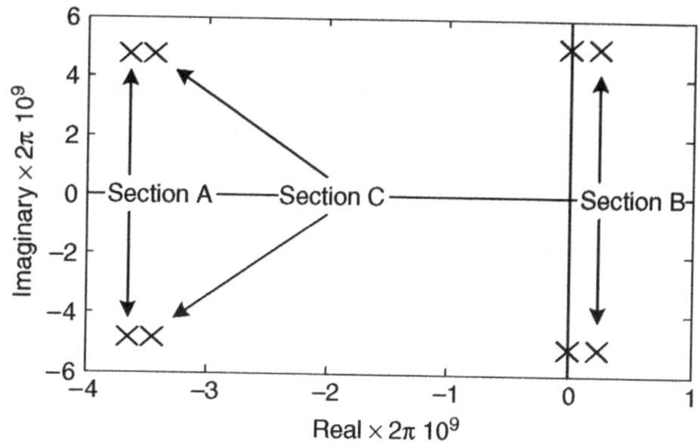

FIGURE 9.7 Self-injected oscillator with a dielectric resonator in the circulator loop. Application of the pole–zero identification technique to the three solutions coexisting for a time delay $\Delta T = 0.1$ ns (Fig. 9.6a).

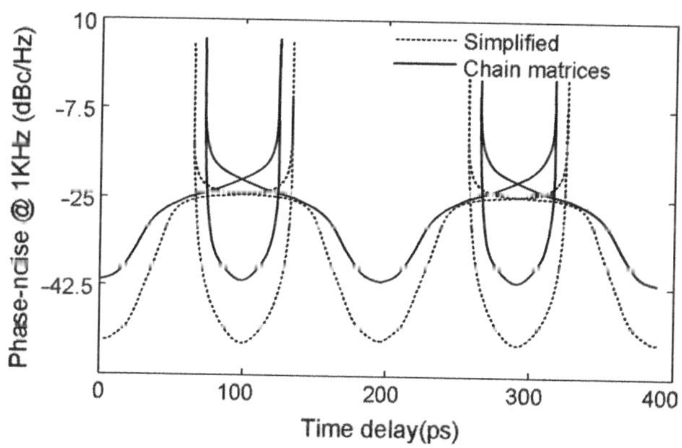

FIGURE 9.8 Self-injected oscillator with dielectric resonator in the circulator loop. Variation versus the time delay ΔT of the phase noise spectral density at $f_m = 1$ kHz. The attenuation value is $A_L = 30$ dB.

Figure 9.8 shows variations in the phase noise spectral density for the attenuation value $A_L = 30$ dB at the constant frequency offset $f_m = 1$ kHz versus the time delay ΔT. Comparing these results with those obtained with the long delay line, higher phase noise reduction is achieved with a resonator of quality factor $Q = 1000$. Again, the phase noise tends to infinity at the turning points of the solution curve, where the analysis is no longer valid. Note that finite values are obtained when the time derivative of the amplitude increment $\Delta\dot{V}(t)$, appearing in the perturbation equation (9.8), is taken into account for calculation of the phase noise spectral density.

9.4 STABILIZATION LOOP

In a proposed feedback topology [5,6], the phase noise reduction is achieved with a stabilization loop containing a frequency discriminator. A schematic of the feedback oscillator is shown in Fig. 9.1b. A fraction of the oscillator output signal is extracted through a coupler and injected into a power divider. One of the divider outputs is delayed by means of a long transmission line, which gives rise to stabilization of the signal frequency. The two signals are introduced in a frequency mixer acting as a phase comparator. The mixer output provides a low-frequency error signal, which after passing through a low-frequency amplifier, is introduced in a varactor diode. Variation in this control voltage corrects the oscillation frequency. Next we present an approximate analytical study of the oscillator with a stabilization loop.

The capacitance of the varactor diode included in the oscillator circuit varies according to $C(v) = C_{jo}/(1 - v_d/\varphi)^{1/2}$, with v_d being the low-frequency feedback voltage and φ, the built-in potential. For an approximate analysis, the steady-state solution is assumed to be sinusoidal, given by $v(t) = V \cos \omega t$ at the point at which the stabilization loop is connected. This signal is introduced into a coupler of L decibels and then split into two signals using the power divider, which further reduces the signal power, to 3 dB. The total attenuation of these two devices reduces the voltage amplitude to γV, with $\gamma < 1$. One of the two signals obtained is introduced directly into the frequency mixer; the other passes through a long delay line before being introduced into this mixer. Neglecting the phase shift due to the coupler and power splitter, the two inputs of the frequency mixer will be $v_1(t) = \gamma V \cos \omega(t)$ and $v_2(t) = \gamma V \cos[\omega(t - \Delta T)]$. Then the dc output signal will be $V_d \infty 1/2\gamma^2 V^2 \cos(\omega \Delta T)$. A low-frequency amplifier of linear voltage gain is also used to enhance the feedback-loop sensitivity to the error signal. The gain contribution from the mixer and amplifier is represented globally as G_A. Then the control voltage at the terminals of the varactor diode is $V_d = 1/2G_A\gamma^2 V^2 \cos(\omega \Delta T)$.

For a simplified analytical study, a parallel connection of the varactor to the oscillator circuit will be assumed, as shown in Fig. 9.1b. The feedback oscillator equations will be formulated in terms of the total admittance function as $Y'_T(V, \omega) = G'_T(V, \omega) + jB'_T(V_d, \omega) = 0$. This admittance includes the parallel susceptance exhibited by the varactor diode, which depends on the bias voltage. The feedback loop will only give rise to a small amplitude and frequency deviation of the oscillator solution. It will be assumed that in free-running operation the dc voltage across the varactor diode is $V_d = 0$. Note that the maximum sensitivity of the varactor capacitance is obtained about this dc value, so it will also be assumed that under feedback conditions, this bias voltage will undergo small variations only. Therefore, it will be possible to linearize the admittance function associated with the original oscillator $Y_T(V, \omega)$ about the original free-running solution given by

$V_o, \omega_o, V_d = 0$. This provides the following equation system:

$$\begin{aligned} G'_T(V,\omega) &= G_{vo}(V-V_o) + G_{\omega o}(\omega-\omega_o) = 0 \\ B'_T(V,\omega) &= B_{vo}(V-V_o) + B_{\omega o}(\omega-\omega_o) + \left.\frac{\partial B_T}{\partial V_d}\right|_o V_d \\ &= B_{vo}(V-V_o) + B_{\omega o}(\omega-\omega_o) + \left.\frac{\partial B_T}{\partial V_d}\right|_o K V_o^2 \cos(\omega\,\Delta T) = 0 \end{aligned} \tag{9.21}$$

where $K = G_A A\gamma^2/2$. It is a nonlinear equation in the frequency ω, due to the sinusoidal dependence on this variable of the second equation. Note that for a simplified analysis, the dc term at the amplifier output has approached $V_{dc} = KV_o^2\cos(\omega\,\Delta T)$, with V_o being the free-running oscillation amplitude. In a more accurate analysis, the voltage amplitude V of the feedback signal should be used, which would give rise to a nonlinear dependence on V. The determinant of the Jacobian matrix associated with the complex system (9.21) can be calculated in a straightforward manner. It is given by

$$\det[JY'_T] = \det[JY_T]_o - \alpha\,\Delta T\sin(\omega\,\Delta T) \tag{9.22}$$

with $\det[JY_T]_o$ being the determinant of the Jacobian matrix associated with the original free-running oscillator and evaluated at the original steady-state solution V_o, ω_o. The newly introduced coefficient is given by $\alpha = KV_o^2G_{vo}\,(\partial B_T/\partial v_d)|_o$.

When analyzing the circuit steady-state solution versus time delay ΔT, turning points will be obtained at the zeros of the determinant function $\det[JY'_T]$, analogous to the one corresponding to the self-injected topology analyzed in Section 9.2. Because the steady-state solution of the original oscillator V_o, ω_o should be stable, its associated determinant will fulfill $\det[JY_T]_o > 0$, and the turning points will be obtained only for sufficiently long ΔT so that the term $\alpha\,\Delta T\sin(\omega\,\Delta T)$ becomes comparable to $\det[JY_T]_o$. When tracing the solution curve versus ΔT, the turning points will give rise to jumps between different sections of this curve.

For illustration, the stabilization technique described has been applied to a parallel topology oscillator with cubic nonlinearity. Figure 9.9 shows the variation in oscillation frequency versus the time delay ΔT when using a low-frequency amplifier with gain $G_A = 20$ dB. As can be seen, turning points are obtained from relatively large ΔT values. These turning points will give rise to a sudden shift in the oscillation frequency. As already discussed, in the approximate stability analysis, limited to a dominant real pole, the stable and unstable sections have been distinguished through the evaluation of the sign of $\det[JY'_T]$. The solution will be stable in curve sections fulfilling $\det[JY'_T] > 0$ and unstable in curve sections with $\det[JY'_T] < 0$. Because this stability analysis is not complete, pole–zero identification has also been used at different points of the multivalue solution curve. No other instabilities have been observed, and the results agree with those based on evaluation of the sign of $\det[JY'_T]$.

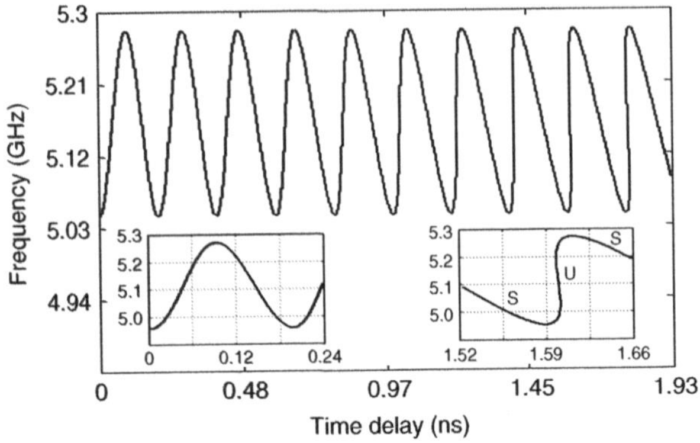

FIGURE 9.9 Parallel topology oscillator with a stabilization loop; variation of the oscillation frequency versus the time delay ΔT.

As in previous cases, for analytical derivation of the phase noise spectral density, only white noise about the oscillator carrier will be considered. Note that the objective of the work is study of the loop action over the oscillator noise, so noise contributions from the loop itself are not considered. Under the noise influence, the perturbed oscillator amplitude is given by $V + \Delta V(t)$, and its frequency becomes $\omega + d\phi(t)/dt - j(d\frac{\Delta V}{V_o}(t)/dt)$. Then the noisy oscillator equations are written

$$\begin{gathered} G_{VO}\Delta V(t) + G_{\omega o}\left(\frac{d\phi(t)}{dt} - j\frac{d\Delta V(t)}{V_o\,dt}\right) = G_n(t) \\ B_{VO}\Delta V(t) + \left(B_{\omega o} - \Delta T\frac{\alpha}{G_{V_o}}\sin(\omega\,\Delta T)\right)\left(\frac{d\phi(t)}{dt} - j\frac{d\Delta V(t)}{V_o\,dt}\right) = B_n(t) \end{gathered} \tag{9.23}$$

By applying the Fourier transform to system (9.23) the phase noise spectral density at relatively low frequency offset from the carrier is given by

$$|\Delta\phi(\Omega)|^2 = \frac{|Y_{V_o}|^2 2S_w}{(\det[JY_T]_o - \alpha\,\Delta T\sin(\omega\,\Delta T))^2 V_o^2\Omega^2} \tag{9.24}$$

where the denominator agrees with the determinant of the Jacobian matrix of system (9.21). For constant offset frequency $\Omega = \Omega_m$, the denominator has a double dependence on ΔT: linear and sinusoidal, as in the case of a self-injected oscillator. In Fig. 9.10, this variation is analyzed at constant frequency offset $f_m = 1$ kHz. The phase noise reduction will be due to the large magnitude reached by the denominator at some ΔT values. The phase noise minima correspond to the positive maxima of $-\alpha\,\Delta T\sin(\omega\,\Delta T)$ obtained from the condition $\sin(\omega\,\Delta T) = -1$. Thus, they are observed at the time-delay values $\Delta T = (3 + 2n)\pi/2\omega$, with n an integer. Because at these phase noise minima the condition $\cos(\omega\,\Delta T) = 0$ is fulfilled, these points will have minimum frequency deviation with respect to the original oscillation frequency ω_o, as can easily be deduced from inspection of (9.21).

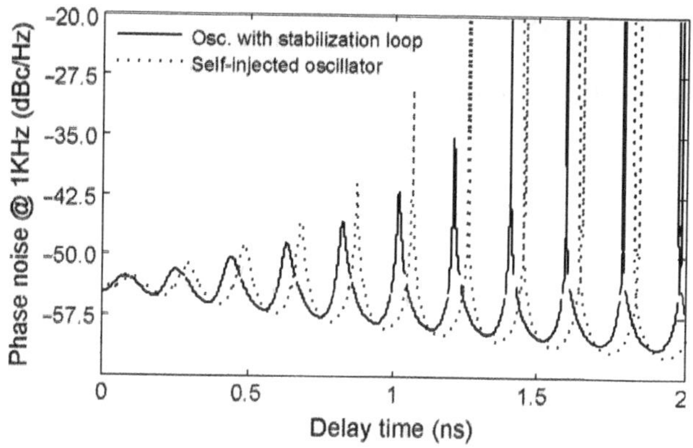

FIGURE 9.10 Self-injection topology and low-frequency stabilization loop applied to the parallel topology oscillator. Comparison of the variation in phase noise spectral density versus ΔT at the constant offset frequency $f_m = 1$ kHz.

Note that the oscillator solution must be stable in the absence of feedback, so it will fulfill $\det[JY_{To}] > 0$. On the other hand, at the phase noise minima, $\sin(\omega\Delta T) = -1$, so the condition $\det[JY'_T] > 0$ is satisfied. Therefore, they should correspond to stable solutions of the feedback oscillator. As soon as the product $\alpha\ \Delta T$ becomes dominant in the denominator of (9.24), the phase noise minima decrease as $-20\ \log(\alpha\ \Delta T)$. For $\alpha\ \Delta T \gg \det[JY_T]_o$, the maximum reduction of the phase noise spectral density with respect to the free-running value tends to $\Delta S = 20\ \log(\alpha\ \Delta T/\det[JY_T])$, with α being directly proportional to the loop gain K. Thus, the reduction increases with the loop gain. The phase noise maxima, tending to infinity, are obtained for $\det[JY'_T] = 0$ and correspond to the turning points of (9.24). These unphysical maxima come from the fact that time derivative $\Delta\dot{V}(t)$ was neglected when deriving an expression for the phase noise spectral density. On the other hand, the linearization used in the amplitude V is no longer valid in the immediate neighborhood of the turning points.

As gathered from all previous results, the behavior of the self-injected oscillator and that of the oscillator with the low-frequency feedback loop are qualitatively similar. In Fig. 9.10, the phase noise variation of the stabilization loop (with the gain $G_A = 20$ dB of the low-frequency amplifier) is compared with that obtained with the self-injection topology (with the attenuation $A_L = 20$ dB between circulator ports 2 and 3). For these particular values, the self-injection topology allows a shorter line length for the same phase noise reduction of about 10 dB with respect to the original free-running oscillation. In both cases the phase noise spectral density tends to the value obtained with the original oscillator for time delay ΔT tending to zero.

In the analytical formulation derived, only white noise perturbations about the oscillator carrier have been taken into account. To consider the influence of flicker noise, the oscillator equations should be written at both dc and the first harmonic term, in a manner similar to (9.15).

9.5 TRANSISTOR-BASED OSCILLATORS

9.5.1 Harmonic Balance Analysis

The preceding analytical study has focused on the effect on the stability and phase noise of the oscillator circuit of the two feedback topologies discussed here. However, accurate application to transistor-based oscillators containing several noise sources would require full harmonic balance techniques [16]. A parameter-switching technique must also be used, since direct parameter sweep will give rise to discontinuous jumps at the turning points. As shown in the preceding sections, the jumps take place between closely spaced turning points. The section between these points, not provided by default harmonic balance simulation, is likely to be unstable, as in all the preceding examples. Thus, use of the continuation technique could be avoided in practice.

When using harmonic balance, phase noise analysis will generally be based on the carrier modulation and conversion matrix approaches. For simplicity, the analysis presented here is limited to the carrier modulation approach. As we know, this analysis technique is based on a mixed harmonic balance formulation of the oscillator circuit. The unknowns are $\overline{X}'$ and ω; $\overline{X}'$ is a vector of harmonic components of the state variables after the arbitrary elimination of the imaginary part of one of them. Thus, the dimension of this vector is $(2N+1)Q-1$, with N the number of harmonic terms considered and Q the number of state variables. The carrier frequency perturbation due to the noise sources is calculated from

$$\begin{bmatrix} \left.\dfrac{\partial \overline{H}}{\partial \overline{X}'}\right|_o & \left.\dfrac{\partial \overline{H}}{\partial \omega}\right|_o \end{bmatrix} \begin{bmatrix} \Delta \overline{X}'(t) \\ \Delta\omega(t) \end{bmatrix} = [G]\overline{N}(t) \tag{9.25}$$

where $\overline{N}(t)$ is the vector of noise sources and $[G]$ is the matrix relating the vector of state variables to these sources. When using a feedback topology, the vector $\partial \overline{H}/\partial\omega|_o$ will contain sinusoidal terms in $(k\omega\,\Delta T)$, with k the harmonic index having amplitude proportional to ΔT, similar to (9.8) and (9.23). These terms are obtained from the frequency derivation of linear embedding matrixes affecting the state-variable and nonlinear element vectors X and Y [see (5.38) and (5.54)]. To solve for $\Delta\omega(t)$, the mixed Jacobian matrix in (9.25) must be inverted, which will give rise to a phase noise variation similar to the one obtained with the analytical formulation, that is, inversely proportional to $[\det[JH]_o + \rho\Delta T(\beta\sin(\omega\,\Delta T) + \gamma\cos(\omega\,\Delta T))]^2$ in the case of the self-injection loop, or to $(\det[JH]_o - \alpha\;\Delta T\sin(\omega\,\Delta T))^2$ in the case of the low-frequency feedback loop, with frequency discriminator. In these expressions, $\det[JH]_o$ is the determinant of the original free-running oscillator and the proportionality constants α, β and γ will, of course, be different from those obtained using the analytical formulation. The fact that the determinant of the feedback oscillator exhibits the same double dependence on the time delay ΔT (linear and sinusoidal) obtained in the analytical expressions explains the good agreement with the harmonic balance results (see Figs. 9.4 and 9.10).

9.5.2 Semianalytical Formulation

The semianalytical formulation for the analysis of feedback oscillator configurations is conceptually similar to that presented for the phase noise analysis of injection-locked oscillators in Section 7.7.2. The analysis technique is based on the use of a perturbation model for the original free-running oscillator extracted from harmonic balance simulations. This perturbation model is given by the derivatives of the oscillator admittance function about the free-running solution. It provides the derivatives $\partial Y_T/\partial V_o$ and $\partial Y_T/\partial \omega_o$ that are used in the formulation of both self-injected and feedback oscillators with a stabilization loop. In the latter case, it will also be necessary to obtain the derivative with respect to the bias voltage of the varactor diode $\partial Y_T/\partial V_d$. The semianalytical approach will require an equivalent model of the oscillator noise sources. This model is given by a current source $i_n(t)$, introduced in parallel at the node where the total admittance $Y_T(V, \omega)$ is calculated [17,18].

When using an auxiliary generator for analysis of the original free-running oscillator, the admittance derivatives required are easily calculated through finite differences. Note that the total admittance function of this oscillator evaluated at its steady-state solution is $Y_T(V_o, \omega_o) = 0$, and Y_T agrees with the admittance function of the auxiliary generator Y_{AG}. The derivative $\partial Y_{AG}/\partial V_{AGo}$, agreeing with $\partial Y_T/\partial V_o$, is obtained by maintaining the auxiliary generator frequency at $\omega_{AG} = \omega_o$ and applying a small sweep (with no optimization) in V_{AG}. Then the derivative $\partial Y_T/\partial V_o$ is obtained as

$$\frac{\partial Y_T}{\partial V_o} = \frac{\partial Y_{AG}}{\partial V_{AG}} = \left.\frac{Y_{AG2} - Y_{AG1}}{V_{AG2} - V_{AG1}}\right|_{\omega_{AG}=\omega_{AGo}} \tag{9.26}$$

Note that Y_{AG1} and Y_{AG2} are different from zero, as the steady-state oscillation conditions are not fulfilled, due to the V_{AG} sweep. The linear variation of Y_{AG} in the interval V_{AG1}, V_{AG2} must be verified to ensure the validity of the derivative calculation (9.26). In most practical cases, it will be possible simply to replace one of the two points considered in (9.26) with the actual steady-state solution $Y_{AG1} = 0$ and $V_{AG1} = \mathrm{V}_{AGo}$, so just one V_{AG} increment is considered.

The derivative $\partial Y_{AG}/\partial \omega_{AG}$, agreeing with $\partial Y_T/\partial \omega_o$, is obtained by maintaining the auxiliary generator amplitude at $V_{AG} = V_o$ and applying a small sweep (with no optimization) in ω_{AG}. Then the derivative $\partial Y_T/\partial \omega_o$ is calculated as

$$\frac{\partial Y_T}{\partial \omega_o} = \frac{\partial Y_{AG}}{\partial \omega_{AG}} = \left.\frac{Y_{AG2} - Y_{AG1}}{\omega_{AG2} - \omega_{AG1}}\right|_{V_{AG}=V_{AGo}} \tag{9.27}$$

Again, the linear variation of Y_{AG} about ω_o when ω_{AG} is swept must be verified.

As already stated, in the case of a feedback oscillator with a stabilization loop, it will also be necessary to determine the derivative $\partial Y_T/\partial V_d$. This is done in an identical manner to the two previous cases. The auxiliary generator amplitude and frequency are maintained at their steady-state values $V_{AG} = \mathrm{V}_o$ and $\omega_{AG} =$

ω_o, performing a small sweep in the bias voltage of the varactor diode V_d. The derivative is given by

$$\frac{\partial Y_T}{\partial V_d} = \frac{\partial Y_{AG}}{\partial V_d} = \frac{Y_{AG2} - Y_{AG1}}{V_{d2} - V_{d1}} \tag{9.28}$$

Note that, in general, the varactor bias voltage corresponding to the original free-running solution will be $V_{d1} = 0$, although the technique can be applied equally for other V_{d1} values.

Once the admittance derivatives have been extracted numerically, they will be introduced in the analytical formulation of the particular feedback oscillator topology. As already stated, accurate phase noise calculation will require a valid equivalent noise model at the analysis node. When considering flicker and white noise, two different equivalent noise current sources will be necessary, and the circuit equations will be formulated taking into account both the dc and fundamental frequency components. Numerical calculation of the nine derivatives appearing in equations (9.14), evaluated at the free-running solution, will be necessary. For the calculation of the derivatives with respect to the dc voltage, a dc auxiliary must be used. Note that this type of formulation can be used only at observation nodes where the dc component is not blocked.

9.5.3 Application to a 5-GHz MESFET-Based Oscillator

The two feedback techniques described above have been applied to a MESFET-based VCO at 5 GHz [16]. The oscillator schematic is shown in Fig. 9.11. For the harmonic balance simulation, the Angelov model has been used, with four nonlinearities, given, respectively, by the gate-to-source current i_{GS}, the capacitance c_{GS}, the drain-to-source current i_{DS}, and the drain-to-gate current i_{GD}. The simulated VCO band has been 4.6 to 5.4 GHz. In the measurements, the oscillation band was 4.7 to 5.30 GHz.

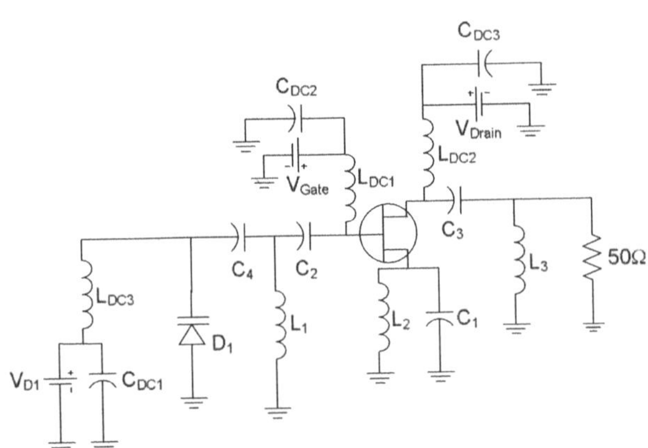

FIGURE 9.11 A 5-GHz VCO to which the two feedback techniques for phase noise reduction have been applied.

For the delay line, the substrate with $\varepsilon_r = 10.2$ has been chosen. In this case, the physical length l of this line will be the analysis parameter instead of the time delay. For each of the two feedback configurations, the initial estimation of the required length and attenuation/gain values has been obtained by means of the semianalytical approach described. The derivatives of the original oscillator have been extracted with harmonic balance. Next, analytical expression (9.11) or (9.24) has been used to estimate the values of the feedback-loop elements required for a phase noise reduction $\Delta S = 10$ dB. For the self-injected oscillator, the estimated values are attenuation $A_L = 20$ dB and line length $l_1 = 1.419$ m. For the stabilization loop, the estimated values are amplifier gain $G_A = 10$ dB and line length $l_2 = 1.5$ m. The line sections l_1 and l_2 were implemented on two separate boards, with multiple line bends, to be of reasonably small size. Loss of the linear structures used has been taken into account in the estimations.

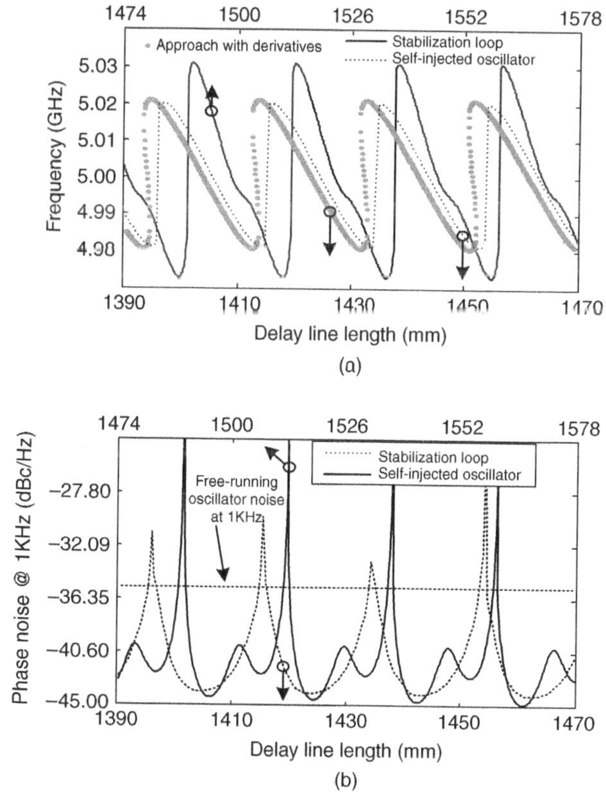

FIGURE 9.12 VCO at 5 GHz with the two types of feedback: a self-injected topology with $A_L = 20$ dB and $l_1 = 1.419$ m and a stabilization loop with discriminator having amplifier gain $G_A = 10$ dB and $l_2 = 1.5$ m. (a) Variation of the oscillation frequency versus the physical line length. (b) Variation of the phase noise spectral density at constant offset frequency $f_m = 1\ kHz$.

Simulations of the two feedback configurations are compared in Fig. 9.12a, where the variation of oscillation frequency with line length is represented. The gray- and black-dotted lines correspond, respectively, to the harmonic balance and semianalytical simulations of a self-injected oscillator. The thin solid line corresponds to the harmonic balance simulation of the stabilization loop with discriminator. As can be seen, a slightly smaller frequency deviation with respect to the free-running value is obtained with the self-injected oscillator (Fig. 9.12a). For this topology, estimation using the analytical model has also been superimposed. This was obtained by introducing into system (9.4) the derivatives Y_{vo} and $Y_{\omega o}$ of the free-running oscillator calculated at the circulator connection node. As shown, there is good qualitative agreement between the analytical model and the harmonic balance results. As expected, turning points are obtained only from sufficiently large values of the line length.

For the phase noise analysis, several noise sources have been considered. A voltage noise source in series with the internal gate terminal accounts for the flicker noise. The spectral density of this source is $S_v = 9 \times 10^{-9}/\Omega\ V^2/\text{Hz}$. White noise sources accounting for the channel thermal noise and thermal noise from all the resistive elements have also been introduced. The phase noise variation versus the delay line length l, calculated using the carrier modulation approach at constant offset frequency $f_m = 1$ kHz, is shown in Fig. 9.12b. The results using the two different configurations can be compared. The sharp maxima at the turning points of the solution curves are in good qualitative agreement with the analytical models. Similar phase noise reduction, of about 9 dB, is obtained with the two configurations. The phase noise spectra measured, confirming this reduction, are superimposed in Fig. 9.13. Simulations using the carrier modulation [11] and conversion matrix approaches [11,19] are also shown. The results with the two phase noise analysis techniques are overlapped up to the frequency offset $f = 4$ MHz for a self-injected oscillator, and up to $f = 3$ MHz for the stabilization loop. Although

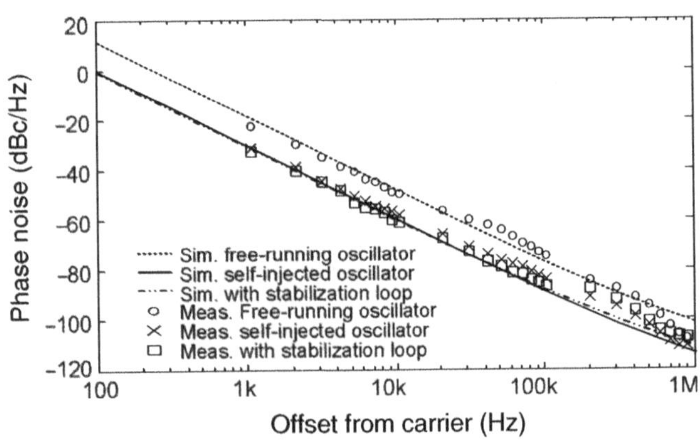

FIGURE 9.13 VCO at 5 GHz; comparison between the simulated and measured phase noise spectra corresponding to the free-running oscillator, the self-injected oscillator, and the stabilization loop with discriminator, respectively.

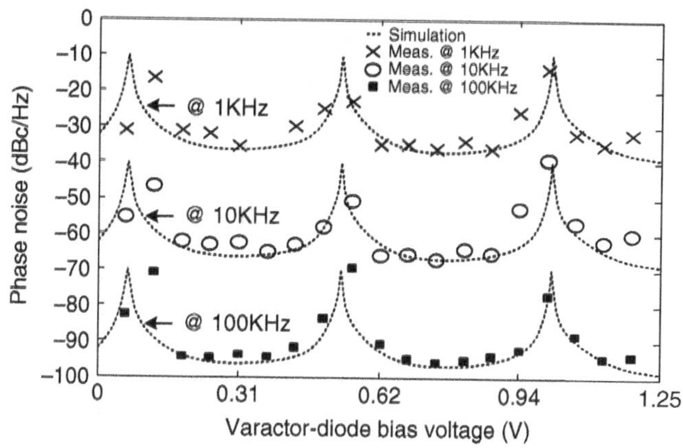

FIGURE 9.14 Self-injected oscillator with $A_L = 20$ dB and $l_1 = 1.419$ m. Showing Variation in phase noise spectral density versus the bias voltage of a varactor diode for three frequency offsets from the carrier. Measurements are superimposed.

not presented here, the conversion matrix approach shows less phase noise improvement as the frequency offset increases, in agreement with the analytical predictions of (9.10) and with the measurements. Finally, the varactor bias voltage has also been used as a parameter. In contrast with the line length, this parameter can be continuously modified in the experiment. The variation in phase noise spectral density using self-injected configuration at the three offset frequencies 1, 10, and 100 kHz is shown in Fig. 9.14. The experimental results are superimposed.

REFERENCES

[1] U. L. Rohde, A. K. Poddar, and G. Bock, *The Design of Modern Microwave Oscillators for Wireless Applications*, Wiley, Hoboken, NJ, 2005.

[2] A. Grebennikov, Noise reduction in transistor oscillators: 1. Resonant circuits, High Freq. Electron., p. 17, 2005.

[3] H. C. Chang, Phase noise in self-injection-locked oscillators: theory and experiment, *IEEE Trans. Microwave Theory Tech.*, vol. 51, pp. 1994–1999, Sept. 2003.

[4] H. C. Chang, Stability analysis of self-injection-locked oscillators, *IEEE Trans. Microwave Theory Tech.*, vol. 51, pp. 1989–1993, Sept. 2003.

[5] J. L. Altman, A technique for stabilizing microwave oscillators, *IEEE Trans. Microwave Theory Tech.*, vol. 2, pp. 16–25, July 1954.

[6] C. E. Free and C. S. Aitchison, Microwave oscillator control using switched delay-line technique, *Proceedings of the IEEE Microwave Theory and Techniques Symposium*, pp. 79–82, May 1995.

[7] C. McNeilage, E. N. Ivanov, P. R. Stockwell, and J. H. Searls, Review of feedback and feedforward noise reduction techniques, *Proceedings of the IEEE Frequency Control Symposium*, pp. 146–155, May 1998.

[8] K. Kurokawa, Some basic characteristics of broadband negative resistance oscillators, *Bell Syst. Tech. J.*, vol. 48, pp. 1937–1955, July–Aug. 1969.

[9] V. Rizzoli and A. Lipparini, General stability analysis of periodic steady-state regimes in nonlinear microwave circuits, *IEEE Trans. Microwave Theory Tech.*, vol. 33, pp. 30–37, Jan. 1985.

[10] J. Jugo, J. Portilla, A. Anakabe, A. Suárez and J. M. Collantes, Closed-loop stability analysis of microwave amplifiers, *IEE Electron. Lett.*, vol. 37, pp. 226–228, Feb. 2001.

[11] V. Rizzoli, F. Mastri, and D. Masotti, General noise analysis of nonlinear microwave circuits by the piecewise harmonic balance technique, *IEEE Trans. Microwave Theory Tech.*, vol. 42, pp. 807–819, 1994.

[12] J. C. Nallatamby, M. Prigent, M. Camiade, and J. Obregón, Phase noise in oscillators: Leeson formula revisited, *IEEE Trans. Microwave Theory Tech.*, vol. 51, pp. 1386–1394, Apr. 2003.

[13] P. F. Combes, J. Graffeuil, and J. F. Sautereau, *Microwave Components, Devices and Active Circuits*, Wiley, Chichester, UK, 1987.

[14] K. Chang, *Microwave Solid-State Circuits and Applications*, Wiley, New York, 1994.

[15] D. J. Vendelin, A. M. Pavio, and U. L. Rohde, *Microwave Circuit Design*, Wiley, New York, 1990.

[16] A. Suárez and F. Ramirez, Analysis of stabilization circuits for phase-noise reduction in microwave oscillators, *IEEE Trans. Microwave Theory Tech.*, vol. 53, pp. 2743–2751, Sept. 2005.

[17] F. Ramirez, M. Ponton, S. Sancho, and A. Suárez, Phase noise of injection-locked oscillators and frequency dividers, *IEEE Trans. Microwave Theory Tech.*, vol. 56, Feb. 2008.

[18] F. Ramirez, A. Suárez, and S. Sancho, Harmonic-balance technique for the shortening of the initial transient of microwave oscillators, *IEEE MTT-S International, Microwave Symposium Digest*, p. 4, 2005.

[19] J. C. Nallatamby, M. Prigent, J. C. Sarkissian, R. Quéré, and J. Obregón, A new approach to nonlinear analysis of noise behaviour of synchronized oscillators and analog-frequency dividers, *IEEE Trans. Microwave Theory Tech.*, vol. 46, pp. 1168–1171, Aug. 1998.

CHAPTER TEN

Coupled-Oscillator Systems

10.1 INTRODUCTION

Coupled oscillator systems are expected to oscillate in a synchronous manner at a single fundamental frequency [1]. In this chapter two different types of multioscillator systems are studied: oscillators with global coupling (Fig. 10.1a), used for power combination at the fundamental frequency [2,3] or at a given harmonic component of the individual oscillation frequency [4], and oscillators with nearest-neighbor coupling (Fig. 10.1b), used for power combination and beam steering [5,6].

The oscillator systems with global coupling considered here are formed by N identical suboscillators connected to a symmetric coupling network (Fig. 10.1a). If the output port is not the same as the coupling part, the total output signal is extracted with a power combination network (not represented in the figure). For N suboscillators combined, a minimum of N different oscillation modes can exist, in correspondence with the N eigenvalues and eigenvectors of the coupling network [2]. Assuming equal suboscillators, each of the N modes corresponds to the phase shift $n2\pi/N$, with $n = 0$ to $N-1$, between adjacent oscillator elements. For power combination at the fundamental frequency ω_o, the in-phase mode ($n = 0$) is used, with all the suboscillators having the same phase values. In turn, for power combination at a multiple $N\omega_o$ of the individual oscillation frequency, we are interested in the mode $n = 1$, providing a $2\pi/N$ phase shift between the oscillator elements. This is the operational mode of N-push oscillators. The suboscillators are in phase at $N\omega_o$ at the coupling ports, whereas the harmonic components $\omega_o, \ldots, (N-1)\omega_o$ are inherently canceled. Thus, the frequency of the output signal

Analysis and Design of Autonomous Microwave Circuits, By Almudena Suárez

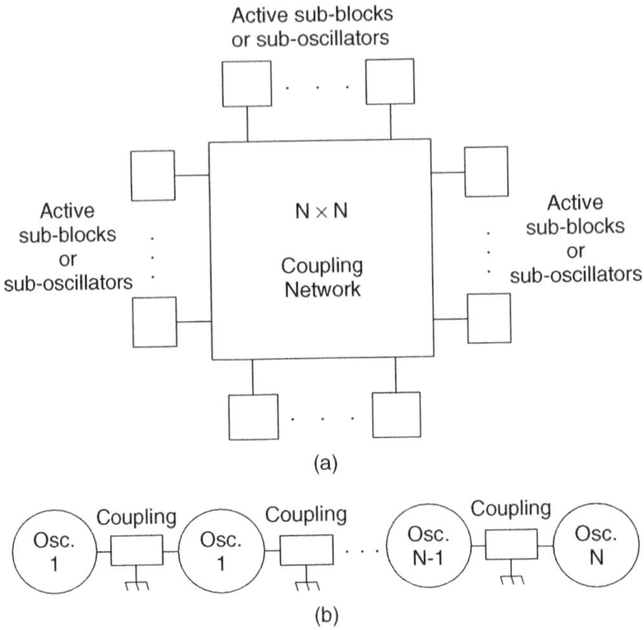

FIGURE 10.1 Coupled-oscillator systems: (a) global coupling with N active subblocks and an $N \times N$ linear coupling network, used primarily for power combination and multiplication by N of the oscillation frequency; (b) nearest-neighbor coupling, used mainly for beam steering in phased-array antennas.

is N times the oscillation frequency of the individual suboscillators. Compared to a single oscillator at the fundamental frequency $N\omega_o$, the N-push oscillator has the advantage of a usually much lower phase noise, since the higher the oscillation fundamental frequency, the lower the quality factor Q and the higher the phase noise spectral density [7]. The power combination at $N\omega_o$ will also be advantageous with respect to the use of an oscillator at ω_o followed by a frequency multiplier by N. This is because the phase noise of the oscillator is inherently increased in 20 log N decibels, due to multiplication by N of the phase perturbations. In contrast, the coupling of N oscillators gives rise to a phase noise improvement at the fundamental frequency of approximately 10 log N decibels with respect to an individual oscillator of the same characteristics. This can be explained roughly by the fact that N resonators are coupled together, which leads to an increase in the oscillator quality factor. Another advantage is the clean cancellation of the frequency components $K\omega_o$ with $k \neq mN$.

The existence of stable oscillations modes other than the desired mode should be prevented. In some cases, this can be done through a suitable design of the individual suboscillator, considering the equivalent impedance exhibited by the coupling. For example, for a two-push ("push–push") oscillator, the common node existing in the coupling network (or in the power combination network, depending on the topology used) is a virtual short-circuit at the desired mode $n = 1$ [7], due to the phase relationship $2\pi/N = \pi$ between the suboscillators. In turn, at the undesired in-phase oscillation, ($n = 0$) will exhibit $N = 2$ times the actual load impedance $NR_L = 2R_L$. Taking these equivalent impedances into account, the individual oscillator will be designed so as to fulfill the oscillation startup conditions at the desired mode only.

The oscillator systems with nearest-neighbor coupling (Fig. 10.1b) considered here are used for beam steering applications [6]. Each element is a full free-running oscillator before its connection to the system. Once in the system, the linear coupling networks introduce relatively low injection signals into each oscillator, coming from its nearest neighbors, and the system gets locked to a single fundamental frequency. When used for the beam steering of an antenna array, each oscillator in the coupled system feeds an array element. The beam is steered electronically by controlling the phase distribution across the aperture elements. Thus, a constant interstage phase shift must be synthesized between the different oscillators. This is achieved by tuning the peripheral system elements only, usually by means of varactor diodes. In the simpler case of a one-dimensional array, the constant phase shift progression is obtained by detuning the two outermost oscillators in opposite directions. In the case of a two-dimensional array, all the peripheral elements must be tuned, which allows beam steering in azimuth and elevation. The vertical and horizontal peripheral elements are detuned, with the same values, in opposite directions, except the corner elements, which are tuned with the addition of the tuning voltages of their neighboring vertical and horizontal elements [5,8]. Beam steering based on coupled oscillator systems allows a smaller system size compared to the traditional topology based on phase shifters, which requires individual control of the polarization and wiring of each phase shifter. Other disadvantages of traditional phase shifters are their high cost and high insertion loss at high frequencies. The power delivered by the oscillator elements mitigates this problem in coupled-oscillator systems. In summary, a smaller and cheaper design can be obtained.

The behavior of oscillator systems with nearest-neighbor coupling involves added difficulties compared to the case of injection-locked oscillators. In an injection-locked oscillator, the synchronizing source acts as an external generator and is not influenced by the circuit oscillation. In a coupled-oscillator system there is mutual synchronization between adjacent oscillator elements, and the overall system behaves in a free-running manner. The oscillation frequency depends on the element values of the individual oscillators and will change with the tuning of peripheral elements used for the beam steering. Note, however, that it is also possible to introduce an injection signal in the coupled-oscillator system, from an independent periodic source, which allows fixing the array oscillation frequency. As we know, for relatively low power in the input-periodic source at a given frequency ω_{in}, a standard injection-locked oscillator exhibits two solutions, with different amplitude and phase. An N-element coupled-oscillator system contains $N-1$ pairs of mutually coupled oscillators. Thus, for given values of its tuning voltages, 2^{N-1} different solutions will coexist [6,9]. Only one of these solutions has a constant interstage phase shift, which, for proper system operation, must be the only stable solution.

This chapter is divided into two main parts. The first part is devoted to multidevice oscillators used for power combination at the fundamental frequency or the Nth harmonic component (N-push). The second part deals with nearest-neighbor coupled-oscillator systems, used for beam steering in phased-array antennas. For each case we study possible oscillation modes as well as mathematical procedures for stability analysis of the solutions and prediction of the phase noise spectrum. Techniques for the harmonic balance design of coupled-oscillator systems are also provided. However, in coupled-oscillator arrays for beam steering the number of oscillator elements is usually very high. Therefore, use of the harmonic balance method can be prohibitive in terms of computational effort and simulation time. To cope with this problem, a semianalytical formulation is presented. This formulation is based on a perturbation model of the individual oscillators in free-running conditions, extracted with harmonic balance simulations. Using this model for each oscillator element and applying Kirchhoff's laws to the coupled structure, a nonlinear system of reduced order is obtained. This will enable a low-cost analysis of arrays with a high number of oscillator elements. It will also allow efficient design of the coupling networks under different criteria, such as the reduction of the frequency and amplitude deviations with respect to the free-running oscillation.

10.2 OSCILLATOR SYSTEMS WITH GLOBAL COUPLING

10.2.1 Simplified Analysis of Oscillation Modes

In this section we present a general study of multidevice oscillators, systems comprised of several suboscillator circuits coupled globally through a multiport linear network. Assuming N identical suboscillators, the system can be split into a passive N-port network, consisting of the coupling network and an active N-port network that consists of suboscillators. Alternatively, the division can be carried out at the transistor input or output terminals, so the linear loading elements connected to these terminals are included in the passive network considered (Fig. 10.1a). Then the active subnetwork is made up of N active subcircuits, each consisting of one transistor, together with its feedback elements and load at the opposite port (the one that is not connected to the passive N-port network). This second option provides more analysis sensitivity, due to the location of the observation ports at the transistor terminals, constituting the "sources" of negative resistance.

Either an impedance analysis (in terms of the branch currents) or an admittance analysis (in terms of the node voltages) can be carried out. Depending on the particular circuit topology, one or another will be more suitable. Here only the admittance analysis is presented. The N-port passive network is represented by its admittance matrix $[Y_L(\omega)]$, which will be frequency dependent. The effect of a particular port voltage on a given branch current will depend only on the relative location of the two ports being considered. Thus, the linear admittance matrix

$[Y_L]$ connecting the various active subcircuits has at any analysis frequency ω the following structure [2,7]:

$$\overline{I} = \begin{bmatrix} I_1 \\ I_2 \\ I_3 \\ \vdots \\ I_N \end{bmatrix} = [Y_L]\overline{V} = \begin{bmatrix} Y_{11} & Y_{12} & Y_{13} & \cdots & Y_{1N} \\ Y_{1N} & Y_{11} & Y_{12} & \cdots & Y_{1,N-1} \\ Y_{1,N-1} & Y_{1,N} & Y_{11} & \cdots & Y_{1,N-2} \\ \vdots & \vdots & \vdots & \vdots & \vdots \\ Y_{12} & Y_{13} & Y_{14} & \cdots & Y_{11} \end{bmatrix} \begin{bmatrix} V_1 \\ V_2 \\ V_3 \\ \vdots \\ V_N \end{bmatrix} \tag{10.1}$$

where the vector $\overline{I}$ contains the currents at the terminals of the linear network, and the vector $\overline{V}$ contains the node voltages. Note that in the case of a reciprocal network, the matrix elements will fulfill the property $Y_{ij} = Y_{ji}$ for $i \neq j$. The admittance matrix in (10.1) will have N eigenvalues and eigenvectors, indexed with $n = 0$ to $N-1$. Due to the matrix properties indicated, the nth complex eigenvalue and eigenvector are given by [2,10]

$$\lambda_n = Y_{11} + Y_{12}e^{jn(2\pi/N)} + Y_{13}e^{j2n(2\pi/N)} + \cdots + Y_{1N}e^{j(N-1)n(2\pi/N)}$$

$$\overline{V}^n = \frac{A}{\sqrt{N}} \begin{bmatrix} 1 \\ e^{jn(2\pi/N)} \\ e^{j2n(2\pi/N)} \\ \vdots \\ e^{j(N-1)n(2\pi/N)} \end{bmatrix} \tag{10.2}$$

where A is a proportionality constant, with voltage dimension, and n varies between 0 and $N-1$. Note that the different complex components of each eigenvector have the same amplitude. The $[Y_L]$ matrix in (10.1) is frequency dependent and so are its eigenvalues and eigenvectors, which fulfill the following relationship:

$$\begin{bmatrix} I_1^n \\ \vdots \\ I_N^n \end{bmatrix} = [Y_L] \begin{bmatrix} V_1^n \\ \vdots \\ V_N^n \end{bmatrix} = \lambda_n \begin{bmatrix} V_1^n \\ \vdots \\ V_N^n \end{bmatrix} \tag{10.3}$$

where $\overline{I}^n$ is the current vector associated with the voltage eigenvector $\overline{V}^n$. From (10.3), the complex scalar ratio $\lambda_n = I_j^n / V_j^n$ is the same at any circuit port, $j = 1$ to N. Thus, if the voltages at the linear network nodes agree with the components of the nth eigenvector, all the network ports will exhibit the same input admittance, equal to the associated eigenvalue λ_n.

In general, the voltage vector $\overline{V}$, which consists of the voltage values at the N ports of the linear network, can differ from any of the eigenvectors in (10.2). However, the vector $\overline{V}$ can be expressed as a linear combination of these eigenvectors [2]

$$\overline{V} = \frac{1}{A^2} \sum_n (\overline{V}_n^+ \overline{V}) \overline{V}_n \tag{10.4}$$

where + indicates an adjoint or conjugate transpose. The scalar coefficients $\overline{V}_n^+\overline{V}$ affecting the vectors $\overline{V}_n$ provide the components of the vector $\overline{V}$ in the orthonormal basis spanned by $\overline{V}_n$, with $n = 0$ to $N-1$. Note the division by the square of the amplitude A of the vector $\overline{V}$.

Next, the admittance matrix $[Y_D]$ accounting for the active part of the circuit is considered. This global active network will have N ports connected directly to the N ports of the linear coupling network. It contains N active blocks, each consisting of a transistor, together with its feedback and terminal elements, assumed identical for the N blocks. The active blocks are connected only through the linear coupling network. Thus, the complex active-admittance matrix $[Y_D]$ will be diagonal. The diagonal elements consist of current-to-voltage functions Y_{Dj}, $j = 1$ to N, depending on the voltage at the corresponding port V_j and the analysis frequency ω. More precisely, the functions Y_{Dj}, $j = 1$ to N depend only on the amplitude of the excitation voltages $|V_j|$, $j = 1$ to N. To understand this it must be taken into account that the response of any nonlinear circuit to an independent sinusoidal source does not depend on the phase of this input source (provided the circuit does not contain any periodic generators operating at the source frequency). Instead, it is the phase shift with respect to the independent source that defines a particular solution. Due to the fact that the N ports of the active matrix are uncoupled, the port voltages behave as independent periodic excitation sources. Thus, the matrix equation describing the coupled system can be written as

$$\left\{[Y_L(\omega)] + \begin{bmatrix} Y_{D1}(|V_1|,\omega) & \cdots & 0 \\ \vdots & \ddots & \vdots \\ 0 & \cdots & Y_{DN}(|V_N|,\omega) \end{bmatrix}\right\} \begin{bmatrix} V_1 \\ \vdots \\ V_N \end{bmatrix} = \overline{0} \tag{10.5}$$

Note that the dependence on the phase values at the various ports is considered in the voltage vector on the right-hand side. For compactness, the system above can be rewritten

$$\{[Y_L] + [Y_D(|\overline{V}|)]\}\overline{V} = [Y_L]\overline{V} + [Y_D(|\overline{V}|)]\overline{V} = \overline{0} \tag{10.6}$$

where, for simplicity, the frequency dependence of the different matrixes is not shown explicitly. Applying the same left multiplication by $\overline{V}_n^+$ and right multiplication by $\overline{V}_n$ used in (10.4) to each of the terms in expression (10.6), the following relationship is obtained:

$$\sum_n \left[\overline{V}_n^+[Y_L]\overline{V} + \overline{V}_n^+\left[Y_D(|\overline{V}|)\right]\overline{V}\right]\overline{V}_n = 0 \tag{10.7}$$

Clearly, each scalar coefficient affecting $\overline{V}_n$ must individually be equal to zero, which makes it possible to write [2]

$$\overline{V}_n^+[Y_L]\overline{V} + \overline{V}_n^+\left[Y_D(|\overline{V}|)\right]\overline{V} = \overline{V}_n^+\left(\lambda_n[I_d] + \left[Y_D(|\overline{V}|)\right]\right)\overline{V} = 0 \tag{10.8}$$

where we have taken into account that $\overline{V}_n^+$ is a left eigenvector of $[Y_L]$ and fulfills $\overline{V}_n^+[Y_L] = \overline{V}_n^+\lambda_n$. As gathered from (10.8), a system solution will be obtained if all the elements of the diagonal active matrix are equal and fulfill

$$\lambda_n = -Y_D(A) \tag{10.9}$$

where A is the common amplitude value at the connection ports $|V_j| = A$. Note that both λ_n and $Y_D(A)$ are complex numbers. On the other hand, the eigenvalue λ_n agrees with the input admittance at the ports of the linear coupling network when the node voltages correspond to the eigenvector associated with λ_n. Thus, equation (10.9) is the steady-state oscillation condition of the oscillator system fulfilled by the solution

$$\overline{V}_n = A_n \begin{bmatrix} 1 \\ e^{jn(2\pi/N)} \\ e^{j2n(2\pi/N)} \\ \vdots \\ e^{j(N-1)n(2\pi/N)} \end{bmatrix} \tag{10.10}$$

Because in principle there are N eigenvalues and eigenvectors, with $n = 0$ to $N-1$, N different oscillation modes are possible, which does not mean that they will actually be observed. Observation of a particular mode n requires fulfillment of the oscillation condition (10.9), with node voltages of the form (10.10). Note that equation (10.9) is, in fact, equivalent to the well-known oscillation condition $Y_N + Y_L = 0$ in a single-oscillator circuit.

We must take into account the fact that the relevant variables determining the different oscillation modes are the phase shifts between the N node voltages, not the absolute phase values. Actually, if $\theta_1, \theta_2, \ldots, \theta_N$ is a solution of the coupled system, shifting the phase of all the node voltages an equal amount provides a new solution: $\theta_1 + \alpha, \theta_2 + \alpha, \ldots, \theta_N + \alpha$, which is due to the system autonomy behaving globally in free-running conditions. Whatever the number of oscillators N, an in-phase solution may, in principle, be possible, given by the n-dimension vector $\overline{V}_o = A_o[1 \quad 1 \quad \cdots \quad 1]^T$. The buildup of a particular mode $\overline{V}_n$ will require fulfillment of the oscillation startup conditions, with a negative-conductance excess under small-signal conditions, so $\text{Re}[\lambda_n + Y_D(0)] < 0$ and $\text{Im}[\lambda_n + Y_D(0)] = 0$, with a positive slope of the imaginary part versus frequency (see Chapter 1, Section 1.3). In the steady state, the condition $\lambda_n + Y_D(A_n) = 0$ must be fulfilled. Both λ_n and $Y_D(A)$ are frequency dependent, so the N possible modes will have different frequencies, in general. For a given circuit, not all of the possible modes will necessarily exist. The oscillation conditions may be fulfilled for none, one, or several modes at different frequencies. In fact, the objective will be to prevent the startup of any undesired oscillation mode from the dc regime.

The analysis above is simplified, but it shows that the potential number of modes of operation is at least equal to the number of devices used in the system. Note that it is assumed that for all the solutions, the active blocks behave in an identical manner, with the same port amplitude. However, other solutions may exist with different amplitudes at the different active blocks, even if these blocks are identical [2]. The full system analysis would require an accurate numerical technique such as harmonic balance with a systematic initialization method, to obtain the various coexisting solutions, and a complementary stability analysis of each of these solutions.

10.2.2 Applications of Globally Coupled Oscillators

Two main applications of globally coupled oscillators are discussed here: the power combination at the fundamental frequency in Rucker's symmetrical oscillator [2,3], and the power combination at the Nth, harmonic component in N-push oscillators.

10.2.2.1 Power Combination at the Fundamental Frequency Rucker's symmetrical oscillator is used to combine the output power of N different suboscillators at the fundamental frequency ω_o. The oscillation mode desired for this power combination is the in-phase mode ($n = 0$), corresponding to the solution $\overline{V}_o = A_o[1 \quad 1 \quad \cdots \quad 1]^T$. The coupling network has a star shape with the output load R_L connected to the middle node, as in the example of Fig. 10.2a, containing four FET-based suboscillators.

The $N \times N$ admittance matrix of the passive part $[Y_L]$ is a particular case of the one considered in (10.1). It contains two different element values only: the diagonal-element values Y_s and the non-diagonal-element values Y_c. Then the expression for linear matrix eigenvalues is simplified to

$$\lambda_n = Y_s + Y_c e^{jn(2\pi/N)} + Y_c e^{j2n(2\pi/N)} + \cdots + Y_c e^{j(N-1)n(2\pi/N)} \tag{10.11}$$

with $n = 0$ to $N-1$. Clearly, whatever N is, only two different eigenvalues exist under the symmetry conditions considered, given by

$$\begin{aligned} \lambda_o &= Y_s + (N-1)Y_c \\ \lambda &= Y_s + Y_c e^{j(2\pi/N)} + Y_c e^{j2(2\pi/N)} + \cdots + Y_c e^{j(N-1)(2\pi/N)} \end{aligned} \tag{10.12}$$

Due to the multiplicity of eigenvalues, the eigenvectors spanning the subspace associated with the repeated eigenvalue are not defined uniquely. Diagonalizable matrixes (such as the ones we are dealing with) have multiple linearly independent eigenvectors in the multidimensional invariant subspace of the multiple eigenvalue [11]. However, vectors of the form (10.10) are still valid eigenvectors of the linear admittance matrix $[Y_L]$. The oscillation mode desired for power combination is $\overline{V}_o = A_o[1 \quad 1 \quad \cdots \quad 1]^T$, and this solution should be unique. Thus, it will be necessary to prevent the excitation and buildup of other modes. This is achieved through a suitable design of the individual suboscillators such that $\text{Re}[\lambda_o] < |\text{Re}[Y_d(0)]| < \text{Re}[\lambda]$. For reasons of symmetry, the in-phase mode is the

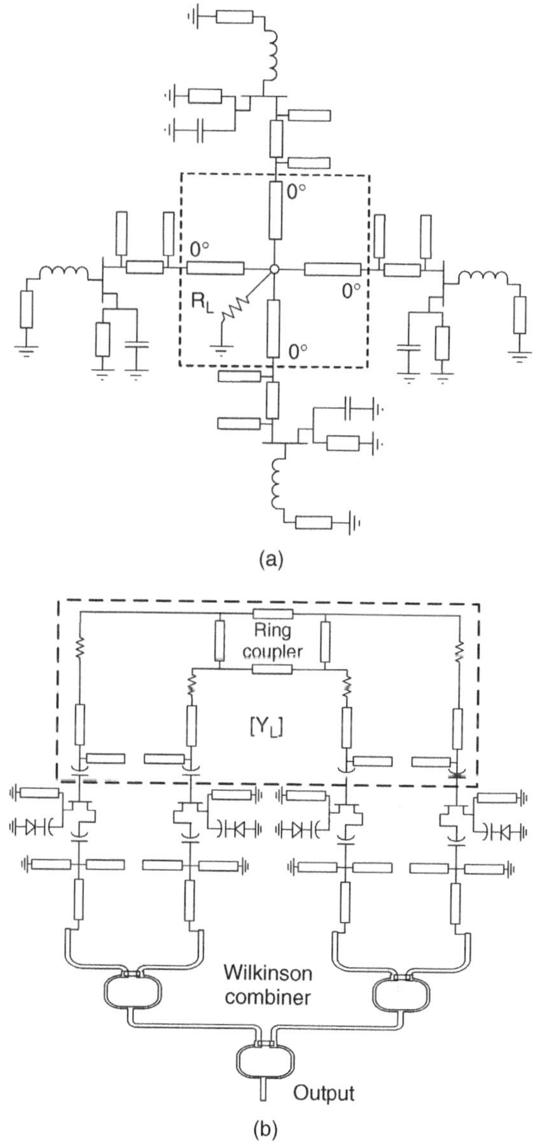

FIGURE 10.2 Topologies with global coupling for power combination: (a) implementation of Rucker's oscillator using four FET transistors for power combination at the fundamental frequency; (b) quadrupled-push oscillator for power combination of the fourth harmonic component. The transistors are coupled at the gate terminals. The output power is extracted using Wilkinson combiners.

most natural one in Rucker's oscillator. On the other hand, the fact that all the undesired modes $n = 1$ to $N-1$ exhibit the same linear admittance $\lambda(\omega)$ will facilitate the global suppression of these modes.

As an example, consider the desired in-phase mode in a system of N equal suboscillators. Applying Kirchhoff's laws, it is easily demonstrated that from the point of view of the oscillator n, the central node of the star network is in virtual open-circuit conditions with respect to the oscillators $1, \ldots, n-1$ and $n+1, \ldots, N$. In turn, the equivalent impedance from this node to ground is NR_L, with R_L being the actual load impedance. For an undesired oscillation mode with a $2\pi/N$ phase shift between the oscillator elements, the central node is a virtual short circuit. This must be taken into account when calculating the length of the transmission line that connects each oscillator to the central node. This length should be determined so that the oscillation startup conditions are fulfilled with the termination NR_L, not with the short-circuit termination. Note that all the oscillators are equal, so a single-oscillator design suffices to obtain the entire coupled topology.

10.2.2.2 Power Combination at the Nth Harmonic Component Ring couplers can be used for power combination at the harmonic component N in the N-push configurations [7,12,13]. When using transistors as active elements, the coupling port may be the same as that from which the output power is extracted, or it may be different. For example, the transistors may be coupled at the gate terminal and the power may be extracted from the drain terminal with Wilkinson combiners. When using the same port for coupling and power extraction, the various nodes of the ring coupler are connected to a common output resistance R_L through a star transmission-line configuration, like the one used in Rucker's oscillator [13].

The admittance matrix of the ring coupler network is a particular case of (10.1). For briefness, only the first row of the linear matrix is written in the following. There are two different cases, depending on the odd or even number of oscillator elements N:

$$Y_{1^{\text{st}}\text{row}} = \begin{cases} \left[Y_{11}, Y_{12}, \ldots, Y_{1,[(N+1)/2]-1}, Y_{1,(N+1)/2}, Y_{1,(N+1)/2}, \right. \\ \qquad \left. Y_{1,[(N+1)/2]-1}, \ldots, Y_{13}, Y_{12}\right] \quad \text{for } N \text{ odd} \\ \left[Y_{11}, Y_{12}, \ldots, Y_{1,\frac{N}{2}}, Y_{1,\frac{N}{2}+1}, Y_{1,\frac{N}{2}}, \right. \\ \qquad \left. Y_{1,\frac{N}{2}-1}, \ldots, Y_{13}, Y_{12}\right] \quad \text{for } N \text{ even} \end{cases} \tag{10.13}$$

The matrix structure of (10.13) is the same for the ring-coupler network and the ring-plus-star network. This matrix will have the form indicated by (10.13) and (10.1), and the structure of its N eigenvalues will be

$$\lambda_o, \lambda_1 = \begin{cases} \lambda_{N-1}, \lambda_2 = \lambda_{N-2}, \ldots, \lambda_{(N-1)/2} = \lambda_{(N+1)/2} & N \text{ odd} \\ \lambda_{N-1}, \lambda_2 = \lambda_{N-2}, \ldots, \lambda_{N/2} & N \text{ even} \end{cases} \tag{10.14}$$

The eigenvector associated to the eigenvalue λ_1 provides the oscillation mode:

$$\overline{V}_1 = A_1 \left[1 \quad e^{j(2\pi/N)} \quad e^{j2(2\pi/N)} \quad \ldots \quad e^{j(N-1)(2\pi/N)}\right]^T \tag{10.15}$$

For this oscillation mode, there is a $2\pi/N$ phase shift between adjacent suboscillators. The phase shift $\Delta\phi = 2\pi/N$ gives rise to the time delay $\tau = (2\pi/N)/(2\pi/T_o) = T_o/N$, with T_o the common oscillation period. At harmonic component m, this delay generates the phase shift $\Delta\phi_m = m2\pi f_o\tau = m(2\pi/N)$. When adding the output signals of the suboscillators, assumed identical, all the harmonic components that differ from $m = kN$, with k an integer, will be zero, as can easily be verified. Thus, the oscillation mode (10.15) allows multiplication of the individual oscillation frequency by N.

To avoid the possible startup of undesired oscillation modes, such as the in-phase mode, the condition $\text{Re}[\lambda_1] < |\text{Re}[Y_d(0)]| < \text{Re}[\lambda_o], \text{Re}[\lambda_2], \ldots, \text{Re}[\lambda_{N-1}]$ must be ensured for all the frequency values. To facilitate oscillation at the desired mode with an $\Delta\phi = 2\pi/N$ phase shift, a transmission line with the electrical length $\beta l = 2\pi/N$ is connected between the oscillator elements, where β is the propagation constant and l the physical length (see, e.g., the quadruple-push configuration of Fig. 10.2b). In that configuration, the output power is extracted with a power combiner connected to the transistor drain terminals. As already stated, it is also possible to extract the output power connecting the different nodes of the ring coupler to a common output resistance R_L with transmission lines [13]. In this kind of design it is taken into account that, at the desired mode with 90° phase shift, the middle node of the power combination network becomes virtually short-circuited at f_o seen from any suboscillator output. In turn, the middle node exhibits the impedance $4R_L$ at the undesired in-phase mode.

In the following, a detailed analysis of the various oscillation modes in the particular case of a quadruple-push oscillator is presented. The quadruple-push oscillator is used to obtain an output signal at four times the oscillation frequency of the individual suboscillators $f_{\text{out}} = 4f_o$. The associated 4×4 linear admittance matrix $[Y_L]$ will have the eigenvalues

$$\begin{aligned}
\lambda_0 &= Y_{11} + 2Y_{12} + Y_{13} \\
\lambda_1 &= Y_{11} + Y_{12}e^{j(2\pi/4)} + Y_{13}e^{j2(2\pi/4)} + Y_{12}e^{j3(2\pi/4)} \\
&= Y_{11} + Y_{12}e^{j(\pi/2)} + Y_{13}e^{j\pi} + Y_{12}e^{j3(\pi/2)} \\
\lambda_2 &= Y_{11} + Y_{12}e^{j2(2\pi/4)} + Y_{13}e^{j2[2(2\pi/4)]} + Y_{12}e^{j2[3(2\pi/4)]} \\
&= Y_{11} + 2Y_{12}e^{j\pi} + Y_{13} \\
\lambda_3 &= Y_{11} + Y_{12}e^{j3[(2\pi/4)]} + Y_{13}e^{j3[2(2\pi/4)]} + Y_{12}e^{j3[3(2\pi/4)]} \\
&= Y_{11} + Y_{12}e^{j(3\pi/2)} + Y_{13}e^{j\pi} + Y_{12}e^{j(\pi/2)}
\end{aligned} \tag{10.16}$$

Note that $\lambda_1 = \lambda_3$, in agreement with the eigenvalue structure of (10.14) for the even case. The oscillation modes associated with the various eigenvalues are determined by the corresponding eigenvectors:

$$\begin{aligned}
\overline{V}_0 &= A_0[1 \quad 1 \quad 1 \quad 1]^{\mathrm{T}} \\
\overline{V}_1 &= A_1\left[1 \quad e^{j(2\pi/4)} \quad e^{j2(2\pi/4)} \quad e^{j3(2\pi/4)}\right]^{\mathrm{T}} \\
&= A_1\left[1 \quad e^{j(\pi/2)} \quad e^{j\pi} \quad e^{j(3\pi/2)}\right]^{\mathrm{T}} \\
\overline{V}_2 &= A_2\left[1 \quad e^{j2[(2\pi/4)]} \quad e^{j2[2(2\pi/4)]} \quad e^{j2[3(2\pi/4)]}\right]^{\mathrm{T}} \\
&= A_2[1 \quad e^{j\pi} \quad e^{j2\pi} \quad e^{j\pi}]^{\mathrm{T}} \\
\overline{V}_3 &= A_3\left[1 \quad e^{j3[(2\pi/4)]} \quad e^{j3[2(2\pi/4)]} \quad e^{j3[3(2\pi/4)]}\right]^{\mathrm{T}} \\
&= A_3\left[1 \quad e^{j(3\pi/2)} \quad e^{j\pi} \quad e^{j(\pi/2)}\right]^{\mathrm{T}}
\end{aligned} \tag{10.17}$$

The oscillation modes of interest for multiplication by 4 of the individual oscillation frequencies are those corresponding to $\overline{V}_1$ and $\overline{V}_3$, which are totally equivalent. In both cases, the phase shift between the oscillator elements is $2\pi/4$. Thus, a quadruple-push oscillator consists of four identical oscillators, with the respective $0^\circ, 90^\circ, 180^\circ$, and 270° phases at the fundamental frequency f_o. When connecting a power combiner to the individual outputs of the four active blocks, assumed identical, the components at $4f_o$, $8f_o$, and so on, are added in phase, whereas the rest are canceled, so the output frequency is the quadruple of the individual oscillation frequency.

The possible startup of the various modes is analyzed by checking the conditions

$$\begin{aligned}
\mathrm{Re}[Y_{Tn}(\omega_o)] &= \mathrm{Re}[\lambda_n + Y_D(0)] < 0 \\
\mathrm{Im}[Y_{Tn}(\omega_o)] &= \mathrm{Im}[\lambda_n + Y_D(0)] = 0 \\
\frac{\partial \mathrm{Im}[Y_{Tn}(\omega_o)]}{\partial \omega} &> 0
\end{aligned} \tag{10.18}$$

with $n = 0, 1, 2$. This mode analysis has been applied to the quadruple-push FET-based oscillator of Fig. 10.2b, with a 12-GHz output frequency and transistor gate and drain bias voltages $V_{GS} = -0.2$ V and $V_{DS} = 2.5$ V. As already stated, the linear admittance matrix $[Y_L]$ has three different eigenvalues $\lambda_0, \lambda_1, \lambda_2$, corresponding to oscillation modes with interoscillator phase shifts $0^\circ, 90^\circ$, and 180°, respectively. Thus, the startup conditions (10.18) should be evaluated individually for each of these three eigenvalues. This analysis, presented in Fig. 10.3, shows that the three possible oscillation modes, with $0^\circ, 90^\circ$, and 180° phase shifts, fulfill the startup conditions. Note that this technique is equivalent to evaluation of the oscillation startup conditions from the dc regime in a single-oscillator circuit using the total admittance function $Y_T = Y_{NL} + Y_L$ and, thus has the same limitations. The additional application of pole–zero identification (or other accurate stability analysis method) to the dc solution is advisable.

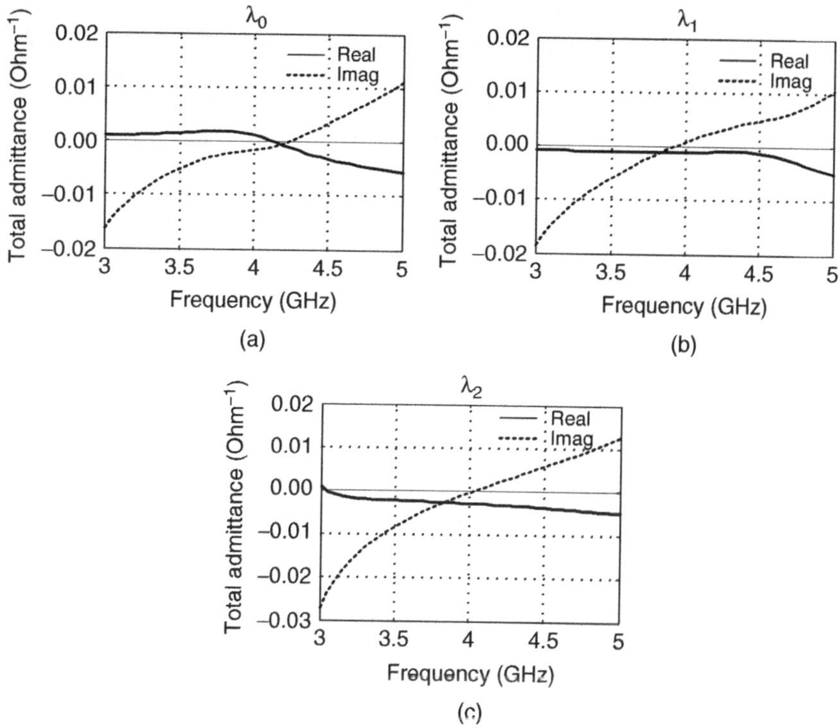

FIGURE 10.3 Frequency variation of the small-signal quantities (a) $Y_{Trmo} - Y_d(0) + \lambda_{rmo}$, (b) $Y_{T1} - Y_d(0) + \lambda_1$, and (c) $Y_{T2} - Y_d(0) + \lambda_2$, respectively, providing the total input admittance to the oscillation modes 0,1, and 2 at the transistor gate terminals. The imaginary part of Y_T is traced with a dashed line in each case.

Let a parameter η capable of quenching all the possible oscillation modes be considered, such as the gate bias voltage V_{GS} in FET-based oscillators. When varying η, the oscillation modes are generated from the dc regime at Hopf bifurcations that take place at different parameter values. At the Hopf bifurcation point, the dc solution gives rise to a degenerate limit cycle (oscillation) of zero amplitude. Because the two solutions are actually the same at the bifurcation point, the stability properties of the dc solution are transferred to the periodic solution as shown in Section 3.3.3. Periodic oscillations generated from stable dc solutions will be stable for a supercritical Hopf bifurcation, or for a subcritical bifurcation, will usually become stable after a turning point in the periodic solution path (see Section 3.3.3). If the rest of oscillation modes are generated from unstable dc solutions, they will be unstable and will have no influence on circuit behavior. The statements above are true provided that no desynchronization occurs in the coupled system, which is relatively unusual. Knowledge of these properties can facilitate the distinction between stable and unstable modes.

As already stated, in FET-based oscillators we can use the gate bias voltage V_{GS} as an analysis parameter. This bias voltage is varied simultaneously in all the suboscillator elements, so it takes the same value V_{GS} in all these elements. Obviously, the dc solution will be stable for V_{GS} below conduction. The mode that is generated from this stable dc solution will be stable or will generally become stable after a turning point. As an example, we have traced the bifurcation diagram of the quadruple-push oscillator versus V_{GS} in Fig. 10.4a. The diagram has been obtained in harmonic balance using the techniques described in Section 10.2.5. In turn, the results of applying pole–zero identification to the dc solution of the coupled system are shown in Fig. 10.4b, where the real part of the various pairs of dominant

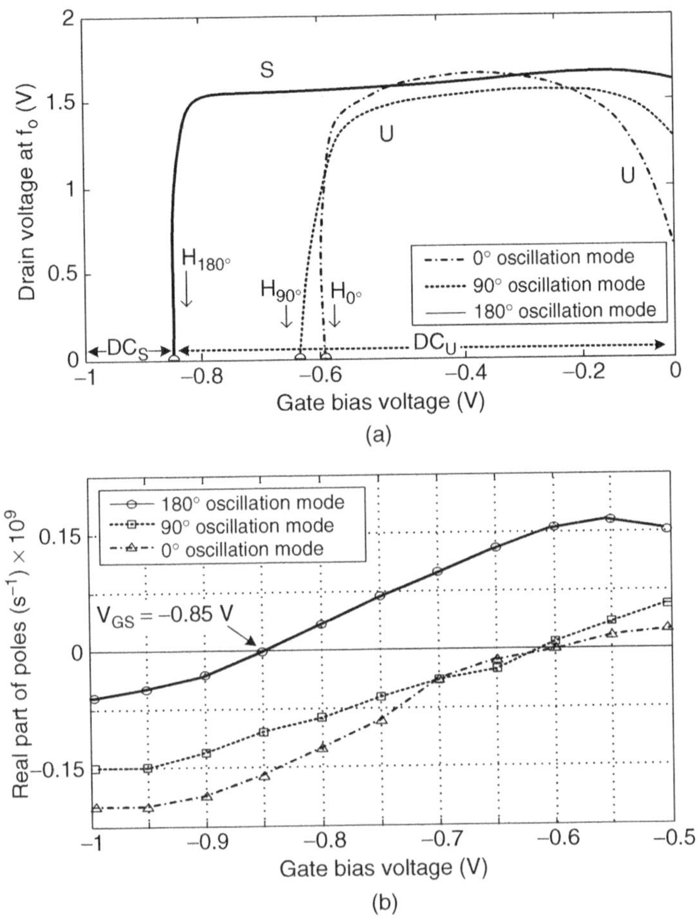

FIGURE 10.4 Bifurcation diagram of the quadruple-push oscillator versus the gate bias voltage V_{GS}: (a) variation of the drain voltage amplitude of the oscillation modes; (b) variation of the real part of the dominant poles versus V_{GS}, obtained through pole–zero identification of the dc solution.

complex-conjugate poles has been represented versus V_{GS}. Increasing V_{GS} from a very low value, the solution with a 180° phase shift between the oscillator elements is generated at the Hopf bifurcation $V_{GS,H180^\circ}$ (Fig. 10.4a). This solution arises from a stable dc regime, so it should also be stable. For $V_{GS,H180^\circ} < V_{GS} < V_{GS,H90^\circ}$, the dc solution is unstable, with a pair of complex-conjugate poles on the right-hand side of the complex plane. At $V_{GS,H90^\circ}$, a second pair of complex-conjugate poles cross the imaginary axis to the right-hand side of the complex plane, in a second Hopf bifurcation (Fig. 10.4b). The solution with a 90° phase shift is generated at this bifurcation point (Fig. 10.4a). It arises from an unstable dc regime, so it should be unstable. For $V_{GS,H90^\circ} < V_{GS} < V_{GS,H0^\circ}$, the dc solution contains two pairs of complex-conjugate poles on the right-hand side of the complex plane. Finally, the in-phase solution is generated at $V_{GS,H0^\circ}$, at a third Hopf bifurcation. This in-phase solution is also generated from an unstable dc regime. For $V_{GS} > V_{GS,H0^\circ}$ the dc solution contains three pairs of complex-conjugate poles on the right-hand side of the complex plane.

With the aim of stabilizing the solution with a 90° phase shift, we have modified the bifurcation behavior of the circuit by connecting interbranch resistances between the gate terminals of the different transistors. We have iteratively calculated the interbranch resistance so that the Hopf bifurcations that lead to the 90° and 180° phase shift solutions fulfill $V_{GS,H90^\circ} < V_{GS,H180^\circ}$. For the interbranch resistance $R_b = 1.4\ \text{k}\Omega$, the desired oscillation mode with 90° phase shift was generated from a stable dc solution at the gate bias voltage $V_{GS,H90^\circ} = -0.78$ V. The solution with a 180° phase shift was generated at $V_{GS,H180^\circ} = -0.32$ V and the in-phase solution was never obtained. Next, the stability of the 90° mode at the desired bias voltage $V_{GS} = -0.2$ V must be confirmed. This will require a technique for the stability analysis of steady-state oscillation modes, presented in the following section.

10.2.3 Stability Analysis of a Steady-State Periodic Regime

In the following, the stability analysis of the periodic oscillation modes of globally coupled oscillators is presented. The analysis is applied to the general mode n, where n can take any of the values $0, 1 \ldots, N-1$. The steady-state oscillation conditions (10.5) are rewritten as

$$[Y_T]\overline{V}^n = \begin{bmatrix} Y_{11} + Y_D(A_n) & Y_{12} & \cdots & Y_{1N} \\ Y_{1N} & Y_{11} + Y_D(A_n) & \cdots & Y_{1,N-1} \\ \vdots & \vdots & \ddots & \vdots \\ Y_{12} & Y_{13} & \cdots & Y_{11} + Y_D(A_n) \end{bmatrix} \begin{bmatrix} A_n \\ A_n e^{jn(2\pi/N)} \\ A_n e^{j2n(2\pi/N)} \\ \vdots \\ A_n e^{j(N-1)n(2\pi/N)} \end{bmatrix} = 0 \tag{10.19}$$

where $n = 0$ to $N-1$ and the matrix $[Y_T]$ has been defined, for compactness. Note that the linear admittance matrix (10.1) has been considered for a generality. In general, the steady-state amplitude A_n will be different for the different oscillation modes (see, e.g., Fig. 10.4a). In the following, the more compact notation $Y_{cd} = Y_{11} + Y_D(A)$ will be used. The Y_{cd} elements in the total admittance matrix depend on the amplitude and frequency. The rest of the matrix elements are frequency dependent only. System (10.11) can be rewritten in terms of the error functions H_k, with $k = 1$ to N, in the following manner:

$$
\begin{aligned}
H_{1o} &= Y_{cd}(A_o)A_o e^{j\theta_1} + \cdots + Y_{1k}A_o e^{j\theta_k} + \cdots + Y_{1N}A_o e^{j\theta_N} = 0\\
&\vdots\\
H_{ko} &= Y_{1,N-(k-1)+1}A_o e^{j\theta_1} + \cdots + Y_{cd}(A_o)A_o e^{j\theta_k}\\
&\quad + \cdots + Y_{1,N-(k-1)}A_o e^{j\theta_N} = 0\\
&\vdots\\
H_{No} &= Y_{12}A_o e^{j\theta_1} + \cdots + Y_{1,k+1}A_o e^{j\theta_k} + \cdots + Y_{cd}(A_o)A_o e^{j\theta_N} = 0
\end{aligned}
\tag{10.20}
$$

where the subscript o indicates a steady state. For simplicity we have introduced the phase variables $\theta_1, \theta_2, \ldots, \theta_N$, with $\theta_m = (m-1)n2\pi/N$, and the steady-state amplitude is denoted A_o. Each H_{ko} can be expressed in a compact manner as the product

$$
H_{ko} = [Y_T(|V_k|, \omega)]_k^T \overline{V}
\tag{10.21}
$$

where $[Y_T]_k^T$ is the row matrix agreeing with the kth row of $[Y_T]$ and $\overline{V}$ is the vector of node-voltage phasors.

For the stability analysis, an instantaneous perturbation will be considered, which will give rise to increments in the system oscillation frequency and the various amplitude and phase values. The perturbed system is obtained by expanding each function H_k in a Taylor series about the steady state $A_o, \theta_1, \ldots, \theta_N, \omega_o$ and taking into account that multiplication by the frequency increment s is equivalent to a time derivation of the variables perturbed. Conceptually, the steps are identical to those used in Section 1.3 for the stability analysis of multiport networks [see (1.18)–(1.20)]. From an inspection of (10.21), the derivation of $[Y_T]_k^T$ with respect to frequency will give rise to terms of the form

$$
\frac{\partial Y_{km}}{\partial \omega}\left(-j\frac{\Delta \dot{A}_m(t)}{A_o} + \Delta\dot{\theta}_m(t)\right)A_o e^{j\theta_m}
$$

with $m = 1$ to N. Taking this into account, the perturbed oscillator system is formulated as

$$\underbrace{\begin{bmatrix} \frac{\partial H_1}{\partial A_1} & \cdots & \frac{\partial H_1}{\partial A_N} \\ \vdots & \cdots & \vdots \\ \frac{\partial H_N}{\partial A_1} & \cdots & \frac{\partial H_N}{\partial A_N} \end{bmatrix}}_{\left([Y_T]+\left[\frac{\partial Y_T}{\partial A}\right]\right)[e^{j\theta_k}]} \begin{bmatrix} \Delta A_1 \\ \vdots \\ \Delta A_N \end{bmatrix} + \underbrace{\begin{bmatrix} \frac{\partial H_1}{\partial \theta_1} & \cdots & \frac{\partial H_1}{\partial \theta_N} \\ \vdots & \cdots & \vdots \\ \frac{\partial H_N}{\partial \theta_1} & \cdots & \frac{\partial H_N}{\partial \theta_N} \end{bmatrix}}_{j[Y_T][e^{j\theta_k}]A_o} \begin{bmatrix} \Delta \theta_1 \\ \vdots \\ \Delta \theta_N \end{bmatrix}$$

$$+ \underbrace{\begin{bmatrix} \frac{\partial Y_{cd}}{\partial \omega} e^{j\theta_1} & \cdots & \frac{\partial Y_{1N}}{\partial \omega} e^{j\theta_N} \\ \vdots & \cdots & \vdots \\ \frac{\partial Y_{N1}}{\partial \omega} e^{j\theta_1} & \cdots & \frac{\partial Y_{cd}}{\partial \omega} e^{j\theta_N} \end{bmatrix}}_{[J_\omega]=\left[\frac{\partial Y_T}{\partial \omega}\right][e^{j\theta_k}]} \begin{bmatrix} -j\frac{\Delta \dot{A}_1}{A_o} + \Delta\dot{\theta}_1 \\ \vdots \\ -j\frac{\Delta \dot{A}_N}{A_o} + \Delta\dot{\theta}_N \end{bmatrix} A_o = 0 \tag{10.22}$$

where $[e^{j\theta_k}]$ is a $N \times N$ diagonal matrix, with $k = 1$ to N. Note that all the derivatives are calculated at the particular steady-state solution of the coupled system, fulfilling (10.20). Splitting system (10.22) into real and imaginary parts, it is possible to obtain a linear time invariant (LTI) system in the $2N$ real variables $(\Delta A_1 \cdots \Delta A_N, \Delta\theta_1 \cdots \Delta\theta_N)^T$:

$$\begin{bmatrix} \Delta\dot{A}_1 \\ \vdots \\ \Delta\dot{A}_N \\ \Delta\dot{\theta}_1 \\ \vdots \\ \Delta\dot{\theta}_N \end{bmatrix} = \frac{[J_\omega^{r,i}]^{-1}}{A_o} \underbrace{\begin{bmatrix} H_{1A}^r & \cdots & H_{1A}^r & H_{1\theta_1}^r & \cdots & H_{1\theta_N}^r \\ H_{1A}^i & \cdots & H_{1A}^i & H_{1\theta_1}^i & \cdots & H_{1\theta_N}^i \\ & & \vdots & \vdots & & \\ H_{NA}^r & \cdots & H_{NA}^r & H_{N\theta_1}^r & \cdots & H_{N\theta_N}^r \\ H_{NA}^i & \cdots & H_{NA}^i & H_{N\theta_i}^i & \cdots & H_{N\theta_N}^i \end{bmatrix}}_{[JH]} \begin{bmatrix} \Delta A_1 \\ \vdots \\ \Delta A_N \\ \Delta\theta_1 \\ \vdots \\ \Delta\theta_N \end{bmatrix} \tag{10.23}$$

where the the superscripts r and i indicate real and imaginary parts, respectively. The real matrix $[J_\omega^{r,i}]$, of $2N \times 2N$ dimension, is obtained directly from the complex $N \times N$ matrix$[J_\omega]$ in (10.22). On the other hand, the subscripts in the elements of $[JH]$ indicate the derivative of each H function with respect to the corresponding variable. As already indicated, these derivatives are calculated at the particular steady-state solution of the coupled system, obtained from (10.20).

Because the coupled system is actually a free-running system, there will be an irrelevance with respect to the phase origin. This means that the phase values at the various nodes can be shifted an equal amount and a new solution is obtained. If a common phase shift $\Delta\alpha$ is applied to all the oscillator elements, the condition

$H_k = 0$ is still fulfilled, so it is possible to write

$$\frac{\partial H_k}{\partial \theta_1}\Delta\alpha + \cdots + \frac{\partial H_k}{\partial \theta_N}\Delta\alpha = 0 \tag{10.24}$$

and the following relationship is satisfied:

$$\frac{\partial H_k}{\partial \theta_1} + \cdots + \frac{\partial H_k}{\partial \theta_N} = 0 \qquad k = 1 \text{ to } N \tag{10.25}$$

Thus, the N columns containing the phase derivatives in the Jacobian matrix $[JH]$ of (10.23) are linearly related, so the matrix is singular and contains a zero eigenvalue $\lambda = 0$. For stability, all the rest of eigenvalues must have a negative real part. Note that due to the Taylor series expansion of the function H_k in terms of the perturbation frequency, the stability analysis above is limited to real poles or complex-conjugate poles $\sigma \pm j\omega$ with small ω. However, this type of instability will be the most common in a globally coupled system. For complex-conjugate poles, the imaginary part ω will generally be on the order of the frequency difference between the oscillation frequencies of the various modes.

The technique above has been applied to the stability analysis of the solution with 90° phase shift in the quadruple-push oscillator of Fig. 10.2b, obtained for the gate bias voltage $V_{GS} = -0.2$ V. Even though the output power is extracted at the fourth harmonic term, the ports of the coupling network are defined at the gate terminal, so the influence of the higher harmonic terms will be rather small. The four dominant poles of the original oscillator (except the zero pole) are shown in Table 10.1. The solution is unstable, as it contains poles on the right-hand side of the complex plane. There are two unstable poles. This result is in agreement with the fact that the oscillation with 90° phase shift was generated from an unstable dc solution with one pair of unstable complex-conjugate poles. Table 10.2 shows the four dominant poles (except the zero pole) obtained from (10.23) after application of the stabilization technique, based on the connection of the interbranch resistance R_b. The solution is stable because all the dominant poles are located on the left-hand side of the complex plane.

TABLE 10.1

Real Poles	Complex Poles
-5.3687×10^6 s^{-1}	2.6524×10^8 s^{-1} $+ j1.1677 \times 10^8$ Hz
-1.2789×10^9 s^{-1}	2.6524×10^8 s^{-1} $- j1.1677 \times 10^8$ Hz

TABLE 10.2

Real Poles	Complex Poles
-3.1975×10^3 s^{-1}	-1.2816×10^7 s^{-1} $+$ j 2.7738×10^6 Hz
-1.5137×10^8 s^{-1}	-1.2816×10^7 s^{-1} $-$ j 2.7738×10^6 Hz

10.2.4 Phase Noise

For the phase noise analysis of the coupled-oscillator system, we should introduce the complex envelopes of the noise perturbations into the perturbed equation system (10.25). Globally coupled oscillators inherently have a phase noise reduction capability compared to a single-oscillator element of the same characteristics. To demonstrate this, a Rucker's oscillator (used for power combination at the fundamental frequency) will be considered, which allows a simpler mathematical derivation. The simplification is due, on the one hand, to the fact that the $N \times N$ admittance matrix $[Y_L]$ of the passive network contains only two element values: the diagonal elements Y_s and the nondiagonal elements Y_c, and on the other hand, to the fact that the steady-state oscillation has equal phase values $\theta_1 = \cdots = \theta_N = 0$ at the various ports. This steady-state solution can be formulated as

$$\begin{bmatrix} H_{1o} \\ H_{2o} \\ \vdots \\ H_{No} \end{bmatrix} = \begin{bmatrix} Y_s + Y_D(A) & Y_c & \cdots & Y_c \\ Y_c & Y_s + Y_D(A) & \cdots & Y_c \\ \vdots & \vdots & \ddots & \vdots \\ Y_c & Y_c & \cdots & Y_s + Y_D(A) \end{bmatrix} \begin{bmatrix} 1 \\ 1 \\ \vdots \\ 1 \end{bmatrix} A_o = 0 \tag{10.26}$$

where the error functions H_k, $k = 1$ to N have been introduced. For the noise analysis, equivalent current noise sources $i_{pk}(t)$ in parallel with the various analysis nodes will be considered. The phase noise analysis will be limited to white noise perturbations about the common oscillation frequency. Note that taking into account the effect of the $1/f$ noise sources will be essential for accurate prediction of the output power spectrum due to phase noise. However, this would require considering the system equations at both the first harmonic component and dc. The large size of the resulting system would make an analytical study very difficult, so only an analysis about the carrier frequency is presented here.

The noise sources will give rise to random increments in the amplitude and phase values of the node voltages and in the common oscillation frequency. For simplicity, it will be considered that the phase shifts $\Delta\theta_1, \Delta\theta_2, \ldots, \Delta\theta_N$ due to the noise sources are small. As we already know, this assumption is valid only at relatively high offset frequency from the carrier (Chapter 2). The vector of perturbed H functions is given by

$$\begin{bmatrix} H_1 \\ H_2 \\ \vdots \\ H_N \end{bmatrix} = \begin{bmatrix} Y_{cd}(A_o + \Delta A_1, j\omega_o + s) & \cdots & Y_c(j\omega_o + s) \\ Y_c(j\omega_o + s) & \cdots & Y_c(j\omega_o + s) \\ \vdots & \vdots \ \ddots & \vdots \\ Y_c(j\omega_o + s) & \cdots & Y_{cd}(A_o + \Delta A_N, j\omega_o + s) \end{bmatrix}$$

$$\begin{bmatrix} (A_o + \Delta A_1(t))e^{j\Delta\theta_1(t)} \\ (A_o + \Delta A_2(t))e^{j\Delta\theta_2(t)} \\ \vdots \\ (A_o + \Delta A_N(t))e^{j\Delta\theta_N(t)} \end{bmatrix} = \begin{bmatrix} I_{p1}(t) \\ I_{p2}(t) \\ \vdots \\ I_{pN}(t) \end{bmatrix} \tag{10.27}$$

where $I_{\text{pk}}(t)$ are the noise current envelopes about the oscillation frequency at the various analysis nodes, obtained by averaging $i_{\text{pk}}(t)\exp(-j\omega_o t)$ over one oscillation period. The Y_{cd} elements in the total admittance matrix depend on the oscillation frequency and amplitude. In turn, the Y_c elements are frequency dependent only. Due to the small amplitude of the noise perturbations, the error functions H_1 to H_N in system (10.27) can be expanded in a first-order Taylor series about the steady-state solution. For this expansion, exponentials of the form $e^{j\Delta\theta}$ are approached as $e^{j\Delta\theta} \cong 1 + j\Delta\theta$. For simplicity, the amplitude dependence of the H_k functions will be neglected, assumed to be much less than the phase and frequency dependencies. This will reduce the accuracy of the calculation but will make it possible to obtain manageable analytical expressions. Next, the values of real perturbation frequencies $\Delta\dot{\theta}_k(t)$ are assumed to be equal at all the suboscillators, which, as shown in Chapters 2 and 7, gives rise to minor accuracy degradation. Then, expansion of the kth equation, corresponding to H_k, is given by

$$\begin{aligned} H_k = H_{ko} &+ \left(A_o \frac{\partial Y_{cd}}{\partial \omega} + (N-1) A_o \frac{\partial Y_c}{\partial \omega} \right) \Delta\dot{\theta}_k + A_o Y_{cd}\, j\Delta\theta_k \\ &+ A_o Y_c\, j\, \underbrace{(\Delta\theta_1 + \cdots + \Delta\theta_N)}_{\text{no } k \text{ term}} \end{aligned} \tag{10.28}$$

In equation (10.28) and in the remainder of this analysis, all the derivatives are calculated at the in-phase steady-state solution of the coupled-oscillator system given by (10.26). After suppressing the steady-state term, it will be possible to approximate the imaginary part of the kth perturbed equation as

$$\begin{aligned} &\left[\frac{\partial B_{cd}}{\partial \omega} + (N-1) \frac{\partial B_c}{\partial \omega} \right] \Delta\dot{\theta}_k + G_c \Delta\theta_1 + G_c \Delta\theta_2 \\ &+ \cdots + G_{cd} \Delta\theta_k + \cdots + G_c \Delta\theta_N = \frac{I^i_{\text{pk}}(t)}{A_o} \end{aligned} \tag{10.29}$$

where the following definitions have been considered: $Y_{cd} = G_{cd} + jB_{cd}$ and $Y_c = G_c + jB_c$. Applying the Fourier transform in the slow time variable associated with the noise sources, the entire set of equations in matrix form is

$$\begin{bmatrix} j\Omega \left.\frac{\partial B_T}{\partial \omega}\right|_o + G_{cd} & G_c & \cdots & G_c \\ G_c & j\Omega \left.\frac{\partial B_T}{\partial \omega}\right|_o + G_{cd} & \cdots & G_c \\ \vdots & \vdots & \ddots & \vdots \\ G_c & G_c & \cdots & j\Omega \left.\frac{\partial B_T}{\partial \omega}\right|_o + G_{cd} \end{bmatrix} \begin{bmatrix} \Delta\theta_1(\Omega) \\ \Delta\theta_2(\Omega) \\ \vdots \\ \Delta\theta_N(\Omega) \end{bmatrix} = \frac{1}{A_o} \begin{bmatrix} I^i_{p1}(\Omega) \\ I^i_{p2}(\Omega) \\ \vdots \\ I^i_{pN}(\Omega) \end{bmatrix} \tag{10.30}$$

where $B_T = B_{cd} + (N-1)B_c$. Note that because of the circuit autonomy, the addition of the phase derivatives is equal to zero, as shown in (10.24)–(10.25). Then the following relationship is fulfilled:

$$(N-1)G_c + G_{cd} = 0 \tag{10.31}$$

Calculation of the phase perturbation $\Delta\theta_k$ requires inverting the matrix in (10.30), which provides the following system:

$$\begin{bmatrix} \Delta\theta_1(\Omega) \\ \Delta\theta_2(\Omega) \\ \vdots \\ \Delta\theta_N(\Omega) \end{bmatrix} = [M(\Omega)] \begin{bmatrix} I^i_{p1}(\Omega) \\ I^i_{p2}(\Omega) \\ \vdots \\ I^i_{pN}(\Omega) \end{bmatrix} \tag{10.32}$$

where $[M(\Omega)]$ is the inverse of the matrix in (10.32), written compactly. Due to the particular topology of the $N \times N$ matrix in (10.32), it will be possible to obtain the inverse matrix $[M(\Omega)]$ in an analytical manner. The process is as follows. Initially, all the matrix terms are divided by G_c:

$$G_c \begin{bmatrix} \frac{j\Omega(\partial B_T/\partial\omega)|_o + G_{cd}}{G_c} & 1 & \cdots & 1 \\ 1 & \frac{j\Omega(\partial B_T/\partial\omega)|_o + G_{cd}}{G_c} & \cdots & 1 \\ \vdots & \vdots & \ddots & \vdots \\ 1 & 1 & \cdots & \frac{j\Omega(\partial B_T/\partial\omega)|_o + G_{cd}}{G_c} \end{bmatrix} \begin{bmatrix} \Delta\theta_1(\Omega) \\ \Delta\theta_2(\Omega) \\ \vdots \\ \Delta\theta_N(\Omega) \end{bmatrix}$$

$$= \frac{1}{A_o} \begin{bmatrix} I^i_{p1}(\Omega) \\ I^i_{p2}(\Omega) \\ \vdots \\ I^i_{pN}(\Omega) \end{bmatrix} \tag{10.33}$$

The resulting normalized matrix has the form [14]

$$[Q] = \begin{bmatrix} a+jb & 1 & \cdots & 1 \\ 1 & a+jb & \cdots & 1 \\ \vdots & \vdots & \ddots & \vdots \\ 1 & 1 & \cdots & a+jb \end{bmatrix} \tag{10.34}$$

with the complex term $a + jb$ being

$$a + jb = \frac{j\Omega(\partial B_T/\partial\omega)|_o + G_{cd}}{G_c} \tag{10.35}$$

The inverse of the general matrix (10.34) is given by [14]

$$[Q]^{-1} = \frac{1}{(a+jb-1)(a+jb+N-1)} \begin{bmatrix} a+jb+N-2 & -1 & \cdots & -1 \\ -1 & a+jb+N-2 & \cdots & -1 \\ \vdots & \vdots & \ddots & \vdots \\ -1 & -1 & \cdots & a+jb+N-2 \end{bmatrix} \tag{10.36}$$

where N is the matrix dimension. Then the inverse matrix $[M(\Omega)]$ can be obtained directly by replacing a and b by their original values. Applying this calculation to (10.32), the perturbed phase values $\Delta\theta_k$, $k = 1$ to N, are given by

$$\begin{bmatrix} \Delta\theta_1 \\ \Delta\theta_2 \\ \vdots \\ \Delta\theta_N \end{bmatrix} = \frac{G_c}{A_o\left(j\Omega\,(\partial B_T/\partial\omega)|_o + G_{cd} - G_c\right)\left(j\Omega\,(\partial B_T/\partial\omega)|_o + G_{cd} + (N-1)G_c\right)}$$
$$\cdot \begin{bmatrix} \frac{j\Omega\,(\partial B_T/\partial\omega)|_o + G_{cd}}{G_c} + N - 2 & \cdots & -1 \\ -1 & \cdots & -1 \\ \vdots & \ddots & \vdots \\ -1 & \cdots & \frac{j\Omega\,(\partial B_T/\partial\omega)|_o + G_{cd}}{G_c} + N - 2 \end{bmatrix} \begin{bmatrix} I_{p1}^i(\Omega) \\ I_{p2}^i(\Omega) \\ \vdots \\ I_{pN}^i(\Omega) \end{bmatrix} \tag{10.37}$$

Solving for any of the phase increments, $\Delta\theta_1$, for example, provides

$$\begin{aligned} \Delta\theta_1 &= \frac{\left(j\Omega\,(\partial B_T/\partial\omega)|_o + G_{cd} + (N-2)G_c\right) I_{p1}^i - G_c I_{p2}^i + \cdots - G_c I_{pN}^i}{A_o\left(j\Omega\,(\partial B_T/\partial\omega)|_o + G_{cd} - G_c\right)\left(j\Omega\,(\partial B_T/\partial\omega)|_o\right)} \\ &= \frac{\left(j\Omega\,(\partial B_T/\partial\omega)|_o - G_c\right) I_{p1}^i - G_c I_{p2}^i - \cdots - G_c I_{pN}^i}{A_o\left(j\Omega\,(\partial B_T/\partial\omega)|_o + G_{cd} - G_c\right)\left(j\Omega\,(\partial B_T/\partial\omega)|_o\right)} \end{aligned} \tag{10.38}$$

where (10.31) has been taken into account. Then, the phase noise spectral density is given by

$$|\Delta\theta_1(\Omega)|^2 = \frac{\left\{\left[\Omega\,(\partial B_T/\partial\omega)|_o\right]^2 + NG_c^2\right\} 2|I_p|^2}{A_o^2\left\{\left[\Omega\,(\partial B_T/\partial\omega)|_o\right]^2 + (G_{cd} - G_c)^2\right\}\left[\Omega\,(\partial B_T/\partial\omega)|_o\right]^2} \tag{10.39}$$

where all the white noise sources of the suboscillators have been assumed uncorrelated and with the same spectral density $|I_p|^2$. Taking into account the relationship (10.31), it will be possible to write $G_{cd} - G_c = -NG_c$. Then the phase noise spectral density is simplified to

$$|\Delta\theta_1(\Omega)|^2 = \frac{\left\{NG_c^2 + [\partial B_T/\partial\omega|_o]^2\Omega^2\right\} 2|I_p|^2}{A_o^2\left\{N^2G_c^2 + [\partial B_T/\partial\omega|_o]^2\Omega^2\right\}[\partial B_T/\partial\omega|_o]^2\Omega^2} \tag{10.40}$$

Relatively close to the carrier (small Ω), the phase noise is given by $|\Delta\theta_1(\Omega)|^2 = 2|I_p|^2/(A_o^2 N[\partial B_T/\partial\omega|_o]^2\Omega^2)$. Thus, it decreases as -20 dB/dec. This is in agreement with the fact that the coupled configuration behaves globally as a free-running oscillator. Compared to a single oscillator with the same total susceptance B_T, the phase noise spectral density is clearly reduced in $10\log N$ decibels. This can be explained roughly by the fact that N resonators exist at the same oscillation frequency that gives rise to a larger equivalent oscillator quality factor. The effect on the noise sources does not increase in the same proportion, due to the fact that the noise sources of the various oscillators are uncorrelated. When increasing the offset frequency Ω, two corner frequencies are obtained, due to the coupled operation of the oscillator system. The values of these corner frequencies will be relatively high, due to the usually large value of the conductance terms in the numerator and denominator compared to the derivative $\partial B_T/\partial\omega|_o$.

The analysis above has been applied to a Rucker's oscillator at 5 GHz, with four suboscillator elements, each consisting of the parallel connection of a cubic nonlinearity and a parallel LC resonator. The nonlinearity is given by $i(v) = -0.037v + 0.021v^3$ A, and the linear element values are $C = 0.26$ pH and $L = 2.12$ nH. The four suboscillator elements are coupled to the output load $R_L = 50\ \Omega$ through transmission-line sections of the same length. In Fig. 10.5 the phase noise obtained with (10.39) is compared to that resulting from harmonic balance simulations using the carrier modulation and conversion matrix approaches. Note that only white noise sources have been considered in both the analytical calculation and the harmonic balance simulations, so the phase noise drops versus this offset frequency as -20 dB/dec. The analysis confirms the phase noise reduction of $10\log N = 10\log 4 = 6$ dB predicted by (10.40). The first corner frequency of the spectrum occurs at the offset frequency Ω_{c1}, for which the two terms in the numerator of (10.39) take

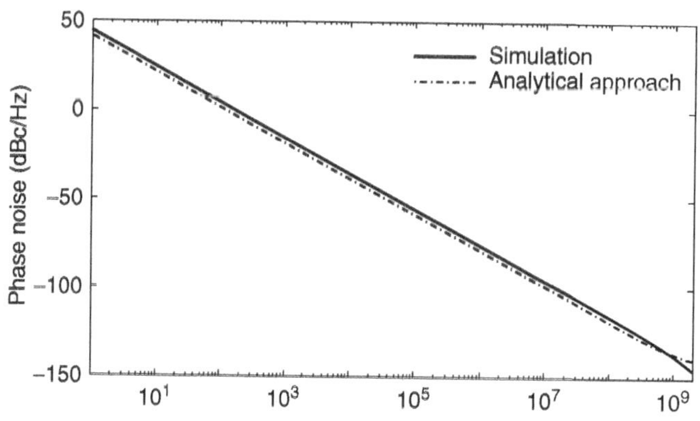

FIGURE 10.5 Phase noise spectrum of a Rucker oscillator consisting of four identical suboscillator elements, each consisting of the parallel connection of a cubic nonlinearity and a parallel LC resonator. The results provided by (10.40) are compared with those obtained with harmonic balance simulations.

identical value. The calculated corner frequency is $f_{c1} = 1$ GHz. The second corner frequency, obtained for the offset frequency at which the two terms in the denominator become equal, is even higher, above 2 GHz.

10.2.5 Analysis and Design Using Harmonic Balance

In the following, harmonic balance tools for the design of push–push and quadruple–push oscillator are provided. The techniques can easily be extended to the design of N-push oscillators with arbitrary positive integer N.

10.2.5.1 Push–Push Oscillator A push–push oscillator is composed of two oscillator elements, oscillating with a 180° phase shift at the fundamental frequency [15,16]. The odd harmonics cancel at the circuit output, whereas the even harmonics add in phase. This type of oscillation is known as *odd-mode oscillation*. The push–push oscillator is used to double the oscillation frequency of the two devices at the circuit output. From the general expressions (10.2), the 2×2 coupling matrix of the push–push oscillator will have the following eigenvalues and eigenvectors:

$$\begin{aligned} \lambda_n &= Y_{11} + Y_{12}\, e^{jn(2\pi/2)} \\ \overline{V}_n &= \frac{1}{\sqrt{2}} \begin{bmatrix} 1 \\ e^{jn\pi} \end{bmatrix} \end{aligned} \tag{10.41}$$

with $n = 0$ or 1. Thus, both an even-mode oscillation, corresponding to $n = 0$, and an odd-mode oscillation, corresponding to $n = 1$, may be possible. The small-signal design should be carried out so as to fulfill the startup conditions at the odd mode ($n = 1$) while preventing start up of the even mode ($n = 0$); that is, $\mathrm{Re}[\lambda_1] < |\mathrm{Re}[Y_d(0)]| < \mathrm{Re}[\lambda_o]$ and $\mathrm{Im}[\lambda_1] + \mathrm{Im}[Y_d(0)] = 0$. Remember that $\lambda_n + Y_d(0)$ agrees with the total small-signal admittance $Y_{T,n}(0, \omega)$ at the connection node of each active subblock of the coupling network. In the design of the individual suboscillator, note that the middle node of this network is a virtual short circuit at the fundamental frequency f_o.

When using harmonic balance software, the startup conditions of the various oscillation modes can be evaluated with no need for an eigenvalue calculation of the passive matrix $[Y_L(\omega)]$. The analysis is performed using two auxiliary generators with small-signal amplitude $A_{AG} = \varepsilon$. The two voltage generators are introduced in parallel at the two equivalent nodes where the active subblocks are connected to the passive network. The even mode is analyzed by setting the phase of the two generators to the same value, which may arbitrarily be made zero: $\phi_{AG1} = \phi_{AG2} = 0$. The odd mode is analyzed by setting $\phi_{AG1} = 0°$ and $\phi_{AG2} = 180°$. For an analysis of startup conditions, the auxiliary generator frequency ω_{AG} is swept from near dc to the maximum oscillation frequency of the device used. The startup conditions should be fulfilled at the desired odd mode only. Due to the small amplitude of the auxiliary generators, this evaluation of the oscillation startup conditions can be performed using a small-signal analysis rather than harmonic balance.

As an example, the design of a bipolar-based push–push oscillator with output frequency 18 GHz is presented below. The oscillator schematic is shown in Fig. 10.6. Two bipolar transistors are coupled at the base terminal through a transmission line of length l. The output power is extracted by connecting the transmission lines at the emitter terminals to a Wilkinson combiner. In the push–push operation expected, the solutions at the two suboscillators have a phase shift of 180°. Because of this, the middle point of the transmission line (of length l) connecting the base terminals of the two transistors is a virtual short circuit at the oscillation frequency. Thus, each suboscillator can be designed independently, with a base transmission line of length $l/2$ ended in a short circuit. For the odd-mode oscillation to start up, the conditions $\mathrm{Re}[Y_{T,1}(\omega_o)] < 0$ and $\mathrm{Im}[Y_{T,1}(\omega_o)] = 0$ must be fulfilled at the oscillation frequency ω_o desired. On the other hand, at the common oscillation mode undesired, the middle point of the coupling line is a virtual open circuit. For this impedance termination, the startup conditions should not be fulfilled at any frequency.

The periodic steady state of a push–push oscillator is analyzed with the same two auxiliary generators as those used for evaluation of the startup conditions. However, their amplitude will no longer be small and will have to be determined by harmonic balance in order to satisfy the steady-state conditions $Y_{AG1} = Y_{AG2} = 0$. An auxiliary generator is connected to each of the two oscillator elements, which will avoid the risk of convergence to a meaningless mathematical solution in which only one of the suboscillators is actually oscillating. In this type of solution, one suboscillator is in an oscillatory state, whereas the second simply responds in a nonautonomous manner to the periodic signal generated by the first suboscillator. The two auxiliary generators will operate at the same frequency, $\omega_{AG} = \omega_o$, with

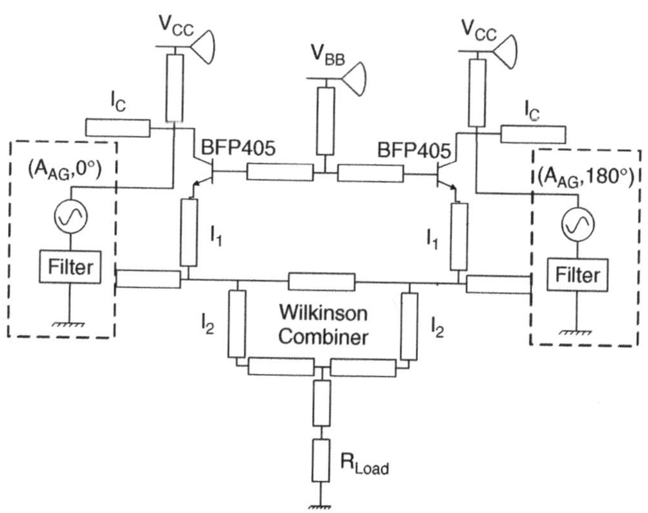

FIGURE 10.6 Bipolar-based push–push oscillator at 18-GHz output frequency. The two auxiliary generators used for the oscillator design are also represented.

the same amplitude, A_{AG}, and phase values $\phi_{AG1} = 0$ and $\phi_{AG2} = 180^\circ$. Use of these two generators will force the harmonic balance convergence to odd-mode oscillation. The frequency $\omega_{AG} = \omega_o$ will have the same amplitude, A_{AG}, and phase, $\phi_{AG1} = 0$ and $\phi_{AG2} = 180^\circ$. As we already know, the nonperturbation condition is given by the zero value of the admittance function (current-to-voltage relationship) associated with each auxiliary generator, $Y_{AG1} = Y_{AG2} = 0$. Due to the circuit symmetry, provided that one of the two conditions is fulfilled, $Y_{AG1} = 0$, the other will necessarily also be fulfilled, $Y_{AG2} = 0$, so it is possible to use one condition only [17], which reduces the computational cost. For analysis of the push–push oscillator, the oscillation frequency $\omega_{AG} = \omega_o$ and common amplitude A_{AG} are calculated or optimized in order to fulfill the nonperturbation condition $Y_{AG1} = 0$. The pair of auxiliary generators can also be used to improve the original circuit design. An example is presented in the following.

In general, the steady-state oscillation frequency of a push–push oscillator will differ from the one predicted by small-signal analysis (due to the nonlinearity of the autonomous circuit) and also differ from the oscillation frequency desired by the designer. Using harmonic balance with the two auxiliary generators, it will be possible to impose the circuit oscillation frequency by doing $\omega_{AG1} = \omega_{AG2} = \omega_o$, maintaining the phase values $\phi_{AG1} = 0^\circ$ and $\phi_{AG2} = 180^\circ$. Next, some circuit elements, together with the common amplitude A_{AG}, should be optimized to fulfill $Y_{AG1} = Y_{AG2} = 0$. It will also be possible to maximize the output power at $2\omega_o$. With this aim, a sweep is performed in the common auxiliary generator amplitude A_{AG}. At each sweep step, two or more circuit elements μ_1 and μ_2 are optimized to fulfill the nonperturbation condition $Y_{AG1}(\mu_1, \mu_2) = 0$. The resulting output power at the second-harmonic component $2\omega_o$ is traced versus A_{AG}, selecting the element values μ_{10} and μ_{20} that maximize this output power. The results of applying this technique to the circuit in Fig. 10.6 are shown in Fig. 10.7. The parameters considered are the line lengths l_1 and l_2. Note that the frequency remains constant during the entire A_{AG} sweep, and each point of the sweep corresponds to a different design. The point selected is the one corresponding to the line lengths $l_1 = 9.98$ mm and $l_2 = 6.48$ mm. The second-harmonic output power obtained is $P_{out} = -2$ dBm.

Next, the stability of the design selected must be verified, together with fulfillment of the oscillation startup conditions. Note that we have changed some element values with respect to the original design, which may have affected the startup conditions at odd and even oscillation modes. This stability analysis can be performed in an accurate manner by applying pole–zero identification to both the dc solution and steady-state oscillation. Figure 10.8 illustrates the synchronization mechanism of the two individual oscillators when starting from different initial values. The simulation has been carried out using the envelope transient method, connecting the two auxiliary generators with different voltage values at the initial time t_o only. This connection of the two auxiliary generators at t_o prevents envelope transient convergence to solutions with either of the two suboscillators in a nonoscillatory state. As can be seen, a stable oscillation is established after a short transient.

Total cancellation of the oscillation fundamental frequency at the push–push output requires perfect symmetry of the two suboscillators. However, this is an

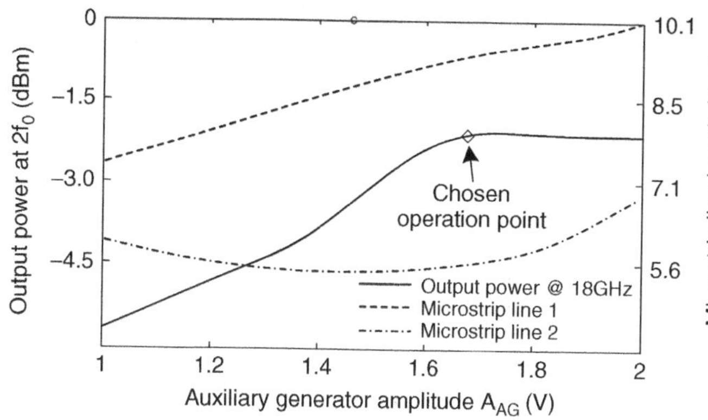

FIGURE 10.7 Selection of circuit element values from an evaluation of the output power at the second-harmonic component versus the oscillation amplitude.

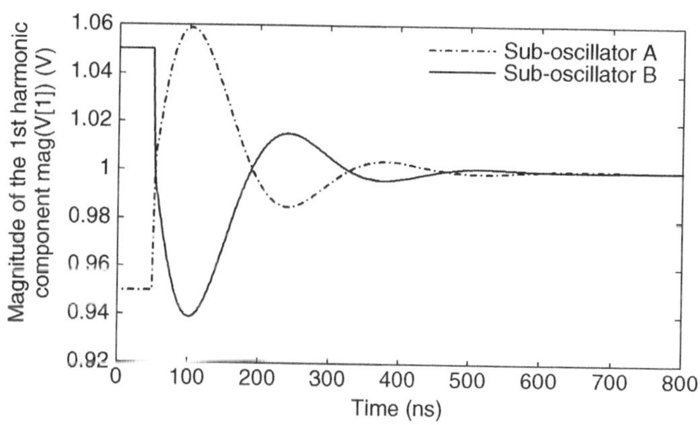

FIGURE 10.8 Envelope transient simulation of the synchronization of two individual oscillations in a push–push oscillator when starting from different initial values. The magnitude represented is the amplitude of the first-harmonic component of the collector voltage of each transistor.

unrealistic situation, as there will necessarily be discrepancies between the oscillator elements, due to tolerances and manufacturing defaults. Even if the oscillators are not symmetrical, they will still be mutually synchronized unless the discrepancies are too large, which may lead to desynchronization, due to significant differences in the individual oscillation frequencies. Assuming that they still behave in a synchronized manner, the solution will be periodic, but the oscillation amplitudes will be different and the phase shift between the oscillator elements will no longer be 180°, due to the loss of symmetry of the coupling network. This will prevent ideal cancellation of the first-harmonic component at the circuit output.

To analyze the effect of circuit element imbalance, discrepancies in line lengths and transistor models will be considered. Assuming synchronized behavior, the harmonic balance analysis of the effect of these asymmetries requires use of two auxiliary generators at the same frequency ω_{AG}. Since the two oscillation amplitudes are different, four variables have to be determined: the oscillation frequency ω_{AG}, the amplitudes A_{AG1} and A_{AG2}, and the phase shift ϕ_{AG2}. Note that the phase of the other auxiliary generator is maintained at $\phi_{AG1} = 0$. The effect of the discrepancy $\Delta\eta$ in a given circuit element is analyzed by means of a sweep in $\Delta\eta$. Here, sweeps in three different elements are considered: the lengths of the collector and emitter lines and the base resistance R_b. The results are shown in Fig. 10.9a.

The demonstrator has been implemented on plastic substrate (Cuclad 2.17) using specific microwave drilling tools instead of chemical processes, in order to match the physical dimensions of the lines. Great accuracy is necessary in the microstrip line used to couple the bases of the transistors, due to the relevant role of its length in circuit behavior, as shown in simulation. For maximum symmetry, one

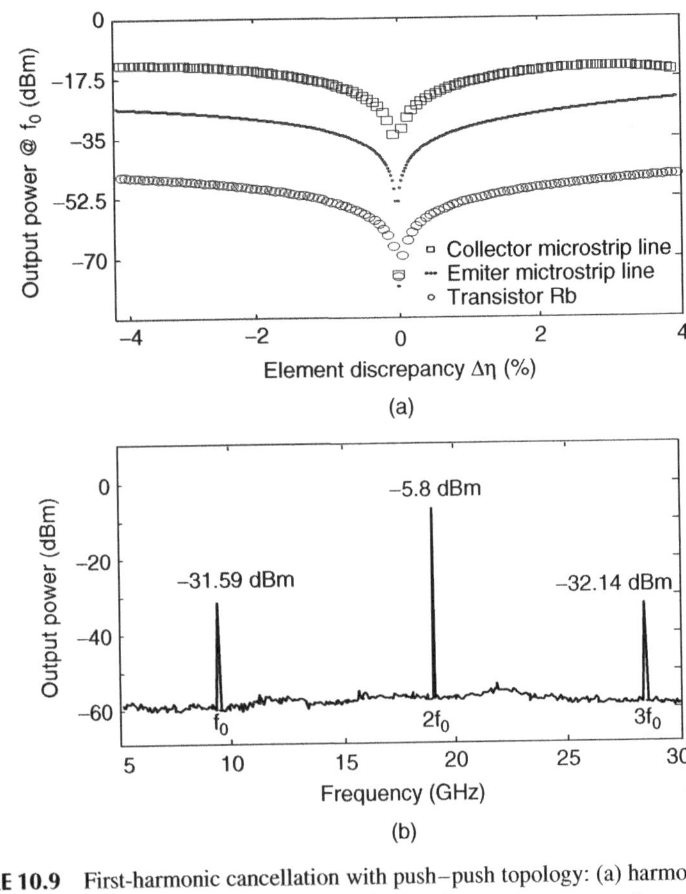

FIGURE 10.9 First-harmonic cancellation with push–push topology: (a) harmonic balance simulation of the output power variations at the first-harmonic component versus discrepancies in various circuit elements: the length of the collector and emitter transmission lines and the base resistance; (b) measured output power spectrum.

of the devices has been back-mounted to allow the same line dimensions at each oscillator. The biasing networks have been implemented using radial stubs along with quarter-wave lines. Some notch filters to eliminate spurious responses have been used. Metallic via holes have been used to minimize the parasitic inductance in ground connections. The spectrum measured, shown in Fig. 10.9b, indicates first- and second-harmonic output powers of −31.59 and −5.8 dBm, respectively.

In the push–push oscillator considered, the phase noise can be estimated from (10.39), considering a matrix order 2×2. Note that although the two suboscillators have a 180° phase shift at the fundamental frequency ω_o, the noise perturbations about this frequency do not cancel out, due to their random nature. This is evidenced by the perturbed system (10.27). If we could design a single-oscillator circuit with characteristics identical to those of the individual suboscillators, phase noise at the fundamental frequency would be about $10 \log 2 = 3$ dB lower than that of an individual suboscillator. On the other hand, the output frequency desired for the push–push oscillator is twice the oscillation frequency, $2f_o$. This frequency multiplication by 2 gives rise to a natural phase noise increase of $20 \log 2 = 6$ dB with respect to the fundamental frequency. Thus, the output noise at $2f_o$ is expected to be 3 dB larger than that of a similar oscillator at the fundamental frequency f_o.

The phase noise of the push–push oscillator of Fig. 10.6 has been analyzed with harmonic balance using the carrier modulation and conversion matrix approaches. The noise contributions in each transistor are thermal noise from the different resistances and two current noise sources, introduced in parallel with the base terminal and the collector terminal, accounting, respectively, for the flicker and base–emitter and base–collector shot noise. The phase noise spectrum obtained is represented in Fig. 10.10, where it may be compared with that of a single oscillator at the fundamental frequency f_o. This individual oscillator was obtained by terminating the middle node of the coupling network in a short circuit. Note

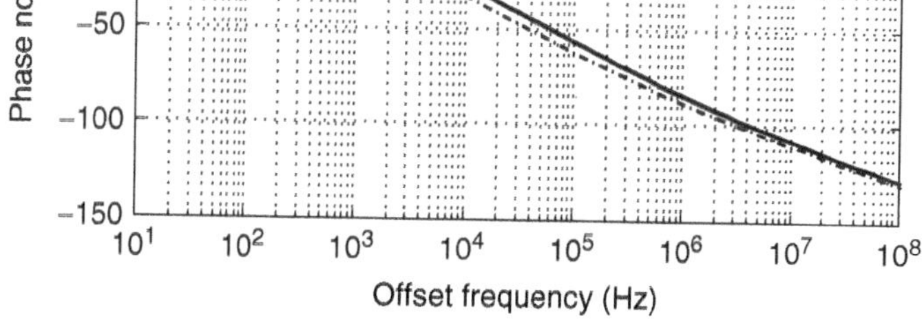

FIGURE 10.10 Phase noise spectral density obtained in harmonic balance with the carrier modulation approach. The noise of a single oscillator (solid line) can be compared with that of the push–push configuration (dashed line). Measurement points are superimposed.

that the individual oscillator is not totally identical to the suboscillator, due to the different loading conditions at the even harmonic terms, which are not short circuits for the push–push suboscillator. Because of that, it was necessary to optimize the circuit elements slightly to fulfill the oscillation condition $Y_{AG} = 0$.

10.2.5.2 Quadruple–Push Oscillator The quadruple–push oscillator [13] consists of four identical oscillators with the phases 0°, 90°, 180°, and 270° at the fundamental frequency f_o. When their outputs are injected into an ideal power combiner, the components at $4f_o$, $8f_o$, and so on, are added in phase while the rest are canceled, so the output frequency is the quadruple of the individual frequency. Due to the in-phase combination of the fourth harmonic component of the individual suboscillator, the output power at $4f_o$ will be 6 dB higher than the power delivered by individual suboscillators at the same harmonic frequency. To increase the output power at $4f_o$, a load–pull technique can be applied to the elementary oscillator. The technique is similar to the one described in Section 8.4.3.

The multiharmonic load consists of four subnetworks (Fig. 10.11), each comprising an ideal bandpass filter at $f_n = nf_o$, with $n = 1$ to 4 and a passive termination, defined by its reflection coefficient Γ_n. The entire set of harmonic loads must sustain self-oscillation at f_o. To guarantee this, a voltage auxiliary generator with amplitude A_{AG} is connected in parallel at the transistor output node. The auxiliary generator operates at the fundamental frequency $f_{AG} = f_o$, so the oscillation frequency is maintained at the desired value during the entire design process. At each variation in the harmonic loads, two or more circuit element values are calculated to fulfill the steady-state oscillation condition $Y_{AG} = 0$, with Y_{AG} being the ratio between the auxiliary generator current and the voltage A_{AG}.

To illustrate, the technique has been applied to optimization of the suboscillator circuit of a quadruple–push oscillator at 20 GHz. To enable circuit oscillation, series feedback is used, consisting of the parallel connection of a capacitor and a

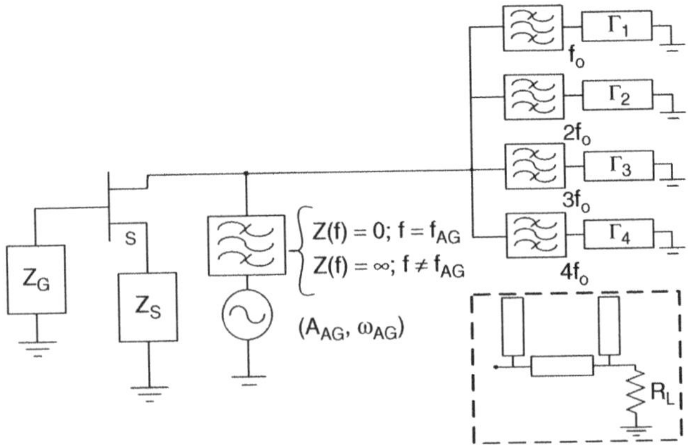

FIGURE 10.11 Load–pull optimization of the individual suboscillator to maximize the output power at $4f_{rmo}$. The harmonic components are terminated with ideal loads Γ_n.

transmission line resonating at $f_o = 5$ GHz at the source terminal. For circuit optimization, reflection coefficients with magnitudes $\rho_1 = \rho_2 = \rho_3 = 1$ are considered. The required magnitude and phase of Γ_4 are determined in two nested sweeps in terms of ρ_4 and ϕ_4. At each sweep step, two or more circuit elements must be optimized to fulfill the oscillation condition $Y_{AG} = 0$. It is possible to optimize the feedback and gate terminal elements or the oscillation amplitude A_{AG} and the ideal multiharmonic phases ϕ_1, ϕ_2, and ϕ_3. The latter option allows maintaining the original values of the small-signal impedance at the transistor output. The output power variation at $4f_o$, calculated as $P_{out4} = \frac{1}{2}\cdot \text{Re}[V_4 I_4^*]$, is shown in Fig. 10.12. The reflection coefficients providing the maximum output power $P_{out4} = 1.25$ dBm are $\Gamma_1 = 1\angle{-89.12^\circ}$, $\Gamma_2 = 1\angle{-52.62^\circ}$, $\Gamma_3 = 1\angle 218^\circ$, and $\Gamma_4 = 0.5\angle 310^\circ$.

The quadruple–push oscillator can be obtained using a ring coupler combiner (Fig. 10.13a), as proposed by Xiao et al. [13], or by using a ring coupler connecting the gate terminals and two Wilkinson combiner stages at the drain loads (Fig. 10.13b). The Wilkinson combiners will allow better port isolation at the expense of a larger circuit size. A good fitting of the entire harmonic load requires two transmission-line sections and three stubs (option 1). A second topology, shown in Fig. 10.13, uses a transmission line and two stubs (option 2). It provides worse fitting at the lower-harmonic terms, but it is smaller and simpler to implement. However, the lower harmonics will inherently cancel in the quadruple–push configuration, so this smaller topology has been chosen here.

For harmonic balance simulation of a quadruple–push oscillator, four auxiliary generators are used. This prevents undesired convergence toward useless mathematical solutions in which one or more of the oscillator elements are not actually oscillating. The phase values of the four auxiliary generators are set to $\phi_{AG1} = 0$, $\phi_{AG2} = 90^\circ$, $\phi_{AG3} = 180^\circ$, and $\phi_{AG4} = 270^\circ$ (Fig. 10.13). Although

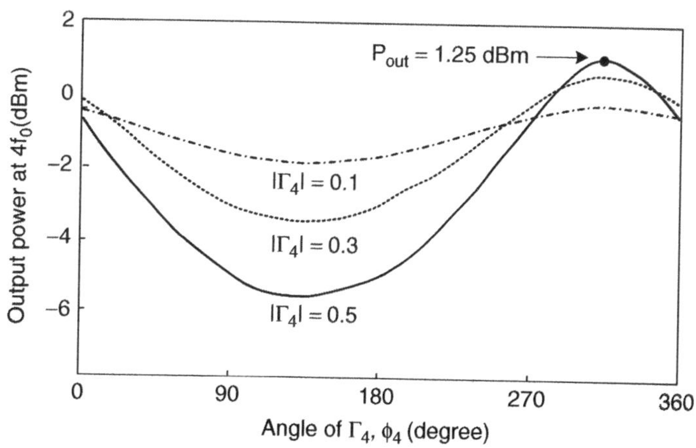

FIGURE 10.12 Maximization of the output power at $4f_{rmo}$. Output power variation versus the amplitude and phase of Γ_4.

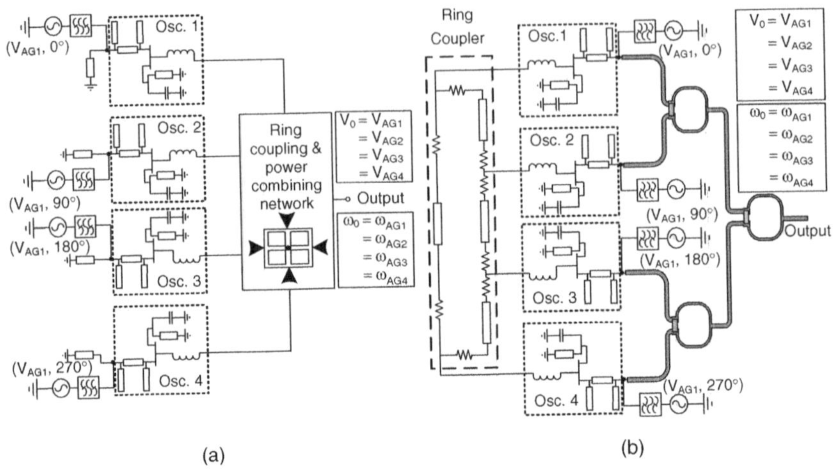

FIGURE 10.13 Quadruple–push configuration: (a) power combination using a ring coupler; (b) treelike configuration with two Wilkinson combiner stages.

four auxiliary generators are introduced in the circuit, a smaller number of non-perturbation conditions $Y_{AGi} = 0$ are needed, due to the circuit symmetry. The simulated output power at $4f_o$ for the optimized configuration in Fig. 10.13b is $P_{\text{out}} = 3.7$ dBm. Note that both startup from the dc regime of the 90° mode and its steady-state stability must be verified.

As shown in Section 10.2.2, the stability analysis of the dc solution versus a parameter capable to quench all possible oscillation modes can be helpful to distinguish between stable and unstable oscillation modes. An example was presented in Fig. 10.4. Results of pole–zero identification of the dc solution (in Fig. 10.4b) were in agreement with the bifurcation diagram of Fig. 10.4a. This bifurcation diagram was obtained through use of a continuation technique in harmonic balance. For each mode n, four auxiliary generators of constant amplitude $A_{AG} = A_n$ and respective phase values $\phi_{AG,k} = nk2\pi/4$ were used. In the flat regions of each curve, V_{GS} was swept, solving $Y_{AG,k} = 0$ for the common amplitude $A_{AG} = A_n$ and the oscillation frequency $\omega_{AG} = \omega_n$. To obtain sections of high slope versus V_{GS}, the amplitude A_{AG} was swept down to zero, solving the equations $Y_{AG,k} = 0$ for the common bias voltage V_{GS} and ω_{AG}.

The use of pole–zero identification for stability analysis of the steady-state modes can be demanding, due to the presence of symmetries in the coupled structure and the usually close values of the pole frequencies. Alternatively, it is possible to use the one-harmonic approach (10.23). This requires obtaining the admittance functions $Y_{NL}(A, \omega)$ that describe the active subblocks. To obtain these functions, a voltage auxiliary generator with amplitude A and frequency ω is connected in parallel with the output terminal of one of the four active subcircuits. A high-value resistance is introduced in parallel, to avoid convergence problems during the double sweep. A double-entry table $Y_{NL}(A, \omega)$ is obtained in this manner, to be used in analyses of (10.19) and (10.23). These numerical models were used to obtain the

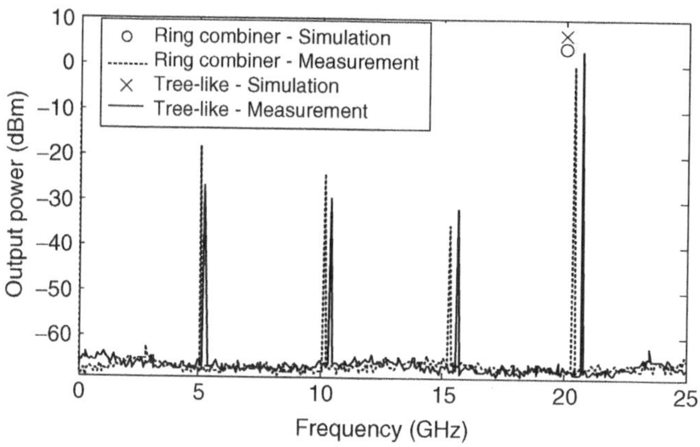

FIGURE 10.14 Simulation and measurements of the output power spectrum of the quadruple–push oscillator.

poles in Tables 10.1 and 10.2, respectively, corresponding to the unstable 90° solution (corresponding to $V_{GS} = -0.2$ V in Fig. 10.4a) and the stabilized solution, after the introduction of interbranch resistances. This solution could actually be measured, with the results shown in Fig. 10.14. The measured phase noise spectral density was −91 dBc/Hz at 100 kHz offset frequency.

10.3 COUPLED-OSCILLATOR SYSTEMS FOR BEAM STEERING

In different applications there is a need to control the direction of the antenna beam Examples are radar surveyance for guidance and tracking, satellite communications for satellite tracking, and remote sensing and cellular communications to steer the beam toward the stations selected [8,18]. Mechanical control of the antenna beam has two main limitations: the scan speed and the natural degradation of the mechanical components. These problems can be overcome with electronic beam steering using array antennas. The array antennas consist of a large number of radiating elements distributed over the area that constitutes the radiating aperture. The beam is steered electronically by controlling the phase distribution across the aperture elements. The overall radiation pattern is the result of the combination in amplitude and phase of waves radiated by the array of elements. This type of antenna is usually called a *phased array*, as signals with appropriate phase values are applied to the discrete antenna elements.

For better understanding, a phased-array operation is depicted in Fig. 10.15. It is assumed that an electromagnetic plane wave reaches the phased array at an angle α with respect to the direction perpendicular to the plane of the array, or "broadside." Due to the angle α, the parallel components of the plane wave have traversed different path lengths l_n when they get to the antenna elements. If the spacing between the antenna elements is d, the path length difference between two adjacent elements is $\Delta l = d \sin\alpha$, as can easily be deduced from Fig. 10.15.

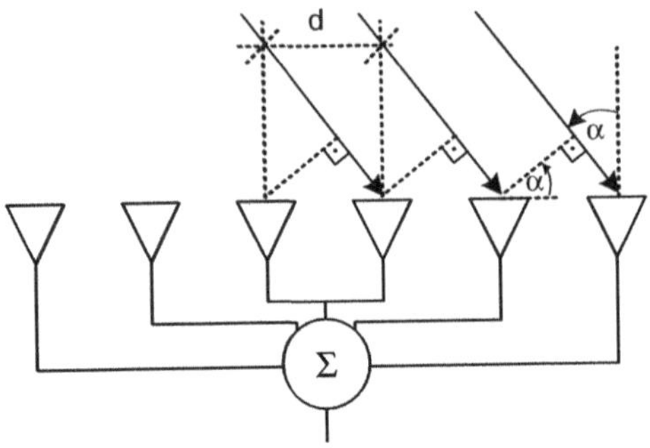

FIGURE 10.15 Phased array with an electromagnetic wave arriving at an angle α off broadside. The distance between adjacent array elements is d, so the interelement phase shift is $\Delta\phi = 2\pi(d/\lambda)\sin\alpha$.

Taking the propagation constant into account, the corresponding phase shift will be $\Delta\phi = 2\pi(d/\lambda)\sin\alpha$, with λ the wavelength. The resulting field received is [19]

$$E(\alpha, t) = \sum_{n=0}^{N-1} E_n e^{-j[(2\pi n d/\lambda)\sin\alpha]} e^{j\omega t} \tag{10.42}$$

Due to the law of reciprocity, the beam transmitted can be steered an angle α off broadside by exciting the antenna elements with a constant interstage phase shift $\Delta\phi$ so that the rays transmitted interfere in a constructive manner. The relationship between the beam angle α and the interelement phase shift is $\Delta\phi = 2\pi(d/\lambda)\sin\alpha$. The distance between the radiating elements is typically on the order of $d = 0.6\lambda$.

The type of radiating elements used in a phased array depends on the application, frequency bands, and polarization and could be dipoles, waveguide slots, horns, or microstrip patches. In transmission, the input power of a conventional array antenna is provided by a power amplifier. In reception, a low-noise amplifier is connected to the array output. Dynamic control of the beam is obtained by using controllable power dividers and phase shifters. Alternatively, a coupled-oscillator system, with a constant interstage phase shift, can be used for beam scanning in phased array, with the output of each oscillator component feeding an antenna element. Beam scanning is achieved by synthesizing a constant phase shift distribution between adjacent oscillators. Then the beam angle is changed by varying the interstage phase shift, which [6,14] can be done by detuning the peripheral oscillators only, leaving the rest of the oscillators unchanged. In other words, the variation in the constant interstage phase shift is achieved without using phase shifters, by varying the control biasing of the peripheral oscillators only. This simplifies the antenna-feeding system, due to the considerable reduction in controlled elements. The technique allows a smaller system size compared with the conventional use of phase shifters, which requires individual control of the biasing and wiring of each phase shifter.

Both one- and two-dimensional oscillator arrays have been proposed for beam steering. In a one-dimensional array, the constant interstage phase shift is varied by detuning the two outermost oscillators of the array in the opposite direction. This allows beam steering in one direction only, either azimuth or elevation. To steer the beam in both azimuth and elevation, a two-dimensional array is necessary with a constant interstage phase shift in each of the two directions ϕ_x and ϕ_y. In a manner similar to the one-dimensional case, this phase distribution can be synthesized by detuning the peripheral oscillator elements only. Note that due to unavoidable technological differences between the oscillator elements, the interoscillator phase shift required will not be achieved with perfect accuracy. Due to the continuity of the equation system, this will simply give rise to a pointing error that must be kept below some tolerance limits [5,6].

10.3.1 Analytical Study of Oscillator-Array Operation

A coupled-oscillator system for beam steering application consists of full oscillator elements, which exhibit a free-running oscillation prior to their insertion in the array. Once in the array, the oscillators are connected through passive coupling elements, typically consisting of a resistively loaded transmission line. Injection signals are introduced through these coupling elements, so the entire array is synchronized at a given oscillation frequency in the order of the original free-running values of the oscillator elements. Note that the situation is different from that of the globally coupled oscillators studied, generally comprised of active blocks, which to oscillate require connection of the passive coupling network. In coupled oscillator systems for beam steering, the array elements are typically equal but tuned differently so as to exhibit different free-running oscillation frequencies. As stated earlier, York [6] has demonstrated that to synthesize a constant interstage phase shift, only the peripheral array elements must be tuned, which considerably simplifies the control technique compared with conventional phase shifters.

In the following, the theoretical basis for the synthesis of a constant phase shift distribution in a coupled-oscillator system is presented [6]. The analytical demonstration will take the familiar case of an oscillator injection-locked to an external source as a starting point. In the absence of a synchronizing source, this circuit is assumed to oscillate at the free-running frequency ω_o, with amplitude V_o, at a given observation port. The total admittance function (current-to-voltage relationship) calculated at this port fulfills $Y_T(V_o, \omega_o) = 0$. For small-signal injection, and once in the synchronized regime, it will be possible to expand the total admittance function $Y_T(V_s, \omega_s)$ in a first-order Taylor series about the free-running solution $Y_T(V_o, \omega_o) = 0$. The periodic synchronized solution can be approached (Section 4.2):

$$\begin{aligned} \frac{I_g}{V_o}\cos\phi &= \left.\frac{\partial G_T}{\partial V}\right|_o (V - V_o) + \left.\frac{\partial G_T}{\partial \omega}\right|_o (\omega_s - \omega_o) \\ -\frac{I_g}{V_o}\sin\phi &= \left.\frac{\partial B_T}{\partial V}\right|_o (V - V_o) + \left.\frac{\partial B_T}{\partial \omega}\right|_o (\omega_s - \omega_o) \end{aligned} \tag{10.43}$$

where the phase origin is taken at the external current generator with amplitude I_g. If the variation in the imaginary part of the total conductance B_T with respect to the amplitude is neglected, it is possible to solve the second equation of (10.43) directly for the synchronization frequency ω_s:

$$\omega_s = \omega_o - \frac{I_g \sin\phi}{(V_o)\partial B_T/\partial\omega|_o} \tag{10.44}$$

In the case of a coupled oscillator system, the current injected into a given oscillator i comes from the two neighbors $i-1$ and $i+1$. It is injected through coupling networks connected between the oscillator elements. These networks, assumed equal, relate the injected current into the oscillator element i to the output voltage at the oscillators $i-1$ and $i+1$. Thus, it is a nearest-neighbor coupling instead of a global coupling. The coupling networks will be described by means of their complex admittance $y_c = |y_c|e^{-j\Phi}$. In phasor notation, the current injected in each oscillator element will be written as

$$I_i = |y_c|e^{-j\Phi}V_{i-1} + |y_c|e^{-j\Phi}V_{i+1} \tag{10.45}$$

where Φ is the phase shift associated with the coupling element. The coupling networks are often implemented with a transmission line ended with series resistances (Fig. 10.16). For simplicity, the electrical length of the line is often taken about $\beta l = 2\pi$ radians at the oscillation frequency.

Two different cases may be distinguished, depending on the magnitude of the coupling admittance $|y_c|$. For small $|y_c|$, which implies relatively large series resistances, the current injected through the coupling networks will be small and will have little influence over the amplitude and frequency of the oscillator element. The situation is similar to that of the injection-locked oscillator with a low input amplitude I_g. As in (10.43), it will be possible to perform an analysis of the coupled oscillator by linearizing the system equations about the free-running solution of the various oscillators. For large $|y_c|$, which implies relatively small series resistances, the current injected through the coupling networks will be larger and a linearized analysis such as the one in (10.43) might not be applicable. Weak coupling is generally preferred for beam scanning applications due to the small variation in the frequency and amplitude of the array elements with respect to their free-running values. However, too weak coupling will give rise to high sensitivity to discrepancies between the oscillator elements. These discrepancies are unavoidable due to the component tolerances. Due to the lack of robustness, the actual system response may differ significantly from the simulated response, with perfectly equal oscillator elements. In the case of weak coupling, undesired nonsynchronized solutions can

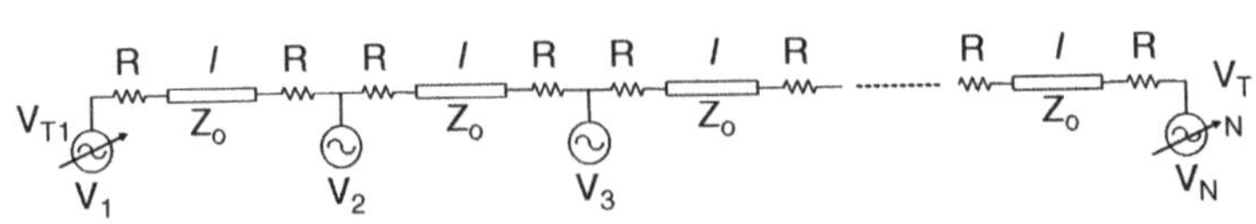

FIGURE 10.16 General configuration of a one-dimensional oscillator array, with coupling networks that consist of resistively ended transmission lines.

also arise when varying the control biasing in opposite directions, which is due to the limited influence of one oscillator over the other.

In the following, the coupling is assumed to be weak enough to allow linearization of the oscillator elements about their free-running regime. Then it will be possible to replace I_g in (10.44) with the current injected through the coupling networks (10.45). It will be considered that the different oscillator elements have different free-running frequencies, given by ω_{oi}. By *free-running frequency* we mean the oscillation frequency of the array element prior to its connection to the coupled system. This system is described in the following equations, obtained from (10.44) and (10.45):

$$\omega_s = \begin{cases} \omega_{oi} - \dfrac{1}{(V_i)\partial B_T/\partial\omega|_o}|y_c|[\sin(\phi_i - \phi_{i-1} - \Phi)|V_{i-1}| \\ \qquad + \sin(\phi_i - \phi_{i+1} - \Phi)|V_{i+1}|] & i = 2 \text{ to } N-1 \\ \omega_{oi} - \dfrac{1}{(V_i)\partial B_T/\partial\omega|_o}|y_c|[\sin(\phi_i - \phi_{i+1} - \Phi)]|V_{i+1}| & i = 1 \\ \omega_{oi} - \dfrac{1}{(V_i)\partial B_T/\partial\omega|_o}|y_c|[\sin(\phi_i - \phi_{i-1} - \Phi)]|V_{i-1}| & i = N \end{cases} \tag{10.46}$$

Note that the phase variables are denoted ϕ_i here instead of θ_i, to make a distinction with respect to the case of globally coupled oscillators. Because the aim is to obtain a constant phase shift distribution, a constant value of the interstage phase shift, $\Delta\phi$, will be imposed. Then the free running oscillation frequencies of the different array elements ω_{oi}, $i = 1$ to N will become the system unknowns. For simplicity, it will be considered that the oscillation amplitude is the same for all oscillator elements; that is, $V_i = V\ \forall i$. The system solution is the following:

$$\omega_s = \begin{cases} \omega_{oi} - \dfrac{1}{\partial B_T/\partial\omega|_o}|y_c|[\sin(\Delta\phi - \Phi) + \sin(-\Delta\phi - \Phi)] & i = 2 \text{ to } N-1 \\ \omega_{o1} - \dfrac{1}{\partial B_T/\partial\omega|_o}|y_c|[\sin(-\Delta\phi - \Phi)] & i = 1 \\ \omega_{oN} - \dfrac{1}{\partial B_T/\partial\omega|_o}|y_c|[\sin(\Delta\phi - \Phi)] & i = N \end{cases} \tag{10.47}$$

From equations (10.47) it is clear that to achieve a constant phase shift distribution, the free-running frequency of the middle oscillators $i = 2$ to $N-1$ must be the same and equal to

$$\omega_{o,m} = \omega_s + \frac{1}{\partial B_T/\partial\omega|_o}|y_c|[\sin(\Delta\phi - \Phi) + \sin(-\Delta\phi - \Phi)] \tag{10.48}$$

In turn, the free-running frequencies of the lower- and upper-edge oscillators must be

$$\omega_{o1} = \omega_s - \frac{1}{\left.\frac{\partial B_T}{\partial \omega}\right|_o} |y_c| [\sin(\Delta\phi + \Phi)] = \omega_{o,m} - \frac{1}{\left.\frac{\partial B_T}{\partial \omega}\right|_o} |y_c| [\sin(\Delta\phi - \Phi)] \quad i = 1$$

$$\omega_{oN} = \omega_s - \frac{1}{\left.\frac{\partial B_T}{\partial \omega}\right|_o} |y_c| [\sin(\Phi - \Delta\phi)] = \omega_{o,m} + \frac{1}{\left.\frac{\partial B_T}{\partial \omega}\right|_o} |y_c| [\sin(\Delta\phi + \Phi)] \quad i = N \tag{10.49}$$

Thus, the two outermost oscillators have to be detuned in opposite directions to achieve a constant interstage phase shift distribution. The constant interstage phase shift $\Delta\phi$ can be varied by changing the bias voltage of the outermost oscillators only, which is a significant advantage with respect to the conventional use of phase shifters, one for every two elements of the array antenna.

The in-phase solution is used for broadside radiation and enables a combination of the power delivered by the various oscillator elements. For the in-phase solution, the free-running frequency distribution should be the following [6]:

$$\begin{aligned}
\omega_{o1} &= \omega_{o,m} + \frac{1}{\partial B_T/\partial\omega|_o} |y_c| \sin\Phi \qquad i = 1 \\
&\vdots \\
\omega_{o,m} &= \omega_s - 2\frac{1}{\partial B_T/\partial\omega|_o} |y_c| \sin\Phi \\
\omega_{oN} &= \omega_{o,m} + \frac{1}{\partial B_T/\partial\omega|_o} |y_c| \sin\Phi \qquad i = N
\end{aligned} \tag{10.50}$$

Unless the phase shift introduced by the coupling network is a multiple of π radians, the outermost oscillators must be detuned with respect to the middle ones.

The technique described for the synthesis of a constant phase distribution between the oscillator elements has been extended to two-dimensional arrays [8]. In these systems there is constant phase shift $\Delta\phi_y$ across the array rows and a different constant phase shift $\Delta\phi_x$ across the array columns. This allows beam steering in both azimuth and elevation. Thus, the following phase distribution must be synthesized:

$$\begin{aligned}
\phi_{m,n\pm1} - \phi_{m,n} &= \pm\Delta\phi_x \\
\phi_{m\pm1,n} - \phi_{m,n} &= \pm\Delta\phi_y
\end{aligned} \tag{10.51}$$

To achieve the phase distribution above, only the peripheral oscillator elements are detuned [5,8]. In an $M \times N$ oscillator array, the horizontal peripheral oscillators $(1,n)$, with $n = 2$ to $N-1$, are detuned in the direction opposite to the oscillators (M,n), with $n = 2$ to $N-1$. In turn, the vertical oscillators $(m,1)$, with $m = 2$ to $M-1$, are detuned in the direction opposite to the oscillators (m,N), with $m = 2$ to $M-1$. The corner elements are tuned with the addition of the tuning voltage values of their neighboring vertical and horizontal elements.

10.3.2 Harmonic Balance Analysis

The harmonic balance analysis of the N coupled-oscillator system requires avoiding unstable nonoscillating solutions in any of the oscillator circuits. Note that together with solutions for which all the coupled circuits oscillate, there exist mathematical solutions in which one or more elements of the coupled system are not actually oscillating. Instead, they are responding to the signal injected through the coupling networks in a nonautonomous manner. Convergence toward these undesired mathematical solutions can be avoided through the use of auxiliary generators, one connected to each oscillator element. All the auxiliary generators operate at the common synchronization frequency $\omega_{AG} = \omega_s$ and must simultaneously exhibit a current-to-voltage relationship equal to zero $Y_{AG,i} = 0$, $i = 1$ to N. When these conditions are fulfilled, the amplitude and phase $V_{AG,i} = V_i$ and $\phi_{AG,i} = \phi_i$ of each auxiliary generator will agree with the first-harmonic component of the corresponding node voltage. The N complex equations, $Y_{AG,i} = 0$, $i = 1$ to N, to be solved in combination with harmonic balance, will allow both analysis and synthesis of the coupled-oscillator system [20].

In the system analysis, the admittance equations are solved for the synchronization frequency $\omega_o = \omega_{AG}$, the interstage phase shift $\Delta\phi_{i+1,i}$, and the oscillation amplitudes $V_i \equiv V_{AG,i}$:

$$\begin{aligned} &Y_{AG,1}(\omega_{AG}, V_{AG,1}, \ldots, V_{AG,N}, \phi_{AG,2}, \ldots, \phi_{AG,N}) = 0 \\ &\vdots \\ &Y_{AG,N}(\omega_{AG}, V_{AG,1}, \ldots, V_{AG,N}, \phi_{AG,2}, \ldots, \phi_{AG,N}) = 0 \end{aligned} \qquad (10.52)$$

where the phase of the first oscillator has been fixed arbitrarily to zero $\phi_{AG,1} = 0$. This is possible, as the entire coupled system operates in a free-running regime.

For the synthesis, the synchronization frequency of the N auxiliary generators is set to the desired value $\omega_{AG} = \omega_s$, and so are the $N-1$ interstage phase-shift values. Then the amplitude of each oscillator $V_i = V_{AG,i}$ and N tuning elements η_i are calculated to fulfill $Y_{AG,1} = 0$ to $Y_{AG,N} = 0$. The tuning elements of the two outermost oscillators η_1 and η_N will generally correspond to bias voltages $\eta_1 = V_{T1}$ and $\eta_N = V_{TN}$. The constant phase shift between the adjacent oscillator elements is imposed using the auxiliary generators (i.e., $\phi_{AG,2} = \Delta\phi$ and $\phi_{AG,N} = \phi_{N-1} + \Delta\phi$). The numerical resolution of system (10.52) shows that the middle parameters η_2 to η_{N-1} take a fixed common value η_o for this constant phase shift progression, in agreement with York's demonstration [6].

To achieve the in-phase condition at the desired synchronization frequency ω_s, the parameters η_2 to η_{N-1}, V_{T1}, V_{TN}, and the amplitudes $V_{AG,1}$ to $V_{AG,N}$ should be calculated, imposing the in-phase condition $\Delta\phi = 0$ at $\omega_{AG} = \omega_o$. The ability to preset the operation frequency prevents, at least in simulation, any frequency deviation with respect to the free-running value. In practice, the frequency variation will be unavoidable, due to the element tolerances. Once this solution has been obtained, the tuning elements of the innermost oscillators are kept fixed to η_o, sweeping $\Delta\phi$ and calculating $\omega_{AG} = \omega_s$, the amplitudes V_1 to V_N, and the two tuning voltages V_{T1} and V_{TN} at each step. The values obtained for ϕ_3 to ϕ_{N-1} correspond to a constant interstage phase shift in agreement with the analytical demonstration by York [6]. The $\Delta\phi$ phase sweep performed in the harmonic-balance simulation provides the required tuning voltages V_{T1}, V_{TN} for each value of the constant interstage phase shift $\Delta\phi$. We also obtain the variation of the common oscillation frequency ω_s and the amplitude of the different oscillator circuits.

For illustration purposes the harmonic balance technique has been applied to the design of a coupled oscillator array with three transistor-based oscillator elements. Each array element is a 6-GHz varactor-tuned oscillator designed using an NE3210S01 NEC HJ-FET and a MACOM MA46H070 varactor. The schematic of this individual oscillator is shown in Fig. 10.17a. The varactor diode is connected to the transistor drain terminal and the oscillator output is taken from the source terminal. The active device is self-biased by dc grounding the gate. An inductive series stub connected at the gate allows one to further tune the center frequency of

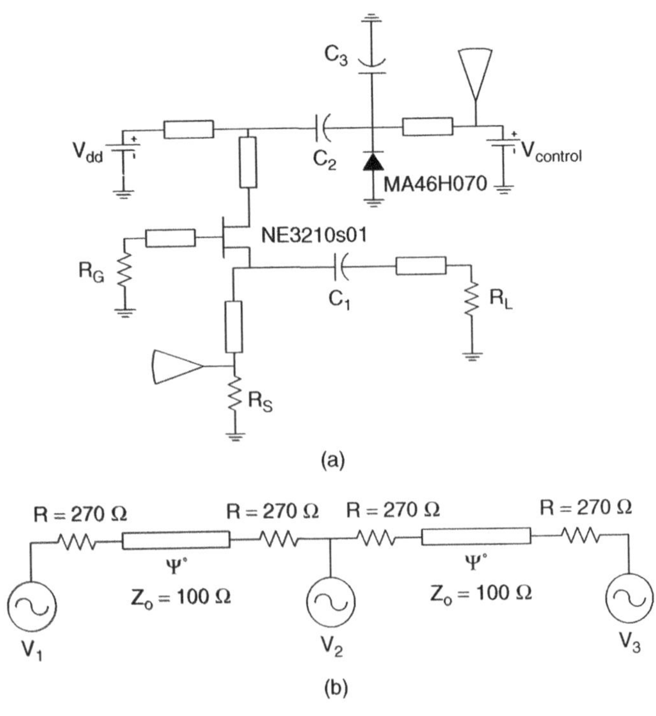

FIGURE 10.17 Coupled-oscillator array. (a) Schematic of the individual oscillator circuit designed using an NE3210S01 NEC HJ-FET and a MACOM MA46H070 varactor. The varactor is connected to the drain of the active device and the oscillator output is taken from the source terminal [21]. (b) The oscillator elements are coupled with resistively loaded transmission lines.

the VCO. The gate also provides an injection port. The circuit was built on CucLad substrate with dielectric constant $\varepsilon_r = 2.17$ and height $H = 0.8$ mm [21].

The oscillator elements are coupled through the transistor source terminals. A constant interstage phase shift is achieved by detuning the two outermost oscillators in opposite directions. The coupling network is composed of a transmission-line section of characteristic impedance Z_o, embedded between two series resistors R, as shown in Fig. 10.17b. This type of coupling network has been used widely in the literature [6], due to its wideband characteristics and fabrication simplicity. It has the following Y-matrix representation:

$$Y_{ij} = \begin{cases} n_i \dfrac{e^{j\psi} + j[(R - Z_0)/Z_0]\sin\psi}{2Re^{j\psi} + j[(R - Z_0)^2/Z_0]\sin\psi} & i = j \\ \dfrac{-1}{2Re^{j\psi} + j[(R - Z_0)^2/Z_0]\sin\psi} & |i - j| = 1 \\ 0 & \text{otherwise} \end{cases} \tag{10.53}$$

where ψ is the electrical length of the transmission line and $n_i = 1$ at the array edge elements ($i = 1$ or N), and 2 otherwise. The coupling network is introduced between every two adjacent oscillator elements. It couples each oscillator element to its two nearest neighbors and also slightly modifies its output load, due to the input admittance introduced.

For the coupled-system implementation, the electrical length of the transmission line of the coupling network has characteristic impedance $Z_c = 100\ \Omega$ and $\beta l = 355^\circ$. The series resistances at both ends of this transmission line have the value $R = 270\ \Omega$. Note that for this relatively large value of this resistance the coupling strength will be weak, meaning that the current injected from the neighboring oscillators will have a low amplitude value.

In the synthesis technique, a constant interstage phase shift is imposed using the auxiliary generators, so the solution necessarily has a constant phase-shift progression. To determine the required tuning voltages for different values of the constant interstage phase shift $\Delta\phi$, this phase shift is varied between -180 and 180°. For each $\Delta\phi$, system (10.52) is solved through error minimization with a two-tier Newton–Raphson or through optimization in harmonic balance. This $\Delta\phi$ sweep is the most convenient way to determine the values required for the two tuning voltages V_{T1} and V_{T3}. This is because no infinite slope points are generally obtained versus $\Delta\phi$, similar to what happens with closed solution curves of injection-locked oscillators. In Chapter 5, Section 5.5.2.2, we saw that the closed solution curves of injection-locked oscillators curves are traced by sweeping the phase shift ϕ_s with respect to the synchronizing source. No tuning points are encountered versus the phase variable, which prevents the harmonic-balance convergence problems associated to these points. The closed curve, with its turning points, is obtained by tracing the resulting amplitude $V_s(\phi_s)$ versus the oscillation frequency $\omega_s(\phi_s)$. In a coupled-oscillator system, the sweep of $\Delta\phi$ allows us to obtain the solution curves in a straightforward manner, even if the solution exhibits turning points versus the actual circuit parameters V_{T1} and V_{T3}.

For a FET-based coupled oscillator, with the coupling networks of Fig. 10.17b, solutions are obtained for the entire phase shift range -180 to 180°. The tuning voltages V_{T1} and V_{T3} of the two outermost oscillators, which take different values for each $\Delta\phi$, are represented in Fig. 10.18a, one versus another. This representation corresponds to a closed curve in the plane (V_{T1}, V_{T3}), which is due to the periodicity of (10.52) in the phase shift $\Delta\phi$. Note that each pair of tuning voltages (V_{T1}, V_{T3}) in this closed curve provides a different value for the constant interstage phase shift $\Delta\phi$. The set of tuning voltages providing constant (and different) $\Delta\phi$ here is called V_T. Note that due to the presence of multiple oscillators in the coupled systems for each pair of tuning voltages $(V_{T1}, V_{T3}) \in V_T$, other oscillation modes with nonconstant interstage phase shift will generally exist. For the proper

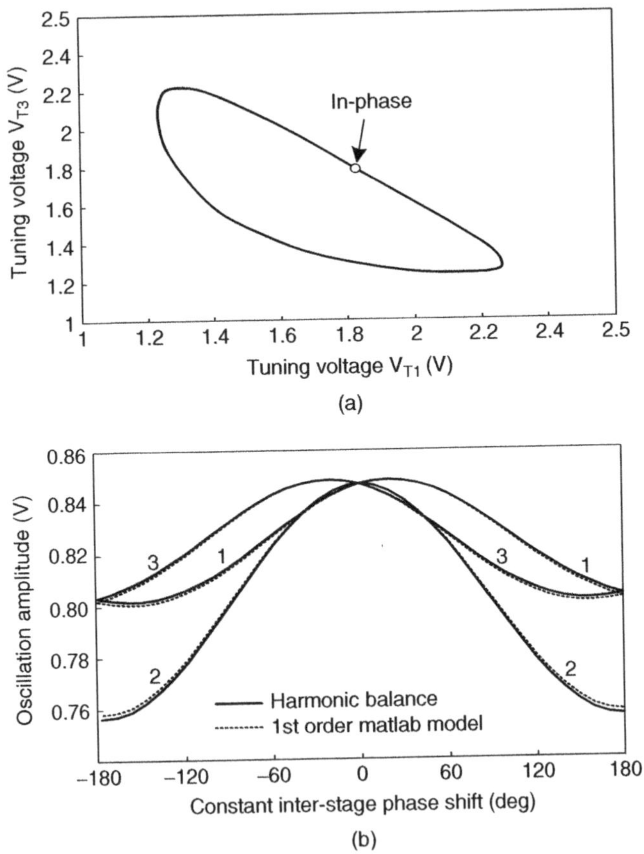

FIGURE 10.18 Solution of the coupled-oscillator system of Fig. 10.17 versus an imposed constant interstage phase shift: (a) required variation of the tuning voltages V_{T1} and V_{T3} of the two outermost oscillator to change $\Delta\phi$ from -180° to 180°; (b) variation of the first-harmonic amplitude in the three oscillator elements versus the imposed phase shift. The results of the semianalytical formulation are superimposed.

system behavior, the solution with constant $\Delta\phi$ should be the only stable solution. The oscillation modes of coupled-oscillator systems are analyzed exhaustively in Section 10.3.5.

As gathered from the downward inclination of the closed curve, the tuning voltages $(V_{T1}, V_{T3}) \in V_T$ must be varied in an opposite sense to obtain the constant phase shift progression from -180 to 180°, in total agreement with the demonstration (10.50). The in-phase solution is obtained at the point with equal values of the tuning voltages $V_{T1} = V_{T3}$. The evolution of the magnitude of the first-harmonic component of the output voltage of the three oscillator elements is represented in Fig. 10.18b. As can be seen, the amplitudes of these three oscillators are not equal. There is a slight variation in the range 0.74 to 0.84 V. Note that the oscillation amplitudes are more similar for weaker coupling strength, tending to the free-running value as this strength decreases.

The behavior of the middle oscillator, 2, is the same for positive or negative values of $\Delta\phi$; thus, the shape of the oscillation magnitude in 2 is symmetrical with respect to $\Delta\phi = 0$. The outermost oscillators, 1 and 3, have different values except at the in-phase solution $\Delta\phi = 0$, for which the two amplitude curves intersect at the middle point. As gathered from the figure, oscillators 1 and 3 exchange their values for positive and negative $\Delta\phi$, which gives the curves their symmetrical shape. This is understood from the fact that the three oscillator elements are identical. Then the positive or negative value of the interstage phase shift simply makes oscillators on the left- or right-hand side of the symmetry axis exchange their roles.

Note that the coupled-oscillator system with no external injection signal behaves in an autonomous manner. The array oscillation frequency will depend on the parameter values, thus on the tuning voltages that are modified to change the constant interstage phase shift. In Fig. 10.19, the array fundamental frequency is represented versus the phase shift imposed. Note that the deviation with respect

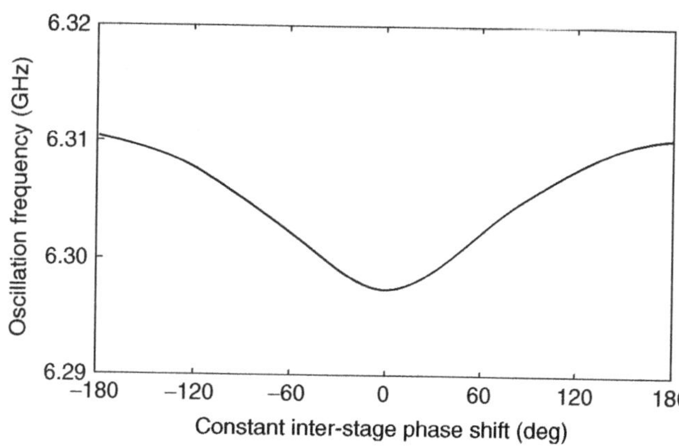

FIGURE 10.19 Variation in the oscillation frequency in a three-element FET-based array versus the imposed constant phase shift between the oscillator elements. Note that the actual system parameters are the tuning voltages of the two outermost elements of the oscillator array.

to the free-running value is undesired and should be minimized in the design. As can be seen, the oscillation frequency varies in symmetrical manner with respect to the value corresponding to the in-phase solution $\Delta\phi = 0^\circ$. This is due to the role exchange between the left- and right-hand-side oscillators for $0^\circ < \Delta\phi < 180^\circ$ and $0^\circ > \Delta\phi > -180^\circ$. The free-running frequency of the array elements is $f_o =$ 6.29 GHz, so, as expected, the maximum deviation is obtained for the tuning voltages that give rise to the maximum phase shift: -180°, 180°.

Note that for weak coupling the constant phase shift solution exists for the entire interval -180° to 180° of constant phase shift values. However, the stable operation range will be smaller. In harmonic balance, the stability analysis can be carried out by applying pole–zero identification to the oscillator chain. Note that the choice of a sensitive circuit node for introduction of the small-signal current source (used for calculation of the closed-loop transfer function) can be demanding in this type of system, due to their symmetry properties.

As already indicated, the solution obtained with constant interstage phase shift $\Delta\phi$ will not be stable for the entire phase interval -180° to 180°. Thus, for the elements of the tuning voltage set V_T at which the constant phase shift solution obtained is unstable, the system will operate in a different regime. When increasing the detuning voltages of the two outermost oscillators, a point is often reached at which the system makes a transition to a periodic solution with nonconstant phase shift. This solution corresponds to one of the periodic oscillation modes coexisting with the constant phase shift solution. These modes are discussed in detail in Section 10.3.3. If the detuning is increased further, the system will evolve to a nonperiodic solution. This is due to the loss of synchronization of the coupled-oscillator system when the free-running frequencies of the outermost oscillators are too different, as a result of the detuning.

For a rough analysis of the variations in system behavior, the envelope transient can be used. As already known, in envelope transient the circuit variables are expressed in a Fourier series with slowly varying harmonic terms: $x(t) = \sum_k X_k(t)e^{jk\omega t}$. Due to the time variation in the harmonic components, this method is more flexible than harmonic balance and can provide both periodic and quasiperiodic solutions. However, for a valid simulation, oscillations of all the oscillator elements must be suitably initialized. This is because the envelope transient may converge to an unstable solution in which one or more of the oscillator elements are not actually oscillating, but responding to the signal injected through the coupling networks in a nonautonomous manner. The oscillation initialization may be carried out with the same auxiliary generators as those used for the harmonic balance analysis of the oscillator array. The auxiliary generators should have the values resulting from a preliminary harmonic balance simulation of the oscillator system. The oscillation frequency ω_{AG} resulting from this preliminary simulation will be the fundamental frequency of the Fourier series expansions $x(t) = \sum_k X_k(t)e^{jk\omega_{AG}t}$. For a clearer explanation, this envelope transient analysis will be illustrated by means of its application to the coupled system of Fig. 10.17.

Only pairs of tuning voltages (V_{T1}, V_{T3}) belonging to the set V_T will be considered. The procedure is the following. The constant $\Delta\phi$ is swept. At each $\Delta\phi$ step the tuning voltage values (V_{T1}, V_{T3}) that provide the constant phase shift solution $\Delta\phi$ are determined with standard harmonic balance, using one auxiliary generator per oscillator element and solving the outer-tier equation (10.52). Once the required pair of tuning voltages (V_{T1}, V_{T3}) has been determined, an envelope transient simulation is carried out. The auxiliary generators with the values resulting from the preceding harmonic balance simulation are connected to the oscillator elements at the initial time instant t_o only. For $t > t_o$, the auxiliary generators are disconnected and the system is allowed to evolve according to its own dynamics. The disconnection can be carried out with a time-varying resistor, fulfilling $R(t_o) = 0$ and $R(t > t_o) = \infty$. The procedure can be summarized as follows: For a sweep of the constant interstage phase shift $\Delta\phi$, at each $\Delta\phi$ step:

1. Calculate the pair (V_{T1}, V_{T3}) that provides the constant phase shift $\Delta\phi$ with harmonic balance, using auxiliary generators to sustain the different individual oscillations.
2. Apply the envelope transient for the pair of tuning voltages (V_{T1}, V_{T3}) obtained, using the harmonic balance solution as a starting point. The auxiliary generators with the values resulting from the preceding analysis are connected to the circuit at the initial time t_o only and disconnected for $t > t_o$.

The advantage of the envelope transient, compared to standard harmonic balance, is its higher flexibility. As already stated, a solution with constant phase shift $\Delta\phi$ becomes unstable when varying the tuning voltages beyond certain values. At a certain pair of tuning voltage values, the system often evolves to a stable solution with nonconstant phase shift. This should be due to an unstable real pole associated with the constant phase shift regime. The reason is that in the transition from one type of solution to the other there is no qualitative variation in the frequency basis: the solution is periodic before and after the transition. That is, there is no generation of a new irrational or subharmonic frequency, which would be due to a pair of unstable complex-conjugate poles. Provided that the fundamental frequencies of the two periodic solutions are close enough, the envelope transient will naturally converge toward the stable solution with nonconstant phase shift. The sampling rate required for the harmonic components $X_k(t)$ will be determined by the frequency difference between these two solutions. On the other hand, for too different tuning voltages V_{T1} and V_{T3}, the free-running oscillation frequencies of the oscillator elements will also be too different and the solution may desynchronize. If this happens, a quasiperiodic solution with two different oscillation frequencies ω and ω' will generally be obtained. In the envelope transient simulation, the magnitudes of the harmonic components $|X_k(t)|$ oscillate at the difference between these two fundamental frequencies, $|\omega - \omega'|$. Note that the harmonic components $X_k(t)$ must be sampled at sufficiently small steps to follow the transient dynamics from a periodic solution to a quasiperiodic solution.

As an example, Fig. 10.20 shows the envelope transient analysis of the FET-based array of Fig. 10.17 versus the imposed constant $\Delta\phi$. The figure shows

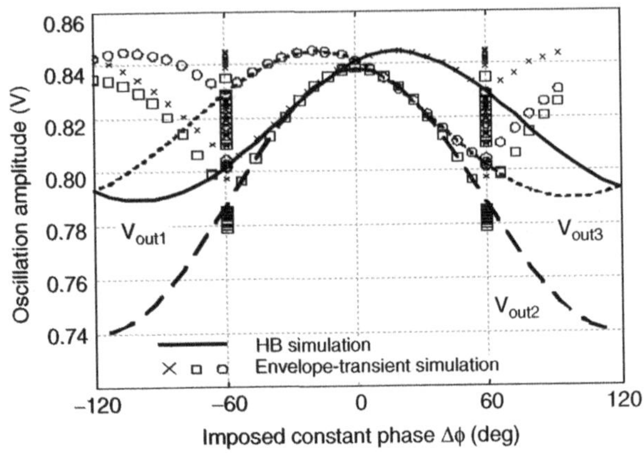

FIGURE 10.20 Envelope transient simulation of the coupled array of three FET-based oscillators. The variation in the magnitude of the first-harmonic component of the output voltage of the three oscillator elements $|V_{1,j}(t)|$, with $j = 1, 2, 3$, has been represented versus the imposed phase shift. The results of the harmonic balance simulation are superimposed for comparison.

the evolution of the first-harmonic amplitude in the three oscillator elements. Although $\Delta\phi$ is used for this representation, the actual circuit parameters are the tuning voltages V_{T1} and V_{T3} that provide the constant $\Delta\phi$ value indicated on the horizontal axis. The results obtained with harmonic balance and with envelope transient are both represented. Six harmonic components have been used in both types of analysis. As already indicated, the harmonic balance solution is used as an initial value for each envelope transient simulation. For the envelope transient, the amplitude of the first-harmonic component of the output voltage of each oscillator element $|V_{1,j}(t)|$, with $j = 1,2,3$ has been represented versus the imposed phase shift $\Delta\phi$. Provided that the integration step is selected accurately, constant values of the steady-state amplitudes $|V_{1,j}(t)|$ indicate a periodic solution. If $|V_{1,j}(t)|$ oscillates, the steady-state solution is quasiperiodic, and when projecting this magnitude in a representation of the type shown in Fig. 10.20, a point distribution is obtained at the particular value of the phase shift imposed.

As shown in Fig. 10.20, the harmonic balance and envelope transient simulations provide the same solution in the phase range -60° to 60°, which indicates that the constant phase shift solution is stable in this interval of imposed interstage phase shift. However, in the intervals $(-180^\circ, -60^\circ)$ and $(60^\circ, 180^\circ)$, the envelope transient converges to a different periodic solution, with nonconstant phase shift. Figure 10.20 provides no clue about the mechanism responsible for the loss of stability of the solution with constant interstage phase shift. The difficulty in understanding the instability mechanism is due primarily to simultaneous variation of the two parameters V_{T1} and V_{T3}. Note that, so far, all the bifurcation analysis presented in the book have been carried out versus variations in one single parameter. Simultaneous variation of the two tuning voltages in the oscillator array makes interpretation of the solution qualitative changes difficult.

10.3.3 Semianalytical Formulation

The harmonic balance method allows very accurate simulation of coupled-oscillator arrays, with complete models of their elements and a certain number of harmonic terms. However, the coupled-oscillator systems used for beam steering generally contain a high number of oscillator elements, so the harmonic balance analysis may be unaffordable, due to its high computational cost. The semianalytical formulation presented in this section may be used instead. In this type of formulation, the oscillators are described with one-port admittance models extracted from harmonic balance simulations. These models are obtained linearizing the oscillator elements about their free-running solution. Thus, the technique is applicable only for relatively weak coupling, that is, for relatively small values of the current injected through the coupling networks. However, this case is the most common in practical applications, since it provides smaller deviations of the oscillation amplitude between the various array elements and a smaller variation of the oscillation frequency with respect to free-running operation. Here we assume that each oscillator originally exhibits a stable free-running oscillation of the same amplitude and frequency. The extension to different oscillator designs or operation conditions is straightforward. The basic assumption is that when coupled to the other identical oscillators, the amplitude and frequency of each oscillator element undergoes small variations only, such that $V_o \rightarrow V_o + \Delta V_k$, $\omega_o \rightarrow \omega_o + \Delta\omega$, $k = 1$ to N. A detailed explanation of the semianalytical formulation is provided below.

Let any of the oscillator elements that compose the coupled array be considered. The oscillator is initially separated from the array and operates in free-running conditions. As shown in previous chapters, the total current-to-voltage ratio of a free-running oscillator is equal to zero at any circuit node. In particular, it will be zero at the node chosen to connect the oscillator to the coupling networks. Thus, prior to their insertion into the array, all the oscillator elements will fulfill the condition $Y = 0$ at the node selected. When using harmonic balance with an auxiliary generator connected to this node, the equation $Y = 0$ is equivalent to the nonperturbation condition $Y_{AG} = 0$. Then, the auxiliary generator amplitude A_{AGo} will agree with the voltage amplitude at this particular node, V_o. Because all the array elements are identical, all of them will fulfill the same equation $Y_{AG}(A_{AGo}, \omega_o) = 0$, with the same values, A_{AGo} and ω_o.

Introduction of the various oscillators into the coupled system will give rise to variation in the amplitude and phase of each oscillator (see Fig. 10.18) as well as the oscillation frequency (see Fig. 10.19). This is due to the current injection through the coupling networks. As a result, the admittance functions associated with each oscillator element will no longer be equal to zero, due to the required fulfillment of Kirchhoff's laws. For small current injected through the coupling networks, the admittance function of each oscillator can be expanded in a Taylor series about the free-running solution V_o, ω_o. As shown in Chapter 5, the use of auxiliary generators allows us to consider an absolute dependence of the admittance function Y on the voltage amplitude at the connection node V and frequency ω. For identical oscillators, the voltage and frequency derivatives of the corresponding admittance functions $\partial Y_o/\partial V$ and $\partial Y_o/\partial\omega$ will also be identical. The two outermost oscillators are tuned with the bias voltages V_{T1} and V_{TN} to modify the constant interstage phase shift. Linear behavior with respect to these voltages will also be assumed. Thus, the two outermost oscillators will be modeled with the three derivatives $\partial Y_o/\partial V$, $\partial Y_o/\partial\omega$, and $\partial Y_o/\partial V_T$. The innermost oscillators will be modeled with only two derivatives: $\partial Y_o/\partial V$ and $\partial Y_o/\partial\omega$.

The foregoing admittance derivatives about the free-running solution constitute the input of the semianalytical formulation. As in other cases, these derivatives are calculated through finite differences in harmonic balance using an auxiliary generator. In the free-running regime, this auxiliary generator fulfills $Y_{AG}(V_o, \omega_o) = 0$. The derivative $\partial Y_{AG}/\partial A_{AGo}$ is calculated through increments in A_{AG} about $A_{AGo} = V_o$ while keeping the auxiliary generator frequency at the free-running value $\omega_{AGo} = \omega_o$. In turn, the derivative $\partial Y_{AG}/\partial\omega_{AGo}$ is calculated through increments in ω_{AG} about $\omega_{AGo} = \omega_o$ while keeping the auxiliary generator amplitude at the free-running value $A_{AGo} = V_o$. The derivative with respect to the tuning voltage V_T that is required in the two outermost oscillators is calculated by keeping the auxiliary generator amplitude and frequency at the free-running values A_{AGo} and ω_{AGo} and considering increments in the tuning voltage V_T.

Once the oscillators are coupled, it is assumed that the coupled array evolves to a periodic steady-state solution. The equation system describing the array behavior is derived from an application of Kirchhoff's laws to the array nodes (Fig. 10.17). For each oscillator element $k = 1$ to N, the product of the perturbed admittance function by the voltage (at the connection node) is equal to the current injected into these nodes from the coupling networks:

$$
\begin{aligned}
&\left(\frac{\partial Y_{AG}}{\partial V_o}(V_1 - V_o) + \frac{\partial Y_{AG}}{\partial\omega_o}(\omega_s - \omega_o) + \frac{\partial Y_{AG}}{\partial V_{To}}(V_{T1} - V_{To})\right) V_1 \\
&\qquad = -y_{c11} V_1 - y_{c12} V_2 e^{j\phi_2} \\
&\vdots \\
&\left(\frac{\partial Y_{AG}}{\partial V_o}(V_k - V_o) + \frac{\partial Y_{AG}}{\partial\omega_o}(\omega_s - \omega_o)\right) V_k e^{j\phi_k} \\
&\qquad = -y_{ck(k-1)} V_{k-1} e^{j\phi_{k-1}} - y_{ck} V_k e^{j\phi_k} - y_{ck(k+1)} V_{k+1} e^{j\phi_{k+1}} \\
&\vdots \\
&\left(\frac{\partial Y_{AG}}{\partial V_o}(V_N - V_o) + \frac{\partial Y_{AG}}{\partial\omega_o}(\omega_s - \omega_o) + \frac{\partial Y_{AG}}{\partial V_{To}}(V_{TN} - V_{To})\right) V_N e^{j\phi_N} \\
&\qquad = -y_{cN(N-1)} V_{N-1} e^{j\phi_{N-1}} - y_{cNN} V_N e^{j\phi_N}
\end{aligned}
\tag{10.54}
$$

where the y_{cjk} are the elements of the admittance matrix $[Y_c]$ of the coupling network, and $V_k e^{j\phi_k}$, with $k = 1$ to N, are the voltage phasors at the oscillator connection nodes. Note that the derivatives with respect to the amplitude and frequency, calculated in free-running conditions at these same nodes, are assumed identical for all the array elements. The derivatives of the admittance function with respect to the tuning voltages are only included in the equation corresponding to the two outermost oscillators [6].

In system (10.54) it is assumed that all the oscillator elements operate in a synchronized manner at the same oscillation frequency ω_s. Note that from the point of view of an external observer, the entire system behaves in an autonomous manner. Thus, the frequency ω_s is an unknown to be determined. The change in the frequency value with respect to free-running conditions gives rise to different phase values at the oscillator elements $\phi_1 = 0, \phi_2, \ldots, \phi_N$, where the phase of the first oscillator has arbitrarily been set to zero, due to the system autonomy. For each pair of values of the tuning voltages V_{T1} and V_{T2}, the unknowns of system (10.54) will be $V_1, V_2, \ldots, V_N, \phi_2, \ldots, \phi_N, \omega_s$. It is a well-balanced system of $2N$ real equations in $2N$ real unknowns. For more clarity and compactness, it is possible to reformulate the system in matrix form:

$$\overline{H}(\overline{V}^s, \overline{\phi}, \omega_s) = \left(\begin{array}{cccc|c} Y_V & 0 & \cdots & 0 & Y_\omega V_1^s e^{j\phi_1} \\ 0 & Y_V & \cdots & \vdots & Y_\omega V_2^s e^{j\phi_2} \\ \vdots & & \ddots & 0 & \vdots \\ 0 & \cdots & 0 & Y_V & Y_\omega V_N^s e^{j\phi_N} \end{array}\right) \begin{pmatrix} V_1^s e^{j\phi_1} & (V_1^s - V_1^0) \\ & \vdots \\ V_N^s e^{j\phi_N} & (V_N^s - V_N^0) \end{pmatrix}$$

$$+ \begin{pmatrix} Y_{V_T^1} & 0 \\ 0 & 0 \\ \vdots & \vdots \\ 0 & 0 \\ 0 & Y_{V_T^1} \end{pmatrix} \begin{pmatrix} V_1^s e^{j\phi_1} & \Delta V_{T1} \\ V_N^s e^{j\phi_N} & \Delta V_{TN} \end{pmatrix} + \begin{pmatrix} Y_{C11}(\omega_s) \cdots Y_{CIN}(\omega_s) \\ \vdots \qquad \vdots \\ Y_{CN1}(\omega_s) \cdots Y_{CNN}(\omega_s) \end{pmatrix}$$

$$\begin{pmatrix} V_1^s e^{j\phi_1} \\ V_2^s e^{j\phi_2} \\ \vdots \\ V_N^s e^{j\phi_N} \end{pmatrix} = \overline{0} \tag{10.55}$$

Note that the matrix $[Y_c]$, associated with the coupling network, will have many zero elements, in agreement with (10.54), as the array behaves under nearest-neighbor coupling only. System (10.55) is a nonlinear system of second order in the amplitudes V_k, which is solved numerically through an error minimization technique such as Newton–Raphson. This resolution is performed externally to the harmonic balance simulator using in-house software. Note that the complexity of the second-order system obtained is considerably lower than that of the original multivariable harmonic balance system. The admittance matrix $[Y_c]$ [considered in (10.55)] can be calculated in the in-house software from passive models of the various network elements, so the matrix is evaluated locally at each frequency value. It is also possible to expand its elements in a Taylor series around ω_0. In the common case of using resistively ended transmission-line coupling networks, which are broadband, the matrix $[Y_c]$ does not appreciably vary over the range of the array frequency values. Due to local calculation of the matrix $[Y_c]$, the semianalytical formulation can also be used for efficient selection of the optimum coupling networks from the characteristics of the oscillator elements [21].

As an example, the semianalytical approach has been used to analyze the same three-element coupled oscillator of Fig. 10.17. As with harmonic balance, a constant interstage phase shift has been imposed between the oscillator elements $\phi_{n+1} - \phi_n = \Delta\phi, n = 1$ to $N - 1$, proceeding to solve the system in terms of $V_1, \ldots, V_N, V_{T1}, V_{TN}, \omega_s$ at each $\Delta\phi$ step. Note that as in the case of injection-locked oscillators, no infinite slope points are found versus $\Delta\phi$, which makes it possible to obtain solution curves in a straightforward manner, even if the solution exhibits turning points versus the actual circuit parameters V_{T1} and V_{TN}. The results are superimposed in Fig. 10.18b, where they can be compared with results obtained using harmonic balance simulations.

10.3.4 Determination of Coexisting Solutions

As has been pointed out, in a coupled array with N oscillator elements, two different synchronized solutions may be expected for each pair of oscillator elements, which gives rise to a total of 2^{N-1} different solutions. The multiple solutions [6,9] can also be understood as the result of the cos and sin terms contained in the exponential functions of (10.54). Because two different angles provide the same cos or sin values, 2^{N-1} steady-state solutions will generally coexist for each pair of tuning voltages V_{T1} and V_{TN}. If the pair of values V_{T1} and V_{TN} belongs to the set V_T containing the pairs of tuning voltages that provide constant interstage phase shift, only one of the 2^{N-1} coexisting solutions will have this constant phase shift. For the $2^{N-1}-1$ remaining solutions, the interstage phase shift will not be constant. As has been shown in Section 10.3.2, the loss of stability of the constant phase shift solution of the coupled-oscillator array is often associated with a transition to a stable periodic solution with nonconstant phase shift. For the reasons mentioned, these nonconstant phase shift solutions are naturally generated in the system. Understanding the mechanism that gives rise to the transition to a nonconstant phase shift solution will require determination of all the periodic solutions that coexist with the constant phase shift solution. This is affordable with the semianalytical formulation due to its low computational cost.

As shown in Section 10.3.3, the continuous set of pairs V_{T1} and V_{TN} providing the progressive constant phase shift $\Delta\phi(V_{T1}, V_{TN})$ is obtained by imposing $\Delta\phi_{n+1,n} = \phi_{n+1} - \phi_n = \Delta\phi$, with $n = 1$ to $N-1$, and sweeping $\Delta\phi$ in system (10.54). For each $\Delta\phi$ value, this system is solved in terms of the vector containing the oscillation amplitudes $\overline{V}$, the frequency ω, and the tuning voltages V_{T1} and V_{TN}. The resulting set of V_{T1} and V_{TN} values had been called V_T and, as already seen, constitutes a closed curve in the plane (V_{T1}, V_{TN}), due to the periodicity of (10.54) in the constant $\Delta\phi$ (Fig. 10.18a). The solution with constant phase shift obtained for each pair of tuning voltages $(V_{T1}, V_{T3}) \in V_T$ is called S_1 here. In Fig. 10.21, the solution S_1 corresponding to the FET-based three-element array has been represented in terms of voltage amplitude in the first oscillator element versus

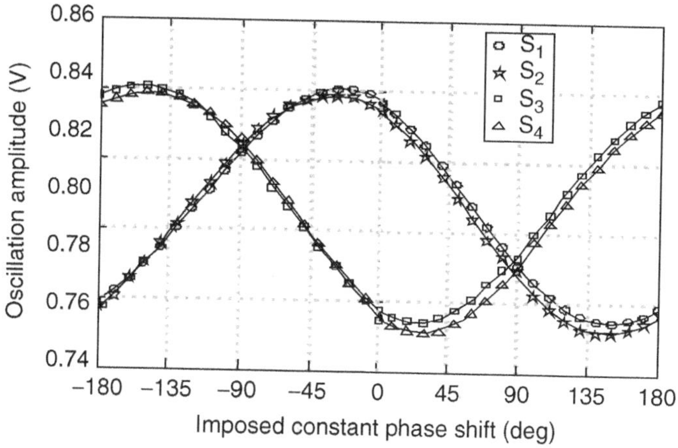

FIGURE 10.21 Variation in the four coexisting solutions of a FET-based system with three elements versus the imposed constant $\Delta\phi(V_{T1}, V_{T3})$.

the imposed constant phase shift $\Delta\phi$. However, as shown in the figure, coexisting with S_1 are three more solutions ($2^{N-1} - 1 = 3$) with nonconstant phase shift. For compactness the phase shift $-180^\circ \le \Delta\phi \le 180^\circ$ is used for the horizontal axis. However, the actual system parameters are the tuning voltages $(V_{T1}, V_{T3}) \in V_T$, which give rise to each constant $\Delta\phi$ in the solution S_1. Thus, the parameters are the tuning voltages that fulfill $\Delta\phi_{S1}(V_{T1}, V_{T3}) = \Delta\phi$. For each pair of tuning voltage values V_{T1} and V_{T3}, the three solutions with nonconstant phase shift that coexist with S_1 will be called S_2, S_3, and S_4. These solutions are difficult to obtain with either harmonic balance or the reduced-order nonlinear system (10.54). This is because each of them must be initialized separately to obtain Newton–Raphson convergence. A systematic initialization technique is described below [22].

Initially, the pair V_{T1o}, V_{TNo}, which provide in-phase operation $\Delta\phi = 0^\circ$ in S_1, is considered. Then 2^{N-1} independent Newton–Raphson resolutions of (10.54) in terms of $(\overline{V}, \overline{\phi}, \omega)$ are carried out for the same pair of tuning voltages (V_{T1o}, V_{TNo}) using 2^{N-1} different sets of initial conditions. These conditions, imposed to the phase values of the oscillator elements, are given by all the 2^{N-1} binary combinations of 0 and π radians:

$$k = \begin{cases} 1 & \overline{\phi}^{1} = 0, \ldots, 0, 0 \\ 2 & \overline{\phi}^{2} = 0, \ldots, 0, \pi, \ldots \\ 2^{N-1} & \overline{\phi}^{2^{N-1}} = \pi, \ldots, \pi, \pi \end{cases} \tag{10.56}$$

Initial conditions (10.56) constitute an extension of the simpler case of an injection-locked oscillator (10.43). This is explained as follows. Assuming for simplicity that the total admittance function of this oscillator has an amplitude and frequency dependence of the form $Y_T(V,\omega) = Y_T^r(V) + jY_T^i(\omega)$, the relationship between the synchronization frequency and the phase shift is given by

$$\omega_s - \omega_o = \frac{I_g}{V_o}\frac{\sin\phi_s}{\partial Y_{\text{im}}/\partial\omega|_o} \tag{10.57}$$

The in-phase solution $\phi_s = 0$ of (10.57) coexists at the same frequency $\omega_s = \omega_o$ with the solution with phase shift $\phi_s = \pi$. The initial values considered in (10.56) are an extension of this simple case to a coupled-oscillator system. They should be close to the actual phase values of the 2^{N-1} solutions that coexist with the in-phase solution.

For compactness, the 2^{N-1} periodic solutions obtained for the tuning voltages $V_{T1}, V_{TN} \in V_T$ will be denoted $S_k(\Delta\phi)$. For each $\Delta\phi$ value, this notation represents solutions that are obtained for the tuning voltage values V_{T1} and V_{TN} that give rise to the constant interstage phase shift $\Delta\phi$ in S_1. It must be emphasized that there is no actual dependence of S_k, $k = 2$ to N, on $\Delta\phi$. Starting from $\Delta\phi = 0^\circ$, a continuation procedure in two stages is followed:

1. At each step $m+1$, small increment of the constant interstage phase shift $\Delta\phi$ is considered: $\Delta\phi^{m+1} = \Delta\phi^m + \delta\Delta\phi$, with $\Delta\phi^1 = 0^\circ$. A constant phase shift solution is obtained by imposing $\Delta\phi_{n+1,n} = \phi_{n+1} - \phi_n = \Delta\phi$, with $n = 1$ to $N-1$, in system (10.55). This system is solved in terms of $\overline{V}$, ω, V_{T1}, and V_{TN} and provides the tuning voltages V_{T1}^{m+1} and V_{TN}^{m+1} required to obtain a constant interstage phase shift $\Delta\phi^{m+1}$ in S_1.
2. The other $2^{N-1} - 1$ solutions coexisting with S_1 for the tuning voltage values V_{T1}^{m+1}, V_{TN}^{m+1}, obtained at stage 1, are independently initialized taking the results $S_k(\Delta\phi_m)$ of the previous $2^{N-1} - 1$ independent Newton–Raphson convergences as initial value. A Newton–Raphson resolution is independently applied for each of them. Remember that the initial phase values used for $m+1$ are given by (10.56).

The procedure described was used to obtain the three additional periodic solutions S_2, S_3, and S_4 of the coupled system of three FET-based oscillators shown in Fig. 10.21. The maximum relative error in the oscillator amplitudes with respect to harmonic balance simulations is about 1%. Due to the virtual impossibility of obtaining multiple solutions with harmonic balance, solutions of the systematic method were provided as initial conditions for the harmonic-balance simulations.

This simultaneous modification of the two parameters V_{T1} and V_{TN} makes an understanding of the circuit behavior quite difficult, so the bifurcation that governs the transition from the constant phase-shift solution to a solution with nonconstant interstage phase shift (Fig. 10.20) has not yet been determined. To gain more insight into the circuit evolution, only one parameter V_{TN} will be varied continuously while considering a variety of constant values of the other parameter, V_{T1}. Note that the pairs of tuning voltages(V_{T1}, V_{TN}) considered are no longer restricted to the set V_T, providing a constant phase shift in S_1. Each solution curve versus V_{TN} for constant V_{T1} is obtained independently, using different initial conditions, through a sweep in one of the oscillator phases ϕ_n (remember that $\phi_1 = 0$), solving (10.55) for the remainder of the system variables. When ϕ_2 is swept, these variables will be $V_1, V_2, \ldots, V_N, \phi_3, \ldots, \phi_N, \omega_s, V_{T1}$.

The technique described has been applied to a FET-based system with three oscillator elements. In Fig. 10.22 the oscillation amplitude V_2 has been traced versus V_{T3} for three different values of V_{T1}. At each V_{T1}, two closed curves, C_1 and C_2, are obtained versus the other parameter, V_{T3}. The turning points of these two curves are located at slightly different values of V_{T3}. Therefore, four solutions for each V_{T1} value will coexist in most of the V_{T3} interval. For each V_{T3} value, the four coexisting solution are located in the upper and lower halves of the two closed curves C_1 and C_2 and correspond to the intersections of these closed curves with a vertical line at a particular V_{T3} value. However, due to the fact that the turning points do not agree exactly, near any of the two edges of the V_{T3} interval there will be only two solutions. It must be noted that for each V_{T1} value, periodic solutions exist inside the intervals delimited by the turning points of the two closed curves. Beyond these points, only desynchronized quasiperiodic solutions are possible.

Consider, initially, the constant value $V_{T1} = 0.4$ V. This value is included in the V_{T1} interval belonging to the set V_T. Therefore, there must be at least one V_{T3} value in the solution curves obtained, providing a constant phase shift. This value is $V_{T3} = 0.6$ V, and the corresponding interstage phase shift is $\Delta\phi(V_{T1} = 0.4, V_{T3} = 0.6) = 50°$. This solution is located in the upper section of C_1 and coexists for the same pair of tuning voltages, V_{T1} and V_{T3}, with a second solution in the lower section of C_1 and two more solutions in the upper and lower sections of C_2. These three additional solutions are determined from the intersection of the vertical dashed line with the two closed curves. For these other three solutions, the interstage phase shift is not constant.

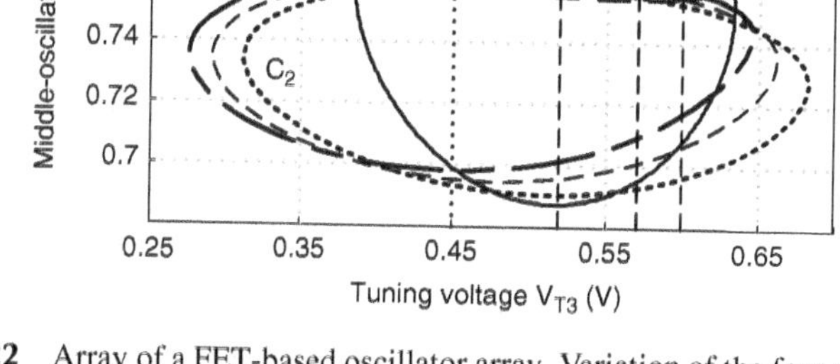

FIGURE 10.22 Array of a FET-based oscillator array. Variation of the four coexisting solutions versus V_{T3} for different values of V_{T1} : $V_{T1} = 0.4$ V (solid dashed line), $V_{T1} = 0.43$ V (dashed line), and $V_{T1} = 0.47$ V (dotted line). The set of constant phase shift solutions is shown by the solid line.

If a slightly different V_{T1} value is now considered, the two closed curves will vary slightly, shifting versus V_{T3} or changing their inclination. This can be seen in Fig. 10.22, where two other V_{T1} values have been considered: $V_{T1} = 0.43$ V and $V_{T1} = 0.47$ V. Increasing V_{T1}, the two closed curves, C_1 and C_2, shift leftward. In curves C_1 and C_2 obtained for $V_{T1} = 0.43$ V there is a V_{T3} value providing a constant interstage phase shift solution. This value is $V_{T3} = 0.6$ V, and the constant phase shift is $\Delta\phi(V_{T1} = 0.43, V_{T3} = 0.56) = 30^\circ$. This solution is located in the upper half of curve C_1, corresponding to $V_{T1} = 0.43$ V. When performing the same analysis for $V_{T1} = 0.47$ V, the point providing a constant phase shift is $\Delta\phi(V_{T1} = 0.47, V_{T3} = 0.5) = 10^\circ$, so the constant $\Delta\phi$ point shifts leftward again in the new C_1 curve. The set of solutions with progressive constant phase $\Delta\phi = 0$ to 2π is given by the envelope of the shifted points.

An array of $N = 4$ coupled FET-based oscillators has also been considered. The two tuning voltages, corresponding to the two outermost oscillators, are V_{T1} and V_{T4}, and the number of coexisting solutions is $2^{N-1} = 8$. In Fig. 10.23, these solutions are traced in terms of the third-oscillator amplitude V_3 versus V_{T3} for constant $V_{T1} = 0.5$. The eight coexisting solutions are arranged in four closed curves. The constant phase shift solution with $\Delta\phi = 0^\circ$ is located in the upper half of C_1. The seven additional solutions with nonconstant phase shift are determined from the intersections of the vertical dashed line with the four closed curves. The results of the harmonic balance simulations with nine harmonic components of the four-element chain are shown for comparison.

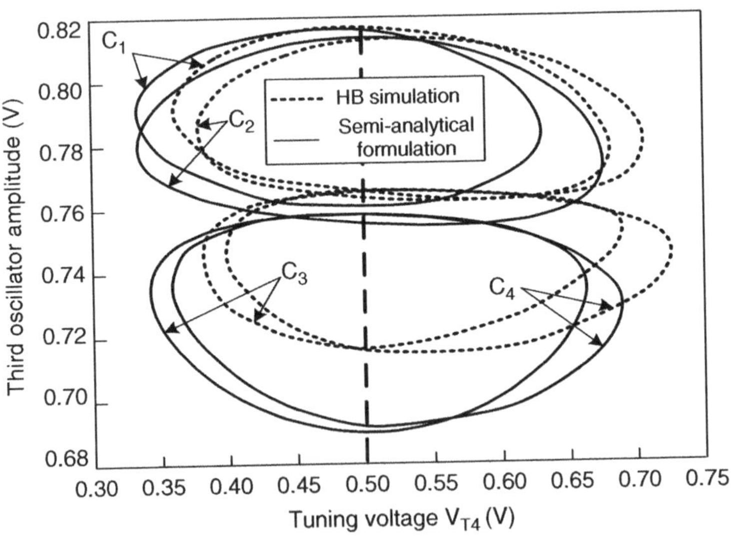

FIGURE 10.23 Array of four FET-based oscillators. Variation of the eight coexisting solutions versus V_{T4} for constant $V_{T1} = 0.5$. The harmonic balance simulations are superimposed for comparison.

10.3.5 Stability Analysis

When simulating any circuit or system in the frequency-domain, the stability analysis of the solution obtained is essential to verify its physical existence. This is even more so when several steady-state solutions coexist, as in the case of the coupled-oscillator systems studied here. Note that the frequency-domain simulation may converge, in principle, to any of these coexisting solutions, regardless of its stability properties. Due to the high number of oscillator elements usually contained in the coupled-oscillator systems, the application of harmonic balance techniques such as pole–zero identification to each of the coexisting solutions would be cumbersome. Instead, the stability analysis technique will be derived from the semianalytical formulation in (10.55). As we already know, this system has been obtained by modeling each oscillator in terms of the amplitude and frequency derivatives of the admittance function Y_V and Y_ω, calculated at the free-running solution. This free-running solution must, of course, be stable. In terms of the admittance derivatives, the approximation used inherently restricts the instability detection to small perturbation frequencies. However, stable oscillators in uncoupled conditions will in coupled operation typically undergo turning points and Hopf bifurcations with small perturbation frequency, due to a loss of synchronization. As shown in Chapter 3, Section 3.3.2 and Section 3.3.3, this perturbation frequency will be in the order of the difference between the unlocked oscillation frequencies.

To derive the formulation for the stability analysis, the variable increments $\delta V_1(t), \ldots, \delta V_N(t)$, $\delta\phi_1(t), \ldots, \delta\phi_N(t)$ are considered in (10.55). Using a development similar to the one in (1.18)–(1.20), it is possible to write

$$\begin{aligned} &H_\omega^1\left(\delta\dot{\phi}_1(t) - j\frac{\delta\dot{V}_1(t)}{V_1^s}\right) + \sum_{i=1}^{N}(H_{V_i}^1\delta V_i(t) + H_{\phi_i}^1\delta\phi_i(t)) = 0 \\ &\cdots \\ &H_\omega^N\left(\delta\dot{\phi}_N(t) - j\frac{\delta\dot{V}_N(t)}{V_N^s}\right) + \sum_{i=1}^{N}(H_{V_i}^N\delta V_i(t) + H_{\phi_i}^N\delta\phi_i(t)) = 0 \end{aligned} \tag{10.58}$$

System (10.58) is a linear time invariant system. It can easily be split into real and imaginary parts, which provides an equation system of the form

$$\begin{bmatrix} \delta\dot{V}_1 \\ \vdots \\ \delta\dot{V}_N \\ \delta\dot{\phi}_1 \\ \vdots \\ \delta\dot{\phi}_N \end{bmatrix} = [JH] \begin{bmatrix} \delta V_1 \\ \vdots \\ \delta V_N \\ \delta\phi_1 \\ \vdots \\ \delta\phi_N \end{bmatrix} \tag{10.59}$$

where $[JH]$ is a real-element matrix composed of the constant derivatives of the real and imaginary parts of the error functions H with respect to the amplitude and phase of the oscillator voltages. The stability or instability of any solution (with either constant or nonconstant interstage phase shift) will be determined by the eigenvalues of the constant Jacobian matrix $[JH]$ calculated at this steady-state solution.

To illustrate the technique described above will be applied along curves C_1 and C_2 in Fig. 10.22 obtained for $V_{T1} = 0.4$ V and containing the constant phase shift solution point $\Delta\phi = 50^\circ$. Each solution curve is obtained through a phase sweep (in either ϕ_2 or ϕ_3) from a different initial point, as indicated in the preceding section. Assuming ϕ_2 is swept, at each phase step the steady-state solution of (10.55) is determined, solving for the variables $V_1, V_2, V_3, \phi_3, \omega_s, V_{T3}$. Then Jacobian matrix $[JH]$ in the linear system (10.59) is calculated at this solution and the corresponding eigenvalues are determined [22]. For a three-element oscillator system, the Jacobian matrix is 6×6 in dimensions. Real and complex eigenvalues are obtained, but the dominant eigenvalues are real for all points in the two solution curves.

In Fig. 10.24, one and two real eigenvalues are traced for C_1 and C_2, respectively. The entire C_2 is unstable, as it always contains a positive eigenvalue describing the closed curve comprised by the two sections C_2^L and C_2^U. The second eigenvalue, which also describes a closed curve, passes through zero at the two turning points of Fig. 10.22, in agreement with the Jacobian matrix singularity at these points. On the other hand, the solution curve C_1 has the dominant eigenvalue shown in Fig. 10.24. The remaining eigenvalues associated with this closed curve have a negative real part, so they do not have any influence on the solution stability. The eigenvalue represented describes a closed curve that passes through zero at the two turning points of C_1. This eigenvalue is positive in the lower section of

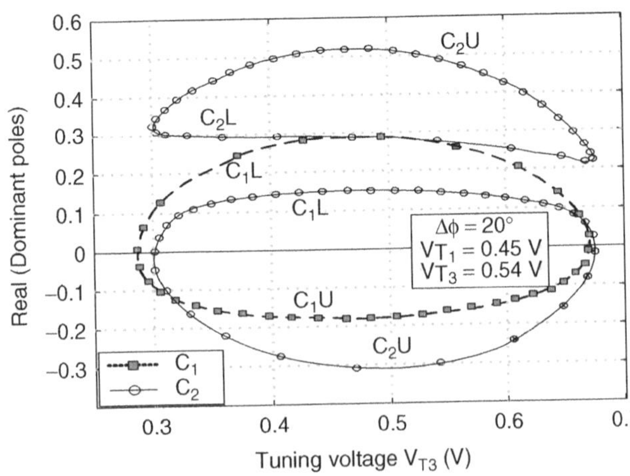

FIGURE 10.24 Stability analysis of two solution curves obtained versus V_{T3} for constant $V_{T1} = 0.45$ V, represented in Fig. 10.22. The upper half of the closed curve C_1 is the only stable section of the two curves. U and L indicate upper and lower sections, respectively.

C_1, so this section is unstable. The eigenvalue is negative in the upper section of C_1, so this section is stable.

From the analysis above, the only stable section of the entire solution diagram of Fig. 10.22 corresponds to the upper section of C_1. The two turning points of C_1 actually correspond to local–global (mode-locking) bifurcations at which the coupled system loses synchronization. Outside the V_{T3} range delimited by these two turning points, the system operates in a quasiperiodic regime, due to the excessive difference between the free-running frequencies of the two outermost oscillators.

Next, simultaneous variation of V_{T1} and V_{T3} to obtain a progressive constant phase shift $\Delta\phi$ will be considered. When increasing $\Delta\phi$, the solution with constant $\Delta\phi$ shifts along the upper section of the various C_1 curves obtained as V_{T3} is varied (Fig. 10.22). For a particular pair of tuning voltages $(V_{T1,M}, V_{T3,M})$, the solution with constant phase shift $\Delta\phi(V_{T1,M}, V_{T3,M}) = \Delta\phi_{\max}$ will be located exactly at the right turning point of C_1. This situation will determine the maximum value of constant interstage phase shift $\Delta\phi$, with stable behavior. After reaching the turning point, the solution with constant $\Delta\phi > \Delta\phi_{\max}$ will be located on the lower section of C_1. This section is entirely unstable, so the constant $\Delta\phi$ solution becomes unstable at the pair of tuning voltages $(V_{T1,M}, V_{T3,M})$ at which this turning point is obtained. Note that it is mathematically possible to obtain solutions with larger $\Delta\phi$, but these solutions will not be stable, as they will lie on the lower section of C_1. Therefore, the right-hand turning point of C_1 delimits the maximum constant phase shift $\Delta\phi_{\max}$ with stable operation. For pairs V_{T1} and V_{T3} providing constant $\Delta\phi$ beyond this maximum stable value, the system will evolve to a stable solution in the upper section of the newly generated C_1 curves, with nonconstant phase shift. The same thing happens when reducing the constant interstage phase shift $\Delta\phi$. The point with constant $\Delta\phi$ will shift leftward along the upper half of C_1. For a particular pair of tuning voltages $(V_{T1,m}, V_{T3,m})$, the solution with constant phase shift $\Delta\phi(V_{T1,m}, V_{T3,m}) = \Delta\phi_{\min}$ will be located exactly at the left turning point of C_1. This situation will provide the minimum stable phase value of the constant phase shift solution. After reaching the left-hand turning point of this curve, the solution with constant phase shift will lie on the lower section of C_1. From this minimum stable value $\Delta\phi_{\min}$, the solutions observed physically for the pairs V_{T1} and V_{T3} will correspond to points in the upper section of the newly generated C_1 curves, with nonconstant phase shift.

Because there is only one stable section in the two closed solution curves of Fig. 10.22, a stable constant phase shift solution which always belongs to the upper section of C_1 never coexists with another stable periodic mode. Note that the transition from a solution with constant interstage phase shift to a solution with nonconstant shift is continuous, in agreement with Fig. 10.20. This is due to the continuous variation of the closed curves versus V_{T1} and V_{T3}. After passing though the turning point, the next constant phase shift solution is located immediately below this turning point, whereas the stable solution with nonconstant phase shift (to which the system evolves) is located immediately above this point.

From the preceding analysis, the two solutions at the edges of the stable constant-phase-shift range will fulfill two different conditions. On the one hand, they must, of course, be constant-phase-shift solutions. On the other hand, its associated Jacobian matrix $[JH]$, in the linearized system (10.59), must be singular. Thus, they must fulfill

$$\begin{aligned} \overline{H}(\overline{V}, \Delta\phi, \omega, V_{T1}, V_{T3}) &= \overline{0} \\ \det\left\{JH(\overline{V}, \Delta\phi, \omega, V_{T1}, V_{T3})\right\} &= 0 \end{aligned} \tag{10.60}$$

where $\Delta\phi$ refers to the constant phase shift between the oscillator elements. Note that the condition above does not imply observation of an infinite-slope point in the constant $\Delta\phi$ curve (the solid line in Fig. 10.22), as the two analysis parameters V_{T1} and V_{T3} are being simultaneously varied along this curve. To see this more clearly, note that the increment $\delta\overline{V}$ in the steady-state variables (amplitudes and phases of the oscillator voltages) versus the small simultaneous variation of the two tuning parameters V_{T1} and V_{TN} can be estimated from

$$\frac{\partial\overline{H}}{\partial\overline{V}}\delta\overline{V} + \frac{\partial\overline{H}}{\partial V_{T1}}\delta V_{T1} + \frac{\partial\overline{H}}{\partial V_{TN}}\delta V_{TN} = 0 \tag{10.61}$$

where $\partial\overline{H}/\partial\overline{V}$ is the same Jacobian matrix $[JH]$ as that associated with system (10.59), and δV_{T1} and δV_{TN} are the increments of the tuning voltages. Then the slope of the variables with respect to one of the parameters (e.g., V_{T1}) is calculated as

$$\frac{\delta\overline{V}}{\delta V_{T1}} = [JH]^{-1} \cdot \left[-\frac{\partial\overline{H}}{\partial V_{T1}} - \frac{\partial\overline{H}}{\partial V_{TN}}\frac{\delta V_{TN}}{\delta V_{T1}}\right] \tag{10.62}$$

An analogous equation would be obtained for a variable slope versus V_{TN}. At the turning points of the solution curve versus either V_{T1} or V_{TN}, the variable slope will tend to infinity. However, unlike what happens in one-parameter systems, there may be two different reasons for this situation. One of them is the singularity of the Jacobian matrix $[JH]$, which would imply a qualitative stability change due to the crossing of a real pole through the origin. The second possible reason would be the infinite value of the ratio $\delta V_{TN}/\delta V_{T1}$, which does not imply any special properties of the Jacobian matrix and thus does not give rise to a qualitative variation of the solution stability. Note that $\partial\overline{H}/\partial V_{T1}$ and $\partial\overline{H}/\partial V_{TN}$ are vectors and cannot be responsible for the infinite value of the derivatives $\lim_{\delta V_{T1}\to 0}\delta\overline{V}/\delta V_{T1}$ and $\lim_{\delta V_{TN}\to 0}\delta\overline{V}_i/\delta V_{TN}$. Thus, a turning point of the solution curve versus either V_{T1} or V_{TN} does not necessarily mean a stability change [21].

10.3.6 Phase Noise Analysis

Due to the high number of oscillator elements usually contained in oscillator arrays, it will be convenient to obtain the phase noise spectrum from the same semianalytical formulation (10.55). This will avoid computationally expensive simulations using harmonic balance. Only white noise sources are considered, as the main purpose will be a qualitative understanding of the phase noise behavior of a coupled-oscillator system. As described in Chapter 7 (Section 7.7.2), this noise model can be obtained by fitting the phase noise spectrum of the free-running oscillator in harmonic balance. The circuit, with all its noise sources, is initially analyzed using the carrier modulation and conversion matrix approaches. Then a different circuit, with a single current noise source is analyzed, fitting this source so as to obtain the same phase noise spectrum. In this manner it is possible to obtain an equivalent noise source $N_j(t) = N_{Rj}(t) + jN_{Ij}(t)$, added in parallel to each oscillator output. In the presence of low-amplitude noise sources, the nonlinear functions $\overline{H}$ of the reduced-order system will be expanded in a first-order Taylor series about the steady-state solution being considered. As in stability analysis, the frequency perturbation will be expressed as $\delta\dot{\phi}_k(t) - j[\delta\dot{V}_k(t)/V_k^s]$, with $k = 1$ to N. This provides the system

$$H_\omega^1\left(\delta\dot{\phi}_1(t) - j\frac{\delta\dot{V}_1(t)}{V_1^s}\right) + \sum_{i=1}^{N}(H_{V_i}^1\delta V_i(t) + H_{\phi_i}^1\delta\phi_i(t)) = N_1(t)$$

$$\vdots \tag{10.63}$$

$$H_\omega^N\left(\delta\dot{\phi}_N(t) - j\frac{\delta\dot{V}_N(t)}{V_N^s}\right) + \sum_{i=1}^{N}(H_{V_i}^N\delta V_i(t) + H_{\phi_i}^N\delta\phi_i(t)) = N_N(t)$$

Note that $N_1(t)$ to $N_N(t)$ are the envelopes of the various noise sources about the oscillation frequency. Thus, only white noise about this frequency is considered. To take the flicker noise sources into account, the system (10.54) should be written in terms of both the dc and fundamental-frequency components, which would imply adding N real equations to (10.54) and (10.63).

Splitting equation (10.63) into real and imaginary parts and applying the Fourier transform in the slow time variable associated with noise perturbations, it is possible to write

$$\{j\Omega[I] - [JH]\}\begin{bmatrix}\delta V_1(\Omega)\\ \vdots\\ \delta V_N(\Omega)\\ \delta\phi_1(\Omega)\\ \vdots\\ \delta\phi_N(\Omega)\end{bmatrix} = \begin{bmatrix}N_1^r\\ \vdots\\ N_N^r\\ N_1^i\\ \vdots\\ N_N^i\end{bmatrix} \tag{10.64}$$

The amplitude and phase noise are calculated by solving for the increment vector $[\delta V_1 \cdots \delta V_N, \delta\phi_1 \cdots \delta\phi_N]^T$. Note that even though the matrix $[JH]$ is singular, the matrix $\{j\Omega[I] - [JH]\}$ is not, so it can be inverted to solve for the variable vector. Multiplication by the adjoint vector $[\delta V_1 \cdots \delta V_N, \delta\phi_1 \cdots \delta\phi_N]^*$ will allow us to determine the phase and amplitude noise spectra $|\delta\phi_j(\Omega)|^2$ and $|\delta V_j(\Omega)|^2$ of each oscillator as well as the various autocorrelation terms.

Phase noise analysis based on (10.64) is validated with the results of the harmonic balance simulation using the carrier modulation approach. The two techniques have been applied to the coupled system of three FET-based oscillators. For comparison, the system (10.64) has been solved in two different manners: assuming a constant value of the coupling network admittance matrix and taking its frequency variation into account. The phase noise spectral density has been evaluated versus the imposed constant phase shift $\Delta\phi$ in the stable range of Fig. 10.20. The phase noise variation at 100 kHz offset from the carrier is represented in Fig. 10.25. It has been normalized to the free-running value because of the lack of an accurate model of the active device. Furthermore, the main goal of this analysis is to compare the array phase noise with that of the free-running circuit. As can be seen, there is very good agreement between the results of the semianalytical approach and those obtained with full harmonic balance simulations. In a frequency-independent admittance matrix, the phase noise reduction for the in-phase solution $\Delta\phi = 0$ with respect to the free-running value is 10 log N, with $N = 3$. However, if the frequency variation of the coupling network is taken into account through a first-order derivative, the phase noise decreases more than 10 log N for this in-phase solution. This is in good agreement with harmonic balance simulations through the carrier modulation approach [23] shown in Fig. 10.25. The phase noise increases sharply as the edges of the stable phase shift range are approached, in agreement with the fact that we are approaching a singularity condition in the system (10.64).

To get more insight into the mechanism for phase noise reduction observed in Fig. 10.25, a simplified phase noise analysis will also be presented. In this analysis the amplitude perturbations δV_j taken into account in (10.63) will be neglected. These perturbations are essential for an accurate calculation of the phase noise spectrum, but the complexity of the resulting complete system allows little intuition as to the general phase noise behavior of the oscillator array. For comparison with the analysis of the globally coupled oscillator in Section 10.2.2, the in-phase solution of a four-element coupled-oscillator array is considered. This in-phase solution fulfills

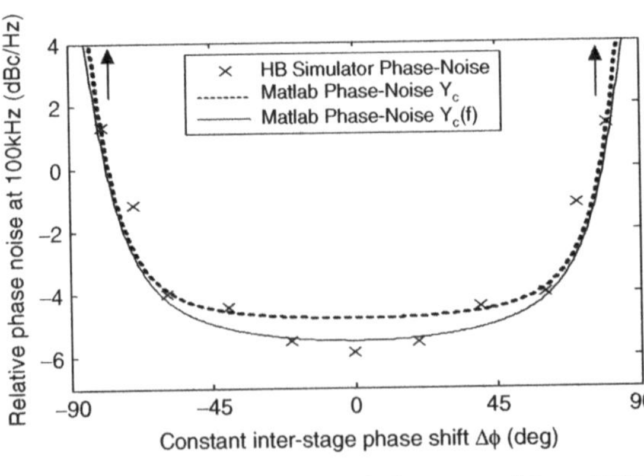

FIGURE 10.25 Phase noise analysis of a coupled system of three FET-based oscillators. The results of the semianalytical approach are compared with harmonic balance simulations based on the carrier modulation approach.

$$\begin{bmatrix} H_{1o} \\ H_{2o} \\ H_{3o} \\ H_{4o} \end{bmatrix} = \begin{bmatrix} Y_s + Y_{o1}(V, \omega_s) & Y_c & 0 & 0 \\ Y_c & Y_s + Y_o(V, \omega_s) & Y_c & 0 \\ 0 & Y_c & Y_s + Y_o(V, \omega_s) & Y_c \\ 0 & 0 & Y_c & Y_s + Y_{o4}(V, \omega_s) \end{bmatrix}$$

$$\times \begin{bmatrix} 1 \\ 1 \\ 1 \\ 1 \end{bmatrix} V_o = 0 \qquad (10.65)$$

where Y_s is the parallel load admittance added by the coupling network to each oscillator circuit and Y_c is the coupling admittance, connecting each oscillator to its nearest neighbors. For simplicity, the same steady-state amplitude V_o is considered in all the oscillator elements. Note that in contrast with (10.26), the admittance functions Y_{o1},Y_o, and Y_{o4} correspond to full oscillator circuits instead of just active subblocks. The functions Y_{o1},Y_o, and Y_{o4} are the perturbed admittance functions of the various oscillator elements, calculated as

$$\begin{aligned} Y_{o1}(V, \omega_s) &= Y_V(V - V_o) + Y_\omega(\omega_s - \omega_o) + Y_{VT}(V_{T1} - V_{To}) \\ Y_o(V, \omega_s) &= Y_V(V - V_o) + Y_\omega(\omega_s - \omega_o) \\ Y_{o4}(V, \omega_s) &= Y_V(V - V_o) + Y_\omega(\omega_s - \omega_o) + Y_{VT}(V_{T4} - V_{To}) \end{aligned} \qquad (10.66)$$

with the subindexes indicating derivatives with respect to the corresponding magnitudes. The noise current sources give rise to small perturbations in equation (10.65). The perturbed system will be obtained by expanding (10.65) in a first-order Taylor series about the steady state given by V, ω_s, $\phi_k = 0\ \forall\ k$. For that we take into account that the first-order Taylor series expansion of $e^{j\delta\phi}$ about $\delta\phi = 0$ is $1 + j\delta\phi$. Then the perturbed equation corresponding to kth oscillator becomes

$$\frac{\partial Y_o}{\partial V}\delta V_k + \frac{\partial Y_o}{\partial \omega}\delta\dot{\phi}_k + jY_{otk}\delta\phi_k + jY_c(\delta\phi_{k-1} + \delta\phi_{k+1}) = \frac{I_{pk}(t)}{V_o} \qquad (10.67)$$

with $\delta\phi_o = \delta\phi_5 = 0$ and

$$Y_{otk} = Y_{ok} + Y_s \qquad \text{with} \quad Y_{o2} = Y_{o3} = Y_o \qquad (10.68)$$

Note that the frequency variation of the coupling networks has been neglected for simplicity. As shown in Fig. 10.25, this frequency dependence has a limited influence on coupled-system behavior. The situation is totally different from the case of the globally coupled oscillator, in which the coupling network (defined from the plane reference of the active subblocks) contributes crucially to the oscillator resonance and is thus frequency dependent. Neglecting, as in (10.26), the amplitude variation of the imaginary part of the admittance function, it is possible to write

$$\frac{\partial B_o}{\partial \omega}\delta\dot{\phi}_k + G_c\delta\phi_{k-1} + G_{otk}\delta\phi_k + G_c\delta\phi_{k+1} = \frac{I^i_{pk}(t)}{V_o} \qquad (10.69)$$

and in matrix form the perturbed system is given by

$$\begin{bmatrix} j\Omega\frac{\partial B_o}{\partial\omega}\Big|_o + G_{ot1} & G_c & 0 & 0 \\ G_c & j\Omega\frac{\partial B_o}{\partial\omega}\Big|_o + G_{ot} & G_c & 0 \\ 0 & G_c & j\Omega\frac{\partial B_o}{\partial\omega}\Big|_o + G_{ot} & G_c \\ 0 & 0 & G_c & j\Omega\frac{\partial B_o}{\partial\omega}\Big|_o + G_{ot4} \end{bmatrix}$$

$$\times \begin{bmatrix} \delta\phi_1(\Omega) \\ \delta\phi_2(\Omega) \\ \delta\phi_3(\Omega) \\ \delta\phi_4(\Omega) \end{bmatrix} = \frac{1}{V_o}\begin{bmatrix} I_{p1,im} \\ I_{p2,im} \\ I_{p3,im} \\ I_{p4,im} \end{bmatrix} \tag{10.70}$$

Determination of the phase noise spectrum requires inverting the matrix in system (10.70). Unlike the case of (10.26), there is no simple analytical expression for the inverse matrix. However, when performing the inversion, some simplifications are possible:

$$\begin{aligned} G_{ot1} + G_c &= 0 \\ G_{ot} + 2G_c &= 0 \\ G_{ot4} + G_c &= 0 \end{aligned} \tag{10.71}$$

The equalities above, from the property (10.25), are due to the invariance of the autonomous system versus a shift in the phase origin. To illustrate, the case of only two oscillator elements is considered in the following. The matrix inversion provides the system

$$\begin{bmatrix} \delta\phi_1(\Omega) \\ \delta\phi_2(\Omega) \end{bmatrix} = \frac{\begin{bmatrix} j\Omega(\partial B_o/\partial\omega)|_o - G_c & -G_c \\ -G_c & j\Omega(\partial B_o/\partial\omega)|_o - G_c \end{bmatrix}}{-\Omega^2(\partial B_o/\partial\omega)\big|_o^2 - 2j\Omega G_c\,(\partial B_o/\partial\omega)|_o} \frac{1}{V_o}\begin{bmatrix} I_{p1,im} \\ I_{p2,im} \end{bmatrix} \tag{10.72}$$

Solving for the phase perturbation in any of the two oscillator elements and multiplying by the adjoint, the phase noise spectrum is given by

$$\begin{aligned} |\delta\phi(\Omega)|^2 &= \frac{\left[2G_c^2 + \Omega^2\,(\partial B_o/\partial\omega)^2\right]|I_p|^2 2}{V_o^2\left[4G_c^2\,(\partial B_o/\partial\omega)^2\,\Omega^2 + (\partial B_o/\partial\omega)^4\,\Omega^4\right]} \\ &= \frac{\left[2G_c^2 + \Omega^2\left(\frac{\partial B_o}{\partial\omega}\right)^2\right]}{V_o^2\left[4G_c^2 + \left(\frac{\partial B_o}{\partial\omega}\right)^2\Omega^2\right]} \frac{2|I_p|^2}{\left(\frac{\partial B_o}{\partial\omega}\right)^2\Omega^2} \end{aligned} \tag{10.73}$$

where it has been taken into account that $|I_{p\_im}|^2 = 2|I_p|^2$. Close to the carrier, the phase noise spectrum tends to

$$|\delta\phi(\Omega)|^2 = \frac{|I_p|^2}{V_o^2\,(\partial B_o/\partial\omega)^2\,\Omega^2} \tag{10.74}$$

For comparison, the case of a single oscillator under free-running conditions is also considered. Neglecting the amplitude variation of the imaginary part of the total admittance, the circuit equation under current noise perturbations is written

$$j\Omega\frac{\partial B_o}{\partial\omega}\delta\phi(\Omega) = \frac{I_{p\_i}}{V_o} \tag{10.75}$$

so the corresponding phase noise spectrum is given by

$$|\delta\phi(\Omega)|^2 = \frac{2|I_p|^2}{V_o^2\,(\partial B_o/\partial\omega)^2\,\Omega^2} \tag{10.76}$$

Thus, under coupled conditions, the phase noise spectral density is reduced by 10 log 2 close to the carrier. In the case of N oscillator elements, a phase noise reduction of $10\log N$ decibels can be expected. In a rough manner, the reduction in oscillator phase noise can be explained by the fact that N resonators exist at the same oscillation frequency, which gives rise to an N-fold higher equivalent oscillator quality factor. The effect on the noise sources does not increase in the same proportion, due to the fact that the noise sources of the various oscillators are uncorrelated. On the other hand, at higher frequency offset, the expression providing the phase noise variation versus the offset frequency depends on the particular number of oscillators N, as demonstrated in [14]. There is not a single general expression as in the case of global coupling (see 10.40). To see this, compare, for instance, system (10.70), corresponding to $N = 4$ oscillator elements, with (10.72), corresponding to $N = 2$.

10.3.7 Comparison Between Weak and Strong Oscillator Coupling

The semianalytical formulation is valid for weak coupling, that is, for a relatively small amplitude of the currents injected through the coupling network. When implementing these networks with resistively loaded transmission lines, the weak coupling will require relatively large resistance values. In turn, strong coupling will be obtained for a relatively small value of the series resistances of the coupling networks. The behavior of the coupled-oscillator array under strong coupling is obviously more nonlinear than under weak coupling, due to the higher current injected through the coupling networks. Larger amplitude and frequency deviations with respect to free-running values can be expected. The oscillation can even be extinguished in one or more of the oscillator elements [21].

Semianalytical formulation based on linearization of the oscillator elements about their free-running solutions cannot deal with an oscillator systems under

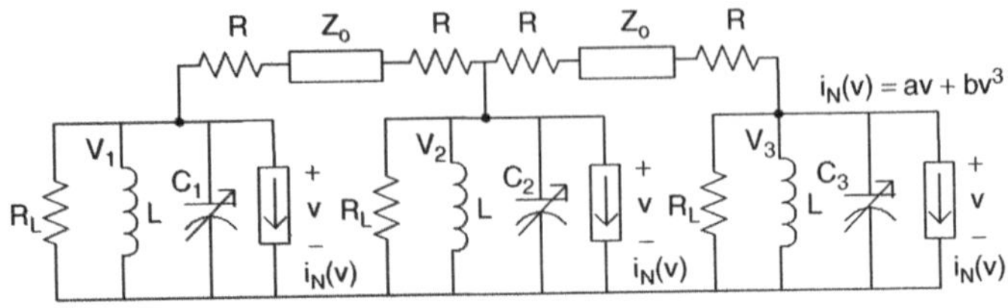

FIGURE 10.26 System of three parallel resonance oscillators. The oscillators are coupled through resistively loaded lines.

strong-coupling conditions. Therefore, the influence of coupling strength will be studied using the harmonic balance technique. For a low computational cost, the comparison will be carried out for a system of three coupled van der Pol oscillators. As shown in Fig. 10.26, they are parallel resonance oscillators with a cubic nonlinear current $i(v)$. The two tuning elements are the two capacitances C_1 and C_3 of the two outermost oscillators. The array has been analyzed for decreasing values of the coupling resistance, which means increasing the coupling strength between oscillator elements.

Figure 10.27a shows the variation in the middle oscillator amplitude versus the imposed constant phase shift for five different values of the coupling resistor. As can be seen, the amplitude variation along the phase shift range increases with the coupling strength. Very small variations are obtained for $R = 200\ \Omega$, whereas for $R = 100\ \Omega$ the variation is so large that the amplitude of this middle oscillator goes to zero at the two phase shift values $\Delta\phi = \pm 90^\circ$. It must also be noted that the curves evolve gradually as the coupling strength increases, exhibiting maxima and minima at similar phase shift values. The amplitude and frequency deviations with respect to the free-running solution and along the constant-phase-shift interval increase with the coupling strength, which is an undesired effect. The oscillation extinction from a certain coupling strength is in good qualitative agreement with the general behavior of injection-locked oscillators, in which the oscillation is typically extinguished from the relatively high amplitude of the synchronizing source. Note that in the three-element system of Fig. 10.26, the middle oscillator circuit is the one that undergoes oscillation extinction, in good agreement with the fact that this oscillator gets the highest current injection. This is because unlike the two outermost oscillators, it is coupled to two oscillator elements, each coupled to only one element.

In Fig. 10.27b, the array oscillation frequency has been represented versus $\Delta\phi$ in the positive range 0 to 180°. Use of the imposed constant phase shift $\Delta\phi$ in the horizontal axis instead of either of the actual parameters C_1 and C_3 gives rise to an apparent discontinuity at the resistance value for $R = 100\ \Omega$. At this R value, the oscillation extinction should prevent $\Delta\phi$ variation beyond $\Delta\phi = \pm 90^\circ$. However, when increasing $\Delta\phi$ beyond these values, the system evolves to a qualitatively different solution with all the array elements in the oscillatory stage. The problem is that this solution has quite a different frequency. A jump is obtained from 5.7 to 4.85 GHz. Therefore, for $|\Delta\phi| \geq 90^\circ$, a constant phase shift solution exists but

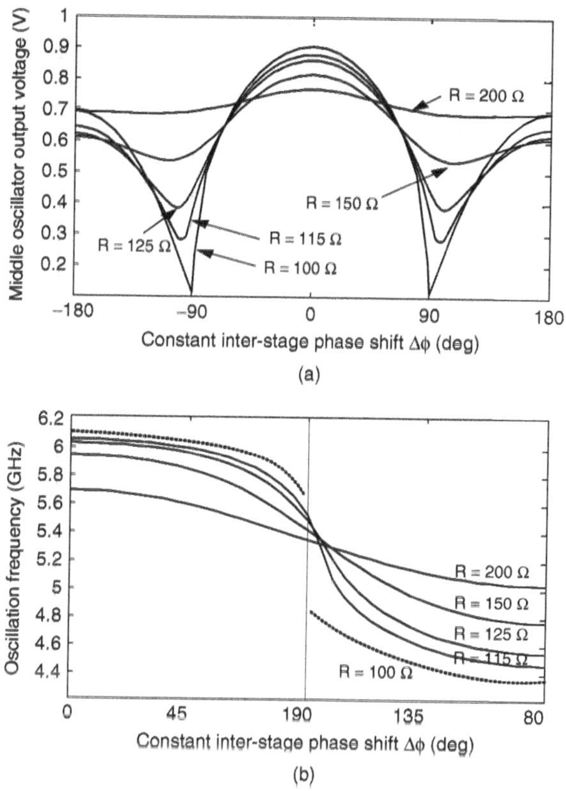

FIGURE 10.27 Coupled-oscillator system with increasing coupling strength. Different values of the coupling resistor are considered. (a) Variation of the output voltage amplitude of the middle oscillator circuit versus the imposed constant phase shift $\Delta\phi$. (b) Variation of the oscillation frequency versus $\Delta\phi$. A discontinuity is obtained for $R = 100\ \Omega$.

with quite a different oscillation frequency. After this analysis it is clear that the solution paths of Fig. 10.27a in the phase intervals $(-180^\circ, -90^\circ)$ and $(90^\circ, 180^\circ)$ correspond to a solution qualitatively different from the one in $(-90^\circ, 90^\circ)$.

Figure 10.28 shows the amplitude variation in three oscillator elements versus the capacitance C_1 and along the constant phase shift solutions for $R = 100\ \Omega$. Note that both C_1 and C_3 vary along the curves to modify the constant phase shift. The first-harmonic amplitude in the middle oscillator is shown by the dashed-dotted line. As expected, this is the amplitude that undergoes the largest variations and the only one in which the oscillation is extinguished. The periodic solutions obtained are not stable for all the phase shift values. The stable and unstable intervals are indicated by S and U. The stability analysis of these solution curves is presented later in this section. The solutions in Fig. 10.27 have also been traced versus the two capacitances C_1 and C_3, which are varied simultaneously to obtain the constant phase shift solution. The representation of the first-harmonic amplitude in oscillator 1 versus C_1 is shown in Fig. 10.29. An identical plot results when representing

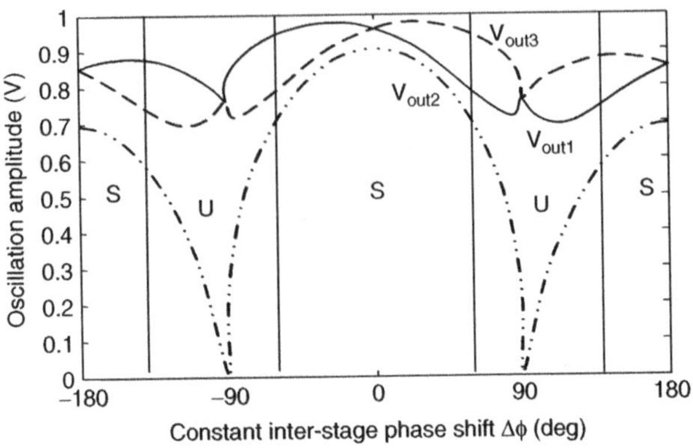

FIGURE 10.28 Coupled-oscillator system under strong coupling. Variation of the output voltage amplitude of the three oscillator circuits versus the constant phase shift imposed.

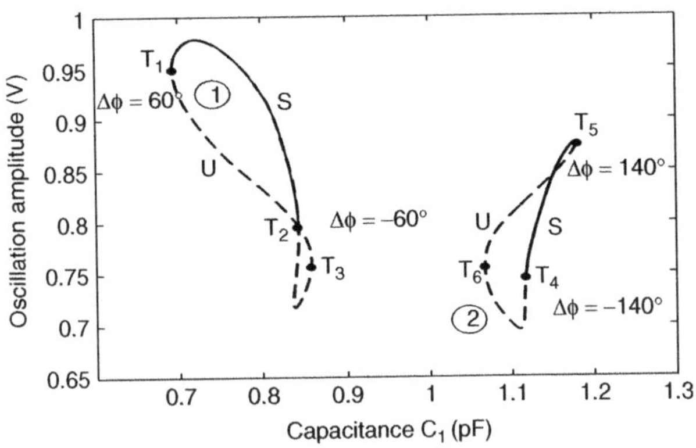

FIGURE 10.29 Coupled-oscillator system under strong coupling. Variation of the output voltage amplitude in the first oscillator versus the tuning capacitance C_1. Two different solution islands are obtained, due to the discontinuity versus the imposed $\Delta\phi$.

the amplitude in oscillator 3 versus C_3. This representation demonstrates the actual continuity of the solution curves of a coupled-oscillator system. It also shows that when increasing $\Delta\phi$ continuously after the extinction of oscillation at $\Delta\phi = \pm 90^\circ$, a different solution is generated. As can be seen, two solution islands are obtained, corresponding to the phase shift intervals $|\Delta\phi| < 90^\circ$ and $|\Delta\phi| > 90^\circ$, respectively. Thus, one island (island 1) is associated with the central curves of Fig. 10.28, whereas the other (island 2) is associated with the outer solution curves.

Next, we apply pole–zero identification to the solution curves of Fig. 10.29. Due to the symmetry of the solution paths, only the positive phase shift range $(0^\circ, 180^\circ)$ is taken into account for this analysis. The pole locus is represented in Fig. 10.30. The capacitance and frequency jump occurring at $\Delta\phi = 90^\circ$ divide the locus into two sections: one at about 6 GHz, corresponding to island 1, and

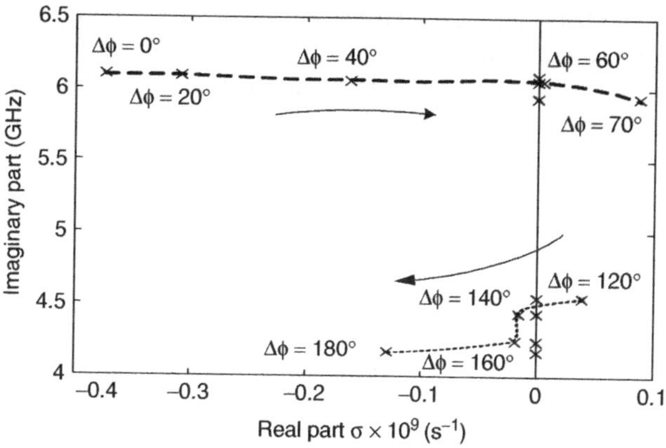

FIGURE 10.30 Coupled-oscillator system under strong coupling. Variation in the pair of poles closest to the imaginary axis along the constant phase shift solution of Fig. 10.28.

another at about 4.5 GHz, corresponding to island 2. Starting from $\Delta\phi = 0°$ in island 1, and increasing $\Delta\phi$, all the solution poles are initially on the left-hand side of the complex plane except the poles on the imaginary axis $\pm j\omega_o$ intrinsic to the solution autonomy. However, at $\Delta\phi = 60°$, a pair of complex-conjugate poles at $\sigma \pm j\omega_o$ cross the imaginary axis to the right-hand side of the complex plane, so the constant phase shift solutions of island 1 become unstable from this $\Delta\phi$ value. As we know, the pair of complex-conjugate poles that cross the imaginary axis at $\pm j\omega_o$ is equivalent to a real pole crossing to zero, which implies a singularity of the system at the crossing points. This is in total agreement with the stability analysis of Section 10.3.5, which showed that the edges of the stable constant phase shift range correspond to solutions at which the system Jacobian matrix $[JH]$ becomes singular. Therefore, the solutions in island 1 or, equivalently, in the middle section of Fig. 10.28, extending from $\Delta\phi = -90°$ to $\Delta\phi = 90°$, are unstable for the phase values $60° < |\Delta\phi| < 90°$.

Next, island 2, with an oscillation frequency at about 4.5 GHz, is considered. Starting from $\Delta\phi = 90°$, the pair of poles $\sigma \pm j\omega_o$ are initially on the right-hand side of the complex plane. With increasing $\Delta\phi$ the poles move rightward, turn, and cross the imaginary axis to the left-hand side again at $\Delta\phi = 140°$, so the solution in island 2 becomes stable at this phase shift value. Again, the frequency of the poles agrees with the oscillation frequency, so a singularity is also obtained at this stabilization point.

As already stated, the behavior is analogous when considering the phase shift interval $(0°, -180°)$, with the solution becoming unstable due to a real pole crossing through zero at $\Delta\phi = -60°$ and becoming stable again for $\Delta\phi = -140°$. Thus, the two phase shift values delimiting the stable range in island 1 are T_1, corresponding to $\Delta\phi = 60°$, and T_2, corresponding to $\Delta\phi = -60°$. The solution is stable between these two phase shift values. The turning point T_3 has no effect on the solution stability. In island 2 the points delimiting the stable range are T_4, corresponding to $\Delta\phi = -140°$, and T_5, corresponding to $\Delta\phi = 140°$. The solution is stable between these two phase shift values. The turning point T_6 has no effect on the solution stability.

Comparing the system behavior for weak and strong coupling, the frequency and amplitude variations versus $\Delta\phi$ become larger for stronger coupling. For the case $R = 100\ \Omega$ considered, there is even an apparent loss of continuity in the required values of the tuning capacitors and the oscillation frequency. In addition, the stable phase shift range is smaller than the one resulting for weaker coupling. On the other hand, the stronger coupling should be less sensitive to modeling inaccuracies and discrepancies between the oscillator elements. More specifically, York [6] and strogatz and Mirollo [24] have shown that the probability of locking in oscillator arrays with nearest-neighbor coupling and randomly distributed free-running frequencies (due to discrepancies in the element values) goes to zero for a number of oscillators N tending to infinity. As explained by York [6], this is due to the difficult communication between oscillator elements at opposite ends when using nearest-neighbor coupling only. This communication is more difficult when increasing the number of oscillator elements. To ensure mutual coupling, the coupling strength must increase at the rate $\sqrt{N}$ [6].

10.3.8 Forced Operation of a Coupled-Oscillator Array

As shown in the previous analyses, the tuning of oscillator-array peripheral elements in order to modify constant interstage phase shift gives rise to an undesired variation in the oscillation frequency ω_s (Fig. 10.19). This variation with respect to the free-running values can be reduced with a suitable design. However, it cannot be eliminated totally, as the entire system behaves in a free-running manner, so the common oscillation frequency is an unknown to be determined when solving system (10.54) versus $\Delta\phi$ or V_{T1}, V_{TN}. The variation in oscillation frequency can be avoided by injection-locking one of oscillator elements of the coupled system, which eliminates the system autonomy. The fundamental frequency of the periodic oscillation is determined by that of the synchronizing source and will not vary with the system parameters [25]. In injection-locked arrays, the constant interstage phase shift can be modified through detuning of the outermost oscillators in opposite directions. The procedure is thus analogous to that used in coupled-oscillator systems with free-running behavior. Note that the response obtained will generally depend on the oscillator selected for connection to an external synchronizing source. This is due to the nearest-neighbour coupling between the oscillator elements, which gives rise to different operation conditions in the array elements depending on the oscillator chosen. It is also possible to use double injection from the same synchronizing source [18]. In a one-dimensional array, the external oscillator signal is split, delayed in one of the two channels, and injected into the two outermost oscillators. Unlike a coupled system under free-running conditions, an injection-locked system has a well-defined phase reference, established by the synchronizing source. In the common case of a synchronizing source with low phase noise, the phase noise will improve with respect to unforced conditions, which is due to the phase relationship between the coupled system and the external source.

The effect is similar to that obtained when injection locking a single free-running oscillator (Chapter 4, Section 4.2.7).

In the following, the semianalytical formulation describing coupled-system behavior under forced conditions is presented. Here the case of injection at a single oscillator k will be considered. As already stated, the synchronizing source now establishes the phase reference of the periodic solution. However, it is possible to maintain the phase of the first oscillator at zero phase value provided that a new phase variable $-\phi_g$ is assigned to the synchronizing source. For simplicity, in the following formulation it is considered that the synchronizing current source $I_g e^{-j\phi_g}$ is connected in parallel at the coupling node of oscillator k:

$$\begin{aligned}
H_1 &= \left(\frac{\partial Y_{AG}}{\partial V_o}(V_1 - V_o) + \frac{\partial Y_{AG}}{\partial \omega_o}(\omega_s - \omega_o) + \frac{\partial Y_{AG}}{\partial V_{To}}(V_{TN} - V_{T1o})\right) V_1 \\
&= -y_{c11} V_1 - y_{c12} V_2 e^{j\phi_2} \\
&\vdots \\
H_k &= \left(\frac{\partial Y_{AG}}{\partial V_o}(V_k - V_o) + \frac{\partial Y_{AG}}{\partial \omega_o}(\omega_s - \omega_o)\right) V_k e^{j\phi_k} \\
&= -y_{ck(k-1)} V_{k-1} e^{j\phi_{k-1}} - y_{ckk} V_k e^{j\phi_k} - y_{ck(k+1)} V_{k+1} e^{j\phi_{k+1}} - I_g e^{-j\phi_g} \\
&\vdots \\
H_N &= \left(\frac{\partial Y_{AG}}{\partial V_o}(V_N - V_o) + \frac{\partial Y_{AG}}{\partial \omega_o}(\omega_s - \omega_o) + \frac{\partial Y_{AG}}{\partial V_{TNo}}(V_{TN} - V_{To})\right) V_N e^{j\phi_N} \\
&= -y_{cN(N-1)} V_{N-1} e^{j\phi_N} - y_{cNN} V_N e^{j\phi_N}
\end{aligned} \tag{10.77}$$

where the injected current $I_g e^{-j\phi_g}$ appears in the equation corresponding to the kth oscillator. To synthesize a constant phase shift solution at the desired frequency ω_s with fixed input amplitude I_g, phase at the different oscillator outputs is set to $\phi_1 = 0, \phi_2 = \Delta\phi, \phi_3 = 2\Delta\phi, \ldots, \phi_N = (N-1)\Delta\phi$. Then $\phi_g, V_{T1}, V_{TN}, V_1, V_2, \ldots, V_N$, as well as the innermost timing voltages (tending to a common value), are calculated to fulfill $H_1 = 0$, $H_2 = 0, \ldots, H_N = 0$. As usual, to obtain the values of V_{T1} and V_{TN} required for each $\Delta\phi$ value, the phase increment $\Delta\phi$ will be swept, solving the system in terms of the variables indicated at each $\Delta\phi$ step. Note that for the entire phase shift interval, the oscillation carrier is fixed to the frequency of the input generator $\omega_g = \omega_s$. Due to the addition of an extra oscillator element, constituted by synchronizing source, the number of coexisting solutions will be 2^{N+1}. As in an unforced oscillator array, solutions with nonconstant phase shift, coexisting with a stable constant phase shift solution (obtained through simultaneous detuning of the outermost oscillators) will generally be unstable. As shown in Section 4.3.2, if the synchronizing source in oscillator k is not connected directly to its coupling node, derivatives of the admittance function at this particular node with respect to

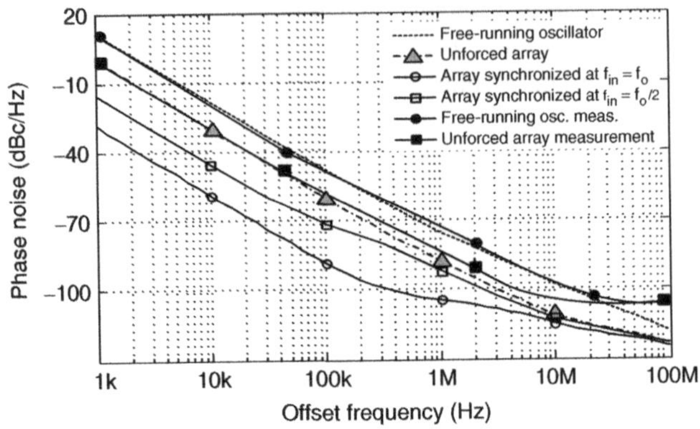

FIGURE 10.31 Oscillator phase noise of a coupled system of four oscillator elements in unforced and injection-locked operation. The spectrum of a single element under free-running conditions is shown as a reference.

the synchronizing source, of either voltage or current type, should be considered [see, e.g., expression (4.70)].

As an example, Fig. 10.31 compares the oscillator phase noise of a coupled system of four oscillator elements in unforced and injection-locked operation. In injection-locked mode, the cases of a synchronizing source at f_o and at the subharmonic frequency $f_{o/2}$ are considered. The spectrum of a single element in free-running conditions is shown, as a reference. As can be seen, synchronization with an external source reduced the phase noise spectral density significantly.

REFERENCES

[1] M. Odyniec, Free, forced, and coupled oscillators' dynamics, *IEEE MTT-S International Microwave Symposium*, Long Beach, CA, pp. 875–878, 2005.

[2] K. Kurokawa, An analysis of Rucker's multidevice symmetrical oscillator, *IEEE Trans.Microwave Theory Tech.*, vol. 18, p. 967, May 1970.

[3] C. T. Rucker, Multiple-diode high- average- power avalanche- diode oscillator, *IEEE Trans. Microwave Theory Tech.*, vol. 17, pp. 1156–1158, 1969.

[4] K. W. Kobayashi, A. K. Oki, L. Tran, et al., A 108-GHz InP-HBT monolithic push–push VCO with low phase noise and wide tuning bandwith, *IEEE J. Solid State Circuits*, vol. 34, pp. 1225–1232, Sept. 1999.

[5] R. J. Pogorzelsky, On the design of coupling networks for coupled oscillator arrays, *IEEE Trans. Antennas Propag.*, vol. 51, pp. 794–801, Apr. 2003.

[6] R. A. York, Nonlinear analysis of phase relationships in quasi-optical oscillator arrays, *IEEE Trans. Microwave Theory Tech.*, vol. 41, pp. 1799–1809, Oct. 1993.

[7] U. L. Rohde, A. K. Poddar, and G. Bock, *The Design of Modern Microwave Oscillators for Wireless Applications*, Wiley, Hoboken, NJ, 2005.

[8] T. Heath, R. R. Kerr, and G. D. Hopkins, Two-dimensional, nonlinear oscillator array antenna, *IEEE Aerospace Conference*, pp. 1104–1115, 2005.

[9] S. Nogi and T. Itoh, Mode analysis and stabilization of a spatial power combining array with strongly coupled oscillators, *IEEE Trans. Microwave Theory Tech.*, vol. 41, pp. 1827–1837, Oct. 1993.

[10] K. Kurokawa, Single-cavity multiple-device oscillator, *IEEE Trans. Microwave Theory Tech.*, vol. 19, pp. 793–801, 1970.

[11] J. H. Wilkinson, *The Algebraic Eigenvalue Problem*, Oxford University Press, New York, 1965.

[12] S. Yen and T. Chu, An *n*th-harmonic oscillator using an *N*-push coupled oscillator array with voltage-clamping circuits, *IEEE MTT-S International Microwave Symposium*, Philadelphia, PA, pp. 2169–2172, 2003.

[13] H. Xiao, T. Tanaka, and M. Aikawa, Basic behavior of quadruple–push oscillator using ring resonator, *IEICE Trans. Electron.*, vol. E88, pp. 1502–1508, 2005.

[14] H. C. Chang, X. Cao, U. K. Mishra, and R. A. York, Phase noise in coupled oscillators: theory and experiment, *IEEE Trans. Microwave Theory Tech.*, vol. 45, pp. 604–615, May 1997.

[15] P. Cortese, A. Sion, D. Baglieri, et al., Low phase noise 2 GHz HBT push–push VCO based on an advanced low frequency noise model, *IEEE Compound Semiconductor Integrated Circuit Symposium (CSIC)*, Monterey, CA, pp. 291–294, 2004.

[16] F. X. Sinnesbichler and G. R. Olbrich, SiGe HBT push–push oscillators for V-band operation, *Proceedings of the IEEE Microwave Theory and Techniques Symposium*, pp. 55–59, June 2000.

[17] F. Ramirez, J. García, T. Fernandez, and A. Suárez, Nonlinear simulation techniques for the optimized design of push–push oscillators, *IEEE MTT-S International Microwave Symposium Digest*, Philadelphia, vol. 3, p. 2157, June 8–13, 2003.

[18] R. A. York and T. Itoh, Injection- and phase-locking techniques for beam control, *IEEE Trans. Microwave Theory Tech.*, vol. 46, pp. 1920–1929, 1998

[19] T. Heath, R. R. Kerr, and G. D. Hopkins, Nonlinear oscillator array antenna development at GTRI, *IEEE Aerospace Conference*, p. 19, 2006.

[20] A. Collado, F. Ramirez, A. Suárez, and J. P. Pascual, Harmonic-balance analysis and synthesis of coupled-oscillator arrays, *IEEE Microwave Wireless Components Lett.*, vol. 14, p. 192, May, 2004.

[21] A. Georgiadis, A. Collado, and A. Suárez, New techniques for the analysis and design of coupled-oscillator systems, *IEEE Trans. Microwave Theory Tech.*, vol. 54, pp. 3864–3877, 2006.

[22] A. Collado, A. Suárez, and S. Sancho, Semi-analytical formulation for the stability analysis of coexisting solutions in coupled-oscillator systems, *IEEE MTT-S International, Microwave Symposium Digest*, pp. 973–976, 2007.

[23] V. Rizzoli, F. Mastri, and D. Masotti, General noise analysis of nonlinear microwave circuits by the piecewise harmonic balance technique, *IEEE Trans. Microwave Theory Tech.*, vol. 42, pp. 807–819, 1994.

[24] S. H. Strogatz and R. E. Mirollo, Phase-locking and critical phenomena in lattices of coupled oscillators with random intrinsic frequencies, *Physica D*, vol. 31, pp. 143–168, 1988.

[25] H. C. Chang, X. Cao, M. J. Vaughan, U. K. Mishra, and R. A. York, Phase noise in externally injection-locked oscillator arrays, *IEEE Trans. Microwave Theory Tech.*, vol. 45, pp. 2035–2042, Nov. 1997.

CHAPTER ELEVEN

Simulation Techniques for Frequency-Divider Design

11.1 INTRODUCTION

The frequency divider is an integral component of frequency synthesizers [1,2]. It is used to divide the output frequency of the voltage-controlled oscillator (VCO) into the frequency of the reference oscillator by an order of 10 to 30 MHz. The phase of the divider output is compared with that of the reference oscillator, and the resulting error signal is used to adjust the VCO output frequency. A frequency divider can also be used to generate a precision I/Q signal from an input signal with a 50% duty cycle [3]. Compared with the traditional resistor and capacitor quadrature generation, the frequency-divider approach is easier to implement, requires lower input power, and gives rise to less phase imbalance. Other applications include frequency counting, phase noise reduction, and bandwidth compression.

Frequency dividers can be designed from either a digital or an analog point of view. The digital dividers can be classified into static, using bistable cells, and dynamic, using capacitances as storage elements. The static approach is based on the transistor implementation of the D-type flip-flop [4–6], which provides a divide-by-2 cell. This cell can be designed in source-coupled logic (SCL) using a CMOS process, or in emitter-coupled logic (ECL) using a bipolar process. The dynamic dividers, using storage capacitances, typically require a much smaller number of transistors. The static dividers are faster and more reliable, whereas

Analysis and Design of Autonomous Microwave Circuits, By Almudena Suárez

the dynamic dividers, with fewer transistors, are easier to implement. The digital dividers enable a broad division bandwidth and nowadays operate up to very high frequency, although the power consumption increases with frequency. Digital dividers operating beyond 60 GHz have been reported [7]. In analog dividers the output frequency is limited only by the maximum oscillation frequency of the devices used. Analog frequency dividers can be obtained with a very small number of devices. They feature lower power consumption and lower phase noise and make possible a versatile and simpler design. Their main drawback is reduced bandwidth. There are different types of analog dividers based on the principle on which the division is based: parametric, regenerative, and harmonic injection. Analytical studies of these three types of divider were described in Chapter 4.

Our aim in this chapter is to show practical simulation tools for the efficient design of frequency dividers. No attempt is made to cover the divider configurations, which have been treated in great depth by other authors [1,6,8,9]. Instead, once a particular type of divider and an initial topology have been selected, the objective will be the optimization of the element values for the best possible performance. Different design goals are considered: for example, the increase in output power at the divided frequency, the reduction in the input power threshold for the frequency division (in regenerative dividers), broadening of the operational bandwidth, or the reduction in settling time. The chapter is organized as follows. In Section 11.2 we present a brief reminder of the various types of frequency dividers, covered in detail in Chapter 4. In Sections 11.3 and 11.4 we describe harmonic balance techniques for the optimized design of regenerative and harmonic injection dividers, respectively. Due to the conceptual similarity in their operational principle, in Section 11.5 we extend some of the design techniques developed for harmonic injection dividers to subharmonically injection-locked oscillators.

11.2 TYPES OF FREQUENCY DIVIDERS

As already stated, frequency dividers can be classified into digital and analog dividers. In turn, analog dividers can be subclassified according to their operational principle into parametric, regenerative, and harmonic injection dividers. Brief descriptions of these dividers are presented next.

1. *Parametric dividers.* The operation of parametric circuits is based on the periodic variation of a nonlinear reactance, usually a capacitance. As shown in Chapter 3, Section 3.3.1.2, a simple circuit comprised of an inductor, a resistor, and a varactor diode, biased at V_b and periodically pumped at the frequency ω_{in}, can operate as a frequency divider by 2 for input frequency about $2\omega_o$, with ω_o being $\omega_o = 1/\sqrt{LC(V_b)}$. The periodic capacitance is able to convert the energy delivered by the input periodic source to the divided frequency $\omega_{in}/2$ [10,11]. There is a constant phase shift between the subharmonic oscillation and the input source, which varies with the input frequency within certain frequency division band. When analyzing the small-signal response of the capacitance, linearized about the periodic

solution at ω_{in}, excess negative resistance is obtained at this divided frequency $\omega_{in}/2$, due to the phase shift between its current and voltage. In addition, the capacitive reactance resonates at this frequency with the inductive element. Thus, the oscillation startup conditions are fulfilled at the subharmonic frequency $\omega_{in}/2$. Note that more efficient topologies with filtering of the input and output frequencies are used in practice. The parametric frequency division requires the nonlinear behavior of the reactance, so it is never observed under a small-signal drive. The parametric frequency division ω_{in}/N with $N \neq 2$ is also possible, but requires an idler frequency component $(N-1)\omega_{in}/N$ inside the nonlinear capacitance to sustain the divided regime through intermodulation. For maximum efficiency, open- or short-circuit terminations are used at frequencies different from the input frequency ω_{in} and the desired subharmonic ω_{in}/N.

2. *Regenerative dividers.* A regenerative divider uses mixing plus subharmonic feedback to achieve frequency division by order N from a certain level of the input power at ω_{in} [12–14]. The division by N is due to an instability at $\omega_o = \omega_{in}/N$ obtained when increasing the input power P_{in}. The regenerative divider can be implemented with a harmonic mixer, providing the frequency component $\omega_{in} - (N-1)\omega_o$, which is subsequently filtered and amplified, plus a feedback loop at $\omega_o = \omega_{in}/N$. It is also possible to introduce a multiplier in the feedback loop, providing $(N-1)\omega_o$, and to use the mixer to get the difference frequency $\omega_{in} - (N-1)\omega_o$. This is the division scheme of Fig. 4.22. Clearly, the regenerative dividers have a nonlinear behavior with respect to the input source. The aim is to obtain instability at $\omega_o = \omega_{in}/N$ from a certain P_{in} value. The gain at this frequency, calculated from the circuit linearization about the periodic solution at ω_{in}, must increase versus P_{in}. The following conditions should be fulfilled: $|H(P_{in}, \omega_{in}/N)| > 1$and $\text{ang}[H(P_{in}, \omega_{in}/N)] = 2k\pi$, with H the open-loop transfer function of the divider circuit and k an integer. This can be achieved by biasing the amplifier device below the conduction threshold voltage. When increasing P_{in}, the device starts conducting for a progressively larger fraction of the period, with the input voltage waveform reaching higher maxima for higher P_{in}.

3. *Harmonic-injection dividers.* A harmonic injection divider by N is basically an oscillator circuit, with the Nth-order harmonic component of the oscillation frequency $N\omega_o$ synchronized to an independent input source at ω_{in} [15–18]. In contrast to parametric and regenerative dividers, harmonic injection dividers exhibit self-oscillation in the absence of input signal. Since oscillation on the order of the desired output frequency $\omega_o \cong \omega_{in}/N$ is already available, they enable frequency division at a very low input power. However, this free-running oscillation of the divider circuit might not be admissible in some systems, which would require the use of either of the other two types of divider.

4. *Digital dividers.* Digital dividers are based on the circuit implementation of logic gates and typically require a large number of transistors. Their main advantage over analog dividers is that they allow broadband frequency division. The basic element for the digital divider is the D flip-flop. A master–slave D flip-flop is commonly used. In a D flip-flop the Q and $\overline{Q}$ outputs change their states only when the clock signal makes a high-to-low transition (or the reverse). When this happens,

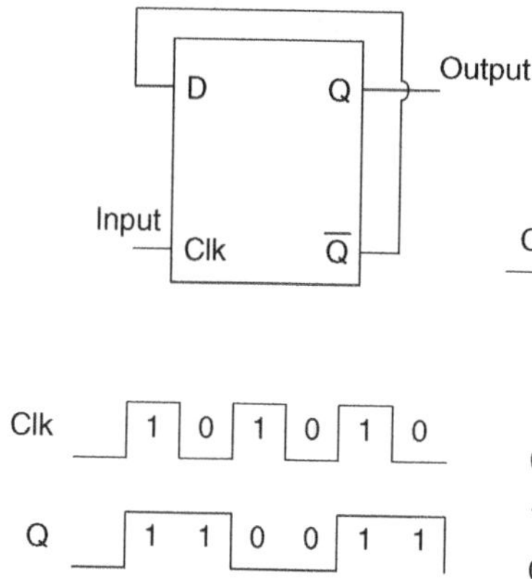

Clk	Q	$\bar{Q} \equiv D$
1	1	0
0	1	0
1	0	1
0	0	1
1	1	0
0	1	0

FIGURE 11.1 *D* flip-flop enabling frequency division by 2 of the clock signal. The table of truth of the *D* flip-flop is also included.

the Q output takes the value of the preceding D input. To obtain the frequency division, the $\bar{Q}$ output is connected to the D input, which clearly leads to the division by 2 of the clock frequency (Fig. 11.1). The D flip-flop is implemented with current mode logic, applying feedback from the differential outputs to the transistor inputs, which corresponds to making the D input equal to $\bar{Q}$ [8]. Inductors are used to compensate the transistor capacitances and reduce the rise and fall times. The clock signal is usually introduced through a frequency mixer based on a Gilbert cell. Even though the digital dividers are designed from a logic gate point of view, they actually behave as nonlinear dynamical systems. Depending on whether or not the circuit self-oscillates, it is possible to obtain an equivalent block diagram corresponding to a regenerative divider or an injection-locked frequency divider, respectively [8].

As can be gathered from the paragraphs above, analog frequency dividers have an inherently autonomous nature, which makes them difficult to synthesize. Here, nonlinear design techniques for analog dividers, based on the use of transistor devices, is presented. Two main types of dividers are considered: the regenerative divider and the harmonic injection divider.

11.3 DESIGN OF TRANSISTOR-BASED REGENERATIVE FREQUENCY DIVIDERS

11.3.1 Frequency-Divided Regime

The regenerative frequency divider can be implemented in two main ways: with a single-ended topology or by means of a Gilbert mixer, in feedback configuration. In a single-ended topology, the mixing function (see the schematic of Fig. 4.22) is enabled by the nonlinearity of the transistor transconductance. The transistor is connected to an output filter, which selects the divided frequency ω_{in}/N. For division by N, a multiplier by $N-1$ must be introduced in the feedback loop. The transistor provides mixing and amplification of the difference frequency ω_{in}/N. As shown by Sarkissian et al. [19], the introduction of a previous amplification stage allows us to increase the division bandwidth. The divider based on a Gilbert mixer has a similar principle. The Gilbert mixer provides amplification by itself and enables much broader division bandwidth, with the drawback of higher power consumption. Another possibility is to use a harmonic mixer to obtain the frequency component $\omega_{in} - (N-1)\omega_{in}/N$, so no multiplier is required in the feedback loop.

For an initial divider design, we can analyze the open-loop transfer function at the divided frequency. A small-signal input source at ω_{in}/N can be introduced into the circuit, calculating the open-loop transfer function at the divided frequency. This transfer function is obtained in the presence of the input source at ω_{in}, with relatively high input power P_{in}. The circuit is analyzed in open-loop conditions at ω_{in}/N and closed-loop conditions at the steady-state frequencies $k\omega_{in}$. When using a transistor, the nonlinear transconductance depends on the voltage signal generated by an external source at ω_{in}/N (instead of its original control voltage). Then, the open-loop transfer function is calculated as the ratio between the voltage drop at the actual control-voltage location and the voltage delivered by the independent source, in similar manner to the return-ratio calculation described in Chapter 6, Section 6.3. If we do not have access to the internal transistor nodes, we can open the loop at a sensitive circuit node. In the example in Fig. 11.2, the small-signal source at ω_{in}/N is introduced into the feedback block and the output of the block is introduced into the mixing element. The open-loop transfer function is calculated as the ratio between the voltage value at ω_{in}/N at the filter output node and the voltage value of the small-signal source. For an accurate design, a good estimation of the circuit impedance at the location of the small-signal source at ω_{in}/N would be necessary. (This impedance has not been taken into account in the figure.) Even though this technique is just approximate, it is very simple to implement on commercial harmonic balance devices and it allows an initial estimation of the circuit element values for the frequency division. Once the loop is closed, the startup conditions of the subharmonic frequency should be verified for accurate small-signal stability analysis, such as pole–zero identification [20].

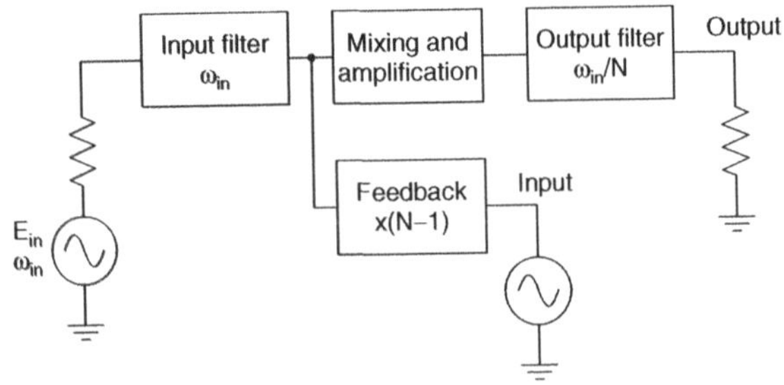

FIGURE 11.2 Design technique of regenerative dividers, based on approximate calculation of the open-loop transfer function at ω_{in}/N.

Using the approximate design technique described in open loop at ω_{in}/N, the circuit elements are tuned or optimized so as to satisfy the conditions $|H(P_{in}, \omega_{in}/N)| > 1$ and $\text{ang}[H(P_{in}, \omega_{in}/N)] = 2k\pi$. Fulfilment of these conditions will probably lead to instability of the periodic solution ω_{in}, which should exhibit a pair of unstable complex-conjugate poles at about the frequency ω_{in}/N. For broad division bandwidth the sensitivity of the phase shift to the input frequency must be minimized, $d(\text{ang}(H))/d\omega_{in} \cong 0$ [19]. The bandwidth will also be very dependent on input and output matching.

We emphasize that the open-loop optimization described is just a procedure to make the periodic regime at ω_{in} exhibit an instability at ω_{in}/N. Because it is based on small-signal analysis of the divider circuit at ω_{in}/N, it does not allow us to predict the output power at the subharmonic frequency ω_{in}/N or accurate determination of the division bandwidth. This will require full harmonic balance analysis of the closed-loop circuit at the divided frequency.

The regenerative divider design is simplified considerably in the case of division by order $N = 2$, since the feedback loop, which should be a multiplier by $N-1$ in the case of a divider by $N = 2$, reduces to a simple passive network. The division is obtained from a flip bifurcation giving rise to the transition from a stable periodic regime at ω_{in} to a regime at $\omega_{in}/2$. For a narrowband design, a free-running oscillator at the desired output frequency $\omega_{in}/2$ can be obtained initially. Next, the device bias voltage is reduced so as to quench the oscillation. For sufficiently high input amplitude, the transistor will conduct for a fraction of the input signal period $2\pi/\omega_{in}$ and division will be obtained from a certain input power P_{ino}. In the case of FET transistors, advantage is taken of the quasiquadratic characteristics of the drain-to-source current i_{DS} versus the gate-to-source voltage v_{GS} near pinch-off for efficient mixing [18]. Note that the frequency division is obtained from the power delivered from the input source.

As shown in Chapter 5, the nonlinear divided regime can be simulated with the aid of an auxiliary generator. The voltage auxiliary generator at the divided frequency $\omega_{AG} = \omega_{in}/N$ will be introduced in parallel at a sensitive circuit node. Besides the nonlinear analysis of the circuit, the auxiliary generator will also allow a correction or improvement in the initial divider design. For instance, it will be possible to increase the output power at the subharmonic frequency $\omega_{AG} = \omega_{in}/N$ for constant input power P_{in}. If a voltage auxiliary generator is connected in parallel with the output load R_L, the output power will be related directly to the auxiliary generator amplitude: $P_L = A_{AG}^2/2R_L$. Thus, to maximize the output power, it will be possible to increase A_{AG} from the original output voltage amplitude $A_{AG} = V_{out}$, by means of a sweep, using a continuation technique. At each sweep step, the values of some circuit elements η_1 and η_2, together with auxiliary generator phase ϕ_{AG}, are calculated or optimized to fulfil $Y_{AG}(\eta_1, \eta_2, \phi_{AG}) = 0$. From a certain A_{AG} value it will no longer be possible to fulfill the nonperturbation conditions $Y_{AG} = 0$, due to inherent limitations of the circuit topology and input power. The optimized circuit elements may be chosen in different manners, taking into account that the elements in the feedback and/or output networks affect conditions for subharmonic oscillation, and the elements of the input networks also have an influence on matching to the input periodic source.

As an example, the technique described will be applied to the FET-based circuit shown in Fig. 4.8, operating as a regenerative divider by $N = 2$. The transistor will be biased at the pinch-off voltage $V_{GS} = -1.8$ V. As already known, in regenerative dividers there is an input-power threshold P_{ino} below which no frequency division is obtained. The division is observed for $P_{in} > P_{ino}$. To know the input voltage threshold E_{ino} for the frequency division by 2 in this original design, we check the flip bifurcation locus of Fig. 4.26. This locus provides variation in the threshold E_{ino} versus the gate bias voltage V_{GS}. For the bias voltage $V_{GS} = -1.8$ V, the amplitude is $E_{ino} = 0.99$ V, corresponding to $P_{ino} = 4$ dBm.

At flip bifurcations, the subharmonic component is generated from zero input power (or extinguished to zero power in the case of an inverse bifurcation). Thus, for $E_{ino} = 0.99V$, the subharmonic amplitude is in the limit of zero value. The divided-solution curve is represented in Fig. 11.3 in terms of the output power at the subharmonic frequency $\omega_{in}/2$ versus the input power. To trace this frequency-divided curve, the harmonic balance techniques described in Chapter 5, Section 5.5.2.2, and Chapter 6, Section 6.4 have been used. As already known, the divided solution coexists with a nondivided solution at ω_{in} (see Chapter 3), which is not represented in Fig. 10.3. When increasing P_{in}, this nondivided solution (at ω_{in}) becomes unstable at the direct flip bifurcation F_1, obtained at $P_{ino} = 4$ dBm. This is because a pair of complex poles $\sigma \pm j\omega_{in}/2$ of the nondivided solution, located initially on the left-hand side of the complex plane, cross the imaginary axis to the right of this plane at $P_{ino} = 4$ dBm. If the input power is increased further, the pair of poles initially moves rightward, then turns and crosses the imaginary axis at the inverse flip bifurcation F_2, where the subharmonic solution is extinguished.

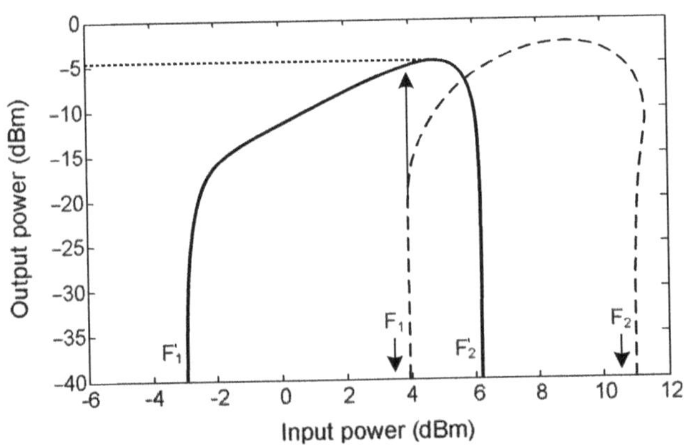

FIGURE 11.3 Auxiliary generator technique to correct an initial frequency-divider design: (a) variation of the subharmonic output power versus the auxiliary generator amplitude; (b) deviation of the input and output inductances with respect to their original values $L_{in} = 0.58$ nH and $L_{out} = 2.89$ nH, respectively. Input inductance, dashed line; output inductance, solid line.

To increase the output power at $P_{ino} = 4$ dBm, the auxiliary generator operating at the divided frequency $\omega_{in}/2$ is connected in parallel at the drain node. The optimization elements considered are the input and output inductive transmission lines and the bias resistance R_{DD} (Fig. 4.8). Then the subharmonic amplitude A_{AG} is increased from the initial near-zero value up to the maximum value that allows fulfillment of the nonperturbation condition $Y_{AG} = 0$. With the technique described, it has been possible to increase the subharmonic output power from 0 W (at the flip bifurcation) to $P_{out} = -5$dBm, indicated by the arrow in Fig. 11.3. Because no additional condition has been imposed on the new design, the change in circuit elements has led to a variation in the entire solution curve. The flip bifurcation determining the lower edge of the division interval is shifted from $P_{in}(F_1) = 4$dBm to $P_{in}(F_1') = -2.9$dBm. In turn, the flip bifurcation determining the upper edge of this interval is reduced from $P_{in}(F_2) = 11$dBm to $P_{in}(F_2') = 6.1$dBm. The design modification has also given rise to a qualitative variation of the solution curve, discussed below.

At the flip bifurcations the slope of the subharmonic amplitude versus the parameter tends to infinite, as explained in Chapter 3, Section 3.3.1.2. However, in the immediate neighborhood of this bifurcation, the slope may be positive, in the case of supercritical flip bifurcation, or negative, in the case of a subcritical bifurcation. In a supercritical bifurcation, the small-amplitude subharmonic solution exists only after the critical parameter value at which the bifurcation takes place. In the subcritical bifurcation, the subharmonic solution exists before this critical value. Note that this definition is local, i.e., it is only valid for small values of the subharmonic amplitude. From the above description, the flip bifurcations $F_1, F_1' F_2'$ are supercritical, whereas the bifurcation F_2 is subcritical. On the other hand, the periodic solution at ω_{in}, corresponding to the original design (not represented in the figure), is unstable between the two flip bifurcations F_1 and F_2. In turn, the periodic solution at ω_{in}, corresponding to the modified design (not represented in the figure), is unstable between the two flip bifurcations F_1' and F_2'.

The subcritical flip bifurcations are characterized by the coexistence in the neighborhood of the bifurcation of the stable nondivided solution at ω_{in} with an unstable divided solution of small amplitude. As shown in Chapter 3, this involves a hysteresis phenomenon in the transformation from a nondivided regime to a divided regime, and vice versa, so the subcritical bifurcations are undesired. The bifurcation F_2 in Fig. 11.3 is subcritical. The design modification has transformed the flip bifurcation delimiting the upper edge of the division band into a supercritical bifurcation. This beneficial transformation is fortuitous, as no mathematical conditions were used to impose the supercritical type of flip bifurcations.

Note that in the way it has been applied to this example, the technique is inherently limited, as only the values of the existing circuit elements are varied, with no modification of the original design topology. In the next section, some changes are introduced by means of the addition of a feedback network or a load circuit at the subharmonic frequency.

11.3.2 Control of Operation Bands in Frequency Dividers by 2

As already shown, the limits of the operation band of a regenerative frequency divider by 2 are determined by flip bifurcations versus the parameter considered, generally the input power or input frequency. Thus, it will be possible to preset the division band by imposing flip bifurcations at the desired parameter values through bifurcation control [21,22]. This will be carried out using the bifurcation control technique described in Section 6.6. The technique is based on the use of an auxiliary generator to impose the bifurcation conditions at the desired parameter value η_b'. Note that some circuit components will have to be modified or added to the original design to satisfy these conditions.

In flip bifurcation control, it is possible, for example, to introduce an ideal load at the subharmonic frequency $\omega_{\text{in}}/2$, which should be implemented at a later stage. The load introduced should have low-pass filter behavior, with a cutoff frequency below ω_{in}, to minimize qualitative changes with respect to the response of the original circuit. However, for bifurcation control, feedback networks are most typically used. The addition of a feedback network, introduced specifically for this control, enables direct action on the stability properties with a very small variation in the original qualitative performance. The feedback network introduced may have a prefixed linear topology or may be defined in terms of ideal impedance boxes $R + jX$ at the divided frequency $\omega_{\text{in}}/2$. The ideal feedback network provides more flexibility for the bifurcation control but requires a later implementation stage of the resulting ideal impedance values.

Regardless of the type of circuit elements selected to perform the control, a vector $\bar{\rho}_c$ containing these element values will be defined. To shift a flip bifurcation from the original parameter value η_b to η_b', we take into account that the subharmonic amplitude tends to zero value at the flip bifurcation points. Thus, an auxiliary generator of very small amplitude ε at the subharmonic component $\omega_{\text{in}}/2$ will be introduced into the circuit. Then the vector of control element values $\bar{\rho}_c$ and the auxiliary generator phase will be calculated so as to fulfill the bifurcation conditions

$$
\begin{aligned}
Y_{AG}(\phi_{AG}, \bar{\rho}_c) &= 0 \\
A_{AG} &= \varepsilon \\
\omega_{AG} &= \frac{\omega_o}{2} \\
\eta &= \eta_b'
\end{aligned}
\tag{11.1}
$$

System (11.1) is solved in combination with harmonic balance at the fundamental frequency $\omega_{\text{in}}/2$ by means of two-level Newton–Raphson or through optimization with the goal $Y_{AG} = 0$.

To avoid a possible hysteresis phenomenon in the transformation from nondivided regime at ω_{in} to divided regime at $\omega_{\text{in}}/2$, and vice versa, the flip bifurcation obtained must be supercritical (see Chapter 3). As stated earlier, the supercritical flip bifurcation is characterized by the positive slope of the subharmonic amplitude versus the parameter when this parameter is varied in the sense that makes the original periodic regime at ω_{in} evolve from stable to unstable. Note that this definition is inherently local only, valid only for small value of the subharmonic amplitude. The supercritical type of bifurcation can be imposed by considering a second copy of the same circuit, operating at $\eta'_b + \Delta\eta$, with an auxiliary generator having the amplitude value $A_{AG2} = \varepsilon + \Delta A$, with $\Delta A > 0$. Therefore, the two circuits, with the same element values, are resolved simultaneously through optimization, with the goals (11.2) and $Y_{AG2} = 0$. The set of optimization elements will include $\overline{\rho}_c$, ϕ_{AG1}, ϕ_{AG1}, and the amplitude ΔA, constrained to small-signal values. The inclusion of ΔA in the set of optimization elements will improve the convergence properties. Once the optimization is concluded, the resulting divided curve should be traced, and an accurate stability analysis must be carried out along the solution curve obtained.

An alternative manner to obtain a supercricial bifurcation is to sweep a circuit parameter η_1 and resolve (11.1) at each η_1 step. Then, pole zero identification is applied to the incipient subharmonic solution with $A_{AG} = \varepsilon$. As shown in Chapter 3, Section 3.3.1.2, for a supercritical bifurcation, this incipient solution will contain a pair of complex-conjugate poles at $\sigma \pm j\omega_{in}/2$ on the left-hand side of the complex plane. For a subcritical bifurcation, it will contain a pair of complex-conjugate poles at $\sigma \pm j\omega_{in}/2$ on the right-hand side of this plane. By performing pole–zero identification of the incipient solution versus a suitable parameter η_1, it will be possible to choose a η_1 value for which the bifurcation at the imposed value η_b is supercritical. In general, the imposed amplitude $A_{AG} = \varepsilon$ of the incipient subharmonic cannot be too small, to ensure sufficient accuracy of the pole–zero identification technique. Despite this, there is little accuracy degradation in the determination of the flip locus, which is due to the high slope of the subharmonic amplitude curve in the neighborhood of the bifurcation.

An immediate application of the control of flip bifurcations will be the reduction of the power threshold for frequency division in regenerative dividers. Remember that this type of divider does not oscillate in the absence of input signal. As a result, they have the drawback of requiring relatively high input power for frequency division. As an example, this technique has been applied to a regenerative divider with input frequency $f_{in} = 10.4$ GHz (Fig. 11.4) in hybrid technology. In this divider, the flip bifurcation leading to frequency division by 2 takes place originally for the input generator power ($P_{inb} = 5$ dBm). The shift in bifurcation can be achieved through modification of the series-feedback network, existing in the original design, so as to fulfill conditions (11.3) at the input power value desired, P'_{inb}. However, this feedback network is an essential part of the divider topology, so its variation may give rise to significant modification of the original circuit response. Instead, a specific feedback network has been added to the circuit for bifurcation control. This feedback network contains three passive elements and acts as a lowpass filter with high insertion loss. With respect to the original circuit, this network increases the feedback amount at the subharmonic frequency, which is higher for a smaller resistance value. In addition, the lowpass characteristic of this network ensures minimum discrepancy, prior to the frequency division, between the original and controlled circuits. Note that prior to this division the solution at ω_{in} is still stable,

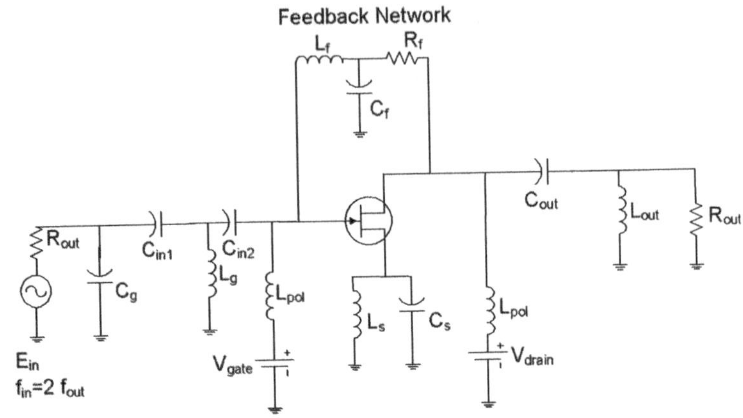

FIGURE 11.4 Regenerative divider by $N = 2$ in hybrid technology, with $f_{in} = 10.4$ GHz.

and thus physically observable, and, in general, we do not want to modify the circuit behavior before division takes place.

In Fig. 11.5a, the values of the feedback inductance (implemented on transmission line) and resistance obtained through optimization of (11.4) are represented versus the input power value P_{inb} at which we impose the flip bifurcation. Harmonic balance simulation has been carried out considering 15 harmonic components. Note that the resistance and inductance both tend to constant values when reducing the input power to small-signal levels. This is due to the lack of sensitivity of the flip bifurcation to the generator power under small-signal conditions. To verify the actual shifting of the bifurcation point to the desired value, we will select particular values of the power threshold and obtain the required feedback element values R_f and L_f from Fig. 11.5a. Then for each pair of feedback element values replaced into the circuit, we perform a simulation of the frequency-divided curve versus the input power. These simulations are shown in Fig. 11.5b, together with that of the original curve, exhibiting the flip bifurcation at $P_{in} = 5$ dBm. The input power of the controlled curves at the flip bifurcation is, in all cases, in excellent agreement with that imposed through the bifurcation control (horizontal axis of Fig. 11.5a). As can be seen, all the curves are qualitatively similar and tend to converge as the input power increases. They are all entirely stable from the flip bifurcation. Measurements are superimposed for the original and controlled curves, corresponding to $P_{inb} = -15$ dBm. Thus, the bifurcation-control technique has allowed reducing the power threshold for frequency division from $P_{inb} = 5$ dBm to $P'_{inb} = -15$ dBm. Further reduction of this threshold would be unreliable, due to the circuit operating under linear conditions with respect to the input source at ω_{in}, as gathered from Fig. 11.5a.

In pure bifurcation control [22,23], the controlled and uncontrolled steady-state solutions must be exactly the same at the bifurcation point. This condition has not been imposed here. However, the feedback network selected, which has a lowpass characteristic, should have a very limited influence on the nondivided regime at

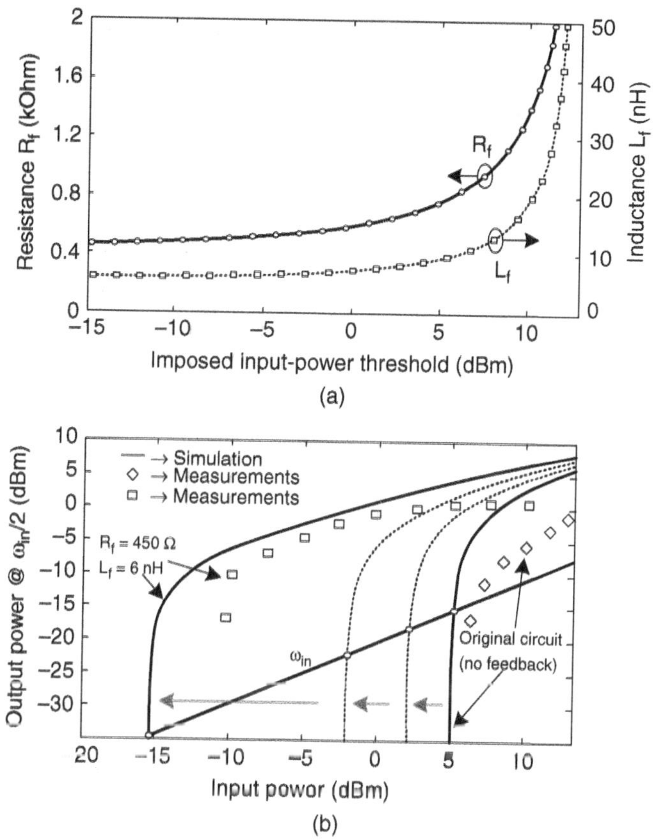

FIGURE 11.5 Shift of the flip bifurcation in a regenerative frequency divider: (a) variation of the feedback network parameters versus the input power for flip-type bifurcation; (b) output power at the divided-by-2 frequency versus the input power for the feedback network values resulting from (a). Measurements for the original and controlled curves corresponding to $P_{inb} = -15$ dBm are superimposed.

ω_{in}. To check this, the default nondivided solution at ω_{in} obtained with each of the controllers considered is also represented in Fig. 11.5b. All the nondivided curves overlap, which confirms the accuracy of the technique. Note that in each case, the curve is stable up to the bifurcation point only. The bifurcation point is shifted, with minimum disturbance of the nondivided solution. According to Fig. 11.5a, to shift the bifurcation point to $P_{inb} = -2$ dBm, the feedback element values must be $R_f = 525\ \Omega$ and $L_f = 7$ nH, as confirmed by the bifurcation diagram of Fig. 11.5b. The pole–zero identification technique [20] has been applied to this particular controlled circuit, with the results shown in Fig. 11.6. As can be seen, a pair of complex-conjugate poles at $f_{in}/2 = 5.2$ GHz shifts rightward versus the input power, as expected in a regenerative divider by 2, and crosses the imaginary axis at exactly $P_{in} = -2$ dBm.

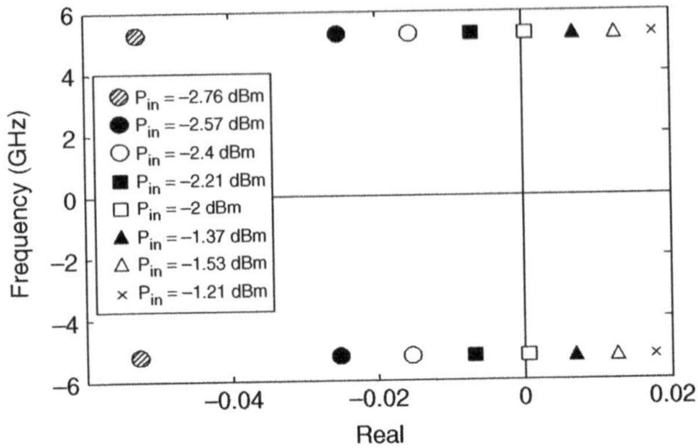

FIGURE 11.6 Evolution of the pair of complex-conjugate poles at $f_{in}/2 = 5.2GHz$ corresponding to the circuit in Fig. 11.4 when the feedback network is synthesized to achieve the bifurcation point at $P_{inb} = -2$ dBm.

11.3.3 Control of Divider Settling Time

High-speed frequency dividers are essential for broadband communication applications [8]. They are characterized by fast switching on when the input signal is introduced. In the following, a technique for shortening of the frequency-divider transient is presented. A divider by 2, with input frequency ω_{in}, is considered. Below the input power threshold P_{ino}, the nondivided solution at ω_{in} is stable and no division takes place. If the input power is now increased to $P_{in} > P_{ino}$, the nondivided solution at ω_{in} will become unstable. This solution must have a pair of dominant complex-conjugate poles $\sigma \pm j(\omega_{in}/2)$, with positive σ, responsible for frequency division by 2. The more positive the σ value, the shorter the transient will be. The real part σ depends on the circuit elements and quality factor, but also on the bias point and input power used. Thus, it will be possible to carry out a rightward shifting of the critical poles $\sigma \pm j(\omega_{in}/2)$ though continuous variation of sensitive elements. This pole-shifting technique is conceptually similar to the one used to shorten the startup transient of free-running oscillators. However, for a frequency divider, the shifted poles correspond to a periodic solution at ω_{in} instead of a dc solution [24]. To avoid altering the main characteristics of the divided solution at $\omega_{in}/2$ (e.g., its output power P_{out}), we perform two consecutive simulations: one corresponding to a divided solution at $\omega_{in}/2$ and the other corresponding to an unstable nondivided solution at ω_{in}, with the dominant poles $\sigma \pm j(\omega_{in}/2)$. The aim is to shorten the transient from an unstable nondivided solution at ω_{in} to a stable frequency-divided solution at $\omega_{in}/2$. Note that the technique applies only to the initial linear stage of the transient, with no aim to control the nonlinear stage.

We assume that the divided solution at $\omega_{in}/2$ has been obtained with an auxiliary generator, fulfilling the nonperturbation conduction $Y_{AG} = 0$, with the amplitude A_{AGo} corresponding to the output power P_{out}. Two copies of this circuit will be analyzed consecutively in harmonic balance. The circuit element values, bias voltage, and input frequency and power are the same for these two copies. The procedure is based on the sweep of a sensitive element η_1 performing, at each step η_1^k the following two consecutive simulations:

1. *First circuit copy, operating at* $\omega_{in}/2$. The first copy is used to maintain the desired output power of the divided steady-state solution at $\omega_{in}/2$. Thus, it will contain an auxiliary generator at the frequency $\omega_{in}/2$ with the same amplitude A_{AGo} of the original solution. At each step η_1^k, two other circuit elements, η_2 and η_3, together with the auxiliary generator phase ϕ_{AG}, are optimized to fulfill $Y_{AG}(\phi_{AG}, \eta_2, \eta_3) = 0$. The parameter values resulting from this simulation are η_1^k, η_2^k, and η_3^k.
2. *Second circuit copy, operating at* ω_{in}. The second circuit copy, with the element values η_1^k, η_2^k, and η_3^k, is used to apply pole–zero identification to the unstable nondivided solution at ω_{in} that coexists with the divided solution. We must emphasize that no auxiliary generator is used for this simulation, as we are analyzing the nondivided solution at ω_{in}. Instead, a small-signal current generator at a frequency ω not harmonically related to ω_{in} is introduced at a sensitive circuit node. This small-signal generator allows us to obtain the closed-loop transfer function $Z_{ss}(\omega) = V_{ss}(\omega)/I_{ss}(\omega)$ through linearization of the large-signal regime at ω_{in} with the conversion matrix approach. A frequency sweep will be carried out in an interval containing $\omega_{in}/2$. The function $Z_{ss}(\omega) = V_{ss}(\omega)/I_{ss}(\omega)$ corresponding to the parameter values η_1^k, η_2^k, and η_3^k is stored.

Once the sweep has been concluded, evolution of the pair of critical poles $\sigma \pm j(\omega_{in}/2)$ versus η_1, η_2, and η_3 is determined by applying pole–zero identification to the impedance function $Z_{ss}(\eta_1, \eta_2, \eta_3, \omega)$. Note that we will have a different frequency function $Z_{ss}(\omega)$ for each of the three-element sets η_1, η_2, η_3 considered in the previous simulation.

For illustration, the technique described to reduce the settling time of frequency dividers has been applied to the regenerative FET-based frequency divider of Fig. 11.7. The input-power value is $P_{in} = -4$ dBm, above the division threshold. The divided frequency is $f_{in}/2 = 5$ GHz and the desired output power value is $P_{out} = 1.5$ dBm. The auxiliary generator is connected in parallel with the output load (Fig. 11.7). The amplitude of this generator is kept fixed at A_{AGo} to maintain the desired output power $P_{out} = 1.5$ dBm during the entire optimization process. The parameter η_1 considered is the capacitance C_S. At each C_S variation, the inductors L_G and L_D, playing the role of η_2 and η_3, are calculated to fulfill $Y_{AG}(L_G, L_D, \phi_{AG}) = 0$ for the desired output power value $P_{out} = 1.5$ dBm. Evolution of the pair of critical poles versus C_S, L_G, and L_D is shown in Fig. 11.8. Note that each point corresponds to a different divider design. As can be seen, the pair of critical poles $\sigma \pm j(\omega_{in}/2)$ initially shifts rightward versus C_S, L_G, and L_D, then turns and starts to move leftward. As shown in Fig. 11.7, the largest positive σ is obtained for $C_S = 1.08$ pF.

To verify the results of pole shifting, the settling time of the frequency divider has been analyzed with an envelope transient (Section 5.6.3) using an auxiliary

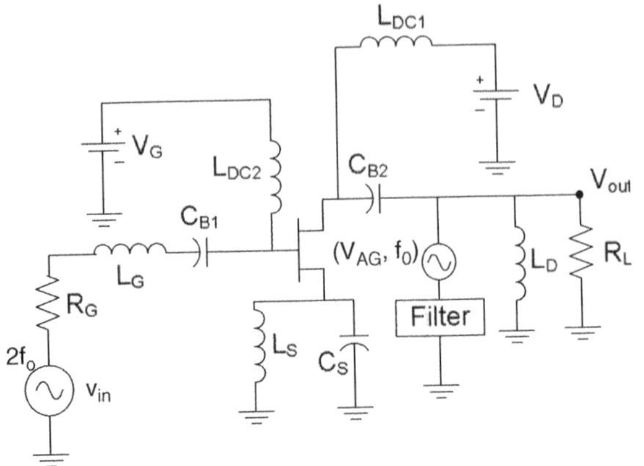

FIGURE 11.7 Hybrid divider with a 5-GHz input frequency, to which the technique for transient shortening is going to be applied.

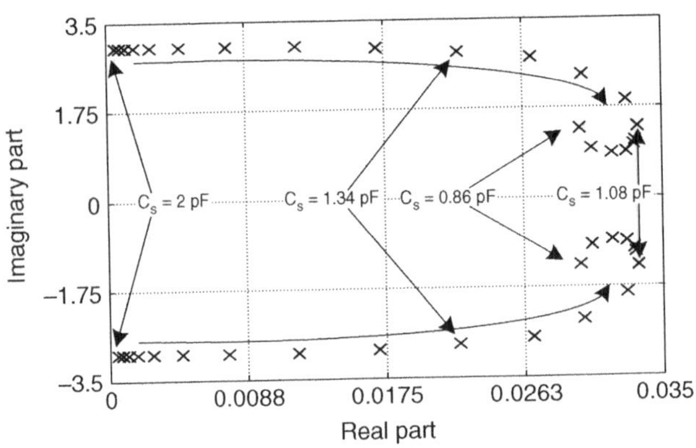

FIGURE 11.8 Pole–zero map of the unstable solution at f_{in}, showing evolution of the critic pair of complex-conjugate poles versus the set C_S–L_G–L_D.

generator at $\omega_{\text{in}}/2$ to initialize the subharmonic component. Figure 11.9 shows the resulting variation amplitude of the first-harmonic component of the output voltage Mag($V_{\text{out}}[1]$), corresponding to the subharmonic frequency $\omega_{\text{in}}/2$. Three different C_S values have been considered. In all cases, the input power is switched from zero to $P_{\text{in}} = -4$ dBm at time $t = 100$ ns. For each C_S value it is possible to distinguish in Fig. 11.9 the transient stage, with an exponential envelope, and the steady-state regime. Note that the three different designs reach the same steady-state amplitude. This is due to the fact that the steady-state amplitude at $\omega_{\text{in}}/2$ has been maintained constant during the entire pole-shifting process through the connection of the auxiliary generator at $\omega_{\text{in}}/2$ with the constant amplitude $A_{AGo} = 0.5$ V. As can be observed, the design corresponding to the largest positive σ, obtained for $C_S = 1.08$ pF, has the shortest transient.

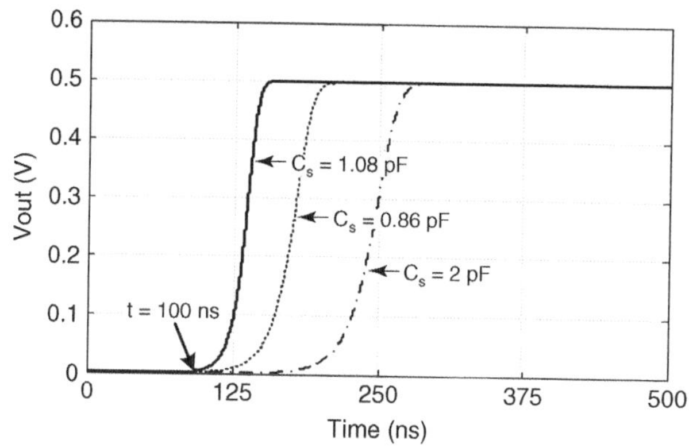

FIGURE 11.9 Envelope transient simulation showing the divider settling time for different locations of the unstable pair of poles in Fig. 11.8.

There is very good agreement between these results and the behavior expected from the pole diagram of Fig. 11.8. The average measured settling time was $t_{s1} \approx 35$ ns and $t_{s2} \approx 100$ ns for $C_{S1} = 1$ pF and $C_{S2} = 2$ pF, respectively. These measurements have been obtained using the HP-83480 digital communications analyzer.

11.4 DESIGN OF HARMONIC INJECTION DIVIDERS

A harmonic injection divider is basically an oscillator with one of its harmonic components synchronized with the input signal [15,18]. In contrast to regenerative dividers, harmonic injection dividers oscillate in the absence of input power. Therefore, the division can be achieved from very low input power. Harmonic injection division can be obtained with topologies simpler to those used for regenerative division. Actually, any oscillator circuit can be used for this type of division, since the harmonic synchronization is a natural nonlinear phenomenon occurring in all oscillatory systems excited by a periodic input. The problem is that unless a careful design is carried out, the division bandwidth, which ordinarily decreases with the order N, will be extremely narrow. In the following, a semianalytical formulation is presented to evaluate the capabilities of a given free-running oscillator for operation as a frequency divider by any order N. Then full harmonic balance design techniques will be shown.

11.4.1 Semianalytical Estimation of Synchronization Bands

Let's consider a free-running oscillator at a frequency ω_o. If an input generator at ω_{in} is connected to this oscillator, it will be possible to obtain synchronization bands for ω_{in} values on the order of $\omega_{in} \cong N\omega_o$. Harmonic balance techniques to obtain these division bands accurately were presented in Chapter 5. However, for a small amplitude in the input generator, the synchronization bands can be determined approximately through linearization of the circuit about its nonlinear free-running oscillator regime. In Section 4.3.2, linearization was applied to the total admittance function of a free-running oscillator, that fulfills $Y_T(V_o, \omega_o) = 0$, with V_o and ω_o being the steady-state oscillation amplitude and frequency, respectively. Under the presence of the input source, this admittance function is expanded in a Taylor series, as shown in Section 4.3.2. In the analysis presented below, the free-running oscillator is simulated through harmonic balance using an auxiliary generator. Then the derivatives of the admittance function required for the Taylor series expansion are calculated through increments in the auxiliary generator amplitude and frequency. The procedure is analogous to the one already used for the semianalytical phase noise analysis of injection-locked oscillators in Chapter 7, and for the analysis of coupled oscillators in Chapter 10. This procedure is described in detail below.

As already stated, the original free-running oscillator is analyzed using harmonic balance. Simulation is carried out by an auxiliary generator connected to the circuit node n, and the resulting values, fulfilling $Y_{AG} = 0$, are $\omega_{AG} = \omega_o$ and $A_{AG} = V_o$ (see Fig. 11.10). Next, the synchronizing source $e_{in}(t)$ is introduced. For frequency division by N, the fundamental frequency and first-harmonic component of the voltage at node n will be given by $\omega_s = \omega_{in}/N$ and V_s. For a small amplitude of $e_{in}(t)$, these two values should be close to ω_o and V_o. Setting the phase origin at node n, the input generator voltage will be expressed: $e_{in}(t) = Re(E_{in}e^{j(\omega_{in}+\phi)t})$, with ϕ being the opposite of phase shift between the frequency component at ω_s at node n and the input generator at $N\omega_s$. The two-tier formulation of the synchronized system is

$$\begin{aligned} \overline{H}\left(\overline{X}(V_s, \omega_s)\right) &= 0 \text{ inner tier} \\ Y(V_s, \omega_s, E_{in}e^{j\phi})V_s e^{j0} &= 0 \text{ outer tier} \end{aligned} \tag{11.2}$$

where Y is the total current-to-voltage relationship at the fundamental frequency ω_s at node n and $\overline{H}(\overline{X}) = 0$ is the pure harmonic balance system, with $\overline{X}$ being the vector comprised of harmonic components of the various state variables. The admittance Y agrees with the admittance function Y_{AG} of the auxiliary generator. For a small amplitude of the input source E_{in}, it is possible to expand the outer tier of (11.2) in a first-order Taylor series about the free-running oscillation

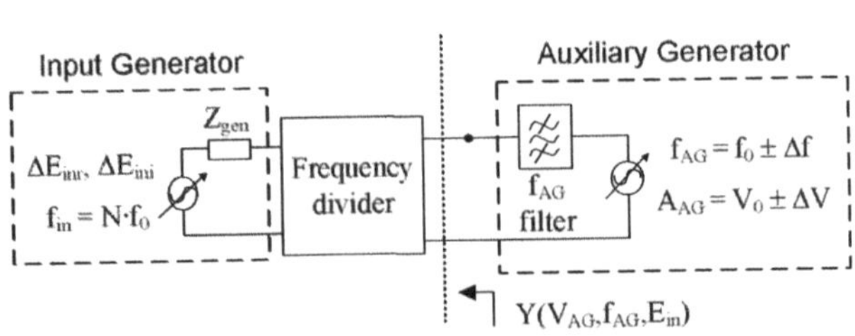

FIGURE 11.10 Circuit arrangement for calculation of the admittance derivatives used in semianalytical estimation of the frequency-divided solution for low input amplitude.

$(V_o, \omega_o, E_{\text{in}} = 0)$:

$$\frac{\partial Y_o}{\partial V_o}\Delta V_s + \frac{\partial Y_o}{\partial \omega_o}\Delta\omega_s = -\frac{\partial Y_o}{\partial E_{\text{in}}^r} E_{\text{in}} \cos\phi - \frac{\partial Y_o}{\partial E_{\text{in}}^i} E_{\text{in}} \sin\phi \tag{11.3}$$

with $\Delta\omega_s = \omega_s - \omega_o$ and $\Delta V_s = V_s - V_o$.

The derivatives in (11.3) can be calculated numerically through increments in the auxiliary generator frequency and amplitude and in the real part of the input generator, as shown in Fig. 11.10. The harmonic balance system must be resolved at each of these increments, according to (11.2). Therefore, use of the auxiliary generator allows us to take into account the entire harmonic content of the free-running steady-state solution. In more detail, the derivative calculation is performed as follows:

$$\begin{aligned}
\frac{\partial Y_o}{\partial V_o} &= \frac{Y_{AG}(V_o + \Delta V, \omega_o) - Y_{AG}(V_o - \Delta V, \omega_o)}{2\Delta V} \\
\frac{\partial Y_o}{\partial \omega_o} &= \frac{Y_{AG}(V_o, \omega_o + \Delta\omega) - Y_{AG}(V_o, \omega_o - \Delta\omega)}{2\Delta\omega} \\
\frac{\partial Y_o}{\partial E_{\text{in}}^r} &= \frac{Y_{AG}(V_o, \omega_o, E_{\text{in}}^r) - Y_{AG}(V_o, \omega_o, -E_{\text{in}}^r)}{2E_{\text{in}}^r}
\end{aligned} \tag{11.4}$$

where E_{in}^r, with the superscript indicating the real part, is a small increment in the real part of the generator, maintaining the imaginary part at zero value, $E_{\text{in}}^i = 0$. Note that it is sufficient to calculate the admittance derivatives with respect to either the real or imaginary part of the input generator, E_{in}^r or E_{in}^i, due to fulfillment of the Cauchy–Riemann relationships, which establish that

$$\frac{\partial Y_o^r}{\partial E_{\text{in}}^r} = \frac{\partial Y_o^i}{\partial E_{\text{in}}^i} \qquad \frac{\partial Y_o^r}{\partial E_{\text{in}}^i} = -\frac{\partial Y_o^r}{\partial E_{\text{in}}^r} \tag{11.5}$$

As shown in Chapter 4, the solution of (11.3) is an ellipse in the plane defined by ω_s and V_s, centered about the free-running oscillation point ω_o and V_o. This ellipse constitutes the divided-solution curve. As shown in Chapter 4, the turning points of the ellipse are, in fact, local–global (mode-locking) bifurcations at which the solution gets or loses synchronization. It must also be taken into account that at the turning point of any solution curve traced versus a given parameter, a real pole passes through zero. Thus, if the upper section of the ellipse is stable, the lower section will be unstable. The division bandwidth $\Delta\omega_{1/N}$ by the order N is given by twice the maximum of the increment $|\Delta\omega_s|$, which is easily calculated from (11.3). This provides the following expression [see (4.73)]:

$$\Delta\omega_{1/N} \equiv 2|\Delta\omega_s|_{\max} = 2E_{\text{in}} \frac{|\partial Y_o/\partial E_{\text{in}}|}{|\partial Y_o/\partial \omega_o|\,|\sin\gamma|} \tag{11.6}$$

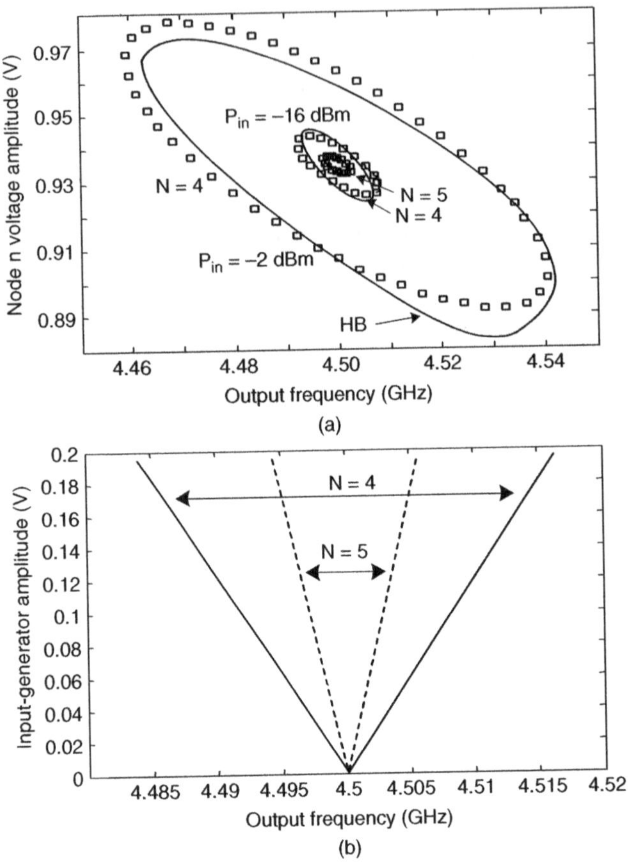

FIGURE 11.11 Estimation of the division capabilities of a free-running oscillator using semianalytical formulation. Two values of the division order, $N = 4$ and $N = 5$, have been tested. (a) Comparison of the synchronization curves obtained with (11.3) and with harmonic balance, considering 15 harmonic components. (b) Variation of the division bandwidth versus the input generator amplitude.

where γ is the angle difference $\gamma = ang(\overline{Y}_{T\omega}) - ang(\overline{Y}_{ov})$. Expression (11.6) enables a simple evaluation of the capabilities of a given oscillator circuit to operate as a harmonic injection divider by order N, as it provides an estimation of the division bandwidth by any order N for low input power at ω_{in}.

The technique has been applied to the FET-based circuit considered in Chapter 4, with free-running frequency $f_o = 4.5$ GHz, operating as a harmonic injection divider by $N = 4$ and $N = 5$. The results of the semianalytical technique are shown in Fig. 11.11. In Fig. 11.11a, the frequency-divided solutions are shown in terms of the voltage amplitude at the observation node versus the divided frequency. The divided-solution curves obtained through the semianalytical approach (11.3) are compared with full harmonic balance using 15 harmonic components.

The harmonic-balance synchronization curves have been obtained by introducing a auxiliary generator at $\omega_{AG} = \omega_{in}/N$, as explained in Chapter 5. The phase of this generator is set to zero value, $\phi_{AG} = 0$, and the phase of the input source at ω_{in} is swept between 0 and 2π. At each step the auxiliary generator amplitude A_{AG} and frequency $\omega_{AG} = \omega_{in}/N$ are optimized so as to fulfill the nonperturbation condition $Y_{AG} = 0$.

The upper sections of the various closed curves correspond to stable behavior. The lower sections are unstable, as has been verified with a complementary stability analysis. For $P_{in} = -16$ dBm, the semianalytical division curves almost overlap with those resulting from harmonic balance analysis for the two division orders-break $N = 4$ and $N = 5$. As the input power increases, the circuit starts to behave nonlinearly with respect to the input source, which reduces the accuracy of the first-order Taylor series expansion (11.3). This is why less agreement is obtained for $P_{in} = -2$ dBm. Clearly, the circuit in Fig. 4.8, operating as a harmonic injection divider, will have a narrower bandwidth for the division order $N = 5$ than for $N = 4$. This is due to the smaller value of the magnitude $|\partial Y_o/\partial E_{in}|$ in the case of $N = 5$, as all the other derivatives in (11.6) are the same for the two division orders.

Figure 11.11b shows the variation in division bandwidth versus the input generator amplitude E_{in} estimated from the linearization (11.6), for two different orders $N = 4$ and $N = 5$. As noted in Chapter 4, these V-shaped synchronization regions obtained for the various division orders are known as *Arnold tongues*. Figure 11.11b provides Arnold tongues corresponding to $N = 4$ and $N = 5$. They have been traced in the plane defined by the output frequency and input amplitude E_{in}, which is why they overlap. Their lower vertexes, obtained for $E_{in} = 0$, correspond to the free-running solution $\omega_{in} = N\omega_o$ and $E_{in} = 0$. As can be seen, a broader bandwidth in obtained for $N = 4$. Changes in the original oscillator to broaden the division bandwidth by $N = 5$ may include a better design of the input filter, centered about $5\omega_o$, and/or increasing the feedback voltage amplitude at the harmonic component $4\omega_o$ at the transistor input. A technique to increase the feedback amount is presented next.

11.4.2 Full Harmonic Balance Design

In the following, a full harmonic balance design of a harmonic injection divider is presented. Two stages are considered: optimization of the initial free-running oscillator and optimization of the frequency divider itself. In the divider optimization, two different aspects are considered: avoiding shifting of the frequency-division band during the design process, and broadening the division band through the introduction of a low-frequency feedback network.

11.4.2.1 Optimization of the Original Free-Running Oscillator The output frequency $\omega_s = \omega_{in}/N$ of a frequency divider is the result of mixing of the input signal at $N\omega_s$ with the feedback signal at $(N - 1)\omega_s$. Increasing the feedback amount at $(N - 1)\omega_s$ with enhance the sensitivity of the input generator. Thus, a suitable feedback network design will be essential for a broad division bandwidth.

In a single-ended divider, mixing is provided by the transistor transconductance, and the increase in harmonic amplitude will be achieved through modification of the transistor operation conditions, varying the bias sources and the circuit-element values.

The technique to increase the feedback amount can be applied to the original free-running oscillator, as this feedback will depend primarily on the circuit topology and bias point of the transistor device. To enhance the harmonic amplitude at $(N-1)\omega_o$, a continuation technique is applied, taking the original oscillator design as the starting point and varying the sensitive circuit parameters. An auxiliary generator at ω_o allows us to maintain the same free-running oscillation frequency during the entire design process. To illustrate, the technique will be applied to the divider circuit considered in Section 11.4.1.

For a FET-based circuit, the feedback voltage $(N-1)\omega_o$ has been analyzed at both $N=4$ for division by $N=4$, and $N=5$ for division by $N=5$. Different gate bias values V_{GG} between -0.5 and -1.5 V have been tested, performing a sweep in the drain bias V_{DD}. For each V_{GG} and V_{DD}, the auxiliary generator (at $\omega_{AG}=\omega_o$, $A_{AG}=V_o$) fixes the desired value of the free-running oscillation frequency ω_o and the voltage amplitude, at this frequency, at the auxiliary generator location. Here this auxiliary generator is connected to the gate terminal. For each point of the V_{DD} sweep (at constant V_{GG}), the values of two circuit elements are determined so as to fulfill the nonperturbation conditions $Y_o(\gamma_1,\gamma_2)=0$. Here, the optimized elements γ_1 and γ_2 are the load resistance R_L and inductance L_L. As already stated, to avoid convergence problems, a continuation technique using the final values R_L^{n-1} and L_L^{n-1} obtained for V_{DD}^{n-1} as the initial guess for V_{DD}^{n}.

The results of the design procedure described above are shown in Fig. 11.12a. The highest harmonic amplitudes at both $3\omega_o$ and $4\omega_o$ obtained in the simulation of Fig. 11.12a correspond to V_{GG} and V_{DD} values at the corners of the characteristic curves, with very nonlinear behavior. Two spectra, at the gate port, are compared in Fig. 11.12b. Both have exactly the same fundamental frequency ($f_o=4.5$ GHz) and voltage amplitude ($V_o=1$ V) at the gate terminal where the auxiliary generator is connected. Bias point 1 is $V_{GG}=-1$ V and $V_{DD}=4$ V, and bias point 2, which has the richest harmonic content, is $V_{GG}=-0.62$ V and $V_{DD}=2$ V. Figure. 11.13 compares the division bandwidth by $N=5$ obtained before and after use of the technique. The input voltage amplitude is $E_{in}=0.1$ V. The closed synchronization curves are traced in terms of the voltage amplitude, at the observation node, versus the output frequency. A significant increase in division bandwidth has been obtained without affecting the free-running frequency $f_o=4.5$ GHz or amplitude. Note that the center point of the ellipse, corresponding to the free-running oscillator, is the same for the two bias points.

11.4.2.2 *Technique to Prevent Frequency Shifting of Division Bands*

In microwave circuit design it is common practice to increase the accuracy of the component models in successive steps, with gradual introduction of parasitic elements and microstrip discontinuities, for example. These changes in the circuit components will lead to an undesired shift of the frequency band unless a complementary design technique is employed. To derive this technique, we will take

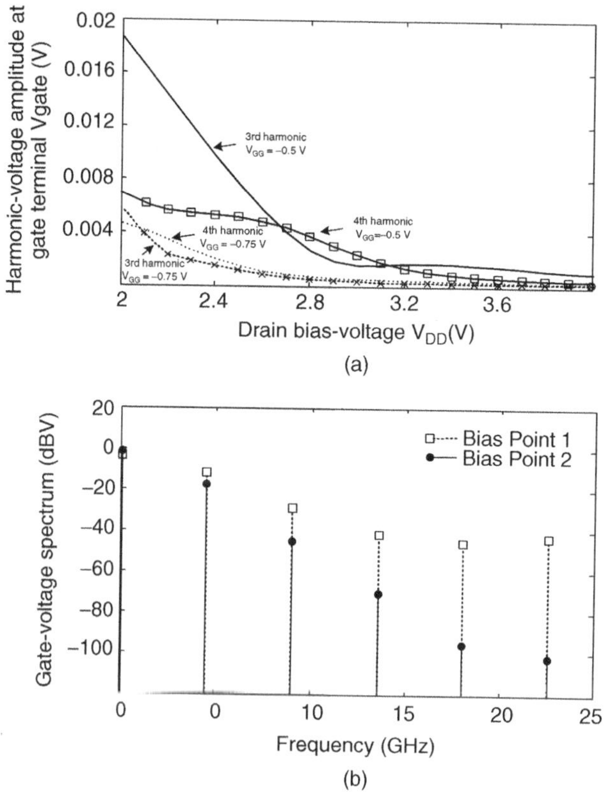

FIGURE 11.12 Free-running oscillator. Technique for increasing the harmonic amplitude of the gate voltage while keeping constant the oscillation frequency at $f_o = 4.5$ GHz. (a) A sweep is performed in the drain bias voltage V_{DD} for two values of the gate bias voltage V_{GG}. Each point of the curves corresponds to a different design (different values of the oscillator load elements). (b) Comparison of the voltage spectrum at the gate terminal for two different designs with the same oscillation frequency.

into account that the linearized analysis of (11.3) predicts a synchronization band centered perfectly about $\omega_{ino} = N\omega_o$. Thus, the objective will be to maintain the free-running oscillation amplitude and frequency at the different design stages. Note that as the divider starts to behave in a nonlinear manner with respect to the input source, the synchronization band will gradually deviate from perfect symmetry about $\omega_{ino} = N\omega_o$. This is due to the usual bending of the V-shaped synchronization regions (Arnold tongues) in the plane defined by the input power and frequency (see Section 4.3.2). If a high input power is required for a nonnegligible division band, the technique may not be applicable, as this division band may not be centered about the free-running solution given by $N\omega_o$ and V_o.

Here it is assumed that the division bandwidth is not negligible for a low input power. Then the frequency shift in the synchronization band can be avoided through reoptimization of the free-running oscillation regime, by setting the frequency of the

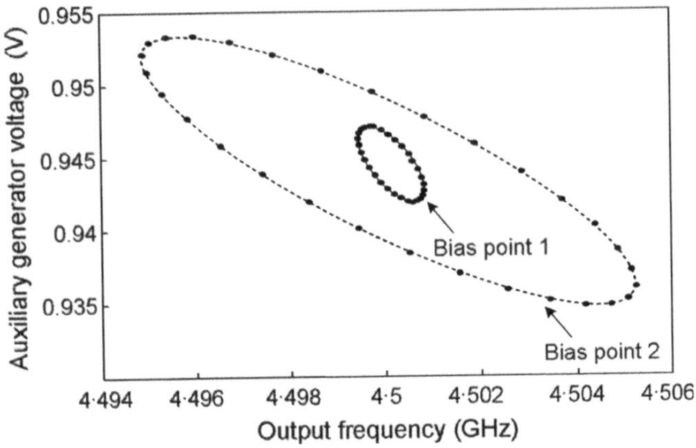

FIGURE 11.13 Comparison of synchronization bands by $N = 5$ before and after use of the technique to increase the feedback harmonic component $(N - 1)\omega_o$.

auxiliary generator to the fixed value $\omega_{AG} = \omega_{ino}/N$ and the auxiliary generator amplitude to $A_{AG} = V_o$. This allows us to maintain the free-running oscillation frequency and amplitude at the desired values ω_o and V_o. Thus, the division bands will keep centered about this point for low input power and there will be no frequency shifting or degradation of the output power. After each new inclusion of parasitic elements or discontinuity models, a couple of variables (e.g., a line length and width) are optimized so as to fulfill the free-running oscillation condition $Y_{AG} = 0$. The stability of the design, together with the oscillation startup conditions, must be checked after each new change.

An example of the use of the technique is shown in Fig. 11.14. The synchronization curves corresponding to the division order $N = 4$ and input power $P_{in} = -2$ dBm are shown at various stages of the circuit design. They have been traced in terms of the output voltage amplitude versus the divided frequency. Initially, only ideal lumped elements are used in the linear embedding network. In the second simulation, some of these elements are implemented on microstrip using the expressions provided in Chapter 8, and the rest of the elements are replaced with full manufacturer models, taking parasitics into account. Microstrip discontinuities associated with steps, tees, and crosses have also been considered. In the third simulation, the biasing networks have been implemented using radial stubs along with quarter-wave lines. Note that all curves obtained through harmonic balance are centered about the values $f_o = 4.5$ GHz and $V_o = 1$ V as a result of free-running optimization of the oscillator. However, as shown in Chapter 4, the slope of the curves changes for the three implementations, depending on the derivatives of the admittance function at the free-running solution (calculated at the observation node). This is because the free-running solution is different in the three cases. The major reduction in the synchronization band has been found to be due to the input radial stub. After noting this, a different implementation of the bias filter must be employed.

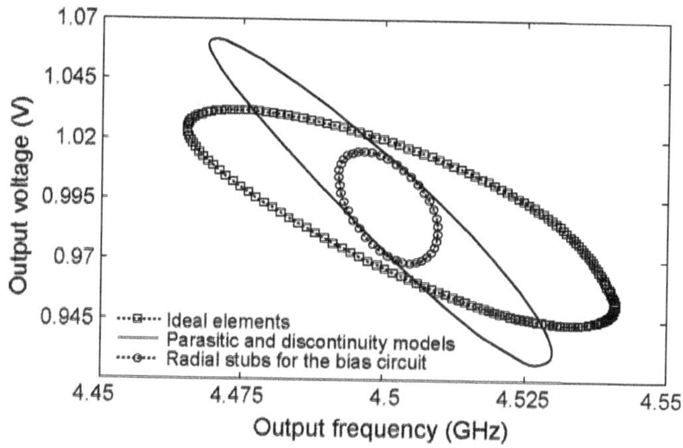

FIGURE 11.14 Method to prevent the frequency shift of the division bands as parasitic elements and discontinuities are introduced in the circuit description, at different stages of the design process. Solution curves corresponding to the division order $N = 4$ have been obtained through harmonic balance with 15 harmonic components. The curves are always centered about the output frequency $f_s = 4.5$ GHz. The input power in all cases is $P_{in} = -2$ dBm.

11.4.3 Introduction of a Low-Frequency Feedback Loop

As already stated, harmonic injection dividers usually have narrow synchronization bandwidth. To broaden this bandwidth, we may add a phase-locking loop to the harmonic injection divider, following the technique presented by Chang et al. [26]. To apply this technique, the low-frequency term $|\omega_{in} - N\omega_a|$, generated by the device transconductance, is extracted from the transistor output terminal with the aid of a choke inductor (see Fig. 11.15). This signal is then introduced into a low-frequency amplifier and the amplifier output is connected to the varactor diode. The low-frequency feedback induces a variation in the oscillation frequency ω_a so as to reduce the error $|\omega_{in} - N\omega_a|$. In a rough manner, this can be seen as a variation of the circuit free-running oscillation frequency, depending on the generator values, since the capacitance of the varactor diode changes with these values. This frequency variation mechanism will be combined with the injection locking for a substantial increase in the division bandwidth. Note that the signal detected at the transistor output should be constant for a periodic steady-state regime. A simplified analytical explanation of the injection-locked, phase-locked divider is presented next.

Input generator values $P_{in}, \omega_{in}, \phi_{in}$ are considered such that the autonomous frequency can be expressed as $\omega_a = \frac{\omega_{in}}{N} + \dot{\phi}(t)$. Using an envelope-type representation with $\frac{\omega_{in}}{N}$ as fundamental, the observation-node voltage at the divided frequency is written as $(V_o + \Delta V(t))e^{j\left(\phi(t)+\frac{\omega_{in}}{N}t\right)}$. Then the detected voltage signal has a low-frequency variation at $\omega_{in} - N\omega_a = -N\dot{\phi}(t)$. This signal is represented as

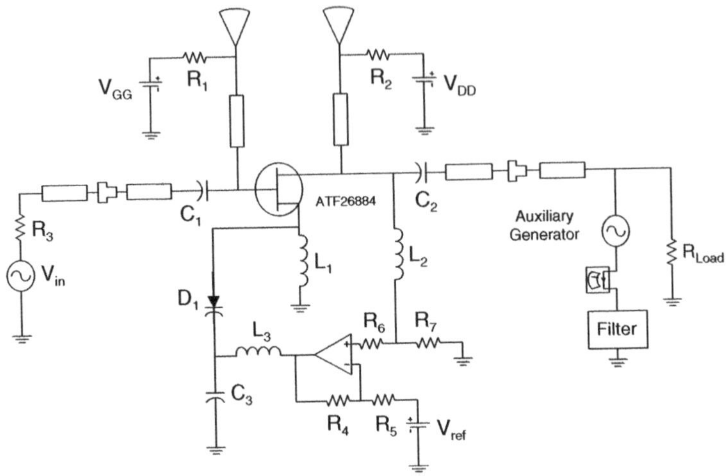

FIGURE 11.15 Harmonic injection divider by order $N = 4$. The external low-frequency loop is introduced to increase the synchronization bandwidth through the combination of injection- and phase-locking effects.

$v_d(t) = V_{d,o}\cos[-N\phi(t) + \phi_d]$. The amplified signal between the varactor terminals is $v_b(t) = G_v v_d(t)$, with G_v the amplifier voltage gain. In these conditions, the free-running frequency is $\omega_o'(t) = \omega_o + KV_{d,o}\cos[-N\phi(t) + \phi_d]$, with K being the oscillator-tuning sensitivity. The admittance function at the observation node will be linearized about this free-running solution, considering the amplitude and frequency increments $\Delta V(t)$ and $\omega_a - \omega_o'$:

$$\frac{\partial Y_o}{\partial V}\Delta V(t) + \frac{\partial Y_o}{\partial \omega}\left(\frac{\omega_{in}}{N} + \dot{\phi}(t) - \omega_o - KV_{d,o}\cos[-N\phi(t) + \phi_d]\right)$$
$$= -\frac{\partial Y_o}{\partial E_{in}} E_{in} e^{j(\phi_{in} - \phi(t))} \qquad (11.7)$$

The incomplete system (11.7) requires an additional model for the prediction of the amplitude $V_{d,o}$ and constant phase shift ϕ_d of the low-frequency signal, which will change with the input generator values. In [26], they use the nonlinear transconductance model of the transistor for the calculation of $V_{d,o}$ and estimate ϕ_d from the phase shift from gate to drain through the device. In spite of its limitations, (11.7) evidences the increased sensitivity to the phase shift $\phi(t)$ through the varactor-bias voltage $v_b(t)$. This sensitivity will be higher for higher tuning sensitivity and higher gain G_V of the dc amplifier. Within the division bandwidth, the phase shift takes a constant value ϕ_o, so equation (11.7) becomes

$$\frac{\partial Y_o}{\partial V}\Delta V_s + \frac{\partial Y_o}{\partial \omega}\left(\frac{\omega_{in}}{N} - \omega_o - KV_{d,o}\cos(-N\phi_o + \phi_d)\right)$$
$$= -\frac{\partial Y_o}{\partial E_{in}^r} E_{in}\cos(\phi_{in} - \phi_o) + \frac{\partial Y_o}{\partial E_{in}^i} E_{in}\sin(\phi_{in} - \phi_o) \qquad (11.8)$$

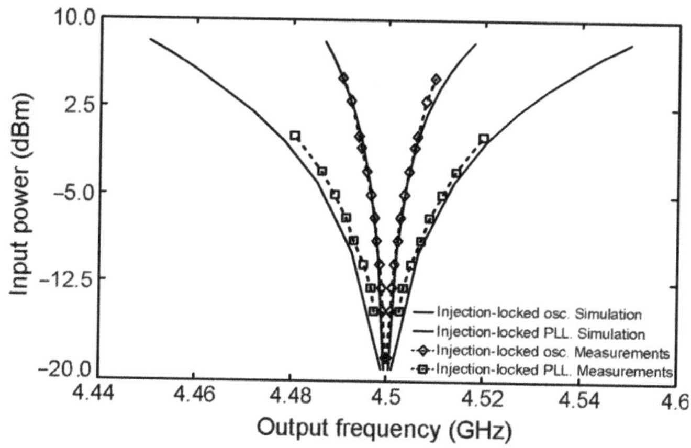

FIGURE 11.16 Connection of the low-frequency feedback loop to the frequency divider by $N = 4$. Turning-point locus delimiting the synchronization bandwidth in the place defined by the input frequency and input power before and after connection of the loop. Measurements are superimposed.

The effect of the use of injection-locking phase locking to the circuit of Fig. 11.15 is shown in Fig. 11.16, where the locus of turning points, delimiting the synchronization bandwidth, has been represented in the plane defined by the output frequency and input power. Note that the frequency point corresponding to zero input power (-∞ dBm) agrees with the free-running oscillation frequency. As can be seen, the technique enables a substantial enlargement of the synchronization bandwidth. Measurements are superimposed.

11.4.4 Control of Turning Points

As shown in previous sections, the edges of the synchronization bands in harmonic and subharmonic injection oscillators are given by local or global bifurcations occurring at turning points of the periodic synchronization curves. Thus, bifurcation control can be used to preset the operation bands of this type of circuit. The objective will be to shift the points of infinite slope of the closed synchronization curves, as illustrated in Fig. 11.17a. As shown in Chapter 6, the two turning points T_L and T_U, determining the lower and upper band edges, respectively, fulfill the condition $\det[JY_{AG}] = 0$, with $[JY_{AG}]$ being the Jacobian matrix associated with the auxiliary generator and defined as follows:

$$[JY_{AG}] = \begin{bmatrix} \dfrac{\partial Y^r_{AG}}{\partial A_{AG}} & \dfrac{\partial Y^r_{AG}}{\partial \phi_{AG}} \\ \dfrac{\partial Y^i_{AG}}{\partial A_{AG}} & \dfrac{\partial Y^i_{AG}}{\partial \phi_{AG}} \end{bmatrix} \tag{11.9}$$

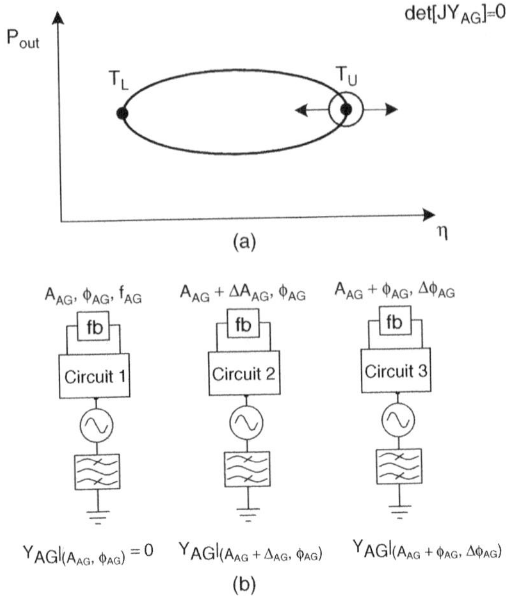

FIGURE 11.17 Control of turning points to preset the operation bands of harmonic injection dividers: (a) the objective of the control technique is the shifting of the curve points fulfilling $\det[JY_{AG}] = 0$; (b) three circuit copies, each containing an auxiliary generator in different operation conditions, for the harmonic balance determination of turning points.

Using the set of control element values $\overline{\rho}_c$, the condition for shifting of a turning point obtained initially for the parameter value η_b to a new value η'_b will be

$$\begin{aligned} Y_{AG}(A_{AG}, \phi_{AG}, \overline{\rho}_c) &= 0 \\ \det[JY_{AG}(A_{AG}, \phi_{AG}, \overline{\rho}_c)] &= 0 \\ \eta &= \eta'_b \end{aligned} \tag{11.10}$$

which is solved in combination with harmonic balance. In system (11.10), three real equations must be solved. Again, no particular constraints are generally imposed on ϕ_{AG}. By setting A_{AG} to a constant value, it is possible to preserve the original subharmonic output power at the band edge. Then the inclusion of at least two optimization elements in the vector $\overline{\rho}_c$ will be necessary. The turning-point condition is given by

$$\det[JY_{AG}(A_{AG}, \phi_{AG}, \overline{\rho}_c)] = \frac{\partial Y^r_{AG}}{\partial A_{AG}} \cdot \frac{\partial Y^i_{AG}}{\partial \phi_{AG}} - \frac{\partial Y^i_{AG}}{\partial A_{AG}} \cdot \frac{\partial Y^r_{AG}}{\partial \phi_{AG}} = 0 \tag{11.11}$$

The derivatives appearing in $[JY_{AG}]$ are calculated through increments, performing a harmonic balance simulation at each particular increment. When using commercial harmonic balance software to shift the turning points, these derivatives must be available for evaluation of the determinant function (11.11) during the optimization process. Note that $\det[JH] = 0$ constitutes a turning-point condition and will be one of the optimization goals of the design process. The derivatives in (11.11) are obtained by resolving three copies of the circuit simultaneously in the same harmonic balance simulation (Fig. 11.17b). The three copies will have the same values of the circuit component, bias generators, and input-source power and frequency P_{in} and ω_{in}. An auxiliary generator is connected to each copy. One of the copies corresponds to the actual operation point, and its associated auxiliary generator must fulfill $Y_{AG} = 0$. The second copy, which operates at A_{AG} and $\phi_{AG} + \Delta\phi_{AG}$, is used to calculate the derivative $\partial Y_{AG}/\partial\phi_{AG}$. The third copy, which operates at $A_{AG} + \Delta A_{AG}$ and ϕ_{AG}, enables calculation of the derivative $\partial Y_{AG}/\partial A_{AG}$.

The three circuit copies are analyzed simultaneously in a single harmonic balance simulation. The optimization goals will be $Y_{AG} = 0$ (to be fulfilled by the first circuit copy) and $\det[Y_{AG}] = 0$. Note that by using the two additional circuit copies with incremented values, the derivatives comprising $[JY_{AG}]$ are available for evaluation of the determinant $\det[Y_{AG}]$. The optimization elements $\overline{\rho}_c$ (which are, of course, common to the three circuit copies) may belong to the original circuit design or may correspond to the element values of an additional feedback network introduced specifically for bifurcation control. Use of this additional feedback network with the parameters $\overline{\rho}_c$ will reduce the disturbance of the original circuit response.

The technique for shifting the edges of the synchronization bands of harmonic injection dividers will be used on a frequency divider by $N = 2$ in hybrid technology. The input power considered is $P_{in} = -10$ dBm. The original frequency-divided curve is represented by the dashed line in Fig 11.18. This curve has been traced in terms of the output power at the divided-by-2 frequency

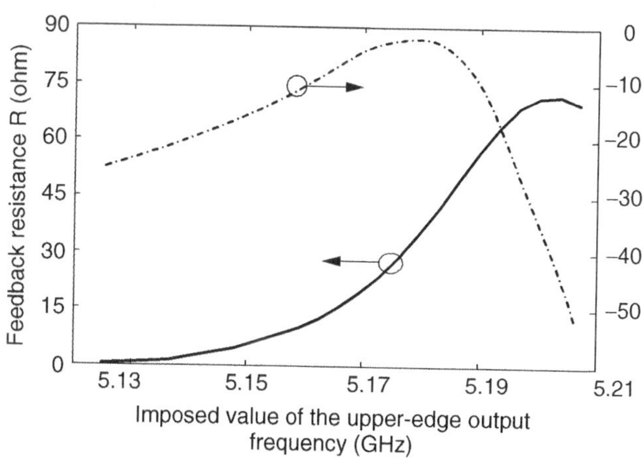

FIGURE 11.18 Variation in the ideal feedback elements versus the frequency imposed at the upper edge of the synchronization band.

versus this divided frequency. The edges of the division band are given by the two turning points obtained for the output frequency and power values f_L, P_L (the lower edge) and f_U, P_U (the upper edge), respectively. For the control of these turning points, an additional parallel feedback network will be introduced into the circuit. As already stated, the feedback network may either have specific topology, as in the example of Section 11.3, or be constituted by an ideal impedance block. To illustrate this second possibility, an ideal impedance block is considered here, defined as

$$Z(\omega) = \begin{cases} R + jX & \omega = \dfrac{\omega_{\text{in}}}{2} \\ \infty & \omega \neq \dfrac{\omega_{\text{in}}}{2} \end{cases} \tag{11.12}$$

The ideal feedback network enables more flexibility for the bifurcation control, but requires a later implementation stage of the resulting ideal impedance function. Using the above ideal feedback network, two different control actions will be performed.

11.4.4.1 Setting the Upper Synchronization Edge to the Desired Frequency and Power Values f'_U and P'_U

The power at the upper edge of the division band will be set to the value $P'_U = 8$ dBm. As shown in Fig. 11.19, this value is higher than the value obtained in the original design, about 2.5 dBm. To set this output-power value, the auxiliary generator is connected in parallel with the output load, with the fixed amplitude value $A_{AG} = \sqrt{2P'_U R_L}$, with $P'_U = 8$ dBm. Next, the value of the imposed upper-edge frequency will be varied between $f'_U = 5.12$ and 5.21 GHz. The values required for the real and imaginary parts, R and X,

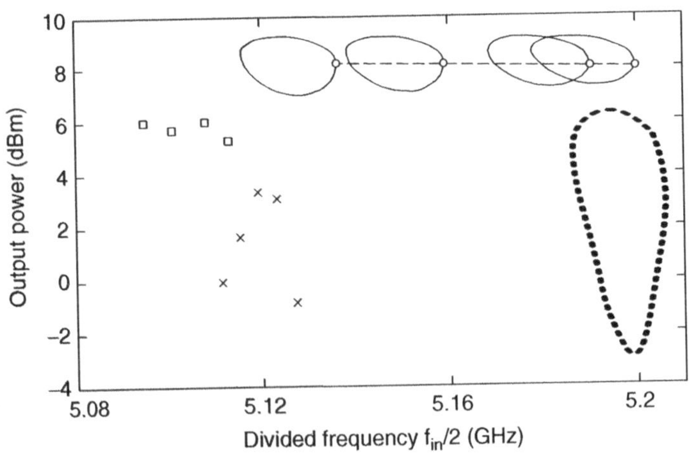

FIGURE 11.19 Technique for the control of turning points in the closed synchronization curves of harmonic injection dividers. The upper edge of the synchronization band has been shifted, maintaining a constant value $P_{\text{out}} = 8$ dBm of the output power at the subharmonic frequency $\omega_{\text{in}}/2$. The original synchronization curve is also represented (dashed line).

of the ideal feedback impedance are determined by solving (11.10) in terms of R, X, and ϕ_{AG}. Evolution of the ideal values R and X versus the upper-edge frequency imposed is shown in Fig. 11.18. Note that each pair of R and X values corresponds, in fact, to a different frequency-divider design. The actual shifting of the upper edge of the division band is verified in Fig. 11.19. This figure shows the frequency-divided curves corresponding to different designs obtained by imposing particular values of the upper-edge frequency, as shown in Fig. 11.18. The identical output power value at the upper edge of all the different synchronization curves can be noted. Measurement points corresponding to the original design and the design with the imposed upper edge frequency 5.17 GHz are superimposed. Disagreements are attributed to discrepancies in the models of the linear and nonlinear circuit elements. Note that general-purpose models were used in the harmonic balance simulations.

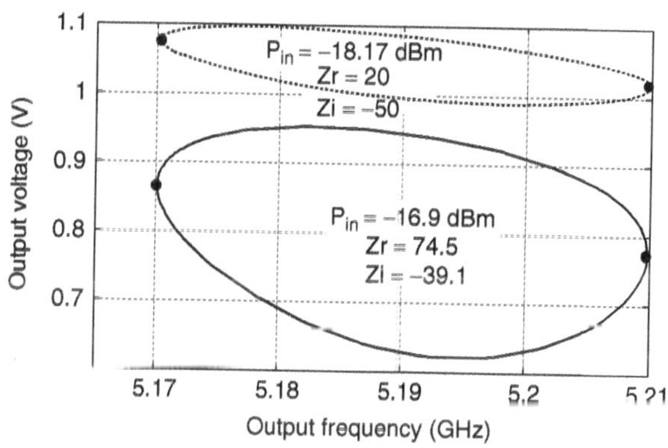

FIGURE 11.20 Simultaneous presetting of the two edges of a synchronization band with the technique for turning-point control.

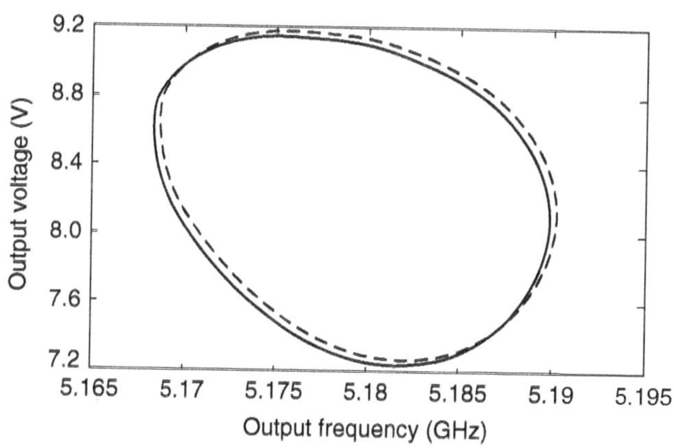

FIGURE 11.21 Comparison of a synchronization curve before and after synthesis of the feedback network.

11.4.4.2 Setting the Frequency Edges of the Synchronization Band, f'_L and f'_U

The second design goal has been to set the two frequency edges of the division band, f'_L and f'_U, simultaneously. To do this, two sets of equations are considered: $Y_{AG1} = 0$ and $\det[JY_{AG1}] = 0$ with $f_{AG1} = f'_L$ and $Y_{AG2} = 0$ and $\det[JY_{AG2}] = 0$ with $f_{AG2} = f'_U$. It is a system of six real equations that must be solved for six unknowns: $A_{AG1}, A_{AG2}, \phi_{AG1}, \phi_{AG2}$, R, and X. Figure 11.20 shows the results of this technique when imposing $f'_L = 5.17$ GHz and $f'_U = 5.21$ GHz, for two different values of input power. As can be seen, the two frequency edges agree closely with the values imposed. The final stage is implementation of the ideal feedback network. In this case, it has been possible to implement the network with just a resistance in series with a capacitance. Note that more complex synthesis techniques will generally be necessary in circuits with broader synchronization bandwidth. Figure 11.21 compares the synchronization curves before and after (dashed line) the synthesis of the ideal network for a particular design.

11.5 EXTENSION OF THE TECHNIQUES TO SUBHARMONIC-INJECTION OSCILLATORS

In a sub-harmonically injected oscillator, the circuit self-oscillation gets locked to the mth harmonic component of the input source at $m\omega_{in}$. Because the source must be able to generate this mth harmonic component, the required input power should be relatively high. Thus, the sub-synchronized oscillator will operate in nonlinear regime with respect to the input source and for reasonable input-power values, the frequency bandwidth is typically much smaller than in harmonically injection-locked oscillators [16]. In similar manner to the harmonic-injection dividers, the edges of the synchronization bandwidth are delimited by two turning points of local/global type.

For the harmonic balance analysis of synchronized solutions, the fundamental frequency considered will be the input-source frequency ω_{in}. In turn, the frequency of the auxiliary generator will agree with $\omega_{AG} = m\omega_{in}$. The nonperturbation condition will be given by $Y_{AG}(A_{AG}, \phi_{AG}) = 0$. The turning points delimiting the synchronization band will fulfil the condition

$$\begin{aligned} &Y_{AG}(A_{AG}, \phi_{AG}, \overline{\rho}_c) = 0 \qquad \text{at } \omega_{AG} = m\omega_{in} \\ &\det[JY_{AG}(A_{AG}, \phi_{AG})] = 0 \end{aligned} \tag{11.13}$$

The operation bands of subharmonic injection oscillators are delimited by the same phenomena as are harmonic injection dividers. Thus, these bands can be imposed by using the same bifurcation control techniques. On the other hand, the combination of injection locking and phase locking enabled by their introduction into the circuit's low-frequency feedback loop, will be very convenient to broaden the narrow synchronization band that is typically obtained in subharmonically injection-locked oscillators.

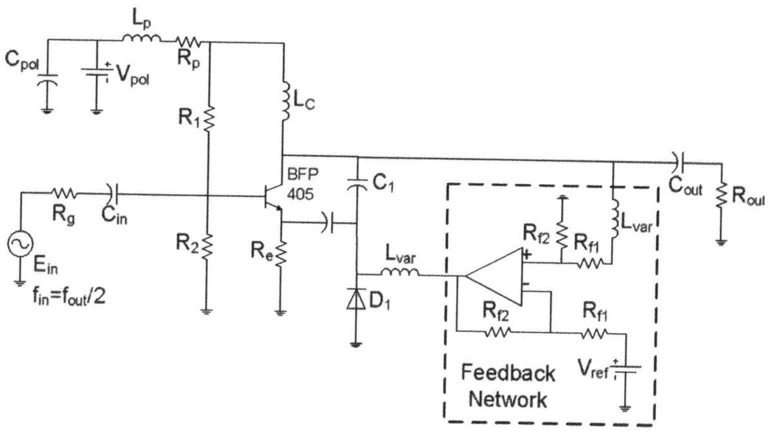

FIGURE 11.22 Subsynchronized oscillator at 4.2 GHz with an injection signal at one-half the oscillation frequency. An external low-frequency feedback loop is used to enlarge the synchronization bandwidth and control its upper edge.

The technique for turning-point control will be used to set the operational bands of a subsynchronized oscillator at one-half the oscillation frequency. The free-running frequency of the oscillator analyzed is $f_o = 4.88$ GHz and the input generator frequency is about $f_{in} = 2.4$ GHz. A BFP405 bipolar transistor from Infineon is used (Fig. 11.22), with a capacitive divider and an inductor at the collector terminal. The input signal at the subharmonic frequency is introduced at the base terminal. The circuit has been manufactured in hybrid technology. For low-frequency feedback, the collector signal is extracted through a choke and introduced into a low-frequency amplifier (Fig. 11.16). The amplifier output is connected to the varactor diode and this modifies the self-oscillation frequency, reducing the frequency error, as in a phase-locked oscillator.

Figure 11.23 shows the global effect of the low-frequency feedback loop on the synchronization region in the plane defined by the oscillation frequency $2\omega_{in}$ and input power. This region is delimited by a set of turning points at which subsynchronization $\frac{1}{2}$ takes place. Its lower vertex corresponds to the free-running oscillation, given by $P_{in} = 0W, 2\omega_o$. All the locus points fulfill the turning-point conditions $Y_{AG} = 0$ and $\det[JY_{AG}] = 0$ at $\omega_{AG} = 2\omega_{in}$. The locus shown by the dashed line corresponds to the original circuit with no feedback. The locus shown by the solid line corresponds to the controlled circuit. Thus, a significant enlargement of the synchronization bandwidth is obtained through the addition of a low-frequency feedback loop.

Condition (11.10) for turning-point control will be used to preset the upper edge of the synchronization band (the turning point on the right-hand side) for the power $P_{in} = -10$ dBm. The control parameters selected have been the reference voltage V_{ref} and one of the resistors (R_{f2}) affecting the gain of the operational amplifier. To verify the general capabilities of the technique, in Fig. 11.24a the specified right edge of the synchronization band has been swept between 4.75 and 5.2 GHz,

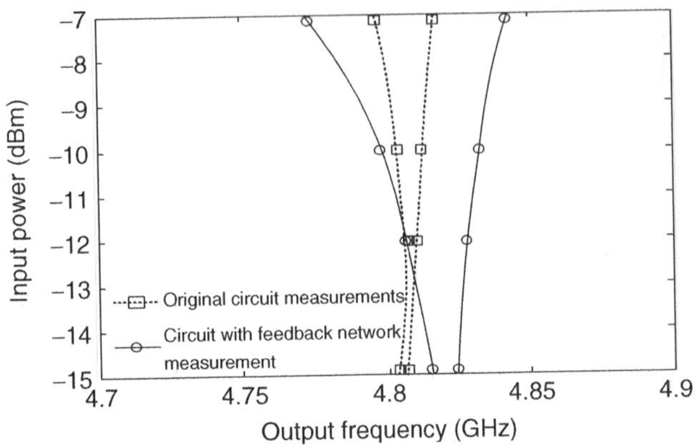

FIGURE 11.23 Enlargement of the synchronization locus of an oscillator at 4.2 GHz, with an injection signal at one-half the oscillation frequency, by means of the connection of a low-frequency feedback loop.

calculating at each step the values required for the feedback network elements through the turning-point control system (11.10).

The collection of shifted synchronization curves is shown in Fig. 11.24b. The synchronization curves have been obtained in a totally independent harmonic balance simulation. For this simulation we use an auxiliary generator at $\omega_{AG} = 2\omega_{\text{in}}$, as explained in Chapter 5. The phase of this generator is set to zero value $\phi_{AG} = 0$ and the phase of the input source at ω_{in} is swept between 0 and 2π. At each step the auxiliary generator amplitude A_{AG} and frequency $\omega_{AG} = 2\omega_{\text{in}}$ are optimized so as to fulfill the nonperturbation condition $Y_{AG} = 0$. Through this independent harmonic balance simulation, it has been verified that for each curve, the upper edge of the synchronization band corresponds exactly to the frequency value specified in the horizontal axis of Fig. 11.24a. This confirms the accuracy of the technique. As shown in Fig. 11.24b, the synchronization bandwidth is barely altered by the shifting. The slight output power decrease (about 1 dB) is unavoidable, due to the intrinsic lowpass nature of the circuit elements. The stability properties of the synchronized solution curve are preserved under the control action. The entire upper section of each closed curve is stable, as has been verified using a complementary stability analysis.

The application presented shows the great frequency-shifting capability compared to the relatively narrow synchronization bandwidth. The fact that the frequency shift is achieved through variation of magnitudes in the low-frequency loop makes possible a simple control of the division band in practical design. Measurements corresponding to the outermost synchronization curves are superimposed in Fig. 11.24b. The experimental curves exhibit the same frequency displacement as that observed in Fig. 11.23. This example shows that it is possible to preset the edge of the synchronization for high input generator power, with the circuit behaving nonlinearly with respect to this power.

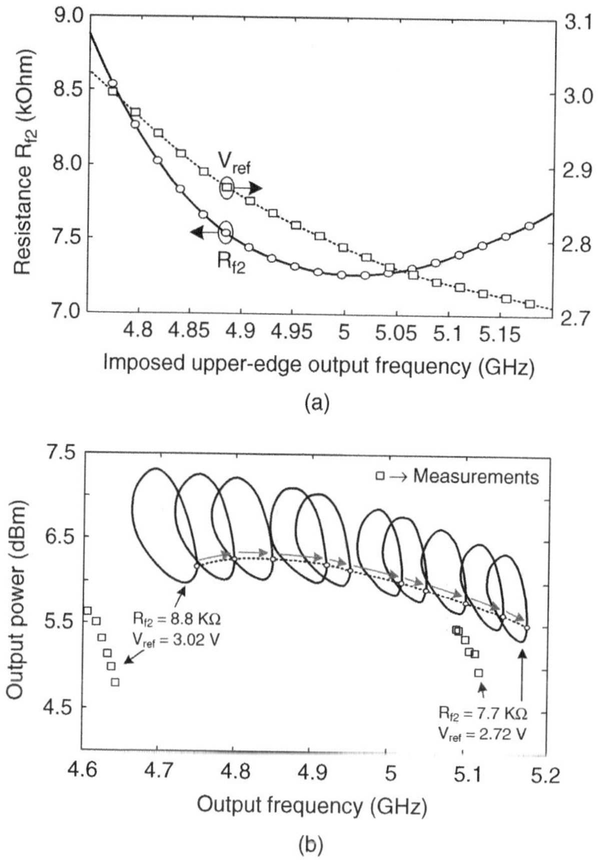

FIGURE 11.24 Subsynchronized oscillator at 4.2 GHz with an injection signal at one-half the oscillation frequency and shifting of the upper edge of the synchronization band through bifurcation control: (a) variation of the control parameters versus the specified frequency of the upper turning point; (b) controlled synchronization curves. The shifted turning point is indicated together with the variation of the control parameter values. Measurements for the outermost curves are superimposed.

REFERENCES

[1] O. Mazouffre, H. Lapuyade, J. Begueret, et al., A 23–24 GHz low power frequency synthesizer in 0.25 μm SiGe, *13th European Gallium Arsenide and Other Compound Semiconductors Application Symposium*, Paris, pp. 533–536, 2005.

[2] U. L. Rohde, *Microwave and Wireless Synthesizers: Theory and Design*, Wiley-Interscience, New York, 1997.

[3] A. Mazzanti, P. Uggetti and F. Svelto, Analysis and design of injection-locked *LC* dividers for quadrature generation, *IEEE J. Solid State Circuits*, vol. 39, pp. 1425–1433, 2004.

[4] S. Yang, C. Lee, and K. Cho, A CMOS dual-modulus prescaler based on a new charge sharing free *D*-flip-flop, *14th Annual IEEE International ASIC/SOC Conference and Exhibit*, Arlington, VA, pp. 276–280, 2001.

[5] Y. Lei, A. Koukab, and M. Declercq, Design and optimization of CMOS prescaler, *PhD Research in Microelectronics and Electronics Conference*, Lausanne, Switzerland, pp. 103–106, 2005.

[6] J. Lee and B. Razavi, A 40-GHz frequency divider in 0.18-μm CMOS technology, *IEEE Symposium on VLSI Circuits*, Kyoto, Japan, pp. 259–262, 2003.

[7] M. Sokolich, C. H. Fields, S. Thomas III, et al., A low-power 72.8-GHz static frequency divider in AlInAs/InGaAs HBT technology, *IEEE J. Solid State Circuits*, vol. 36, pp. 1328–1334, 2001.

[8] U. Singh and M. M. Green, High-frequency CML clock dividers in 0.13-μm CMOS operating up to 38 GHz, *IEEE J. Solid State Circuits*, vol. 40, pp. 1658–1661, 2005.

[9] K. J. Wong, A. Rylyakov, and C. K. Yang, A broadband 44-GHz frequency divider in 90-nm CMOS, *IEEE Compound Semiconductor Integrated Circuit Symposium*, Palm Springs, CA, pp. 196–199, 2005.

[10] A. D'Ambrosio and A. Tattanelli, Parametric frequency dividers: operation and applications, *3rd European Microwave Conference*, pp. 1–5, 1973.

[11] G. Sarafian and B. Z. Kaplan, Dynamics of parametric frequency divider and some of its practical implications, *IEEE Convention of Electrical & Electronics Engineers*, Jerusalem, Israel, pp. 523–526, 1996.

[12] M. Mossammaparast, C. McNeilage, P. Stockwell, and J. H. Searls, Phase noise of X-band regenerative frequency dividers, *IEEE International Frequency Control Symposium*, Kansas, MO, pp. 531–535, 2000.

[13] J. Muellrich, W. Klein, R. Khlifi, and H. Rein, SiGe regenerative frequency divider operating up to 63 GHz, *Electron. Lett.*, vol. 35, pp. 1730–1731, 1999.

[14] C. Rauscher, Regenerative frequency division with a GaAs FET, *IEEE Trans. Microwave Theory Tech.*, vol. 32, pp. 1461–1468, 1984.

[15] F. Ramirez, E. de Cos, and A. Suárez, Nonlinear analysis tools for the optimized design of harmonic-injection dividers, *IEEE Trans. Microwave Theory Tech.*, vol. 51, June 2003.

[16] Y. Yang and T. Chu, Locking performance analysis of MESFET subharmonically injection-locked oscillator, *IEEE Trans. Microwave Theory Tech.*, vol. 47, pp. 1014–1020, 1999.

[17] P. Dorta and J. Perez, On the design of MESFET harmonic injection frequency dividers using the harmonic balance technique, *20th European Microwave Conference*, Budapest, Hungary, pp. 1730–1735, 1990.

[18] R. Quéré, E. Ngoya, M. Camiade, A. Suárez, M. Hessane, and J. Obregón, Large signal design of broadband monolithic microwave frequency dividers and phase-locked oscillators, *IEEE Trans. Microwave Theory Tech.*, vol. 41, pp. 1928–1938, Nov. 1993.

[19] J. C. Sarkissian, M. Camiade, P. Savary, A. Suárez, R. Quéré, and J. Obregón, 60 GHz HEMT-MMIC analog frequency divider by two, *Gallium Arsenide Integrated Circuit Symposium*, Philadelphia, PA, pp. 104–107, 1994.

[20] J. Jugo, J. Portilla, A. Anakabe, A. Suárez, and J. M. Collantes, Closed-loop stability analysis of microwave amplifiers, *IEE Electron. Lett.*, vol. 37, pp. 226–228, Feb. 2001.

[21] A. Collado and A. Suárez, Application of bifurcation control to practical circuit design, *IEEE Trans. Microwave Theory Tech.*, vol. 53, pp. 2777–2788, 2005.

[22] E. H. Abed and J. Fu, Bifurcation control and feedback stabilization, *American Control Conference*, Seattle, WA, pp. 387–391, 1986.

[23] X. S. Luo, G. Chen, B. H. Wang, and J. Q. Fang, Hybrid control of period-doubling bifurcation and chaos in discrete nonlinear dynamical systems, *Chaos Solitons Fractals*, vol. 18, pp. 775–783, 2003.

[24] F. Ramirez, A. Suárez, and S. Sancho, Stabilization techniques for frequency dividers, *Proceedings of the International Workshop on Integrated Nonlinear Microwave and Millimeter-Wave Circuits*, pp. 120–123, 2006.

[25] H. C. Chang, A. P. Yeh, and R. A. York, Analysis of oscillators with external feedback loop for improved locking range and noise reduction, *IEEE Trans. Microwave Theory Tech.*, vol. 47, pp. 1535–1543, Aug. 1999.

[26] A. Collado and A. Suárez, Application of bifurcation control for the optimized design of frequency dividers, *Integrated Non-linear Microwave and Millimetre-wave Circuits (INMMiC) Workshop*, Aveiro, Portugal, Jan. 30–31, 2005.

CHAPTER TWELVE

Circuit Stabilization

12.1 INTRODUCTION

Microwave circuits are often analyzed using the harmonic balance method, due to its efficiency in dealing with distributed elements. As already known, this method provides only steady-state solutions, with no information about the stability or instability of these solutions. Unstable solutions cannot be observed physically. Thus, if the simulated solution is unstable, a qualitatively different solution will be obtained in the measurements. The physical solution may contain undesired oscillations at incommensurate or subharmonic frequencies [1] or exhibit unexpected jumps versus the variation of a parameter such as the input power or bias voltage [2]. The instability may degrade or invalidate the design, increasing the production cycles and the final circuit cost. Undesired oscillations give rise to unexpected spectral components, which can generate interference or damage some circuit elements. On the other hand, a circuit exhibiting jumps or hysteresis is generally unreliable, since different stable solutions coexist for the same values of the circuit parameters. The observation of one or another depends on the form of variation of these parameters, and undesired transitions may be obtained in some cases.

The suppression of instability phenomena through trial-and-error variation of the circuit components is costly and inefficient. Stabilization is often achieved at the expense of a significant alteration of the expected circuit response. A different procedure will be applied here. The simulation techniques presented in this chapter allow determination of the instability mechanisms that lead the circuit to the undesired behavior. From a knowledge of these mechanisms and the characteristics of the undesired solution, the values of the stabilization components will be calculated applying dedicated stabilization techniques, which respect the original circuit specifications. The stabilization techniques will enable suppression of the most common instability phenomena in nonlinear circuits, such as spurious oscillations, hysteresis, chaos, and sideband noise amplification. This suppression will have to be tackled in a different manner depending on the type of circuit. In the case of forced circuits, such as power amplifiers or frequency multipliers, the objective will be circuit stabilization with minimum degradation of the output power or efficiency. In the case of free-running oscillators, the solution autonomy will constitute an additional difficulty because the oscillation frequency changes under any variation of the circuit elements [3]. In particular, it will vary with the introduction of a stabilization network. In this chapter a technique is presented to avoid undesired variations in the oscillation power or frequency during the stabilization process.

The chapter is organized as follows. In Sections 12.2 to 12.4 we present examples of stability analysis and stabilization of power-combining class AB, class E/F, and class E amplifiers, respectively. In Section 12.5 we present different techniques for the stabilization of unstable oscillator circuits. In Section 12.6 a MMIC multifunction chip is considered, showing a systematic technique to determine the origin of unstable behavior in complex multiloop systems.

12.2 UNSTABLE CLASS AB AMPLIFIER USING POWER COMBINERS

Power amplifiers are often implemented with multistage topologies based on power combiners [4,5]. In these topologies, the input power is divided between 2^N different amplification branches using N two-way power dividers, such as Wilkinson dividers (Fig. 12.1). After amplification, the outputs of the various branches are combined successively using two-way adders, which, again, can be implemented through Wilkinson combiners. Each amplification branch operates at the input power P_{in}/N. Thus, the 1-dB compression point of the combined topology increases as $10 \log N$ compared to a single branch with identical design. The transistors are usually operated in class AB, which provides a good compromise between efficiency and linearity. However, in multitransistor amplifiers, multidevice configuration and the symmetry of the power-combining blocks may give rise to different undesired oscillation modes. In a manner similar to the case of globally coupled oscillators (Chapter 10, Section 10.2), these modes are associated with the different eigenvalues of the symmetric power-combining networks. A study of these possible oscillation modes, together with stability analysis and stabilization techniques, is presented below.

12.2.1 Oscillation Modes

In this section the various oscillation modes of power-combining amplifiers are analyzed at the reference plane that connects the output of the transistors to the passive output network (see Fig. 12.1). The analysis will be an extension of the verification oscillation startup conditions in a single-transistor oscillator, based on

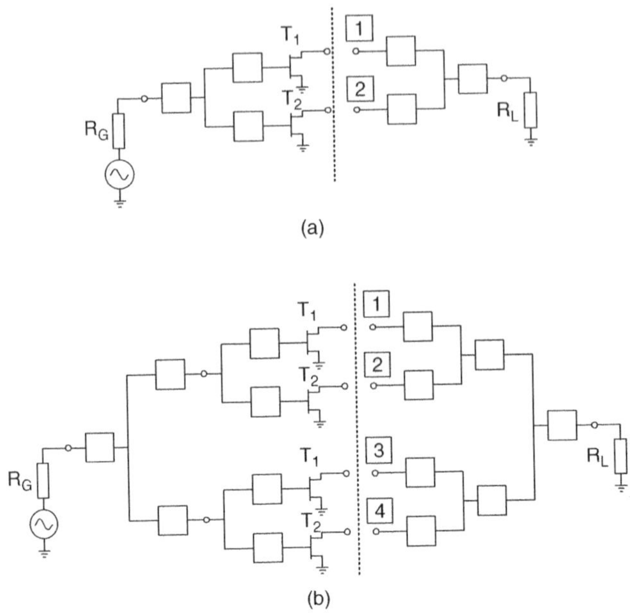

FIGURE 12.1 General topology of power-combining amplifiers: (a) basic cell with two transistors combined; (b) four transistors combined.

evaluation of its total admittance function at a single observation port [conditions (6.4)]. Here total admittance analysis will be carried out at the multiple ports that connect the transistor outputs to the passive network. Note that an analogous impedance analysis, in terms of the branch current, can be more suitable for some specific topologies. As in the case of a single-transistor oscillator, this analysis has limited accuracy, due to the possible inobservability of some internal feeback loops from the transistor outputs. Despite this, analysis of the power-combining amplifier at the transistor output reference plane allows us to relate the symmetry properties of the power-combining networks to different possible oscillation modes. For more complete stability analysis of the entire amplifier circuit, either pole–zero identification or the Nyquist criterion, described in Chapter 6, should be used.

Taking the transistor output as the reference plane (Fig. 12.1), for N transistors combined there will be N possible modes of oscillation. Each mode is associated with one of the N eigenvalues and eigenvectors of the $N \times N$ impedance or admittance matrix of the linear power-combining network [6]. An example corresponding to two transistors combined is shown in Fig. 12.1a. The ports of the 2×2 matrix considered are defined at the two different inputs of the output power-combining network [7]. The output load R_L is not taken as a port but is assumed to be inside the network. For the usual case of a symmetric power-combination network, the associated admittance matrix has the form

$$[Y_L] = \begin{bmatrix} Y_{11} & Y_{12} \\ Y_{12} & Y_{11} \end{bmatrix} \tag{12.1}$$

Note that an ideal network symmetry is considered. The matrix in (12.1) has two different eigenvalues and eigenvectors. One of the eigenvalues is given by $\lambda_1 = Y_{11} + Y_{12}$. The associated eigenvector is $\overline{V}_1 = [1, 1]$. The second eigenvalue is $\lambda_2 = Y_{11} - Y_{12}$, and the corresponding eigenvector is $\overline{V}_2 = [1, -1]$. The basic fact is that for equal voltage phasors $\overline{V}$ at the two inputs of the power-combining network, the input admittance looking from each port of this network to the output R_L agrees with the eigenvalue λ_1; that is, $Y_{\text{in}} = \lambda_1$ [6,8,9]. This is easily derived from

$$\left.\begin{aligned} I_1 &= Y_{11}V + Y_{12}V \\ I_2 &= Y_{12}V + Y_{11}V \end{aligned}\right\} \qquad Y_{\text{in}1} = \frac{V}{I_1} = Y_{11} + Y_{12} = \frac{V}{I_2} = Y_{\text{in}2} \tag{12.2}$$

In a similar manner, for voltage phasors $\overline{V}_1$ and $\overline{V}_2$ such that $\overline{V}_2 = \overline{V}_1 e^{j180^\circ}$, the input admittance at any of the two ports is given by $Y_{\text{in}1} = Y_{\text{in}2} = Y_{11} - Y_{12} = \lambda_2$.

A second multiport network, containing the transistor devices, will also be defined. For the stability analysis under small-signal conditions, the input drive generator can be made zero, so the active multiport network will extend from the grounded generator resistance R_g to the two ports to which the passive network is connected (see Fig. 12.1a). These output ports will be the two ports (1 and 2) of the active network. Thus, ideally, the corresponding small-signal admittance matrix $[Y_A]$ will have the same form as (12.1), with different element values Y_{ij} and eigenvalues λ_{ai}, $i = 1, 2$. Because the transistor terminal voltages, constituting the port voltages, must be the same for both the active and passive subnetworks, two different operation modes may naturally arise: the in-phase or even mode $\overline{V}_1 = \overline{V}_2$ and the out-of-phase or odd mode $\overline{V}_2 = \overline{V}_1 e^{j180^\circ}$ [10]. The input admittance looking into the transistor output terminals will agree, for each mode, with the corresponding eigenvalue $Y_{\text{in,i}} = \lambda_{ai}$, $i = 1, 2$.

The next objective will be to find out whether even- or odd-mode oscillations will actually startup from the dc regime. This would give rise to unstable amplifier behavior under small-signal conditions. For this verification it must be taken into account that the eigenvalues of the admittance matrixes will change with the small-signal frequency f. The possible oscillation modes will be checked with the aid of the oscillation startup conditions derived by Kurokawa [11]. It is taken into account that for each operation mode, the input admittance looking into each active and passive network port agrees with the corresponding eigenvalues. Then the possible oscillation modes $i = 1, 2$ will fulfill

$$\begin{aligned} &\text{Re}(\lambda_{ai}(f) + \lambda_i(f)) < 0 \\ &\text{Im}(\lambda_{ai}(f) + \lambda_i(f)) = 0 \\ &\frac{\partial[\text{Im}(\lambda_{\text{ai}}(f) + \lambda_i(f))]}{\partial f} > 0 \end{aligned} \tag{12.3}$$

The conditions above may never be fulfilled, be fulfilled for one single mode at a particular frequency f_o, or be fulfilled for different modes at different frequencies. For instance, a given amplifier may fulfill the oscillation startup conditions corresponding to mode 1 at the frequency f_o and the conditions corresponding to mode 2 at f_o'. The situation is conceptually analogous to the situation encountered for globally coupled oscillators (see Fig. 10.4 in Chapter 4).

Power-combining amplifiers are usually made up of several cells of the form of the one shown in Fig. 12.1a. Then the number of possible oscillation modes is equal to the number of transistors contained in the power-combining structure. To see this, the case of a two-stage amplifier such as the one shown in Fig. 12.1b is considered. The four ports of the output passive network are defined between the output terminals of the transistors and ground. Assuming identical blocks at levels 1 and 2 of the output network, the admittance matrix of the four-port passive-subcircuit matrix will have the general form

$$[Y_L] = \begin{bmatrix} Y_{11} & Y_{12} & Y_{13} & Y_{13} \\ Y_{12} & Y_{11} & Y_{13} & Y_{13} \\ Y_{13} & Y_{13} & Y_{11} & Y_{12} \\ Y_{13} & Y_{13} & Y_{12} & Y_{11} \end{bmatrix} \tag{12.4}$$

The corresponding eigenvalues and eigenvectors are

$$\begin{aligned} \lambda_1 &= Y_{11} + Y_{12} + 2Y_{13} & \overline{V}_1 &= \begin{bmatrix} 1 & 1 & 1 & 1 \end{bmatrix}^{\mathrm{T}} \\ \lambda_2 &= Y_{11} - Y_{12} & \overline{V}_2 &= \begin{bmatrix} 1 & -1 & -1 & 1 \end{bmatrix}^{\mathrm{T}} \\ \lambda_3 &= Y_{11} + Y_{12} - 2Y_{13} & \overline{V}_3 &= \begin{bmatrix} 1 & 1 & -1 & -1 \end{bmatrix}^{\mathrm{T}} \\ \lambda_4 &= Y_{11} - Y_{12} & \overline{V}_4 &= \begin{bmatrix} 1 & -1 & 1 & -1 \end{bmatrix}^{\mathrm{T}} \end{aligned} \tag{12.5}$$

According to (12.5), there are four possible oscillation modes: one mode with the output voltage of four transistors in phase (even-mode oscillation), two different modes with the two transistors in the same subcombiner having 180° phase shift (odd-mode oscillation in each subcombiner) and one mode with the two transistors in each subcombiner in phase and 180° out of phase with the transistors in the other subcombiner (odd-mode oscillation between the subcombiners). At each mode the input admittance at all four ports of the linear network agrees with the corresponding eigenvalue of the matrix (12.4): $Y_i = \lambda_i, i = 1$ to 4.

A four-port network is now defined with the active part of the circuit, that is, looking into the transistor outputs. This matrix will have exactly the same structure as that of the linear part given by (12.4), although the element values will, of course, be different. The four eigenvalues $\lambda_{ai}, i = 1$ to 4, of this matrix will agree with the input admittance looking into the transistor output at the four different modes. Due to the symmetry properties of the two subcircuits and the necessary fulfillment of Kirchhoff's laws, the possible oscillation will take place in any of the four modes shown in (12.5). The oscillation will generally start up for the negative real part of the total port admittance $\mathrm{Re}[\lambda_{ai}(f) + \lambda_i(f)] < 0$ and the resonance of the imaginary part of this total admittance with positive slope versus the frequency f, as indicated in (12.3).

The eigenvalues of both the active and passive subcircuits will vary with the small-signal frequency. The oscillation startup conditions in (12.3) may be fulfilled at different frequencies for the different oscillation modes. As an example, for the same amplifier circuit, operation mode 1 may fulfill the oscillation startup conditions at the frequency f_o. The operation mode 2 may fulfill the startup conditions at the frequency f_o', and the operation modes 3 and 4 may never fulfill these conditions at any frequency. The previous analyses used for two- and four-transistor cells can be generalized to any number of transistors combined.

In the preceding discussion, approximate conditions were derived to detect undesired oscillation modes of power-combining amplifiers in small-signal operation. These possible oscillations are due to the instability of the dc solution. However, power amplifiers are often stable in small-signal mode, and the oscillation is generated from a certain level of the input power only. This oscillation is generally due to the nonlinear capacitances contained in active devices, which under high pumping amplitude from the input source at the frequency f_{in} exhibit negative resistance in some frequency intervals (see Section 3.3). This type of oscillation is parametric or due to periodic variation of the nonlinear capacitances contained in the circuit. Actually, the current through a nonlinear capacitance, with the voltage $v(t)$ between its terminals, is given by

$$\text{i(t)} = c(v(t))\frac{dv}{dt} \tag{12.6}$$

In the large-signal regime of the power amplifier, $v(t)$ is a periodic signal at the input-source frequency f_{in} and gives rise to a significant periodic variation of $c(v(t))$. To analyze the small-signal admittance exhibited by the capacitance at a frequency f, a small perturbation $\Delta v(t) = \Delta V \cos 2\pi f t$ is introduced at this frequency. Then it is possible to linearize equation (12.6) about the steady-state regime at f_{in}, through the conversion matrix approach:

$$\Delta\overline{I}(kf_{\text{in}} + f) = j2\pi(kf_{\text{in}} + f)\left[\frac{\partial\overline{Q}}{\partial\overline{V}}\right]_o \Delta\overline{V}(kf_{\text{in}} + f) \tag{12.7}$$

where the vectors $\Delta\overline{I}, \Delta\overline{V}, \overline{Q}$, and $\overline{V}$ contain the harmonic components of $\Delta i(t), \Delta v(t), q(t)$, and $v(t)$, respectively. The input admittance at a given frequency $kf_{\text{in}} + f$ is obtained by calculating the ratio between the increments ΔI and ΔV at this frequency. If negative conductance plus resonance is obtained at a certain perturbation frequency f_o, an oscillation will start up at this frequency. The negative conductance (in a certain frequency band about f_o) is the result of the phase-shifting between the small signal current and voltage under the periodic variation of the nonlinear capacitance at f_{in}. As shown in Chapter 3, Section 3.3.1.2, frequency division by 2 may occur if the input frequency is about twice the resonance frequency of the averaged capacitance with an inductive element. The parametric oscillation will be extinguished from a certain input power, because for physical reasons the negative resistance exhibited by any active device decreases with the current amplitude. Note that although most oscillations obtained from a certain power of the input drive source have a parametric origin, they can also be due to gain expansion plus feedback at a given frequency $kf_{\text{in}} + f$.

The same study of the possible oscillation modes from dc regime, based on the admittance matrixes, can be extended to the case of oscillations generated from a large-signal periodic regime. However, the complexity increases significantly due to the fact that we have to linearize the amplifier about its large signal periodic regime. The admittance matrix that corresponds to the passive subnetwork will be the one obtained in the small-signal regime, although for this analysis it will be analyzed at the sidebands $[Y_L(kf_{\text{in}} + f)]$. The admittance matrix $[Y_A(kf_{\text{in}} + f)]$ corresponding to the active subnetwork should be calculated by linearizing the transistors about the steady-state regime at the input generator frequency f_{in}, through the conversion matrix approach. For two transistors combined, the two matrixes $[Y_L(kf_{\text{in}} + f)]$ and $[Y_A(kf_{\text{in}} + f)]$ will have the order $(2N + 1)$ 2 x $(2N + 1)$ 2, with N the number of considered harmonic terms. Due to the periodicity of the poles it will be sufficient to limit the f variation to the interval $0 < f < f_{\text{in}}$. For two transistors combined, even and odd oscillations are possible, arising from the steady-state solution at kf_{in}. Even the signal at the f_{in} keeps in phase at the two different transistor branches, the oscillation at these different branches can be in phase (at f_a) or 180° out of phase. For four transistors combined, equivalent modes to the four oscillation modes indicated in (12.5) are possible. These modes can give rise to either a self-oscillating mixer regime or to a subharmonic oscillation.

In power-combining amplifiers, the odd-mode instability is often associated to division by 2 of the input drive frequency f_{in}. This is due to the natural generation of the 180° phase shift in the frequency division by 2. Although an oscillation at an incommensurable frequency is also possible, frequency division by 2 seems to be more common. Instabilities at incommensurable frequencies will be studied in great depth in Section 12.3, so the following analysis is limited to frequency divisions by 2.

12.2.2 Analytical Study of the Mechanism for Frequency Division by 2

This section presents an approximate analytical study of the mechanisms for the amplifier instability, leading to a frequency division by two. A perfect symmetry of the two circuit branches in the basic cell of Fig. 12.1a will be assumed [12]. Because the objective is to analyze the instability that leads to a frequency division by 2, the input generator frequency will be expressed as $f_{\text{in}} = 2f_o$. The odd harmonic components $(2n + 1)f_o$ exist only in the presence of odd-mode instabilities. Under these conditions, a virtual ground will be obtained at the odd components $(2n + 1)f_o$ at the each of the two connection nodes between the two amplification branches. In turn, a virtual open circuit is obtained at $2nf_o$ at these connection nodes.

The following analysis is totally analogous to that carried out for the detection of a division by 2 in an *RL*-diode circuit in Chapter 4. Assuming the existence of a frequency division by 2, the gate-to-source voltage will be written $v_{gs}(t) = V_{gs0} +$

$V_{gs1}\cos(\omega_o t+\phi_1)+V_{gs2}\cos(2\omega_o t+\phi_2)$, with an analogous expression for the drain-to-source voltage. For each transistor, a simplified model is used containing two nonlinear elements. One is a quadratic current source $i_{ds}(v_{gs})$, depending on the pinch-off voltage V_p. The other is the gate-to-source charge $q(v_{gs})=q_0+q_1 v_{gs}+q_2 v_{gs}^2$. The basic cell is solved at dc, f_o, and 2 ω_o. As already stated, input generator conditionsE_{in} and ω_{in} for unstable behavior are assumed, so $V_{gs1}\neq 0$, $V_{ds1}\neq 0$. For this subharmonic solution to exist, the circuit total admittance at ω_o, evaluated at the transistor gate port of any of the two circuit branches, must be zero:

$$\begin{aligned}0 = {} & \frac{\omega_o C_{gd}}{C_T\omega_o + B_2(\omega_o)}\left(\frac{I_{dss}}{V_p^2}V_{gs2}\cos\Delta\phi + \frac{2I_{dss}}{V_p^2}V_{gs0} - \frac{2I_{dss}}{V_p}\right) - q_2\omega_o V_{gs2}\sin\Delta\phi \\ & + j(B_1(\omega_o) + \omega_o(C_{gd} + q_1 + 2q_2 V_{gd0} + q_2 V_{gs2}\cos\Delta\phi)) \\ & + j\frac{\omega C_{gd}}{\omega C_T + B_2(\omega_o)}\left(\frac{I_{dss}}{V_p^2}V_{gs2}\sin\Delta\phi - \omega_o C_{gd}\right)\end{aligned} \tag{12.8}$$

where $\Delta\phi=\phi_2-2\phi_1$, $C_T=C_{gd}+C_{ds}$, and $B_1(\omega_o)$ and $B_2(\omega_o)$ are the susceptances associated with the input and output distributed elements, respectively. The complex equation 12.8 is coupled with a second equation at the input frequency, $f_{in}=2f_o$, with an explicit dependence on the input generatorE_{in}, which for briefness has not been included. As gathered from the approximate equation (12.8) and Fig. 12.1a, the instability is favored by the symmetry of the power-combining topology, since the virtual ground at the connection node greatly reduces the resistive loading at the subharmonic frequency. The increase in the power delivered by the input generator gives rise to a higher amplitude V_{gs2}. As a result, for $\pi<\Delta\phi<2\pi$, there will be a negative conductance term in (12.8), due to the nonlinear capacitance. Feedback through C_{gd} can also contribute a negative conductance term, for $0<\Delta\phi<\pi$. This term will generally be small due to the low value of C_{gd}. On the other hand, the dc voltage V_{gs0} also affects the total conductance of (12.8).

An additional difficulty in the detection of this type of instability is that the subharmonic component at $f_{in}/2$ generated can be measured only if the two branches of the amplification cell are not totally symmetrical. Then a low-power subharmonic component is detected in the spectrum analyzer. However, the transistors are often damaged due to the significant change in the power consumption when this frequency division takes place, leading to amplifier failure. The various stabilization networks of power-combining amplifiers proposed in the literature [10,13,14] aim at increasing the resistive loading at the subharmonic component $f_o=f_{in}/2$. This means adding a positive conductance term to (12.8), which can be done by connecting a resistance R between the two amplifier branches. Due to the symmetry of the structure, the equivalent load resistance at the subharmonic frequency will be $R/2$. Ideally, this resistance between the two amplifier branches will have no effect at the input frequency f_{in}, as no current will flow through it, due to the circuit symmetry and the in-phase operation of the two branches.

The analytical explanation above provides insight into the causes for the common frequency division by 2 in power-combining amplifiers and the way this instability can be suppressed. However, accurate detection and elimination of subharmonic oscillations will be carried out with harmonic balance techniques. These techniques are presented in the next section.

12.2.3 Global Stability Analysis with Harmonic Balance

The stability analysis of the power-combining amplifier using harmonic balance will be limited to the common case of flip bifurcations. These bifurcations lead to the division by 2 of the input generator frequency. The generalization to Hopf-type bifurcations leading to an oscillation at an incommensurate frequency is straightforward. This case is considered in the class E/F power amplifier of Section 12.3.

As an example, a GaAs FET-based amplifier at 4 GHz with the topology of Fig. 12.2 will be considered. The various microstrip lines have been calculated to enable transistor input and output matching. Using stability analysis based on pole–zero identification, a frequency division by 2 instability had initially been detected for the input generator conditions $P_{in} = -4$ dBm and $f_{in} = 4$ GHz. Figure 12.3 shows the bifurcation diagram versus P_{in}. One of the solution curves is the periodic solution at the input generator frequency f_{in}. This solution is physically observable only up to the flip bifurcation point F, at which the subharmonic component at $f_{in}/2$ is generated. The subharmonic solution generated has been analyzed versus P_{in} with the aid of an auxiliary generator (see Chapter 5, Section 5.5.2.2). This subharmonic solution arises at the flip bifurcation F. From this bifurcation, the output power at the generator frequency f_{in} is smaller than the one provided by the unstable periodic solution. This power reduction is due to the presence of the subharmonic component.

The power-combining amplifier will most often be expected to operate in a certain input power interval and a certain frequency band. The set of input frequency and input power values (f_{in}, P_{in}) for which the amplifier exhibits the undesired frequency division by 2 will be delimited by the flip bifurcation locus, at which the solution at the divided frequency $f_o = f_{in}/2$ is generated or extinguished. Thus, for an accurate and global determination of the input generator conditions with unstable

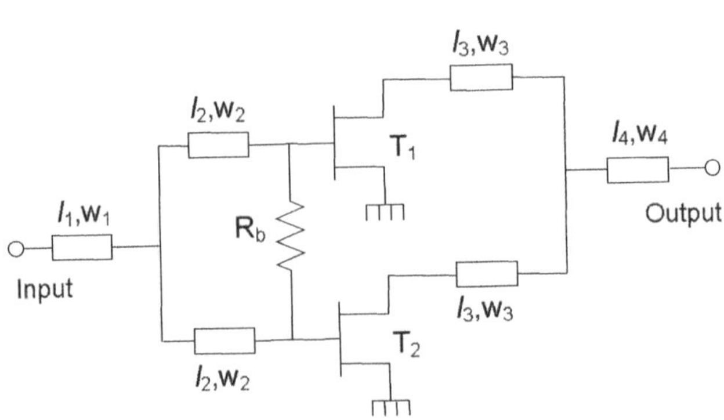

FIGURE 12.2 GaAs FET-based amplifier at 4 GHz.

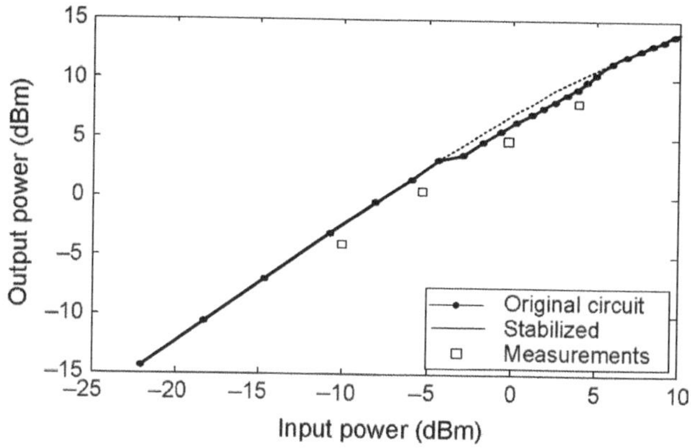

FIGURE 12.3 Power transfer curve $P_{in} - P_{out}$ of a single-cell power-combining amplifier before and after the introduction of a stabilization resistance. Measurements have been superimposed.

behavior, bifurcation analysis tools can be applied. The flip bifurcation that delimits the pairs of values f_{in} and E_{in} providing unstable behavior is obtained using the auxiliary generator technique. A voltage auxiliary generator at the divided-by-2 frequency $f_{AG} = f_{in}/2$ of very small amplitude $A_{AG} = \varepsilon$ and phase ϕ_{AG} is connected to the output of one of the two transistors. To determine the flip bifurcation locus in the plane defined by f_{in} and E_{in}, the following system is solved (Chapter 6):

$$\begin{aligned} Y_{AG}^r(E_{in}, f_{in}, \phi_{AG}) &= 0 \\ Y_{AG}^i(E_{in}, f_{in}, \phi_{AG}) &= 0 \\ A_{AG} &= \varepsilon \end{aligned} \tag{12.9}$$

It is a system of two equations in three unknowns, which provides a curve in the plane defined by f_{in} (horizontal axis), E_{in}. At low slope sections of the locus, f_{in} is swept, solving $Y_{AG} = 0$ for E_{in} and ϕ_{AG}. At high slope sections of the locus, E_{in} is swept, solving $Y_{AG} = 0$ for f_{in} and ϕ_{AG}. As already indicated, the method can easily be generalized to other types of instability; for instance, to obtain the boundary of unstable behavior due to the onset of an oscillation at an incommensurate frequency $f_a \neq (m/n)f_{in}$. Then determination of the unstable operation region in the plane defined by f_{in} and E_{in} will require obtaining the Hopf bifurcation locus in this plane, by solving $Y_{AG} = 0$ in terms of E_{in}, f_{in}, and f_{AG}.

For the power amplifier of Fig. 12.2, the resolution of (12.9) provides the closed curve in Fig. 12.4, which constitutes the flip bifurcation locus in the plane f_{in}, E_{in}. Comparison with the results of pole–zero identification indicates that the frequency division is obtained for f_{in} and E_{in} inside the locus. The locus provides at a glance the circuit operation conditions for unstable behavior. Measurements have been superimposed with good agreement. The locus indicates that a relatively large input generator amplitude is necessary for the nonlinear capacitances to exhibit the

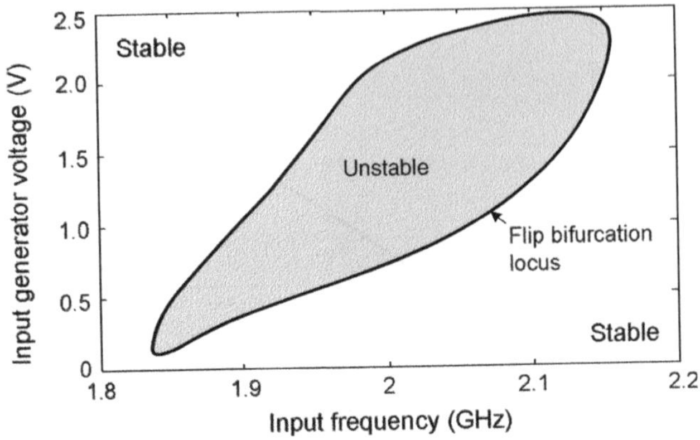

FIGURE 12.4 Locus of unstable behavior of the basic amplifier cell of Fig. 12.1.

negative resistance that gives rise to the instability. On the other hand, once the circuit is in the divided-by-2 regime, the negative resistance decreases from certain E_{in} value. This is in agreement with the physical reduction of the negative resistance of any nonlinear element with the amplitude of the current that transverses this element (Chapter 1, Section 1.3). It explains the oscillation extinction from a certain value of the input generator amplitude.

12.2.4 Amplifier Stabilization

As already stated, the odd-mode instability of power-combining amplifiers can be eliminated through the connection of a resistance R_b between the input (or output) terminals of transistors in the two amplifying branches (Fig. 12.2). Due to the odd-mode subharmonic oscillation, each amplifier branch sees an equivalent resistance $R_b/2$ at $f_{in}/2$, in parallel between the gate terminal and the transistor and a virtual ground. The smaller the value of the resistance introduced, the higher its damping effect will be. The next objective will be development of a technique to obtain directly the resistance value that stabilizes the amplifier globally for all the operation conditions f_{in} and E_{in}, that is, a technique that makes the unstable region in Fig. 12.4 disappear. The technique, described below, is based on a bifurcation analysis in terms of the stabilization resistance R_b.

A pair of constant input amplitude and input frequency values E_{ino} and f_{ino}, belonging to the unstable region of Fig. 12.4, will be considered. The resistance R_b is connected between the two amplifier branches. Next, variations in R_b from very small to very large values will be assumed. For a very small value of R_b, no frequency division will take place, due to the high damping introduced at the subharmonic frequency. By increasing R_b continuously, a particular value R_{bo} will be reached, from which the added damping is not enough to prevent subharmonic oscillation. Thus, when reducing the resistance, a flip bifurcation will take place at the value R_{bo}. A bifurcation detection technique can be used for direct determination of the threshold resistance value R_{bo}. Note that the circuit is not expected to operate for a single pair of values P_{ino} and f_{ino}, but in certain ranges of input power and input frequency. In the technique proposed the flip bifurcation locus is traced in the plane defined by P_{in} and R_b for different values of f_{in}, covering the entire operation band. For each f_{in} value, the flip locus is obtained by solving

$$\begin{aligned} Y_{AG}^r(E_{in}, R_b, \phi_{AG}) &= 0 \\ Y_{AG}^i(E_{in}, R_b, \phi_{AG}) &= 0 \\ A_{AG} &= \varepsilon \end{aligned} \tag{12.10}$$

System (12.10) contains two real equations in three unknowns, which provides a curve in the plane defined by E_{in} and R_b. Application of this technique to the cell of Fig. 12.1 provides, for $f_{in} = 3.85$ GHz, the solid line curve of Fig. 12.5a. This curve indicates the evolution of the threshold value of the stabilization resistance R_b versus the input generator amplitude E_{in}. It constitutes the flip bifurcation locus in the plane defined by E_{in} and R_b. The dashed line curve corresponds to the amplitude of the generated subharmonic component. As can be seen, smaller stabilization resistance is needed at those E_{in} values with higher subharmonic amplitude. The stabilization resistance tends to infinity at the edges of the unstable range, as resistance is no longer necessary from these edges.

As already stated, a locus must be traced for each of the f_{in} values considered, covering the entire frequency band. Each locus will have a minimum resistance value providing stable behavior for the entire input power interval. Figure 12.5b shows the bifurcation loci obtained for two different values of the input frequency $f_{in} = 1$ GHz and $f_{in} = 2$ GHz for the amplifier of Fig. 12.2. To guarantee amplifier stability in the entire frequency band, the value of the interbranch resistance R_{stab} should be equal to the minimum R_b value of all the loci.

Due to the symmetry of the power-combining cell of Fig. 12.2, the interbranch resistance used for amplifier stabilization should have little effect at the input frequency f_{in}, as signals at the amplification branches are in phase at this frequency. To verify this, the $P_{in} - P_{out}$ curve obtained after introduction of the interbranch resistance R_{stab} that stabilizes the amplifier globally has been superimposed in Fig. 12.3. The curve is nearly overlapped with the curve obtained in the original analysis of the unstable periodic regime. Provided that the two amplifying braches are perfectly symmetric, the resistance does not degrade the power transfer curve of the unstable amplifier; it only stabilizes this solution. Measurements of the stabilized amplifier are also shown.

The same stabilization techniques can be to multiple-stage amplifiers. As an example, Collado et al. [12] applied the technique to an amplifier with the topology proposed by Anakabe et al. [10] containing three cells, like the one depicted in Fig. 12.1. Even with large circuits, the harmonic balance technique presented is computationally very efficient, which is due in part to the inherent symmetry of the power-combining structures.

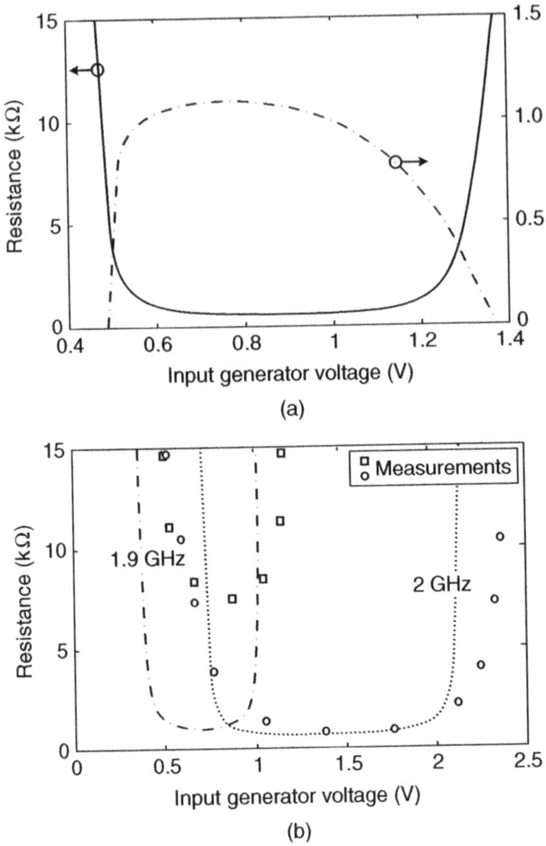

FIGURE 12.5 Stabilization of a power amplifier with interbranch resistance: (a) threshold value of the stabilization resistance obtained with a bifurcation locus in the plane E_{in}, R_b, with the amplitude variation of the subharmonic component $f_o = f_{in}/2$ superimposed; (b) bifurcation loci for two values of input frequency.

12.3 UNSTABLE CLASS E/F AMPLIFIER

12.3.1 Class E/F Operation

The power amplifier efficiency can be maximized by reducing overlapping between the voltage and current waveforms at the transistor output, responsible for the transistor resistive loss. In the standard class E design the transistor operates basically as a switch instead of a voltage-controlled current source [15]. The design is achieved using a resonant output network with precisely calculated element values [16]. A capacitance is connected in parallel to the transistor output contributes to the total capacitance of the resonant circuit, so it must be taken into account in the calculation of the capacitance value. The resonant circuit must be tuned at about the output frequency. The total capacitance, together with that of the inductive load, is determined so as to fulfill the zero-voltage switching conditions [17], which imply the zero value of both the voltage across the capacitance and its time derivative when the transistor starts to conduct. If this voltage is different from zero when it goes from the "off" to the "on" state, the energy stored in the capacitance will be discharged through this device, giving rise to overlap of the current and voltage waveforms. To avoid the discharge, the voltage of the switching device is made zero before transition to the "on" state. This is achieved through suitable tuning of the output resonant network. This circuit must ensure conduction of the switching device for about 50% of the period of the input signal.

On the other hand, in class F amplifiers, the transistor is biased in class B and a harmonic load is tuned so as to constitute an open circuit at several odd harmonics of low order and a short circuit at even harmonic terms [17]. In this way, the voltage waveform is compressed, resembling a square waveform, since the odd harmonic components tend to flatten the upper and lower sections of the waveform. The flattening of the waveform avoids the voltage peaks of the transistor while it is conducting. The current waveform takes the shape of half a sinusoid, and the efficiency increases with the number of odd harmonics. Class F^{-1} is the dual of class F. The even harmonic components are open-circuited and the odd harmonic terms are short-circuited. Then the square waveform corresponds to the current, and the voltage waveform takes the shape of half a sinusoid.

Class E and class F both have advantages and drawbacks. The switching operation of the device in class E design can be achieved with simple load circuits. If the device has a large value of the output capacitance, the total capacitance required for voltage-switching conditions may be smaller than the capacitance exhibited by the transistor itself, so a modified design technique must be used [18]. On the other hand, class F requires a complex synthesis of harmonic loads but has the advantage of providing smaller values of peak voltage and rms current. Taking the consideration above into account, D. Rutledge and his group at the California Institute of Technnology (Caltech) proposed a new class of operation which combines two principles: zero-voltage switching, used in class E, and harmonic tuning, used in class F.

In class E/F operation the output load inductance is tuned as in class E to fulfill the zero-voltage switching condition. On the other hand, the load impedances at a certain number of harmonic terms are fixed to values that would correspond to class F^{-1}. The high circuit complexity associated with load harmonic tuning is avoided by using a push–pull configuration, which allows taking advantage of symmetries at even and odd harmonics. An example is shown in Fig. 12.6. This circuit was designed by Rutledge's group at Caltech. The power amplifier consists of two push–pull transistor pairs in a double-differential drive. The four transistors are vertical double-diffused MOSFETs (VDMOS), operated in class E/F_{odd} mode at 29 MHz. The outputs of the four VDMOS transistors are combined using a distributed active transformer [19]. The transistors are self-biased at the gate terminal. The drain bias circuit is connected at the middle point of each transformer primary. The secondary circuits of the different transformers are connected in series, which

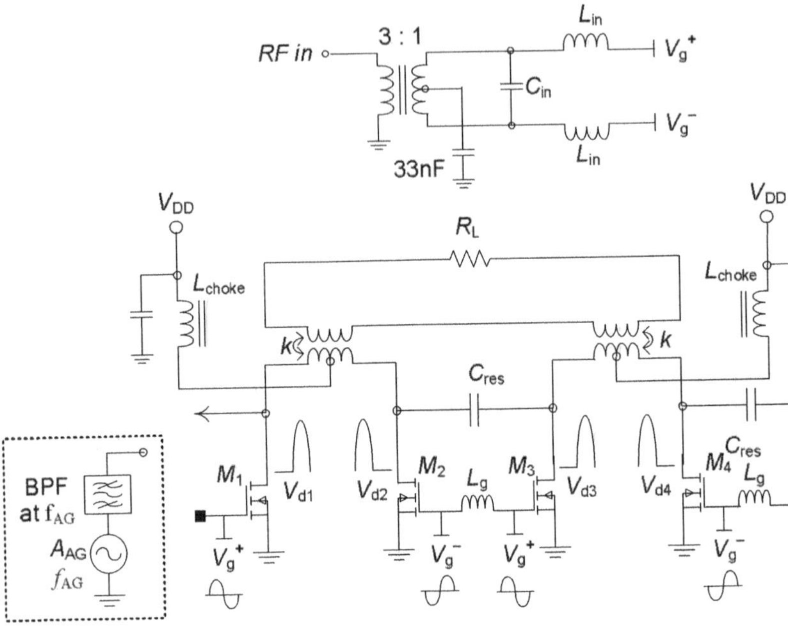

FIGURE 12.6 Class E/F_{odd} power amplifier, showing voltage waveforms at the main circuit nodes. (Reprinted with permission of IEEE.)

gives rise to a distributed transformer enabling the output power combination. The transformer also performs the impedance change 2 : 1.

By means of the input balun, the input signals at a same push–pull pair have a 180° phase shift. Due to the symmetry, the odd harmonics at the outputs of the two transistors have the same amplitude and opposite phase. Thus, there is a virtual ground node at the center of the differential impedance, which short-circuits the impedance exhibited by the output bias network, Z_{vdd}, at the odd harmonic terms. At the even harmonic terms, there is a virtual open circuit at the symmetry line. Applying Kirchhoff's laws [17] it is easily demonstrate that the load impedance (between the transistor output and ground) at the odd harmonic terms is half the impedance $Z_L/2$ between the output nodes of the two transistors. In turn, the even harmonics will see load impedance equal to $Z_L/2 + 2Z_{vdd}$.

The load seen by each transistor at the first and second harmonic components must be the one required for zero-voltage switching operation (class E), which is achieved by tuning the parallel resonant circuit connected between the two drain nodes of the two transistors. This resonant circuit consists of a capacitor C_{res} and transformer magnetization inductance L_k. If, as in this case, the design contains more than a push–pull pair, the capacitance of the resonant network is not connected between the transistors in the same pair, as very long lines would be required for this connection. It is connected between the output terminals of the adjoint transistors, which due to the symmetry, does not give rise to any change in the circuit behavior. If the quality factor of the resonant circuit $C_{res} - L_k$ is sufficiently high, approximate short circuits will be obtained at the odd harmonic component with order higher than 1, which is the condition for class F^{-1} operation.

The input power must be sufficiently large for each transistor to operate basically as a switch between the triode and cutoff regions. Due to the input balun, the two transistors in the push–pull pair have 180° phase shift. When one transistor is in the triode region, the other will be in the cutoff region, and vice versa. Thus, the waveforms v_1 and v_2 are zero for approximately half the input period. The output signal of each push-pull pair is taken between the drain terminals of the two corresponding transistors. Note that the even harmonic terms are inherently canceled in this output signal, due to the odd symmetry of the push–pull topology. For a sufficiently high quality factor, the output voltage waveform should be nearly sinusoidal [17].

12.3.2 Anomalous Experimental Behavior in a Class E/F_{odd} Power Amplifier

The class E/F amplifier of Fig. 12.6 exhibited anomalous behavior in the intermediate range of input power. The research objective has been twofold: to investigate the origin of various undesired phenomena and to suppress these phenomena without disturbing the original circuit performance (at higher input power) in terms of output power and efficiency. The research work presented here was carried out at Caltech by Jeon et al. [19].

Figure 12.7 presents variation in the output spectrum versus the input power drive for the constant drain-bias voltage $V_{DD} = 72$ V. For low input power, only a leakage signal at the input frequency is obtained (Fig. 12.7a). The amplifier periodic solution is stable. However, at $P_{in,1} = 5.5$ W, the output spectrum suddenly turns into the spectrum shown in Fig. 12.7b. The continuity of this spectrum suggests chaotic (nonperiodic) behavior. As the input power is increased further, this type of spectrum continues to be observed until the input power reaches $P_{in,2} = 13.0$ W. Starting with this value, the spectrum becomes the proper one (Fig. 12.7c), with high drain efficiency and output power. Next, a reduction in the input power from above $P_{in,2}$ will be considered. When reducing the input power, the periodic spectrum becomes chaotic at $P_{in,2}$, so no hysteresis is observed at the upper edge of the unstable interval. If we continue to reduce the input power, the chaotic spectrum is observed up to the power value $P_{in,3} = 5.3$ W. At this power value a mixer-like spectrum is generated (Fig. 12.7d) that is never obtained when increasing the power. It is a self-oscillating mixer regime in which the input signal at 29 MHz (f_{in}) mixes with self-oscillation at about 4 MHz (f_a). If the input power continues to be reduced, the oscillation vanishes at the input power $P_{in,4} = 5.0$ W.

All the qualitative changes in the spectrum observed correspond to bifurcations. When increasing the input power, the amplifier undergoes bifurcations at $P_{in,1} = 5.5$ W, where the periodic solution is transformed into a chaotic regime, and at $P_{in,2} = 13.0$ W, where the chaotic regime turns into a periodic solution. When reducing the input power, the periodic spectrum turns into a chaotic regime at the same power value, $P_{in,2} = 13.0$ W. It we continue to reduce the input power, two other bifurcations occur at $P_{in,3} = 5.3$ W, where the chaotic solution becomes

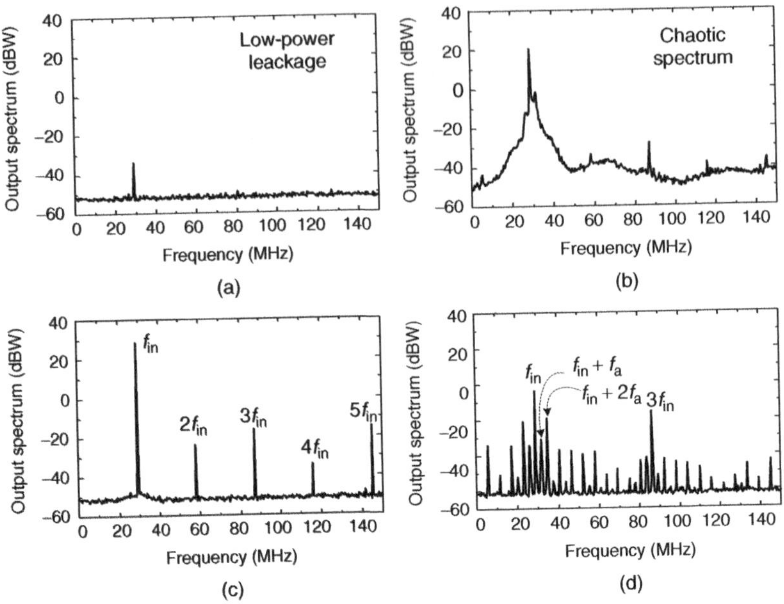

FIGURE 12.7 Variation in the output spectrum for constant drain bias voltage $V_{DD} = 72$ V: (a) $P_{in} = 4$ W; (b) $P_{in} = 10$ W; (c) $P_{in} = 16.5$ W; (d) power reduction; $P_{in} = 5.2$ W. (Reprinted with permission of IEEE.)

a quasiperiodic solution, and at $P_{in,4} = 5.0$ W, where the self-oscillation is extinguished.

12.3.3 Stability Analysis of a Class E/F$_{odd}$ Power Amplifier

In the following, various stability analysis techniques are applied to a class E/F$_{odd}$ amplifier [19]. One objective is to fully understand and obtain in simulation all the phenomena observed experimentally. This will give insight into the instability mechanisms leading to the various types of anomalous behavior. Because the amplifier is expected to operate under different drain bias and input power conditions, another objective will be to determine the unstable behavior intervals when varying for these two parameters.

For the circuit analysis, the VDMOS transistor is modeled primarily as a voltage-controlled current source with two nonlinear capacitances [20,21]. One is the drain-to-source capacitance C_{ds}. This is modeled as a reverse-biased diode in which the parameters of the junction capacitance are fitted to match the capacitance as a function of the drain bias voltage. The other is the feedback capacitance between the gate and the drain. Values of the feedback capacitance are extracted from the data sheet of the transistor, and a junction capacitance model is also used to fit the values. The gate-to-source capacitance is assumed to be constant as a first-order approximation. The parasitic resistances and inductances at both gate and drain are also incorporated in the model as linear elements.

12.3.4 Stability Analysis with Pole–Zero Identification

Initially, we carried out a stability analysis of the periodic solution at f_{in} provided by default by the harmonic balance simulator using pole–zero identification. For this analysis, the constant drain bias voltage $V_{DD} = 72$ V was considered, sweeping the input power from $P_{in} = 4$ W to $P_{in} = 16.5$ W as was done in the measurements. Stability analysis through pole–zero identification of a given steady-state solution requires obtaining a closed-loop transfer function of the circuit linearized about this particular solution. For that, a small-signal current generator $I_n(f)$ is connected in parallel to the drain node of one of the four transistors. The input power P_{in} is swept, and at each P_{in} step, the default periodic harmonic balance solution is determined using a standard analysis. Then the closed-loop transfer $Z_{in}(f)$ is calculated through the conversion matrix approach. This requires sweeping the frequency f to obtain the complex function $Z_{in}(P_{in}, f)$. Pole–zero identification will be applied to this function. The data $Z_{in}(P_{in}, f)$ are stored for later application of pole–zero identification versus the input power P_{in}.

Note that a circuit such as the one in Fig. 12.6 contains many poles. At each P_{in} step, we are interested only in the poles with largest σ, as these poles will dominate the transient response of the power amplifier. Figure 12.8 shows the evolution versus P_{in} of the pair of complex-conjugate poles that are closer to the imaginary axis. The periodic solution is initially stable, but at $P_{in,Hl} = 6.1$ W, the pair of complex-conjugate poles crosses the imaginary axis in a direct Hopf bifurcation (Chapter 3, Section 3.3.1.2). The periodic solution is not physically observable from this power value, as it contains a pair of poles with positive σ which will give rise to the onset of an oscillation at a corresponding frequency of about 5 MHz. This frequency will mix with the one delivered by the input source. As P_{in} continues to be increased, the poles move farther to the right, turn, and again cross the imaginary axis, to the left-hand side, at the power value $P_{in,Hu} = 13.5$ W, for which the oscillation vanishes in an inverse Hopf bifurcation. Note that this analysis provides only stability of the periodic solution at the input frequency f_{in}. It does not provide any information on the behavior or stability properties of the quasiperiodic solution generated at the Hopf bifurcation, which could also be unstable.

12.3.5 Hopf Bifurcation Locus

Because the amplifier is expected to operate under a variety of bias conditions, the next objective will be efficient determination of a set of (V_{DD}, P_{in}) values with unstable behavior. In the preceding analysis it has been observed that the instabilization of the periodic regime at f_{in} is due to a Hopf bifurcation occurring in this regime. Thus, the instability contour in the plane defined by (V_{DD}, P_{in}) is given by the Hopf bifurcation locus, comprised of the points at which the oscillation is generated or extinguished. To obtain this locus, it is taken into account that the oscillation amplitude tends to zero at the Hopf bifurcation points, as was shown in Chapter 6, Section 6.5.2. The oscillation condition, given by circuit total admittance equal to zero at any circuit node, should be fulfilled for oscillation amplitude tending

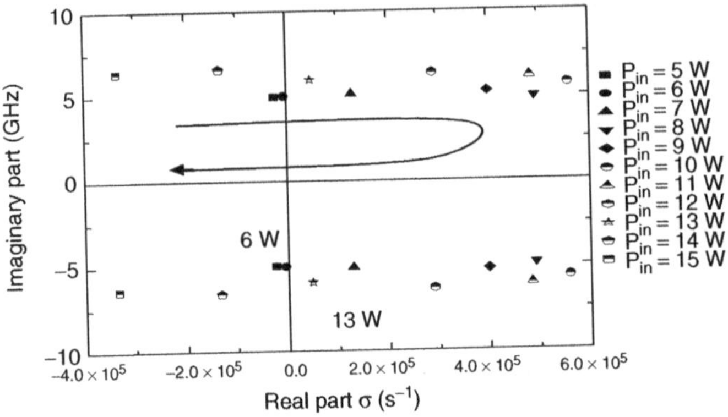

FIGURE 12.8 Pole locus versus P_{in} for $V_{DD} = 72$ V. The input power has been varied from 5 to 15 W in 1-W steps. The pair of poles crosses to the right-hand side of the complex plane for $P_{\text{in},Hl} = 6.1$ W and return to the left-hand side for $P_{\text{in},Hu} = 13.5$ W. (Reprinted with permission of IEEE.)

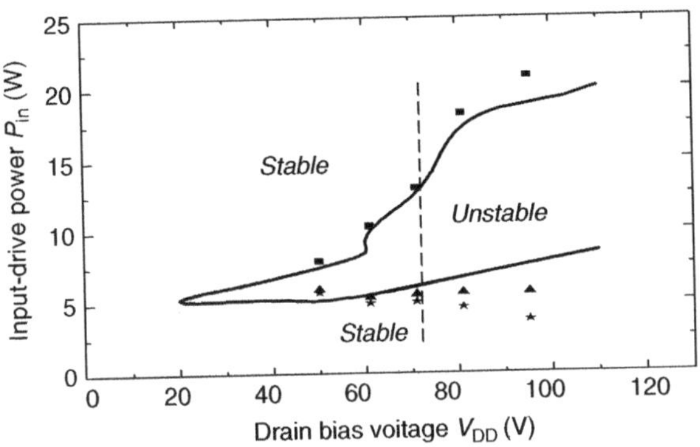

FIGURE 12.9 Instability contour in the plane defined by the drain bias voltage V_{DD} and the input power P_{in}. (Reprinted with permission of IEEE.)

to zero. Thus, at the Hopf bifurcation, the linear admittance $Y_{\text{in}} = 1/Z_{\text{in}}$ must be zero at the oscillation frequency $f = f_a$ generated. This frequency agrees with the frequency of the pair of critical complex-conjugate poles when they cross the imaginary axis. The Hopf bifurcation locus, which delimits the region of unstable behavior in terms of V_{DD} and P_{in}, is obtained by solving $Y_{\text{in}}(f_a, V_{DD}, P_{\text{in}}) = 0$. Note that the oscillation frequency f_a must be included in the calculation, as this frequency is autonomous and thus varies along the locus. It is a two-equation system in three unknowns, providing a curve in the plane defined by V_{DD} and P_{in}. This curve, constituting the Hopf locus, is represented in Fig. 12.9.

The bifurcation points resulting from this analysis are consistent with the pole–zero identification. The periodic solution of the power amplifier is unstable inside the locus and stable outside the locus. Thus, the Hopf locus traced in

Fig. 12.9 provides an instability contour analogous to the one shown in Fig. 12.3, corresponding to the power-combining amplifier. The shape of the locus can be reasoned in a similar manner. For low input power, the nonlinear capacitances do not undergo sufficient pumping to exhibit negative resistance. A negative resistance (and resonance) is obtained from a certain value of this power. As the input power continues to increase, the signal amplitude at the nonlinear capacitances also increases, so, beyond certain value, the negative resistance must decrease, in agreement with the general behavior of all physical devices. As can be seen, the locus is in very good agreement with the results of the pole–zero identification, performed for $V_{DD} = 72$ V (dashed line) and shown in Fig. 12.8. Experimental points have been superimposed. In the lower border, two different sets of experimental points are represented because of the hysteresis phenomenon observed. The locus represented delimits the points for which the periodic solution is unstable, so it should agree with the higher power measurements (triangles). Actually, when increasing P_{in} from a low value, the periodic solution is observable in the interval between the stars and the triangles. Thus, the periodic solution existing within this section must be stable. On the other hand, a chaotic regime is obtained immediately (see Fig. 12.7b) when the periodic solution becomes unstable for most of the measurement points represented, which is not explained by the locus.

12.3.6 Analysis of an Undesired Oscillatory Solution

To devise an efficient stabilization technique, the characteristics of the undesired self-oscillating mixer regime will be analyzed. The analysis is carried out for V_{DD} and P_{in} values inside the instability locus, where the oscillation amplitude is not small. Note that the oscillation amplitude tends to zero only at points of the instability contour. Inside the contour we will have a fully grown oscillation. The analysis is carried out with the aid of an auxiliary generator. Even though the amplifier contains four transistors, only one auxiliary generator is necessary to determine the oscillating steady state. The voltage auxiliary generator is connected in parallel at the drain terminal of one of the four transistors. For given V_{DD} and P_{in}, the amplitude V_{AG} and frequency f_{AG} of the auxiliary generator are calculated to fulfill the condition $Y_{AG}(V_{AG}, f_{AG}) = 0$. Note that this solution is the undesired self-oscillating mixer solution of the power amplifier at the two fundamental frequencies f_{in} and f_{AG}, the latter agreeing with the oscillation frequency f_a, so two-tone harmonic balance must be used. The technique for the harmonic-balance analysis of self-oscillating mixer solutions was presented in Chapter 5, Section 5.5.2.3.

After obtaining the steady-state self-oscillating mixer solution for a particular P_{in} value, the phase of the intermodulation products of f_{in} and f_{AG} at the drain terminal of each transistor has been analyzed. At the oscillation frequency f_a, the two transistors in the same push–pull pair are in phase, whereas the two pairs are 180° out of phase. However, at the input frequency f_{in}, the original phase shift relationships, induced by the push–pull drive, are maintained (i.e., 180° phase

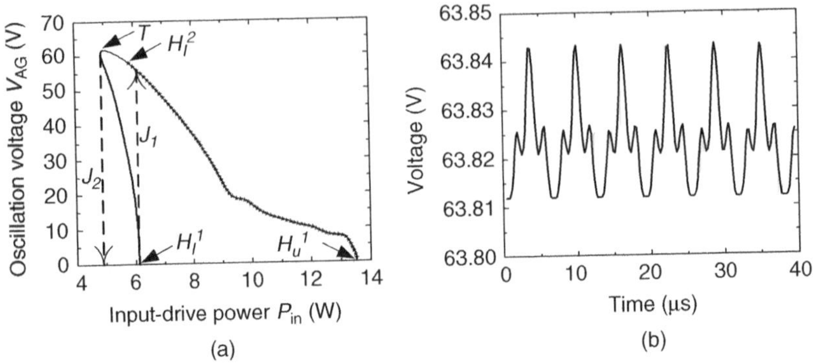

FIGURE 12.10 Quasiperiodic solution at f_{in} and f_a: (a) variation in the oscillation amplitude versus P_{in}; (b) time-varying envelopes obtained in the input power interval indicated by stars in (a). The time-varying envelopes indicate a solution with three incommensurate fundamentals, very unlikely to be observed in practice. (Reprinted with permission of IEEE.)

shift between the two transistors in the same push–pull pair and 180° phase shift between the two pairs as well). Because the two transistors of a same push–pull pair are in phase at the oscillation frequency, the middle point of the pair becomes a virtual open at this frequency. Thus, this point will be ideally suited for the introduction of a stabilization network, as this network will not be seen by the input drive frequency f_{in} (at which the middle node is a virtual short circuit) but will have an influence at the oscillation frequency, at which the middle node is a virtual open circuit.

To find reasons for the observation of hysteresis in the experimental characterization of the amplifier, the self-oscillating mixer solution at the two fundamental frequencies f_{in} and f_a has been analyzed versus the input power P_{in}. The analysis has been carried out for constant $V_{DD} = 72$ V, which is the value considered in the measurements. The equation $Y_{AG}(V_{AG}, f_{AG}) = 0$ is solved at each P_{in} step through optimization in two-tone harmonic balance. The resulting variation of the oscillation amplitude at the drain terminal, agreeing with V_{AG}, is represented in Fig. 12.10a. The curve exhibits a turning point T related to the experimentally observed hysteresis phenomenon. Actually, when increasing P_{in}, the transition from stable amplifier behavior to the self-oscillating mixer regime (J_1 in Fig. 12.10a) is due to the Hopf bifurcation H_l^1 occurring in the amplifier solution. When P_{in} is decreased, the jump back to the stable amplifier solution (J_2) should be due to turning point T, in the solution curve corresponding to self-oscillating mixer operation. Note that in terms of the input power, the simulated hysteresis interval is in good correspondence with the experimental interval, presented in Fig. 12.7.

The hysteresis phenomenon is well predicted by Fig. 12.10a. However, in the measurement, an abrupt transition from stable amplifier behavior to the chaotic regime occurred when increasing the input power. So far we have only detected a transition from a periodic solution to a self-oscillating mixer regime. However, direct transition to the chaotic regime was found in the measurements. Note that the harmonic balance analysis does not allow the simulation of chaotic solutions with a continuous spectrum. For more analysis flexibility, the envelope transient will be used. As already seen, in this analysis technique, the circuit variables are expressed in a Fourier series with time-varying coefficients (see Section 5.6, Chapter 5). When introducing these time-frequency expressions of the circuit variables into the modified nodal equations that describe the circuit behavior, we obtain a system of differential algebraic equations in the time-varying harmonic coefficients. The system is integrated in the slow time variable associated with these harmonic terms. In the particular case of the unstable power amplifier, the circuit variables are represented as $\overline{x}(t) = \sum_{k,m} \overline{X}_{k,m}(t)e^{j(k2\pi f_{in}+m2\pi f_a)t}$, with f_{in} the input drive frequency and f_a the frequency of the already detected oscillation at about 4 MHz. The application of envelope transient to autonomous circuits requires proper initialization of the self-generated oscillation. Initialization of the power amplifier will be carried out with the same auxiliary generator at $f_{AG} = f_a$ as that used for an harmonic-balance analysis of the undesired self-oscillating mixer regime. An auxiliary generator with the values resulting from this analysis and fulfilling $Y_{AG}(A_{AG}, f_{AG}) = 0$ is connected to the circuit at the initial time t_0 only. From this time, the circuit is allowed to evolve according to its own dynamics. The disconnection of the auxiliary generators for $t > t_0$ can be done with the aid of a time varying resistor [see equation (5.78), Chapter 5].

The described technique has been applied using the solution points in the curve of Fig. 12.10a (corresponding to self-oscillating mixer operation) as initial value. The envelopes are constant in the section $T - H_l^2$. This indicates that the self-oscillating mixer regime at the two fundamental frequencies f_{in} and f_a should be stable at this section. From point H_l^2 in Fig. 12.10a, the magnitude of the harmonic components $X_{k,m}(t)$ becomes time-varying (Fig. 12.10b). The envelopes oscillate at a few hundred kilohertz. Therefore, from H_l^2 there is a second oscillation, in addition to the previous oscillation at about 4 MHz (f_a). The bifurcation H_l^2 is a tertiary Hopf bifurcation, leading to instability of the self-oscillating mixer solution at f_{in} and f_a. In a manner analogous to Hopf bifurcations from dc and periodic regimes, it leads to the onset of an additional incommensurate oscillation frequency f_a'. Thus, a three-fundamental regime is obtained from H_l^2. According to the Ruelle–Takens theory [30], the three-fundamental solution generated is likely to break into chaos, which explains the chaotic spectrum observed in the experiment (Fig. 12.7b). The interval in which the self-oscillating mixer solution at f_{in} and f_a is unstable has been indicated by stars in Fig. 12.10a. Thus, the regime at f_{in} and f_a is stable only in the curve section between the turning point T and the Hopf bifurcation H_l^2. To validate the results, a detailed measurement has been carried out for input power slightly above that corresponding to point T of Fig. 12.10a. This measurement is shown in Fig. 12.11 and confirms the existence of a second oscillation at about 500 kHz, which agrees with the results of the envelope transient simulation.

The bifurcation diagram obtained explains all the phenomena observed in the measurements. In Fig. 12.12, examples of the experimental spectra obtained in the input power intervals are shown for a better understanding. Increasing the input

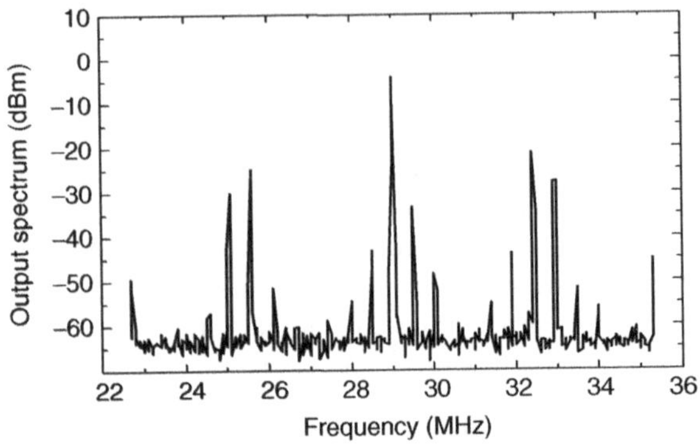

FIGURE 12.11 Experimental solution with three fundamental frequencies. Measured spectrum for $V_{DD} = 72$ V and $P_{in} = 5.15$ W. The coexistence of two oscillations can be observed at about 4 MHz and 500 kHz. (Reprinted with permission of IEEE.)

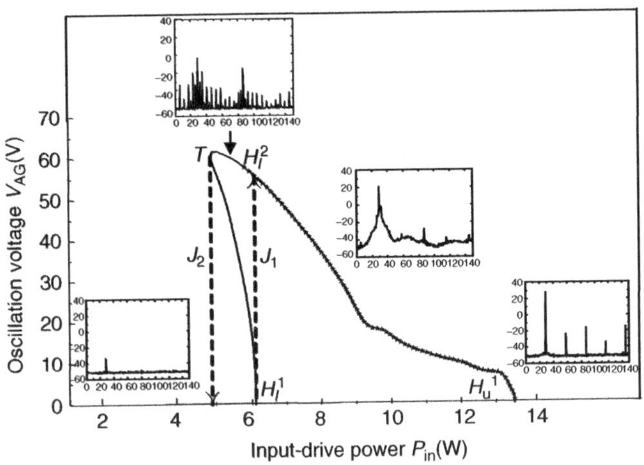

FIGURE 12.12 Measured spectra for various input power values. Stars have been superimposed in sections of the oscillatory curve with periodic harmonic components. (Reprinted with permission of IEEE.)

power from a very low value, periodic solution of the amplifier is initially stable and turns into chaos at the Hopf bifurcation H_l^1. This dramatic change is due to the fact that the input power value at which the instabilization of the desired periodic solution takes place, corresponding to the bifurcation H_l^1, is higher than the one at which the self-oscillating mixer regime becomes unstable, corresponding to the bifurcation H_l^2. Thus, the system goes directly from a periodic regime to a chaotic solution. The chaotic regime persists until the input power reaches the inverse Hopf bifurcation H_u^1, at which the amplifier periodic solution becomes stable again. When the input power is decreased, the second oscillation is extinguished at the Hopf bifurcation H_l^2. Thus, the self-oscillating mixer regime at the frequencies f_{in}

and f_a is stable for a very small interval of input power. At the turning point T, the system jumps down to the stable periodic regime.

12.3.7 Circuit Stabilization

After an understanding of the various instability phenomena, the goal will be the global stabilization of the power amplifier for all expected operational conditions in terms of V_{DD} and P_{in}. From phase analysis of the various frequency components of the undesired self-oscillating solution, it was gathered that the middle node of the push–pull pair behaves as a virtual open circuit at the oscillation frequency. To determine the type of circuit resonance taking place at the oscillation frequency, a simplified equivalent of a quarter of the amplifier can be derived, considering the equivalent impedances seen by the output of each transistor at the oscillation frequency (Fig. 12.13). It is easily seen that each transistor sees a parallel resonator between the drain terminal and ground. Both the load resonator and the bias network contribute to this equivalent resonator, which is obtained by taking into account the symmetry of the push–pull pairs. The output capacitance C_{res} connected between two transistor pairs is doubled due to the virtual ground developed at the center of the capacitance, and the magnetization inductance L_{res} in the output transformer is divided by 2 due to the virtual open at the center of the transformer. The RF choke inductance L_{choke} is also a critical element of the equivalent circuit. Taking this inductance into account, the resonance frequency is 5.3 MHz, quite close to the oscillation frequency obtained in both measurement and simulation. This

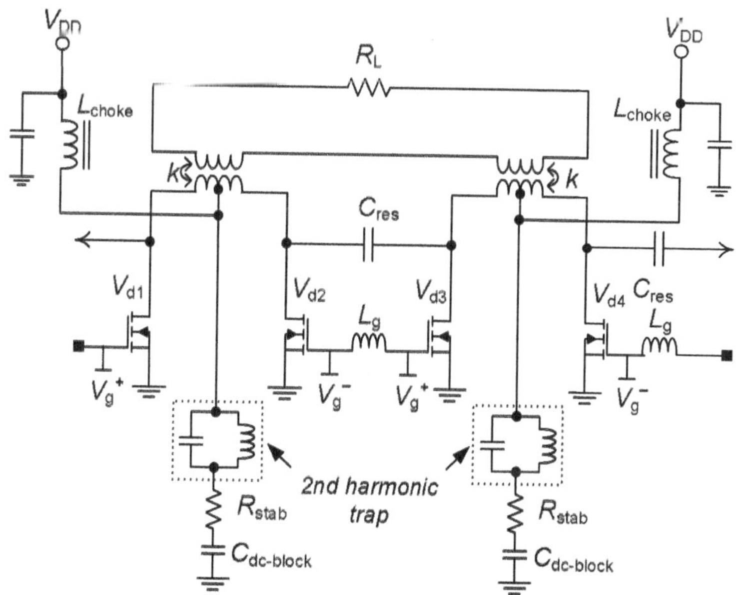

FIGURE 12.13 Class E/F amplifier containing one stabilization network connected to the middle point of each of two push–pull pairs. (Reprinted with permission of IEEE.)

confirms the assumption of a parallel resonance. Thus, a simple means to stabilize the amplifier will be to add a parallel resistor to the middle node of each push-pull. The aim is to increase the total positive resistance of the parallel circuit. This resistance will have little effect at the input drive frequency because the middle node is a virtual short circuit at this frequency. Because the middle node should be an ideal open circuit at all even harmonic terms, a second harmonic trap at $2f_{in}$, the most influential even harmonic component, is also connected in series.

Once the topology and location of the stabilization network have been determined, the next step will be accurate determination of the stabilization resistance value in order to globally stabilize the amplifier for all possible values of V_{DD} and P_{in}. Two different techniques have been considered. In the first, the threshold value of this stabilization resistance is obtained through a bifurcation analysis, similar to what was done in Section 12.2. The second technique is graphical, based on an analysis of the small-signal input admittance at $f \neq f_{in}$, seen from a sensitive circuit node.

12.3.7.1 Bifurcation Condition Let constant values of the drain bias voltage V_{DD} and input power P_{in} be considered initially. For a very small value of the stabilization resistance R, no oscillation can take place, due to the high damping introduced by this parallel resistance. As R increases, a particular value R_{stab} will be reached from which the undesired oscillation is generated in a direct Hopf bifurcation. Thus, the threshold value for stable behavior will correspond to a Hopf bifurcation obtained versus variations in this resistance. As already known, this bifurcation condition can be imposed by introducing into the circuit an auxiliary generator with very small amplitude and solving the nonperturbation condition $Y_{in}(R_{stab}, f_a) = 0$. The amplifier will be unstable for $R > R_{stab}$ and stable for $R < R_{stab}$.

The resistance value R^o_{stab} that stabilizes the circuit globally will be the minimum of all the bifurcation resistance values R_{stab} obtained for all possible operation conditions, in terms of V_{DD} and P_{in}. To determine R^o_{stab}, different V_{DD} values will be considered in an interval covering the entire operational range in V_{DD}. For each V_{DD} value, the Hopf bifurcation locus is obtained in the plane defined by P_{in} and the stabilization resistance R_{stab}. This is done by solving the complex equation $Y_{in}(R_{stab}, P_{in}, f_a) = 0$ that determines the Hopf bifurcation locus. At flat sections of the locus, P_{in} is swept, calculating $Y_{in}(R_{stab}, f_a) = 0$. At high slope sections, R_{stab} is swept, calculating $Y_{in}(P_{in}, f_a) = 0$. This technique has provided the results shown in Fig. 12.14. A different Hopf locus is obtained for each V_{DD} value. As can be seen, the value required for the stabilization resistance decreases with this bias voltage. On the other hand, as P_{in} approaches the values for which the periodic solution becomes stable, the resistance required tends to infinity. This is because the stabilization resistance is no longer necessary from these power values. From the results of Fig. 12.14, global stabilization of the amplifier requires a resistance smaller than $R^o_{stab} = 17\ \Omega$.

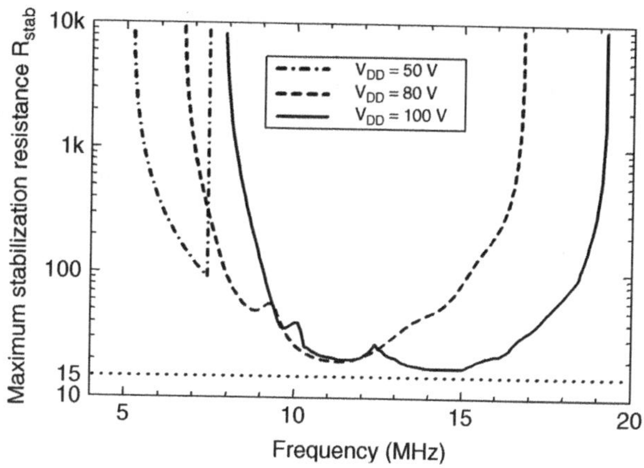

FIGURE 12.14 Variation of the resistance required for circuit stabilization versus the input power P_{in} and bias voltage V_{DD}, obtained through Hopf bifurcation analysis. (Reprinted with permission of IEEE.)

12.3.7.2 Admittance Analysis As an alternative to the bifurcation analysis described above, the threshold value of the stabilization resistance can be determined using a graphical technique. This technique is based on an analysis of the circuit small-signal admittance (or impedance) at $f \neq f_{in}$, calculated from the circuit linearization about its steady state regime at kf_{in}. This analysis is described in Section 6.2.2, of Chapter 6. This requires the previous location of a sensitive observation node, such that whenever the pair of complex-conjugate poles is located on the right-hand side of the complex plane, the oscillation startup conditions in terms of admittance (or impedance) are fulfilled at this observation node. The usual fulfillment of the oscillation startup conditions when a pair of complex-conjugate poles is located on the right-hand side of the complex plane was discussed in Chapter 1, Section 1.3. In the particular case of the class E/F amplifier considered, the oscillation startup condition is fulfilled at the drain node of any of the four transistors in terms of the admittance function $Y_{in} = 1/Z_{in}$. In fact, comparing the frequency variation of Y_{in} with the pole–zero identification results, it has been possible to associate the instability with the existence of negative conductance at the oscillation frequency, together with a zero crossing of the susceptance with positive slope. The graphical technique is used as described below.

Several discrete values of the stabilization resistance R are considered. For each R value, a double sweep is carried out in V_{DD},P_{in}. At each pair of values V_{DD},P_{in} a default harmonic balance simulation is carried out, which provides a corresponding periodic amplifier solution, stable or not. Once the solution for V_{DD},P_{in} has been calculated, a second simulation for the same V_{DD},P_{in} values is performed. This simulation should provide the input admittance $Y_{in}(f)$. It is otained with the same small-signal current source $I_n(f)$ used for pole–zero identification. The input admittance $Y_{in}(f)$ seen by the current source is calculated using the conversion matrix approach. In this approach, the circuit is linearized about the periodic

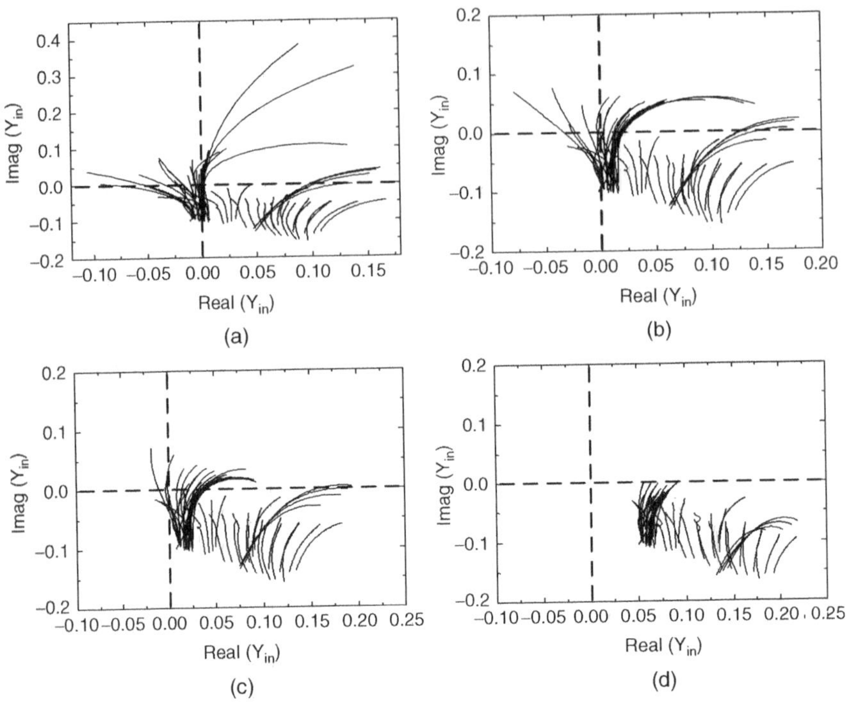

FIGURE 12.15 Determination of the stabilization resistance obtained through a double sweep in the two circuit parameters V_{DD} and P_{in}: (a) original circuit; (b) $R = 100\ \Omega$; (c) $R = 50\ \Omega$; (d) $R = 15\ \Omega$. Global stability of the power amplifier. (Reprinted with permission of IEEE.)

solution at f_{in} obtained using default harmonic balance simulation. In this conversion matrix analysis, the frequency f of the small-signal source is swept in a sufficiently large interval to cover all possible values of the unstable frequency.

Proceeding as described above, for each resistance value the double sweep in V_{DD},P_{in} provides a family of polar plots. Each plot is obtained for a particular pair of values V_{DD},P_{in}, by sweeping the frequency of the small-signal current source and calculating the total input admittance $Y_{in}(f)$ using the conversion matrix approach. For global stability, no crossing of the negative real semiaxis with positive increase in the susceptance must be obtained. Figure 12.15 illustrates the application of this technique for three resistance values: 100, 50, and 15 Ω. The same frequency sweep range has been considered for all the curves. For the sake of clarity in the representation, this range has been limited to 3 to 5.5 MHz. As expected, larger stability ranges are achieved as the stabilization resistance is reduced. For $R_{stab} = 15\ \Omega$, the amplifier becomes stable for all possible values of V_{DD} and P_{in}. The results of the two stabilization techniques described are in good agreement.

A new version of the amplifier was manufactured, with a stabilization resistance of 15 Ω in series with the second harmonic trap. We never observed any oscillation or chaotic regimes. The drain efficiency of the amplifier has been measured, and the only significant degradation is 1.3% for 240 W of output power. This small effect of the stabilization network on overall performance is due mainly to the fact that the middle point of the push–pull pairs, where the stabilization network is connected, is a virtual ground at the input frequency f_{in}. The second harmonic trap prevented the influence of the stabilization network at the second harmonic frequency. The influence of even harmonic terms of higher order turned out to be negligible.

12.4 UNSTABLE CLASS E AMPLIFIER

In this section, a class E amplifier showing anomalous behavior in the intermediate range of input power is analyzed and stabilized. The undesired phenomena and underlying instability mechanisms observed are different from those of the first example. As will be shown, they involve the effect of the noise perturbations. The research work presented here was carried out at Caltech by Jeon et al. [2].

A schematic of a class E amplifier is shown in Fig. 12.16. The output capacitance C_{out} together with the resonant tank composed by C_{res} and L_{res} allows fulfillment of the zero-voltage switching condition, which prevents overlap of the transistor output voltage and current waveforms. Note that the switching mode requires a relatively high input power value, with the transistor operating between the triode and cutoff regions [15]. The resonant circuit is slightly detuned from the operation frequency $f_{in} = 7.4$ MHz, to achieve the zero-voltage switching condition at the drain terminal. A second harmonic trap guarantees a second-harmonic level below −40 dBc. This subcircuit allows us to obtain a transistor load of about 8 Ω at the fundamental frequency, which is the required value for the switching operation. The lowpass filter consisting of C_{lpf} and L_{lpf} eliminates the VHF harmonic components from the output power spectrum. In saturated operation the amplifier provides an output power of 360 W with a gain of 16.1 dB and drain efficiency 86.1% at 7.4 MHz.

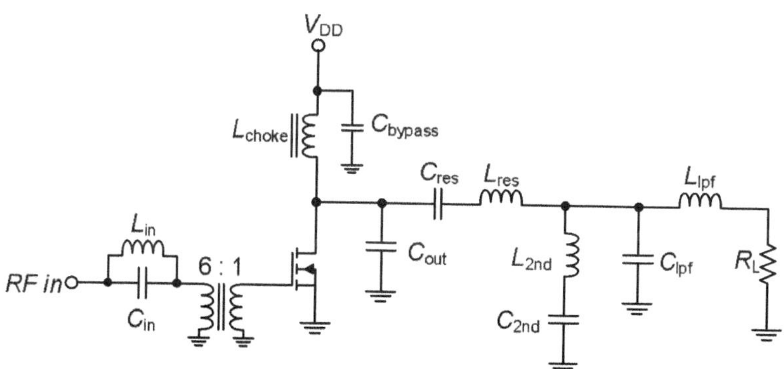

FIGURE 12.16 Class E power amplifier operating at 7.4 MHz. (Reprinted with permission of IEEE.)

12.4.1 Amplifier Measurements

In measurement of the power amplifier, various anomalous phenomena were observed. For small input power, only leakage output power is obtained at the input drive frequency f_{in}. This is due to transistors in near-cutoff conditions below this input power. From $P_{in} = 0.5$ W, noise bumps appear in the spectrum, with relatively high power. As already noted, these noise bumps, also known as "noisy precursors," (Section 2.5.5, Chapter 2), are due to parametric sideband noise amplification. There are three bumps centered at about $f_c = 560$ kHz and $f_{in} \pm f_c$ (Fig. 12.17a). If the input power is increased further, the central bump frequency f_c decreases and the bumps around f_{in} approach each other. The bump power increases (Fig. 12.17b) until for $P_{in} = 0.83$ W, an oscillation arises at the frequency $f_a = 1$ MHz (Fig. 12.17c). This frequency is, surprisingly, quite a bit higher than the central bump frequency f_c, which was actually decreasing versus the input power. From $P_{in} = 0.83$ W, the circuit operates in a self-oscillating mixer regime at the two frequencies f_{in} and f_a (Fig. 12.17c). The oscillation frequency is close to $f_{in}/7$, which gives rise to spectral lines at small offset from the input frequency. For $P_{in} = 0.89$ W there is a frequency division by 7, due

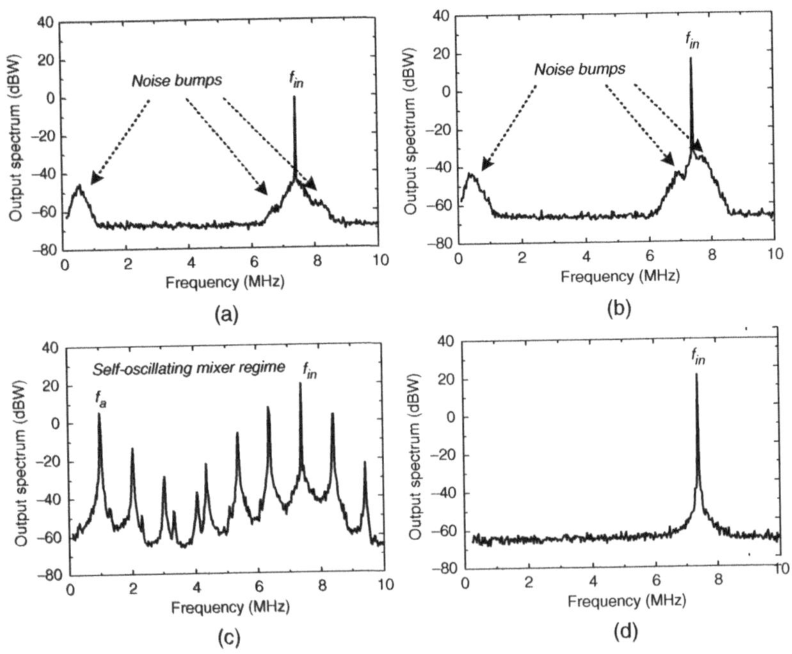

FIGURE 12.17 Measurements of the output power spectrum of a class E amplifier for different input power values. The resolution bandwidth is 3 kHz. (a) $P_{in} = 0.5$ W. Pronounced noise bumps about the frequencies $f_c = 560$ kHz and $f_{in} \pm f_c$. (b) $P_{in} = 0.8$ W. The bump frequency f_c is reduced and the bump power increases. (c) $P_{in} = 0.84$ W. An oscillation is generated at the frequency $f_a = 1$ MHz. (d) Proper switching mode operation of the amplifier. (Reprinted with permission of IEEE.)

to synchronization of $7f_a$ to the input frequency f_{in}. This synchronization is maintained within the input power interval 0.89 to 0.92 W. The synchronization capability is in agreement with the high phase noise of the oscillation prior to synchronization and suggests a low quality factor for this self-generated oscillation. For an input power higher than $P_{in} = 0.92$ W, the circuit again behaves as a self-oscillating mixer. Finally, at $P_{in} = 1.7$ W, the oscillation is extinguished (Fig. 12.17d) and the amplifier starts to operate in the periodic regime desired. For reasons to be given later, when reducing the input power, no hysteresis phenomenon was noted in the initial measurements of the amplifier.

As already known, the noise bumps at frequencies $kf_{in} \pm f_c$, with k an integer, are due to a pair of canonical complex-conjugate poles $\sigma \pm j2\pi f_c$ of the periodic solution at f_{in} located on the left-hand side of the complex plane, but relatively close to the imaginary axis. Thus, one would expect to get the spectral lines generated at $P_{in} = 0.83$ W at approximately $kf_{in} \pm f_c$, with k an integer, which are the central frequencies of the bumps. However, in the class E amplifier there is a significant difference between the frequency of the bumps, f_c, and the oscillation frequency, f_a. To understand this phenomenon it will be necessary to combine various analysis techniques, which will be shown in the next sections.

12.4.2 Stability Analysis of the Power Amplifier

The initial objective was to analyze the amplifier stability versus the input power P_{in} for the same bias conditions considered in the measurements. Thus, a P_{in} sweep will be carried out, performing at each P_{in} step a default harmonic balance simulation of the periodic solution at the input frequency f_{in}, plus a conversion matrix analysis using the small-signal current source in order to determine the impedance function $Z_{in}(f)$. However, when performing this P_{in} sweep, a discontinuous jump is observed in the solution curve, which could be due to a hysteresis phenomenon. The hysteresis is associated with the existence of turning points or infinite slope points in the solution curve at which the Jacobian matrix of the harmonic balance system becomes singular. In Chapter 6, Section 6.4, it was shown how turning points in the solution curves of forced circuits, in a periodic regime, could be circumvented by sweeping the amplitude of the auxiliary generator operating at the fundamental frequency.

To obtain a multivalued nonoscillatory solution, the auxiliary generator frequency is made equal to the frequency of the input source $f_{AG} = f_{in}$. The variables to determine in order to fulfill the nonperturbation condition $Y_{AG} = 0$ will be auxiliary generator amplitude A_{AG} and its phase ϕ_{AG}. Note that because it is a nonautonomous regime, the first-harmonic voltage at the node at which the auxiliary generator is connected will have a certain phase shift with respect to the input generator which provides the phase reference. Remember that when the condition $Y_{AG} = 0$ is satisfied, the auxiliary generator amplitude agrees with the voltage at the node at which the auxiliary generator is connected. Thus, sweeping the auxiliary generator amplitude will be equivalent to sweeping the voltage amplitude at f_{in} at the connection node of the auxiliary generator. The continuation technique

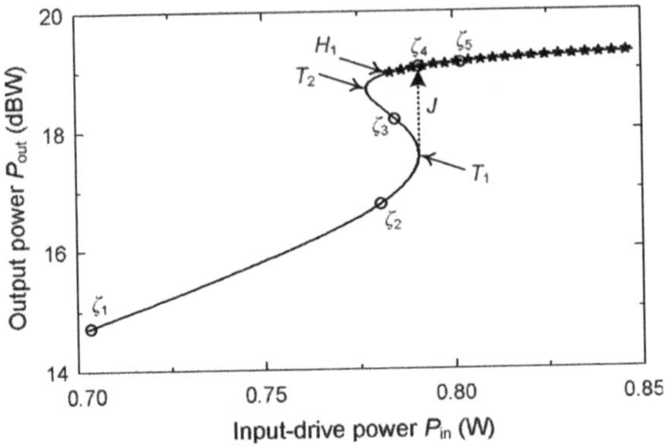

FIGURE 12.18 Multivalued power transfer curve of a class E power amplifier obtained using a parameter-switching technique based on an auxiliary generator. The curve section in which the amplifier operates in a self-oscillating mixer regime is indicated by stars. (Reprinted with permission of IEEE.)

proceeds as follows. In curve sections with low slope with respect to the input power, this input power will be swept, as in a standard harmonic balance sweep. In sections with high slope versus P_{in}, the parameter to sweep will be the amplitude of the auxiliary generator A_{AG}. In the power amplifier analyzed, the auxiliary generator is connected to the transistor drain terminal. The technique described has allowed us to obtain the multivalued $P_{in} - P_{out}$ curve of Fig. 12.18.

In a second step, the stability of the various sections of the multivalued periodic solution curve obtained is going to be determined. The input power is swept and at each P_{in} step the amplifier periodic solution at f_{in} is determined. Then the closed-loop transfer $Z_{in}(f)$ is calculated using the conversion matrix approach. To obtain $Z_{in}(f)$, a small-signal current source $I_n(f)$ is connected to the transistor drain terminal. The incommensurate frequency f is swept in a sufficiently large interval. Pole–zero identification is applied externally to the stored functions $Z_{in}(f)$ obtained for the various P_{in} values.

Initially, the input power has been varied from $P_{in} = 0.70$ W to $P_{in} = 0.78$ W in a direct sweep, that is, without passing through the turning points, with a constant step of 0.01 W. By doing so, we follow the section $\zeta_1 - \zeta_2 - \zeta_4$ of Fig. 12.18, by a jump between ζ_2 and ζ_4. For the calculation of Z_{in}, the frequency f of the small-signal current source has been swept between 1 kHz and 2 MHz. The resulting pole locus is represented in Fig. 12.19a. For most of the P_{in} values considered there are two pairs of complex-conjugate poles at a small distance from the imaginary axis. These two pairs of poles will be denoted $\sigma_1 \pm j2\pi f_1$ and $\sigma_2 \pm j2\pi f_2$. The pair of poles $\sigma_1 \pm j2\pi f_1$ is initially much closer to the imaginary axis than is the pair $\sigma_2 \pm j2\pi f_2$. As shown below, these two pairs of poles will have a decisive influence on the power amplifier behavior versus the input power.

As the input power increases, the two pairs of poles $\sigma_1 \pm j2\pi f_1$ and $\sigma_2 \pm j2\pi f_2$ shift rightward at very different velocity $d\sigma/dP_{in}$ versus the parameter. The movement of $\sigma_2 \pm j2\pi f_2$ is quicker than that of $\sigma_1 \pm j2\pi f_1$, which remains in

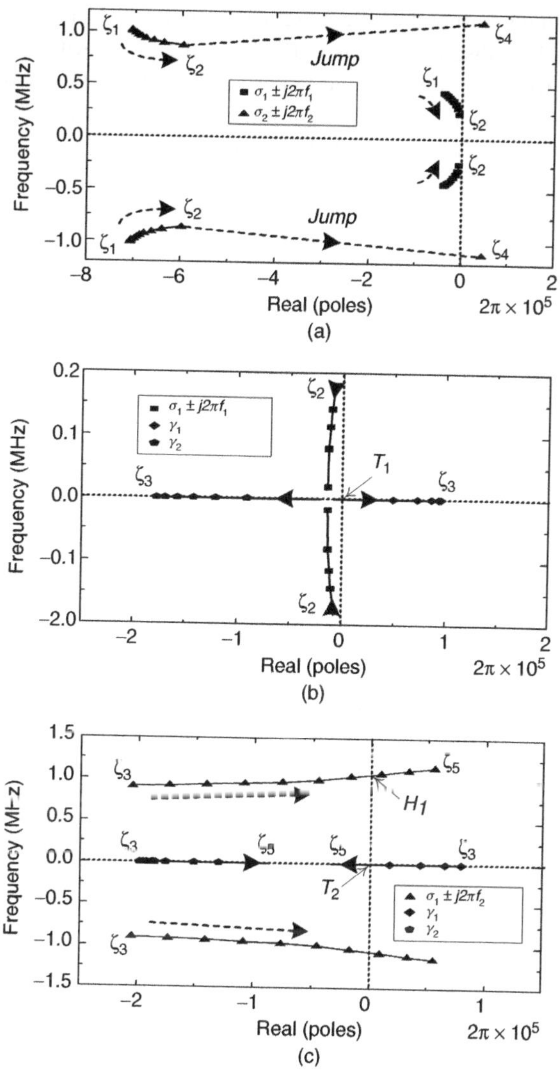

FIGURE 12.19 Pole evolution along the curve $P_{in} - P_{out}$ of Fig. 12.18. (a) Section $\zeta_1 - \zeta_2 - \zeta_4$. The two pairs of poles $\sigma_1 \pm j2\pi f_1$ and $\sigma_2 \pm j2\pi f_2$ closest to the imaginary axis have been represented. (b) Section $\zeta_2 - T_1 - \zeta_3$. The two complex-conjugate poles $\sigma_1 \pm j2\pi f_1$ approach each other until they meet on the imaginary axis. Thy turn into two real poles and from the parameter value at which the merging takes place, follow opposite directions. The real pole γ_1 crosses the imaginary axis at $P_{in} = 0.791$ W, in correspondence with the turning point in the periodic solution curve. (c) Section $\zeta_3 - T_2 - H_1 - \zeta_4 - \zeta_5$. At the input power $P_{in} = 0.781$ W, the pole γ_1 crosses the imaginary axis again to the right-hand side of the complex plane, which gives rise to the turning point T_2. The pair of complex-conjugate poles $\sigma_2 \pm j2\pi f_2$ crosses the imaginary axis to the right-hand side of the complex plane at the power value $P_{in} = 0.781$ W, at which the inverse Hopf bifurcation H_1 takes place. (Reprinted with permission of IEEE.)

the immediate neighborhood of the imaginary axis for all values of P_{in} considered. At $P_{in} = 0.79$ W, a discontinuity is obtained in the pole locus associated with the solution jump from the lower to the upper section of the power transfer curve in Fig. 12.18. This jump is due to the turning point T_1 of this solution curve. The discontinuity gives rise to a relatively large shift of $\sigma_2 \pm j2\pi f_2$, whereas the second pair of poles, $\sigma_1 \pm j2\pi f_1$, seems to disappear. From the power value $P_{in} = 0.79$ W, the pair $\sigma_2 \pm j2\pi f_2$ is located on the right-hand side of the complex plane. Thus, the periodic solution is unstable, as the poles $\sigma_2 \pm j2\pi f_2$ should give rise to the startup of an oscillation at about the frequency $f_a = 1$ MHz.

For a detailed study of the pole evolution, the curve sections about the two turning points have also been considered, that is, the sections $\zeta_2 - T_1 - \zeta_3$ and $\zeta_3 - T_2 - H_1 - \zeta_4 - \zeta_5$. Instead of a simple P_{in} sweep, the continuation technique based on the use of the auxiliary generator will be applied to pass through the turning points. Therefore, the pole zero identification, applied sequentially after each harmonic balance analysis, will also pass through these turning points. Then it will be possible to get into the section between T_1 and T_2 that has not yet been analyzed. Figure 12.19b shows pole evolution along the section $\zeta_2 - T_1 - \zeta_3$. The pair of complex-conjugate poles $\sigma_2 \pm j2\pi f_2$ approaches the imaginary axis without crossing it. For clarity, only evolution of the pair of complex-conjugate poles $\sigma_1 \pm j2\pi f_1$ has been represented. The frequency of the small-signal current source is swept in the interval 0 to 300 kHz. As the input power increases, the two complex-conjugate poles keep close to the imaginary axis, and from 0.78 to 0.791 W move nearly vertically, approaching each other until they merge on the real axis. This merging gives rise to a qualitative variation of the pole structure, since the two complex-conjugate poles become two real poles, γ_1 and γ_2. If the input power is increased further, the two real poles move in opposite direction. The pole γ_1 moves rightward and crosses the imaginary axis at $P_{in} = 0.791$ W. This is the power value at which the turning point T_1 occurs in the solution curve $P_{in} - P_{out}$ of Fig. 12.18. It is in agreement with the fact that a pole at zero corresponds to a singularity of the Jacobian matrix associated to the harmonic balance system. Thus, it gives rise to infinite slope of the solution curve versus the parameter.

Figure 12.19c shows the evolution of a pair of complex-conjugate poles $\sigma_2 \pm j2\pi f_2$ and the pair of real poles γ_1 and γ_2 along the section $\zeta_3 - T_2 - H_1 - \zeta_4 - \zeta_5$ of the $P_{in} - P_{out}$ curve. The frequency interval considered for the small-signal current source is 0 to 1.5 MHz. After passing through zero, the real pole γ_1 moves farther right, turns, and crosses the origin again at the power value $P_{in} = 0.777$ W, corresponding to the turning point T_2. For the entire section $T_1 - T_2$ of the $P_{in} - P_{out}$ curve, the pole γ_1 is on the right-hand side of the complex plane, so this curve section is instable. Along the same section $\zeta_3 - T_2 - H_1 - \zeta_4 - \zeta_5$ of Fig. 12.19c, the pair of poles $\sigma_2 \pm j2\pi f_2$ moves rightward, approaching the imaginary axis. At $P_{in} = 0.781$ W, this pair of poles crosses the imaginary axis to the right-hand side of the complex plane. At the crossing point, a Hopf bifurcation H_1 is obtained, which gives rise to an oscillation at the frequency $f_a = 1$ MHz. If the input power P_{in} is increased further, the poles move rightward, turn, and cross the imaginary axis again at $P_{\text{in}} = 1.45$ W. At this power value, an inverse Hopf bifurcation H_2 takes place, which gives rise to the oscillation extinction.

The stability analysis is in good agreement with the measurements of Fig. 12.18. For the power interval 0.5 to 0.79 W, the pair of complex-conjugate poles $\sigma_1 \pm j2\pi f_1$ is very close to the imaginary axis and the small stability margin is one of the causes for observation of the noisy precursors. In fact, the pole frequency f_1 agrees with the central frequency of the bumps f_c. As the input power increases, the pole frequency decreases, which explains the reduction in bump frequency observed in the measurements. At $P_{\text{in}} = 0.791$ W, the turning point T_1 is obtained, which should give rise to a jump to the upper section of the periodic curve. But this upper section is already unstable when the jump takes place, since the pair of complex-conjugate poles $\sigma_2 \pm j2\pi f_2$, with $f_2 = 1$ MHz, is on the right-hand side of the complex plane. Thus, the system evolves directly into a self-oscillating mixer regime. In the measurements, the sudden variation in the amplifier output spectrum was attributed to the onset of the oscillation. On the other hand, the hysteresis interval was too small, so the difference between the input power values at which the oscillation was generated and extinguished (when sweeping the parameter in opposite direction) was not noticed initially.

12.4.3 Analysis of Noisy Precursors

The noisy precursors are associated with the amplitude perturbation of the steady-state solution due to the noise sources. In Chapter 7, Sections 7.3.3 and 7.7.2, a detailed explanation of this phenomenon in oscillator circuits was presented, together with frequency-domain techniques for its analysis. The phenomenon can also occur in forced circuits exhibiting no oscillation, as shown in the previous example. Here the mathematical model derived in Chapter 7 is adapted to the analysis of noisy precursors in nonautonomous circuits. Then the conditions required for the experimental observation of the precursors are discussed.

12.4.3.1 Mathematical Model As we already know, once a given circuit has reached a stable periodic steady state $\overline{x}_{sp}(t)$, the perturbation transient will be dominated by the real pole or pair of complex-conjugate poles with the smallest magnitude of negative real part. Assuming complex-conjugate dominant poles, $\sigma_c \pm j2\pi f_c$, the smaller the absolute value $|\sigma_c|$, the longer will be the transient at the frequency f_c. Under the continuous noise perturbations, noise bumps will arise in the output power spectrum about the frequencies $kf_{\text{in}} \pm f_c$ [2].

If a parameter η is varied and the critical poles $\sigma_c \pm j2\pi f_c$ approach the imaginary axis, the noise bumps will become higher and narrower (Chapter 7). If the poles cross the imaginary axis, they will become distinct spectral lines: thus the term *noisy precursors*. The phenomenon can also be understood to indicate the result of sideband noise amplification. Under the action of a relatively large pumping signal, the nonlinear capacitances exhibit negative resistance about the sideband frequencies $kf_{\text{in}} \pm f_c$. Before the bifurcation takes place, the absolute value of the negative at the resonance frequency will be smaller than the positive resistance of the linear embedding circuit. At the bifurcation point, the positive and negative resistances are balanced at the resonance frequency. From this point onward, negative resistance will be dominant and the solution will be unstable. In negative-resistance amplifiers, the product $\sqrt{G}\Delta f$, with G and Δf the maximum gain and 3-dB bandwidth, respectively, increases for a smaller quality factor Q of the resonant circuit [22]. In parametric sideband amplification about $kf_{\text{in}} \pm f_c$, a larger bump power should be expected for a smaller quality factor of the resonance at the frequency f_c.

In a forced regime such as the periodic regime of a power amplifier, there is no irrelevance with respect to the phase origin, as this is established by the input generator at the drive frequency f_{in}. Thus, there is no common phase noise and the solution noise observed is due to perturbations in amplitudes and phases of the various harmonic components. The Jacobian matrix associated with the harmonic balance system perturbed by the noise sources and linearized about the steady-state solution is not singular and can be inverted (without numerical difficulty) to solve for the amplitude noise. Thus, the circuit output noise spectrum can be predicted accurately using the conversion matrix approach. Following a procedure similar to the one explained in Chapter 7, it will be possible to relate the noisy precursor power to the stability margin and the offset frequency from the central bump frequencies $kf_{\text{in}} \pm f_c$. Assuming white noise perturbations, the output power spectrum can be expressed as

$$S_k(f) \cong \frac{M_k^{\text{lsb}}}{\sigma_c^2 + 4\pi^2(f + f_c)^2} + \frac{M_k^{\text{usb}}}{\sigma_c^2 + 4\pi^2(f - f_c)^2} \tag{12.11}$$

where only the dominant poles $\sigma_c \pm j2\pi(f_c + kf_{\text{in}})$ have been taken into account. As shown in Section 7.3.3, the coefficients M_k^{lsb} and M_k^{usb} depend on the Jacobian matrix of the harmonic balance system and the input-noise sources. The model (12.11) corresponds to pairs of Lorentzian lines about the frequencies $kf_{\text{in}} \pm f_c$. As can be gathered from (12.11), the height of the Lorentzian lines increases for lower $|\sigma_c|$, which implies a shorter distance of the critical poles to the imaginary axis [2]. Larger precursor power will also be obtained for lower offset frequency from the critical values $kf_{\text{in}} \pm f_c$. Note that the linearization about the steady-state regime becomes invalid in the immediate neighborhood of the bifurcation point. Thus, use of the conversion matrix approach in the noisy precursor analysis is limited to relatively low sideband power. For higher power, the nonlinearities contained in the circuit will give rise to the sideband gain saturation and other effects. A general noise analysis based on the envelope transient method can be used to study the noisy precursors under these conditions.

When using the envelope transient technique to analyze the noisy precursors of a steady-state solution at the frequency f_{in}, the circuit variables will be expressed in a Fourier series at this fundamental frequency f_{in}, with slowly time-varying harmonic terms:

$$\overline{x}(t) = \sum_k \overline{X}_k(t) e^{j2\pi k f_{\text{in}} t} \tag{12.12}$$

Prior to the occurrence of bifurcation, the time variation of the harmonic terms $\overline{X}_k(t)$ will be due exclusively to noise perturbations. For an accurate analysis, the integration step of the differential equation system in the harmonic terms $\overline{X}_k(t)$ must be small enough to allow taking into account the entire precursor band about the critical frequency f_c in the frequency domain. The integration time will typically be much smaller than that required for a full time-domain simulation. Accurate determination of the output spectrum will require a long simulation interval that will be divided in subintervals with a windowing method and averaged according to the periodogram technique.

12.4.3.2 Application to the Class E Amplifier For an analysis of the noisy precursors in the power amplifier, different noise contributions have been taken into account. Only white noise sources have been considered, as the white noise dominates at the frequencies of the nearly critical poles, $\sigma_c \pm j2\pi(f_1 + kf_{in})$. In particular, the noise sources considered are the channel thermal noise, the shot noise, the thermal noise associated with all the circuit resistive elements, and the input generator noise. To study the sideband amplification, the conversion matrix approach will be used to evaluate the gain from the various noise sources to the circuit output about the nearly critical frequencies f_1, $f_{in} - f_1$, and $f_{in} + f_1$. The gain from the channel noise source to the circuit output has initially been evaluated. The original noise current source has been replaced with a small-signal determinist source of the same type, introduced to evaluate the gain from the noise source to the output using the conversion matrix approach. The frequency of this source is swept about f_1, $f_{in} + f_1$, and $f_{in} - f_1$, in three different analyses. These analyses allow us to calculate the conversion gain at the three sideband frequencies.

As a representative analysis, Fig. 12.20 shows the variation of the current conversion gain from the noise source (about f_1) to the output current at the three sidebands f_1, $f_{in} + f_1$, and $f_{in} - f_1$. To obtain the gain curves, the frequency of the small-signal current source is swept about f_1. Three different values of input power P_{in} have been considered. Note that for each input power value P_{in}, the circuit is linearized about a different steady-state solution with the conversion matrix approach, as a different solution is obtained for each P_{in} value. All the gain curves have the same qualitative form, corresponding to the two Lorentzian lines of expression (12.11). The central frequency of the gain sidebands changes with P_{in}, due to the variation in the near-critical-pole frequency f_1. Extremely high gains are not physical, as small changes in P_{in} cause large gain variation. It must also be taken into account that in the immediate neighborhood of the bifurcation, nonlinear effects occur which have not been accounted for in the calculation. When varying the current source used for the gain analysis about $f_{in} + f_1$ or $f_{in} - f_1$, a qualitatively similar result is obtained.

At $P_{in} = 0.65$ W, the gain curves are centered about the offset frequency 490 kHz from the main spectral lines, in agreement with the pole frequency f_1 at this particular power value. The highest gain corresponds to the upper sideband $f_{in} + f$, in agreement with experimental observations. For higher input power ($P_{in} = 0.70$ W and $P_{in} = 0.76$ W), there is reduction of the central (offset) frequency of the

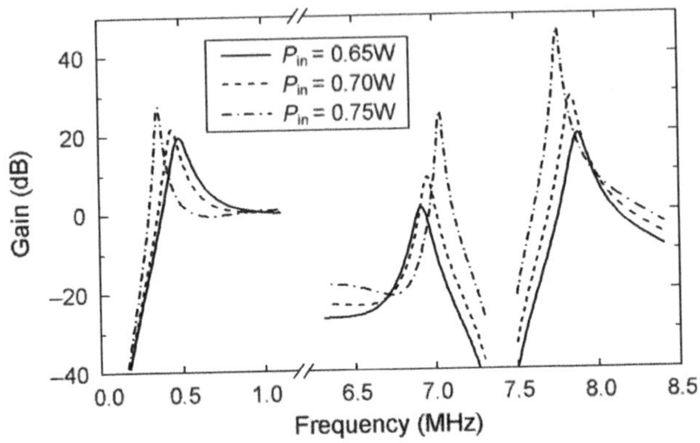

FIGURE 12.20 Frequency variation in the current gain from terminals of the channel noise source to the amplifier output, calculated using the conversion matrix approach. Three different input power values have been considered. (Reprinted with permission of IEEE.)

bumps, in correspondence with the pole variation and the measurements. The gain at the two sidebands $f_{in} \pm f$ increases more quickly than does the low-frequency gain, also in agreement with the measurements. Applying the same gain analysis to other noise sources yields very similar qualitative results.

For an analysis of the amplifier output spectrum, all existing noise sources must be considered simultaneously. The phase noise from the input source has been modeled from experimental measurements. The available noise source with the power required for the switching mode has a high phase noise, which constitutes the largest noise contribution. Results obtained using the conversion matrix approach and using an envelope transient are compared in Fig. 12.21. The good agreement indicates that no significant nonlinear effects are taking place in the noise amplification. The upper sideband power about f_{in} is higher than the lower sideband power, in correspondence with the gain analysis of Fig. 12.20 and with the measurements.

The poles of a power amplifier exhibiting parametric instability will cross the imaginary axis at a certain input power value. Therefore, and due to the circuit continuity, they must be close to the imaginary axis for a certain input power interval. However, the noisy precursors are not always observable. The class E amplifier analyzed exhibits three characteristics that contribute to the practical observation of the noisy precursors. The quality factor at the resonance frequency of the near-critical poles f_1 is low, as gathered from the high phase noise of the observed oscillation. The near-critical poles evolve slowly versus the parameter (i.e., they have a small derivative $d\sigma/dP_{in}$), so they remain close to the axis in a relatively large input power interval. Finally, the gain from the noise sources to the circuit output is high, as shown in Fig. 12.21. This gain is essential for practical observation of the noisy precursors. In other amplifiers, this gain is low and the precursor power is below the noise floor of the measurement equipment. Understanding the circuit characteristics that contribute to the observation of noisy precursors will be useful

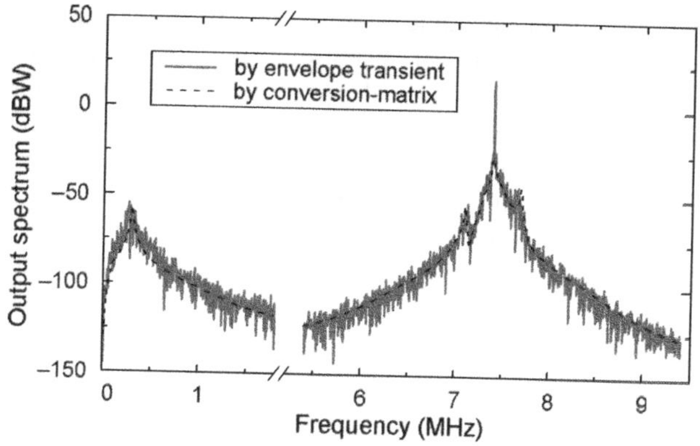

FIGURE 12.21 Analysis of noisy precursors. Comparison of the results obtained using the conversion matrix approach and using an envelope transient. Higher output power is obtained at the upper sideband, in agreement with the gain analysis. (Reprinted with permission of IEEE.)

to devise a technique for their efficient suppression from the output spectrum. This technique is described in Section 12.4.5.

12.4.4 Elimination of the Hysteresis Phenomenon from the Power Transfer Curve $P_{in} - P_{out}$

Hysteresis is a commonly observed phenomenon in both forced and autonomous nonlinear circuits. As has been shown, its accurate analysis with harmonic balance requires the use of a continuation technique such as the one described in Section 12.4.2. The technique described, based on the introduction of an auxiliary generator into the circuit, has the advantage of being externally applicable by users of commercial harmonic balance software, as it requires only the use of standard library elements and a switch of the sweep parameter. This analysis is essential for the prediction of hysteresis phenomena. However, the actual design objective will be the suppression of this phenomenon from the power transfer curve $P_{in} - P_{out}$. Instead of a trial-and-error procedure, a systematic technique is presented here to eliminate this phenomenon with a single harmonic balance simulation.

The hysteresis in a class E power amplifier is due to the existence of turning points in the power transfer curve $P_{in} - P_{out}$, thus can be eliminated by suppressing these turning points. For the curve in Fig. 12.18, the two turning points occur for $P_{in} = 0.791$ W and 0.777 W, with a power difference of 0.014 W. These turning points can be eliminated by making them approach each other until they merge in a single turning point, corresponding to a cusp bifurcation (Chapter 4, Section 4.2.4). This task is not as difficult as it may seem. Actually, the nonlinear equation system that describes the circuit behavior is continuous, so its solutions will evolve continuously versus P_{in}. In a similar manner, entire solution curves, traced versus P_{in}, also evolve continuously versus the variation of an additional parameter, μ. Thus, provided that we find a sensitive parameter μ, the two turning points of the $P_{in} - P_{out}$ curve, corresponding to P_{inT1} and P_{inT2}, will approach each other continuously versus a decrease or increase in μ. Again, provided μ possesses enough sensitivity, the two points will merge at a cusp bifurcation, obtained for $\mu = \mu_{cp}$, so that $P_{in,cp} = P_{inT1} = P_{inT2}$. Once the cusp point is met, if μ is varied further in the same direction, the two turning points will disappear from the solution curve. The system continuity also prevents an excessive disturbance of the original response of the power amplifier.

As we already know, the Jacobian matrix of a harmonic balance system becomes singular at the turning points, due to the existence of a pole at zero $\gamma = 0$. Similarly, the Jacobian matrix associated with the admittance function of the nonperturbation condition $Y_{AG} = 0$ also becomes singular at these points, as was shown in Chapter 6, Section 6.5. When simulating the periodic solution of the oscillator, the auxiliary generator frequency f_{AG} agrees with the input drive frequency f_{in}, and the 2×2 real Jacobian matrix associated with the admittance function is given by

$$[JY_{AG}] = \begin{bmatrix} \dfrac{\partial Y_{AG}^r}{\partial A_{AG}} & \dfrac{\partial Y_{AG}^r}{\partial \phi_{AG}} \\ \dfrac{\partial Y_{AG}^i}{\partial A_{AG}} & \dfrac{\partial Y_{AG}^i}{\partial \phi_{AG}} \end{bmatrix} \tag{12.13}$$

where Y_{AG}^r and Y_{AG}^i are the real and imaginary parts of the admittance function Y_{AG}, respectively. The derivatives are calculated through finite differences in harmonic balance, as explained in Chapter 6.

For the original values of the circuit parameters, the power amplifier exhibits two turning points; that is, there are two solutions of the singularity condition $\det[JY_{AG}(P_{in}, A_{AG}, \phi_{AG})]$. These two turning points will correspond to the input power values P_{inT1} and P_{inT2}. As already explained, if a stabilization parameter μ is now varied continuously from its original value μ_0, the two turning points will shift to $P_{inT1}(\mu)$ and $P_{inT2}(\mu)$, due to the continuity of the circuit equations. Considering the plane defined by μ and P_{in}, we can trace the locus of all these turning points, fulfilling $\det[JY_{AG}] = 0$. The locus will contain two sections: a section that for constant μ provides the turning point obtained for a smaller P_{in} value (e.g., T_1 in Fig. 12.18), and a section that for constant μ provides the turning point obtained for a larger P_{in} value (e.g., T_2 in Fig. 12.18). Note that both sections of the curve can be obtained globally by tracing the turning-point locus in the plane defined by μ and P_{in}. This locus is defined by the three real equations

$$\begin{aligned} Y_{AG}(A_{AG}, \phi_{AG}, P_{in}, \mu) &= 0 \\ \det[JY_{AG}(A_{AG}, \phi_{AG}, P_{in}, \mu)] &= 0 \end{aligned} \tag{12.14}$$

System (12.14) contains four real unknowns in three real equations, so it provides a curve in the plane (μ, P_{in}). All the points in this curve have a real pole at zero $\gamma = 0$. The parameter μ may correspond to a circuit element of the original circuit or may be added to the circuit for hysteresis suppression.

The locus defined by (12.14) can be obtained in commercial harmonic balance software. This requires simultaneous harmonic balance simulation of the original circuit and two other copies, used for derivative calculation. This technique was shown in Chapter 6, Section 6.5, but is recalled here briefly, for completeness. An auxiliary generator, with different amplitude and phase values, is connected to each of the three circuit copies. The auxiliary generator in the first circuit copy operates at nominal auxiliary generator values A_{AG}, ϕ_{AG}, and must fulfill the nonperturbation condition $Y_{AG}(A_{AG}, \phi_{AG}, P_{\text{in}}, \mu) = 0$. The two other copies allow calculation of the derivatives contained in the Jacobian matrix through finite differences. Thus, one of the copies operates at $A_{AG} + \Delta A_{AG}$ and ϕ_{AG} and is used for calculation of the derivative $\partial Y_{AG}/\partial A_{AG}$. The second copy operates at A_{AG} and $\phi_{AG} + \Delta\phi_{AG}$ and is used for calculation of the derivative $\partial Y_{AG}/\partial\phi_{AG}$. The three circuits are solved simultaneously in a single harmonic balance simulation. The expression for $\det[JY_{AG}]$ in terms of the four different real derivatives is introduced symbolically (programmed) in the circuit page of the harmonic balance simulator. To obtain the turning point locus efficiently, the phase ϕ_{AG} is swept, optimizing A_{AG}, P_{in}, and μ (common to the 3 circuits), so as to satisfy $Y_{AG} = 0$ and $\det[JY_{AG}] = 0$ [conditions (12.14)] simultaneously.

The method described has been used to remove the hysteresis phenomenon from the solution curve of the class E power amplifier [2]. The choice of the parameter μ was limited to circuit element values that should not significantly affect the drain efficiency or output power delivered. This is the case for the reactive elements L_{lpf} and C_{lpf} in the lowpass filter that constitutes the final stage of the output network. Note that it is also necessary to determine through simulation whether the turning points of the curve $P_{\text{in}} - P_{\text{out}}$ are sensitive to L_{lpf} and C_{lpf} and can be removed through a variation of these parameters. The capacitance C_{lpf} was taken as the stabilization parameter $\mu = C_{\text{lpf}}$, considering also several values of L_{lpf}. To analyze the influence of these elements over the hysteresis phenomenon, the turning point locus has been traced in the plane defined by C_{lpf} and P_{in} for three different values of the inductance L_{lpf}.

The results are shown in Fig. 12.22. Considering any of the three loci, the two input power values ($P_{\text{in}T1}$ and $P_{\text{in}T2}$) obtained for each C_{lpf} provide the turning points of the corresponding power transfer curve $P_{\text{in}} - P_{\text{out}}$. In agreement with the power transfer curve of Fig. 12.18, the original amplifier, with $L_{\text{lpf}} = 257$ nH and $C_{\text{lpf}} = 100$ pF, has the two turning points indicated as black points in Fig. 12.22. For each L_{lpf} value, the two turning points approach each other when reducing the capacitance C_{lpf} until they meet at the cusp point CP. For a capacitance value smaller than C_{lpf} there are no turning points, so no hysteresis phenomenon will be observed in the circuit measurements. The same behavior is obtained for the three L_{lpf} values. On the other hand, the range of C_{lpf} values for which the power transfer curve $P_{\text{in}} - P_{\text{out}}$ exhibits turning points decreases with L_{lpf}.

The results of Fig. 12.22 have been verified tracing the power transfer curves $P_{\text{in}} - P_{\text{out}}$ for $L_{\text{lpf}} = 257$ nH and different values of C_{lpf} between 80 and 100 pF (Fig. 12.23). For $C_{\text{lpf}} = 100$ pF, two turning points exist at the input power values predicted by the locus of Fig. 12.22, so the hysteresis phenomenon will be observed.

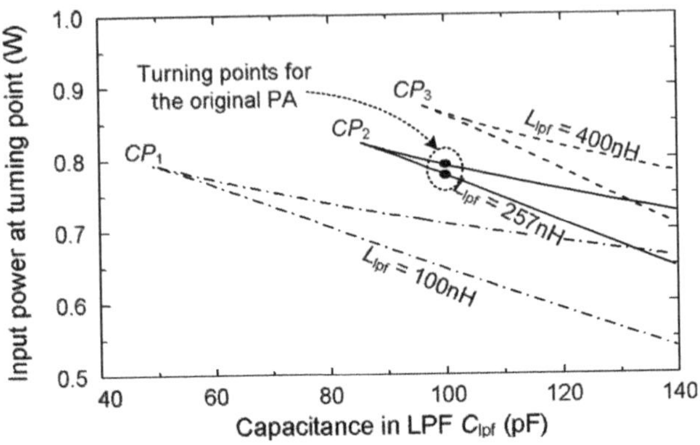

FIGURE 12.22 Turning-point locus in a plane defined by the stabilization parameter C_{lpf} and the input power P_{in}. Three different values of the output network inductance L_{lpf} have been considered; thus, three different loci are represented. (Reprinted with permission of IEEE.)

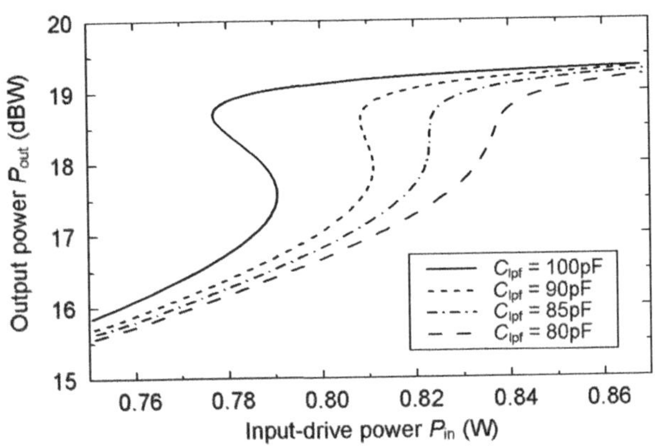

FIGURE 12.23 Elimination of the hysteresis phenomenon. The lowpass filter inductance is maintained at $L_{\text{lpf}} = 257$ nH. The hysteresis interval decreases with the capacitance C_{lpf}, in agreement with the loci of Fig. 12.22. For capacitance values smaller than $C_{\text{lpf}} = 85$ pF, corresponding to the cusp point, no hysteresis is observed. (Reprinted with permission of IEEE.)

For $C_{\text{lpf}} = 90$ pF, the two turning points have closer P_{in} values, in agreement with Fig. 12.22, so the hysteresis interval in terms of P_{in} will be narrower. For $C_{\text{lpf}} = 85$ pF, the two turning points meet at the cusp point CP_2. For $C_{\text{lpf}} = 80$ pF, no hysteresis is observed. As can be seen, except for the turning point suppression, there is only a minimum alteration of the shape of the original power transfer curve $P_{\text{in}} - P_{\text{out}}$. This similarity to the original curve is due to the continuity of the system equations, which will make the solution curve evolve in a gradual manner versus the stabilization parameter, regardless of the particular choice of μ.

As stated earlier, all the points of the turning-point locus contain a real pole at zero $\gamma = 0$. The cusp point μ_{cp}, belonging to this locus, fulfills the two conditions

$$\begin{aligned} \gamma(P_{in,cp}, \mu_{cp}) &= 0 \\ \frac{\partial \gamma}{\partial P_{in}}(P_{in,cp}, \mu_{cp}) &= 0 \end{aligned} \tag{12.15}$$

The second condition, $\partial\gamma/\partial P_{in} = 0$, indicates that, in fact, no qualitative stability change actually takes place at the cusp point. This is because the unstable section between the turning points shrank to zero and has disappeared. The rest of the locus points fulfill $\partial\gamma/\partial P_{in} \neq 0$ and give rise to a qualitative stability change in the solution curve obtained for fixed μ.

If we analyze the evolution of the real pole along the power transfer curve $P_{in} - P_{out}$ exhibiting the cusp point (obtained for fixed $\mu = \mu_{cp}$), we find that this real pole moves rightward versus P_{in}, touches the imaginary axis at the cusp point $P_{in,cp}$, and turns at this same point, moving leftward again without ever crossing to the right-hand side of the complex plane. Thus, the pole is tangent to the imaginary axis at μ_{cp}. For either $\mu < \mu_{cp}$ or $\mu > \mu_{cp}$, depending on the sense of variation considered of this parameter, the solution curve will no longer exhibit turning points and the hysteresis phenomenon will be suppressed. The cusp point is a codimension 2 bifurcation, which requires the fine tuning of two different circuit parameters. In the case described, the two tuned parameters are P_{in} and μ.

For illustration, pole zero identification has been applied along the solution curves of Fig. 12.23 [2,23]. Figure 12.24 shows the variation of the real pole γ_1 versus the input power P_{in} along each of the three solution curves. Note that the P_{in} interval considered is about the hysteresis region alone. As can be seen, for the C_{lpf} values that give rise to hysteresis, the γ_1 curve crosses the horizontal axis $\gamma_1 = 0$ twice, at each of the turning points T_1 and T_2. At the cusp point, the γ_1 curve is tangent to the axis $\gamma_1 = 0$, in agreement with conditions (12.15). For C_{lpf} values showing no hysteresis, there is no crossing of the real pole through the axis $\gamma_1 = 0$.

For $C_{lpf} = 80$ pF, without hysteresis, a larger P_{in} interval has been considered in the pole analysis. For low P_{in} there are two pairs of complex-conjugate poles, $\sigma_1 \pm j2\pi f_1$ and $\sigma_2 \pm j2\pi f_2$, as in the original circuit. As P_{in} increases, the two poles $\sigma_1 \pm j2\pi f_1$ approach each other, and at $P_{in} = 0.83$ W they split into two real poles. These real poles never cross the imaginary axis to the right-hand side of the complex plane. On the other hand, for C_{lpf} values smaller than 50 pF, instability due to the pair of complex-conjugate poles $\sigma_2 \pm j2\pi f_2$, with $f_2 \cong 1$ MHz, is not obtained for any input power value. Extinction of the oscillation is another beneficial effect of modification of the output lowpass filter. Thus, to have a solution curve $P_{in} - P_{out}$ without hysteresis and without oscillation, the capacitance C_{lpf} of the output lowpass filter must be lower than 50 pF. Thus, it is possible to eliminate both the hysteresis and the oscillation by choosing $C_{lpf} < 50$ pF. However, the noisy precursors due to the pair of poles $\sigma_1 \pm j2\pi f_1$ are not suppressed with these design modifications. Although the distance of these poles

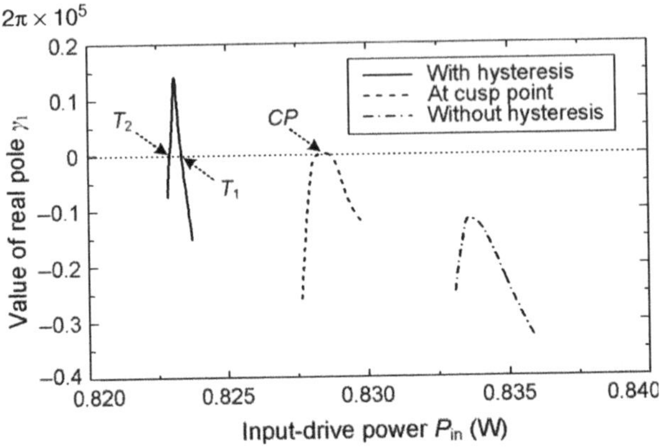

FIGURE 12.24 Variation in the real pole γ_1 versus the input power for different values of the capacitance C_{lpf}. The curve is tangent to the horizontal axis $\gamma_1 = 0$ at the cusp point. (Reprinted with permission of IEEE.)

to the imaginary axis has increased, they are still relatively close to this axis and the precursors are observable in the output power spectrum both in simulation and in actual measurements.

The suppression of the hysteresis phenomenon has been verified experimentally. Keeping the output inductance constant at $L_{\text{lpf}} = 257$ nH, various C_{lpf} values, all smaller than 50 pF, have been tested. The experimental power transfer curves $P_{\text{in}} - P_{\text{out}}$ of the stabilized amplifier, corresponding to the output capacitance values $C_{\text{lpf}} = 20$, 10, and 0 pF, are shown in Fig. 12.25, where they can be compared with the original curve, at which a jump phenomenon takes place. It must be noted that only curves without oscillation are represented. In the measurements, this oscillation could be suppressed for capacitance values smaller than 30 pF. Although this value is smaller than that obtained in simulation (50 pF), there is good qualitative agreement with the predictions of Fig. 12.23. The reduction in C_{lpf} suppresses the hysteresis phenomenon with minimum alteration of the power transfer curve. When reducing the capacitance, the intermediate zone around the two turning points becomes smoother and shifts to the right, in agreement with the prediction of Fig. 12.23. For the capacitance value selected, $C_{\text{lpf}} = 20$ pF, all the harmonic components have power levels 50 dB below the output power at the fundamental frequency, so the filter still fulfills the required suppression of VHF frequencies. No oscillation was observed when varying the input power, but the noisy precursors were still noticeable in the output spectrum. Thus, an additional technique is needed to eliminate the noisy precursors. This technique is presented in the following section.

12.4.5 Elimination of Noisy Precursors

As noted earlier, noisy precursors are due to noise amplification when a steady-state solution has a small stability margin. Its practical observation also depends on the

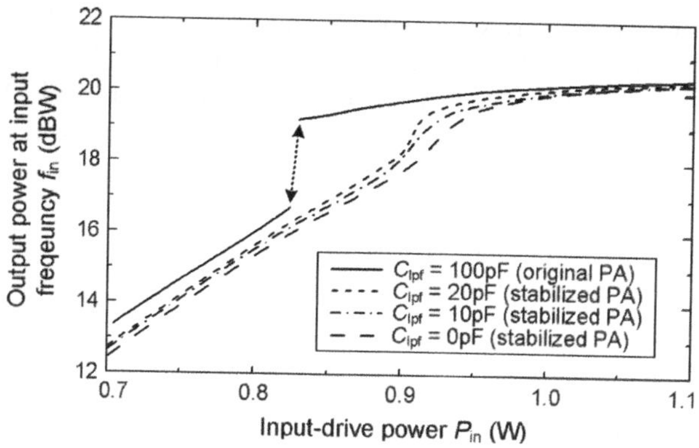

FIGURE 12.25 Experimental power transfer curve $P_{in} - P_{out}$ before and after the elimination of noisy precursors. (Reprinted with permission of IEEE.)

gain values from the various noise sources in the circuit output. To eliminate noisy precursors, both aspects must be taken into account.

From the analysis described in Section 12.4.3, it was concluded that the noisy precursors of a class E power amplifier were due to proximity to the imaginary axis of the pair of complex-conjugate poles $\sigma_1 \pm j2\pi f_1$, with $\sigma_1 < 0$. According to (12.11), one way to reduce noisy-precursor power could be to shift the pair of poles $\sigma_1 \pm j2\pi f_1$ leftward, away from the imaginary axis. This is equivalent to increasing the magnitude $|\sigma_1|$. The increase in $|\sigma_1|$ can be achieved by adding a resistive load at the resonance frequency f_1. The effect of the resistive load can be analyzed by calculating the total input admittance with the same small-signal current source i_s as that used for pole–zero identification. Provided that the observation node is sensitive enough, it will be possible to detect the critical resonance by analyzing the total admittance $Y_n(f)$ from this observation node. The addition of the resistive load in parallel will increase the total positive conductance at the observation node and will shift the pair of poles leftwards in the complex plane.

This small-signal current source at the frequency f, related nonharmonically to f_{in}, is connected in parallel to the drain terminal. The input admittance $Y_{in}(f)$ is calculated, using the conversion matrix approach, as the ratio between the current delivered and the node voltage. When tracing the input admittance Y_{in} versus the source frequency f, a clear resonance is obtained at $f_c = 400$ kHz, with positive real part Re[Y_{in}] and positive slope of the imaginary part Im[Y_{in}] versus the frequency f. The positive sign of Re[Y_{in}] > 0 is in agreement with the fact that the amplifier is actually stable. The parallel connection of a resistance R_d to the drain node should increase the real part of the input admittance Re[Y_{in}]. In the pole locus, it should give rise to the rightward shift of the pair of poles $\sigma_1 \pm j2\pi f_1$, which would increase the stability margin. However, this resistance connected to the transistor output will considerably reduce the drain efficiency and output power of a class E amplifier. To avoid this performance degradation, an inductor L_d with relatively high value is introduced in series with the resistance R_d so as to increase the parallel impedance at the frequency of the input source. The values selected for the elements of the series subnetwork are $R_d = 33\ \Omega$ and $L_d = 4\ \mu\text{H}$. A dc block capacitance $C_b = 80$ nF is also connected in series. As already stated, the series subnetwork is connected in parallel to the drain terminal [2]. Because of the series inductance $L_d = 4\ \mu\text{H}$, the subnetwork introduced will exhibit a large impedance at the input drive frequency f_{in}. However, it must also be taken into account that the resonance frequency f_c is relatively close to f_{in}, so the value of the series inductance L_d cannot be too large. Otherwise, the resistance would have little effect on the near-critical poles. A compromise value of the inductance L_d should be chosen.

To analyze the influence of the series subnetwork on the circuit poles, pole–zero identification has been applied to the amplifier periodic solution, considering a sweep in the input power P_{in}. For low input power P_{in}, introduction of the inductance L_d transforms the original pair of complex-conjugate poles $\sigma_1 \pm j2\pi f_1$ into two real poles γ_1' and γ_2', where the prime refers to the modified circuit. As P_{in} increases, the poles approach each other and become a pair of complex-conjugate poles $\sigma_1' \pm j2\pi f_1'$, with f_1' being significantly smaller than f_1 for the entire P_{in} interval. The reduction of the resonance frequency is due to the effect of the inductance L_d, connected in parallel at the drain node. The corresponding real part σ_1' is only slightly smaller than σ_1. However, as seen in previous sections, observation of the noisy precursors depends strongly on the gain from the noise sources to the amplifier output at the near-critical frequencies f_1', $f_{\text{in}} - f_1'$, $f_{\text{in}} + f_1', \ldots$. As shown in Section 12.4.3, this gain can be analyzed by replacing the particular current noise source with a deterministic small-signal source and obtaining the current gain from the location of the current source to the transistor output [2]. An analogous technique is used in the case of voltage noise sources.

Here the gain from the current source associated with the channel noise has been analyzed. The current source used for the gain calculation is initially swept about f_1'. The resulting gain bands are shown in Fig. 12.26. In comparison with

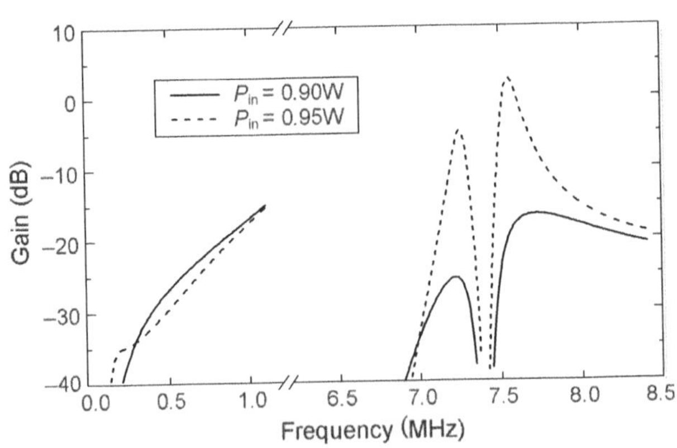

FIGURE 12.26 Corrected amplifier. Frequency variation of the current gain from the terminals of the channel noise current source to the circuit output, calculated using the conversion matrix approach. (Reprinted with permission of IEEE.)

Fig. 12.20, with the introduction of the series subnetwork there is a significant gain reduction for all the P_{in} values. This is the result of the high attenuation of the embedding network at the frequency f_1', much smaller than f_1, which is due mainly to the blocking capacitor. The sideband gain curves about the input frequency maintain a Lorentzian shape. Low gain values are also obtained when sweeping the current source (used for the gain calculation) about $f_{in} + f_1'$ and $f_{in} - f_1'$. The output power spectrum simulated by both the envelope transient and conversion matrix approaches is shown in Fig. 12.27. No noise bumps are observed.

The elimination of noisy precursors has also been verified experimentally. No noisy precursors or instability were obtained in the entire range of input power and drain bias voltage, in agreement with simulations. Figure 12.28 shows a representative measurement of the output power spectrum, corresponding to $P_{in} = 0.95$ W.

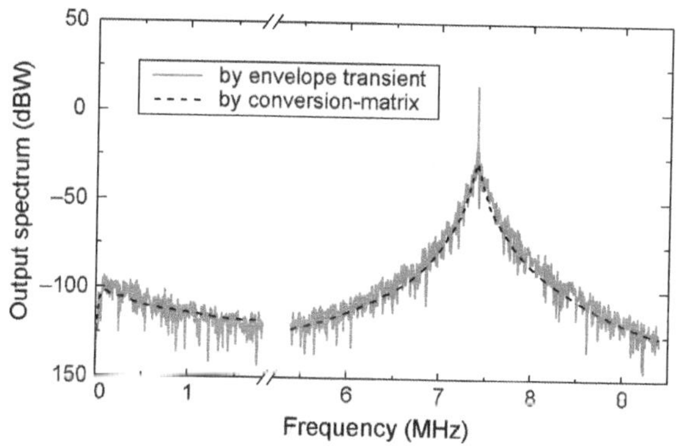

FIGURE 12.27 Simulated output power spectrum after introduction of the correction subnetwork $R_d - L_d$, connected in parallel to the transistor drain terminal. No noise bumps are observed. (Reprinted with permission of IEEE.)

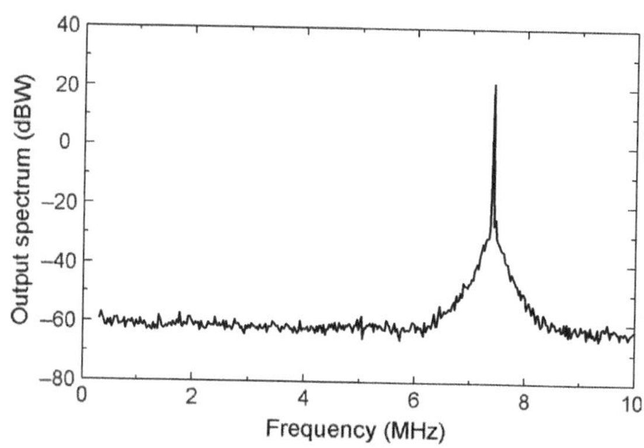

FIGURE 12.28 Experimental output power spectrum of the corrected amplifier for $P_{in} = 0.95$ W. The noise from the input source is still present about f_{in}. However, the noise bumps are no longer observed, in agreement with simulations. (Reprinted with permission of IEEE.)

The phase noise from the input source is still present about the spectral line at the input frequency f_{in}, but unlike the spectrum of Fig. 12.18b, there are no noise bumps either about this spectral line or at low frequency. Compared with the original amplifier, the drain efficiency degradation is below 1.4% at all output power values [2].

12.5 STABILIZATION OF OSCILLATOR CIRCUITS

Free-running oscillator circuits are expected to provide a self-sustained periodic oscillation at a frequency f_o. However, it is not uncommon to observe additional oscillations at either incommensurate or subharmonic frequencies. The undesired oscillations can be suppressed through modification of the circuit parameters or by introducing a stabilization network. However, this will surely give rise to an undesired variation of the primary oscillation frequency f_o and output power. This variation is due to the fact that the fundamental frequency of a free-running oscillator depends on the values of all the circuit elements, and the oscillation frequency and power may change significantly after stabilization. The undesired spurious oscillation must be eliminated using a technique that prevents any variation of the main characteristics of the primary oscillation. Thus, the technique will have to be different from those used for power amplifiers or other forced circuits.

The technique presented here enables systematic stabilization of free-running oscillators exhibiting spurious oscillations at incommensurate or subharmonic frequencies. The technique relies on the fact that the periodic solution desired by the designer is still a valid mathematical circuit solution, even though it is unstable, and this is why it is not observed physically. The objective will be to change the stability properties of the periodic solution without modifying its fundamental frequency f_o and output power P_o. As already stated, the oscillator may be stabilized by changing the values of its components or by introducing a stabilization network. The advantage of adding a stabilization network is that the original oscillator design remains unchanged. The stabilization technique is illustrated by means of its application to an unstable push-push oscillator at 18 GHz.

12.5.1 Stability Analysis of an Oscillator Circuit

For stabilization of the periodic solution of a free-running oscillator, some circuit elements will be modified in order to shift the unstable poles of this solution to the left-hand side of the complex plane. As in the technique for the shortening of the oscillator transient (Section 8.5.1), the stabilization technique is based on a two-stage simulation of the same oscillator circuit. A sweep in one of the circuit elements, μ_1, will be considered. This element may belong to the original oscillator circuit or be part of an additional stabilization network. At the first simulation stage, an auxiliary generator is used to maintain the desired values of the oscillation amplitude and frequency, setting $A_{AG} = V_o$ and $f_{AG} = f_o$. The values of two other elements, μ_2 and μ_3, are calculated to fulfill the steady-state oscillation condition $Y_{AG}(\eta_2, \eta_3) = 0$. At the second simulation stage, pole–zero identification is applied to the circuit resulting from the preceding optimization process, with parameter values μ_1, μ_2, and μ_3. For this analysis, a small-signal current generator $I_n(f)$ is connected in parallel to a circuit node. The closed-loop transfer function $Z_n(f) = V/I_n(f)$ is obtained, linearizing the circuit about its periodic steady-state solution using the conversion matrix approach. The pole–zero identification is performed for each steady-state solution, obtained through the μ_1 sweep. If the poles shift rightward, the sense of the sweep in μ_1 should be changed. If they barely move, there is little sensitivity, so variations in a different circuit parameter should be considered. In case of using a stabilization network, a different network parameter or a different topology of this network should be considered.

As an example, the technique will be applied to a version of the push–push oscillator considered in Chapter 10. The circuit (Fig. 12.29) comprises two identical suboscillators at the frequency $f_o = 9$ GHz which are coupled with a microstrip line that connects the two base terminals [24]. The oscillation frequency specified is obtained by suitably tuning the series resonator connected to the emitter terminal of each of two transistors. The circuit output is extracted from the emitter terminals of the transistors using a Wilkinson combiner. In the odd-oscillation mode, the odd-harmonic components are 180° out of phase, whereas the even-harmonic components are in phase. Then the circuit output frequency is twice the oscillation frequency of each suboscillator, $2f_o = 18$ GHz.

As already seen, for efficient harmonic balance analysis of the periodic solution of the push–push oscillator, two auxiliary generators at $f_{AG} = f_o$ with a 180° phase shift are introduced at equivalent nodes of the two suboscillators. The aim is to ensure convergence to the odd-mode oscillation. Amplitude A_{AG} and frequency f_{AG}

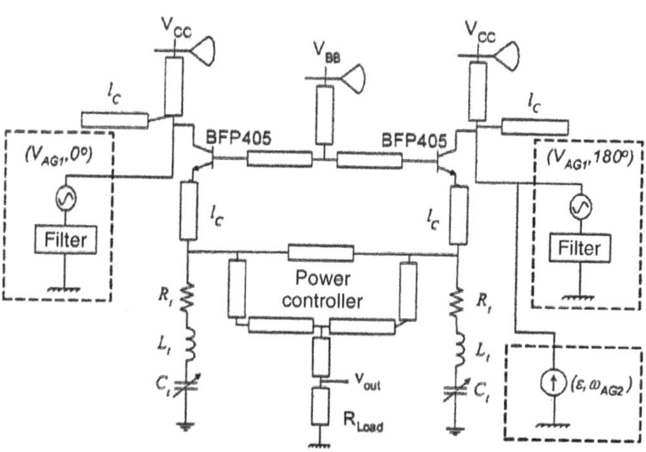

FIGURE 12.29 Push–push oscillator. The auxiliary generator is used to determine the steady-state periodic solution. The small-signal current generator at the incommensurate frequency f allows calculation of a closed-loop transfer function modeled using pole–zero identification.

of the two auxiliary generators must be calculated to satisfy $Y_{AG}(A_{AG}, f_{AG}) = 0$. Due to the circuit symmetry, it is generally sufficient to apply this condition to one suboscillator only.

The push–push oscillator will operate as a voltage-controlled oscillator with a varactor diode connected to the emitter terminal of each suboscillator. Thus, the steady-state oscillation has been analyzed versus the bias voltage of the two varactor diodes V_{dc}. The circuit should behave in stable manner for the entire range of dc bias variation. Thus, the stability of the oscillatory solution will be analyzed initially versus this bias voltage. After obtaining each new steady-state solution by solving $Y_{AG}(V_{AG}, f_{AG}) = 0$, a stability analysis with pole–zero identification is carried out. The simulation procedure is summarized as follows:

For $k = 1$ to N

$V_{dc} = V_{dc}^k$

First analysis : Determination of the steady - state solution with an auxiliary generator

Harmonic balance optimization $Y_{AG}(A_{AG}, f_{AG}) = 0$

Final values: $V_{dc}^k, A_{AG}^k, f_{AG}^k$

$V_{dc} = V_{dc}^k, A_{AG} = A_{AG}^k, f_{AG} = f_{AG}^k$

Second analysis :

The auxiliary generator is kept fixed at

$A_{AG} = A_{AG}^k, f_{AG} = f_{AG}^k$

$I_s(f)$ is introduced in parallel with the auxiliary generator

Conversion matrix approach; sweep in the I_n frequency f

Calculation of the transfer function $Z_{in}(f) = V_n(f)/I_s(f)$

Function $Z_{in}(V_{dc}^k, f)$

Due to the autonomous behavior of the free-running oscillator, circuit linearization about the steady-state periodic oscillation will have a pair of imaginary poles at the oscillation frequency located on the imaginary axis $\pm j2\pi f_o$. When varying the varactor bias voltage V_{dc}, these imaginary poles will slide along the imaginary axis. When reducing the bias voltage from $V_{dc} = 4.5$ V, an additional pair of complex-conjugate poles at the frequency $f_a \approx 4$ MHz crosses the imaginary axis to the right-hand side of the complex plane at $V_{dc} = 2.5$ V. Figure 12.30 shows the variation in the real part of this critical poles and the variation in the oscillation frequency f_o versus the bias voltage V_{dc}. As shown, the solution becomes unstable at the bifurcation bias voltage $V_b = 2.5$ V. At V_b, a secondary Hopf bifurcation takes place in the circuit since a pair of poles crosses the imaginary axis at a frequency related nonharmonically to that of the main oscillation about 9 GHz. This gives rise to the onset of a second oscillation frequency at $f_a \cong 4$ MHz, so for

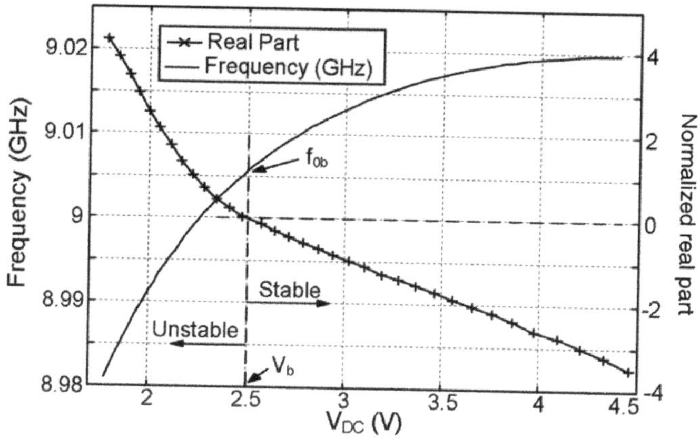

FIGURE 12.30 Variation in oscillation frequency and stability properties of a periodic solution of the push–push oscillator versus the bias voltage V_{dc}. The real part of the critical poles is represented. The periodic solution becomes unstable for $V_{dc} < V_b = 2.5$ V.

bias voltage $V_{dc} < V_b$, the circuit behaves in a self-oscillating mixer regime with two autonomous fundamental frequencies. The objective here is stabilization of the voltage-controlled push–push oscillator for most of the varactor bias range, with minimum variation of the original characteristics of the periodic oscillation [25].

12.5.2 Stabilization Technique for Fixed Bias Voltage

The push–push oscillator described can be stabilized by varying the values of the bias network, due to the low frequency of its spurious oscillation. Thus, there is no need to introduce an additional stabilization network. However, for generality, use of a stabilization network will be considered here, as the frequency of the additional undesired oscillation will not always be low. Initially, the stabilization technique is applied to a particular value of the varactor bias voltage, $V_{dco} = 2.0$ V, belonging to an interval with unstable behavior. As shown in Fig. 12.30, this bias voltage provides the periodic oscillation frequency $f_o(V_{dco}) = 9.013$ GHz, which is unstable. To stabilize the periodic solution of a push–push oscillator, a compensation network can be introduced between the emitter terminals of the two bipolar transistors (Fig. 12.31). This compensation network consists primarily of a lowpass filter connecting the two emitter terminals. In control theory this is equivalent to the insertion of a lead–lag network [26] to modify location of the critical poles.

The stabilization procedure is implemented as follows. An auxiliary generator is introduced at the same oscillation frequency as that of the unstable solution $f_{AG} = f_o(V_{dco}) = 9.013$ GHz and with the same value of the first-harmonic amplitude $V_{AG} = V_o$. Then the inductance in the compensation network L_c is swept, at each sweep step calculating the values of the other two network elements C_c and R_c, so as to fulfill the steady-state oscillation condition $Y_{AG} = 0$. For each L_c value, the first-harmonic amplitude and the oscillation frequency will be the same. However,

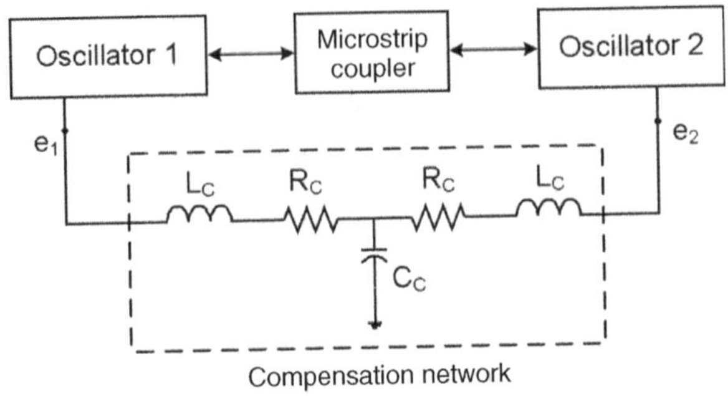

FIGURE 12.31 Push–push oscillator with compensation network.

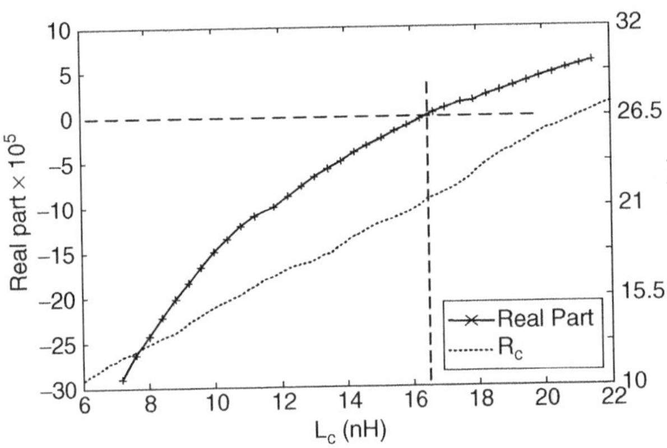

FIGURE 12.32 Stability analysis of push–push oscillator versus inductance L_c of the compensation network for a fixed value of the varactor bias voltage $V_{dc} = 2.0$ V. The resistance value required to maintain the original oscillation frequency f_o and first-harmonic amplitude is also shown.

the actual solution (with different harmonic content) and its associated characteristic matrix [see expression (6.15), Chapter 6] will be different, so its poles can shift to the left-hand side of the complex plane. The results obtained when varying L_c are shown in Fig. 12.32. As can be seen, for inductance values smaller than $L_c = 16.5$ nH, the solution is stable. The R_c value resulting from the optimization $Y_{AG} = 0$ is superimposed. It must be noted that this pole shifting has been carried out without modifying the original oscillator design. The only modified elements are the resistance R_c and capacitance C_c of the compensation network.

A second technique allows stabilizing the free-running oscillator solution, also imposing a certain stability margin. The technique is based on use of the small-signal admittance function $Y_{in}(f)$ instead of the closed-loop transfer function $Z_{in}(f)$. This is because the use of $Z_{in}(f)$ requires subsequent application of pole–zero identification in an external program. In contrast, when using $Y_{in}(f)$ it is possible to deal with the instability–stability conditions in the same harmonic balance program. As already known, the admittance function $Y_{\text{in}}(f)$ is given by the ratio between the injected current and the node voltage $Y_{\text{in}} = I_s(f)/V_n(f)$, so it is exactly the inverse of $Z_{\text{in}}(f)$. If there is enough sensitivity at the observation node, at bias voltages with unstable behavior this admittance function will exhibit a resonance at about $f_a \cong 4$ MHz with a negative real part. This real part of the input admittance will be zero at the bifurcation point and positive at bias voltages with stable behavior. In fact, there is a linear relationship between $\text{Re}[Y_{\text{in}}]$ at the resonance frequency and the real part of the complex-conjugate poles at this same frequency σ (see Section 1.2, Chapter 1).

The objective here will be to stabilize the solution with a sufficient stability margin. With this aim, the original negative value $\text{Re}[Y_{\text{in}}] < 0$ will be changed to a positive value $\text{Re}[Y_{\text{in}}] = G_o$, which is imposed in simulation. The higher G_o is, the larger the stability margin will be. Note that the objective is not to eliminate the resonance, as this is considered to be intrinsic to the oscillator circuit, but to impose a positive conductance, ensuring a sufficient stability margin. Thus, a given stability margin can be imposed by solving

$$
\begin{aligned}
\text{Re}[Y_{\text{in}}(f_a, R_c, C_c, L_c)] &= G_o \\
\text{Im}[Y_{\text{in}}(f_a, R_c, C_c, L_c)] &= 0 \\
Y_{AG}(R_c, C_c, L_c) &= 0 \\
V_{AG} &- V_o \\
f_{AG} &= f_o
\end{aligned}
\tag{12.16}
$$

The three last conditions on the auxiliary generator allow us to keep constant the values of the oscillation frequency and first-harmonic amplitude desired. These were the values of the original unstable periodic oscillation at the considered bias point V_{dco}. The conditions $\text{Re}[Y_{\text{in}}] > 0$ and $\text{Im}[Y_{\text{in}}] = 0$ at the frequency f_a allow us to impose a certain stability margin over the pair of complex-conjugate poles at that particular frequency. Solving (12.16) requires the combined use of harmonic balance and the conversion matrix approach, since the complete equation system involves the small-signal admittance function Y_{in}, obtained by linearizing the harmonic balance equation about the periodic steady-state regime. The harmonic balance solution changes at each optimization step. Thus, the harmonic balance system is linearized using the conversion matrix approach about a different solution at each iteration. The optimization variables are incremented according to the gradient of a total error given by the summation of a harmonic balance error and the conversion matrix error.

To facilitate the resolution of system (12.16), a continuation technique can be used, increasing G_o gradually from the initial negative value $G_o = \text{Re}[Y_{\text{in}}] < 0$ to the final desired value $G_o = G_{of}$, which sets the stability margin. System (12.16) is solved at each G_o step. Figure 12.33 shows the variation of L_c versus the G_o value imposed. Only L_c is represented, although C_c has also been varied [see (12.16)]. For validation, the pole–zero identification technique has been applied

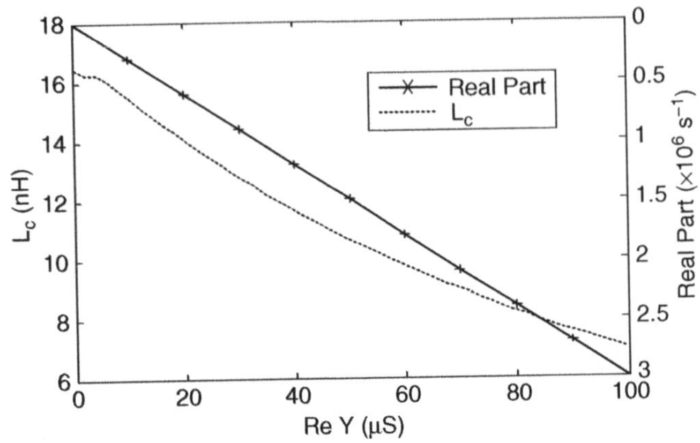

FIGURE 12.33 Variation in the inductance L_c of the compensation network versus the stability margin, imposed by presetting the value of the real part of the input admittance G_o. The real part σ of the critical poles is also represented and shows a linear relationship with G_o.

to the resulting set of steady-state solutions versus G_o. As already known, this identification requires exporting $Z_{in}(f)$ and fitting this function with a quotient of polynomials in a separate program. Figure 12.33 shows the variation in the real part of the complex-conjugate poles at f_a versus the conductance G_o, used in system (12.16). Note the linear relationship between G_o and σ. In agreement with the simulations of Fig. 12.31, the stability margin increases when reducing the inductance L_c of the compensation network. Note that the other two network elements, R_c and C_c, also vary at each inductance step. This is because they have to be optimized in order to maintain periodic oscillation at the oscillation amplitude and frequency values desired, $A_{AG} = V_o$ and $f_{AG} = f_o$. In Fig. 12.34 the waveform of the originally unstable periodic solution is compared with the waveform of the stabilized solution for the element values of the compensation network $R_c = 6.18\ \Omega$, $L_c = 6$ nH, and $C_c = 3.5$ pF. There is a remarkable resemblance between the waveforms. This is due to the fact that both the oscillation frequency and first-harmonic amplitude are being fixed with the auxiliary generator. Also, the compensation network is lowpass in nature, thus has little influence at the higher harmonic terms.

The stabilization procedure illustrated in Figs. 12.32 and 12.33 has been carried out for a particular value of the varactor bias voltage $V_{dc} = 2.0$ V. For other values of this bias voltage the periodic solution of the voltage-controlled push-push oscillator may still be unstable. To check whether this is the case, a new stability analysis has been carried out for fixed values of the compensation network versus the bias voltage V_{dc}. The resulting pole locus is shown in Fig. 12.35. As can be seen, for a bias voltage below $V_{dc} = 0.75$ V, the oscillator periodic solution is unstable, with a pair of unstable complex-conjugate poles at about 4 MHz.

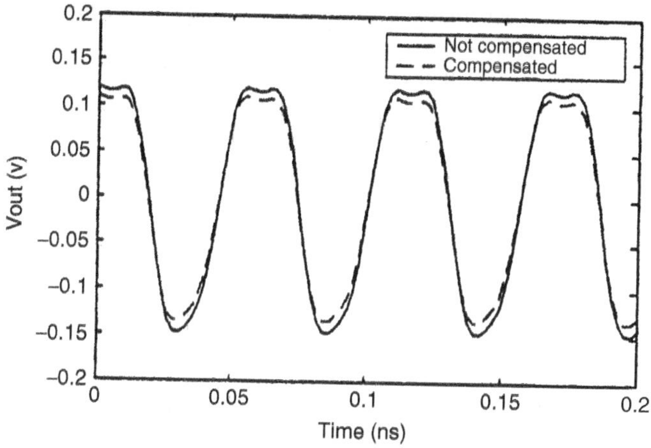

FIGURE 12.34 Comparison of the waveforms of the originally unstable periodic solution of a push–push oscillator for the bias voltage $V_{dc} = 2.0$ V and the stabilized solution for the same bias voltage value.

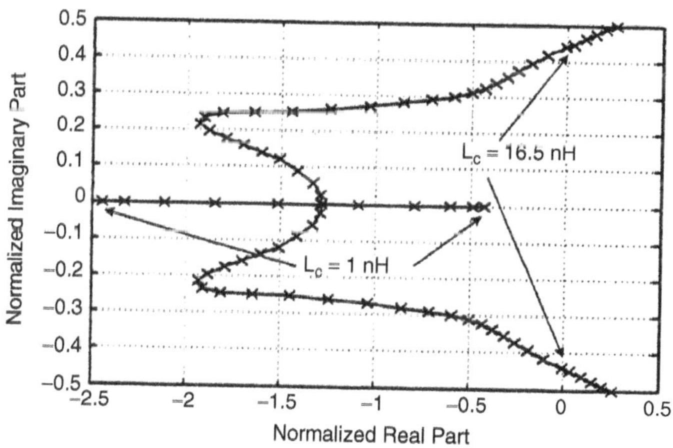

FIGURE 12.35 Pole locus versus the periodic solution of a push–push oscillator versus the bias voltage V_{dc} for fixed values of the compensation network $R_c = 6.18\ \Omega$, $L_c = 6$ nH, and $C_c = 3.5$ pF.

12.5.3 Stabilization Technique for the Entire Tuning Voltage Range

To extend the stabilization technique to the entire bias voltage range, a bifurcation detection technique is used. The procedure is similar to that used for the global stabilization of power amplifiers. We will obtain a bifurcation locus in the plane defined by the actual circuit parameter, in this case the bias voltage V_{dc}, and a stabilization parameter such as the inductance L_c of the stabilization network. The locus delimits the pairs of values L_c and V_{dc} giving unstable behavior. Next, we chose a value of L_c such that the oscillator remains in the stable region for the entire range of V_{dc} values.

In the free-running oscillator analyzed, the bifurcation that leads to unstable behavior is of Hopf type, since the oscillator instability is due to a pair of complex-conjugate poles at a frequency f_a, related nonharmonically to the primary oscillation frequency, f_o. To obtain the Hopf bifurcation locus in the plane defined by L_c and V_{dc}, we proceed as follows. For each bias voltage value V_{dc}, we determine the compensation network inductance L_{cb} that constitutes the border between stable and unstable behavior. This is done by setting to zero both the real and imaginary parts of the small-signal input admittance Y_{in} at the incommensurable frequency f_a. Note that this is the condition for Hopf bifurcation, as the steady-state oscillation amplitude tends to zero at Hopf bifurcation points (see Section 6.5.2, Chapter 6). The network elements C_o and R_o are constant during this calculation, as in the stabilized design, the network will, of course, be fixed during V_{dc} variations. The Hopf bifurcation locus in the plane defined by L_c and V_{dc} is given by

$$\begin{aligned} \text{Re}[Y_{in}(f_a, V_{AG}, f_{AG}, L_{cb}, V_{dc})] &= 0 \\ \text{Im}[Y_{in}(f_a, V_{AG}, f_{AG}, L_{cb}, V_{dc})] &= 0 \\ Y_{AG}(V_{AG}, f_{AG}, L_{cb}, V_{dc}) &= 0 \end{aligned} \tag{12.17}$$

System (12.17) must be solved by combining the harmonic balance and conversion matrix approaches, in a manner similar to system (12.16). This is because the small-signal admittance Y_{in} is calculated by linearizing the harmonic balance system about the periodic solution that fulfills $Y_{AG} = 0$. The system comprises four real equations in five real unknowns, so a curve is obtained in the plane defined by L_{cb} and V_{dc}, corresponding to the Hopf locus. This locus is represented in Fig. 12.36. Note that the admittance function Y_{AG} does not depend on the resonance frequency f_a. It depends only on the periodic solution at the frequencies kf_o, with k an integer, obtained with harmonic balance. In contrast, the small-signal admittance Y_{in} depends on f_a and on the steady-state solution V_{AG}, f_{AG}. On the other hand, both admittance functions depend on the inductance value L_{cb}. Actually, the linearized frequency response versus small-signal current $I_n(f)$ changes with the steady-state solution about which the harmonic balance system is linearized.

The Hopf bifurcation locus in Fig. 12.36 delimits the pairs of L_c and V_{dc} values, providing either stable or unstable behavior. For $V_{dc} = 2.5$ V, corresponding to the limit of stable operation in the original push–push design, L_{cb} tends to infinity. This is in agreement with the fact that a stabilization network is no longer necessary for $V_{dc} > 2.5$ V. The accuracy of the bifurcation locus has been verified using an envelope transient. As is already known, in this analysis method the circuit variables are represented in a Fourier series with slowly varying harmonic terms $\overline{x}(t) = \sum_k \overline{X}_k(t)e^{j2\pi kf_ot}$. For a stable periodic oscillation, the envelope amplitudes $|\overline{X}_k(t)|$ should have a constant value. Furthermore, if the frequency of the Fourier basis f_o used agrees with the actual oscillation frequency, both the real and imaginary parts of the complex harmonic components will have constant values. In the case of

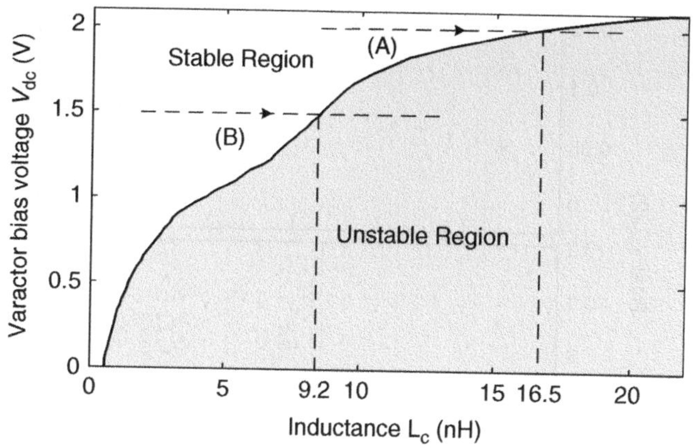

FIGURE 12.36 Hopf bifurcation locus in a plane defined by the inductance of the compensation network L_c and the tuning voltage V_{dc}.

unstable behavior, the steady-state solution will be quasiperiodic and the amplitude $|\overline{X}_k(t)|$ will oscillate at the low frequency f_a. This analysis has been carried out versus continuous variation of L_c for two different values of V_{dc}. For each V_{dc} value, L_c is swept in a large range. The two auxiliary generators with values resulting from harmonic balance optimization are connected only at the initial instant in time. Remember that it is a push–push oscillator, so two auxiliary generators are necessary. This allows oscillation initialization, avoiding trivial nonoscillatory solutions at either of the two suboscillators. In Fig. 12.37 the magnitude of the real part of the second harmonic of the output voltage $\text{Re}[V_{out}][2]$ has been traced versus L_c. (Remember that the output frequency of the push–push oscillator is $2f_o$). A constant value of $\text{Re}[V_{out}][2]$, after reaching the steady-state regime, implies a periodic regime at the same fundamental frequency as that considered in the Fourier series representation of the circuit variables $f_{AG} = f_o$. When represented versus L_c, this gives rise to a single point. Thus, a sequence of single points versus L_c is obtained until reaching the Hopf bifurcation, from which the periodic solution becomes unstable. From this Hopf bifurcation, a vertical line corresponding to projection of $\text{Re}[V_{out}][2]$ is obtained. The stability border is represented by a vertical line. This border agrees with the prediction of Fig. 12.34 for the two different V_{dc} values. As V_{dc} decreases, a smaller L_C value is required for stabilization of the periodic solution, in agreement with Fig. 12.36.

Finally, an experimental validation of the stabilization technique has been carried out. Figure 12.38 compares the measured spectra obtained before and after the connection of the compensation network at the bias voltage $V_{dc} = 2.0$ V. As can be seen, before introduction of the network, we obtain a mixer-like spectrum. It corresponds to a quasiperiodic solution in which the output frequency $2f_o = 18$ GHz mixes with a low-frequency oscillation. When a compensation network is introduced, a periodic spectrum is obtained at the same oscillation frequency $2f_o = 18$ GHz.

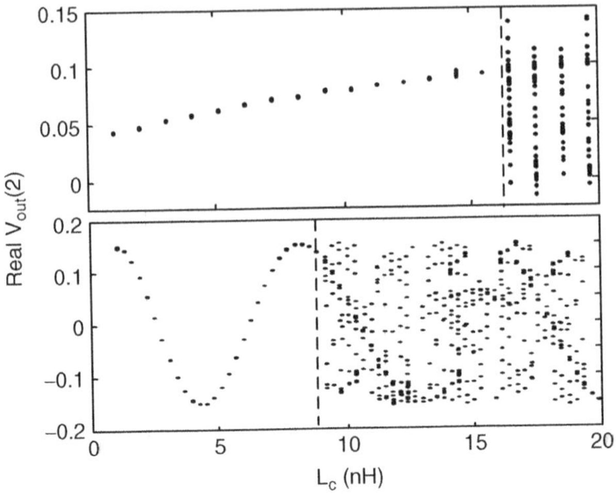

FIGURE 12.37 Use of an envelope transient to verify the accuracy of the stability border provided by the Hopf bifurcation locus of Fig. 12.36. Variation in the real part of the second-harmonic component of the output voltage has been represented versus the inductance L_c for two different values of the bias voltage, $V_{dc} = 2$ V and $V_{dc} = 1.5$ V.

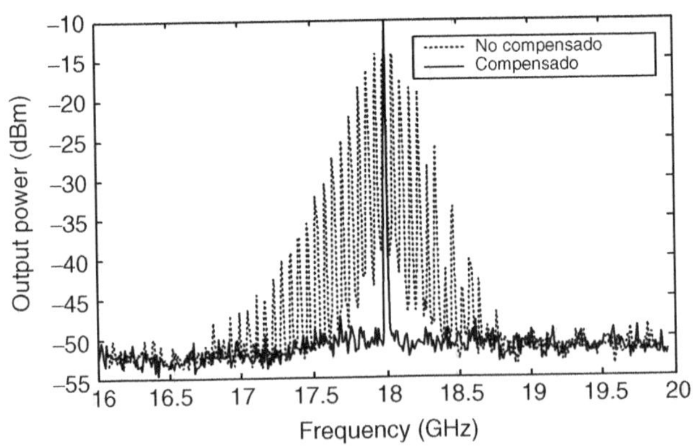

FIGURE 12.38 Measured output power spectrum of an oscillator circuit about the harmonic $2f_o$ with and without a compensation network. The network parameters are $R_c = 5\ \Omega$, $L_c = 20$ nH, and $C_c = 3$ pF.

12.6 STABILIZATION OF MULTIFUNCTION MMIC CHIPS

MMIC multifunctional chips allow reducing the physical dimensions of electronic equipment while maintaining, or even improving, system performance [27]. Such circuits include several functions optimized together in the same chip, allowing a high level of integration and a minimization of chip surface and routing interfaces. The success of the design relies on aspects such as selection of the most suitable technology process, which should provide accurate measurement-based models for the circuit components and the use of appropriate analyses and design techniques. Unfortunately, the use of multifunction circuits increases circuit complexity, requiring the development of systematic and agile design techniques. Due to the multiple active elements contained in a circuit and the existing feedback loops (e.g., due to common bias circuitry), stability analysis will also be essential in order to predict undesired behavior. The numerous difficulties involved in stability analysis in complex circuits have been highlighted by experts in MMIC design [28]. In the following, a specific design strategy is proposed to predict and correct unstable behavior in multifunction MMIC. The study is combined with analysis of the small-signal admittance $Y_{in}(f)$ at an observation node, for a quick and simple test, and pole–zero identification of the closed-loop transfer function $Z_{in}(f) = 1/Y_{in}(f)$, for better accuracy. Due to the complexity of a multifunction circuit, a systematic method must be used for the selection of observation nodes, which will be done at the lumped-element schematic level and the layout level. The technique will be illustrated by means of its application to a multifunction MMIC chip in S-band [27].

The complete multifunction MMIC chip is an S-band frequency down-converter including five functions: RF, LO, and IF amplification, phase shifting, and doubly balanced frequency conversion. The mixer is implemented using a symmetric ring-FET resistive configuration [29], with differential amplifiers used at both RF and LO ports to balance the signal and increase the power level. Finally, a push-pull amplifier is implemented at the IF port in order to combine the 180° out-of-phase signals at the mixer outputs, improving the gain and efficiency of the converter. This IF amplifier included a first-order *RC* parallel filter connected in series at its output. This filter was used to stabilize the amplifier unconditionally. The resulting MMIC is a complex structure containing 12 nonlinear transistors. The circuit is shown in Fig. 12.39.

The first version of the circuit exhibited oscillation at 23.5 GHz. Note that the frequency at which instability is observed is much higher than the IF frequency of operation, around 170 MHz. Oscillation was observed even without employing drive signals at either the RF or the LO port, which means that it was caused by instability in the dc solution of the multifunction circuit. The oscillation persisted at the nominal LO power value required ($P_{LO} = -5$ dBm), as shown in the spectrum in Fig. 12.40. The oscillation degraded the converter performance significantly, especially the output power at 1-dB gain compression, as only −5 dBm was obtained instead of the expected value of +2 dBm.

The multifunction chip oscillates in the absence of input signal from the local oscillator or the RF input. Thus, its dc solution is unstable. For the stability analysis of this dc solution, a small-signal current source is introduced into the circuit. The input admittance at the injection node is calculated using ordinary small-signal analysis. As already known, oscillation at a frequency f_o will start up if the conditions $\text{Re}[Y_{in}(f_o)] < 0$ and $\text{Im}[Y_{in}(f_o)] = 0$ with $\partial\text{Im}[Y_{in}(f_o)]/\partial f > 0$ are fulfilled (see Chapter 1, Section 1.3). The advantage of the function $Y_{in}(f)$ is that it can be evaluated directly at the harmonic balance simulator used, with no need to

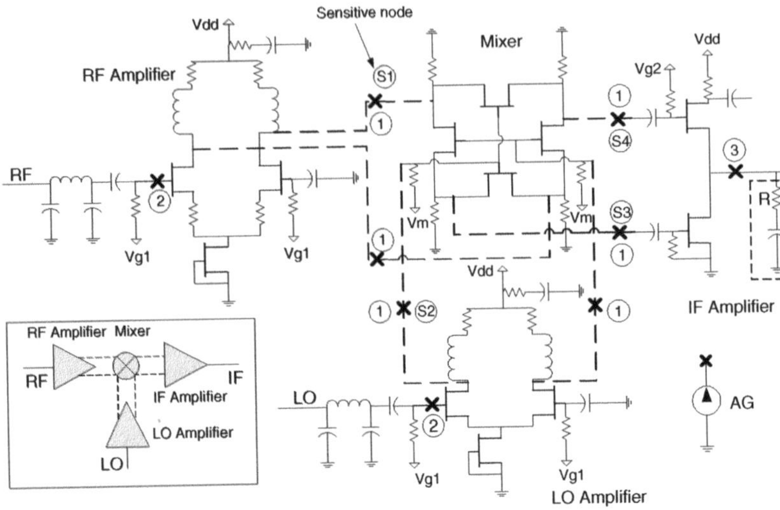

FIGURE 12.39 Multifunction MMIC converter developed using stability design techniques. (Reprinted with permission of IEEE.)

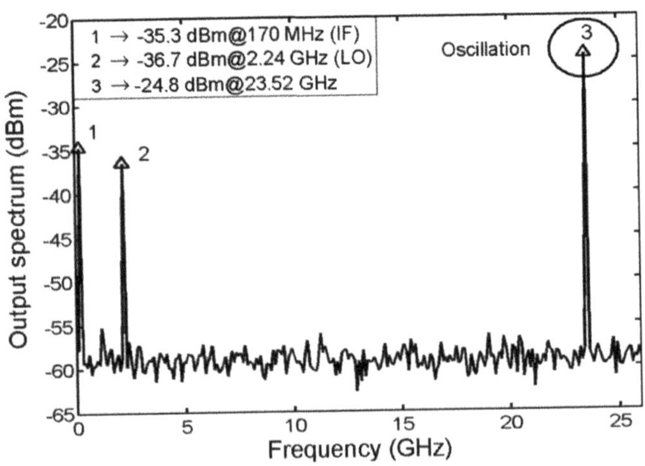

FIGURE 12.40 Output spectrum of the first version of a multifunction MMIC converter, with an undesired oscillation at 23.5 GHz. (Reprinted with permission of IEEE.)

export data. For a more rigorous analysis, pole–zero identification is applied to the closed-loop transfer function provided by the inverse of the defined admittance $Z_{\text{in}}(f) = 1/Y_{\text{in}}(f)$.

Due to the complexity of multifunction circuits, a systematic method must be used for selection of the observation nodes, which is done at the lumped-element schematic level and the layout level. The methodology [27] is summarized below.

12.6.1 Analyses at the Lumped-Element Schematic Level

Initially, the multifunction converter was analyzed at the lumped-element schematic level, without including the effects of the microstrip lines contained by the complete MMIC. In this way, the complexity of the circuit was reduced and, consequently, the computation time required. The nodes in the schematic of Fig. 12.39 were classified into three groups, indicated as 1, 2, and 3. First, observation nodes classified as 1 were considered, which correspond to the six terminals of the mixer. These nodes are, in fact, the external terminals of the mixer transistors. It should be pointed out that the internal transistor nodes were not accessible. At type 1, external nodes multiloop effects can be obtained due to the connection of three circuit amplifiers used for the RF, LO, and IF signals. This may lead to feedback through signal or dc bias paths. However, when using these analysis techniques, none of the mixer nodes exhibited instability at any frequency f [27].

In the second stage, the nodes classified as 2 in Fig. 12.39 were analyzed. These nodes correspond to the gate terminals of the transistors used in the RF and LO amplifiers. Typical instabilities occur at these locations due to feedback through the gate–source capacitance of the transistors. However, no instabilities were obtained at any of these observation nodes. Finally, the output node of the IF amplifier, classified as 3, was analyzed, and no instability was detected.

In conclusion, the oscillation observed in the measurement could not be detected at the lumped-element schematic level. Therefore, it was likely to be due to additional feedback effects introduced by the distributed elements.

12.6.2 Analyses at the Layout Level

As stated previously, all the analyses at the schematic level predicted stable behavior, so new analyses were performed at the layout level. This type of simulation is demanding, since the multifunction circuit consists of more than 800 elements. Moreover, the large number of control-bias and radio-frequency signal paths contained in a multifunction MMIC circuit leads to much higher layout complexity than that in single-function circuits. Normally, bridges are necessary between transmission lines in different layers of the MMIC in the layout. These bridges give rise to signal coupling between the layers, as represented in the equivalent circuit in Fig. 12.41. The capacitance value C_p depends on the line width. The wider the lines, the higher the value of C_p. Typically, it will be around picofarads or femtofarads. This capacitance results in coupling signals between layers, which gets higher as the frequency increases. Such coupling can introduce feedback and cause unstable behavior. In our application, in addition to the number of control bias and radio-frequency signal paths, symmetry is required for the signals at the mixer. This fact increases the complexity of the layout, so various bridges were needed, especially in the area of the mixer to which the three amplifiers were connected. Therefore, for analyses at the layout level, the nodes near bridges between lines had to be considered in addition to the three sets of nodes discussed in Section 12.6.1. Bridges between two signal paths, between dc paths and signal paths, and between two dc paths were taken into account.

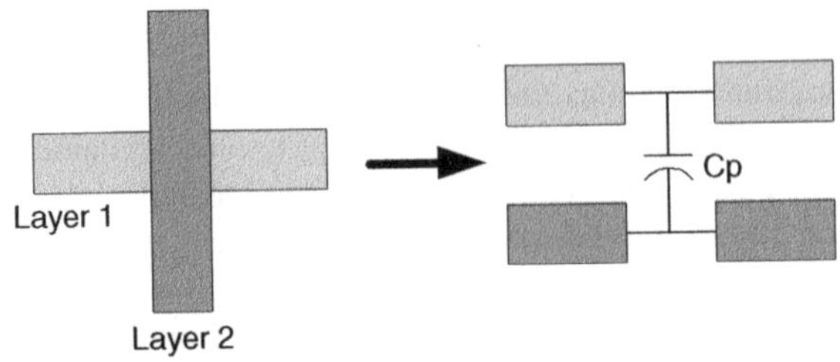

FIGURE 12.41 Equivalent circuit model of a bridge between lines in different layers.

Various nodes were analyzed until instability was found at the oscillation frequency obtained in the measurements. The evaluation of $Y_{in}(f)$ at the RF input node of the mixer (S1 in Fig. 12.39), depicted in Fig. 12.42a, showed fulfillment of the oscillation condition around 26.5 GHz. For a more rigorous verification, identification techniques were also applied to the transfer function $Z_{in}(f) = 1/Y_{in}(f)$. The results are shown in Fig. 12.42b. This analysis provided a pair of complex-conjugate poles with positive real part around 26.5 GHz, which confirmed the instability observed in the measurements. Other converter nodes, designated as S2, S3, and S4 in Fig. 12.39, were also sensitive, confirming the unstable frequency obtained at the S1 node. Note that all four nodes were in the area of the mixer, thereby ratifying our assumption regarding the most sensitive area of a multifunction circuit.

To investigate the influence of the various sections of a multifunction circuit on the oscillation detected, the analysis was repeated by replacing different parts of the layout with an ideal schematic connection. In this way, the impact of these parts on the transfer function could be evaluated. The parts selected were the most likely to give rise to unstable behavior. First, bridges in the mixer area were studied. Due to the complexity of the layout, that area included various bridges between signal paths, specifically between the lines containing nodes S1–S2, S1–S3, and S3–S4. Although these bridges were substituted consecutively for ideal connections, unstable behavior was obtained in all cases. Second, the remaining bridges in the layout were changed, with unsuccessful results. Third, the microstrip lines between the amplifiers and the mixer were replaced, but unstable behavior continued in all cases. Finally, the microstrip lines between components and transistors were substituted for ideal connections. After replacing the microstrip line at point 3 in Fig. 12.39, the circuit became stable. Hence, the most critical part of the multifunction circuit turned out to be the output of the IF amplifier.

To devise a stabilization technique, the influence of various parameters on the observed oscillation was analyzed using a continuation technique. For this analysis, no LO power is considered because, as indicated in all the previous study, in terms of bias voltages and other parameters, the instability is more relevant when no power is injected. Actually, the circuit becomes stable at a certain LO power. The instability was found to be very sensitive to the length L of the microstrip line between the transistors of the amplifier and the output RC filter. Figure 12.43a shows the variation in the oscillation amplitude versus the line length. For lengths

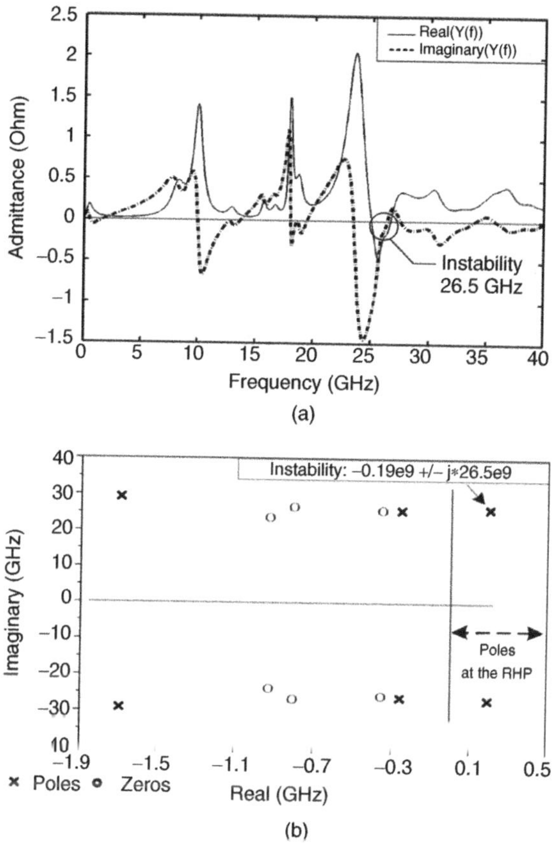

FIGURE 12.42 Stability analysis at the RF input node of the mixer (S1 in Fig. 12.39): (a) real and imaginary parts of the small-signal admittance $Y(f)$; (b) pole–zero map obtained at the S1 node of the first version of the multifunction MMIC converter, showing unstable behavior around 26.5 GHz. (Reprinted with permission of IEEE.)

shorter than 180 μm, the oscillation is extinguished in an inverse Hopf bifurcation. On the other hand, the oscillation amplitude is nearly flat for lengths longer than 250 μm. These results are confirmed through small-signal analysis. For each L value, the closed-loop function $Z_{in}(f)$ was obtained by sweeping the frequency of the small-signal current generator introduced at node S1. The results obtained are depicted in Fig. 12.43b. The dependence of the unstable frequency on the length L is confirmed by comparing the results of Fig. 12.43a and b. Therefore, the multifunction circuit becomes stable for lengths shorter than 177 μm. Note that in comparison to the direct sweep of Fig. 12.11a, pole–zero identification of Fig. 12.43b requires postprocessing of the data obtained in the frequency sweep.

Taking the earlier results into account, the second version of the multifunction MMIC converter was designed minimizing the critical length L. Moreover, bridges between transmission lines in different layers were avoided as much as possible

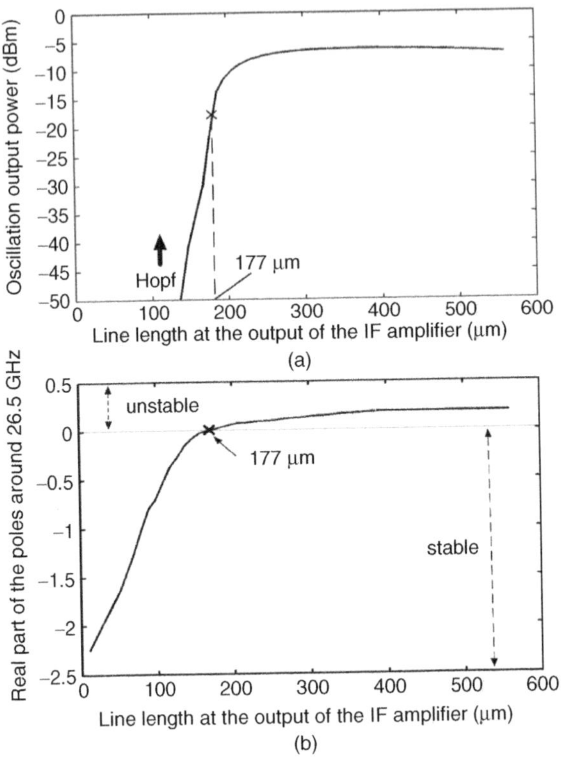

FIGURE 12.43 Stability analysis versus line length at the output of an IF amplifier: (a) evolution of the oscillation output power at 26.56 GHz obtained by steady-state analysis; (b) evolution of the real part of poles versus the length L of the transmission line between the output of the IF amplifier and the RC filter. The circuit becomes stable for $L < 177\ \mu m$. (Reprinted with permission of IEEE.)

during the development of the layout so as to reduce the number of locations prone to feedback. The complexity of this layout should be noted since the circuit involved several functions, required symmetry for the mixer, and included a large number of control-bias and signal paths. Finally, the values of the RC circuit were also optimized according the following general rules:

1. Minimum disturbance to the converter performance
2. Sufficient stability margin for $\mathrm{Re}[Y(f)] > 0$, to ensure stable performance given the technological dispersion of the circuit parameters and the very strict environmental requirements (-30° to $+65^\circ$)
3. Minimum chip area

The stability analyses were integrated into our design cycle for the second version. These analyses were carried out at various potentially sensitive nodes of the circuit at both the lumped-element schematic and layout levels. The circuit was also

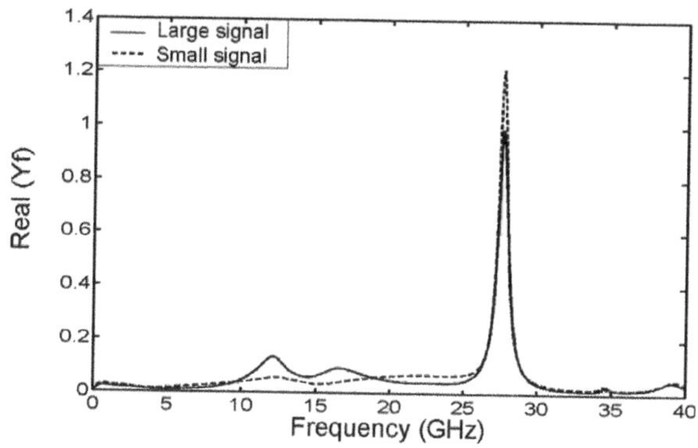

FIGURE 12.44 Real part of the admittance $Y(f)$ for the second version of the multi-function MMIC converter, obtained by small- and large-signal analysis with $P_{LO} = 0$ dBm, showing stable performance and a large stability margin. (Reprinted with permission of IEEE.)

analyzed for several mixer voltages, V_m and P_{LO}. All these techniques helped to achieve successful chip design. After simulations, an RC output filter of $R = 15\ \Omega$ and $C = 2$ pF was implemented next to the output of the push-pull amplifier, obtaining a stable admittance function with no degradation in the circuit performance. The real part of the admittance $Y_{in}(f)$ obtained by means of small- and large-signal analysis with $P_{LO} - 0$ dBm are depicted in Fig. 12.44. As can be seen, the second version of the chip showed stable performance and presented a considerable stability margin for $\text{Re}[Y_{in}(f)] > 0$. This stable operation was confirmed by pole–zero identification of $Z_{in}(f)$.

REFERENCES

[1] A. Anakabe, J. M. Collantes, J. Portilla, et al., Analysis and elimination of parametric oscillations in monolithic power amplifiers, *IEEE MTT-S International Microwave Symposium*, Seattle, WA, pp. 2181–2184, 2002.

[2] S. Jeon, A. Suárez, and D. B. Rutledge, Analysis and elimination of hysteresis and noisy precursors in power amplifiers, *IEEE Trans. Microwave Theory Tech.*, vol. 54, pp. 1096–1106, 2006.

[3] F. Ramirez, A. Suárez, and S. Sancho, Stabilization techniques for microwave oscillators, *Analyse de stabilité avec l'outil STAN. Organizador: CNES (Centre National d'Etudes Spatiales)*, Institut Aéronautique et Spatial, Toulouse, France, 2006.

[4] A. K. Ezzeddine and H. A. Hung, Novel power combiner for MIC and MMIC amplifiers, *IEEE MTT-S International Microwave Symposium Digest*, vol. 1, pp. 479–482, 1990.

[5] D. J. Vendelin, A. M. Pavio, and U. L. Rohde, *Microwave Circuit Design*, Wiley, New York, 1990.

[6] K. Kurokawa, An analysis of Rucker's multidevice symmetrical oscillator, *IEEE Trans. Microwave Theory Tech.*, vol. 18, p. 967, May 1970.

[7] U. L. Rohde, A. K. Poddar, and G. Bock, *The Design of Modern Microwave Oscillators for Wireless Applications*, Wiley, Hoboken, NJ, 2005.

[8] R. G. Freitag, A unified analysis of MMIC power amplifier stability, *IEEE MTT-S International Microwave Symposium*, Albuquerque, NM, pp. 297–300, 1992.

[9] A. Mortazawi, H. D. Foltz, and T. Itoh, A periodic second harmonic spatial power combining oscillator, *IEEE Trans. Microwave Theory Tech.*, vol. 40, pp. 851–856, 1992.

[10] A. Anakabe, J. M. Collantes, J. Portilla, S. Mons, and A. Mallet, Detecting and avoiding odd-mode parametric oscillations in microwave power amplifiers, *Int. J. RF Microwave Comput. Aided Eng.*, vol. 15, pp. 469–478, 2005.

[11] K. Kurokawa, Some basic characteristics of broadband negative resistance oscillators, *Bell Syst. Tech. J.*, vol. 48, pp. 1937–1955, July–Aug. 1969.

[12] A. Collado, F. Ramirez, and A. Suárez, Analysis and stabilization tools for microwave amplifiers, *IEEE MTT-S International Microwave Symposium Digest*, Fort Worth, (TX), vol. 2, pp. 945–948, June 2004.

[13] D. Teeter, A. Platzker, and R. Bourque, Compact network for eliminating parametric oscillations in high power MMIC amplifiers, *IEEE MTT-S International Microwave Symposium Digest*, vol. 3, pp. 967–970, 1999.

[14] M. Mochizuki, M. Nakayama, Y. Tarui, Y. Itoh, S. Tsuji, and T. Takagi, Nonlinear analysis of $f_0/2$ loop oscillation of high power amplifiers, *IEEE MTT-S International Microwave Symposium*, Orlando, FL, pp. 709–712, 1995.

[15] J. F. Davis and D. B. Rutledge, Low-cost class-E power amplifier with sine-wave drive, pp. 1113–1116, 1998.

[16] N. O. Sokal, Class-E switching-mode high-efficiency tuned RF/microwave power amplifier: improved design equations, *IEEE MTT-S International Microwave Symposium Digest*, vol. 2, pp. 779–782, 2000.

[17] S. D. Kee, I. Aoki, A. Hajimiri, and D. Rutledge, The class-E/F family of ZVS switching amplifiers, *IEEE Trans. Microwave Theory Tech.*, vol. 51, pp. 1677–1690, 2003.

[18] N. O. Sokal, Class E high-efficiency power amplifiers, from HF to microwave, *IEEE MTT-S International Microwave Symposium*, Baltimore, MD, pp. 1109–1112, 1998.

[19] S. Jeon, A. Suárez, and D. B. Rutledge, Global stability analysis and stabilization of a class-E/F amplifier with a distributed active transformer, *IEEE Trans. Microwave Theory Tech.*, vol. 53, pp. 3712–3722, 2005.

[20] C. E. Cordonnier, P. Rossel, R. Maimouni, H. Tranduc, D. Allain, M. Napieralskadouard, Spice model for TMOS power MOSFETs, *Motorola Application Note AN1043*, Phoenix, AZ, 1989.

[21] Y. Kim and J. G. Fossum, Physical DMOST modeling for high-voltage IC CAD, *IEEE Trans. Electron Devices*, vol. 37, pp. 797–803, 1990.

[22] R. E. Collin, *Foundations for Microwave Engineering*, 2nd ed., Wiley, New York, 2001.

[23] J. Jugo, J. Portilla, A. Anakabe, A. Suárez, and J. M. Collantes, Closed-loop stability analysis of microwave amplifiers, *IEE Electron. Lett.*, vol. 37, pp. 226–228, Feb. 2001.

[24] F. Ramirez, J. García, T. Fernandez, and A. Suárez, Nonlinear simulation techniques for the optimized design of push–push oscillators, *IEEE MTT-S International Microwave Symposium Digest*, Philadelphia, vol. 3, p. 2157, June 8–13, 2003.

[25] S. Sancho and F. Ramírez, General stabilization techniques for microwave oscillators, *IEEE Microwave Wireless Components Lett.*, vol. 15, pp. 868–870, Dec. 2005.

[26] K. Ogata, *Modern Control Engineering*, Prentice-Hall, Englewood Cliffs, NJ, 1980.

[27] C. Barquinero, A. Suárez, A. Herrera, and J. L. García, Complete stability analysis of multifunction MMIC circuits, *IEEE Trans. Microwave Theory Tech.*, vol. 55, pp. 2024–2033, 2007.

[28] J. Cazaux, MMIC's for space-borne systems: status and prospectives, pp. 3–6, 1994.

[29] S. A. Maas, *Nonlinear Microwave Circuits*, Artech House, Norword, MA, 1988.

[30] J. M. T. Thompson and H. B. Stewart, *Nonlinear Dynamics and Chaos*, 2nd ed., John Wiley & Sons, 2002.

Index

WILEY SERIES IN MICROWAVE AND OPTICAL ENGINEERING

KAI CHANG, Editor
Texas A&M University

FIBER-OPTIC COMMUNICATION SYSTEMS, Third Edition • *Govind P. Agrawal*

ASYMMETRIC PASSIVE COMPONENTS IN MICROWAVE INTEGRATED CIRCUITS • *Hee-Ran Ahn*

COHERENT OPTICAL COMMUNICATIONS SYSTEMS • *Silvello Betti, Giancarlo De Marchis, and Eugenio Iannone*

PHASED ARRAY ANTENNAS: FLOQUET ANALYSIS, SYNTHESIS, BFNs, AND ACTIVE ARRAY SYSTEMS • *Arun K. Bhattacharyya*

HIGH-FREQUENCY ELECTROMAGNETIC TECHNIQUES: RECENT ADVANCES AND APPLICATIONS • *Asoke K. Bhattacharyya*

RADIO PROPAGATION AND ADAPTIVE ANTENNAS FOR WIRELESS COMMUNICATION LINKS: TERRESTRIAL, ATMOSPHERIC, AND IONOSPHERIC • *Nathan Blaunstein and Christos G. Christodoulou*

COMPUTATIONAL METHODS FOR ELECTROMAGNETICS AND MICROWAVES • *Richard C. Booton, Jr.*

ELECTROMAGNETIC SHIELDING • *Salvatore Celozzi, Rodolfo Araneo, and Giampiero Lovat*

MICROWAVE RING CIRCUITS AND ANTENNAS • *Kai Chang*

MICROWAVE SOLID-STATE CIRCUITS AND APPLICATIONS • *Kai Chang*

RF AND MICROWAVE WIRELESS SYSTEMS • *Kai Chang*

RF AND MICROWAVE CIRCUIT AND COMPONENT DESIGN FOR WIRELESS SYSTEMS • *Kai Chang, Inder Bahl, and Vijay Nair*

MICROWAVE RING CIRCUITS AND RELATED STRUCTURES, Second Edition • *Kai Chang and Lung-Hwa Hsieh*

MULTIRESOLUTION TIME DOMAIN SCHEME FOR ELECTROMAGNETIC ENGINEERING • *Yinchao Chen, Qunsheng Cao, and Raj Mittra*

DIODE LASERS AND PHOTONIC INTEGRATED CIRCUITS • *Larry Coldren and Scott Corzine*

RADIO FREQUENCY CIRCUIT DESIGN • *W. Alan Davis and Krishna Agarwal*

MULTICONDUCTOR TRANSMISSION-LINE STRUCTURES: MODAL ANALYSIS TECHNIQUES • *J. A. Brandão Faria*

PHASED ARRAY-BASED SYSTEMS AND APPLICATIONS • *Nick Fourikis*

FUNDAMENTALS OF MICROWAVE TRANSMISSION LINES • *Jon C. Freeman*

OPTICAL SEMICONDUCTOR DEVICES • *Mitsuo Fukuda*

MICROSTRIP CIRCUITS • *Fred Gardiol*

HIGH-SPEED VLSI INTERCONNECTIONS, Second Edition • *Ashok K. Goel*

FUNDAMENTALS OF WAVELETS: THEORY, ALGORITHMS, AND APPLICATIONS • *Jaideva C. Goswami and Andrew K. Chan*

HIGH-FREQUENCY ANALOG INTEGRATED CIRCUIT DESIGN • *Ravender Goyal (ed.)*

ANALYSIS AND DESIGN OF INTEGRATED CIRCUIT ANTENNA MODULES • *K. C. Gupta and Peter S. Hall*

PHASED ARRAY ANTENNAS • *R. C. Hansen*

STRIPLINE CIRCULATORS • *Joseph Helszajn*

THE STRIPLINE CIRCULATOR: THEORY AND PRACTICE • *Joseph Helszajn*

LOCALIZED WAVES • *Hugo E. Hernández-Figueroa, Michel Zamboni-Rached, and Erasmo Recami (eds.)*

MICROSTRIP FILTERS FOR RF/MICROWAVE APPLICATIONS • *Jia-Sheng Hong and M. J. Lancaster*

MICROWAVE APPROACH TO HIGHLY IRREGULAR FIBER OPTICS • *Huang Hung-Chia*

NONLINEAR OPTICAL COMMUNICATION NETWORKS • *Eugenio Iannone, Francesco Matera, Antonio Mecozzi, and Marina Settembre*

FINITE ELEMENT SOFTWARE FOR MICROWAVE ENGINEERING • *Tatsuo Itoh, Giuseppe Pelosi, and Peter P. Silvester (eds.)*

INFRARED TECHNOLOGY: APPLICATIONS TO ELECTROOPTICS, PHOTONIC DEVICES, AND SENSORS • *A. R. Jha*

SUPERCONDUCTOR TECHNOLOGY: APPLICATIONS TO MICROWAVE, ELECTRO-OPTICS, ELECTRICAL MACHINES, AND PROPULSION SYSTEMS • *A. R. Jha*

OPTICAL COMPUTING: AN INTRODUCTION • *M. A. Karim and A. S. S. Awwal*

INTRODUCTION TO ELECTROMAGNETIC AND MICROWAVE ENGINEERING • *Paul R. Karmel, Gabriel D. Colef, and Raymond L. Camisa*

MILLIMETER WAVE OPTICAL DIELECTRIC INTEGRATED GUIDES AND CIRCUITS • *Shiban K. Koul*

MICROWAVE DEVICES, CIRCUITS AND THEIR INTERACTION • *Charles A. Lee and G. Conrad Dalman*

ADVANCES IN MICROSTRIP AND PRINTED ANTENNAS • *Kai-Fong Lee and Wei Chen (eds.)*

SPHEROIDAL WAVE FUNCTIONS IN ELECTROMAGNETIC THEORY • *Le-Wei Li, Xiao-Kang Kang, and Mook-Seng Leong*

ARITHMETIC AND LOGIC IN COMPUTER SYSTEMS • *Mi Lu*

OPTICAL FILTER DESIGN AND ANALYSIS: A SIGNAL PROCESSING APPROACH • *Christi K. Madsen and Jian H. Zhao*

THEORY AND PRACTICE OF INFRARED TECHNOLOGY FOR NONDESTRUCTIVE TESTING • *Xavier P. V. Maldague*

METAMATERIALS WITH NEGATIVE PARAMETERS: THEORY, DESIGN, AND MICROWAVE APPLICATIONS • Ricardo Marqués, Ferran Martín, and Mario Sorolla

OPTOELECTRONIC PACKAGING • *A. R. Mickelson, N. R. Basavanhally, and Y. C. Lee (eds.)*

OPTICAL CHARACTER RECOGNITION • *Shunji Mori, Hirobumi Nishida, and Hiromitsu Yamada*

ANTENNAS FOR RADAR AND COMMUNICATIONS: A POLARIMETRIC APPROACH • *Harold Mott*

INTEGRATED ACTIVE ANTENNAS AND SPATIAL POWER COMBINING • *Julio A. Navarro and Kai Chang*

ANALYSIS METHODS FOR RF, MICROWAVE, AND MILLIMETER-WAVE PLANAR TRANSMISSION LINE STRUCTURES • *Cam Nguyen*

FREQUENCY CONTROL OF SEMICONDUCTOR LASERS • *Motoichi Ohtsu (ed.)*

WAVELETS IN ELECTROMAGNETICS AND DEVICE MODELING • *George W. Pan*

OPTICAL SWITCHING • Georgios Papadimitriou, Chrisoula Papazoglou, and Andreas S. Pomportsis

SOLAR CELLS AND THEIR APPLICATIONS • *Larry D. Partain (ed.)*

ANALYSIS OF MULTICONDUCTOR TRANSMISSION LINES • *Clayton R. Paul*

INTRODUCTION TO ELECTROMAGNETIC COMPATIBILITY, Second Edition • *Clayton R. Paul*

ADAPTIVE OPTICS FOR VISION SCIENCE: PRINCIPLES, PRACTICES, DESIGN AND APPLICATIONS • *Jason Porter, Hope Queener, Julianna Lin, Karen Thorn, and Abdul Awwal (eds.)*

ELECTROMAGNETIC OPTIMIZATION BY GENETIC ALGORITHMS • *Yahya Rahmat-Samii and Eric Michielssen (eds.)*

INTRODUCTION TO HIGH-SPEED ELECTRONICS AND OPTOELECTRONICS • *Leonard M. Riaziat*

NEW FRONTIERS IN MEDICAL DEVICE TECHNOLOGY • *Arye Rosen and Harel Rosen (eds.)*

ELECTROMAGNETIC PROPAGATION IN MULTI-MODE RANDOM MEDIA • *Harrison E. Rowe*

ELECTROMAGNETIC PROPAGATION IN ONE-DIMENSIONAL RANDOM MEDIA • *Harrison E. Rowe*

HISTORY OF WIRELESS • *Tapan K. Sarkar, Robert J. Mailloux, Arthur A. Oliner, Magdalena Salazar-Palma, and Dipak L. Sengupta*

PHYSICS OF MULTIANTENNA SYSTEMS AND BROADBAND PROCESSING • *Tapan K. Sarkar, Magdalena Salazar-Palma, and Eric L. Mokole*

SMART ANTENNAS • *Tapan K. Sarkar, Michael C. Wicks, Magdalena Salazar-Palma, and Robert J. Bonneau*

NONLINEAR OPTICS • *E. G. Sauter*

APPLIED ELECTROMAGNETICS AND ELECTROMAGNETIC COMPATIBILITY • *Dipak L. Sengupta and Valdis V. Liepa*

COPLANAR WAVEGUIDE CIRCUITS, COMPONENTS, AND SYSTEMS • *Rainee N. Simons*

ELECTROMAGNETIC FIELDS IN UNCONVENTIONAL MATERIALS AND STRUCTURES • *Onkar N. Singh and Akhlesh Lakhtakia (eds.)*

ANALYSIS AND DESIGN OF AUTONOMOUS MICROWAVE CIRCUITS • *Almudena Suárez*

ELECTRON BEAMS AND MICROWAVE VACUUM ELECTRONICS • *Shulim E. Tsimring*

FUNDAMENTALS OF GLOBAL POSITIONING SYSTEM RECEIVERS: A SOFTWARE APPROACH, Second Edition • *James Bao-yen Tsui*

RF/MICROWAVE INTERACTION WITH BIOLOGICAL TISSUES • *André Vander Vorst, Arye Rosen, and Youji Kotsuka*

InP-BASED MATERIALS AND DEVICES: PHYSICS AND TECHNOLOGY • *Osamu Wada and Hideki Hasegawa (eds.)*

COMPACT AND BROADBAND MICROSTRIP ANTENNAS • *Kin-Lu Wong*

DESIGN OF NONPLANAR MICROSTRIP ANTENNAS AND TRANSMISSION LINES • *Kin-Lu Wong*

PLANAR ANTENNAS FOR WIRELESS COMMUNICATIONS • *Kin-Lu Wong*

FREQUENCY SELECTIVE SURFACE AND GRID ARRAY • *T. K. Wu (ed.)*

ACTIVE AND QUASI-OPTICAL ARRAYS FOR SOLID-STATE POWER COMBINING • *Robert A. York and Zoya B. Popović (eds.)*

OPTICAL SIGNAL PROCESSING, COMPUTING AND NEURAL NETWORKS • *Francis T. S. Yu and Suganda Jutamulia*

SiGe, GaAs, AND InP HETEROJUNCTION BIPOLAR TRANSISTORS • *Jiann Yuan*

ELECTRODYNAMICS OF SOLIDS AND MICROWAVE SUPERCONDUCTIVITY • *Shu-Ang Zhou*

Printed and bound by CPI Group (UK) Ltd, Croydon, CR0 4YY
16/09/2026
14963457-0008